Computational Fluid Dynamics 2002

Springer
Berlin
Heidelberg
New York
Hong Kong
London
Milan
Paris
Tokyo

S. Armfield P. Morgan K. Srinivas (Eds.)

Computational Fluid Dynamics 2002

Proceedings
of the Second International Conference
on Computational Fluid Dynamics,
ICCFD, Sydney, Australia,
15-19 July 2002

Springer

Professor Steve W. Armfield
Professor Patrick Morgan
Professor Karkenahalli Srinivas
University of Sydney
School of Aerospace
Building J07
2006 Sydney
Australia

Cataloging-in-Publication Data applied for

Bibliographic information published by Die Deutsche Bibliothek Die Deutsche Bibliothek lists this publication in the Deutsche Nationalbibliografie; detailed bibliographic data is available in the Internet at <http://dnb.ddb.de>.

ISBN-13:978-3-642-63938-8 e-ISBN-13:978-3-642-59334-5
DOI: 10.1007/978-3-642-59334-5

Springer-Verlag Berlin Heidelberg New York
a member of BertelsmannSpringer Science+Business Media GmbH
http://www.springer.de

Softcover reprint of the hardcover 1st edition 2003

Typesetting: Data prepared by the author using a Springer TEX macro package
Cover design: *design & production* GmbH, Heidelberg

Printed on acid-free paper 57/3141/xo 5 4 3 2 1 0

Preface

We are pleased to present the Proceedings of the Second International Conference on Computational Fluid Dynamics held at the University of Sydney, Australia, from July 15 to 19, 2002. The conference was a productive meeting of scientists, mathematicians and engineers involved in the computation of fluid flow. Keynote lectures were presented in the areas of optimisation, algorithms, turbulence and bio-fluid mechanics.

Two hundred and fifty abstracts from many countries were received for consideration. The executive committee, consisting of A. Lerat, M. Napolitano, J.J. Chattot, N. Satofuka and myself, were responsible for the selection of papers. Each of the members had a separate subcommittee to carry out the evaluation. One hundred and seventy papers were selected of which one hundred and fifty two were presented at the conference. All papers that appear in the proceedings have been peer reviewed by a panel of experts (with a minimum of two for every paper) before publication.

The conference was attended by 160 delegates with a minimum of late withdrawals. The informal and friendly atmosphere provided by the university surroundings was highly appreciated, and the technical aspects of the conference were stimulating.

It is appropriate here to thank Alain Lerat, the retiring secretary of the international scientific committee of the conference. We also wish to welcome J. J. Chattot who is the incoming secretary.

I would also like to thank the members of the local organising committee who processed the abstracts and the manuscripts, the referees for their invaluable comments, the sponsors for their generous support and the many students and spouses for their help.

Sydney, Australia
September 2002

K. Srinivas
chairman

Sponsors

NASA, Ames, USA
Dassault Aviation, Paris, France
Defence Science & Technology Organisation, Australia
Aerospace, Mechanical & Mechatronic Engineering, University of Sydney, Australia
The Institution of Engineers, Australia.

Local Organising Committee

K Srinivas,	University of Sydney, Chairman, Editor
S W Armfield,	University of Sydney, Co-Chairman, Editor-in-Chief
P Morgan,	University of Sydney, Secretary, Editor
M Behnia,	University of New South Wales
D J Auld,	University of Sydney
J Madani,	University of Sydney
B Dally,	University of Adelaide
L Cheng,	University of Western Australia
J P Denier,	University of Adelaide
G Bain,	Defence Science & Technology Organisation, Australia

International Scientific Committee

C H Bruneau,	France	R MacCormack,	USA
J J Chattot,	USA	K Nakahashi,	Japan
H Deconinck,	Belgium	Y Nakamura,	Japan
S Dennis ,	Canada	M Napolitano,	Italy
S M Deshpande,	India	J Periaux,	France
D X Fu,	China	R Radespiel,	Germany
K Fujii,	Japan	P L Roe,	USA
M Giles,	UK	F Sabetta,	Italy
M Hafez,	USA	N Satofuka,	Japan
Y Hussaini,	USA	Y Shokin,	Russia
R Klein,	Germany	K Srinivas,	Australia
D Kwak ,	USA	H M Wu,	China
A Lerat,	France	Y Yang,	Taiwan
P Le Quere,	France	F G zhuang,	China

Honorary Members

H Cabbannes,	France
M Holt,	France
E Krause,	Germany
K W Morton,	UK
K Oshima,	Japan
R Peyret,	France
P Zandbergen,	Netherlands

Contents

Part VI Direct Numerical Simulation

Part VII Fluid Structure Coupling and Heat Transfer

Part XVIII Unsteady Flow

Part XIX Vortical Flows

Part XX Technical Notes

Part I

Invited Lectures

A Newton-Krylov Algorithm for Aerodynamic Analysis and Design

D.W. Zingg, M. Nemec, and T.T. Chisholm

Institute for Aerospace Studies, University of Toronto,
4925 Dufferin St., Toronto, Ontario M3H 5T6 Canada

Abstract. A Newton-Krylov algorithm is presented for analysis and optimization of single- and multi-element airfoil configurations. The algorithm provides fast and robust convergence for the compressible Navier-Stokes equations with a one-equation turbulence model. Objective function gradients are efficiently computed using the preconditioned generalized minimum residual method to solve the discrete adjoint equations. A number of key features of the algorithm are highlighted. Several examples are given showing the effectiveness of the algorithm in high-lift system design, multi-point design, and generation of a Pareto front.

1 Introduction

Engineering design is a creative process that involves innovative thinking, setting and balancing priorities, and weighing trade-offs. Challenges are present that require designer intervention and cannot be fully automated. In contrast, optimization is a mathematical task that can be performed by a computer. Given constraints and objectives defined by the designer, an optimal design can hopefully be found within a specified design space. Whether the optimal design is actually useful depends on the skill and ingenuity of the designer in posing the optimization problem.

Optimization tools can be of great assistance in the design process. Instead of searching the design space, the designer can concentrate on modifying the objectives and constraints until the desired goals are achieved. The optimization tool is unlikely to provide the most useful design based on the initial problem specification. Rather, the objectives and constraints will evolve as the design progresses, based on feedback from the optimization tool.

In aerodynamic design, efficient performance must be achieved over a range of operating conditions, including Mach number, Reynolds number and angle of attack. This leads to both multi-objective and multi-point optimization problems. For example, a wing must provide efficient take-off, climb, and landing performance with high-lift devices extended and efficient cruise performance with such devices retracted. The multi-objective nature of the problem stems from the individual objectives associated with take-off, climb, landing, and cruise performance. These can be combined into a composite objective function based on an overriding objective, such as direct operating cost, or they can be considered as competing objectives. In the latter case, it can be instructive to generate a non-inferior solution set or Pareto front. The multi-point nature of the optimization problem arises from the fact that a single evaluation of a composite objective function requires multiple flow solutions for different geometric configurations, e.g. slat and flap settings, and different operating conditions, e.g. Reynolds and Mach numbers.

Once an aerodynamic optimization problem has been posed by defining an objective, constraints, and design variables, there are several different ways to formulate and solve the problem. Constrained optimization problems are often solved using sequential quadratic programming methods [1]. Alternatively, the constraints can be added as a penalty to the objective function, leading to an unconstrained formulation [1]. Constraints can also be lumped using the Kreisselmeier-Steinhauser function [2]. Unconstrained problems can be solved efficiently using a quasi-Newton algorithm together with a line search [3].

Optimization algorithms can be classified as derivative-free or gradient-based. Genetic algorithms [4] are a popular example of the former, which are often slow to converge but potentially provide a global optimum. Gradient-based algorithms typically converge more rapidly but to a local optimum. The gradient can be calculated using finite differences or the complex step approach [5] at a cost of at least one nonlinear flow solution per design variable. Computing the gradient using flow sensitivities requires the solution of one linear system per design variable [6]. The adjoint approach [7,5] is even more efficient, requiring the solution of one linear system independent of the number of design variables. In the continuous adjoint method, the adjoint equations are determined from the governing partial differential equations and then discretized. In the discrete adjoint method, the adjoint equations are determined from the discrete flow equations. Gunzburger [6] provides an excellent description of the various strategies. Giles [9] concludes that the discrete adjoint approach has a number of advantages over the continuous approach. Linearization of the discrete equations can be performed by hand or using automatic differentiation [10]. Refs. 11-14 provide examples of adjoint methods applied to aerodynamic optimization.

The efficiency of an adjoint-based optimization technique is dependent on fast, reliable algorithms for flow solution and gradient computation. Although several suitable algorithms exist for solving the compressible Navier-Stokes equations, Newton-Krylov algorithms [15,16] are particularly well-suited to aerodynamic optimization, since the flow Jacobian and the Krylov solver can be used in the solution of both the flow equations and the adjoint equation. Furthermore, during the design iterations, the flow solution is typically initialized from a nearby state, thus bypassing the sometimes problematic start-up phase of the Newton-Krylov algorithm.

In this paper, we describe an efficient Newton-Krylov algorithm for aerodynamic analysis and optimization based on the discrete adjoint approach. The governing equations are the compressible Navier-Stokes equations with the one-equation turbulence model of Spalart and Allmaras [17]. Aspects of the algorithm that are particularly critical to its performance are emphasized. Examples are shown demonstrating the efficiency of the flow solver and the gradient computation. Optimization examples are presented that demonstrate the capabilities of the algorithm for high-lift system design, multi-point design, and generation of a Pareto front. Further details can be found in Refs. 14 and 18-20.

2 Newton-Krylov Algorithm for the Compressible Navier-Stokes Equations

In this section, we describe our Newton-Krylov algorithm for the solution of the two-dimensional thin-layer compressible Navier-Stokes equations. The reader is referred to Zingg et al. [16] for a discussion of various strategies that are possible within a Newton-Krylov framework. The present algorithm is an extension of that of Pueyo and Zingg [15] to multi-block grids, matrix-based numerical dissipation, and a field-equation turbulence model. While the basic implementation is relatively straightforward, there are a number of subtle aspects of the algorithm that are critical to achieving fast convergence.

2.1 Inexact-Newton Strategy

The coupled system of nonlinear algebraic equations arising from a spatial discretization of the governing equations can be written as

$$R(Q) = 0, \tag{1}$$

where Q is the vector of conservative variables at every node of the grid. The residual vector, $R(Q)$, includes the steady Navier-Stokes equations, the turbulence model equation, the boundary condition equations and the interface conditions for multi-block grids. An inexact-Newton iteration is given by

$$||R(Q_n) + A(Q_n)\Delta Q_n|| \leq \eta_n ||R(Q_n)||, \tag{2}$$

where A is the Jacobian of R, $\Delta Q_n = Q_{n+1} - Q_n$, Q_n is the current solution, and Q_{n+1} is the updated solution. The Jacobian includes a linearization of boundary and interface conditions, which should be ordered within the blocks such that the diagonal entries are as large as possible. Although some functions, such as the absolute value, cannot be differentiated, it is important that the residual be linearized as accurately as possible. The generalized minimal residual (GMRES) algorithm [21] is used to solve the linear system to the tolerance prescribed by the parameter η_n at each Newton (or outer) iteration. A matrix-free implementation of the matrix-vector products needed by GMRES avoids the issue of nondifferentiable functions and produces an accurate linearization [15]. Although quadratic convergence can be obtained with an appropriate sequence of η_n values, this is not particularly important in order to achieve fast convergence in terms of computing time. It is more critical to avoid oversolving. A value of η_n equal to 0.1, which means that the residual of the linear problem is reduced by one order of magnitude, is typically appropriate. The inexact-Newton approach is very effective once the solution is within the region of convergence of Newton's method. During the iterations prior to reaching the region of convergence, which we refer to as the start-up phase, various strategies can be used to ensure convergence. The convergence history of a Newton-Krylov algorithm can thus be divided into two distinct phases, the start-up phase, which is dominated by nonlinear issues, and the inexact-Newton phase, which is dominated by linear issues.

2.2 Solving the Linear System

Due to the ill-conditioning of the Jacobian matrix, preconditioning is needed in order to obtain reliable convergence of the GMRES algorithm. The preconditioning strategy is a key component of an efficient algorithm. A preconditioner based on a block incomplete lower-upper factorization with some fill (ILU(p), where p controls the level of fill) is very effective. Note that the factorization can be based on an approximation to the exact Jacobian matrix, A, without affecting the outer iterations, as long as the exact Jacobian is retained in Eq. 2. In order to achieve at least second-order accuracy with a dissipative scheme, the spatial discretization must involve a stencil of at least five points, i.e. the Jacobian matrix must include entries associated with distance-2 neighbours. In contrast, the Jacobian matrix associated with a first-order upwind spatial discretization includes only entries corresponding to distance-1 neighbours. In order to save memory, some researchers use a preconditioner based on the approximate Jacobian corresponding to a first-order spatial discretization. However, the resulting preconditioner is less effective, and both speed and robustness are adversely affected [13,22].

Our ILU(p) factorization is based on neither the exact Jacobian nor the approximate Jacobian corresponding to a first-order upwind spatial discretization. Instead an intermediate approximate Jacobian is used to form the preconditioner. The spatial discretization uses second-order centered differences together with second and fourth-difference matrix artificial dissipation using a pressure switch to turn off the second-difference dissipation everywhere except near shocks (see Nemec and Zingg [23] for details). In our intermediate Jacobian, the coefficient of the second-difference dissipation, ϵ_2^l, is calculated from the following expression, which combines the second- and fourth-difference coefficients in the residual function, ϵ_2^r and ϵ_4^r:

$$\epsilon_2^l = \epsilon_2^r + \sigma\epsilon_4^r, \tag{3}$$

where σ is a user-defined parameter. The coefficient of fourth-difference dissipation is set to zero in the approximate Jacobian. Hence the approximate Jacobian matrix includes only entries associated with distance-1 neighbours. Fig. 1 shows the effect of the parameter σ on convergence for a transonic flow over the RAE 2822 airfoil. The angle of attack is 2.31°, the Mach number is 0.729, the Reynolds number is 6.5 million, and the grid has roughly 15,000 nodes with an off-wall spacing of 2×10^{-6} chords. The residual of the mass conservation equation is plotted versus CPU time on an Alpha 667 MHz EV67 processor. Based on these and other results, we typically use $\sigma = 5$ or 6. Note that $\sigma = 12.5$ roughly corresponds to a first-order upwind Jacobian and is much less effective than lower values of σ. Pueyo and Zingg [15] have shown that the present approach with $\sigma = 5$ leads to a much better preconditioner than the exact Jacobian, because

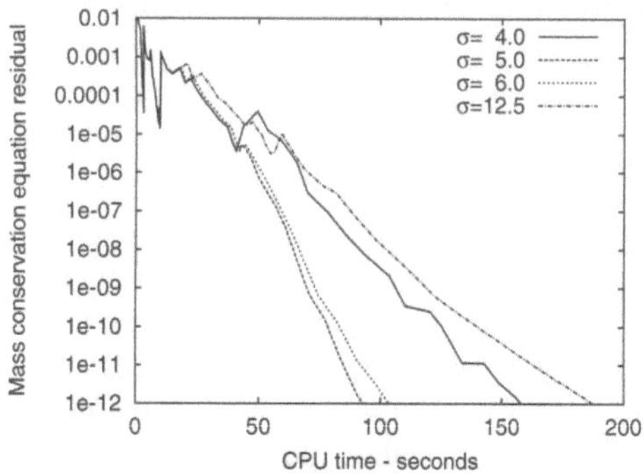

Fig. 1. Effect of the parameter σ

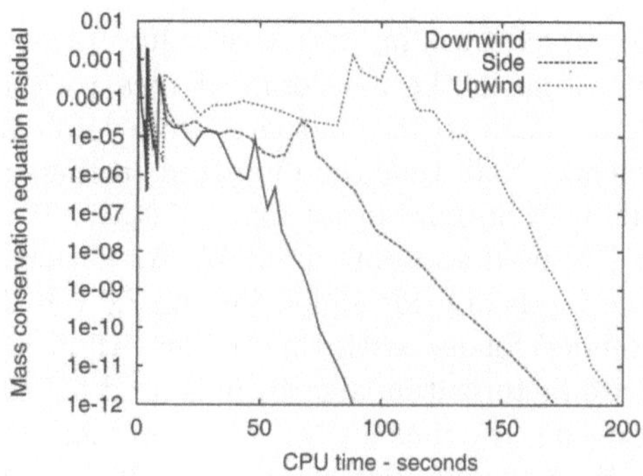

Fig. 2. Effect of the root node location in the RCM ordering

the latter produces a poorly conditioned factorization. Fill values of 2 to 4 in the incomplete factorization are optimal, with $p = 4$ somewhat more robust.

The ordering of the unknowns can have a significant impact on the convergence of the preconditioned GMRES algorithm, primarily through its effect on the incomplete factorization [24]. The reverse Cuthill-McKee (RCM) reordering, originally developed for direct methods, has been shown by numerous authors to be effective in this context as well. However, it is less widely recognized that the choice of the root node in the RCM ordering can also be important. If the RCM ordering runs from downstream to upstream, convergence can be hampered. This is illustrated in Fig. 2, which shows three convergence histories, resulting from RCM orderings obtained with root nodes lying on the downwind, upwind, and side boundaries of the computational domain for a subsonic flow over the NACA 0012 airfoil. The angle of attack is 6°, the Mach number is 0.3, the Reynolds number is 9.0 million, and the grid has roughly 17,000 nodes with an off-wall spacing of 10^{-6} chords. Note that the root node becomes the last node in an RCM ordering. Hence the use of a root node on the downwind boundary is effective because it ensures that an ordering which runs against the primary flow direction is avoided.

2.3 The Start-Up Phase

In order to ensure global convergence of the inexact-Newton algorithm, a start-up strategy is required. Several effective strategies have been devised, including damping the inexact-Newton updates [25], introducing a finite time step, i.e. using a time-linearized implicit Euler method[1] [26], using an approximate-Newton approach in which an approximate Jacobian is used to drive the outer iterations [28], grid sequencing, and using another solver for the initial iterations. For inviscid and laminar flows, the start-up phase is generally not problematic. Field-equation turbulence models introduce an additional degree of difficulty. The Spalart-Allmaras turbulence model can become unstable when the eddy viscosity becomes negative. Avoiding negative values of eddy viscosity is thus an important consideration in the development of a robust start-up strategy. Failure of the boundary-layer to trip properly from laminar to turbulent flow can also cause divergence. Note that the increased stiffness associated with the k–ω turbulence model is even more problematic [29].

Pueyo and Zingg [15] used an approximate-factorization algorithm in diagonal form [30] with two levels of grid sequencing during the start-up phase. This approach can be extended for use with the Spalart-Allmaras turbulence model [31]. If the approximate-factorization algorithm is used to reduce the residual by three orders of magnitude, the inexact-Newton algorithm can be initiated and converges reliably [14]. Although this strategy is effective and reliable, it has some drawbacks. It requires considerable extra programming and is not applicable to unstructured grids. Hence we also seek a start-up strategy that is more consistent with the overall Newton-Krylov framework.

Our alternative start-up strategy is motivated primarily by the difficulties associated with the Spalart-Allmaras turbulence model. In order to avoid negative values of the eddy viscosity, the turbulence model Jacobian is modified such that it becomes an M matrix. This precludes the use of a matrix-free implementation of the matrix-vector products needed by the GMRES algorithm. However, this is not detrimental in the context of optimization, since the full Jacobian must be formed and stored in order to solve for the adjoint variables. A finite time step is used which increases as the residual decreases. The time step for the mean flow is spatially varying based on the grid metric Jacobian using the formula given by Pulliam [30]. The time step for the turbulence model is proportional to the square of the distance to the nearest wall, since the destruction term in the turbulence model, which is the source of most instabilities, is inversely proportional to the square of this distance. The residual of the linear problem is reduced by four orders of magnitude during the start-up phase. This is not as detrimental as it would appear, since the modification to the turbulence model Jacobian and

[1] Note that Newton's method is recovered from the time-linearized implicit Euler method in the limit as the time step goes to infinity [27].

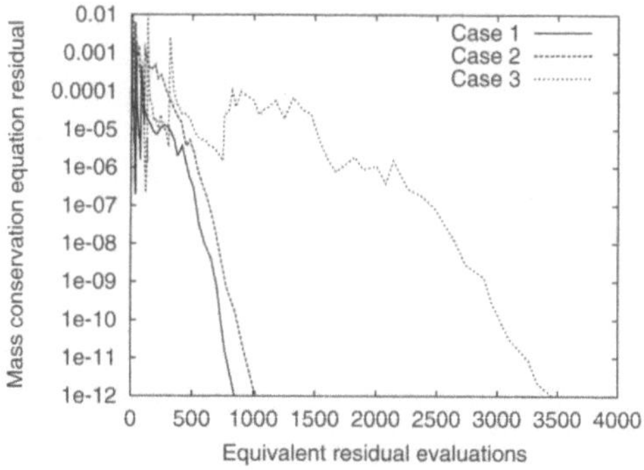

Fig. 3. Convergence histories for three cases

the finite time step produce a well-conditioned linear problem that converges rapidly.

Grid sequencing is an important component of the start-up strategy. Using the trip term in the Spalart-Allmaras turbulence model, a number of iterations can be required before tripping actually occurs. This can be expensive on a fine grid, but can be accomplished rapidly on a twice-coarsened grid, which has sixteen times fewer nodes. In order to increase efficiency and robustness, scalar artificial dissipation is used on the coarse grids. The residual is reduced by two orders of magnitude on each grid. Once the fine grid residual is reduced by two orders, the start-up phase is complete, and the inexact-Newton phase driven by the exact Jacobian begins. Since the pressure switch used to add first-order numerical dissipation near shock waves is difficult to linearize, matrix-vector products are calculated using the matrix-free approach in order to provide an accurate linearization for transonic cases.

2.4 Flow Solver Convergence Results

Fig. 3 shows the convergence histories obtained using the Newton-Krylov algorithm for three cases. Cases 1 and 2 are the subsonic flow over the NACA 0012 airfoil and the transonic flow over the RAE 2822 airfoil with the conditions and grids described previously. Case 3 is a three-element high-lift configuration, configuration A2 in Moir [32]. The angle of attack is 20.18°, the Mach number is 0.197, and the Reynolds number is 3.52 million. The grid has roughly 72,000 nodes and an off-wall spacing of 10^{-6} chords. The accuracy and grid dependence of the numerical solution for this high-lift configuration is discussed by De Rango and Zingg [33] and Rumsey and Ying [34]. In Fig. 3, the norm of the residual of the mass conservation equation is plotted versus "equivalent residual evaluations," which is the CPU time normalized by the CPU time required for a single evaluation of the complete residual vector, $R(Q)$. The equivalent residual evaluation provides a work unit that permits comparisons of different algorithms independent of the computing hardware used. The CPU times for a

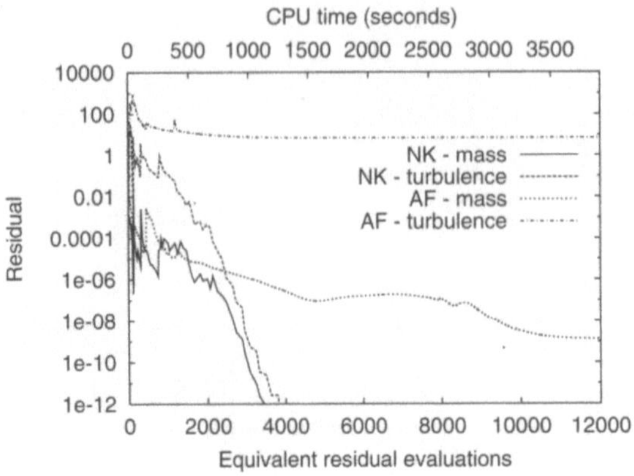

Fig. 4. Convergence histories for case 3: Newton-Krylov algorithm (NK) and approximate factorization algorithm (AF)

residual evaluation on the Alpha 667 MHz EV67 processor are 0.11, 0.091, and 0.325 seconds for the three cases, respectively. Therefore, the time required to converge the single-element cases is about 90 seconds, while the three-element case converges in under 20 minutes on a single processor.

Fig. 4 compares the convergence of the Newton-Krylov algorithm with that obtained using the approximate factorization algorithm in diagonal form [30]. CPU time is shown in both seconds and equivalent residual evaluations. The conservation of mass and turbulence model residuals are shown separately. The convergence of the turbulence model stalls using the approximate factorization algorithm as a result of a few points near the edges of recirculation bubbles behind the slat, the cove on the main body, and at the trailing edge of the flap. Nevertheless, the flow is converged elsewhere, and the two solvers give the same converged solution. The Newton-Krylov algorithm converges further and more rapidly for this complex turbulent flow.

3 Application to Aerodynamic Optimization

A gradient-based algorithm for aerodynamic optimization must include the following components: a geometry representation in terms of a finite number of design variables, a means of perturbing the grid to conform to a modified geometry, a flow solver, an algorithm for computing the objective function gradient, a method for incorporating constraints, and an optimizer. In order to represent the geometry, we use a B-spline parameterization. Due to the local nature of these curves, one can freeze some portions of the airfoil geometry while modifying other portions. Design variables are the vertical co-ordinates of the B-spline control points. Angle of attack and the relative position of flaps and slats can provide additional design variables. Details of the geometry representation as well as the grid perturbation strategy for single- and multi-block grids can be found in Nemec and Zingg [14] and Nemec et al. [20].

3.1 Formulation

Constraints, such as minimum thickness requirements, are cast as inequality constraints and added to the objective function as a penalty term. Thus the objective function, J, becomes

$$J = J_{\mathrm{d}} + J_{\mathrm{p}}, \tag{4}$$

where J_{d} is the design objective, and J_{p} the penalty term arising from the constraints. The penalty term is given by

$$J_{\mathrm{p}} = \omega_{\mathrm{p}} \sum_{j=1}^{N_{\mathrm{c}}} C_j, \tag{5}$$

where N_{c} is the number of constraints, and ω_{p} is a user-specified weight for the penalty term. The individual constraints are written as

$$C_j = \begin{cases} [1 - h(z_j)/h^*(z_j)]^2 & \text{if } h(z_j) < h^*(z_j) \\ 0 & \text{otherwise} \end{cases} \tag{6}$$

where $h^*(z_j)$ is a minimum allowable thickness at a location given by z_j. Similar constraints are applied to limit the movement of individual elements in a high-lift system.

Using the penalty approach, the optimization problem is unconstrained and can be solved using the BFGS quasi-Newton method with a backtracking line search [3]. Each step of the line search requires both a flow solution and a gradient calculation. A property of the unconstrained formulation is that the gradient vanishes at the optimum solution. Thus if the gradient can be reduced by several orders of magnitude, this provides a quantitative indication that the solution has converged to at least a local optimum.

3.2 Computing the Gradient

The variables defining the geometry, including the B-spline control points, coordinates of individual elements, and the angle of attack, are collected into the vector of design variables, X. With the introduction of the design variables, the residual equation, Eq. 1, becomes

$$R(Q, X) = 0. \tag{7}$$

This equation defines the flow variables, Q, for specified values of the design variables, X, thus implicitly defining $Q(X)$. Note that the grid is perturbed each time the design variables are modified. Hence the grid point locations are functions of the design variables, and the grid sensitivities are implicitly included in the formulation.

The gradient of the objective function, $J(X, Q(X))$, is computed from the following equation:

$$\frac{\mathrm{d}J}{\mathrm{d}X} = \frac{\partial J}{\partial X} - \psi^{\mathrm{T}} \frac{\partial R}{\partial X}, \tag{8}$$

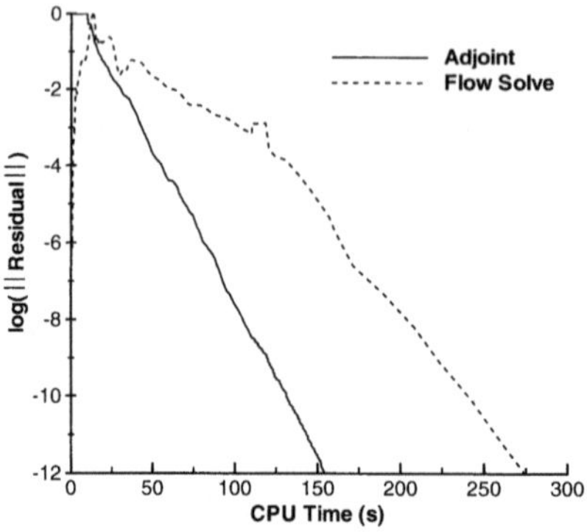

Fig. 5. Comparison of adjoint and flow solver convergence

where ψ is the vector of adjoint variables. The adjoint variables are determined by solving the linear system

$$\frac{\partial R}{\partial Q}^{\mathrm{T}} \psi = \frac{\partial J}{\partial Q}^{\mathrm{T}}. \tag{9}$$

The matrix $\frac{\partial R}{\partial Q}^{\mathrm{T}}$ is the transpose of the Jacobian of the discrete flow equations. As a result of the transpose, the matrix-free implementation of GMRES cannot be used. The partial derivative terms in Eqs. 8 and 9, $\frac{\partial J}{\partial X}$, $\frac{\partial R}{\partial X}$, and $\frac{\partial J}{\partial Q}$, are generally straightforward and inexpensive to compute, either analytically or using finite differences.

The adjoint equation is solved using the same preconditioned GMRES algorithm used for the flow solution with $\sigma = 6$ and $p = 6$. Fig. 5 shows that the adjoint solution converges roughly twice as fast as the flow solution for a two-element optimization problem. This contrasts markedly with other solution approaches in which the adjoint problem converges more slowly than the flow problem. In order to provide sufficiently accurate gradients and flow solutions, we typically converge the adjoint problem three orders of magnitude and the flow problem ten orders of magnitude, so the gradient computation cost is about one-fifth of the flow solution cost.

Gradient accuracy can be assessed by comparison with centered-difference gradient approximations. Nemec and Zingg [14] show that the gradients calculated using the discrete adjoint method are very accurate for subsonic flows, but a small error is introduced for transonic flows due to difficulties in differentiating the pressure switch. The error in the discrete-adjoint gradients does not affect the converged solution and has a minimal effect on convergence of the optimization.

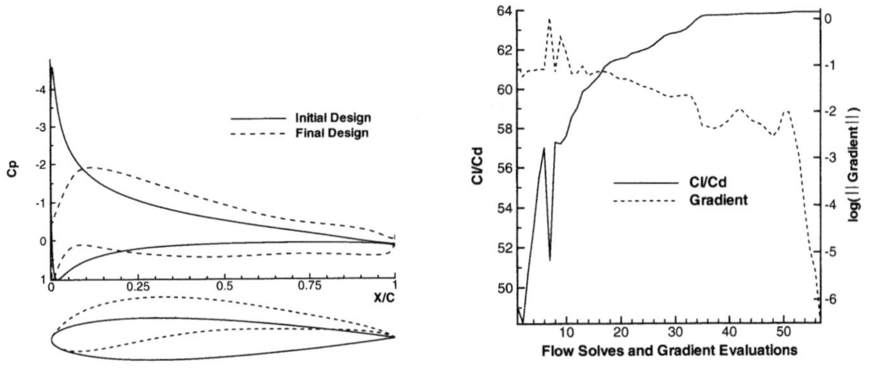

(a) Airfoils and C_p

(b) Convergence history

Fig. 6. Maximization of lift-to-drag ratio

3.3 Optimization Examples

Maximization of Lift-to-Drag Ratio: Figs. 6(a) and 6(b) show the results of an optimization in which a single-element airfoil is calculated that maximizes the lift-to-drag ratio. There are ten design variables, including nine B-spline control points and the angle of attack. The airfoil thickness is constrained to be at least 11% at the 35% chord location and at least 1.6% at the 90% chord location. The Mach number is 0.25 and the Reynolds number is 2.88 million. The initial condition is the NACA 0012 airfoil at an angle of attack of 9°, which produces a lift-to-drag ratio of 49. As shown in Fig. 6(b), after 57 flow solves and gradient evaluations, the gradient is reduced by six orders of magnitude, and the lift-to-drag ratio is roughly 64. The angle of attack is 4.3°. For practical purposes, the optimization is converged after 40 flow solves and gradient evaluations. Each flow solution is initialized from the preceding solution and requires roughly 80 seconds to meet the convergence criterion. Hence the optimization converges in just over one hour. This particular design is quite heavily constrained, not so much by the thickness constraints, but by the limited number of design variables used to define the shape. Increasing the number of design variables leads to more optimal designs but possibly with undesirable features not conducive to efficient off-design performance.

Optimizing a Two-Element Configuration: Aspects of high-lift system design are reviewed by van Dam [35]. Here we consider a relatively simple problem in which we modify the position of a flap to increase the lift coefficient without increasing the drag coefficient. The design objective function is given by

$$J_{\mathrm{d}} = \begin{cases} \omega_{\mathrm{L}} \left(1 - \frac{C_{\mathrm{L}}}{C_{\mathrm{L}}^{*}}\right)^2 + \omega_{\mathrm{D}} \left(1 - \frac{C_{\mathrm{D}}}{C_{\mathrm{D}}^{*}}\right)^2 & \text{if } C_{\mathrm{D}} > C_{\mathrm{D}}^{*} \\ \omega_{\mathrm{L}} \left(1 - \frac{C_{\mathrm{L}}}{C_{\mathrm{L}}^{*}}\right)^2 & \text{otherwise} \end{cases} \tag{10}$$

Table 1. Gap-overlap optimization summary

Design	C_L	C_D	Gap	Overlap
Initial	2.145	0.04720	2.40	-5.31
#1	2.165	0.04687	1.99	-4.28
#2	2.173	0.04677	1.95	-3.30
Final	2.175	0.04675	2.02	-2.68
Target	2.180	$\leq$ 0.0472		

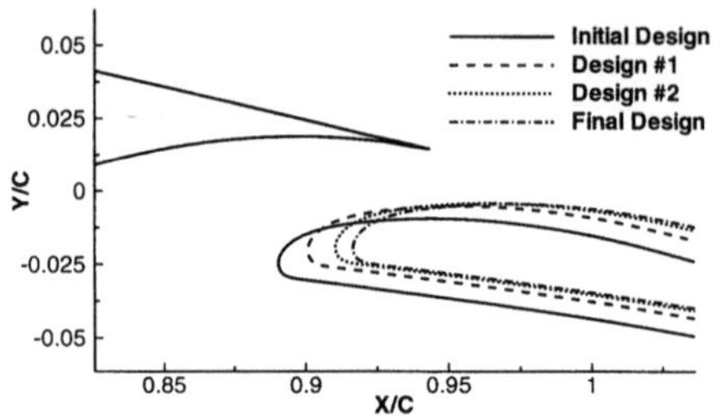

Fig. 7. Evolution of flap position

with $C_L^* = 2.180$, $C_D^* = 0.04720$, and weights, ω_L and ω_D, equal to unity. The angle of attack is 4°, the Mach number is 0.25, and the Reynolds number is 2.51 million. The initial values of C_L and C_D are 2.145 and 0.04720, respectively. Both the angle of attack and the flap deflection angle are held fixed. The design variables are the horizontal and vertical displacement of the flap. In order to avoid excessive grid distortion for the multi-block grids, the flap movement is limited to maximum changes in gap and overlap of 0.5% chord and 1.0% chord, respectively.

Table 1 shows the progress of the optimization. From the initial gap and overlap positions, the optimization converges to Design #1 with the flap movement constraint active. A new grid is generated, and the process is repeated, leading to Design #2, again with the flap movement constraint active. A new grid is again generated and the final converged design is achieved with the flap movement constraint inactive. The drag is actually lower than the original value, so the drag component of the objective function is inactive. Fig. 7 shows the progress of the flap position from the initial design to the final design.

The final design is well converged, with a reduction in the gradient of over five orders of magnitude. This ensures that a local minimum has been found, but a global minimum is not guaranteed. In order to examine the uniqueness of this optimum, we repeat the optimization with a completely different initial flap position. The results are shown in Fig. 8, which shows the previously described optimization labelled G24-O53 and the second optimization labelled G29-O05. The identical optimum flap position is found in both cases.

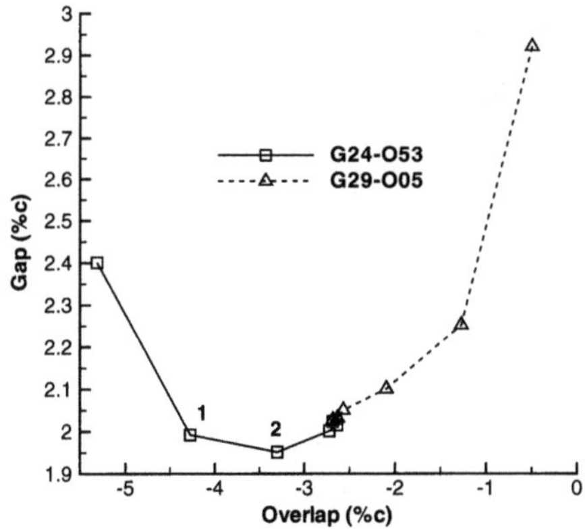

Fig. 8. Convergence to optimal flap position from distinct initial conditions

A Pareto Front: Next we consider a simple example involving competing objectives given by

$$J_{\mathrm{L}} = \left(1 - \frac{C_{\mathrm{L}}}{C_{\mathrm{L}}^{*}}\right)^{2}, \qquad J_{\mathrm{D}} = \left(1 - \frac{C_{\mathrm{D}}}{C_{\mathrm{D}}^{*}}\right)^{2} \tag{11}$$

with $C_{\mathrm{L}}^{*} = 0.55$, and $C_{\mathrm{D}}^{*} = 0.0095$. The design variables are 10 free B-spline control points and the angle of attack. The Mach number is 0.7; the Reynolds number is 9 million. Three thickness constraints are imposed. Within this design space, the value of C_{L}^{*} can be attained, but the value of C_{D}^{*} cannot. These two objectives are competing since increasing the lift coefficient generally increases the drag coefficient, and vice-versa.

A design is said to be *Pareto-optimal* if there exists no other feasible design in the design space which is better or equal with respect to all of the objectives [4]. For convex fronts, the weighted-sum method can be used to generate the Pareto front. The objective function is defined as

$$J = \omega_{\mathrm{L}} J_{\mathrm{L}} + (1 - \omega_{\mathrm{L}})\, J_{\mathrm{D}} + J_{\mathrm{p}} \tag{12}$$

The Pareto front shown in Fig. 9 is found by minimizing J for several different values of the weight ω_{L}. Also shown are two airfoils corresponding to different points on the front. Due to the use of a gradient-based optimization algorithm, this is a local Pareto front. However, we have compared with the front generated using a genetic algorithm, which finds a global minimum, and the two fronts are identical [20]. The Pareto front provides a clear picture of the trade-off between lift and drag for this particular set of design variables and constraints.

A Multi-Point Design Example: Finally, we consider transonic drag minimization at fixed lift [36]. The design objective function is given by Eq. 10, with $C_{\mathrm{L}}^{*} = 0.733$, $C_{\mathrm{D}}^{*} = 0.013$, $\omega_{\mathrm{L}} = 1$, and $\omega_{\mathrm{D}} = 0.1$. The Reynolds number is 9 million. The angle of attack is free. There are 19 free control points

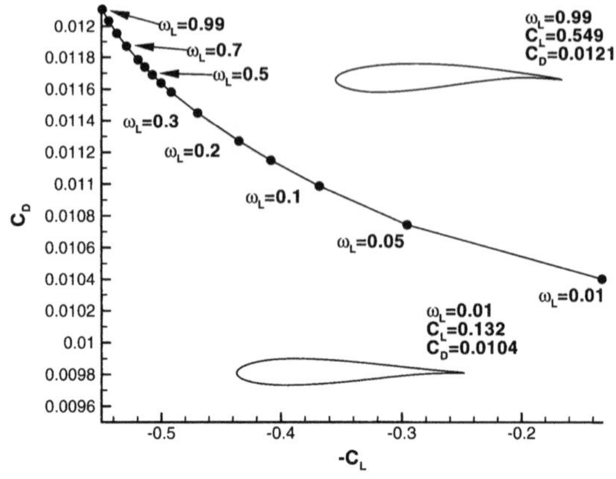

Fig. 9. Pareto front for two competing objectives

and three thickness constraints. Results of a single-point drag minimization at a Mach number of 0.74 are shown in Fig. 10(a) in comparison with the RAE 2822 airfoil. Fig. 10(b) shows the drag coefficient as a function of the Mach number at fixed lift. The drag coefficient has been greatly reduced at the design Mach number, but it has been increased at Mach numbers below 0.71.

Using multi-point optimization, the drag-divergence Mach number can be increased without a penalty at lower Mach numbers. Fig. 11 shows the results of an optimization performed using a four-point composite objective function based on the weighted-sum method. The design Mach numbers are 0.68, 0.71, 0.74, and 0.76. Although the resulting airfoil has a somewhat higher drag coefficient at a Mach number of 0.74 than that achieved by the single-point optimization, the drag at Mach numbers below 0.74 is greatly reduced. Although the airfoil is still not suitable for practical use, since off-design angles of attack also need to be incorporated into the composite objective function, this example demonstrates the applicability of the Newton-Krylov algorithm to practical multi-point optimization.

4 Conclusions and Recommendations

Several examples have been presented demonstrating the efficiency and applicability of the Newton-Krylov algorithm for practical aerodynamic optimization. Further work is needed in two key areas. The validity of the computed optimum airfoils and high-lift configurations needs to be assessed. A quantitative understanding is required of the sensitivity of the optimum shapes to numerical resolution and accuracy of physical models. For example, is the turbulence model sufficiently accurate that the optimum flap position determined in our two-element example can be confirmed experimentally? Further work is also needed in applying the algorithm to practical design problems, which are characterized by

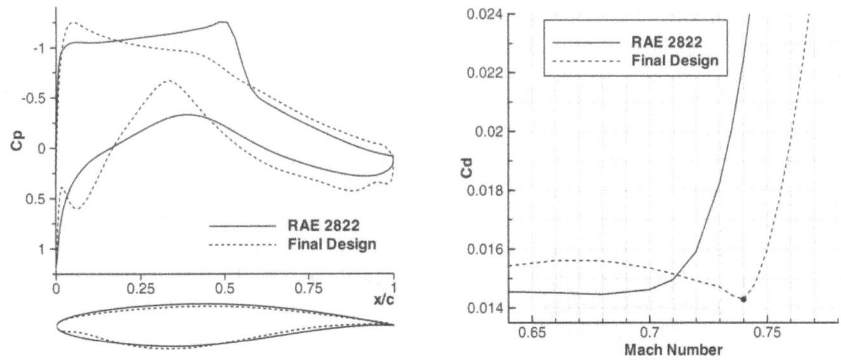

(a) Airfoils and C_p, $M = 0.74$ (b) Drag coefficient

Fig. 10. Single-point drag minimization at fixed lift

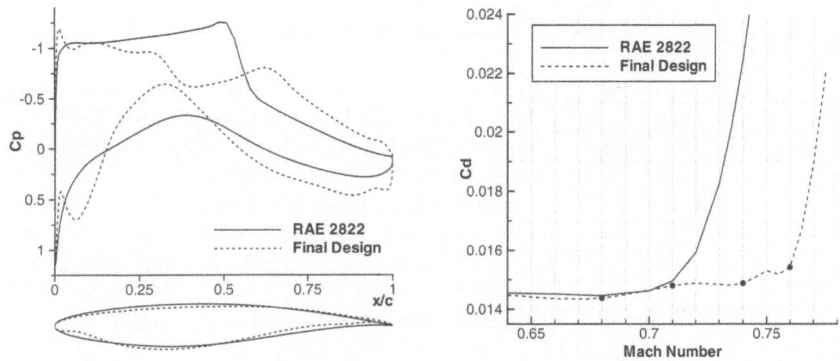

(a) Airfoils and C_p, $M = 0.74$ (b) Drag coefficient

Fig. 11. Four-point drag minimization at fixed lift

complex constraints and a large number of operating points. Assessment of the algorithm in this context is necessary to guide further development.

References

1. J. Nocedal, S.J. Wright: *Numerical Optimization*, Springer-Verlag, New York, 1999.
2. G.A. Wrenn: "An Indirect Method for Numerical Optimization Using the Kreisselmeier-Stienhauser Function," NASA CR-4220, March 1989.
3. J.E. Dennis Jr., R.B. Schnabel: *Numerical Methods for Unconstrained Optimization and Nonlinear Equations*, Prentice-Hall, Englewood Cliffs, NJ, 1983.
4. D.E. Goldberg: *Genetic Algorithms in Search, Optimization, and Machine Learning*, Addison-Wesley, 1989.

5. W.K. Anderson, J.C. Newman, D.L. Whitfield, E.J. Nielsen: AIAA J., **39**(1):56-63, 2001.
6. M.D. Gunzburger: "Introduction into Mathematical and Physical Aspects of Flow Control and Optimization," von Karman Institute for Fluid Dynamics Lecture Series 1997-05, Belgium, 1997.
7. O. Pironneau: J. Fluid Mech., **64**:97–110, 1974.
8. A. Jameson: J. Sci. Comput., **3**:233–260, 1988.
9. M.B. Giles, N.A. Pierce: Flow, Turbulence and Combustion **65**:393–415, 2000.
10. B. Mohammadi: "Practical Applications to Fluid Flows of Automatic Differentiation for Design Problems," von Karman Institute for Fluid Dynamics Lecture Series 1997-05, Belgium, 1997.
11. J. Reuther, A. Jameson, J.J. Alonso, M.J. Remlinger, D. Saunders: Parts 1 and 2, J. Aircraft, **36**(1):51–74, 1999.
12. J. Elliott, J. Peraire: Aeronautical J., **102**:365–376, 1998.
13. E.J. Nielsen, W.K. Anderson: AIAA J., **40**(6):1155–1163, 2002.
14. M. Nemec, D.W. Zingg: AIAA J., **40**(7):1146–1154, 2002.
15. A. Pueyo, D.W. Zingg: AIAA J., **36**(11):1991–1997, 1998.
16. D.W. Zingg, S. De Rango, A. Pueyo: "Advances in Algorithms for Computing Aerodynamic Flows," in *Frontiers of Computational Fluid Dynamics 2002*, D.A. Caughey, M.M. Hafez, eds., World Scientific, Singapore, 2002.
17. P.R. Spalart, S.R. Allmaras: "A One-Equation Turbulence Model for Aerodynamic Flows," AIAA 92-0439, Jan. 1992.
18. M. Nemec, D.W. Zingg: "From Analysis to Design of High-Lift Configurations Using a Newton-Krylov Algorithm," Paper 173, ICAS 2002, September 2002.
19. T. Chisholm, D.W. Zingg: "A Fully-Coupled Newton-Krylov Solver for Turbulent Aerodynamic Flows," Paper 333, ICAS 2002, September 2002.
20. M. Nemec, D.W. Zingg, T.H. Pulliam: "Multi-Point and Multi-Objective Aerodynamic Shape Optimization," AIAA 2002-5548, September 2002.
21. Y. Saad, M.H. Schultz: SIAM J. on Sci. and Stat. Comput., **7**(3):856–869, 1986.
22. T.J. Barth, S.W. Linton: "Unstructured Mesh Newton Solver for Compressible Fluid Flow and Its Parallel Implementation," AIAA 95-0221, Jan. 1995.
23. M. Nemec, D.W. Zingg: AIAA J., **38**(3):402–410, 2000.
24. M. Benzi, D.B. Szyld, A. Van Duin: SIAM J. Sci. Comput. **20**(5):1652–1670, 1999.
25. P.R. McHugh, D.A. Knoll: AIAA J., **12**(12):2394–2400, 1994.
26. W.A. Mulder, B. van Leer: J. Comp. Phys., **59**:232-246, 1985.
27. H. Lomax, T.H. Pulliam, D.W. Zingg: *Fundamentals of Computational Fluid Dynamics*, Springer-Verlag, Berlin, Heidelberg, 2001.
28. M. Blanco, D.W. Zingg: AIAA J., **36**(4):607–612, 1998.
29. P. Guezaine: AIAA J., **39**(3):528–531, 2001.
30. T.H. Pulliam: "Efficient Solution Methods for the Navier-Stokes Equations," von Karman Institute for Fluid Dynamics Lecture Series, Belgium, Jan. 1986.
31. P. Godin, D.W. Zingg, T.E. Nelson: AIAA J., **35**(2):237-243, 1997.
32. I.R.M. Moir: "Measurements on a Two-Dimensional Aerofoil with High-Lift Devices," AR 303, AGARD, Aug. 1994.
33. S. De Rango, D.W. Zingg: AIAA J., **39**(7):1296–1304, 2001.
34. C.L. Rumsey, S.X. Ying: Prog. in Aero. Sciences, **38**: 145-180, 2002.
35. C.P. van Dam: Prog. in Aero. Sciences, **38**: 101-144, 2002.
36. M. Drela: "Pros & Cons of Airfoil Optimization," in *Frontiers of Computational Fluid Dynamics 1998*, D.A. Caughey, M.M. Hafez, eds., World Scientific, Singapore, 1998.

Studying Bypass Transition to Turbulence by Computer Simulation

P. A. Durbin and G. Kalitzen

Mechanical Engineering Department
Stanford University
Stanford, CA 94305-3030, U.S.A.

1 Introduction

Transition from laminar to turbulent flow is one of the more intriguing of fluid phenomena. It has been approached by instability theory. When transition proceeds from inflectional shear flow velocity profiles that is a sound conception. The breakdown to turbulence is the endpoint that evolves from initial instability waves. However, in boundary layers that scenario seems uncommon. Tollmein-Schlicting waves are rarely the precursors of transition, except in carefully controlled wind tunnel experiments. Tollmein-Schlichting waves have very small growth rates and are usually preempted by other mechanisms. External disturbances often initiate such processes. This scenario is then called bypass transition.

A transition Reynolds number can be defined by a plot of skin friction (C_f) versus R_θ as in figure 1. At low Reynolds number the data and DNS follow the laminar, Blasuis curve. As the Reynolds number increases, they rise above the laminar level, eventually transitioning to the turbulent curve. The transition Reynolds number Re_t is defined by the minimum of the actual skin friction curve.

Transition Reynolds number has been measured experimentally as a function of free-stream turbulence intensity. The data correlation $R_\theta = 163 + e^{6.91-Tu}$ [1] is displayed in figure 2. Tu here is the turbulent intensity in percent: $Tu = 0.01u'/U_\infty$. Tollmein-Schlichting wave growth rates combined with e^N transition estimates suggest that when Re_t is less than about 10^3 bypass transition is predominant. Thus when $Tu \lesssim 0.5\%$ Tollmein-Schlichting waves probably play a role in transition; but at higher levels they cannot be important. When $Tu \gtrsim 3\%$ transition occurs near the the Reynolds number (200) at which Tollmein-Schlichting instability just occurs, which is well upstream of where it could cause transition. Hence something else is at play.

The term 'bypass' is non-specific. Experiments have shown that elongated, contours of negative and positive velocity perturbation occur in the boundary layer beneath free-stream turbulence [6]. These have been called Klebanoff modes. Experiments also show that in the transition region (see figure 1) the flow contains patches of turbulence within a laminar background. The patches are called turbulent spots — although their form differs from the classic Em-

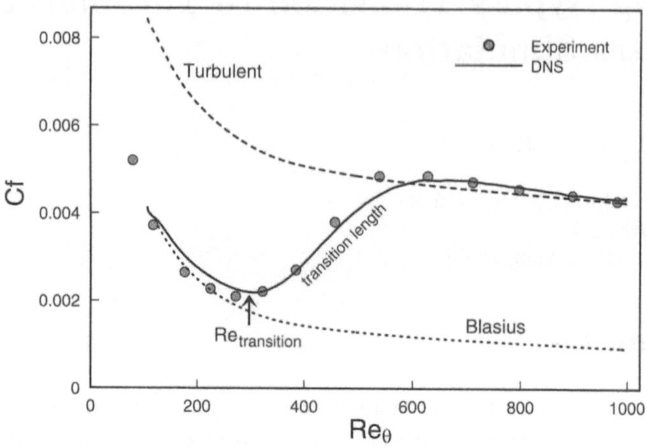

Fig. 1. Skin friction curve, illustrating transition terminology.

mons, kidney shaped, spots. Hence, the transition region is not an intermediate state, rather it is an intermittent zone within which fully turbulent patches exist.

A further understanding of bypass transition has been obtained by computer simulations [4], [7]. These showed that elongated low speed regions lift away the wall to form isolated, backward jets in the perturbation velocity field. To the extent that they are isolated and rise to the top of the boundary layer, they are distinct from Klebanoff modes. The elevated jets are perturbed by the free-stream turbulence. They develop instabilities that grow quickly, fill the boundary layer, and eventually are observed next to the wall as turbulent spots.

Several aspects of bypass transition make it well suited to direct numerical simulation (DNS). By definition, it occurs at the lowest Reynolds number at which turbulence is observed: the computational expense of DNS rises nearly as the cube of Reynolds number, so it is best suited to lower Reynolds phenomena. Even then, on the order of 50 million grid points are needed for the present problem.

The transition length is much shorter than that needed for Tollmein-Schlichting waves to grow, become three-dimensional and break down. Hence the full range, from laminar to fully turbulent flow fits into a reasonable computational domain. We have also found that an outflow sponge region is not needed, unlike stability wave studies. A convective outflow condition is used.

In this article we briefly describe the computational algorithm, then we will discuss current simulations of a turbine passage, followed by results from a fundamental study of transition on a flat plate boundary layer caused by incident wakes.

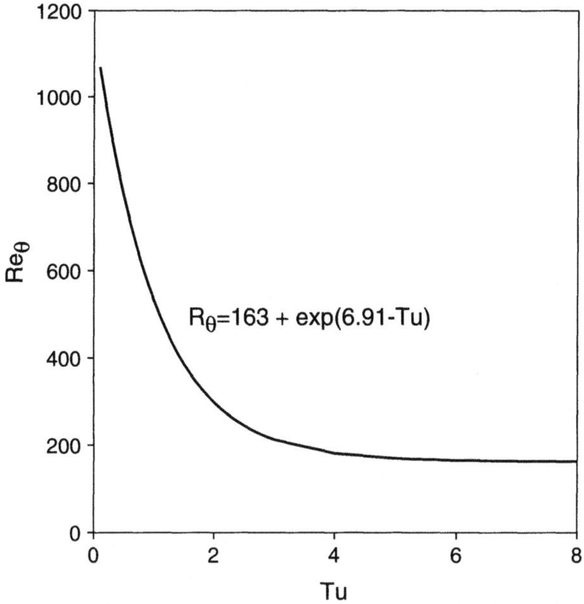

Fig. 2. Empirical formula for transition Reynolds number in the presence of free-stream turbulence.

2 Computational method

The incompressible Navier-Stokes equations were integrated by the method of [5]. The grid is curvilinear in two directions and uniformly spaced, Cartesian in the third. The latter is because the flow is statistically homogeneous in that direction. A direction of homogeneity increases the sample size and makes the DNS less costly.

A staggered grid, finite volume, semi-discrete form of the momentum equations is

$$V\partial_t \boldsymbol{u} = \sum_{faces} \boldsymbol{S}^l \cdot (\boldsymbol{T}_l - \boldsymbol{u}\boldsymbol{u}) \qquad (1)$$

where $\boldsymbol{S}^l$ is the area vector of face l and V is the elemental cell volume. $\boldsymbol{T}$ is the stress tensor. Following [5], define

$$\boldsymbol{u} = \sum_{l=1,2,3} \boldsymbol{S}_l U^l \qquad (2)$$

where U^l are the volume fluxes and $\boldsymbol{S}_l$ is the inverse area vector. In two dimensions $\boldsymbol{S}^1 = (S_x^1, S_y^1)$ and

$$\boldsymbol{S}_1 = (S_y^2, -S_x^2)/(S_x^1 S_y^2 - S_y^1 S_x^2)$$

where the superscripts indicate the face number, not exponentiation. Hence $\boldsymbol{S}_l \cdot \boldsymbol{S}^k = \delta_{kl}$. Substituting (2) into (1) and dotting with $\boldsymbol{S}^1$ gives the evolution equation

$$V\partial_t U^1 = \sum_{l=1,2,3} \boldsymbol{S}^l \cdot \boldsymbol{T}_l \cdot \boldsymbol{S}^1 - U^l U^1 \tag{3}$$

for the 1-component of the flux vector. Similar formulas can be derived for the other component. Thus the dependent variables are the flux vectors in the face normal direction. Cartesian velocities are computed when post-processing the results.

Each volume flux is stored on the appropriate grid of face centers; this is the motive for working with flux vectors. The i-direction flux is stored at the points $(i \pm {}^1/_2, j, k)$ of the cell with center (i, j, k); and similarly for the other components. Staggered grids remain popular for incompressible DNS.

The stress tensor in (3) contains the pressure, $-P\sum_{l=1,2,3} \boldsymbol{S}^l \cdot \boldsymbol{S}^1$ as well as the linear, viscous stress. The fractional step method consists of evaluating P at the nth time level when U^1 is advanced provisionally to the $n+1^{st}$ level. Then the pressure is corrected to enforce solenoidality and the velocity is corrected [2]. In the present case, the viscous stress is treated by Crank-Nicholson and the convective flux on the right side of (3) is evaluated at $n + 1/2$ by the 2-level Adams-Bashforth formulation. This provides second order temporal accuracy to the velocity field.

The pressure correction is obtained from a Poisson equation. In the Cartesian direction it is solved by fast Fourier transform. This produces a set of Helmholtz equations that are solved by line Gauss-Seidel, with multi-grid acceleration.

3 Turbine passage

Boundary conditions were imposed by adding ghost cells. In the turbine passage (figure 3) a portion of the boundary consists of the solid, no-slip walls. The portions upstream and downstream of the blade are periodic. The latter is implemented by copying data from the first interior cell on the opposite side of the passage into the ghost cell. Although the mean flow should satisfy this periodicity, the turbulence should not. As is usual in DNS, we assume that the correlation length of the eddies is short enough for the artifice of periodicity to be innocuous.

Although these simulations are still in progress, we will present initial results for three inlet conditions. The first is turbulence free. In that case a simulation with incident wakes was restarted with no inlet disturbance; hence, it started with a disturbance in the free-stream, that washed out, leaving behind a flow with no free-stream turbulence, but with a self-sustained transition near the trailing edge.

The second inflow consists of periodically passing wakes, as explained in [7] and shown in figure 4. These consist of fully-developed, turbulent wakes obtained from a separate simulation. The wakes are swept across the inlet at velocity $-V_c$. Let the velocity in the frame of the wake generator be $(U, 0) + (u_w, v_w)$ where U

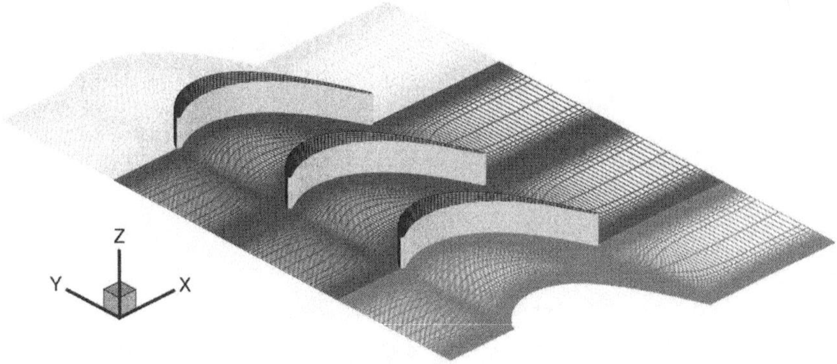

Fig. 3. Grid for turbine simulation, shown at coarsened scale and extended periodically for display.

is a uniform oncoming flow and $\boldsymbol{u}_w$ is the turbulence plus the mean wake defect. Then if the free-stream velocity in the blade frame is (U_∞, V_∞) the inlet velocity is

$$(U_\infty, V_\infty) + (u_w \cos\beta - v_w \sin\beta, v_w \cos\beta + u_w \sin\beta) \qquad (4)$$

where $\tan\beta = (V_\infty + V_c)/U_\infty$. (u_w, v_w) are obtained from a time-evolving wake simulation. Time-dependent inflow turbulence is obtained by sweeping through the frozen field in the streamwise direction.

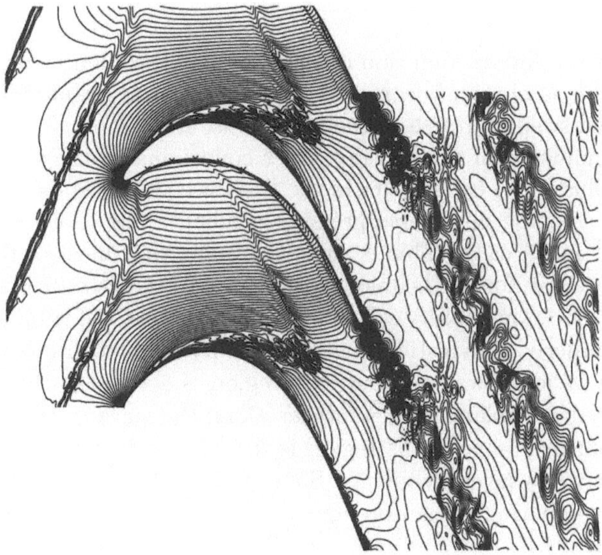

Fig. 4. Magnitude of instantaneous velocity from DNS.

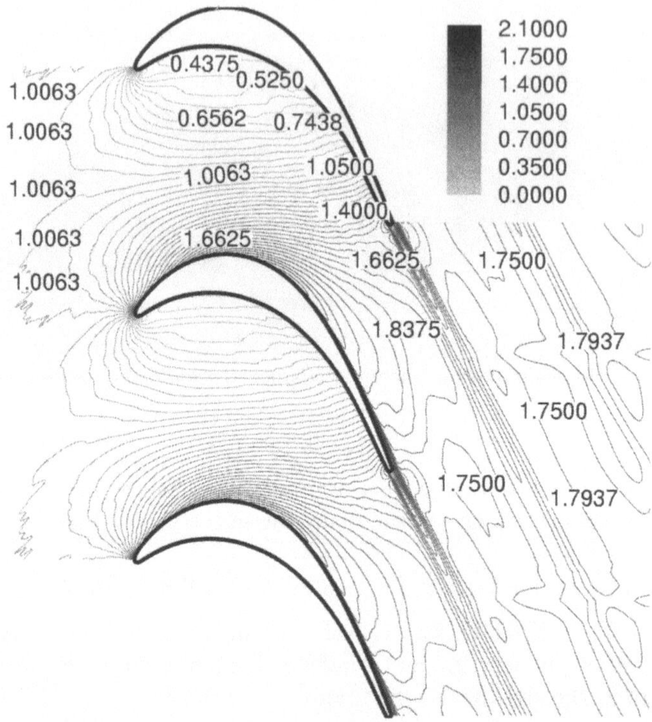

Fig. 5. Turbulent intensity with grid turbulence at inlet.

The third inflow condition is a field of grid turbulence, carried with the flow. Again, this is obtained from a separate, temporal simulation that is frozen in time, then swept through in the streamwise direction.

The influence of the different inlet conditions on the state of the boundary layer is illustrated by figure 6. In the first case transition occurs near the trailing edge, in a region of adverse pressure gradient. The chord Reynolds number is 1.5×10^5 in all cases. This is sufficiently low that most of the blade would be laminar were it not for the inlet disturbance. In the absence of free-stream disturbance (case 1), instability waves are observed near the trailing edge, suggesting that an orderly, Kelvin-Helmholtz instability leads to transition.

With incident wakes a buffeted region is seen near the leading edge. The boundary layer relaminarizes further along the surface; indeed, it is likely that the turbulence seen in this region is driven by the incident wake, rather than being boundary layer turbulence. Shortly past midspan boundary layer transition begins. Comparing to the top figure, it is clear that the instability wave route is bypassed.

It can be seen that bypass transition occurs in the lower most case of figure 6 as well. The incident grid turbulence seems to produce a straighter transition line. At this stage it is unclear what causes the qualitative difference between the second and third cases. Clearly the intermittent wake lets the turbulent zone

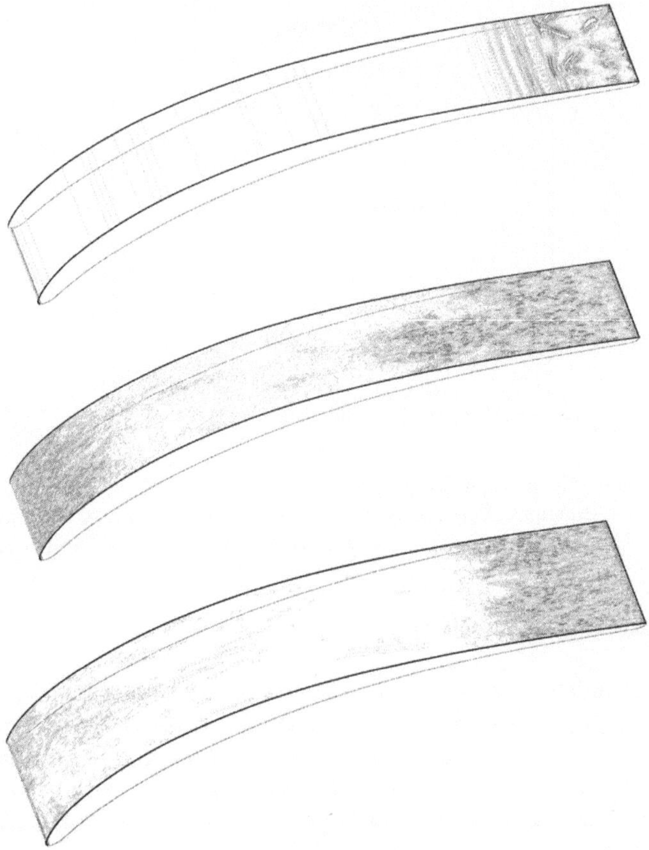

Fig. 6. Contours of normal velocity component, showing transitition. Top: no inlet disturbance; middle, incident wakes; lower incident grid turbulence.

convect toward the trailing edge before being reinvigorated by the subsequent wake. This might be why the transition region has a different appearance when the blade is continuously perturbed by grid turbulence.

Skin friction distributions are presented in figure 7. They show that the middle section of the suction side is unaffected by the incident disturbances, but transition moves away from the trailing edge ($x = 1$ in the figure) when the wake or grid turbulence is introduced.

The irregularities in the solid curve in figure 7 are due to non-uniformities in the grid. DNS has been found to be quite sensitive to grid stretch in other studies [3]. Here the grid cells are skewed in the passage and seem to give rise the an analogous mesh error. Mesh errors peculiar to DNS are associated with distortion of vortices convected through the grid; from a computational perspective mesh stretch errors are wave-number dependent [3].

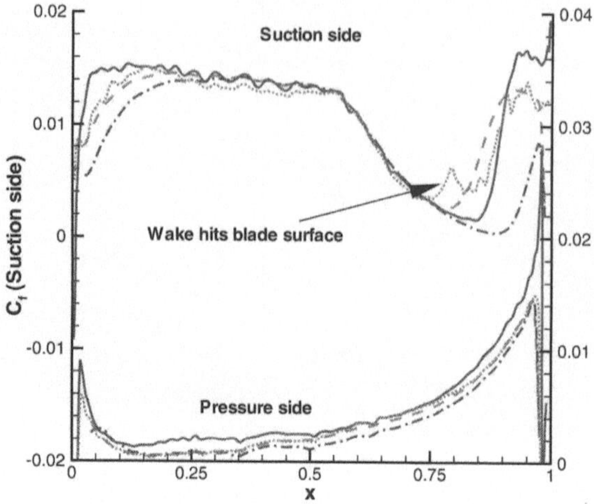

Fig. 7. Averaged skin friction around blade for the three cases: — · —, no incident disturbance; - - -incident wake turbulence; ——grid turbulence. An extra curve is included that shows the instantaneous skin friction for the wake impingement.

4 Fundamental studies

A theoretical understanding of some portion of the bypass transition problem comes from consideration of the continuous spectrum of the Orr-Sommerfeld equation [4]. These modes form a basis for representing free-stream disturbances. They have the remarkable property of nearly vanishing within the boundary layer, as illustrated in figure 8. That figure is a full Navier-Stokes simulation of a spatially evolving mode. It is seen to be oscillatory in the free-stream, hence is similar to a Fourier mode.

The paradox raised by figure 8 is that the lack of disturbance inside the boundary layer implies that external disturbances have no path to trigger transition. However, very low frequency modes can enter the boundary layer. That is illustrated by figure 9. Two modes are superimposed in the free stream. The mean shear seems to filter out the higher frequency component and only the low frequency mode penetrates the boundary layer.

Direct numerical simulation has furthered our fundamental understanding of the process of bypass transition. Because of the simple geometry and relatively low Reynolds number, the number of grid points and computational time required to capture the development of eddies is not prohibitive. The simulations in this section were on a 70 million point grid and required about 350 CPU hours on 64 processors of an SGI origin 2000.

A full field of turbulence was simulated by extension of figure 9: a sum of three-dimensional modes, with random phases and amplitudes chosen from a turbulence spectrum was imposed at the inlet to a computational box: see [4] for further details. The low frequency components induce long streaks of low or high speed flow near the wall, caused by displacement of mean momentum. The

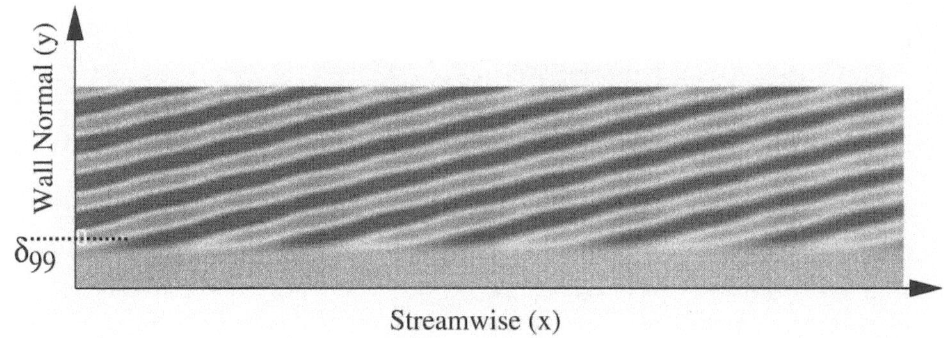

Fig. 8. A mode of the continuous spectrum

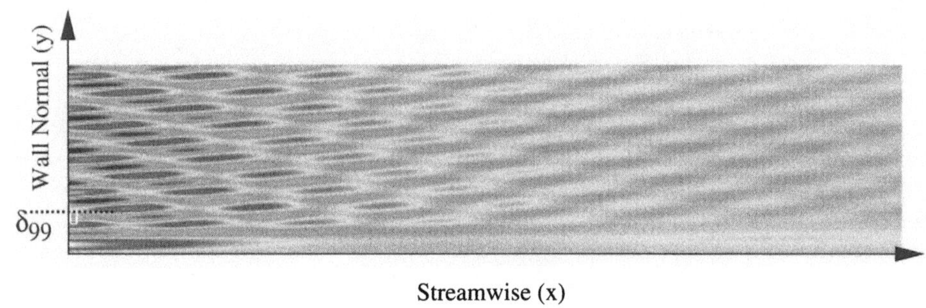

Fig. 9. Two modes of the continuous spectrum, one being of very low frequency

velocity in these streaks is in the streamwise direction; so they can be though of as forward and backward jets in the perturbation field. Their existence in perturbed laminar flow has been known for a long time, originally being referred to as breathing modes by G.I. Taylor. They do not directly cause transition.

Figure 10 shows contours of the perturbation u field in the upper frame. Long streaks are seen in the laminar region. They extend upstream. This particular view is in the transitional region, with the fully turbulent boundary layer starting at the right. An isolated patch of turbulence, upstream of the fully turbulent zone, is captured in the view; it is seen more clearly in the v-component contours in the lower portion of the figure.

The observation of turbulent patches next to the wall has been a source of some confusion. They do not appear to be due to instability of the jets; indeed, the v contours suggest that they appear spontaneously.

Direct numerical simulations allow us to track these patches back to their origin by examining data at earlier times. The times in figure 11 are prior to figure 10. They identify the origin of the turbulent patch as an instability near the top of the boundary layer. It appears that backward jets, associated with lift-up of low momentum fluid, approach the top of the boundary layer, where

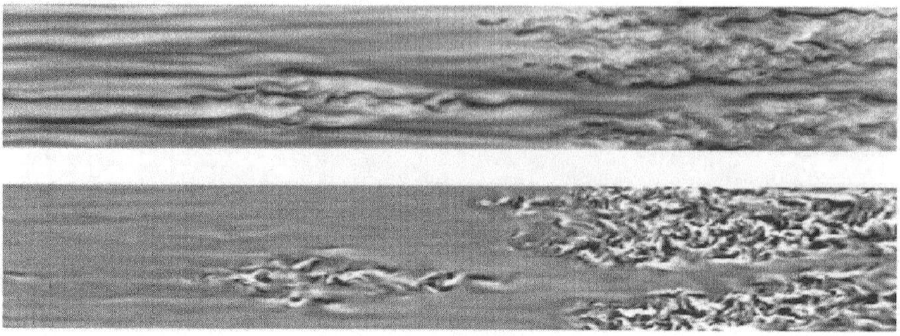

Fig. 10. Contours of u (above) and v (below) in an $x - z$ plane near the wall.

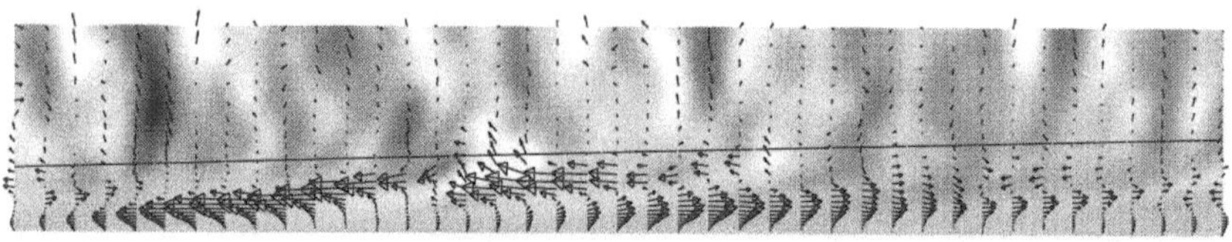

t/T=-0.35

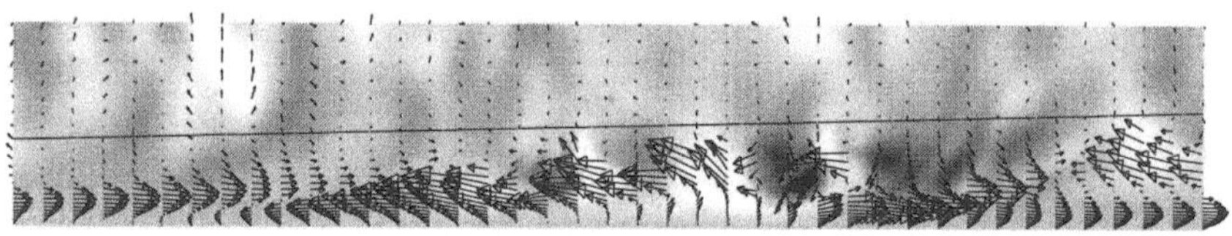

t/T=-0.05

Fig. 11. Velocity vectors in a $x - y$ plane at two times. The field of view shifts right with time.

they are perturbed by shorter wavelength, free-stream turbulence. The long jets are unstable and develop undular perturbations that grow rapidly, filling the boundary layer, and finally being seen as a turbulent spot near the wall. This process is discussed fully in [4] and [7].

Acknowledgement Sponsorship by the Department of Energy ASCI-ASAP program under contract B341491, by NSF, and by the Office of Naval Research is gratefully acknowledged.

References

1. B.J. Abu-Gannam, R. Shaw: J. Mech. Eng. Sci. **22**, 213 (1980)
2. J. Ferziger, M. Peric: *Computational methods for fluid dynamics*, (Springer-Verlag, 2000)

3. T. Ikeda: *Direct simulation of a rough wall channel*, Ph. D. thesis, Aerospace Engineering, Stanford University
4. R.G. Jacobs, P.A. Durbin: J. Fluid Mech. **428**, 185 (2000)
5. M. Rosenfeld, D. Kwak, M. Vinokur: J. comp. Phys. **94**, 102 (1991)
6. K.J Westin, A.V. Boiko, B.G. Klingmann, V.V Kozlov, P.H. Alfredsson: J. Fluid Mech. **281**, 193 (1994)
7. X.Wu, R. Jacobs, J.C.R. Hunt, P.A. Durbin: J. Fluid Mech. **398** 109 (1999)

A Parallel Non-Staggered Navier-Stokes Solver Implemented on a Workstation Cluster

S. Armfield[1], S. Norris[2], P. Morgan[1] and R. Street[3]

[1] The University of Sydney, Sydney, NSW 2006, Australia
[2] Bioengineering Institute, University of Auckland, New Zealand
[3] Environment Fluid Mechanics Laboratory, Stanford University, USA

Abstract. Non-staggered, fractional step Navier-Stokes solvers are presented and analysed in comparison to a standard staggered solver. It is shown analytically that the non-staggered schemes have the same order of accuracy as the staggered scheme, although one of the schemes is non-elliptic in the pressure field at the grid scale. Numerical results are used to validate the analysis, with only the non-elliptic non-staggered scheme showing a reduced accuracy and efficiency compared to the staggered scheme. Details of a computer cluster built using Pentium 4 processors with a fast ethernet switch are presented. A serial benchmark demonstrates the excellent performance of the P4 for large array size compute intensive jobs, while both MPI and HPF coded parallel benchmarks show the effective parallel performance of the cluster.

1 Introduction

The simulation of complex three-dimensional flows requires substantial computer resources and accurate and efficient solvers. The purchase of large scale turn-key machines is beyond the ability of many small Computational Fluid Dynamics (CFD) facilities. A far cheaper alternative is the use of work-station clusters. Such facilities are comprised of standard off the shelf desk top computers linked via a switched network. Work-station clusters are effectively distributed memory parallel computers, and some work is required to convert a standard serial code to make efficient use of such a facility. In this paper details of a cluster constructed from Pentium 4 machines and a 100MBits/second fast ethernet network are presented. Both message passing and data parallel approaches are considered and it is shown that for an array based structured code both approaches provide approximately the same performance, with the data parallel approach requiring considerably less code modification.

As well as suitable computer facilities three-dimensional flow simulation also requires efficient and accurate Navier-Stokes solvers. In recent years one of the more popular methods for the solution of the Navier-Stokes equations is the fractional step method, which is a single step time-integrator, typically implemented in second-order form [1]. Details of a non-staggered Navier-Stokes solver are presented below and it is shown that the non-staggered solver provides equivalent accuracy and efficiency to a standard staggered grid fractional step code. Converting the non-staggered code to run on the workstation cluster using the data parallel approach is relatively straightforward and benchmark results are presented showing that excellent results are achieved.

2 Navier-Stokes Code

The Navier-Stokes equations in non-dimensional and incompressible form are as follows,

$$u_t + (u \cdot \nabla)u = -\nabla P + \tfrac{1}{Re}\nabla^2 u, \tag{1}$$
$$\nabla \cdot u = 0, \tag{2}$$

where u is the velocity, P the pressure and Re is the Reynolds number. Using a fractional step method the momentum equations are first integrated in time with Adams-Bashforth for the advective terms and Crank-Nicolson for the viscous terms, giving

$$\frac{v^* - v^n}{\Delta t} + \left[\frac{3}{2}H(v^n) - \frac{1}{2}H(v^{n-1})\right] = -Gp^{n-1/2} + \frac{1}{2Re}L(v^{n+1} + v^n), \tag{3}$$

with superscript n representing the time step, and v and p the discrete velocity and pressure, respectively, H the discrete advection operator, G the discrete gradient and L the discrete Laplacian. The pressure is located in time at the $n - 1/2$ level, obtained from the previous time-step integration, and v^* will therefore not be divergence free and is an approximation to the correct divergence free v^{n+1} velocity. The v^* field is corrected to provide a divergence free v^{n+1} as,

$$v^{n+1} = v^* - \Delta t G\pi,$$

with π a pressure correction obtained from the following Poisson equation,

$$L\pi = Dv^*/\Delta t,$$

where D is the discrete divergence. The pressure is then updated as

$$p^{n+1/2} = p^{n-1/2} + \pi$$

and the integration continues to the next time-step.

The fractional step method, as described above, has been shown to be second order in time on both a conventional MAC staggered grid and on a non-staggered grid, [2–4]. Figure 1 contains sketches of the staggered and non-staggered grids, where it is seen that on the staggered grid the velocity storage locations are offset a half grid spacing in their respective directions from the pressure nodes, while on the non-staggered grid they are coincident. The staggered grid approach has been widely used and has some advantages in the discretisation of the pressure correction equation. However it has a number of disadvantages, particularly for curvilinear grids, where non-staggered grids are commonly used [6]. On staggered grids each of the momentum equations and the continuity equation will have their own finite volumes, leading to a larger and more complex code. A non-staggered code is will therefore require less work to parallelise.

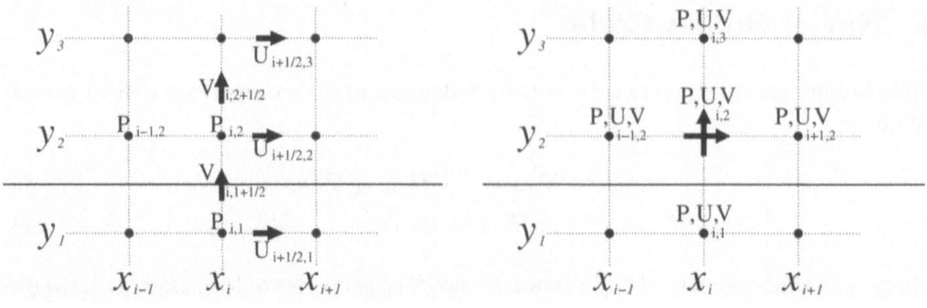

Fig. 1. Staggered grid (left) and non-staggered grid (right). Solid circles indicate pressure storage locations, horizontal arrows indicate horizontal (U) velocity locations, vertical arrows indicate vertical (V) velocity locations. The index i represents location in the x direction, the index j represents location in the y direction and a wall is located at $y_{j+1/2}$

2.1 Spatial Discretisation

The primary differences between the discretisations on the staggered and non-staggered grids occur in the pressure gradient and divergence terms, and these, together with the associated errors, will be considered in detail below. For brevity the analysis will be presented in two dimensional form on a uniform cartesian mesh.

Staggered Grid The pressure gradient terms are located at the center of their respective finite-volumes for the U and V momentum equations, that is at $(x^{i+1/2}, y^j)$ for the U-momentum equation and $(x^i, y^{j+1/2})$ for the V-momentum equation, giving,

$$p_y^{i,j+1/2} \sim \left(\frac{p^{i,j+1} - p^{i,j}}{\Delta y}\right),$$

$$p_x^{i+1/2,j} \sim \left(\frac{p^{i+1,j} - p^{i,j}}{\Delta x}\right),$$

which are $\Delta y, \Delta x$ spaced centered differences.

The velocity corrections are made using the same discrete gradient for the scalar π,

$$U^{i+1/2,j} = U^{i+1/2,j*} - \Delta t\left(\frac{\pi^{i+1,j} - \pi^{i,j}}{\Delta x}\right),$$

$$V^{i,j+1/2} = V^{i,j+1/2*} - \Delta t\left(\frac{\pi^{i,j+1} - \pi^{i,j}}{\Delta y}\right).$$

A centered $\Delta y, \Delta x$ spaced scheme is used for the divergence, that is,

$$D(U,V)^{i,j} \sim \left(\frac{U^{i+1/2,j} - U^{i-1/2,j}}{\Delta x} + \frac{V^{i,j+1/2} - V^{i,j-1/2}}{\Delta y}\right).$$

The discrete Poisson equation for π is then obtained directly by applying the discrete divergence to the velocity corrections, yielding,

$$\left(\frac{\pi^{i+1,j} - 2\pi^{i,j} + \pi^{i-1,j}}{\Delta x^2}\right) + \left(\frac{\pi^{i,j+1} - 2\pi^{i,j} + \pi^{i,j-1}}{\Delta y^2}\right) =$$
$$\frac{1}{\Delta t}\left(\frac{U^{i+1/2,j*} - U^{i-1/2,j*}}{\Delta x} + \frac{V^{i,j+1/2*} - V^{i,j-1/2*}}{\Delta y}\right). \tag{4}$$

Thus for the staggered scheme all the discretisations for the pressure and divergence terms are second order in space with $\Delta x, \Delta y$ spacing.

Non-Staggered Grid For the non-staggered grid the finite volumes for the U- and V-momentum equations and for the divergence are all located at (x^i, y^j). A $2\Delta x, 2\Delta y$ spaced centered difference is used for the pressure gradients as,

$$p_x \sim \frac{p^{i+1,j} - p^{i-1,j}}{2\Delta x},$$
$$p_y \sim \frac{p^{i,j+1} - p^{i,j-1}}{2\Delta y}$$

and similarly for the π gradient in the velocity correction

$$U^{i,j} = U^{i,j*} - \Delta t\left(\frac{\pi^{i+1,j} - \pi^{i-1,j}}{2\Delta x}\right),$$
$$V^{i,j} = V^{i,j*} - \Delta t\left(\frac{\pi^{i,j+1} - \pi^{i,j-1}}{2\Delta y}\right)$$

and for the divergence,

$$D(U,V)^{i,j} \sim \left(\frac{U^{i+1,j} - U^{i-1,j}}{2\Delta x} + \frac{V^{i,j+1} - V^{i,j-1}}{2\Delta y}\right).$$

Applying the discrete divergence to the velocity correction, as for the staggered scheme, gives a Poisson π equation of the form,

$$\left(\frac{\pi^{i+2,j} - 2\pi^{i,j} + \pi^{i-2,j}}{4\Delta x^2}\right) + \left(\frac{\pi^{i,j+2} - 2\pi^{i,j} + \pi^{i,j-2}}{4\Delta y^2}\right) =$$
$$\frac{1}{\Delta t}\left(\frac{U^{i+1,j*} - U^{i-1,j*}}{2\Delta x} + \frac{V^{i,j+1*} - V^{i,j-1*}}{2\Delta y}\right), \tag{5}$$

where it is seen that the Laplacian of π has a $2\Delta x$ and $2\Delta y$ spacing. This approach will be denoted the sparse scheme.

An alternative to obtaining the π equation via a direct application of the discrete divergence to the velocity correction is to derive the π equation first in continuous form and then discretise it using a compact $\Delta x, \Delta y$ spacing for the Laplacian, as follows,

$$\left(\frac{\pi^{i+1,j} - 2\pi^{i,j} + \pi^{i-1,j}}{\Delta x^2}\right) + \left(\frac{\pi^{i,j+1} - 2\pi^{i,j} + \pi^{i,j-1}}{\Delta y^2}\right) =$$
$$\frac{1}{\Delta t}\left(\frac{U^{i+1,j*} - U^{i-1,j*}}{2\Delta x} + \frac{V^{i,j+1*} - V^{i,j-1*}}{2\Delta y}\right), \quad (6)$$

which will be denoted the compact scheme. This approach introduces an additional discretisation error into the corrected velocity as a divergence error which has the following form,

$$Error = \Delta t \left[\Delta x^2 \left(\frac{\pi^{i+2,j} - 4\pi^{i+1,j} + 6\pi^{i,j} - 4\pi^{i-1,j} + \pi^{i-2,j}}{4\Delta x^4}\right) + \right.$$
$$\left.\Delta y^2 \left(\frac{\pi^{i,j+2} - 4\pi^{i,j+1} + 6\pi^{i,j} - 4\pi^{i,j-1} + \pi^{i,j-2}}{4\Delta y^4}\right)\right], \quad (7)$$

which is to second order,

$$Error \sim \Delta t \left[\Delta x^2 \pi_{xxxx} + \Delta y^2 \pi_{yyyy}\right].$$

As $\pi = p^{n+1/2} - p^{n-1/2} \sim \Delta t$ the compact divergence error is expected to behave as,

$$Error \sim \Delta t^2 (\Delta x^2, \Delta y^2).$$

Thus for the non-staggered scheme, as for the staggered scheme, all the pressure gradient and divergence terms are discretised using second order centered schemes.

2.2 Pressure Accuracy

As shown above the use of a non-staggered grid leads to either a sparse pressure correction equation, or to a pressure correction equation that is not derived directly from the discrete forms of the velocity correction and divergence operators. The effect of this on the accuracy and behaviour of the pressure is investigated by deriving the implied pressure equation. The pressure equation is obtained by taking the divergence of the momentum equations,

$$\frac{v^{n+1} - v^n}{\Delta t} + \left[\frac{3}{2}H(v^n) - \frac{1}{2}H(v^{n-1})\right] = -Gp^{n+1/2} + \frac{1}{2Re}L(v^{n+1} + v^n)$$
$$+ \frac{\Delta t}{2Re} LG\pi. \quad (8)$$

This is the equation that is satisfied by the v^{n+1} and $p^{n+1/2}$ obtained using the fractional step method. As can be seen it includes an additional term when compared to the discretised momentum equation (3) above, which represents the error resulting from the use of a fractional step method. As $\pi \sim \Delta t$ this error is second order in time, as has been shown previously [2–4]. Given that second order schemes are used for the advection operator H and the viscous Laplacian L, and that the velocity v^{n+1} is divergence free to second order, the overall fractional step method as described here will therefore produce a velocity field that is second order accurate in space and time.

Staggered Grid The pressure for the staggered scheme will satisfy the equation,

$$Lp^{n+1/2} = -D(\frac{3}{2}H(v^n) - \frac{1}{2}H(v^{n-1}) + \frac{\Delta t}{2Re}LG\pi). \tag{9}$$

The first and second terms on the right hand side of this equation may be represented as,

$$D(\frac{3}{2}H(v^n) - \frac{1}{2}H(v^{n-1})) \sim D(H(v^{n+1/2})) + O(\Delta t^2).$$

The third term on the right hand side, $\frac{\Delta t}{2Re}LG\pi$, is an order Δt^2 error term, as $\pi \sim \Delta t$. The pressure obtained using the fractional step method is therefore second order accurate in time and space.

Non-Staggered Grid The pressure equation for the sparse non-staggered scheme is as follows,

$$\frac{(p^{i+2,j} - 2p^{i,j} + p^{i-2,j})}{4\Delta x^2} + \frac{(p^{i,j+2} - 2p^{i,j} + p^{i,j-2})}{4\Delta y^2} = -D(\frac{3}{2}H(v^n) - \frac{1}{2}H(v^{n-1}) + \frac{\Delta t}{2Re}LG\pi), \tag{10}$$

where the $2\Delta x, 2\Delta y$ spaced discretisation of the pressure gradient and divergence operator on the non-staggered grid, gives a sparse pressure Laplace operator, as for the π pressure correction equation above in equation (5). The right hand side is again, $D(H(v^{n+1/2}))$ to second order in time and the pressure is again second order accurate in space and time. However the sparse $2\Delta x, 2\Delta y$ spaced discretisation of the pressure Laplacian means that the grid scale pressure is un-coupled, and the discrete operator is non-elliptic [7,8]. This means that grid scale error in the pressure field can accumulate without bound, leading to a grid scale oscillation that may cause the solution to break down completely.

The non-staggered scheme with compact Poisson π equation has the following pressure equation,

$$\frac{(p^{i+2,j} - 2p^{i,j} + p^{i-2,j})}{4\Delta x^2} + \frac{(p^{i,j+2} - 2p^{i,j} + p^{i,j-2})}{4\Delta y^2} = -D(\frac{3}{2}H(v^n) - \frac{1}{2}H(v^{n-1}) + \frac{\Delta t}{2Re}LG\pi) + (\Delta t^2\left[\Delta x^2 p_{xxxx} + \Delta y^2 p_{yyyy}\right])_{tt} + \frac{1}{Re}L(\Delta t^2\left[\Delta x^2 p_{xxxx} + \Delta y^2 p_{yyyy}\right])_t. \tag{11}$$

The $2\Delta x, 2\Delta y$ spaced discretisation of the pressure gradient and divergence again gives a sparse Laplace operator for p. The additional right hand side terms, when compared to equation (10) above, arise from the compact scheme divergence error. These terms are all of the form $(\Delta t^2 \Delta x^2, \Delta t^2 \Delta y^2)$ and thus the compact scheme pressure will still be second order accurate, although the magnitude of the error may have been increased. However these additional error terms have the effect of re-coupling the grid scale pressure field, as the form of the pressure fourth derivatives is as given above equation (7), which are elliptic at the grid scale. When compared to the sparse scheme above the compact scheme will therefore have the advantage of not allowing the grid scale pressure error to grow without bound, and this is the reason for using such an approach, in spite of the additional error that is introduced.

Elliptic Interpolation A further modification to the non-staggered compact scheme has frequently been used. In this approach the divergence is defined as,

$$D(U,V)^{i,j} \sim \left(\frac{U^{i+1/2,j} - U^{i-1/2,j}}{\Delta x} + \frac{V^{i,j+1/2} - V^{i,j-1/2}}{\Delta y} \right),$$

with the inter-node cell face velocities obtained via interpolation as,

$$V^{i,j+1/2} = \left(\tfrac{V^{i,j+1}+V^{i,j}}{2}\right) + \Delta t \left[\left(\tfrac{p^{i,j+2}-p^{i,j}}{2\Delta y}\right) + \left(\tfrac{p^{i,j+1}-p^{i,j-1}}{2\Delta y}\right) - \left(\tfrac{p^{i,j+1}-p^{i,j}}{\Delta y}\right) \right],$$

and

$$U^{i+1/2,j} = \left(\tfrac{U^{i+1,j}+U^{i,j}}{2}\right) + \Delta t \left[\left(\tfrac{p^{i+2,j}-p^{i,j}}{2\Delta x}\right) + \left(\tfrac{p^{i+1,j}-p^{i-1,j}}{2\Delta x}\right) - \left(\tfrac{p^{i+1,j}-p^{i,j}}{\Delta x}\right) \right].$$

This approach, which was first suggested by Rhie and Chow [6], has been used in equivalent forms in a number of schemes and has been defined as elliptic interpolation as it has effect of recovering the grid scale pressure ellipticity of the continuous equations [7,8]. The use of elliptic interpolation adds a divergence error of the form,

$$\Delta t \left[\Delta x^2 p_{xxxx} + \Delta y^2 p_{yyyy} \right],$$

which is expected to behave as $(\Delta t \Delta x^2, \Delta t \Delta y^2)$. The corresponding pressure equation will be,

$$\begin{aligned} \frac{(p^{i+1,j} - 2p^{i,j} + p^{i-1,j})}{\Delta x^2} + \frac{(p^{i,j+1} - 2p^{i,j} + p^{i,j-1})}{\Delta y^2} = \\ -D(\frac{3}{2}H(v^n) - \frac{1}{2}H(v^{n-1}) + \frac{\Delta t}{2Re}LG\pi) \\ +(\Delta t^2 \left[\Delta x^2 p_{xxxx} + \Delta y^2 p_{yyyy} \right])_{tt} + \frac{1}{Re}L(\Delta t^2 \left[\Delta x^2 p_{xxxx} + \Delta y^2 p_{yyyy} \right])_t \\ -(\left[\Delta x^2 p_{xxxx} + \Delta y^2 p_{yyyy} \right]) + \frac{1}{Re}L(\Delta t \left[\Delta x^2 p_{xxxx} + \Delta y^2 p_{yyyy} \right]). \end{aligned} \tag{12}$$

When compared to the pressure equation for the compact Poisson π equation scheme, equation (11) above, using elliptic interpolation for the cell face velocities is seen to produce a compact discrete pressure Laplace operator. This means that the grid scale pressure is fully coupled, however additional error of the form, $\Delta x^2 \Delta y^2, \Delta t \Delta x^2, \Delta t \Delta y^2$ has been introduced.

Both the compact and elliptic interpolation non-staggered schemes effectively couple the grid scale pressure into the solution, however there is a difference. The elliptic interpolation scheme was originally developed for use with iterative solvers, rather than non-iterative fractional step solvers. In the context of the present pressure correction scheme an iterative solver would repeat the momentum equation/pressure correction equation cycle at each time step until the divergence of the v^* field was less than a preset convergence criterion. This effectively means that the pressure correction π becomes very small, and the additional error terms in the compact scheme, associated with π, in equation (11), become negligible. As it is these terms that act to couple the grid scale pressure into the solution, the grid scale coupling will be lost. The elliptic interpolation approach is therefore necessary for iterative schemes as the grid scale coupling is then not dependent on the pressure correction π.

Pressure Boundary Conditions For the staggered scheme no explicit boundary conditions are required, and this is one of the advantages of this approach. For the non-staggered scheme a boundary condition is required to provide the pressure at the boundary exterior point, and this is obtained by fourth order extrapolation from the interior. For the pressure correction π in all cases the boundary condition is obtained directly from the velocity boundary condition.

3 Results

The accuracy and efficiency of the schemes described above are compared for the solution of start-up natural convection flow in a square cavity. The simulation of natural convection flow requires the solution of a temperature equation in addition to the momentum and Poisson pressure correction equations described above, together with a buoyancy term added to the vertical momentum equation. The temperature equation is discretised using the same time and space schemes as the momentum equations, and is solved prior to the momentum equations at each time step. The advective operator H uses a third order QUICK discretisation, and the diffusion Laplace operator L uses a standard second order compact discretisation. For both the staggered and non-staggered schemes the temperature values are stored at the pressure node locations. The control parameters for the natural convection flow are the Prandtl number $Pr = \nu/\kappa$ and the Rayleigh number $Ra = g\alpha\Delta T H^3/\nu\kappa$, with g gravity, α the coefficient of thermal expansion, H the height of the cavity, ν the kinematic viscosity and κ the thermal diffusivity.

Initially the fluid in the square cavity is stationary and isothermal at temperature $T = 0$. At time $t = 0$ the left and right walls are instantaneously heated and cooled to $\Delta T/2$ and $-\Delta T/2$ respectively, with the top and bottom boundaries adiabatic. All boundaries are no-slip. The results presented were obtained with $Ra = 6 \times 10^5$ and $Pr = 7.5$.

To test the behaviour of the methods the flow was integrated from $t = 0$ to $t = 2$ for a series of grid and time steps sizes as follow: $[20 \times 20, \Delta t = 0.1]$, $[40 \times 40, \Delta t = 0.05]$, $[80 \times 80, \Delta t = 0.025]$ and $[160 \times 160, \Delta t = 0.0125]$ grids. The 'error' is expressed as the L^2 norm of the difference between the test solution obtained at a given Δt and grid and a benchmark solution also integrated from $t = 0$ to $t = 2$ on a 640×640 grid with $\Delta t = 3.125 \times 10^{-3}$. Times have been non-dimensionalised using the boundary layer start-up time for the natural convection cavity. The time-step selected in each case is the largest time step at which the spatial error totally dominates the temporal error, that is a further reduction in time-step on a fixed grid leads to no change in the error, which gives a Courant number of approximately 0.1. Both the staggered and non-staggered schemes have been previously shown to have second order accuracy in time, [2–4]. The results shown below will be used to validate the spatial accuracy of the schemes.

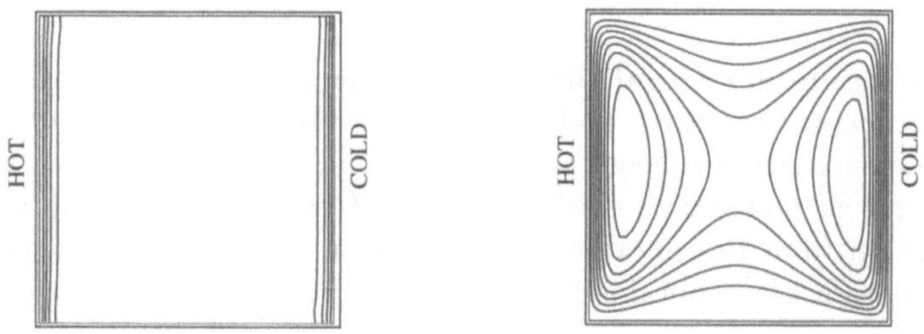

Fig. 2. Temperature contours(left) and streamfunction contours(right) for the natural convection flow

Figure 2 shows the streamfunction and temperature contours respectively for the natural convection flow at time $t = 2$. The streamfunction contours show the recirculation's that are associated with each of the thermal boundary layers that form on the heated and cooled walls. The flow continues to develop from this stage with hot and cold intrusions ejected from the boundary layers travelling across the horizontal boundaries and stratifying the interior of the domain, and the two recirculation's coalescing to form a single cavity scale recirculation.

Figure 3 shows the pressure, temperature, U and V velocity errors for each scheme, together with the difference between the staggered and the non-staggered compact schemes. For the pressure all but the non-staggered sparse scheme show second order convergence. The non-staggered sparse scheme shows a low order convergence which is assumed to be a result of the accumulation of grid-scale error. The temperature results again show that all but the non-staggered sparse scheme are exhibiting second order behaviour. The non-staggered sparse scheme, while having approximately the same error, is showing inconsistent behaviour at the finest grid size. The U-velocity results show all schemes exhibiting second

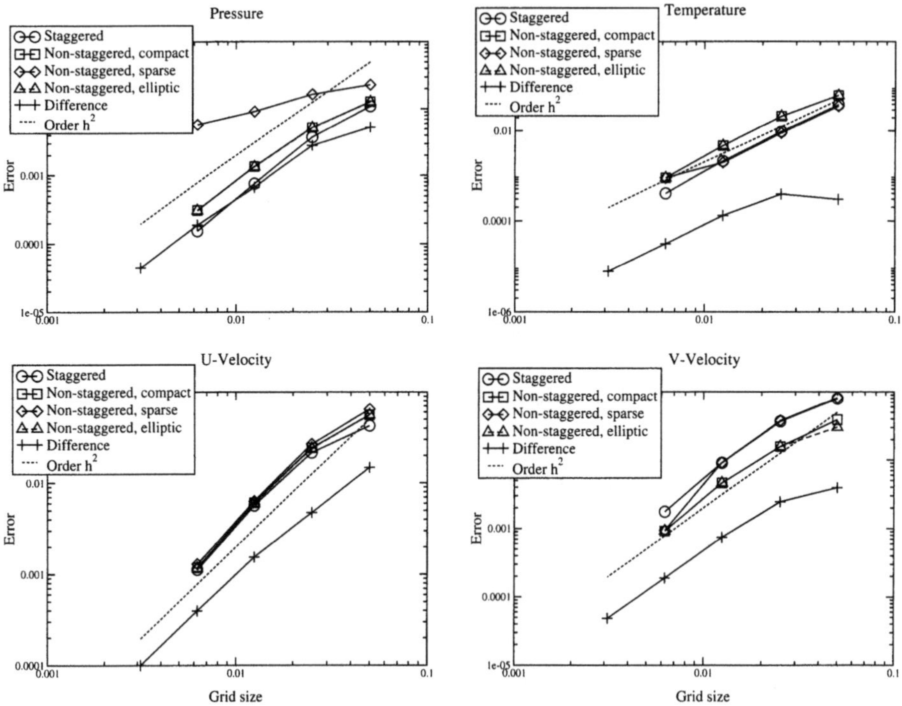

Fig. 3. Pressure accuracy(top left), temperature accuracy(top right), U-velocity accuracy(bottom left) and V-velocity accuracy(bottom right)

order behaviour, while the V-velocity results again show inconsistent behaviour for the non-staggered sparse scheme, while all other schemes are converging at second order.

Figure 4 contains the average error plotted against compute time for each of the schemes, allowing the efficiency to be compared. The non-staggered sparse scheme is clearly the least efficient of those tested, while all others are approximately equal. Overall therefore the non-staggered schemes have performed equivalently to the staggered scheme, with very similar efficiency and accuracy. The non-staggered scheme with elliptic interpolation performed approximately the same as the non-staggered compact scheme, demonstrating that the additional error terms did not significantly affect the solution. However it is also apparent that the stronger grid scale coupling of the elliptic interpolation is not necessary when using the compact scheme.

The poor performance of the non-staggered sparse scheme is attributed to the lack of grid scale pressure coupling. Figure 5 shows the pressure contours for sparse and compact non-staggered schemes at time $t = 2$, where it is clearly seen that a grid scale error component is present in the sparse scheme.

In summary the non-staggered scheme provides equivalent accuracy and efficiency to the staggered code when the compact pressure correction equation

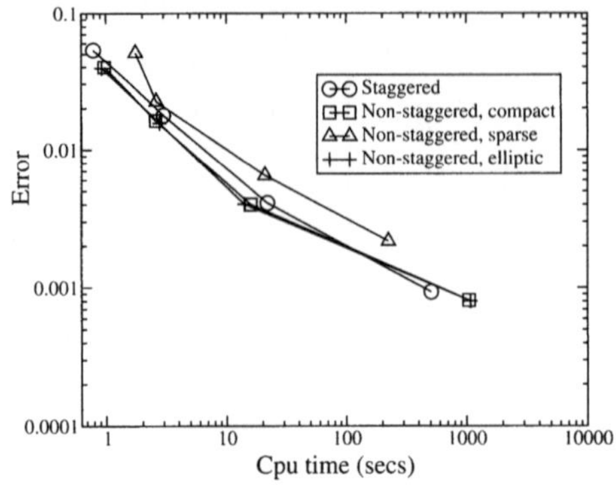

Fig. 4. Average error versus compute time.

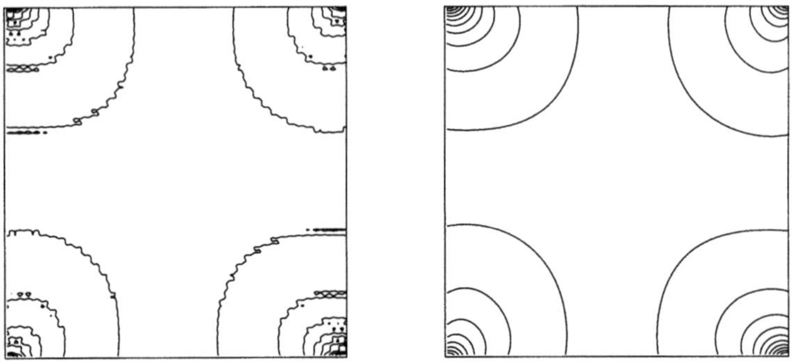

Fig. 5. Pressure contours for the non-staggered sparse scheme (left) and the non-staggered compact scheme (right)

is used. The non-staggered sparse pressure correction equation scheme leads to an inaccurate oscillatory pressure and inconsistent grid convergence for temperature and V velocity, and should not be used. The use of elliptic interpolation for the cell face velocities does not significantly detract from the accuracy and efficiency of the non-staggered compact scheme, however it is not necessary for the fractional step method used here.

4 Computer Cluster

4.1 Serial Performance

The cluster is based on Pentium 4 1.6 Ghz Processor workstations, with the Intel 850 chipset, 400 Mhz Front Side Bus and 500MBytes PC-800 RDRAM. These workstations cost approximately US$700 each in March, 2002, and provide an excellent base for the construction of a work station cluster dedicated to large scale computational fluid dynamics. The 400Mhz front side bus in combination with the PC-800 memory provides a peak memory bandwidth of 3.2 GBytes/second using a dual channel architecture with simultaneous 1.6Gbyte/second read and 1.6Gbytes/second write. The high main memory bandwidth of these machines provides good performance for large array jobs that run primarily out of cache.

The serial performance is demonstrated using the following benchmark loop,

$$x(i) = a(i) * y(i-1) + b(i) * x(i) + c(i) * y(i+1).$$

Each iteration of this loop requires five floating point operations, with five reads and one write. The Pentium 4 in single precision can simultaneously perform 400 MWords/second read and 400 MWords/second write. The out of cache floating point performance will therefore be limited by the read to approximately 400MFlops/second.

Figure 6 presents the results obtained for this benchmark loop for the Pentium 4 compared to a Pentium 3 running at 1.2Ghz and a Compaq DS10 running at 466Mhz. The code for the Pentiums was compiled using the Intel Fortran compiler, version 5.0, running on the Redhat Linux 7.2 operating system, while that for the Compaq used the Compaq Fortran V5.4 compiler running under Compaq OSF1 V5.0.

The floating point speed for each of the machines has been obtained for a range of array sizes to demonstrate the effect of array size on performance and to determine the asymptotic large array performance. For all the machines it is seen that the best performance is obtained for small array sizes. This clearly demonstrates the effect of the faster cache-memory. The Pentium 4 has a 256Kbyte level two cache, which corresponds to a vector length of 12,800 for this loop. It is seen that for arrays less than 12,800 the Pentium 4 is providing about 1.2GFlops/second performance, while for larger array sizes the performance is approximately 400MFlops/second, as predicted based on the main memory band-width. The Pentium 3 shows a similar cache/main memory variation in performance, although with the asymptotic large array performance reduced to 130MFlops/second, corresponding to the memory bandwidth for that machine. The Compaq DS10 is also a cached memory architecture machine, with a 2 MByte level two cache, corresponding to vectors of length 100,000. The out

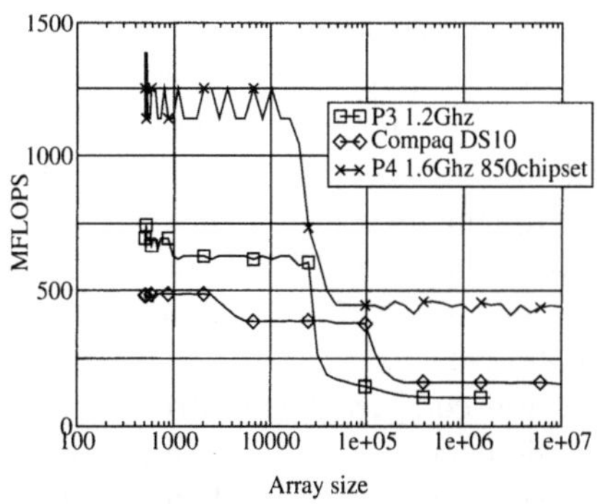

Fig. 6. Serial performance for Pentium 4, Pentium 3 and Compaq DS10

of cache performance of the DS10 is approximately 180MFlops/second, again corresponding to its main memory bandwidth.

It is clear from this simple benchmark that main memory bandwidth is the determining factor in the out of cache performance for these representative workstations. Large scale three-dimensional fluid simulations are very memory intensive and both in serial form and on moderate sized work station clusters the codes will be running primarily out of cache. The very high main-memory bandwidth and corresponding out of cache performance of the Pentium 4 clearly make it an excellent machine on which to base a computational fluid dynamics parallel cluster.

4.2 Parallel Performance

The workstation cluster consists of 16 Pentium 4 machines, as described above, connected by a 100 MBits/second fast ethernet switch. The switch cost approximately US$200 and thus the cost of the cluster is $11,400.

To run a code in parallel on a workstation cluster it is generally necessary to make some modifications. Parallelisation may be carried out using either message passing or data parallel methods. Both these approaches have been tested for the Pentium 4 cluster, with the MPICH library [9,10], together with the Intel compiler, used for the message passing code and the Portland Group High Performance Fortran compiler for the data parallel code [11].

The cluster was tested for a 3D finite volume discretisation of Laplace's equation upon a structured mesh with Dirichlet boundary conditions, and with the equations and solution field being stored in three dimensional arrays corresponding to the nodes of the mesh. The arrays were partitioned across the processors in their third axis. The discretised equations were solved using an unpreconditioned conjugate gradient solver, [12][13].

To provide a comparison of the different coding required for the MPI and HPF implementations of the conjugate gradient solver code fragments for a matrix vector multiply are shown below in Figure 7 for the HPF implementation and Figure 8 for the MPI implementation. As can be seen the implementation of the matrix vector multiply is straight forward using HPF. With the MPI code the matrix-vector operation can be implemented in the same way as the HPF code, with the additional step of a swap operation between neighbouring processors to share elements at the array boundaries. The code is simplest when written with a dummy element at the boundaries that is used in these swaps. The neighbour swap operation can be implemented in a number of ways. Here the code uses a persistent communication request, or a "half channel", with the communication pattern being initialised with the `MPI_*_INIT` calls, and then executed with the `MPI_STARTALL` and `MPI_WAITALL` calls.

The code sizes for these code fragments are representative of those of the complete conjugate gradient codes. The message passing code is larger than the HPF code. Extra care must be taken when writing message passing codes to prevent deadlocks caused by mismatched send and receive operations. Code

```
INTEGER nx,ny,nz,i,j,k,l,noit
REAL, DIMENSION(nx,ny,nz) :: s,p,ae,aw,an,as,at,ab,ap
!hpf$ DISTRIBUTE (*,*,block) :: s,p,ae,aw,an,as,at,ab,ap
do l = 1,noit
  forall(i=2:nx-1,j=2:ny-1,k=2:nz-1)                          &
      s(i,j,k) = ae(i,j,k)*p(i+1,j,k) + aw(i,j,k)*p(i-1,j,k) &
               + an(i,j,k)*p(i,j+1,k) + as(i,j,k)*p(i,j-1,k) &
               + at(i,j,k)*p(i,j,k+1) + ab(i,j,k)*p(i,j,k-1) &
               + ap(i,j,k)*p(i,j,k)
enddo
```

Fig. 7. The HPF implementation of a matrix vector multiply in an iterative loop.

```
INCLUDE 'mpif.h'
INTEGER :: nx,ny,nn,i,j,k,l,noit,id_previous, &
           id_next,stat,req(4),mtag = 15
REAL, DIMENSION(nx,ny,nn) :: s,p,ae,aw,an,as,at,ab,ap
call MPI_RECV_INIT( p(1,1,nn),nx*ny,MPI_REAL,id_next,mtag, &
                    MPI_COMM_WORLD,req(3),stat )
call MPI_RECV_INIT( p(1,1,1),nx*ny,MPI_REAL,id_previous,mtag, &
                    MPI_COMM_WORLD,req(4),stat )
call MPI_SEND_INIT( p(1,1,2),nx*ny,MPI_REAL,id_previous,mtag, &
                    MPI_COMM_WORLD,req(1),stat )
call MPI_SEND_INIT( p(1,1,nn-1),nx*ny,MPI_REAL,id_next,mtag, &
                    MPI_COMM_WORLD,req(2),stat )
do l = 1,noit
   call MPI_STARTALL( 4,req,stat )
   call MPI_WAITALL( 4,req,status,stat )
   forall(i=2:nx-1,j=2:ny-1,k=2:nn-1)                          &
        s(i,j,k) = ae(i,j,k)*p(i+1,j,k) + aw(i,j,k)*p(i-1,j,k) &
                 + an(i,j,k)*p(i,j+1,k) + as(i,j,k)*p(i,j-1,k) &
                 + at(i,j,k)*p(i,j,k+1) + ab(i,j,k)*p(i,j,k-1) &
                 + ap(i,j,k)*p(i,j,k)
enddo
do i = 1,4
   call MPI_REQUEST_FREE( req(i),stat )
enddo
```

Fig. 8. The MPI implementation of a matrix vector multiply in an iterative loop.

must also be written to partition the data across the nodes and to initialise the multiple processes – tasks which are done transparently by HPF.

The performance of each of the conjugate gradient implementations is shown in Figure 9 for a 100^3 problem size. As can be seen both the message passing and data parallel codes provide good parallel speedups, with approximately 60% parallel efficiency for up to ten processors. While both approaches are providing approximately the same performance it is clear that parallelising an existing structured Fortran 90 coded Navier-Stokes solver will be much simpler using

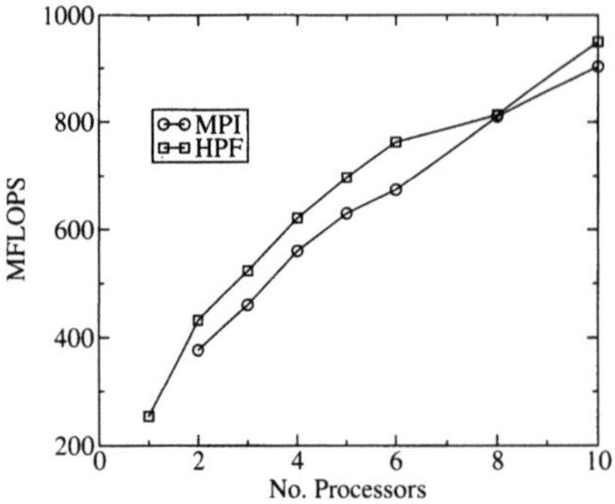

Fig. 9. Laplace equation solver parallel speed with MPI and HPF

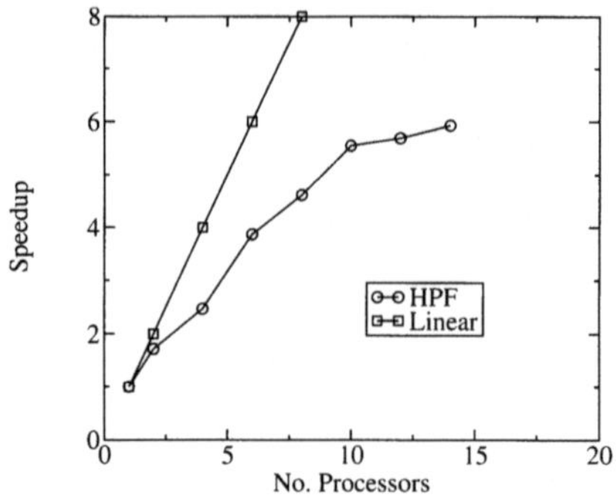

Fig. 10. Parallel Navier-Stokes code performance on 3D natural convection flow with HPF compared to linear speedup

HPF, and this approach was chosen. It is noted however that HPF is limited to structured array based codes; MPI has much greater flexibility with respect to data distributions.

4.3 Parallelisation of CFD Code

The Navier-Stokes solver was written using Fortran 90 array operations and forall structures. Parallelisation of such a structured code using High Performance Fortran is very straightforward, only requiring the inclusion of the compiler directives specifying the layout of the arrays on the processors, no modification of the code itself is required. The code was again tested on the natural convection start up problem in three-dimensional form on a 100^3 grid. The parallel performance on up to fourteen processors is shown in Figure 10, where speedup is plotted against number of processors. As can be seen the speedup ratio is approximately constant up to 10 processors, when a 5.5 times speedup is obtained, an efficiency of 55%. The speedup ratio then reduces with a 6 times speedup obtained on 14 processors, an efficiency of 43%. The roll-off in performance with increased number of processors seen here is typical of parallel codes and can result from the un-parallelised fraction of the code or the inter-processor communication overhead.

5 Discussion and Conclusions

The fractional step non-staggered scheme with compact pressure correction equation is second order with equivalent accuracy and efficiency as the staggered scheme. The non-staggered scheme with sparse pressure correction equation performs poorly and provides inconsistent convergence. Elliptic interpolation of cell face velocities is not required with the non-staggered fractional step method if a compact pressure correction scheme is used, however use of elliptic interpolation does not degrade accuracy. This is in contrast to iterative time integration schemes where the pressure correction goes to zero with the iteration and elliptic interpolation is then required to control the growth of grid scale pressure error [7,8]

The high main memory bandwidth of the Pentium 4 with the 850 chipset provides excellent floating point performance and is well suited for large scale fluid mechanics simulations. Both message passing (MPI) and data parallel (HPF) codes were tested on a P4 cluster to demonstrate the parallel capability of such a facility. In particular it was found that both approaches gave equivalent performance, and given the relative ease of applying HPF to an existing structured Navier-Stokes solver written in Fortran 90, this approach was chosen. The parallel code was found to have good parallel speedup for up to ten processors, with efficiencies approximately the same as the benchmark conjugate gradient solver. For more than 10 processors the parallel efficiency dropped, and it is clear that for a 100^3 grid communication overheads are then becoming dominant. In the future a faster Gigabit network will be tested which will allow more processors to be efficiently utilised before communications overheads begin to degrade the performance.

Finally it is clear that relatively inexpensive off the shelf work station clusters can provide Gflops/second speeds on Navier-Stokes solvers at costs less than US$10.00/Megaflop.

Acknowledgements: The authors wish to acknowledge the support of the Office of Naval Research Grant N00014-99-1-0413, NSF Grant ATM-0073395, Australian Research Council Grant A00000777 and the University of Sydney Major Equipment Fund.

References

1. Kim J, Moin P.: J. Comp. Physics **59**, 308 (1985)
2. Armfield S.W. and Street R.: Journal of Computational Physics **153**, 660 (1999)
3. Armfield S.W. and Street R.: ANZIAM J. **42**, C13 (2000)
4. Armfield S.W. and Street R.: Int. J. for Numerical Methods in Fluids **38**, 255 (2002)
5. Zang Y, Street R, Koseff J.: J. Comp. Physics **114**, 18 (1994)
6. Rhie C.M. and Chow W.L.: AIAA-82-0988 (1988)
7. Armfield S.W.: Computers and Fluids **20**, 1 (1991).
8. Armfield S.W.: J. of Comp. Physics **114**, 176 (1994)
9. William Gropp, Ewing Lusk, and Anthony Skjellum: *Using MPI: Portable Parallel Programming with the Message-Passing Interface.* (MIT Press, Cambridge, Massachusetts, 1994.)
10. Marc Snir, Steve Otto, Steven Huss-Lederman, David Walker, and Jack Dongarra. *MPI–The Complete Reference: Volume 1, The MPI Core.* (MIT Press, Cambridge, Massachusetts, second edition, 1998.)
11. Charles H. Koelbel, David B. Loveman, Robert S. Schreiber, Guy L. Steele Jr, and Mary E. Zosel. *The High Performance Fortran Handbook.* (Scientific and Engineering Computation Series. MIT Press, Cambridge, Massachusetts, 1994.)
12. R. Barrett, M. Berry, T. F. Chan, J. Demmel, J. Donato, J. Dongarra, V. Eijkhout, R. Pozo, C. Romine, and H. van der Vorst: *Templates for the Solution of Linear Systems.* (SIAM, Philadelphia, 1994.)
13. Magnus R. Hestenes and Eduard Stiefel: *Journal of Research of the National Bureau of Standards*, 49(6):409–437, 1952.

Towards Computational Biomechanics Based Cardiovascular Medical Practice

Takami Yamaguchi[1], Tomoaki Hayasaka[2], Daisuke Mori[1], Hiroyuki Hayashi[1], Kouichiro Yano[3], Fumio Mizuno[1], and Makoto Harazawa[1]

[1] Tohoku University, Department of Mechatronics and Precision Engineering Sendai 980-8579 Japan
[2] The Physical and Chemical Research Institute (RIKEN), Division of Advanced Computing, Wako 351-0198, Japan
[3] Nagoya Institute of Technology, Department of Mechanical and Systems Engineering, Nagoya 466-8555, Japan

Abstract. Some of our recent fundamental computational biomechanical studies of the cardiovascular system, from modeling aortic flow to the microscale dynamics of the platelet, are discussed. It is emphasized that a living system inherently needs multi-scale, multi-physics analysis, and the rapid advancement of the biological sciences must also be incorporated in computational biomechanics studies. Based on these efforts, it is now expected that a computational biomechanics-based software system that can be applied to real clinical situations will be developed for the cardiovascular system. This will help medical practitioners to accumulate the fundamental data and clinical experience that they obtain from real patients, and use it to make quantitative predictions.

1 Introduction

Cardiovascular diseases, particularly ischemic heart disease, and especially myocardial infarction, are the leading cause of death in the industrialized world. Cerebral strokes, another vascular disorder, are also a major cause of death. These vascular diseases share a common background, atherosclerosis. They also share a common final event, the failure or destruction of the vascular wall structure. It is well known that atherosclerosis preferentially occurs at specific arterial sites, which are suspected to be related to hemodynamics, particularly the wall shear stress (WSS) distribution. The final vascular events that induce fatal outcomes, such as acute coronary syndrome, are now strongly suspected of being triggered by the sudden mechanical disruption of an atherosclerotic plaque on an arterial wall [1]. Subarachnoid hemorrhage, which is another very sudden, fatal disease, is the direct result of rupture of a cerebral aneurysm. Therefore, both the onset and final consequences of tragic fatal vascular diseases are inseparably connected to mechanical events that occur on the vascular wall, which are strongly suspected of being influenced by blood flow alterations.

Consequently, the fluid-solid mechanical interactions between blood and the vascular wall must be analyzed in order to predict, diagnose, and prevent the fatal results of these vascular diseases. This is clearly not a task that can be performed by conventional experimental or theoretical methods used in physiological mechanics. It is usually prohibitively difficult to measure detailed mechanical properties in a living system, and they are too complicated to be studied elegantly in a theoretical framework alone. Therefore, we need to use computational studies to elucidate the mechanism of such disease, to refine the diagnostic measures, and to develop invasive and non-invasive therapies [2] [3] [4]. This article discusses our preliminary studies of various aspects of the cardiovascular system using computational mechanics, and focuses on why computational studies are necessary, how a computational model is built, the prerequisites for computation, and the pitfalls of interpreting computational studies.

2 Examples of Computational Mechanical Studies of the Cardiovascular System

2.1 Parametric and Realistic Modeling of the Large Arteries

Real human arteries, particularly, the aortic arch, have a significant characteristic configuration. One of the characteristics of the aortic arch is that its centerline does not lie in a plane (non-planarity) [5] Another is that there are major branches at the top of the arch. Individual variants in these geometrical factors are suggested to induce serious diseases. Yoshii et al. [6] constructed 3-D images of human aortas by stacking a series of computed tomographical (CT) slices, and indicated that the angle between the ascending and descending legs of the arch in the cranial view tends to be sharper in aortic aneurysm cases. However, no systematic investigations from a fluid dynamics point of view have been conducted. Kilner et al. [7] observed a characteristic helical blood flow pattern in the human aortic arch using magnetic resonance (MR) measurement, and showed the qualitative flow structure in the arch. The human aortic flow was found to consist of very complex vortices. However, it is very difficult to collect precise hemodynamics information, such as the WSS, which has long been suspected of playing an important role in various disorders, particularly atherosclerosis, even from state-of-the-art clinical image-based measurements, because the spatial resolution is insufficient for precise evaluation of near-wall phenomena, such as the WSS. Instead of experimental methods, we studied the flow in the human aortic arch in more detail by constructing realistic three-dimensional (3-D) vessel models using computational fluid dynamics (CFD) methods, paying full attention to the curvature and torsion of the aortic trunk and the major branches that emerge from the top of the arch [8].

Aortic Branching Model We proposed the so-called differential geometrical method [11] of modeling arterial trees as a method that combines these two approaches to modeling as shown in Fig. 1. In our method, vascular trees are represented using their centerlines and their normal planes defined by a combination of normal and binormal vectors [9]. Each branch is modeled separately and connected using an overset (or a chimera) mesh technique when fluid mechanical computations are made. To evaluate this method, we constructed an aortic arch model for CFD simulations, which incorporates both non-planarity and the major branches, based on a set of MR images, and discuss their combined effects on blood flow.

Our basic idea is that the branching structure of the aortic arch can be crudely simplified as a main pipe, i.e., the aortic arch, with three (or more, as necessary) major branches connected to it. Using the differential geometry modeling method that was devised and applied in our previous study, the aortic trunk and its three branches were constructed individually, based on their centerlines, using 3-D reconstructed MR images. There were regions in each model in which the corresponding computational grids overlapped, and we used these to combine the models to form a single computational domain consisting of the arch and its branches. To model each vessel geometrically, we used the interactive clinical interface for MR images developed by Hayasaka et al. [10]. Figure 2 shows an MR image, and the CFD model constructed from it. We assumed that the cross-sectional shape is circular for simplicity. It is important to note that any kind of irregularity of the wall, such as a non-circular cross-section, localized stenosis, or aneurysmal expansion, can easily be incorporated into the model using the proposed method. The diameter of the aortic arch in our preliminary study was 2 cm at the inlet, and it decreased with the distance from the inlet, in order to compensate for the effect of the decrease in the blood flow in the descending aorta caused by the escape of a large proportion of the blood flow into the three branches. The aortic diameter was assumed to decrease in an exponential manner. The diameters of the brachiocephalic, left carotid, and left subclavian arteries were uniformly 1, 0.7, and 0.7 cm, respectively.

Conditions of Analysis To use the overset grid method, the velocity vectors and pressure for each grid in the overlapping regions had to be interpolated and transferred from one region to another at every computational step. The transferred data were always regarded as part of the boundary conditions in the recipient region. The CFD computations used a commercial program, SCRYU

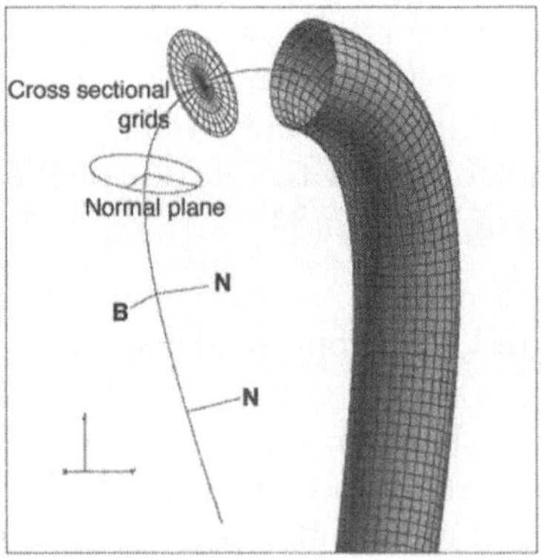

Fig. 1. Schematic representation of the differential geometrical modeling of the aorta

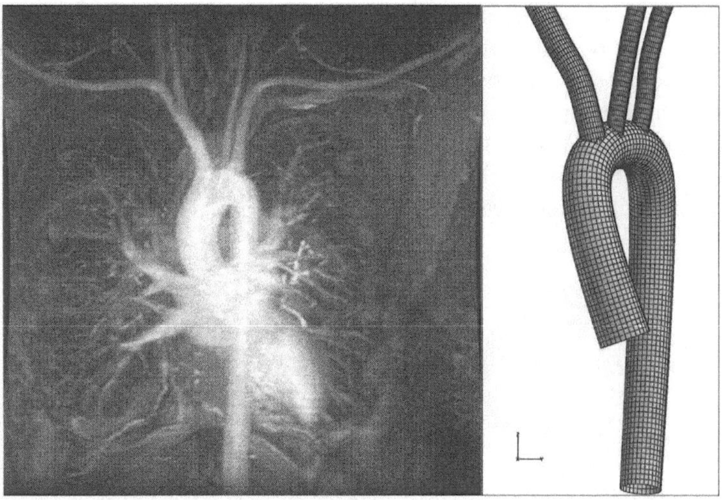

Fig. 2. Left: A 3D reconstructed MRA (magnetic resonance angiography) Image of a voluntary adult. Right: A CFD model based on the left image

version 1.4 (Software Cradle, Osaka, Japan). The 3-D unsteady Navier-Stokes equations for incompressible flow were solved and the solution was advanced in time using a finite volume method of discretization. The boundary conditions were as follows:

- A steady uniform velocity perpendicular to the cross section at the aortic inlet.
- No-slip at the wall.
- Zero pressure and zero velocity gradient at all the outlets.

The Reynolds number defined using the diameter and the inflow velocity was 1,600, which was deemed an appropriate value for a resting adult human.

Computational Results and Discussion of the Effect of Geometry Figure 3 shows the secondary flow at several cross-sections under a steady uniform inflow condition with a Reynolds number (Re) of 1600. Figures 3 (a) - (c) and (d) - (f) illustrate the flow along the arch, and the flow in the three branches, respectively. As shown in Figures 3 (a) - (c), the flow along the arch consisted of a pair of nearly symmetrical circulations at the top of the arch, a large right-handed rotational flow in the descending part of the arch, and a large left-handed rotational flow at the end of the arch. These characteristics of the global flow were similar to the results obtained using our previous arch model without branches [11]. Although we found a complex pattern of vortices occupying the whole aorta, no single vortex predominated at any one cross-section; there was always at least a pair of vortices, and one vortex sometimes became much smaller than the other. This is physically plausible, because angular momentum is conserved in the computation. It may be different in the real aorta, where the inflow from the left ventricle may be fully rotational due to the 3D curvature and torsion of the flow path. By contrast, the secondary flow differed in each branch, where unique structures were seen, probably due to the asymmetric 3-D structure of the arch. Figure 4 shows the WSS distribution under the same conditions. There was a region of high WSS around the branches at the top of the arch. Particularly, the distal part of each branch was a region of significantly higher WSS.

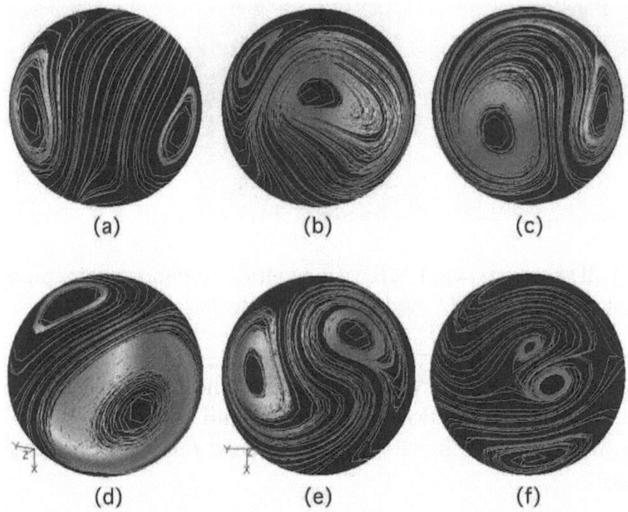

Fig. 3. Stream line patterns at cross sections of the aorta; (a) at the top of the aorta, (b) at the descending part, (c) at the end of the arch, and three branches; (d) at the brachiocephalic, (e) at the left common carotid, and (f) at the left subclavian arteries.

This study demonstrated that the combined method of differential geometrical modeling based on the vascular centerline, with overset meshing for the branches, is a powerful means of conducting a CFD analysis of the blood flow in the aortic arch. Further work is necessary to construct a more realistic CFD model with arbitrary cross-sectional shapes, which probably affect the blood flow, using medical images, such as MRI.

2.2 Cellular Mechanics of the Arteries and Blood Constituents

Blood is not a Newtonian fluid when the shear rate is very low. In the larger arteries that we usually model and analyze, the fluid shear rate can be safely assumed to be high enough to justify Newtonian assumptions. It is, however, necessary to treat blood as a multi-phase fluid in some cases, such as in very small arteries, and even in large arteries with very complex shapes, where flow retardation or stagnation may occur, and when we are interested in the cellular

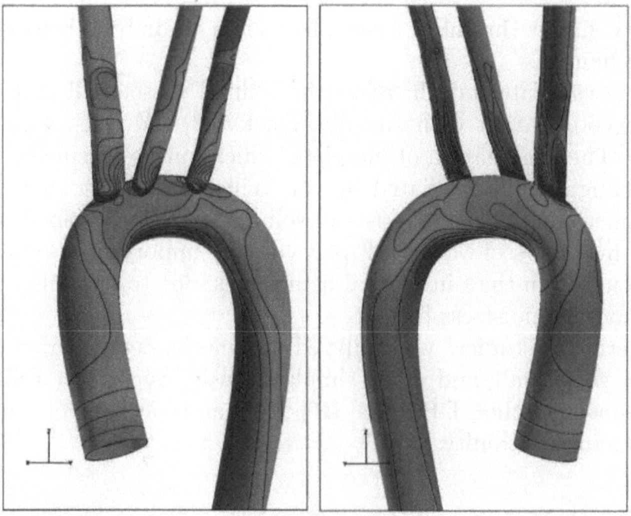

Fig. 4. WSS distribution pattern on the aortic luminal surface (Left: front view, Right: rear view)

response of the vascular system. We have already conducted some studies to model and compute the spontaneous change in the shapes of cultured endothelial cells subject to flow. We recently reported a series of studies of the mechanical effects of different modes of cyclic stretching on cultured cells. They were studies of the vascular wall. In the following section, we introduce a series of studies on the behavior of platelets in flowing blood. The motion of the platelets was modeled using a combination of the Discrete Element Method (DEM) and the Stokesian Dynamics Method (SDM).

Computational mechanical simulation of the aggregation and dissociation of platelets mediated by von Willebrand factor Platelets play the most important role in blood coagulation, particularly in the formation of primary thrombi. With respect to the mechanical interactions between the blood flow and the aggregated platelets forming a primary thrombus, we can distinguish completely different mechanisms. In normal hemostatic thrombus formation, the vascular wall injury that induces the aggregation of platelets is usually transmural; that is, an aggregation of platelets forms at the site of wall injury to stop hemorrhage. Once the hemorrhage stops, the formed thrombus does not affect the blood flow in the blood vessel. By contrast, in pathological thrombus formation, the blood vessel injury or dysfunction occurs on or close to the luminal surface of the vessel wall. The formed thrombus is thought to protrude from the luminal surface towards the blood flow; therefore, it mechanically interacts with the blood flow. For a pathological thrombus to persist on an arterial wall, its precursor or the primary thrombus must survive for a certain span of time resisting the shear and drag forces exerted by the blood flow before formation of the secondary, firmer thrombus takes place via a thrombin-fibrinogen mediated chemical mechanism.

In both cases, platelets adhere to the injured vessel wall and then form a thrombus in cooperation with the fibrin network, red cells, and other blood constituents. The aggregation of platelets, which initiates primary thrombi formation, is thought to be mediated by von Willebrand factor (vWF). vWF is a long-chain macromolecule that exists in soluble form in the blood and in an insoluble form in the vessel wall. vWF plays a more important role in pathological thrombosis formation than in normal hemostasis, due to its ability to sense and react to hemodynamic stress [12].

In this series of studies, we analyzed the mechanical interactions between platelets, the vessel wall, and vWF, simultaneously, using both a SDM [13]. and a discrete element method DEM [14] [15] that can treat solid elements in a flow without assuming continuity.

Computational Method A two-dimensional (2D) mechanical model of a blood vessel, platelets, and vWF was constructed. The mechanical interactions between individual platelets, and the interactions between the platelets and vessel walls that were mediated by vWF, were modeled, as shown in Figure 5. Here, vWF was assumed to act at a relatively remote distance from the direct collision of components, and a vWF molecule was assumed to produce only a tensile force at a certain distance from the other elements (platelets or the wall). Furthermore, it was assumed that only repulsive forces were generated in the direct interactions. Details of the assumptions have been reported elsewhere [16].

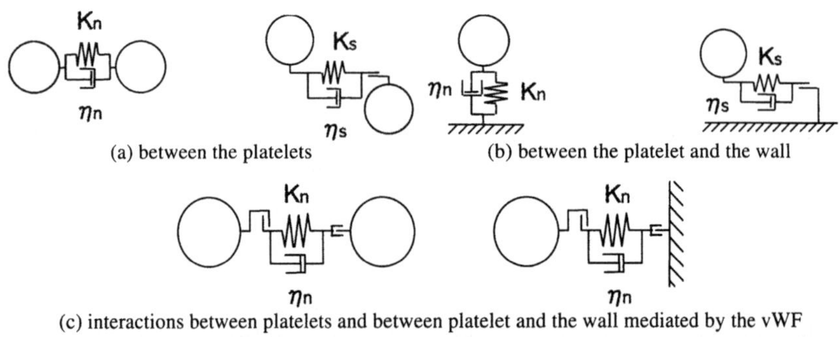

Fig. 5. Mechanical models of the direct interactions between platelets (a) and between the platelet and the wall (b), and indirect interactions mediated by the vWF (c).

A 2D Couette flow model was used in this study, and the influence of other blood constituents, including red blood cells, which were proven to be very important when we consider the very onset of the initial thrombus formation, was neglected. The fluid force acting on the particles was calculated using the SDM. In the SDM, the fluid-solid mechanical interactions adjacent to each particle are evaluated using the Stokes flow assumptions and are summed to calculate the localized flow field around a particle. Adjacent particles are assumed to act on a particle via the velocity of the particle (induced velocity method). This method focuses on the contribution of the movement of surrounding particles within a truncated range of the flow field, when the velocity of the individual particle is computed to evaluate the total flow field.

Computed Results of the Effect of the Strength of the Chemical Binding of vWF to Platelet Aggregation vWF has numerous receptors on it, which have different characteristics and cause chemical binding with the corresponding receptors on other blood constituents, such as platelets. Some of the receptors on vWF are activated by physical factors, including WSS. Considering the nature of the receptors on vWF, we simulated the variable strength of the chemical binding between platelets and vWF by changing the number of receptors (Nre) on vWF. The destruction of the primary thrombus under the influence of the blood flow was then modeled.

Figure 6 (a) shows the results for Nre = 0, i.e., no activation of the vWF receptors was assumed. Consequently, there were no interactions mediated by vWF. Since the only force acting between the platelets and between the platelets and the vessel wall in this model is the repulsive force due to direct collisions, the simulated platelets separated from the aggregation individually. Figure 6 (b) shows the results for Nre = 5. When there were both repulsive forces due to direct collisions and tensile forces due to vWF binding between platelets or between platelets and the vessel wall, the modeled platelets moved together to form aggregations in the blood flow. Figure 6 (c) shows the results for Nre = 10. Larger platelet aggregations formed, since there were twice as many receptors, resulting in a much stronger binding force. The temporal evolution of the number of aggregations is shown in Figure 7. The number of receptors, and hence the total binding force between the platelets and vWF, significantly affects the destruction of the primary thrombus.

2.3 Fluid-Solid Interaction Studies

Analyses of the cardiovascular system must examine two types of fluid-solid interactions. One involves an actively moving solid boundary, such as the contracting ventricular wall propelling blood from the left ventricle into the aorta. The other type involves passive wall motion, such as the deformation of arterial walls due to pulsatile blood flow and pressure. From a computational mechanics viewpoint, these two phenomena require completely different approaches. In the former analysis, solid mechanical modeling is unnecessary and the wall motion can be modeled as a voluntary movement of the solid boundary from a fluid mechanics viewpoint. A moving boundary and the addition of some mesh-updating formulation, such as the arbitrary Lagrangian-Eulerian method, suffices. The latter problem is very difficult, and is still a very challenging target for computational mechanics. We have conducted several studies of each type of mechanical

(a) Nre=0

(b) Nre=5

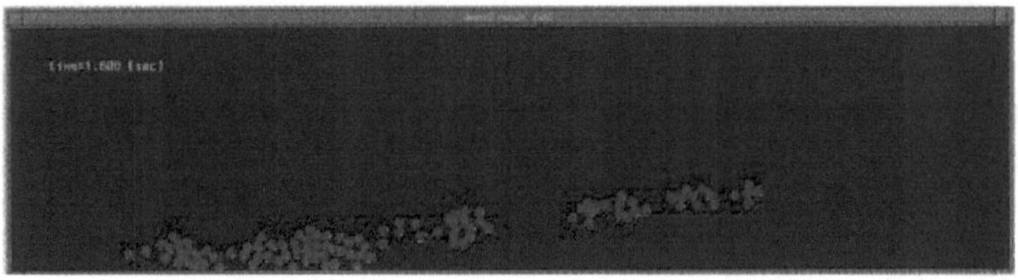

(c) Nre=10

Fig. 6. Transient arrangement of particles simulating platelets with various strengths of chemical binding induced by the vWF. Top: Nre=0, i.e.,. No bindings, Middle: Nre = 5, i.e.,. Modcorrespcorrespondencesnding, Bottom: Nre=10, the strongest binding force between the platelets and between the platelet and the wall was assumed.

interaction. In this article, we present preliminary results from recent studies of right coronary artery flow.

Model of the Right Coronary Artery that Moves with the Ventricular Contraction The beating motion of the heart wall, to which the major coronary arteries are tethered, mechanically influences the flow inside the coronary arteries, and hence affects atherogenesis. In this study, we conducted a CFD simulation using a simplified model of the right coronary artery, which deforms with the contraction and dilatation of the heart. The results are discussed with respect to the local hemodynamic characteristics, particularly the streamline pattern and WSS distribution.

The CFD model and Computation The CFD model of the right coronary artery consisted of a coiled pipe with an ordinary helix as a centerline, surrounding a simple body that simulated the heart as shown in Fig 8 [17]. The diameter

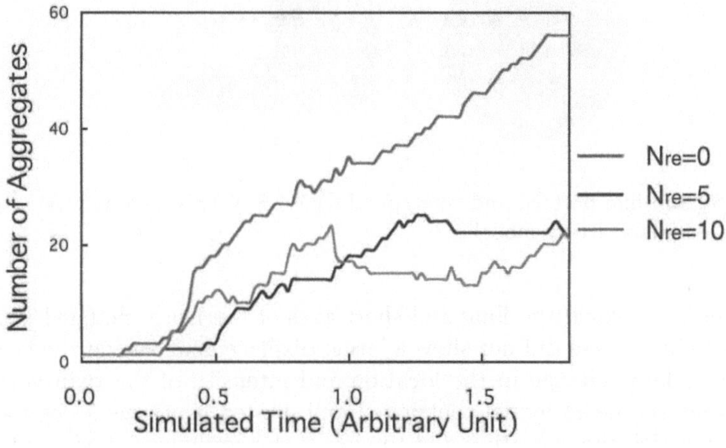

Fig. 7. Time course of the dissociation of primary thrombus with respect to the strength of the chemical bindings

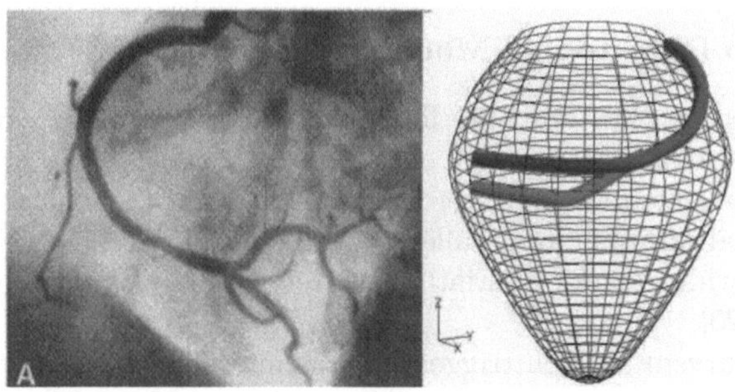

Fig. 8. Angiogram of the right coronary artery (left) and a simple model of the heart and the right coronary artery (right)

and height of the cylinder changed over time at a given rate. The volume reduction in one beat of the cylinder was 75

Results and Discussion Representative results are given by the streamline patterns and corresponding WSS distribution at a cross-section in the middle of the right coronary artery model at a phase between early systole and late diastole, as shown in Figure 9.

Many interesting features of the fluid dynamics were evident. The most characteristic changes in the streamline and WSS distribution appeared when the flow decelerated. Striking differences in the formation and change in the shape of the twin vortices corresponded to the mode of heart contraction and dilatation.

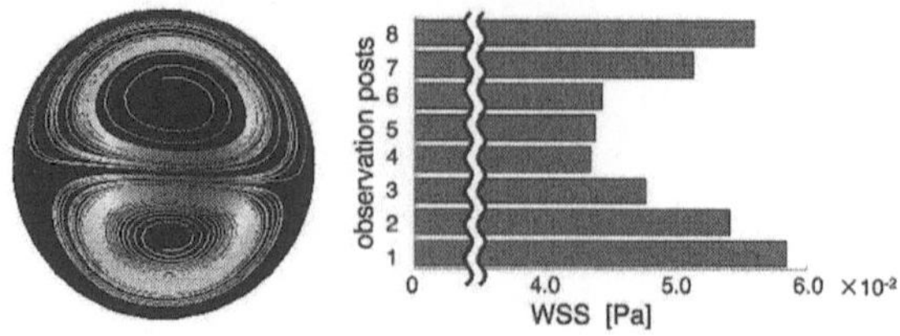

Fig. 9. Stream line pattern and corresponding WSS distribution at a cross section of the right coronary artery model.

In the mode in which the long and short axes of the heart changed proportionally, the twin vortices did not show a larger displacement during the heart cycle. However, a large change in the location and intensity of the twin vortices was found when the heart model contracted and dilated using the second and third modes, in which the curvature and torsion of the coronary artery model behave in opposite directions. Useful information about the specific mechanisms for the formation, growth, and breakdown of coronary plaque can be obtained from an analysis of the results under various physiological conditions.

2.4 New Two Dimensional Model of the Circle of Willis

Blood is supplied to the brain via four arteries: two internal carotid arteries (ICA) and two vertebral arteries. To ensure the reliability of the supply of blood to the brain, these arteries are interconnected at the base of the brain, where they form a closed arterial circle, called the circle of Willis. It is well known that there is a very wide range of variation in the cerebral arteries, especially the circle of Willis [20].

Cerebral aneurysms, which trigger subarachnoid hemorrhage (SAH), a serious cerebrovascular disease, are known to preferentially occur at or adjacent to the circle of Willis, and the complex blood flow pattern in the circle of Willis is suspected of influencing the development and sometimes fatal rupture of aneurysms. This is suspected to be dependent on the complex geometry of the circle. Therefore, we conducted a CFD simulation of the blood flow in the circle of Willis, focusing on the relationship between the geometric configuration and the redistribution of blood flow by parametrically changing inlet flow conditions.

The CFD model and Computation The circle of Willis consists of multiple inlet and outlet vessels. They have a complex 3D geometry, with a very wide variety of anomalies with respect to diameter, bifurcation pattern, etc. It is extremely difficult to model the real 3D geometry of the circle of Willis, and previously reported 1D models are not sufficient to include the complex shape factors of the circle of Willis [21] [22]. Since 1D models do not seem to be effective for determining the real fluid mechanics of the circle, we built a generic 2D model of the circle of Willis using a schematic drawing in an established monograph.

First, by sampling the outline points from the 2D schematic model (Figure 10 (left)), and interpolating these points with a cubic spline curve, the outline boundary of the CFD model was obtained. Then, we made a computational model by defining computational grids inside the outline (Figure 10 (right)) [23].

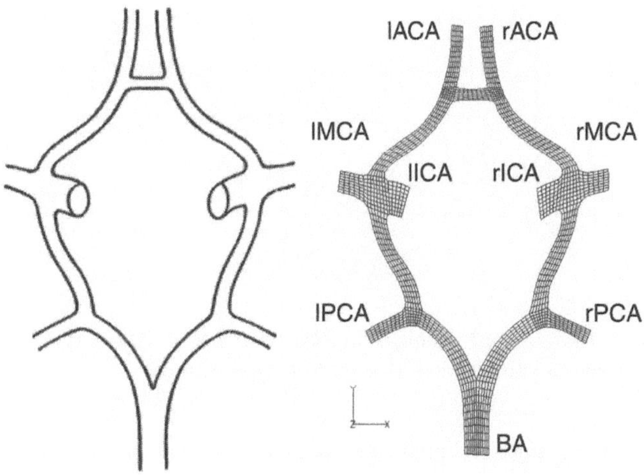

Fig. 10. An outline of the circle of Willis (left) and the 2D CFD model based on it (right).

Governing Equations and Boundary Conditions The boundary conditions at every inlet consisted of a steady uniform inflow velocity perpendicular to the cross-section. It is known that carotid artery atherosclerosis induces a gradual decrease in the cerebral blood flow supply over the long term. In addition to this gradual alteration to the blood flow that may be compeKilnerby remodeling the anastomotic patterns of the cerebral arteries, a sudden disruption of the carotid plaque and its subsequent acute embolization producing unilateral obstruction of crucial cerebral arteries is of serious clinical importance. To analyze the effect of a unilateral decrease or even the complete obstruction of the blood flow supply via the internal carotid and basal arteries, we parametrically changed the inflow rate at one (in this case the left) internal carotid artery in combination with a decrease in the basal artery blood flow rate. At the right internal carotid artery (rICA), the inflow rate was always kept constant. At all of the outlets, a zero-pressure, zero-velocity gradient was given as the boundary condition. No-slip on the wall was assumed.

RESULTS and Discussion We performed CFD computations for a wide range of combinations of blood flow rates. While the rICA inflow was fixed, the inflows at the left internal carotid artery (lICA) and basilar artery (BA) were reduced.

The value of the inflow Reynolds number at the lICA was set at 500, 375, 250, 125, and 0. The basilar artery inflow was controlled using 3 steps: 500, 250, and 0. In total 3x5 = 15 cases were examined. Representative results are shown in Figure 11.

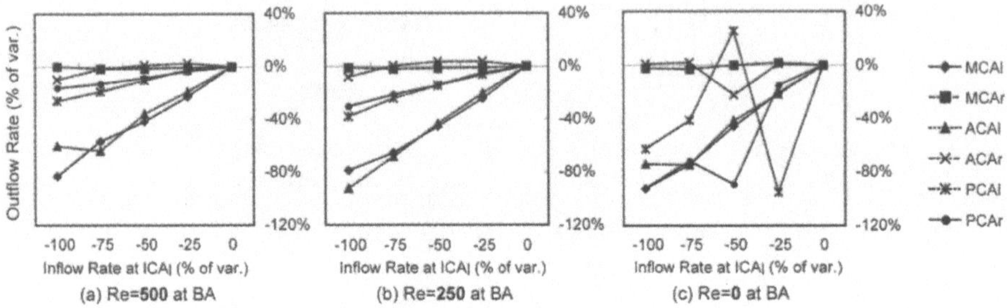

Fig. 11. Blood flow output changes in the branches of the circle of Willis due to unilateral decrease or obstruction of the blood flow supply.

In brief, the change in outflow, which is directly connected to an internal carotid artery, was proportional to the change in the inflow from the internal carotid artery. At the outlets on the contralateral side, relatively little change in the flow was observed. This influence was considered to be diminished by bypass flow from the vertebral artery. Therefore, the outflow on the contralateral side was relatively spared.

Flow Pattern in the Circle of Willis Representative results are shown in Figure 12. The inflow conditions were lICA: Re=250, rICA: Re=500, and BA: Re=500. Significant vortex formation was observed at the posterior and anterior communicating arteries.

Discussion and Remarks for Future Study It is well known that there is autoregulation of CBF. Our results show that the fluid dynamics in the circle of Willis strongly depend on its shape. When the balance between blood flow on the right and left sides was destroyed, the outflow rates from the outlet nearest the inlet where the inflow rate was changed were greatly affected. The necessity of analyzing the blood flow in the circle of Willis using 2D or 3D models is obvious. This result strongly suggests that CBF is affected by the geometry of the circle of Willis, not just by the connectivity of the component arteries.

3 Development of Clinical Applications

Due to the rapid advances in invasive and non-invasive imaging technology in clinical medicine, we can now obtain images of target organs to use in computational analysis modeling. However, the resolution and reproducibility of

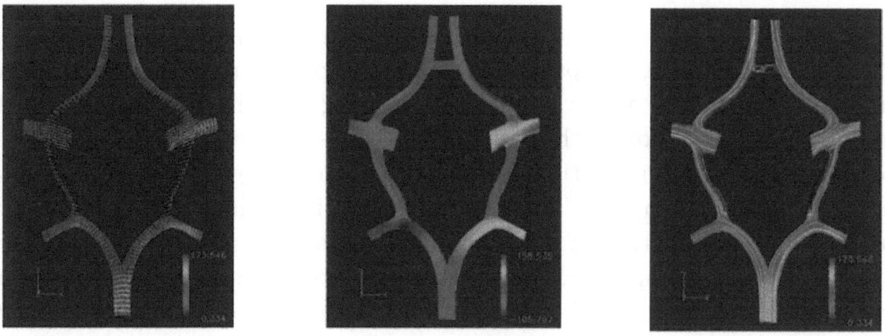

(a) Velocity Vector (b) Pressure Distribution (c) Streamline

Fig. 12. Blood flow output changes in the branches of the circle of Willis due to unilateral decrease or obstruction of the blood flow supply.

non-invasive methods, such as magnetic resonance imaging (MRI), are still insufficient for automatic modeling. It is also difficult to automatically eliminate contamination due to spurious information and noise from the clinical images. Therefore, we believe that any clinical application should fully utilize human pattern recognition and image reconstruction abilities. This is also a practical and efficient way of processing images. Consequently, we have been developing a comprehensive computational analysis support system that includes the preprocessing of medical images, segmentation, structure registration, database and human interfaces, mesh generation, and the final evaluation of the computed results.

Since our ultimate goal is to build a system that can assist clinicians in diagnosis, treatment planning, and following-up patients with cardiovascular diseases, and since patients differ in terms of anatomical configuration and disease condition, a wide variety of patient data must be accumulated, not only for statistical analysis, but also to improve the processing system. Therefore, a comprehensive database that includes patient data, medical images, computational models, and computed results must be constructed within the scope of our current projects. To make the database more usable, a novel method of searching and finding patient datasets using morphological and physiological queries must be developed. It is also necessary to make this database accessible through high-speed Internet connections, so that medical practitioners can obtain information easily. Quantitative evaluation of the benefits and risks of treatment is one of most demanding aspects of clinical medicine, since it is based on experience and intuition, which are difficult to quantify. Although evidence-based medicine (EBM) is currently strongly recommended in medical practice, most clinical evidence still seems to consist of accumulated statistics and clinical experiences that are not rigorously expressed in a quantitative way. By contrast, computational mechanics analysis gives us truly quantitative assistance in diagnosis, treatment planning, assessing the prognosis, etc.

The operations necessary to perform computational mechanics analyses, particularly CFD analysis, comprise modeling, computation, and post-processing, including visualization of the results and storing data in an ordered way that enables later retrieval for further analysis. Of these stages of CFD analysis, it is recognized that modeling is still the most time-consuming process, and is heavily dependent on human power. There is very wide individual-to-individual structural and functional variation in real clinical cases, so that any modeling should necessarily be patient specific. Again, structural differences cover a wide range, from macroscopic anatomical ones to minute irregularities in the wall structure of small arteries.

When a computational analysis is utilized in a clinical situation, there are several additional requirements. They are the exactness and speed of all processes. In some circumstances, clinical diagnosis and treatment decisions must be made very quickly. This is particularly true if the system to be developed is aimed at emergency treatment, such as for unstable angina, which sometimes leads to the acute coronary syndrome and sudden cardiac death. Subarachnoid hemorrhage (SAH) is another example. Clinical decisions should be made within an hour. Although there is a trade off between exactness and speed, we need to find an optimal compromise between them, otherwise the developed system cannot be widely used.

This is why we started the project called CREAM (Computational Risk Estimation And Management for Cardiovascular Disease) with the support of the Japan Science and Technology Cooperation (a subsidiary of the Ministry of Education, Culture, Sports, Science, and Technology). This project will develop a comprehensive system to support clinical decisions in cardiovascular medicine using computational biomechanical methodology. By the end of the project, we should have a system composed of several subsystems, including medical image retrieval and storage, analysis to extract the information necessary to build computational models, a modeling system with properly defined boundary conditions, a multi-scale computation system using GRID computers over a nationwide high-speed computer network, a sophisticated visualization system, and a comprehensive tool to support clinical decisions in real situations. The results will be reported elsewhere in the near future.

4 Concluding Remarks

This article discussed some of our computational biomechanical studies of the cardiovascular system, from aortic flow to the microscale dynamics of blood constituents. Many subjects were not discussed here, including intracardiac flow, cellular mechanics, and the flow adjacent to endothelial cells. Since the living system inherently needs multi-scale, multi-physics analysis, new technology should be continuously introduced to this field from various fields of scientific research. In addition, computational biomechanics studies must incorporate the tremendous advances in the biological sciences.

References

1. P.Libby: Circulation**104** 365 (2001)
2. T.Yamaguchi: 'Computational visualization of blood flow in the cardiovascular system'. In: *Biological Flows*, ed. by M.Y.Jaffrin, C.G.Caro (Plenum Press, New York, 1996) pp. 115–136.
3. T.Yamaguchi: 'Computational visualization of blood flow'.. In: *Computational Biomechanics*, ed. by K.Hayashi, H.Ishikawa (Springer-Verlag, Tokyo, 1996) pp. 165–184.
4. T.Yamaguchi, *Clinical Application of Computational Mechanics to the Cardiovascular System* (Springer-Verlag, Tokyo, 2000)
5. C.G.Caro, D. J. Doorly, MTarnawski, K. T. Scott, Q. Long, C. L. Dumoulin: Proc. R. Soc. Lond., A **452**,185 (1996)
6. S. Yoshii, K. Kamiya, T. Matsukawa, A. Ueno: J. Jpn. Surg. Soc. **89** 972 (1988)
7. P. J. Kilner, G. Z. Yang, R. H. Mohiaddin, D. N. Firmin, D. B. Longmore: Circulation **88** (part 1) 2235 (1993)
8. D. Mori, T. Hayasaka, T. Yamaguchi: J.S.M.E. Intl. J. (C), submitted
9. Y. C. Pao, J. T. Lu, E. L. Ritman: J. Biomech. **25** 287 (1992)
10. T. Hayasaka, H. Liu, R. Himeno,T. Yamaguchi: Proceedings of the 2001 Bioengineering Conference, ASME, **BED-Vol.50** 563 (2001)
11. D. Mori, T. Yamaguchi: Computer Methods in Biomechanics and Bioengineering (2002) in print
12. S. Goto, et al.: J. Biol. Chem. **270** 22352 (1995)
13. S. Kamiyama, A. Satou: *Microsimulation of Colloidal Dispersions* (Baifukan, Tokyo 1997) pp. 43–51
14. J. Hidaka, A. Simosaka: *Funtai Simulation Nyuumon* (Sangyo-Tosho Tokyo 1998) pp. 29–82.
15. P.A. Cundall, O. D. L. Strack: Geotechnique **29** 47 (1979)
16. H. Miyazaki, T. Yamaguchi: Biorheology (2002) in print
17. H. Hayashi, T. Yamaguchi: Biorheology (2002) in print
18. T. W. Taylor, H. Okino, T. Yamaguchi: J. Biomech. Eng **116** 127 (1994)
19. T. W. Taylor, H. Suga, Y. Goto, H. Okino, T. Yamaguchi: J. Biomech. Eng. **118** 106 (1996)
20. B. Adachi: *Das Arteriensystem der Japaner* Band1. (Maruzen, Japan 1928) pp. 116–134
21. B. Hillen, B. A. Drinkenburg, H. W. Hoogstraten, L Post: J. Biomech..**21** 807 (1988)
22. C.A. Lodi, M. Ursino: Ann. Biomed. Eng. **27** 257 (1999)
23. M. Harazawa, T. Yamaguchi, 'Computational fluid dynamics model analysis of the blood flow in the circle of Willis'. In *Proceedings of the Fifth JSME-KSME Fluids Engineering at Nagoya, Nov.22, 2002*

Part II

Algorithms

Uncertainty Quantification in CFD Simulations: A Stochastic Spectral Approach

Lionel Mathelin and M. Yousuff Hussaini

Florida State University, Tallahassee, FL 32306-4120

Abstract. The purpose of this work is to study the propagation of uncertainty in numerical simulations. For this purpose, we employ a methodology that couples the polynomial chaos technique with Karhunen-Loeve decomposition. This method is applied to the problem of a quasi- one-dimensional nozzle flow with uncertainty in inlet conditions and nozle shape. It is shown that the polynomial chaos approach efficiently handles uncertainty propagation and establishes confidence intervals for the simulation results. Furthermore, such an approach enables computation of statistical moments of arbitrary order in a much more effective way than other usual techniques such as the Monte Carlo method or the perturbation method.

1 Background

Scientific context. With the continuous improvement in both accuracy and reliability of computational fluid dynamic (CFD) simulations, further developments toward physically realistic and reliable results require accounting for variablity in operating conditions. Most simulations treat boundary conditions, initial conditions, geometry of the problem, and physical properties as deterministic. However, the actual conditions are not known precisely or are stochastic in nature, and one must account for a certain level of uncertainty in the simulations. It is indeed of critical importance to quantify uncertainty (e.g. [1,8,13]) and establish confidence intervals for the simulation-based predictions or design ([3]). See [1] and [7] for a review of different sources of uncertainty.

Several techniques have been proposed to study the propagation of uncertainty due to the boundary conditions, to the flow field and to the output of the simulation. They are not without some drawbacks. For instance, the perturbation technique is limited to low variance, the Monte Carlo method has very slow convergence, and the inverse stochastic operator method (expansion of the inverse of the stochastic operator in a Neumann series) is complex and limited to low variance. An alternative and relatively more effective approach is the polynomial chaos technique. Originally due to Wiener ([10]) and mainly applied in the field of structural mechanics, this technique is now making its way into the field of fluid mechanics ([3,11,12]).

Polynomial chaos. The basic idea of polynomial chaos is to project the variables of the problem onto a stochastic space spanned by a complete set of orthogonal polynomials, Ψ, functions of a random variable $\xi(\theta)$ where θ is a random event. A random variable can have different distributions. For example, $\xi(\theta)$ can have a Gaussian distribution, which is appropriately represented by Hermite polynomials; random variables with Gamma and Beta distributions are usually represented by Laguerre polynomials and Jacobi polynomials, respectively. Obviously, the convergence rate, and thus, the number of terms required for a given accuracy, depends both on the random process being approximated and on the random variables used. This is explained in [11]. In this approach, each variable of the problem is expanded, say for the velocity u, as

$$u(\boldsymbol{x},t,\theta) = \sum_{i=0}^{\infty} u_i(\boldsymbol{x},t)\, \Psi_i(\xi(\theta)). \tag{1}$$

In practice, the series is truncated to a finite number of terms, hereafter denoted by Npc. This framework is also valid for partially correlated or non-correlated variables. In that case, the variables of the problem are functions of several independent random variables and $\boldsymbol{\xi}$ is a vector. In the case of Hermite polynomials, the zero*th* order term represents the mean value, the first order term the Gaussian part, and higher order terms account for non-Gaussian contributions.

The statistics of a variable u can be determined from its polynomial chaos expansion. in addition to the mean value, given by the zero*th* term of the series, and the variance, the spatial covariance C is obtained from:

$$C_{uu}(x_1;x_2) \;=\; \sum_{i=1}^{P} u_i(x_1)\, u_i(x_2) \;< \Psi_i^2 >, \tag{2}$$

where $< . >$ denotes the stochastic scalar product.

Karhunen-Loeve expansion. When the structure of the covariance function is known, the most effective representation of a random process is through the Karhunen-Loeve decomposition. The spectral representation of the covariance function is :

$$C(x_1,t_1;x_2,t_2) \;=\; \sum_{i=0}^{\infty} \lambda_i\, f_i(x_1,t_1)\, f_i(x_2,t_2), \tag{3}$$

where λ_i and $f_i(x,t)$ are the eigenvalue and eigenfunction of the covariance kernel, respectively. This leads the random process $u(\boldsymbol{x},t,\theta)$ being expressed as (see derivation in [2]):

$$u(\boldsymbol{x},t,\theta) \;=\; \overline{u}(\boldsymbol{x},t,\theta) + \sum_{i=1}^{\infty} \sqrt{\lambda_i}\, f_i(\boldsymbol{x},t)\, \varphi_i(\theta), \tag{4}$$

where φ_i are random variables to be determined.

Depending on the stochastic spectrum of the random process, the rate of decay of the eigenvalues λ_i varies and thus affects the number of terms retained in the Karhunen-Loeve expansion and also governs the number of stochastic dimensions of the problem.

2 Stochastic solution of the nozzle flow problem

Fundamentals. To illustrate the polynomial chaos approach, we consider a quasi-one-dimensional nozzle flow. The solution is well known, and the simple flow configuration makes it an appropriate choice for focusing on the proposed stochastic treatment.

The system of compressible Euler equations is solved in conservative form:

$$\boldsymbol{Q}_t + \boldsymbol{F}_x = \boldsymbol{S}, \tag{5}$$

$$\text{where} \quad \boldsymbol{Q} = \begin{pmatrix} \varrho\, A \\ \varrho\, u\, A \\ \varrho\, E\, A \end{pmatrix} \quad \boldsymbol{F} = \begin{pmatrix} \varrho\, u\, A \\ \varrho\, u^2\, A + P\, A \\ \varrho\, u\, E\, A + P\, u\, A \end{pmatrix} \quad \boldsymbol{S} = \begin{pmatrix} 0 \\ P\, \partial A/\partial x \\ 0 \end{pmatrix} \tag{6}$$

Here, ϱ is the density, u the velocity, A the cross-sectional area, P the static pressure, and E the total energy.

The problem considers supersonic flow in a divergent nozzle. Uncertainty arises from inlet boundary conditions (say pressure, velocity and density) and also from "fuzzy" nozzle shape. The Euler equations are spatially discretized by the spectral element method and temporally discretized by a fourth-order Runge-Kutta scheme. This spatial discretization is consistent with the polynomial chaos expansion.

The solution to Equation (5) is expanded in terms of spectral and stochastic basis functions, and a Galerkin technique is applied. Specifically, the spectral basis functions are the Chebyshev polynomials, and the stochastic basis functions are Hermite polynomials (Eq. (1)).

Manufacturing limitations introduce uncertainty in the shape of the nozzle. Let $A(x)$ be Gaussian and partially correlated along the x direction:

$$C_{AA}(x_1, x_2) = \sigma^2\, e^{-\frac{|x_1 - x_2|}{b}}, \tag{7}$$

where b is the correlation length, possibly determined from manufacturing constraints, and σ^2 is the associated variance. Here, we consider $b = 20$ and $\sigma = 0.05$. As the covariance structure of the cross-sectional area A is known, the Karhunen-Loeve decomposition is used to represent the stochastic process.

Since A is assumed to be Gaussian, the corresponding random variables φ_i are also Gaussian and form an orthonormal basis, avoiding the need to project onto the space of polynomial chaos. Thus, we have $\varphi_i(\theta) = \xi_i(\theta)$, and A can be directly expressed as

$$A(x) = \overline{A}(x) + \sum_{i=1}^{\infty} \sqrt{\lambda_i}\, \Phi_i(x)\, \xi_i(\theta) \tag{8}$$

Validation of polynomial chaos results. The polynomial chaos (PC) approach is validated by comparison with regular Monte Carlo (MC) simulations (which do not take advantage of any acceleration procedures). All the inlet conditions are assumed Gaussian and fully correlated while the cross-sectional area

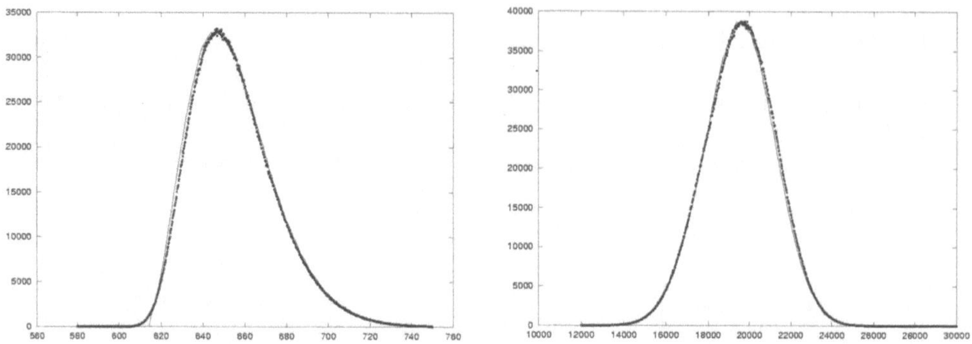

(a) Velocity. (b) Pressure.

Fig. 1. Outlet probability density function. 1-D *2nd* order Polynomial Chaos (*solid line*) vs MonteCarlo (*dots*). Arbitrary units for y-axis.

is assumed deterministic. MC simulations are carried out with over 9 millions independent samples, while the PC approach is 1-D (i.e., a single stochastic parameter) second order approximation (three-term expansion). Figure 1 shows the comparison between the probability density distributions (PDF) obtained from PC and MC techniques for the pressure and velocity at the nozzle exit. The pressure is found to remain approximately Gaussian at the nozzle exit while the velocity PDF becomes skewed due to the non-linearity of the Euler equations. The agreement between the results of the two techniques is noted to be excellent, in particular, for the pressure where the two curves match almost perfectly. For the velocity, the agreement is very good except in the left tail region. Higher order terms in the PC expansion are required to improve the approximation in the tail region. This example clearly demonstrates the ability of the PC technique to accurately propagate uncertainty through flow evolution, even with a limited number of terms in the expansion (3 terms in the present case). Considering the difference in the CPU time required ($1.7\,10^6$ s for MC, 70 s for PC), the speed-up ratio is approximately 24000, hence making the PC a much more effective alternative to the MC technique for this case, even if the current convergence acceleration procedures improve MC performance by two orders of magnitude.

Results. *1-D stochastic simulation.* We call the simulation with a single parameter as a random variable a 1-D stochastic simulation. The impact of uncertainty is best understood with a non-correlated random variable. In the present case, the inlet velocity is considered to be stochastic while all the remaining boundary conditions are deterministic. The covariance of pressure is easily determined and plotted in Fig. 2(a) for the pressure. An even more graphic representation is the evolution of the PDF of a variable along the nozzle. Figure 2(b) shows

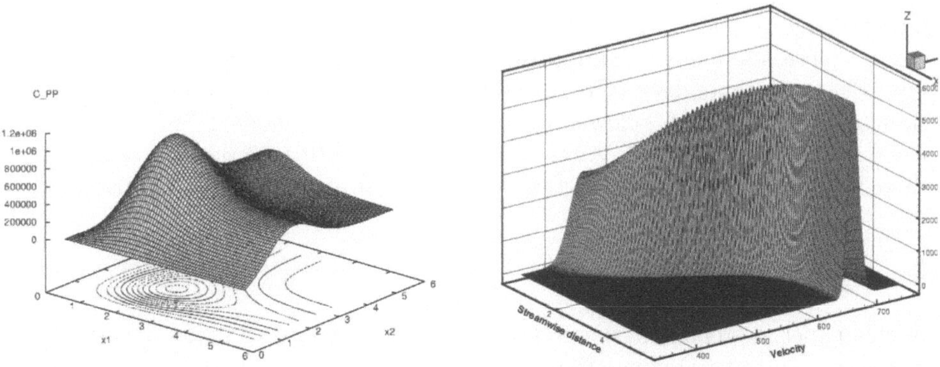

(a) Pressure covariance. (b) Velocity PDF along the nozzle. Arbitrary units for y-axis.

Fig. 2. 1-D stochastic simulation.

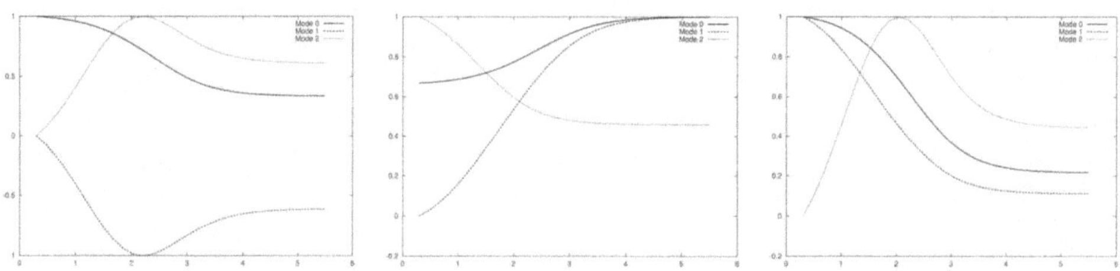

(a) Density modes. (b) Velocity modes. (c) Pressure modes.

Fig. 3. Distribution of the first three normalized modes for a 2-D *2nd* order PC.

the velocity PDF. It is to be noted that the variance decreases with the streamwise distance. This is due to the fact that in the configuration considered here (supersonic flow in a divergent nozzle), we have $\partial\, u_{\text{exit}} \,/\, \partial\, u_{\text{inlet}} < 1$.

2-D stochastic simulation. In this simulation, the inlet pressure and velocity are both assumed stochastic with no correlation while the inlet density is considered to be deterministic. The uncertainty in inlet pressure is represented by the random variable ξ_1 while velocity uncertainty is represented by ξ_2. The nozzle shape, A, is assumed uncertain as well and is expanded in a Karhunen-Loeve series. Figure 3 shows the first three modes of the variables along the nozzle.

Mode 0 is the mean value, mode 1 is the variance in stochastic space dimension ξ_1, and mode 2 is the variance in stochastic dimension ξ_2. Higher order modes account for coupling between the two stochastic dimensions and thus bring some skewness to the resulting PDF. They are not plotted to avoid cluttering of the figure.

As the inlet pressure and velocity are assumed to be independent, the evolution of their modes 1 and 2 are little affected by cross coupling. The velocity mode 1 increases due to a finite pressure variance. As the inlet density is deterministic, both its modes 1 and 2 start from 0. As pressure and density behave similarly, modes 1 and 2 downstream evolve similarly: mode 1 first decreases, while mode 2 first increases. After they peak, they both asymptote to a limiting value as the velocity variance and the uncertainty in stochastic space ξ_2 decrease.

3 Conclusions

In this paper, a stochastic spectral approach is developed to study uncertainty quantification in CFD simulations. It is applied to a quasi-one-dimensional supersonic flow problem. The results demonstrate that this approach is effective and holds promise to become a viable technique, although computationally intensive, even for realistic configurations. Furthermore, it yields complete statistics of arbitrary order.

Further development and improvement of this methodology in treating uncertainty in simulations could be achieved with a stochastic collocation scheme, which could enhance the area of its applicability (e.g. to highly nonlinear and complex models) and could reduce the computational time in some cases. It is critically relevant to extending the popular discontinuous Galerkin spectral method (where a Riemann problem has to be solved) to include stochastic partial differential equations. A preliminary study of the stochastic Riemann problem and the development of a stochastic discontinous spectral element method can be found in [6].

Acknowledgements: The authors thank Thomas A. Zang for suggesting this problem and for his support.

References

1. H.W. Coleman, F. Stern: J. Fluids Eng. **119**, 795 (1997)
2. R.G. Ghanem, P.D. Spanos: *Stochastic finite elements: a spectral approach*, (Springer-Verlag 1991)
3. L. Huyse: ICASE Report 2001-18 (2001)
4. L.D. Landau, E.M. Lifchitz: *Cours de physique théorique. Vol.6. Mécanique des fluides*, 2nd edition (Mir Moscow 1989) [in French]
5. H.W. Liepmann, A. Roshko: *Elements of gas dynamics*, (Wiley, New York 1957)
6. L. Mathelin, M.Y. Hussaini, preprint (2002)
7. U.B. Mehta: J. Fluids Eng. **113**, 538 (1991)
8. P.J. Roache: J. Fluids Eng. **116**, 405 (1994)
9. G.A. Sod, J. Comp. Phys. **27**, 1 (1978)
10. N. Wiener: Amer. J. Math. **60**, 897 (1938)
11. D. Xiu, G.Em. Karniadakis: preprint (2002*a*)
12. D. Xiu, G.Em. Karniadakis: J. Comput. Phys.(2002*b*)
13. H.C. Yee,P.K. Sweby: AIAA J. **36**(5), 712 (1998)

A Computational Method for Combustion Using Conditional Moment Closure

M.J. Cleary, J.H. Kent and R.W. Bilger

School of Aerospace, Mechanical and Mechatronic Engineering, The University of Sydney 2006, Australia

Abstract. Two numerical methods are compared to solve the steady-state Conditional Moment Closure equations for a combustion application. A time stepping scheme for stiff equations, VODE, is compared with a modified Newton-Raphson method. The results show that the modified Newton-Raphson has a significant advantage under CMC conditions where the solution is mostly close to the starting estimates and resort to time-stepping method is not needed.

1 Introduction

A computationally costly part of numerical codes for combustion is the solution of the differential equations containing the chemical kinetics. Due to the widely varying rates between the different reactions the equations tend to be stiff and the solution poses a significant challenge. For combustion problems such as fires in large spaces very many computational fluid dynamics (CFD) grid cells are required and computational efficiency becomes very important.

This paper compares two numerical methods employed to solve the stiff set of equations governing the flow and chemistry in an elliptic, axisymmetric CFD code for turbulent flow to predict carbon monoxide formation in a methane hood-fire [1]. Turbulent combustion is modelled using steady-state Conditional Moment Closure (CMC) [2]. A comparison is made here between a scheme employing a modified Newton-Raphson (N-R) solver [3], and a time-stepping scheme which employs the variable-coefficient ODE solver, VODE [4,5]. VODE is widely used by others [6] for this type of application, whereas N-R is not normally considered useful for chemical kinetics, because of stiffness. Newton-Raphson is used by Grcar [7] in combination with a time-stepping routine for similar two-point boundary value problems whereas here the modified N-R is used alone.

2 Equations

The fundamental hypothesis in CMC is that the fluctuating reaction rates in a turbulent, reacting flow can be more accurately modelled by considering species conditioned upon the mixture fraction, ξ, having a particular value η. The steady state form of the CMC equation is given by

$$\left[\langle \mathbf{v}|\eta\rangle - D_t \frac{\nabla(\rho\tilde{P}(\eta))}{\rho\tilde{P}(\eta)}\right] \cdot \nabla Q_i - \nabla \cdot D_t \nabla Q_i = \frac{1}{2}\langle\chi|\eta\rangle \frac{\partial^2 Q_i}{\partial \eta^2} + \langle w_i|\eta\rangle \quad (1)$$

where Q_i and $\langle w_i|\eta\rangle$ are the conditional average mass fraction and conditional average chemical source term of species i, and $\langle \mathbf{v}|\eta\rangle$ and $\langle \chi|\eta\rangle$ are the conditional average velocity and conditional average scalar dissipation respectively. $\tilde{P}(\eta)$ is the Favre probability density function (PDF) of the mixture fraction about its mean. D_t is the turbulent diffusivity. Equation(1) is solved for I species, J nodes in mixture fraction space and K nodes in physical space.

The stiffness of this set of equations is due to large differences between the conditional chemical source terms, which for first-order closure are taken as

$$\begin{aligned}\langle w_i(\mathbf{Y},T)|\eta\rangle &= w_i(\mathbf{Q},Q_T)\\ &= \sum_{l=1}^{L}(\nu''_{i,l}-\nu'_{i,l})\left[k_{f,l}\prod_{i=1}^{I}[X_i|\eta]^{\nu'_{i,l}} - k_{r,l}\prod_{i=1}^{I}[X_i|\eta]^{\nu''_{i,l}}\right].\end{aligned} \qquad (2)$$

Here $k_{f,l}$ and $k_{r,l}$, the forward and reverse rate constants of reaction l, are functions of the conditional average temperature, Q_T, and $\nu'_{i,l}$ and $\nu''_{i,l}$ are the stoichiometric coefficients on the reactant and product sides, respectively, of each reaction. $[X_i|\eta]$ is taken here to be the conditional molar concentration of species i.

The spatial advection and diffusion terms on the left hand side (LHS) of (1) distinguish the CMC method from the Stationary Laminar Flamelet Model (SLFM) [8]

$$\frac{1}{2}\langle \chi|\eta\rangle\frac{\partial^2 Q_i}{\partial\eta^2} + \langle w_i|\eta\rangle = 0. \qquad (3)$$

The transport terms on the LHS of (1) allow spatial effects to be properly included such as for example the transport of carbon monoxide from the flame to a cooled smoke layer where reaction is frozen [1]. However, these terms are small compared with the chemical source terms when a solution is being sought, which has not yet converged. Therefore (3) is used here as a simpler model to evaluate the numerical methods.

Equation (3) is discretised using a second-order central difference for the mixture fraction diffusion term and the chemical source term is linearised so that

$$\langle w_i|\eta\rangle = S_{U,i} + S_{P,i}Q_{P,i} \qquad (4)$$

Obtaining a solution to (3) amounts to satisfying the set of equations

$$\mathbf{F}(\mathbf{Q}) = [F_1, F_2, ..., F_N]^T = \mathbf{0} \qquad (5)$$

at each of the J nodes in mixture fraction space, where

$$F_i = -a_{P,i}Q_{P,i} + a_R Q_{R,i} + a_L Q_{L,i} + S_{U,i} \qquad (6)$$

$$a_R = \frac{1}{2}\langle\chi|\eta\rangle\frac{1}{\frac{1}{2}(\eta_R-\eta_L)(\eta_R-\eta_P}) \qquad (7)$$

$$a_L = \frac{1}{2}\langle\chi|\eta\rangle\frac{1}{\frac{1}{2}(\eta_R-\eta_L)(\eta_P-\eta_L}) \qquad (8)$$

$$a_{P,i} = a_R + a_L - S_{P,i} \qquad (9)$$

Subscripts L and R indicate values at adjacent nodes to the "*left*" and "*right*" of the point P in mixture fraction space.

3 Newton-Raphson solver

The Newton-Raphson method as applied to reacting systems is described by Turns [9] and the Fortran code is adapted from Press *et al.* [3]. The Jacobian matrix $J_{ij} \equiv \frac{\partial F_i}{\partial Q_i}$ is required at each iteration. For chemical mechanisms with elementary reactions this can be determined analytically.

The solution is applied to solve for all species at a particular conditioned $\xi = \eta$. It is repeated for all ξ using the previous iteration values for the neighbouring terms in mixture fraction space.

An advantage of solving for species mass fractions is that all species are bounded by values of zero and unity. However, there is a range of many orders of magnitude between the major species (percentage range) and some minor species (parts per million range). The performance of the Newton-Raphson method is modified here by setting appropriate limits to the

- allowed range of values
- allowed movement between iterations.

The allowed value upper limit sets scaled upper bounds for all species and is taken from known solutions. The iteration-to-iteration movement limit of Q_i in the upward direction is set to a fraction of the scaled upper limit (typically 0.1). A simple under-relaxation factor is not useful because the chemical source term may predict a value orders of magnitude greater than unity. For downward movement the implicit formulation prevents large excursions into negative values and a conventional under-relaxation factor can be used (here 0.9).

The convergence criterion is that for all species and all mixture fractions, the maximum residual of (5) normalised by the maximum conditional average mass fraction Q, is less than some specified tolerance.

4 VODE solver

VODE is a publicly available code from the Lawrence Livermore National Laboratory [4]. It is a variable-coefficient ODE solver that solves unsteady equations of the form

$$\frac{\partial \mathbf{Q}}{\partial t} = \mathbf{F}(\mathbf{Q}) \tag{10}$$

where for a steady-state solution $\partial \mathbf{Q}/\partial t \to \mathbf{0}$.

A solution is obtained simultaneously for all species and mixture fractions so that $\mathbf{F}$ and $\mathbf{Q}$ have length $I \times J$. This solver also requires the Jacobian matrix, which is determined analytically as described above.

User inputs include the output time, the number of internal time steps, and a tolerance, which is scaled according to the upper mass fraction limit of each species. For stiff problems VODE develops an interpolation function, up to order five, from time zero to the output time within the prescribed number of internal time steps. If the output time is not attained within constraints then, the solution is produced at an earlier time. Successive calls to VODE can be made to approach a steady-state solution.

5 Initial and boundary conditions

The initial conditions are obtained from a simple infinite-rate, one-step reaction of methane fuel in air to produce complete combustion. In mixture fraction space this is characterised by no fuel on the lean side of stoichiometry, and no oxygen on the rich side, with maximum products at stoichiometry. Nitrogen is inert and minor species are initially zero.

The two-point boundary conditions are at $\xi = 0$ and $\xi = 1$. Here the conditioned species have the values found in the pure air stream (oxygen and nitrogen only) and the pure fuel stream.

Temperature, which strongly influences reaction rates is obtained from transport of conditioned enthalpy. For unity Prandtl and Schmidt numbers and adiabatic conditions the conditioned standardised enthalpy has a simple algebraic relation to mixture fraction

$$h = \xi h_{\xi=1} + (1 - \xi) h_{\xi=0}. \tag{11}$$

The boundary values of standardised enthalpy are obtained from the inlet stream temperatures.

6 Results and discussion

The flames modelled here are adiabatic flames for methane fuel. The chemical mechanism is a truncated version from D'Anna *et al.* [10] with 26 species and 107 reactions. Scalar dissipation is taken to be constant, $\tilde{\chi}$, across mixture fraction. Solutions are obtained over a range of scalar dissipation from $\tilde{\chi} = 1$ to $\tilde{\chi} = 40$. The lower value is similar to values at the base of buoyant fire plumes, and the upper value is at about the value which causes flame extinguishment. Fuel ($\xi = 1$) and air ($\xi = 0$) streams are at $20^0 C$.

Figure 1 shows solutions of selected species in mixture fraction space for initial conditions and conditions at $\tilde{\chi} = 1$ and $\tilde{\chi} = 40$. At low scalar dissipation rate the major species distributions are similar to the initial values for the one-step complete reaction. Minor species such as the toxicant CO and the radical OH are initially zero but are evident at $\tilde{\chi} = 1$. The temperature at stoichiometry is $2084^0 C$. At the high scalar dissipation rate the flame is extinguished and non-reacting mixing at room temperature is shown.

Figure 2 shows the ratio of the time to convergence of VODE to N-R over the range of scalar dissipation rates. The convergence criteria is 5×10^{-5}. For VODE the maximum number of internal time steps at each solver call is set to 50 for a time interval of 0.1s at each call.

These results show that N-R achieves a significant advantage over VODE at low scalar dissipation rate. As the scalar dissipation rate increases the steady-state solution departs further from the initial conditions (the infinite-rate one-step reaction) and the advantage decreases. At about $\tilde{\chi} = 30$ as the flame nears extinction there is little advantage between the two methods and for extinguishment the Newton-Raphson solver is about half the speed of VODE.

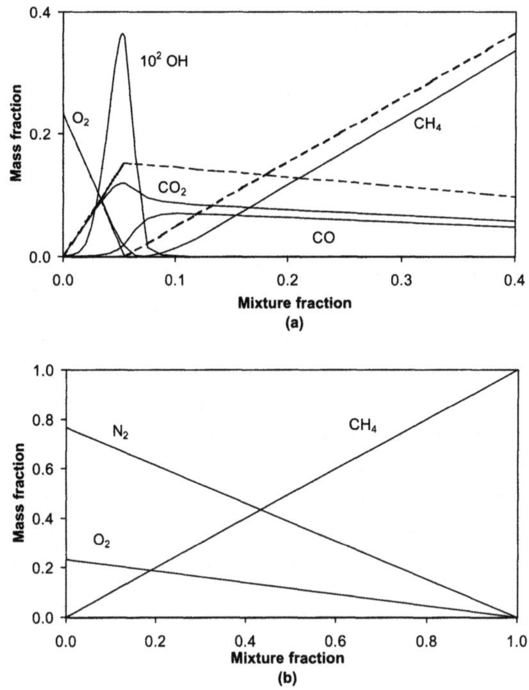

Fig. 1. Selected species distribution versus mixture fraction for an adiabatic methane flame: **(a)** broken lines: initial conditions, full lines: $\tilde{\chi} = 1$, and **(b)** $\tilde{\chi} = 40$.

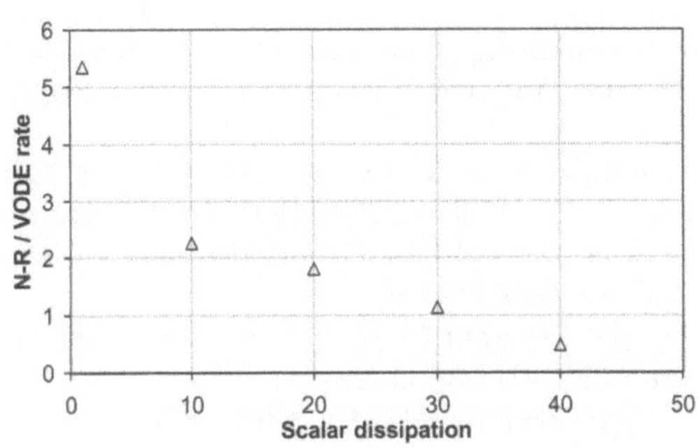

Fig. 2. Rate of steady-state convergence of Newton-Raphson to VODE versus scalar dissipation.

The convergence time for VODE is not strongly dependent on the difference between the initial condition and final solution varying from 36s at $\tilde{\chi} = 1$ to 58s at $\tilde{\chi} = 40$. By contrast Newton-Raphson ranges from 7s to 120s showing its great speed when near to a solution.

7 Conclusion

The comparison of the convergence times of two methods to solve a model for the steady-state CMC equation shows that the modified Newton-Raphson method can have a clear advantage over the time stepping method, VODE. The advantage is manifest when the initial conditions are in the vicinity of the solution. This is the case for the CMC equations where the transport terms tend to vary mildly due to spatial coupling. Under these conditions the solution is usually close to the starting estimates and the modified Newton-Raphson method described here is the preferred choice. The modifications, which set scaled limits to the values and limits on movement of values per iteration allow N-R to be used alone to solve stiff equations without resort to time stepping.

References

1. M.J. Cleary, J.H. Kent, R.W. Bilger: 'Prediction of Carbon Monoxide in Fires by Conditional Moment Closure'. Submitted to *The 29th Symposium (International) on Combustion at Sapporo, Japan, July 21–27,2002*
2. A.Y. Klimenko, R.W. Bilger: Prog. Energy Combust. Sci. **25**, 595 (1999)
3. W.H. Press, S.A. Teukolsky, W.T. Vetterling, B.P. Flannery: *Numerical Recipes in Fortran 77*, 2nd edn. (Cambridge University Press, Cambridge 1992)
4. P.N. Brown, G.D. Byrne, A.C. Hindmarsh: SIAM J. Sci. Stat. Comput. **10**, 1038 (1989)
5. G.D. Byrne, A.C. Hindmarsh: ACM Trans. Math. Softw. **1**, 71 (1975)
6. R.J. Kee, F.M. Rupley, J.A. Miller: 'CHEMKIN-II: A Fortran Chemical Kinetics Package for the Analysis of Gas-Phase Chemical Kinetics'. Sandia National Laboratories Report SAND89-8009B (1996)
7. J.F. Grcar: 'The Twopnt Program for Boundary Value Problems'. Sandia National Laboratories Report SAND91-8230 (1992)
8. N. Peters: Prog. Energy Combust. Sci. **10**, 319 (1984)
9. S.R. Turns: *An Introduction to Combustion: Concepts and Applications*, 2nd edn. (McGraw Hill, USA 2000)
10. A. D'Anna, A. D'Alessio, J. Kent: Combust. Flame **125**, 1196 (2001)

Building-Cube Method for Flow Problems with Broadband Characteristic Length

Kazuhiro Nakahashi

Tohoku University, Aoba 01, Sendai 980-8579, JAPAN
email: naka@ad.mech.tohoku.ac.jp

Abstract. In this paper, requirements for the next generation CFD in the near-future computational environment is discussed and a new approach named Building-Cube Method is proposed aimed for large-scale, high resolution computations around a complex geometry. In the present method, a flow field is divided into a number of cubes (squares in 2D) of various sizes. Each cube is a computational sub-domain with Cartesian mesh of equal spacing and equal number of nodes. The geometrical size of each cube is determined by adapting to geometry and flow features so as to cope with broadband characteristic length of the flow. Equal spacing and equal number of Cartesian grid in each cube make it easy to parallelize the flow solver and to handle huge data output. The approach is tested for a shock/boundary layer interacting problem.

1 Introduction

CFD has achieved a significant progress in its algorithms and the applications in the last 30 years. Among the various approaches developed, computational methods using unstructured grids are becoming popular in the last decade because of their advantages over conventional structured grids in treating complex geometry (for example, Fig.1 [1]).

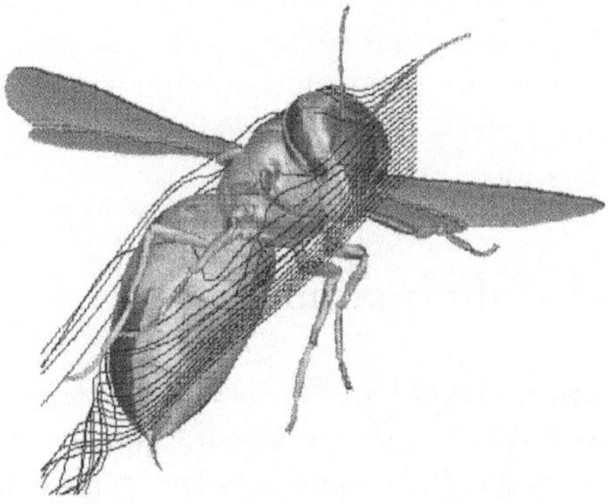

Fig. 1. Flow computation around a hornet by unstructured mesh

However the current CFD is not powerful enough for flows that require high resolution not only near wall boundaries but also in space. Flow with large separations is one of them. Accurate prediction of airfoils/wings at stall is still one of the remaining problems [2]. Direct numerical simulation (DNS) of turbulent flow around a real geometry is in the far distant future because of the broadband characteristic flow lengths. Aeroacoustics simulation also requires high resolution with highly dense grid near the sound source and higher-order scheme in space for sound propagation where the mean-flow amplitude scales are typically orders of magnitude larger than those of the generated acoustics.

In this paper, requirements for the next generation CFD algorithm that will enable the turbulent flow simulations around a complex geometry in the near-future computational environment are discussed at first. Then a new approach named Building-Cube Method is proposed aimed for the large-scale, high resolution computations for flow around a complex geometry.

2 Requirements for next generation CFD

The requirements for the next generation CFD method are; (1) capability to treat complex geometries, (2) capability to use locally dense grid fitted to the local characteristic flow length, (3) easy adaptive refinement, (4) easy construction of higher-order scheme, (5) efficient use of parallel computers, (6) easy post processing for extremely large-scale computations.

Unstructured mesh is suitable to resolve various characteristic flow lengths by use of adaptive mesh refinements. However, construction of higher-order spatial accuracy on unstructured grid is painfully expensive in computational time and memory.

Higher-order scheme in space can be easily constructed on a constant-spacing Cartesian mesh, which has been demonstrated its capability for internal flows such as automobile engine rooms. The drawback of this approach, however, is the difficulty for outer flows where the large difference in characteristic flow lengths exists. A variant of the Cartesian grid method is to use adaptive mesh refinement in space and cut cells on the wall boundaries. However, the introduction of irregular subdivisions and cells into the Cartesian grid complicates the algorithm and increases the memory requirement. Therefore the advantages of the Cartesian mesh over the unstructured grid, such as simplicity and less memory requirement, disappear.

3 Building-Cube method

To satisfy the requirements for the next generation CFD mentioned above, basic strategies employed here are; (a) subdivisions of a flow field by cubes (squares in 2D) of various sizes, (b) equally-spacing, highly dense Cartesian mesh in each cube, (c) adaptive refinement method for cube construction, and (d) same grid size for all cubes for easy parallel computation.

The Cartesian mesh is easy to extend to the higher-order scheme. The most critical problem of the conventional Cartesian mesh is how to fit to the local characteristic flow length. Here, a flow field is divided into a number of sub-domains, named Cube. The geometrical size of each cube is determined by adapting to geometry and flow features like an adaptive refinement Cartesian mesh [3].

The outline of the method is shown for 2D computation around a multi-element airfoil in Fig.2. Solid lines in Fig.2 show the cube boundaries. Gray lines show the Cartesian mesh inside of each cube. In this case each cube is equally divided by 16x16 mesh. Even with this relatively coarse grid in cube, the staircase representation approximates the original surface well as shown in the figure. For the Navier-Stokes computations, each cube will be divided by 64x64 mesh or 128x128 to resolve the boundary layers and wakes accurately. By use of this fine grid in cube, the original body surface is accurately reconstructed by staircase representation of 10^{-4} scale to the airfoil chord length.

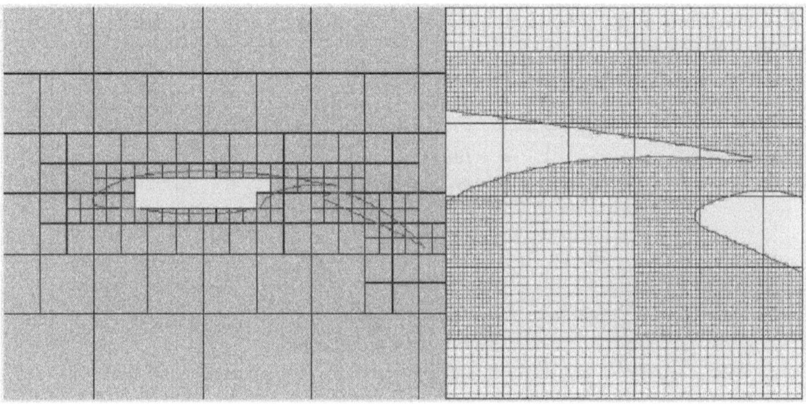

Fig. 2. Building cube mesh around a multi-element airfoil. Total number of cubes is 228 and each cube is divided by 16x16 Cartesian mesh. The airfoil surface is approximated by staircase representation

4 Test case

The present method was tested for an interacting flow between an incident shock wave and a laminar boundary layer on a flat plate (Fig.3). Freestream Mach number is 2 and the Reynolds number is 2.96×10^5. The flow field is divided into 92 cubes. Cartesian mesh in each cube is 64x64 and the minimum spacing on the wall is about 6×10^{-4} which is small enough to resolve the laminar boundary layer.

The compressible Navier-Stokes equations were solved by the third-order upwind method. The computed result reproduces a separation bubble on a flat plate caused by the incident shock wave.

The parallel computation using the OpenMP was tested on the SGI Origin 2000 parallel computer at the Institute of Fluid Science in Tohoku University.

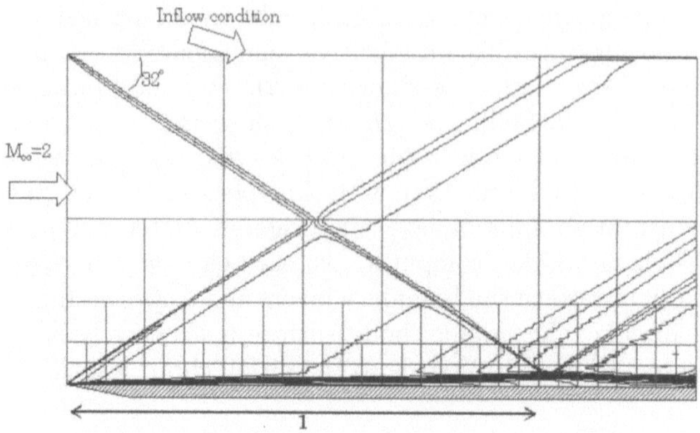

Fig. 3. Computed density contours of a shock/boundary layer interacting problem and the cube boundaries.

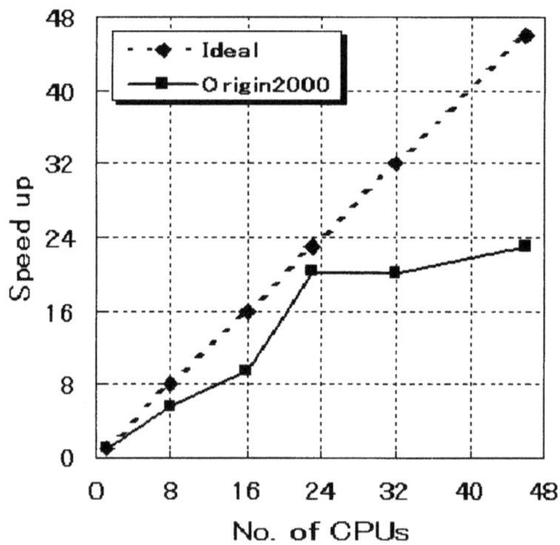

Fig. 4. Speedup in SGI Origin 2000. Number of cubes is 92.

Figure 4 shows the parallel performance of this computation. The number of the cubes is 92 so that the result with 23 CPUs is very close to the ideal performance. The implementation of the method on a parallel computer is straightforward and the efficient computation can be expected for large scale computations.

5 Conclusion

In this paper, an approach aimed for large-scale computation on a parallel computer, named Building-Cube Method, was proposed. The method is based on the Cartesian mesh, and the local grid density is adapted to the flow characteristic length by changing the cube size. Equal spacing and equal number of Cartesian grid in each cube make it easy to parallelize the flow solver and to handle huge data output.

References

1. K. Nakahashi : 'Progress in Unstructured- Grid CFD'. Proc. 1st ICCFD, Computational Fluid Dynamics 2000, Ed. by N. Satofuka, Springer, pp.3-13, (2000)
2. N. Hasuike, A. Oyama, K. Nakahashi, S. Obayashi: 'CFD Stall Prediction of 2D Airfoils'. J. Japan Society for Aeronautical and Space Sciences, **50**, 577 (2002)
3. D.D.Zeeuw, K. G. Powell : 'An Adaptively Refined Cartesian Mesh Solver for the Euler Equation'. J. Comp. Phys., 104, pp.56-68, (1993)

A Discrete Sensitivity Solver for an Industrial CFD Code via Automatic Differentiation

David W. F. Standingford[1] and Shaun A. Forth[2]

[1] BAE SYSTEMS Advanced Technology Centre, Sowerby Building, PO Box 5, FPC 267 Filton, Bristol BS34 7QW, UK, email: `David.Standingford@baesystems.com`

[2] Applied Mathematics & Operational Research, ESD, Cranfield University (RMCS Shrivenham), Swindon SN6 8LA, UK, email: `S.A.Forth@rmcs.cranfield.ac.uk`

Abstract. We report on development and validation of a discrete sensitivity solver for the BAE SYSTEMS/Airbus UK CFD code FLITE3D. We used the Odyssée automatic differentiation (AD) tool to create a discrete forward sensitivity version of FLITE3D. Validation is via comparing sensitivities of integrated forces (lift, drag, side force) with respect to (w.r.t) angle-of-attack (AoA) α, calculated using the sensitivity solver and central-differencing. Validation, w.r.t. changes in the surface and field mesh, is performed by setting sensitivities of all mesh related quantities as if rotated by an infinitesimal angle $\Delta\alpha$. Such sensitivities correspond to those calculated w.r.t AoA α. We investigate calculating discrete sensitivities under 2 approximations regarding so-called mesh sensitivities, concluding that they are not applicable to FLITE3D. We present results of a wing geometry optimisation using forward sensitivities.

1 Introduction

Industrial use of CFD is now routine and the current challenge is to incorporate large scale CFD codes into the engineering design cycle via design optimisation. The associated use of gradients is non-trivial and is the subject of the monograph [8]. Aerodynamic designs may have 100's of design variables making the cost of finite–differencing prohibitive. This prompted interest in adjoint methods which compute gradients in time proportional to the (small) number of design objectives [5]. Such methods may be implemented at the continuous level (e.g. [1]) in which adjoint PDEs are formulated, discretised and solved. This requires mathematical analysis, both initially and as the CFD solver is updated, and may be perceived by industry as hard to maintain. Discrete adjoint methods, in which the discretised flow solver is adjoined, are therefore of interest. A discrete adjoint solver may be generated and maintained by hand–coding [9], again with maintenance problems, or via automatic differentiation (AD) [6] tools. We are using the AD tool Odyssée [4] to develop both forward and adjoint sensitivity solvers for the BAE SYSTEMS/Airbus UK compressible flow solver FLITE3D. Here we report on development and validation of the forward sensitivity solver.

2 The Flite3D CFD Package

The CFD package FLITE3D consists of a number of separate programs.

Surface mesh For geometric parameters $\underline{\beta}$ generates surface mesh $\boldsymbol{x_S}$.

Volume mesh Generate volume mesh $\boldsymbol{x}$ using advancing front algorithm.

Solver preprocessor Produces agglomeration data for multigrid acceleration and edge–based data for the solver.

Flow Solver Uses unstructured, node–based finite volumes with second–order spatial accuracy, Runge–Kutta (RK) time integration and multigrid acceleration to solve the compressible Euler equations. It is similar to [3], by default uses a pseudo-Laplacian numerical dissipation, and features many complex boundary conditions associated with aeronautical design.

A 2-D mesh, close to a geometric boundary, is depicted in Fig. 1. Away from the boundary, an area-weighted normal to a typical control volume is labelled $\boldsymbol{w}$. Control volumes for boundary vertices are cut by the boundary. An area weighted normal for a boundary face is denoted $\boldsymbol{n}$ and it contributes to several control–volumes. A unit point normal for a boundary vertex is denoted $\boldsymbol{a}$.

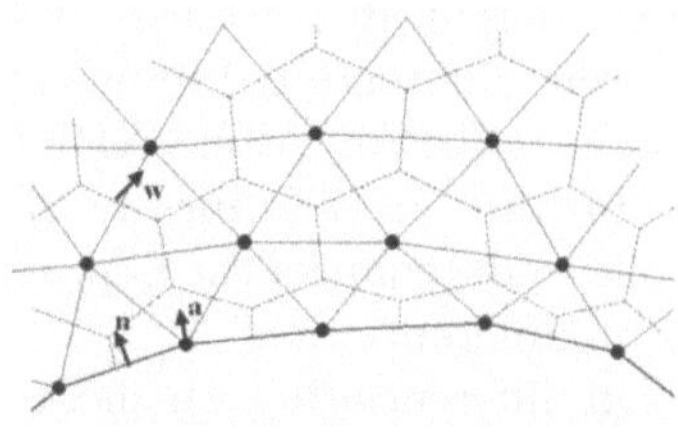

Fig. 1. Flite3D grid showing: actual mesh (unbroken lines); node points (black dots); and dual mesh (dashed lines) defining the flow–solver control–volumes.

Neglecting RK time-stepping and multigrid, FLITE3D's solver is described by (1–3).

$$\left.\begin{array}{l}\textbf{Surface Mesh Generator}\\ \text{Given design parameters } \underline{\beta}\\ \text{Surface mesh generation } \boldsymbol{x_S} = \boldsymbol{X_S}(\underline{\beta})\\ \text{Calculate point and face surface normals } \boldsymbol{a}(\boldsymbol{x_S}, \underline{\beta}), \boldsymbol{n}(\boldsymbol{x_S})\end{array}\right\} \quad (1)$$

$$\left.\begin{array}{l}\textbf{Volume Mesh Generator}\\ \text{Volume Mesh generation } \boldsymbol{x} = \boldsymbol{X}(\boldsymbol{x_S})\\ \text{Calculate Volume Mesh normals } \boldsymbol{w}(\boldsymbol{x}, \boldsymbol{x_S})\end{array}\right\} \quad (2)$$

$$\left.\begin{array}{ll}\multicolumn{2}{l}{\textbf{Flow Solver}}\\ \multicolumn{2}{l}{\text{Initialise flow variables } \boldsymbol{u}^0}\\ \texttt{Do} & n = 0, n_{max} - 1\\ & \text{on surface set momentum } \rho\boldsymbol{u} \text{ s.t. } \rho\boldsymbol{u}.\boldsymbol{a} = 0\\ & \boldsymbol{r}^n = \boldsymbol{R}(\boldsymbol{u}^n, \boldsymbol{x}, \boldsymbol{w}, \boldsymbol{n})\\ & \text{on surface set momentum residual } \boldsymbol{R_M} \text{ s.t. } \boldsymbol{R_M}.\boldsymbol{a} = 0\\ & \texttt{If (converged) Exit Do Loop}\\ & \boldsymbol{u}^{n+1} = \boldsymbol{u}^n - \boldsymbol{P_L}(\boldsymbol{u}^n, \boldsymbol{x})\boldsymbol{r}^n\\ & \text{Aerodynamic Coefficients } [C_L, C_D, C_S] = C(\boldsymbol{u}^n, \boldsymbol{x}, \boldsymbol{n})\\ \texttt{EndDo} & \end{array}\right\} \quad (3)$$

For the flow–solver (3), $\boldsymbol{P_L}(\boldsymbol{u}^n, \boldsymbol{x}) = -\Delta t/V$ is the diagonal matrix of negative time steps $-\Delta t_i^n$ divided by control volumes V_i for each cell i in the mesh. In (3) wall zero–mass–flux boundary conditions are applied in 2 stages [3].

Before residual evaluation, the momentum $\boldsymbol{\rho u}_j$ for boundary vertices j is *chopped* removing its normal component $\boldsymbol{\rho u}_j \to \boldsymbol{\rho u}_j - (\boldsymbol{a}_j \cdot \boldsymbol{\rho u}_j)\boldsymbol{a}_j$. After residual evaluation, the corresponding momentum residuals $\boldsymbol{R}_{M\,i}$ are similarly chopped $\boldsymbol{R}_{M\,i} \to \boldsymbol{R}_{M\,i} - (\boldsymbol{a} \cdot \boldsymbol{R}_{M\,i})\boldsymbol{a}_i$ to ensure stability.

3 Forward Sensitivities via AD Tool Odyssée

From the flow solver we extracted 75 subroutines, comprising 21,000 lines of code (including comments), concerned with calculating normals ($\boldsymbol{w}$, $\boldsymbol{n}$, $\boldsymbol{a}$), flow residuals and conservative variable updates. Twelve of these routines were individually differentiated, either by hand or with Odyssée's aid. The remaining 63 were differentiated exclusively by Odyssée. Forty–three, concerning the flow solver, were differentiated in one application of Odyssée taking 20 minutes CPU time on a COMPAQ Alpha XP1000 workstation (RAM: 1GB, clock–speed 667 MHz). In total 38,500 lines of differentiated Fortran code were produced. To ease code-maintenance, the C pre–processor and makefiles are used to automate nearly all differentiation.

When differentiating we chose free-stream Mach number M, AoA α, lateral angle of incidence β, together with all mesh points $\boldsymbol{x_S}$, $\boldsymbol{x}$ as independent variables. Sensitivities of lift C_L, drag C_D and side force C_S coefficients may then be calculated.

We validated the AD generated sensitivity solver first w.r.t. changes in AoA α and then w.r.t. changes in the surface $\boldsymbol{x_S}$ and volume $\boldsymbol{x}$ meshes.

3.1 Angle-of-Attack (AoA) Sensitivities

As is standard, flows of speed U at incidence α are calculated by rotating the far-field velocity $u = U\cos(\alpha)$, $v = U\sin(\alpha)$. To validate our sensitivity solver we calculated $dC_L/d\alpha$ using both AD code and central-differencing (CD). Convergence criteria were, at iteration n, $|C_L^i - C_L^{i-10}| \le 10^{-5} \times \max\left(1, \left|C_L^i\right|\right)$ for $i = n$ and $i = n - 10$. For the sensitivity solver we ensured,

$$\left|\frac{dC_L^i}{d\alpha} - \frac{dC_L^{i-10}}{d\alpha}\right| \le 10^{-5} \times \max\left(1, \left|\frac{dC_L^i}{d\alpha}\right|\right),$$

for $i = n, n - 10$. Identical convergence criteria were used for all further calculations. Results are in Table 1, and show good agreement of the two techniques. Table 2(a), gives AD results for a succession of finer meshes.

3.2 Differentiation with respect to all Mesh Quantities

To validate the sensitivity solver to changes in surface and volume meshes, we set sensitivities of all mesh points as for a rotation α about the y (wing span–wise) axis. This unconventional approach calculates AoA sensitivities as if by rotating the mesh by the infinitesimal $\Delta\alpha \to 0$. We calculated $d(\boldsymbol{w}, \boldsymbol{n}, \boldsymbol{a})/d\alpha$ using differentiated versions of volume, surface and point surface normal calculation routines. We zeroed far-field flow derivatives and the resulting sensitivities of Table 2(b) show excellent agreement with Table 2(a).

Table 1. AoA α sensitivities for M6 wing, AoA $\alpha = 5^o$, Mach no. 0.84 calculated by AD and approximated by CD (perturbations $\pm\Delta\alpha$) on 14,148 vertex mesh.

Method	$\Delta\alpha$	$dC_L/d\alpha$	$dC_D/d\alpha$	$dC_S/d\alpha$
AD	$\to 0$	6.5743×10^{-2}	1.1317×10^{-2}	9.8631×10^{-4}
CD	0.1	6.5729×10^{-2}	1.1318×10^{-2}	9.8373×10^{-4}
CD	0.05	6.5736×10^{-2}	1.1322×10^{-2}	9.8329×10^{-4}
CD	0.01	6.5887×10^{-2}	1.1321×10^{-2}	9.6541×10^{-4}

Table 2. Grid convergence of force sensitivities to AoA α calculated by 3 forward mode AD enabled techniques.

(a) No approximation and freestream flow derivatives set by rotation.

Mesh pts.	$dC_L/d\alpha$	$dc_d/d\alpha$	$dc_s/d\alpha$
28, 244	7.3186×10^{-2}	1.1625×10^{-2}	1.8982×10^{-3}
47, 177	7.3712×10^{-2}	1.1495×10^{-2}	2.0702×10^{-3}
70, 102	7.4098×10^{-2}	1.1752×10^{-2}	2.0775×10^{-3}

(b) No approximation and derivatives of all grid quantities set by rotation.

Mesh pts.	$dC_L/d\alpha$	$dc_d/d\alpha$	$dc_s/d\alpha$
28, 244	7.3186×10^{-2}	1.1626×10^{-2}	1.8980×10^{-3}
47, 177	7.3711×10^{-2}	1.1495×10^{-2}	2.0702×10^{-3}
70, 102	7.4098×10^{-2}	1.1752×10^{-2}	2.0776×10^{-3}

(c) Approximated by neglecting volume mesh sensitivities.

Mesh pts.	$dC_L/d\alpha$	$dC_D/d\alpha$	$dC_S/d\alpha$
28, 244	2.1455×10^{-1}	2.6997×10^{-2}	3.9774×10^{-3}
47, 177	2.1970×10^{-1}	2.6838×10^{-2}	4.4372×10^{-3}
70, 102	2.2079×10^{-1}	2.7964×10^{-2}	4.2330×10^{-3}

4 Eliminating Volume Mesh Sensitivities

In Sect. 3.2 we used sensitivities of all mesh related quantities in calculating surface mesh sensitivities. For design sensitivities of solution algorithm (1-3), this necessitates differentiating both surface (1) and volume (2) mesh generators. We would anticipate improved efficiency if we only differentiated the surface grid generation and flow solver stages. This approach is motivated by noting that in continuous sensitivity formulations there are no volume mesh sensitities [2].

4.1 Neglecting Volume Mesh Normals $\boldsymbol{w}$

We set all sensitivities of volume mesh normals $\boldsymbol{w}$ to zero, retaining sensitivies of the surface $\boldsymbol{n}$ and point surface $\boldsymbol{a}$ normals of Fig. 1. Results are in Table 2(c). We might expect grid convergence to the results of tables 2(a,b). Instead, sensitivities $dC_L/d\alpha$, $dC_D/d\alpha$ and $dC_S/d\alpha$ converge, but to different values.

Others [2], have obtained a consistent continuous sensitivity solver from a discrete (AD) sensitivity solver on neglecting variation of interior mesh points but, crucially, by also augmenting the boundary conditions. Extending their analysis to the Euler equations is straightforward.

Consider the zero mass flux boundary condition,

$$\boldsymbol{\rho u} \cdot \boldsymbol{a} = 0. \tag{4}$$

Application of an AD tool to (4)'s discretisation produces code equivalent to,

$$\frac{\partial \boldsymbol{\rho u}}{\partial \alpha} \cdot \boldsymbol{a} + \boldsymbol{\rho u} \cdot \frac{\partial \boldsymbol{a}}{\partial \alpha} = 0. \tag{5}$$

Now consider $\boldsymbol{\rho u}(\boldsymbol{x}, \alpha)$ to be a function of $\boldsymbol{x}$ and α, as when deriving a continuous sensitivity solver. Differentiating (4) gives,

$$\frac{\partial \boldsymbol{\rho u}}{\partial \alpha} \cdot \boldsymbol{a} + \boldsymbol{\rho u} \cdot \frac{\partial \boldsymbol{a}}{\partial \alpha} + \sum_{i=1}^{3} \left(\mathbf{grad}(\rho u_i) \cdot \frac{\partial \boldsymbol{x}_S(\alpha)}{\partial \alpha} \right) a_i = 0, \tag{6}$$

with $\boldsymbol{x}_S$ the location of the boundary mesh point. Obviously (6) differs from (5) only in its final term. We have hand-coded this term, and a similar modification of the surface pressure integration for C_L, C_D and C_S, into our sensitivity solver. This was facilitated by FLITE3D's option of a second–order, high resolution AUSM solver [7] which necessarily calculates gradients of momentum and pressure fields. The resulting sensitivity solver did not converge. From (3) and [3], we see that to maintain stability of the flow solver we must enforce $\boldsymbol{R}_M.\boldsymbol{a} = 0$ where $\boldsymbol{R}_M$ is the momentum residual at a boundary vertex. Comparison with (6) indicates the need for the gradient of the momentum residual to enforce the appropriate condition in the differentiated solver. Such a quantity is not calculated within FLITE3D and it would not seem sensible to pursue this further.

4.2 Incomplete Sensitivities

Mohammadi and Pironneau [8, p146-153] considered use of *incomplete sensitivities*, arguing, for example that the boundary condition (4), which on differentiation is (5), be approximated as $\boldsymbol{u}.\nabla \boldsymbol{n} = 0$. We implemented this by using only our differentiated force calculation routine after convergence of the flow solver. The results (not shown) were seriously in error with $dC_L/d\alpha$ of incorrect sign. This indicates that incomplete sensitivities, at best, must be used under restricted conditions (e.g. fixed wing leading and trailing edges).

5 Optimisation

We employ a quadratic programming method to minimise the drag C_D of an ONERA M6 wing at a fixed lift coefficient $C_L = 0.5$. We employ a single design variable: the linear twist about the leading-edge, θ. The AoA, α is also an input. The optimal design (θ=5.9^o,α=9.7^o,C_L=0.5,C_D=0.0509) is achieved in 12 steps by supplying the sensitivity matrix J=($dC_L/d\alpha$, $dC_D/d\alpha$; $dC_L/d\theta$, $dC_D/d\theta$) at each stage. This represents a drag saving of 2.8%

Conclusions

Using the AD tool Odyssée we have developed forward sensitivity version of the industrial unstructured mesh flow solver FLITE3D. Sensitivities w.r.t. changes in AoA and the field and volume mesh have been validated. Techniques to eliminate the need for volume mesh sensitivites have been unsuccessful in line with previous work [1,9]

Acknowledgements: We thank the UK Department of Trade and Industry and Murray Cross of Airbus UK for supporting this project under the CAST program. We thank Laurent Hascoët and INRIA for use of the Odyssée AD tool.

References

1. W. K. Anderson, V. Venkatakrishnan: Computers and Fluids **28** (1999)
2. J. Borggaard, A. Verma: SIAM Journal on Scientific Computing **22** (2000)
3. P. I. Crumpton, P. Moinier, M. B. Giles: 'An Unstructured Algorithm for High Reynolds Number Flows on Highly-Stretched Grids.' In: *10th International Conference on Numerical Methods for Laminer and Turbulent Flow, 1997.*
4. C. Faure, Y. Papegay: 'Odyssée User's Guide. Version 1.7' Rapport Technique RT–0224, INRIA, Sophia-Antipolis, France (1998)
5. M. B. Giles, N. A. Pierce: An introduction to the adjoint approach to design. *Flow, Turbulence and Combustion,* to appear.
6. A. Griewank: *Evaluating Derivatives: Principles and Techniques of Algorithmic Differentiation* (SIAM, Philadelphia 2000)
7. M. S. Liou and C. J. Steffen: Journal of Computational Physics **107** (1993)
8. B. Mohammadi, O. Pironneau: *Applied Shape Optimization for Fluids.* (Oxford Science Publications 2001)
9. E. J. Nielsen and W. K. Anderson: AIAA Journal **37** (1999)

Conservative Multidimensional Upwind Residual Distribution Schemes for Arbitrary Finite Elements

Tiago Quintino[1], Mario Ricchiuto[1], Arpi Csík[1,2], Herman Deconinck[1] and Stefaan Poedts[2]

[1] Dept. Aeronautics and Aerospace, von Karman Institute for Fluid Dynamics
[2] Center Plasma Astrophysics, Katholieke Universiteit Leuven

Abstract. We introduce monotone first order fluctuation splitting schemes for solving hyperbolic systems on arbitrary finite elements, thereby generalizing the N-scheme previously proposed for linear P1 triangles. Conservation is retained by relaxing on strict monotonicity, using a simple method based on contour integration over the element boundaries. Numerical examples are given for the Euler equations solved on Q1 elements for applications ranging from transonic to hypersonic regimes.

1 Introduction

Over recent years, multi-dimensional upwind fluctuation splitting or residual distribution schemes (RDS) have gained some momentum with the development of the *matrix* schemes for the solution of systems of conservation laws [1,2]. These schemes have been mainly constructed for linear P1 elements, although some generalization to Q1 quadrilaterals have already been discussed in [4,6].

Multi-dimensional upwind schemes strongly rely on the quasilinear form of the governing equations, for the computation of the so called *upwind parameters*. For first order schemes, this adds a conservation constraint, which is normally treated by an appropriate linearization of the Jacobians [2]. In the case of the Euler equations on P1 elements a conservative linearization is available, based on the Roe parameter vector [10,8]. This linearization depends on two properties, the linearity of the finite element space (P1 elements) and the requirement to have quadratic fluxes in some variable [10] ($\mathbb{LRD}$ approach). Such a set of variables fails to exist for *arbitrary* systems of conservation laws and/or for nonlinear finite elements. Some remedies have been investigated [6,9,11], but the solutions are much too costly or complicated to be of practical use.

In [5], Csík et al. proposed a new formulation of the positive N-scheme, based on a conservative contour flux integration ($\mathbb{CRD}$ approach). By relaxing on strict positivity, this formulation is conservative independently from the averaged state used for evaluation of the Jacobians of the system. Due to this property, the conflicting issues of conservation and upwinding along local averaged advection speeds are decoupled, allowing the extension of the schemes to more complex systems and nonlinear elements.

In this paper we address the generalization of the first order monotone N-scheme for P1 elements to *arbitrary* finite elements. This is important since the existence of a monotone, shock capturing, first order scheme is the basis for the construction of nonlinear higher order schemes, that automatically follow using standard procedures [2,3].

2 Residual distribution schemes

Consider the Euler equations for a perfect gas written as:

$$\boldsymbol{U}_t + \nabla \cdot \boldsymbol{F} = \boldsymbol{0}, \quad \boldsymbol{U} = \big(\rho, \rho \boldsymbol{v}, E\big)^T, \quad \boldsymbol{F} = \big(\rho \boldsymbol{v}, \rho \boldsymbol{v}\boldsymbol{v} + \boldsymbol{I}p, (E+p)\boldsymbol{v}\big)^T \tag{1}$$

where $\boldsymbol{v}$ is the velocity vector, E the total energy density and for closure, the pressure p is computed $p = (\gamma - 1)\big(E - \frac{1}{2}\rho \boldsymbol{v} \cdot \boldsymbol{v}\big)$. Rewriting the system in the quasilinear form (2), where d is the number of dimensions, we obtain:

$$\frac{\partial \boldsymbol{U}}{\partial t} + \boldsymbol{A}_j \frac{\partial \boldsymbol{U}}{\partial \boldsymbol{x_j}} = 0, \quad j = 1 \ldots d \tag{2}$$

Where $\boldsymbol{A}_j = \partial F_j / \partial \boldsymbol{U}$, is the jacobian in terms of $\boldsymbol{U}$. In this framework, the solution is approximated in a compact piecewise finite element space over an unstructured mesh composed of elements $\boldsymbol{\Omega}$ with m number of nodes.

$$\boldsymbol{U}^h(\boldsymbol{x}, t) = \sum_{l=1}^{m} \boldsymbol{U}_l(t) N_l(\boldsymbol{x}), \tag{3}$$

Where $\boldsymbol{U}_l(t)$ is the time dependent nodal value of the solution at node l and $N_l(\boldsymbol{x})$ is the shape function with properties of interpolation (4a), constant summation and conservation (4b):

$$N_l(\boldsymbol{x}_k) = \delta_{kl} \; : \; k, l = 1 \ldots m \tag{4a}$$

$$\sum N_k(\boldsymbol{x}) = 1, \sum \nabla N_k(\boldsymbol{x}) = 0 : \; k = 1 \ldots m, \forall \boldsymbol{x} \in \Omega \tag{4b}$$

Integrating (2) over an element $\boldsymbol{\Omega}$ we get the total cell residual ϕ^{Ω}:

$$\phi^{\Omega} = \int_{\Omega} \boldsymbol{A}_j \frac{\partial \boldsymbol{U}^h}{\partial \boldsymbol{x_j}} \, \mathrm{d}\Omega \tag{5}$$

Making use of (3) and (4a) we can write (5) as:

$$\phi^{\Omega} = \int_{\Omega} \boldsymbol{A}_j \frac{\partial \boldsymbol{N_l}}{\partial \boldsymbol{x_j}} \boldsymbol{U}_l \, \mathrm{d}\Omega = \boldsymbol{U}_l \int_{\Omega} \boldsymbol{A}_j \frac{\partial \boldsymbol{N_l}}{\partial \boldsymbol{x_j}} \, \mathrm{d}\Omega = \sum_{l=1}^{m} \boldsymbol{K}_l \boldsymbol{U}_l \tag{6}$$

where the $\boldsymbol{K}_l$ are defined by

$$\boldsymbol{K}_l = \int_{\Omega} \boldsymbol{A}_j \frac{\partial \boldsymbol{N_l}}{\partial \boldsymbol{x_j}} \, \mathrm{d}\Omega \tag{7}$$

Integral (7) can be approximated by any particular numerical integration for which we require for consistency $\sum \boldsymbol{K}_l = 0$. This is automatically satisfied by the property (4b) if the integration is exact for polynomials of sufficient high order, considering the order of $\boldsymbol{N_l}$ and the jacobian $\boldsymbol{A}_j$. The system being hyperbolic,

$\boldsymbol{K}_l$ have a complete set of real eigenvalues and eigenvectors and considering the positive and negative eigenvalue matrices, we write (8):

$$\boldsymbol{K}_l = \boldsymbol{R}_l \boldsymbol{\Lambda}_l \boldsymbol{L}_l \qquad \boldsymbol{K}_l^+ = \boldsymbol{R}_l \boldsymbol{\Lambda}_l^+ \boldsymbol{L}_l, \qquad \boldsymbol{K}_l^- = \boldsymbol{R}_l \boldsymbol{\Lambda}_l^- \boldsymbol{L}_l \tag{8}$$

We now distribute the residual in fractions to the nodes that compose the element, using some distribution rule, which defines the numerical scheme. We require for consistency that these fractions sum up to the overall residual:

$$\phi^\Omega = \sum_{l=1}^{m} \phi_l^\Omega \tag{9}$$

This finally leads to the following semi-discrete scheme for each node l, where S_l is the dual cell area and the summation is performed over the cells surrounding the node.

$$\left(\frac{\partial \boldsymbol{U}}{\partial t}\right)_l = -\frac{1}{S_l} \sum_{\Omega, l \in \Omega} \phi_l^\Omega \tag{10}$$

The choice of the scheme to compute these fractions ϕ_l^Ω determines the overall properties of the method: multidimensional upwinding, conservativity, monotonicity, k-exactness for polynomials of order k. Refer to [1] for a review on the properties of the matrix schemes.

2.1 The standard N-scheme

For the N-scheme the fractions ϕ_l^Ω are computed as (11). Then we combine (9) and (6) to yield (12) and enforce consistency by defining $\boldsymbol{U}^*$ as (13). Clearly the scheme is positive and linear, hence first order.

$$\phi_l^N \triangleq \boldsymbol{K}_l^+ (\boldsymbol{U}_l - \boldsymbol{U}^*) \tag{11}$$

$$\phi^\Omega = \sum_{l=1}^{m} \phi_l^N \triangleq \sum_{l=1}^{m} \boldsymbol{K}_l^+ (\boldsymbol{U}_l - \boldsymbol{U}^*) \equiv \sum_{l=1}^{m} \boldsymbol{K}_l \boldsymbol{U}_l \tag{12}$$

$$\Longrightarrow \boldsymbol{U}^* = \left(\sum_{l=1}^{m} \boldsymbol{K}_l^-\right)^{-1} \sum_{l=1}^{m} \boldsymbol{K}_l^- \boldsymbol{U}_l \tag{13}$$

2.2 Conservation of the N-scheme

The above scheme is not conservative. To gain back conservation we have to ensure (14). This is obtained by redefining $\boldsymbol{U}^*$ as (15), thereby relaxing on the positive property ($\mathbb{CRD}$ approach [5]).

$$\phi^\Omega = \sum_{l=1}^{m} \phi_l^N \equiv \phi^c = \oint_{\partial\Omega} \boldsymbol{F}_j \cdot \boldsymbol{n}_j \mathrm{d}\partial\Omega \tag{14}$$

$$\boldsymbol{U}^* = \left(\sum_{l=1}^{m} \boldsymbol{K}_l^+\right)^{-1} \left(\left(\sum_{l=1}^{m} \boldsymbol{K}_l^+ \boldsymbol{U}_l\right) - \phi^c\right) \tag{15}$$

Here ϕ^c is the conservative residual computed with any integral approximation that satisfies the order of the fluxes involved. In theory any method of integration for the integrals in (14) can be chosen as long as the *telescopic* property is retained. Consequently, a contour integration is chosen with an Simpson rule. Using this approach, the variables chosen for the averaged state to compute the $\boldsymbol{K}_l$ do not affect conservation.

2.3 Application to particular Finite Elements

The P1 elements: Assuming in (7) linear triangular elements and the jacobian $\boldsymbol{A}_j$ constant per cell and computed in an averaged state of $\bar{\boldsymbol{U}}_l$, the general upwind parameters have the form (16).

$$\boldsymbol{K}_l = \frac{1}{2}\bar{\boldsymbol{A}}_j \boldsymbol{n}_{jl}, \quad \bar{\boldsymbol{A}}_j = \boldsymbol{A}_j(\bar{\boldsymbol{U}}) \tag{16}$$

Where $\boldsymbol{n}_{jl}$ are the inward pointing scaled normals of the triangle, as shown in figure (1). The choice of this averaged state, as mentioned, is at the source of the conservation property. With $\mathbb{LRD}$ approach, a linearization with Roe parameter vector is imposed, but $\mathbb{CRD}$ allows any choice of variables to linearize.
The Q1 elements: For quadrilateral elements, with bi-linear shape-functions, and the jacobian evaluated again at an averaged state, we get the same expression (16), but where the normals $\boldsymbol{n}_{jl}$ are defined as in figure (1). In this case, the $\mathbb{LRD}$ with Roe linearization is not accurate, due to the non linearity of the shape-functions. In the computation of (14), the shape-function variation along the edges of arbitrary quadrilaterals is no longer linear. Tests have shown that this error is rather small for roughly regular elements.

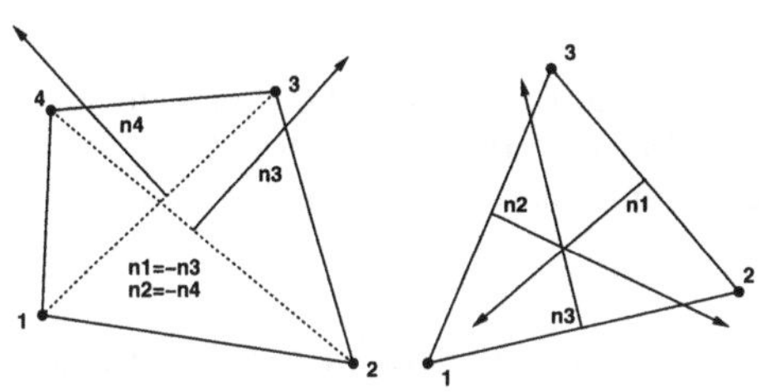

Fig. 1. Quadrilateral and Triangle $\boldsymbol{n}_l$ normals

General Elements: For general higher-order elements, including three dimensional ones, the upwind parameters (7) are computed by averaging the jacobian in the element Ω and computing (17) to satisfy $\sum \boldsymbol{K}_l = 0$.

$$\boldsymbol{K}_l = \bar{\boldsymbol{A}}_j \cdot \int_\Omega \partial_j \boldsymbol{N}_l \, \mathrm{d}\Omega \tag{17}$$

2.4 Higher Order schemes: a boundedness issue

Once a first order monotone scheme is available as discussed above, a high order scheme preserving exact solutions belonging to the Finite Element space can be obtained. An example for the case of a scalar conservation law follows:

$$\beta_l^N = \frac{\phi_l^N}{\phi^\Omega} \qquad \phi_j^\Omega = \frac{\max(0,\beta_l^N)}{\sum_{l=1}^{m}\max(0,\beta_l^N)}\phi^\Omega \tag{18}$$

3 Results and Discussion

3.1 Transonic flow, *Ma* = 0.675 circular arc bump channel

This transonic test-case, also known as GAMM channel, consists of a 3×1 channel, with an unity chord circular arc in the middle with 10% height. This results in a sonic pocket that culminates in a transonic shock at $\approx 72\%$ of the chord, reaching $Ma \approx 1.32$, [7]. The reference solution was calculated with a finer 300×90 quadrilateral mesh with the $\mathbb{CRD}$ N scheme. The triangle schemes perform better in edge aligned flows, where the cross diffusion is minimal. The quadrilateral schemes, although conservative, have shown to be very diffusive, as their cross-diffusion minimizes in diagonal aligned flows.

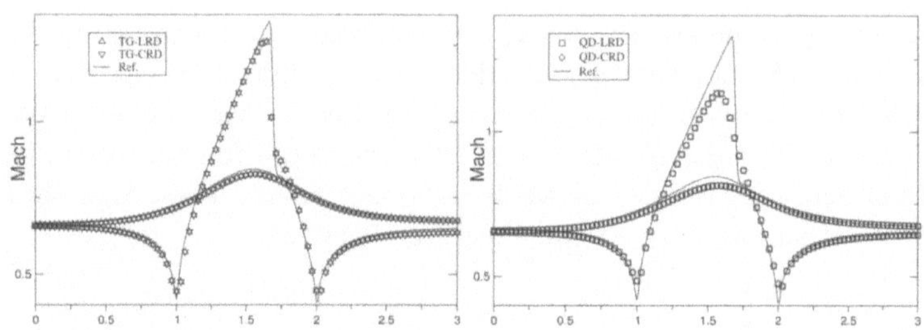

Fig. 2. Mach distribution along the top and bottom walls in the GAMM channel. In the left, the results of the isotropic triangle mesh (TG) ≈ 3400 nodes, in the right the computed results with the quadrilateral (QD) 100×30 mesh.

3.2 Hypersonic flow, *Ma* = 6 bow shock in cylinder

To demonstrate the robustness of the N-scheme for Q1 elements, we present the results of a simulation of a hypersonic bow shock in front of a cylinder of unity radius. The flow Mach number is 6. Plots along the stagnation line are presented, fig.(3). In this case, a strong normal shock is observed in the stagnation line, and the scheme yielded monotone evolutions. The computation grid had 60×70 Q1 elements. The reference solution was computed in a grid with approximately 10 times more elements.

4 Conclusion

The general definition of the upwind parameters $\boldsymbol{K}_l$ for *arbitrary* finite elements is presented and applied in conjunction with the $\mathbb{CRD}$ approach to retain conservation. This separates the conflicting issues of multidimensional upwinding and

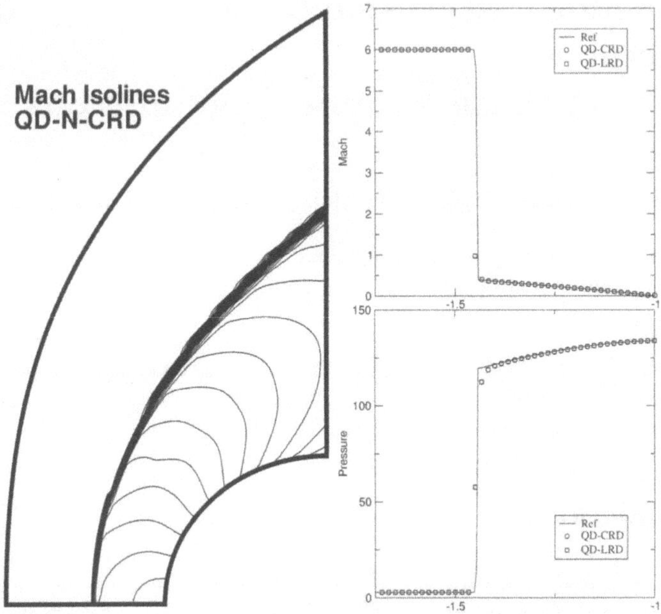

Fig. 3. Left: Q1 flow field results, Mach distribution with conservative N-scheme. Right: Q1 Stagnation line results for N-scheme with both 𝕃ℝ𝔻and ℂℝ𝔻. Top Right: Mach number evolution. Bottom Right: pressure distribution.

linearization of the jacobians. It shows potential for application to non-linear or higher order finite elements.

References

1. E. van der Weide: Compressible Flow Simulation on Unstructured Grids using Multi-dimensional Upwind Schemes.PhD thesis, ULB, Belgium (1998)
2. H. Deconinck, K.Sermeus, R. Abgrall: Status of Multidimensional Upwind Residual Distribution Schemes and Applications in Aeronautics. AIAA CP 2000-2328(2000).
3. R. Abgrall: J. Computational Physics, **167**, pp. 277-315 (2001)
4. E. van der Weide, H. Deconinck: Fluctuation Splitting Schemes for the Euler Equations on Quadrilaterals Grids. ICFD Proceedings (1995)
5. A. Csík, M. Ricchiuto, H. Deconinck: J. Comp. Physics **179**, pp. 286-312 (2002)
6. R. Abgrall, T. Barth: Residual distribution schemes for conservations laws via adaptive quadrature. Submitted to SIAM Journal (2001)
7. S. Eidelman, P. Colella, R. Shreeve: AIAA Journal, **22**, pp.1609-1615 (1984)
8. H. Deconinck, P.L. Roe, R. Stuijs: Computers & Fluids **22**, pp. 215-222 (1993)
9. B. Bruinsma: Conservative linearization of the Euler equations for fluctuation splitting schemes on hybrid grids. von Karman Institute Stagiaire Report (1996-05)
10. P.L. Roe: J. Computational Physics **43**, pp. 357-372 (1981)
11. T. Quintino: Multidimensional Upwind Residual Distribution schemes for Quadrilaterals in Hybrid Grids. Graduation Thesis, IST Lisbon (2001)

An Embedded Grid Adaptation Strategy for Unstructured Data Based Finite Volume Computations

Nikhil Shende[1], Umesh Garg[2], D. Karthikeyan[2], and N. Balakrishnan[1]

[1] Indian Institute of Science, Bangalore, India
[2] Former Graduate Students

Abstract. A grid adaptation strategy for unstructured data based codes, employing a combination of hexahedral and prismatic elements, generalizable to tetrahedral and pyramidal elements has been developed.

1 Introduction

One of the important advantages of the unstructured data based cell centre finite volume code is the ease with which h-adaptation can be effected, without significantly increasing the grid size. It is further advantageous to employ a mesh embedding procedure allowing for the hanging nodes [1], because, the *goodness* of the initial grid is retained even after adaptation. The flow solvers employing face based algorithm capable of handling arbitrary polyhedral volumes are most suitable to be used in conjunction with such a mesh embedding procedure.

Different procedures have been employed in the past to handle grids with hanging nodes. Particularly in the context of structured mesh computations, a zonal mesh approach is employed [2]. Here each of the embedded regions is identified as a separate zone and flux transfer across the volume interface is performed akin to the procedure employed in a multi-block grid. Computations with hanging nodes are also common in more recent Cartesian mesh calculations [3]. Here an hierarchical division based on a quad-tree/oct-tree structure is employed. While the use of former strategy is limited to quadrilateral/hexahedral volumes, the latter has severe limitations as applied to viscous flow computations. At this stage, it is worthwhile to remark that use of unstructured data based computation is a natural option for such an adaptation procedure, because of its ability to handle resulting arbitrary polyhedral volumes [4]. Also, the procedure itself can be expected to be relatively less expensive, making it suitable for dynamic mesh adaptation. In this paper, we discuss an adaptation strategy, which can handle two basic mesh elements, namely, hexahedral and prismatic elements. The procedure is general enough and can be extended to tetrahedral and pyramidal elements commonly encountered in an unstructured mesh computation. The details of the algorithm are presented in section 2 and results are discussed in section 3.

2 Algorithm

To start with, definitions of different terminologies and the data requirements are presented. The cell *type* refers to the basic mesh element in the initial grid. In the present case, *type one* refers to the hexahedral element and *two* refers to the prismatic element. The *flag* assigned to a cell indicates if the cell is a candidate for further division or not. The *level* assigned to a cell refers to the number of times a cell in the initial grid has been divided leading to the formation of the cell under consideration. In the present algorithm the maximum level difference between two cells sharing a face is *one*. The *mark* is another important parameter which indicates if the cell has hanging nodes on its faces or not. At this stage, it should be remarked that the flow solver does not require any of the parameters described above and these parameters are exclusively used in the adaptation stage alone. All that is input to the flow solver by the adaptation routine is the information pertaining to the *inside* and the *outside* cells for a given face; inside cell being a cell for which the surface normal defined for the face is an outward normal and the other cell being the outside cell. This distinguishes the present methodology from the more commonly used *hierarchical* structures employed for effecting the cell division. Three categories of data, namely, the cell, the face and the point are input to the adaptation routine. The cell data involves *type, level, flag* and *mark* assigned to the cell. Apart from this, the nodes constituting a cell are also stored in a certain order. The nodes constituting a face, inside and outside cells for a given face, the boundary type form the face data. The coordinates of the nodes form the point data.

The grid adaptation algorithm can be understood in two stages, namely, the flagging and the cell division.

2.1 Flagging

In the present work, the cell division employs a face based algorithm (interestingly, it is recalled that the flow solver also employs a face based algorithm). Therefore, flagging of the cells (or feature detection) is followed by flagging of the faces.

To start with, all the cells in the initial grid are at *level 0.* When a cell is divided, the level of its offspring is incremented by *one.* To avoid repeated refinement, either a minimum cell volume criterion or a volume weighted sensor parameter may be used [3]. The steps involved are:
(1) Flag the cell based on a suitable sensor parameter. We assign *flag one* to the cell which is a candidate for division and *zero* otherwise. (2) Flag the cell to ensure that the maximum level difference between the resulting offsprings of the cell and those of its face based neighbours is at most *one.* (3) Perform smoothing by flagging and de-flagging the cells to ensure that there are no *pits and islands* in the flagged surface front. The inherent assumption in this subjective procedure is that the feature to be detected exists in *patches.* This is a valid assumption for most of the flows encountered in the nature. (4) Repeat steps 2 and 3 until there are no more cells to be flagged. (5) During the cell division, an iterative

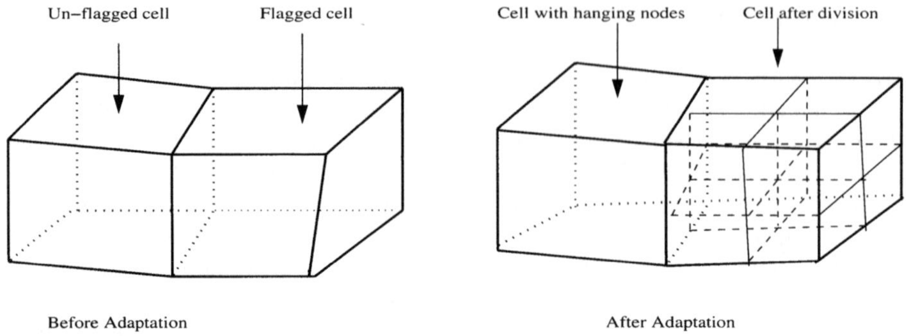

Fig. 1. Division of an hexahedron

procedure, wherein, out of two flagged cells (at different levels) sharing a face, the lower level cell is divided first and the higher level cell is divided in the subsequent iterations. For this purpose, flag of the higher level cell is reset to *two*.

The flagging procedure is pictorially depicted in figure 1 for a hexahedral element. As a consequence of the flags received by the cells, the faces also receive suitable flags. Possibilities those may arise during the adaptation procedure and the resulting face flags are summerized in Table 1. A boundary face to be divided is flagged 11.

2.2 Cell division

In this procedure, the cell division is effected by dividing a hexahedral cell into eight hexahedra and a prismatic cell into four hexahedra and four prisms. The sequence in which the *nodes* and the *new nodes* are stored for a hexahedral and a prismatic volume is depicted in figure 2. In addition to the nodes, the faces constituting a given volume are also stored in a specific sequence. The *nodes* and the *new nodes* associated with the face are related to the volume data as presented in the Table 2 for the case of an hexahedral volume. Similar data

Table 1. Flags received by the faces based on the flag and level of the cells

Inside cell	Outside cell	Level diff.	Face flag
Not flagged	Not flagged	0	00
Flagged	Flagged	0	10
Flagged	Not flagged	0	10
Not flagged	Flagged	0	10
Not flagged	Not flagged	1	00
flagged	flagged	1	20
Flagged	Not flagged	1	21
Not flagged	Flagged	1	21

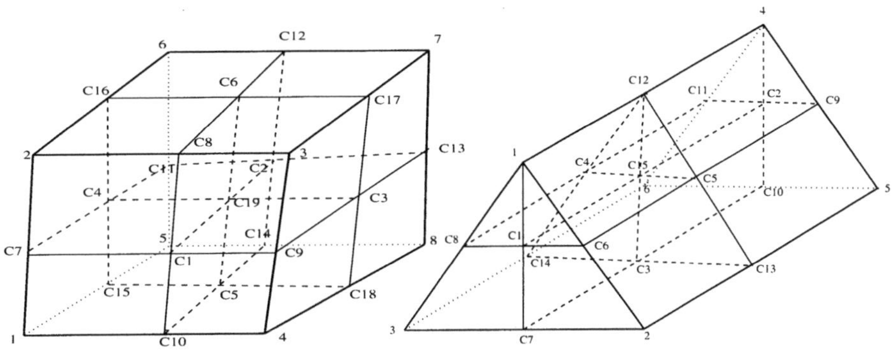

Fig. 2. Adaptation of hexahedron and prism

can be written for a prismatic volume also. A node based data, wherein, the neighbours of a given node connected by an edge carrying a hanging node is also maintained. This array suitably gets updated during the division. The aforesaid data obviate any global search during the cell division and at most involve a search in a given volume.

The steps involved in the cell division are as follows:
(1) Resetting the flags: During the iterative division, the flag of any intermediate level cell has to be reset to *one* from *two* after effecting the division of its lower level neighbour. Corresponding faces flagged 20 are reset to flag 10. (2) Marking the cells: During an iteration, if two face sharing cells are at same level and one of them is flagged, the resulting lower level cell has hanging nodes at least on one of its faces. Such a cell should be marked 1. (3) Sweep over the cells: If a cell is flagged, a new node is created at its centroid and is assigned to an appropriate location corresponding to the cell. (4) Sweep over the faces: If a face has received one of the flags 10 or 11, new nodes have to be created at the face centroids and edge mid points. These nodes are appropriately assigned to the inside and outside cells. Any other non zero face flag (such as 20) corresponds to the case where the *new node* slots associated with the volume are already filled. While the creation of the new node on the face centroid is straight-forward, the creation of the new nodes on the edge mid points requires a search to determine if a node already exists on the edge mid-point or not. Now that all the new

Table 2. Face topology for hex element

Face no.	Nodes forming the face	Edge midpoints	Face center
1	1, 4, 3, 2	C10, C9, C8, C7	C1
2	5, 6, 7, 8	C11, C12, C13, C14	C2
3	3, 4, 8, 7	C9, C18, C13, C17	C3
4	1, 2, 6, 5	C7, C16, C11, C15	C4
5	1, 5, 8, 4	C15, C9, C18, C10	C5
6	2, 3, 7, 6	C8, C17, C12, C16	C6

Table 3. Prism element: connectivity for new volumes

Number	Nodes forming cell	cell type
1	1,C6,C1,C12,C5,C15	prism
2	C7,C1,C6,2,C3,C15,C5,C13	hex
3	3,C8,C1,C7,C14,C4,C15,C3	hex
4	1,C1,C8,C12,C15,C4	prism
5	C12,C5,C15,4,C9,C2	prism
6	C3,C15,C5,C13,C10,C2,C9,5	hex
7	C14,C4,C15,C3,6,C11,C2,C10	hex
8	C12,C15,C4,4,C2,C11	prism

nodes have been created, the new volumes are formed as per the connectivity information presented for a prism in Table 3. Similar data for an hexahedral cell can be easily derived. (5) New faces are created both on the surface and inside of the flagged cells. The inside and outside cells for new faces are appropriately assigned. (6) Proceed to the next iteration if the cells flagged for division still exist. Otherwise, exit.

3 Results and Discussion

The capability of the adaptation algorithm is demonstrated using the following two cases, wherein, divergence of velocity ($\nabla . \vec{u}$) is used as a sensor parameter. (1)**Flow past ONERA M6 wing, $M_\infty = 0.84$, $\alpha = 3.06$ deg.** : The adapted grid and the results are presented in the figure 3. The initial grid contains only hexahedral elements. For this test case, a lambda shock is formed on the upper surface of the wing. The results clearly show the mesh enrichment in the region of shock. The initial grid has 48000 cells and the adapted grid has 139861 cells. (2)**Flow past a rectangular wing with NACA 0012 section, $M_\infty = 0.85$, $\alpha = 1.00$ deg.** : The adapted grid and the results are presented in the figure 4. The initial grid contains only prismatic elements. The grid, after adaptation, contains both hexahedral and prismatic elements. The initial grid has 53340 cells and the adapted grid has 91693 cells.

References

1. Y. Kallinderis, 'Grid adaptation by redistribution and local embedding', VKI Lecture series 1996-06.
2. V. Venkatakrishnan, 'Viscous Computations using a Direct Solver', Computers & Fluids, Vol. 18, No. 2, pp 191-204, 1990
3. W. J. Coirier: An Adaptively-Refined, Cartesian, Cell Based Scheme for the Euler and Navier-Stokes Equations, Ph.D. Thesis, University of Michigan (1994)
4. M. Delanaye,J. A. Essers, 'An accurate finite volume scheme for Euler and Navier-Stokes equations on Unstructured Adaptative Grids', AIAA-95-1718-CP.

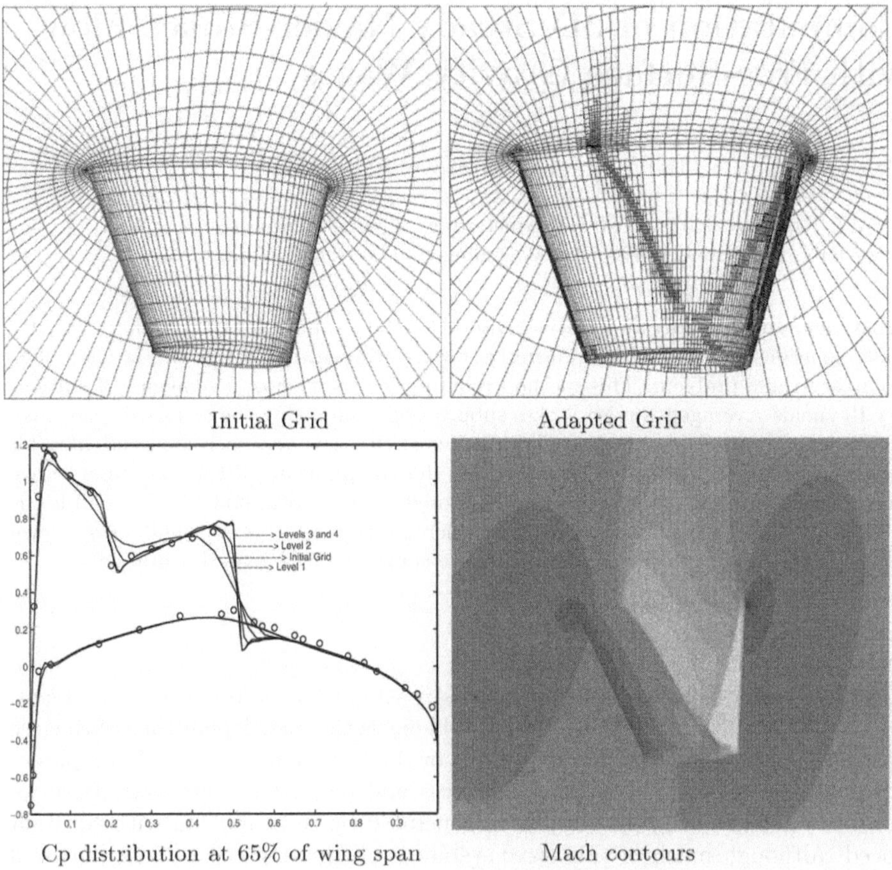

Fig. 3. The grids and the results obtained for ONERA M6 wing

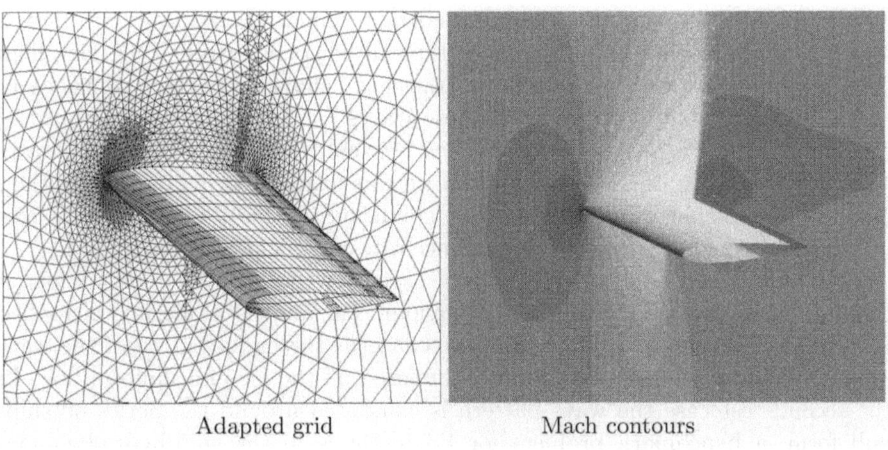

Fig. 4. The grid and the result for rectangular wing with NACA 0012 section

Computation of 3D Steady Navier-Stokes Flow with Free-Surface Gravity Waves

Mervyn R. Lewis[1], Barry Koren[1], and Hoyte C. Raven[2]

[1] CWI, P.O. Box 94079, 1090 GB Amsterdam, The Netherlands
[2] MARIN, P.O. Box 28, 6700 AA Wageningen, The Netherlands

Abstract. In this paper an iterative method for the computation of stationary gravity-wave solutions is investigated, using a novel formulation of the free-surface (FS) boundary-value problem. This method requires the solution of a sequence of stationary Reynolds-Averaged Navier-Stokes subproblems employing the so-called quasi free-surface condition. The numerical performance of this new approach is investigated for two test cases. The first test case involves the computation of the 3D gravity-wave pattern due to a pressure perturbation imposed on a uniform flow. The second is the computation of the gravity-wave pattern generated by a realistic ship-hull form, known as Series 60. Results of the ship-hull case are compared with experimental data.

1 Introduction

In fluid dynamics a large class of problems exists in which a free surface is present. The inherent difficulty of this class of problems is the interdependence of the free surface (FS) location and the unknowns of the bulk-flow problem. Examples of FS flow problems in the physical sciences and engineering are vast. Here we consider the stationary gravity waves generated by ships moving with constant speed. Although much of this wave system is adequately described by inviscid theory, certain areas in the ship-hydrodynamics problem, such as the boundary layer and the wake, require a viscous description. Therefore, present-day developments are primarily directed towards computing stationary FS Navier-Stokes flow problems. Current numerical methods for solving the stationary viscous FS flow problem often employ a time-dependent formulation and integrate until a steady state is reached. This approach typically displays two defects, viz., high computational costs due to persistent transient behaviour of the gravity waves and substantial spatial damping of these waves. To reduce the computational effort of solving FS Navier-Stokes flow an efficient iterative method, employing a novel formulation of the FS flow problem was introduced in [1]. This method employs the so-called quasi free-surface condition (QFSC). The method shows mesh-independent asymptotic convergence rates for a 2D test case.

The emphasis of this paper lies on the extension to 3D of this method and on the investigation of its convergence behaviour. The numerical method is applied to two 3D test cases. The first test case concerns the computation of the wave system due to a pressure perturbation imposed on the FS of a uniform flow. In the second test case the wave pattern is computed around the Series 60 ship-hull form, a benchmark problem for FS methods in the ship-hydrodynamics community.

2 Mathematical Model

In this section a short description is given of the mathematical model which is involved and, secondly, a short summary of the free surface conditions from which the QFSC can be derived.

Let $\Omega(\boldsymbol{x}) \subset R^3$ denote the physical domain which is occupied by the fluid and $\partial\Omega(\boldsymbol{x}) = \Gamma_{\mathrm{FS}} \cup \Gamma_0$ the boundary of $\Omega(\boldsymbol{x})$, where Γ_{FS} denotes the FS and Γ_0 the remaining (fixed) part of the boundary. The dynamics of the viscous, incompressible flow subjected to a constant gravitational force is described by the Navier-Stokes equations, which read

$$\nabla \cdot \boldsymbol{u}\boldsymbol{u}^T + \nabla\varphi - \nabla \cdot \boldsymbol{\tau}(\boldsymbol{u}) = 0, \quad \forall \boldsymbol{x} \in \Omega, \tag{1a}$$

$$\nabla \cdot \boldsymbol{u} = 0, \quad \forall \boldsymbol{x} \in \Omega, \tag{1b}$$

where $\boldsymbol{\tau}(\boldsymbol{u})$ is the viscous stress tensor and $\nabla\varphi(\boldsymbol{x}) = \nabla p(\boldsymbol{x}) + \mathrm{Fr}^{-2}\boldsymbol{e}_z$ the gradient of the hydrodynamic pressure containing the Froude number $\mathrm{Fr} = U_\infty/\sqrt{g\ell}$ with g the acceleration of gravity and ℓ the reference length.

The dynamics of the free surface boundary, Γ_{FS}, is governed by the following conditions. Assuming that the FS can be described as a single-valued function, denoted by $\Gamma_{\mathrm{FS}} = \{(\boldsymbol{x}) : z = \zeta(x, y)\}$, the kinematic condition can be written as

$$\boldsymbol{u} \cdot \nabla\zeta(x, y) = \boldsymbol{u} \cdot \boldsymbol{e}_z, \quad \forall \boldsymbol{x} \in \Gamma_{\mathrm{FS}}, \tag{2}$$

where $\boldsymbol{e}_z$ is the unit vector directed in the opposite direction of the gravitational force. The FS has to adhere to the three dynamic conditions imposed by the assumption of vanishing interfacial stresses. The dynamic condition in the direction normal to the FS reads

$$p(\boldsymbol{x}) - 2\mathrm{Re}^{-1}\frac{\partial \boldsymbol{u}_n}{\partial \boldsymbol{n}} = p_{\mathrm{FS}}(\boldsymbol{x}), \tag{3}$$

whereas the conditions in the two tangential directions are

$$\boldsymbol{t}^{(\alpha)} \cdot \boldsymbol{\tau}(\boldsymbol{u}) \cdot \boldsymbol{n} = 0, \quad \alpha = 1, 2. \tag{4}$$

Here $p_{\mathrm{FS}}(\boldsymbol{x})$ is the specified pressure distribution along the FS. In our first and second test case, $p_{\mathrm{FS}}(\boldsymbol{x}) \neq 0$ and $p_{\mathrm{FS}}(\boldsymbol{x}) = 0$, respectively. For the practical application envisaged here the viscous contribution in (3) will be neglected. Also, the effects of surface tension can be safely ignored.

Many modern FS-iteration methods apply a formulation in which they solve (1a), (1b), in time-dependent form, subjected to the normal dynamic condition applied at an approximate location of the FS, followed by an update of the FS using the kinematic condition in time-dependent form. This approach is disadvantageous because it decouples the two FS conditions. It is the combination of the kinematic and (normal) dynamic condition which is responsible for wave-like solutions. Therefore it was proposed in [1] to combine these conditions resulting in the quasi free-surface condition (QFSC)

$$\mathrm{Fr}^2\boldsymbol{u} \cdot \nabla\varphi - \boldsymbol{u} \cdot \boldsymbol{e}_z = \mathrm{Fr}^2\boldsymbol{u} \cdot \nabla p_{\mathrm{FS}}(\boldsymbol{x}), \quad \forall \boldsymbol{x} \in \Gamma_{\mathrm{FS}}. \tag{5}$$

Note the nonlinearity of the boundary condition.

3 Computational Method

It has been shown that the time-integration method to solve the steady-state FS flow problem is computationally inefficient. This is due to the fact that the attenuation of the surface-gravity waves behaves like $\mathcal{O}(t^{(1-d)/2})$ in R^d, see e.g. [1,2]. To circumvent this slow transient behaviour we adopt an iterative method for the stationary formulation of the FS flow problem. This iterative method requires the solution of a number of stationary Navier-Stokes subproblems employing the QFSC. The solution of the FS flow problem can be found by iterating the following two steps:

1. For a given boundary Γ_{FS}, solve $(\boldsymbol{u}(\boldsymbol{x}), \varphi(\boldsymbol{x}))$ from (1a), (1b) with on Γ_{FS} the boundary conditions (4), (5) and on Γ_0 other appropriate boundary conditions.
   If $\|p - p_{\text{FS}}\|$ is still larger than a chosen small tolerance (assume that at convergence $p = p_{\text{FS}}$), then do step 2, else stop.
2. Use the solution $(\boldsymbol{u}(\boldsymbol{x}), \varphi(\boldsymbol{x}))$ of step 1 to obtain a (better) approximation of Γ_{FS} according to

$$\{(x, y, z = \text{Fr}^2(\varphi(\boldsymbol{x}) - p_{\text{FS}}(\boldsymbol{x})) : (x, y, z) \in \Gamma_{\text{FS}}\},$$

   next return to step 1.

The nonlinear stationary Navier-Stokes subproblem is discretized by replacing the differential operators by second and third-order finite-difference schemes and next solved by Newton's method. The resulting linear system of algebraic equations is solved using a space-marching strategy whereby the marching is performed in the main flow direction. As a result a number of smaller systems has to be solved, which is done by preconditioned GMRES. More details on this solution procedure can be found in [3].

4 Numerical Results

4.1 Pressure Perturbation

The first test case considers the computation of the gravity-wave pattern due to a pressure perturbation imposed on a uniform flow. The perturbation has the Gaussian distribution

$$p_{\text{FS}}(x, y) = Pe^{-\alpha r^2}, \quad r^2 = (x - x_0)^2 + (y - y_0)^2, \quad \forall (x, y) \in \Gamma_{\text{FS}}, \qquad (6)$$

where the parameter values $P = 0.05$, $\alpha = -4$ and $(x_0, y_0) = (0, 0)$ correspond to the values set in [4]. The Froude number is $\text{Fr} = 0.6$ based on unit length. The solutions are computed on a mesh containing $161 \times 41 \times 41$ grid points in the x, y and z direction, respectively, with $z = 0$ as initial estimate for the FS. The correction to the FS which results from the solution of the first Navier-Stokes subproblem is shown in Fig. 1. The convergence behaviour of the FS

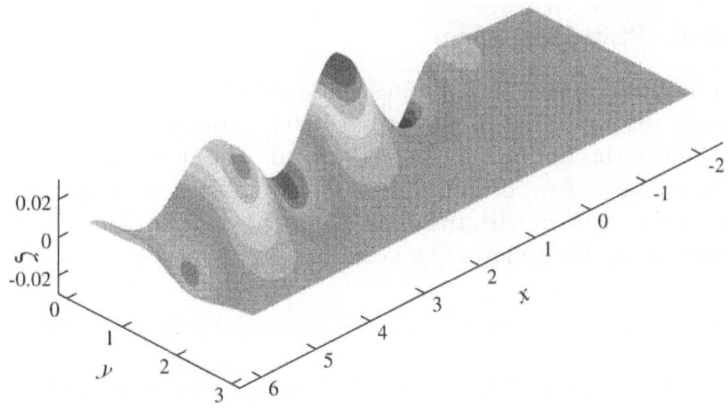

Fig. 1. FS for pressure-perturbation problem, after solving first Navier-Stokes subproblem

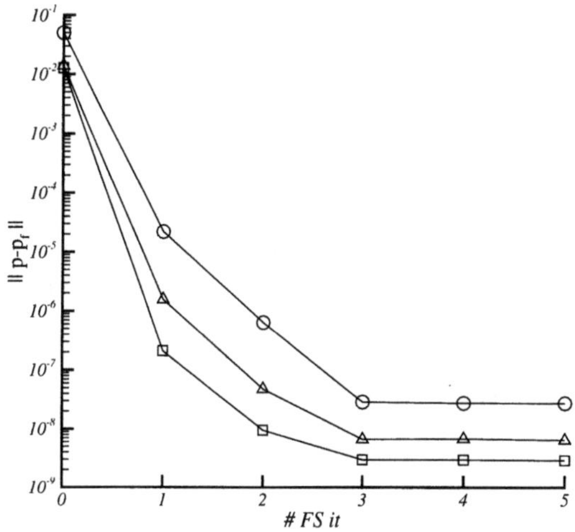

Fig. 2. FS pressure defect measured in $\|\cdot\|_\infty(\circ)$, $\|\cdot\|_1(\square)$ and $\|\cdot\|_2(\triangle)$

iteration process, described in the previous section, is shown in Fig. 2. Correspondence of the computed solution with that from [4] is good although small differences in amplitude and wave length are present. Increasing the amplitude of the perturbation, entering the range of the mildly nonlinear wave systems, shows a reduction of the convergence rate. Computing highly nonlinear waves is not yet possible with the FS iteration method due to robustness problems of the Navier-Stokes subproblem solver.

4.2 Realistic Ship-Hull Form

The computational domain for this test case contains $321 \times 121 \times 41$ points. Fr and Re are set at 0.316 and 10^6, respectively, both based on the length of the ship hull. For the initial Γ_{FS} we take $z = 0$. The wave pattern of the first Navier-Stokes subproblem, obtained after 255 iterations, is shown in Fig. 3. A comparison of this pattern with the experimental data from [5] is given in Fig. 4. For this test case no FS updates were computed yet.

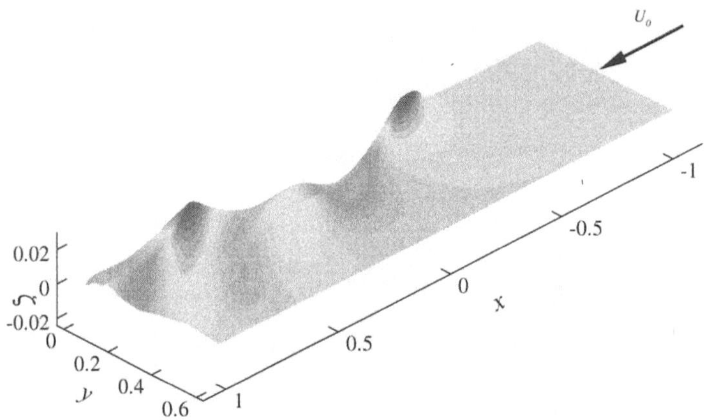

Fig. 3. Wave pattern Series 60 ship hull at Fr = 0.316 after solving first Navier-Stokes subproblem. The flow enters the domain at $x = -1$ with speed U_0. The hull is located at $-0.5 \leq x \leq 0.5$

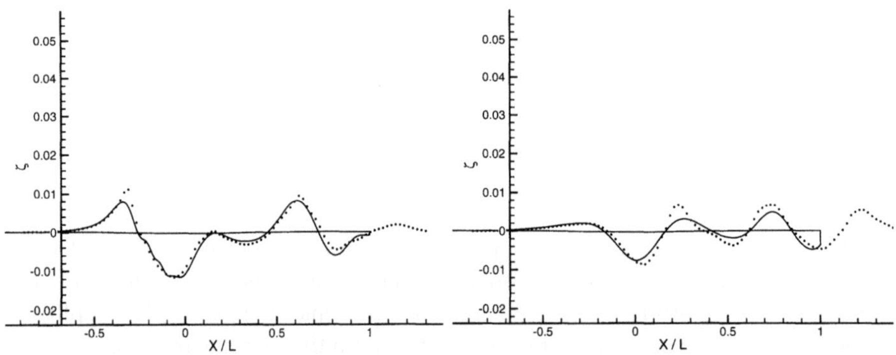

Fig. 4. Comparison of computed longitudinal wave cuts (-) with experimental results (...), Series 60 hull at Fr = 0.316 and Re = 10^6. Left graph : $\frac{y}{\ell} = 0.0755$, right graph :$\frac{y}{\ell} = 0.2067$, $\frac{y}{\ell} = 0$ corresponds to the ship's plane of symmetry

5 Conclusions

The proposed FS iteration method, employing the QFSC, shows very fast convergence in the case of the pressure perturbations for linear to mildly nonlinear wave systems. For both the pressure-perturbation problem and the ship-hull problem, the solution of the first Navier-Stokes subproblem already reveals a large portion of the final wave system. The present method does not suffer from the slow transient effects encountered in time-dependent formulations. So, based on the results of the two test cases considered here, we can conclude that the new formulation of the FS flow problem possesses the proper 3D wave physics and shows very fast convergence.

Acknowledgments: This work was supported by the Dutch Technology Foundation STW.

References

1. E.H. van Brummelen, H.C. Raven, B. Koren: J. Comput. Phys. **174**, 120 (2001)
2. M.J. Lighthill: *Waves in Fluids* (Cambridge University Press, Cambridge, 1958)
3. M. Hoekstra: Numerical Simulation of Ship Stern Flows with a Space-Marching Navier-Stokes Method. PhD Thesis, Delft University of Technology, Delft (1999)
4. D. C. Wyatt: Journal of Ship Research **44**, 96 (2000)
5. Y. Toda, F. Stern, J. Longo: Report 352, Iowa Institute of Hydraulic Research, Iowa city (1991)

Profile-modifiable Conservative Transport Schemes and a Simple Multi-integrated Moment Formulation for Hydrodynamics

Feng Xiao

Tokyo Institute of Technology, 4259 Nagatsuta, Midori-ku, Yokohama, 226-8502 Japan

Abstract. This paper introduces a class of conservative transport schemes (CIP-CSL), which combines an Eulerian conservative formulation with a semi-Lagrangian solution procedure. Emphasis is particularly placed on the implementations of the methods to various fluid dynamic problems. One of the schemes can be directly used as an interface tracking method for multi-fluid computation by modifying the slope of the reconstructed interpolation. A simple methodology based on multi-integrated moment, namely VSIA, is also presented. VSIA (Volume Surface Integrated Average) method employs two integrated moments which are called VIA (Volume Integrated Average) and SIA (Surface Integrated Average), and gives a completely conservative formulation for the VIA. The numerical model is based on the CIP-CSL schemes and can be constructed under a lucid guideline. Some numerical aspects of the VSIA for general evolution equations will be briefly discussed, and implementations for inviscid compressible flow and incompressible viscous flow will be then presented.

1 Introduction

A class of transport schemes, namely CIP-CSL (Constrained Interpolation Profile - Conservative Semi-Lagrangian) schemes, has been recently devised. The CSL type schemes are designed to solve the scalar conservation equation in a form as

$$\frac{\partial f}{\partial t} + \frac{\partial}{\partial x}(uf) = 0, \tag{1}$$

where t refers to the time, x the spatial coordinate, u the characteristic speed and f the transported quantity.

Given the cell-integrated averages $\overline{f}_i^n$ as well as the cell interface values $f_{i-\frac{1}{2}}^n$ and $f_{i+\frac{1}{2}}^n$ at the nth step, a piece-wise interpolation function $\mathcal{F}_i(x)$ with a freedom of three can be constructed over the ith cell $[x_{i-\frac{1}{2}}, x_{i+\frac{1}{2}}]$. The interpolation function $\mathcal{F}_i(x)$ is essential in determining the reconstructed profile and consequently the numerical solution. Some practical candidates of $\mathcal{F}_i(x)$ are:

i. CSL2 reconstruction [6],

$$\mathcal{F}_i(x) = c_{0i} + c_{1i}(x - x_{i-\frac{1}{2}}) + c_{2i}(x - x_{i-\frac{1}{2}})^2. \tag{2}$$

ii. CSLR0 reconstruction [5],

$$\mathcal{F}_i(x) = \frac{c_{0i} + 2c_{1i}(x - x_{i-\frac{1}{2}}) + \beta_i c_{1i}(x - x_{i-\frac{1}{2}})^2}{[1 + \beta_i(x - x_{i-\frac{1}{2}})]^2}. \quad (3)$$

iii. CSLR1 reconstruction [5],

$$\mathcal{F}_i(x) = \frac{c_{0i} + 2c_{1i}(x - x_{i-\frac{1}{2}}) + (3c_{2i} + \tilde{\beta}_i b_i)(x - x_{i-\frac{1}{2}})^2 + 2\beta_i c_{2i}(x - x_{i-\frac{1}{2}})^3}{[1 + \beta_i(x - x_{i-\frac{1}{2}})]^2}. \quad (4)$$

iv. CSL3 reconstruction [4],

$$\mathcal{F}_i(x) = c_{0i} + c_{1i}(x - x_{i-\frac{1}{2}}) + c_{2i}(x - x_{i-\frac{1}{2}})^2 + c_{3i}(x - x_{i-\frac{1}{2}})^3. \quad (5)$$

The coefficients $\{c_i\}$ are determined by using $\overline{f}_i^n$, $f_{i-\frac{1}{2}}^n$ and $f_{i+\frac{1}{2}}^n$ through continuity constraints. In the CSL3 reconstruction, the first-order derivative of the interpolation polynomial at each cell center is used as an extra constraint to control the shape of reconstructed profile. It works, for example, as a limiter to enforce the monotonicity in some variants of the scheme. Once the all pieces of interpolation function $\mathcal{F}_i(x)$ are determined, the numerical solution of the interface value $f_{i-\frac{1}{2}}$ at time step $n+1$ is updated through a semi-Lagrangian solution, while the cell-integrated average $\overline{f}_i$ is advanced by a flux formulation (consult the corresponding references for details).

Our numerical experiments show that regarding to dispersion accuracy, the CSL type schemes are superior to conventional numerical schemes, which use only either the cell-averaged value or the point value for reconstruction. In addition, a CSL scheme requires less grid stencils for reconstruction and appears spatially compact. In the rest part of this paper, we further explore the possibility to utilize the CIP-CSL type methods to a wider spectrum of applications in fluid dynamics.

Section 2 shows that an interface tracking scheme can be easily constructed by just modifying the slope limiter in the CSL3 scheme. A new formulation, namely VSIA, using the volume integrated average (VIA) and the surface integrated average (SIA) of the physical field to solve general dynamical problems defined in time-space domain, is described in section 3. The applications of VSIA to Euler conservation law and viscous incompressible flow will be briefly introduced in section 4 with some preliminary numerical results. The paper ends with a short summary in section 5.

2 A simple interface tracking scheme

Since CSL3 provides a formulation that allows modifying the interpolation profile without violation of mass-conservation, we can easily design an algorithm that maintains sharp transition jumps.

There are many ways to determine the first-order derivative at cell center $d^n_{i-\frac{1}{2}}$ (see [4] for details). For capturing a Heaviside function, we firstly compute $\hat{d}_{i-\frac{1}{2}} = \text{minmod}(\tilde{d}_{i-\frac{1}{2}}, 2S_i, 2S_{i-1})$ [3], with S_i and $\tilde{d}_{i-\frac{1}{2}}$ being approximated by the central differencing. A modification can be made by determining $d^n_{i-\frac{1}{2}}$ with a function in the form as $\Phi(\alpha, \hat{d}_{i-\frac{1}{2}})$. One of the simplest forms is $d^n_{i-\frac{1}{2}} = \alpha \hat{d}_{i-\frac{1}{2}}$. The parameter α functions as a modification to the slope. It is obvious that any α larger than one will enforce the local sharpness of each piece-wisely constructed profile. When applied to a VOF function valued between 0 and 1, we used $\alpha = (1 + \gamma|f^n_i - f^n_{i-1}|/\Delta_{i-\frac{1}{2}})$ with $0 \le \gamma \le 1$.

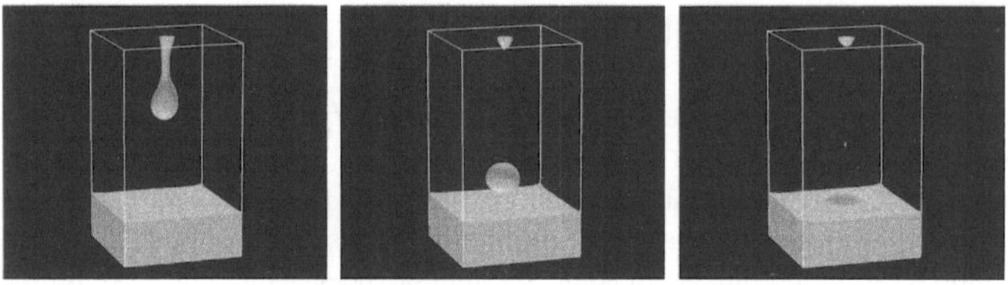

Fig. 1. Simulation of the interaction between a falling drop and liquid.

We have applied the interface tracking method to the simulation of multi-fluid dynamics. Fig.1 shows the snapshots of the interaction between a falling drop and a mass of liquid. The drop was initially attached onto the top of a sealed tank. It was torn under the gravity, and the major part fell downward to the liquid surface while the remainder adhered to the ceiling due to the capillary force. The impact of the drop caused a splashing of the liquid. All process was reasonably re-produced. The total mass of both the drop and the liquid was conserved in this simulation.

3 The general VSIA formulation for evolution equation

Let $\phi(x_1, x_2, \cdots, x_m, t)$ be any physical variable expressed in m dimensional coordinate and time t $(x_1, x_2, \cdots, x_m, t)$. We denote a fixed control volume (CV) by $\mathcal{V} \subset \mathcal{R}^m$ and its surface by $\mathcal{S} \subset \mathcal{R}^{m-1}$. In practice, the surface is approximated by a set of L connected planes, i.e. $\mathcal{S} = \cup_{l=1}^{L} \mathcal{S}_l$.

The volume-integrated average (VIA) of $\phi(x_1, x_2, \cdots, x_m, t)$ can be defined as

$$\overline{^{V}\phi} = \frac{1}{|\mathcal{V}|} \int_{\mathcal{V}} \phi(x_1, x_2, \cdots, x_m, t) dV \tag{6}$$

and the surface-integrated average (SIA) of $\phi(x_1, x_2, \cdots, x_m, t)$ on each segment of the surface planes is computed as

$$\overline{^{S_l}\phi} = \frac{1}{|\mathcal{S}_l|} \int_{\mathcal{S}_l} \phi(x_1, x_2, \cdots, x_m, t) dS, \quad \text{for} \quad l = 1, 2, \cdots, L; \tag{7}$$

where $|\mathcal{V}|$ and $|\mathcal{S}_l|$ denote the volume of the CV and the area of the lth surface segment.

Choosing the control volume and the surface segment is arbitrary. From practical consideration, the simplest control volume can be a hyper cube in $\mathcal{R}^m$ with all surface segments being respectively normal to the coordinate axes. The surface segments whose external normal directions are coincident with or opposite to the l axis are respectively denoted as $\mathcal{S}_l$ and $\mathcal{S}_{-l}$. Thus, we have the dependent variables as $\overline{^V\phi}$, $\overline{^{S_1}\phi}$, $\overline{^{S_2}\phi}, \cdots, \overline{^{S_m}\phi}$. The VIA $\overline{^V\phi}$ is inside the CV and staggeringly located in respect to SIA $\overline{^{S_l}\phi}$ which are on the surface segments. The grid is designed so that $\overline{^V\phi}$ and $\overline{^{S_l}\phi}$ are always on the same straight line for all l.

We consider the evolution equation for ϕ

$$\frac{\partial \phi}{\partial t} + \nabla \cdot (\phi \mathbf{u}) = \psi \tag{8}$$

with $\nabla = (\frac{\partial}{\partial x_1}, \frac{\partial}{\partial x_2}, \cdots, \frac{\partial}{\partial x_m})^T$, $\mathbf{u} = (u_1, u_2, \cdots, u_m)$ and ψ representing all the rest terms except the advection. The corresponding equations for the VIA and the SIA of ϕ are written as

$$\frac{\partial \overline{^V\phi}}{\partial t} + \frac{1}{|\mathcal{V}|} \sum_{l=-m}^{m} \left[\int_{\mathcal{S}_l} (\phi \mathbf{u} \cdot \mathbf{n}_l) dS \right] = \frac{1}{|\mathcal{V}|} \int_{\mathcal{V}} (\psi) dV \tag{9}$$

and

$$\frac{d\overline{^{S_l}\phi}}{dt} + \overline{^{S_l}\phi} \nabla \cdot \mathbf{u} = \frac{1}{|\mathcal{S}_l|} \int_{\mathcal{S}_l} (\psi) dS, \quad \text{for} \quad l = 1, 2, \cdots, m; \tag{10}$$

where $\mathbf{n}_l$ is the unit external vector normal to $\mathcal{S}_l$ and $\frac{d}{dt}$ is the material derivative. From (9) we know that the conservation of the VIA is automatically guaranteed.

The left hand sides of eqs.(9) and (10) are so-called the advection part and can be solved by a CSL scheme. The effects of the right hand sides are considered by firstly updating the VIAs based on the all forcing terms, and then advancing the SIAs through the following evolution equations with $\frac{\delta(\overline{^V\phi}}{\delta t})^{\text{forcing}}$ known

$$\mathsf{H}_{S_l}(\frac{\delta \overline{^{S_l}\phi}}{\delta t}) = \mathsf{H}_V(\frac{\delta \overline{^V\phi}}{\delta t})^{\text{forcing}}, \quad \text{for} \quad l = 1, 2, \cdots, m; \tag{11}$$

where $\mathsf{H}_{S_l}()$ and $\mathsf{H}_V()$ are 1D approximation which can be a high order interpolation, a spline or even an ENO approximation. Our experiments show that a simple averaging also works well.

4 Implementations to various hydrodynamic system

In this section, we present VSIA models for some typical fluid dynamic systems. All the formulations can be constructed according to a simple principle, i.e.(i) solving the advection phase by a CSL scheme, (ii) computing the non-advection phase for part of the VIAs or SIAs by conventional discretization and (iii) updating the rest VIAs and SIAs by the known time variations of VIAs or SIAs obtained in (ii) using the relations like (11).

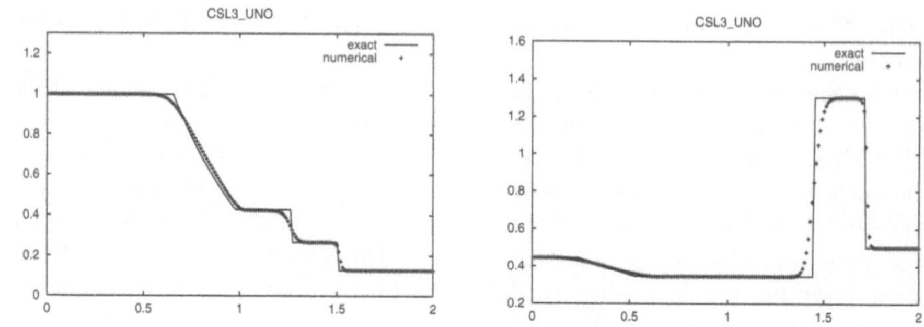

Fig. 2. Numerical results of test 1 (left) and test 2 (right). Displayed are the density profiles.

4.1 The 1D Euler equations

Suppose that the dependent variables in an Euler conservation law are the density ρ, the momentum ρu and the total energy E, we denote the VIAs and the SIAs by $\overline{{}^{x}\phi}$ and $\overline{{}^{0}\phi}$. Given VIA and SIA at $t = t^n$ as $\overline{{}^{x}\rho}^{n}$, $\overline{{}^{x}\rho u}^{n}$, $\overline{{}^{x}E}^{n}$, $\overline{{}^{0}\rho}^{n}$, $\overline{{}^{0}\rho u}^{n}$ and $\overline{{}^{0}E}^{n}$, we can use any CIP-CSL scheme to solve the advection part and get $\overline{{}^{x}\rho}^{\diamond}$, $\overline{{}^{x}\rho u}^{\diamond}$, $\overline{{}^{x}E}^{\diamond}$, $\overline{{}^{0}\rho}^{\diamond}$, $\overline{{}^{0}\rho u}^{\diamond}$ and $\overline{{}^{0}E}^{\diamond}$. The provisional values after the advection phase is then used for computing the non-advection part. Following [7], a simple flux vector splitting across cell boundary $x = x_{1+\frac{1}{2}}$ is adopted for the non-advection flux, and an Euler forward time integration is then used to update the VIAs. Finally, the SIA quantities are advanced by using the time variation of the VIAs.

We computed two widely used 1D shock tube tests. Shown in Fig.3, shock, contact discontinuity and rarefaction wave were well reproduced except that the low-order accuracy in the non-advection calculations seems to have caused an extra viscosity to the solutions.

4.2 The 2D incompressible Navier-Stokes equations

The 2D (x-y plane) incompressible Navier-Stokes equations are solved. A collocated grid is used. All the VIAs of the velocity components $\overline{{}^{xy}u}$ and $\overline{{}^{xy}v}$ as well as the pressure p are located at the regular grid points (cell-point). The SIAs of u and v, i.e. $\overline{{}^{x}u}$, $\overline{{}^{y}u}$, $\overline{{}^{x}v}$ and $\overline{{}^{y}v}$ are respectively defined at the middle points (interface-point) between the two neighboring cell-points. A projection approach is utilized. We computed the widely used benchmark test of the cavity flow. A simulation with the Reynolds number being 1000 is conducted on a 60×60 grid. Fig.3 shows the stream lines of the steady solution and the vertical velocity component v along the x axis passing through the geometric center against that in [1]. The typical flow patterns are adequately captured.

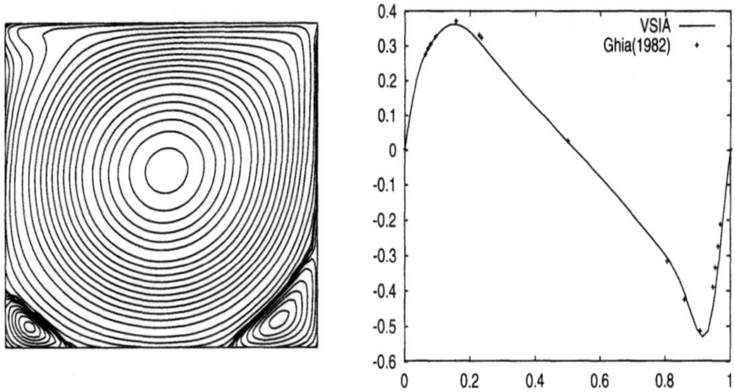

Fig. 3. Cavity flow computed on a 60×60 grid with the Reynolds number being 1000. Displayed are the stream line (left, the contours of the secondary vortices are densely plotted.) and the y component of the velocity (right).

5 Summary

CSL type schemes have been implemented to various fluid dynamic problems. A method for tracking free boundary is constructed by just modifying the slope of the constructed profile. We have also devised a general formulation based on both the VIA and the SIA of prognostic variable for an evolution equation. The simple VSIA models for some typical governing equations were designed and tested. The numerical models appear to be computationally stable, and the numerical results are reasonable.

Using both VIA and SIA of a physical variable enable us to make high order reconstruction with less mesh stencils and leads to a formulation that automatically conserves the VIA. Our previous numerical experiments showed that a VSIA model is superior to the conventional numerical method which uses only one 'moment' in respect of reconstructing the subgrid structures as well as preserving the dispersion properties of the original transport equation. Although much work has to be done before some well established numerical models of this sort come out, we can expect the VSIA methodology as a promising solution to many hydrodynamic problems.

References

1. U.Ghia, K.N.Ghia, C.T.Shin: J. Comput. Phys. , **48**, 387(1982)
2. S.Osher: SIAM J. Numer. Anal. , **31**, 217(1984)
3. B.van Leer: J. Comput. Phys. , **32** 101(1979)
4. F.Xiao, T. Yabe: J. Comput. Phys. , **170**, 498(2001)
5. F.Xiao, T.Yabe, X.Peng, H.Kobayashi: to be published.
6. T.Yabe, R. Tanaka, T. Nakamura, F.Xiao: Mon. Wea. Rev. , **129**, 332(2001)
7. G.C.Zha, E.Bilgen: Int. J. Numer. Meth. in Fluids , **17**, 115(1993)

Preconditioned Implicit Flux-splitting Scheme for Condensate Flows

Satoru Yamamoto[1] and Byeong Rog Shin[2]

[1] Dept. of Aeronautics and Space Engineering, Tohoku University
Sendai 980-8579, Japan
[2] Institute of Fluid Sciences, Tohoku University, Sendai 980-8577, Japan

Abstract. A numerical method for simulating high- and very low-speed condensate flows in atmospheric flow conditions is proposed. This is based on the preconditioning method, Roe's scheme, the 4th-order compact MUSCL scheme, and the LU-SGS scheme. A preconditioned implicit flux-splitting form is newly derived. This form is easily applied to the Roe scheme and the LU-SGS scheme. Condensate flows around the NACA0012 and a cylinder in several atmospheric flow conditions are calculated using the present preconditioned implicit scheme.

1 Introduction

Condensation and evaporation are fundamental physics dominating the global atmosphere in the earth. We also know several engineering problems which is deeply dependent to these physics, for examples, condensate flows in steam turbine, vapor trails over and behind airplane, chemical vapor deposits, and so on. Therefore, condensation and evaporation are of quite important and key physics not only in science but also in engineering. Recently, we have proposed a numerical method for simulating condensate transonic flows in atmospheric wind tunnel and flight conditions[1,2]. The continuity equation of mixed gas, the momentum equations, the energy equation, and the continuity equations of vapor, liquid and the number density of atmospheric small particles are solved by the high-resolution finite-difference method based on the fourth-order compact MUSCL scheme[3], the Roe's approximate Riemann Solver[4], and the LU-SGS scheme[5]. Then, vapor trails around the ONERA M6 wing and over the single-delta wing in humid air have been successfully visualized by the numerical simulations[1,2].

In this paper, this method is extended to a method for simulating condensate flows at very low Mach number by introducing the preconditioning algorithm developed by Weiss and Smith[6]. The preconditioning method has been originally proposed by Turkel[7], and Choi and Merkle[8]. Compressible Navier-Stokes equations are transformed to a set of incompressible equations with the pseudo-compressibility. A pseudo-speed of sound is also defined to avoid the stiff problem. It is known that the preconditioning algorithm is somewhat hard to apply the existing compressible flow code because of the difficulty of the code remaking according to preconditioned Jacobian matrices. This paper may relieve you from such a coding stress. We have developed a preconditioned implicit flux-splitting form in general curvilinear coordinates. This form has been applied to Roe's scheme and the LU-SGS scheme. Since the preconditioned Roe's scheme and the preconditioned LU-SGS scheme have been derived as almost similar forms with the previous forms we developed, we spent only a few weeks for making a new code as a preconditioning version.

In this paper, some preliminary calculations, such as condensate flows around the NACA0012 airfoil at high- and very low-Mach number, and condensate natural convections around a horizontal tube, are presented to show the potential of the present method.

2 Preconditioned flux-splitting form

The numerical flux $(F_i)_{\ell+1/2}$ defined at the interface between the control volume ℓ and $\ell+1$ in each coordinate $i (i = 1, 2, 3)$ in the present fundamental equations can be written by a flux-splitting form as

$$(F_i)_{\ell+1/2} = (A_i^+)_{\ell+1/2} Q_{\ell+1/2}^L + (A_i^-)_{\ell+1/2} Q_{\ell+1/2}^R \tag{1}$$

where, $Q = J[\rho \; \rho u_1 \rho u_2 \; \rho u_3 \; e \; \rho_v \; \rho\beta \; \rho n]^T$. $\rho, u_i, e, \rho_v, \beta$, and n are the density of mixed gas, the physical velocities, the total internal energy, the density of water vapor, the condensate mass fraction, and the number density of atmospheric small particles. $A_i^\pm$ are the Jacobian matrices split by the sign of eigenvalues. L and R mean the MUSCL extrapolation[3] for primitive variables according to the direction of characteristics.

Preconditioned flux-vectors $(\hat{A}_i^\pm)_{\ell+1/2} \hat{Q}^M = (A_i^\pm)_{\ell+1/2} Q^M$ ($M = L$ or R) are newly derived as

$$(\hat{A}_i^\pm)_{\ell+1/2} \hat{Q}^M = \hat{\lambda}_{i1} \Gamma \hat{Q}^M + \frac{\hat{\lambda}_{ia}}{\hat{c}_i \sqrt{g_{ii}}} \hat{Q}_{ia} + \frac{\hat{\lambda}_{ib}}{\hat{c}_i^2} \hat{Q}_{ib} \tag{2}$$

where

$$\begin{aligned}
\hat{\lambda}_{i1} &= U_i, \quad \hat{\lambda}_{i4} = \frac{(1+\alpha)U_i}{2} + \hat{c}_i\sqrt{g_{ii}}, \quad \hat{\lambda}_{i5} = \frac{(1+\alpha)U_i}{2} - \hat{c}_i\sqrt{g_{ii}} \\
\hat{\lambda}_{ia} &= \frac{1}{2}(\hat{\lambda}_{i4} - \hat{\lambda}_{i5}), \quad \hat{\lambda}_{ib} = \frac{\ell_i^- \hat{\lambda}_{i4} - \ell_i^+ \hat{\lambda}_{i5}}{\ell_i^- - \ell_i^+} - \hat{\lambda}_{i1} \\
\hat{c}_i &= \frac{1}{2}\sqrt{U_i^2(1-\alpha)^2/g_{ii} + 4U_r^2}, \quad \ell_i^\pm = \frac{\rho U_r^2}{U_i(\alpha-1)/2 \pm \hat{c}_i\sqrt{g_{ii}}} \\
\alpha &= U_r^2(\rho_p + \frac{\rho_T}{\rho C_p})
\end{aligned}$$

$\hat{Q} = J[p \; u_1 \; u_2 \; u_3 \; T \; \rho_v/\rho \; \beta \; n]^T$. p and T are the static pressure and the static temperature. U_i are the contravariant velocities, $g_{ii} = \nabla\xi_i \cdot \nabla\xi_i$. Γ and U_r are the preconditioning matrix and the preconditioning parameter. They are fundamentally the same with those by Weiss and Smith[6]. The vectors $\hat{Q}_{ia}$

and $\hat{Q}_{ib}$ are almost the same with those already derived for condensate flow equations. They are defined by

$$\begin{aligned}\hat{Q}_{ia} &= \hat{q}_1^M Q_{ic} + \rho\hat{U}_i Q_d, \quad \hat{Q}_{ib} = (\rho\hat{U}_i\hat{c}_i^2/g_{ii})Q_{ic} + (\hat{q}_1^M\hat{c}_i^2/U_r^2)Q_d \\ Q_{ic} &= [0\ \partial\xi_i/\partial x_1\ \partial\xi_i/\partial x_2\ \partial\xi_i/\partial x_3\ U_i\ 0\ 0\ 0]^T \\ Q_d &= [1\ u_1\ u_2\ u_3\ (e+p)/\rho\ \rho_v/\rho\ \beta\ n]^T\end{aligned}$$

$\hat{U}_i = (\partial\xi_i/\partial x_j)\hat{q}_{j+1}^M$ $(j=1,2,3)$. $\hat{q}_j$ $(j=1,2,\cdots,8)$ is the j-th element in $\hat{Q}$.

3 Preconditioned Roe's scheme

The flux-splitting form can be applied to Roe's approximate Riemann solver[4]. The preconditioned numerical flux $(F_i)_{\ell+1/2}$ is defined by

$$(F_i)_{\ell+1/2} = \frac{1}{2}[F_i(\hat{Q}^L_{\ell+1/2}) + F_i(\hat{Q}^R_{\ell+1/2}) - \mid(\hat{A}_i)_{\ell+1/2}\mid(\hat{Q}^R_{\ell+1/2} - \hat{Q}^L_{\ell+1/2})] \quad (3)$$

Each flux-difference splitting in the right hand side of Eq.(3) can be calculated by using a modified form of Eq.(2) as

$$\mid(\hat{A}_i)_{\ell+1/2}\mid\hat{Q}^M = \mid\hat{\lambda}_{i1}\mid\Gamma\hat{Q}^M + \frac{\mid\hat{\lambda}_{ia}\mid}{\hat{c}_i\sqrt{g_{ii}}}\hat{Q}_{ia} + \frac{\mid\hat{\lambda}_{ib}\mid}{\hat{c}_i^2}\hat{Q}_{ib} \quad (4)$$

4 Preconditioned LU-SGS scheme

The present flux-splitting form can be also applied to the maximum second-order LU-SGS scheme[9] as

$$\begin{aligned}\Gamma D\Delta\hat{Q}^{*m} &= RHS^m + \theta_L\Delta t G^+(\Delta\hat{Q}^{*m}) && (5)\\ \Delta\hat{Q}^m &= \Delta\hat{Q}^{*m} - \Gamma^{-1}D^{-1}\theta_L\Delta t G^-(\Delta\hat{Q}^m) && (6)\\ G^+(\Delta\hat{Q}^{*m}) &= (\hat{A}_1^+\Delta\hat{Q}^{*m})_{i-1,j,k} + (\hat{A}_2^+\Delta\hat{Q}^{*m})_{i,j-1,k} + (\hat{A}_3^+\Delta\hat{Q}^{*m})_{i,j,k-1} && (7)\\ G^-(\Delta\hat{Q}^m) &= (\hat{A}_1^-\Delta\hat{Q}^m)_{i+1,j,k} + (\hat{A}_2^-\Delta\hat{Q}^m)_{i,j+1,k} + (\hat{A}_3^-\Delta\hat{Q}^m)_{i,j,k+1} && (8)\end{aligned}$$

where D is the diagonal matrix approximated by the spectral radius of the preconditioned Jacobian matrices. m is the number of Newton iterations in each time step. The delta-form fluxes in Eqs.(7) and (8) can be calculated by using Eq.(2). Then, $\hat{Q}^M$ is replaced by the time derivative $\delta\hat{Q}^M$. The script of each time-derivative flux in the right hand side of Eqs.(7) and (8) denoted by i, j, k indicates the grid location of the flux.

5 Numerical Examples

5.1 Transonic condensate flows around NACA0012 airfoil

Condensate flows around NACA0012 airfoil at transonic speed in atmospheric flight conditions have been calculated. The uniform Mach number is 0.75. The

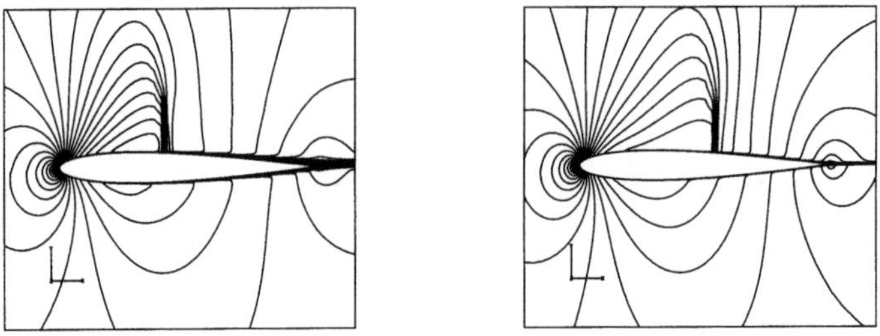

(a) dry air (b) 90% relative humidity

Fig.1 Calculated Mach number contours

angle of attack is 2°. The uniform temperature is 293K. At first, the calculated Mach number contours of the flow assuming dry air and specifying the chord length $L = 0.1$m is shown in Fig.1(a). The flow pattern in the well-known benchmark conditions is well demonstrated in this figure. Figure 1(b) shows the calculated Mach number contours of the flow specifying the relative humidity $\phi = 90\%$ and $L = 1.5$m. Atmospheric small particulates are assumed to be uniformly floating and to be same spheres with a radius. The number density is fixed at $1 \times 10^{12}/\text{m}^3$. These figures indicate that the normal shock on the upper wing surface tends to be shifted downward if condensation occurs. It is due to the release of latent heat when water vapor changes into water liquid as explained in the previous studies[1,2].

5.2 Condensate flows at very low Mach number

Condensate flows around a very-cooled NACA0012 airfoil at very low Mach number in atmospheric conditions are next taken as a virtual flow problem into consideration. The uniform Mach number and the uniform temperature are fixed at 0.01 and 293K. The angle of attack and the chord length are 0° and 0.1m. The relative humidity is $\phi = 90\%$. The airfoil surface is assumed to be cooled at 253K. The Reynolds number in this flow is corresponding to 2000. The atmospheric small particulates are assumed to be the same with those in the previous calculations. Another flow at the same Reynolds number, where the uniform Mach number is 0.1 and the chord length is 0.01m, is also calculated. Figures 2(a)-(c) show that the calculated Mach number contours, the calculated temperature contours, and the condensate mass fraction contours for the former case. Figures 3(a)-(c) show the same contours for latter case. These figures are well compared with each other except for the condensate mass fraction contours. The Mach number contours and the temperature contours in Figs.2 and 3 are almost the same with each other. Since the Reynolds number is equivalent in two cases, a flow similarity is preserved between two cases. However, the difference of the condensate mass fraction between two cases is quite large. The similarity law based on the Reynolds number is broken. It is a nature that the growth of condensed water particles depends on the time and space. The latter case results

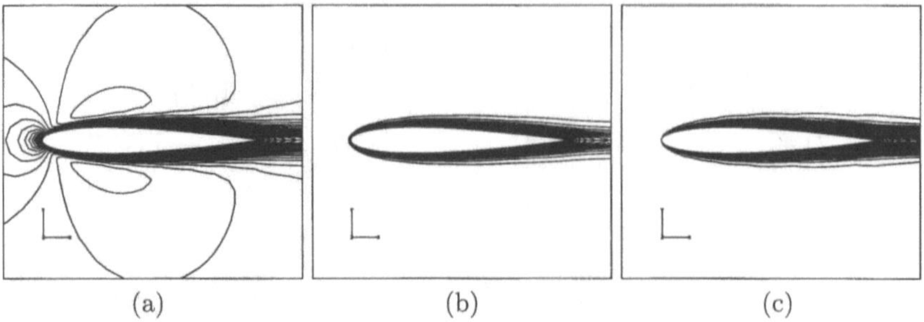

Fig.2 Calculated results of condensate flow(Mach number: 0.01, chord length:0.1m). (a) calculated Mach number contours, (b) calculated temperature contours, (c) calculated condensate mass fraction contours.

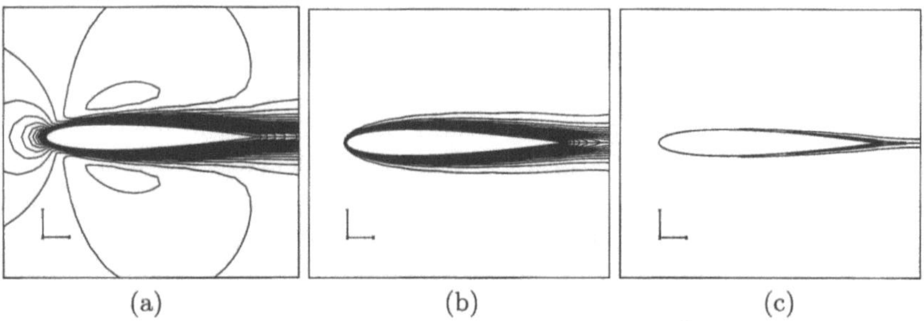

Fig.3 Calculated results of condensate flow(Mach number: 0.1, chord length:0.01m). (a) calculated Mach number contours, (b) calculated temperature contours, (c) calculated condensate mass fraction contours.

in that the particles assuming a smaller chord length and a higher speed than those of former case don't grow sufficiently. This result can explain the nature of the present condensation.

5.3 Laminar film condensation around horizontal tube

A preliminary study for simulating a condensate convection around an atmospheric horizontal tube in zero gravity and 1g environments is lastly introduced. The uniform Mach number is 0. The uniform temperature is 293K. The relative humidity is $\phi = 70\%$. The wall temperature of the tube is 253K. The atmospheric small particulates are assumed to be the same with those in the previous calculations. At first, the calculated temperature contours assuming zero gravity is shown in Fig.4(a). Since no convection generates, only thermal diffusion due to the temperature difference between the atmospheric field and the tube dominates the field in the zero gravity environment. Figures 4(b) and 4(c) show the calculated temperature contours and the condensate mass fraction contours when the Rayleigh number is Ra=10^5 in 1g environment. The temperature contours indicate that the atmospheric mixed gas cooled by the tube moves

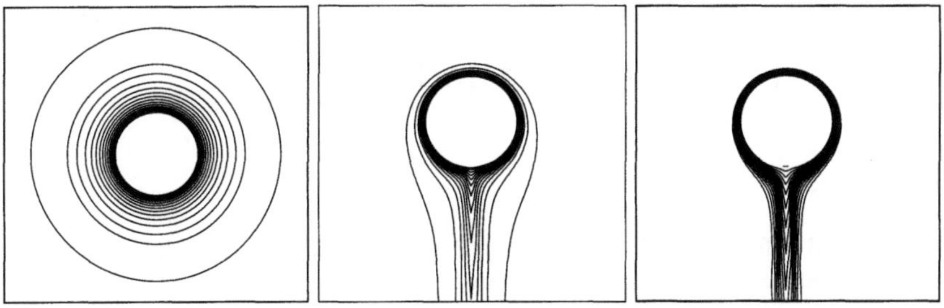

(a) (b) (c)

Fig.4 Calculated results of laminar film condensation. (a) calculated temperature contours at zero gravity, (b) calculated temperature contours at Ra=10^5, (c) calculated condensate mass fraction contours at Ra=10^5.

downward due to the gravitational effect. According to the thermal cooling, a laminar condensate film is produced around the tube and the condensed water particles move downward, too.

6 Conclusions

A numerical method for simulating high- and very low-speed condensate flows in atmospheric flow conditions has been proposed. This method is based on a preconditioned implicit flux-splitting method applied to Roe's scheme and the LU-SGS scheme. The calculated results suggest that the present method would be widely applicable to a lot of condensate flow problems such as condensate flows with homogeneous and heterogeneous nucleations at zero gravity and 1g environments, at high- and low-Mach number, and in science and engineering.

References

1. S. Yamamoto, H. Hagari and M.Murayama: Trans. of the Japan Society for Aeronautical and Space Sciences, **42**, (2000), pp.182-189.
2. S.Yamamoto: Proc. 15th AIAA CFD Conference, AIAA Paper 2001-2651, (2001).
3. S.Yamamoto and H.Daiguji: Computers & Fluids, **22**,(1993), pp.259-270.
4. P.L.Roe: J. Comp. Phys., **43**, (1981), pp.357-372.
5. S. Yoon and A. Jameson: AIAA Journal, **26**,(1988), pp.1025-1026.
6. J.M. Weiss and W.A. Smith: J. Comp. Phys., **33**,(1995), pp.2050-2056.
7. E.Turkel: J. Comp. Phys., **72**(1987), pp.277-298.
8. Y.-H. Choi and C.L. Merkel: J. Comp. Phys., **105**, (1993), pp.207-223.
9. S. Yamamoto, S. Kano and H. Daiguji: Computers & Fluids, **27**, (1998), pp.571-580.

Numerical Model for Flows Including a Solid Body by Level Set Method

Kensuke Yokoi

Computer and Information Division, RIKEN (The Institute of Physical and Chemical Research), Wako Saitama 351-0198, Japan

Abstract. We propose a numerical method to capture and track complex moving boundaries advected in flows. Our method is based on the level set method, but it overcomes the the problem of accumulation of re-initialization error in the level set method. The method can capture the shape of the solid body well.

1 Introduction

Fluid phenomena with complex moving boundaries (gas-liquid, liquid-liquid, gas-liquid-solid, etc.) appear everywhere, such as flows in rivers, kitchens, bodies and plants. Studies on these phenomena are considered challenging and are important for various fields such as physics, industry and medicine. However, numerical studies on these phenomena are difficult because two different scales of a phenomena and a complex moving boundary must be treated simultaneously, and boundary conditions of complex shapes must be considered. Therefore, numerical studies for these problems have not been done hardly. In this work, a numerical algorithm for these problems is proposed.

In our approach, a Cartesian fixed grid is used. One advantage of using such a grid is that a grid for complex moving boundaries need not be generated. Large deformation and topological change of interfaces can be treated because the grid is never distorted. However, there are some problems such as numerical diffusion at the interface and influence of the grid. Various techniques have been proposed to overcome such problems. In these techniques, we consider the level set method [1,2] to be suitable for complex moving boundary problems. The level set method has been applied to various complex moving boundary problems [1–5].

The level set method is an interface-capturing method. The method expresses the surface of an $N-1$ dimension as the zero-contour of an N dimensional function. The function and the zero-contour are called the level set function and the zero-level set, respectively. In the level set method, the interface (zero-level set) is not tracked directly but implicitly through the level set function. In this paper, the function that satisfies Eq. (1) is used as a level set function f_{ls}:

$$\begin{aligned} f_{ls} &= 0 \text{ at the interface} \\ |\nabla f_{ls}| &= 1 \text{ for the whole region.} \end{aligned} \tag{1}$$

The function is also called a distance function that represents the distance from the interface. The time evolution of the level set function is calculated by the advection equation

$$\frac{\partial f_{ls}}{\partial t} + \mathbf{u} \cdot \nabla f_{ls} = 0. \tag{2}$$

The level set method has various features. The method can automatically treat topological change in interfaces without special treatment. To precisely capture interfaces in a Cartesian grid, smoothing of the interface must be introduced. In the level set formulation, the thickness of the interface can be easily maintained because the distance from the interface can be known from the distance function. The level set method is useful for estimating the curvature κ. To compute the curvature, the divergence of a normal vector to the interface is required. In the formulation, the normal vector is always well defined from the level set function. The curvature is calculated as follows

$$\kappa = \nabla \cdot (\frac{\nabla f_{ls}}{|\nabla f_{ls}|}). \tag{3}$$

From the curvature, the surface tension force can be calculated easily by using the CSF (Continuum Surface Force) model [6].

Although the level set method is a useful interface-capturing method as mentioned above, there is a problem of re-initialization error, which is discussed below. The initial level set function f_{ls} satisfies the property of the distance function. However, after calculation of the advection equation Eq. (2), f_{ls} does generally not satisfy the feature of the distance function as shown in Fig. 1(2). Therefore, the feature of the distance function must be recovered while maintaining the position of the zero-level set; here, it is assumed that the zero-level set has represented the interface. This procedure is called a re-initialization [3]. However, the position of the zero-level set moves slightly in this procedure (Fig. 1(3)). This error is the re-initialization error. Although the error is small, the error accumulates in each time step because the next time evolution is calculated from f_{ls} with the error. Therefore, the accumulation of errors is not negligible in some cases [for example Fig. 4 and Fig. 5 (lower side)].

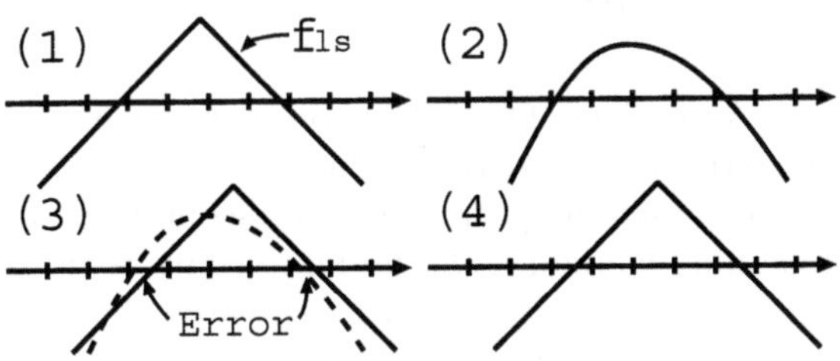

Fig. 1. Schematic figure of re-initialization by the original level set method.

In this paper, we propose a method to prevent the accumulation of re-initialization errors. First, an outline of the numerical method is given. After the explanation of the numerical implementation of the method, a comparison of results obtained using the present method and those obtained using the original method is presented.

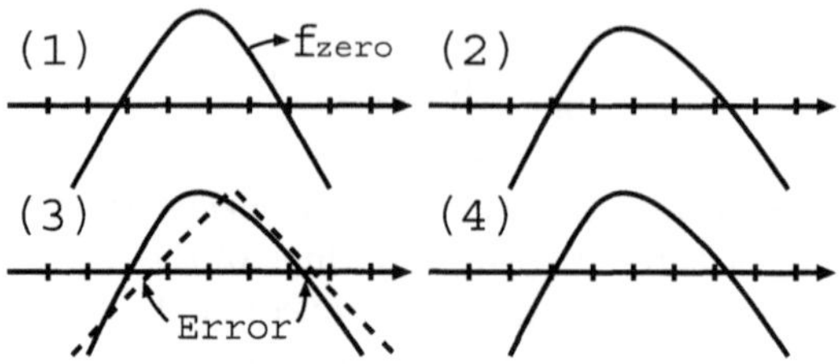

Fig. 2. Schematic figure of the present method.

2 Numerical method

An outline of our method is presented using Fig. 2. In our method, a function, f_{zero}, that is not re-initialized is used to track the interface. The interface is tracked through the zero-contour of f_{zero}. The time evolution of f_{zero} is calculated by the advection equation ((1) → (2) in Fig. 2). To capture the interface (i.e., to compute curvature, smoothing etc.), the level set function f_{ls} is constructed from the zero-contour of f_{zero} (Fig. 2(3)). This gives rise to error similar to the error that occurs in the original level set method. However, there is no accumulation of errors in the present method because the time evolution is calculated from f_{zero} without the error (Fig. 2(4)).

We will now explain the numerical procedure. The time evolution of f_{zero} to track the zero-level set is calculated by the advection equation:

$$\frac{\partial f_{zero}}{\partial t} + \mathbf{u} \cdot \nabla f_{zero} = 0, \tag{4}$$

((1) → (2) in Fig. 2). As a solver of the advection equation, we use the CIP (Cubic Interpolated Propagation) method [7,8], which is less diffusive and robust. To construct the level set function f_{ls} from the zero-contour of f_{zero} (Fig. 2 (3)), we use the fast marching method [2,9] and an iterative re-initialization scheme proposed by Sussman et al. [4] (referred to hereafter as Sussman's method). Before using these methods, the position of the zero-contour of f_{zero} is calculated by interpolating between grids. In this time, linear interpolation is used. From the zero-contour, f_{ls} within Δh (where Δh is the grid spacing) from the zero-contour of f_{zero} are computed by the fast marching method, solving the Eikonal equation:

$$|\nabla f_{ls}| = 1. \tag{5}$$

Other f_{ls} are calculated by Sussman's method. In the calculation, f_{ls} computed by the fast marching method is fixed. Sussman's method solves the following problem to a steady state

$$\frac{\partial f_{ls}}{\partial \tau} = S(f_{ls})(1 - |\nabla f_{ls}|), \tag{6}$$

where, τ is artificial time and S is a smoothed sign function:

$$S(f_{ls}) = \frac{f_{ls}}{\sqrt{f_{ls}^2 + \varepsilon^2}}. \tag{7}$$

In the calculation of (6), a narrow band approach [3] is used. The procedure is repeated using f_{zero} without the re-initialization error as the initial state of the next time step.

We give supplementary explanations of the procedure to construct the level set function f_{ls} from the zero-contour of f_{zero}. f_{ls} developed by Eq. (2) should be used as an initial condition of the iterative calculation of Eq. (6) in each time step. If the value of the former time step or $1, -1$ is used as an initial condition of the iteration, the number of iterations will become large. For the calculation of Eq. (2), high-order schemes such as the CIP method may be not needed because the error in the advection calculation disappears in the iterative calculation of Eq. (6). In this paper, we used a first-order upwind scheme. The fast marching method can also be used for the calculation of f_{ls} beyond Δh from the interface. In this time, we did not use this method because it requires more calculation time than that required in the present method. Although the calculation time in Sussman's method is generally short, the re-initialization error is large. However, in our method, the error does not arise because the value around the zero-level set computed by the fast marching method has been fixed in the iteration.

3 Numerical examples

As a testing problem, we applied the present method to a two-dimensional rigid body rotation problem, called a Zalesak problem [10]. Fig. 3 shows the results after one revolution and ten revolutions by the present method. The interfaces have been captured well after ten revolutions as well as one revolution.

As a comparison, the original method was also applied to the Zalesak problem. The results are shown in Fig. 4. Deviation of the zero-level set can be

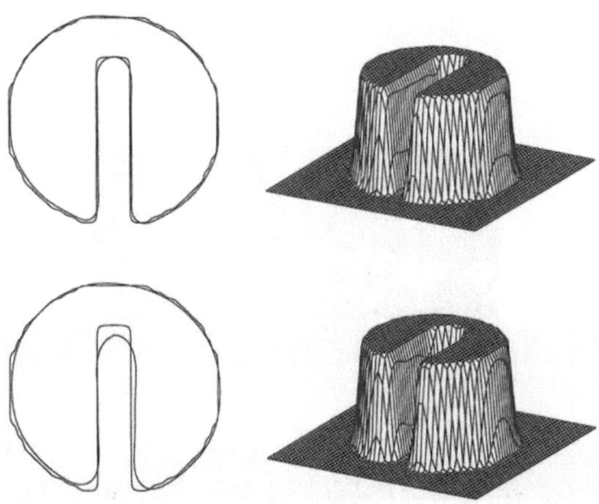

Fig. 3. Results after one revolution (upper figure) and ten revolutions (lower figure) by using the present method. Left and right figures show the zero-level set with the exact solution (initial condition) and the profiles of f_{color}, respectively.

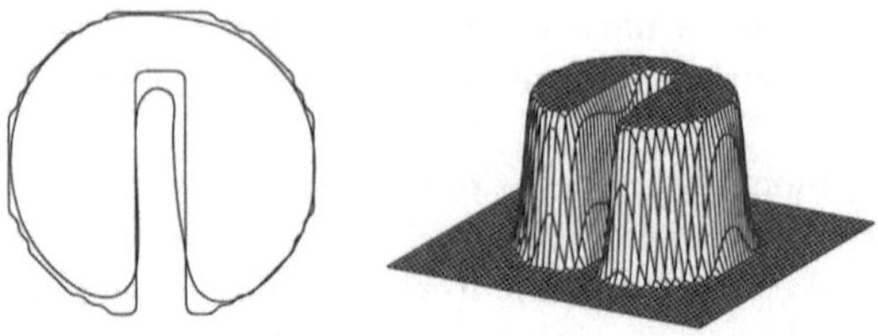

Fig. 4. Zero-level set and f_{color} after one revolution by Sussman's re-initialization with the CIP method.

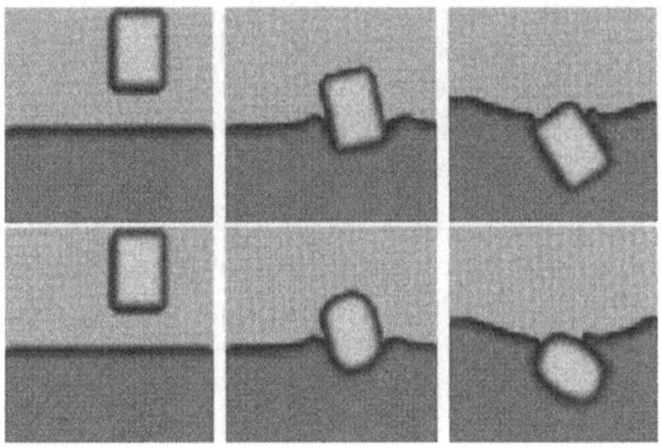

Fig. 5. The rectangular rigid body falling into the liquid. The figures show the time series by the modified level set method (upper side) and the original level set method (lower side). 40×40 computational grids were used. Animation files are available on http://hellfire.riken.go.jp/%7Ekensuke.

seen. The error is the re-initialization error because the same advection equation solver, the CIP method, was used. The error increases with time.

The present method was applied to a hydrodynamic problem including a rigid body. In this calculation, we used the CIP-CUP method [11] based on a Poisson equation of the pressure as a solver of multi fluid flows, and a rigid body

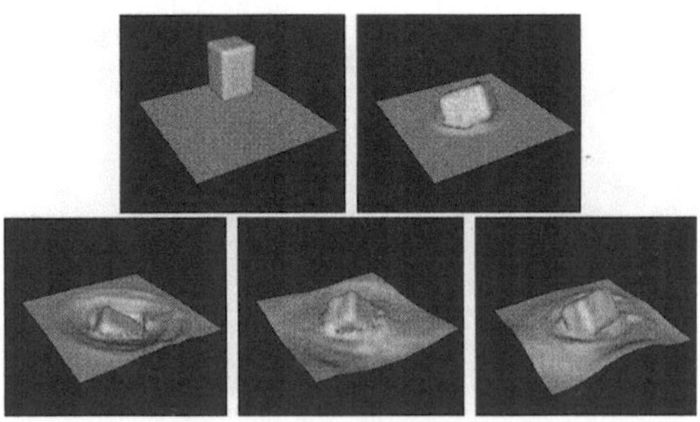

Fig. 6. An example of three dimensional calculation. $40 \times 40 \times 40$ computational grids were used. The rigid body is represented by $8 \times 8 \times 12$ grids.

dynamics model [12,13]. The results are shown in Fig. 5 (upper side). The density ratio of gas:liquid:solid is 1.25 : 1000 : 500. 40×40 computational grids were used. The rectangular rigid body is represented by 8×12 grids. The present method captured the shape of the rigid body well, though the grid number is small. Fig. 5 (lower side) shows the results obtained by using the original method for treatment of the interface of the rigid body. The results show that the shape can not be maintained by using the original method because of the re-initialization error. Fig. 6 shows an example of three dimensional calculation.

4 Summary

We have proposed a numerical method based on the level set method for capturing moving interfaces in a Cartesian fixed grid. This method can avoid the problem of accumulation of re-initialization error in the original method.

References

1. S. Osher and J.A. Sethian: J. Comput. Phys. **79**, 12 (1988).
2. J.A. Sethian: Level Set Methods and Fast Marching Methods, Cambridge University Press (1999).
3. D.L. Chopp: J. Comput. Phys. **106**, 77 (1993).
4. M. Sussman, P. Smereka and S. Osher: J. Comput. Phys. **114**, 146 (1994).
5. K. Yokoi and F. Xiao: Physica D **161** 202 (2002).
6. J.U. Brackbill, D.B. Kothe and C. Zemach: J. Comput. Phys. **100**, 335 (1992).
7. T. Yabe and T. Aoki: Comput. Phys. Commun. **66** 219 (1991).
8. T. Nakamura and T. Yabe: Comput. Phys. Commun. **120** 122 (1999).
9. D. Adalsteinsson and J.A. Sethian: J. Comput. Phys. **148**, 2 (1999).
10. S.T. Zalesak: J. Comput. Phys. **31**, 335 (1979).
11. T. Yabe and P.Y. Wang: J. Phys. Soc. Jpn. **60** 2105 (1991).
12. F. Xiao et al.: Comput. Phys. Commun. **102** 147 (1997).
13. F. Xiao: J. Comput. Phys. **155**, 348 (1999).

Designing Adaptive Low Dissipative High Order Schemes

H.C. Yee[1] and B. Sjögreen[2]

[1] NASA Ames Research Center, Moffett Field, CA, 94035, USA
[2] NADA, The Royal Institute of Technology, 100 44 Stockholm, Sweden

1 Motivation

Proper control of the numerical dissipation/filter to accurately resolve all relevant multiscales of complex flow problems while still maintaining nonlinear stability and efficiency for long-time numerical integrations poses a great challenge to the design of numerical methods. The required type and amount of numerical dissipation/filter are not only physical problem dependent, but also vary from one flow region to another. This is particularly true for unsteady high-speed shock/shear/boundary-layer/turbulence/acoustics interactions and/or combustion problems since the dynamics of the nonlinear effect of these flows are not well-understood. Even with extensive grid refinement, it is of paramount importance to have proper control on the type and amount of numerical dissipation/filter in regions where it is needed.

2 Objective

The objective of this paper is to propose an integrated approach for the control of the numerical dissipation in high order schemes. Among other design criteria, the key idea consists of automatic detection of different flow features as distinct sensors to signal the appropriate type and amount of numerical dissipation/filter where needed and leave the rest of the region free of numerical dissipation contamination. These scheme-independent detectors are capable of distinguishing shocks/shears, flame sheets, turbulent fluctuations and spurious high-frequency oscillations. In addition, these sensors are readily available as an improvement over existing grid adaptation indicators. The detection algorithm is based on an artificial compression method (ACM) [2] (for shocks/shears), and redundant multiresolution wavelets (WAV) [1] (for all types of flow feature). The ACM and wavelet filter schemes using a second-order nonlinear filter with sixth-order spatial central scheme for both the inviscid and viscous flux derivatives and a fourth-order Runge-Kutta method are denoted by ACM66-RK4 and WAV66-RK4. The scheme without any filter is denoted by CEN66-RK4 and the filter scheme using the dissipative portion of the fifth order WENO scheme (WENO5) as the nonlinear filter is denoted by WAVweno66-RK4. Computations blending a second-order nonlinear filter with an eighth-order linear filter using the wavelet as the sensor are denoted by WAV66-RK4-D8. For smooth flows, one of the design criteria to improve nonlinear stability is the use of the entropy splitting of the inviscid flux derivatives [3]. These schemes are indicated by appending the letters “ENT”.

3 Numerical Examples

Extensive grid convergence studies using WAV66-RK4 and ACM66-RK4 for a complex viscous shock/shear/boundary-layer interaction, and a viscous supersonic combustion with four species (a planar Mach 2 shock in air interacting with a circular zone of hydrogen bubble) were conducted (see e.g., Fig. 1). More accurate solutions were obtained with WAV66-RK4 and ACM66-RK4 than with WENO5-RK4, which is nearly three times more expensive. Figure 2 shows an inviscid vortex convection problem at 150 periods on a 81 × 81 grid. The second-order MUSCL and the Tam & Web DRP method are extremely diffusive whereas WENO5 is more diffusive than the rest of the filter schemes using RK4. The CEN66-ENT-RK4-D8 linear filter scheme preserves the vortex convection beautifully. Figure 3 shows the highly accurate computation by blending of two different filters WAV66-RK4-D8 for a 1-D shock-turbulence problem.

For a comprehensive study of this work, see [4].

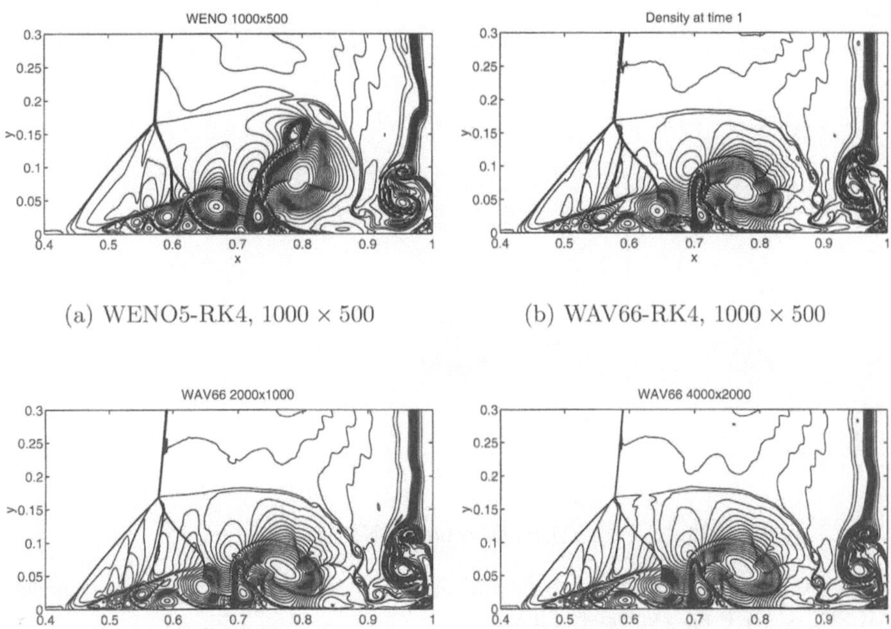

(a) WENO5-RK4, 1000 × 500

(b) WAV66-RK4, 1000 × 500

(c) WAV66-RK4, 2000 × 1000

(d) WAV66-RK4, 4000 × 2000

Fig. 1. Density contours of viscous shock/shear/boundary-layer interaction, Re = 1000.

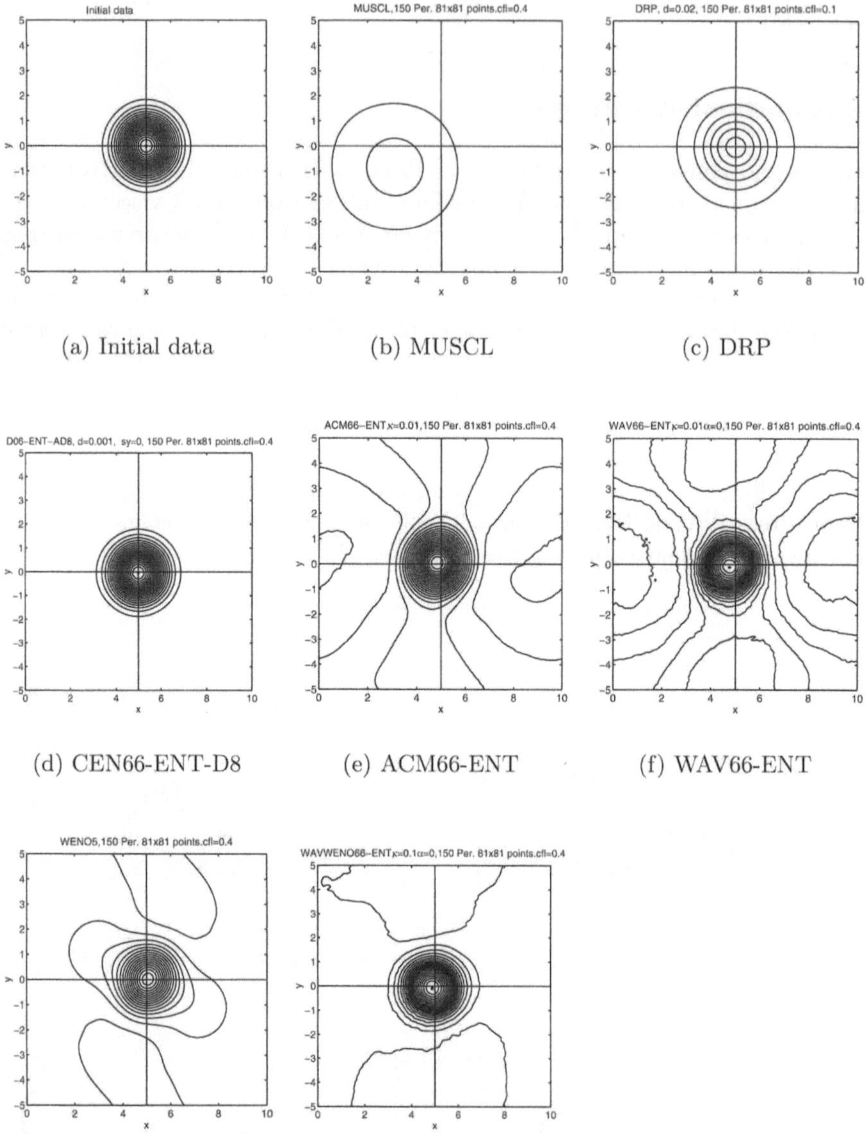

(a) Initial data (b) MUSCL (c) DRP

(d) CEN66-ENT-D8 (e) ACM66-ENT (f) WAV66-ENT

(g) WENO5 (h)WAVweno66-ENT

Fig. 2 Vortex convection at 150 Periods, cfl=0.4 ($\Delta t \approx 0.01$; DRP 4 times smaller). MUSCL & DRP are extremely diffusive.

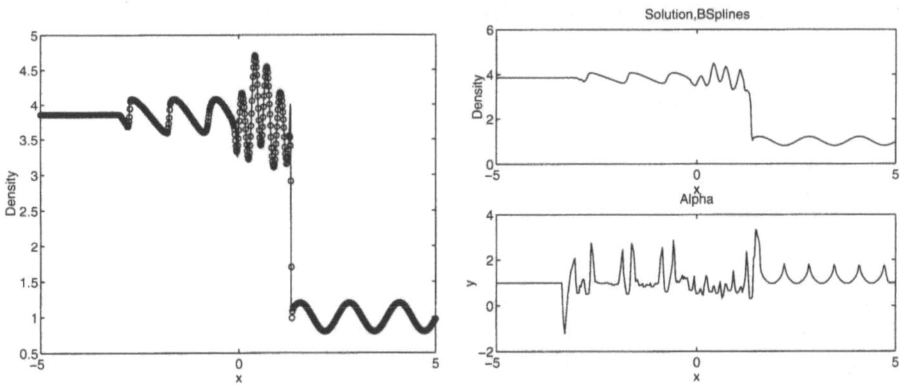

(a) 2nd-order UNO scheme on a 8000 grid (solid), WAV66-RK4-D8 on a 800 grid (circle)

(b) Density & wavelet detector (Alpha), WAV66-RK4-D8, 400 grid

Fig. 3. Blending of two different filters by WAV66-RK4-D8 for a 1-D shock-turbulence problem.

References

1. B. Sjögreen and H. C. Yee: 'Multiresolution Wavelet Based Adaptive Numerical Dissipation Control for Shock-Turbulence Computations'. In: RIACS Report 01.01, NASA Ames research center (Oct 2000).
2. H. C. Yee, N. D. Sandham, and M. J. Djomehri: 'Low Dissipative High Order Shock-Capturing Methods Using Characteristic-Based Filters'. In:J. Comput. Phys., **150**, 199-238 (1999).
3. H.C. Yee, M. Vinokur, and M.J. Djomehri: 'Entropy Splitting and Numerical Dissipation'. In: J. Comput. Phys., **162**, pp. 33-81 (2000).
4. H. C. Yee and B. Sjögreen: 'Designing Adaptive Low-Dissipative High Order Schemes for Long-time Integrations'. In: **Turbulent Flow Computation**, Fluid Mechanics & its Applications, **66**, D. Drikakis & B.J. Geurts, editors, Kluwer Academic, pp. 141-198 (2002).

A New Grid-free Method for Conservation Laws

Praveen Chandrashekar and S M Deshpande

ARDB CFD Center, Department of Aerospace Engg.,
Indian Institute of Science, Bangalore 560012, INDIA
Email: praveen@aero.iisc.ernet.in

Abstract. We present a grid-free or meshless approximation called the Kinetic Meshless Method (KMM), for the numerical solution of hyperbolic conservation laws that can be obtained by taking moments of a Boltzmann-type transport equation. The meshless formulation requires the domain discretization to have very little topological information; a distribution of points in the domain together with local connectivity information is sufficient. For each node i, the connectivity consists of a set of nearby nodes $\mathcal{C}_i$ which are used to evaluate the spatial derivatives appearing in the conservation law. The derivatives are obtained using a modified form of the least squares approximation. The method is applied to the Euler equations for inviscid flow and results are presented for some 2-D problems.

1 Introduction

Meshless methods are being increasingly proposed as an alternative to conventional grid-based methods for the solution of problems involving complex geometry and moving boundaries. They only require a distribution of points in the domain which can be obtained from either one or more of the available grid generation tools. For complex domains, the point distribution can also be obtained from overlapping chimera meshes and there is no need for interpolation at the boundaries of the overlap regions. Another option is to use quadtree/octree based grids which are relatively easier to generate and we do not have to worry about badly deformed or cut-cells near the boundaries. Thus a single (meshless) solver can be used on all types of grids or point distributions [1]. Recently some effort has been directed towards point generation techniques (Lohner & Onate, Lee) also.

Meshless methods for compressible flows where the governing equations are hyperbolic have been developed by Ghosh and Deshpande [2,3], Batina [4], Morinishi [5], and Balakrishnan [6], all of which make use of a least squares formulation. They differ from one another in the way they introduce upwinding. They solve the governing equations in the pde form unlike many other meshless methods (Belytschko et. al., Junk et. al., Liu et. al.) which use a weak formulation and require the evaluation of many integrals over the computational domain. The LSKUM [3] of Ghosh and Deshpande is a kinetic meshless scheme and is obtained by taking moments of a meshless upwind discretization of the Boltzmann equation, and the same approach is taken in the present work. The present method differs from LSKUM in that it uses a different upwind-type discretization of the Boltzmann equation which requires less number of points in each connectivity. Kinetic schemes are attractive due to the fact that they are free of all tunable parameters and artificial fixes, and are known to be very robust in terms of positivity, entropy consistency and convergence history.

2 Kinetic Representation

We are interested in the numerical solution of hyperbolic conservation laws which can be obtained by taking suitable moments of a Boltzmann-type transport equation. The system of conservation laws is written in differential form as

$$\frac{\partial U}{\partial t} + \sum_{\alpha=1}^{d} \frac{\partial F^\alpha}{\partial x^\alpha} = 0 \tag{1}$$

where U is the vector of conserved variables and F^α are the flux vectors. We assume that the system of equations (1) has a kinetic representation, that is, it can be obtained by taking moments of a Boltzmann equation

$$\frac{\partial f}{\partial t} + \sum_{\alpha=1}^{d} v_\alpha \frac{\partial f}{\partial x^\alpha} = J(f) \tag{2}$$

where the moments are defined as

$$\langle \cdot \rangle = \int_{\mathbb{R}^d \times \mathbb{R}_+} \Psi(\boldsymbol{v}, I)(\cdot) \mathrm{d}\boldsymbol{v} \mathrm{d}I \tag{3}$$

Equation (1) is obtained by taking moments of (2) provided $\langle f \rangle = U$, $\langle v_\alpha f \rangle = F^\alpha$ and $\langle J(f) \rangle = 0$. The variable I is sometimes necessary to get the correct moment closure and represents non-translational degrees of freedom. The kinetic representation gives a particle picture of the system under consideration, where the particles are moving with velocity $\boldsymbol{v}$, with the function $f(\boldsymbol{v}, I; \boldsymbol{x}, t)$ giving the statistical distribution of molecular velocity. The macroscopic behaviour is obtained by taking average/moments of the microscopic behaviour. It has been shown by Harten, Lax and van Leer that (1) has a kinetic representation if either $F^\alpha(U) = A^\alpha(U)U$, with all the A^α having only real eigenvalues and a complete set of eigenvectors, or, if the system of equations (1) has a convex entropy. Conservation laws modeling a physical system will always be associated with an entropy condition due to the second law of thermodynamics. In such a case we can use the *moment method strategy* of Deshpande: first obtain an upwind discretization of (2) and take moments to obtain a discretization for the conservation law (1). Note that we do not actually solve equation (2) but merely use it in an intermediate step to obtain a discretization of (1).

3 Dual Least Squares Approximation

A variant of the least squares approximation, which we call the *dual least squares approximation*, has been used previously by some authors [5,6]. Let $f : \mathbb{R}^d \to \mathbb{R}$

be a sufficiently smooth function. If $j \in C_i$ then let f_{ij} denote the value of f at the mid-point of the line segment $\boldsymbol{ij}$. Using Taylor's formula about the point $\boldsymbol{x}_i$, we obtain

$$f_{ij} = f_i + \frac{1}{2}\Delta \boldsymbol{x}_{ij} \cdot \nabla f_i + O(|\Delta \boldsymbol{x}_{ij}|^2), \quad \Delta \boldsymbol{x}_{ij} = \boldsymbol{x}_j - \boldsymbol{x}_i$$

The dual least squares approximation for ∇f_i is obtained from the following weighted minimization:

$$\text{minimize} \sum_{j \in C_i} w_{ij} \left[f_{ij} - f_i - \frac{1}{2}\Delta \boldsymbol{x}_{ij} \cdot \nabla f_i \right]^2, \quad \text{w.r.t. } \nabla f_i \tag{4}$$

This gives a linear system of $d \times d$ equations, $\mathbb{A}_i \nabla f_i = \mathrm{b}_i$, where $\mathbb{A}_i$ is a square matrix that depends only on the coordinates of node i and its neighbours. In the present work the weight function is chosen as $w_{ij} = |\boldsymbol{x}_i - \boldsymbol{x}_j|^{-2}$. The matrix $\mathbb{A}$ is symmetric and positive definite as long as the connectivity is non-degenerate; hence a unique solution always exists. Moreover all the elements of $\mathbb{A}$ have the same dimensions which leads to good condition numbers. The solution of the above linear system can be written as

$$\text{Dual Least Squares (DLS):} \quad \left.\frac{\partial f}{\partial x^\alpha}\right|_i = \sum_{j \in C_i} \mu_{ij}^\alpha (f_{ij} - f_i) \tag{5}$$

In order to get first order accurate estimate of ∇f, Balakrishnan [6] has shown that we must use atleast a second order estimate of f_{ij}, i.e., $f_{ij} = f\left((\boldsymbol{x}_i + \boldsymbol{x}_j)/2\right) + O(|\boldsymbol{x}_i - \boldsymbol{x}_j|^2)$. We also mention the least squares approximation of the derivative,

$$\text{Least Squares (LS):} \quad \left.\frac{\partial f}{\partial x^\alpha}\right|_i = \frac{1}{2} \sum_{j \in C_i} \mu_{ij}^\alpha (f_j - f_i) \tag{6}$$

Equations (5), (6) show the difference between DLS and LS; in LS we make use of known nodal values of the function whereas in DLS we use the function values at the mid-points which are not known a priori and have to be reconstructed in some way. An important issue in DLS is to get suitable mid-point values for satisfying specified accuracy and upwinding.

4 Kinetic Meshless Method using DLS

The present method is built on the DLS approximation and the moment-method strategy. The advantage of DLS is that the value of f_{ij} is not known a priori, and we can introduce upwinding into the scheme by defining f_{ij} in a consistent and upwind manner. Substituting the DLS estimate given by equation (5) into the Boltzmann equation and assuming that the collision term is identically zero, we obtain

$$\frac{\mathrm{d}f_i}{\mathrm{d}t} + \sum_\alpha \sum_{j \in C_i} \mu_{ij}^\alpha (v_\alpha f_{ij} - v_\alpha f_i) = 0 \tag{7}$$

Taking moments we obtain a semi-discrete form of the conservation law at node i

$$\frac{\mathrm{d}U_i}{\mathrm{d}t} + \sum_{\alpha} \sum_{j \in C_i} \mu_{ij}^{\alpha} (F_{ij}^{\alpha} - F_i^{\alpha}) = 0 \tag{8}$$

In order to upwind-bias the approximation given by equation (8), we construct f_{ij} from the particle distribution functions of nodes i and j, depending on whether the particle velocity has a component towards node i or towards node j. If V is a complete set of macroscopic variables which uniquely define f, then this approximation is given by

$$f_{ij}(\boldsymbol{v}, I) = \begin{cases} f(\boldsymbol{v}, I; V_{ij}^{+}), & \boldsymbol{v} \cdot \hat{e}_{ij} \geq 0 \\ f(\boldsymbol{v}, I; V_{ij}^{-}), & \boldsymbol{v} \cdot \hat{e}_{ij} \leq 0 \end{cases} \tag{9}$$

where $\hat{e}_{ij}$ is the unit vector which is directed from node i to node j. The $V_{ij}^{\pm}$ must be atleast second-order accurate interpolants [6,7] of V from $\boldsymbol{x}_i$, $\boldsymbol{x}_j$ to the mid-point of $\boldsymbol{ij}$. This is achieved by using linear interpolation

$$V_{ij}^{+} = R_{ij}(V_i + \frac{1}{2}\Delta \boldsymbol{x}_{ij} \cdot \nabla V_i), \quad V_{ij}^{-} = R_{ij}(V_j - \frac{1}{2}\Delta \boldsymbol{x}_{ij} \cdot \nabla V_j) \tag{10}$$

The matrix R_{ij} effects a rotation of the coordinate system such that the x^1-axis becomes parallel to $\hat{e}_{ij}$ upon rotation. This transformation makes it easier to use equation (10) in (9). The derivatives of V are calculated using the least squares approximation given by equation (6). We can show [7] that the scheme given by (8), (9), (10) is rotationally invariant, i.e., if U^{n+1}, $\bar{U}^{n+1}$ are the updated values in two Cartesian coordinate frames, then they are related by $\bar{U}^{n+1} = RU^{n+1}$, where R is the coordinate transformation operator. Numerical experiments on a scalar conservation law indicate an effective accuracy of two even on highly non-uniform grids. Higher order schemes can be easily derived by including more terms in the Taylor's formula but higher order limiting for discontinuous flows is still an unsolved problem, especially on non-structured grids. It must be mentioned that we use the word upwinding in a somewhat general sense (upwind biased) and not in the usual sense that the update at a node is influenced by the data upwind of the node. We obtain mid-point values based on sign of signal propagation and then use full stencil in DLS. This hopefully gives enough dissipation to obtain a stable scheme and the large amount of numerical experience corroborates this expectation.

5 Numerical Results

We have applied KMM to solve the 2-D Euler equations governing inviscid compressible flows. These can be obtained by taking moments of the Boltzmann equation together with the Maxwellian distribution function. For interpolation, the primitive variables $V = [\rho,\ \boldsymbol{u},\ p]^{\top}$ are used; in the case of discontinuous flows, the gradients of V are limited using Venkatakrishnan limiter. The wall boundary conditions are based on the strong formulation [8]. The point discretization is obtained from standard triangular grid generation techniques and the average number of points in the connectivity is six. The coefficients μ in (5) are calculated once in the beginning and stored (average of 12 floating point numbers per node in 2D) which reduces the cost of each time integration step. The first case is flow over the two-element Williams airfoil at $M_\infty = 0.15$, $\alpha = 0^o$ and is computed on a discretization of 6415 points with 233 points on the main airfoil and 120 on the flap. Figure (1a)-(1b) show comparison of the pressure coefficient between present results and the exact incompressible potential solution. The minimum and maximum entropy in the domain are $S_{\text{min}} = 3.45724$, $S_{\text{max}} = 3.46488$ while the free-stream value is $S_\infty = 3.45776$, the maximum change being 0.22% of the freestream value. The lift and drag coefficients are 3.7608 and -0.0069 while the potential solution gives 3.736 and -0.0001 respectively. The second test case is flow over cylinder at $M_\infty = 0.38$ on a discretization of 4111 points with 250 points on the cylinder. Figure (2) shows the Mach number variation on the cylinder which is seen to have good left-right and top-bottom symmetry. The entropy production is very less with $S_{\text{min}} = 1.59731$, $S_{\text{max}} = 1.59886$ while the free-stream value is $S_\infty = 1.59869$, the maximum change being 0.097% of the freestream value. The lift and drag coefficients are 0.0006 and 0.0012 respectively.

The third case is transonic flow over NACA-0012 at $M_\infty = 0.85$, $\alpha = 1^o$ which is computed on a discretization of 4385 points and another computation is performed with five levels of solution based adaptation, with the final grid having 6938 points. Figure (3) shows the pressure coefficient on the un-adapted and adapted grids and figure (4) shows the pressure contours on the final level of adapted grid. Finally, supersonic flow over a cylinder at $M_\infty = 3.0$ is computed with 12857 points. The interior scheme is applied on the wall boundary points also and no special treatment is done except to kill the normal component of the velocity. The Mach contours are shown in figure (5) which show the bow shock on the front and the slip line on the leeward side of the cylinder. At the stagnation point behind the shock, the computed total pressure and total temperature are 0.3234 and 1.0006, while the exact values are 0.3283 and 1.0000 respectively.

6 Summary

A new meshless method for numerical solution of hyperbolic conservation laws derivable from kinetic equations is described and applied to 2-D inviscid compressible flows. The results show good resolution of the flow features over a range of Mach numbers and are comparable to second order methods, with a numerical order of accuracy of two for a scalar conservation law. Combined with the strong wall boundary condition, it produces little spurious entropy as shown in the subsonic cases.

References

1. Ramesh V and Deshpande SM, ICCFD-1, Kyoto, 2000.
2. Mandal, Ghosh and Deshpande, *FM Report* 1989-04, Dept. of Aerospace Engg., IISc, Bangalore, 1989.

3. Ghosh AK and Deshpande SM, AIAA Paper 95-1735, 1995.
4. Batina JT, AIAA Paper 93-0333, 1993.
5. Morinishi K, Int. Symp. on CFD, Bremen, 1999.
6. Sridar D and Balakrishnan N, *FM Report* 2001-01, Dept. of Aerospace Engg., IISc, Bangalore, 2001.
7. Praveen C and Deshpande S M, *FM Report* 2001-08, Dept. of Aerospace Engg., IISc, Bangalore, 2001.
8. Balakrishnan N and Fernandez G, *Int. Jl. Num. Meth. Fluids*, Vol. 28, pp. 1481-1501, 1998.

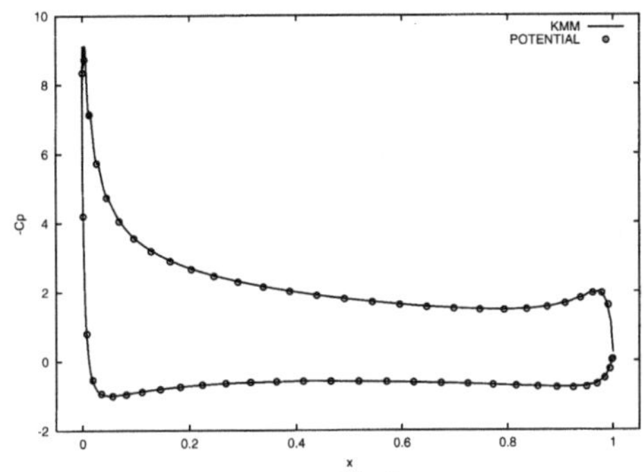

(1a)

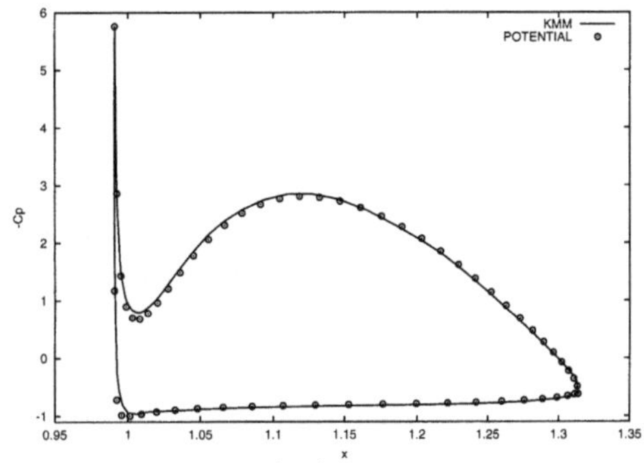

(1b)

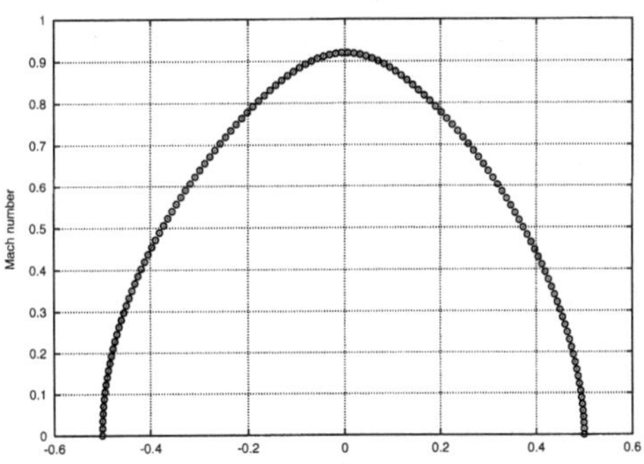

(2)

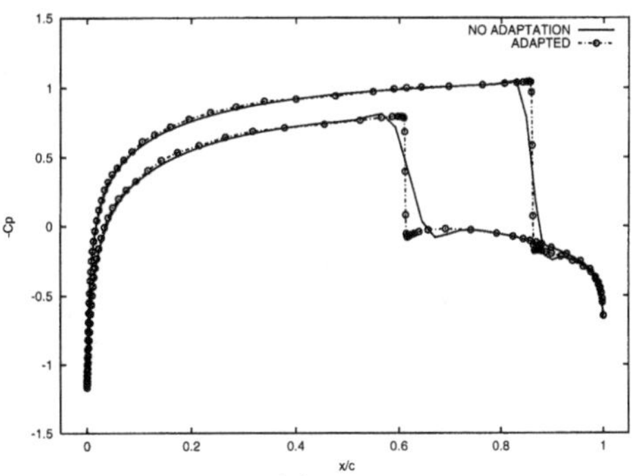

(3)

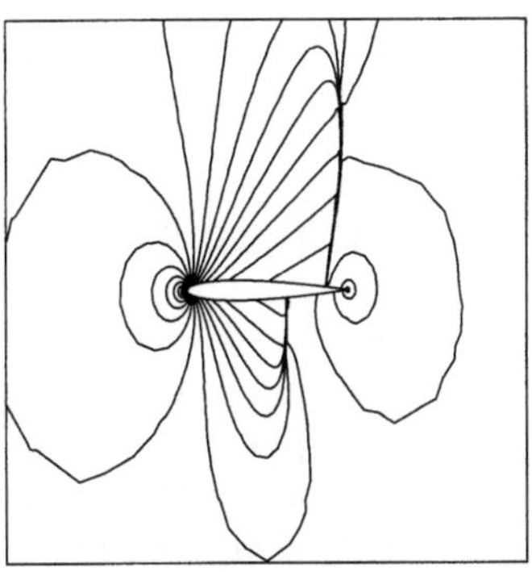

(4)

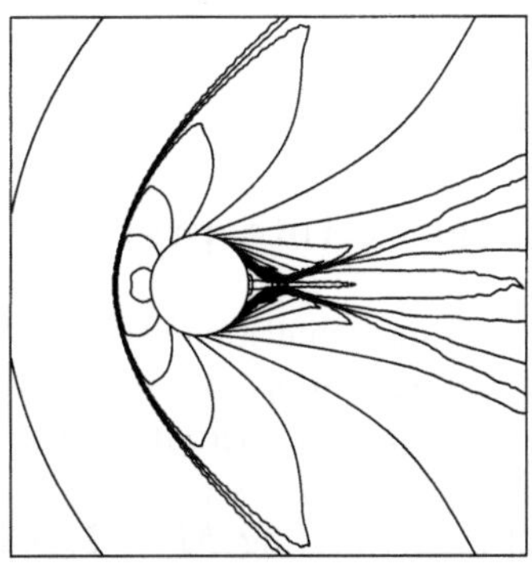

(5)

CFD Newton Solvers with EliAD: An Elimination Automatic Differentiation Tool

Shaun A. Forth and Mohamed Tadjouddine

Applied Mathematics & Operational Research Group, Cranfield University (RMCS Shrivenham), Swindon, SN6 8LA, UK, email `S.A.Forth@rmcs.cranfield.ac.uk`

Abstract. We present a matrix interpretation of standard forward and reverse modes of automatic differentiation (AD) in terms of forward- and back-substitution of the extended Jacobian system. We then show how efficiency improvements for Jacobian calculation are achieved by performing Gaussian elimination on the extended Jacobian. We introduce the ELIAD tool, developed to enable such elimination AD and present results demonstrating significant run-time improvements both for individual finite-volume flux Jacobian calculations and for a 2-D parabolised Navier-Stokes (PNS) flow solver.

1 Introduction

There is increased interest in Newton solvers for CFD due to their quadratic convergence. In [11], the Jacobian $\boldsymbol{J}$ associated with the residual vector $\boldsymbol{R}(\boldsymbol{q})$ for conservative flow variables $\boldsymbol{q}$ was hand-coded. Others [10] automated this process using symbolic algebra tools but after much refinement to control the growth of symbolic expressions and account for all dependencies. This approach is severely limited for code with branches, loops and sub-programs. Attention has shifted to Newton-Krylov solvers requiring calculation of Jacobian-vector products $\boldsymbol{J} \cdot \boldsymbol{d}$. Local flux Jacobians are still required for preconditioning and in [5], finite-differencing was used.

The field of Automatic Differentiation (AD) [3] comprises tools and techniques enabling numerically exact differentiation of functions defined by computer code. State of the art AD tools for Fortran code are ADIFOR [1] and TAMC [2], which will produce local flux Jacobian code. Recently, the elimination approach to AD [4] via source transformation has been implemented in the ELIAD tool [9]. Here, we present its use for local flux Jacobian code.

2 Matrix Interpretation of Automatic Differentiation

Source transformation AD tools take as input computer code defining dependent variables (outputs) $\boldsymbol{y} \in \mathbb{R}^m$ in terms of independent variables (inputs) $\boldsymbol{x} \in \mathbb{R}^n$ via intermediate variables $\boldsymbol{v} \in \mathbb{R}^p$. They output code to calculate Jacobian×matrix[matrix×Jacobian] products $\boldsymbol{J} \cdot \boldsymbol{M}$ [$\bar{\boldsymbol{M}} \cdot \boldsymbol{J}$] via the forward [reverse] mode. Setting $\boldsymbol{M} = \boldsymbol{I}_n$ [$\bar{\boldsymbol{M}} = \boldsymbol{I}_m$] the $n \times n$ [$m \times m$] identity, enables Jacobian calculation. This process is described mathematically via the extended Jacobian system.

2.1 The Extended Jacobian System

Consider an example code comprising 5 floating point operations (FLOPs) and 1 nonlinear operation (NLOP),

$$\left.\begin{aligned} v_1 &= x_1 * x_2 \\ v_2 &= v_1^2 \\ v_3 &= 2 * v_2 \\ v_4 &= \sin(v_3) \\ y_1 &= v_4 * x_3 \\ y_2 &= v_4 * x_1 \end{aligned}\right\} , \tag{1}$$

with $n = 3$ independent variables x_1, x_2, x_3, $p = 4$ intermediate variables $v_1, \ldots, v_4$ and $m = 2$ dependent variables y_1, y_2. We wish to calculate the dense Jacobian $\boldsymbol{J} = [\partial y_i / \partial x_j], i = 1, 2, j = 1, 2, 3$. We define the gradient operator $\nabla = (\partial/\partial x_1, \ldots \partial/\partial x_n)$, differentiate the 3 identities $-x_i = -x_i, i = 1, \ldots, 3$ and each line of (1) to give the *extended Jacobian System* $\boldsymbol{J_E} \nabla \boldsymbol{X} = \boldsymbol{R}$,

$$\left[\begin{array}{ccc|cccc|cc} -1 & 0 & 0 & 0 & 0 & 0 & 0 & 0 & 0 \\ 0 & -1 & 0 & 0 & 0 & 0 & 0 & 0 & 0 \\ 0 & 0 & -1 & 0 & 0 & 0 & 0 & 0 & 0 \\ \hline x_2 & x_1 & 0 & -1 & 0 & 0 & 0 & 0 & 0 \\ 0 & 0 & 0 & 2*v_1 & -1 & 0 & 0 & 0 & 0 \\ 0 & 0 & 0 & 0 & 2 & -1 & 0 & 0 & 0 \\ 0 & 0 & 0 & 0 & 0 & \cos(v_3) & -1 & 0 & 0 \\ \hline 0 & 0 & v_4 & 0 & 0 & 0 & x_3 & -1 & 0 \\ v_4 & 0 & 0 & 0 & 0 & 0 & x_1 & 0 & -1 \end{array}\right] \left[\begin{array}{c} \nabla x_1 \\ \nabla x_2 \\ \nabla x_3 \\ \hline \nabla v_1 \\ \nabla v_2 \\ \nabla v_3 \\ \nabla v_4 \\ \hline \nabla y_1 \\ \nabla y_2 \end{array}\right] = \left[\begin{array}{ccc} -1 & 0 & 0 \\ 0 & -1 & 0 \\ 0 & 0 & -1 \\ \hline 0 & 0 & 0 \\ 0 & 0 & 0 \\ 0 & 0 & 0 \\ 0 & 0 & 0 \\ \hline 0 & 0 & 0 \\ 0 & 0 & 0 \end{array}\right] , \tag{2}$$

at a cost of 1 FLOP and 1 NLOP. In general, we may write the computational cost, cost($\boldsymbol{J}$), of calculating the Jacobian $\boldsymbol{J}$ of a function $\boldsymbol{F}$ by AD as cost($\boldsymbol{J}$) = cost($\boldsymbol{F}$) + cost($\boldsymbol{J_E}$) + cost(linear solve), with cost($\boldsymbol{F}$) that of calculating the original function and cost($\boldsymbol{J_E}$) that of calculating entries of the extended Jacobian, both of which are fixed for a particular function $\boldsymbol{F}$. However, different modes of AD are defined in terms of how the system $\boldsymbol{J_E} \nabla \boldsymbol{X} = \boldsymbol{R}$ is solved and this cost, cost(linear solve), therefore varies.

2.2 Conventional Forward and Reverse Mode AD

By writing the solution of (2) as $\nabla \boldsymbol{X} = \boldsymbol{J_E}^{-1} \boldsymbol{R}$, the Jacobian is expressed as

$$\boldsymbol{J} = \boldsymbol{P}^T \boldsymbol{J_E}^{-1} \boldsymbol{R} , \tag{3}$$

where $\boldsymbol{P}^T = [\boldsymbol{0}_{m \times (p+n)}, \boldsymbol{I}_m]$ and $\boldsymbol{0}_{m \times (p+n)}$ is the $m \times (p+n)$ zero matrix. We may group the operations in (3) as $\boldsymbol{J} = \boldsymbol{P}^T (\boldsymbol{J_E}^{-1} \boldsymbol{R})$ and solve $\nabla \boldsymbol{X} = \boldsymbol{J_E}^{-1} \boldsymbol{R}$ via forward substitution ($\boldsymbol{J_E}$ is lower triangular). Premultiplying by $\boldsymbol{P}^T$ extracts the last m rows of $\nabla \boldsymbol{X}$ so requires no FLOPs. This is the matrix interpretation of forward mode AD as in ADIFOR and TAMC. Such tools exploit the sparsity of $\boldsymbol{J_E}$ and $\boldsymbol{R}$ but not of the derivatives $\nabla \boldsymbol{X}$[1]. This approach solves (2) to give

[1] N.B. Run-time exploitation of $\nabla \boldsymbol{X}$'s sparsity via SparsLinC's [1] sparse derivatives is potentially costly due to the manipulation of non-zero entry indecies [3, p134]

$\boldsymbol{J}$ using cost(linear solve) $= 27$ multiplications. Generally, for forward mode AD cost(linear solve) is bounded above by $n \times N_{c\neq 0}$ multiplications with $N_{c\neq 0}$ the number of non-zero off-diagonal entries in $\boldsymbol{J_E}$.

Alternatively, group (3) as $\boldsymbol{J} = (\boldsymbol{P}^T \boldsymbol{J_E}^{-1})\boldsymbol{R}$ and solve the upper triangular system $\boldsymbol{J_E}^T \boldsymbol{A} = \boldsymbol{P}$ for the *adjoints* $\boldsymbol{A}$ via back substitution. This is a matrix interpretation of reverse mode AD. Tools like TAMC exploit the sparsity of $\boldsymbol{J_E}^T$ but not of $\boldsymbol{A}$ and give cost(linear solve) bounded above by $m \times N_{c\neq 0}$ multiplications. For our example this requires cost(linear solve) $= 18$ multiplications.

2.3 EliAD and AD Elimination Techniques for Jacobians

Computing $\boldsymbol{J}$ is equivalent to eliminating all intermediate variables ∇v_k from (2) to obtain the ∇y_i in terms of the ∇x_j. The ∇v_k may be eliminated in any order, but the cost may be order dependent. In [4], it was shown that monotonic increasing[decreasing] pivot orders are equivalent to conventional forward[reverse] mode AD with compile time exploitation of the sparsity of $\nabla \boldsymbol{X}[\boldsymbol{A}]$. For our example this costs 10[10] multiplications, substantially less than conventional AD.

Efficiency is also boosted, though to a lesser extent, by choice of a good elimination ordering. For our example, eliminating ∇v_3 first costs one multiplication. Then eliminating ∇v_2, ∇v_4 and finally ∇v_1 costs $1, 2$ and 4 multiplications respectively. This so called *cross country elimination* [3] obtains $\boldsymbol{J}$ from $\boldsymbol{J_E}$ at a cost of only 8 multiplications. Though this example is contrived, more than halving the number of multiplications compared to the best of forward/reverse modes, we will demonstrate (Sect. 3) great efficiency gains for functions taken from a CFD solver. For the results in this paper, all orderings are obtained by choosing, at each step of the elimination, the ∇v_j with the lowest *Markowitz cost* [4], defined as the product of the number of non-zero off-diagonal entries in ∇v_j's row and column in the partially eliminated extended Jacobian. It is an upper bound on both the number of multiplications to eliminate the ∇v_j, and on the associated fill-in.

Our tool ELIAD reads in a Fortran subroutine, determines the entries of $\boldsymbol{J_E}$, obtains a pivot order for elimination and then writes a new Fortran subroutine comprising the original statements augmented by code to calculate the Jacobian.

3 Results

We apply ELIAD (Markowitz ordering) to flux calculations in a finite-volume PNS space-marched flow-solver. Cell-centred primitive variables (pressure , density and velocity components) are extrapolated, utilising the twice differentiable Van Albada limiter, to either side of a cell-face. Inviscid fluxes are then calculated by Roe's flux [7]. We calculate inviscid fluxes in the streamwise direction via the Vigneron technique [12]. This ensures stable space-marching by multiplying the stream-wise pressure gradient by a factor $\min(\omega(M), 1)$, with M the stream-wise Mach number. We make the thin-layer approximation for viscous fluxes, set the viscosity by Sutherland's law and discretise gradients by finite-differencing. We used a COMPAQ Alpha DS20E workstation with 667 MHz CPU and 1GB of RAM.

3.1 Flux Jacobian Calculation

Table 1 shows the ratio of Jacobian to function CPU times used to calculate, for randomised independent variables (inputs), the Jacobian of (a) the Roe flux ($n = 20$, $m = 5$), (b) the Vigneron Flux ($n = 5$, $m = 5$) and (c) the viscous flux ($n = 10$, $m = 4$). All calculations were performed for 10,000 random sets of inputs repeated 10, 50 and 50 times respectively. We use the forward and reverse modes of TAMC [2], 1-sided finite-differencing and ELIAD with Markowitz ordering. The approximate linearisations: assume constant eigen-vector/value matrices and neglect TVD interpolation (Roe fluxes); assume constant $\omega(M)$ (Vigneron fluxes); neglect temperature dependence of viscosity (viscous fluxes). The ELIAD generated code is twice as fast as the next fastest numerically exact technique (TAMC reverse) for Roe and viscous fluxes and significantly faster than TAMC(forward) for the Vigneron flux. We also show the maximum error in the Jacobians, assuming TAMC(forward) is exact. The ELIAD Jacobians are correct to machine precision. The error in the 1-sided finite-difference approximation is in line with its truncation error. The error in the approximate

Table 1. Efficiency of flux Jacobian calculations

(a) Roe flux, CPU($\boldsymbol{F}$) = 7.31×10^{-7} s

technique	CPU($\boldsymbol{J}(\boldsymbol{F})$)/CPU($\boldsymbol{F}$)	Error($\boldsymbol{J}(\boldsymbol{F})$)
TAMC(forward)	18.5	0.0
TAMC(reverse)	9.1	3.6×10^{-16}
Finite Difference	24.1	4.0×10^{-6}
ELIAD	4.7	5.6×10^{-15}
Approximate	15.0	3.6

(b) Vigneron flux CPU($\boldsymbol{F}$) = 1.93×10^{-7} s

technique	CPU($\boldsymbol{J}(\boldsymbol{F})$)/CPU($\boldsymbol{F}$)	Error($\boldsymbol{J}(\boldsymbol{F})$)
TAMC(forward)	3.8	0.0
Finite Difference	9.8	1.3×10^{-6}
ELIAD	2.5	4.4×10^{-16}
Approximate	1.1	1.1

(c) Viscous flux Jacobian, CPU($\boldsymbol{F}$) = 4.22×10^{-7} s

technique	CPU($\boldsymbol{J}(\boldsymbol{F})$)/CPU($\boldsymbol{F}$)	Error($\boldsymbol{J}(\boldsymbol{F})$)
TAMC(forward)	5.8	0.0
TAMC(reverse)	3.6	0.0
Finite Difference	13.4	4.5×10^{-8}
ELIAD	1.8	5.6×10^{-17}
Approximate	1.5	2.8×10^{-1}

linearisations is high, presumably due to the approximations made and use of randomised inputs (a flow-field would have a smooth variation in inputs except at discontinuities).

Mach 2 Supersonic Diffuser After incorporating the ELIAD and TAMC Jacobian code into our PNS solver we calculated the inviscid flow-field into a diffuser of initial width 0.4 units, final width 0.26 units and length 0.8 units. The half geometry grid uses 25 cells cross-stream and 160 evenly spaced stream-wise cells with a symmetry boundary condition on the centre-line. Within the space-marching algorithm we enforced a fall in residual by a factor of 10^{-6} for each successive cross-flow solution. Table 2 gives overall performance details. Excepting the approximate linearisation, quadratic convergence was observed and use of ELIAD generated code improved overall performance.

Table 2. CPU Times and Speed-Up for Supersonic Diffuser

Linearisation	total No. iterations	run-time (s)	speed-up
Approximate	1335	1.03	1
Sparse FD	497	0.74	1.4
TAMC	497	0.37	2.8
ELIAD	497	0.33	3.1

Mach 2 Laminar Boundary Layer As a viscous test case we calculated the laminar boundary layer over a 0.93 units long flat-plate with temperature $T_\infty = 221.6K$ at inflow and a Reynolds number of 1.65×10^6 per unit length. Far-field free-stream and no-slip, isothermal ($T_{wall} = T_\infty$) wall boundary conditions are applied. We used a grid of 40 cells cross-stream and 93 evenly spaced stream-wise cells. The residual fell by a factor of 10^{-10}, or reached an absolute value less than 1×10^{-10}, at each successive cross-flow solution plane. Table 3 gives performance data. Excepting the approximate linearisation, quadratic convergence was again observed and the ELIAD generated code greatly improved overall performance.

Table 3. CPU Times and Speed-Up for Laminar Boundary Layer

Linearisation	total No. iterations	run-time (s)	speed-up
Approximate	4906	2.86	1
Sparse FD	383	1.10	2.6
TAMC	383	0.43	6.7
ELIAD	385	0.34	8.4

4 Conclusions

Local Jacobian code generated by the elimination AD tool ELIAD executes up to twice as fast as that generated by standard AD techniques and its use within Jacobian assembly for a 2-D PNS Newton solver gave increased performance.

The Roe flux results are our first application of ELIAD to a code with branching (Harten's entropy fix) and the implementation is sketched in [8]. ELIAD is to be extended (nested branching, subprograms) and new pivot order heuristics developed to take account of all FLOPs and re-use of arithmetic registers [9].

Acknowledgements: We wish to thank the UK MOD and EPSRC for funding this work under grant GR/R21882/01 and, particularly, our ELIAD co-authors Dr John Pryce and Prof. John Reid who made many useful suggestions for improving this paper.

References

1. C.H. Bischof, A. Carle, P.D. Hovland, P. Khademi, and A. Mauer: 'ADIFOR 2.0 User's Guide (Revision D)'. Mathematics and Computer Science Division, Rice University, Technical Memorandum no. 192 (1998).
2. R. Giering: 'Tangent Linear and Adjoint Model Compiler, Users Manual'. Center for Global Change Sciences, Department of Earth, Atmospheric, and Planetary Science, (MIT, Cambridge, MA,1997).
3. A. Griewank: *Evaluating Derivatives: Principles and Techniques of Algorithmic Differentiation* (SIAM, Philadelphia 2000).
4. A. Griewank and S. Reese: 'On the calculation of Jacobian matrices by the Markowitz rule'. In: *Automatic Differentiation of Algorithms: Theory, Implementation, and Application*, ed. by A. Griewank, G. F. Corliss (SIAM, Philadelphia, 1991) pp.126–135
5. P.D. Hovland, L.C. McInnes: Parallel Computing **27**, 4 (2001)
6. U. Naumann: Efficient Calculation of Jacobian Matrices by Optimized Application of the Chain Rule to Computational Graphs. PhD Thesis, Technical University of Dresden (1999)
7. P.L. Roe: Journal of Computational Physics **43** (1981)
8. M. Tadjouddine, S.A. Forth, J.D. Pryce: 'AD Tools and Prospects for Optimal AD in CFD Flux Jacobian Calculations'. In *Automatic Differentiation: From Simulation to Optimization.* ed. by. G. Corliss, C. Faure, A. Griewank, L. Hascoët, U. Naumann (Springer, New York 2001) pp. 247–252.
9. M. Tadjouddine, S.A. Forth, J.D. Pryce, J.K. Reid. 'Performance Issues for Vertex Elimination Methods in Computing Jacobians using Automatic Differentiation'. In *Proceedings of the Second International Conference on Computational Science, Volume 2.* ed. by P.M. Sloot (Lecture Notes in Computer Science, Springer-Verlag, Amsterdam, 2002) pp. 1077–1086,
10. K.J. Vanden and P.D. Orkwis: AIAA Journal **34**, 6 (1996)
11. V. Venkatakrishnan: AIAA Journal **27**, 7 (1989)
12. Y.C. Vigneron, J.V. Rakich, and J.C. Tannehill: 'Calculation of Supersonic Viscous Flow over Delta Wings with Sharp Subsonic Leading Edges'. AIAA Paper 78-1137 (1978)

Parallelized, Solution Adaptive, Multi-Grid Hybrid Unstructured Methods for the Navier-Stokes Equations on Complex Configurations

S. J. Zhang, J. Liu, Y. S. Chen, D. Godavarty, and P. Mallapragada

Engineering Sciences, Inc., Huntsville, AL 35802, USA

Abstract. This paper presents parallelized, solution adaptive, multi-grid hybrid unstructured methods to the turbulent flows past complex configurations. The numerical techniques are accomplished by parallel computing algorithm with multi options of serial, PVM or MPI to accommodate any type of machines, hanging node solution adaptation, algebraic multi-grid speeding-up convergence method. With the enhancement of higher order scheme data reconstruction, higher order upwind-biased differencing, multi-pressure-correction, and Bi-CGSTAB and GMRES matrix solvers, the proposed methods can efficiently and accurately handle flow problems in several cases including benchmark and practical. Very good convergence and accurate solutions have been demonstrated even for complex 3D configurations under high Reynolds number turbulent flow conditions.

1 Introduction

Computational Fluid Dynamics (CFD) has evolved rapidly as a discipline and is increasingly being used as valuable tool for engineering design. Based largely on the mathematical foundations laid by Lax [1] and Godunov [2] among others, the field has come into its own in the last two decades. Great advances have been made in the areas of spatial discretization, grid generation, and solution strategies. Many early computations of the flow over complex geometries were developed on structured grids with body-fitted coordinates which take the advantage of automatic indexing and are efficient for many applications. However, the task of generating structured grids about complex configurations presented a serious challenge. The widely used multi-block structured grid approach solves the problem by tessellating the domain between the body and other boundaries into simple logically rectangular blocks so that structured grids can be generated easily with each block. The automation of blocking and ensuring grid smoothness at the block interfaces are difficult tasks, which have prompted intensive research and development in unstructured grid algorithms in the past one-and-half decades. Unstructured grid methods are characterized by their ease in handling completely unstructured meshes and have been widely used in computational fluid dynamics. The relative increase of computer memory and CPU time with unstructured grid methods is quite much, but can be offset by using parallel techniques in which many processors are put together to work on the same problem. Furthermore, when unstructured grid method is facilitated with the multi-grid technique its convergence can be greatly speed up to reach its solution. Specifically, flexible mesh adaptation is another attraction for unstructured grid method where high resolution can be obtained by mesh refinement in high gradient region such as shock waves. Solution adaptive algorithms have become quite popular in CFD. They provide flexibility to adjust the grid during the solution to achieve high flow field resolution during the solution procedure without any intervention by the user. Those adaptive grid algorithms have been developed for sequential execution and have reached a high level of maturity. However, the area of parallel adaptive algorithms still remains wide open to explore.

In this study, a cell-centered unstructured finite volume method enhanced by parallelization, anisotropic adaptation and multi-grid techniques was developed to predict the turbulent flows over complex configurations.

2 Solution Algorithms

The transport equations can also be written in integral form as

$$\frac{\partial}{\partial t}\int_{\Omega} \rho\phi d\Omega + \oint_{\Gamma} \boldsymbol{F}\cdot\boldsymbol{n} d\Gamma = \int_{\Omega} S_{\Omega} d\Omega \tag{1}$$

where Ω is the domain of interest, Γ the surrounding surface, $\boldsymbol{n}$ the unit normal of Γ in outward direction. The flux function $\boldsymbol{F}$ consists of the inviscid and the viscous parts:

$$\boldsymbol{F} = \rho\vec{V}\phi - \mu_{\phi}\nabla\phi \tag{2}$$

The viscous flux for the face e between control volumes P and E as shown in Fig. 1 can be approximated as:

$$(\nabla\phi\cdot\boldsymbol{n})_e \approx \frac{\phi_E - \phi_P}{|\boldsymbol{r}_E - \boldsymbol{r}_P|} + \nabla\phi_e\cdot\left(\boldsymbol{n} - \frac{\boldsymbol{r}_E - \boldsymbol{r}_P}{|\boldsymbol{r}_E - \boldsymbol{r}_P|}\right) \tag{3}$$

The inviscid flux is evaluated through the value at the upwind cell and a linear reconstruction procedure to achieve second order accuracy

$$\phi_e = \phi_u + \Psi_e\nabla\phi_u\cdot(\boldsymbol{r}_e - \boldsymbol{r}_u) \tag{4}$$

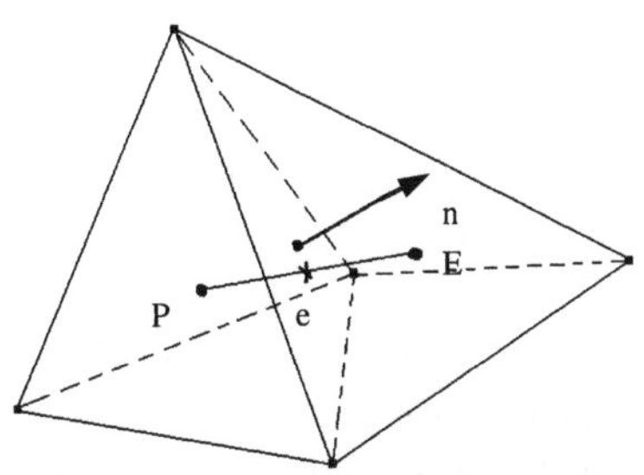

Fig. 1. Unstructured control volume

where the subscript u represents the upwind cell and Ψ_e is a flux limiter used to prevent from local extrema introduced by the data reconstruction. The flux limiter proposed by Barth [3] is employed. Defining $\phi_{\max} = \max(\phi_u, \phi_j)$, $\phi_{\min} = \min(\phi_u, \phi_j)$, the scalar Ψ_e associated with the gradient at cell u due to face e is:

$$\Psi_e = \begin{cases} \min(1, \frac{\phi_{\max}-\phi_u}{\phi_e^0-\phi_u}) \text{ if } \phi_e^0 - \phi > 0 \\ \min(1, \frac{\phi_{\min}-\phi_u}{\phi_e^0-\phi_u}) \text{ if } \phi_e^0 - \phi < 0 \\ \\ 1 \end{cases} \tag{5}$$

where ϕ_e^0 is computed without limiting (i.e. $\Psi_e = 1$).

In this work the preconditioned Bi-CGSTAB [4] and GMRES [5] matrix solvers are used to efficiently solve transport equations and pressure-correction equation, respectively. Both of Recursive Coordinate Bisection (RCB) and Recursive Graph Bisection (RGB) were employed to decompose the sub-domains. Both of PVM and MPI were implemented for data communication among processors. Specifically, algebraic multi-grid (AMG) [6] method was implemented to enhance the stability and capability of the solver to handle complex problems. The details of parallel algorithm and domain decomposition can be found in [7,8]. Unstructured meshes are particularly flexible for local grid refinement and coarsening. The primary idea behind adaptation is to modify the grids to better resolve the features in the flow fields and hence to achieve accurate numerical solutions. The modification of the grids is carried out by a hanging node approach. The adaptive techniques such as adaptation function , criteria for cell refinement or coarsening and smoothing strategy were detailed in [9]. Almost all-iterative solution methods converge more slowly on finer grids. It has been demonstrated by many authors, including the present ones [10], that solution of the Navier-Stokes equations by multi-grid methods is very efficient. AMG method [6] was implemented to solve the algebra equation resulting from the pressure correction equation. The details of implementation procedures are referred to [7].

3 Results and Discussion

The present approach has been applied to many external and internal flow problems for validation. In this study, the issues of accuracy and efficiency of the proposed methods, and robustness of the flow solver are addressed for NACA 0012 airfoil and Boeing 747 airplane. Two kinds of two-equation turbulence models are employed here for N-S equation computations, one is the standard k-ε model with wall function, the other the extended k-ε model taking the effect of compressibility into account with wall function.

The transonic flow (M = 0.799) around a NACA 0012 profile at an angle of attack (AoA) of 2.26 is considered,first. Both of Euler and N-S computations were carried out of the case. The initial coarse grid for numerical computations contains 4399 nodes and 8666 triangles.

Fig. 2 illustrates the partitioned domains, pressure contours, solution adaptive grids with Euler computations. It clearly demonstrates the capability of

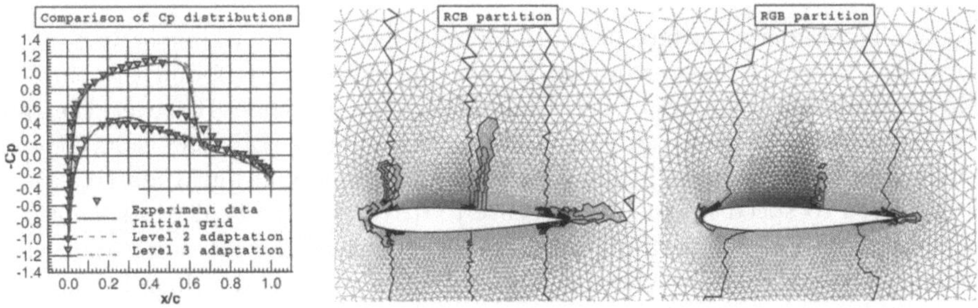

Fig. 2. Parallelized, solution adaptation and multi-grid Euler solutions

solution adaptation, the validity of developed parallel/adaptive scheme and the feasibility of employed domain decomposition methods. Furthermore, we found that the present parallelized algebraic multi-grid method is quite robust and highly effective. The accuracy of the present method with the turbulence models dealing with some special flow features such as boundary and shear layers, recirculation zones and shock waves was assessed. In this study, two-equation models with wall functions are employed because they could provide good predictions at low cost. Fig. 3 shows the N-S computational results with extended k-ε model. It can be seen the predicted results are in reasonable agreement with observations. Therefore the present turbulence models will be used to predict turbulent flow on more complex configuration in the following considerations, i.e. Boeing 747.

Unstructured transonic computations are also presented for a complete Boeing 747 airplane configuration as a demonstration and assessment of the present methodology for a very complex geometry and flow field. The 747 geometry includes the fuselage, the wing, horizontal and vertical tails, underwing pylons and flow-through engine nacelles. The unstructured mesh for the 747 contains 105,372 tetrahedra and 19,698 nodes for the half-span airplane. Also, there are 8,402 triangles and 4,371 nodes on the boundaries of the mesh that include the airplane, the symmetry plane, and the far field. Euler and Navier-Stokes (N-S) flow solutions were obtained for angle of attack (AoA) of 2.7 at Mach 0.84. The flow solver was executed under sequential and parallel modes, respectively. Figs. 4 and 5 illustrate partitioned domains along with adaptive grids and cor-

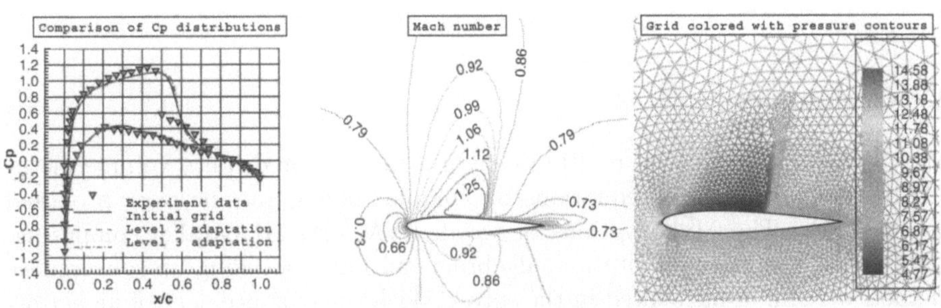

Fig. 3. N-S computational results

responding predicted surface pressure contours based on RCB, respectively. The parallel efficiency is also examined for the Euler case. It is found the parallel efficiency under the modes of PVM and MPI is almost same. However, the efficiency for 6 and 12 processor run on network computers (LINUX) is 90.7% and 84.2%, respectively. In addition, parallelized algebraic multi-grid matrix solver is employed to speed up the convergence to drive the flow field to its solution. Furthermore these 3D computations further confirmed the robustness and efficiency of the present parallelized algebraic multi-grid method to handle complex flow problems.

Fig. 4. Partitioned grids with RCB together with solution adaptation

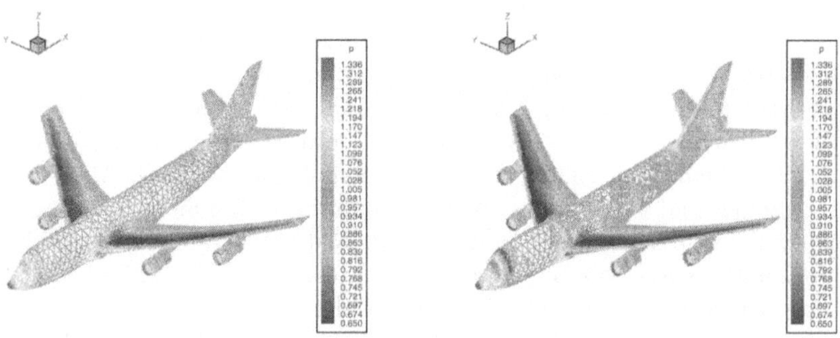

Fig. 5. Pressure contours with combining parallelization and adaptation

4 Conclusions

The parallelized, adaptive, multi-grid hybrid unstructured finite volume methods to the Navier-Stokes equations on complex configurations were proposed in this study. The Multi options of serial, PVM or PMI parallel computing, hanging node solution adaptation, algebraic multi-grid speeding-up convergence method were all implemented in the solver. With the enhancement of higher order scheme data reconstruction, higher order upwind-biased differencing, multi-pressure-correction, and Bi-CGSTAB and GMRES matrix solvers, the present approach can efficiently and accurately handle flow problems in several cases including benchmark and practical. Very good convergence and accurate solutions have been demonstrated even for 3D very complex geoemtries under high Reynolds number turbulent flow conditions.

References

1. Lax, P., "Hyperbolic Systems of Conservation Laws and the Mathematical Theory of Stoke Waves," 2nd ed., Vol. 11, SIAM Regional Series on Applied Mathematics, SIAM Philadelphia, PA, 1973.
2. Godunov, S. K., "A Difference Scheme for Numerical Computation of Discontinuous Solution of Hydrodynamic Equation," *Math. Sbornik*, 47, 1959, 271-306 (in Russian). Translated US Joint Publ. Res. Service, JPRS 7226 (1969).
3. Barth, T.J., "Recent Development in High Order K-Exact Reconstruction on Unstructured Meshes," AIAA Paper 93-0668, 1993.
4. Van Der Vorst, H.A., "Bi-CGSTAB: A Fast and Smoothly Converging Variant of Bi-CG for the Solution of Nonsymmetric Linear Systems," SIAM J. Sci. Stat. Comput., Vol. 13, pp. 631-644, 1992.
5. Saad, Y. and Schultz, M.H., "GMRES: A Generalized Minimal Residual Algorithm for Solving Nonsymmetric Linear Systems," SIAM J. Sci. Stat. Comput., Vol. 7, pp. 856-869, 1986.
6. Raw, M., "Robustness of Coupled Algebric Multigrid for the Navier-Stkoes Equations," AIAA Paper 96-0297, 1996.
7. Zhang, S.J., Liu, J., Chen, Y.-S., Godavarty, D., Mallapragada, P. and Wang, T.-S., "A Parallelized, Adaptive, Multi-Grid Hybrid Unstructured Solver for All-Speed Flows," AIAA paper 2002-0109, 2002.
8. Mallapragada, P, Zhang, S.J., Liu, J., Chen, Y.-S. and Godavarty, D., "A Parallel Implementation of General Purpose Unstructured Flow Solver," AIAA paper 2002-2979, 2002.
9. Zhang, S.J., Liu, J., Chen, Y.-S. and Wang, T.-S., "Adaptation for Hybrid Unstructured Grid with Hanging Node Method," AIAA Paper 2001-2657, 2001.
10. Zhang, S.J., Yu, A.B. and Zulli, P., "Preconditioned Multigrid Method for Fluid Flow in Porous Media, Second International Conference on CFD in the Minerals & Process Industries, Dec. 1999, pp. 461-466, Melbourne.

Uncertainty of Solution in Euler Equations and Numerical Instability

Kyu Hong Kim[1], Chongam Kim[2], Oh-Hyun Rho[2], and Kyung-Tae Lee[1]

[1] Sejong-Lockheed Martin Aerospace Research Center, Sejong University, SEOUL 143-747, KOREA
[2] Department of Aerospace Engineering, Seoul National University, SEOUL 151-742, KOREA

Abstract. Numerical instability is studied based on numerics and mathematics. It is frequently observed in the region where velocity is zero. In that region, Euler equations have numerous solutions and, thus, it is impossible to determine an unique solution with only governing equations. The unique solution can be determined by additional outer flow conditions or outer numerical disturbances. This peculiarity of Euler equations can cause numerical instability in a shock discontinuity since information on the unique solution under undisturbed conditions is lost by disturbances. As a result, shock instability cannot be removed completely in the scheme which is consistent with Euler equations.

1 Introduction

The one of importance in developing numerical schemes is the removal of numerical instability. Since numerical instability is closely related to a converged solution or solution accuracy, there have been so many studies for successful and robust computations, and as by-products, so many numerical schemes have been developed. Shock instability has been analyzed by Quirk [1], Liou [2], and Xu [3] and so on. For the first time, Quirk explored and suggested the defect and the failure of Roe's FDS throughout many numerical test cases [1]. Liou identified the dissipative terms that are responsible for shock instabilities, such as the odd-even decoupling and the so-called "carbuncle phenomenon" and propose a conjecture stating the connection of the pressure difference term to these multi-dimensional shock instabilities and hence a cure for those difficulties. He reached the conclusion that the scheme, which has the numerical dissipation term expressed by pressure difference in a continuity equation, always has the probability of numerical shock instability [2]. Xu explained the shock instability in introducing the streamline converging or diverging, such as quasi one-dimensional nozzle. He concluded that the odd-even decoupling and the carbuncle phenomena are due to the numerical discretization of the inviscid Euler equations in the shock region and the special dissipative mechanism of the Godunov method in the projection stage. In order to stabilize normal to shock front, a direct way is to add dissipation through entropy fix, such as to solve the viscous shear layer equation [3]. Rhbinet said that the development of the instability has been proven to be ruled by a linear instability mechanism and though carbuncle phenomenon has

been thought to be the purely numerical problem for many years, this is not true [4]. Pandolfi explained and predicted the generation of such instabilities using the analysis of the linearized form of the numerical scheme [5]. In order to cure shock instability and instability in contact discontinuity, AUSMPW [6] and AUSMPW+ [7] schemes has been developed. During the development, Kim et al. showed the intrinsic problem by Euler equations in Ref. [8].
In this paper, numerical instability, including instability in a contact discontinuity as well as in a shock discontinuity, will be explained by the uncertainty of Euler equation. Since Euler equation has the characteristic that cannot determine density distribution in the region of zero velocity, the property across discontinuity cannot be determined uniquely. In computations this uncertainty acts as a source of numerical disturbance continuously. This disturbance may increase, in some conditions, i.e., numerical instability increases, and as a result, unwanted solutions are introduced.

2 Characteristics of Euler Equations

In Euler equations, velocity of zero is the one of solutions. It is commonly treated as a trivial solution and has no importance in mathematics. However, in computations, it leads to many numerical problems. If velocity becomes zero asymptotically, Euler equations have the form of $0 = 0$, i.e., have numerous solutions. Thus, solutions, including density and energy and so on, cannot be determined uniquely by only Euler equations and boundary conditions. Unique solution can be determined with the additional restriction equation besides given equations and boundary conditions.
There are so many schemes developed to solve Euler equations. The scheme, completely consistent with Euler equations such as Roe's FDS and AUSM-type schemes, also has the uniqueness problem like Euler equations.

3 Numerical instability in slip flows

In case of a slip flow, the velocity normal to a control surface, v, is zero and the governing equations should be always satisfied independent to u, which is parallel velocity to control surface. Providing that AUSM+ is applied to a slip flow problem, solutions is $p_{i-1} = p_{i+1}$ and $p_i = p_{i+2}$. It means that AUSM+ shows even-odd decoupling of pressure. Different from Euler equations, AUSM+ cannot determine pressure uniquely as well as density. Uncertainty of pressure is the direct cause that AUSM+ shows pressure oscillations in a shear layer. On the other hand, in AUSMPW+, $v_{i-1} = v_i = v_{i+1} = v_{i+2} = 0$ and $p_{i-1} = p_i = p_{i+1} = p_{i+2}$. Different from AUSM+, the solution of AUSMPW+ never shows even-odd decoulping characteristic, and as a result, AUSMPW+ does not show pressure oscillations near boundary layers. It is noticeable that the numerical dissipation term by pressure difference in continuity equation removes oscillations and provides a restriction equation to settle the uncertainty problem of pressure.

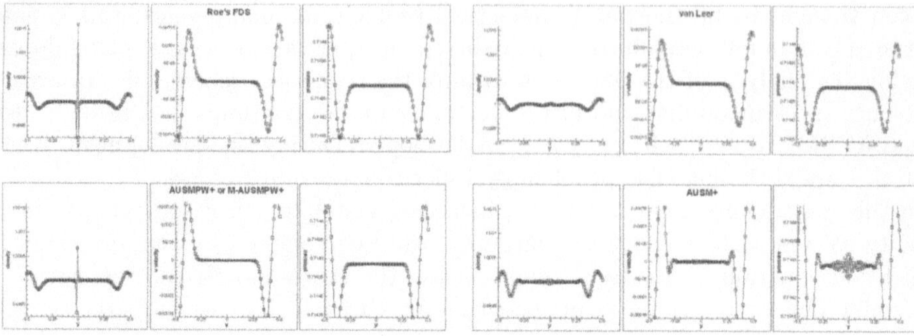

Fig. 1. Density, velocity and pressure distributions influenced by the pressure disturbance; (**a**) Roe's FDS, (**b**)van Leer, (**c**) AUSMPW+, (**d**) AUSM+

Table 1. Uncertainty and instability in a contact discontinuity of each scheme

Numerical scheme	Uncertainty of density	Uncertainty of u	Uncertainty of pressure	Pressure oscillation
Euler equation	yes	yes	no	no
Roe's FDS	yes	yes	no	no
AUSM+	yes	yes	yes	yes
AUSMPW+	yes	yes	no	no
M-AUSMPW+	yes	yes	no	no
van Leer FVS	no	no	no	no

The followings are numerical experiments for slip flows. Initial converged solutions are disturbed by pressure. Initial conditions are $(\rho, u, v, p)_L = (1, 0.25, 0.0.714)$ and $(\rho, u, v, p)_R = (1, 0.5, 0.0.714)$. Pressure disturbances, $(\rho, u, v, p)_D = (1, 0.5, 0.0.775)$ are entered right behind the contact discontinuity.

As shown in Fig. 1, AUSM+ shows pressure oscillations because of uncertainty of pressure and also shows density and velocity oscillations. Moreover, the solution was not converged in this test case. It is caused by the fact that AUSM+ cannot determine density and pressure uniquely in a contact discontinuity. On the other hand, in case of Roe's FDS or AUSMPW+, after disturbed and converged to a new solution, the newly converged pressure and velocity are the same as initial converged values, However, density are seen to be converged to a completely different value. This shows that, once disturbances are introduced, it is impossible for results to be restored to the original state, when AUSM+, Roe's FDS, and AUSMPW+ which are consistent with Euler equations are applied. This characteristic, sometimes disadvantage, is caused by the fundamental characteristic of Euler equations, not by the problem of the numerical scheme itself.

4 Numerical instability in a shock discontinuity

Figure 2 shows that pressure is disturbed in converged normal shock discontinuities. In a flow direction, flows are governed by a shock relation. In the normal direction to a flow direction, flows are governed by slip flows which may have different velocities across a discontinuity. Thus, the uncertainty of density in a slip discontinuity may lead to physically unacceptable density distributions in a shock discontinuity. The difference between Euler equations and the consistent numerical scheme is as follows. Euler equations can determine density behind a shock discontinuity uniquely from a shock relation. So, even if a solution is disturbed, information of original state could be preserved by a shock relation, and after all, the results are restored to the original state. However, in computations, a shock relation can not be used any more since there is no information to indicate what a shock is and where a shock is locate on. Thus, it is impossible to preserve information of original state. As a result, once a solution is disturbed, the information of original state is lost and there is no way to restore to the original state. It is a main cause of numerical instability from a shock discontinuity.

Figure 2 shows pressure distributions after the shock is converged to the machine accuracy. In case of (a), the pressure disturbance which is 10 percentage larger than the converged value is put into the cell which is located right in front of a sonic transition position. In case of (b), the pressure disturbance is put into the cell which is located right behind a sonic transition position. Figure 3 shows that pressure is disturbed in converged normal shock discontinuities. In the case (b), all upwind schemes do not show the increase of numerical instability since the numerical dissipation of test schemes are so enough to remove the disturbance. On the other hand, in the cases (a) where disturbance is put into in front of the sonic transition position, the disturbance is easily increased in crossing shock discontinuity. Roe's FDS shows severe numerical instability. AUSM+ suppresses the increase of the numerical instability somehow.

Although AUSM+ has been known as robust scheme, numerical instability may increase in some conditions, and after all, severe oscillations occur. Figure 4 and 5 are good examples. Figure 4 is the result of moving shock with Mach number 25. The numerical disturbance generated in shock discontinuity grows

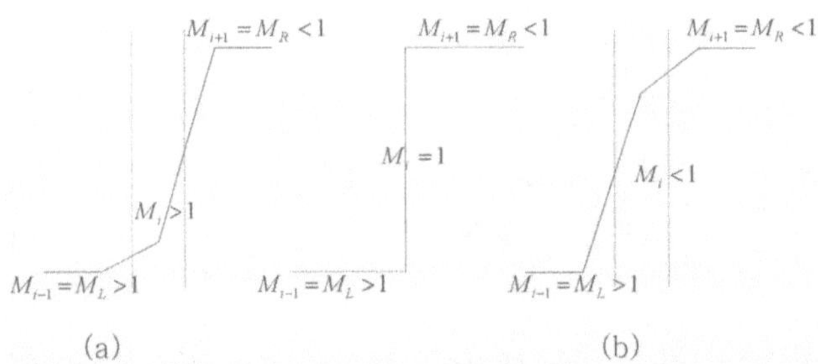

Fig. 2. Schematics of shock discontinuities according to the sonic transition position

Table 2. Uncertainty and instability in a shock discontinuity of each scheme

Numerical scheme	Uncertainty of density	Uncertainty of u	Uncertainty of pressure	Pressure oscillation
Euler equation	no	no	no	no
Roe's FDS	yes	yes	no	yes
AUSM+	yes	yes	yes	yes
AUSMPW+	yes	yes	no	yes
M-AUSMPW+	no	no	no	no
van Leer FVS	no	no	no	no

larger and larger and oscillation is amplified along the flow. Figure 5 shows that the numerical disturbance also grows larger along the flow and unwanted vortex flow is produced.

5 Conclusions

Numerical instability of AUSM-type scheme is studied and the causes and cures of instability are proposed. Numerical instability occurs in the region of velocity of zero where Euler equation has numerous solutions and cannot determine unique solution fundamentally. Thus, the scheme consistent with the Euler equation cannot determine unique solutions and as a result, numerical instability is induced. The cause of numerical instability from a slip flow and a shock discontinuity could be explained with the uniqueness of solution. If a solution could

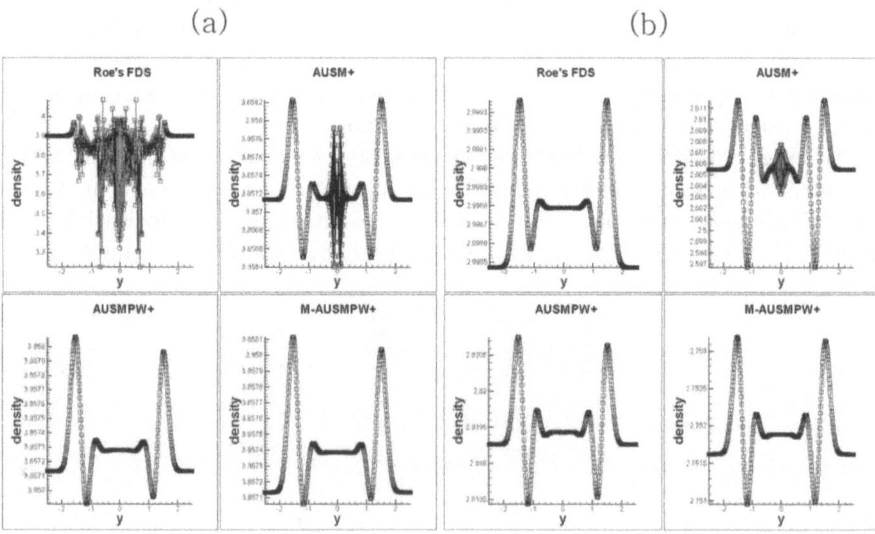

Fig. 3. Density distributions when the sonic transition position is located; (**a**) between i^{th} and $i+1^{th}$ cells, (**b**) between i^{th} and $i-1^{th}$ cells

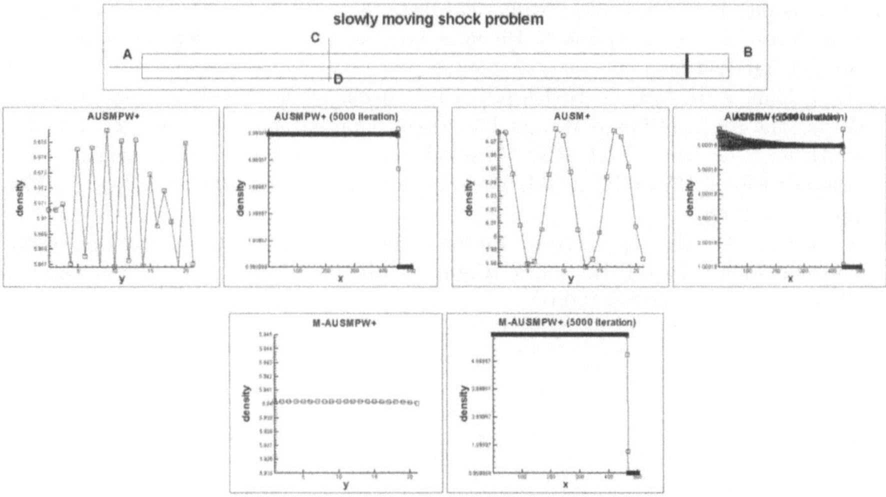

Fig. 4. Density distributions of moving shock discontinuity along AB and CD

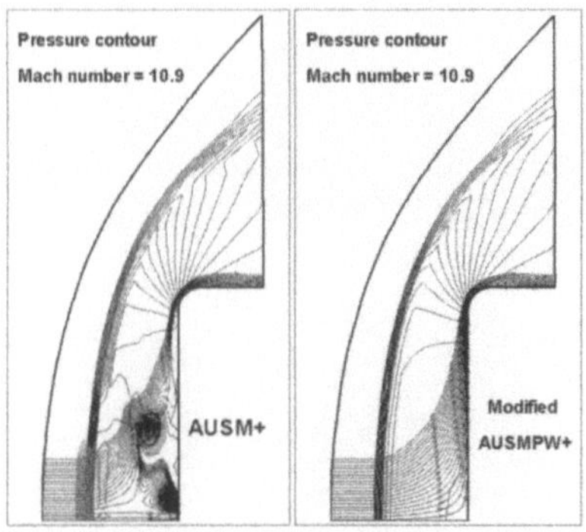

Fig. 5. Density distributions along the line AB and CD

be determined uniquely, all numerical instability from a slip flow and a shock discontinuity can be removed.

References

1. J. J. Quirk: 'A Contribution to the Great Riemann Solver Debate'. Int. J. for Numerical Methods in Fluids, **18**, 555 (1994)
2. M. S. Liou: 'Mass Flux Schemes and Connection to Shock Instability'. J. of Computational Physics, **160**, pp. 623-648 (1994)
3. Kun Xu: 'Gas Evolution Dynamics in Godunov-type Schemes and Analysis of Numerical Shock Instability'. ICASE Report No. 99-6 (1999)
4. J.-Ch. Robinet, J. Gressier, G. Vasalis, J.-M. Moschetta: 'Shock Wave Instability and Carbuncle Phenomenon: Same intrinsic Origin?'. under consieration for publication in J. Fluid Mech.

5. M. Pandolfi, D.D'Ambrosio: 'Numerical Instabilities in Upwind Methods: Analysis and Cures for the "Carbuncle" Phenomenon'. J. of Computational Physics, **166**, pp. 271-301 (2001)
6. K. H. Kim, J. H. Lee, O. H. Rho: 'An Improvement of AUSM Schemes by Introducing the Pressure-based Weight Functions'. Computers and Fluids, **27**, 3 (1998)
7. K. H. Kim, C. Kim, O. H. Rho: 'Methods for the Accurate Computations of Hypersonic Flows, PART I: AUSMPW+ Scheme'. J. of Computational Physics, **174**, pp. 38-80 (2001)
8. K. H. Kim, C. Kim, O. H. Rho: 'Methods for the Accurate Computations of Hypersonic Flows, PART II: Shock Aligned Grid Technique'. J. of Computational Physics, **174**, pp. 81-119 (2001)

Discontinuous Galerkin Method for the Time-domain Maxwell's Equations

Adour V. Kabakian, Vijaya Shankar, and William F. Hall

HyPerComp, Inc.
31255 Cedar Valley Dr., Suite 327
Westlake Village, CA 91362, USA

Abstract. A parallel, unstructured, high-order discontinuous Galerkin method is developed for the time-dependent Maxwell's equations, using simple monomial polynomials for spatial discretization and a fourth-order Runge-Kutta scheme for time marching. Scattering results for a number of validation cases are computed employing polynomials of up to third order. Accurate solutions are obtained on coarse meshes and grid convergence is achieved, demonstrating the capabilities of the scheme for time-domain electromagnetic wave scattering simulations.

1 Introduction

The more traditional and predominant frequency-domain approach in Computational Electromagnetics (CEM) remains restricted to single frequency solutions, making them unsuitable for applications that require broadband frequency response. Such problems are most naturally solved by the time-domain approach which can model broadband frequency response from a single pulse calculation. Maintaining accurate phase and amplitude of waves over long propagation distances, however, remains a challenge for current time-domain methods. Numerical experiments have shown that employing a high-order scheme rather than performing mesh refinement with a low-order method is computationally more efficient for such problems. In this article, the development of an unstructured high-order time-domain solver for Maxwell's equations based on the discontinuous Galerkin method [1] is presented.

2 Governing Equations

Electromagnetic wave propagation is governed by Maxwell's equations, a set of quasi-linear hyperbolic equations which are written in vector differential form as

$$\frac{\partial \boldsymbol{D}}{\partial t} - \nabla \times \boldsymbol{H} = -\boldsymbol{J}\,, \qquad \frac{\partial \boldsymbol{B}}{\partial t} + \nabla \times \boldsymbol{E} = \boldsymbol{0}\,, \tag{1}$$

$$\nabla \cdot \boldsymbol{D} = \rho\,, \qquad \nabla \cdot \boldsymbol{B} = 0\,, \tag{2}$$

where $\boldsymbol{E}$ and $\boldsymbol{H}$ are the electric and magnetic field intensities, respectively, $\boldsymbol{D}$ and $\boldsymbol{B}$ are the electric and magnetic flux densities, respectively, $\boldsymbol{J}$ is the electric current density, and ρ is the electric charge density.

If an electromagnetic field initially satisfies the divergence equations (2), it will continue doing so when evolved in time via the curl equations (1). Thus, only the curl equations are typically solved for time domain simulations.

For ordinary materials, the field intensities $\boldsymbol{E}$ and $\boldsymbol{H}$, the flux densities $\boldsymbol{D}$ and $\boldsymbol{B}$, and the current $\boldsymbol{J}$ are related through the constitutive relations

$$\boldsymbol{D} = \epsilon \boldsymbol{E}\,, \qquad \boldsymbol{B} = \mu \boldsymbol{H}\,, \qquad \text{and} \qquad \boldsymbol{J} = \sigma \boldsymbol{E}\,, \tag{3}$$

where ϵ is the electric permittivity, μ is the magnetic permeability, and σ is the conductivity. The material parameters μ and ϵ determine the speed of propagation c through a medium by the relation $c^2 = 1/\epsilon\mu$.

The boundary conditions on the surface of a perfect electric conductor (PEC) are $\hat{\boldsymbol{n}} \times \boldsymbol{E} = \boldsymbol{0}$ and $\hat{\boldsymbol{n}} \cdot \boldsymbol{B} = 0$. The linearity of Maxwell's equations can be exploited to solve for the scattered fields directly. For this scattered field formulations, the PEC surface boundary conditions become

$$\hat{\boldsymbol{n}} \times \boldsymbol{E}^s = -\hat{\boldsymbol{n}} \times \boldsymbol{E}^i \qquad \text{and} \qquad \hat{\boldsymbol{n}} \cdot \boldsymbol{B}^s = -\hat{\boldsymbol{n}} \cdot \boldsymbol{B}^i. \tag{4}$$

3 Numerical Scheme

3.1 Discontinuous Galerkin Method

Consider the generic hyperbolic conservation law for a vector quantity $\boldsymbol{U}(\boldsymbol{x})$ in a domain $\mathcal{D}$

$$\frac{\partial \boldsymbol{U}}{\partial t} + \nabla \cdot \boldsymbol{F}(\boldsymbol{U}) = 0\,, \tag{5}$$

where $\boldsymbol{F}(\boldsymbol{U})$ is a flux vector. The space domain $\mathcal{D}$ is partitioned into non-overlapping subdomains, or elements $\mathcal{D}_i$. A semi-discrete solution is sought in an approximation space V_h which is discontinuous across element faces

$$V_h = \left\{ v \in L^2(\mathcal{D}) : v\,|\,\mathcal{D}_i \in \mathcal{P}(\mathcal{D}_i) \right\}\,,$$

where $\mathcal{P}(\mathcal{D}_i)$ is a polynomial space defined on $\mathcal{D}_i$.

Let $\mathcal{B}_i = \{v_l^i\}_{l=0,\ldots,N-1}$ be a polynomial basis set local to $\mathcal{D}_i$ such that

$$\operatorname{Span}(\mathcal{B}_i) = \mathcal{P}(\mathcal{D}_i) \qquad \text{and} \qquad \operatorname{Supp}(v_l^i(\boldsymbol{x})) = \mathcal{D}_i\,, \quad l = 0, \ldots, N-1.$$

The approximate solution $\boldsymbol{U}_h$ is obtained by requiring that a weak Galerkin formulation of (5) be satisfied over each element $\mathcal{D}_i$. Multiplying (5) by $v_l^i(\boldsymbol{x})$, integrating over $\mathcal{D}_i$, and applying the divergence theorem, we obtain

$$\int_{\mathcal{D}_i} v_l^i \frac{\partial \boldsymbol{U}_h}{\partial t}\, d\boldsymbol{x} = -\int_{\partial\mathcal{D}_i} v_l^i\, \boldsymbol{F}(\boldsymbol{U}_h) \cdot \hat{\boldsymbol{n}}\, ds + \int_{\mathcal{D}_i} \nabla v_l^i \cdot \boldsymbol{F}\left(\boldsymbol{U}_h\right) d\boldsymbol{x}\,, \tag{6}$$

where $\partial\mathcal{D}_i$ is the boundary $\mathcal{D}_i$ and $\hat{\boldsymbol{n}}$ is the outward unit normal vector.

Since the solution is discontinuous across the interface of any two contiguous elements $\mathcal{D}_i$ and $\mathcal{D}_j$, a numerical flux $\boldsymbol{F}_\nu(\boldsymbol{U}_h^i, \boldsymbol{U}_h^j)$ is constructed for the flux term $\boldsymbol{F}(\boldsymbol{U}_h)$ in the surface integral of (6). The numerical flux $\boldsymbol{F}_\nu$ provides the coupling mechanism between elements, and it is typically an exact Riemann flux for linear systems or an approximation for non-linear problems.

3.2 Formulation of Scheme for Maxwell's Equations

The curl equations (1) are cast in the conservation law form of (5) by setting

$$\boldsymbol{U} = \begin{bmatrix} D_x \\ D_y \\ D_z \\ B_x \\ B_y \\ B_z \end{bmatrix}, \ \boldsymbol{F}(\boldsymbol{U}) = \begin{bmatrix} \mathcal{E} \\ \mathcal{F} \\ \mathcal{G} \end{bmatrix}; \quad \mathcal{E} = \begin{bmatrix} 0 \\ H_z \\ -H_y \\ 0 \\ -E_z \\ E_y \end{bmatrix}, \ \mathcal{F} = \begin{bmatrix} -H_z \\ 0 \\ H_x \\ E_z \\ 0 \\ -E_x \end{bmatrix}, \ \mathcal{G} = \begin{bmatrix} H_y \\ -H_x \\ 0 \\ -E_y \\ E_x \\ 0 \end{bmatrix}.$$

It should be noted that

$$\boldsymbol{F}(\boldsymbol{U}) \cdot \hat{\boldsymbol{n}} = \begin{bmatrix} -\hat{\boldsymbol{n}} \times \boldsymbol{H} \\ \hat{\boldsymbol{n}} \times \boldsymbol{E} \end{bmatrix}.$$

The numerical flux $\boldsymbol{F}_\nu$ is obtained by solving exactly the one-dimensional Riemann problem on each elemental interface. Accordingly, Shankar *et al.* have derived jump conditions from which the flux is obtained [2]:

$$\hat{\boldsymbol{n}} \times \boldsymbol{E}^* = \hat{\boldsymbol{n}} \times \frac{(c\epsilon \boldsymbol{E} - \hat{\boldsymbol{n}} \times \boldsymbol{H})_L + (c\epsilon \boldsymbol{E} + \hat{\boldsymbol{n}} \times \boldsymbol{H})_R}{(c\epsilon)_L + (c\epsilon)_R} \tag{7}$$

$$\hat{\boldsymbol{n}} \times \boldsymbol{H}^* = \hat{\boldsymbol{n}} \times \frac{(c\mu \boldsymbol{H} + \hat{\boldsymbol{n}} \times \boldsymbol{E})_L + (c\mu \boldsymbol{H} - \hat{\boldsymbol{n}} \times \boldsymbol{E})_R}{(c\mu)_L + (c\mu)_R} \tag{8}$$

where subscripts L and R refer to the left and right sides of the interface, superscript $*$ refers to field values on the interface, and $\hat{\boldsymbol{n}}$ is the normal pointing from the left side to the right side.

3.3 Numerical Boundary Conditions

On boundary and interior element faces alike, we seek to compute the flux terms $\hat{\boldsymbol{n}} \times \boldsymbol{E}$ and $\hat{\boldsymbol{n}} \times \boldsymbol{H}$. The same jump conditions employed to compute the numerical flux $\boldsymbol{F}_\nu$ for interior elements are used for the boundary condition update procedure. For a perfect conductor, $\hat{\boldsymbol{n}} \times \boldsymbol{E}$ follows directly from the analytical boundary condition (4), while the the jump conditions are used to obtain

$$\hat{\boldsymbol{n}} \times \boldsymbol{H} = \hat{\boldsymbol{n}} \times \boldsymbol{H}_e - \frac{\hat{\boldsymbol{n}}}{(c\mu)_e} \times (\hat{\boldsymbol{n}} \times \boldsymbol{E} - \hat{\boldsymbol{n}} \times \boldsymbol{E}_e) \tag{9}$$

where subscript e refers to the field and material parameter values inside the element, and $\hat{\boldsymbol{n}}$ is the outer normal of the element face.

Since the computational domain has to be truncated a finite distance away from the surface of the scatterer, absorbing boundary conditions have to be imposed to allow waves to exit the domain without numerically induced reflections of appreciable size that can contaminate the near field solution. Setting the incoming Riemann grouping R to zero in (7) and (8), we obtain

$$\hat{\boldsymbol{n}} \times \boldsymbol{E}^* = \frac{1}{2}\, \hat{\boldsymbol{n}} \times (\boldsymbol{E} - \hat{\boldsymbol{n}} \times \boldsymbol{H}) \quad \text{and} \quad \hat{\boldsymbol{n}} \times \boldsymbol{H}^* = \frac{1}{2}\, \hat{\boldsymbol{n}} \times (\boldsymbol{H} + \hat{\boldsymbol{n}} \times \boldsymbol{E}),$$

where it is assumed that $c\epsilon = 1$ and $c\mu = 1$.

3.4 Basis Function Sets

For this initial study, the monomials of multidimensional Taylor expansions about cell centroids are chosen for the local basis function sets. When a polynomial basis of order n is used, we refer to the scheme as Galerkin-p_n. The spatial discretization order of accuracy is $n+1/2$ for arbitrary grids and $n+1$ for uniform meshes. While the scheme has been implemented with an architecture that could support arbitrary n, the code has been exercised for n=1, 2, and 3 only.

The basis set of element $\mathcal{D}_i$ for the general case of Galerkin-p_n is

$$\mathcal{B}_i(p_n) = \left\{ (x - x_i)^a \, (y - y_i)^b \, (z - z_i)^c \, : \, 0 \leq a + b + c \leq n \right\}, \tag{10}$$

where x_i, y_i, and z_i are the centroid coordinates of $\mathcal{D}_i$. For the familiar linear expansion, $\mathcal{B}_i(p_1) = \{1, \, x - x_i, \, y - y_i, \, z - z_i\}$.

It is well known that as n increases, the matrix operators arising from these basis functions quickly become ill-conditioned, preventing spectral-like convergence for arbitrary n. Our choice of basis functions is merely a consequence of our initial interest in and implementation of a p_1-only scheme. The code was subsequently extended to order n. For $n >> 3$, orthonormal basis sets and expansions in computational domains rather than in physical space may be considered.

3.5 Implementation Details

The standard fourth-order Runge-Kutta method is used for time integration. The experimentally observed stability limits for the one-dimensional case, given by $\Delta t \leq \nu \, \Delta x/c$, are $\nu = 1/2$, $1/4$, and $1/5$, for p_1, p_2, and p_3, respectively.

Since $\boldsymbol{F}(\boldsymbol{U}_h)$ is a linear function of $\boldsymbol{U}_h$ here, we pre-compute and store the surface and volume integrals of the basis function combinations, i.e. the moments, that arise in (6). These moment matrices are unique for to each element for the discontinuous Galerkin formulation used in this work.

The discontinuous Galerkin CEM solver has been parallelized with the MPI interface. Given the very compact nature of the scheme, highly efficient parallelization of better than 90% is easily achieved.

4 Numerical Results

4.1 Sphere (P_1 Results)

A sphere of radius $a = 1$ is used to demonstrate broadband capability. On the surface of the sphere, the resolution of the hexahedral mesh is five cells per wavelength for an electrical diameter of $ka = 20$, where $k = 2\pi/\lambda$ is the wavenumber and λ is the wavelength. The outer boundary is a sphere of radius four. The sphere is illuminated by a Gaussian derivative pulse tailored to encompass frequencies corresponding to a range of ka's from one to 20. With this single pulse calculation, the backscatter and bistatic RCS are computed for 192 frequencies. The monostatic RCS as a function of ka is shown in Fig. 1. It agrees with the exact solution to within 0.1 dB. The E-plane bistatic cross sections obtained from the same calculation for $ka = 10$ is shown in Fig. 2

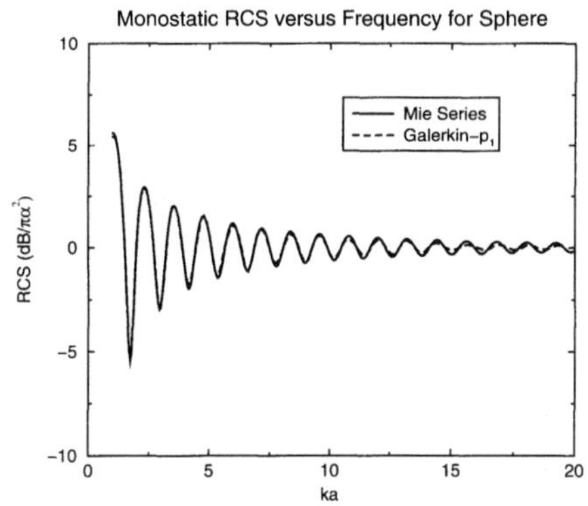

Fig. 1. Monostatic RCS versus frequency from single pulse calculation.

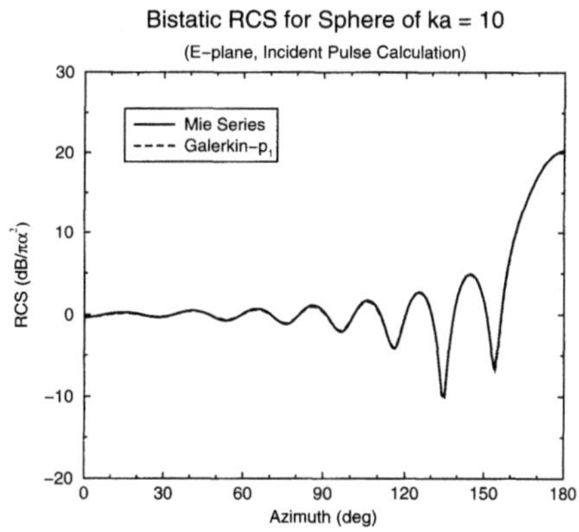

Fig. 2. Bistatic RCS for sphere of ka=10 from pulse calculation.

4.2 EMCC Flat Plate 1 (P_1)

The EMCC flat plate 1 geometry [4], shown in Fig. 3, has a tip where the electromagnetic fields are singular. The plate is illuminated at a 10 degree grazing angle. Monostatic RCS convergence is achieved with Galerkin-p_1 at a surface resolution of six tetrahedral cells per wavelength, with 12 cells per wavelength clustering along the edges. The RCS is shown in Fig. 3 along with the measured RCS for both the left and right sides of the target. The two sets of measured data exhibit discrepancies of up to 4 dBs near 40 degrees, even though the geometry is perfectly symmetric along the centerline. In that region, the computed RCS curve lies between the two experimental RCS curves, and elsewhere, there is overall good agreement between experiment and computation.

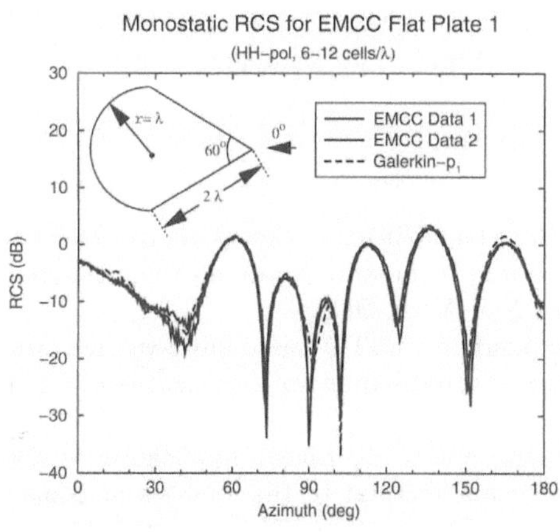

Fig. 3. EMCC flat plate 1.

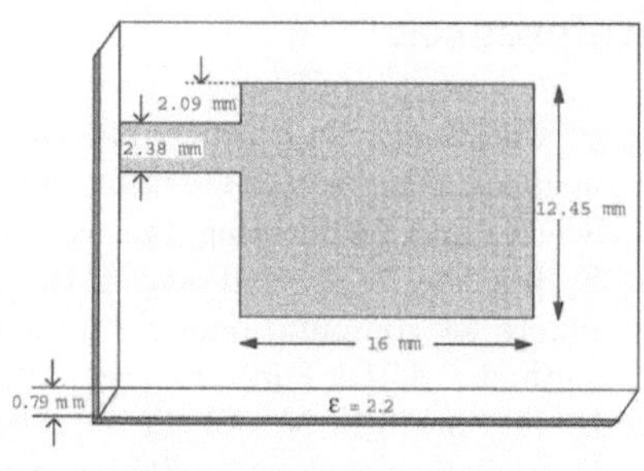

Fig. 4. Microstrip patch antenna detail.

4.3 Microstrip Patch Antenna (P_3)

The microstrip patch antenna, shown in Fig. 4, consists of a flat rectangular PEC patch atop a 0.79 mm thick layer of material of relative permittivity $\epsilon_r = 2.2$, laying over a PEC sheet. A Gaussian pulse in the space between the narrow 2.38 mm wide strip and the PEC sheet underneath is launched and allowed to propagate towards the patch antenna. The field values at an observation point near the entrance to the antenna are recorded and then Fourier transformed to compute the return loss, or the fraction of energy returned by the antenna.

Results are computed by Galerkin-p_3 for two hexahedral meshes of different resolutions. The field time response at the antenna entrance is shown in Fig. 5. It consists of the Gaussian pulse entering the antenna followed by the return of the energy that has not radiated out. The loss return computed from this time response is shown in Fig. 6. There is overall good agreement with the experimental data reported in [3], and the resonances at about 7.5 and 18 GHz are captured accurately. The under-prediction of the valleys near 12.5, 14.5, and 20 Ghz is eliminated by increasing the number of cells from 8,064 to 20,706.

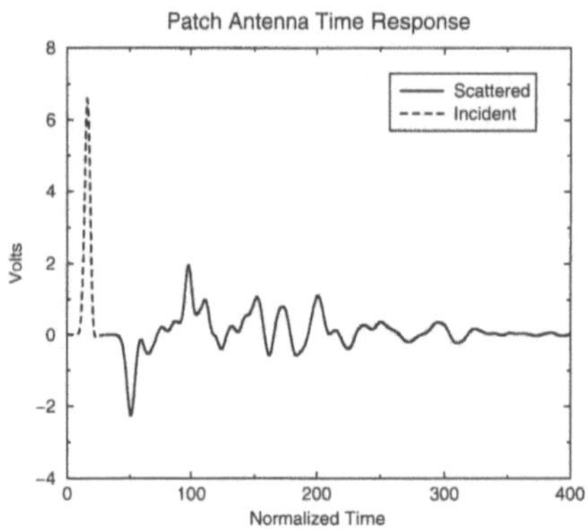

Fig. 5. Time response at antenna input.

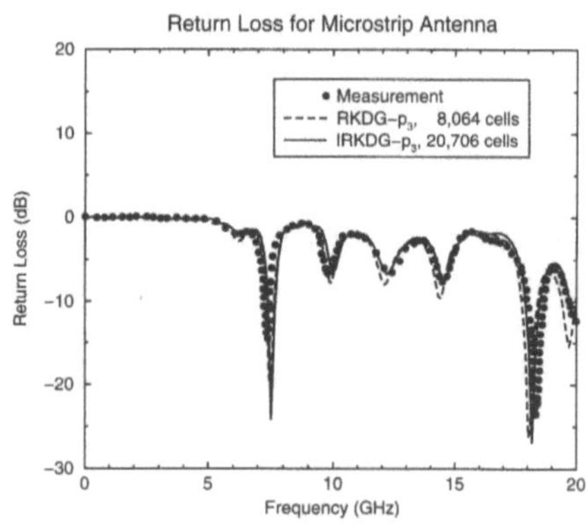

Fig. 6. Patch antenna return loss.

References

1. B. Cockburn, G. Karniadakis, and C.-W. Shu, editors. *Discontinuous Galerkin methods: Theory, Computation, and Applications.* Lecture Notes in Computational Science and Engineering. Springer-Verlag, New York, 2000.
2. V. Shankar, W. F. Hall, and A. H. Mohammadian. A CFD-based finite-volume procedure for computational electromagnetics - interdisciplinary applications of CFD methods. *AIAA Paper 89-1987*, 1989.
3. D. M. Sheen, S. M. Ali, M. D. Abouzahara, and J. A. Kong. Application of the three-dimensional finite-difference time-domain method to the analysis of planar microstrip cuircuits. *IEEE Trans. Microwave Theory Tech.*, 38(7):849–856, 1990.
4. A. C. Woo, H. T. G. Wang, M. J. Schuh, and M. L. Sanders. Benchmark plate radar targets for the validation of computational electromagnetics programs. *IEEE Antennas and Propag.*, 34(6), Dec 1992.

Controllable and Observable Subspaces in Computational Fluid Dynamics

Karen Willcox

Massachusetts Institute of Technology, Cambridge, MA, 02139, USA

Abstract. The concepts of controllability and observability are explored for computational fluid dynamic (CFD) applications. Identification of a balanced subspace which accounts for both controllability (inputs) and observability (outputs) can be performed using the proper orthogonal decomposition (POD) and the concepts of primal and dual state-space systems. Conventional model reduction approaches in CFD consider only controllability and do not take account of engineering outputs of interest. For a two-dimensional, inviscid flow problem, the balanced reduced-order model is shown to have superior performance, implying that the underlying CFD model is unbalanced with respect to the chosen input and output.

1 Introduction

Consider a general linearised CFD model, which can be written as

$$\frac{d\mathbf{x}}{dt} = A\mathbf{x} + B\mathbf{u}, \qquad \mathbf{y} = C\mathbf{x}, \tag{1}$$

where $\mathbf{x}(t)$ is the state vector containing the n unknown flow perturbation quantities at each point in the computational grid. The vectors $\mathbf{u}(t)$ and $\mathbf{y}(t)$ contain the system forcing inputs and outputs respectively. The linearisation matrices A, B and C in (1) are real, and are evaluated at steady-state conditions.

The CFD model typically contains tens of thousands of unknowns, even if just two dimensions are considered. The equations are expensive to time-march, even under the assumption of linearity, and the system is prohibitively large for many applications such as aeroelasticity, optimisation and active flow control. Model reduction is a powerful technique for obtaining low-order, high-fidelity models for complex systems. Reduction techniques have been developed and applied in many contexts, including fluid dynamics, structural dynamics and controls.

The fundamental concept of model reduction is to identify a low-order, dominant subspace. By projecting the high-order system onto this subspace, a low-order model is obtained which accurately captures the dynamics of interest. A number of reduction techniques have been developed and applied to CFD models, including the POD, which uses empirical data to determine a basis which is optimal in the sense that it minimises the error between the exact and projected data. The POD has been used extensively for identifying dominant subspaces in fluid dynamic applications (see [1] for a review). In this paper, it is shown that the POD can be applied using either system inputs (the conventional approach), system outputs or, as shown in [2], a balanced combination of the two.

2 Model Order Reduction

Consider the large n^{th}-order linear system defined by (1). The objective of the reduction procedure is to determine an n_r^{th}-order reduced-space basis onto which the state vector can be projected, that is $\mathbf{x} = V\mathbf{x_r}$, and an orthonormal set $\tilde{V}$, so that $\tilde{V}V = I$. This basis is chosen appropriately so that the reduced-order system

$$\dot{\mathbf{x}}_r = \tilde{V}AV\mathbf{x}_r + \tilde{V}B\mathbf{u} \qquad \mathbf{y}_r = CV\mathbf{x}_r \tag{2}$$

accurately reproduces the desired dynamics of the original system (1) with many fewer states ($n_r << n$). While the basis vectors are determined mathematically, the space they span often has physical significance. While many model reduction techniques have been developed in a controls setting, the challenge has been to find techniques which can be used for the very large systems encountered in CFD applications. One widely used, efficient reduction technique will now be described, and its connection to classical control concepts will be outlined.

2.1 Proper Orthogonal Decomposition

The POD has been widely used to determine efficient bases for dynamic systems. It can be shown that an optimal set of basis vectors, $\mathbf{\Psi}$, are the eigenfunctions of the kernel, K. Sirovich introduced the method of snapshots as a way of determining the eigenfunctions $\mathbf{\Psi}$ without explicitly calculating the kernel [3]. The kernel can be approximated as

$$K(\theta, \theta') = \frac{1}{m}\sum_{i=1}^{m} \mathbf{x}_i(\theta)\mathbf{x}_i^T(\theta'), \tag{3}$$

where $\mathbf{x}_i(\theta)$ is the instantaneous system state or "snapshot" at a time t_i and the number of snapshots m is sufficiently large. It can be shown that the eigenvectors of K are linear combinations of the snapshots, and can be computed by solving an $m \times m$ eigenvalue problem.

Rather than performing a set of time domain simulations to obtain the snapshots $\mathbf{x}_i$, a better approach is to use linearity and evaluate the POD basis vectors in the frequency domain. Any general forcing function can be considered as a superposition of sinusoidally time-varying components each at a frequency ω. Because the system is linear, the component of forcing at frequency ω induces a response which is also harmonic with frequency ω, that is $\mathbf{x}(t) = \mathbf{X}e^{j\omega t}$ and $\mathbf{y} = \mathbf{Y}e^{j\omega t}$. Considering a single temporal harmonic, ω_k, the state-space system (1) can be written in the frequency domain as

$$\mathbf{X}_k = (j\omega_k I - A)^{-1} B\mathbf{U}_k, \qquad \mathbf{Y}_k = C\mathbf{X}_k. \tag{4}$$

The POD snapshots can therefore be obtained by choosing a set of sample frequencies $\{\omega_k\}$ based on the frequency content of problems of interest and solving the frequency domain system (4) to obtain the responses $\{\mathbf{X}_k\}$ [4].

2.2 Controllability

The POD process is often explained in the context of a least squares fit through a set of data obtained from the high-order system; the POD vectors minimize the error between the actual snapshots and their reconstruction in the reduced space. Another insightful interpretation of the reduction process can be gained by considering the concept of controllability, borrowed from state-space theory.

The controllable subspace is that set of states that can be obtained in a finite time with zero initial state and a given input $\mathbf{u}(t)$ (also called the set of reachable states). The controllability grammian is an $n \times n$ matrix whose eigenvectors span the controllable subspace and is defined for the linear system (1) as

$$W_c = \int_0^\infty e^{At} B B^T e^{A^T t} dt. \tag{5}$$

Note that W_c does not depend on the definition of the output. The eigenvalues of W_c are real, non-negative numbers and denote the "reachability" of the corresponding eigenvector, or the amount of control "energy" required to obtain it. For a state $\mathbf{x}_0$, the controllability function is defined as

$$L_c(\mathbf{x}_0) = \min_{\mathbf{u} \in L_2(-\infty,0)_{\mathbf{x}(-\infty)=0, \mathbf{x}(0)=\mathbf{x}_0}} \frac{1}{2} \int_{-\infty}^0 ||\mathbf{u}(t)||^2 dt \tag{6}$$

where $||.||$ denotes the L_2 norm. The lower the value of $L_c(\mathbf{x}_0)$, the more reachable the state $\mathbf{x}_0$. If $\mathbf{x}_0$ is the kth eigenvector of W_c, $L_c(\mathbf{x}_0) = 1/2\lambda_k^c$, where λ_k^c is the kth eigenvalue of W_c. The eigenvectors of W_c, ordered by increasing eigenvalue, therefore describe the most reachable states of the system.

Further insight can be gained by thinking about the problem in the frequency domain. The controllability grammian can also be written as

$$W_c = \frac{1}{2\pi} \int_{-\infty}^{\infty} (j\omega I - A)^{-1} B B^T \left(-j\omega I - A^T\right)^{-1} d\omega. \tag{7}$$

The term $\mathbf{X}_\omega = (j\omega I - A)^{-1} B$ is the magnitude of the response of the linear system to sinusoidal forcing at a frequency ω (the actual response is given by $\mathbf{x}(t) = \mathbf{X}_\omega e^{i\omega t}$).

Although the kernel is never explicitly computed in the POD frequency domain analysis, by choosing a finite set of discrete frequencies for the snapshots, (3) can be written

$$K = \frac{1}{m} \sum_{i=1}^{m} (j\omega_i I - A)^{-1} B B^T \left(-j\omega_i I - A^T\right)^{-1}. \tag{8}$$

By comparing equations (7) and (8), we can see that, within a scaling factor, the POD kernel is therefore an approximation to the controllability grammian over a chosen, restricted frequency range. The subspace spanned by the POD basis vectors approximates the most controllable subspace.

2.3 Observability

While controllability considers only the inputs to the state-space system, an analogous concept exists which considers system outputs. The observable subspace comprises those states which as initial conditions could produce a non-zero output with no external input, and the observability grammian is defined as

$$W_o = \frac{1}{2\pi} \int_{-\infty}^{\infty} \left(j\omega I - A^T\right)^{-1} C^T C \left(-j\omega I - A\right)^{-1} d\omega. \tag{9}$$

The observability function L_o is defined as

$$L_o(\mathbf{x}_0) = \frac{1}{2} \int_0^{\infty} ||\mathbf{y}(t)||^2 dt \tag{10}$$

where $\mathbf{y}(t)$ is the output generated with $\mathbf{x}_0$ as an initial condition and no external forcing. The higher the value of $L_o(\mathbf{x}_0)$, the more observable the state $\mathbf{x}_0$. If $\mathbf{x}_0$ is the kth eigenvector of W_o, $L_o(\mathbf{x}_0) = \lambda_k^o/2$, where λ_k^o is the kth eigenvalue of W_o. The eigenvectors of W_o, ordered by increasing eigenvalue, therefore describe the most observable states of the system.

Controllability and observability are dual concepts. Consider the state-space system

$$\frac{d\mathbf{z}}{dt} = A^T \mathbf{z} + C^T \mathbf{u}_d, \qquad \mathbf{y}_d = B^T \mathbf{z}, \tag{11}$$

which is the dual of the primal system defined by (1). Here, $\mathbf{z}$ is the dual state vector, and $\mathbf{u}_d$ and $\mathbf{y}_d$ contain the dual inputs and outputs. Comparing (7) and (9), it can be seen that the controllability grammian of the primal system is equal to the observability grammian of the dual system, and vice versa.

We have seen that the POD basis vectors approximate the most reachable states of the system. Similarly, a set of vectors could be generated which approximate the most observable states of the system by applying the POD technique to the dual state-space system. The question remains: which basis is the better one to use for the purposes of model reduction? If a state is very reachable, then it is likely to play an important role in the system dynamics. However, if this state generates little or no output, then it will not be important for predicting the output of interest accurately. Truly what is of importance is the way in which inputs are transferred to outputs; the state itself is an intermediate mechanism, which is moreover dependent on the arbitrary choice of state coordinates.

This dilemma can be resolved through the concept of a balanced realization, first introduced by Moore [6]. The basic idea is that a balanced representation of the system is one in which the controllability grammians are equal and diagonal. Model reduction can then be performed using the Hankel singular values of the system. These values are independent on the choice of system realization, and represent the importance of a state for transmitting input to output. As described in [2], using classical balanced realization theory, the POD-based grammian approximations can be combined in an efficient way to obtain an approximately balanced subspace for projection. In the following section, results will be presented that compare the three subspaces, controllable, observable and balanced, and the reduced-order models they produce.

3 Results

Results will be presented for a two-dimensional NACA 0012 airfoil operating in unsteady plunging motion with a steady-state Mach number of 0.755. A finite volume CFD formulation for the Euler equations is used, which is described in [5]. The CFD mesh has 3482 grid points, which corresponds to a total of $n = 13928$ unknowns in the linear state-space system. The input to this system is the rigid plunging motion (vertical motion of the airfoil), while the output of interest is the lift force generated.

POD was applied to the primal and dual systems to generate two sets of basis vectors approximating the most controllable and most observable subspaces for the input and output selected. The approximate balancing technique from [2] was also used to create a set of balanced basis vectors. The snapshots of the primal and dual systems were taken over a frequency range from $\omega = 0$ to $\omega = 2$ in steps of $\Delta\omega = 0.1$. From these snapshots, 15 POD basis vectors were computed to perform the balancing. Reduced-order models were constructed using each of the resulting bases. Figure 1 shows the transfer function from plunging motion to lift force. The results from the CFD model are compared to results from each of the reduced-order models with four and ten states.

From Figure 1 it is apparent that the primal POD vectors form a much better basis than the dual POD vectors, although the approximately balanced basis yields the best results. Although a wider range is plotted, the reduced-order models are strictly only valid over the frequency range sampled ($\omega \leq 2$). With just four states, the balanced reduced-order model gives a very accurate representation of the dynamics. The controllability basis shows some error for

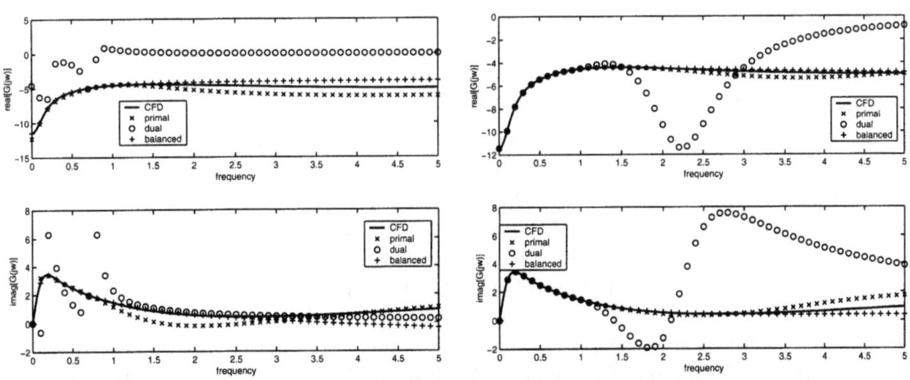

Fig. 1. Transfer function from plunging motion to lift force for subsonic airfoil. Results from CFD model ($n = 13,928$) are compared to reduced-order models derived using POD for primal, dual and balanced models. Left: $k = 4$. Right: $k = 10$.

$\omega > 1$, while the observability basis is very inaccurate for all frequencies. If the size of the model is increased to ten states, results for all three bases improve. The dual POD model gives good results for low frequencies, but still shows significant error at higher frequencies. In order to obtain an accurate representation using this basis, more states would be required.

This result yields interesting insight to the CFD formulation. For this case, choosing vectors from the most controllable subspace yields a much better approximation than choosing vectors from the most observable subspace, which suggests that the flow states capable of generating the most lift force are not typically excited by a plunging motion. The better performance by the balanced basis suggests that the CFD coordinate system is unbalanced with respect to the chosen input and output. An interesting avenue to explore is whether a rescaling of the CFD model could lead to improved convergence properties.

4 Conclusions

Controllability and observability subspaces have been investigated for a CFD model in the context of model reduction. While the results will vary depending on the specific CFD formulation and the inputs and outputs of interest, the case presented demonstrates some interesting aspects. The problem studied considers the lift force generated on an airfoil due to a plunging motion input. For this case, the most controllable subspace provides a much better approximation to the system dynamics than the most observable subspace. This suggests that the flow states capable of producing the most lift force are not those excited by plunging motion. Conversely, the states reached during a typical plunging motion are important for generating lift. However, the approximately balanced model yields the best results, suggesting that the CFD formulation is unbalanced with respect to the chosen input and output.

References

1. E.H. Dowell and K.C. Hall: 'Modeling of fluid-structure interaction', *Annual Review of Fluid Mechanics*, 33, 2001, pp. 445-90
2. K. Willcox and J. Peraire: 'Balanced Model Reduction via the Proper Orthogonal Decomposition', AIAA Paper 2001-2611, 2001, to appear *AIAA Journal.*
3. L. Sirovich: 'Turbulence and the Dynamics of Coherent Structures. Part 1 : Coherent Structures', *Quarterly of Applied Mathematics*, 45(3), Oct. 1987, pp. 561-571
4. T. Kim: 'Frequency-Domain Karhunen-Loeve Method and Its Application to Linear Dynamic Systems', *AIAA Journal* 36(11), 1998, pp. 2117-2123
5. K. Willcox: *Reduced-Order Aerodynamic Models for Aeroelastic Control of Turbomachines*, PhD Thesis, Dept. of Aeronautics and Astronautics, MIT, Feb. 2000
6. B.C. Moore: 'Principal Component Analysis in Linear Systems: Controllability, Observability, and Model Reduction', *IEEE Transactions on Automatic Control*, AC-26(1), 1981, pp. 17-31

A Compact Residual Scheme for Unsteady Compressible Flow Problems

Rémi Abgrall[1,2] and Mohamed Mezine[1]

[1] Mathématiques Appliquées de Bordeaux Université Bordeaux I, 351 cours de la Libération, 33 405 Talence Cedex, France
[2] Institut Universitaire de France

Abstract. The aim of this paper is to present the construction of upwind *compact* residual distribution schemes for the time accurate solution of hyperbolic conservation laws problems when using triangular unstructured meshes. Several examples demonstrate the efficiency of our technique.

In the past decades, many numerical schemes have been devoted to the simulation of unsteady compressible flows which might contain discontinuities. In most cases, this is done with methods that are extension of the so-called high order upwind schemes. There is at least two important exceptions to this, the finite element type schemes of the stream-line diffusion method type [1,2], and the Discontinuous Galerkin schemes [3]. The schemes constructed following the classical upwind high order philosophy are, for unstructured meshes, most of the time quite disappointing because they suffer from excessive numerical diffusion. To overcome this problem, one has try to adapt ideas forming from the ENO/WENO [4,5] philosophy. Examples of such attempts are [6–8]. If these extensions had some success, the price to pay is very high : the coding is rather complex, and more disturbing, these schemes do not have a compact stencil, so parallel implementation becomes difficult.

In order tackle these two problems – compacity of the stencil, effective accuracy of the solution–, at least two classes of methods have emerged in the recent years, the Discontinuous Galerkin schemes, and the Residual Distributive schemes (RD). In this paper, we focuss on the RD schemes. Though different in spirit, they have a common core at least in their "unstabilised" versions, the residual property. This property allows to show the formal accuracy of the scheme on a very general mesh. The RD schemes use a pointwise representation of the solution, like in finite difference schemes. The unknowns are updated by evaluation of the amount of residual sent to the vertices, and the stabilization mechanism can be similar to artificial viscosity, as in the SUPG finite element type methods [1,2], or inspired by the nonlinear techniques of the so-called high resolution schemes as in [9]. One can also take into account the genuinely multidimensional structure of the problem [10]. Here, we present an extension of the work in [9] to unsteady problems.

1 Review of residual distributive schemes and motivations

In the paper, $\mathcal{T}_h$ is a triangular type mesh, the triangles are $\{T_j\}_{j=1,n_t}$, the vertices are $\{M_j\}_{j=1,n_s}$. The parameter h stands for the maximum diameter of the triangles of $\mathcal{T}_h$. Since there is no ambiguity in the following, we denote by 1, 2, 3 the three vertices of the triangle T. We consider the PDE

$$\frac{\partial W}{\partial t} + \text{div } F(W) = 0 \qquad W(x, t=0) = W_0(x) \tag{1}$$

on a domain $\Omega \subset \mathbb{R}^2$. In the applications, W is the vector of conserved variables in compressible fluid dynamics ($W = (\rho, \rho\boldsymbol{u}, E)^T$), and F is the related flux. We assume a perfect gas equation of state, $\gamma = 1.4$.

If W_i^n is an approximation of W at the vertex M_i and time t_n, the residual scheme writes

$$|C_i| \frac{W_i^{n+1} - W_i^n}{\Delta t} + \sum_{T, M_i \in T} \Phi_i^T = 0, \tag{2}$$

where C_i is the medial dual cell at M_i. In (2) , Φ_i^T is the residual sent by T at M_i ; the three residuals defined on T satisfy

$$\sum_{M_j \in T} \Phi_j^T = \int_{\partial T} F^h \cdot \boldsymbol{n} d\sigma := \Phi^T \tag{3}$$

where F^h is an approximation of the exact flux. In [11], we show that if F^h is continuous accross the edges of T and is a second order accurate approximation of F, the solution of (2) converges to a weak solution of (1) under the standard assumptions of the Lax Wendroff theorem.

Under the linear preserving (LP) condition, i.e. $\Phi_i^T = \beta_i^T \Phi^T$ with β_i^T uniformly bounded when $h \to 0$, one can show [10,11] that (2) is second order accurate *at steady state* under rather general assumptions on the mesh. However, this scheme provides very poor results for unsteady problems. As an example, we consider on Figure. 1 the problem of advection by the solid velocity

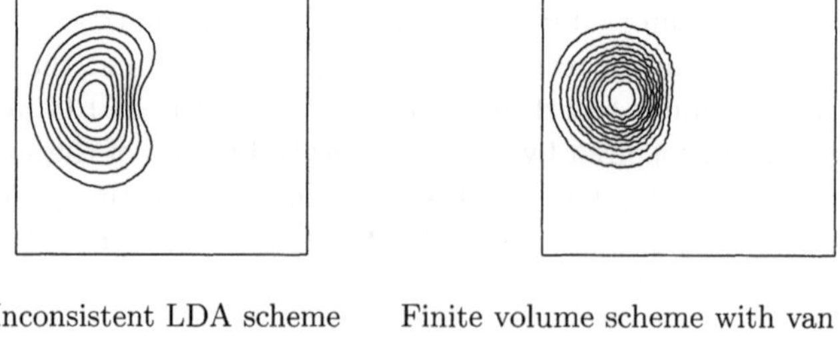

Inconsistent LDA scheme   Finite volume scheme with van Leer limiter

Fig. 1. Solution for the rotating cylinder after one revolution with the N scheme field $\boldsymbol{\lambda} = (-y, x)$ centered at the origin, the initial condition is the characteristic function of a cylinder obtained. We use the LDA scheme [10] which is an LP scheme, and a standard finite volume scheme with a standard limiter. The results are very disapointing. However, in the steady case, the quality of the LP residual distributive scheme is higher than those of the high order finite volume scheme [9–11]. The reason for this behavior is that (2) discretises the time and space have been discretised *unconsistantly* using a splitting technique.

2 A scheme for unsteady flow problems

The complete description of the scheme is given in [12]. The construction of RD schemes need three ingredients. First a partition of the computational domain, here $\Omega \times [0, T]$ where $\Omega \subset \mathbb{R}^2$ and $T > 0$; then a second order approximation of the time-space fluxes for reaching second order accuracy. The last ingredient is a first order monotonicity preserving scheme and a technique to increase the accuracy without losing the monotonicity.

The main idea is to consider the unsteady problem (1) as a steady problem in a subset of $\mathbb{R}^2_{\mathbf{x}} \times \mathbb{R}^+_t \times \mathbb{R}^+_\tau$. In other words, the fictitious time τ is an iteration parameter. The computational domain is discretised by a triangular mesh in space. The space times elements are prisms, $K = T \times [t_n, t_{n+1}]$. Following [10], we interpolate linearly in space and linearly in time the Roe parameter vector $Z = (\sqrt{\rho}, \sqrt{\rho}\boldsymbol{u}, \sqrt{\rho}H)^T$ where H is the total enthalpy,

$$Z^h(\mathbf{x}, t) = \frac{t_{n+1} - t}{\Delta t} Z^n(\mathbf{x}) + \frac{t - t_n}{\Delta t} Z^{n+1}(\mathbf{x}) \tag{4}$$

where Z^n and Z^{n+1} represent the piecewise linear interpolant of $\{Z_j\}_{j=1,n_s}$ at times t_n and t_{n+1}. Using the fact that $W = \frac{1}{2}\mathcal{D}(Z)\cdot Z$ and the flux $F = \frac{1}{2}\mathcal{R}(Z)\cdot Z$ are quadratic in Z, we see that

$$\begin{aligned} \Phi_{x,t} &:= \int_K \left(\frac{\partial W^h}{\partial t} + \text{ div } F(W^h) \right) dx dt \\ &= \frac{|T|}{4} \sum_{M_j \in T} \left[(\frac{1}{3} W_j^{n+1} + \frac{1}{2} \widetilde{W}_j^{n+1}) - (\frac{1}{3} W_j^n + \frac{1}{2} \widetilde{W}_j^n) \right] \\ &+ \frac{\Delta t}{2} \sum_{M_j \in T} K_j^{n+1} \widetilde{W}_j^{n+1} + K_j^n \widetilde{W}_j^n, \end{aligned} \tag{5}$$

In (5), n_ℓ^T denotes the inward scaled normal to the side of T opposite to the vertex ℓ, $\overline{Z} = \frac{Z_1 + Z2 + Z_3}{3}$ is the average of the linear interpolant of Z on T. For $m = n, n+1$, we have also set $K_j^m = \frac{1}{2}\mathcal{R}(\overline{Z}^m) D^{-1}(\overline{Z}^m) \cdot n_j^T$, $\widetilde{W}_j^m = D(\overline{Z}^m) Z_j^m$. The matrices K_j^m are diagonalisable in $\mathbb{R}$ with the standard eigenvalues $\overline{\mathbf{u}} \cdot \mathbf{n}_j^T \pm c||\mathbf{n}_j^T||$ and $\overline{\mathbf{u}} \cdot \mathbf{n}_j^T$. We set $K_j = L_j \Lambda_j R_j$ where Λ_j is the diagonal matrix of the eigenvalues, R_j (resp. L_j) are the right (resp. left) eigenvectors of K_j. This decomposition enable to consider the matrices $K_j^\pm = L_j \Lambda_j^\pm R_j$ as usual.

We need to define six residuals Φ_j^m (for $j = 1, 2, 3$ and $m = n, n+1$) per prism K. They have to sum up to $\Phi_{x,t}$ defined in (5). Causality suggests that K does not send any signal to the vertices located at time t_n, since they are in the past for $K : \Phi_j^n \equiv 0$ for $j = 1, 2, 3$.

Extension of the system N scheme. We define the residual Φ_j^{n+1} by an extension of the standard N scheme of Roe,

$$\begin{aligned}\Phi_i^T = & \frac{|T|}{4}\left[\left(\frac{1}{3}W_i^{n+1} + \frac{1}{2}\widetilde{W}_i^{n+1}\right) - \left(\frac{1}{3}W_i^n + \frac{1}{2}\widetilde{W}_i^n\right)\right] \\ & + \frac{\Delta t}{2}(K_i^{n+1})^+(\widetilde{W}_i^{n+1} - W_{n+1}^*) + \frac{\Delta t}{2}(K_i^n)^+(\widetilde{W}_i^n - W_n^*).\end{aligned} \tag{6}$$

The states W_{n+1}^* and W_n^* are computed to make the scheme conservative and defined by $W_m^* = N^m\left(\sum_j K_j^{m-}\widetilde{W}_j^m\right)$ with $N^m = \left(\sum_j K_j^{m-}\right)^{-1}$. Experiments show that this scheme, where the summation in (2) is replaced by a summation on all the prisms that share M_i as a vertex at time t_{n+1}, provide non oscillatory results.

Getting second order accuracy. A deeper analysis, as in [9,12], shows that this N scheme can be proved to be non oscillatory on special solutions which mimic simple waves. It is also shown that any function defined by (4) restricted on any prism K can be decomposed into a sum of $3 \times 2 \times 4$ simple waves. This gives an indication about the nonoscillatory behavior of the scheme. Hence, we adopt the following procedure to get second order accuracy. All the calculations are local and refers to a single prism K.

First we pick a direction $\boldsymbol{n}$, and consider the eigenvectors of the matrix K_n denoted by $\{\mathbf{r}_p\}_{p=1,4}$ We denote by $\{\ell_p\}$ the associated left eigenvectors.

Next we consider, for each left eigenvector ℓ_p, the residuals φ_i^p, $i = 1, \cdots, 3$ defined by $\varphi_i^p = \ell_p(\Phi_i)$. Since $\sum_{i=1}^{3} \varphi_i^p = \ell_p(\Phi) := \varphi^p$, we can introduce, as in the scalar case [13], the parameters $x_i^p = \frac{\varphi_i^p}{\varphi^p}$. which sum up to unity. Then , the problem reduces to the following. For any $p = 0, \cdots, 3$ find a mapping $(x_1^p, x_2^p, x_3^p) \mapsto (\hat{x}_1^p, \hat{x}_2^p, \hat{x}_3^p)$ such that the three properties hold true

$\sum_{i=1}^{3} \hat{x}_i^p = 1$	Conservation
for all i, $\hat{x}_i^p x_i^p \geq 0$	Preservation of the monotony property
for all i, $\left\lvert\hat{x}_i^p\right\rvert$ is bounded	LP property

Several examples of such mapping are presented in [13]. The simplest one is $\hat{x}_i^p = (x_i^p)^+\left[\sum_{j=1}^{3}(x_j^p)^+\right]^{-1}$ where $x^+ = \max(0., x)$. This mapping is used to defined the scalar PSI scheme [10]. The modified residual is defined by $\Phi_i^{LP} = \sum_{\ell=0}^{3} \hat{x}_i^\ell \varphi_\ell \mathbf{r}_\ell$. From this relation , it can be seen that $\Phi_i^{LP} = \mathbf{B}_i \Phi$ where the matrix $\mathbf{B}_i$ depends on $Z_i^m, i = 1, \cdots, 3$, $m = n, n+1$ and is uniformly bounded : the scheme is LP. This results in a non linear system which is solved by Newton iterations.

3 Numerical experiments

To show that there is no degradation of accuracy compared to a good one dimensional scheme (Roe's scheme with superbee limiter [14]), we have computed Shu's test case [15] with our scheme with meshes of similar resolution (800 points for the 1D scheme, an unstructured mesh of 10×800 points for the multiD scheme. We display the density at time $t = 1.8$ on Figure 2. We see that there is *no* degradation of the solution, contrarily to what happens in most unstructured schemes.

Among the numerous test cases we have run, we select the Collela and Woodward Mach 3 case. The same density isolines are plotted on Figure 3. We note the improvement of the quality of the slip lines, the occurence of the compression shock that hits the first reflected shock wave. Its trace is clearly visible, and the slip line that comes out is also clean. We have not done any special treatment at the corner where the (unphysical) entropy production is clearly visible. However, the entropy production for the LP scheme is much smaller that for the N scheme. An effect of that is that the second reflected shock wave has almost no lambda structure. A closer look at the picture shows that there is no degradation of the thickness of the shock waves (1-2 mesh points), even at the exit of the computational domain.

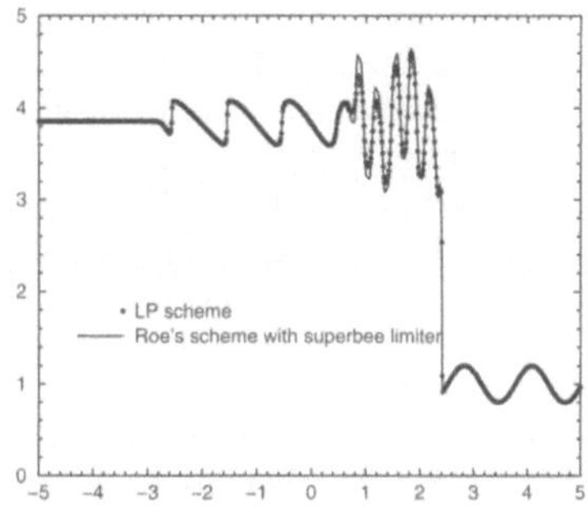

Fig. 2. Density on Shu's case

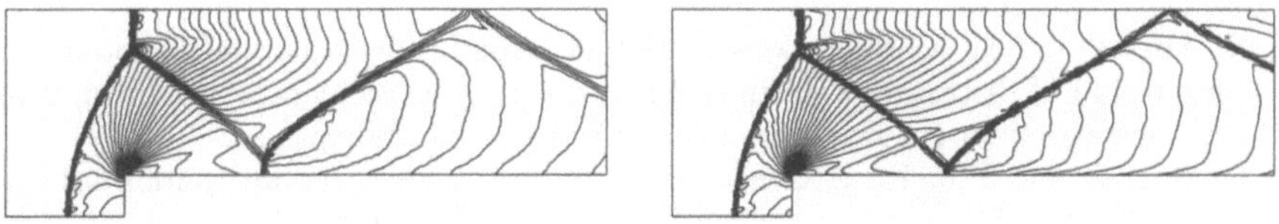

N scheme LP scheme

Fig. 3. Density isolines on Collela and Woodward Mach 3 case.

4 Conclusions

We have presented a new scheme that is genuinely second order accurate for unsteady problems using unstructured meshes. We have shown there is no degradation of the quality of the solution for one dimensional problems computed with the genuinely multidimensional compared with a very good one dimensional solver. The scheme is robust, and compact, so well adapted to parallel computers. Higher order schemes are beeing now considered, along the lines of [13].

References

1. C. Johnson, U. Nvert, and J. Pitkranta. Finite eleMent methods for linear hyperbolic problems. *Computer methods in applied mechanics and engineering*, pages 285–312, 1984.
2. Th.J.R. Hughes and M. Mallet. A new finite element formulation for Computational Fluid Dynamics : III the generalized streamline operator for multidimensional advective–diffusive systems. *Computer Methods in Applied Mechanics and Engineering*, 58:305–328, 1986.
3. B. Cockburn and C.-W. Shu. The local discontinuous Galerkin method for time-dependent convection-diffusion systems. *SIAM J. Numer. Anal.*, 35(6):2440–2463 (electronic), 1998.
4. A. Harten, S. Osher, B. Engquist, and S.R. Chakravarthy. Uniformly high–order accurate non–oscillatory schemes iii. *J. Comput. Phys.*, 71:231–303, 1987.
5. X.L. Liu, S.J. Osher, and T. Chan. Weighted essentially non-oscillatory schemes. *J. Comput. Phys.*, 115(1):200–212, 1994.
6. R. Abgrall. On Essentially Non-Oscillatory Schemes on Unstructured Meshes : Analysis and Implementation. *J. Comput. Phys.*, 114(1):45–54, September 1994.
7. O. Friedrich. Weighted essentially non-oscillatory schemes for the interpolation of mean values on unstructured grids. *J. Comput. Phys.*, 144(1):194–212, 1998.
8. C. Hu and C.W. Shu. Weighted essentially non-oscillatory schemes on triangular meshes. *J. Comput. Phys.*, 150(1):97–127, 1999.
9. R. Abgrall and M. Mezine. Construction of second order accurate monotone and stable residual schemes : the steady case. *submitted*, 2002. http://www.math.u-bordeaux/ abgrall/Publi/steady.ps.gz.
10. R. Struijs, H. Deconinck, and P. L. Roe. Fluctuation Splitting Schemes for the 2D Euler equations. *VKI LS 1991-01, Computational Fluid Dynamics*, 1991.
11. R.Abgrall. Toward the ultimate conservative scheme : Following the quest. *J. Comput. Phys*, 2001.
12. R. Abgrall and M. Mezine. Construction of second order accurate monotone and stable residu al distributive schemes for unsteady flow problems. *submitted*, 2002. http://www.math.u-bordeaux.fr/ abgrall/Articles/unsteady.ps.gz.
13. R. Abgrall and P.L. Roe. Construction of very high order fluctuation scheme. *submitted*, 2002. http://www.math.u-bordeaux/ abgrall/Publi/high.ps.gz.
14. P.L. Roe. Fluctuations and signal. a framework for numerical evolution problems. In K.W. Morton and M.J. Baines, editors, *Numerical Methods for Fluid Dynamics*, pages 208–242, New York, 1982. Academic Press.
15. C.W. Shu and S.J. Osher. Efficient implementation of eno schemes. *J. Comput. Phys.*, 1, 1989.

Coordinate Transformation for CFD

Zi-Niu Wu[1] and Jing Shi[1]

Department of Engineering Mechanics, Tsinghua University,
Beijing 100084, P.R. China

Abstract. In this paper we consider rather general coordinate transformation including given, flow driven (such as the unified coordinate and the generalized characteristic coordinates proposed here), and space-time transformation. We discuss the equivalence of solutions (such as simple waves, weak solutions, and hyperbolicity) in different coordinate systems.

1 Introduction

Coordinate transformation is largely used in CFD for treating complex geometries. For nonlinear conservation laws with discontinuous solutions, it is well-known that both the original and the transformed equations must be put in conservation form in order to have (possibly) correct weak solutions[3].

By forced coordinate transformation we mean the new coordinate system only depends on the old coordinate system. There is another kind of coordinate transformation, called passive or flow driven transformation, in which the transformation depends on the flow properties. We know there are two traditional approaches to describe fluid flows: Eulerian approach and Lagrangian approach. The switch between the Lagrangian approach and the Eulerian approach can be considered as flow driven transformation, since in the Lagrangian approach the coordinate moves at the velocity of fluid particles. Recently, Hui and his co-workers proposed a unified coordinate system which unifies both approaches [1] and in which the coordinate follows the direction of the flow. Unification has been also studied under the name of arbitrary Lagrangian-Eulerian approach [2].

For the Eulerian approach, severe numerical diffusion may smear the contact discontinuities badly. The lagrangian or unified coordinate system follows the fluid particle or flow direction, so that the contact discontinuity is tracked automatically within the framework of a shock-capturing scheme. No method actually follows the speed of an expansion fan which moves at the speed $u \pm a$ (where u is the convective speed and a is the sound speed). Hence we propose a more general coordinate system, called generalized characteristic coordinate system, which moves locally at the speed $\alpha u + \beta a$, so that all the characteristic speeds of a compressible flow can be followed through the choice of α and β.

There are quite a number of theoretical questions that remain to be answered.

(1) For continuous forced transformation, Viviand[3] proved the equivalence of weak solutions between the original coordinate system and the transformed system. But the question remains unresolved for discontinuous transformation.

(2) Wagner[4] established the equivalence of Euler and Lagrangian equations of gas dynamics for weak solutions. It is yet unknown whether simple waves and weak solutions are equivalent between the original coordinate system and the flow driven coordinate system, especially for the unified coordinate system and for the GCCS.

(3) Almost all the transformations studied before are purely spatial. The problem of space-time transformation was rarely addressed. What would happen when both the space coordinate and the time axis are allowed to change?

We are interested in the unresolved problems stated above, and in their significance in CFD. Some results will appear in [5,6].

2 Coordinate Transformation

In order to avoid complexity, we consider the 1D case for presentation. The transformation is defined by

$$dt = d\tau \tag{1}$$

$$dx = Ad\xi + Bd\tau \tag{2}$$

For dx to be a full differential, the Cauchy-Riemann relation

$$\frac{\partial A}{\partial \tau} = \frac{\partial B}{\partial \xi} \tag{3}$$

must hold. It can be proven that the frame derivative

$$\frac{D_B \xi}{Dt} = \xi_t + B\xi_x$$

vanishes so that B is the speed of the moving frame.

Still using (1)-(2), we have

$$A = \frac{\partial x}{\partial \xi} = \xi_x$$

Hence A is the Jacobian of the coordinate transformation.

If B is a given function of x and t only, then the new frame (ξ, τ) is called a forced moving frame.

If B is a function of the solution w, then the new frame (ξ, τ) is called a passively moving frame or flow driven flame.

In the unified coordinate system of Hui et al [1],

$$B = hu \tag{4}$$

where h is a constant. The traditional Eulerian approach and Lagrangian approach correspond to $h = 0$ and $h = 1$, respectively.

The generalized characteristic coordinate system is defined by

$$B = \alpha u + \beta a \tag{5}$$

The compatibility relation and the physical conservation law should be written in the following conservative form to have correct weak solution:

$$W_\tau + F(W)_\xi = 0 \tag{6}$$

where

$$W = \begin{pmatrix} \rho A \\ \rho u A \\ \rho E A \\ A \end{pmatrix}, \quad F(W) = \begin{pmatrix} \rho u - \rho B \\ \rho u^2 + p - \rho u B \\ \rho u(E + \frac{p}{\rho}) - \rho E B \\ -B \end{pmatrix}$$

The eigenvalues of the Jacobian $N = \frac{dF}{dW}$ are

$$\mu_1 = \frac{u - B}{A}, \mu_2 = \frac{u - B + a}{A}, \mu_3 = \frac{u - B - a}{A}, \mu_4 = 0$$

In 3D, let (x, y, z) be the coordinates in the physical space and t be the physical time. The inviscid flow equations are known to be linearly well-posed in the physical frame (t, x, y, z). Now introduce the coordinate transformation $(x, y, z, t) \to (\xi, \eta, \varsigma, \tau)$. In traditional coordinate transformations we have $\tau = t$. But in the present paper τ may depend on both t and (x, y, z). For convenience we call τ the pseudo time. Precisely, the coordinate transformation we consider is defined by

$$d\xi = \xi_t dt + \xi_x dx + \xi_y dy + \xi_z dz \tag{7}$$
$$d\eta = \eta_t dt + \eta_x dx + \eta_y dy + \eta_z dz \tag{8}$$
$$d\varsigma = \varsigma_t dt + \varsigma_x dx + \varsigma_y dy + \varsigma_z dz \tag{9}$$
$$d\tau = \tau_t dt + \tau_x dx + \tau_y dy + \tau_z dz \tag{10}$$

3 Properties of the GCCS

Let us denote $\nabla = (\frac{\partial}{\partial \rho}, \frac{\partial}{\partial u}, \frac{\partial}{\partial p}, \frac{\partial}{\partial A})$ and study the four characteristic fields. Let R_i be the right eigenvector for the eigenvalue μ_i. Then we have

$$\nabla \mu_1 \bullet R_1 = 0, \nabla \mu_2 \bullet R_2 = \frac{2 - \beta(\gamma + 1)}{2A}, \nabla \mu_3 \cdot R_3 = \frac{1 + \gamma + 2\beta\gamma}{2A}, \nabla \mu_4 \cdot R_4 = 0$$

Proposition 1. *The first characteristic field is linearly degenerate. The second characteristic field is genuinely nonlinear if* $\beta \neq \frac{2}{\gamma+1}$ *and linearly degenerate if* $\beta = \frac{2}{\gamma+1}$. *The third characteristic field is genuinely nonlinear if* $\beta \neq -\frac{\gamma+1}{2\gamma}$ *and linearly degenerate if* $\beta = -\frac{\gamma+1}{2\gamma}$. *The fourth characteristic field is linearly degenerate.*

Now let us consider simple waves.

Proposition 2. *There are possibly four types of simple waves. The 1-simple wave is a generalized contact discontinuity across which only ρ and A are discontinuous. The 2 and 3-simple waves are exactly right and left-going expansion fans as in the Eulerian system. The 4-simple wave is a motionless A-discontinuity across which only A is discontinuous (and this discontinuity is motionless).*

For convenience, we use (ρ_0, u_0, p_0, A_0) to denote either (ρ_l, u_l, p_l, A_l) or (ρ_r, u_r, p_r, A_r) and consider shock waves. Let s' be the shock speed in the physical space and σ be the shock speed in the GCCS.

Proposition 3. *If (and only if) A satisfies the jump relation*

$$-(B_l - B_r) = \sigma(A_l - A_r), \tag{11}$$

then the Rankine-Hugoniot relation in the physical space and the Rankine-Hugoniot relation in the transformed space are equivalent, with σ given by

$$\sigma = \frac{s' - B_l}{A_l} \tag{12}$$

Proposition 4. *Consider the GCCS. The function $S = A_r/A_l$ becomes negative for α satisfying*

$$\min(\alpha_a, \alpha_b) < \alpha < \max(\alpha_a, \alpha_b)$$

with

$$\alpha_a = \frac{(\gamma+1)s(M-s) - \beta(\gamma+1)(M-s)\sqrt{\left(\frac{2\gamma(M-s)^2-(\gamma-1)}{\gamma+1}\frac{(\gamma-1)(M-s)^2+2}{(\gamma+1)(M-s)^2}\right)}}{((\gamma-1)M+2s)(M-s)+2}$$

$$\alpha_b = \frac{s-\beta}{M}$$

where M is the Mach number upstream of the shock and s is the shock speed relative to the upstream sound speed. In other words, there is a parameter range for α such that the GCCS is not invertible.

Consider the equivalence of expansion waves in GCCS and in the Eulerian system.

Proposition 5. *Let the initial data for $A(\xi, \lambda)$ be constant, i.e., $A(\xi, 0) = Const$, then, in the GCCS, the physical expansion fan remains to be an expansion wave if*

$$\alpha < \frac{\gamma+1}{2} + \frac{\beta(\gamma-1)}{2},$$

degenerates to a linear wave if

$$\alpha = \frac{\gamma+1}{2} + \frac{\beta(\gamma-1)}{2},$$

and changes to a compression wave if

$$\alpha > \frac{\gamma+1}{2} + \frac{\beta(\gamma-1)}{2}.$$

Now consider the solution of the Riemann problem with the initial discontinuity $(\rho_l, u_l, p_l, A_l)/(\rho_r, u_r, p_r, A_r)$.

Proposition 6. *The solution of the Riemann problem is generally composed of a left-going expansion fan (or shock wave), a motionless A-discontinuity (or generalized contact discontinuity), a generalized contact discontinuity (or motionless A-discontinuity), a right-going shock wave (or expansion fan). The solution for the physical properties ρ, u, and p is exactly the same as in the Eulerian system.*

4 Properties of Space-time Transformtion

First we are interested in whether the space-time transformation will change the hyperbolicity of the gas dynamic equations.

Proposition 7. *If $\tau_t + \tau_x u \neq 0$, $\tau_t + (u + a)\tau_x \neq 0$ and $\tau_t + (u - a)\tau_x \neq 0$, then the transformed one-dimensional inviscid compressible flow equations are hyperbolic with respect to the pseudo time and thus strongly well-posed. If the pseudo time follows a characteristics of the original system so that $\tau_t + \theta_i \tau_x = 0$ for some $i \in (1, 2, 3)$, then the transformed system becomes elliptic. The same holds in two dimensions and in three dimensions.*

Hence space time transformation may alter the hyperbolicty of the system. Consider the simplest case with $\xi_t = 0,\ \xi_x = 1,\ \tau_t = 1,$ so that

$$x = \xi, dt = d\tau + \frac{dx}{\theta}$$

for $\frac{d\tau}{dt} |_\theta = 0$. Let $\theta = u$. In this case we have $\tau_t + u\tau_x = 0$ and $\xi = x$, $d\tau = dt - \frac{dx}{u}$. Then the transformed system becomes

$$(R - \frac{1}{u}A)W_\tau + AW_\xi = 0$$

or

$$\begin{pmatrix} 0 & -\frac{1}{u}\rho & 0 \\ 0 & 0 & -\frac{1}{u\rho} \\ 0 & -\frac{1}{u}\rho a^2 & 0 \end{pmatrix} W_\tau + \begin{pmatrix} u & \rho & 0 \\ 0 & u & \rho^{-1} \\ 0 & \rho a^2 & u \end{pmatrix} W_\xi = 0$$

or

$$\begin{aligned} -\frac{\rho}{u}\frac{\partial u}{\partial \tau} + u\frac{\partial \rho}{\partial \xi} + \rho\frac{\partial u}{\partial \xi} \qquad &= 0 \\ -\frac{1}{u\rho}\frac{\partial p}{\partial \tau} + u\frac{\partial u}{\partial \xi} + \frac{1}{\rho}\frac{\partial p}{\partial \xi} \qquad &= 0 \\ -\frac{1}{u}\rho a^2\frac{\partial u}{\partial \tau} + \rho a^2\frac{\partial u}{\partial \xi} + u\frac{\partial p}{\partial \xi} &= 0 \end{aligned}$$

Hence we have two equations for u, one equation for p, and no equation for ρ.

Acknowledgements: This work was supported by Chinese National Natural Science Foundations Contract No. 10025210).

References

1. W.H. Hui, W.P. Li, Z.W. Li, *A unified coordinate system for solving the two-dimensional Euler equations*, J. Comput. Phys., **153**, 596 (1999)
2. L.G. Margolin, *Introduction to "An arbitrary Lagrangian-Eulerian computing method for all speeds"*, J Comput Phys, **135**, 198 (1997)
3. H. Viviand, *Formes conservatives des equations de la dynamique des gaz*, La Recherche Aerospatiale, **1**, 65 (1974)
4. D.H. Wagner, *Equivalence of Euler and Lagrangian equations of gas dynamics for weak solutions*, J. Differential Equations, **68**, 118 (1987)
5. Z.N. Wu, *Expansion fans in forced and passively moving frames*, Acta Mechanica, **156**, 235 (2002)
6. Z.N. Wu, *A note on the unified coordinate system for computing shock flows*, J. Comput. Phys., **180**, 110 (2002)

Part III

Application

3-Dimensional SPH Simulations of High Pressure Die Casting

Joseph Ha and Paul W. Cleary

CSIRO Mathematical and Information Sciences, Private Bag 10, Clayton South MDC, Victoria 3168, Australia.

Abstract. Three dimensional modelling of high pressure die casting (HPDC) using SPH is presented. The SPH method used to simulate three dimensional fluid flow is described. Three examples of HPDC are shown. The first is a generic component, the second is a real automotive piston head and the third one is a rack and pinion steering column. In each case the filling of the runner, gate and die are shown. The importance of the order of the die filling is shown as is the role of flow separation from corners and even moderately curved surfaces. The degree of surface fragmentation, droplet formation and the strongly transient nature of the voidage are also shown.

1 Introduction

High pressure die casting (HPDC) is an important process for manufacturing high volume and low cost components. Examples from the automotive industry include automatic transmission housings, piston heads and gear box components. In this process liquid metal (generally an aluminium alloy) is injected at high speed (around 50 to 100 m/s) and under very high pressures through complex gate and runner systems and into the die. The geometric complexity of the dies leads to strongly three dimensional fluid flow with significant free surface fragmentation. Crucial to forming homogeneous cast components with minimal entrapped voids is the order in which the various parts of the die fill and the positioning of the gas exits. This is determined by the design of the gating system and the geometry of the die. Numerical simulation offers powerful insights into the effect of changes in die design and the filling process.

Smoothed particle hydrodynamics (SPH) is a special type of numerical method for simulating fluid flows. It uses only particles to represent the fluid and these are the computational framework on which the fluid equations are solved. SPH automatically follows complex flows and is particularly suited for high speed fluid flows that involve droplet formation, splashing and complex free surface motion such as those found in HPDC.

Recently, [1] and [3] reported on the application of SPH to HPDC in 2-dimensions and the favourable comparison of these SPH results with water analogue experiments. Note that particularly good resolution of the small scale fluid structures was obtained as well as accurate predications of voids in the flow. In [2] SPH simulations were compared with water analogue experiments of gravity die casting (GDC) for a complex die in two different orientations. In [4] SPH simulations were compared with experimental results from X-ray imaging of GDC.

The SPH results were able to capture the free surface wave behaviour and the fine details of the flow. In this paper, we present SPH results for the simulation of detailed fluid flows in a generic and two automotive components.

2 The SPH Methodology

SPH is a Lagrangian method that uses an interpolation kernel of compact support to represent any field quantity in terms of its values at a set of disordered points (the particles). The fluid is discretised, and the properties of each of these elements are associated with its centre, which is then interpreted as a particle. A particle b has mass m_b, position $\boldsymbol{r}_b$, density ρ_b and velocity $\boldsymbol{v}_b$. In SPH, the interpolated value of any field A at position $\boldsymbol{r}$ is approximated by:

$$A(\boldsymbol{r}) = \sum_b m_b \frac{A_b}{\rho_b} W(\boldsymbol{r} - \boldsymbol{r}_b, h) \tag{1}$$

where W is an interpolating kernel, h is the interpolation length and the value of A at $\boldsymbol{r}_b$ is denoted by A_b. This allows smoothed approximations to the physical properties of the fluid to be calculated from the particle information. The smoothing formalism also provides a way to find gradients of fluid properties. In this way, the SPH representation of the hydrodynamic governing equations can be built. These equations of motion are given in [1]. The simulation progresses by explicitly integrating this system of ordinary differential equations. The form of these equations are the same regardless of the dimension of the governing equations. There is no reference to a computational grid. The particle position is the only geometric information in the equations. The computation of the sums in the equations requires only the identification of the particles' neighbours.

3 Numerical Simulation

In this section, SPH simulations of a generic, transmission piston head and rack and pinion steering column are presented. In all cases, fluid initially fills the cylindrical shot sleeve and is forced downward by a piston at the top of the fluid so as to give average gate speed of 50 m/s. The visualisation of the SPH particles is by rendering the free surfaces constructed from the particle data. The fluid surface conveys much more information to the viewer than do the particles, in particular depth perception and variations in the surfaces are much easier to identify.

3.1 Filling of a Generic Component

Here, we simulate the filling of a relatively simple component involving a slotted annular extension of a base plate with a series of cylindrical holes and a long rounded slot at one end. The component itself measures around 200 mm long. Figure 1 shows two perspective views of the filling pattern at selected times.

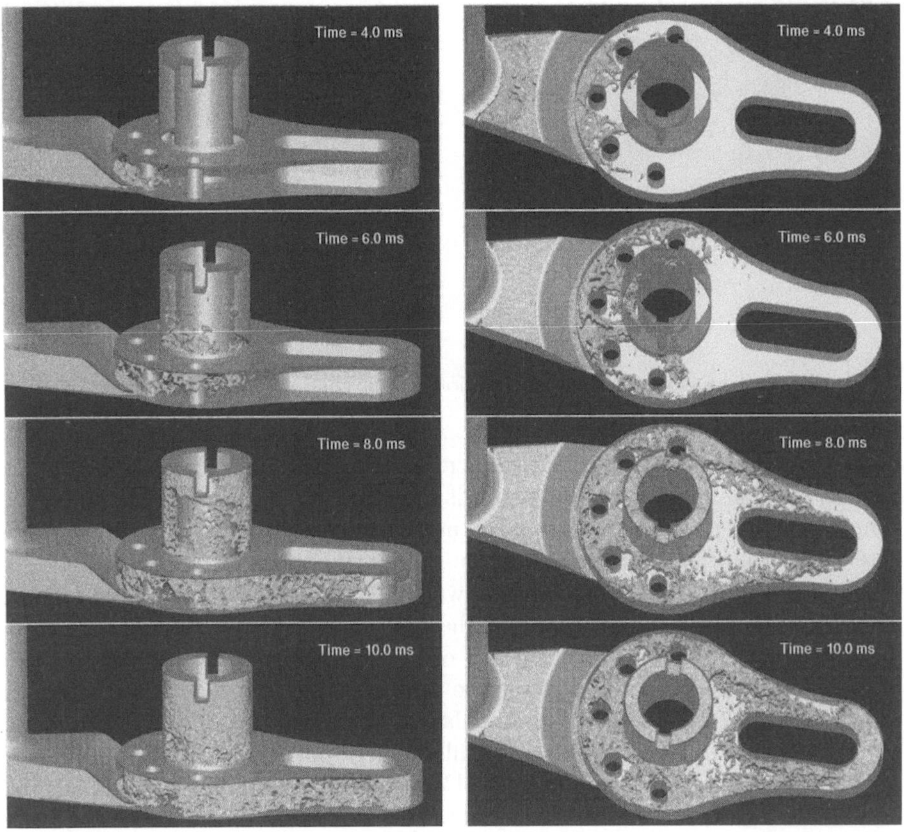

Fig. 1. Two perspective views of the filling process of a generic component.

The frames show that progressively larger volumes of the die cavity are filled. There is preferential flow around the curved outsides of the die initially with substantial voidage created by the free surface fragmentation. The voids behind the central cylindrical obstacle and the voids behind the ones to either side are long lived. Splashing droplets are visible in most of the non-filled parts of the die. The filling of the die is almost completed at 10 ms. The final voids to fill are behind the second and fourth of the cylindrical cutouts at the front of the die and small spaces on either side of the gate, just at the point of entry into the die. Overall, the filling process is complex, non-front filling with significant splashing and fragmentation. It clearly demonstrates that the order of fill does not occur according to proximity to the gate.

3.2 Filling of a Piston Head

Here, the filling of a transmission piston head is considered. It is the circular part in the centre of the casting configuration shown in Figure 2. At one end

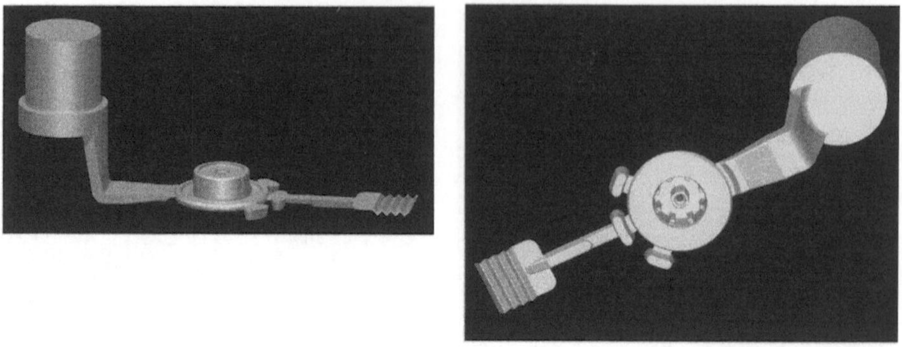

Fig. 2. Two views of the transmission piston head casting configuration.

is the cylindrical shot sleeve and runner. At the other end of the casting is the venting system. The top of the piston is cylindrically symmetric, but relatively simple. The underneath view reveals a more complex geometry with eight ridge extensions from the side wall and a non-symmetric central attachment.

Figure 3 shows two perspective views of the filling process. It takes around 26 ms for the leading fluid to reach the gate from the shot sleeve. When the fluid strikes the vertical inner surface of the annular upward part of the die, it is deflected up onto the top surface of the piston and also flows sideways around the vertical walls of the annular section. Initially, the free surface is quite fragmented. By 40 ms, the fluid fills a large portion of the piston die and the amount of voidage has diminished substantially, with main voids still to be filled being on either side of the gate. This is a result of the increasing back pressure from the increasingly well filled die cavity forcing fluid back towards the gate. This allows the voids to be progressively filled. At 56 ms, the die and side vents are completely filled and fluid flows only down the central vent towards the corrugated chill vent.

3.3 Filling of a Rack and Pinion Steering Column

Here, we present the SPH simulation of the filling of a rack and pinion steering column shown in Figure 4. The filling process is shown in Figure 5. The runner system is much more complex than for the previous parts. It splits into two branches with one bending to the left and entering an annular gate at the left end of the section at the bottom. The upper branch continues straight for awhile and then bends gently to the left and splits into two smaller symmetric branches that enter chisel gates in the center of the right section of the component. Highly fragmented fluid is sprayed from all three gates and flies down the lengths of the two cylindrical sections. The fluid collects at the closed ends of each of the sections and begins to back fill. Some fluid is trapped in the regions protruding from the lower cylindrical section. As time passes and the die fills, the part starts to take shape. By 58 ms, the die is around half filled and the shape of the cast component is becoming discernible.

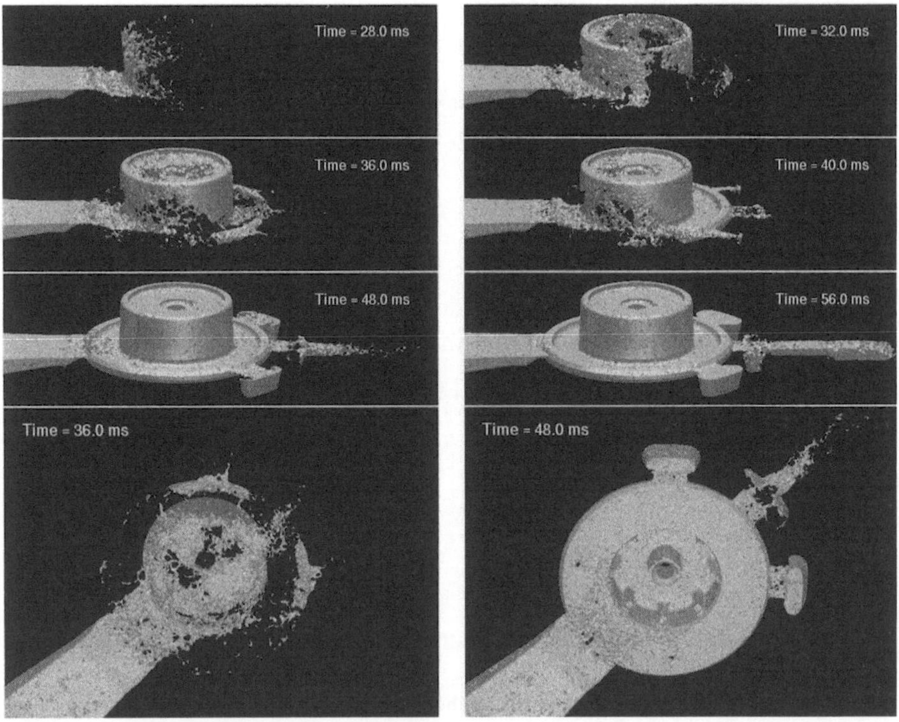

Fig. 3. Two perspective views of the filling of the piston head die.

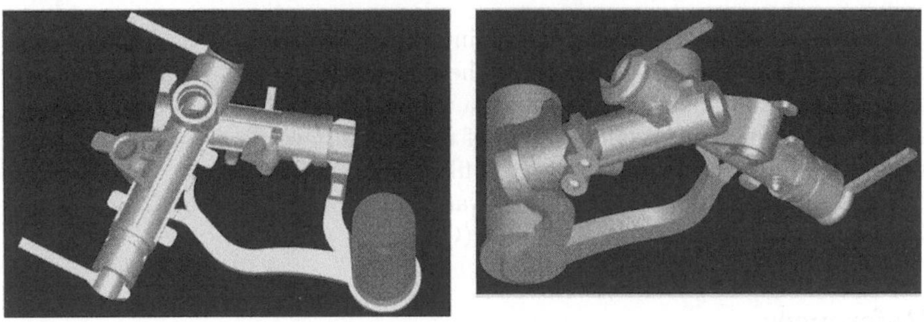

Fig. 4. Two views of the geometry of the steering column configuration.

Similarly to the previous two examples, the filling is initially highly fragmented and the order of fill does not occur according to proximity to the gates. The regions that filled first are those furthest away from the gates. Then the fill takes place progressively back towards the gates. This again has clear implications for the venting design.

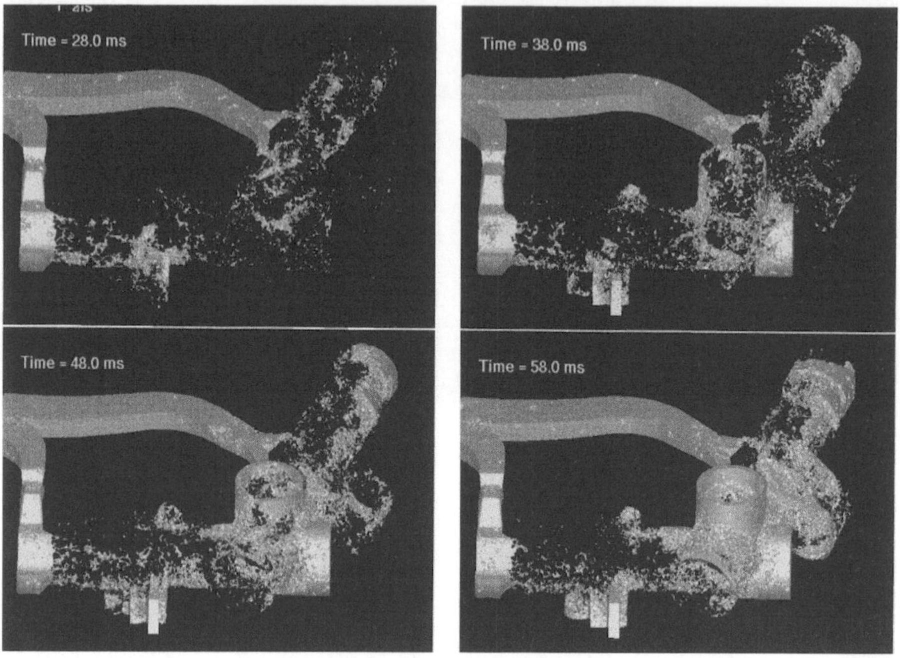

Fig. 5. Filling of the steering column.

4 Conclusion

The modelling of high pressure die casting using SPH has been described. Three dimensional isothermal simulation of the filling of three industrial components demonstrates that die filling is far from a uniform front fill. In each case, flow separation and significant fragmentation of the free surfaces lead to large transient void formation with much of the parts filled by back flows.

Acknowledgements: This work was funded by the Cooperative Research Centre for Cast Metals Manufacturing (CAST).

References

1. P.W. Cleary, J. Ha, V. Ahuja: Int. J. Cast Metals Res., **12**, 335 (2000).
2. J. Ha, P.W. Cleary, V. Alguine, T. Nguyen: 'Simulation of die filling in gravity die casting using SPH and MAGMAsoft'. In: *Proc. 2nd Int. Conf. on CFD in Minerals & Process Industries, Melbourne, Australia*, 1999, pp 423–428.
3. J. Ha, P.W. Cleary: Int. J. Cast Metals Res., **12**, 409 (2000).
4. J. Ha, R. Schuhmann, V. Alguine, P.W. Cleary, T. Nguyen: 'Real-time X-ray imaging and numerical simulation of die filling in gravity die casting'. In: *Proc. Modeling of Casting, Welding and Advanced Solidification Processes IX*, 2000, pp 151–158.

Numerical Simulations of Blood Flow at the Distal Anastomosis between a Prosthetic Graft and a Native Popliteal Artery

Siro Kitamura[1], Yoshinori Inoue[2], Takehisa Iwai[2], and Junichi Ooida[1]

[1] Institute of Computational Fluid Dynamics (iCFD), 1-22-3 Hara-machi Meguro-ku 152-0011 Tokyo, Japan
[2] Tokyo Medical and Dental University (TMDU), 1-5-45 Yushima Bunkyo-ku 133-0034 Tokyo, Japan
[3] The Institute of Space and Astronautical Science (ISAS), 3-1-1 Yoshinodai Sagamihara-shi 229-8510 Kanagawa, Japan

Abstract. The blood flow fields at distal anastomosis between a prosthetic graft and a native popliteal artery are simulated numerically under two different models and three wave patterns. The models are normal and protrusion type. The protrusion type was developed through surgery. Three wave patterns represent the different graft failure rates. In these calculations, the Reynolds number is fixed, and the wall of the blood vessel is assumed to be rigid. The different distributions of shear stress and velocity among some conditions are discussed.

1 Introduction

Recently, vascular reconstructions using the prosthetic graft have been performed to treat femoral artery occlusions [1]. Blood flow could be sufficiently restored by these vascular reconstruction. After operations, graft occlusions occur in the distal anastomosis between prosthetic graft and native popliteal artery under some conditions. It is found that the graft occlusion is related to the shape of distal anastomosis and the pulsative wave pattern in the prosthetic graft. Some numerical simulations for this flow field have been done [2],[3]. In this study, numerical simulation of the flow fields at the distal anastomosis are performed for various cases.

2 Flow model

Figure 1 shows the flow models, the normal type(left one) and the protrusion type (right one). These models are based on the clinical experience of vascular reconstructions. It is known that the graft failure rate of the protrusion type is better than that of the normal type.

The pulsative wave pattern models, which are applied to the inlet velocity profile, were based on the results measured after the operation by the Doppler velocimeter.

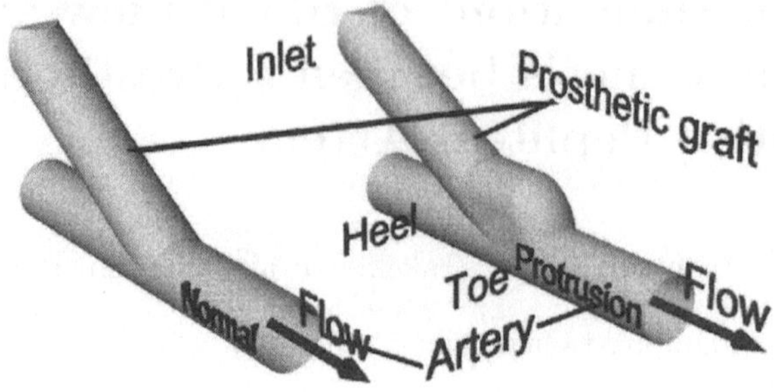

Fig. 1. Flow models.

3 Calculation method

The governing equations are the unsteady incompressible Navier-Stokes equations and the equation of continuity;

$$\mathrm{div}\boldsymbol{u} = 0 \tag{1}$$

$$\frac{\partial \boldsymbol{u}}{\partial t} + \boldsymbol{u} \cdot \mathrm{grad}\boldsymbol{u} = -\mathrm{grad}p + \frac{1}{\mathrm{Re}}\Delta \boldsymbol{u} \tag{2}$$

where $\boldsymbol{u}$ is the velocity vector, Re is Reynolds number, p is pressure. The pressure field is obtained by solving the Poisson equation. The multi-directional upwind method is used. A third-order upwind scheme[4] is applied for the discretization of the non-linear convection terms, and the second-order central difference for the diffusive terms. For the temporal integration, the Crank-Nicolson implicit scheme is utilized.

The blood flow is assumed to be Newtonian. The Reynolds number is set to 480. The blood vessel wall is treated as rigid in this paper. An uniform mesh size of $97 \times 49 \times 17$ is used for the computations.

4 Inlet velocity profile

In this study, three typical wave patterns are used, as shown in Fig.2.(a). They model the typical cases, based on the measured velocity profile in the prosthetic graft after the vascular reconstruction. They are classified according to the graft failure rate. Wave 1 represents the case in which the graft failure rate is the highest, that has only one peak and no reverse flow. Wave 3 represents the lowest graft failure rate, which has three peaks and one reverse zone. And wave 2 represents middle case, which has two peaks and one reverse zone. In all wave patterns, the last half part of the cycle is flat, that is, no stream at the inlet. The wave 3 is similar to a healthy wave pattern.

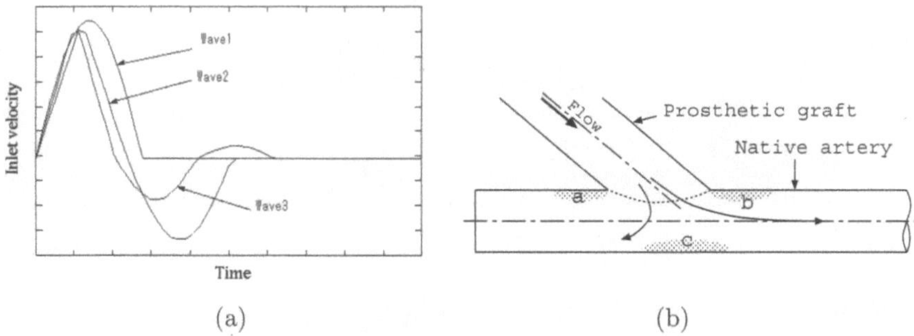

Fig. 2. (a)Wave patterns and (b)intimal thickening areas.

5 Reocclusion

Figure 2.(b) shows the area a,b,c that are liable to the intimal thickening. During the process of the reocclusion, the intimal thickens in one or more areas, and finally the artery is closed.

It was found from the clinical study that the occurrence of the constant small shear stress on the vessel wall is one of the causes of thickening. The intense variation of shear stress also causes the intimal thickening.

6 Calculation results

Figures 3 and 4 show the shear stress distributions on the center plane at four points in time for both models, at the wave pattern 3 (see Fig.2.(a)). The curves on the right-upper corner in each figure express the inlet velocity profile. In this paper, "shear stress" is defined as follows :

$$\left|\frac{\partial u}{\partial y}\right| + \left|\frac{\partial v}{\partial x}\right| + \left|\frac{\partial v}{\partial z}\right| + \left|\frac{\partial w}{\partial y}\right| + \left|\frac{\partial w}{\partial x}\right| + \left|\frac{\partial u}{\partial z}\right|$$

where u, v, w are the velocity compoents in each direction.

Comparing Fig.3.(a) and Fig.4.(a), it is found that the shear stress near the toe for the protrusion type is greater than that for the normal type, i.e., the variation of shear stress for the protrusion type is greater than that for the normal type because the lower limit of shear stress is nearly zero in both type. This fact is related to the intimal thickening in the area "b" in Fig.2.(b). And, it stimulates the intimal thickening. This result will be contrary to the fact that the graft failure rate of the protrusion type is better than that of normal type. (the prosthetic graft and native artery are anastmosed smoothly to prevent the intimal thickening near the toe(region b) in clinic.)

The shear stress along the base wall (area "c" in Fig.2.(b)) in Fig.4.(b) is less than that in Fig.3.(b), i.e, the variation of shear stress in area "c" for the normal type is greater than that for the protrusion type. This implies that the intimal

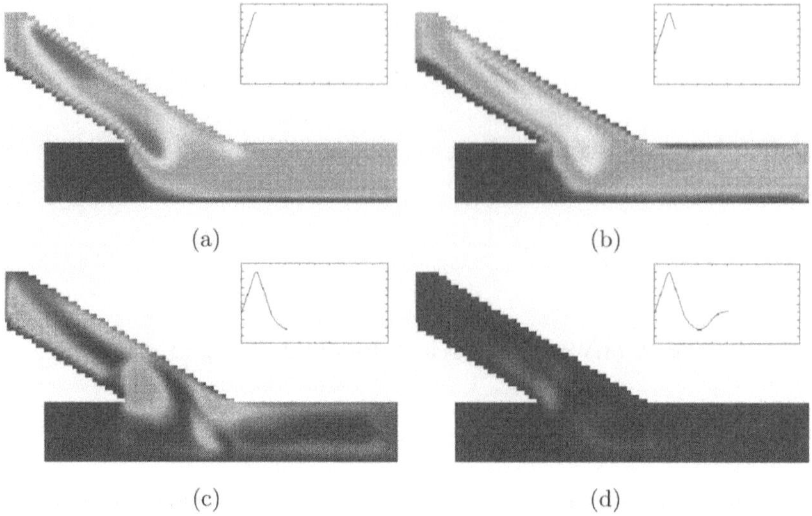

Fig. 3. Distribution of absolute value of shear stress. (Normal type, wave pattern 3 in Fig.2.(a)) (a)->(b)->(c)->(d)

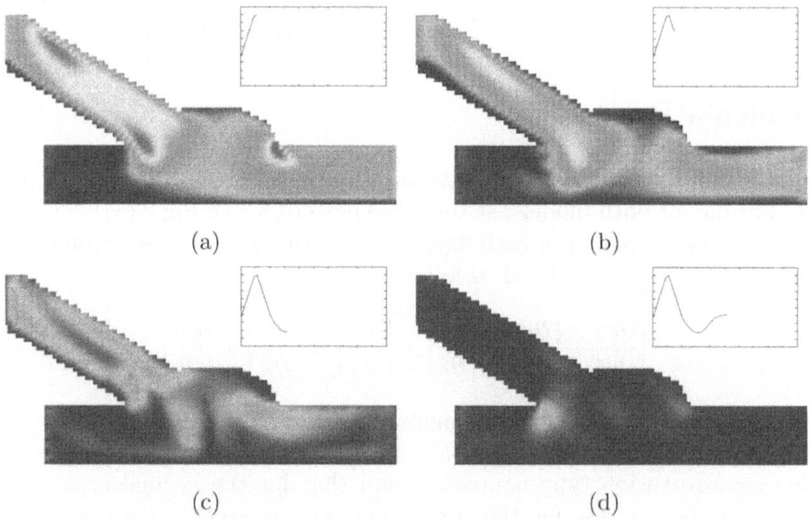

Fig. 4. Distribution of absolute value of shear stress. (Protrusion type, wave pattern 3 in Fig.2) (a)->(b)->(c)->(d)

thickening on the base wall is suppressed in the protrusion type. In other word, the intimal thickening is encouraged in the normal type.

Figure 5 shows the variation of the stagnation area along the base wall. The abscissa represents the position in flow direction, and the ordinate the time. The black area expresses the stagnation area, and white no stagnation area along

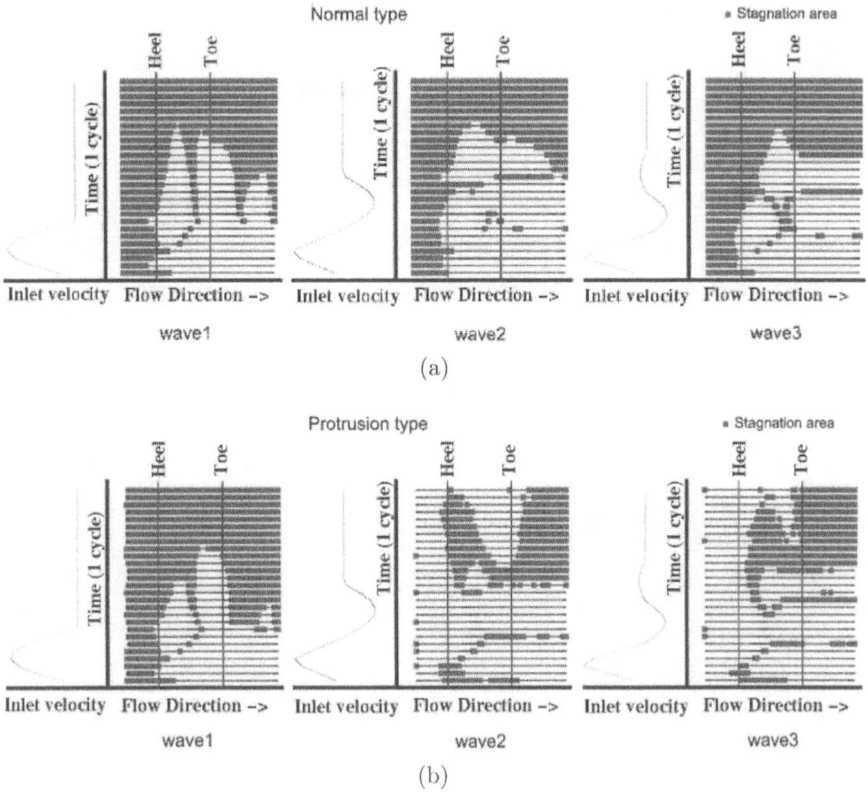

Fig. 5. Variation of the stagnation area along the base wall in three waves. (a) Normal type (b) Protrusion type

the base wall. The two vertical lines shows the position of heel and toe. In the stagnation areas, the absolute value of velocity is less than 2.5% of the half maximum velocity.

In the normal type, it is seen that the stagnation profile in all wave patterns are similar. In the protrusion type, the stagnation profile in wave 2 and wave 3 are also similar to each other, however, it is smaller in the last half part of the cycle. Wave 1 differs from the others in which a greater stagnation area exitst. Therefore, Wave 1 is inappropriate for the improvement of the graft failure rate even for the protrusion type because the low shear stress exist in long time.

Figure 6 shows the variations of the shear stress along the base wall. The dark area express the no shear stress area. We can only identify the explicit path of no shear stress point movement for wave 1 in the both types and during the last half part of the cycle for wave 3 (normal type). Especially, it is difficult to follow the path of no shear stress point movement when the inlet velocity is not zero.

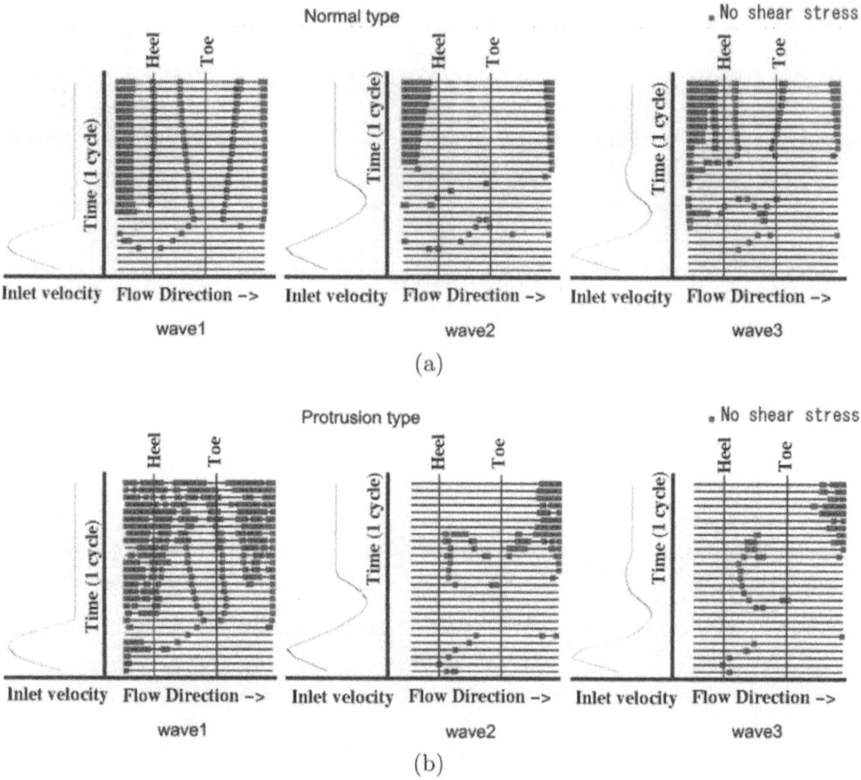

Fig. 6. Variation of the shear stress along the base wall in three waves. (a) Normal type (b) Protrusion type

7 Conclusions

The blood flow fields at distal anastomosis between a prosthetic graft and a native popliteal artery are simulated numerically under two models and three wave patterns. Wave 1 is inappropriate for the improvement of the graft failure rate in both models. Wave 2 and 3 for the protrusion model give better results in comparing with that for the normal model. The difference of intimal thickening between the two types is due to the thickening in area c in Fig.2.(b)

References

1. Inoue, Y., Iwai, T., Kure, N., Muraoka, Y., Endo, M., Surg. Today, (1997).
2. Moore, J.A., Steinman, D.A. et al., ASME J. Biomech. Eng., Vol.121, pp.265-271, 1999.
3. Lei, M., Giddens, D.P. et al., ASME J. Biomech. Eng., Vol.123, pp.81-87, 2001.
4. Kuwahara, K., AIAA paper, 99-3405, (1999).

Effect of Ship Frontal Variation on the Flow Field in the Flight-Deck Region

R. Toffoletto[1], K.R. Reddy[1], and J. Lewis[2]

[1] Defence Science and Technology Org, Fishermans Bend 3207, Vic, Australia
[2] Vacation Student, University of Melbourne, Vic, Australia

Abstract. Pilot workload and helicopter flight dynamic characteristics are influenced by the airwake in the flight deck region of a ship. Determination and inclusion of this complex flow field in helicopter mathematical modelling presents a considerable challenge. In this paper, results of the research conducted to investigate the effects of the ship frontal geometry on the flow in the flight deck region are presented for a range of on–set flow conditions. Calculated results, using Computational Fluid Dynamics (CFD), are compared with water tunnel flow visualisation data.

1 Introduction

The flight deck of a small ship contains regions of strong vorticity and associated severe velocity gradients. One of the main reasons why the operation of helicopters from naval ships is (Fig. 1) so hazardous is the interaction between the helicopter rotors and vortices [1] in the airwake. This is especially true during a landing, when the aerodynamic loading of the rotor system can be considerably altered by the passage through an eddy or a region of crossflow [2]. The changes in rotor loading may affect the aircraft control margins, pilot workload, and power margin [3] for safe landing.

Fig. 1. Helicopter approaching the flight deck of a ship

Because of this, and the civilian demand for the safe operation of helicopters from oil rigs and heliports in built-up areas, there is a growing research interest in the interaction between a bluff body airwake and the helicopter [4–7].

In this paper, we describe recent progress at the Air Operation Division of the Defence Science and Technology Organization using Computational Fluid Dynamics (CFD) as a tool to study the ship airwake problem. Our aim is to gain an understanding of the typical flow field structures around the flight deck and how they affect the performance and handling characteristics of rotorcraft.

The work described was done as part of a collaborative program under the auspices of The Technical Cooperation Program (TTCP). In the program, the US, UK, Canada, and Australia are using different CFD codes to study the airwake over an agreed geometrical representation of a simple three-dimensional frigate-like shape. Imperial units are being used in the study.

2 Generic Frigate Model

The Simple Frigate Shape (SFS) has a block superstructure consisting of a forward funnel followed by a backward facing step to an aft flight deck (Fig. 2a), and a second backward facing step at the stern. The flight deck region has dimensions similar to those of the ANZAC and FFG–7 class frigates in use by the Royal Australian Navy. The reference point, (X=0, Y=0, Z=0), used in this paper is at the hangar, along the centreline of the ship, and at water–level. Our aim is to gain an understanding of the typical flow field structures around the flight deck and how they affect the performance and handling characteristics of rotorcraft.

At a TTCP panel meeting held in Australia during May 2001, it was decided to improve the realism of the SFS by adding a bow. This model, designated the SFS 2, shown in Fig. 2b, is based on the SFS with an extended superstructure and a triangular shaped bow that is the same height as the flight deck.

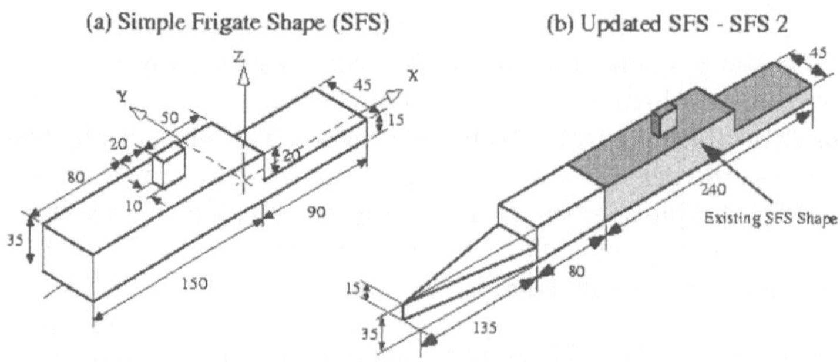

Fig. 2. General dimensions of the simple frigate shapes – dimensions in feet

3 Basis of CFD Modelling

The flow simulations reported here were performed using the commercially available Navier–Stokes code FLUENT Version 6.0 [8]. This code supports unstructured grids and a variety of models including laminar flow, the k–ε turbulence model (where k is the kinetic energy of the turbulent fluctuations and ε is the dissipation per unit mass), a model based on the Renormalization Group (RNG) method, and a Large Eddie Simulation model (LES). This paper concentrates on results for the RNG k–ε model. This provides a good balance between the numerical instabilities of simple laminar flow simulations for a highly irregular body and the greater physical fidelity, but more expensive computational demands, of LES methods. The simple shape chosen for the ship avoids curved surfaces and allows a grid to be easily defined. The model will produce the main features of the vortex flow field similar to those found on a typical frigate. Using this shape, other researchers in the US, UK, and Canada are collaborating with Australia in conducting parallel numerical studies, as well as conducting wind and water tunnel tests. This collaboration will provide an excellent opportunity to compare different CFD code predictions with experimental data and establish confidence in numerical simulation of this complex flow problem.

The water surface is treated as a non–slip wall to eliminate the formation of a boundary layer prior to the ship. The forward (inlet) boundary is treated as a velocity inlet condition. All other boundaries (sides, rear, and top) are treated as pressure far–fields. The ship is a viscous non–slip boundary (wall), which allows for the formation of boundary layers. However, apart from a denser grid in the vicinity of the ship, no special emphasis has been given to the development of boundary layers around the ship. A more rigorous treatment of boundary layers was not considered necessary as all separation points are defined by sharp corners. For this problem, more emphasis is given to resolving the vortical flow in the landing-deck region. The turbulence intensity of 1% of the freestream kinetic energy at a length scale of 1 ft was assumed at sea level and standard atmosphere conditions.

The key requirement of the numerical grid is an adequate resolution of all separated flow structures, so the detailed flow characteristics can be calculated in the flight deck region. The grid developed for this ship was a compromise between grid fidelity in the regions required (flight deck), an outer boundary far enough away from the ship so as not to affect the flow, and an overall grid size that is not too large (due to computer memory limitations). The domain is a regular hexahedron with the dimensions 3000 ft $\times$ 1400 ft $\times$ 700 ft. The cells near the ship surface are in the order of 1.0 ft^3 and the number of cells in the domain is in the order of 1.0×10^6. As mentioned previously, no specialised 'boundary layer' cells were used in this model.

4 Results and discussion

4.1 Unsteady Flow Calculations

The results presented here are a follow–on of work conducted previously [9,10] and concentrate mainly on the differences in the wake caused by the change in bow shape with the SFS 2. The majority of the differences in the flight deck between the two ship models observed were for the 0 deg yaw case (wind from directly ahead) and the majority of results in this paper concentrate on this case.

The on–set flow velocity for all cases was a constant 70 ft/s at the required yaw angle. Also, the initial condition for all cases was a uniform 70 ft/s at the appropriate yaw angle for the entire domain. In general, the transient flow dissipated after 6–7 s of flow time. The cases were then run to about 25 s of flow time with a time step of 0.02 s (50 Hz).

The resultant flow solution could then be used to obtain time dependant data for any point in the flow. Figure 3 shows a typical variation of velocity at

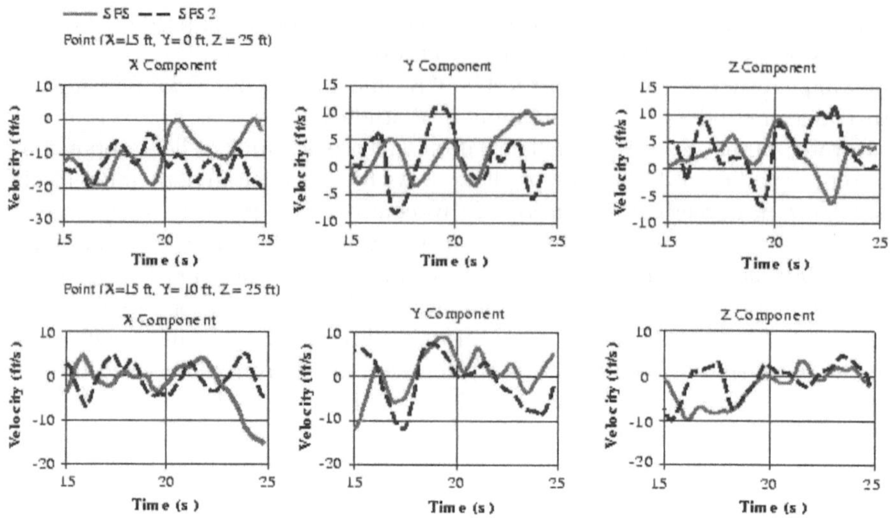

Fig. 3. Velocity profile at various points over the flight deckfor 0 deg yaw

two points 15 ft aft of the hangar and 25 ft above water–level. The first point is along the centreline of the ship and the other 10 ft off the centreline. The data is shown for flow time between 15 and 25 s. Data shows that in the case of the SFS 2, velocity at a reference point in the hangar region varies more frequently compared to SFS, however the magnitudes of the variations are generally smaller.

Also of interest for helicopter flight dynamics analysis is the comparison of time averaged data for the two ship models. The time averaged data is obtained by averaging the velocity data between 6 and 25 s for a given set of data points. An example of this is given in Fig. 4. In this figure, comparisons of the time averaged velocities are made along a line leading from the hangar (X=0) to 150 ft aft of the hangar along the ship centreline, 25 ft above the water–level. From this figure, it can be seen that there are only small differences in the velocity field for the two ship models. The geometry changes to the front of the ship produce considerable flow differences in that area. However, as can be seen in Fig. 4, these changes do not substantially influence the time averaged velocity field in the flight deck region.

4.2 Water Tunnel Flow Visualisation

Recently, a flow visualisation experiment of the SFS was conducted in the DSTO water tunnel at Fishermans Bend. The model is a 300:1 replication of the SFS model that can be yawed between 0 and 45 deg in 5 deg increments. The technique used to visualise the flow was a simple dye (sodium fluorescein) injection technique where the dye was mixed with toothpaste and placed on the model surface. Using this technique, the flow visualisation data was obtained for a range of operating conditions, including 0, 30, and 45 deg. Figure 5a shows results for the 0 deg case. The flow is turbulent with a strong recirculation zone close to

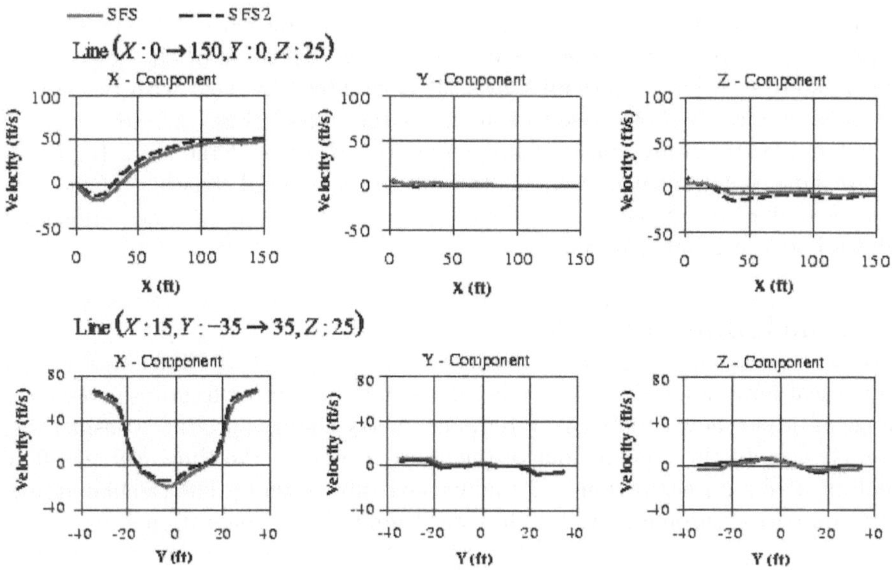

Fig. 4. Time averaged velocity profile at various points in the flight deck region for 0 deg yaw

the hangar. Figure 5b shows the CFD data for the same on–set flow conditions. The reattachment points are indicated by arrows for both CFD and water tunnel results.

Figure 6a shows a vortex pair originating at the leading bow of the ship when yawed at 30 deg. These vorticies were quite strong and stable at 30–

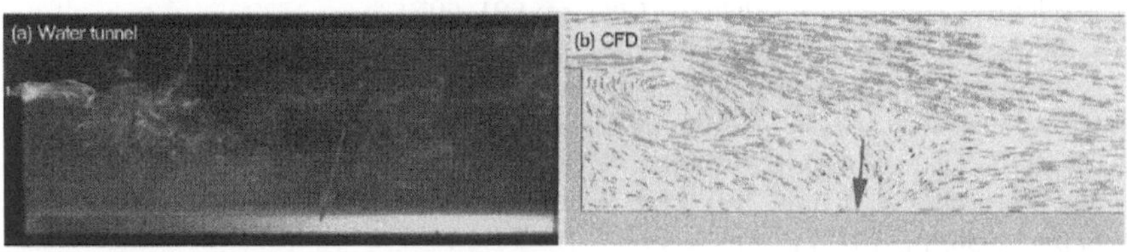

Fig. 5. Comparison between water tunnel and CFD for flow over the flight deck region for 0 deg yaw

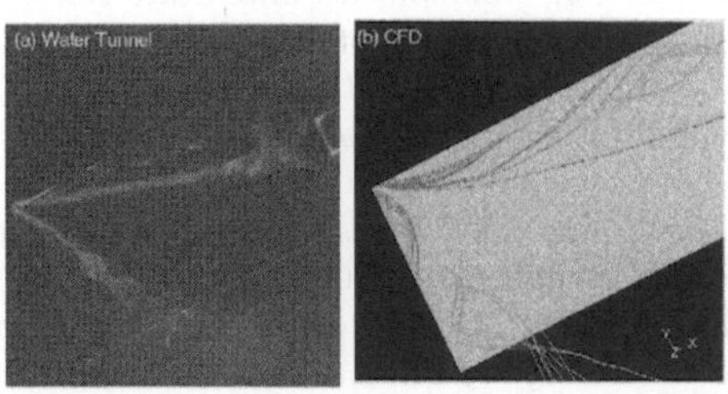

Fig. 6. Vortex pair originating from the leading bow of the SFS for 30 deg yaw

45 deg yaw. To obtain a CFD solution for comparison with the experimental results, streamlines were released at the leading edge bow of the ship in the steady state solution. Figure 6b shows the calculated flow field with two strong vorticies similar to those observed in the water tunnel (Fig. 6a), however, the calculated vorticies are considerably closer to the edges of the ship. Figures 5 and 6 show that the calculated results capture the general structure of the flow field including severe velocity gradients and large recirculation zones associated with a backward facing step.

5 Concluding Remarks

Numerical flow simulations were carried out for two simple frigate shapes for a range of on–set flow conditions. Frontal geometry changes produce considerable flow changes to the flow in front of the stack. However, the time averaged flow field in the flight deck region is not influenced substantially. The calculated flow field compares reasonably well with water tunnel flow visualisation data.

References

1. J. Val Healey, 'The Aerodynamics of Ship Superstructures.' In: *Proceedings of AGARD conference - Aircraft Ship Operations France, AGARD-CP-509, 1991*, pp 4.1–4.14
2. Azuma, A., Saito, S., and Kawachi, K., 'Response of a Helicopter Penetrating the Tip Vortices of a Large Airplane,' In: *Vertica, 1987, Vol 11, No 1/2*, pp 65–76
3. Reddy, R. and Truong, T., 'Effect of Ship Airwake Interference on Helicopter Trim and Control Positions,' In: *Proceedings of the Twelfth Australasian Fluid Mechanics Conference, Sydney, 1995*, pp 691–695
4. Landsberg, A.M., Boris, J.P., Sandberg, W., and Young, Jr., T.R., 'Analysis of the Nonlinear Coupling Effects of a Helicopter Downwash with an Unsteady Airwake,' In: *American Institute of Aeronautics and Astronautics, 1995, Paper 95-0045*
5. Tai, T.C., 'Effect of Ship Motion on DD–963 Ship Airwake Simulated by Multizone Navier–Stokes Solution,' In: *21st Symposium on Naval Hydrodynamics, Trondheim, 1996*, pp 107–117
6. Zan, S.J., 'Experimental Determination of Rotor Thrust in a Ship Airwake,' Journal of the American Helicopter Society, April 2002, pp 100–108
7. Xin, H., He, C., and Lee, J., ' Combined Finite State Rotor Wake and Panel Ship Deck Models for Simulation of Helicopter Operations,' In: *57th Annual Forum Proceedings, Washington, DC, May 9–11 2001*
8. FLUENT User's Guide, Version 5.2., *Fluent Inc., Centra Resource Park, Lebanon, NH 03766, 1999*
9. KR Reddy, R Toffoletto, and KRW Jones, 'Numerical Simulation of Ship Airwake,' Computers and Fluids, Vol **29**, No **4**, 2000, pp 451–465
10. R Toffoletto and KR Reddy, 'Helicopter Operating Environment in the Vicinity of a Ship,' In: *Proceedings of 3rd Australian Pacific Vertiflite Conference on Helicopter Technology, Canberra, 2000*

Part IV

Compressible Flow

Numerical Solution of the Euler Equations with a Multiorder Discontinuous Finite Element Method

Francesco Bassi[1] and Stefano Rebay[2]

[1] Università di Bergamo, Facoltà di Ingegneria
via Marconi 5, 24044 Dalmine, Italy. e-mail: fbassi@unibg.it
[2] Università di Brescia, Dipartimento di Ingegneria Meccanica
via Branze 38, 25123 Brescia, Italy. e-mail: rebay@ing.unibs.it

Abstract. Discontinuous Galerkin methods have proven to be very well suited for the construction of robust high order accurate numerical schemes on arbitrary unstructured and possibly non conforming grids for a wide variety of applications. These accurate, flexibile and robust methods are however rather expensive in terms of computational cost. In this paper we propose to address the issue of computational efficiency of discontinuous finite element methods by introducing a "multiorder" discontinuous Galerkin solution strategy which is similar to a multigrid techinque but uses progressively lower order polynomial DG approximations on the same grid instead of the same discretization on progressively coarsened grids. We present the results obtained in the numerical solution of the Euler equations for the subsonic flow around a circle which give some indication of the effectiveness ot the proposed multiorder solution strategy.

1 Introduction

Discontinuous Galerkin (DG) methods, originally introduced for purely advective problems, have been successively extended to advection-diffusion and to purely elliptic problems [3,1], and are presently finding use in a wide variety of applications. They are in fact accurate and robust methods very well suited for the construction of high order accurate schemes on arbitrary unstructured and possibly non conforming grids. They can handle complex geometries and boundary conditions. They are compact and consequently highly parallelizable. Elements of different order of accuracy can be easly accommodated in the same grid, thus allowing a straightforward implementation of h-p adaptive methods.

The price to pay for the robustness, accuracy and flexibility offered by DG methods is their relatively high computational cost. Several options may be considered in order to improve the computational efficiency of DG methods, such as for example the quadrature-free implementation described in [2] or the use of implicit time discretization schemes and the PGMRES iterative linear system solver, see e.g. [4]. Another possible way to improve the efficiency of DG methods is to take advantage of multigrid techniques [6].

In this paper, however, we introduce a DG solution approach which uses the multigrid idea in a nonstandard way, in that, instead of using the same discretization on progressively coarsened grids, we use progressively lower order polynomial DG approximations on the same grid. To distinguish between the standard multigrid algorithm and the one here proposed we will call the latter the "multiorder" approach. More precisely we will consider a "full multiorder" (FMO) algorithm (analogous to the standard full multigrid technique). The FMO solution approach seems to be perfectly suited for DG discretizations for many of the reasons previously outlined. Notice in particular that DG methods have proved to be very effective for both low and high-order polynomial approximations, and that, unlike standard FEM methods, they include the case of piecewise constant approximations. As an additional advantage, the use of a single grid avoids all the complications of multiple grids in the unstructured case.

In the following we first describe the DG-FMO algorithm and then give a brief report of the results obtained in the solution of the subsonic compressible flow around a circle.

2 Discontinuous Galerkin Full Multiorder Algorithm

The DG space discretized equations, see e.g. [5], can be advanced in time using different time integration schemes. Runge-Kutta methods are a popular choice for DG space discretizations. These schemes have in fact proved to be very effective for the solution of difficult unsteady problems characterized by strong discontinuities and/or fast space-time oscillations, and can match in time the high-order accuracy in space of the DG discretization while retaining total variation stability (provided that the simple forward Euler step does). See e.g. [5].

However, for steady state computations these schemes may be far from ideal since in this case time accuracy is of no relevance. Implicit schemes are potentially much more efficient in this case, see e.g. [4]. Unfortunately, despite the compactness of the DG approximation, the solution of the linear systems arising at each time step in implicit methods requires considerable computational resources, expecially for 3-dimensional problems when a high-order accurate DG space discretization is employed.

In an attempt to improve the computational efficiency of DG methods, we here consider a DG-FMO solution strategy. As in the full multigrid case, we have found that the DG-FMO efficiency is crucially dependent on the many ingredients which are to be choosed in this type of algorithms, such as the shape of the cycle (V, W, etc.), the number of pre and post smoothing sweeps at the various levels, and the type of smoother employed.

As a general rule, the above mentioned ingredients must be selected so that, at each level, the errors in the high frequency range at that level are effectively damped. In the context of a DG-FMO algorithm this means that at each level the smoother must be designed in order to quickly damp the high-order component of the error.

The first smoother considered is the explicit multistage Runge-Kutta (EXPRK) scheme

$$
\begin{aligned}
&W^0 = U^n;\\
&\textbf{For } k = 1, \ldots, m \textbf{ Do:}\\
&\qquad W^k = W^0 - \alpha_k \Delta t M^{-1} R\left(W^{k-1}\right);\\
&\textbf{End;}\\
&U^{n+1} = W^m.
\end{aligned} \tag{1}
$$

where U^n denotes the vector of the discrete unknowns at time step n (i.e. the expansion coefficients of the solution for all the elements $e \in \mathcal{T}$), M the block diagonal mass matrix, $R(U)$ the residual of the spatial part of the equations, Δt the time increment, and α_k, $k = 1, \ldots, m$ a set of scalar coefficients which completely define the scheme. Optimized coefficients α_k for usage in multigrid algorithms are available in the literature for the most common space discretization schemes, but, unfortunately, they are not directly applicable to the case of DG space discretizations. The result obtained by using the "standard" coefficients α_k (i.e. those that maximise the order of accuracy of the RK scheme) were not very satisfactory. Work is underway to calculate the optimized coefficients in the case of DG space approximations using Fourier analysis applied to the simple linear advection equation.

On the other hand, a scheme which treats implicitely the unknowns within an element should have better chances to be a good smoother for a DG-FMO algorithm and in fact we found by numerical experiment that a rather effective smoother in this respect is the semi-implicit m-stage RK-like scheme

$$
\begin{aligned}
&W^0 = U^n;\\
&\textbf{For } k = 1, \ldots, m \textbf{ Do:}\\
&\qquad \left[M + \alpha_1 \Delta t D\left(W^0\right)\right] \delta W^k = -M\left(W^{k-1} - W^0\right) - \alpha_k \Delta t R\left(W^{k-1}\right);\\
&\qquad W^k = W^{k-1} + \delta W^k\\
&\textbf{End;}\\
&U^{n+1} = W^m.
\end{aligned} \tag{2}
$$

where $D(U)$ denotes a block diagonal matrix. We will refer to the above scheme as to the IBDRK smoother. The $pq \times pq$ diagonal blocks of $D(U)$, where p denotes the number of expansion coefficients used in the DG approximation and q the number of equations to be solved ($q = 4$ for the 2-dimensional Euler equations), are equal to the block diagonal part of corresponding shape of the matrix

$$
A(U) = \frac{\partial R(U)}{\partial U}.
$$

The above scheme has been obtained starting from the scheme (1) by replacing the residual $R(W^{k-1})$ with the residual at the new level $R(W^k)$, which is then approximated by linear extrapolation as $R(W^k) \approx R(W^{k-1}) + A(W^{k-1})(W^k -$

W^{k-1}). For the sake of computational efficiency, the matrix $A(W^{k-1})$ is then replaced with the block diagonal part D and "freezed" to $D(W^0)$ as indicated in (2). The above scheme displayed promising performance in the test cases attempted, as documented in the next section.

3 Numerical Results

For reasons of space we will only report the results obtained on the simple but well documented test case of the compressible subsonic flow around a circle. The flow impinges on the circle at a Mach number $M_\infty = 0.38$, which is close to the limiting value to have a completely subsonic and therefore shockless flow. For this reason we could compute high-order accurate DG solution with no need for an oscillation control procedure which would otherwise be necessary to cope with the discontinuities of the solution.

Three grids have been considered, namely the triangulation of a regular distribution of 16×4, 32×8 and 64×16 points, where the first figure refers to the number of points distributed along the circle, and the latter to the number of points in the radial direction, respectively. All the grids extend to about twenty diameters away from the circle centre.

All the solutions have been marched to a L^2 residual error less than 10^{-8} by means of the V-cycle DG-FMO algorithm described in the previous section. We here present the results obtained using the 5 stage IBDRK solver with "standard" coefficients $a_k = 1/(6-k)$ for $k = 1, \ldots, 5$. The operators needed to transfer solutions of different order of accuracy (usually referred to as the "restriction" and the "prolongation" operators in the multigrid literature) are simply L^2 projection operators.

Table 1 shows the performance of the DG-FMO algorithm on the three grids with piecewise constant (P0), linear (P1), and quadratic (P2) elements. Table 2 shows the results of the same algorithm in which we used a single full implicit Newton step for the P0 elements instead than the IBDRK smoother. This test was performed in order to check if it is possible to achieve a grid independent computational cost by combining the IBDRK scheme for high-order accurate

Table 1. Number of iterations and work required to reduce the L^2 norm of the residual error to 10^{-8}. DG-FMO algorithm, V-Cycle, one pre and one post smoothing iterations at each level, five stage IBDRK smoother.

	P0		P1		P2	
Grid	Iter	Work	Iter	Work	Iter	Work
16×4	68	225	54	212	81	286
32×8	129	452	139	646	82	298
64×16	305	1160	415	2450	154	600

Table 2. Number of iterations and work required to reduce the L^2 norm of the residual error to 10^{-8}. DG-FMO algorithm, V-Cycle, one pre and one post smoothing iterations at each level, five stage IBDRK smoother and Full implicit Newton step for P0 elements.

	P0		P1		P2	
Grid	Iter	Work	Iter	Work	Iter	Work
16×4	9	115	49	305	77	290
32×8	11	170	77	480	81	430
64×16	11	511	123	1200	139	650

Table 3. Number of iterations and work required to reduce the L^2 norm of the residual error to 10^{-8} with the Newton method. Linear systems solved with the PGMRES algorithm, ILU preconditioner.

	P0		P1		P2	
Grid	Iter	Work	Iter	Work	Iter	Work
16×4	10	127	10	518	8	265
32×8	10	140	9	205	7	170
64×16	15	674	11	1464	7	726

elements and the relatively cheap full implicit scheme for piecewise constant elements. Table 3 shows instead the performance of the Newton method as a comparison. The fully coupled linear systems have been solved by means of the PGMRES algorithm and the incomplete LU factorization preconditioner. Notice that the P1 and P2 Newton computation initial guesses were the previously computed lower-order solutions.

The columns labeled "Work" appearing in the three tables give the CPU time of the various computations in terms of "working units," defined as the time needed to compute the residual $R(U)$ and the block diagonal matrix $D(U)$ appearing in eq. (2) at that level.

The performance of the DG-FMO algorithm is very similar to that of the Newton method, even on the relatively small grids here employed. This is in the authors' opinion a very interesting result by itself. Notice in addition that the DG-FMO algorithm requires substanatially less memory than the full implicit Newton algorithm and that for larger problems (expecially in three dimensions) the solution of fully coupled systems should have a much greater impact on the overall computational cost than that displayed in the above tables.

The use of a single Newton step for the P0 elements reduces somewhat the number of iterations required to reach convergence on the finer grids. The cost of each iteration is however increased by the need to solve a full linear system for the

P0 elements. It appears that this amount to an overall saving in computational work only for the case of P1 elements. A possible explanation of this behavouir is that, since the P1 solution is already quite accurate, expecially on the finer 64×16 grid, the work required to compute the P2 solution is artificially low and not really relevant. Moreover, the work required is not grid independent even if, when passing from coarser to finer grids, the increase is less rapid than in the plain IBDRK case. More work is howevere needed to try to address these issue.

4 Conclusions

We have presented a discontinuous Galerkin full multiorder algorithm for the solution of the steady state Euler equations. The performance of the algorithms is strongly dependent on the various ingredients that typically appear in this kind of multilevel algorithms. We have tested three smoothers and found that a block diagonal implicit modification of the standard RK time integration scheme seems to give very promising results, at least for the test cases attempted. In order to gain more insight in the DG-FMO algorithm, work is underway to analize theoretically the method in the case of the one dimensional linear advection equation.

As a final remark, it is worth to mention that our attempts to use piecewise cubic elements with the proposed methodology have been a complete failure since the computation appears to be unconditionally unstable. Due to the lack of a theoretical analysis of the method it is still unclear at this moment if this unexpected behaviour is simply due to an error in our program or if it is a real limitation of the proposed methodology.

References

1. D.N. Arnold, F. Brezzi, B. Cockburn and D. Marini, "Unified analysis of discontinuous Galerkin methods for elliptic problems", *Siam J. Numer. Anal.*, to appear.
2. H. Atkins, and C.-W. Shu, "Quadrature-free implementation of discontinuous Galerkin methods for hyperbolic equations", AIAA Journal **36**, (1998), pp. 775–782.
3. F. Bassi and S. Rebay, "A high-order accurate discontinuous finite element method for the numerical solution of the compressible Navier–Stokes equations", *J. Comput. Phys.*, **131**, (1997), pp. 267–279.
4. F. Bassi and S. Rebay, "GMRES discontinuous Galerkin solution of the compressible Navier–Stokes equations", Lecture Notes in Computational Science and Engeneering, **11**, Springer-Verlag, 2000.
5. B. Cockburn and C.-W. Shu "TVB Runge-Kutta local projection discontinuous Galerkin finite element method for conservation laws III: One dimensional systems" *J. Comput. Phys.*, **84**, (1989), pp. 90–113.
6. J.J.W. van der Vegt, and H. van der Ven "Space-time discontinuous Galerkin finite element method with dynamic grid motion for inviscid compressible flows I. General formulation" Memorandum No. 1599, MST, Faculty of Mathematics, University of Twente, December 2001.

A Positivity Preserving Finite Element Method for Hyperbolic Partial Differential Equations

Matthew Hubbard and Martin Berzins

School of Computing, University of Leeds, Leeds, LS2 9JT, U.K.

Abstract. This paper describes a framework in which flux and slope limiters, commonly used in the finite volume community, can be applied in the context of finite element methods, both to the spatial discretisation and the mass matrix. This gives a nonlinear finite element method which uses different basis functions for discretisation of the time and space derivatives and is inherently positivity preserving for hyperbolic partial differential equations. The procedure can be carried out on irregular triangular meshes and is applied here to the two-dimensional scalar advection equation. A number of alternative methods are possible, but the end results do not differ enormously and one representative scheme is picked out to be compared with other schemes for a simple test case involving constant advection at an oblique angle to the mesh. Other cases, not discussed here, show similar qualities.

1 Introduction

The aim of this paper is to suggest links between limiting techniques, which are used as a matter of course in the finite volume community for the modelling of hyperbolic partial differential equations, and finite element methods, which are notoriously poor at approximating problems of this type, particularly in more than one space dimension. Limiters are used to combine a low order but positive (and usually upwind) scheme with a high order scheme to create an accurate method which avoids spurious, numerically induced, oscillations. They are an important component of many of today's most successful finite volume methods but have yet to be employed in the traditional finite element framework: it has proved difficult to construct upwind and positive finite element methods, although SUPG schemes have been a partial success, and the more recent developments in Discontinuous Galerkin and Fluctuation Splitting methods have illustrated the strength of the relationship between finite volume and finite element methods.

This work essentially employs a cell vertex finite volume approach, but considers how it might be recast as a mass-lumped finite element scheme with nonlinear basis functions. It then considers how a mass matrix might be incorporated, what it might look like, and how it might be modified to give a positive scheme when it is inverted. This follows the approach of Cardle [5] in which the basis functions are modified differently for the space and time derivatives, and extends earlier, one-dimensional work of Berzins [2]. Results are presented for a standard test case to illustrate the accuracy of the new method.

2 A Mass-Lumped Scheme

The system which will be studied here is the two-dimensional scalar advection equation, given by

$$u_t + \boldsymbol{\lambda} \cdot \boldsymbol{\nabla} u = 0 ,$$

approximated on triangular meshes. This represents the transport of a quantity u with velocity $\boldsymbol{\lambda}$. Applying the mass-lumped, linear Galerkin finite element method to this equation leads, after some simple algebraic manipulation, to an edge-based form of the scheme, given by

$$\frac{\Omega_i}{3}\,\dot{u}_i = \sum_{j \in \cup \Delta_i} \frac{1}{6}\,(\boldsymbol{\lambda}_{ij} \cdot \boldsymbol{n}^i_{j-1,j+1})(u_j + u_i) . \tag{1}$$

Here the sum is over the nodes j adjacent to node i, Ω_i is the area of the patch of cells surrounding node i and $\boldsymbol{n}^i_{j-1,j+1}$ are the 'inward' pointing normals (see Fig. 1) scaled by the distance between nodes $j-1$ and $j+1$ (j is incremented and decremented modulo N_i, the number of nodes adjacent to node i, and increases anticlockwise).

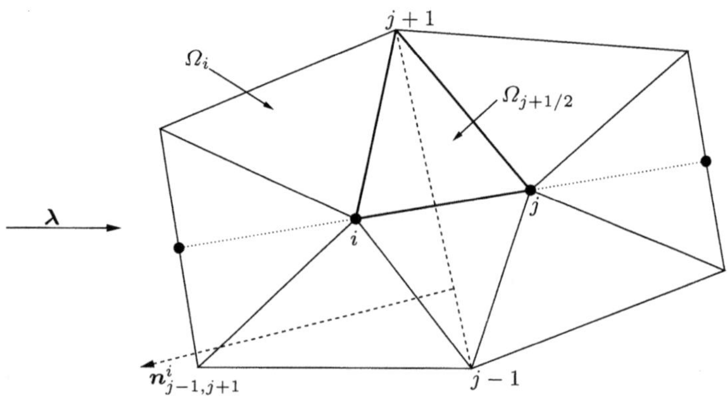

Fig. 1. The geometry associated with the edge adjoining nodes i and j.

This scheme is unconditionally unstable, but can be adjusted to introduce an upwind bias along the grid edges. A diffusive component is added (the difference between the contribution of an edge to the central scheme (1) and to a purely upwind scheme), and a limiter, denoted here by $V(\cdot)$, applied to give

$$\frac{\Omega_i}{3}\,\dot{u}_i = \sum_{j \in \cup \Delta_i} \frac{1}{6}\,(\boldsymbol{\lambda}_{ij} \cdot \boldsymbol{n}^i_{j-1,j+1}) \left[2u_{\mathrm{m}} + \frac{V(r_j)}{r_j}(u_{\mathrm{d}} - u_{\mathrm{m}}) \right] , \tag{2}$$

in which

$$r_j = \left[\frac{u_{\mathrm{d}} - u_{\mathrm{m}}}{u_{\mathrm{m}} - u_{\mathrm{u}}} \times \frac{d_{\mathrm{u}}}{d_{\mathrm{d}}} \right]_j .$$

The value u_{d} is taken from the downwind vertex of edge ij (node i if $\boldsymbol{\lambda}_{ij} \cdot \boldsymbol{n}^i_{j-1,j+1} \geq 0$, node j otherwise), u_{m} is taken from the upwind vertex of edge ij, and u_{u} is taken from the intersection of the extension of edge ij beyond the upwind vertex with the opposite edge of the triangle into which it extends (see Fig. 1). Linear interpolation is used to evaluate this value, which is consistent with the Galerkin method originally considered and guarantees that the value of u_{u} remains bounded by the local solution values. $\boldsymbol{\lambda}_{ij}$ is evaluated at the midpoint of edge ij. d_{d} is invariably the length of edge ij, while d_{u} is the distance between the upwind vertex and the intersection point with the opposite edge.

It can be shown that, as long as $0 \leq V(r)/r \leq 2$ and $0 \leq V(r) \leq 2$, the scheme (2), combined with forward Euler time-stepping, is positive (and hence stable) for δt satisfying

$$\frac{\delta t}{\Omega_i} \sum_{j \in \cup \triangle_i} \left[1 + \frac{d_{\mathrm{d}}}{d_{\mathrm{u}}}\right] |\alpha^i_j| \leq 1$$

for all nodes i, where $\alpha^i_j = \boldsymbol{\lambda}_{ij} \cdot \boldsymbol{n}^i_{j-1,j+1}$. In this work, two limiters have been used, both of which satisfy the above conditions:

- $V(r) = \max(0, \min(2r, \min(0.25 + 0.75r, 4)))$, a third order limiter derived by Gaskell and Lau [6].
- $V(r) = (r + |r|)/(1 + \max(1, |r|))$, a modified form of van Leer's limiter [4].

The relationship between the scheme (2) and finite elements is not immediately obvious. However, it can be replicated by augmenting the standard linear basis function $\phi_i(x,y)$, used in the Galerkin approach, with a nonlinear term, based around the local grid edges:

$$\tilde{\phi}_i(x,y) = \phi_i(x,y) + 4 \sum_{j \in \cup \triangle_i} \mathrm{sgn}(\alpha^i_j) \left[1 - \frac{V(r_j)}{r_j}\right] \phi_i(x,y)\phi_j(x,y).$$

3 Including a Mass Matrix

The consistent linear Galerkin scheme on triangles gives a system of ordinary differential equations defined by

$$\sum_{j \in \cup \triangle_i} \frac{\Omega_{j+1/2}}{12} [2\dot{u}_i + \dot{u}_j + \dot{u}_{j+1}] = G_i(\underline{u}) \qquad (3)$$

where G_i is simply the right hand side obtained from the chosen spatial discretisation, $\underline{u}$ is the vector of all solution values at the given time level and $\Omega_{j+1/2}$ is the area of the triangle with vertices i, j, $j+1$ (see Fig. 1).

The lumped scheme is positive, but this property will normally be lost when the mass matrix, introduced in (3), is inverted. The aim here is to manipulate the mass matrix in a manner which will retain the positivity property of the scheme. The approach is similar to that used in spatial limiting: the lumped matrix is taken as a starting point, and to this is added a component proportional to the difference between the lumped matrix and the consistent matrix derived from linear basis functions. It follows the philosophy of Cardle [5], in that the time and space derivatives are treated independently.

A variety of schemes of this form have been implemented [2,3] and another alternative is presented here. In fact, most of the procedures produce similar results: there is far more sensitivity to the discretisation of the spatial terms. Forward Euler time-stepping is used in all the schemes, but a cell-based 'limiting' is chosen here rather than an edge-based one. As in [3], the time derivative is discretised first, giving

$$\sum_{j \in \cup \triangle_i} \frac{\Omega_{j+1/2}}{12} (2u_i^{n+1} + u_j^{n+1} + u_{j+1}^{n+1}) = \delta t\, G_i(\underline{u}^n) + \sum_{j \in \cup \triangle_i} \frac{\Omega_{j+1/2}}{12} (2u_i^n + u_j^n + u_{j+1}^n)\,, \tag{4}$$

in which n indexes the time level. The mass matrix can be modified *before* discretising the time derivative, and a nonlinear method constructed based on limiting ratios of differences in time derivatives so that the modified mass matrix is an M-matrix (whose inverse contains only non-negative entries) [2]. However, this only enforces positivity of $\dot{u}$. It doesn't automatically impose positivity on the solution.

Equation (4) can be rewritten as

$$\sum_{j \in \cup \triangle_i} \frac{\Omega_{j+1/2}}{12} \left[4u_i^{n+1} + (u_j^{n+1} - u_i^{n+1}) + (u_{j+1}^{n+1} - u_i^{n+1})\right] = \delta t\, G_i(\underline{u}^n) + \sum_{j \in \cup \triangle_i} \frac{\Omega_{j+1/2}}{12} \left[4u_i^n + (u_j^n - u_i^n) + (u_{j+1}^n - u_i^n)\right]$$

Cumulatively, the latter terms on each side give the difference between lumped and consistent Galerkin schemes. These differences are limited in a manner which guarantees that the matrix on the left hand side is diagonally dominant with non-positive off-diagonal entries (and hence an M-matrix), so its inversion will retain the positivity of the scheme. The resulting method takes the form

$$\begin{aligned} \sum_{j \in \cup \triangle_i} \frac{\Omega_{j+1/2}}{12} \Big[4u_i^{n+1} &+ \max(r_{j+1/2}^{n+1} - 1, 0)(u_i^{n+1} - u_{j+1}^{n+1}) \\ &+ \max(s_{j+1/2}^{n+1} - 1, 0)(u_i^{n+1} - u_j^{n+1})\Big] \\ = \delta t\, G_i(\underline{u}^n) + \sum_{j \in \cup \triangle_i} \frac{\Omega_{j+1/2}}{12} \Big[4u_i^n &+ \max(r_{j+1/2}^n - 1, 0)(u_i^n - u_{j+1}^n) \\ &+ \max(s_{j+1/2}^n - 1, 0)(u_i^n - u_j^n)\Big] \end{aligned} \tag{5}$$

in which

$$r_{j+1/2} = \frac{u_j - u_i}{u_i - u_{j+1}} \qquad \text{and} \qquad s_{j+1/2} = \frac{1}{r_{j+1/2}}$$

The right hand side of (5) can be shown to be a positive combination of local solution values for an appropriate limit on the time-step, so the overall scheme should be positive.

Note that, throughout this discussion, j and $j+1$ refer to consecutive nodes around a patch of cells centred on node i. In [3] this cell-based pairing of edges is replaced by associating each edge with its projection across the patch, in the manner of the above spatial discretisation.

Equations (5) are solved iteratively (indexed below by m), using a Jacobi-type method to give

$$\begin{aligned}
&\left(\sum_{j\in\cup\triangle_i}\left[\max(r^{m+1}_{j+1/2}-1,0)+\max(s^{m+1}_{j+1/2}-1,0)\right]\right)u_i^{m+1}\\
&= \frac{3\delta t}{\Omega_i}G_i(\underline{u}^n)\\
&+ \sum_{j\in\cup\triangle_i}\left[\max(r^n_{j+1/2}-1,0)(u_i^n-u^n_{j+1})+\max(s^n_{j+1/2}-1,0)(u_i^n-u_j^n)\right]\\
&+ \sum_{j\in\cup\triangle_i}\left[\max(r^m_{j+1/2}-1,0)u^m_{j+1})+\max(s^m_{j+1/2}-1,0)u_j^m)\right] .
\end{aligned}$$

The initial estimate at time level n is taken to be $\underline{u}^n$ and, in the following results, ten iterations were always enough to reach convergence. In this work, r and s have been replaced by $V(r)$ and $V(s)$ (using the modified van Leer limiter) because this improves the accuracy of the scheme and improves the restriction on the time-step. The positivity property remains intact.

4 Results

The test case shown here represents the doubly periodic advection of a double sine wave profile, given initially by

$$u(x,y,0) = \sin(2\pi x)\sin(2\pi y),$$

with a constant velocity of $\boldsymbol{\lambda} = (1,2)^{\mathrm{T}}$, across the domain $[0,1]\times[0,1]$. All of the results are obtained on uniform triangular grids consisting of squares divided by diagonals from bottom left to top right corners. This means that the advection velocity is *not* aligned with grid edges.

The results indicate that the new scheme matches the performance of a good cell centred scheme (the MLG scheme of Batten *et al.* [1]) on a similar type of mesh, having smaller L_∞ error but larger L_1 error, but still lags behind a cell vertex Lax-Wendroff scheme to which Flux-Corrected Transport (FCT) has been applied [7]. This is also seen in other cases approximated.

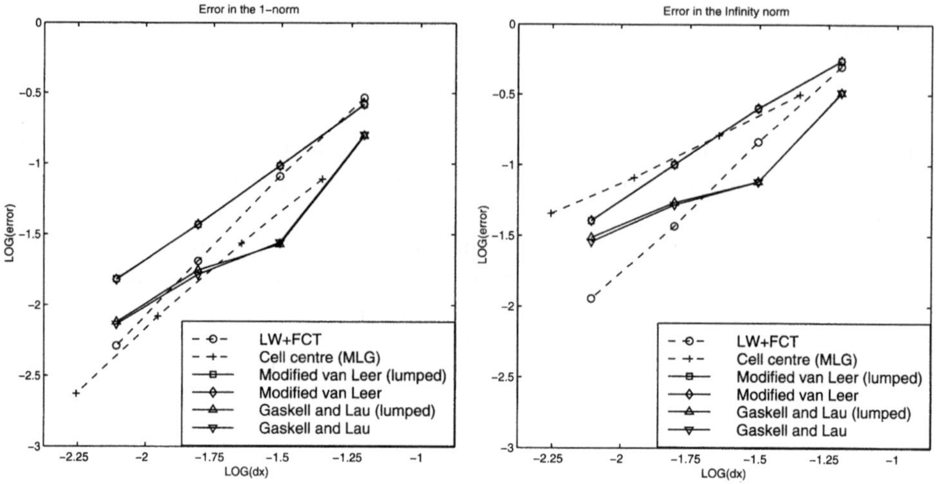

Fig. 2. Approximations of the L_1 and L_∞ norms of the errors for the double sine wave test case at $t = 1$.

5 Conclusions

The method presented above is still in the very early stages of development. The 'lumped' method is very accurate, particularly when the third order limiter is used, but this is unsurprising since it is effectively a flux limited finite volume method applied on the dual of the triangular grid: the novelty being the method used to calculate the ratios to be limited. As presented it is only first order accurate in time. The consistent finite element method is, as yet, less successful. This may be because the Galerkin method is used as the basis, and the nonlinear approach, while guaranteeing positivity, appears to add a diffusive term to the lumped scheme. It would be more interesting to apply it to other methods such as Taylor-Galerkin, or a fluctuation splitting method which have more leeway for improvement. At the moment, the more restrictive the limiting is on the mass matrix, the better the solution.

References

1. P. Batten, C. Lambert, D.M. Causon: Int. J. Numer. Methods Eng. **39**, 1821 (1996).
2. M. Berzins: Com. Num. Meth. Eng. **17**, 659 (2001)
3. M. Berzins, M.E. Hubbard: submitted to J. Comput. Phys. (2001)
4. M. Berzins, J.M. Ware: Appl. Numer. Math. **16**, 417 (1995)
5. J.A. Cardle: Int. J. Numer. Methods Eng. **38**, 171 (1995)
6. P.H. Gaskell, A.C. Lau: Int. J. Numer. Methods Eng. **8**, 617 (1988)
7. M.E. Hubbard, P.L. Roe: Int. J. Numer. Methods Fluids **33**, 711 (2000)

A Residual-Based Scheme for Solving Unsteady Compressible Viscous Flows

Grégoire Hanss, Christophe Corre and Alain Lerat

SINUMEF Laboratory, ENSAM,
151 Bd de l'Hopital, 75013 Paris, France

Abstract. A residual-based compact scheme initially developed for steady compressible flows computation is extended to the unsteady case. Within a dual time step framework, the full residual, including the physical time derivative, is used to obtain high-accuracy and build the numerical dissipation in a compact way. The resulting scheme is compared with a conventional high-order method for a model advection problem and a complex viscous compressible flow.

1 Introduction

A residual-based compact (RBC) scheme has been recently proposed [1] for computing multidimensional inviscid and viscous steady flows governed by conservation laws of the form $w_t + f_x + g_y = 0$; it makes a systematic use of the residual $r = f_x + g_y$ vanishing at steady-state to obtain high-accuracy and to build dissipation in a compact way. In the present paper the RBC scheme is extended to the unsteady case within a dual time step (DTS) framework; aiming at simplicity, one basically retains the design principles derived for the steady case with a residual now defined with respect to the dual time, which includes the physical time-derivative : $r = w_t + f_x + g_y$. The results provided by a second-order in time and third-order in space RBC scheme are compared with those of a conventional high-order scheme for a model circular advection test-problem and the computation of vortex pairing in a time-developing mixing layer.

2 High-order within a DTS framework

Let us consider in a first step the 1D hyperbolic system $w_t + f_x = 0$ approximated at point $j\delta x$ of a uniform grid by an implicit scheme of the form :

$$\mathcal{T}(\phi, \psi, w_j^{n+1}, w_j^n, w_j^{n-1}, w_j^{n-2}) + \frac{\delta h_j}{\delta x}^{n+1} = 0, \tag{1}$$

where $\mathcal{T}$ is a pth order linear multistep approximation of w_t at time $(n+1)\Delta t$:

$$\mathcal{T}(\phi, \psi) = \frac{(1+\phi+\psi)\Delta w^n - \phi \Delta w^{n-1} - \psi \Delta w^{n-2}}{\Delta t} = (w_t)^{n+1} + O(\Delta t^p), \tag{2}$$

with $\Delta w^n = w^{n+1} - w^n$; δ denotes the difference operator over one grid cell $(\delta\phi_j = \phi_{j+\frac{1}{2}} - \phi_{j-\frac{1}{2}})$ and h is some high-order numerical flux such that $\frac{\delta h_j^{n+1}}{\delta x} =$

$(f_x)_j^{n+1} + O(\delta x^q)$. For instance, the third-order ($q = 3$) directional non-compact (DNC) scheme proposed in [2] reads :

$$h^{DNC} = (Id - \frac{1}{6}\delta^2)\mu f + \frac{1}{12}|A^R|\delta^3 w, \tag{3}$$

with μ the average operator over one grid cell ($\mu\phi_j = \frac{1}{2}(\phi_{j+\frac{1}{2}} + \phi_{j-\frac{1}{2}})$) and A^R the Roe-average of the Jacobian $A = df/dw$. The numerical flux (3) is similar to that of the well-known Roe-MUSCL scheme. Clearly, the truncation error of scheme (1)-(2)-(3) is $\epsilon^{DNC} = O(\Delta t^p, \delta x^3)$ and $p = 2$ if we choose a 3-level formula with ($\phi = 1/2, \psi = 0$) for (2), which is preferred to the less robust third-order 4-level formula with ($\phi = 7/6, \psi = -1/3$).
In order to solve efficiently the non-linear system (1), formally written $R^*(w) = 0$, one computes its solution w as the steady-state of

$$w_\tau + R^*(w) = 0, \tag{4}$$

where τ is the so-called dual time [3]. The known physical state at time-level n is set as an initial intermediate state $w^{n,0}$, which is evolved through M intermediate states $w^{n,m}$ using (4) until the steady-state of (4) is considered to be reached, at which point $w^{n,M}$ is taken as the new physical state w^{n+1}. Thus (1) is computed in practice as :

$$\frac{\Delta w_j}{\Delta\tau}^{n,m} + \mathcal{T}(\phi, \psi, w_j^{n,m}, w_j^n, w_j^{n-1}, w_j^{n-2}) + \frac{\delta h_j}{\delta x}^{n,m} = 0, \tag{5}$$

with $\Delta w^{n,m} = w^{n,m+1} - w^{n,m}$. Scheme (5) is made implicit with respect to τ in order to reach steady-state after a limited number of subiterations and ensure the high efficiency of the method. To keep the scheme simple and cheap, a Harten-Roe type 3-point implicit stage is built :

$$(Id + \lambda + \dot{A}\delta\mu - \frac{1}{2}|\dot{A}|\delta^2)\Delta w_j^{n,m} = -\sigma\delta h_j^{n,m} - \Delta\tau\mathcal{T}(\phi, \psi), \tag{6}$$

where $\sigma = \Delta\tau/\delta x$, $\dot{A} = \sigma A$ and the term $\lambda = (1 + \phi + \psi)\Delta\tau/\Delta t$ comes from the implicitation of the physical time-derivative approximation $\mathcal{T}(\phi, \psi)$.
For the $2D$ problem $w_t + f_x + g_y = 0$, with $A = df/dw$, $B = dg/dw$, the DNC scheme is extended on a $(j\delta x, k\delta y)$ grid in a straightforward manner :

$$\begin{aligned}(Id + \lambda + \dot{A}\delta_1\mu_1 - \frac{1}{2}|\dot{A}|\delta_1^2 + \dot{B}\delta_2\mu_2 - \frac{1}{2}|\dot{B}|\delta_2^2)\Delta w_{j,k}^{n,m} =& - \sigma_1\delta(h_1)_{j,k}^{n,m} \\ &-\sigma_2\delta(h_2)_{j,k}^{n,m} - \Delta\tau\mathcal{T}(\phi, \psi),\end{aligned} \tag{7}$$

with δ_p, μ_p the difference and average operators in the pth space direction, $\sigma_1 = \Delta t\tau/\delta x$, $\dot{A} = \sigma_1 A$, $\sigma_2 = \Delta\tau/\delta y$, $\dot{B} = \sigma_2 B$ and the DNC fluxes given by :

$$h_1^{DNC} = (Id - \frac{1}{6}\delta_1^2)\mu_1 f + \frac{1}{12}|A^R|\delta_1^3 w \ , \ h_2^{DNC} = (Id - \frac{1}{6}\delta_2^2)\mu_2 g + \frac{1}{12}|B^R|\delta_2^3 w. \tag{8}$$

In the viscous case, the physical fluxes read $f = f^E(w) - f^V(w, w_x, w_y)$, $g = g^E(w) - g^V(w, w_x, w_y)$ and (8) is used to approximate at third-order the inviscid fluxes derivatives f_x^E, g_y^E while f_x^V and g_y^V are approximated at second-order only using simple centered differences.

3 Design of the RBC scheme

Building on the ideas developed in [1] for steady problems, the following RBC scheme is proposed to approximate $w_t + f_x = 0$:

$$\frac{\Delta w_j}{\Delta \tau}^{n,m} + \mathcal{T}(\phi_1,\psi_1) + \frac{1}{6}\delta^2\mathcal{T}(\phi_3,\psi_3) + \frac{\delta\mu f_j}{\delta x}^{n,m} = \frac{1}{2}\delta(sgn(A)(\frac{\delta f}{\delta x} + \mu\mathcal{T}(\phi_2,\psi_2)))_j^{n,m}. \tag{9}$$

Since

$$\mathcal{T}(\phi_1,\psi_1) + \frac{1}{6}\delta^2\mathcal{T}(\phi_3,\psi_3) + \frac{\delta\mu f}{\delta x} = r + \frac{\delta x^2}{6}r_{xx} + O(\Delta t^{p_1}, \delta x^2\Delta t^{p_3}, \delta x^4),$$

with $r = w_t + f_x$, and for the "dissipation-like" term on the RHS

$$\frac{1}{2}\delta(sgn(A)(\frac{\delta f}{\delta x} + \mu\mathcal{T}(\phi_2,\psi_2))) = \frac{\delta x}{2}(sgn(A)r)_x + O(\delta x\Delta t^{p_2}, \delta x^3),$$

the truncation error of (9) at steady-state in dual-time τ, when $r = 0$, is given by $\epsilon^{RBC} = O(\Delta t^{p_1}, \delta x\Delta t^{p_2}, \delta x^2\Delta t^{p_3}, \delta x^3)$ and third-order space accuracy is ensured with a first-order formula for $\mathcal{T}(\phi_3,\psi_3)$ and a second-order one for $\mathcal{T}(\phi_2,\psi_2)$. In what follows however, we will use the same 3-level second-order formula ($\phi_p = 1/2, \psi_p = 0$) for all the time-derivatives, which eventually yields $\epsilon^{RBC} = O(\Delta t^2, \delta x\Delta t^2, \delta x^2\Delta t^2, \delta x^3)$ on a compact 3-point stencil. Note the necessary dissipation for compressible flow computations is brought in (9) by the higher-order terms remaining in the RHS operator at steady-state; the first-order brings robustness in the transient phase.

When applied to the 2D problem $w_t + f_x + g_y = 0$, the RBC scheme reads :

$$(Id + \frac{\delta_1^2}{6})(Id + \frac{\delta_2^2}{6})\mathcal{T} + \frac{\delta_1\mu_1}{\delta x}(Id + \frac{\delta_2^2}{6})f_{j,k}^{n+1} + \frac{\delta_2\mu_2}{\delta y}(Id + \frac{\delta_1^2}{6})g_{j,k}^{n+1} = \tilde{P}, \tag{10}$$

where the dissipation operator $\tilde{P}$ is given by :

$$\begin{aligned}\tilde{P} =& \frac{1}{2}\delta_1[\Phi sgn(A)(\mu_1\mathcal{T} + \frac{\delta_1 f}{\delta x} + \frac{\delta_2\mu_1\mu_2 g}{\delta y})]_{j,k}^{n+1} \\ &+ \frac{1}{2}\delta_2[\Psi sgn(B)(\mu_2\mathcal{T} + \frac{\delta_1\mu_1\mu_2 f}{\delta x} + \frac{\delta_2 g}{\delta y})]_{j,k}^{n+1}.\end{aligned} \tag{11}$$

The matrix dissipation coefficients Φ, Ψ have been designed in the steady case [1]. The second-order correction $\frac{\delta_1^2}{6} + \frac{\delta_2^2}{6}$ applied to the temporal approximation $\mathcal{T}$ in (10) to form a derivative of the residual in the truncation error has to be completed with the higher-order term $\frac{\delta_1^2\delta_2^2}{36}$ to ensure the scheme's unconditional stability.

The RBC scheme is made implicit with respect to the dual time τ with the following simple 3-point per direction implicit stage :

$$\begin{aligned}&(Id + \lambda + \epsilon_3\lambda\tfrac{1}{6}(\delta_1^2 + \delta_2^2) - \tfrac{1}{2}\epsilon_2\lambda(sgn(A)\delta_1\mu_1 + sgn(B)\delta_2\mu_2) \\ &+ \dot{A}\delta_1\mu_1 - \tfrac{1}{2}|\dot{A}|\delta_1^2 + \dot{B}\delta_2\mu_2 - \tfrac{1}{2}|\dot{B}|\delta_2^2)\Delta w_{j,k}^{n,m} \\ &= -\sigma_1\delta_1(h_1^{RBC})_{j,k}^{n,m} - \sigma_2\delta_2(h_2^{RBC})_{j,k}^{n,m} - \Delta\tau(Id + \tfrac{\delta_1^2}{6})(Id + \tfrac{\delta_2^2}{6})\mathcal{T}(\phi,\psi)\end{aligned} \tag{12}$$

where the RBC fluxes are easily deduced from (10)-(11). Ideally the parameters ϵ_p should be taken equal to 1 in order to ensure a maximum efficiency; in practice, the above implicit stage is solved by a line-relaxation method which requires a diagonal dominance condition that is satisfied with $\epsilon_p = 0$. The amplification factor (with respect to the dual time) of the corresponding implicit stage is compared in Fig. 1 with that of (7) : interestingly, it is observed the efficiency of RBC improves when λ tends to 0, that is, for a fixed value of the dual timestep $\Delta\tau$, when the physical time Δt increases, while a reverse behaviour, in accordance with common DTS practice is observed with DNC.
For the Navier-Stokes equations, the third-order dissipation (11) retains the same form with r now including the viscous fluxes; approximation (10) is used with $f = f^E - f^V$, $g = g^E - g^V$ and an extra corrective term is added to form the full residual [1], yielding a globally third-order space-accurate scheme.

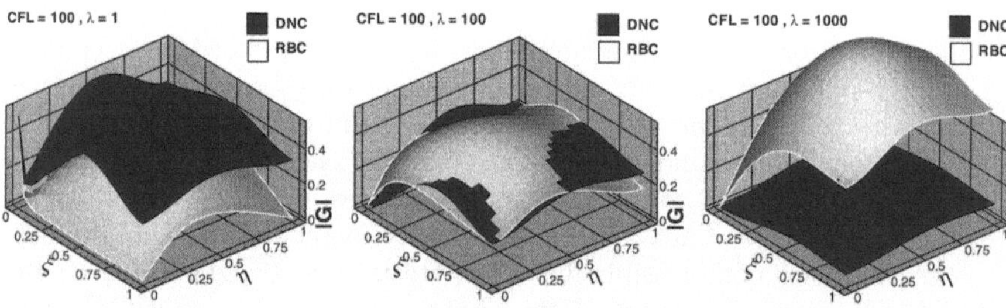

Fig. 1. Efficiency analysis of DNC / RBC in 2D. Amplification factor in the wavenumbers plane for increasing values of λ.

4 Applications

DNC and RBC are first applied to a model scalar advection problem taken from [4] : a Gaussian distribution initially centred at point $(-0.5, 0)$ is rotated around $(0, 0)$. Both schemes are used on a 80×80 uniform grid, with $\Delta t = 2.5 \times 10^{-3}$ and a CFL number based on the dual timestep equal to 100 in order to ensure fast convergence to steady-state at each pysical time-level (less than 10 iterations are required to reduce the residual by 4 orders of magnitude). When comparing the solution after one turn in the $[-1, 1]^2$ box (see Fig. 2), the RBC scheme is found to yield less dissipative and dispersive error. Next the schemes are applied to the computation of vortex pairing in a time-developing mixing layer at a convective Mach number equal to 0.8 and Reynolds number based on velocity jump across the shear layer and the vorticity thickness equal to 1000 (see [5] for a detailed description of the test case set-up). The flow has been computed with Δt varying from 0.1 up to 4 and $\Delta t = 0.5$ has been found to provide a good trade-off between solution accuracy and efficiency in the time-marching process. For this time-step, λ is of order 100 in the flow and, as expected, DNC and RBC have roughly the

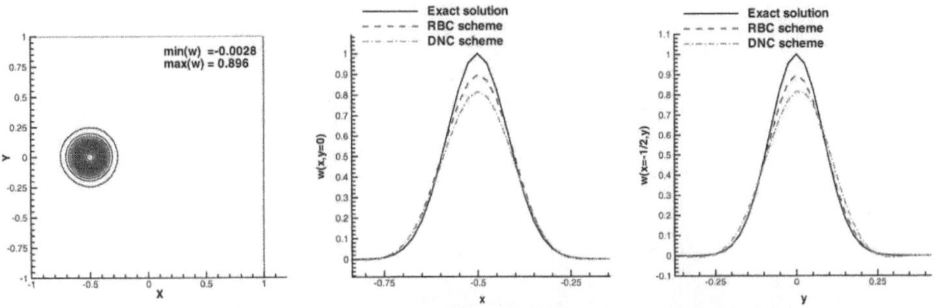

Fig. 2. Circular advection on a 80 × 80 regular grid. Left : contours of RBC solution after one turn. Center and right : horizontal ($y = 0$) and vertical ($x = -1/2$) cutline.

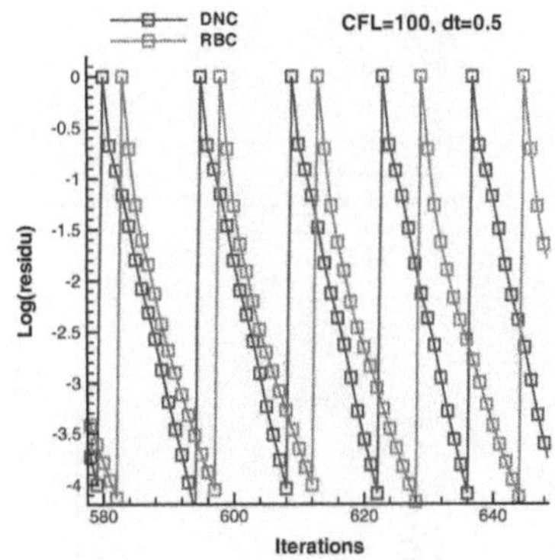

Fig. 3. Convergence to dual steady-state at several time-levels for DNC and RBC on the 181 × 181 grid, with $\Delta t = 0.5$ and $CFL_\tau = 100$.

same efficiency (see Fig. 3). The computed solutions on a series of increasingly refined grids are plotted in Fig. 4. The solutions obtained with DNC and RBC on the finest 181×181 grid are very similar; however, RBC provides better shock resolution (almost oscillation-free) and is less sensitive to grid coarsening, with in particular better preserved vortex structures on the coarsest 61×61 grid. Since the residual-based dissipation can be accurately computed on curvilinear grids thanks to its compactness, the present work is currently extended to the computation of unsteady flows over airfoil geometries.

References

1. A. Lerat A. and C. Corre, *J. Comput. Phys.* **170**, 642-675 (2001).
2. A. Rezgui *et al*, *Computers and Fluids* **30**, 875-901 (2001).
3. R. Peyret and T.D. Taylor, *Comput. methods for fluid flow*, Springer, NY, 1983.
4. P. De Palma *et al*, in Notes Numer. Fluid Mech. **78**, 103-116 (2001).
5. H.C. Yee *et al*, *J. Comput. Phys.* **162**, 33-81 (2000).

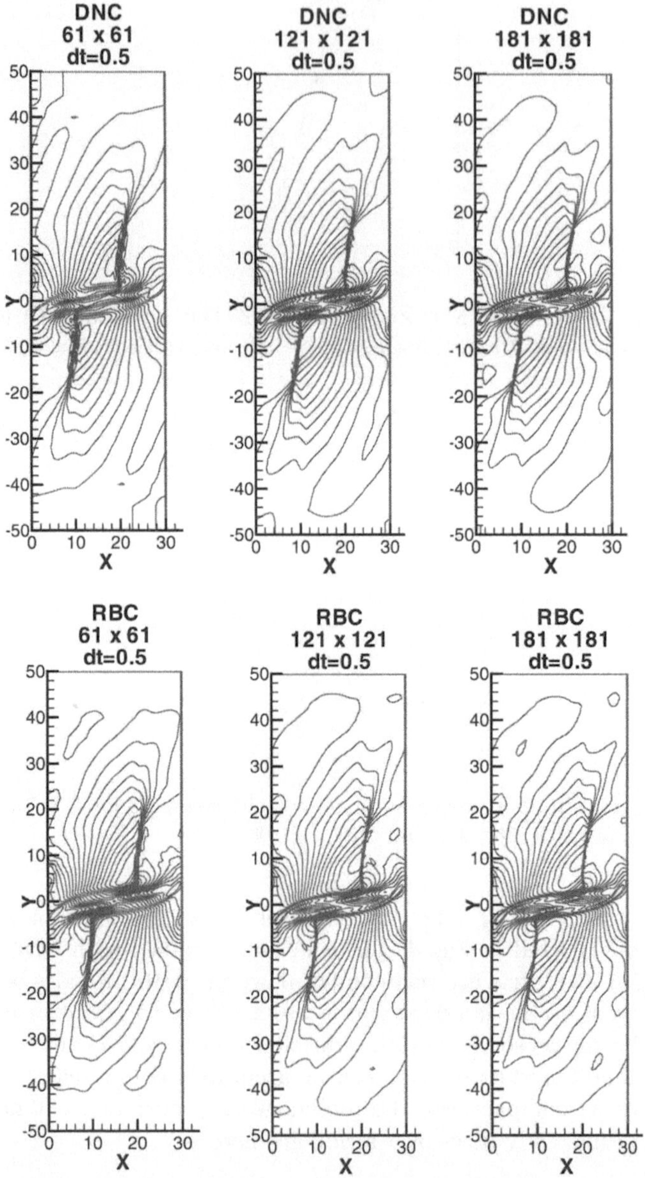

Fig. 4. Vortex pairing in a time-developing mixing layer. Temperature contours at $t = 160$. Grid-convergence for the DNC (top) and RBC (bottom) schemes.

A Multi-Block Shock-Fitting Technique to Solve Steady and Unsteady Compressible Flows

Francesco Nasuti

University of Rome "La Sapienza", Dip. Meccanica Aeronautica, Via Eudossiana 18, Rome 00184, Italy

Abstract. A numerical technique based on the coupling of shock fitting and lambda scheme has been extended to solve unsteady inviscid and viscous flows around complex geometries. This goal has been achieved by a multi-block approach, which allows the shocks to float through the different blocks. The result is that many complex flow fields can be handled while preserving the main features of the lambda scheme, of the shock fitting and of orthogonal grids.

1 Introduction

In the last decades computational fluid dynamics (CFD) has grown its predicting capabilities becoming more and more an indispensable design tool. The most successful numerical approaches have undergone both a refining process, eventually leading to robust codes able to treat a wide range of complex problems, and an analysis of the limits connected to their own nature, that encourages the continuous study of less popular approaches [1]. One of them is the shock fitting technique, left aside after its success in the sixties and seventies [2]. The main merit of shock fitting is its capability of considering shocks as discontinuities, that is as they actually are in all scales larger than the molecular scale. Moreover, it allows discretizing the equations written in their primitive variables without passing through the formulation in "conservation" variables. The latter merit greatly reduces the numerical error because of the "quasi-linear" form of the equations and especially because of the possibility of naturally describing wave propagation phenomena, respecting the domains of dependence. These merits yielded the solution of the inviscid blunt body flows by just a few dozens of cells [3].

While shock fitting was considered the best approach to solve blunt body flows, where the shock was taken as a boundary of the computational domain, it enjoyed less success for complex problems featuring multiple shocks. Nevertheless, attempts were done to make the approach as general as possible for two-dimensional problems including inviscid and viscous flows in both steady-state and unsteady regimes, and also to solve specific three-dimensional problems [4]. The most general technique was that of letting independent shocks move over a fixed grid: it was referred to as floating shock fitting [5]. In particular, a recent version of the latter technique has been used to study a wide range of problems, including some cases of shock interaction [6].

Aiming to solve more complex problems, a further generalization of the technique is provided in the present paper. The goal is achieved by a multi-block approach, as widely done to solve complex flow fields by other techniques. Indeed, multi-block techniques allow solving whatever two-dimensional problem by orthogonal grids, exploiting their properties of accuracy and easy implementation. The peculiarity of the present approach is that both the information relevant to the model for the solution of the equations and that relevant to the shock fitting must be shared among blocks. In particular, to preserve floating shock fitting properties, shocks must be allowed to float through the whole computational domain divided in as many blocks as desired.

2 Outline of the Technique

The governing equations for the present study of steady and unsteady turbulent compressible flows are the twodimensional (linear and axisymmetric) Reynolds-Averaged Navier-Stokes equations, written in nonconservative form. They are integrated by an explicit second-order time and space accurate scheme that follows Moretti's "lambda" formulation for the convective terms, and uses central differencing for the viscous terms [8]. Turbulence is treated according to the one-equation model of Spalart and Allmaras, including a correction that takes into account the effect of compressibility in shear-layers [9,10]. The present technique features less nonlinearity than the conservative form and thus it yields the same error with less cells in the inviscid regions of the flow field. Moreover, it does not require any flux limiter like those needed in conservative approaches to capture shocks, because it is capable of treating shock discontinuities as discontinuities by shock-fitting.

2.1 Shock Fitting

The shock fitting technique models shocks as real discontinuities, governed by their own algebraic equations [5]. Therefore, the states upstream and downstream the shock and its position are determined by coupling the solutions of the Rankine-Hugoniot jump relations and those of the flow field equations. In particular, in the present floating shock fitting technique the shock position is defined by its intersection with the computational grid (Fig. 1). Each intersection point, where the solution of the shock is computed at each time step, is called shock point. The field variables at each side of the shock point are computed from the neighboring mesh nodes respecting the physical domain of dependence, then the jump relations are enforced and finally the updated solution at the shock point is used to update the neighboring mesh nodes.

The shock-fitting procedure can be divided in three main phases. In the first phase the formation of a shock is detected by looking for a possible coalescence of compression waves in the neighborhood of each mesh node. In the second phase, the position of each shock point is compared with that of the neighboring shock points to determine the shock slope. Consequently the direction normal to the shock is obtained, which permits to compute the local shock velocity and to update high-pressure side variables according to Rankine-Hugoniot relations. In the final phase, the value of the shock velocity is used to displace the shock point accordingly.

2.2 Multi-Block

In the present paper the validated CFD approach described in [6] and based on floating shock fitting is extended to overlapping multi-block mesh structures, that permit a suitable distribution of cells throughout the flow field while preserving a structured distribution and reasonable cell aspect ratio. The equations are treated at the block boundary as usual by introducing ghost nodes and by using bilinear interpolation, whereas a peculiar interpolation procedure must be introduced to correctly treat shocks.

Equations At boundaries some of the spatial derivatives cannot be computed by differences among values at nodes inside the block, because some signal will propagate from outside the block, except for the case of supersonic outflow. In general, signals propagating from outside can be obtained by enforcing the proper boundary condition, whereas in case of boundary between blocks the missing derivatives must be inferred from the solution in the neighboring block. A possible approach that preserves the accuracy of the method, while maintaining a general easy-to-implement coding, is that of using two rows of ghost nodes outside the block. In general, considering all of the boundaries of the block, two rings of ghost nodes are defined as shown in Fig. 2, whose position is obtained by smoothly prolonging outside of the block the inner node distribution.

To make possible that the solution at the ghost nodes is interpolated from that of the neighboring block, such nodes are required to lie inside the neighboring block, or in other words, they have to be such that there are no mesh-size holes between neighboring blocks. A possible configuration is drawn in Fig. 3, where only the ghost nodes of the left block (block 1) are displayed for the sake of clarity. The generic ghost point A lies inside a real cell BCED of the neighboring block (block 2) and the solution of the generic function $\mathcal{F}_\mathcal{A}$ at the point A is obtained by the bilinear interpolation

$$\mathcal{F}_A = \mathcal{A} + \mathcal{B}\hat{x}_A + \mathcal{C}\hat{y}_A + \mathcal{D}\hat{x}_A\hat{y}_A \tag{1}$$

where the coordinates of the node A in the frame of reference $(\hat{x}, \hat{y})$ having origin in B and axes directed as BC and BD are known, whereas $\mathcal{A}$, $\mathcal{B}$, $\mathcal{C}$ and $\mathcal{D}$ are computed by solving at each step and for each variable the 4×4 linear system obtained applying (1) in B, C, D and E, respectively.[1] Then, at the next iteration, the solution in the boundary nodes can be updated as done in the inner nodes because the missing derivatives are obtained from the solution in the inner and ghost nodes, the latter bringing information from the neighboring blocks.

[1] Bilinear interpolation is second order accurate as can be easily shown.

Shocks The use of ghost nodes allows introducing also ghost shock points needed to follow the general shock fitting procedure to transfer information between blocks. In particular, the introduction of ghost shock points is necessary to compute the local shock slope of real shock points at the block boundary and to avoid interpolating across shocks. The procedure used to find ghost shock points is illustrated by the example shown in Fig. 4, where the problem is to find the ghost shock points in the prolongation of block 1, when the shock position is known by the real shock points in block 2. A possible approach is to consider for each ghost cell $A_1A_2A_4A_3$ a quadrilateral B_1C_2ED, that is the smallest group of cells large enough to include the whole ghost cell. Once the quadrilateral is defined, it is possible to look for shock points along its boundaries, and, in case some are found, connect them by straight lines and find the intersection of these lines with the boundaries of the ghost cell. In the example of Fig. 4 the procedure finds the shock points S_1 and S_4, and the intersection of S_1S_4 with A_1A_3 and A_3A_4 provides the ghost shock points T_1 and T_2, respectively.

3 Results and Conclusions

A first example of application of the present technique is the computation of aerospike nozzle flow fields (Fig. 5). This axisymmetric rocket nozzle is of interest in aerospace propulsion since the external expansion adapts itself to the ambient pressure with the consequent performance benefit [7,11]. The resulting flow field is particularly interesting because of the presence of two supersonic afterbody regions, one behind the nozzle external shroud, the other behind the truncated spike, characterized by turbulent vortices, mixing layers and shocks. The study has been carried out by partitioning the whole computational domain in 22 blocks and considering a coarse grid made of 28828 cells and two further grids obtained by multiplying the number of cells in each direction by 1.5 and 2.0, respectively. Figure 5 shows that the three main fitted shocks smoothly cross the boundaries of blocks, some of which are overlapping, and the enlargement shows an example of crossing, together with real and ghost nodes and shock points. It is also interesting to observe that good grid convergence is obtained, as shown by the wall pressure behavior along the upper afterbody base (shroud, Fig. 6).

Another example of application is that of a dual-bell nozzle operating in overexpanded condition, that yields a large separated region (Fig. 7). The relevant flow field must be carefully studied because the main feature of this rocket nozzle is its capability of safely operating with separated flow. The computed solution is clearly unaffected by the decomposition in blocks as shown by the supersonic jet region, the subsonic recirculating region, and the fitted shock. The flow evolution during transition between operating modes with separated and attached flow is also of great practical interest for dual bell nozzles. To this goal an unsteady simulation obtained by doubling with a linear ramping chamber pressure in time has been carried out. Figure 8 shows the capability of the method of studying such flows by the expected smooth evolution of the shock (left), and the displacement of the separation point towards the nozzle exit as chamber pressure rises in time (right).

These results indicate that although more cumbersome to develop than common CFD methods widespread in commercial and scientific communities, shock fitting approaches retain their own merits that, if fully exploited, can yield the best result for some class of problems of interest, like flow transients with shocks moving through the flowfield. The use of multi-block techniques allows extending the number of applications to more and more complex problems. Further extensions are presently being achieved by parallelization of the code.

Acknowledgements: The applications of the technique discussed in the last section are being studied in the framework of European Space Agency (ESA) Technology Research Programme contracts. Partial support of Italian Ministry of University (MIUR) and Italian Space Agency (ASI) is also acknowledged. The author thanks Prof M Onofri and Dr R Paciorri for the many valuable discussions.

References

1. M. Arora, P.L. Roe: J. Comput. Phys. **130** (1997) pp. 25–40
2. G. Moretti: Computer & Fluids **31** (2002) pp. 719–723
3. G. Moretti, M. Abbett: AIAA J. **4** (1966) pp. 2136–2141
4. F. Marconi, M.D. Salas: Computer & Fluids **1** (1973) pp. 185–195
5. G. Moretti: Computer & Fluids **15** (1987) pp. 59–75
6. F. Nasuti, M. Onofri: AIAA J. **34** (1996) pp. 1428–1434
7. F. Nasuti, M. Onofri: J. Propulsion & Power **14** (1998) pp. 318–326
8. G. Moretti, F. Marconi, M. Onofri: M. 'Shock Boundary Layer Interactions Computed by a Shock Fitting Technique'. In: *Lecture Notes in Physics 414* (Springer, Berlin 1993) pp. 345–349
9. P.R. Spalart, S.R. Allmaras: La Recherche Aerospatiale **1** (1994) pp. 5–21
10. R. Paciorri, F. Nasuti, F. Sabetta: AIAA Paper 2001–3039 (AIAA, Washington 2001)
11. F. Nasuti, M. Onofri: J. Propulsion & Power **17** (2001) pp. 809–817

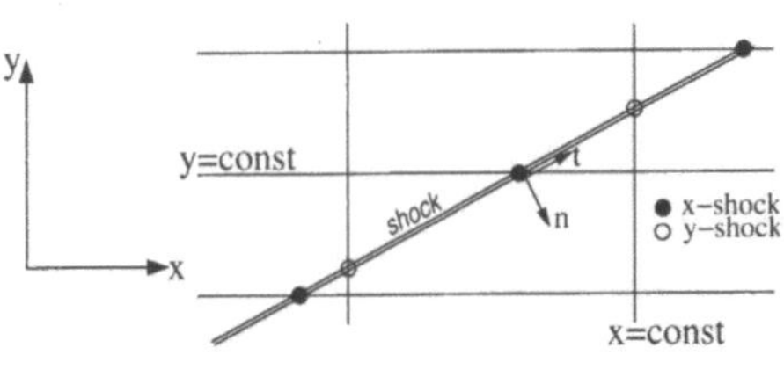

Fig. 1. Shock points

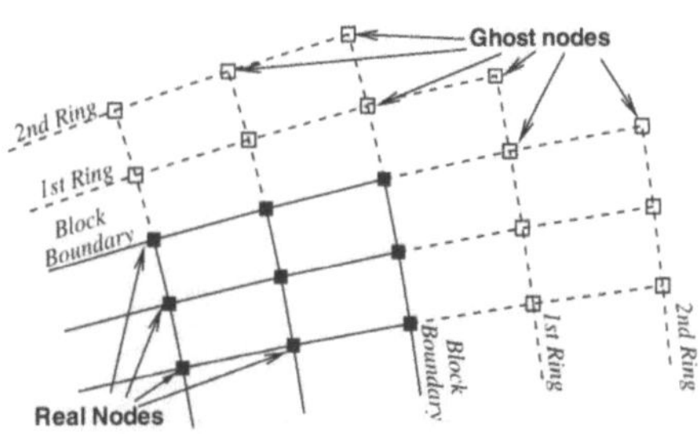

Fig. 2. Ghost Points

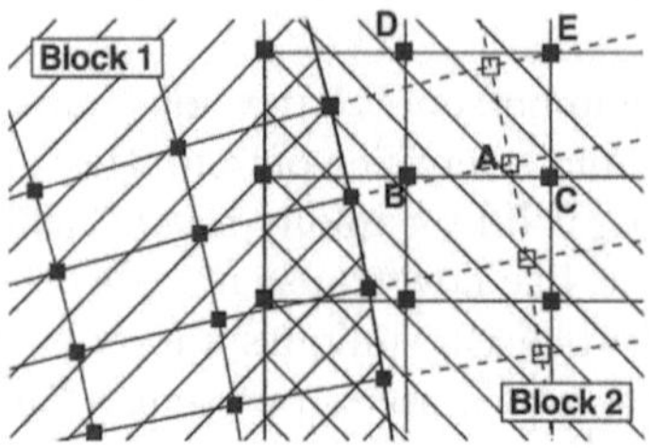

Fig. 3. Interpolation at ghost points

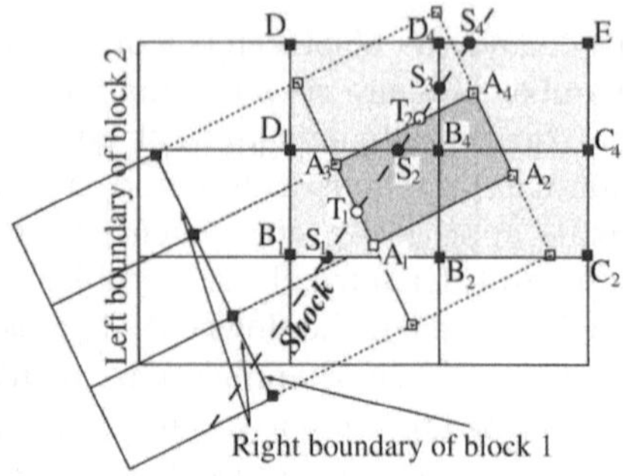

Fig. 4. Ghost shock points

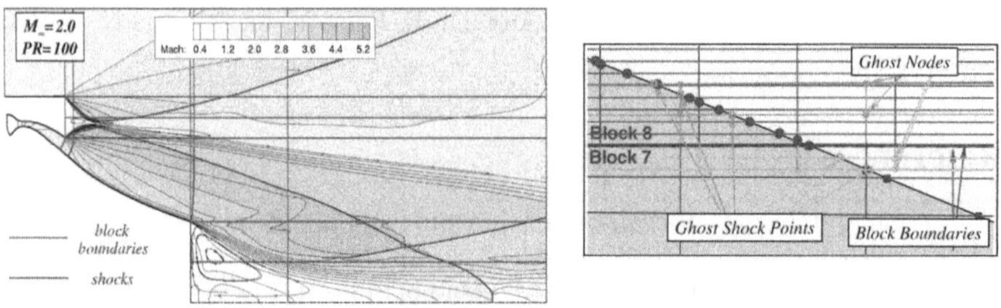

Fig. 5. Aerospike nozzle: Mach number contours and shock-fitting details across blocks

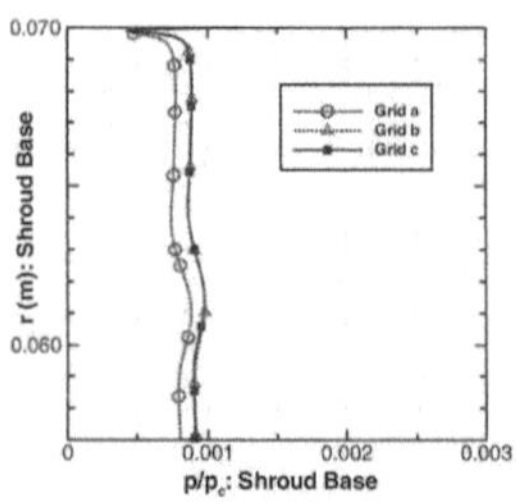

Fig. 6. Grid convergence for aerospike nozzle

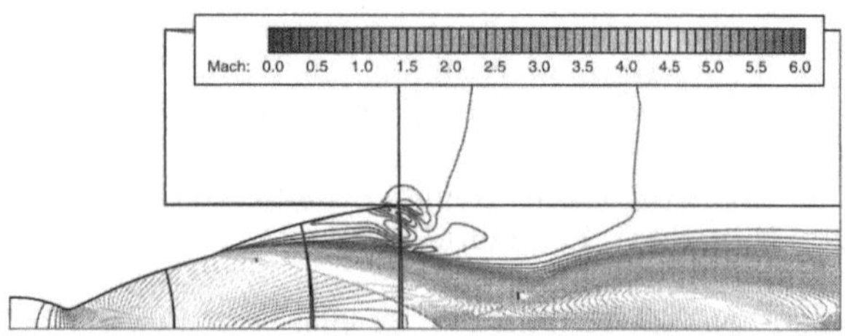

Fig. 7. Computed dual-bell nozzle flow field

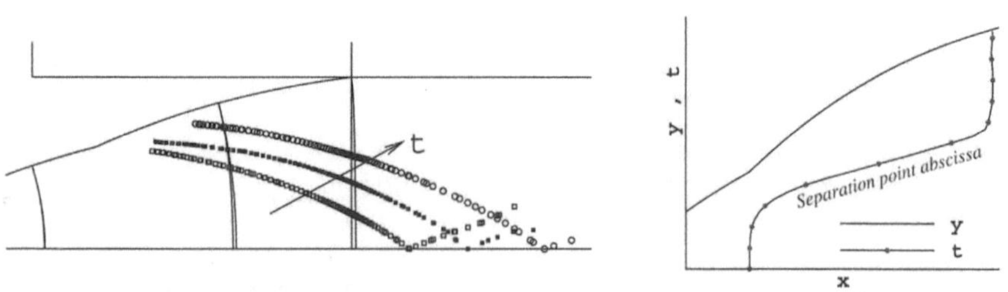

Fig. 8. Unsteady computation of dual bell nozzle: evolution of the shock (left) and of the separation point (right)

3D Numerical Simulation of a Planar Compressible Free Mixing Layer

Song Fu and Qibing Li

Department of Engineering Mechanics, Tsinghua University, Beijing 100084, PRC

Abstract. In this paper Xu's gas-kinetic BGK scheme is simplified based on the conservative constraint and then applied in the simulation of 3-D planar compressible mixing layer with $M_c = 0.4, 0.8$ and 1.2. The numerical results here provide the flow-field structure, the characteristic of velocity fluctuation. The velocity fluctuation intensity and shear stresses are in good agreement with experimental and existing numerical results. The normalized growth rate agrees well with experimental results.

1 Introduction

Turbulent mixing layer is commonly observed in various engineering applications such as combustion, propulsion and environmental flows. The rich flow physics thus attracted numerous experimental [1,6,7,14,16] and computational [5,9,10,13] studies till these days. These results showed that with increased convective Mach number M_c, the ratio of the free-stream velocity difference to the sum of the sound speeds, the mixing layer become highly three dimensional and the normalized growth rate decreased rapidly. The normalized turbulent quantities also decreased with the increasing M_c.

Numerical simulation of compressible mixing layer where discontinuity may exist requires numerical methods with good quality. Recently [19,21] developed a Gas-kinetic BGK scheme, which is based on the BGK model of Boltzmann equation, satisfies the entropy condition inherently and can be proved to be identical to Navier-Stokes equations at the condition of continuously varying flow field. For the discontinuous flow field, it is able to deliver better results than those based on the Riemann solver.

In present work Xu's BGK scheme [19] was simplified and applied to the numerical study of spatially evolving compressible mixing layers. Of particular interest is the compressibility effect on the mixing layer with variable M_c.

2 Numerical method–Simplified BGK Scheme

The 3D BGK-Boltzmann equation can be written as

$$f_t + uf_x + vf_y + wf_z = (g - f)/\tau \tag{1}$$

where f is the gas distribution function and g is the equilibrium state approached by f. They are both functions of space (x, y, z), time t, particle velocities (u, v, w), and internal variable ξ. τ is the particle collision time. The equilibrium state is a Maxwellian distribution. Since mass, momentum and energy are conserved during particle collisions, f and g satisfy the conservation constraint

$$\int (g-f)\psi_\alpha d\Xi = 0, \quad \alpha = 1,2,3,4,5, \tag{2}$$

at any point in space and time. Here $d\Xi = dudvdwd\xi$ is the volume element in the phase space, ψ_α is component of the vector of moments $\psi_\alpha = (1, u, v, w, (u^2 + v^2 + w^2 + \xi^2)/2)^T$.

Take (1) in constraint (2), one can get the finite volume formulation

$$\begin{aligned} Q_{ijk}^{n+1} = Q_{ijk}^{n} &+ \frac{1}{\Delta x}\int_{t^n}^{t^n+\Delta t} (F_{i-1/2,j,k} - F_{i+1/2,j,k})dt \\ &+ \frac{1}{\Delta y}\int_{t^n}^{t^n+\Delta t} (G_{i,j-1/2,k} - G_{i,j+1/2,k})dt \\ &+ \frac{1}{\Delta z}\int_{t^n}^{t^n+\Delta t} (H_{i,j,k-1/2} - H_{i,j,k+1/2})dt \end{aligned} \tag{3}$$

where the relation between macro quantities and the distribution function f is

$$Q = (\varrho, \varrho U, \varrho V, \varrho W, \varrho\varepsilon)^T = \int f\psi_\alpha d\Xi, \tag{4}$$

$$F = \int uf\psi_\alpha d\Xi, \quad G = \int vf\psi_\alpha d\Xi, \quad H = \int wf\psi_\alpha d\Xi. \tag{5}$$

If τ is a local constant, (1) has the integral solution

$$\begin{aligned} f(x,y,z,t,u,v,w,\xi) = &\frac{1}{\tau}\int_0^t g(x',y',z',t',u,v,w,\xi)e^{-(t-t')/\tau}dt' \\ &+ e^{-t/\tau} f_0(x-ut, y-vt, z-wt), \end{aligned} \tag{6}$$

where $x' = x - u(t-t'),\ y' = y - v(t-t'),\ z' = z - w(t-t')$ is the trajectory of a particle motion and f_0 is the initial gas distribution function at $t = 0$. If f_0 and g are obtained, one can easily get f through above equation, then the fluxes across the cell interface can be calculated through (5). To simplify the above method, a directional splitting method is adopted. The following derivation is in the x direction.

f_0 and g can be constructed as

$$\begin{aligned} f_0(x,0) = &g^l(1 + a^l x - \tau(a^l u + A^l))(1 - \mathrm{H}[\mathrm{x}]) \\ &+ g^r(1 + a^r x - \tau(a^r u + A^r))\mathrm{H}[\mathrm{x}], \end{aligned} \tag{7}$$

$$g(x,t) = g_0(1 + (1-\mathrm{H}[\mathrm{x}])\overline{a}^l x + \mathrm{H}[\mathrm{x}]\overline{a}^r x + \overline{A}t), \tag{8}$$

where subscripts l, r denote the local quantity at the left and right sides of the cell interface respectively. H[x] is the Heaviside function. g^l, g^r, g_0 are local and initial Maxwellians respectively. The terms a^l, a^r, A^l and A^r are from the Taylor expansion of a Maxwellian and have the similar form such as $a^l = a^l_\alpha \psi_\alpha$.

To determine $\overline{A}$, [19] using the integration of the conservation constraint (2) at $(x = 0)$ over the whole time step, which leads to a equation with eleven terms at the right side. In this article, a new simple method is given out through the differentiation of (2) at $(x = 0, t = 0)$,

$$\frac{\partial}{\partial t} \int [g(0,t) - f(0,t)] \psi_\alpha d\Xi \bigg|_{t=0} = 0, \tag{9}$$

from which a simplified equation

$$M^0_{\alpha\beta} \overline{A}_\beta = \frac{1}{\varrho_0} \int \left[a^l \mathrm{H}[\mathrm{u}] u g^l + a^r (1 - \mathrm{H}[\mathrm{u}]) u g^r \right] \psi_\alpha d\Xi \tag{10}$$

is obtained. Here only two terms need to be calculated in the present method, leading to a remarkable decrease in CPU time. The new scheme appears similar to previous scheme of [20] or SBGK scheme of [11], but takes the effect of nonequilibrium into account and shows more clear physical meanings.

Details of the scheme can be found in [12].

3 Computational Parameters

All variables are nondimensionallized by the initial vorticity thickness $\delta_\omega(0)$ the free stream parameters $\varrho_1, a_1, \varrho_1 a_1^2$. The initial momentum thickness $\delta_m(0) = 0.25$. The free stream sound speed is set to $a_1 = a_2 = 1$, and pressure $p_1 = p_2 = 1/\gamma$, where gas constant $\gamma = 1.4$. The kinetic viscosity $\nu = 0.001$ is used for all of the simulations.

Table 1 provides the computational parameters. The domain is defined as $(L_x \times L_y \times L_z)$ with $(N_x \times N_y \times N_z)$ grid points. The grid mesh is uniformly spaced in the streamwise x and spanwise z directions. In the transverse y direction the grid mesh is stretched outwards. The local cell stretching ratio, $A_j = \Delta y_{j+1}/\Delta y_j$, is a in the interior of the domain, $-L_y/6 \le y \le L_y/6$, and b in the other region. The minimal cell size in the y direction is 0.2 for all the cases.

At the inflow boundary, the flow quantities are prescribed, but a broadband forcing, whose fluctuation intensity at $y = 0$ is $0.03U_c$, is also superimposed [18]. The periodic condition at spanwise boundaries, Thompson's nonreflecting

Table 1. Computational parameters

Case	M_1	M_2	M_c	Re	$L_x \times L_y \times L_z$	$N_x \times N_y \times N_z$	a	b
3D1	1.9	1.1	0.4	200	350 × 120 × 30	875 × 200 × 80	1.015	1.04
3D2	2.9	1.3	0.8	400	350 × 120 × 30	875 × 140 × 80	1.04	1.18
3D3	3.9	1.1	1.2	600	350 × 120 × 30	875 × 200 × 80	1.015	1.04

boundary conditions at the outflow and two lateral side boundaries are employed. For the outflow boundary, the pressure correction is applied [15],

The simulation is started with an initialized flow-field in which the transverse and spanwise velocity are fixed by zero value and the streamwise velocity profiles are fixed by the hyperbolic tangent function, i.e.

$$u = 0.5(U_1 + U_2) - 0.5(U_1 - U_2)\tanh\left(0.5y/\delta_m(0)\right) \tag{11}$$

where U_1 and U_2 are the velocities of the high- and low-speed free streams, respectively. $\delta_m(0)$ is the momentum thickness of the shear layer that is unchanged at $x = 0$. The density and pressure fields are uniform throughout the flow field.

4 Results

Figure 1 shows the isosurface of Q, the second invariant of $\nabla \boldsymbol{u}$, in the mixing layer, from which we can easily find the large structure in the flow field. The flow lose its stability upstream, and then vortex occur, roll up and interact with each other while moving downstream. The flow instability is dominated by the spanwise rollers when $M_c = 0.4$, similar to the incompressible results in [8], while oblique structures play a much more important role when $M_c = 1.2$. From the isosurface of Q, one can also find the rib structures between rollers in low M_c mixing layer. But complex Mach waves can not be found in high M_c case which can be seen easily from the isosurface of pressure.

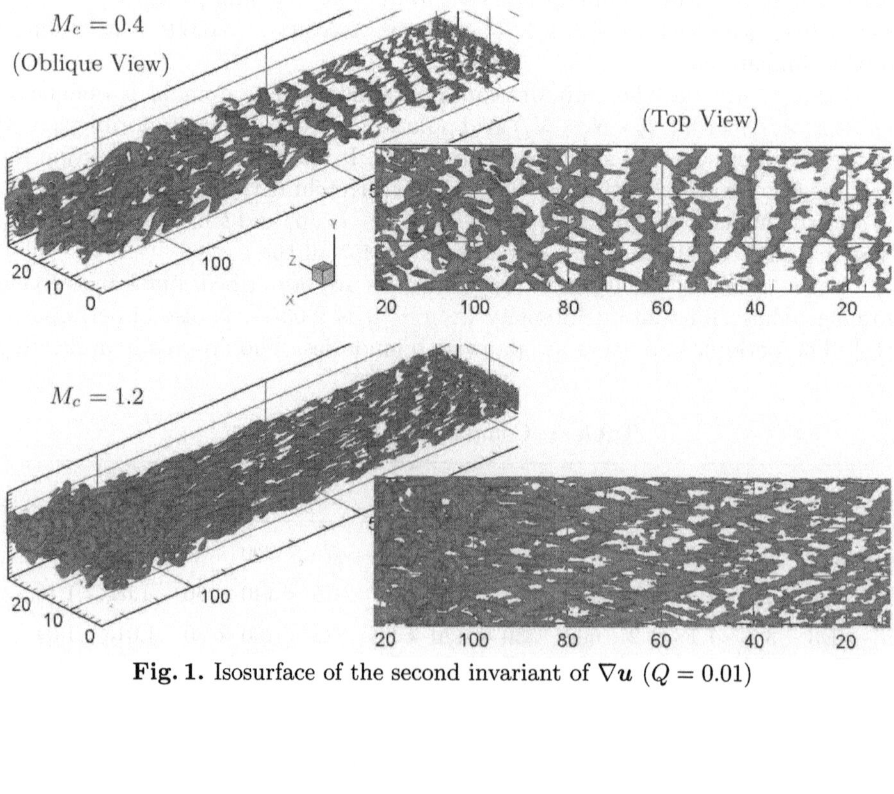

Fig. 1. Isosurface of the second invariant of $\nabla \boldsymbol{u}$ ($Q = 0.01$)

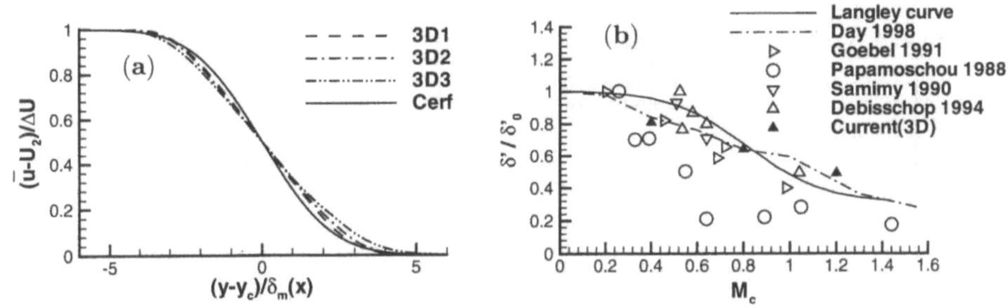

Fig. 2. Profiles of mean streamwise velocity (**a**), growth rate of mixing layers (**b**)

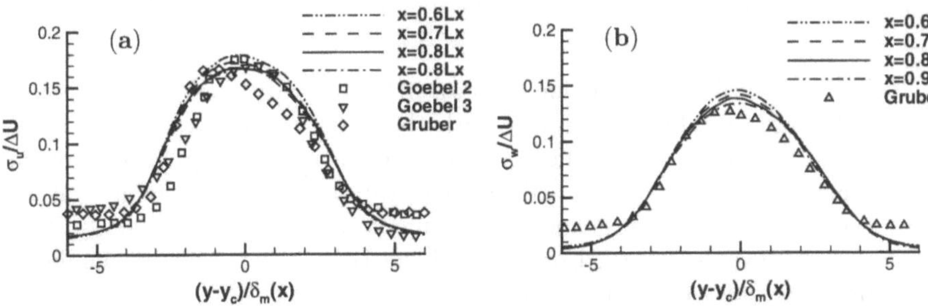

Fig. 3. Profiles of velocity fluctuation intensity: (**a**) streamwise fluctuation, (**b**) spanwise fluctuation, where σ is the r.m.s. quality of the fluctuation ($M_c = 0.8$)

Figure 2(a) shows the profiles of the mean streamwise velocity at the full developing region downstream. In low M_c mixing layer, the profile agrees very well with the error function which is the first order approximation to the mean streamwise velocity in incompressible case. The distribution becomes more and more gently with increasing M_c, when scaled by the thickness of the layer, which means rapid velocity changes near the layer boundary in high M_c flow. The normalized growth rate decreases with increasing M_c (Fig. 2(b)), which agrees well with experimental data ([3,6,14,17]) and theoretic curve ([2]).

Figure 3 shows the profiles of the streamwise and spanwise fluctuation intensity where "Goebel 2" and "Goebel 3" represent the results of [6] with $M_c = 0.46, 0.8$ respectively. "Gruber" is from [7] where M_c is 0.8. One can also find the similarity easily. Comparing with the experimental results, the velocity fluctuation intensities agree well (Fig.3(a,b);4(a)). Figure 4(b) shows that the ratio between the shear stress and the turbulent energy keeps constant (≈ 0.35)in the inner regime of the mixing layer, which agree well with the experimental result in [3]. The increase with M_c in the outer regime maybe come from the complex Mach waves outside the mixing layer.

5 Conclusion

In present paper Xu's gas-kinetic BGK scheme is simplified and applied in the simulation of 3D planar compressible mixing layer with $M_c = 0.4, 0.8$ and 1.2.

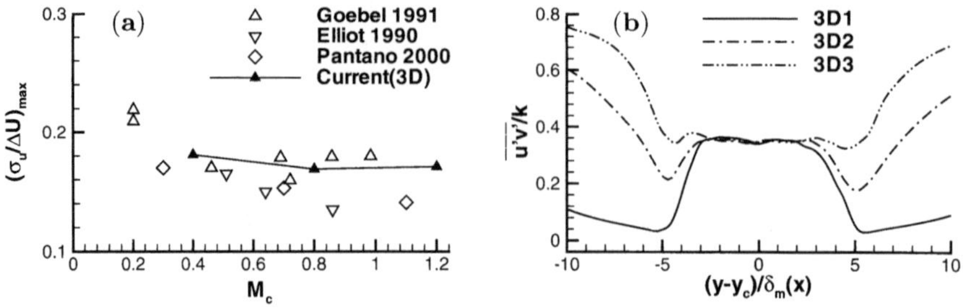

Fig. 4. Maximum of fluctuation versus M_c: (**a**) $\sigma_u/\Delta U$, (**b**) $\sigma_v/\Delta U$ and (**c**) $\overline{u'v'}/(\Delta U)^2$, and the distribution of the ratio in different cases: (**d**) $\overline{u'v'}/k$

The numerical results here provide the flow-field structure, the characteristic of velocity fluctuation which agree well with the experimental results and other numerical study. With increasing M_c, the streamwise velocity fluctuation intensity varies little; the normalized growth rate decrease remarkable; but the ratio between the shear stress and the turbulent energy keeps almost constant. Present study reveals the good property of the BGK scheme in the simulation of compressible flows that ensures its wide applications.

References

1. G.L. Brown, A. Roshko: J. Fluid Mech. **64**, 775 (1974)
2. M.J. Day, W.C. Reynolds, N.N. Mansour: Phys. Fluids **10**, 993 (1998)
3. J.R. Debisschop, O. Chambres, J.P. Bonnet: Exp. Therm. Fluid Sci. **9**, 147 (1994)
4. J.B. Freund, P. Moin, S.K. Lele: J. Fluid Mech. **421**, 229 (2000)
5. N.J. Georgiadis, J.I.D. Alexander, E. Reshotko: AIAA Paper 2001-0289 (2001)
6. S.G. Goebel, J.C. Dutton: AIAA J. **29**, 538 (1991)
7. M.R. Gruber, N.L. Messersmith, J.C. Dutton: AIAA J. **31** (11), 2061 (1993)
8. S. Iwase, M. Tanahashi, T. Miyauchi: Large Scale Structure and Coherent Fine Scale Eddies in High Reynolds Number Turbulent Mixing Layers. In *Turbulence and Shear Flow Phenomena (2nd international symposium)*, ed. by E. Lindborg, A. Johansson, J. Eaton et al. (KTH, Stockholm 2001) pp. 131–136
9. L.J. Leep, J.C. Dutton, R.F. Burr: AIAA J. **31**, 2039 (1993)
10. S.K. Lele: AIAA Paper 89-0374
11. Y.S. Lian, R.Q. Wang: Chinese J. Comput. Phys. **16** (3), 275 (1999)
12. Q.B. Li: Numerical Study of Compressible Mixing Layer with BGK Scheme. PhD Dissertation, Tsinghua University, Beijing (2002)
13. C. Pantano, S. Sarkar: J. Fluid Mech. **451**, 329 (2002)
14. D. Papamoschou, A. Roshko: J. Fluid Mech. **197**, 453 (1988)
15. T.J. Poinsot, S.K. Lele: J. comput. Phys. **101**, 104 (1992)
16. R. Rossmann, M.G. Mungal, R.K. Hanson: J. Turbulence **3** (009) (2002)
17. M. Samimy, G.S. Elloit: AIAA J. **28**, 439 (1990)
18. S. Stanley, S. Sarkar: Theoret. Comput. Fluid Dynamics **9**, 121 (1997)
19. K. Xu: J. Comput. Phys. **171** (1), 289 (2001)
20. K. Xu, A. Jameson: AIAA Paper 95-1973 (1995)
21. K. Xu, K.H. Prendergast: J. Comput. Phys. **114** (1), 9 (1994)

A Blunt Body Analogy for Bow Shock Characteristics in Front of a Supersonic Liquid Jet

Sam Zakrzewski, Kulachate Pianthong, Masud Behnia, and Brian E. Milton

School of Mechanical and Manufacturing Engineering, University of New South Wales, Sydney, NSW 2052, Australia

Abstract. The unique characteristics of high-speed liquid fuel jets suggest that they have the potential to be applied to high performance diesel engines, SCRAM jets and other combustion applications where small combustion residence times are required. When a liquid jet is injected at supersonic speed, with respect to the surrounding medium, a bow shock wave is produced. The ability of such jets to increase atomisation rates and decrease ignition times has led to interest in the liquid jet head - bow shock region. The use of CFD to study this region is advantageous as experimental studies are unable to take many measurements without being intrusive. This paper investigates the effect of the dynamic jet shape on the bow shock stand-off distance and shape using contemporary CFD methods. The shock wave characteristics in the shock layer are also presented.

1 Introduction

It has been hypothesised that increasing fuel injection speeds to a supersonic level (with respect to air or other oxidant) can positively aid the combustion process. In order to fully understand this it is essential to comprehend the shock wave characteristics and dynamics around the leading edge of a high-speed liquid jet.

The description 'high-speed liquid jet' in this case refers to a jet of liquid that is ejected from a nozzle orifice into air. The speed of the jet is above the supersonic range with respect to the air. The air is considered quiescent and atmospheric. These initial setup criterion produce unique characteristics as the jet is ejected from the nozzle orifice into the atmospheric medium.

The first and most obvious characteristic is the shock wave that precedes the liquid jet head. When the quiescent air is accelerated up to the jet's supersonic velocity a shock wave is formed. The shock is either attached or detached from the head depending on the speed of the jet. The shock gives rise to some interesting jet dynamics as shown in Fig. 1. Other characteristics are the increased shear layer intensity that improves the atomisation process and the unique jet head shape, which is dictated by the conditions in the shock layer [1].

The dynamics of a high-speed liquid jet are such that their future introduction into high performance engines is foreseeable. They are able to increase atomisation rates by the intense free shear layers and decrease ignition time. The latter is achieved by the increase in temperature and pressure in the liquid head

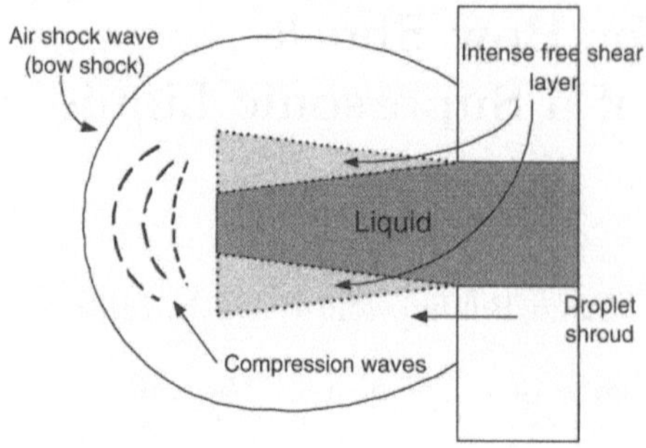

Fig. 1. Characteristics of a high-speed liquid jet

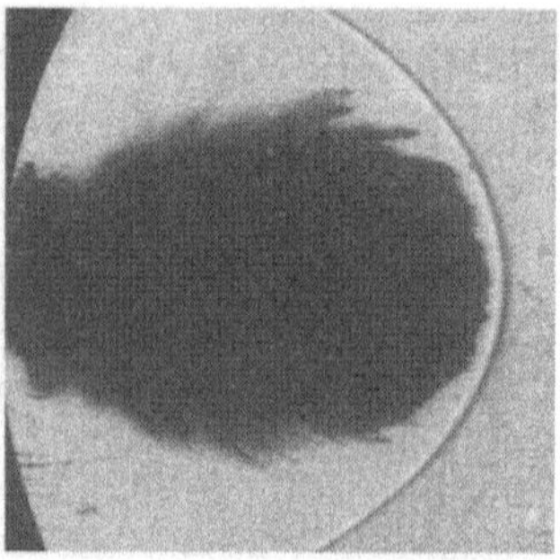

Fig. 2. Experimental visualisation of a high-speed liquid jet

caused by the preceding shock wave. Possible applications include Supersonic Combustion RAM (SCRAM) jets, high performance diesel engines and the improvement of conventional combustion technologies for emissions reduction [2]. Some negative aspects of the high-speed liquid jets are that they are not easily created, the overall energy consumption may be increased by their creation and there is a possibility of internal engine damage.

2 Analogy

It is known that a body travelling at supersonic speed produces a shock wave, in the medium it is travelling in, ahead of the body. Examples of this are shown from early 20th century experiments in supersonic flows [3], through to state-of-the-art injection systems in high-speed liquids jets. The complicated nature of the liquid jet has made it hard to predict shock strength, location and shape using experimental and numerical methods. It is reasonable to suggest that there may be some analogous characteristics of a solid body travelling in a supersonic flow to a high-speed liquid jet. This reasoning was initially studied by Bloor [4]. His findings were limited by the numerical techniques of the time.

When injected into quiescent air, a liquid jet takes on a shape that is influenced by the characteristics of the surrounding gas. The shape of the jet is dynamic as it moves through the gas. Although these effects are transient, an analogy between a solid blunt body and the jet head shape at a specific time frame can be made. The liquid jet shape and its geometry, such as that shown in Fig. 2 obtained by experiment, can be transferred to an axi-symmetrical computational grid on the assumption that the liquid-gas interface will disrupt the flow similarly to a blunt body. The primary simplification is that the jet core shape is assumed non-porous (i.e. a wall). The jet speed is represented by the air velocity over the blunt body. The empirical nature of this comparative analogy is due to the inability to take quantitative experimental readings without being intrusive.

3 Experimental and Computational Procedure

High-speed liquid jets have been experimentally generated for this study using the momentum exchange by impact technique [5]. The nozzle was designed to give a liquid jet velocity of approximately Mach 1.8. Visualisation was undertaken using a shadowgraph apparatus [6]. The experimental results at these speeds show that the jet is axi-symmetrical and that the shock wave stands away from the jet head.

The computational model employs the general purpose code FLUENT. It incorporates a coupled, time-marching algorithm for compressible flow with shock capturing capabilities [7]. To capture the effects of compressibility and the shock wave, the ideal gas law is used for a calorifically perfect gas. To initialise the solution, the whole domain has a free stream Mach number equal to the jet injection speed. An axi-symmetrical boundary is used to reduce computational time.

As the Mach number of the flow is increased, the shock layer decreases and the shock wave occurs closer to the body. The pressure, temperature and density in the shock layers increase as indicated by the Rankine-Hugoniot relationships. There is a decrease in area for a larger property gradient. Pre-solution node-bunching and post-solution grid adaption was undertaken to ensure that the high gradients are fully resolved and a grid dependent solution was attained.

The code was validated between Mach 1 and 10 [8] using the standard supersonic results of shock stand-off distance over a sphere. This fully illustrates the applicability of this numerical scheme.

4 Results and Discussion

4.1 Comparisons with Previous Work

Initial comparisons were made against the low-speed ($470\,\mathrm{ms}^{-1}$) jet interferometry visualisation produced by Shi [9]. They suggest a typical conical, subsonic jet shape with a preceding shock wave. When comparing these to a relevant blunt body grid configuration the shock stand-off distance is over estimated. This is attributed to the presences of acute corners that initiate the formation of expansion waves. The unsatisfactory comparison shows that a traditional conical jet shape cannot be assumed.

4.2 Comparisons with Current Work

As the assumed jet shape is important in predicting a realistic shock location and shape, experimental high resolution liquid jet photographs were used to review the jet shape geometry. An example is shown in Fig. 2. The realistic shape suggested by visualisation is a hemispherical, bulbous, mushroom shape. This is confirmed by the jet dynamics first described by Bloor [4]. The actual physical jet shape is hard to determine from visualisation due to the droplet cloud around it.

Fig. 3. Comparison of Mach 1.8 jet

Figure 3 shows a typical experimental, numerical comparison. The experimental section shows a water jet travelling at Mach 1.8. The computational section illustrates the density contours of a blunt body in air of free stream Mach number 1.8. The comparison shows that the computational shock stand-off predicted is in an acceptable range even though the ambiguous placement of the jet head means that the actual interface between the liquid and the gas is open to interpretation. The blunt body stagnation point properties are comparable to those determined by normal shock relations. For a Mach number of 1.8 the static temperature is 450 K and the density is $2.94\,\mathrm{kg\,m^{-3}}$.

An inadequacy in the comparison between the experimental and computational results is the oblique shock angle away from the geometry, the experimental results having a more acute angle than the computational data.

This shortfall may be caused by the presence of downstream cavities also described in the work of Bloor [4]. The incorporation of these cavities is modelled and compared in Fig. 4. The figure shows that the oblique shock angle is slightly reduced but not to a realistic level. The remaining discrepancy can be attributed to assumptions made by the modelling analogy. Firstly, the computational results show a steady state solution. Physically the jet is moving; therefore the shock is a transient phenomenon, which is evolving as the jet geometry changes. Secondly, the interface between the liquid and the gas is not a solid wall. As the liquid jet is atomised, the core becomes increasingly porous so that some airflow may pass through the quasi-porous mediums. The effect of transience and variable body density in full liquid jets simulations are shown to provide some answers [10].

Fig. 4. Comparison of jet with cavities

4.3 Higher Velocity Studies

As an extension of this work, the jet speed has been increased to a hypersonic level. At these speeds the shock wave is attached to the jet head and further influences its shape. Figure 5 shows the comparison of a liquid diesel jet injected into quiescent atmospheric air at Mach 5.5 [5] and the density contours from a CFD simulation for a geometrically similar body in a free stream with a velocity of 2000 ms^{-1}. The shock angle and stand-off distance show good correspondence.

5 Conclusions

A knowledge base for the understanding of high-speed liquid jets is a crucial stepping-stone for the development and implementation of next generation air-breathing propulsion systems. The current CFD and experimental studies are aimed at expanding that base.

The jet has been generated experimentally by the impact momentum transfer method and subsequently captured by a visualisation technique. The numerical solver used here has proven able to solve blunt body, axi-symmetrical flow with success when compared to previous numerical and experimental findings. The use of grid improvement techniques has allowed for grid independent solutions and the ability to resolve large gradients and discontinuities.

Overall, the blunt body analogy performs relatively well for shock layer data. This method shows a good correlation between experimental and computational

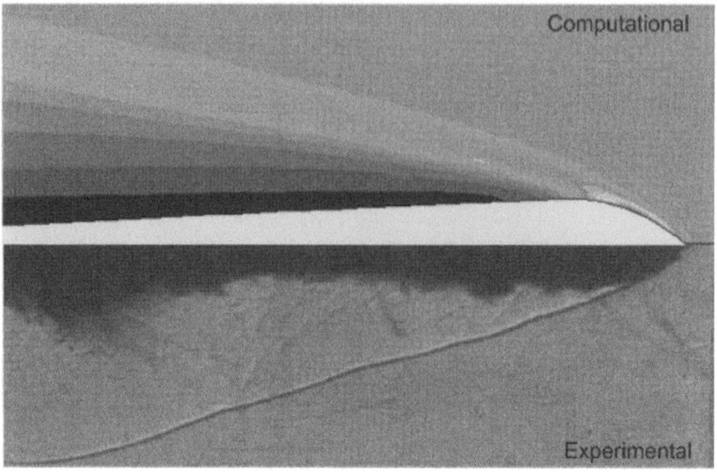

Fig. 5. Comparison of jet at hypersonic velocity

results. The shock strength, location and characteristics are represented well. Shock angles away from the body at low Mach numbers are less realistic. This analogy has helped to develop a more realistic idea of the conditions between the liquid jet head and the bow shock wave.

References

1. J.E. Field, M.B. Lesser: Proceedings of the Royal Society of London. **A, 357**, (1977) pp. 143-162
2. T.F Su, P.V. Farrell: Atomization and Sprays. **8**, (1998) pp. 83-107
3. J.D. Anderson: *Modern Compressible Flow with Historical Perspective*, 2nd edn. (McGraw-Hill, 1990)
4. M.I.G. Bloor: Journal of Fluid Mechanics. **84**, 2 (1978) pp. 375-384
5. K. Pianthong, S. Zakrzewski, M. Behnia, B.E. Milton: Shock Waves. **11**, 6 (2002) pp. 457-466
6. K. Pianthong, S. Zakrzewski, M. Behnia, B.E. Milton: 'Visualization of Supersonic Liquid Jets Using Shadowgraph Technique'. In: *14th Australasian Fluid Mechanics Conference. Adelaide, Australia, December, 2001*
7. Fluent Inc.: *Fluent 5.2 Manual* (Fluent Inc., Lebanon, 1999)
8. S. Zakrzewski: A Numerical and Experimental Investigation of High-Speed Liquid Jets - Their Characteristics and Dynamics. PhD Thesis, UNSW, Sydney, Australia (2002)
9. H.H. Shi: Study of Hypersonic Liquid Jets. PhD Thesis, Tohoku University, Sendai, Japan (1994)
10. S. Zakrzewski, K. Pianthong, M. Behnia, B.E. Milton: 'Predictions of Bow Shock Characteristics in Front of a High-Speed Liquid Jets Using Current CFD Methods'. In: *23rd International Symposium on Shock Waves (ISSW23) Conference. Fort Worth, Texas, USA, July, 2001*

Spectral (Finite) Volume Method for the One Dimensional Euler Equations

Z.J. Wang[1] and Yen Liu[2]

[1] Michigan State University, East Lansing, MI 48824, U.S.A., zjw@egr.msu.edu
[2] NASA Ames Research Center, Moffett Field, CA 94035, U.S.A., liu@nas.nasa.gov

Abstract. In this paper, the spectral volume (SV) method is extended to one-dimensional systems — the quasi-1D Euler equations. The focus of this paper is to study the performance of the SV method on non-linear systems, and to verify that high order solution accuracy can be achieved for hyperbolic systems of conservation laws. Both steady state and time accurate problems are used to demonstrate the overall capability of the SV method.

1 Introduction

The spectral volume (SV) method is a new efficient finite volume method for unstructured grids recently developed by Wang, Wang and Liu in [11][12]. The SV method belongs to a general class of Godunov-type finite volume method [5], which has been under development for several decades, and has become the state-of-the-art for the numerical solution of hyperbolic conservation laws. For a more detailed review of the literature on the Godunov-type method, refer to [11][12], and the references therein. Some of the widely used numerical methods for conservation laws such as the essentially non-oscillatory (ENO)[6][1], weighted ENO [7], and discontinuous Galerkin (DG) [3][4] methods also belong to the Godunov-type methods. Similar to the Godunov method, the SV method has two key components. One is data reconstruction, and the other is the (approximate) Riemann solver [8]. What distinguishes the SV method from the k-exact finite volume (FV) method [2] is the data reconstruction. Instead of using a (large) stencil of neighboring cells to perform a high-order polynomial reconstruction, the unstructured grid cell — called a spectral volume — is partitioned into a "structured" set of sub-cells called control volumes (CVs), and cell-averages on these sub-cells are then the degrees-of-freedom (DOFs). These DOFs are used to perform a high-order polynomial reconstruction inside the SV. All the spectral volumes are partitioned in a geometrically similar manner, and thus a universal resonstruction is obtained for all simplexes. Next, the DOFs are updated to high-order accuracy using the usual Godunov method. Numerical tests with scalar conservation laws in both 1D and 2D have verified that the SV method is indeed highly accurate, conservative, and geometrically flexible [12]. In this paper, we further extend the SV method to one-dimensional systems.

2 Spetral Volume Method for the Quasi-1D Eular Equations

2.1 Basic Formulation

The unsteady quasi-1D Euler equation in conservative form can be written as

$$\frac{\partial Q}{\partial t} + \frac{\partial F}{\partial x} = G, \tag{1}$$

where Q is the vector of conserved variables, F is the inviscid flux vector, and G is the vector for the source terms. Assume we solve (1) in the computational domain of $[a, b]$ subject to proper initial and boundary conditions. Given a partition of the domain $[a, b]$, $\{x_{i+1/2}\}_{i=0}^{N}$, the domain is then divided into N non-overlapping spectral volumes (SVs)

$$[a, b] = \bigcup_{i=1}^{N} S_i, \quad S_i = [x_{i-1/2}, x_{i+1/2}],$$

with $x_{1/2} = a$, and $x_{N+1/2} = b$. Let $h_i = x_{i+1/2} - x_{i-1/2}$, and denote the quantity $\max_{1 \le i \le N} h_i$ by h. Given a desired order of accuracy k for (1), each spectral volume S_i is then partitioned into k control volumes (CVs) using the following partitioning $\{x_{i,j+1/2}\}_{j=0}^{k}$ with $x_{1,1/2} = x_{i-1/2}$ and $x_{i,k+1/2} = x_{i+1/2}$. The j_{th} CV of S_i is then $C_{i,j} = (x_{i,j-1/2}, x_{i,j+1/2})$, $j = 1, \ldots, k$. Denote $h_{i,j} = x_{i,j+1/2} - x_{i,j-1/2}$. Let q denote any of the conserved variables in Q. The cell-averaged conserved variable q at time t for control volume $C_{i,j}$ is defined as

$$\overline{q}_{i,j}(t) = \frac{\int_{x_{i,j-1/2}}^{x_{i,j+1/2}} q(x,t)\mathrm{d}x}{h_{i,j}}, \quad j = 0, \ldots, k, \quad i = 1, \ldots, N.$$

Given the cell-averaged conserved variables for all the CVs in S_i, a polynomial $p_i(x) \in P^{k-1}$ (the space of polynomials of degree at most $k-1$ can be reconstructed such that it is a *k-th* order accurate approximation to the function $q(x)$ inside S_i:

$$p_i(x) = q(x) + O(h^k), \quad x \in S_i, \quad i = 1, \ldots, N. \tag{2}$$

In particular, this gives approximations to the function $q(x)$ at the CV boundaries

$$q_{i,j+1/2} \equiv p_i(x_{i,j+1/2}) = q(x_{i,j+1/2}) + O(h^k), \quad j = 0, \ldots, k, \quad i = 1, \ldots, N. \tag{3}$$

The reconstruction can be conveniently expressed as

$$p_i(x) = \sum_{j=1}^{k} L_j(x)\overline{q}_{i,j}, \tag{4}$$

where $L_j(x) \in P^{k-1}$ are the "shape" functions which satisfy

$$\frac{\int_{x_{i,j-1/2}}^{x_{i,j+1/2}} L_m(x)\mathrm{d}x}{h_{i,j}} = \delta_{jm}. \tag{5}$$

The high-order reconstruction is then used to generate high-order updates for the *CVs*. Integrating (1) in $C_{i,j}$, we obtain the following intergral equation for the *CV*-averaged mean

$$\frac{d\overline{Q}_{i,j}}{dt} h_{i,j} + (F_{i,j+1/2} - F_{i,j-1/2}) = \int_{x_{i,j-1/2}}^{x_{i,j+1/2}} G dx, \tag{6}$$

where $\overline{Q}_{i,j}$ is the vector of the *CV*-averaged conserved variables in $C_{i,j}$. The flux vectors are replaced with a Riemann flux. It is the Riemann solver which introduces "upwinding", and dissipation into the *SV* method such that the *SV* method is not only high-order accurate, but also stable. In this paper, we employ Roe approximate Riemann solver. Equation (6) is then intergrated in time with a third order TVD Runge–Kutta scheme [9]

2.2 TVB and TVD Limiters

For the non-linear Euler equations, it is necessary to perform data limiting to maintain stability if the solution contains discontinuities. TVD limiters enforce strict monotinicity by sacrificing solution accuracy at local extrema, while TVB [10] limiters relax the monotonicity requirement to achieve uniform accuracy away from discontinuities. In this paper, both limiters are presented in a uniform TVB limiter form, with TVD being a special case. And we choose to limit the conserved variables. To this end, we first express the reconstruction in the following form

$$Q^L_{i,j+1/2} = \overline{Q}_{i,j} + \Delta Q^+_{i,j}, \qquad Q^R_{i,j-1/2} = \overline{Q}_{i,j} - \Delta Q^-_{i,j}, \tag{7}$$

Then the quantities $\Delta Q^+_{i,j}$ and $\Delta Q^-_{i,j}$ are limited in the following manner to achieve TVB

$$\Delta\widetilde{Q}^+_{i,j} = \begin{cases} \Delta Q^+_{i,j} & \text{if } |\, \Delta Q^+_{i,j} \,| \le M h^2_{i,j} \\ minmod(\Delta Q^+_{i,j}, \overline{Q}_{i,j+1} - \overline{Q}_{i,j}, \overline{Q}_{i,j} - \overline{Q}_{i,j-1}) & \text{otherwise,} \end{cases} \tag{8}$$

$$\Delta\widetilde{Q}^-_{i,j} = \begin{cases} \Delta Q^-_{i,j} & \text{if } |\, \Delta Q^-_{i,j} \,| \le M h^2_{i,j} \\ minmod(\Delta Q^-_{i,j}, \overline{Q}_{i,j+1} - \overline{Q}_{i,j}, \overline{Q}_{i,j} - \overline{Q}_{i,j-1}) & \text{otherwise,} \end{cases} \tag{9}$$

where M represents some measure of the second derivative of the solution, and the *minmod* function is defined as:

$$minmod(x,y,z) = \begin{cases} s \text{ MIN}(|\, x \,|, |\, y \,|, |\, z \,|) \text{ if } s = sign(x) = sign(y) = sign(z) \\ 0 \qquad\qquad\qquad\qquad\qquad \text{otherwise} \end{cases}$$

In the case of $M = 0$, the limiter becomes TVD.
The availability of cell-averaged data on the *CVs* inside a *SV* makes this *CV*-based data limiting possible, whereas in the *DG* method, one can only do an element based data limiting. Due to the increased local resolution, the *SV* method is expected to have better resolutions for discontinuities than the *DG* method.

3 Numerical Tests

3.1 Accuracy Study with a Steady 1D Flow through a Nozzle

In this accuracy study, we compute the numerical order of accuracy of the SV method using a steady subsonic quasi-1D flow through a converging nozzle with the following area variation

$$A(x) = 1.5 - 0.5tanh(x) \quad -5 \leq x \leq 5.$$

The flow condition is so defined that an exit Mach number of 0.8 is produced. An analytical solution can be computed based on the iso-entropic flow assumption. To compute the errors in the numerical solution, the CV-averaged densities are used to reconstruct the densities at the SV boundaries, which are then compared with the analytical solutions to determine the L_1 and L_∞ error norms. The SVs were partitioned with the Gauss–Lobatto points. The computed L_1 and L_∞ errors with SV schemes of various orders of accuracy on different grids are plotted in Figures.1 and 2. Note that the expected orders of accuracy for all the SV schemes are achieved in both the L_1 and L_∞ norms in this case.

3.2 Shock-Acoustic Wave Interaction Problem

This case was suggested by Shu and Osher to demonstrate the advantages of high-order schemes in capturing both discontinuities and complex smooth structures, such as those occurring in shock-acoustic wave interactions. The initial conditions are

$$Q(x,0) = \begin{cases} Q_L, & x \leq -4, \\ Q_R, & x \geq -4, \end{cases}$$

where

$$\{\rho_L, u_L, P_L\} = \{3.857143, 2.629369, 10.333333\}$$
$$\{\rho_R, u_R, P_R\} = \{1 + 0.2sin(5x), 0, 1.0\}$$

A converged solution using a second-order MUSCL scheme on a grid of 3,200 cells is used as the "exact" solution. Two uniform grids with 200 and 400 SVs were adopted. In the first set of tests, a TVD limiter was used, and the results are plotted in Figures 3 and 4. On the coarse mesh, the second order SV scheme heavily smeared the complex smooth structure, while the fourth-order SV scheme gave excellent results. The solution computed with the third-order SVscheme is between the results of second and fourth-order SV schemes. On the fine mesh, both the third and fourth-order schemes yielded excellent results, while the second-order scheme still smeared some of the density waves. We also tested the SV schemes with the same DOFs, and a TVB limiter using $M = 250$, and the results are shown in Figure 5. Note that the smooth structures are much better resolved with the TVB limiter than with a TVD limiter. The results computed with the third and fourth-order SV schemes are excellent.

4 Conclusions

The spectral volume method has been successfully extended to one-dimensional hyperbolic systems of conservation laws. The extension to the system has been performed in a component-wise manner. Therefore, the mean conserved variables at the *CVs* (sub-cells) are used in the polynomial reconstruction. Both TVD and TVB limiters are also implemented in a component-wise manner. This approach is very efficient, and has been shown to produce very good numerical results for both steady and unsteady, smooth and non-smooth flow problems. An accuracy study with a steady subsonic flow through a converging nozzle has verified that the designed high-order accuracy can be achieved in the system setting. The *SV* method performed very well for the unsteady cases with discontinuities, and both discontinuities and complex smooth structures. In particular, TVB limiters performed better than TVD limiters for smooth structures. The extension to 2D Euler equations is now under way, and will be reported in a future publication.

References

1. R. Abgrall, *J. Comput. Phys.*, **114**, 45-58(1994).
2. T.J. Barth and P.O. Frederichson, AIAA Paper No. 90-0013, 1990.
3. B. Cockburn and C.W. Shu, *Mathematics of Computation*, **52**, 411-435 (1989).
4. B. Cockburn, S.Y. Lin and C.W. Shu, *J. Comput. Phys.*, **84**, 90-113 (1989).
5. S.K. Godunov, *Mat. Sb.*, **47**, 271(1959).
6. A. Harten, B. Engquist, S. Osher and S. Chakravarthy, *J. Comput. Phys.*, **71**, 231 (1987).
7. C. Hu and C.W. Shu, *J. Comput. Phys.*, **150**, 97-127 (1999).
8. P.L. Roe, *J. Comput. Phys.*, **43**, 357-372 (1981).
9. C.W. Shu, *SIAM Journal on Scientific and Statistical Computing*, **9**, 1073-1084 (1988).
10. C.W. Shu, *Math. Comp.*, **49**, 105-121 (1987).
11. Z.J. Wang, *J. Comput. Phys.*, **178**, 210 (2002).
12. Z.J. Wang and Yen Liu, *J. Comput. Phys.*, **179**, 665 (2002).

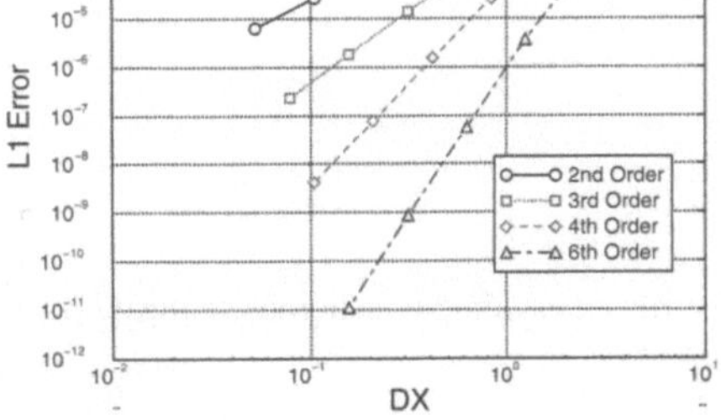

Fig. 1. L_1 errors for nozzle case

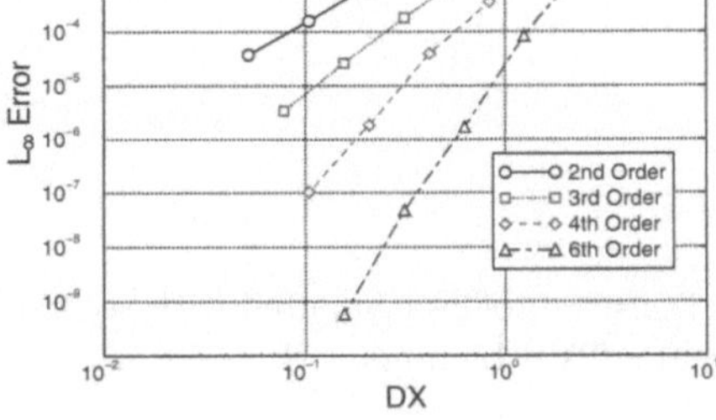

Fig. 2. L_∞ errors for nozzle case

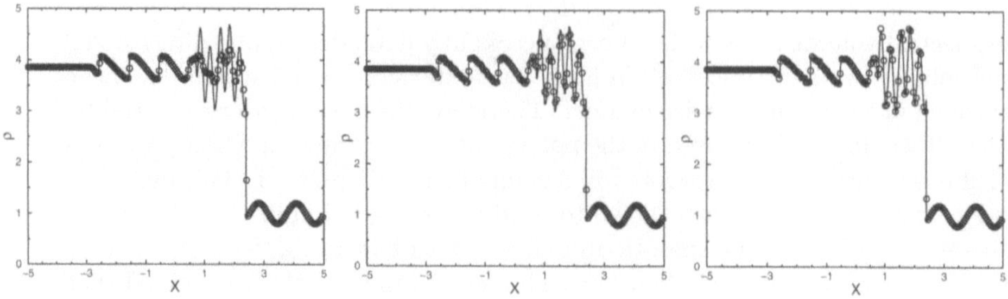

(a) 2nd Order (b) 3rd Order (c) 4th Order

Fig. 3. Density profiles computed with second to fourth order SV schemes on 200 SVs with TVD Limiters. One data point from a SV is shown

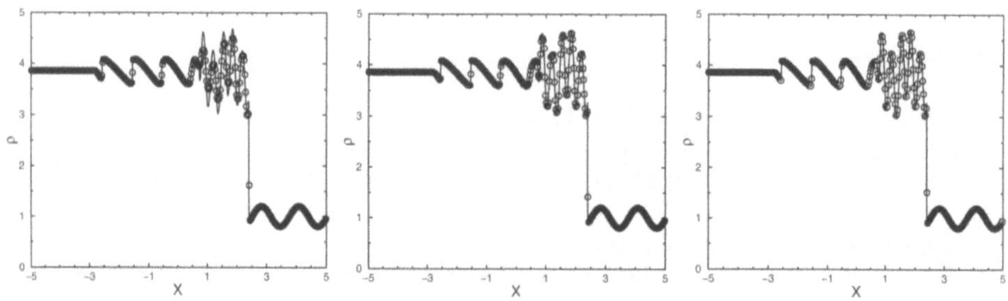

(a) 2nd Order (b) 3rd Order (c) 4th Order

Fig. 4. Density profiles computed with second to fourth order SV schemes on 400 SVs with TVD Limiters. One data point from a SV is shown

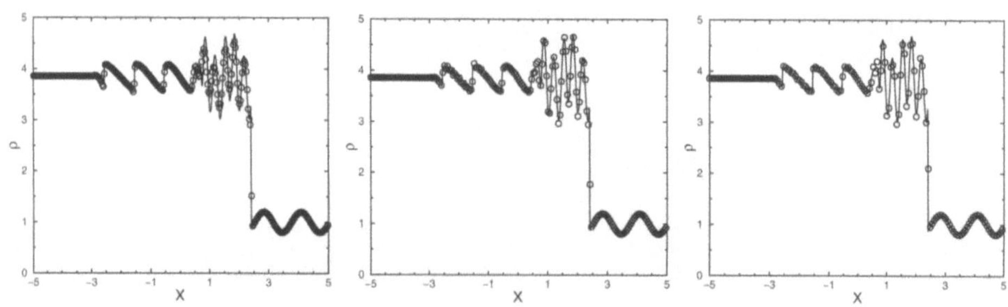

(a) 2nd Order (b) 3rd Order (c) 4th Order

Fig. 5. Density profiles computed with second to fourth order SV schemes using 600 DOFs and a TVB Limiters. ($M = 250$)

High Resolution Monotonicity-Preserving Schemes for Unsteady Compressible Flows

Virginie Daru[1,2] and Christian Tenaud[2]

[1] ENSAM, 151 Boulevard de l'Hôpital 75013 Paris, France
[2] LIMSI - CNRS UPR 3251, BP.133, 91403 Orsay Cedex, France

Abstract. This paper deals with the development of high order accurate schemes for the numerical simulation of unsteady compressible flows. It is shown that very accurate monotonicity preserving schemes can be constructed using a combined time-space discretization approach.

1 Introduction

LES is a powerful tool for performing fine analysis of flow dynamics in turbulent flows. However, in such an approach, the intrinsic dissipation of the scheme must be low enough to minimize the interaction with the subgrid scale model. When dealing with compressible flows involving shock waves, one must use a numerical scheme which can both represent small structures with the minimum of numerical dissipation, and capture discontinuities with robustness. To achieve this dual objective, Total Variation Diminishing (TVD) schemes are generally considered as well suited for the capture of shocks but too diffusive in smooth regions, due to the limitation of the accuracy to the first order near extrema. More recent schemes like the ENO/WENO family are very accurate in smooth regions but show a diffusive behavior in the vicinity of discontinuities. Another approach has been developed in [1], based on high order polynomial fixed stencil reconstruction, and Runge-Kutta time stepping. The reconstructed values at the interfaces of the mesh cells are then limited such as to preserve both monotonicity and high order accuracy. This is achieved by using local geometrical considerations to relax the monotonicity constraints near extrema.
We rather here follow the Lax-Wendroff approach to develop high order TVD combined time-space (one-step) schemes. In the linear case, TVD schemes accurate up to the 7th order in time and space away from extrema are obtained. The implementation of these schemes is very simple as the increase of accuracy can be obtained *via* a change of the limiting function applied to a classical TVD scheme. We then relax the TVD constraint following [1], and obtain a monotonicity preserving scheme which gives excellent results at low cost.

2 Construction of high order one-step schemes

1D scalar case. Let us first consider the 1D scalar linear transport equation :

$$u_t + f(u)_x = 0 \tag{1}$$

with $f(u) = au$, a being the velocity that is supposed to be constant and positive (the case $a < 0$ can be treated by symmetry relatively to each cell interface).

As regards the discrete solution, in the following we denote δt the time step and δx the cell width, u_j^n refering to the discrete solution at time $t = t_0 + n \cdot \delta t$ and location $x = x_0 + j \cdot \delta x$. To access to a high order approximation, the present scheme, based on the Lax-Wendroff approach, is obtained by applying successive correction on the error terms of the modified equations. Following this way, we obtain high order accurate schemes in both time and space. As an example, let us consider the Lax-Wendroff scheme, that we write :

$$u_j^{n+1} = u_j^n - \frac{\delta t}{\delta x}(F_{j+1/2}^{lw} - F_{j-1/2}^{lw}), \quad F_{j+1/2}^{lw} = f_j^n + \frac{(1-\nu)}{2}(f_{j+1}^n - f_j^n) \quad (2)$$

where $F_{j+1/2}^{lw}$ is the numerical flux and ν is the CFL number $\nu = a\delta t/\delta x$. By removing to that Lax-Wendroff scheme an upwind discretization of the 2nd order error term ($a\frac{\delta x^2}{6}(\nu^2 - 1)u_{xxx}$) appearing in the equivalent equation, one obtains the classical third order upwind-biased scheme with the following numerical flux:

$$F_{j+1/2}^3 = f_j^n + \Phi_{j+1/2}^3 \frac{(1-\nu)}{2}(f_{j+1}^n - f_j^n), \quad \Phi_{j+1/2}^3 = 1 - \frac{1+\nu}{3} \cdot (1 - r_j) \quad (3)$$

with $r_j = (f_j^n - f_{j-1}^n)/(f_{j+1}^n - f_j^n)$. Following such successive corrections of the higher order error terms, one can construct schemes with arbitrary order of accuracy, which can be expressed by only modifying the Φ function in the numerical flux. We have then performed the calculations until the seventh order. As an example, we present the successive Φ functions for the 4th and 5th order:

$$\Phi_{j+1/2}^4 = \Phi_{j+1/2}^3 + \frac{1+\nu}{3} \cdot \frac{\nu-2}{4} \cdot (1 - 2r_j + r_j r_{j-1}) \quad (4)$$

$$\Phi_{j+1/2}^5 = \Phi_{j+1/2}^4 - \frac{1+\nu}{3} \cdot \frac{\nu-2}{4} \cdot \frac{\nu-3}{5} \cdot (1/r_{j+1} - 3 + 3r_j - r_j r_{j-1}) \quad (5)$$

Let us underline that this kind of one-step scheme uses only a eight point stencil to recover a seventh order of accuracy in both time and space, which is much less than in a separate time-space discretization using fractional step time integration. Let us also note that the present schemes have the advantage to get the exact solution if the CFL number is equal to 1, which seems desirable.

Following the criteria developed by Harten on the TVD property, to recover a TVD one-step scheme, the Φ function must be limited as follows:

$$\Phi^{o-TVD} = \max\left(0, \min\left(\frac{2r_j}{\nu}, \Phi_{j+1/2}^o, \frac{2}{1-\nu}\right)\right) \quad (6)$$

where o denotes the order of accuracy. In [3], such a third order TVD scheme was proposed, and was used to compute a viscous flow in a shock tube in [2]. In the present paper, the principle is naturally extended up to $o = 7$. The resulting scheme is then seventh order accurate almost everywhere, except around extrema and discontinuities where it becomes first order accurate, as it is the case for all TVD schemes. This is a serious drawback that needs correction for our purpose.

As it is well known, the TVD constraint is active close to an extremum. This constraint leads to a lack of accuracy near this extremum because, respectively, the first ($0 \leq \Phi_{j+1/2} \leq \frac{2r_j}{\nu}$, see (6)) and the second ($0 \leq \Phi_{j+1/2} \leq \frac{2}{1-\nu}$, see (6)) TVD constraints ensure that the numerical flux $F_{j+1/2}$ must belong to the intersection of the two intervals, respectively, $[f_j^n, f_{j+1}^n]$ and $[f_j^n, f_j^{ul}]$ (where $f_j^{ul} = f_j^n + \frac{1-\nu}{\nu}(f_j^n - f_{j-1}^n)$). To get a better representation of the numerical solution in the vicinity of the extrema, it is necessary to enlarge this interval. Following the work in [1], near an extremum, the TVD constraint should not be activated and then be replaced by a monotonicity preserving constraint. We denote by $[f^1, f^2, ..., f^k]$ the interval $[min(f^1, f^2, ..., f^k), max(f^1, f^2, ..., f^k)]$. The first TVD interval is then enlarged to $[f_j^n, f_{j+1}^n, f_j^{md}]$, where

$$f_j^{md} = \frac{1}{2}(f_j^n + f_{j+1}^n) - \frac{1}{2}d_{j+1/2} \tag{7}$$

with $d_{j+1/2} = \text{minmod}(d_j, d_{j+1})$ and $d_j = f_{j+1}^n - 2f_j^n + f_{j-1}^n$ is the local curvature measurement of the flux function. Similarly, we enlarge the second TVD interval $[f_j^n, f_j^{ul}]$ to $[f_j^n, f_j^{ul}, f_j^{lc}]$, where the value f_j^{lc} (lc stands for large curvature):

$$f_j^{lc} = f_j^n + \frac{1}{2}(f_j^n - f_{j-1}^n) + \frac{\beta}{3}d_{j-1/2} \tag{8}$$

where β is a coefficient which allows a degree of freedom on the local curvature (typical values of β are between 1.5 and 2.5). We then define the intersection $[f^{min}, f^{max}]$ of the two previously enlarged intervals $[f_j^n, f_{j+1}^n, f_j^{md}]$ and $[f_j^n, f_j^{ul}, f_j^{lc}]$. Finally, the monotonicity preserving constraint, which gives room for the numerical flux to keep an accurate value, assures that $F_{j+1/2} \in [f^{min}, f^{max}]$. When this constraint is fulfilled, the TVD constraint should not be applied and the numerical flux $F_{j+1/2}$ is computed by using the $\Phi_{j+1/2}^o$ function. Otherwise, $\Phi_{j+1/2}^{o-TVD}$ is used.

Extension to the Euler equations. To extend the high order one step scheme to the non linear 1D system case, we use a classical direct extension of the Φ formula (6) relatively to each eigenvalue of the system. The numerical flux of the scheme is written :

$$F_{j+1/2} = F_{j+1/2}^{Roe} + \frac{1}{2}\Sigma\Phi_k(1 - |\nu_k|)\delta\alpha_k|\lambda_k|r_k \tag{9}$$

where F^{Roe} is the numerical flux of the first order Roe scheme, λ_k and r_k are respectively the eigenvalues and eigenvectors of the Roe-averaged jacobian matrix $A = \frac{df}{dw}$, $\delta\alpha_k$ is the contribution of the k-wave to the variation $\delta w = w_{j+1} - w_j$ and $\nu_k = \frac{\delta t}{\delta x}\lambda_k$. The monotonicity constaints are extended by working on the flux difference components $\delta F_{j+1/2,k} = \frac{1}{2}\Phi_{j+1/2,k}(1 - |\nu_{j+1/2,k}|)\delta\alpha_{j+1/2,k}|\lambda_{j+1/2,k}|$ and all values defining the monotonicity intervals are directly extended from the scalar case using flux differences components $\lambda_{j+1/2,k}\delta\alpha_{j+1/2,k}$. In this way, the schemes are formally only second order accurate, but in practice they show a considerable improvement over classical second order accurate schemes.

The extension to the 2D case makes use of a Strang directional splitting. Due to the cross-derivatives terms, a non-splitted extension cannot preserve monotonicity easily.

3 Numerical results

1D scalar test case. We will consider the classical test case of the advection of an initial profile composed of a Gaussian wave, a square wave, a triangular wave and a ellipse. The initial condition is defined on the interval $[-1, 1]$ and periodic boundary conditions are imposed. The calculations are performed on a uniform mesh with 200 cells, by using CFL=0.5. The results are discussed on the solution obtained at $t = 100$ (after fifty periods) (Fig. 1). We present the results obtained through the use of the schemes previously described: the one-step seventh order scheme (named OS7) and its TVD and MP variants (respectively named OSTVD7 and OSMP7). These schemes are compared to a RK3/WENO5 scheme [6] using a multi-step third order Runge-Kutta time integration coupled with a WENO spatial approximation (r=5). The OS7 gives very good results, nevertheless spoiled by numerical oscillations near discontinuities. The TVD correction does a very good job by eliminating the oscillations, but tends to smooth extrema although it is very localized. The MP correction is almost perfect, keeping the properties of the TVD correction while leaving unchanged the extrema compared to the original scheme. We can underline the quality of the discontinuities representation after 20,000 time iterations. By comparison, the RK3/WENO5 scheme, while very accurate for smooth extrema, is very diffusive around discontinuities and tends to round off all the "corners".

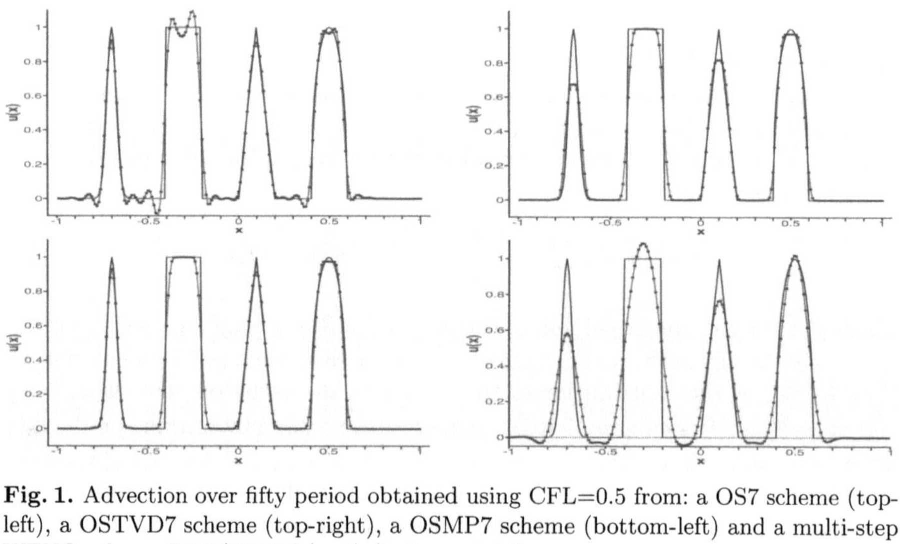

Fig. 1. Advection over fifty period obtained using CFL=0.5 from: a OS7 scheme (top-left), a OSTVD7 scheme (top-right), a OSMP7 scheme (bottom-left) and a multi-step WENO scheme RK3/WENO(r=5) (bottom-right)

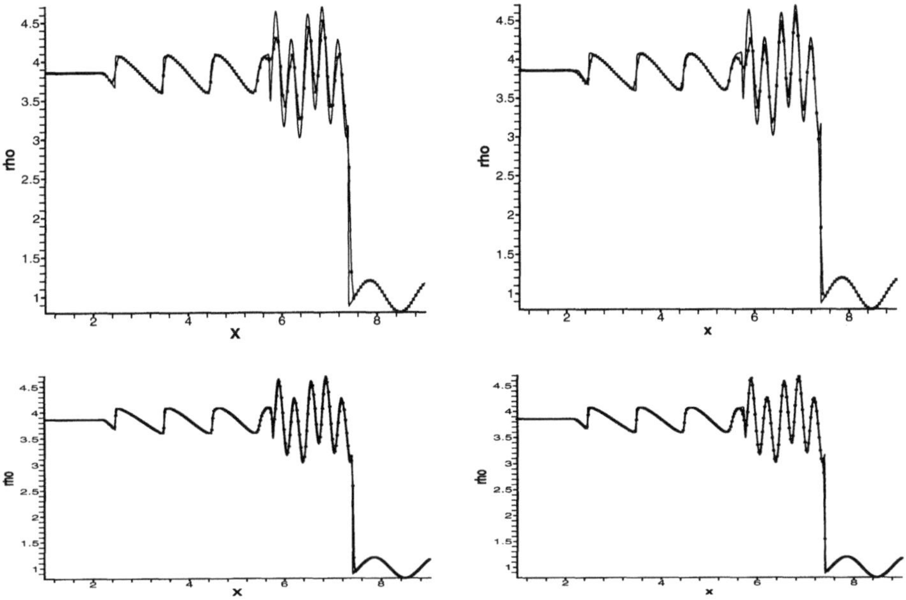

Fig. 2. Density distribution for the Shu-Osher test case a t=1.8, CFL=0.5. Top : 200 grid points, bottom : 400 grid points, left : OSMP7 scheme ($\beta = 2.25$), right : RK3/WENO5 scheme. Solid line : 12800 grid points RK3/WENO5 reference solution.

1D Euler test case. The 1D Euler test case chosen is the one proposed in [4] concerning a moving Mach=3 shock interacting with sine waves in density. The numerical solution obtained by the OSMP7 and RK3/WENO5 schemes are compared in Fig. 2, using two grids of 200 and 400 cells. Inspecting these results, we can see that while the one-step scheme is only formally second order accurate, it allows a very impressive improvement compared to a classical second order TVD scheme, and gives results comparable to those given by the more expensive WENO scheme (the OSMP7 is about six times less expensive in this case), even in coarse grids.

2D viscous shock-vortex interaction. This test case, treated in [5], consider the viscous ($Re = 2000$) interaction of a plane weak shock with a single isentropic vortex, during which acoustic waves are produced and transported. The pressure field contours obtained by using the two schemes are shown in Fig. 3, using a 100x100 cells grid for the OSMP7 scheme and a 200x200 cells grid for the RK3/WENO5 scheme, considered as the reference. One can notice that the WENO scheme exhibits a slightly oscillatory behavior, and that the OSMP7 result is quite accurate even in a coarser grid. This can be seen more precisely on the pressure curve along the line $y = 1$, which shows that the OSMP7 scheme gives even better results than the WENO scheme in representing the acoustic wave in the coarsest grid.

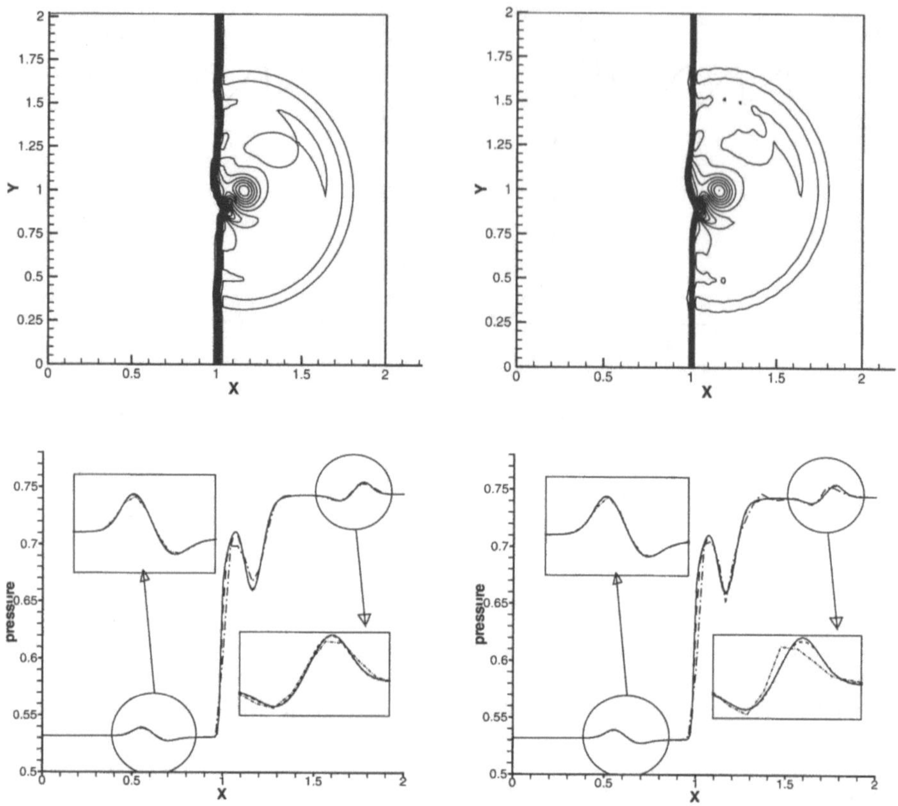

Fig. 3. Top : pressure contours for the 2D shock-vortex interaction a t=0.7, CFL=0.5; left : OSMP7 scheme 100x100 grid points, right : RK3/WENO5 scheme 200x200 grid points. Bottom : pressure along the line $y = 1$, 50x50 grid (dash-dotted line), 100x100 grid (dashed line), 200x200 grid (solid line); left : OSMP7 scheme, right : RK3/WENO5 scheme.

References

1. A. Suresh, H.T. Huynh: 'Accurate Monotonicity-Preserving Schemes with Runge-Kutta Time Stepping' Journal of Computational Physics, **136**, 83–99 (1997).
2. V. Daru, C. Tenaud: 'Evaluation of TVD high resolution schemes for unsteady viscous shocked flows' Computers and Fluids, **30**, 89–113 (2001).
3. M. Arora, P.L. Roe: 'A well behaved TVD limiter for high-resolution calculations of unsteady flows' Journal of Computational Physics, **132**, 2–11 (1997).
4. C.W. Shu, S. Osher: 'Efficient implementation of essentially non-oscillatory shock-capturing schemes, II', Journal of Computational Physics, **83**, 32–79 (1989).
5. C. Tenaud, E. Garnier, P. Sagaut: 'Evaluation of some high-order shock capturing schemes for direct numerical simulation of unsteady two-dimensional free flows', International Journal for Numerical Methods in Fluids, **126**, 202–228 (1996).
6. G.S. Jiang, C.W. Shu: 'Efficient implementation of weighted ENO schemes', Journal of Computational Physics, **83**, 32–79 (1989).

Arbitrarily Accurate Non-Oscillatory Schemes for Nonlinear Scalar Conservation Laws with Source Terms II

Yoko Takakura[1] and Eleuterio F. Toro[2]

[1] Tokyo Noko University, 2-24-16 Nakacho, Koganei, Tokyo 184-8588, Japan
[2] University of Trento, 38050 Mesiano de Povo, Trento, Italy

Abstract. The ADER approach for constructing non-oscillatory explicit one-step schemes with very high order of accuracy in space and time has been extended to nonlinear scalar conservation laws with source terms by using both the ways of state-series expansion and direct expansion. Numerical verification has shown that the ADER schemes work well.

1 Introduction

Modern shock-capturing schemes called Godunov-type schemes[1] rely on exact or approximate solutions of Riemann problem (RP). The conventional RP (Fig. 1(a)) is the initial value problem for the homogenious conservation laws with the initial conditions (ICs) consisting of two constant states separated by a discontinuity. For this problem the similarity solution exists. In many of modern

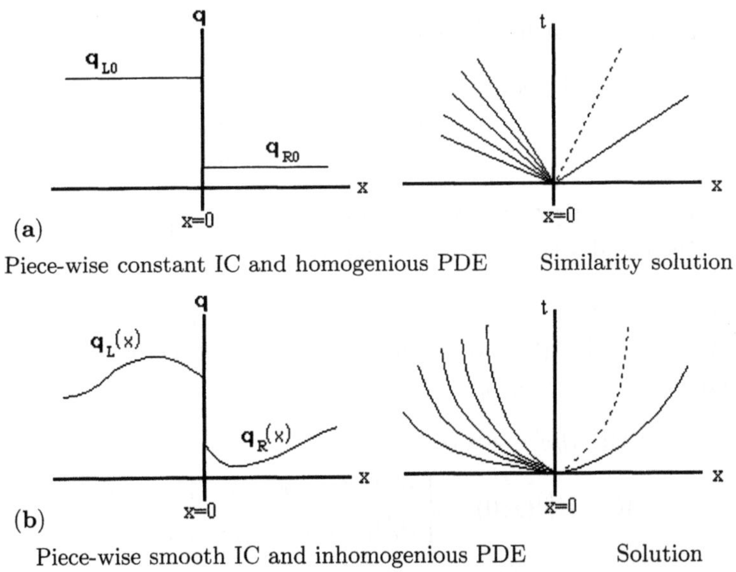

Fig. 1. Concept of RP. (**a**) Conventional RP. (**b**) Generalized RP.

schemes, cell-averaged solution data are reconstructed to polynomials, but for such ICs the problem does not have a similarity solution. Furthermore, in the case of inhomogenious conservation laws including source terms, even if numerical flux functions retaining high accuracy for homogenious conservation laws are used, high accuracy is not achieved.

In this research, the concept of generalized RP (GRP) (Fig. 1(b)), is shown to resolve the problems stated above and achieve high accuracy. Generalization covers two respects: the first is that ICs admit piece-wise polynomial functions, and the second is that governing equations include source terms.

In the approach called ADER [5], Toro et al. [6], carried out the above generalization of RP by using the method of state-series expansion, and Takakura, et al. [3], showed the whole perspective of ADER schemes in the extension to nonlinear homogenious problems by including both newly developed ways of state-series expansion and direct expansion. The aim of this paper is to extend the ADER further to a nonlinear scalar conservation laws with source terms.

2 ADER Approach

The ADER approach[5] is the conservative scheme satisfying

$$q_i^{n+1} = q_i^n - \frac{\Delta t}{\Delta x}\left[f_{i+\frac{1}{2}}^{ader} - f_{i-\frac{1}{2}}^{ader}\right] + \Delta t\ s_i^{ader}, \tag{1}$$

where q_i^n is the spatial average of q for cell $I_i = [x_{i-\frac{1}{2}}, x_{i+\frac{1}{2}}]$ at time $t = t_n$.

2.1 Method Based on State-Series Expansion

Consider the GRP at each cell interface $i + \frac{1}{2}$:

$$\text{PDE: } \partial_t q + \partial_x f(q) = s(q), \tag{2}$$

$$\text{IC: } \quad q(x,0) = \begin{cases} p_i(x + x_{i+\frac{1}{2}}) & \text{if } x < 0 \\ p_{i+1}(x + x_{i+\frac{1}{2}}) & \text{if } x > 0 \end{cases}, \tag{3}$$

where $p_i(x)$ are the ENO/WENO polynomial functions[2] for cell I_i . Notice that the PDE includes the source term and the IC consists of piece-wise smooth functions

As Eq. (2) can be written as $\partial_t q + \lambda(q)\ \partial_x q = s(q)$ where $\lambda(q) = df/dq$, differentiating the GRP (2) and (3) on x sequentially k times gives the derivative GRP (DGRP) for $v^{(k)} \equiv \partial_x^{(k)} q$:

$$\text{PDE: } \partial_t v^{(k)} + \lambda\ \partial_x v^{(k)} = rhs^{(k)}\ , \tag{4}$$

$$\text{IC: } \quad v^{(k)}(x,0) = \begin{cases} \partial_x^{(k)} p_i(x + x_{i+\frac{1}{2}}) & \text{if } x < 0 \\ \partial_x^{(k)} p_{i+1}(x + x_{i+\frac{1}{2}}) & \text{if } x > 0 \end{cases}, \tag{5}$$

The solutions at $x = 0$ and $t \to +0$ of the DGRPs above are given by similarity solutions at $x/t = 0$ of m homogeneous derivative RPs (DRPs) for

$k = 0, 1, \cdots, m-1$ with the conventional RP for $k = 0$ included:

$$\text{PDE: } \partial_t v + \lambda\, \partial_x v = 0, \tag{6}$$

$$\text{IC}: \; v(x,0) = \begin{cases} q_{L\ i+\frac{1}{2}}^{(k)} \equiv \lim\limits_{x \to x_{i+\frac{1}{2}} - 0} \partial_x^{(k)} p_i(x) & \text{if } x < 0 \\ q_{R\ i+\frac{1}{2}}^{(k)} \equiv \lim\limits_{x \to x_{i+\frac{1}{2}} + 0} \partial_x^{(k)} p_{i+1}(x) & \text{if } x > 0 \end{cases} \tag{7}$$

because the influence of the source terms may be neglected as $t \to +0$.

As the k-th time derivatives of q can be replaced by space derivatives of q up to the k-th order plus extra terms, i.e., $q_x^{(l)}$, $\lambda_q^{(l-1)}(q)$, and $s_q^{(l-1)}(q)$ for $l = 1, \cdots, k$, it can be described as:

$$\partial_t^{(k)} q = \alpha^{(k)}(\, q_x^{(0)}, q_x^{(1)}, \cdots, q_x^{(k)}\,). \tag{8}$$

Therefore, from a time Taylor series expansion near $x = 0$ and $t \to +0$, the solution of the GRP (2)- (3) is given by:

$$q_{i+\frac{1}{2}}^{GRP}(0,\tau) = q_{i+\frac{1}{2}}^{(0)} + \sum_{k=1}^{m-1} \frac{\tau^k}{k!}\, \alpha^{(k)}(q_{i+\frac{1}{2}}^{(0)}, q_{i+\frac{1}{2}}^{(1)}, \cdots, q_{i+\frac{1}{2}}^{(k)}) + O(\,\tau^m), \tag{9}$$

where $q_{i+\frac{1}{2}}^{(j)}$ for $j = 0, \cdots, k$ are the solutions at $x/t = 0$ of DRPs (6) and (7).

Thus the ADER numerical flux is obtained by:

$$f_{i+\frac{1}{2}}^{ader} = \frac{1}{\Delta t} \int_0^{\Delta t} f(\, q_{i+\frac{1}{2}}^{GRP}(0,\tau)\,)\, d\tau\,, \tag{10}$$

and the ADER numerical source is expressed by:

$$s_i^{ader} = \frac{1}{\Delta t \Delta x} \int_{x_{i-\frac{1}{2}}}^{x_{i+\frac{1}{2}}} \int_0^{\Delta t} s(\, \bar{q}_i^n(x, t_n + \tau)\,)\, d\tau\, dx, \tag{11}$$

where, with $p_i^{(k)}(x) = \partial_x^{(k)} p_i(x)$,

$$\bar{q}_i^n(x, t_n + \tau) = p_i(x) + \sum_{k=1}^{m-1} \frac{\tau^k}{k!}\, \alpha^{(k)}(\, p_i^{(0)}(x), p_i^{(1)}(x), \cdots, p_i^{(k)}(x)\,). \tag{12}$$

2.2 Method Based on Direct Expansion

Another approach relies on the Taylor series expansion directly for flux and source functions.

Consider the GRP for f, that is obtained by multiplying (2) by $\lambda(q)$

$$\text{PDE: } \partial_t f + \lambda\, \partial_x f = \lambda s, \tag{13}$$

$$\text{IC: } \; f(x,0) = \begin{cases} g_i(x + x_{i+\frac{1}{2}}) & \text{if } x < 0 \\ g_{i+1}(x + x_{i+\frac{1}{2}}) & \text{if } x > 0 \end{cases}, \tag{14}$$

As is similar to the case of state-series expansion, through DGRPs, consider m DRPs of f for $k = 0, 1, \cdots, m-1$:

$$\text{PDE: } \partial_t v + \lambda\, \partial_x v = 0, \tag{15}$$

$$\text{IC:} \quad v(x,0) = \begin{cases} f_L^{(k)}{}_{i+\frac{1}{2}} \equiv \lim_{x \to x_{i+\frac{1}{2}}-0} \partial_x^{(k)} g_i(x) & \text{if } x < 0 \\ f_R^{(k)}{}_{i+\frac{1}{2}} \equiv \lim_{x \to x_{i+\frac{1}{2}}+0} \partial_x^{(k)} g_{i+1}(x) & \text{if } x > 0 \end{cases} \tag{16}$$

To obtain the values of $f_L^{(k)}{}_{i+\frac{1}{2}}$ and $f_R^{(k)}{}_{i+\frac{1}{2}}$, two methods for reconstructing f are possible:

Rf-1) from cell-averaged values of f, i.e., $\bar{f}_i = \dfrac{1}{\Delta x} \displaystyle\int_{x_{i-\frac{1}{2}}}^{x_{i+\frac{1}{2}}} f(\ p_i(x)\ )\ dx$, construct ENO/WENO interpolation polynomials $g_i(x)$ for cells I_i;

Rf-2) from definition of flux, i.e., $f(x,t) = f(\ q(x,t)\ )$, represent $\partial_x^{(k)} f(x,t)$ by terms of $\partial_x^{(j)} q(x,t)$ for $j = 0, \cdots, k$.

In the ADER approach the k-th time derivatives of f should be replaced by space derivatives of f up to the k-th order plus extra terms, i.e., $q_x^{(l)}$ for $l = 1, \cdots, k-1$ and $f_x^{(l)}$, $\lambda_q^{(l-1)}(q)$, and $s_q^{(l-1)}(q)$ for $l = 1, \cdots, k$, it can be described as:

$$\partial_t^{(k)} f = \beta_I^{(k)}(\ q_x^{(0)}, q_x^{(1)}, \cdots, q_x^{(k-1)},\ f_x^{(1)}, \cdots, f_x^{(k)}\ ). \tag{17}$$

A time Taylor series expansion near $x = 0$ and $t \to +0$ brings the solution of the GRP (13)- (14) as:

$$f_{i+\frac{1}{2}}^{GRP}(0,\tau) = f_{i+\frac{1}{2}}^{(0)} + \sum_{k=1}^{m-1} \frac{\tau^k}{k!}\, \beta_I^{(k)}(q_{i+\frac{1}{2}}^{(0)}, q_{i+\frac{1}{2}}^{(1)}, \cdots, q_{i+\frac{1}{2}}^{(k-1)}, f_{i+\frac{1}{2}}^{(1)}, \cdots, f_{i+\frac{1}{2}}^{(k)}) + O(\ \tau^m), \tag{18}$$

where $q_{i+\frac{1}{2}}^{(j)}$ and $f_{i+\frac{1}{2}}^{(j)}$ are solutions at $x/t = 0$ of DRPs for q and f, respectively. Thus the ADER numerical flux is obtained by the integral average

$$\begin{aligned} f_{i+\frac{1}{2}}^{ader} &= \frac{1}{\Delta t} \int_0^{\Delta t} f_{i+\frac{1}{2}}^{GRP}(\ 0, \tau\ )\ d\tau \\ &= f_{i+\frac{1}{2}}^{(0)} + \sum_{k=1}^{m-1} \frac{(\Delta t)^k}{(k+1)!}\, \beta_I^{(k)}(q_{i+\frac{1}{2}}^{(0)}, q_{i+\frac{1}{2}}^{(1)}, \cdots, q_{i+\frac{1}{2}}^{(k-1)}, f_{i+\frac{1}{2}}^{(1)}, \cdots, f_{i+\frac{1}{2}}^{(k)}). \end{aligned} \tag{19}$$

Regarding the form of $\partial_t^{(k)} f$ in (17), different expressions are possible by eliminating $f_x^{(j)}$ or $q_x^{(j)}$ from $f = f(q)$:

$$\partial_t^{(k)} f = \beta_{II}^{(k)}(\ q_x^{(0)}, f_x^{(1)}, \cdots, f_x^{(k)}\ ), \tag{20}$$

$$\partial_t^{(k)} f = \beta_{III}^{(k)}(\ q_x^{(0)}, q_x^{(1)}, \cdots, q_x^{(k)}\ ). \tag{21}$$

Thus five ways are possible by combination of function forms for $\partial_t^{(k)} f$ with reconstructing methods for f.

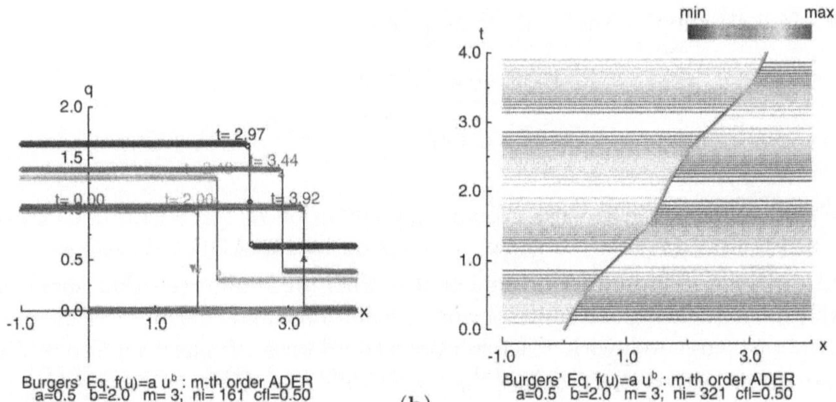

(a) (b)

Fig. 2. Solution of ADER-WENO $m = r = 3$ with $\partial_t^{(k)} f = \beta_I^{(k)}$ and a CFL number 0.5. (**a**)Exact (solid lines) and numerical (symbols) solutions: 160 cells. (**b**) $x - t$ diagram: 320 cells

As the k-th time derivatives of s should be expressed by space derivatives of q up to the k-th order by using $s(q)$ and the governing equation:

$$\partial_t^{(k)} s = \gamma^{(k)}(\ q_x^{(0)}, q_x^{(1)}, \cdots, q_x^{(k)}\ ), \tag{22}$$

from time Taylor series near x and $t = t_n$, the distribution of s in $[t_n, t_{n+1}] \times [x_{i-\frac{1}{2}}, x_{i+\frac{1}{2}}]$ is approximated by using the ENO/WENO polynomial $p_i(x)$

$$s_i^n(x, t_n + \tau) = s(\ p_i(x)\ ) + \sum_{k=1}^{m-1} \frac{\tau^k}{k!}\ \gamma^{(k)}(\ p_i^{(0)}(x), \cdots, p_i^{(k)}(x)\ ), \tag{23}$$

Then the ADER numerical source is obtained as:

$$\begin{aligned} s_i^{ader} &= \frac{1}{\Delta t \Delta x} \int_{x_{i-\frac{1}{2}}}^{x_{i+\frac{1}{2}}} \int_0^{\Delta t} s_i^n(x, t_n + \tau)\ d\tau\ dx \\ &\approx s(\ q_i^n\ ) + \sum_{k=1}^{m-1} \frac{(\Delta t)^k}{(k+1)!}\ \gamma^{(k)}(\ p_i^{(0)}(x_i), \cdots, p_i^{(k)}(x_i)\ ). \end{aligned} \tag{24}$$

It is proved that the ADER schemes with $r = m$ are of order m in time and space[4].

3 Numerical Verification

In Reference[3] a verification problem to study convergence rate shows that ADER schemes achieve the designed order of accuracy with CFL number close to unity. In Reference[4] a verification problem shows that the ADER schemes well capture shock and expansion waves without spurious oscillations.

Here we adopt is the verification problem:

$$\text{PDE: } q_t + (q^2/2)_x = sin(\pi t) \tag{25}$$

$$\text{IC} : q_0(x) = \begin{cases} 1 \text{ if } x < 0 \\ 0 \text{ if } x > 0 \end{cases} . \tag{26}$$

defined on $x \in [-1, 4]$ with x boundary condition of the 0-th order extrapolation. Figure 2 (a) and (b) show the solution by the ADER-WENO $m = r = 3$ with $\partial_t^{(k)} f = \beta_I^{(k)}$ and CFL number 0.5, where it is observed that shock waves are captured clearly without spurious oscillations, and that numerical solutions (symbols) agree well with exact solutions (solid lines). It has been also confirmed that the ADER-WENO with $\partial_t^{(k)} q = \alpha^{(k)}$, $\partial_t^{(k)} f = \beta_{II}^{(k)}$ or $\partial_t^{(k)} f = \beta_{III}^{(k)}$ give the same solution displayed here.

4 Conclusions

The ADER approach for constructing non-oscillatory explicit one-step schemes with very high order of accuracy in space and time has been extended to non-linear scalar conservation laws with source terms by using both the ways of state-series expansion and direct expansion. Numerical verification has shown that the ADER schemes are hopeful.

In addition, the authors would like to express their sincere gratitude to Dr. T. Takahashi, NAL, for his fruitful discussion.

References

1. S. K. Godunov: 'Finite Difference Methods for the Computation of Discontinuous Solutions of the Equations of Fluid Dynamics'. Mat. Sb., Vol. 47 (1959) pp. 271-306
2. C. W. Shu: 'Essentially Non-Oscillatory and Weighted Essentially Non-Oscillatory Schemes for Hyperbolic Conservation Laws'. Technical Report, NASA/CR-97-206253, ICASE Report No. 97-65 (1997)
3. Y. Takakura and E. F. Toro: 'Arbitrarily Accurate Non-oscillatory Schemes for Nonlinear Conservation Law'. CFD Journal Vol.11, No.1 (2002) pp.6-17
4. Y. Takakura and E. F. Toro: 'Arbitrarily Accurate Non-oscillatory Schemes for Nonlinear Scalar Conservation Laws with Source Terms '. AIAA paper 2002-2736 (2002)
5. E. F. Toro, R. C. Millington and L. A. M. Nejad: 'Towards Very High Order Godunov Schemes'. In: *Godunov Methods: Theory and Applications, Edited Review.* ed. by E. F. Toro (Kluwer/Plenum Academic Publisher 2001) pp. 905-938
6. E. F. Toro and V. A. Titarev: 'Solution of the Generalized Riemann Problem for Advection-Reaction Equations'. Proc. Roy. soc. London. Vol.458 No.2018 (2002) pp. 271-281

Parallel Preconditioners with Mesh Adaptation for Compressible Flow Computations

Marzio Sala[1], Pénélope Leyland[2], Luca Formaggia[1], Farid Moussaoui[2], and Yannick Savoy[3]

[1] IMA-MOCS, SB, Ecole Polytechnique Fédérale de Lausanne, Switzerland
[2] LIN-ISE, STI, Ecole Polytechnique Fédérale de Lausanne, Switzerland
[3] APCO, Vevey, Switzerland

Abstract. Parallel domain decomposition techniques and auto-adaptive meshing are combined to solve some large scale problems in compressible flow simulation for aeronautics. In particular, Schwarz type preconditioners are investigated as well as integrated parallel mesh adaptivity and optimisation with the two-fold objective of enhancing solution accuracy and reduce computational costs.

1 Introduction

The simulation of complex compressible flows is required in many industrial applications. The large computational demand of realistic problems (e.g. a complete aircraft simulation), makes parallel implementation mandatory. On the other hand, robust and accurate methods are necessary to treat a large variety of regimes. In this scenario, parallel solvers can take advantage of domain decomposition based preconditioners and grid adaptation methods in an effort to optimise the computational time and solution accuracy.

The European Union founded project IDeMAS has grouped three aircraft industrials: EADS (Germany), Dassault Aviation (France), and Alenia (Italy), and three academic partners: EPFL (Switzerland), INRIA (France) and VKI (Belgium). The goal of this project was to demonstrate the applicability in a parallel context of new upwind unstructured grid-based scheme technology for complete aerodynamic configurations. Within this project, the authors investigated steady-state compressible solutions using domain decomposition techniques and parallel mesh adaptation. In particular, an algebraic multilevel Schwarz-type domain-decomposition technique has been developed, as well as a a-posteriori adaptation scheme for unstructured grids. Here, the two techniques are combined in a coordinate fashion.

The paper setup is as follows: after a brief introduction to the discretisation schemes in Section 2, in Section 3 the iterative solution of the linear system arising from an implicit discretisation scheme is addressed. A 2-level Schwarz preconditioner with an agglomeration-based coarse correction discussed. In Section 4 the main concepts of parallel grid adaptivity are presented. Section 5 presents the results for an X29 aircraft obtained on an SGI Origin 3800.

2 Space and time discretisation

Even when looking for the steady-state solution, it is common practise to seek for the stationary point of the time-dependent version of the Euler equations, that can be written in conservation form as

$$\frac{\partial W}{\partial t} + div\mathcal{F}(\mathcal{W}) = 0, \tag{1}$$

where $W = (\rho, \rho u, \rho v, \rho w, \rho e)$ is the vector of conservative variables, with ρ the density, u, v and w are the components of the velocity vector, e is the specific total energy. $\mathcal{F}$ denotes the generalised flux which is homogeneous in W. The equation system is completed by a state equation, usually relating pressure, p to ρ, e or h - the specific total enthalpy. System (1) is discretised within a bounded domain $\Omega \subset R^3$, and completed by appropriate boundary conditions. Here the domain is discretised by a tessellation Ω_h of tetrahedra T of a mesh $\mathcal{T}_h$.

Many are the ingredients needed for an efficient numerical solution of (1). The method here chosen follows the so-called *multidimensional upwind residual distribution schemes.* Distribution schemes were introduced by P.-L. Roe in the beginning of the 80's for scalar advection laws and have been then developed by several authors and extended to systems [2]. They exploit the semi-linear equivalence of the Euler system due to the homogeneous flux, which can be written as

$$\frac{\partial W}{\partial t} + \mathcal{A}_i \frac{\partial W}{\partial \boldsymbol{x}_i} = 0, \quad \text{where } \mathcal{A} = \frac{\partial \mathcal{F}(W)}{\partial W} \quad \text{or equivalently } \frac{\partial U}{\partial t} + \tilde{\mathcal{A}}_i \frac{\partial U}{\partial \boldsymbol{x}_i} = 0, \tag{2}$$

where U is a set of characteristic variables obtained by a similarity transformation S and $\tilde{\mathcal{A}}$ the matrix $S^{-1}\mathcal{A}S$. The Jacobian matrices $\mathcal{A}$ are not-commuting.

Multidimensional upwind schemes correspond to the way the discretisation computational cell residuals given by the numerical flux Φ_T at time t^n are distributed at time t^{n+1} to the vertexes i of $T \in \mathcal{T}_h$:

$$W_i^{n+1} = W_i^n + \frac{\triangle t_i^n}{|C_i|} \sum_{\{i \in T\}} \beta_T^i \Phi_T, \text{ with } \Phi_T = \oint_{\partial T} \mathcal{F}(W(t^n, x))\boldsymbol{n} dx \approx \sum_{l=1}^{d+1} K_l W_l \tag{3}$$

β_T^i are the distribution coefficients, $\beta_T^i \Phi_T = \Phi_T^i$, K_l are upwind parameters, $|C_i|$ is the volume of the topological dual cell around the node i, $\boldsymbol{n}$ the normal to the cell, W_i^n represents the average of W at time $t^n = n\triangle t_i^n$ within the cell (usually assuming linear variation over each cell); $\beta_T^i \Phi_T = \Phi_T^i$, Conservation requires that $\Phi_T^i + \Phi_T^j + \Phi_T^k = \Phi_T \equiv \sum_k \beta_T^{i(k)} = 1$.
Equations (3) yield a system of ordinary differential equations in time, of the form

$$\frac{dW}{dt} = R(W) = -\mathcal{M}^{-1} \sum_T \beta_T^l \Phi_T, \; \mathcal{M} \approx C_l \delta_{lk}, \tag{4}$$

where $W = \{W_l\}^T$ is the vector of nodal states. The way the coefficients β_T^i are constructed define the type of scheme. Standard Galerkin is given by $\beta_T^i =$

1/3, while the first order positive N-scheme sends flux contributions only to downstream vertexes. More details and other possibilities are given in [2,1].
To integrate Equation (4), in time an implicit scheme is chosen. As $R(W)$ is non-linear in W, an outer Newton method is used, leading to the resolution of an inner large sparse linear system at each time step, $A^n \delta W^n = b^n$, where $A^n = \left[\frac{I}{\Delta t^n} - \frac{\partial R}{\partial W}(W^n)\right]$, $\delta W^n = (W^{n+1} - W^n)$, and $b^n = R(W^n)$. Memory requirements and parallel performances suggest the use of an iterative solver like GMRES. For the kind of problems considered in this paper, a preconditioner is mandatory. One that is particularly adapted to a parallel environment is the Schwarz preconditioner, presented in the next Section.

3 A 2-level Schwarz preconditioner

The Schwarz method is a well known parallel technique based on a domain decomposition strategy for the solution of PDE problems. The procedure is as follows. First, the computational domain Ω is subdivided into M subdomains $\Omega_i, i = 1\ldots, M$, such that $\cup_{i=1}^{M} \bar{\Omega}_i = \Omega$ and $\Omega_i \cap \Omega_j \neq \emptyset$ for some i and j. Then, the solution of the original problem is replaced by an iterative procedure that requires, at each step, the solution of local problems in each Ω_j. In particular, on $\partial\Omega_i \cap \Omega_j$ we apply Dirichlet conditions by imposing the latest values available from the neighbouring sub-domains. See [3] for more details.

Although rather slow as a solver, the Schwarz method is a quite popular parallel preconditioner, because of its generality and simplicity of implementation. The Schwarz preconditioner P_S can be expressed as

$$P_S^{-1} = \sum_{i=1}^{M} \mathcal{R}_i^T A_i^{-1} \mathcal{R}_i, \qquad (5)$$

where $\mathcal{R}_i$ is a restriction operator $\Omega \to \Omega_i$ and A_i corresponds to the discretisation of the differential problem on subdomain Ω_i. This preconditioner is parallel since we can assign each subdomain Ω_i to a different processor. However, the scalability of the Schwarz preconditioner is hindered by the weak coupling between far away sub-domains. A good scalability may be recovered by the addition of an operator defined on a coarse grid, $\mathcal{T}_0$. In this latter case, the preconditioner reads

$$P_{C,add}^{-1} = \mathcal{R}_0^T A_0^{-1} \mathcal{R}_0 + P_S, \qquad (6)$$

where A_0 represents the discretisation of our problem on the coarser mesh. This preconditioner is *additive*, since it can be shown that the correction on the coarse grid space is added to the corrections on the fine grid spaces defined by the subdomains Ω_i. An *hybrid* preconditioner may be used as well, and it reads

$$P_{C,hybrid}^{-1} = P_S^{-1} + \mathcal{R}_0^T A_0^{-1} \mathcal{R}_0 - P_S^{-1} A \mathcal{R}_0^T A_0^{-1} \mathcal{R}_0.$$

In particular, we have found that the hybrid version has a more favourable behaviour for our class of problems.

In general, the construction of the coarse grid may reveal to be difficult or computationally expensive for problems defined on 3D unstructured grids. Therefore, an algebraic procedure similar to the agglomeration technique which has been implemented in the context of multigrid, is devised, see [4] and the references therein. The computational domain is divided into N_p *aggregates*. Each aggregate represents a set of neighbouring cells or nodes. Then, the column of $\mathcal{R}_0$ corresponding to an aggregate will have 1's for all nodes internal to the aggregate, 0's otherwise. The coarse matrix can now be obtained as $A_0 = \mathcal{R}_0 A \mathcal{R}_0^T$. In this way a 2-level method is obtained without constructing a coarse grid.

4 The mesh adaptivity procedure

Parallel mesh adaptation techniques were based on previous work developed at EPFL using a non-hierarchical data structure, because of its greater flexibility [6]. In 3D, the techniques were re-formulated, maintaining the possibilities of local refinement and de-refinement and grid improvement algorithms. The methods developed here are all *local*, so that their influence is limited within a so-called *shell* which is a sub-mesh isolated around the node or segment in consideration, i.e. the neighbouring cells, see [8]. Theses techniques remain valid for arbitrary types of cells, tetrahedra, hexahedra, prisms or hybrid. Mesh optimisation techniques concern many other algorithmic changes such as stretching, smoothing adaptation by displacement of the nodes and re-meshing adaptation by redistribution of the nodes, as well as structural changes such as collapsing and diagonal swapping. These optimisation techniques are primordial since although grid adaptation can provide highly accurate solution meshes, the goal is not only to solve accurately the physical phenomena of the particular problem, but also to maintain this accuracy on as coarse and as regular mesh as possible. Such techniques require a precise description of the boundaries in order to be able to re-project correctly.

The parallelisation uses the packages Metis and ParMETIS[7] for the domain partitioning and redistribution. The ParMETIS library is implemented directly within the software for dynamic mesh adaptation and dynamic load balancing. In parallel mesh adaptation the modes of partitioning, de-refinement, refinement and optimisation require fast and efficient multiple renumbering techniques for the mesh entities: cells, segments, faces, nodes. For this a fast graph/tree technique was used; [5], based on a balanced binomial search tree algorithm AVL (Adelson-Velskii and Landis). Pairs of subtrees differ in height by at most 1. The AVL tree is implemented as a recursive structure of interlinked nodes. When a new node is inserted into the tree, it appears at the root, then moves along the branches until it finds an attachment to the tree. Once the node is inserted, the tree balance is checked. If no imbalance is found, another node is inserted and the process continues. If an imbalance is found, the heights of some nodes are fixed and the process repeated. When a node is deleted, the root becomes unbalanced. The look-up is performed to balance again. Look-up, insertion and deletion operations are of $O(logN)$, where N is the number of nodes in the tree.

The mesh adaptation system acts dynamically during the solution process; i.e. the code advances in global time iterations towards steady state solutions and the mesh is automatically de-refined, refined, optimised and smoothed at user's defined stages (convergence or iterations).

5 Numerical results

The test case retained for demonstrating the efficiency of the parallel preconditioner and mesh adaption techniques was an X29 experimental aircraft geometry, flying at $M_\infty = 0.75$ and with angle of attack $\alpha = 5$. Several levels of solution adaptive meshes were obtained. We have used the following notation: `iters` refers to the sum of the iterations needed by GMRES at each time step of the backward Euler method, while `CPU-time` is the sum of the (elapsed) CPU-time in seconds required by the solution of the linear system. `N_procs` represents the number of processors (and hence also subdomains) used in the computation. The linear system has been solved up to a tolerance $||r||/||r_0|| < 10^{-2}$, and for the Schwarz preconditioner we have used ILU(0) factorisations on each subdomain, with an overlap of 1 element. The coarse problem was solved using an (exact) LU decomposition. The number of aggregates in each subdomain $N_p = 4$ for all test cases using $P_{C,hybrid}$.

Table 1. Comparison of P_S and $P_{C,hybrid}$ using different number of processors. The grid is made up of 136767 nodes and 726713 cells. The N scheme is used. Both CPU-time and iterations refer to the first 14 time iterations of the backward Euler method.

	`N_procs` = 8	`N_procs` = 16	`N_procs` = 32	`N_procs` = 64
		`iters`		
P_S	270	310	369	347
$P_{C,hybrid}$	-	199	210	171
		`CPU-time`		
P_S	4995.1	1461.2	451.6	202.9
$P_{C,hybrid}$	-	1253.3	434.3	189.1

Table 1 shows the effectiveness of the agglomeration coarse operator, especially when the number of processors gets larger. Table 2, instead, reports the CPU-time and iterations to converge using the original mesh, and two adapted meshes. The first adaptation cycle required, on 8 processors, 78 seconds for mesh refinement, 690 seconds for mesh regularisation and optimisation, and 132 seconds of overhead. The second cycle required, on 16 processors, 854 seconds for the mesh refinement, 2860 seconds for mesh regularisation and optimisation, and 1002 seconds of overhead.

6 Conclusions

Results obtained by combining efficient parallel solvers with parallel mesh adaption procedure for external aerodynamics computations are presented. The scal-

Table 2. Performance of $P_{C,hybrid}$ with different number of processors and with different meshes. Starting from an initial solution the mesh was adapted. After again reaching steady-state convergence, a further step of mesh optimisation was performed. "range": refers to the time iterations considered for time measures and iterations count.

	N_procs = 8	N_procs = 16	N_procs = 32	N_procs = 64
starting mesh: N_nodes=136767, N_cells=726713, range=70:100				
iters	90	84	86	91
CPU-time	11402.7	3992.3	1535.4	738.7
first refinement: N_nodes=407479, N_cells=2371914, range=50:100				
iters	-	89	162	226
CPU-time	-	17730.4	7783.2	3570.1
further mesh optimisation: N_nodes=407362, N_cells=2383243, range=70:80				
iters	25	25	25	25
CPU-time	10565.6	6165.3	1999.8	803.2

Fig. 1. X29: Mach number distribution and detail of large intermediate refinement

ability of the parallel technique and the effectiveness of the adaption algorithm have been assessed on a realistic three-dimensional geometry.

References

1. Final Technical Report : European Community BRITE/EURAM III BE 97-4162 Project, VKI, 2002.
2. AGARD R-787. *Unstructured Grid Methods for Advection Dominated Flows'*; May 1992.
3. A. Quarteroni, A. Valli. *Domain Decomposition Methods for Partial Differential Equations*; Oxford University Press, Oxford, 1999.
4. M. Sala, L. Formaggia. *Parallel Schur and Schwarz Based Preconditioners and Agglomeration Coarse Correction for CFD problems*; EPFL-DMA Report 15-2001. Submitted to *Comp. Methods in Appl. Mech. Engng.*.
5. M. A. Weiss. *Data Structure and Algorithm Analysis*; Benjamin/Cummings Publishing Compagny, 1992.
6. P. Leyland, R. Richter. *Completely Parallel Compressible Flow Simulations using Dynamic Adaptive Meshes*; Comp. Methods in Appl. Mech. and Engng., Vol. 184, 2000.
7. K. Schloegel, G. Karypis, and V. Kumar. *Parallel Multilevel Distribution Algorithms for Repartitioning of Adaptive Meshes*; University of Minnesota, Department of Computer Science Technical Report 97-014.
8. G. Carey. *Computational Grids – Generation, Adaptation and Solution Strategies*; ed. Taylor & Francis, 1992.

A New Grid-free Relaxation Scheme for Conservation Laws

S. Balasubramanyam[1] and S.V. Raghurama Rao[2]

[1] M.Sc.(Engg.) student
[2] Assistant Professor
Centre of Excellence for Aerospace CFD
Department of Aerospace Engineering
Indian Institute of Science, Bangalore-560012, India

Abstract. Relaxation Schemes of Jin and Xin (Comm. Pure App. Math., 48, 235-276, 1995) are simple alternatives to Riemann Solvers and complicated flux splittings. With this framework, non-linear conservation laws are converted to a linear system of convection equations with non-linear source terms, which can be easily solved by a splitting method. In this study, a new Grid-free Relaxation Scheme for solving Euler equations of Gas Dynamics is developed by coupling the strategy of Relaxation Schemes with Least Squares Upwinding. The new grid-free scheme is tested on some standard bench-mark problems and its capabilities in resolving the flow features are demonstrated.

1 Introduction

The upwind methods for solving the equations of inviscid compressible flows can be broadly classified as (i) Riemann solvers (exact or approximate), (ii) flux splitting methods and (iii) Kinetic Theory based methods. Jin and Xin [1] introduced a fourth category of methods called as *Relaxation Schemes* which are much simpler than the other upwind methods. The simplicity of the *Relaxation Schemes* is due to the fact that the original non-linear conservation laws are replaced by a linear system of equations with stiff non-linear source terms. The stiff non-linear parts are separated by a *splitting method* and one has to just solve a set of linear convection equations instead of the original non-linear conservation laws.

Grid generation is an important aspect of numerical simulation of fluid flows. Grid Generation around complex configurations is often a difficult and time-consuming task. To reduce the efforts required for the grid generation, there has been a search for numerical algorithms that are *grid-free* in the recent past, in the sense that the they should be able to operate on any arbitrary distribution of points. The *Least Squares Kinetic Upwind Method* (LSKUM) [2] is an important development in this research. In this work, we present a new *grid-free* Euler solver by combining the *Least Squares Upwinding* with the framework of a *Relaxation System* for non-linear conservation laws.

2 Relaxation Systems

Consider a non-linear conservation law in 1-D

$$\frac{\partial u}{\partial t} + \frac{\partial g(u)}{\partial x} = 0 \tag{1}$$

where $g(u)$ is a non-linear function of u. With $g(u) = \frac{1}{2}u^2$, we recover the inviscid Burgers equation. The Relaxation System of Jin and Xin [1] for the above equation is

$$\begin{aligned} \frac{\partial u}{\partial t} + \frac{\partial v}{\partial x} &= 0 \\ \frac{\partial v}{\partial t} + \lambda^2 \frac{\partial u}{\partial x} &= -\frac{1}{\epsilon}\left[v - g(u)\right] \end{aligned} \tag{2}$$

with $\epsilon \to 0$. We can rearrange the second equation of the *Relaxation System* (2) as

$$\epsilon\left(\frac{\partial v}{\partial t} + \lambda^2 \frac{\partial u}{\partial x}\right) = -\left[v - g(u)\right] \tag{3}$$

With $\epsilon \to 0$, we obtain $v = g(u)$ from the above equation, which, when substituted into the first equation of the *Relaxation System* (2), yields the original conservation law (1). Hence, solving the *Relaxation System* (2) in the limit of $\epsilon \to 0$ is equivalent to solving the original conservation law (1). Chapman–Enskog type expansion for the *Relaxation System* results in (see [1] and [3] for details)

$$\frac{\partial u}{\partial t} + \frac{\partial v}{\partial x} = \epsilon \frac{\partial}{\partial x}\left[\frac{\partial u}{\partial x}\left\{\lambda^2 - \left(\frac{\partial g(u)}{\partial u}\right)^2\right\}\right] + O\left[\epsilon^2\right] \tag{4}$$

The right hand side of the above equation contains a second derivative and represents a viscous term. Thus, the *Relaxation System* provides a vanishing viscosity model to the original conservation law. For the dissipation to be positive, we obtain the condition on the positive constant λ as $\lambda^2 \geq \left(\frac{\partial g(u)}{\partial u}\right)^2$. For a vector conservation law

$$\frac{\partial U}{\partial t} + \frac{\partial G(U)}{\partial x} = 0 \tag{5}$$

modeled by the *Relaxation System*

$$\begin{aligned} \frac{\partial U}{\partial t} + \frac{\partial V}{\partial x} &= 0 \\ \frac{\partial V}{\partial t} + D \frac{\partial U}{\partial x} &= -\frac{1}{\epsilon}\left[V - G(U)\right] \end{aligned} \tag{6}$$

the Chapman–Enskog type analysis gives the condition $D \geq \left(\frac{\partial G(U)}{\partial U}\right)^2$ where D is a diagonal matrix. For the Euler equations with

$$U = \begin{bmatrix} \rho \\ \rho u \\ \rho E \end{bmatrix} \; ; \; G(U) = \begin{bmatrix} \rho u \\ p + \rho u^2 \\ pu + \rho u E \end{bmatrix} \tag{7}$$

and with the choice $D = \lambda^2 I$, following Jin and Xin [1], we can choose

$$\lambda^2 = max\left[|u-a|, |u|, |u+a|\right] \tag{8}$$

3 Relaxation System as a Discrete Boltzmann Equation

Introducing the characteristic variables

$$\mathbf{f} = \begin{bmatrix} f_1 \\ f_2 \end{bmatrix} = \begin{bmatrix} \frac{1}{2}U - \frac{1}{2\lambda}V \\ \frac{1}{2}U + \frac{1}{2\lambda}V \end{bmatrix} \tag{9}$$

the *Relaxation System* can be rewritten as a Discrete Boltzmann Equation (see Droillet and Natalini [2]) as

$$\frac{\partial \mathbf{f}}{\partial t} + \Lambda \frac{\partial \mathbf{f}}{\partial x} = \frac{1}{\epsilon}\left[\mathbf{F} - \mathbf{f}\right] \tag{10}$$

Here Λ is a diagonal matrix with $-\lambda$ and λ as the elements and the local Maxwellian is defined by

$$\mathbf{F} = \begin{bmatrix} F_1 \\ F_2 \end{bmatrix} = \begin{bmatrix} \frac{1}{2}U - \frac{1}{2\lambda}G(U) \\ \frac{1}{2}U + \frac{1}{2\lambda}G(U) \end{bmatrix} \tag{11}$$

The original variables U and V can be recovered from

$$U = P\mathbf{f} = P\mathbf{F} \text{ and } V = P\Lambda\mathbf{f} \text{ where } P = [1\ 1] \tag{12}$$

Note that the left hand side of the discrete Boltzmann equation is just a linear convection equation and the non-linear source term on the right hand side can be separated by a simple splitting method. It is preferable to work with the Discrete Boltzmann equation as the multi-dimensional Relaxation System will be in diagonal form, unlike the multi-dimensional Relaxation System given by Jin and Xin [1].

4 Least Squares Upwinding and the Grid–Free Relaxation Scheme

Consider a set of arbitrarily distributed points as shown in Fig.1. Taylor expansion at any point P gives for any neighbouring point i

$$f_i = f_P + (x_i - x_P)(f_x)_P + (y_i - y_P)(f_y)_P + O\left[(x_i - x_P)^2, (y_i - y_P)^2\right] \tag{13}$$

Defining the error e as

$$e = \Delta f_i - \Delta x_i f_x - \Delta y_i f_y \text{ where } \Delta f_i = f_i - f_P\text{ , } \Delta x_i = x_i - x_P \text{and } \Delta y_i = y_i - y_P \tag{14}$$

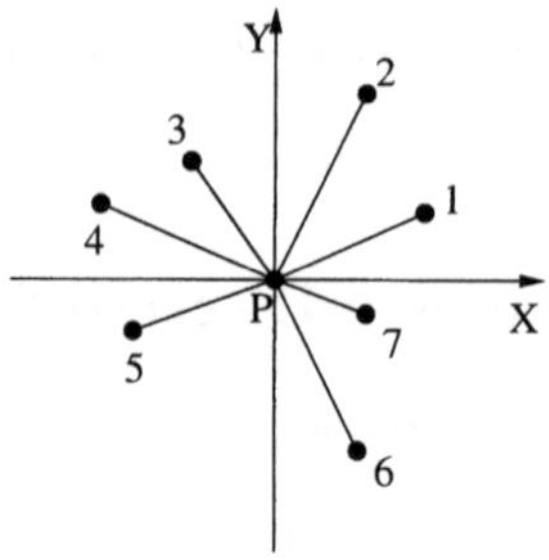

Fig. 1. Arbitrary distribution of points in 2 dimensions

and minimizing the sum of squares of the errors, we obtain the expressions for the derivatives as

$$f_x = \frac{\sum \Delta y_i^2 \sum \Delta x_i \Delta f_i - \sum \Delta x_i \Delta y_i \sum \Delta y_i \Delta f_i}{\sum \Delta x_i^2 \sum \Delta y_i^2 - \left(\sum \Delta x_i \Delta y_i\right)^2} \tag{15}$$

$$f_y = \frac{\sum \Delta x_i^2 \sum \Delta y_i \Delta f_i - \sum \Delta x_i \Delta y_i \sum \Delta x_i \Delta f_i}{\sum \Delta x_i^2 \sum \Delta y_i^2 - \left(\sum \Delta x_i \Delta y_i\right)^2} \tag{16}$$

where $\sum$ stands for $\sum_{i=1}^{n}$. The above two expressions (15) and (16) give approximations for space derivatives for an arbitrary distributions of n points, w.r.t. x and y respectively. These approximations can be used for discretizing the derivatives to obtain an upwind scheme with suitable weights for determining the upwind points. Second order accuracy on a compact stencil of points can be obtained by a 2-step procedure. We couple this framework with the Relaxation Approximation for the hyperbolic systems to obtain a *Grid–Free Relaxation Scheme* for solving hyperbolic conservation laws. The 2-D discrete Boltzmann equation is given by

$$\frac{\partial \mathbf{f}}{\partial t} + \Lambda_1 \frac{\partial \mathbf{f}}{\partial x} + \Lambda_2 \frac{\partial \mathbf{f}}{\partial x} = \frac{1}{\epsilon}\left[\mathbf{F} - \mathbf{f}\right] \tag{17}$$

where

$$\Lambda_1 = \begin{bmatrix} -\lambda & 0 & 0 \\ 0 & 0 & 0 \\ 0 & 0 & \lambda \end{bmatrix} \text{ and } \Lambda_2 = \begin{bmatrix} 0 & 0 & 0 \\ 0 & -\lambda & 0 \\ 0 & 0 & \lambda \end{bmatrix} \tag{18}$$

We use the modified Strang splitting introduced by Jin [5] along with the Least Squares Upwinding for the above discrete Boltzmann equation to obtain a new *Grid–Free Relaxation Scheme.* The details are available in [2,3,6].

5 Results

This new grid-free scheme is tested on some bench-mark problems in 1-D and 2-D. The first order and second order accurate results for the 2-D Burgers equation, on a point distribution obtained from a triangular grid-generator, are shown

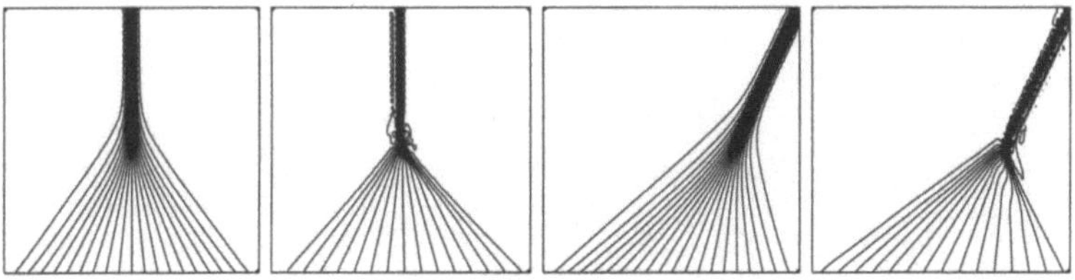

Fig. 2. I order and II order accurate results for 2-D Burgers equation

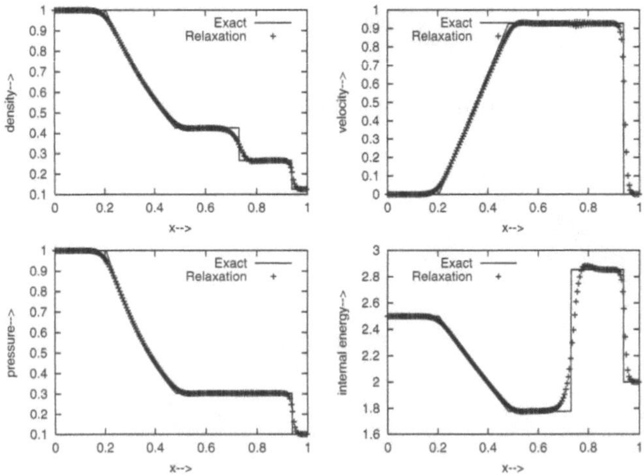

Fig. 3. II order accurate results for 1-D Euler equations on points obtained by a random number generator

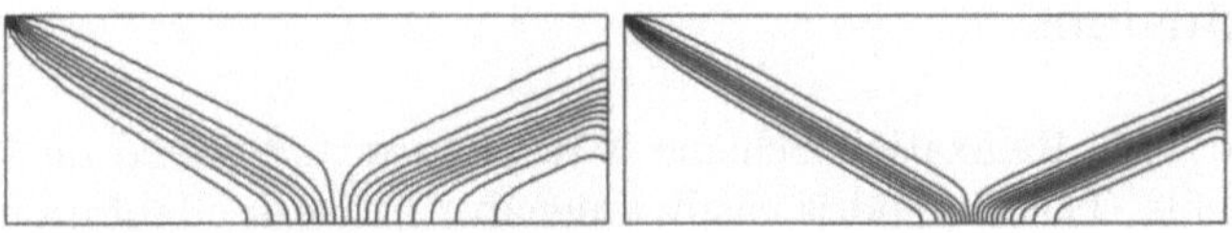

Fig. 4. II order accurate results for 2-D Euler equations on 1452 points and 5667 points obtained by a triangular mesh generator

in Fig.2. The II order accurate results for the 1-D shock-tube problem are shown in Fig.3. The second order accurate results obtained for the 2-D shock reflection problem on 1452 points and 5667 points obtained from a triangular mesh generator are shown in Fig.4. The Grid–Free Relaxation Scheme is easily amenable to mesh adaptation and this is demonstrated with the test case of a supersonic flow around a ramp in a channel Fig.5 and Fig.6. In all the cases, the point distribution obtained from the triangular mesh generator is used as an arbitrary distribution of points for this scheme.

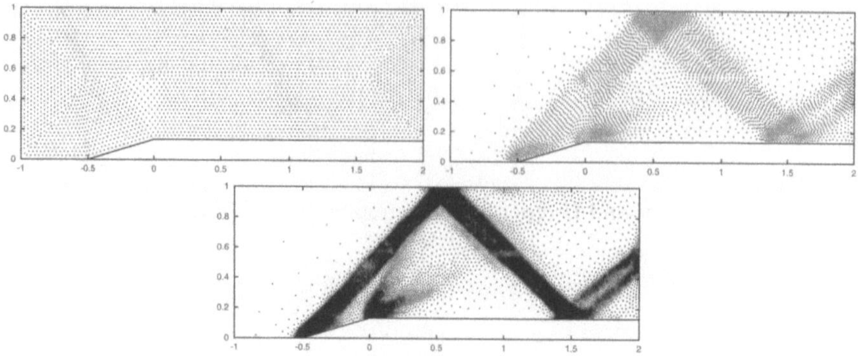

Fig. 5. Point distributions for different levels of mesh adaptation

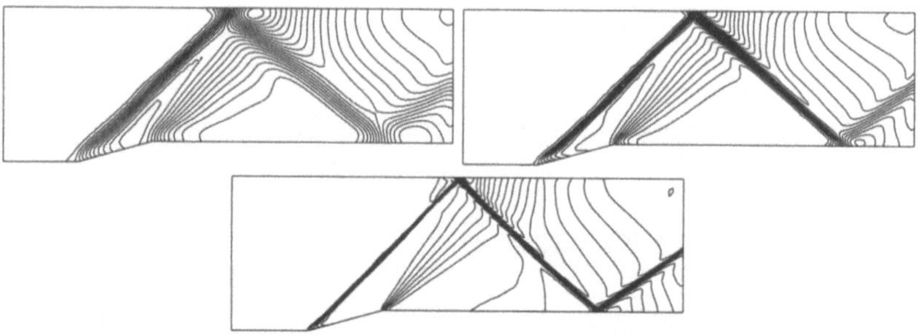

Fig. 6. II order accurate results for flow over a ramp in a supersonic channel with different levels of mesh adaptation

6 Conclusions

A new Grid–Free Relaxation Scheme is developed and tested on some benchmark problems. The method is easily amenable to mesh adaptation and avoids Riemann solvers and complicated flux splittings.

References

1. S. Jin and Z. Xin: Comm. Pure App. Math., 48, 235-276 (1995)
2. A.K. Ghosh and S.M. Deshpande : AIAA Paper-95-1735
3. S.V. Raghurama Rao : Berichte 249, Arbeitsgruppe Technomathematik, Universitaet Kaiserslautern, Germany (2002)
4. D. Aregba-Droillet and R. Natalini : SIAM J. Num. Anal., 37, 6, 1973-2004 (2000)
5. S. Jin : J. Comp. Phy., 122, 51-67 (1995)
6. S. Balasubramanyam and S.V. Raghurama Rao : Report 2001 FM 07, Dept. of Aerospace Engg., Indian Institute of Science, Bangalore

Part V

Computational Aerodynamics

On the Potential of Numerical Optimization of High-Lift Multi-Element Airfoils based on the Solution of the Navier-Stokes-Equations

Jochen Wild

DLR German Aerospace Center, Institute of Aerodynamics and Flow Technology, 38108 Braunschweig, Germany

Abstract. The objective of the presented work is to demonstrate the benefits of using numerical optimization of high-lift multi-element airfoils based on RANS-simulation in practical design. It is first shown, that the design methodology chosen in the presented framework is able to detect optimal configurations determined by wind tunnel tests. Beneath this basic statement sensitivities for the validity of the obtained results are outlined. Finally the method is applied to the drag reduction of a 3-element airfoil.

1 Introduction

Recent development in computational flow simulation has shown, that todays flow solvers based on solving the Reynolds-averaged Navier-Stokes equations are able to predict aerodynamic behavior of single and multi-element airfoils quite well, even under high-lift conditions up to stall [1–5].

Although the reached level of accuracy of RANS-simulation has been demonstrated, this type of simulation has not yet found its application in aerodynamic design methods. Only few attempts are recently published [6,7] but without stating the reliability of the result to be optimal. Also design methods based on coupled panel-boundary layer methods have already been reported [8,9]. Nevertheless the application of CFD for the design of high-lift multi-element airfoils has not been established in industry yet [10].

It has already been shown, that RANS-simulation is able to capture sensitivities of the flow on geometric changes [11]. Now to demonstrate the reliability and potential of aerodynamic design by numerical optimization based on advanced CFD methods for high-lift multi-element airfoils the presented work shows the accuracy together with the benefit of applying this method to practical high-lift design.

2 Optimization System

The optimization system CHAeOPS is designed as a modular system consisting of a main handling module, a geometry-generation tool, the grid-generator and the flow solver. The optimization strategy chosen is the downhill-simplex algorithm SUBPLEX [13]. Previous investigations had shown [14] that the simplex-method provides the best compromise of robustness, efficiency and globality when applied to high-lift design. The geometry generation module is able to handle single element airfoil design variables as well as the typical variables for multi-element airfoils shown in **Fig. 1**. For the latter the first set of design variables consist of the translational and rotational degrees of freedom for each element. The second set describes the design parameters used for varying the curves for splitting up the airfoil into multi elements. The grid generation for the spatial discretization is done by the DLR MegaCads system [15]. Due to the parametric concept the rerun of grid generation is also possible for a changed geometry with conserved grid quality. In **Fig. 2** the generated grid is shown along with two very different flap settings. It can be seen, that the grid generation process is able to follow these dramatic changes in input geometry without loosing the quality of the grid cells. The used flow solver is the DLR FLOWer-code [16], which is a structured multi-block Jameson-type finite-volume method with commonly used convergence acceleration techniques. For turbulence modeling the Spalart-Allmaras with Edwards-Modification (SAE) and the Wilcox k-ω (k-ω) are used within this framework.

In the following the accuracy of RANS-simulation for high-lift flows will be shown by calculations for the so called L1T2-configuration [17]. The flow conditions are given with $Ma_\infty = 0.197$ and $Re_\infty = 3.52 \cdot 10^6$. The calculations are performed both fully turbulent and with prescribed transition. The grid used contains 77.000 cells arranged in 6 blocks so that each surface part is covered within a C-type grid and another outer C-type grid is surrounding the others. A first view to the overall pressure distribution in **Fig. 3** shows nearly no effect of turbulence model or consideration of transition. It can be seen, that the pressure distribution is captured by the calculation with high accuracy. Only a close zoom to the suction peak on the slat shows differences. Another point of view is looking at the dependencies of the aerodynamic coefficients from the angle of attack (**Fig. 4**). It is observed, that the angle of attack for maximum lift is evidently higher for fully turbulent calculations. On the other hand the sensitivity found for the previously selected angle of attack is retained as long as the calculation results in an attached flow.

For the validation of the optimization contour-lines of $C_{L,max}$ for the positioning of the slat at two different deflection angles $\delta_S = 15°$ and $\delta_S = 25°$ are published for the L1T2 airfoil [18]. **Fig. 5** now shows the experimentally obtained contour-lines together with the search-path for the $\delta_S = 25°$ case. It can be easily observed that the presented optimization system is able to detect properly the optimal location for the slat.

3 Application

In the following the described optimization system is applied for increasing the performance of the described 3-element airfoil. The selected design task can be described as minimizing the drag coefficient at a given constant lift coefficient. For optimization the objective function can then be simply formulated as $F_{obj} = 100 \cdot C_D$ in order to get values of the objective function in the order of 1.

For a first attempt the 6 degrees of freedom for the deflection variables are used. The optimization, which is summarized in **Table 1**, yields in a benefit in drag coefficient of about 20%. **Figure 6** shows the changes of the geometry in the aerodynamic coordinate system. It can be seen that the angle of attack for the targeted lift coefficient is significantly reduced by the optimization. This is achieved mainly by increasing the flap deflection angle. The decrease in drag coefficient is consequently an outcome of the reduced loading of the main element resulting in thinner boundary layers and less friction drag. But the shown optimization leads to a non-practical design, since the pitching down moment is increased dramatically by approximately 40%. Regarding a complete aircraft this would result in a reduction of the overall lift coefficient due to an increased negative lift of the trimming tail.

Regarding this it is a good idea to apply a constraint on the pitching down moment. The results of this run in **Table 2** show a decrease of drag coefficient in the range of 5%. As expected the change in the flap deflection angle now stays moderate, as can be observed in **Fig. 7**. But the mechanism of reducing the drag coefficient stays the same: the reduction of the angle of attack to reduce the load of the main element. In contrast to the first optimization now the slat load is increased to retain the target lift.

In a last optimization additionally the sizing of the elements is added, so that the number of design variables is increased. Now again the drag coefficient is decreased by the optimization by 20% (**Table 3**). But contrary to the first optimization the pitching moment is retained. In this case the reduced loading of the main element is compensated by an increased loading of the high-lift devices, which is achieved mainly by extending the element sizes. **Figure 8** shows the differences in the geometry. An interesting result of this optimization is that the maximum lift coefficient was nearly not affected by the geometrical changes, although it was not part of the optimization objective.

4 Concluding remarks

The work described the application of a design tool for the optimization of multi-element airfoils based on RANS-simulation. It has been shown first that the DLR FLOWer-code is able to predict accurately the flow around a multi-element airfoil. For multi-element airfoils mainly the turbulence model and the consideration of laminar/turbulent transition are identified to have a major influence on the result. Further on it has been outlined that the presented optimization tool is able to detect an optimal configuration accurately. Finally the presented optimization system was used to increase the take-off performance of a 3-element airfoil. In a loop, that revealed additional aspects of using the optimization, the drag of the initial configuration was decreased by 20%. In this context it was observed that applying constraints is a necessary requirement for practical optimization. It should be noted here, that the presented design method still needs the expertise of the user, since the identified sensitive parameters can lead to wrong results if they are misused.

References

1. S. Rogers, N. Wiltberger and D. Kwak: Efficient Simulation of Incompressible Viscous Flow Over Single and Multi-Element Airfoils. AIAA Paper 92-0405 (1992)
2. I. Lindblad and K. de Cock: CFD Prediction of Maximum Lift of a 2D High Lift Configuration. AIAA Paper 99-3180 (1999)
3. R. Rudnik: Untersuchung der Leistungsfähigkeit von Zweigleichungs-Turbulenzmodellen bei Profilumströmungen. Dissertation, TU Berlin, DLR-FB 97-49 (1998)
4. I. Fejtek: Summary of Code Validation Results for a Multiple Element Airfoil Test Case. AIAA Paper 97-1932 (1997)
5. F. Lynch, R. Potter and F. Spaid: Requirements for Effective High Lift CFD. In: *ICAS Proceedings 1996* (1996) pp. 1479–1492
6. S. Eyi, K. Lee, S.E. Rogers and D. Kwak: High-Lift Design Optimization Using the Navier-Stokes Equations. AIAA Paper 95-0447 (1995)
7. J. Lacasse and O. Baysal: Shape Optimization of Single- and Two-Element Airfoils on Multiblock Grids. AIAA Paper 94-4273 (1994)
8. E. Besnard, A. Schmitz, E. Boscher, N. Garcia and T. Cebeci: Two-Dimensional Aircraft High Lift System Design and Optimization. AIAA Paper 98-0123 (1998)
9. K.-W. Bock: 'Aerodynamic Design by Optimization'. In: *AGARD CP 463* (1989) pp. 20–1 to 20–12
10. D. Reckzeh: 'Design Work for the A3XX High-Lift-Wing'. In: *Notes on Numerical Fluid Mechanics 77*, ed. by S. Wagner, U. Rist, J. Heinemann, and R. Hilbig, (Springer-Verlag, Berlin/Heidelberg/NewYork 2002) pp. 3–10
11. J. Wild: 'Numerical Investigation of the Reynolds-Number-Effect on the Setting Sensitivity of a 2D High-Lift Multi-Element Airfoil'. In: *Notes on Numerical Fluid Mechanics 77*, ed by S. Wagner, U. Rist, J. Heinemann, and R. Hilbig, (Springer-Verlag, Berlin/Heidelberg/NewYork 2002) pp. 35–41
12. J. Wild: Numerische Optimierung von zweidimensionalen Hochauftriebskonfigurationen durch Lösung der Navier-Stokes-Gleichungen. Dissertation, TU Braunschweig, DLR-FB 2001-11 (2001)
13. T. Rowan: Functional Stability Analysis of Numerical Algorithms. Ph.D. thesis, Department of Computer Sciences, University of Texas at Austin, USA, (1990)
14. J. Wild: 'Direct Optimization of Multi-Element Airfoils for High-Lift using Navier-Stokes Equations'. In: *Computational Fluid Dynamics '98* (Wiley & Sons, Chechester, UK 1988) pp. 383–390
15. O. Brodersen, M. Hepperle, A. Ronzheimer, C.-C. Rossow and B. Schöning: 'The Parametric Grid Generation System MegaCads'. In: *5th International Conference on Numerical Grid Generation in Computational Field Simulation*, ed. by B. Soni, J. Thompson, J. Häuser, and P. Eiseman (National Science Foundation (NSF) 1996) pp. 353–362
16. DLR (Ed.), Institut für Entwurfsaerodynamik, Braunschweig: *FLOWer Installation and User Handbook, Release 115* (1997)
17. I. Moir: 'Measurements on a Two-Dimensional Airfoil with High-Lift Devices'. In: *AGARD AR 303* (1994) pp. A2–1 to A2–12
18. D.S. Woodward, and D.E. Lean: 'Where is High-Lift Today'. In: *AGARD CP 515* (1993) pp. 1–1 to 1–45

Table 1. Results of the drag optimization of the device deflections

$F_{obj}(\boldsymbol{x}) = 100 \cdot C_D$
$C_L = 3.75$
$\boldsymbol{x} = (x_S, y_S, \delta_S, x_F, y_F, \delta_F)$

	initial	*optimized*	Δ
C_L	3.7509	3.7514	+0.02%
C_D	0.08271	0.06586	-20.37%
C_m	-0.3706	-0.5157	-39.17%
α [°]	19.66	13.20	-6.46

Table 2. Results of the drag optimization of the device deflections with constrained pitching moment

$F_{obj}(\boldsymbol{x}) = 100 \cdot C_D$
$C_L = 3.777,\ C_m > -0.3640$
$\boldsymbol{x} = (x_S, y_S, \delta_S, x_F, y_F, \delta_F)$

	initial	*optimized*	Δ
C_L	3.7776	3.7786	+0.03%
C_D	0.08590	0.08123	-5.43%
C_m	-0.3640	-0.3612	+0.77%
α [°]	20.11	19.34	-0.77

Table 3. Results of the drag optimization of the device deflections and sizes with constrained pitching moment

$F_{obj}(\boldsymbol{x}) = 100 \cdot C_D$
$C_L = 3.777,\ C_m > -0.3640$
$\boldsymbol{x} = (x_S, y_S, \delta_S, x_{o,S}, x_{n,S}, x_{u,S},$
$x_F, y_F, \delta_F, x_{o,F}, x_{n,F}, x_{u,F})$

	initial	*optimized*	Δ
C_L	3.7767	3.7780	+0.03%
C_D	0.08637	0.06840	-20.81%
C_m	-0.3624	-0.3639	-0.41%
α [°]	20.16	17.93	-2.23
$C_{L,max}$	4.1782	4.2567	+1.88%

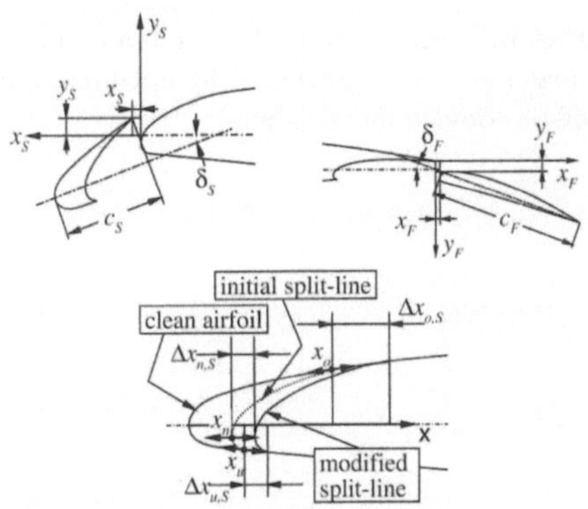

Fig. 1. Typical design variables for the design of a multi-element airfoil

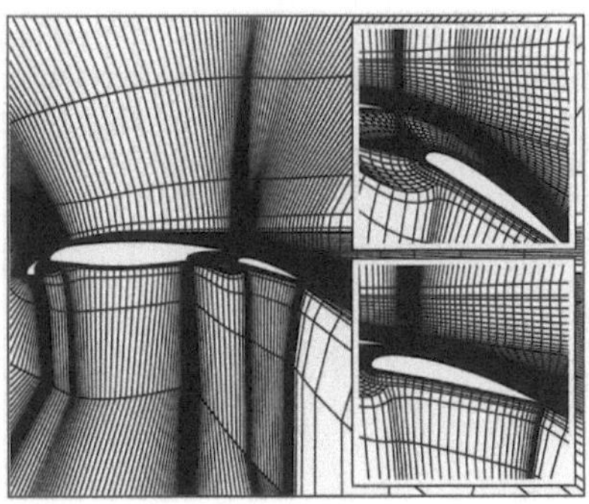

Fig. 2. Variable grid around a 3-element airfoil with two different flap settings (only every 4th grid line displayed)

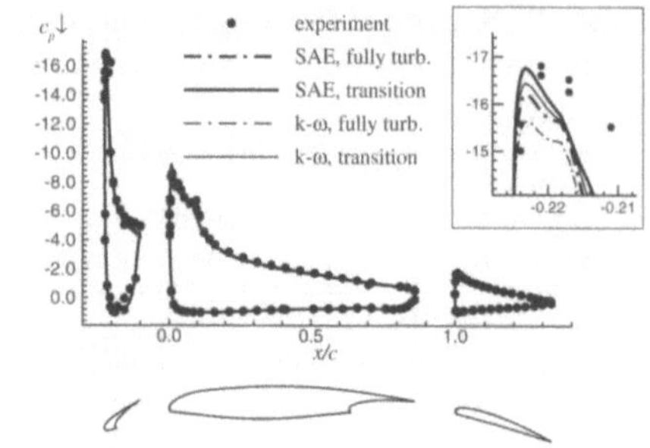

Fig. 3. Calculated pressure distributions compared with experimental data

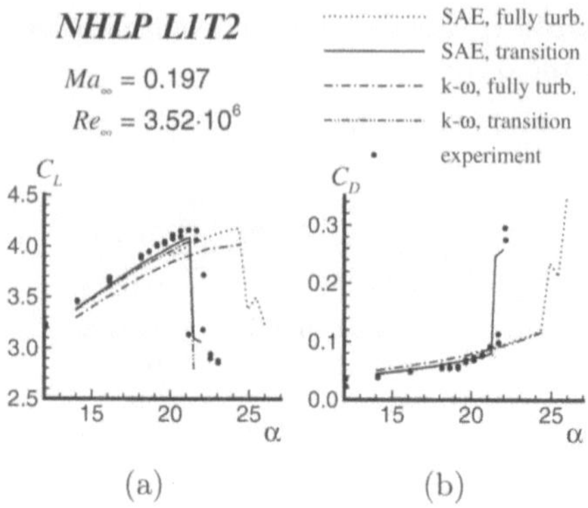

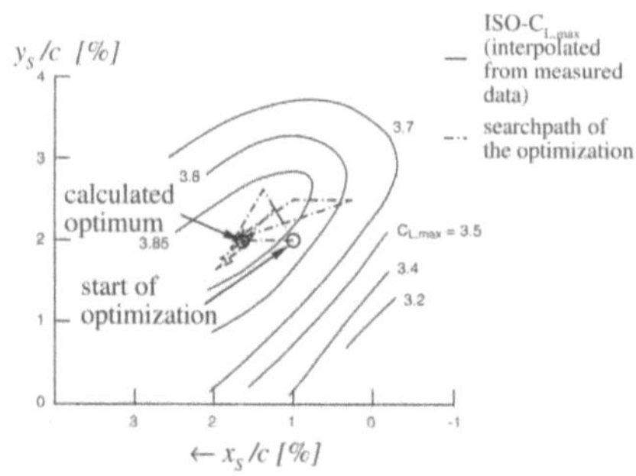

Fig. 5. Comparison of calculated optimum with experimentally determined contour-lines of $C_{L,max}$ for the L1T2-configuration

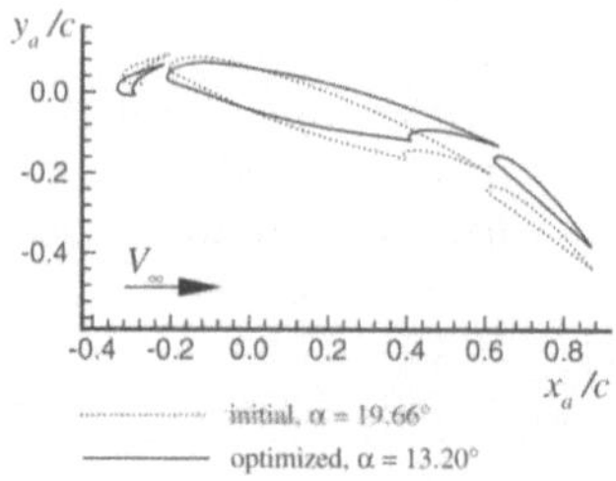

Fig. 6. Comparison of initial and optimized geometry for the drag optimization of the device deflections

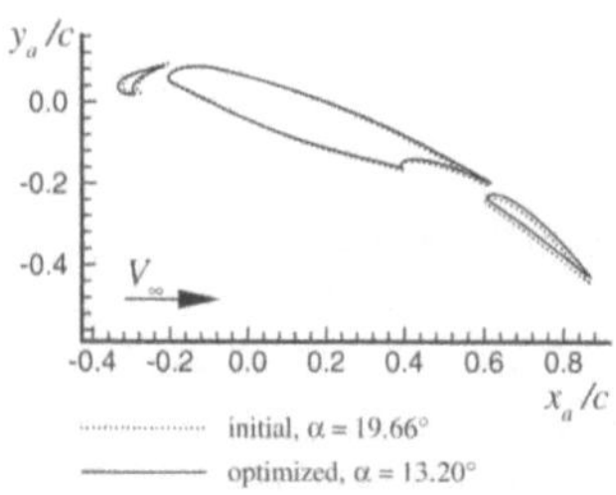

Fig. 7. Comparison of initial and optimized geometry for the drag optimization of the device deflections with constrained pitching moment

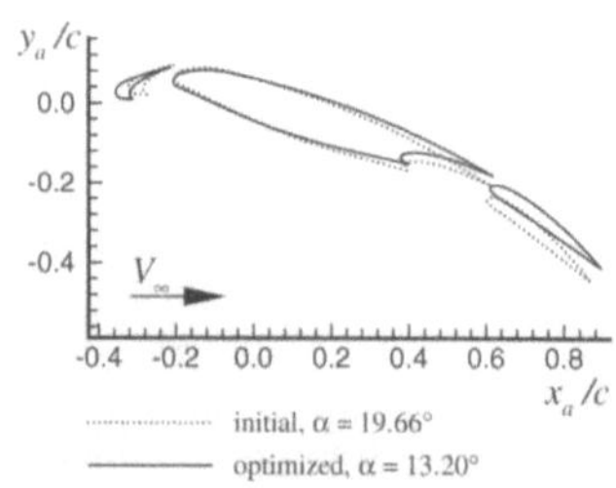

Fig. 8. Comparison of initial and optimized geometry for the drag optimization of the device deflections and sizes with constrained pitching moment

Thrust and Lift Dependence of a Plunging Airfoil on Oscillation Frequency and Amplitude

J. Young and J.C.S. Lai

School of Aerospace and Mechanical Engineering, UNSW@ADFA, Australian Defence Force Academy, Canberra, ACT, 2600, Australia

Abstract. The flow around a plunging airfoil at Reynolds number Re = 20,000 is simulated with a 2D Navier Stokes solver. At high frequency and low amplitude, lift and thrust of the airfoil are found to agree closely with potential flow solutions produced by an unsteady panel method. At low frequency and larger amplitude, the thrust is reduced significantly by leading edge separation effects.

1 Introduction

Oscillating airfoils have been studied in a variety of contexts such as dynamic stall (eg. Choudhuri and Knight [1]), fish and marine mammal propulsion (eg. Lighthill [2], Triantafyllou *et al* [3]), and lift and thrust generation for Micro Air Vehicles using flapping wings (eg. Shyy *et al* [4]). This work uses a Navier Stokes CFD solver to explore the dependency of thrust and lift for a plunging airfoil on frequency and amplitude. In a 2D flow the parameters of interest are the reduced oscillation frequency $k = \omega c/2U_\infty$, non-dimensional plunging amplitude $h = a/c$, and their product kh (equivalent to Strouhal number St).

2 Numerical Method

2.1 Navier Stokes (NS) solver

The NS code solves the strong-conservation form of the 2D unsteady compressible Navier Stokes equations, in generalised coordinates $\partial\hat{Q}/\partial\tau + \partial\hat{F}/\partial\xi + \partial\hat{G}/\partial\eta = \partial\hat{R}/\partial\xi + \partial\hat{S}/\partial\eta$ with $\hat{Q}$ the vector of conserved variables $(\rho, \rho u, \rho v, e)/J$, $\hat{F}$ and $\hat{G}$ the inviscid fluxes and $\hat{R}$ and $\hat{S}$ the viscous flux terms.

These equations may be re-written to show the variation of the Jacobian of the coordinate transformation J with time, $\partial Q/\partial\tau + J\partial\hat{F}/\partial\xi + J\partial\hat{G}/\partial\eta = J\partial\hat{R}/\partial\xi + J\partial\hat{S}/\partial\eta + (Q/J)\partial J/\partial\tau$, with $Q = J\hat{Q}$.

The equations are solved on a structured C-grid wrapped around the airfoil. The plunging motion of the airfoil is introduced by deforming the grid adjacent to the airfoil, resulting in time-dependent grid metrics and Jacobian. The method generally follows Tuncer and Platzer [5], but with modification. The numerical method uses Crank-Nicolson 2nd-order time discretisation; viscous flux terms are evaluated using 2nd-order central differences in space, and inviscid fluxes with a 3rd-order accurate Osher upwind scheme. The resulting semi-implicit equations are non-linear in Q^{n+1}, and this solution at the next time step is solved for using Newton sub-iteration. Typically, 2 iterations were found sufficient.

$$\left[I + \frac{\Delta\tau}{2} J\partial_\xi \hat{A}\right]^p \left[I + \frac{\Delta\tau}{2} J\left\{\partial_\eta \hat{B} - \partial_\eta \hat{N}\right\}\right]^p \left(Q^{p+1} - Q^p\right) =$$
$$-\left(Q^p - Q^n\right) + \frac{\Delta\tau}{2} J^p \left[-\partial_\xi \hat{F} - \partial_\eta \hat{G} + \partial_\xi \hat{R} + \partial_\eta \hat{S}\right]^p$$
$$+\frac{\Delta\tau}{2} J^n \left[-\partial_\xi \hat{F} - \partial_\eta \hat{G} + \partial_\xi \hat{R} + \partial_\eta \hat{S}\right]^n + \frac{Q^p + Q^n}{J^p + J^n}\left(J^p - J^n\right) \quad (1)$$

The Newton sub-iteration process involves formulation of the inviscid and viscous flux-Jacobian matrices $\hat{A} = \partial\hat{F}/\partial Q$, $\hat{B} = \partial\hat{G}/\partial Q$, $\hat{M} = \partial\hat{R}/\partial Q$ and $\hat{N} = \partial\hat{S}/\partial Q$. The method in Tuncer and Platzer uses the thin-layer approximation and ignores the streamwise viscous derivative $\partial\hat{R}/\partial\xi$, any cross-derivatives in the $\partial\hat{S}/\partial\eta$ term, and the flux-Jacobian $\hat{M}$. The assumption of this approximation is that streamwise derivatives in the viscous terms are negligible in comparison to normal derivatives, and flow separation is moderate. To more fully account for all viscous effects without sacrificing computational speed, the current method retains all viscous terms on the right-hand side of the equations. However $\hat{M}$ is still ignored on the left-hand side. It may be seen that with sufficient sub-iterations at each time-step, Q^{p+1} approaches Q^p and the left-hand side vanishes, and the full viscous equations are recovered. Inflow and outflow boundary conditions are implemented using second order Riemann invariant extrapolations. The method is thus overall 2nd-order in time and space.

2.2 Unsteady Panel Method (UPM) code

To isolate and compare elements of the controlling physics of the plunging airfoil flow, an unsteady panel method code was also developed, following Basu and Hancock [6] and Jones *et al* [7]. The method assumes potential flow (ie. inviscid and incompressible). The method enforces the Kutta condition at the trailing edge, which does not allow any separation over the airfoil surface. As the airfoil translates and the lift changes, vorticity is shed into the wake at each time step, and this is allowed to evolve due to self-induced velocities. Thus the non-linear character of the wake, and any consequent effect on lift and thrust, is preserved. The UPM code was tested against, and accurately reproduced, both wake shapes and thrust coefficient results for the panel method used in Jones *et al.*

3 Code Verification and Validation

Results were calculated using the NS solver for a NACA0012 airfoil pitching about the quarter-chord point, at *Re* = 12,000. The NS code was run at M_∞ = 0.2 and 0.05, with pitch amplitudes of ±2° and ±4°, and reduced pitching frequencies of 1 to 12. Fig. 1 shows the comparison with calculations by Ramamurti and Sandberg [8] using an incompressible finite-elements solver, and experiments by Koochesfahani [9]. We see that with $M_\infty = 0.05$, the NS code reproduces the same trend observed by Ramamurti and Sandberg, in that thrust is approximately dependent on the product $k\alpha$ rather than k or α individually. The NS code is also closer to the experimental values observed by Koochesfahani, although it still under-predicts the thrust. Potential sources of error in the Koochesfahani observations are discussed in Ramamurti and Sandberg. It is worth noting that for $M_\infty = 0.2$ the NS code shows very different behaviour. It is clear that at this free-stream mach number, the flow cannot be considered incompressible due to the large velocities induced over the nose of the airfoil during the pitching cycle. As further validation, a stationary NACA0012 airfoil at $Re = 12{,}000$ was considered. Koochesfahani measured a natural vortex shedding frequency in the wake equivalent to $k = 8.7$. The NS code reproduced this natural vortex shedding, at a frequency $k = 8.7$.

NS calculations were performed on a 541×61 structured C-grid, with 189 streamwise points on the airfoil. The first normal point was at a distance of 8.75×10^{-6} chords from the airfoil surface, with approximately 25-30 points within the boundary layer. The boundaries were located approximately 20 chords from the airfoil in all directions. Grid refinement studies were performed by doubling and quadrupling the number of normal and streamwise points in turn, and by doubling both normal and streamwise points together, for selected frequency and amplitude combinations. Maximum variations of 1% on lift coefficients, 2% on drag coefficients and 5% on stationary shedding frequency, were observed.

4 Thrust and Lift Dependence on Plunging Frequency and Amplitude

Fig. 2 shows the mean (time-averaged) thrust coefficient predicted by the Garrick flat plate potential flow analysis [10], and by the UPM code for a NACA0012 airfoil. It also shows the thrust predicted by the NS code for a NACA0012 airfoil, for free stream Mach numbers $M_\infty = 0.2$ and 0.05 and Reynolds number $Re = 20{,}000$. The flow is assumed to be completely laminar, which is supported by qualitative comparison of the wake structure with experiment (see Young and Lai [11]). The plunging frequency k and plunge amplitude (non-dimensionalised with respect to the chord) h are varied so as to keep their product kh constant at a value of 0.3. Note that kh is equivalent to the Strouhal number St based on the total excursion of the trailing edge of airfoil, differing only by the factor $2/\pi$. Previous studies (eg. Triantafyllou *et al* [3]) have postulated that kh (or St) is the sole parameter that determines thrust for both pitching and plunging, although the recent work by Ramamurti and Sandberg suggests there may be some frequency dependence at higher pitching frequencies.

Here we ask what we would expect to see with kh held constant across a range of plunging frequencies and amplitudes. The Garrick analysis, based on a plunging flat plate in a potential flow, is useful as a first approximation. The mean thrust may be calculated by $C_{Tmean} = 4\pi(kh)^2(F^2 + G^2)$, with $F(k)$ and $G(k)$ the Theodorsen functions.

We see that indeed the mean thrust coefficient does asymptote to a constant value dependent on kh alone, but only for $k > 4$ to 5. Below this frequency there is significant variation with both k and h. Next we can determine the effect of some of the assumptions inherent in the Garrick analysis. Firstly the wake, consisting of vorticity shed from the trailing edge, is 'frozen' in that it is not allowed to evolve in response to the self-induced velocities created by the wake vorticity. The UPM code removes this restriction by allowing wake evolution and non-linear vortex wrap-up. It also simulates a finite-thickness airfoil rather than a flat plate. However the flow is assumed still to be incompressible and inviscid, and the Kutta condition is still enforced at the trailing edge, so that no separation is allowed anywhere on the airfoil surface.

We can see from Fig. 2 that the UPM results fairly closely match the Garrick analysis, including the decrease in thrust as frequency is increased, and an asymptotic approach to a constant value, but with a slightly lower asymptotic value. This may be attributed to the fact that the wake becomes increasingly non-linear with increasing frequency, in that the vortex wrap-up takes place faster and closer to the airfoil trailing edge, so having more effect on the velocities over the airfoil surface.

The NS code is used to add viscosity and the possibility of flow separation. Results were calculated for free stream Mach numbers of $M_\infty = 0.2$ and 0.05. The results at $M_\infty = 0.05$ showed a reasonable match with the asymptotic trend of the Garrick and UPM results at higher frequency, indicating that for high enough frequency, the thrust does indeed become approximately independent of k and h separately, and dependent only on kh. However the trend at low frequency is the opposite of that for Garrick and UPM, in that the thrust is increasing with frequency rather than decreasing. This is to be expected, as the small amplitude assumption inherent in the Garrick analysis is increasingly in error as the frequency decreases with kh held constant.

It was determined from the code validation presented earlier, that as the pitching frequency increased, the velocities induced over the airfoil nose became too high for the flow to be assumed incompressible with $M_\infty = 0.2$. This effect was also observed for plunging as shown here by the dramatic loss of thrust for $k > 4$.

The plunging motion of the airfoil is sinusoidal about zero displacement, so the downstroke and upstroke produce the same magnitude of vertical force. Fig. 3 shows the peak lift coefficient achieved during the plunging cycle, as predicted by the Garrick analysis $C_{Lpeak} = 4\pi kh(F^2 + G^2 + kG + k^2/4)^{1/2}$, as well as the UPM code and the NS code for $M_\infty = 0.2$ and 0.05. The Garrick result shows some initial decrease in lift with increasing frequency, then an asymptotic approach to a linear variation with frequency. The UPM code reproduces both these trends, but with marginally less peak lift predicted at a given frequency. The NS results at $M_\infty = 0.05$ also reproduce these trends, although with higher lift predicted at the higher frequencies.

Thus the NS code (at $M_\infty = 0.05$) predicts much lower thrust, but approximately the same lift, at the low plunging frequencies. Figures 4, 5 and 6 show the pressure distribution on the airfoil surface for $k = 2$, 4 and 8 respectively (with kh held constant at 0.3). Shown are UPM and NS $M_\infty = 0.05$ results at the top of the upstroke in each case. These figures show a sharp dip in pressure near the leading edge, caused by vortex separation, which grows less pronounced as the plunging frequency increases. For the $k = 2$ case, the vortex still exhibits a significant influence on the airfoil as it passes over the point of maximum thickness, at which time the low pressure will cause drag rather than thrust, but with little change in lift. In this way LE separation appears to cause the reduction in thrust coefficient for $k < 4$. The reasons for this dependence of LE separation on plunging frequency remain to be determined.

5 Conclusion

The thrust of a plunging airfoil, in an incompressible flow at low Reynolds number, is shown to depend on both the frequency k and amplitude h of the airfoil motion. At higher frequencies the mean thrust becomes approximately a function of kh only. Here the thrust is dominated by inviscid effects, as shown by comparison with unsteady panel method results and linearised analysis. At lower frequencies, leading edge separation causes a marked reduction in mean thrust. Peak lift appears to be relatively unaffected by leading edge separation.

References

1. P.G. Choudhuri, D.D. Knight: Two-Dimensional Unsteady Leading-Edge Separation on a Pitching Airfoil, AIAA J. **32**(4), 1994, pp. 673–681
2. M.J. Lighthill: A Note on the Swimming of Slender Fish, J. Fluid Mech. **9**, 1960, pp. 305–317
3. G.S. Triantafyllou, M.S. Triantafyllou, M.A. Grosenbaugh: Optimal Thrust Development in Oscillating Foils with Application to Fish Propulsion, J. Fluids and Structures **7**, 1993, pp. 205–224
4. W. Shyy, M. Berg, D. Ljungqvist: Flapping and Flexible Wings for Biological and Micro Air Vehicles, Prog. Aerospace Sci. **35**, 1999, pp.455–506
5. I.H. Tuncer, M.F. Platzer: Computational Study of Flapping Airfoil Aerodynamics, J. Aircraft **37**(3), 2000, pp. 514–520
6. B.C. Basu, G.J. Hancock: The Unsteady Motion of a Two-Dimensional Aerofoil in Incompressible Inviscid Flow, J. Fluid Mech. **87**(1), 1978, pp. 159–178
7. K.D. Jones, C.M. Dohring, M.F. Platzer: Wake Structures Behind Plunging Airfoils: A Comparison of Numerical and Experimental Results, AIAA Paper 96-0078, 1996
8. R. Ramamurti, W. Sandberg: Simulation of Flow About Flapping Airfoils Using Finite Element Incompressible Flow Solver, AIAA J. **39**(2), 2001, pp. 253–260
9. M.M. Koochesfahani: Vortical Patterns in the Wake of an Oscillating Airfoil, AIAA J. **27**(9), 1989, pp. 1200–1205
10. I.E. Garrick: Propulsion of a Flapping and Oscillating Airfoil, NACA Report 567, 1937
11. J. Young, J.C.S. Lai: A Computational Study of the Wake Structure of a Plunging Airfoil, Proc. 1st ICCFD, Kyoto, 2000, pp. 151–156

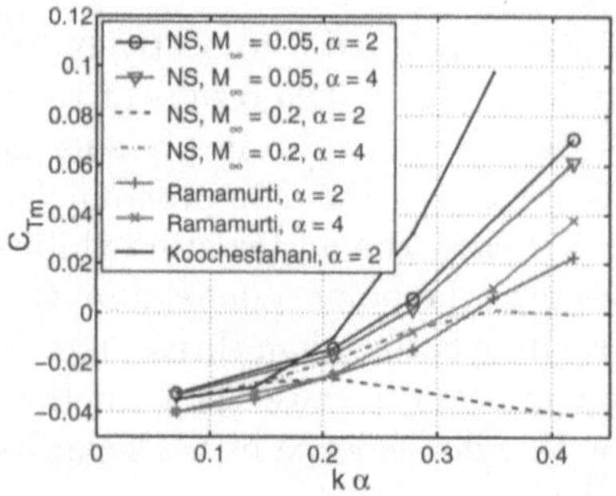

Fig. 1. Mean thrust coefficient, pitching airfoil, Re = 12,000

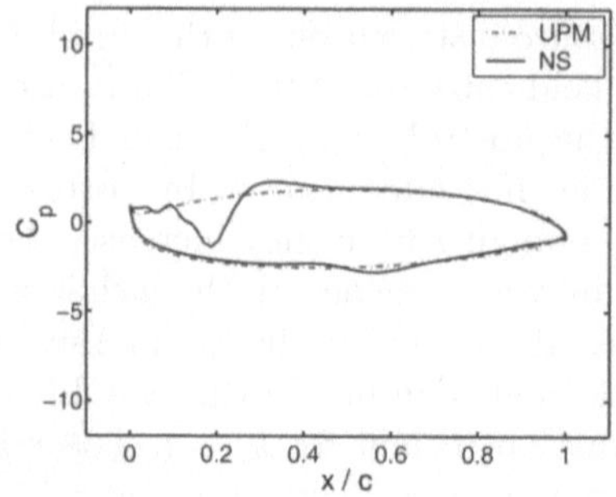

Fig. 4. Airfoil surface pressure distribution, $k = 2$

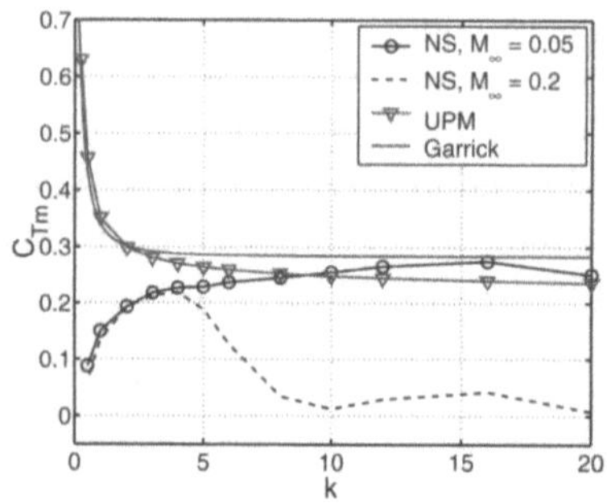

Fig. 2. Mean thrust coefficient, plunging airfoil, Re = 20,000

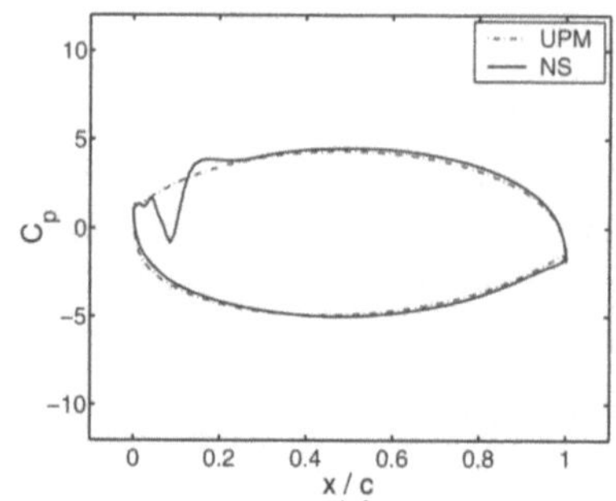

Fig. 5. Airfoil surface pressure distribution, $k = 4$

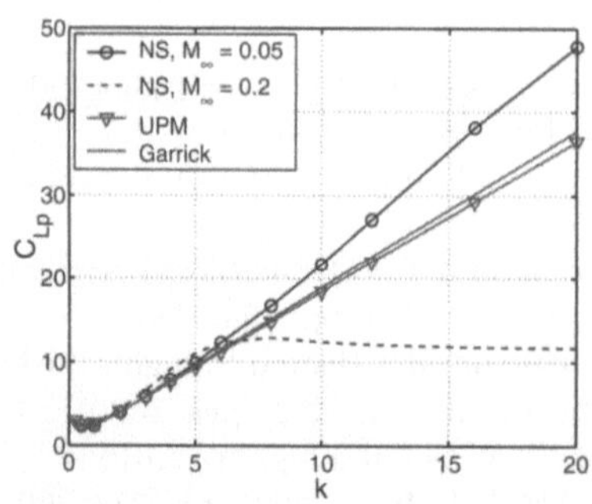

Fig. 3. Peak lift coefficient, plunging airfoil, Re = 20,000

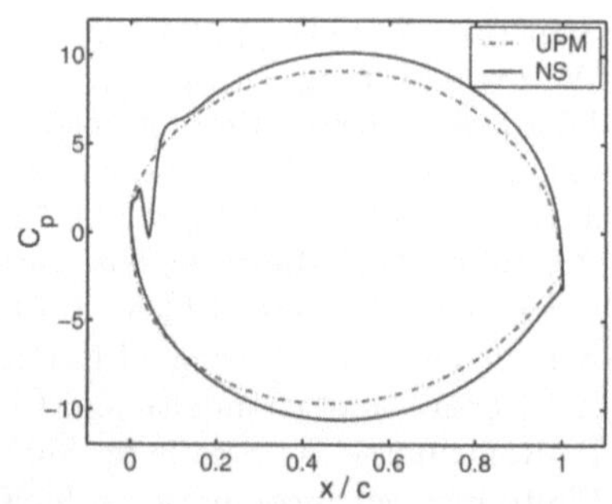

Fig. 6. Airfoil surface pressure distribution, $k = 8$

A New Set of Boundary Conditions for Aerodynamic Design

Michele Ferlauto, Angelo Iollo, Roberto Marsilio, and Luca Zannetti

DIASP, Politecnico di Torino, c.so Duca degli Abruzzi, 24, Torino, Italy

Abstract. The reflectiveness of various boundary conditions is here investigated in the contest of Godunov methods and measured by an *ad hoc* introduced parameter τ. A new set of boundary conditions, for which is always $\tau = 0$, i.e. totally non-reflecting, is then proposed. The approach is also integrated in a novel procedure of aerodynamic design, in the framework of a distributed boundary control based on the governing equations adjoint.

1 Introduction

A common practice in CFD is to set virtual boundaries to limit the computational domain and to obtain a tractable problem. A crucial consideration is related to how the boundary conditions (BC) reflect the signals coming from the interior domain [1,2]. In general a new perturbation, which can be stronger or weaker than the original one, will arise at the boundary. In external flow computations for example, the presence of such reflections is typically observed when the boundary condition is intended to simulate asymptotic flow-field conditions. This behavior is highly pernicious in simulations of systems over-relaxed to the steady state, since in this case one is interested in the steady flow field and any undamped perturbation is merely CPU-time consuming and delaying convergence.

In the present paper we review the reflecting/non-reflecting properties of several boundary conditions typically used in CFD and present a class of non-reflecting boundary conditions derived in the framework of a distributed boundary control based on the governing equations adjoint. The Riemann invariants are taken as boundary controls and a functional representing the convergence to given boundary conditions is defined. Minimizing such functional with respect to the controls we are able to obtain totally non-reflecting boundaries, while ensuring the prescribed physical conditions.

2 Boundary condition reflectiveness

In this section we analyze the reflecting features of the most commonly adopted boundary conditions. Dealing with compressible flows, it is natural to work with characteristic variables, which also give to the analysis a straightforward physical interpretation. Whenever a characteristic, approaching the boundary, is not in accord with the imposed boundary conditions, it reflects to the interior, giving rise to a new perturbation traveling backwards into the flow. This reflection is governed by the coupling of boundary condition and traveling signal. In a 1-D isentropic sketch, the set of BCs can be represented by the functional form

$$\mathcal{B}(a^*(\boldsymbol{U}), u^*(\boldsymbol{U})) = 0 \tag{1}$$

where a is the sound velocity, u the flow velocity and $\boldsymbol{U}$ the conservative variables as defined in (13). The signals, or Riemann invariants, conveyed along characteristics are

$$R_3 = a/\kappa + u, \qquad R_1 = a/\kappa - u \tag{2}$$

Let consider that R_3 is the signal approaching the boundary. The reflected signal will travel as R_1. A measure of the strength of the new perturbation is given by

$$\delta R_1 = \frac{\partial R_1}{\partial R_3} \delta R_3 = \tau \delta R_3 \tag{3}$$

where $\kappa = (\gamma - 1)/2$. The derivative $\tau = \partial R_1 / \partial R_3$ plays here the role of an amplification/damping factor. If the coupling of (1) and (2) results in values of $|\tau| < 1$, then the boundary condition is able to reduce the initial gap between flow state and the incoming signal. For values of $|\tau| > 1$ the perturbation is somehow amplified by boundary condition. When $|\tau| = 1$ the perturbation is reflected without changes in strength. Low values of τ are advantageous because they enhance convergence by damping perturbations that eventually arise during unsteady flow computations.

2.1 Inlet Boundary

At inlet the total temperature, total pressure and flow angles are imposed. In our 1D sketch, only total temperature is imposed, while the Riemann invariant $R_1 = R_1^*$ is given, coming from the interior domain (see fig.1b)

$$\gamma T^0 = a^2 + \kappa u^2 \tag{4}$$

recalling (2), sound and flow velocities in (4) can be expressed in terms of R_1 and R_3. By differentiating (4) with respect to R_1, we have

$$\tau = \frac{\partial R_3}{\partial R_1} = \frac{1-\kappa}{1+\kappa} - \frac{2\kappa}{1+\kappa} \frac{1/\kappa - M}{\sqrt{\frac{1+\kappa}{\kappa}(1+\kappa M^2) - \kappa(1/\kappa - M)}} \tag{5}$$

The τ function is plot versus the mach number in figure 1. We observe that the inlet boundary reflects exactly an incoming perturbation for $M = 0$ and then it has a dumping effect ($\tau < 1$) as the mach increases. The proposed model, although its simplicity, retains most of its effectiveness even for 2D or 3D configurations, as a basic model for the approximate solution of the Riemann problem [3]. Supersonic inlets are not considered, since they are inherently non-reflecting boundaries. In fact a reflection is impossible since no one of the signals can flow back to the inlet.

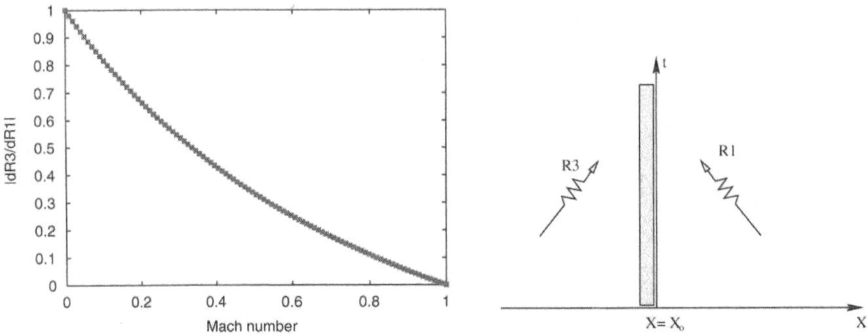

Fig. 1. Inlet reflectiveness versus inlet Mach number.

2.2 Outlet Boundary

One boundary condition must be prescribed at subsonic outlets. In general, the static pressure level p_{ex} is imposed, while $R_3 = R_3^*$ is coming from the inner domain. We can express the sound velocity at outlet as

$$a^* = a_{ex} = \sqrt{\gamma p_{ex}^{\frac{\gamma-1}{\gamma}}} \tag{6}$$

which lead to

$$R_1^* = 2a_{ex}/\kappa - R_3, \qquad \tau = \frac{\partial R_1}{\partial R_3} = -1 \tag{7}$$

The static pressure is not the only variable that can be adopted. One can, for instance, set the outlet mach number M_{ex}. In this case we have

$$u^* = a^* M_{ex} \tag{8}$$

and the expression for τ is

$$\tau = \frac{\partial R_1}{\partial R_3} = \frac{(1 - \kappa M_{ex})}{(1 + \kappa M_{ex})} \tag{9}$$

Another possibility is to set the mass flow rate $\dot{m} = (\rho u)^*$. Again, reformulating the density as function of sound velocity and by using (2), from $d\dot{m} = 0$ is

$$\tau = \frac{(1 + M_{ex})}{(1 - M_{ex})} \tag{10}$$

The τ-diagram as from (9) and (10) are shown in figure 2. As visible, a constraint on the mass flow rate results in a severe and strongly reflecting BC.

3 A new set of boundary conditions

Boundary conditions which result always in $\tau = 0$ can be obtained by decoupling Riemann invariants from the boundary condition. This can be done by setting

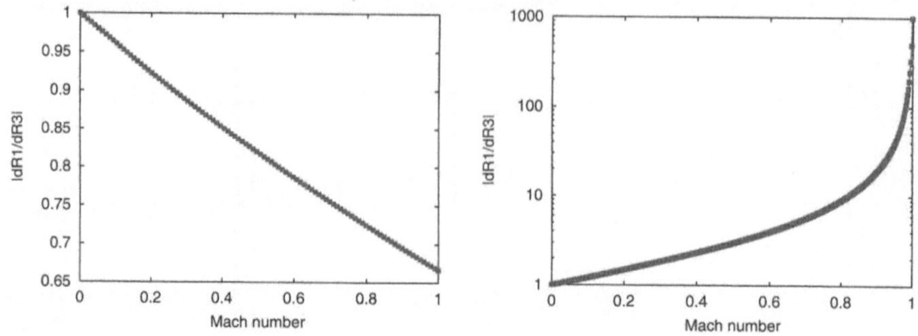

Fig. 2. Outlet reflectiveness versus Mach number. BC on the exit mach number (*left*) BC on the mass flow rate (*right*).

explicitly the value of the Riemann invariant virtually coming from the border. This value is then updated during the computation in order to satisfy the original boundary condition. In other words we consider an optimization process where Riemann invariants are some of the control parameters and the original BCs are accounted by penalization in the objective function to be minimized. Whatever the boundary condition to be satisfied, we set both R_3 and R_1. Depending on the flow velocity direction, one invariant is imposed, the other is coming from the domain interior. The flow state at boundary is given by

$$a^* = \frac{\kappa(R_1^* + R_3^*)}{2}, \qquad u^* = \frac{(R_3^* - R_1^*)}{2} \tag{11}$$

The second step of the procedure refers to the integration of the proposed boundary conditions, e.g. the inlet BC, in the framework of a tool for optimal design [4–6]. Here we recall briefly the mathematical model.

In a cylindric frame of reference, the compressible Euler equations with volume forces acting on the fluid are

$$\frac{\partial \boldsymbol{U}}{\partial t} + \frac{\partial \boldsymbol{A}}{\partial x} + \frac{\partial \boldsymbol{B}}{\partial r} + \boldsymbol{Q} = 0 \tag{12}$$

where $\boldsymbol{U} = \{\rho, \rho u, \rho v, \rho w, e\}^T$ and

$$\begin{aligned}
\boldsymbol{A} &= \left\{\rho u, p + \rho u^2, \rho uv, \rho uw, u(p+e)\right\}^T \\
\boldsymbol{B} &= \left\{\rho w, \rho uw, \rho vw, p + \rho w^2, w(p+e)\right\}^T \\
\boldsymbol{Q} &= \left\{\frac{\rho w}{r}, \frac{\rho uw}{r} - F^x, 2\frac{\rho vw}{r} - F^\theta, \frac{\rho(v^2 - w^2)}{r} - F^r, \frac{w(p+e)}{r} - \boldsymbol{F}\cdot\boldsymbol{q}\right\}^T
\end{aligned} \tag{13}$$

as usual ρ is density, p pressure, e total internal energy per unit volume. System (12) is integrated in time using a finite volume formulation based on an approximate Riemann solver to compute the fluxes at cell interfaces with second order spatial accuracy. Regarding the optimization step, we derive the functional differential using the adjoint technique. The objective functional is

$$\mathcal{I} = \int_{\Gamma_o} (p + \rho u^2) d\Gamma - \int_{\Gamma_i} (p + \rho u^2) d\Gamma - \frac{\chi_o}{2} \int_{\Gamma_i} (T_i^o - \overline{T}_i^o)^2 d\Gamma \tag{14}$$

the first two terms express the conventional thrust and the last term is pertinent to the new BC at the inlet. The technique is general and can be adapted to other functionals and BCs. The maximum of $\mathcal{I}$ is constrained by the steady Euler equations

$$\boldsymbol{E}(F^\vartheta) = \boldsymbol{A}_x + \boldsymbol{B}_r + \boldsymbol{Q} = 0 \tag{15}$$

and, in this particular case, by the kinematic constraint on the blades

$$\boldsymbol{g}_t(\boldsymbol{U}(F^\vartheta)) = u g_x + w g_r - \frac{v - \omega r}{r} = 0 \tag{16}$$

The constrained maximization problem is solved by introducing the Lagrangian

$$\mathcal{L}(\boldsymbol{U}, g, F^\vartheta, \Lambda, \mu, \xi_i) = I + \int_\Omega {}^t\boldsymbol{\Lambda}\, \boldsymbol{E}\, d\Omega + \int_{\Omega_b} \mu G d\Omega + \int_{\Gamma_i} \xi_i (R_{3i} - \overline{R_{3i}}) d\Gamma \tag{17}$$

where ${}^t\boldsymbol{\Lambda}(x,r) = (\lambda_1, \lambda_2, \lambda_3, \lambda_4, \lambda_5)$, $\mu = \mu(x,r)$ and $\xi_i = \xi_i(r)$ are Lagrange multipliers. A stationary point is found when the variation of $\mathcal{L}$ vanish with respect to all of its arguments. We have

$$\delta\mathcal{L} = \delta\mathcal{L}_U + \delta\mathcal{L}_{F^\vartheta} + \delta\mathcal{L}_g + \delta\mathcal{L}_\Lambda + \delta\mathcal{L}_\mu + \delta\mathcal{L}_{\xi_i} + \delta\mathcal{L}_{\overline{R_{3i}}} \tag{18}$$

In order to have $\delta\mathcal{L} = 0$, all the single contributions to $\delta\mathcal{L}$ must vanish at the maximum, so that we enforce $\delta\mathcal{L}_U = 0$, $\delta\mathcal{L}_\Lambda = 0$, $\delta\mathcal{L}_\mu = 0$, $\delta\mathcal{L}_g = 0$, $\delta\mathcal{L}_{\xi_i} = 0$. From $\delta\mathcal{L}_U = 0$ we obtain the so called *adjoint* of the Euler equations, that is

$${}^t\boldsymbol{\Lambda}_x\, \boldsymbol{A}_U + {}^t\boldsymbol{\Lambda}_r\, \boldsymbol{B}_U - {}^t\boldsymbol{\Lambda}\, \frac{\partial \boldsymbol{Q}}{\partial \boldsymbol{U}} - \mu \frac{\partial G}{\partial \boldsymbol{U}} = 0 \tag{19}$$

on the domain Ω, and also the related conditions

$$\left[\frac{\partial \mathcal{I}}{\partial \boldsymbol{U}} + {}^t\boldsymbol{\Lambda} (\boldsymbol{A}_U\, n_x + \boldsymbol{B}_U\, n_r) + \xi_i \frac{\partial R_3}{\partial \boldsymbol{U}} \right] \delta \boldsymbol{U} = 0 \tag{20}$$

on its boundary Σ with unit normal vector $\boldsymbol{n} = (n_x, n_r)$. From ((20)) also the value of ξ_i is deduced. In general (18) results in $\delta\mathcal{L}_{F^\vartheta} \neq 0$ and $\delta\mathcal{L}_{\overline{R_{3i}}} \neq 0$. To reach the maximum δF^ϑ and $\delta\overline{R_{3i}}$ are taken such that $\delta\mathcal{L}_{F^\vartheta} > 0$, $\delta\mathcal{L}_{\overline{R_{3i}}} > 0$. for example using a conjugate gradient method [8].

As numerical examples, we discuss two paradigmatic cases arising from actual turbomachine design. In the first test a stator design is concerned. It has been shown in [7] that the related inverse problem may become either ill-posed or unstable depending on the adopted boundary conditions. Figure 6 shows the convergence history of the gradient residual (curve A) for a configuration unstable to conventional BC. In the same plot a second curve (B) is drawn, which refers also to simultaneous imposition of the new BC and thrust optimization. The second example refers to the design of a three-stage compressor via inverse problem. As allowed by the new approach, the global, nor the local, mass flow has been imposed at the exit. The final blade geometry is presented in Figure 6.

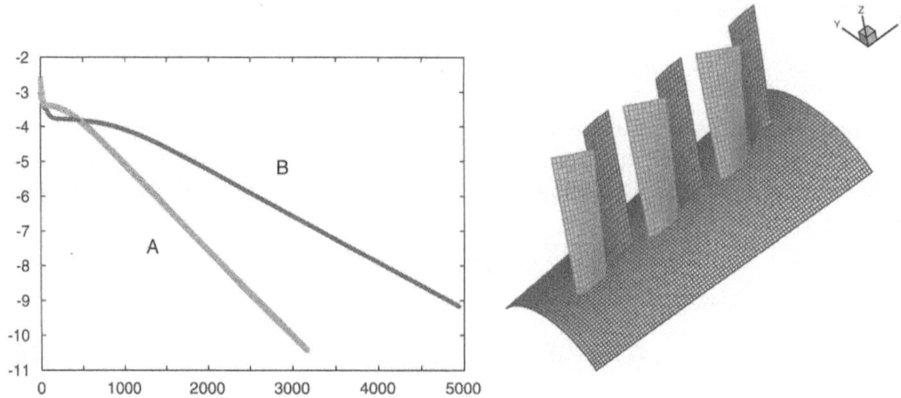

Fig. 3. Numerical experiments. L^2-norm of the gradient residual for the NASA CR 3836 test-case (*left*); blade geometry of the three-stage compressor (*right*).

4 Conclusions

New boundary conditions for aerodynamic design have been investigated and integrated in a design procedure which combines inverse problem solution and adjoint optimization. Numerical experiments have shown that such conditions are quite effective in the inverse problem solution, while the related penalitation is delaying convergence of the adjoint procedure due to the increased number of optimization parameters. Nevertheless, the convergence of the whole numerical procedure is affected by both phenomena and the advantages against conventional BC should be evaluated case by case. The effectiveness of the new set of BC is evident when conventional BCs fail or when integral variables, as the global mass flow, are imposed at the boundary.

References

1. P.A. Thompson. Time Dependent Boundary Conditions for Hyperbolic Systems. *J. Comp. Phys.* 68:1–24, 1987.
2. M.B. Giles. Non-reflecting Boundary Conditions for Euler Equations Calculations, *AIAA J.*, 28:2050–58, 1990.
3. M. Pandolfi, 'A Contribution to the Numerical Prediction of Unsteady Flows', AIAA J. 22, pp. 602–610, (1984).
4. A. Iollo, M. Ferlauto, and L. Zannetti. An aerodynamic Optimization Method Based on the Inverse Problem Adjoint Equations. *J. Comp. Phys.*, 173:87–115, 2001.
5. M. Ferlauto, A. Iollo, and L. Zannetti. Propfan Optimization Based on Inverse Problem Adjoint Equations. *4th Eur. Conf. on Turbomachinery*, Florence, 2001.
6. C. Bena, F. Larocca, L. Zannetti, Design of Multistage Axial Flow Turbines and Compressors, *IMech-E 3rd Eur. Conf. on Turbomachinery*, London, 1999.
7. L. Zannetti, M. Pandolfi. Inverse Design Technique for Cascades. *NASA CR 3836*, 1984.
8. R. Fletcher, *Practical Methods of Optimization*, John Wiley & Sons, (1980)

Numerical Analysis of the Clustered Type Aerospike Nozzle Flow: Flow Structures and Thrust Performance

Takashi Ito[1] and Kozo Fujii[2]

[1] University of Tokyo (ISAS graduate student), 3-1-1 Yoshinodai, Kanagawa, 229-8510, Japan
[2] The Institute of Space and Science, 3-1-1 Yoshinodai, Kanagawa, 229-8510, Japan

Abstract. Fundamental characteristics of the clustered type aerospike nozzle is analyzed by the computational fluid dynamics approach. The base thrust of the aerospike nozzle plays an important role to maintain the thrust even in high spike-length truncations. The pressure at the tip of the ramp increases and stronger expansion occurs toward the nozzle axis for lower module number nozzles. The stagnation pressure increases and consequently the base pressure increases for lower module number nozzles. The thrust loss at the ramp is the major reason for the performance decrease for the lower module number aerospike nozzles.

1 Introduction

Discussions on the development of the SSTO (Single Stage To Orbit) launch vehicles has become active throughout the world by the requirement of the launch cost reduction. One of the key technologies indispensable to make those vehicles in practical use is the highly performanced propulsion device.

The conventional bell-type nozzle produces thrust loss during most of its trajectory. The inherent altitude compensational feature of the aerospike nozzle allows its nozzle to perform better than the bell-type nozzle in the whole range of the altitude. [1] Furthermore, it is believed that the length of the aerospike nozzle (nozzle cut off) does not influence to the total performance and the total vehicle's weight for the annular type aerospike nozzles can be saved. The authors analyzed the fundamental characteristics of the annular-type aerospike nozzle and both the qualitative and the quantitative aspects of the above features were confirmed. [2,3] However, the throat area of the annular-type aerospike nozzle requires forming accuracy and often faces a heat deformation problem which leads to the difficulty in designing larger-scale annular-type aerospike nozzles. Therefore, clustered-type aerospike nozzles are considered to be more reliable for the actual flight even though the deficiency exists compared to the annular-type nozzles. In this study, the fundamental characteristics of the clustered aerospike nozzles are investigated based on the computational results of the numerical simulation using the three-dimensional equations.

2 Geometry of the Aerospike Nozzle and Flow Conditions

Six, twelve and twenty-four module aerospike nozzle configurations are considered. As an example, twelve module aerospike nozzles with different spike-length truncation are shown in Fig. 1. Each of the aerospike nozzle contours are designed under the assumption that the expansion process over the annular-type aerospike nozzle surface is isentropic and the nozzle contains two parts (i.e. module part and the external expansion region).

The length of the aerospike nozzle is believed not to influence the thrust performance and it was confirmed by the former studies for the annular-type aerospike nozzle. [2,3] However, it may not be true for the clustered-type aerospike nozzles. Therefore, 20%, 30% and 100% length aerospike nozzle, shown in Fig. 1 are considered.

To make the performance comparison easier, total flow rate for all the nozzle configurations are set to be the same. Therefore, all types of nozzle has the same module exit area and has 1.7 for its area ratio.

Optimum expansion for this aerospike nozzle is achieved at a pressure ratio (PR) of 71 under the assumption that the flow is isentropic. The flow field and the thrust characteristics, at PR's from 5 to 1000, are studied. The Re number is set to be 5×10^5 based on the length between the nozzle axis and cowl lip. The pressure is normalized by the sea level pressure. In the current study, the external flow is not considered and the ideal gas is assumed for the exhaust flows.

3 Numerical Method

The governing equations are three-dimensional thin-layer Navier-Stokes equations. The convective terms are evaluated by the SHUS [4] scheme which is a family of AUSM-type schemes. High-order space accuracy is obtained using the MUSCL, the primitive variable interpolation. The viscous terms are evaluated by the central differencing and the eddy viscosity is modeled by the Baldwin-Lomax turbulence model with the Degani and Schiff model implemented. Only the steady-state solutions are considered, and the LU-ADI factorization time integration algorithm[5] is used. The computational code used in the present study has been used for a wide variety of flow field and validated in the past.[6,7] Although there is no experimental data available for the comparison, the code has been used for the annular aerospike nozzle flow fields and showed good agreement with the experiment.[2]

4 Truncation and the effect on the nozzle performance

The pressure contour maps for the six module clustered-type aerospike nozzles of the 20%, 30%, and 100% length under Pressure Ratio 200 are shown in Fig. 2(a)-(c). Three high pressure regions appear where the exhaust flow interact each other on the ramp. One is created by the exhaust flow interaction where high pressure region appears along the intermodular symmetry line. The other two high pressure region is created by the interaction of the oblique shock wave which develops toward module centerline. Small peak can be observed for 100% aerospike nozzle where the oblique shock waves interact each other downstream the module centerline.

For the 30% length aerospike nozzle, the nozzle truncation occurs where the oblique shock waves interact each other on the module centerline. For the 20% length aerospike nozzle, the nozzle truncation occurs where the high pressure region between the oblique shock waves appear. Therefore, as the nozzle length becomes shorter, high pressure region on the ramp disappears and the thrust at the ramp becomes smaller.

Figure 2(d) shows the proportion of the thrust produced by each component of the annular type and six module aerospike nozzle at the pressure ratio of 200. The vertical axis shows the thrust and the horizontal axis shows the nozzle length. From our previous studies for the annular type aerospike, it was found that the truncation of the nozzle has little influence on the total nozzle performance. The same feature can be clearly observed for the clustered aerospike nozzle presented in Fig. 2(d).

When the spike is truncated, the nozzle length becomes shorter and the ramp area for the nozzle decreases. Therefore, the pressure thrust produced at the ramp reduces. On the other hand, the thrust generated by the base pressure increases due to the increase of the base area. It compensates the total thrust loss caused by the decrease of the ramp pressure thrust. Accordingly total nozzle thrust becomes almost the same for any nozzle truncation.

This feature observed in annular aerospike nozzle can also be observed for the six-module aerospike nozzle shown in Fig. 2(d). However, the ramp thrust loss are significant and the thrust slightly decrease.

From the discussions above, it is clear that the base pressure thrust plays an important role to keep the thrust even in high truncation.

5 Flowfield of the aerospike nozzle

5.1 Flow features and pressure distributions

The flowfield for the six module 20% aerospike nozzle is investigated as an example. The pressure distributions on the nozzle surface and the Mach number and the pressure distributions on one of the cross sections of the module centerline and intermodular symmetry line are shown in Figs. 3(a)-(b). The pressure distributions on the nozzle surface and on the nozzle axis is also shown in the same figure (Figs. 3(c)(d)).

At the truncated portion of the ramp region, the intermodular symmetry area has higher pressure than that of the module centerline area due to the exhaust flow interactions. Therefore, stronger expansion of the exhaust flow occurs at the truncated point for the flow on the intermodular symmetry line as shown in Fig. 3(b). The shear layer from the ramp tip impinges on the nozzle axis and creates high pressure region on the nozzle axis. This high pressure region on the nozzle axis is the place where the reverse base flow stagnates.

The exhaust flow on the module centerline at the truncated portion does not expand so much and does not converge to the nozzle axis as shown in Fig. 3(a). The pressure on the base and the exhaust flow at the truncation point on the ramp is almost the same.

The pressure distributions on the nozzle surface and nozzle axis is shown in Fig. 3(c) and Fig. 3(d), respectively.

From the discussions above, it is found that the exhaust flow which expands from the high pressure region at the tip of the ramp and impinges on the nozzle axis creates the stagnation. Therefore, the pressure distributions on the nozzle surface of one of the cross section which has higher pressure at the tip of the ramp is selected for different types of nozzles in Fig. 3(c). The pressure distributions on the intermodular symmetry line is shown for six module aerospike nozzle and that of module centerline is shown for twelve and twenty-four module aerospike nozzles, respectively.

At the truncated point on the ramp, six module aerospike nozzle has the highest pressure level. Therefore, the stronger expansion occurs for six module aerospike nozzle and consequently the exhaust flow impinges on the nozzle axis closer to the base surface and creates higher stagnation pressure compared to the other nozzles as shown in Fig.3(d). The reverse flow from the stagnation impinges the base surface. The base pressure near the nozzle axis becomes high as the module number decreases as the stagnation pressure becomes higher.

From the discussions above, the base pressure (thrust) increases as the number of the module decreases.

6 Location of the thrust loss on aerospike nozzles

Figure 4 shows the proportion of the thrust for 20% aerospike nozzles. To make the total flow rate the same, the module exit area is set to be the same for all the configurations. Therefore, the pressure thrust and the momentum thrust produced at the module exit is about the same for all the configurations.

The aerospike nozzle used in this study has axisymmetric configuration. When comparing the axisymmetric area, the projection area toward propulsive direction becomes larger as the distance from the nozzle axis becomes long. Therefore, although the base pressure increases as the number of the module decreases, the thrust loss at the ramp region which is located at larger distance from the nozzle axis influences the total thrust more and consequently the total performance decreases as the module number decreases.

7 Conclusion

The flow structure and thrust performance of the 6, 12 and 24 module aerospike nozzles which has various nozzle length were numerically investigated. The truncation will be carried out at various portions of the high pressure region on the ramp and significant thrust loss occurs at the ramp region. However, the base thrust increases as the length and module number decreases and compensate the thrust loss produced at the ramp. The lower module number aerospike nozzle has higher base thrust because the stagnation pressure on the axis is higher. However, the broad thrust loss area at the ramp causes total thrust to decrease when the module number decreases.

References

1. G. Hagemann, H. Immichi, M. Terhardt: *AIAA Paper*, AIAA 98-3522, July (1998)
2. T. Ito, K. Fujii: J. of Propulsion and Power, 18, No.2, Mar (2002), pp254-260
3. T. Ito, K. Fujii: *AIAA Paper*, AIAA 2001-1051, Jan (2001)
4. E. Shima, T. Jounouchi: *Proceedings of the 14th NAL Symposium on Aircraft Computational Aerodynamics*, SP-34, pp7-12, Jan (1997)
5. K. Fujii: *AIAA Paper*, AIAA 86-0513, Jan (1986)
6. K. Fujii , S. Obayashi: AIAA J., 25, No.12, Dec (1987), pp.1587-1596
7. K. Fujii , Y. Tamura: Inter. J. Comp. Fluid. Phy., 8, No.4, (1997), pp.235-246

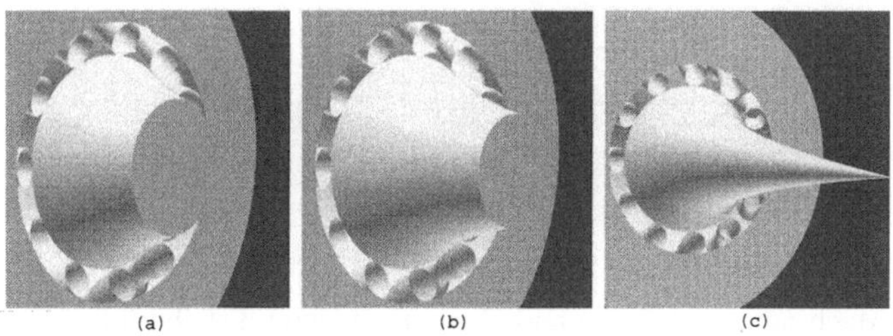

Fig. 1. (**a**) 12 modules 20% aerospike nozzle. (**b**) 12 modules 30% aerospike nozzle. (**c**) 12 modules 100% aerospike nozzle

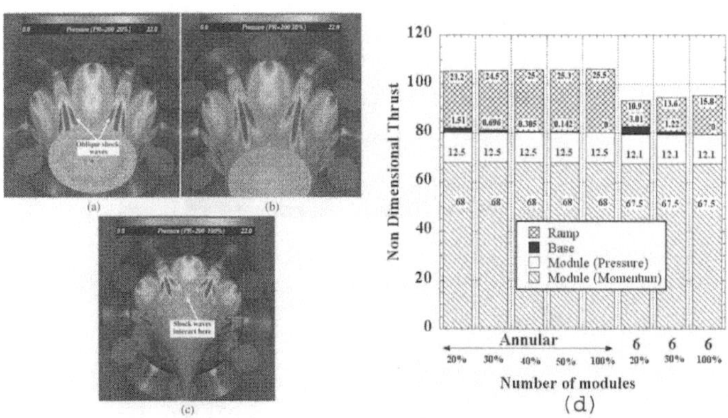

Fig. 2. (**a**) Pressure contour plots for 6 modules 20% aerospike nozzle. (**b**) Pressure contour plots for 30% aerospike nozzle. (**c**) Pressure contour plots for 100% aerospike nozzle. (**d**) Proportion of the thrust for each aerospike nozzles

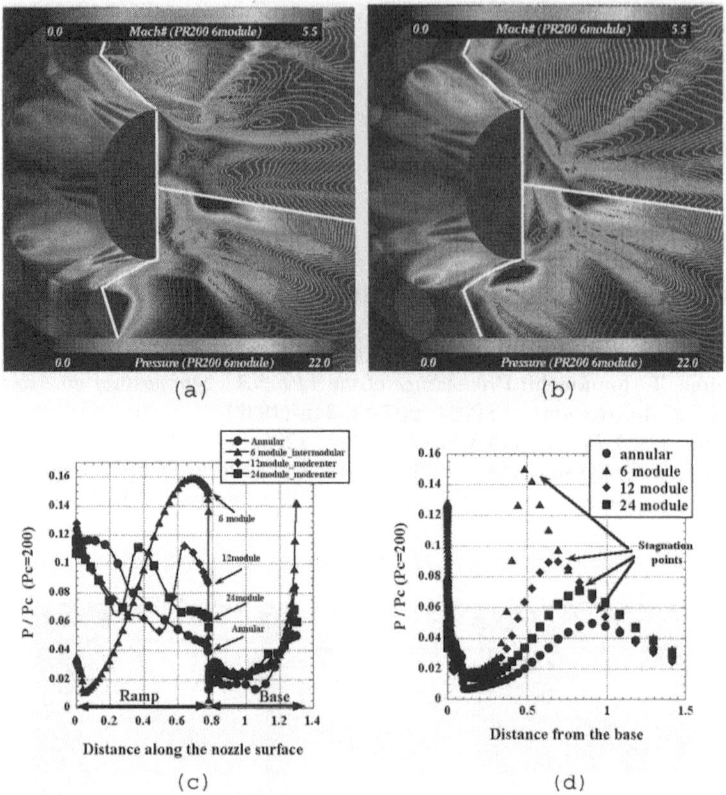

Fig. 3. (**a**) Mach number (above) and Pressure (below) distributions - cross section of the module centerline (**b**) Mach number (above) and Pressure (below) distributions - cross section of the intermodular symmetry line (**c**) Pressure distribution on the nozzle surface (**d**) Pressure distribution on the nozzle axis.

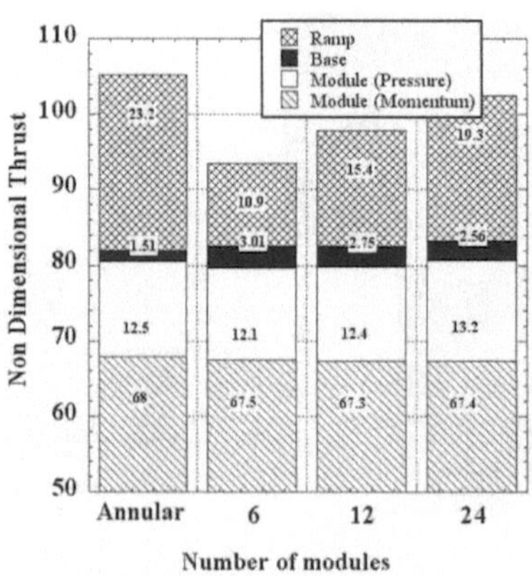

Fig. 4. Proportion of the thrust for each aerospike nozzles

Part VI

Direct Numerical Simulation

A Comparison of 2-D and 3-D Solutions for Incompressible Bifurcating Jets in Stratified Environments

P. L. Morgan, M. Kirkpatrick and S. W. Armfield

School of Aerospace, Mechanical and Mechatronic Engineering,
The University of Sydney 2006, Australia

Abstract. The 3-D results are compared with 2-D results obtained previously and it is observed that in both cases, at jet initialisation a jet intrusion front is formed with associated trailing vortex pairs. The intrusion front transits the domain and for low Reynolds number/low Richardson number the unsteadiness associated with the initialisation eventually decays and a steady state flow results. At a critical combination of Reynolds and Richardson numbers the jet becomes unstable and bifurcates to an unsteady flow. The 3-D simulations of both neutrally buoyant and positively buoyant jets discharged into a stratified environment were found to exhibit complex flow structures with a range of instabilities. It will be shown that although the spanwise wall boundary layers have some affect on the flow field the near field is not significantly affected and has similar characteristics to that observed in the 2-D simulations.

All jets exhibit evolving coherent structures, and their development stems from an initial perturbation, that of the fundamental frequency f_o [2,1,3,7]. This is also the case for incompressible jets discharged into stratified environments. When a planar incompressible jet enters a density stratified medium with the jet density equal to the ambient density at the jet entrance, the occurrence of flow bifurcation will depend on both the jet Reynolds number and the degree of ambient stratification, which is parameterised by a Richardson number Ri [6]. Morgan and Armfield formulated this conclusion from results obtained using a two dimensional time accurate second order fractional step scheme [6]. Recent experiments of a planar water jet discharged into a salt stratified environment indicate that although the flow structure generally is in good agreement with the 2-D solutions obtained numerically, a three dimensional investigation is necessary to identify the role and significance three dimensional flow features have on the flow field. Detailed results will be presented for both neutral and positively buoyant jets discharged into a stably stratified environment.

1 Numerical Method

The 3-D computational domain is a rectangular box with a length $x = 0.3$ meters, height $y = 0.23$ meters and width $z = 0.05$ meters. The governing equations are therefore solved in the domain $0 \leq x \leq 0.3$, $-0.115 \leq y \leq 0.115$ and $-0.025 \leq z \leq 0.025$ with the jet inlet at $x = 0$, $-0.005 \leq y \leq 0.005$,

$-0.025 \leq z \leq 0.025$ with all dimensions here and below in metres. The governing equations solved in this domain are the Navier-Stokes equations, which are written in incompressible and dimensional form,

$$\frac{\partial \varrho}{\partial t} + \frac{\partial(\varrho u_j)}{\partial x_j} = 0, \tag{1}$$

$$\frac{\partial \varrho u_i}{\partial t} + \frac{\partial(\varrho u_i u_j)}{\partial x_j} = -\frac{\partial P}{\partial x_i} + \frac{\partial}{\partial x_j}(2\mu S_{ij}) + f_i, \tag{2}$$

$$\frac{\partial \varrho c}{\partial t} + \frac{\partial(\varrho u_j c)}{\partial x_j} = \frac{\partial}{\partial x_j}\left(\frac{\mu}{Pr}\frac{\partial c}{\partial x_j}\right), \tag{3}$$

where c is the salt concentration, f_i the buoyancy and the diffusion coefficient is written in terms of the dynamic viscosity μ and Prandtl number Pr, and the non-Boussinesq effects are retained.

These equations are solved using a fractional step Navier-Stokes solver, with the domain discretised on a staggered grid using finite volumes. Standard second order spatial differencing is used for all terms in the momentum equations. The concentration equation employs second order differencing for the diffusive terms and the SHARP [4] scheme is used for the advection terms. Time integration is achieved using a third order hybrid Adams-Bashforth/Adams-Moulton scheme for the momentum equations and an implicit Crank-Nicolson scheme for the scalar equation.

The three dimensional planar jet is discharged into the sudden expansion box with non-slip impervious boundary conditions applied to the walls above and below the jet inlet and on the top, bottom and side walls of the expansion. A zero normal gradient boundary condition is applied to the streamwise outlet which is the side of the domain opposite the inlet.

2 Grid and Time Dependency Tests

The computational mesh used for this investigation consist of $250 \times 250 \times 50$ nodes in the x, y and z directions respectively, with 50 nodes spanning the inlet in the y direction. The smallest cell size in the x and y directions were set to $\Delta x = 4.0 \times 10^{-2}$ and $\Delta y = 2.0 \times 10^{-2}$ with the smallest cell size in the z direction set to $\Delta z = 2.0 \times 10^{-2}$. Grid dependency tests were carried out on fine and coarse mesh configurations and their solutions compared with the solution obtained for the mesh outlined above. A fine and coarse mesh configuration was tested by doubling and halving the nodes above in the x, y and z directions. Three time steps were tested and compared for all mesh configurations, $\Delta t = 5.0 \times 10^{-2}$, $\Delta t = 1.0 \times 10^{-3}$ and $\Delta t = 5.0 \times 10^{-4}$ seconds with $\Delta t = 1.0 \times 10^{-3}$ seconds used for this investigation. Grid and time step dependency results show negligible variation in the flow pattern characteristics between the grids and the time step tested.

3 Results

Figure 1(a) illustrates a typical 3-D solution for $Re = 90$ and $Ri = 4.0 \times 10^{-2}$, where the Reynolds number is based on H, the jet inlet vertical thickness, and V_0, the inlet velocity, and $Ri = \frac{gH\Delta\varrho\prime}{V_0^2}$ with $\Delta\varrho\prime = \Delta\varrho\prime_{jet} + \Delta\varrho\prime_b$, $\Delta\varrho\prime_{jet}$ the ratio of density difference between the inlet jet and the ambient at that height to that of water at STP (ϱ_w), and $\Delta\varrho\prime_b$ the ratio of the variation in density of the linear ambient density stratification over height H to ϱ_w. In this case the jet is neutrally buoyant, that is the jet density is equal to the ambient density at the jet inlet height and thus $\Delta\varrho\prime_{jet} = 0$. The figure shows the $c = 13$ isosurface at three times shortly after flow initiation. Some 3-D effects are visible adjacent to the $z = -0.025$ non-slip spanwise boundary, although the overall flow structure is two dimensional. Figure 1(b) contains equivalent 2-D results in the form of c contours. It is seen that the overall structure and rate of development is very

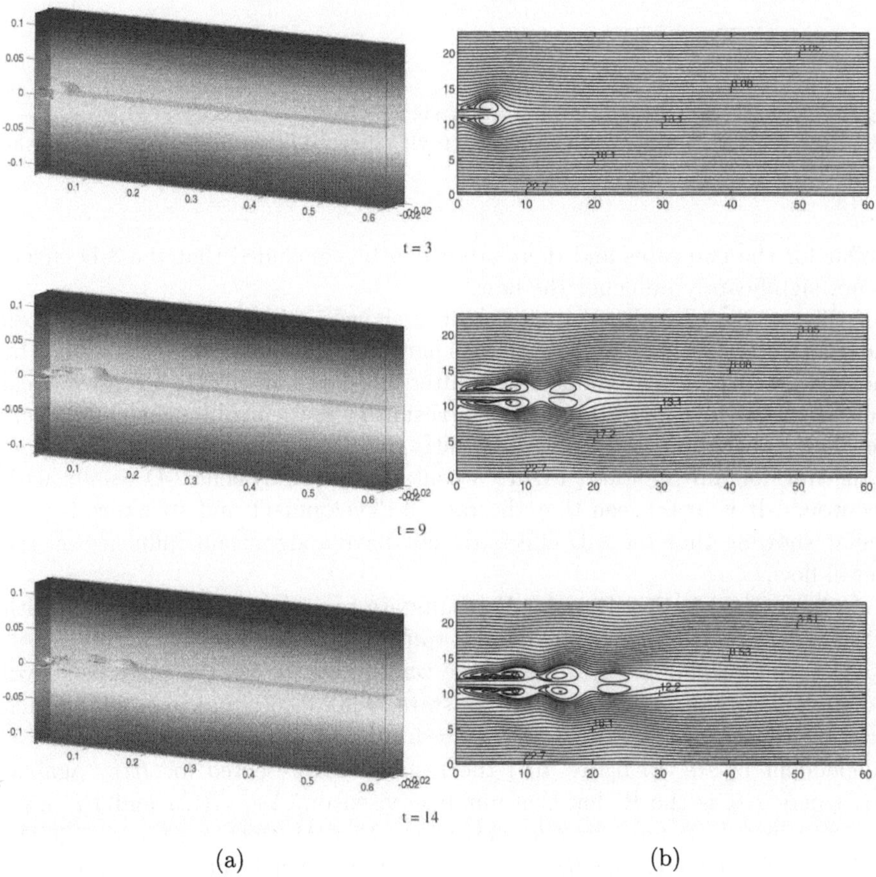

Fig. 1. (a) Typical 3-D neutrally buoyant jet with $Re = 90$, $Ri = 4.0\times10^{-2}$, $t = 3, 9, 14$ seconds with c isosurface shown. (b) Equivalent 2-D results with c contours.

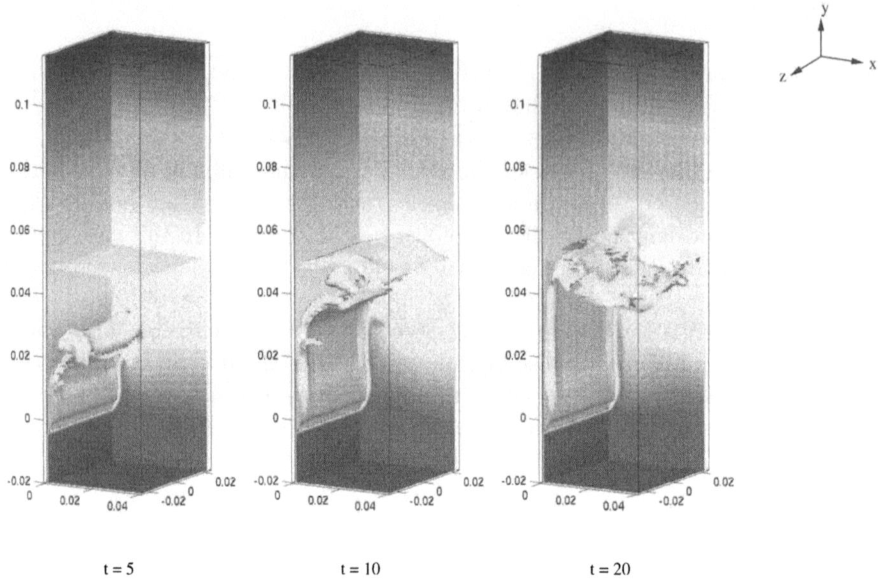

Fig. 2. Typical 3-D solution showing a c isosurface for $Re = 25.6$, $Ri = 11.89$, $Ri_{jet} = 9.8$, at $t = 5, 10, 20$ seconds for a buoyant jet with neutral height 0.04 metres above the jet inlet.

similar for the two cases and therefore it can be concluded that the 3-D effects do not significantly influence the flow.

In figure 2 a typical 3-D buoyant jet is shown with $Re = 25.6$, $Ri = 11.89$ and $Ri_{jet} = 9.8$, where Ri_{jet} is the Richardson number based on $\Delta\varrho\prime_{jet}$. The buoyant jet rises rapidly immediately after discharge, attaching to the vertical wall above the inlet and forming an intrusion at its neutral height. 3-D effects are clearly visible in the turning flow/intrusion region. In this case only a part of the full domain is shown. Figure 3 contains the equivalent 2-D results with c contours. It is again seen that the rate of development and structure is very similar showing that the 3-D effects do not have a significant influence on the overall flow.

In figure 4(a) both 2-D and 3-D results are presented for buoyant jets with $0.05 \leq Ri \leq 24$, $10 \leq Re \leq 200$ and a range of positive jet buoyancies. Cases in which the buoyant jet attaches to the vertical wall above the inlet, as shown in figures 2 and 3, are represented as stars, while those which do not attach are shown as circles. It is seen that only for $Ri \geq 1.0$ does attachment occur, independent of Re. In figure 4(b) the results are presented for Ri_{jet} against Ri_b, where Ri_b is the Richardson number based on $\Delta\varrho\prime_b$. Ri_{jet} and Ri_b have the opposite effect. For example increasing Ri_{jet} will increase the rise height of the jet whilst increasing Ri_b will reduce the rise height of the jet. From the figure, as Ri_b goes to zero the critical Ri_{jet} goes to 1.0 and thus for small Ri_b the jet attachment is determined predominantly by the jet buoyancy. As Ri_b is increased the attachment is prevented by the stabilising effects of Ri_b. A range

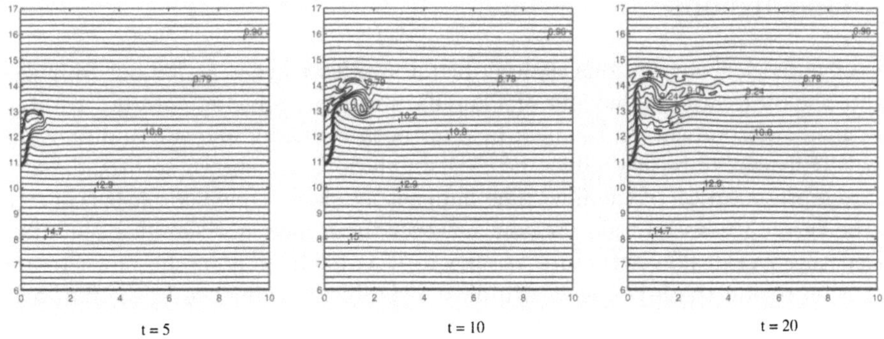

Fig. 3. 2-D buoyant jet with $Re = 25.6$, $Ri = 11.89$, $t = 5, 10, 20$ seconds with a neutral height 0.04 metres above the jet inlet.

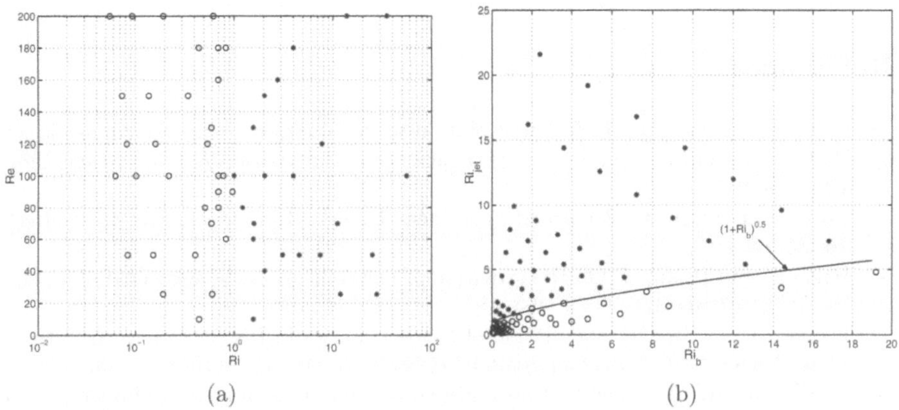

Fig. 4. (a) Results for $0.05 \leq Ri \leq 24$ with $10 \leq Re \leq 200$, $\circ$ unattached, $\star$ attached. (b) Same results plotted as Ri_{jet} against Ri_b, $\circ$ unattached, $\star$ attached.

of functions were tested to obtain a functional form for the critical Ri_{jet}, Ri_b relation and it was found that $Ri_{jet} \sim (1 + Ri_b)^{0.5}$ provided a good empirical fit to the data, as shown in figure 4(b).

Detailed comparison of the 3-D results and the 2-D results obtained by Morgan and Armfield [6] have shown that in both 2-D and 3-D cases, at jet initialisation a jet intrusion front is formed with associated trailing vortex pairs. The intrusion front transits the domain and for low Reynolds number/low Richardson number the unsteadiness associated with the initialisation eventually decays and a steady state flow results. At a critical combination of Reynolds and Richardson numbers the jet becomes unstable and bifurcates to an unsteady flow. Although the 3-D simulation showed that some 3-D features are generated, particularly in the turning flow/intrusion region for the buoyant jet, these do not significantly affect the major flow features, such as rate of development and attachment as shown.

4 Conclusions

Direct numerical simulations as described above have been carried out to investigate the bifurcation behaviour of a laminar incompressible jet entering a stably stratified environment. In the near field for both neutrally buoyant and buoyant jets 3-D features are negligible and 2-D simulations are shown to provide adequate representation of the flow. The buoyant jet shows considerable 3-D effects in the turning flow/intrusion region, however these have not significantly influenced the overall flow structure as shown. It is likely that in the downstream intrusion region the 3-D effects will have a significant influence in the flow and thus 2-D simulations will therefore only reveal a partial representation of the overall flow structure.

Acknowledgements: This project was supported by the Australian Research Council through grant No. A00104152.

References

1. Cohen J.and Wygnanski I. The evolution of instabilities in the axisymmetric jet. Part 1. The linear growth of disturbances near the nozzle. *J. Fluid Mech.*(1987) **176**, 191–219.
2. Ho C. and Huerre P. Perturbed Free Shear Layers. *Ann. Rev. Fluid Mech.*(1984) **16**, 365–424.
3. Hsiao F. and Huang J. On the evolution of instabilities in the near field of a plane jet. *Phys. Fluids A*(1989) **2**, 400–412.
4. Leonard B. P. SHARP Simulation of Discontinuities in Highly Convective Steady Flow. *NASA Technical Memorandum 100240, ICOMP-87-9*(1987), 1–33.
5. B. P. Leonard, A stable and accurate convective modelling procedure based on quadratic upstream interpolation. *Comput. Meth. Appl. Mech. Eng.*(1979) **19**, 59–98.
6. Morgan P. L. and Armfield S. W. Bifurcation of Neutrally Buoyant Jets in Stratified Environments. *Fourteenth Australasian Fluid Mechanics Conference (AFMC 14)*(2001), 761–764.
7. Thomas F. O. and Prakash K. M. K. An experimental investigation of the natural transition of an untuned planar jet. *Phys. Fluids A*(1991) **3**, 90–105.

Simulations of Turbulent and Gravity Interfaces by Lagrangian Block Method

Vincent H. Chu and Wihel Altai

McGill University, Montreal, Quebec H3A 2K6, Canada

Abstract. Computations for the turbulent interfaces in gravity-stratified flows are conducted using a system of contiguous blocks as the computational elements. The blocks are containers of the flow properties. Salinity, temperature, vorticity, turbulence kinetic energy and dissipation rate are carried by the blocks that move in the direction of the velocity following the Lagrangian procedure. The blocks are re-generated at each time increment. The flow properties in the blocks are transferred from the initial system of blocks to the re-generated system of blocks by conserving the flow properties and their moments. This method known as the Lagrangian Block Method (LBM) was introduced in the first ICCFD conference by the authors. Further development of the method and application of the method to turbulence in gravity stratified flows are reported in this paper.

1 Introduction

In Large Eddy Simulations (LES) of turbulent flow, the required viscosity for energy dissipation at the sub-grid scales is quite small. The sub-grid viscosity is comparable in order of magnitude to numerical dissipation. Accurate simulation of the flow depends on ability of the numerical scheme to compute the advection process. The need for numerical accuracy is even greater when the buoyancy force is involved in the numerical simulation. Buoyancy is carried by the turbulence motion, which is driven by the force of the buoyancy. Because the process is advection dominated, the numerical method must be relatively free from unphysical diffusion and oscillations.

In this paper, the turbulent flows are simulated by an advection/diffusion scheme known as the Lagrangian Block Method (LBM). The computational elements of the LBM are systems of contiguous blocks that carry the flow properties and move in the direction of the velocity following the Lagrangian procedure. Flow and fluid properties, such as salinity, temperature, vorticity, turbulence kinetic energy and dissipation rate, are each carried by separated system of blocks. The blocks are re-generated at each time increment so that the flow property in a block can be exchanged with the masses of its neighboring blocks. Fig. 1 shows the movement of a block in the direction of the velocity and the expansion in the size of the block due to diffusion. The computation details of the LBM have been given in a series of papers by Chu and Altai [3] [4] and Altai and Chu [1] [2].

The LBM is non-negative, mass conserving and unconditional stable [3] [1] [4]. These are outstanding attributes of the LBM. Using the method, simulations

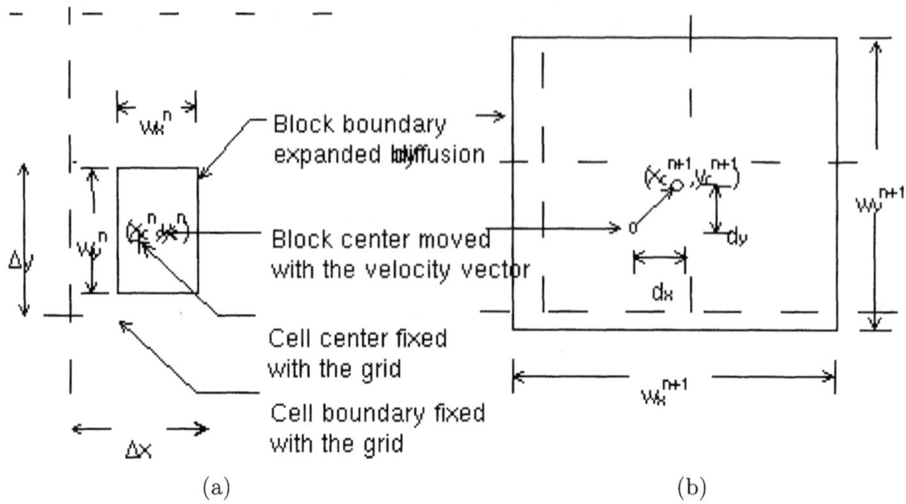

Fig. 1. Lagrangian Block Method. (a) The block in the cells at time $t = n\Delta t$. (b) The block at time $t = (n+1)\Delta t$, after the displacement and enlargement of the block by advection and diffusion.

have been conducted for lock exchanged flow, plume in stratified flow and density plume. Some of the results are presented and compared with experimental observation.

2 Lock Exchange Flow

Simulations using the LBM were conducted for the lock exchange flows. Figure 2 shows one of the lock exchanged flow obtained in a 192x96 grid. A block of densed fluid (e.g., saline in fresh water), initially occupied the height of one side of the tank, is released under the force of gravity. The dense block is narrow. The initial width is equal to one-eighth of its height as marked in (a) and (b) of the figure, The gravity currents at subsequent dimensionless times (a) t = 0.06, (b) 1.35, (c) 2.27, and (d) 3.87, are shown in the figure. In this simulation, the subgrid viscosity was set to zero and the computation was conducted for a pure advection process. The LBM was able to produce sharp density interfaces with considerable details of the turbulent motion. Fluids entrained into the gravity current head maintain its identity in layers as thin as one computational grid thickness.

Figure 3 shows the perfomance of the LBM compared with a third-order upwind-bias scheme known as QUICKEST [6]. The QUICKEST method tends to produce small unphysical oscillation. To suppress the oscillation and improve the stability of the QUICKEST scheme, a procedure known as universal flux limiter is included in the QUICKEST algorithm [7]. For the present problem, we were able to carry out the calculations by the QUICKEST scheme using a maximum Courant number Co = 0.5. Therefore, the comparison with the LBM

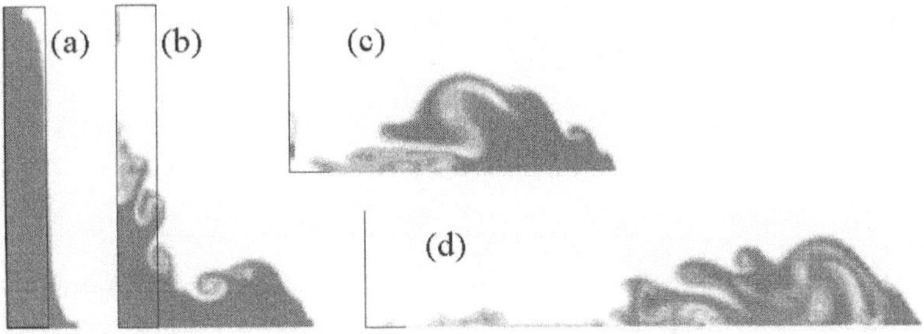

Fig. 2. Gravity interfaces simulated by the LBM at dimensionless time (**a**) $t = 0.06$, (**b**) 1.35, (**c**) 2.27, (**d**) 3.87. The dense fluid occupy initially an area as marked by the black rectangular frame in (a) and (b) of the figure. Courant number Co = 0.95

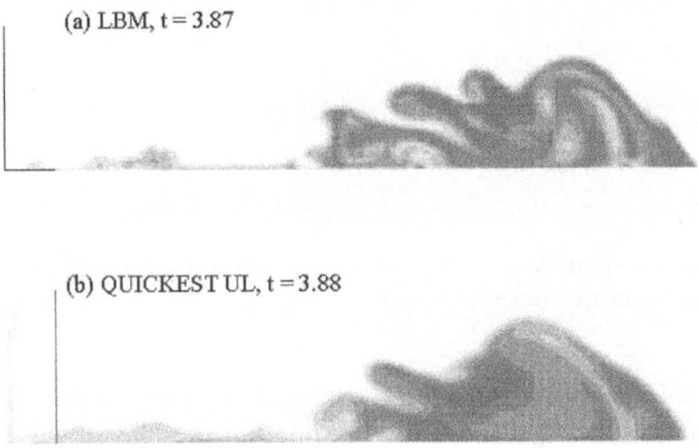

Fig. 3. (**a**) Gravity interfaces by the LBM; dimensionless time $t = 3.87$. (**b**) Gravity interfaces by QUICKEST, with a correction by Universal Flux Limiter; dimensionless time $t = 3.88$. Courant number Co = 0.5

is made using Co = 0.5. Figure 3(a) shows the simulated results when salinity and buoyancy were computed by the LBM. Figure 3(b) shows the result when salinity/buoyancy were computed by the QUICKEST method. Both simulations used LBM to calculate the vorticity.

The LBM not only is capable to re-produce sharp interfaces. It also is more stable. The method is in fact unconditionally stable [4]. Stable computation can be carried out with the LBM using a Courant number that is greater than unity. The numerical results of course would not be correct if the time step is too large.

3 Negatively Buoyant Plume

Simulations using the LBM have been conducted also for a negatively buoyant plume produced by an upward discharge of fluid that is heavier than its

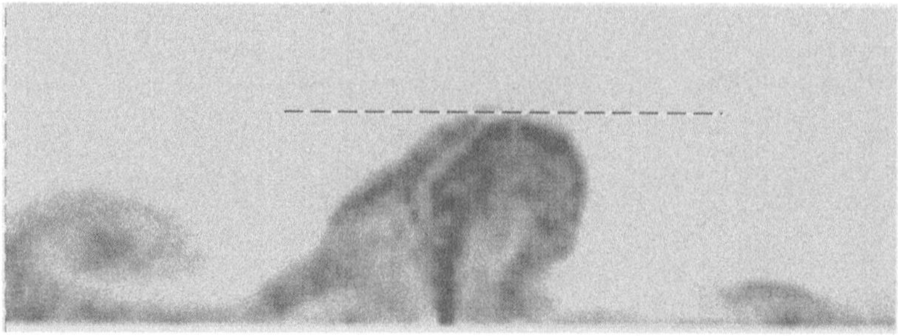

Fig. 4. LES of a negatively buoyant plume produced by upward discharge of fluid heavier than its surrounding. The dash line marks the maximum height of the plume obtained from a formula derived from laboratory data

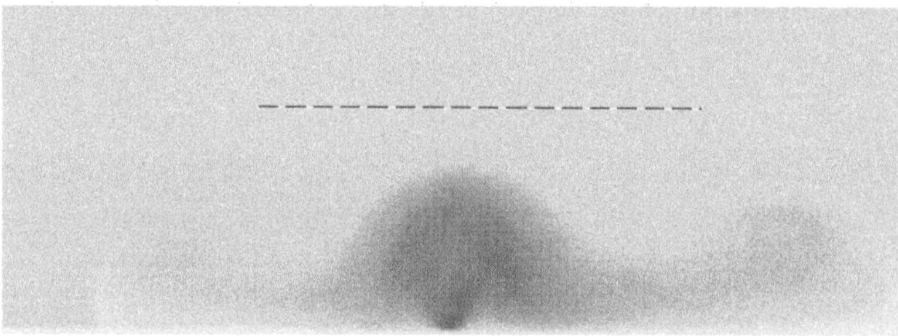

Fig. 5. K-ϵ model simulation of the same negatively buoyant plume with the same buoyant and momentum fluxes as the plume in Fig. 4.

surrounding. Fig. 4 shows the results obtained by Large Eddy Simulation and Fig. 5 the results obtained by a K-ϵ turbulence model. As the plume rises to its maximum height, it falls back and re-entrains into the rising plume. The recirculation of the turbulence energy in the rising and falling fluid have led to very high and unrealistic intensity of turbulence by the K-ϵ model. The maximum height of the plume is significantly under predicted by the K-ϵ model because the high diffusion rate produced by the model. The plume has been observed in the laboratory experiment to be flappying from side to side in a nearly periodic manner [12]. This highly unsteady process are correctly reproduced by the LES using the LBM. We do not have experimental data for the concentration field but the maximum rise of the negative plume is correctly reproduced. The LES were successful because the details of the laminar and turbulent interface of the plume were precisely defined by the LBM.

4 Plume in Stratified Flow

By far the most impressive results were obtained in the simulation of a buoyant plume in a stably stratified fluid. Figure 6 shows the plume rise in the stratified fluid obtained by LES using the LBM. As the plume rises through its stratified

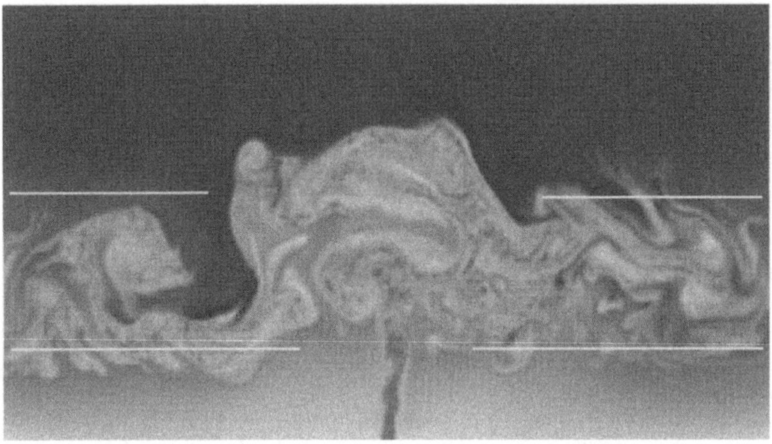

Fig. 6. Plume in stably stratified flow. The lines shows the upper and lower interfaces of the plume as predicted by the formulae derived from experimental observation.

surrounding, eddies with variable buoyancy are produced. The eddies over shot the stratification due to the inertia effect but eventually seek the level of equal density with its surrounding. Only correct simulation of the turbulence could have led to the correct equilibrium level of the plume. The gravity stratification is characterized by the Brunt-Väisälä frequency

$$N = \sqrt{-\frac{g}{\rho}\frac{d\rho}{dz}} \tag{1}$$

and the buoyancy flux at the source

$$F_o = g_o' Q_o. \tag{2}$$

Based on reviews of laboratory data, Wright [11] and Robert [9] recommended the following formulae for the top and bottom elevations of the plume

$$\frac{z_{\text{top}}}{\ell_s} = 3.7, \quad \frac{z_{\text{bottom}}}{\ell_s} = 1.5 \tag{3}$$

where the length scale $\ell_s = F_o^{\frac{1}{3}} N^{-1}$. The lines in Fig. 6 mark the top and bottom of the plume according to these formulae. It is clear that the LES by LBM has been able to re-produce both the top and the bottom level of the plume in nearly perfect agreement with the experimental data.

5 Conclusion

Turbulence driven by the forces of buoyancy is difficult to compute. Accurate advection scheme for the scalar transport is required so that the buoyancy forces are accurately determined. The advection of buoyancy in the gavity-driven turbulent flow is a highly unsteady process. The simulation of this process have provided the opportunity to test and evaluate the performance of numerical schemes. The present LES of the turbulence in gravity driven fluid have demonstrated clearly the advantage of the LBM for its ability to reproduce the turbulent interface and the gravity interface equally with high precision.

References

1. Altai, W. and Chu, V. H.: 'Large Eddy Simulation by Lagrangian Block Method'. In: *Computational Fluid Dynamics 2000.* ed. by N. Satofuka (Springer-Verlag, New York 2001) pp.467-472
2. Altai, W. and V. H. Chu: 'K-ϵ model simulation by Lagrangian Block Method'. In: *Proc. First MIT Conf. on Computational Solid and Fluid Mechanics,* Vol.2 (Elsevier Science Ltd., Oxford, England, 2001) pp. 767-771
3. Chu, V. H. and Altai, W.: 'Lagrangian Block Method'. In: *Computational Fluid Dynamics 2000.* ed. by N. Satofuka (Springer-Verlag, New York 2001) pp.771-772
4. Chu, V. H. and Altai, W.: 'Simulation of shallow transverse shear flow by generalized second moment method,' *J. Hydraulic Research,* 39:6 (2001) pp. 575-582
5. Egan, B. A. and Mahoney, J. R.: 'Numerical modeling of advection and diffusion of urban area source pollutants'. *J. Appl. Meteorology,* 11 (1972) pp. 312-322
6. Einstein, A.: *Ann. d. Physik,* 17 (1905) p. 549. This and other articles of Einstein is collected in a book entitled *Investigations on the Theory of the Brownian Movement,* edited by R. Füth, translated by A. D. Cowper (Dutton, New York).
7. Leonard, B. P.: 'A stable and accurate convection modelling procedure based on quadratic upstream interpolation'. Comput. Meth. Appl. Mech. Engrg., 19 (1979) pp.59-78
8. Leonard, B. P. and Mokhtari, S.: 'Beyond first-order upwinding: the ultra-sharp alternative for non-oscillatory steady-state simulation of convection'. Int'l J. Num. Meth. Engrg., 30 (1990) pp.729-766.
9. Robert, P.J.W., Snyder, W.H., and Baumgartner, D.J.: 'Ocean outfalls, I, II and III'. *J. Hydr. Engrg.,* 115:HY1 (ASCE, 1989) pp. 1-70
10. Taylor, G. I.: 'Diffusion by continuous movements'. *Proc. Lond. Math. Soc.,* 20:2 (1921) pp.196-212
11. Wright, S. J., Wong, D. R., Zimmerman, K. E. and Wallace, R. B.: 'Outfall diffuser behavior in stratified ambient fluid'. *J. Hydr. Div.,* 108:HY4 (ASCE,1982) pp. 483-501
12. Zhang, H. and Baddour, R. E.: 'Maximum vertical penetration of plane turbulent negatively buoyant jets'. *J. Eng. Mech.,* 123:10 (ASCE, 1997) pp. 973-977

Parallel DNS of Homogeneous Isotropic Turbulence Using Variable Order Method of Lines

Hidetoshi Nishida[1], Satoshi Nakai[2], and Nobuyuki Satofuka[1]

[1] Department of Mechanical & System Engineering, Kyoto Institute of Technology, Matsugasaki, Sakyo-ku, Kyoto 606-8585, JAPAN
[2] Daikin Industries, Ltd., Nakazakinishi, kita-ku, Osaka 530-8323, JAPAN

Abstract. The DNSs of 3-D homogeneous isotropic turbulence are carried out on parallel platforms by using a variable order method of lines. The variable order method of lines shows the 91% parallel efficiency on Pilot-3 system and gives 46.3 times speed up on SSCMPP system. In comparison with the spectral solutions, the present higher order method is as accurate as the spectral method at comparable resolution. The large-scale DNSs up to 1024^3 grid resolution are carried out successfully on the parallel platform with 256 PEs.

1 Introduction

In the recent practical flow simulations, the more complicated 3-D flows are considered. These flow simulations, especially the direct numerical simulations (DNSs) of turbulence, need the enormous processing time and memory in order to obtain the reliable solutions. Then, it is difficult to compute on the single CPU platform, so that these large-scale numerical simulations must be carried out on the parallel platforms. Moreover, the parallel computing will shift to the massively parallel computing or grid computing. Under these situations, the computational scheme applicable to the different type of parallel architecture with high parallel efficiency is indispensable. We have presented the variable order method of lines (VOMOL)[1] for the DNS of turbulence. The VOMOL gives the arbitrary order of spatial accuracy by changing only one parameter. In this paper, we try to implement the VOMOL on the different parallel platforms, to validate the parallel efficiency, and to be carry out the parallel DNSs of 3-D homogeneous isotropic turbulence.

2 Variable Order Method of Lines

2.1 Governing Equations

In this paper, we consider the incompressible flows. Then, the incompressible Navier-Stokes equations can be written by

$$\frac{\partial u_i}{\partial x_i} = 0, \tag{1}$$

$$\frac{\partial u_i}{\partial t} + u_j \frac{\partial u_i}{\partial x_j} = -\frac{\partial p}{\partial x_i} + \frac{1}{Re}\frac{\partial^2 u_i}{\partial {x_j}^2}. \tag{2}$$

Equation (1) denotes the continuity equation and eq.(2) the momentum equations. The velocity components and the pressure are expressed by u_i and p, respectively. Re denotes the Reynolds number defined by $Re = UL/\nu$, where ν is the kinematic viscosity. The above equations are nondimensionalized by the reference length L, the reference velocity U and the reference pressure ρU^2.

2.2 Variable Order Method of Lines

The solution procedure of the incompressible Navier-Stokes equations (1) and (2) is based on the fractional step approach on the collocated grid system.

In the method of lines approach, the spatial derivatives are discretized by the appropriate scheme, so that the partial differential equations (PDEs) in space and time are reduced to the system of ordinary differential equations (ODEs) in time. The resulting ODEs are integrated by the Runge-Kutta type time integration scheme.

In the spatial discretization, the convective terms are approximated by the variable order proper convective scheme[2], because of the consistency of the discrete continuity equation, the conservation property, and the variable order of spatial accuracy. This scheme is the extension of the proper convective scheme proposed by Morinishi[3] to the variable order. The variable order proper convective scheme can be described by

$$u_j \frac{\partial u_i}{\partial x_j}|_{\mathbf{x}} = \sum_{\ell=1}^{M/2} c_{\ell'} \overline{\overline{u_j}^{x_j} \frac{\delta_{\ell'} u_i}{\delta_{\ell'} x_j}}^{\ell' x_j}|_{\mathbf{x}}, \quad (\ell' = 2\ell - 1), \tag{3}$$

where M denotes the order of spatial accuracy, and the operators in eq.(3) are defined by

$$\overline{f}^{\ell' x_j}|_{\mathbf{x}} = \frac{1}{2}(f_{\mathbf{x}_j+\ell'/2} + f_{\mathbf{x}_j-\ell'/2}), \tag{4}$$

$$\overline{u_j}^{x_j}|_{\mathbf{x}_j \pm \ell'/2} = \sum_{m=1}^{M/2} \frac{c_{m'}}{2}[u_j|_{\mathbf{x}_j \pm \{\ell'+m'\}/2} + u_j|_{\mathbf{x}_j \pm \{\ell'-m'\}/2}], \tag{5}$$

$$\frac{\delta_{\ell'} u_i}{\delta_{\ell'} x_j}|_{\mathbf{x}_j \pm \ell'/2} = \frac{\pm 1}{\ell' \Delta x_j}(u_i|_{\mathbf{x}_j \pm \ell'} - u_i|_{\mathbf{x}_j}), \tag{6}$$

where $m' = 2m - 1$. In this technique, the arbitrary order of spatial accuracy can be obtained automatically by changing only one parameter M. The coefficients $c_{\ell'}$ and $c_{m'}$ are the weighting coefficients and Δx_j denotes the grid spacing in the x_j direction.

On the other hand, the diffusion terms are discretized by the modified differential quadrature (MDQ) method[4] as

$$\frac{\partial^2 u_i}{\partial {x_j}^2}|_{\mathbf{x}} = \sum_{m=-M/2}^{M/2} \Phi_m''(\mathbf{x}) u_i|_{\mathbf{x}_j+m}, \tag{7}$$

where $\Phi_m''(\mathbf{x})$ is the second derivative of the function $\Phi_m(\mathbf{x})$ defined by

$$\Phi_m(\mathbf{x}) = \frac{\Pi(\mathbf{x})}{(\mathbf{x} - \mathbf{x}_{j+m})\Pi'(\mathbf{x}_{j+m})}, \tag{8}$$

$$\Pi(\mathbf{x}) = (\mathbf{x} - \mathbf{x}_{j-M/2}) \cdots (\mathbf{x} - \mathbf{x}_j) \cdots (\mathbf{x} - \mathbf{x}_{j+M/2}). \tag{9}$$

The coefficients of the variable order proper convective scheme, $c_{\ell'}$, can be computed automatically by using the MDQ coefficients. Then, the incompressible Navier-Stokes equations are reduced to the system of ordinary differential equations (ODEs) in time. This system of ODEs is integrated by the Runge-Kutta type scheme. The pressure equation described by

$$\frac{\partial^2 p^{n+1}}{\partial x_i^2} = \frac{1}{\alpha \Delta t} \frac{\partial \overline{u_i^*}^{x_i}}{\partial x_i}, \tag{10}$$

is solved by the variable order multigrid method[5]. In eq.(10), u_i^* denotes the fractional step velocity, α is the parameter determined by the time integration scheme. The overbar denotes the interpolation from the collocated location to the staggered location. In the variable order multigrid method, the unsteady term is added to the pressure equation. Then the pressure (elliptic) equation is transformed to the parabolic equation in space and pseudo-time. This can be solved by the variable order method of lines.

In the pseudo-time integration, the rational Runge-Kutta scheme[6] is adopted, because of its wider stability region. In addition, the multigrid technique[7] is incorporated into the method in order to accerelate the convergence. Then, the same order of spatial accuracy as the momentum equations can be specified.

3 Parallel Efficiency of Variable Order Method of Lines

3.1 Parallel Platforms

In this paper, the DNSs of homogeneous isotropic turbulence are carried out on two parallel platforms. One is the Pilot-3 system in the Tsukuba University. Another is the SSCMPP system in Japan Atomic Energy Research Institute (JAERI).

In the Pilot-3 system, CPU architecture is PA-RISC 1.1 with PVP-SW feature and clock cycle of 150MHz. PVP-SW means the pseudo vector processor based on slide window. The number of nodes is 128, the total memory is 32GB. And the network topology is 2-D crossbar with 300MB/sec data fransfer throughput. Then, the total performance is 38.4GFLOPS (0.3GFLOPS/PE).

On the other hand, in the SSCMPP, which means the super simulation center / massively parallel processing, system, the CPU architecture is Alpha EV68 with clock cycle of 833MHz. The number of nodes is 900, the total memory is 1.8TB. The network topology is QSW interconnect with 200MB/sec data transfer throughput. Then, this system has the total performance of 1.5TFLOPS (1.65GFLOPS/PE)

3.2 Computational Conditions

In this paper, we consider the free decaying turbulence. The computational region is $(0,0,0) \leq (x,y,z) < (2\pi,2\pi,2\pi)$. The initial conditions are given by the random velocity field with the energy spectrum defined by $E(k) = 8{u_0}^2{k_p}^{-5}k^4\exp\{-2(k/k_p)^2\}/(3\sqrt{\pi})$, where ${u_0}^2 = 1/3$ and $k_p = 2$. The periodic boundary conditions are imposed in all directions. The uniform grids with 128^3, 256^3, 512^3, and 1024^3 points are specified, and the corresponding kinematic viscosity ν and time step Δt are $\nu = 4.5 \times 10^{-3}$, 1.8×10^{-3}, 7.6×10^{-4}, 2.9×10^{-4}, $\Delta t = 1/200$, $1/1000$, $1/1000$, $1/1600$, respectively.

3.3 Parallel Efficiency

The parallel computational code is made by using the message passing interface (MPI) library. The present parallel computational code is implemented on Pilot-3 system and SSCMPP system. Figure 1 shows the total speed up of the present method on Pilot-3 system. The speed up is defined by $S_p = T(base)/T(parallel)$, where $T(base)$ denotes the CPU time on based processor element (PE) and $T(parallel)$ the CPU time on each PE. In this case, T(base) is the CPU time on 4 PEs with 128^3 grid points. The 91% parallel efficiency is obtained in the 10th order of spatial accuracy. Figure 2 shows the speed up on the SSCMPP system. In this case, $T(base)$ is the CPU time on 2PEs. The number of grid points is 256^3, and the order of spatial accuracy is specified by 8th order. The speed up on the SSCMPP system is lower than on the Pilot-3 system. The data transfer throughput of SSCMPP system is 200MB/sec. This is more slower than the processor performance, 1.65GFLOPS/PE, so that the high parallel efficiency can not be obtained. However, it is confirmed that the present result shows the linear speed up, and the CPU time on 256 PEs is 46.3 times faster than the CPU time on 2 PEs.

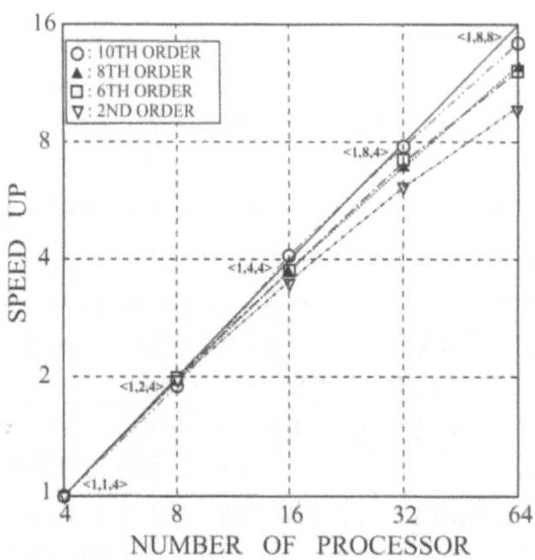

Fig. 1. Speed up on Pilot-3 system

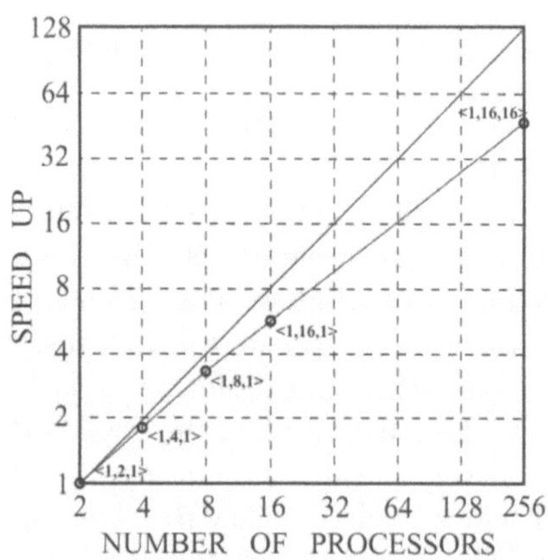

Fig. 2. Speed up on SSCMPP system

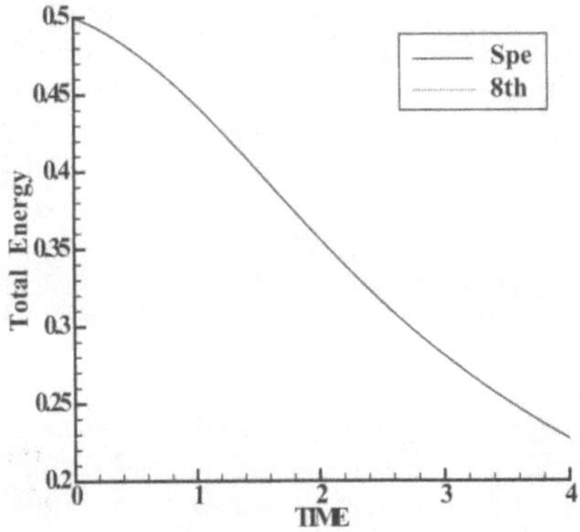

(a) Total energy

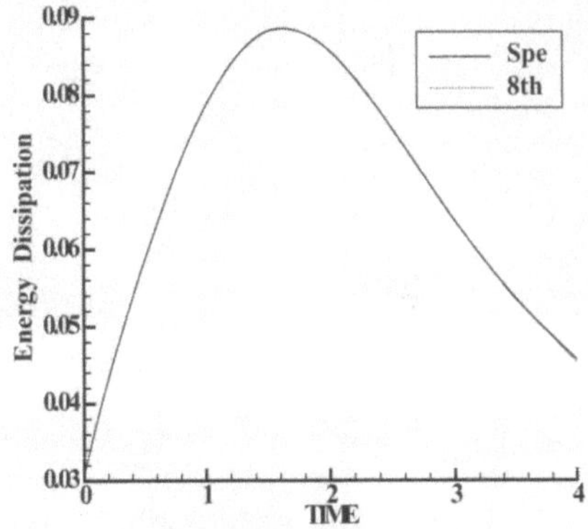

(b) Energy dissipation

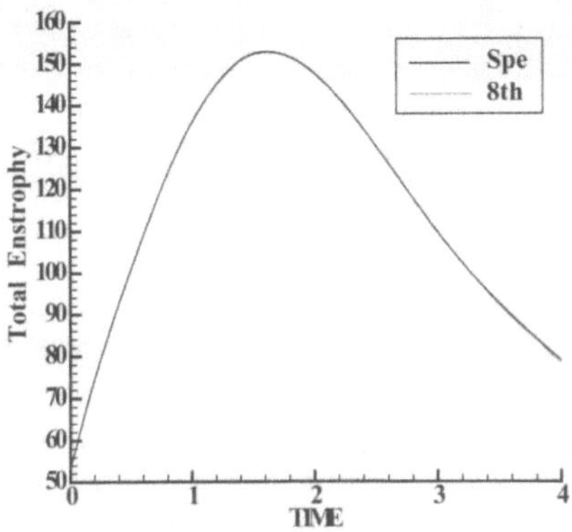

(c) Enstorophy

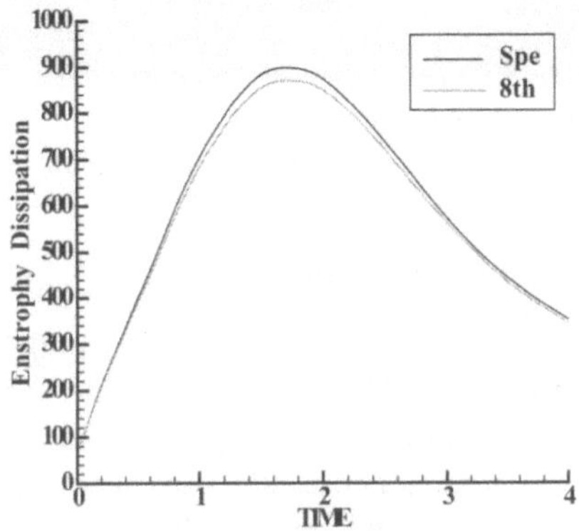

(d) Enstorophy dissipation

Fig. 3. Comparison of statistical quantities (1024^3, $\nu = 2.9 \times 10^{-4}$)

4 Parallel DNS of Homogeneous Isotropic Turbulence

We try to carry out the parallel DNSs on SSCMPP system with 256 PEs. At 512^3 and 1024^3 grid resolutions, the 8th order solutions are compared with the spectral solutions. The convergence criterion of the pressure equation is 1.0×10^{-6}. The averaged CPU time per time step is 23.19sec for 512^3 grid resolution, and 182.78sec for 1024^3 grid resolution. Figure 3 shows the statistical quantities, i.e., the total energy, the energy dissipation, the enstrophy, and the enstrophy dissipation. In the total energy, energy dissipation, and enstrophy, the present solution gives almost the same results as the spectral solution. The present enstrophy dissipation profile shows the slight difference from the spectral solution at peak time. Figure 4 shows the time history of the enstrophy field with 1024^3 grid resolution. These draw the 1/512 of the computational domain. The transition of vortices is clearly presented.

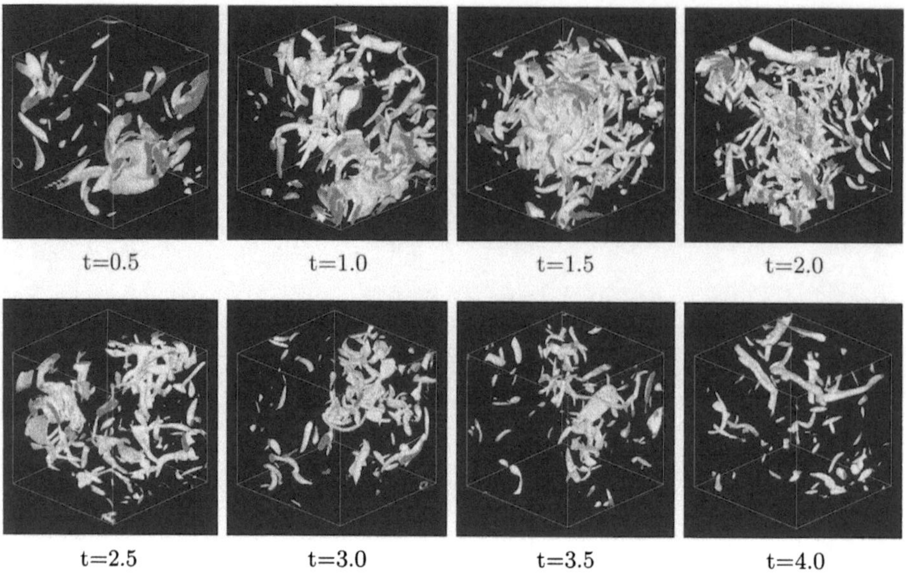

Fig. 4. Time history of enstrophy field (1024^3, $\nu = 2.9 \times 10^{-4}$, $\Omega_{plot} = 500$)

5 Concluding Remarks

In this paper, the variable order numerical method with high parallel efficiency is developed. The present variable order method of lines gives the 91% parallel efficiency on Pilot-3 system, and 46.3 times speed up on SSCMPP system. DNSs of 3-D homogeneous isotropic turbulence can be carried out on parallel platform successfully. The results show that the present method is as accurate as the spectral method at comparable resolution. Then, the present method is very promising for the parallel DNS of the incompressible turbulent flows.

Acknowledgements: This work was supported in part by the Research for the Future Program (97P01101) from Japan Society for the Promotion of Science. We would like to thank Prof. Kaneda at Nagoya University for providing the initial data and spectral solutions.

References

1. H. Nishida: Computational Fluid Dynamics journal, special number, pp.598–604 (2001)
2. H. Nishida, N. Satofuka: Lecture Notes in Computational Science and Engineering, **21**, pp.167–174 (2002)
3. Y. Morinishi, T.S. Lund, O.V. Vasilyev, P. Moin: J. Comput Phys. **143** pp.90–124 (1998)
4. N. Satofuka, K. Morinishi: NASA TM-81339 (1982)
5. H. Nishida, N. Satofuka: Finite Elements in Analysis and Design, **16**, pp.285–297 (1994)
6. A. Wambecq: Computing, **20**, pp.333–342 (1978)
7. A. Brandt: Mathematics of Computation, **31-138**, pp.333–390 (1977)

Direct Numerical Simulation of a Turbulent Shear Flow Beneath a Wind-Driven Water Surface

Wu-ting Tsai

Department of Civil Engineering, National Chiao Tung University, Hsin Chu 300, Taiwan and Institute of Hydrological Sciences, National Central University, Chung Li 320, Taiwan

Abstract. Direct numerical simulation of a turbulent shear flow beneath a wind-driven water surface is performed to gain a better understanding of the hydrodynamic processes governing the vertical exchanges of heat and gases across the air-water interface. Coherent flow structures are identified which serve as the renewal eddies that pump the submerged fluids toward the water surface and bring down the upper fluids, and therefore enhance the scalar exchange between the air and the water body. The computed vertical distributions of the mean velocity, temperature and gas concentration compare well the parameterized profiles derived based on the surface renewal model.

1 Introduction

The exchanges of gases, heat and momentum between the atmosphere and the ocean are determined by the turbulent shear flow beneath the air-water interface, of which the dominant generation mechanism is the wind stress acting on the interface. The transfer rates of these exchange processes are traditionally parameterized by the wind speed through empirical relations although it is well known that other parameters, such as viscoelastic properties of surface films, also significantly influence the processes [8]. The situation is improved recently with the advent of the novel remote-sensing instruments which measure the interfacial properties such as surface temperature and roughness. These measurement techniques raise the questions on the correlation between the remotely-sensed surface signatures and the underlying hydrodynamic processes. One example is the use of infrared camera to estimate the heat transfer velocity from the spatially resolved temperature measurement [4]. In this technique, heat flux across the air-water interface is estimated from the vertical conduction within the thermal sublayer immediately beneath the water surface which undergoes cyclic growth and disruption by the underlying "surface renewal eddies". The parameters of the renewal processes, such as the probability distribution of the renewal frequency, are then derived from the infrared images.

To elucidate the transfer processes across the air-water interface, turbulent shear flow driven by a wind stress acting at the interface is simulated numerically in this study. The numerical simulation primarily focuses on providing further insight into the hydrodynamic mechanism, particularly the coherent flow structures that govern the process of scalar exchange across the air-water interface. Here, the emphasis is on educing the surface features resulting from the interaction between a wind-sheared interface and the subsurface turbulence structures.

2 Numerical Simulation

The approach adopted herein is a direct numerical simulation of the flow without using any subgrid turbulence parameterization. The three-dimensional Navier-Stokes equations subject to the solenoidal condition are solved in a computational domain closed by imposing periodic conditions in both streamwise (x) and spanwise (y) directions, a free-slip condition at the bottom, $z = -h$, and boundary conditions at the interface, $z = \eta(x, y, t)$. The physical conditions that there be no mass flux and balance of the normal and tangential stresses at the interface give rise to the kinematic condition and the dynamic conditions. A constant streamwise stress is applied at the interface, which is balanced by the local tangential stresses. These interfacial conditions are linearized and satisfied at the mean interface, $z = 0$, based on the analyses of [9].

A second-order Runge-Kutta scheme is used for time integrating the Navier-Stokes equations for the velocities u, v and w, and the kinematic condition for the interfacial deformation η. Solenoidal condition is enforced by solving a pressure Poisson equation in a vertically staggered grid system. The pressure Poisson equation is obtained by taking the divergence of the time-discretized Navier-Stokes equations and imposing the solenoidal condition at the updated time interval step. The normal-stress condition at the interface is used as a Dirichlet condition for the pressure equation, while the tangential-stress conditions are satisfied implicitly in evaluating the diffusion terms in the Navier-Stokes equations. Pseudo-spectral and finite difference discretizations of the equations are used in the horizontal and vertical dimensions, respectively.

Numerical simulation of the flow is initiated by adding a three-dimensional fluctuating velocity field to a mean streamwise velocity distribution: $U(z) = At_0(1 + 2z^2)\mathrm{erf}(-z) + 2\pi^{-1/2}\exp(z^2)$, corresponding to the velocity profile of a parallel shear layer driven by a constant surface stress which induces a linear increase in the surface velocity for a period of t_0. Accordingly, the simulated flow corresponds to the subsequent development of a surface shear current which has been driven by a wind field accelerating constantly from rest to a final speed of u_a in a period of t_0 as in the experiment of [7]. The constant increasing rate A depends on the wind speed and is determined to simulate the experiments in [7].

The transport of scalar within the bulk fluid is also simulated by integrating the advection-diffusion equation for the temperature field θ of heat or the concentration c of a dissolved gas. To simulate the distributions of the temperature and the dissolved gas concentration, different boundary conditions are applied at the water surface [3]. For the physical situation in which the water surface is being cooled from above by evaporation, the heat flux Q across the interface can be considered as a constant, $Q = -k(\partial\theta/\partial z)_{z=0}$, where k is the thermal conductivity, assuming the humidity and wind speed in the atmosphere remain constant. Such a uniform heat flux across the water surface results in a Neumann condition for the temperature field at the interface, $(\partial\theta/\partial z)_{z=0} = -Q/k$. As for the transfer of a sparingly soluble gas, such as carbon dioxide, across the water surface, the resistance is dominated by the aqueous flow. In this case, the atmosphere can be considered as an infinite reservoir of the gas with constant concentration c_a. This gives rise to a Dirichlet condition for the concentration at the water surface, $(c)_{z=0} = sc_a$, where s is the solubility of the gas in water.

3 Mean Profiles

For temporal mean current beneath a wind-blown air-water interface, a two-layer velocity profile, analogous to that of a turbulent wall layer, has been found from the observations in the laboratory [11] and field experiments [1]. The current near but not immediately below the interface is found to be characterized by a logarithmic velocity profile,

$$\frac{u_0 - \overline{u}(z)}{u_*} = \frac{1}{\kappa} \ln z^+ + \psi^+ = \frac{1}{\kappa} \ln \frac{z}{z_0}, \tag{1}$$

where $\overline{u}(z)$ represents temporal or horizontal-spatial averaging of u, $u_0 = \overline{u}(0)$ is the mean surface velocity, the von Karman constant $\kappa \approx 0.41$, the friction velocity $u_* = (\tau_0/\rho)^{0.5}$, the wall coordinate $z^+ = zu_*/\nu$, the mean shear stress at the surface $\tau_0 = \rho\nu\overline{u}'(0)$, ρ and ν are the density and kinematic viscosity of water, and the constant of integration ψ^+ which is related to surface roughness length z_0 by $\kappa\psi^+ = -\ln(u_* z_0/\nu)$.

Immediately below the interface, the turbulence is largely suppressed so that a thin viscous sublayer exists where molecular viscosity dominates the momentum transport. This viscous sublayer was found to be thinner than the analogous sublayer in close proximity to a solid no-slip wall [2], and undergoes cyclic growth and disruption by the turbulent "renewal" eddies randomly arise from the underneath well-mixed turbulent region. Similar flow structures also exist for the transfer of heat and dissolved gas. Within these sublayers, thermal conduction and molecular diffusion predominate the transfer processes of heat and dissolved gas, and consequently resist the direct vertical exchange between the atmosphere and the ocean. The exchange rates are enhanced when renewal eddies impinge the interface and temporarily destruct the sublayers. Adopting the surface renewal model and assuming the renewal of sublayer elements is Gaussian, [2] derived an exponential profile for the mean temperature distribution within the thermal sublayer, given a constant heat flux. In a similar manner, the mean velocity profile within a viscous sublayer driven by a surface stress τ_0 can be expressed in the wall coordinate as

$$\frac{u_0 - \overline{u}(z)}{u_*} = \xi_u \left[1 - \exp\left(-\frac{z^+}{\xi_u}\right)\right], \tag{2}$$

where $\xi_u = \delta_u u_*/\nu$ is the nondimensional sublayer parameter, δ_u is a scaling depth equivalent to the thickness of the viscous sublayer which is related to the average residence duration of renewal fluid elements t_θ through $\delta_u = (vt_u)^{1/2}$. The asymptotic limit of (2) as $(z^+/\xi_u) \to 0$ becomes the familiar linear profile $(u_0 - \overline{u})/u_* = z^+$. The matching of the exponential profile of laminar sublayer to the logarithmic profile requires that the right-hand sides of the logarithmic profile (1) and the exponential profile (2) and their derivatives are equal at $z^+ = \zeta_u^+$ [6]. This yields two nonlinear equations for the parameter ξ_u and the matching coordinate ζ_u^+. For heat and gas transports, similar exponential profiles can be derived for temperature and concentration distributions within the sublayers.

The profiles of the mean streamwise velocity $(u_0 - \overline{u}(z^+))/u_*$, temperature $(\overline{\theta}(z^+) - \theta_0)/\theta_*$ and gas concentration $(\overline{c}(z^+) - c_0)/c_*$ are shown in Fig. 1 for the wind speed $u_a = 5$ m s^{-1}, where $\theta_* = -Q/u_*$ and $c_* = -J/u_*$, θ_0 and c_0 are the mean surface temperature and concentration, Q and J are the mean heat and gas fluxes, respectively. Also plotted in Fig. 1(a) are the experimental measurements of [12] for the wind speed of 4.8 m s^{-1}. The experimental data scatter but the mean values vary close to the present simulated velocity profiles. The simulated mean distributions of velocity, temperature and gas concentration agree well with the exponential profiles derived using the surface renewal model of [2]. The experimental measurements, the numerical simulations and the parameterized profiles are comparative for the whole sublayer range extending to the buffer zone.

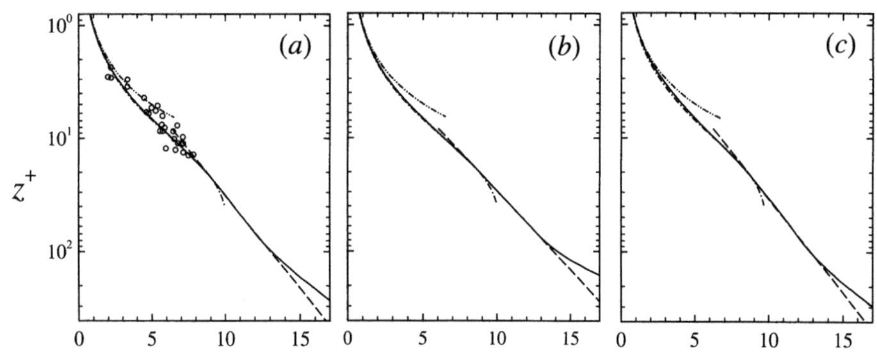

Fig. 1. Vertical distributions of the mean streamwise velocity $(u_0 - \overline{u}(z^+))/u_*$, temperature $(\overline{\theta}(z^+) - \theta_0)/\theta_*$ and gas concentration $(\overline{c}(z^+) - c_0)/c_*$ for the wind speed $u_a = 5$ m s^{-1}. The present numerical results are represented by solid curves. The dash-dot-dotted curves are the traditional linear profiles of viscous sublayer and the dash-dotted curves are the exponential profiles developed based on the surface renewal model [2]. The dashed curves are the logarithmic distributions. The circles in (a) are measurement data of [12] for the wind speed of 4.8 m s^{-1}.

4 Flow Structures

Although the initial turbulent field has been constructed lacking any prescribed spatial structures, organized surface signatures emerge at the interface as the turbulent flow evolves. Two distinct flow features, which are qualitatively identical to those observed in the experiment of [7], appear at the water surface in the simulated results. The first characteristic feature is surface streaming, which consists of elongated high-speed streaks aligned with the wind stress (see bright regions in Fig. 2(a)). These surface streaks consist of major fast-moving strips with several narrower and shorter ones between the major streaks. Comparing the distributions of velocity and temperature in Fig. 2, surface temperatures along these streaks are lower (indicated by higher grey values in Fig. 2(b)) than the average temperature.

The second characteristic surface signatures are localized low-speed spots which emerge randomly at the water surface (dark spots in Fig. 2(a)). A high-speed streak bifurcates and forms a dividing flow when it encounters a low-speed surface spot [10]. In contrast to the cooler surface regions along the high-speed streaks, the water around a low-speed spot is warmer (bright area in Fig. 2(b)).

To reveal flow structures beneath the water surface, contours of the high and low streamwise velocities as well as the regions of warm and cool waters on a cross-wind vertical plane are plotted in Fig. 3. It can be seen, from the vertical distribution of the streamwise velocity, that the high-speed surface streaks appear to be formed by an array of near-surface streamwise jets (bright regions in Fig. 3(a)) carrying cooler waters (dark regions in Fig. 3(b)).

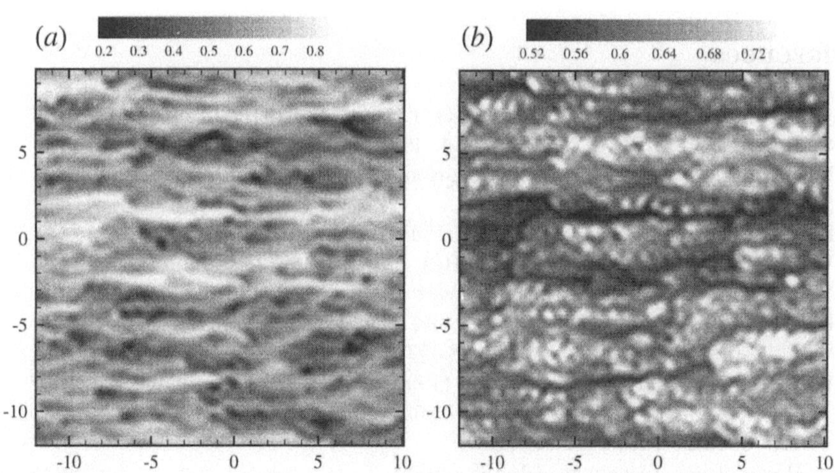

Fig. 2. Contours of instantaneous (a) streamwise velocity and (b) temperature at the interface showing predominant structures of elongated streaks and localized spots. In the figures, the flow travels from left to right. The length and width of the computational domain are both 21.5 cm. The contour values of the velocity and the temperature have been normalized by the maximum velocity and the bulk temperature, respectively.

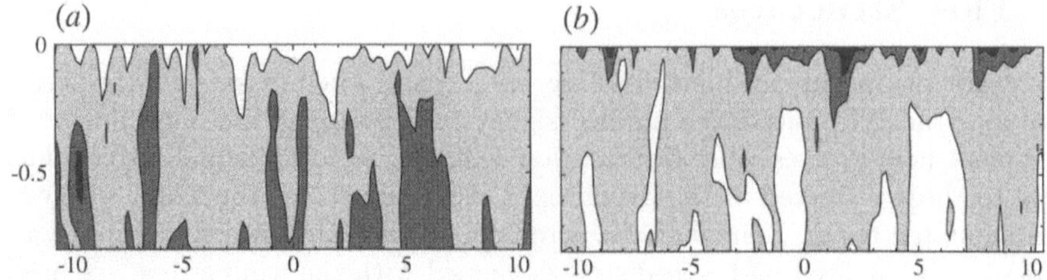

Fig. 3. Contours of instantaneous (a) streamwise velocity and (b) temperature on a vertical plane perpendicular to the wind direction. The direction of wind is out of the page. The contours values of the bright and dark regions in (a) are half and one tenth of the maximum streamwise velocity, respectively. While in (b), the contours levels of the bright and dark regions are 80% and 60% of the bulk temperature, respectively.

Beneath the fast-moving jets, slowly-moving "tongues" are observed (dark regions in Fig. 3(a) and bright areas in (b)). These tongues are associated with upwellings which bring the slowly-moving and warmer underneath fluids up towards the water surface where they consequently induce localized low-speed spots. Such ascendant flows are attributed to the formation and upward movement of coherent Ω-shaped horseshoe vortices in the turbulent shear layer [9]. The underlying coherent horseshoe vortices move upwards and impinge on the water surface. During impingement, these vortices induce upwelling flows and consequently enhance the renewal of water in contact with the surface.

Acknowledgments: This work was supported by grants from the National Science Council of Taiwan (NSC 89-2611-M-009-002 and 90-2611-M-009-001).

References

1. J.H. Churchill, G.T. Csanady: J. Phys. Oceanogr. **13**, 1669 (1983)
2. G.T. Csanady: J. Phys. Oceanogr. **14**, 402 (1984)
3. R.A. Handler, J.R. Saylor, R.I. Leighton, A.L. Rovelstad: Phys. Fluids, **11**, 2607 (1999)
4. B. Jähne, H. Haußecker: Annu. Rev. Fluid Mech. **30**, 443 (1998)
5. W.T. Liu, J.A. Businger: Geophys. Res. Lett. **2**, 403 (1975)
6. W.T. Liu, K.B. Katsaros, J.A. Businger: J. Atmos. Sci., **36**, 1722 (1979)
7. K. Melville, R. Shear, F. Veron: J. Fluid Mech. **364**, 31 (1998)
8. W.T. Tsai: J. Geophys. Res. **101**, 28,557 (1996)
9. W.T. Tsai: J. Fluid Mech. **354**, 239 (1998)
10. W.T. Tsai: Geophys. Res. Lett. **28**, 3959 (2001)
11. J. Wu: J. Fluid Mech. **68**, 49 (1975)
12. J. Wu: J. Phys. Oceanogr. **14**, 138 (1984)

New Bench-mark Results for the 2D Lid-driven Cavity Problem

Charles-Henri Bruneau[1] and Mazen Saad[1]

Mathématiques Appliquées de Bordeaux, Université Bordeaux 1
351 cours de la Libération, 33405 Talence, France
bruneau@math.u-bordeaux.fr, saad@math.u-bordeaux.fr

Abstract. The 2D lid-driven cavity problem is revisited for a wild range of Reynolds numbers. Accurate bench-mark results are provided for steady solutions as well as for periodic solutions around the critical Reynolds number and turbulent solutions at high Reynolds number. Data are given for $Re = 10^3$, $Re = 5 \times 10^3$, $Re = 10^4$ and $Re = 10^5$. In addition, the first Hopf bifurcation is localised by a study of the linearized problem and the computation of the first Lyapunov exponent.

1 Introduction

The 2D lid-driven cavity problem has been intensively studied for more than twenty years. In a first period, the aim of the researchers was to develop robust enough methods to compute nice solutions up to Reynolds numbers $Re = 10^4$. At this value, some authors could find stable steady solutions whereas others were faced to some instabilities [7] [14]. One motivation of this work comes from the following questions – Is the steady solution stable at $Re = 10^4$ or is the stability due to the numerical diffusion of the approximation that is strong enough to delay the transition ? – Are the observed instabilities due either to the approximation or to the presence of the first Hopf bifurcation ? Another motivation is to find out a robust and accurate approximation that gives solutions close to the best solutions computed for instance by spectral methods for low Reynolds numbers, let say $Re = 10^3$, and also allows to compute transient solutions up to $Re = 10^5$ or more. We propose in addition to the well-known $Re = 10^3$ and $Re = 5 \times 10^3$ bench-mark tests, three other bench-mark tests one related to the evaluation of the critical Reynolds number and the two others to describe the flow at $Re = 10^4$ and $Re = 10^5$.

Let $\Omega = (0,1) \times (0,1)$ be a 2D square cavity and $T > 0$ the simulation time, the governing equations of an evoluting incompressible flow are the unsteady Navier-Stokes equations written in primitive variables as:

$$\begin{cases} \partial_t U - \dfrac{1}{Re}\Delta U + (U.\nabla)U + \nabla p = 0 & \text{in }]0,T[\times\Omega \\ \nabla.U = 0 & \text{in }]0,T[\times\Omega \end{cases} \tag{1}$$

where $U = (u,v)$ and p are respectively the velocity and the pressure. The boundary condition is chosen so that the primary vortex is positive. Most of the time, the flow is assumed to start from rest.

2 Outline of the numerical method

The system of equations (1) is discretized by means of a first order Euler scheme. The linear terms are treated implicitly whereas the convection terms are treated explicitly. We denote by U^n the approximation of U at time $t_n = n\delta t$ where $n \in \mathbb{N}$ and δt is the time step. Thus, the semi-discretized equation reads:

$$\frac{U^n}{\delta t} - \frac{1}{Re}\Delta U^n + \nabla p^n = \frac{U^{n-1}}{\delta t} - (U^{n-1}.\nabla)U^{n-1} \text{ in } \Omega \qquad (2)$$

These equations are discretized by finite differences on uniform staggered grids. The discrete values of the pressure are located at the center of each cell and those of the velocity field are located at the middle of the sides. The discretization of the left hand side of equations (2) is achieved using second order centered finite differences. A third order Murman-like scheme is used for the convection terms. For instance, the discretization of the term $v^{n-1}\partial_y u^{n-1}$ is done as:

$$\frac{1}{3}v^{n-1}_{i-\frac{1}{2},j+\frac{1}{2}}\frac{\Delta_{i-\frac{1}{2},j+\frac{1}{2}}u^{n-1}}{\delta y} + \frac{5}{6}v^{n-1}_{i-\frac{1}{2},j-\frac{1}{2}}\frac{\Delta_{i-\frac{1}{2},j-\frac{1}{2}}u^{n-1}}{\delta y} - \frac{1}{6}v^{n-1}_{i-\frac{1}{2},j-\frac{3}{2}}\frac{\Delta_{i-\frac{1}{2},j-\frac{3}{2}}u^{n-1}}{\delta y} \quad \text{if } v^{n-1}_{i-\frac{1}{2},j-\frac{1}{2}} > 0$$

$$\frac{1}{3}v^{n-1}_{i-\frac{1}{2},j-\frac{1}{2}}\frac{\Delta_{i-\frac{1}{2},j-\frac{1}{2}}u^{n-1}}{\delta y} + \frac{5}{6}v^{n-1}_{i-\frac{1}{2},j+\frac{1}{2}}\frac{\Delta_{i-\frac{1}{2},j+\frac{1}{2}}u^{n-1}}{\delta y} - \frac{1}{6}v^{n-1}_{i-\frac{1}{2},j+\frac{3}{2}}\frac{\Delta_{i-\frac{1}{2},j+\frac{3}{2}}u^{n-1}}{\delta y} \quad \text{if } v^{n-1}_{i-\frac{1}{2},j+\frac{1}{2}} < 0$$

Finally, we solve the discrete linear system by means of the FAS multigrid solver to improve the efficiency of the numerical simulation (see [3] for more details).

3 Numerical bench-marks at *Re* = 1000 and *Re* = 5000

We first compare our solution at $Re = 10^3$, on the one hand, on a coarse grid to the classical results in [6], [13] and [14] and, on the other hand, on a fine grid to the very accurate results obtained in [1] with 160 Chebyshev polynomials (table 1). We can see that the present results compare very well. In addition to local quantities it is interesting to compare global quantities as the total kinetic energy $E = \frac{1}{2}\int_\Omega \|U\|^2 dx$, the enstrophy $Z = \frac{1}{2}\int_\Omega \|\omega\|^2 dx$ and the palinstrophy $P = \frac{1}{2}\int_\Omega \|\nabla\omega\|^2 dx$ where $\omega = \partial_x v - \partial_y u$ is the vorticity. As the solution is stationary, the kinetic energy converges to $E = 0.044503$ whereas the two other quantities do not converge. Indeed, because of the singularity at the corner, it is not possible to get convergence as there is a jump of the velocity that induces infinite derivatives. Thus the enstrophy increases slowly to $Z = 22.424$ and the palinstrophy goes to infinity as the mesh size goes to zero.
The solution at $Re = 5000$ is also compared to the solutions available in the literature. Unfortunately, there is no very accurate results as in the previous case and we choose our results obtained on a fine 2048×2048 grid as a reference. The comparison with the results of the literature (table 2) show in particular that our results are coherent with the results given in [6] and [7]. The energy reaches the value $E = 4.7255 \times 10^{-2}$ on the finest grid and the enstrophy increases to $Z = 3.8643 \times 10^3$.

Table 1. Comparison on the two primary vortices at $Re = 1000$.

Scheme	Grid	ψ_{max}	ω	x	y
Present	128×128	0.11786	2.0508	0.46875	0.5625
Ghia [6]	128×128	0.117929	2.04968	0.4687	0.5625
Schreiber [13]	140×140	0.11603	2.026	0.47143	0.56429
Vanka [14]	320×320	0.1173	–	0.4562	0.5625
Present	1024×1024	0.11892	2.0674	0.46875	0.56543
Botella [1]	N = 160	0.1189366	2.067753	0.4692	0.5652
Scheme	Grid	ψ_{min}	ω	x	y
Present	128×128	-1.7003×10^{-3}	-1.1304	0.14063	0.10938
Ghia [6]	128×128	-1.75102×10^{-3}	-1.15465	0.14062	0.1094
Schreiber [13]	140×140	-1.700×10^{-3}	-0.999	0.13571	0.10714
Vanka [14]	320×320	-1.74×10^{-3}	–	0.1375	0.1063
Present	1024×1024	-1.7292×10^{-3}	-1.1120	0.13672	0.11230
Botella [1]	N = 160	-1.729717×10^{-3}	-1.109789	0.13602	0.1118

Table 2. Comparison on the two primary vortices at $Re = 5000$.

Scheme	Grid	ψ_{max}	ω	x	y
Present	256×256	0.12064	1.9125	0.48438	0.53516
Ghia [6]	256×256	0.118966	1.86016	0.4883	0.5352
Huser [8]	80×80 stretched	0.1219	2.001	–	–
Kim-Moin [10]	96×96 stretched	0.112	1.812	–	–
Goodrich [7]	256×256	0.118	–	0.48438	0.53516
Vanka [14]	160×160	0.0920	–	0.4875	0.5313
Present (reference)	2048×2048	0.12197	1.9327	0.48535	0.53516
Scheme	Grid	ψ_{min}	ω	x	y
Present	256×256	-3.0348×10^{-3}	-2.6330	0.19141	0.074219
Ghia [6]	256×256	-3.0835×10^{-3}	-2.66354	0.1914	0.07422
Goodrich [7]	256×256	-3.13×10^{-3}	–	0.1953	0.07422
Vanka [14]	160×160	-5.49×10^{-3}	–	0.15	0.0813
Present (reference)	2048×2048	-3.0699×10^{-3}	-2.7240	0.19434	0.073242

4 Linear stability and periodic solutions

To study the stability of the steady solution, we compute the first Lyapunov exponent of the linearized system. A small perturbation (V, q) is added to the steady solution (U_S, p_S) of system (1). The first Lyapunov exponent defined by

$$\mu_1 = \lim_{t \longrightarrow +\infty} \frac{\mathrm{Log}\|V(t)\|}{t} \quad \text{is approximated by } \mu_1^n = \frac{\sum_{i=0}^{n-1} \mathrm{Log}\|V_0^i\|}{n\delta t}$$

where $V_0^i = V_0^i/\|V_0^{i-1}\|$ is the normalized solution at each time iteration (see [3] for more details).

For low Reynolds numbers the convergence is very fast and monotonous. On the contrary, for high Reynolds numbers the convergence is very slow and not monotonous. Actually the sequence μ_1^n oscillates around the limit for a long time.

Table 3. Evolution of the Lyapunov exponent with Reynolds number.

Re	10	100	1000	5000	7500	7800	8000	8050
T	38	110	242	1386	2087	2243	7442	-
μ_1	-5.2	-0.56	-0.076	-0.013	-0.0088	-0.0082	-0.0026	-0.0011

For example at $Re = 8000$ the value $\mu_1 = -0.0026$ (see table 3) is reached at time $T = 2400$ and then μ_1^n oscillates slightly around this value. We can see that μ_1 is quite small at $Re = 8050$ and we beleive that the critical Reynolds number is very close to this number. The localization of the critical Reynolds number is also confirmed by the few results available in the literature, in particular in [5] where the first eigenvalues are computed. Due to numerical diffusion, we can conclude that the critical Reynolds number for the 2D lid-driven cavity problem is $8000 \leq R_c \leq 8050$ within less than 1% of error.

After the first Hopf bifucation ($Re = 8100$ and $Re = 8200$) there is a stable periodic solution with frequency $f = 0.45$. Then we compute the solution at $Re = 10000$. Some computations on several consecutive grids show that the stable solution is mainly periodic with small variations in the amplitude. The time evolution at the monitoring points and the Fourier analysis as well as the phase portraits show that the variations in amplitude yield a solution which is not purely periodic and exhibits some low frequencies (see figure 1). On figure 2 the vorticity field is plotted along a main period from time $t = 0$ to time $t = 1.64$ to show the evolution of the secondary vortices that deform slightly the primary vortex. The flow patterns of figure 2 are very close to those shown in [4].

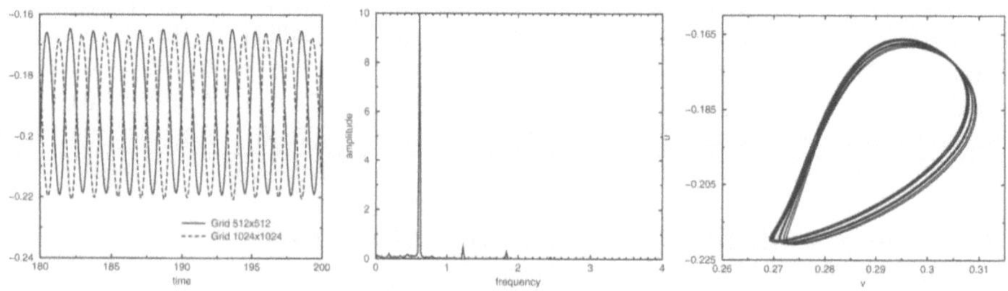

Fig. 1. Horizontal velocity (left), power spectrum (middle) and phase portrait (right) on grid 1024 × 1024 at monitoring point (14/16,13/16) for $Re = 10000$.

Fig. 2. Vorticity field at times $t = 0$, $t = 0.984$ and $t = 1.64$ on grid 1024 × 1024.

5 Solutions at *Re* = 100000

To get reliable results the numerical simulations are performed on a fine grid of 1024 × 1024 points which ensures that there are a few grid points in the boundary layer. The first numerical test is to compute the solution from rest, and the second numerical test is performed initializing the solution at time $t = 0$ with the steady solution obtained at $Re = 5000$.

On the one hand, the plots of figure 3 reveal the presence of numerous eddies coming from the walls inside the whole cavity. A carefull study indicate that the effects of viscosity are confined to a thin layer close to the solid boundaries.

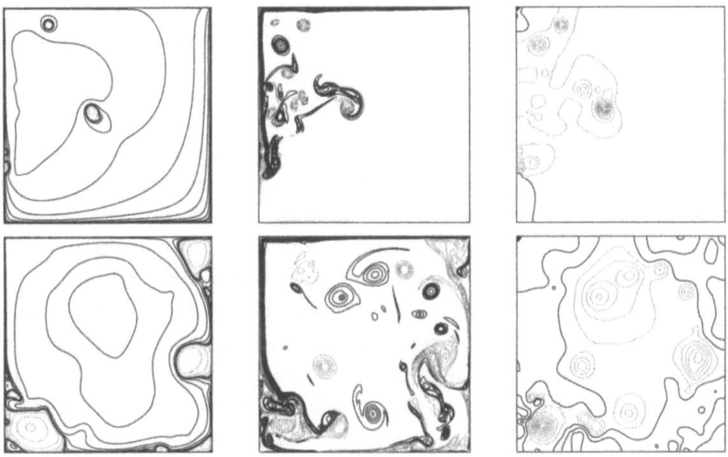

Fig. 3. Solution at $Re = 100000$ computed from initialization at rest. Streamfunction, vorticity and presure fields at times $t = 10$ and $t = 120$.

Fig. 4. Solution at $Re = 100000$ computed from initialization by the steady solution at $Re = 5000$. Streamfunction, vorticity and presure fields at times $t = 10$ and $t = 120$.

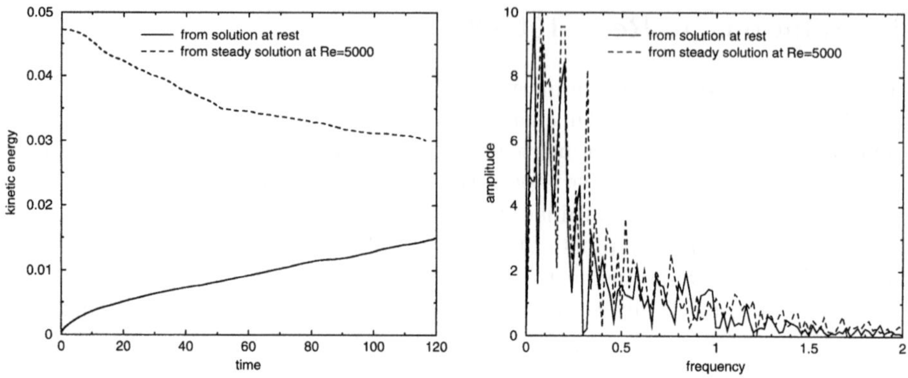

Fig. 5. Comparison of the energy (left) and power spectrum of the vertical velocity (right) at $Re = 100000$ with the two initializations.

Small eddies develop in the vicinity of the walls and the two lower corners. On the other hand, the figure 4 show that the strong primary vortex is progressively reduced as many eddies coming from the wall spread into the whole domain. This is confirmed by the energy history. Indeed the energy of the first simulation increases with time from zero as the energy of the second simulation decreases (figure 5). In addition the power spectrum of the vertical velocity of the two signals cut down to the time interval $(80, 120)$ are very similar in frequency and amplitude. The energy curves suggest that a time of the order of 300 is needed to get the same energy and consequently the solutions converge towards the same asymptotic stage, confirming the presence of an attractor.

References

1. O. Botella, R. Peyret, Comput. Fluids **27** n° 4, 1998.
2. Ch.-H. Bruneau, C. Jouron, J. Comput. Phys. **89** n° 2, 1990.
3. Ch.-H. Bruneau, M. Saad, *The 2D lid-driven cavity problem revisited*, submitted.
4. F. Dexun, M. Yanwen, L. Hong, Proc. 5th Int. Symp. CFD, Sendai 1993.
5. A. Fortin, M. Jardak, J.J. Gervais, R. Pierre, Int. J. Numer. Methods Fluids **24** n° 11, 1997.
6. U. Ghia, K.N. Ghia, C.T. Shin, J. Comput. Phys. **48**, 1982.
7. U. Goodrich, Proc. IMACS 1st Int. Conf. Comp. Physics, Boulder 1990.
8. A. Huser, S. Biringen, Int. J. Numer. Methods Fluids **14** n° 9, 1992.
9. T. Kawamura, H. Takami, K. Kuwahara, Lect. Notes Phys. **218**, 1985.
10. J. Kim, P. Moin, J. Comput. Phys. **59**, 1985.
11. K. Kuwahara, http://www2.icfd.co.jp
12. B.P. Leonard, Comput. Methods Appl. Mech. Eng. **88** n° 1, 1991.
13. R. Schreiber, H.B. Keller, J. Comput. Phys. **49**, 1983.
14. S.P. Vanka, J. Comput. Phys. **65**, 1986.

Numerical Simulation of Viscoelastic Flow past a Linear Array of Cylinders by Parallel Computation

Hua-Shu Dou and Nhan Phan-Thien

Fluid Mechanics Laboratory
Department of Mechanical Engineering
The National University *of* Singapore
Singapore 119260

Abstract. Parallelized pressure-based finite volume method (FVM) with unstructured meshes is used to simulate the large scale problems of viscoelastic flows in distributed environment. The Discrete Elastic Viscous Split Stress vorticity (DEVSS-omega) formulation and the SIMPLER iteration algorithm are employed for the numerical process. An Oldroyd-B fluid past an array of circular cylinders in a channel is simulated for the spacing parameter L=4, 6, and 28. The mechanism of flow instability and the effect of parameter L on the flow behaviour are discussed. The results show that the shear motion in the gap between the wall and the cylinder dominates the flow at high Deborah number.

1 Introduction

Simulations of viscoelastic fluid flows consumes large computer resources. Effects of viscoelasticity on the wall flows concentrate in very thin boundary layers. Only by using fine meshes could one obtain reliable solutions within these thin layers. Thus, a large-scale computational problem is encountered which requires large CPU time and memory. Here, we use a finite volume method (FVM) with unstructured meshes to deal with these problems in the distributed computing environment. FVM with unstructured meshes has several advantages over structured mesh: the ability to better model the geometry, flexibility in domain decomposition and therefore better load balancing under a distributed computing environment [1].

Viscoelastic flow past an array of cylinders has received much attention in the last decade due to its implicative applications in flows through porous media [2] [3] [4] [5]. In this study, the flow of an Oldroyd–B fluid past a linear array of cylinders in a channel with geometry $h/R = 2$ (Figure 1) is simulated using very fine meshes. The computing results are discussed for the flow instability of viscoelastic flows. The results agrees qualitatively with the experimental behaviour reported in literature.

2 Governing Equations

The equations of mass and momentum conservation, and the constitutive equation for the Oldroyd-B model are [6]:

$$\nabla \cdot \mathbf{u} = 0 \quad \text{(incompressibility)}, \tag{1}$$

$$\rho(\frac{\partial \mathbf{u}}{\partial t} + \mathbf{u} \cdot \nabla \mathbf{u}) = -\nabla p + \nabla \cdot \tau + \eta_N \nabla^2 \mathbf{u} \quad \text{(momentum)}, \tag{2}$$

$$\tau + \lambda(\frac{\partial \tau}{\partial t} + \mathbf{u} \cdot \nabla \tau - \nabla \mathbf{u}^T \cdot \tau - \tau \cdot \nabla \mathbf{u}) = 2\eta_{m0} \mathbf{D} \quad (\textit{constitutive}), \tag{3}$$

where ρ is the fluid density, t the time, $\mathbf{u}$ the velocity vector, p the hydrodynamic pressure, η_N the solvent viscosity, $\mathbf{D} = (\nabla \mathbf{u} + \nabla \mathbf{u}^T)/2$ the rate-of-strain tensor, η_{m0} a "polymer-contributed" viscosity, λ the (constant) relaxation time, and τ the "extra" stress tensor, not necessarily traceless. The total viscosity of the Oldroyd-B fluid is $\eta_0 = \eta_N + \eta_{m0}$. By defining $\beta = \eta_{m0}/\eta_0$ as the retardation ratio, we have $0 < \beta < 1$ for Oldroyd-B fluid. For the UCM model, $\beta = 1$, and $\eta_0 = \eta_{m0}$. For the Newtonian fluid, $\beta = 0$ and $\eta_0 = \eta_N$.

To stabilize the solution, the identity

$$\nabla^2 \mathbf{u} = -\nabla \times \omega + \nabla(\nabla \cdot \mathbf{u}) \tag{4}$$

is employed, where the vorticity $\omega = \nabla \times \mathbf{u}$. With this, the momentum equations (2) are re-written as

$$\rho(\frac{\partial \mathbf{u}}{\partial t} + \mathbf{u} \cdot \nabla \mathbf{u}) = -\nabla p + \nabla \cdot \tau + (\varepsilon + \eta_N)\nabla^2 \mathbf{u} + \varepsilon \nabla \times \omega - \varepsilon \nabla(\nabla \cdot \mathbf{u}), \tag{5}$$

where the value of ε can be adapted to the stress level in the DAVSS-ω scheme [7] [8]. In this paper, we use a constant value of ε, $\varepsilon = \beta\eta_0$, and this formulation is named as DEVSS-ω scheme [8].

The vorticity vector ω can be calculated using the vorticity transport equation. This equation is obtained by applying a curl operator to Eq.(2),

$$\rho(\frac{\partial \omega}{\partial t} + \mathbf{u} \cdot \nabla \omega) = \eta_N \nabla^2 \omega + \nabla \times (\nabla \cdot \tau), \tag{6}$$

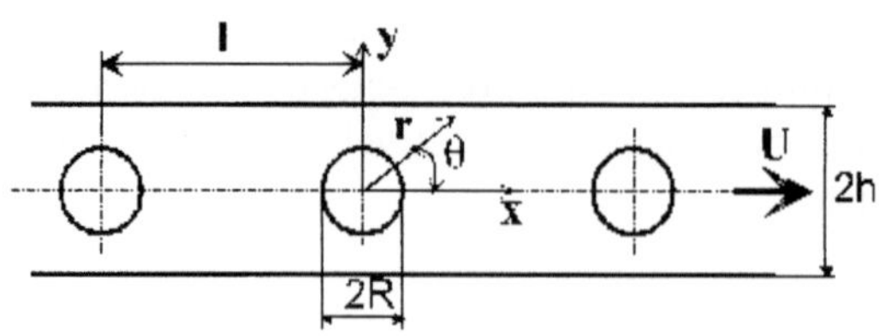

Fig. 1. Flow geometry.

All the dependent variables in the governing equations (3), (5) and (6) can be written in the form of the general transport equations as

$$\frac{\partial(\Lambda\phi)}{\partial t} + \nabla \cdot (\Lambda \mathbf{u}\phi - \Gamma\nabla\phi) = S_\phi = S_P\phi + S_c, \tag{7}$$

where ϕ is one of the dependent variables, Γ denotes the diffusivity of ϕ, S_ϕ is the source term, and $(\Lambda\mathbf{u}\phi - \Gamma\nabla\phi)$ is the "flux" of the variable ϕ. The source term is arbitrarily split in the manner shown on the right side of equation (7).

3 Numerical Method

An unstructured FVM using pressure correction with triangular meshes (control volume-based finite-element method) is used to discretize the governing equations with a co-located mesh arrangement [9]. Triangular meshes are generated in the computing domain (unstructured triangles of irregular size and shape). All the dependent variables such as velocities and pressure as well as stresses are stored at the vertices of the triangular elements. Equation (7) (for velocities, vorticity and stresses) is discretized by using the same interpolation function (8), the pressure and pressure correction are linearly interpolated over the elements as in Prakash and Patankar [9]. A set of local coordinates (X, Y) is defined in the element, as shown in Fig. 2. The origin of this coordinate system (X, Y) is located at the centroid O, and the X axis is aligned with the element-average velocity vector $\mathbf{U}_{avg}$, $\mathbf{U}_{avg} = (\mathbf{U_1} + \mathbf{U_2} + \mathbf{U_3})/3$, $U_{avg} = |\mathbf{U}_{avg}|$. Here, ρ, Γ, and S are assumed to be uniform over any triangular element, and the convection-diffusion flux is calculated by assuming that over any element ϕ varies exponentially in the direction of the average velocity vector and linearly in the normal direction. The shape function is

$$\phi = A \exp\left(\frac{\rho U_{avg} X}{\Gamma}\right) + BY + C \tag{8}$$

where A, B and C are the coefficients in the interpolation formula.

The discretization of equation (7) is carried out using Eq.(8) in all the elements, then integration is performed over a control volume surrounding a node,

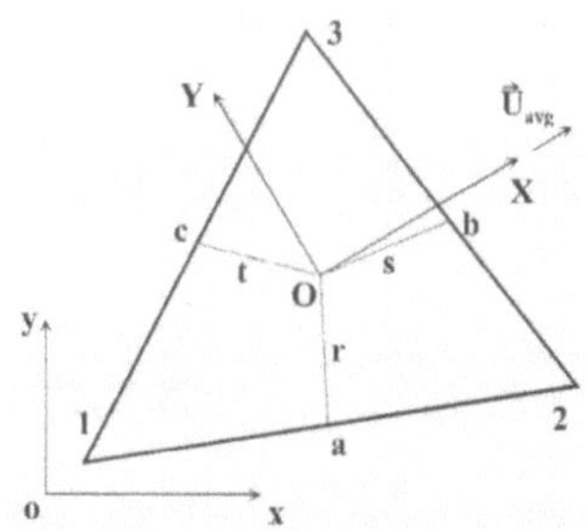

Fig. 2. Local coordinate system within an element.

the final discretization equation for the dependent variable can be expressed by a linear algebraic equation. The discretization equations for pressure and pressure correction are obtained by applying the mass conversation principle over the surfaces of the control volume of a node and utilizing the momentum equations. The equal-order method proposed by Prakash and Patankar [9] is used in this study. All of these algebraic equations are solved sequentially by using the SIMPLER iteration procedure with the Gauss-Seidel point solver.

The computing domain is shown in Fig. 1, which is taken as a unit cell around a cylinder. Periodic conditions for velocities, stresses and pressure are exerted on the inlet and outlet. No-slip conditions are imposed for u, v on solid surfaces. The boundary conditions on the solid surface for are given by analytical solutions such as at the channel walls and the cylinder surface.

The parallel algorithm is detailed in [8]. The parallelization of the computations is implemented by means of a domain decomposition (DD) technique, using the master-slave communication model. In brief, the program was run on a distributed computing environment (called a workstation farm) consisting of 28 DEC Alpha 500/256. Communication is through a fast ethernet of 100 MBits per second. The message passing is supported by PVM software libraries version 3.4.1 from Oak Ridge National Laboratory.

4 Results and Discussion

The dimensionless space parameter between the cylinders is $L = l/R = 4, 6$ and 28, as shown in Fig.1. The width of the passage is $2h = 4R$. The viscosity ratio for the Oldroyd-B fluid used is $\beta = 0.41$. The Reynolds and the Deborah numbers are defined as: $Re = \rho UR/\eta$ and $De = \lambda U/R$, where U is the average velocity in the channel. The steady inertialess flow is considered here. We computed the flow field for the three geometries with the same meshes around the cylinders,

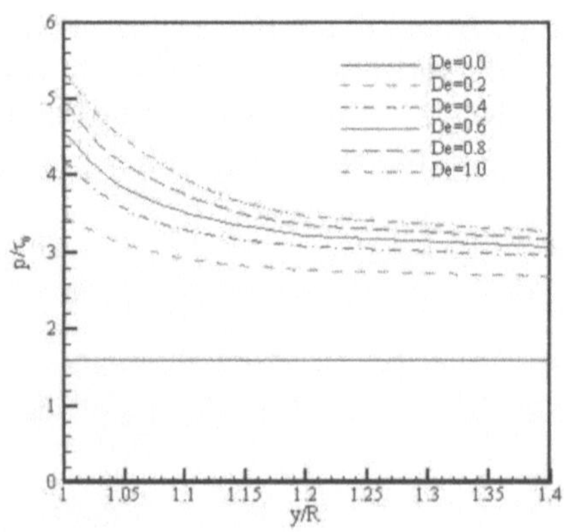

Fig. 3. (a) Pressure contour for the geometry L=4 at Deborah nmber of 0.6. (b) The pressure distribution at the plane x=0 in the gap versus the De number for L=28.

and compared the effect of cylinder spacing. The mesh convergence property of the numerical method has been proved in our previous work. In this study, the mesh size on the surface of the cylinder in the radial direction is $\Delta r/R = 0.0048$. The degree of freedom of the biggest sized mesh is 132368. The attainable maximum Deborah numbers for steady solution are 1.0, 1.1, and 1.2 for the spacing parameter $l = 28, 6$ and 4. However, the appearance of the velocity reflection on the top of the cylinder is about 0.60, 0.65 and 0.75 for $l = 28, 6$ and 4.

Experiments showed that this type of flow exhibits instability at high Deborah number[2] [3] [11]. Liu et al's experiments [2] [3] indicated that the critical *De* number for instability decreases with the increasing of the dimensionless parameter *L*. Extension in the wake has been nominated as the mechanism for the instability initiation [3] [11] because the extension in the wake of cylinders of array is reduced when *L* decreases. However, Dou and Phan-Thien [10] found that the mechanism for instability of viscoelastic flows past a single cylinder in a channel is generated by the strong shear on the cylinder rather than the elongation in the wake. Linear stability analysis also revealed the evident shear stress gradients near the walls and the cylinder surface [4] [5].

Fig.3 (a) shows the pressure contour at $De = 0.60$ for $L = 4$. Fig.3(b) shows the pressure distribution along the plane $x = 0$ with the increasing *De* number for $L = 28$. Fig. 4(a) shows the velocity gradient at the plane $x = 0$ in the gap. A velocity inflection near the cylinder is clearly seen with increasing *De*. Fig. 4(b) shows the velocity gradient at the plane $x = 0$ in the gap for the three values of *L*. Dou and Phan-Thien [10] think that the velocity inflection in the shear layer is generated by the cross-streamline pressure gradient and resultant normal stress. This will lead to the instability of the shear flow and be convected to the downstream wake. Byars' [11] experiments showed that the instability

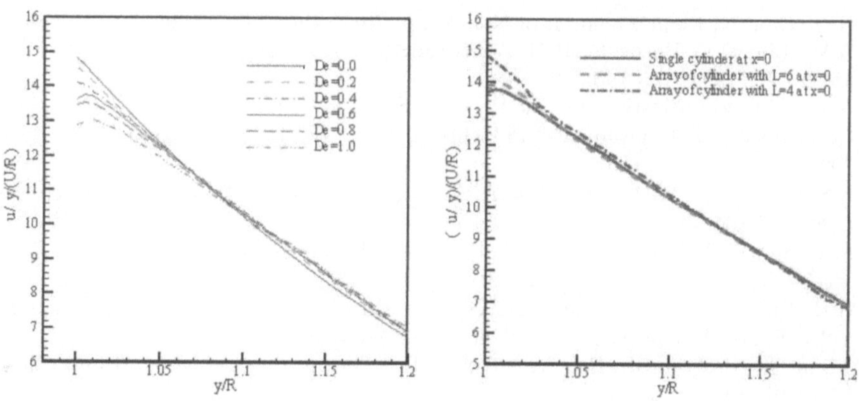

Fig. 4. (a) Velocity gradient near the cylinder at the plane x=0 in the gap versus the De number for L=28. (b) Velocity gradient near the cylinder at the plane x=0 in the gap at De=0.6.

exists in the shear region at shoulder of cylinder upstream the wake at high *De*. From Fig.4(b), it can be seen that the cylinder approaching increases the velocity gradient near the cylinder surface and delays the velocity inflection. This contributes to the stability extension to high *De* for reduced *L*. This means that the shear flow on the top of the cylinder is a dominate factor for flow instability.

The simulation results also show that with the reducing *L*, the stress shear layer on the cylinder becomes thinner, the value of normal stress within the layer becomes low. The distortion of pressure in the shear layer on the cylinder is relaxed for closing cylinder space. This reduces the secondary flow near the cylinder surface and delays the occurrence of velocity inflection on the cylinder. Thus, the approaching cylinders can increase the *Dec* for instability.

5 Conclusions

A detailed simulation is performed using unstructured meshes for the flow of an Oldroyd-B fluid past an array of cylinder in a channel with $h/R = 2$. The results showed that parallel computing is a very useful means for the large scale problems. The unstructured FVM displayed its advantage dealing with complex geometry and distributed computing.

The instability mechanism of viscoelastic flows past an array of cylinders is dominated by shear flow in the gap between the wall and the cylinder. The pressure distortion near the cylinder due to the development of normal stress induces an inflection in the velocity and leads to the flow instability. With the approaching of cylinders, the bahaviour of the shear layer on the cylinder changes. This change delays the occurrence of the velocity inflection on the cylinder and increases the *Dec* number for instability.

References

1. H.-S. Dou, N. Phan-Thien: J. of Non-Newt. Fluid Mech., 77, (1998), 21-51.
2. A.W. Liu, D.E. Bornside, R.C. Armstrong and R.A. Brown: J. non-Newt. Fluid Mech., 77 (1998) 153-190.
3. A.W. Liu: Viscoelastic Flow of Polymer Solutions Around Arrays of Cylinders: Comparisons of Experiments and theory, Ph.D thesis, MIT, Cambridge, USA, 1997.
4. R.Sureshkumar, M.D.Smith, R.C.Armstrong, et al.: J. non-Newt. Fluid Mech., 82 (1999) 57-104.
5. M.D. Smith, R.C. Armstrong, R.A. Brown and R. Sureshkumar: J. non-Newt. Fluid Mech., 93 (2000) 203-244.
6. R.B. Bird, R.C. Armstrong, and O. Hassager, Dynamics of Polymeric Liquids, Vol.1: Fluid Mechanics, 2nd ed.,Wiley, New York, 1987.
7. J.Sun J, M.D.Smith, R.C.Armstrong, R.A.Brown: J Non-Newt. Fluid Mech 86(1999) 281-307
8. H.-S. Dou, N. Phan-Thien: J. Non-Newt. Fluid Mech., 87 (1999) 47-73.
9. C. Prakash, S.V. Patankar: Num. Heat Trans., 8 (1985) 259–280.
10. H.-S. Dou, N. Phan-Thien: J. non-Newt. Fluid Mech., 2001, submitted.
11. J. A. Byars: Experimental Characterization of Viscoelastic Flow Instabilities, Ph.D thesis, MIT, Cambridge, MA 02139, USA, 1996.

Numerical Investigations of Near-Wall Turbulence Structures

Zinedine Khatir

Fluid Dynamics Research Centre, School of Engineering
University of Warwick, Coventry CV4 7AL, UK
Email: es2184@eng.warwick.ac.uk / zkhatir@hotmail.com

Abstract. Near-wall turbulence structure is investigated by a vortex method simulation. The results presented show the evolution of hairpin vortices in the near-wall region and are in good agreement with experimental observations and previous numerical studies of this kind. They suggest that the hairpin vortices being generated coalesce together and induce complicated processes of vortex amalgamation, as well as breaking and reconnection. This complex process is an additional mode of the growth in larger scale and highlights the role of hairpin vortices in the production of turbulence.

1 Introduction

For the past 20 years, considerable advances in flow visualization, measurements and instrumentation have led to an ever-growing knowledge and insight into boundary layers flows. As suggested by Cantwell [1], there is an increasing belief that turbulent flows are not totally random. It has been found that turbulence has quasi-deterministic, repeatable patterns, as well as stochastic elements, that are characterized by a remarkable degree of order. This apparently deterministic sequence is usually referred to as coherent structures or the coherent behaviour of the wall-layer flow, where the wall layer is understood to imply that portion of turbulent flow between the wall and locations where the mean profile is logarithmic in z, the distance from the wall. Thus, the discovery of coherent structure in turbulence has encouraged the hope that the study of vortex dynamics will lead to better models and understanding of turbulent flow, thereby solving or at least making less mysterious one of the great unresolved problem of classical physics. A comprehensive review on structures in turbulent flow and self-sustaining mechanisms of wall turbulence is available in [2] and [3].

The hairpin vortex is postulated to be the basic flow structure of turbulent boundary-layer flows. It is believed that the central features of near-wall flow can be explained in terms of how hairpin vortices interact with the shear flow, with each other, and with the surface layer near the wall.

Hama [4] was the first to model and study numerically turbulent structures in three-dimensional shear flows with his 'Localised-Induction Approximation' (LIA). Hama's original motivation was to attempt to extent the vortex modelling from free-shear flows to boundary-layer flows. This LIA vortex calculation has been used sucessfully by various authors. For instance, Moin *et al* [5] analysed numerically the deformation of hairpin-shaped vortex filament and Hon

& Walker [6] used the LIA to investigate numerically the evolution of hairpin vortices in a shear flow.

The previous aforementioned studies neglected the interaction between the hairpin vortices and the background vorticity. As mentioned above, it is desirable to study the interaction of hairpin vortices with the shear flow in the near-wall regions. In this paper, the generation of hairpin vortices is simulated numerically using a discrete vortex method description of the vorticity field. The attempt is to provide additional insight into the structure of the near-wall regions in turbulent flows.

2 Mathematical Formulation

The evolution of the vorticity field, $\mathbf{\Omega}(\mathbf{x},t) = (\Omega_x, \Omega_y, \Omega_z)$, in three-dimensional flow is governed by the transport equation

$$\frac{\partial \mathbf{\Omega}}{\partial t} + (\mathbf{u}\cdot\nabla)\mathbf{\Omega} = (\mathbf{\Omega}.\nabla)\mathbf{u} + \frac{1}{Re}\Delta\mathbf{\Omega} \tag{1}$$

where Re is the Reynolds number, $\mathbf{u} = (u, v, w)$ is the velocity field and $\nabla\mathbf{u}$ is the velocity gradient tensor with components $(\nabla\mathbf{u})_{ij} = \partial u_i/\partial x_j$. In the present scheme, which is part of a newly developed approach [7,8,9], the velocity field (and its gradients) are made up of several contributions. They are decomposed into a rotational vortical field, $\mathbf{u}'_\Omega(\nabla\mathbf{u}'_\Omega)$, and a potential component, $\mathbf{u}'_p(\nabla\mathbf{u}'_p)$, according to the Helmholtz decomposition rule:

$$\mathbf{u} = \underbrace{U(z)\,\vec{\imath}}_{meanflow} + \underbrace{\mathbf{u}_\Omega{}'}_{Rotational\,disturbance} + \underbrace{\mathbf{u}_p{}'}_{Irrot.\,disturbance} \tag{2}$$

with $\mathbf{u}'_\Omega = (u'_\Omega, v'_\Omega, w'_\Omega)$ and with $\mathbf{u}'_p = (u'_p, v'_p, w'_p)$.
The rotational velocity perturbation is calculated using discrete vortex elements, the rotational disturbance takes the form: $\mathbf{u}'_\Omega = (u'_\Omega = u_\Omega - u^*_\Omega, v'_\Omega, w'_\Omega)$ where $u^*_\Omega\ \vec{\imath}$ is the velocity induced by the undisturbed vortex elements. In a perfect approximation $u^*_\Omega = U(z)$ and it is determined as the initial approximation to the required mean flow field.
The rotational velocity field is formally given by

$$\mathbf{u}_\Omega(\mathbf{x},t) = U(z)\ \vec{\imath} + \mathbf{u}'_\Omega = \int_D \mathbf{K}(\mathbf{x}-\mathbf{z})\ \mathbf{\Omega}(\mathbf{z})\ d\mathbf{z}$$

where $\mathbf{K}$ is the Biot-Savart kernel. Under the discrete-vortex approximation this takes the form

$$\mathbf{u}_\Omega(\mathbf{x},t) = \sum_{n=1}^{\infty} \mathbf{K}_\delta(\vec{r_n})\,\mathbf{\Omega}_n \tag{3}$$

where $\vec{r_n} = (x - X_n, y - Y_n, z - Z_n)$ and r_n is its magnitude with (X_n, Y_n, Z_n) the Lagrangian coordinates of the centre of the nth vortex element.

The desinguralised kernel $\mathbf{K}_\delta(\vec{r_n})$ is given by:

$$\mathbf{K}_\delta(\vec{r_n}) = \frac{f_\delta(r_n)}{4\pi {r_n}^3} \begin{pmatrix} 0 & z - Z_n & -(y - Y_n) \\ -(z - Z_n) & 0 & x - X_n \\ y - Y_n & -(x - X_n) & 0 \end{pmatrix} . \quad (4)$$

Following [10], the irrotational $\mathbf{u}'_p$ is found by writing $\mathbf{u}'_p = \nabla\phi'$ and solving $\nabla^2\phi' = 0$ using a boundary-element method and subject to the kinematic boundary condition:

$$\left(\nabla\phi' + U(z)\,\vec{\imath} + \mathbf{u}'_\Omega\right) \cdot \hat{\mathbf{n}} = 0 . \quad (5)$$

Finally, in limit of infinite *Re*, we have the following closed system of equations for the evolution of the perturbation vorticity and the evolution of the N perturbed vortex elements

$$\begin{cases} \frac{d\mathbf{x}}{dt} = U(z)\,\vec{\imath} + \mathbf{u}'_\Omega + \mathbf{u}'_p \\ \frac{d\mathbf{\Omega}}{dt}(t) = (\mathbf{\Omega}(t) \cdot \nabla) \left\{U(z)\,\vec{\imath} + \mathbf{u}'_\Omega + \mathbf{u}'_p\right\} . \end{cases} \quad (6)$$

3 Numerical Results and Discussion

The evolution of near-wall turbulent flow structures past a flat plate is investigated based on experimental data given in [11]. The growth of hairpin vortices is shown and their contibution towards the development of turbulence structures and streak flow is demonstrated. The results are in good agreement with experimental observations [12, 13]. In 1, the hairpin vortices being generated coalesce together and induce complicated processes of vortex amalgamation, as well as breaking and reconnection. As stated by Smith *et al* [14], this complex process is an additional mode of the growth in larger scale. This highlights the role of hairpin vortices in the production of turbulence as emphasized in Fig. 2.

The coherent structure in the near-wall regions was vizualised by subtracting different values of constant streamwise velocity U_c from the instantaneous

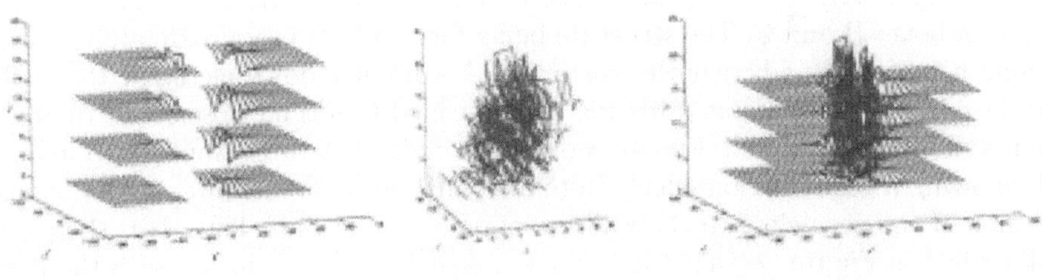

a) b) c)

Fig. 1. Three-dimensional view of the flow field at $t^+ = 1$, showing the different regions forming the disturbance and a snap-shot of a bursting event: (*a*) Region delimited by $-\infty \leq x^+ \leq -20$ and $20 \leq x^+ \leq \infty$ indicating the creation of hairpin vortices, (*b*) Region delimited by $-20 < x^+ < 20$ showing a "break-up" and (*c*) Entire region consisting of the regions as described in (*a*) and (*b*).

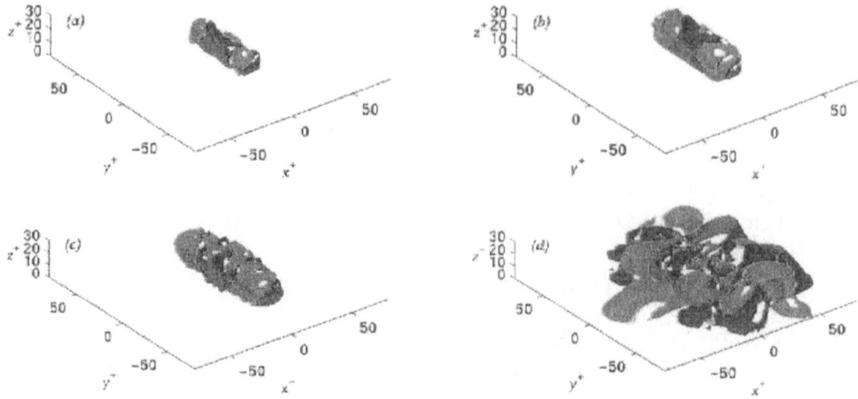

Fig. 2. Isosurface of the spanwise vorticity perturbation ω'_y ($\omega'_y = 0.75$ (light gray) and $\omega'_y = -0.75$ (dark gray)) at time: (*a*) $t^+ = 0.18$, (*b*) $t^+ = 0.45$, (*c*) $t^+ = 0.72$ and (*d*) $t^+ = 1$.

vector fields. This decomposition of the vector field allows us to view the instantaneous spatial relationship between relative motions, without introducing artificial distortion which would be imposed by other decomposition techniques, such as Reynolds-averaged decomposition ([15]). Such vortical and eddy structures were completely obscured if a velocity unequal to their centre velocity was superposed. This kind of realisation is visualised in Figure 3, in terms of a two-dimensional velocity vector plot that occurs in a streamwise wall-normal plane of symmetry at $t^+ = 1$. A constant velocity $U_c = 0.35\,U_\infty$ has been subtracted from the streamwise velocity in order to bring out the packet of hairpins and eddies marked in the 6 boxes A, B, C, D, E and F. A magnified version of those 6 boxes can be seen in more detail in Figures 3 *a-f* respectively. The signature of a hairpin vortex appears to be identifiable in the main plot of 3 and in 3 *a*. It seems that the vortex identified in box A is the primary vortex. The vortex in box F appears to be the head of a relatively young hairpin that is probably being formed upstream of A. Note also the creation of vortical structures downstream of A in boxes B and C. The structure being formed in B is approximately at the same height of A, whereas the vortex in C is further down closer to the wall. In box D, another hairpin vortex is being formed this time closer to A than C, but still close to the wall. Finally, look at the eddy structures emphasized in box F showing a recirculation effect. Note also, in Figure 3, the flow vectors in the region of the very close to the wall are parallel to the wall. Thus, it is there, as described in Figure 2, that the low-speed streaks of [11] coincide with the flow that is strongly parallel to the wall.

As pioneered by Cantwell *et al* [16], another way to identify the structure of a numerically simulated turbulent flow over a flat plate is to use the invariants of the velocity gradient tensor $(\nabla \mathbf{u})_{ij} = \partial u_i / \partial x_j$ (Q and R) and a related scalar quantity, the cubic discriminant $D = 27R^2/4 + Q^3$. (The eigen values λ of the characteric equation of the tensor is given by $\lambda^3 + P\lambda^2 + Q\lambda + R = 0$.) This

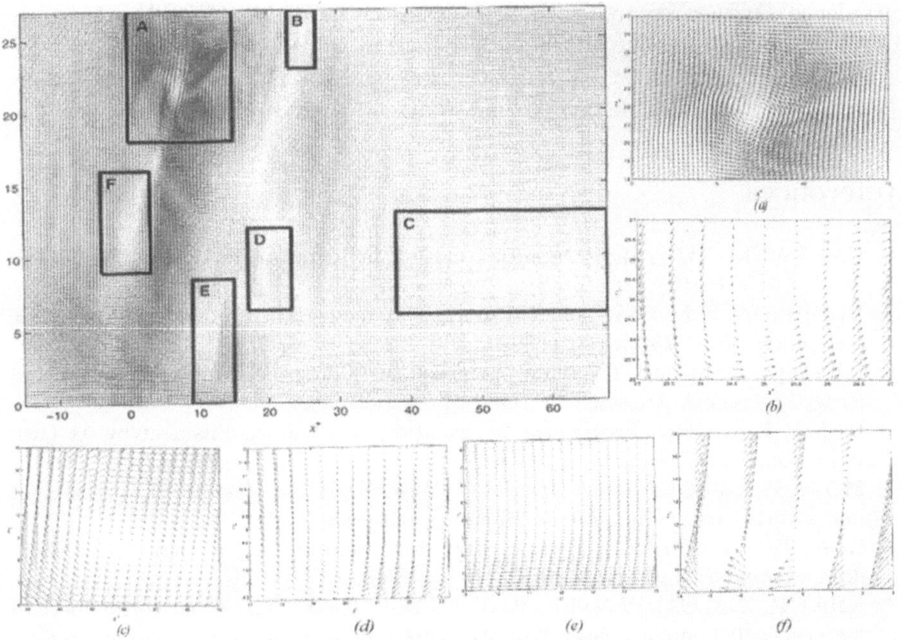

Fig. 3. Velocity vector plot on a streamwise - wall normal plane of symmetry at $t^+ = 1$. A constant velocity $U_c = 0.35\,U_\infty$ has been subtracted from the streamwise velocity in order to bring out the packet of hairpins and eddies marked A, B, C, D, E and F clearly.

Fig. 4. Isosurface of the discriminant $D = 27R^2/4 + Q^3 > 0$.

approach was used to produce Figure 4. Structures are clearly identified and inclined in the streamwise direction. This demonstrates the tendency for those near-wall region structures to retain their identity, as explained by Cantwell *et al* [16].

4 Conclusion

Three-dimensional vortex method simulation of turbulence flow structures has been presented. The investigation strenghtens the idea of coherent structure in turbulence and the key role played by hairpin vortices in the production of turbulence. Further research of those structures is needed and is still under study to improve our knowledge on the underlying mechanisms of turbulence.

Acknowledgements: The author wishes to thank Dr. Srinivas from Sydney University for his valuable help for me to attend the ICCFD2 Conference.

References

1. CANTWELL, B. J. 1981 Organised motion in turbulent flow. *Ann. Rev. Fluid Mech.* **13**, 457–515.
2. ROBINSON, S. K. 1991. Coherent motions in the turbulent boundary layer, *Ann. Rev. Fluid Mech.* **23**, pp. 601–639.
3. PANTON, R. L. 2001. Overview of the self-sustaining mechanisms of wall turbulence, *Progress in Aerospace Sciences* **37**, pp. 341–383.
4. HAMA, F. R. 1962. Progressive Deformation of a Curved Filament by its Own Induction, *Phys. Fluids* **5**, pp. 1156–1162.
5. MOIN, P. & LEONARD, A. & KIM, J. 1986. Evolution of a curved of a filament into a vortex ring, *Phys. Fluids* **29**(4), pp. 955–963.
6. HON, T,-L & WALKER, J. D. A. 1991. Evolution of hairpin vortices in a shear flow, *Computers and Fluids* **20**(3), pp. 343–358.
7. KHATIR, Z. & CARPENTER, P. W. C. 2001. Three-dimensional vortex method for near-wall boundary-layer flow structures, in *Computational Fluid Dynamics 2000*, N. Satofuka (Ed.), Springer, 589-594
8. KHATIR, Z. & CARPENTER, P. W. C. 2000. Three-dimensional vortex/boundary element method for the numerical study of near-wall boundary-layer flow structure, 4th EUROMECH Fluid Mechanics Conference (EFMC2000), 19-23 November.
9. KHATIR, Z. 2002. A Vortex Method Simulation of Near-Wall Turbulence, Proceeding of the Ninth European Turbulence Conference (ETC9), and to apper in *Advances in Turbulence IX*, I. P. Castro & P. E. Hancock (Eds.), ©CIMNE, Barcelona 2002
10. KHATIR, Z. 2001. An Algorithm for Velocity Calculation in Close Proximity to Body Surfaces in Panel Methods", *International Journal of Computers Mathematics* **78**(3), pp. 471- 480.
11. KLINE, S. J. & REYNOLDS, W. C. & SCHRAUB, F. A. & RUNSTADLER, P. W. 1967. The structure of turbulent boundary layers, *J. Fluid Mech.*, **30**, pp. 741–773.
12. HEAD, M. R. & BANDYOPHADYAY, P. 1981. New aspects of turbulent boundary layer structure, *J. Fluid Mech.*, **107**, pp. 297-338.
13. ACALAR, M. S. & SMITH, C. R. 1987. A study of hairpin vortices in a laminar boundary layer, Part 1. & 2., *J. Fluid Mech.* **175**, pp. 1–83.
14. SMITH, C. R., WALKER, J. D. A., HAIDIRI, A. H. & SOBRUN, U. 1991. On the dynamics of near wall turbulence, *Phil. Trans. Roy. Soc. London Ser. A* **336**, pp. 131-175.
15. ZHOU, J., MEINHART, C. D., BALACHANDAR, S. & ADRIAN, R. J. 1997. Formations of Coherent Hairpin Packets in Wall Turbulence, in *self-Sustaining Mechanisms of Wall Turbulence* (ed. R. L. PANTON), pp. 109-134.
16. CANTWELL, B. J., CHACIN, J. M. & BRADSHAW, P. 1997. On the Dynamics of Turbulent Boundary Layers, in *self-Sustaining Mechanisms of Wall Turbulence* (ed. R. L. PANTON), pp. 365-385.

Part VII

Fluid Structure Coupling and Heat Transfer

Optimization of a Thermal Coupled Flow Problem

Günter Bärwolff

TU Berlin, Institut of Mathematics, D-10623 Berlin, Str. d. 17. Juni 135, Germany

1 Introduction

During the growth of crystals crystal defects were observed under some conditions of the growth device. A transition from the twodimensional flow regime of a crystal melt in axisymmetric zone melting devices to an unsteady threedimensional behavior of the velocity and temperature field was found experimentally. This behavior leads to striations as undesirable crystal defects. To avoid such crystal defects it is important to know the parameters, which guarantee a stable steady twodimensional melt flow during the growth process. There are several possibilities for parameter finding. In this paper optimization problems will be discussed. It is known that an unsteady behavior of the melt and vorticies near the fluid-solid-interphase decrease the crystal quality. Thus it makes sense to look for example for flows, which are nearly steady or equal to a given flow regime in the melt zone. This leads to tracking type optimization problems with a functional like

$$J(\boldsymbol{u}, \theta_c) = \frac{1}{2} \int_0^T \int_\Omega |\boldsymbol{u} - \overline{\boldsymbol{u}}|^2 \, d\Omega dt + \frac{1}{2} \int_0^T \int_{\Gamma_c} (\theta_c^2 + \theta_{c_t}^2) \, d\Omega dt \,. \tag{1}$$

$\boldsymbol{u}$ is the velocity vector field in the melt and $\overline{\boldsymbol{u}}$ is the state, which we want to have, θ_c is the control temperature on the control boundary Γ_c.

2 Mathematical model

The crystal melt is described by the Navier-Stokes equation for an incompressible fluid using the Boussinesq approximation coupled with the convective heat conduction equation and the diffusion equation. Heat conductivity and viscosity depend on the temperature. Because of the axisymmetric situation of the melting zone we write down the equations in cylindical coordinates. Thus we have a Boussinesq equation system in cylindrical coordinates for the velocity $\boldsymbol{u} = (u, v, w)$, the pressure p and the temperature θ.
For the velocity no slip boundary conditions are used. At the interfaces between the solid material and the fluid crystal melt we have for the temperature inhomogenous Dirichlet data, i.e. the melting point temperature. On the heated coat (control boundary) of the ampulla the experimentators gave us measured temperatures and we will look further for optimal profiles during the optimization. The initial state was assumed as the neutral position of the crystal melt ($\boldsymbol{v} = \boldsymbol{0}$)

and a temperature field, which solves the non convective heat conduction equation with the given temperature boundary conditions. The material properties and the dimensionless parameters for the investigated crystal close the initial boundary value problem for the description of the melt flow.

3 Optimization and Numerical Method

For the calculus of optimization and the derivation of an optimization system we use the mathematical model in cartesian coordinates, which reads as

$$\boldsymbol{u}_t + (\boldsymbol{u}\cdot\nabla)\boldsymbol{u} - \Delta\boldsymbol{u} + \nabla p - \rho(\theta)\boldsymbol{g} = 0 \quad \text{on } \Omega_T \tag{2}$$

$$-div\,\boldsymbol{u} = 0 \quad \text{on } \Omega_T \tag{3}$$

$$\theta_t + (\boldsymbol{u}\cdot\nabla)\theta - \frac{1}{Pr}\Delta\theta - q = 0 \quad \text{on } \Omega_T\ . \tag{4}$$

$\boldsymbol{u}$ is the velocity vector and $\Omega_T = \Omega \times (0,T)$ is the considered time cylinder. For the boundary conditions we have

$$\boldsymbol{u} = \boldsymbol{0} \quad \text{on } \Gamma \times (0,T), \quad \theta = \theta_c \quad \text{on } \Gamma_{cT}, \quad \text{and} \quad \theta = 0 \quad \text{on } \Gamma_d \times (0,T), \tag{5}$$

where Γ is the boundary of the spatial region $\Omega \subset \mathbb{R}^3$, on which the problem lives, and Γ_c is the control boundary, Γ_d is the Dirichlet part of the boundary and $\Gamma_{cT} = \Gamma_c \times (0,T)$. For $t = 0$ we have the initial condition $\boldsymbol{u} = \boldsymbol{0}$ and a temperature field, which solves the non convective heat conduction equation with the given temperature boundary conditions $\theta = \theta_0$ on Ω.
The use of formal Lagrange parameters technique with respect to the functional of type (1) means the consideration of the Langrange functional

$$L(\boldsymbol{u},p,\theta,\theta_c,\boldsymbol{\mu},\xi,\kappa,\chi) = J(\boldsymbol{u},\theta_c) + <\boldsymbol{\mu}, mo>_{\Omega_T}$$
$$- <\xi, div\,\boldsymbol{u}>_{\Omega_T} + <\kappa, en>_{\Omega_T} + <\chi, \theta-\theta_c>_{\Gamma_{cT}}\ .$$

mo and en stand for the left sides of the equations (2) and (4), and for example for $<\boldsymbol{\mu}, mo>_{\Omega_T}$ we have

$$<\boldsymbol{\mu}, mo>_{\Omega_T} = \int_{\Omega_T} [\boldsymbol{u}_t + (\boldsymbol{u}\cdot\nabla)\boldsymbol{u} - \Delta\boldsymbol{u} + \nabla p - \rho(\theta)\boldsymbol{g}]\cdot\boldsymbol{\mu}\, d\Omega\, dt\ . \tag{6}$$

$\boldsymbol{\mu}$, ξ, κ and χ are Lagrange parameters. We will not discuss the functional analytical aspects of the used Lagrange method, i.e. function spaces, smoothness properties etc. A very good overview over the functional analytical background and the fundation of the optimization of Navier-Stokes problems is developed in [2].
To find candidates $\boldsymbol{u}(\theta_c)$ and θ_c, which minimize the functional (1) we have to analyze the necessary conditions

$$L_{\boldsymbol{u}}\tilde{u} = J_{\boldsymbol{u}}\tilde{u} + <\boldsymbol{\mu}, mo_{\boldsymbol{u}}>_{\Omega_T} - <\xi, div\tilde{u}>_{\Omega_T} + <\kappa, en_{\boldsymbol{u}}>_{\Omega_T} = 0, \tag{7}$$

$$L_p\tilde{p} = <\nabla\tilde{p}, \boldsymbol{\mu}>_{\Omega_T} = 0, \tag{8}$$

$$L_\theta\tilde{\theta} = <-\rho_\theta \boldsymbol{g}\tilde{\theta}, \boldsymbol{\mu}>_{\Omega_T} + <\kappa, en_\theta>_{\Omega_T} + <\chi, \tilde{\theta}>_{\Gamma_{cT}} = 0, \tag{9}$$

$$L_{\theta_c}\tilde{\theta}_c = J_{\theta_c}\tilde{\theta}_c + <-\chi, \tilde{\theta}_c>_{\Gamma_{cT}} = 0\ . \tag{10}$$

Let us have a closer look at the condition (7). For $J_{\boldsymbol{u}}\tilde{u}$ we find

$$J_{\boldsymbol{u}}\tilde{u} = \int_{\Omega_T} (\boldsymbol{u} - \overline{\boldsymbol{u}}) \cdot \tilde{u}\, d\Omega dt \,. \tag{11}$$

The term $< \boldsymbol{\mu}, mo_{\boldsymbol{u}} >_{\Omega_T}$ means the derivative of the Navier-Stokes equation, i.e.

$$< \boldsymbol{\mu}, mo_{\boldsymbol{u}} >_{\Omega_T} = \int_{\Omega_T} [\tilde{u}_t - \Delta\tilde{u} + (\boldsymbol{u}\cdot\nabla)\tilde{u} + (\tilde{u}\cdot\nabla)\boldsymbol{u}] \cdot \boldsymbol{\mu}\, d\Omega dt \,. \tag{12}$$

The discussion of the term $< \kappa, en_{\boldsymbol{u}} >_{\Omega_T}$ gives

$$< \kappa, en_{\boldsymbol{u}} >_{\Omega_T} = \int_{\Omega_T} [(\tilde{u}\cdot\nabla)\theta]\kappa\, d\Omega dt \,. \tag{13}$$

Using the rules of integration by parts from (11)-(13) and (7) we get for all test vector functions $\tilde{u}$

$$L_{\boldsymbol{u}}\tilde{u} = \int_{\Omega_T} [-\boldsymbol{\mu}_t - \Delta\boldsymbol{\mu} + (\nabla\boldsymbol{u})^t\boldsymbol{\mu} - (\boldsymbol{u}\cdot\nabla)\boldsymbol{\mu} + \nabla\xi + (\boldsymbol{u} - \overline{\boldsymbol{u}}) + \kappa\nabla\theta] \cdot \tilde{u}\, d\Omega dt = 0 \,,$$

or

$$-\boldsymbol{\mu}_t - \Delta\boldsymbol{\mu} + (\nabla\boldsymbol{u})^t\boldsymbol{\mu} - (\boldsymbol{u}\cdot\nabla)\boldsymbol{\mu} + \nabla\xi = -(\boldsymbol{u} - \overline{\boldsymbol{u}}) - \kappa\nabla\theta \quad \text{in} \quad \Omega_T \,, \tag{14}$$

with the boundary condition and the final condition

$$\boldsymbol{\mu} = \boldsymbol{0} \quad \text{on} \quad \Gamma \times (0,T), \quad \text{and} \quad \boldsymbol{\mu}(T) = \boldsymbol{0} \quad \text{in} \quad \Omega \,. \tag{15}$$

The necessary condition (8) gives for all test functions $\tilde{p}$ the equation

$$-div\, \boldsymbol{\mu} = 0 \quad \text{in} \quad \Omega_T \,. \tag{16}$$

On the same way we get from $L_\theta\tilde{\theta} = 0$ for the adjoint temperature κ after integration by parts the equation

$$-\kappa_t - \frac{1}{Pr}\Delta\kappa - (\boldsymbol{u}\cdot\nabla)\kappa = -\rho_\theta \boldsymbol{g}\cdot\boldsymbol{\mu} \quad \text{in} \quad \Omega_T \,, \tag{17}$$

with the boundary condition and the final condition

$$\kappa = 0 \quad \text{on} \quad \Gamma \times (0,T) \,, \quad \text{and} \quad \kappa(T) = 0 \quad \text{in} \quad \Omega \,, \tag{18}$$

and the choice of χ as $\chi = \frac{1}{Pr}\frac{\partial\kappa}{\partial\mathbf{n}}$ on Γ_{cT}. The evaluation of the condition (10) finally gives

$$-\theta_{c_{tt}} + \theta_c = \frac{1}{Pr}\frac{\partial\kappa}{\partial\mathbf{n}} \quad \text{on} \quad \Gamma_{cT} \,, \tag{19}$$

with the time boundary conditions

$$\theta_{c_t}(0) = \theta_{c_t}(T) = 0 \,. \tag{20}$$

Now we can summarize and the fully optimization system consists of

- the forward model with the Boussinesq equations (2),(3),(4), the boundary conditions (5) and the given initial state for the velocity field $\boldsymbol{u}$, the pressure p and the temperature θ, and
- the adjoint model with the equations (14),(16),(17),(19), and the conditions (15),(18),(20) for the adjoint variables $\boldsymbol{\mu}$, ξ, κ and the control θ_c.

The global existence of a solution of the forward problem is well known (see [4], [5]). In three dimensions only the local uniqueness of the forward solution could be shown. Hinze [2] has shown the existence and uniqueness of a solution of the adjoint model. For the minimization functional (1) Hinze has shown the positive definiteness of the Hessian $\hat{J}''(\theta_c)$ of

$$\hat{J}(\theta_c) := J(\boldsymbol{u}(\theta_c), \theta_c) ,$$

and with this result we have a sufficient second order optimality condition.

The optimization system (2)-(5) and (14)-(20) is now under consideration for a numerical solution. The Boussinesq equation system is solved with a finite volume method [1].
If we have axisymmetric conditions we can transform the adjoint equations into a cylindrical coordinate system. Using the adjoint divergence condition $div\, \boldsymbol{\mu} = 0$ we can write the adjoint equations in the following quasi conservative form. For adjoint velocity $\boldsymbol{\mu} = (\mu, \nu, \omega)$ in the cylindrical coordinate system with the radial component μ, the azimuthal component ν and the z-component ω we get from (14) for example for μ

$$\begin{aligned} &-\mu_t - ((r\mu)_r/r)_r - \mu_{\varphi\varphi}/r^2 + 2\mu_\varphi/r^2 - \mu_{zz} + \mu u_r + \nu v_r + \omega w_r \\ &-(ru\mu)_r/r - (v\mu)_\varphi/r - (w\mu)_z + v\nu/r + \xi_r = -(u - \overline{u}) - \kappa\theta_r \, . \end{aligned} \tag{21}$$

From equation (17) we get for the adjoint temperature κ

$$-\kappa_t - \frac{1}{Pr}[(r\kappa_r)_r/r + \kappa_{\varphi\varphi}/r^2 + \kappa_{zz}] - (ru\kappa)_r/r - (v\kappa)_\varphi/r - (w\kappa)_z = -\rho_\theta g w \, . \tag{22}$$

Equation (22) is a convective heat conduction equation and the discretization can be done like those in [1]. In the equation (21) and the equations for ν, ω the terms $(\nabla \boldsymbol{u})^t \boldsymbol{\mu}$ and $\kappa\nabla\theta$ are not known from the classical Navier-Stokes equations and they will by discretized in a canonical way for staggered grids. The solution of the discretized system (2)-(5) and the equations for ν, ω and ξ (14) is difficult and expensive, because of the opposite time direction of the forward system and the adjoint system. That means we know the forward solution $\boldsymbol{u}, \theta$ on the whole time interval $[0, T]$ to get the adjoint solution $\boldsymbol{\mu}, \kappa$ and vice versa. If we have discretized the time interval $[0, T]$ by Z timesteps and the dimensions of the spatial discretizations are N, M and P a direct solution of the whole system means the solution of an algebraic equation system with $2Z \times N \times M \times P \times 10$ equations. Iterative methods of the form

i) choose a suitable start value of $\boldsymbol{u}, \theta$,
ii) solve the adjoint problem and get $[\boldsymbol{\mu}, \kappa, \theta_c](\boldsymbol{u}, \theta)$

iii) solve the forward problem and get $[\boldsymbol{u}, \theta](\theta_c)$
iv) until convergence, go to ii),

are under consideration, but such algorithms are also very expensive if one needs some iteration steps. Quite another and a realizable algorithm will be discussed in the next section.

Suboptimal control The starting point for suboptimal or instantanous control is a time discretization of the Boussinesqu equation system, i.e. in the case of an Euler backward time discretization with the time step parameter τ

$$\boldsymbol{u} - \tau \Delta \boldsymbol{u} + \tau \nabla p = \tau \rho(\theta) \boldsymbol{g} - \tau (\boldsymbol{u}^o \cdot \nabla) \boldsymbol{u}^o + \boldsymbol{u}^o \,, \quad -div\, \boldsymbol{u} = 0 \quad \text{in} \quad \Omega, \tag{23}$$

$$\boldsymbol{u} = \boldsymbol{0} \quad \text{on} \quad \Gamma, \tag{24}$$

where the upper index o means the values at the actual time level. Quantities without an index are considered at the new time level. The Euler backward time discretization of the heat conduction equation leads to

$$\theta - \tau \frac{1}{Pr} \Delta \theta + \tau (\boldsymbol{u}^o \cdot \nabla) \theta = \tau q^o + \theta^o \quad \text{in} \quad \Omega, \tag{25}$$

$$\theta = \theta_s \quad \text{on} \quad \Gamma_c, \qquad \theta = 0 \quad \text{on} \quad \Gamma_d \,. \tag{26}$$

Now we look for a control θ_s, which minimizes the functional

$$J_s(\boldsymbol{u}, \theta_s) := \frac{1}{2} \int_{\Gamma_c} \theta_s^2 \, d\Gamma + \frac{1}{2} \int_{\Omega} |\boldsymbol{u} - \overline{\boldsymbol{u}}|^2 \, d\Omega \,. \tag{27}$$

With $\hat{J}_s(\theta_s) := J_s(\boldsymbol{u}(\theta_s), \theta_s) = min!$ for $\boldsymbol{u}$ as a solution of the boundary value problem (23)-(26) for a control θ_s we have a stationary optimization problem per time step and with a sequence of such problems we will get a suboptimal control θ_s over the time period $[0, T]$. The optimality system per time step we get on the same way, which we used in the above discussed time-dependend case.
For the adjoint variables $\boldsymbol{\mu}$, ξ. κ and the control θ_s we get for the Lagrange function

$$\begin{aligned} L(\boldsymbol{u}, p, \theta, \theta_s, \boldsymbol{\mu}, \xi, \kappa, \chi) &= J_s(\boldsymbol{u}, \theta_s) + < \boldsymbol{\mu}, mo >_\Omega \\ &- < \xi, div\, \boldsymbol{u} >_\Omega + < \kappa, en >_\Omega + < \chi, \theta - \theta_s >_{\Gamma_c} \,. \end{aligned} \tag{28}$$

analyzing the nessecary condition $\nabla L = \boldsymbol{0}$ the adjoint system

$$\boldsymbol{\mu} - \tau \Delta \boldsymbol{\mu} + \nabla \xi = -(\boldsymbol{u} - \overline{\boldsymbol{u}}) \,, \quad -\tau div\, \boldsymbol{\mu} = 0 \quad \text{in} \quad \Omega, \tag{29}$$

$$\boldsymbol{\mu} = \boldsymbol{0} \quad \text{on} \quad \Gamma, \tag{30}$$

$$\kappa - \frac{\tau}{Pr} \Delta \kappa - \tau (\boldsymbol{u}^o \cdot \nabla) \kappa = -\tau \rho_\theta g \omega \quad \text{in} \quad \Omega, \tag{31}$$

$$\kappa = 0 \quad \text{on} \quad \Gamma \,, \tag{32}$$

$$\theta_s = \frac{\tau}{Pr} \frac{\partial \kappa}{\partial \mathbf{n}} \quad \text{on} \quad \Gamma_c \,. \tag{33}$$

The advantage of this technique is obvious, because we have to solve per time step only a small stationary optimization problem. The results of [3] showed the efficiency of the suboptimal or instantanous control strategy in the case of isothermic flows and it could be shown, that suboptimal controls are very effective compared to optimal controls, i.e. the value of the $\hat{J}(\theta_s)$ was only 10% higher than $\hat{J}(\theta_c)$ in the case of a boundary controlled backward facing step.

4 Conclusion

With the Langrange parameter technique it's possible to derive an optimization system for a given functional, which solution gives an optimal control. The numerical solution of the fully time-depend optimization system is not possible for realistic configurations yet.
Suboptimal strategies with the used linearizations of (23) and (25) lead to a sequence of time-independend stationary optimization problems, which bring suboptimal results near the optimal control. The developed strategies are now applied to the above discussed crystal melt problem in two and three dimensions.

Acknowledgements: The presented techniques and the resulting optimization systems are based on the results of a very close cooperation with Prof. Dr. Micheal Hinze, Dresden/Berlin, and I have to thank him very much.

References

1. Bärwolff, G., König, F. and G. Seifert: ZAMM 77 (1997) 10
2. Hinze, M.: Optimal and instantaneous control of the instationary Navier-Stokes equations, habilitation thesis, Berlin, August 2000 (available on the webpage `http:\\www.math.tu-dresden.de\~hinze`)
3. Hinze, M.: Optimization of the Navier-Stokes equation, Adjoints workshop, Decin/Czech Republic, September 2001
4. Ladyzhenskaya, O.: The Mathematical Theory of Viscous Incompressible Flows, Gordon & Breach, 1969
5. Constantin, P. and C. Foias: Navier-Stokes Equations, The University of Chicago Press, 1988

Stabilization of a Linearized Navier-Stokes Solver for Turbomachinery Aeroelasticity

M. Sergio Campobasso[1] and Michael B. Giles[2]

[1] DIMEG, Politechnic of Bari, Via Re David 200, 70125 Bari, Italy
[2] Oxford University Computing Laboratory, Parks Road, OX1 3QD Oxford, UK

Abstract. The linear analysis of turbomachinery aeroelasticity relies on the assumption of small level of unsteadiness and requires the solution of both the nonlinear steady and the linear unsteady flow equations. The objective of the analysis is to compute a complex flow solution which represents the amplitude and phase of the unsteady flow for the frequency of unsteadiness of interest. The solution procedure of the linear harmonic Euler/Navier-Stokes solver often consists of a preconditioned fixed-point iteration which in some circumstances may become numerically unstable. The paper summarizes the use of the Recursive Projection Method and the Generalized Minimum Residuals algorithm to provide stabilization and presents a realistic application of both approaches.

1 Introduction

Blade flutter and forced response of turbomachinery rotors [1] are aeroelastic phenomena which may both lead to dramatic mechanical failures if not properly accounted for in the design of the engine. Flutter occurs if the working fluid feeds energy into the vibrating blades, while forced response vibrations are due to an external source of excitation such as the wakes shed by an upstream blade-row. The estimate of both the mean energy flux between fluid and structure in the flutter case and the unsteady forces acting on the blades in the forced response problem requires knowledge of the unsteady flow field. A number of methods have been developed to carry out the analysis of turbomachinery aeroelasticity [9], ranging from uncoupled linearized potential flow solvers [4] to fully-coupled nonlinear three-dimensional unsteady viscous methods [5] and within this range the uncoupled linear harmonic Euler and Navier-Stokes (NS) methods have proved to be a successful compromise between accuracy and cost. This approach relies on the fact that the level of unsteadiness in turbomachinery flows is small and hence views it as a small perturbation of a space-periodic mean steady flow. The unsteady flow field can be linearized about the steady one and due to linearity can be decomposed into a sum of harmonic terms, each of which can be computed independently. The analysis considers a single blade passage rather than the whole blade-row thanks to the cyclic periodicity of both the steady and unsteady flows, thus leading to a great reduction of computational costs. The periodic boundary condition of the linear harmonic equations introduces a phase-shift between the two periodic boundaries, known as *Inter-Blade Phase Angle* (IBPA). The small amplitude of the unsteady aerodynamic forces with respect to the mechanical forces often allows one to neglect the aerodynamic coupling of structural modes and the investigation can be carried out considering one mode at a time. The complete analysis [2] consists of two phases: *a*) calculation of the nonlinear steady flow field about which the linearization is performed and *b*) solution of the linear harmonic equations.

The *HYDRA* suite of parallel codes developed at the *Oxford University Computing Laboratory* includes both a nonlinear (*hyd*) [6] and a linear harmonic (*hydlin*) [3] solver of the inviscid and viscous equations and the solution procedure of both codes is a preconditioned fixed-point iteration. Usually the linear code converges without difficulty, but numerical instabilities have been encountered in situations in which the steady flow calculation itself failed to converge to a steady state but instead finished in a low-level limit cycle, often related to some physical phenomenon such as corner stalls. The main objectives of this paper are to: 1) investigate the numerical instabilities of *hydlin* and 2) demonstrate its stabilization achieved by means of two methods: the *Recursive Projection Method* (RPM) and the *Generalized Minimal Residuals* (GMRES) algorithm.

2 Linear equations

The discrete linear harmonic Euler and NS equations [2] can be viewed as a complex linear system $Ax = b$ of dimension k equal to the product of the number of grid nodes and flow variables. The matrix A depends on the sensitivity of the nonlinear residuals to flow perturbations and the right-hand-side vector b is due to incoming perturbations through the inflow or outflow boundary in the forced response case and to the harmonic deformation of the grid in the flutter problem. The unknown complex vector x represents the amplitude and phase of the unsteady flow for the frequency of unsteadiness of interest. The linear solver *hydlin* can be regarded as the fixed-point iteration:

$$x_{n+1} = (I - M^{-1}A)x_n + M^{-1}b \tag{1}$$

in which M^{-1} is the preconditioning operator resulting from the Runge-Kutta time-marching algorithm, the Jacobi preconditioner and the multigrid scheme [3]. It should be noted that M^{-1} depends on several numerical parameters such as the number of iterations on each grid level and neither M^{-1} nor A are built explicitly, as *hydlin* only uses the matrix-vector products $M^{-1}Ax$. Linear stability analysis of (1) shows that necessary condition for its convergence is that all the eigenvalues of $M^{-1}A$ lie in the unit disc centred at $(1,0)$ in the complex plane. For most aeroelastic problems of practical interest, this condition is fulfilled and *hydlin* converges without difficulty. However an exponential growth of the residual has been encountered for some turbomachinery test-cases caused by a few complex conjugate eigenvalues lying outside the unit disc (*outliers*). In these circumstances the steady flow calculation itself usually failed to converge to a steady-state but instead finished in a small-amplitude limit cycle, related to some physical instability such as flow separations or vortex shedding. The solution procedure of *hyd* is not time-accurate but it nevertheless reflects some physical properties of the flow field due to the pseudo time-marching strategy associated with the Runge-Kutta algorithm. Small-amplitude limit cycles do not prevent the steady solver from converging (their effect is sometimes visible in small oscillations of the residual around a constant low level), but they result in a small number of complex conjugate outliers causing the exponential growth of the residual in the linear calculation. Two different solutions to this problem have been achieved implementing RPM and GMRES in *hydlin*.

3 RPM

RPM is an iterative algorithm for the solution of linear and nonlinear systems [8] and is based on the projection of $M^{-1}A$ onto the orthogonal subspaces $\mathcal{P}$ and $\mathcal{Q}$ of $\mathcal{R}^k$ associated respectively with the subset of m_{out} outliers and that of the remaining $(k-m_{out})$ eigenvalues lying in the unit disc. At each RPM iteration the projection of the linear equations on the low-dimensional subspace $\mathcal{P}$ is solved with Newton's method and that on the subspace $\mathcal{Q}$ with the standard fixed-point iteration (1). Denoting by Z an orthonormal basis of $\mathcal{P}$, the orthogonal projectors P and Q on the subspaces $\mathcal{P}$ and $\mathcal{Q}$ are defined respectively as $P=ZZ^T$ and $Q=I-P$. The basis Z is augmented with the current dominant eigenmode each time the calculation is diverging or converging very slowly. The projections f and g of (1) on $\mathcal{P}$ and $\mathcal{Q}$ are defined respectively as

$$f = P[(I - M^{-1}A)x + M^{-1}b] \qquad g = Q[(I - M^{-1}A)x + M^{-1}b]$$

and the outline of the RPM loop is:

$$p_{init} = Px_{init}, \qquad q_{init} = Qx_{init}$$

Do until convergence:

$$i.\ \ p_{\nu+1} = p_\nu + (I - f_p)^{-1}(f(p_\nu, q_\nu)) - p_\nu)$$
$$ii.\ \ q_{\nu+1} = g(p_\nu, q_\nu))$$

$$x_* = p_* + q_* = p_{\nu_{final}} + q_{\nu_{final}}$$

where $p = Px$, $q = Qx$ and $f_p = P(I - M^{-1}A)P$. It is easily verified that

$$(I - f_p)^{-1} = Z[I - Z^T(I - M^{-1}A)Z]^{-1}Z^T = ZH^{-1}Z^T$$

where H is a small matrix of size m_{out}, whose inversion requires minimum computational effort. The stability analysis of this algorithm shows that its spectral radius is smaller than 1, that is the stabilized RPM iteration is stable. The implementation of RPM in *hydlin* has required only minor changes to the existing code, as q is determined using the core-part of the code performing the standard fixed-point iteration (1) and the remaining computationally cheap operations are performed at the top routine-level.

4 GMRES

GMRES is an iterative method for the solution of linear systems which belongs to the family of Krylov subspace methods [7] and is guaranteed to converge even in the presence of outliers. The Krylov subspace of dimension m generated by $M^{-1}A$ and b is the vectorial space spanned by the powers $((M^{-1}A)^j b,\ j = 0, \dots m-1)$, that is

$$\mathcal{K}_m = < b, M^{-1}Ab, \dots, (M^{-1}A)^{m-1}b > .$$

The GMRES algorithm is based on the progressive reduced Arnoldi factorization [7] of $M^{-1}A$:

$$M^{-1}AQ_m = Q_{m+1}\tilde{H}_m \qquad (2)$$

where $\tilde{H}_m$ is a Hessemberg matrix of size $((m+1) \times m)$, Q_m is a matrix whose m columns form an orthonormal basis for $\mathcal{K}_m$ and Q_{m+1} is Q_m augmented with a new Krylov vector. The length k of each column of Q_m is equal to that of the complex linear flow field. At the mth GMRES iteration the solution x is approximated by the linear combination of the available m Krylov vectors which minimizes the 2-norm of the residual.

The preconditioned GMRES algorithm implemented in *hydlin* uses its core part as a black-box to determine the Krylov vectors which are preconditioned in the existing way (multigrid+Runge-Kutta+Jacobi preconditioner) and the computationally cheap optimization is carried out at the top routine level. The *restart* option [7] is used in order to limit the required memory. Using between 10 and 30 GMRES iterations per restarted cycle makes the computation affordable even for large problems and a good convergence level is usually achieved within 20 restarted cycles.

5 Results

The considered test-case is a three-dimensional fan rotor whose geometry and surface grid are shown in fig. 1-a. This grid has only 157441 nodes and is quite coarse, but all the phenomena discussed in this section have been also observed with finer computational meshes and for other test-cases. The complete flutter analysis of this rotor is reported in [2] and shows that the rotor is aeroelastically stable for all considered steady working conditions for $IBPA = 180^0$. However all linear calculations using the standard fixed-point iteration (1) do not converge. Figure 1-b provides the residual histories of *hydlin* for the near-stall mean steady conditions and for $IBPA = 180^0$ obtained using the RPM and the GMRES solvers with different numerical parameters. The solid line refers to the RPM solver which adds the current dominant eigenmode to the subspace $\mathcal{P}$ only if the calculation diverges. The iterations at which a new partitioning of $M^{-1}A$ is carried out are labelled from 1 to 4. Before the first dominant mode is added to $\mathcal{P}$ this convergence history is that of the standard preconditioned iteration (1) which therefore does not converge. Conversely the stabilized RPM iteration converges (branch $4_0 - E_0$) once all the unstable modes have been included in $\mathcal{P}$. The subset of the spectrum of $M^{-1}A$ with the first 150 dominant eigenvalues is provided in fig. 2, which reveals the presence of 4 complex conjugate outliers (eigenvalues labelled from 1 to 4). The complex conjugate eigenpair in the unit disc labelled with 5 determines the asymptotic convergence rate of

RPM (slope of the branch $4_0 - E_0$). The dotted line in fig. 1-b refers to the convergence of RPM obtained adding to $\mathcal{P}$ also the eigenpair 5 at the iteration labelled with 5. The slope of the branch $5 - E_1$ is steeper than that of $4_0 - E_0$, since in the former case the asymptotic convergence rate is determined by the eigenpair 6, which is closer than 5 to the centre of the unit disc. The residual of the calculation with restarted GMRES performing 10 iterations per restarted cycle and one multigrid cycle per GMRES iteration stagnates (dashed line in fig. 1) and an acceptable convergence rate can be retrieved only by using 30 GMRES iterations per restarted cycle and three multigrid cycle per GMRES iteration (dashed-dotted line). The analysis of the dominant eigenmodes [2] has shown that these numerical instabilities are due to small physical unsteadiness of the nonlinear flow field such as the hub corner stall.

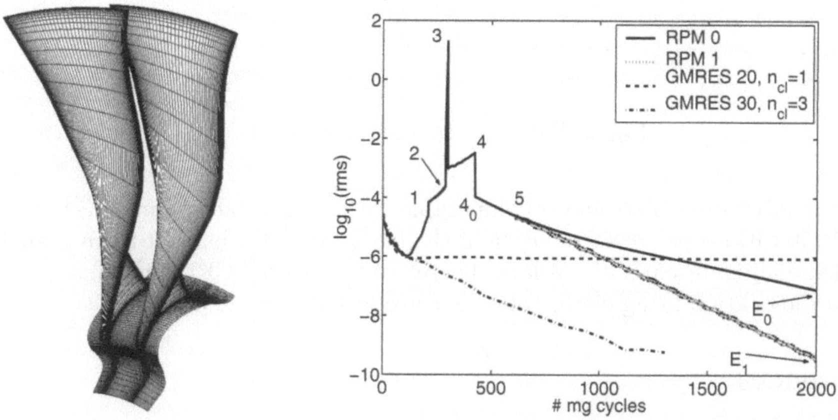

(a) (b)

Fig. 1. **(a)** Fan geometry and surface mesh and **(b)** convergence plots of *hydlin*

6 Conclusions

The implementation of the RPM and GMRES algorithms in the existing linear solver based on a preconditioned fixed-point iteration has stabilized the code. This allows one to carry out the aeroelastic analysis even in the presence of small unsteady phenomena in the mean flow, which are believed not to affect significantly the aeroelastic behaviour of the component. The asymptotic convergence rate of the restarted GMRES algorithm depends on the spectrum of the linear operator, on the number of GMRES iterations per restarted cycle and the number of multigrid cycles per GMRES iteration. The extra memory allocation for storing the Krylov vectors depends only on the number of GMRES iterations per restarted cycle and not on the number of outliers. The asymptotic convergence rate of RPM depends on the spectral radius of the projection of the linear operator onto the stable space $\mathcal{Q}$. The required extra memory allocation depends

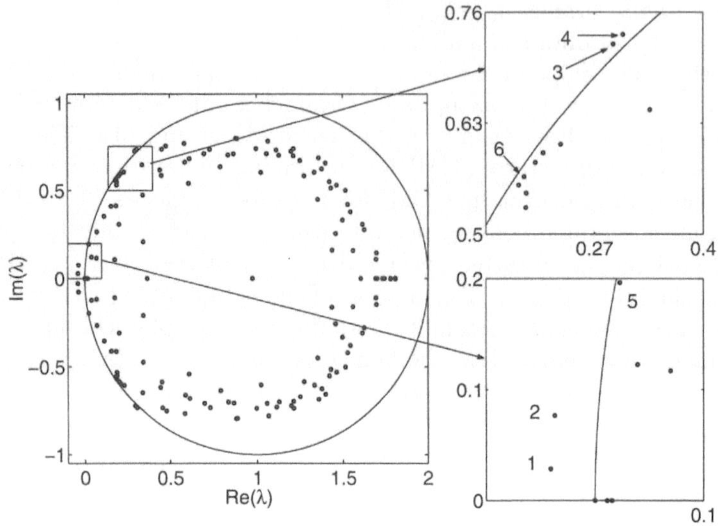

Fig. 2. First 150 eigenvalues of $M^{-1}A$

on the number of outliers and is comparable with that of the restarted GMRES with 10 iterations per restarted cycle if the linear operator has not more than 4 complex conjugate pairs of outliers. Therefore the overall CPU-time and extra memory allocation using either solver is considerably case-dependent.

References

1. M.S. Campobasso, M.B. Giles: 'Analysis of the effect of mistuning on turbomachinery aeroelasticity'. In: *Unsteady Aerodynamics, Aeroacoustics and Aeroelasticity of Turbomachines, 2nd International Symposium in Lyon, France, september 4-8, 2000*, ed. by Pascal Ferrand, Staphane Aubert (Presses Universitaires de Grenoble) pp. 885-896
2. M.S. Campobasso, M.B. Giles: 'Effect of Flow Instabilities on the Linear Analysis of Turbomachinery Aeroelasticity', AIAA paper 2002-4085, Indianapolis 2002
3. M.C. Duta: The use of the Adjoint Method for the Minimization of Forced Response, PhD Thesis, Oxford University, Oxford (2001)
4. K.C. Hall: *Deforming grid variational principle for unsteady small disturbance flows in cascades*, AIAA J. **5**, 31 (1993)
5. J.G. Marshall and M. Imregun: *A review of aeroelasticity methods with emphasis on turbomachinery applications*, J. of Fluid and structures **3**, 10 (1996)
6. P. Moinier: Algorithm developments for an unstructured viscous flow solver, PhD Thesis, Oxford University, Oxford (1999)
7. Y. Saad: *Iterative methods for sparse linear systems* (PWS Publishing Company, Boston 1996)
8. G.M. Shroff and H.B. Keller: *Stabilization of Unstable Procedures: the Recursive Projection Method*, J. of Applied Mathematics **4**, 30 (1993)
9. J.M. Verdon: *Review of Unsteady Aerodynamic Methods for Turbomachinery Aeroelastic and Aeroacoustic Applications*, AIAA J. **2**, 31 (1993)

Diffusive Mass Transfer and Its Effect upon Boundary-Layer Flows

Iordan A. Halatchev and James P. Denier

Department of Applied Mathematics, The University of Adelaide, South Australia 5005, AUSTRALIA

Abstract. The boundary-layer flow above a permeable plate is considered. Fluid transport through the slot occurs as a result of species diffusion; the fluid contained below the plate having a higher concentration than the fluid in the body of the flow. The diffusive mass transfer serves to reduce the skin-friction, but not to the point at which flow separation can occur. The decelerated flow is, however, highly susceptible to inviscid Rayleigh waves indicating a predisposition for the flow to undergo a transition to turbulence.

1 Introduction

Early work on the problem of boundary-layer separation due to slot injection (or blowing), summarised in [5], was mainly concerned with self-similar flows in which the free-stream velocity is proportional to x^m and the blowing velocity is proportional to $x^{1/2(m-1)}$. In this case the boundary-layer streamwise velocity $u = f'(\eta)$ is governed by the classical Falkner-Skan equation

$$f''' + ff'' + \beta(1 - f'^2) = 0, \tag{1}$$

subject to boundary conditions

$$f'(0) = 0, \quad f(0) = -C; \quad f'(\infty) = 1. \tag{2}$$

Here $\beta = 2m/(m+1)$, $C = O(1)$ is a parameter which measures the level of blowing and $\eta = y/x^m$ is the similarity variable. Numerical solutions of this equation with $\beta = 0$ were obtained by [4] who found that, as $C \to C_0 = 0.876$, the boundary-layer flow approaches a state of zero wall shear and is blown off the plate; no solutions were found for values of $C > C_0$. Uniform injection was considered in [3], the boundary-layer equations being examined numerically and analytically. In the numerical study the streamwise location x_s of the point of separation was computed to high accuracy and found to be $x_s = 0.7456$. In the accompanying analytical study the structure of the singularity at separation was determined.

The problem of boundary-layer flow under conditions of interfacial heat and mass transfer, governed by the classical Prandtl equations, the laminar boundary-layer convection-diffusion equation and the steady heat transfer equation was first treated in [8]. The governing equations were solved subject to boundary conditions taking account of foreign fluid injection. It was shown that in the case when $V(x,0) \sim x^{-1/2}$ the flow develops in a self-similar form; the solutions demonstrate that the velocity, concentration and temperature profiles are greatly influenced by either "suction" or "blowing". For instance, the skin-friction coefficient demonstrated a significant decrease with increasing level of blowing.

Here we wish to consider the problem of diffusion-driven boundary-layer separation and instability. As we wish to focus our attention upon the diffusion process we will exclude from our study one of the dominant causes for flow separation, namely the effect of an adverse pressure gradient. We focus on a boundary-layer flow under conditions of interfacial mass transfer with zero pressure gradient. In contrast to previous work on boundary-layer flows in the presence of intense interfacial mass transfer (see [8], [2] and [7]) we consider a finite region of active mass-transfer, in an attempt to provide a more realistic model of many such diffusion dominated flows.

2 Formulation

Consider then the boundary-layer flow of a fluid containing a binary mixture of chemical species A and B over a rigid flat plate. Let C_∞ denote the concentration of the species A in the incipient flow and let U_∞ denote the fluid speed. Suppose that at the surface of the plate the concentration of species A, denoted by C_0, differs from the free-stream value C_∞. If the flat plate is impermeable, so that there is no mechanism for diffusion of the molecular species into the flow, then a concentration boundary layer will develop along with the momentum boundary layer (see [9]). Suppose however that the flat plate is not uniformly impermeable. To be precise, let us consider the case when there is a region of the plate, a finite distance from the leading edge, that is permeable to the molecular species A but otherwise rigid. Situated under this permeable "slot" is a reservoir of fluid containing species A. Because the plate is permeable at this point, mass (i.e. fluid) will diffuse into the flow from the reservoir due to the concentration gradient at the surface. The concentration gradient induces a secondary flow normal to the plate and the resulting mass transfer has the effect of "blowing" (or "suction") of the fluid into (or from) the flow depending upon the sign of the concentration gradient (positive for blowing and negative for suction). The rate v_n^* of the induced flow can be defined in terms of the mass flux through the surface as

$$v_n^* = -\frac{MD}{\rho^*}\frac{\partial c^*}{\partial y^*},$$

where M is the molecular mass, D is the diffusion coefficient, ρ^* the density of the fluid, c^* the concentration and $\partial/\partial y^*$ denotes the derivative normal to the permeable surface.

2.1 The Basic Boundary-Layer Flow

In the limit of large Reynolds number ($Re = U_\infty L/\nu \gg 1$) the flow develops a boundary layer of thickness $O(Re^{-1/2})$ attached to the leading edge of the plate.

We therefore introduce the boundary-layer variables $y^* = LRe^{-1/2}\bar{y}$,$u^* = U_\infty \bar{u}$, $v = U_\infty Re^{-1/2}\bar{v}$, $c^* = C_\infty + (C_0 - C_\infty)\bar{c}$ and $p^* = \rho U_\infty^2 \bar{p}$. The full Navier-Stokes equations reduce to the classical boundary-layer equations coupled to a concentration equation which describes the change in the species concentration within the boundary layer:

$$\frac{\partial \bar{u}}{\partial x} + \frac{\partial \bar{v}}{\partial \bar{y}} = 0, \qquad \bar{u}\frac{\partial \bar{u}}{\partial x} + \bar{v}\frac{\partial \bar{u}}{\partial \bar{y}} = \frac{\partial^2 \bar{u}}{\partial \bar{y}^2},$$

$$\frac{\partial \bar{p}}{\partial \bar{y}} = 0, \qquad \bar{u}\frac{\partial \bar{c}}{\partial x} + \bar{v}\frac{\partial \bar{c}}{\partial \bar{y}} = \frac{1}{Sc}\frac{\partial^2 \bar{c}}{\partial \bar{y}^2},$$

which must be sovled subject to the boundary conditions

$$\bar{u} = 0,\ \bar{c} = 1,\ \bar{v} = -\theta Sc^{-1}\bar{c}_{\bar{y}} \quad \text{on} \quad \bar{y} = 0;$$

$$\bar{u} \to 1,\ \bar{c} \to 0 \ \text{ as } \ \bar{y} \to \infty.$$

Here $\theta = M(C_0 - C_\infty)/\rho^*$ and $Sc = \nu/D$ is the Schmidt number. We will take $Sc = 1$, in what follows, in which case the thickness of the momentum and concentration boundary layers are comparable.

2.2 Linear Inviscid Instability

To study the inviscid instability of the boundary-layer flow we express the total flow, pressure and concentration fields as a superposition of the basic flow and a small amplitude wave-like disturbance

$$(u, v, p, c) = \bar{\mathbf{u}} + \epsilon\left(\frac{\partial \psi}{\partial \bar{y}}, -i\alpha\psi, P, C\right)\exp\left(\mathrm{i}Re^{1/2}\alpha(x - \omega t)\right),$$

where $0 < \epsilon \ll 1$ is the infinitesimally small wave amplitude. The streamfunction ψ is determined from the solution of the classical Rayleigh equation

$$(\bar{u} - \omega)(\psi'' - \alpha^2\psi) - \bar{u}''\psi = 0, \tag{3}$$

which must be solved subject to the boundary conditions

$$\psi(0) = 0, \qquad \psi \to 0 \quad \text{as} \quad \bar{y} \to \infty. \tag{4}$$

Equation (3) subject to boundary conditions (4) constitutes an eigenvalue problem for complex ω as a function of α. The flow is deemed to be unstable if $\alpha \mathrm{Im}(\omega) > 0$.

3 Results

In order to facilitate a numerical solution of the boundary-layer equations the Levy-Lees transformation $\xi(x) = x$, $\eta(x, \bar{y}) = \bar{y}/x^{1/2}$ is employed (see Blottner [1] for full details). New dependent variables are defined according to $F = \bar{u}$,

$V = -\frac{1}{2}\eta F + \xi^{1/2}\bar{v}$ and $G = \bar{c}$. The resulting system is parabolic in ξ and can be solved using a standard marching scheme. We chose to employ second-order accurate centred finite differences in the $\bar{y}$ direction together with a semi-implicit Crank-Nicholson scheme to implement the marching in the ξ-direction (full details of the algorithm can be found in Ref. [6]).

We monitor the state of the boundary-layer flow by calculating the skin-friction coefficient

$$c_f = \frac{1}{\xi^{1/2}} \left(\frac{\partial F}{\partial \eta} \right)_{\eta=0}$$

at each streamwise location. Separation occurs when $c_f = 0$.

3.1 Flow Separation

We follow the methodology of [10] and chose the mass-transfer parameter to have the form

$$\theta = \begin{cases} 0 & \text{for } \xi < \xi_{le}, \\ \epsilon^{-1}\vartheta(\xi - \xi_{le}) & \text{for } \xi_{le} \leq \xi \leq \xi_{le} + \epsilon, \\ \vartheta & \text{for } \xi_{le} + \epsilon < \xi < \xi_{le} + \epsilon + l, \\ -\epsilon^{-1}\vartheta(\xi - \xi_{le} - 2\epsilon - l) & \text{for } \xi_{le} + \epsilon + l \leq \xi \leq \xi_{le} + 2\epsilon + l, \\ 0 & \text{for } \xi > \xi_{le} + 2\epsilon + l. \end{cases}$$

The scaled-skin friction coefficient, as a function of the streamwise coordinate ξ, is plotted in Fig. 1 for different values of the magnitude ϑ of the mass transfer parameter θ. As a test of the accuracy of our code we first calculated the case for a simple Blasius flow (corresponding to $\theta \equiv 0$). This is represented by the horizontal line in Fig. 1 and corresponds to the classical value of $\xi^{1/2}c_f = 0.33206$.

The effect of diffusive mass transfer on the boundary layer is readily observed from Fig. 1. When the mass transfer is active (i.e. for non-zero values of ϑ) the skin friction is reduced from its corresponding Blasius value. The skin-friction

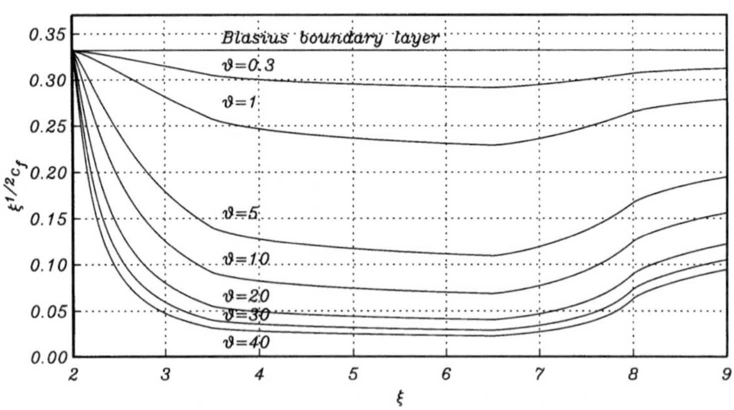

Fig. 1. The scaled skin-friction coefficient as a function of ξ, for the case of diffusion driven flow at $\xi_{le} = 2$, $l = 3$ and $\epsilon = 1.5$ for different values of ϑ

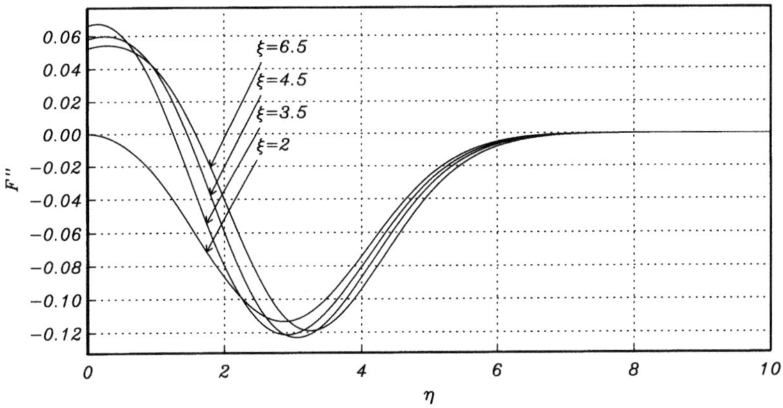

Fig. 2. Plot of $F''(\xi, \eta)$ at different downstream locations $\xi = 2$, 3.5, 4.5 and 6.5, for the case of $\vartheta = 1$

values are identical at the start of the slot; for the geometry used in producing the data for Fig. 1 we have the leading edge of the slot situated at $\xi = 2$. The skin friction reduces as the flow evolves over the slot region, attaining a minimum value at the point where the diffusion parameter θ begins to decrease from its maximum value; for the slot studied here this occurs at $\xi = 6.5$. Note from Fig. 1 that the flow does not attain a point of separation; for a value of $\vartheta = 40$ the flow is still attached, with the minimum value of the skin friction small but non-zero. Higher values of ϑ were studied, however convergence becomes increasingly difficult to achieve with increasing values of ϑ.

In Fig. 2 we present plots of $F''(\xi, \eta)$ for the case $\vartheta = 1$. Although the flow is still attached it now shows evidence of an inflection point at which $F''(\xi, \eta) = 0$. This occurs for all values of ξ lying in the slot regime. Classical inviscid stability theory suggests that the flow will then become susceptible to short Rayleigh waves. We therefore turn our attention to the question of flow instability.

3.2 Growth Rates Curves

The growth rate $\sigma_i = \alpha \mathrm{Im}(\omega)$ was calculated by solving the eigenvalue problem (3) with boundary conditions (4). This was done at different values of the mass-transfer parameter θ, at several streamwise locations over the slot. Representative results are shown Fig. 3 corresponding to a streamwise location of $\xi = 3.5$. These results demonstrate that the flow is unstable at the location $\xi = 3.5$ and that the magnitude of the growth rate increases with an increase in the magnitude of the diffusion parameter θ. Although the growth rates appear relatively small (a maximum of approximately 5.0×10^{-3} for $\vartheta = 1$) it must be remembered that the true growth rate of an inviscid disturbance is multiplied by the factor $Re^{1/2}$ which is large (as $Re \gg 1$ for the boundary layer to exist). In this respect the onset of an inviscid instability can be used to give quantitative predictions of the effect of diffusion on transition to turbulence in such flows.

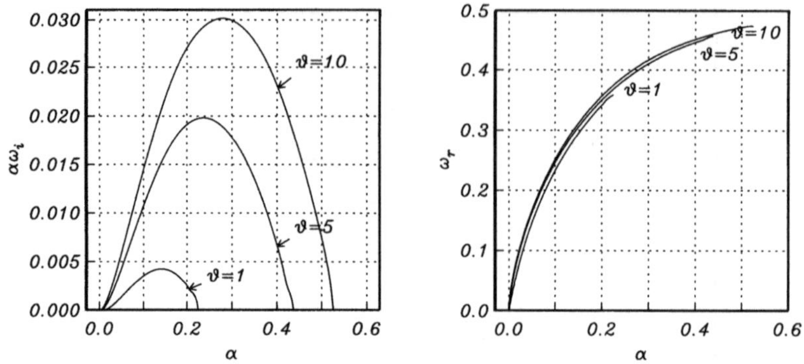

Fig. 3. Graphs of the dimensionless growth rate σ_i and wave-speed ω_r with respect to the wave number α in the unstable regime for a diffusion driven boundary-layer flow for several values of the magnitude of the mass-transfer parameter θ over the "slot" at $\xi = 3.5$

4 Conclusion

This paper presents results of a study of the effect of diffusive mass transfer on the dynamics of a class of boundary-layer flows. Diffusion serves to reduces skin-friction but is unable, at physically realistic levels, to promote flow separation. It does however serve to promote rapidly growing short wave instabilities in the flow and thus may have important implications to transition and mixing in chemical and biological processing systems.

Acknowledgements: Parts of this work were supported through an Australian Research Council Postdoctoral Fellowship awarded to IAH.

References

1. F. G. Blottner: Comp. Meth. Appl. Math. Eng. **6**, 1 (1975).
2. C. B. Boyadjiev, I. A. Halatchev, B. T. Chavdarov: Int. J. Heat Mass Trans. **39**, 2571 (1996).
3. D. Catherall, K. Stewartson, P .G. Williams: Proc. R. Soc. Lond. A **284**, 370 (1965).
4. H.W. Emmons, D. Leigh: *Aero. Res. Counc. Current Paper* No. 157 (1954).
5. G. E. Gadd, C.E. Jones, E. J. Watson: in *Laminar boundary layers* (Oxford University Press, Oxford, 1963).
6. I. A. Halatchev, J. P. Denier: Theor. Comp. Fluid Dyn. (submitted).
7. I. A. Halatchev, J. P. Denier: Int. J. Heat Mass Trans. (submitted).
8. J.P. Hartnett, E.G. Eckert: Mass-transfer cooling in a Trans. ASME **79** 247 (1957).
9. F. P. Incropera, D. P. DeWitt: *Fundamentals of Heat and Mass Transfer* (Wiley, New York, 1990).
10. S. M. Roy: Acta Mechanica, **139**, 43 (2000).

Numerical Estimations of Transonic Flutter Calculation with the Fluid/Structure Coupling Method

Hiroshi Terashima[1] and Kozo Fujii[2]

[1] Graduate student-University of Tokyo, 3-1-1 Yoshinodai Sagamihara kanagawa, 229-8510, Japan
[2] The Institute of Space and Astronautical Science, 3-1-1 Yoshinodai Sagamihara kanagawa, 229-8510, Japan

Abstract. Flutter phenomena for the AGARD445.6 standard aeroelastic wing in the transonic flow are simulated using the fluid/structure coupling method. The effect of time accuracy, the number of inner-iteration, and comparison of a fully implicit with a loosely coupling approch are examined. Improving the time accuracy and increasing the number of the inner-iteration lead to the reduction of the amplitude of time responses. The one-step time-lag error that is generated in a loosely coupling method doesn't have an effect on the flutter predictions in the transonic flow.

1 Introduction

Aeroelastic phenomena such as flutter or buffet, which are caused by the interaction between aerodynamic forces and elastic forces become important factors on the design and development of the aircraft. In recent years, since nonlinear aerodynamics on transonic flow regions could be well predicted by using CFD, time-marching methods solving Navier-Stokes equations coupled with structural equations have been popular for the transonic aeroelastic analysis. In these methods, loosely coupling methods where aerodynamic and structural equations are solved by turns are used. An advantage of using the loosely coupling method is being able to use current optimum aerodynamic or structural algorithms. However, the loosely coupling method produces a one-step time lag error between both solutions. Therefore, only first-order time accuracy is achieved as a whole solver even if the individual solver of aerodynamic and structural equations can achieve a higher-order time accuracy.

Morton et al. [1] developed a fully implicit coupling method with the second-order time accuracy to overcome this limitation. By introducing the inner-iteration method to both solvers, this method could achieve a fully implicit coupling between aerodynamic and structural equations and allow higher-order time accurate schemes. This method was applied to the computations of two-dimensional elastic mounted cylinder where low Mach number flow and showed the importance of the fully implicit coupling, the higher-order time accuracy, and the implementaion of the inner-iteration. For the practical application of this method, the AGARD445.6 aeroelastic wing [2] known as the standard test model for aeroelastic simulations was computed by Gordnier et al. and Yang et al.. [3][4] Their results showed the sensitivity of numerical factors such as the grid resolution and the number of modes used in the structural solver for flutter simulations. In their reports, however, the effect of the time accuracy, the number of the inner-iteration and the difference between coupling methods as was discussed in Ref. [1] were not showed.

The purpose of this study is to investigate the time accuracy and coupling issues of aeroelastic solvers through AGARD445.6 wing computaions. Transonic flow case under $M_\infty = 1.141$ is computed and the effect of the time accuracy and the number of the inner-iteration for flutter predictions are examined. The comparison of a fully implicit coupling method with a loosely coupling method is also discussed.

2 Governing Equations

Aerodynamic governing equations are the unsteady three-dimensional compressible Navier-Stokes equations written in the generalized coordinate system. All variables in these equations are normalized by the combination of freestream density ρ_∞, freestream sound of speed a_∞, and a characteristic length L (in this calculation, the wing root chord length).

For a flexible structure, second-order linear structural dynamic gorverning equations in Eq.(1) are solved.

$$[M]\{\ddot{d}\} + [K]\{d\} = \{F\} \tag{1}$$

where $[M]$ and $[K]$ are the mass and stiffness matrices, respectively. $\{d\}$ and $\{F\}$ are the displacement and aerodynamic force vectors, respectively. In order to solve Eq.(1), it is assumed that the deformed shape of the wing can be represented by a finite set of discrete displacements at selected nodes. From the modal analysis, the displacement vector $\{d\}$ can be expressed as

$$\{d\} = [\phi]\{q\} \tag{2}$$

where $[\phi]$ is the modal matrix which consists of natural mode vectors $\{\phi_i\}$ and $\{q\}$ is the generalized displacement vector. Putting the solution expressed by Eq. (2) into Eq. (1) and multiplying by $[\phi]^T$ yields

$$[\phi]^T[M][\phi]\{\ddot{q}\} + [\phi]^T[K][\phi]\{q\} = [\phi]^T\{F\} \tag{3}$$

If the orthogonal $\{\phi_i\}$ and $\{\phi_i\}^T$ with respect to both $[M]$ and $[K]$ are employed, the final form of structural equations for the generalized coordinate is

$$M_i\ddot{q}_i + \omega_i^2 M_i q_i = f_i \tag{4}$$

where ω_i is the natural frequency of the ith mode. The generalized forces of the ith mode, f_i in Eq.(4) is defined as $1/2\rho_\infty U_\infty^2 \iint_S [\phi]^T \Delta C_p dxdy$. In this study, the data of natural mode vectors and frequencies are directly taken from the Ref. [2], which were calculated by the finite-element analysis. Finally, all variables in Eq.(4) are normalized by the same values with the aerodynamic equations.

3 Numerical Method

For the aerodynamic governig equations, the LU-SGS algorithm [5] is used for the time integration with the convective terms discretized by the SHUS scheme [6] that is a family of the AUSM-type schemes. Third-order accuracy is acheved by the MUSCL interplolation using primitive variables. Viscous terms are discretized by the central differencing and the eddy viscosity is modeled by the Baldwin-Lomax turbulence model.[7] For the time integration, the time accuracy for aerodynamic equations is obtained by utilizing the inner-iteration method [8], which could vanish the factorization and diagonalization error of left hand side operator by the minor modification of base-line solver. The numerical algorithms for deforming grids can be deduced as (written in the one-dimensional form)

$$\left[I - \frac{\Delta t}{1+\phi} J^m \frac{\partial R(Q^m)}{\partial Q^m}\right] \Delta Q^m \tag{5}$$
$$= \frac{\Delta t}{1+\phi} J^m R(Q^m) - (Q^m - Q^n) + Q^n\Big(1 - \frac{J^m}{J^n}\Big) + \frac{\phi}{1+\phi} J^m \Big(\frac{Q^n}{J^n} - \frac{Q^{n-1}}{J^{n-1}}\Big)$$

where the superscript n denotes the time step and m refers to the number of the inner-iteration within each time step. As $m \to \infty$, $Q^m \to Q^{n+1}$. First- or second-order time accuracy is obtained by specifying either $\phi = 0$ or $\phi = 1/2$, respectively. In the present scheme, the ratio of the metric derivatives to the Jacobians is taken to be the appropriate projected area of cell faces and the reciprocal of the Jacobians is taken to be the cell volumes. The metric derivatives and the Jacobinas obtained by such a approach then satisfy the geometric conservation law.[9] For the inflow boundary conditions, the freestream values are specified. For the outflow boundary conditions, the pressure is fixed to the freestream value and the other physical variables are extrapolated.

The structural equations are solved by the similar mannar with the aerodynamic equations.[1] Firstly, Eq.(4) can be rewritten as a first-order system by defining $S = [q_i, \dot{q}_i]^T$

$$\begin{bmatrix} 1 & 0 \\ 0 & M_i \end{bmatrix} \dot{S} + \begin{bmatrix} 0 & -1 \\ \omega_i^2 M_i & 0 \end{bmatrix} S = \begin{Bmatrix} 0 \\ f_i \end{Bmatrix} \tag{6}$$

The resulting scheme becomes as follows

$$\begin{bmatrix} 1 & -\phi' \Delta t \\ \phi' \Delta t \omega_i^2 & 1 \end{bmatrix} \Delta S^m = -\phi' \{(1+\phi)S^m - (1+2\phi)S^n + \phi S^{n-1}$$
$$+\Delta t \begin{bmatrix} 0 & -1 \\ \omega_i^2 & 0 \end{bmatrix} S^m - \Delta t \begin{Bmatrix} 0 \\ f_i^m \end{Bmatrix}\} \tag{7}$$

where $\phi' = 1/(1+\phi)$. As $m \to \infty$, a fully implicit second-order time accuracy scheme is formed by the coupling solutions of Eq.(5) and (7).

4 Interfacing and Grid Deformation Approach

In this study, the CVT(Constant Volume Tetrahedron Transformation) method proposed by Goura et al. [10] is used as the interpolation method for transferring information between the aerodynamic grids and the structural grids. This method defines tetrahedra out of triangles on the structural surface grids and one point from the aerodynamic surface grids. The motion of the aerodynamic grids is derived from the known motions of the structural grids by conserving the volume of the tetrahedron. The grid deformation scheme developed by Nakamichi [11] is employed to generate a time-varying grid system. In this method, the grid points near the surface are moving completely with the surface motion, while the far-field grid points is at rest. The initial grid conditions generated so as to satisfy the orthogonality is maintained near the body surface though the far-field orthogonality might be violated.

5 Results and Discussion

The AGARD445.6 weakened wing model at the transonic flow condition, $M_\infty = 1.141$ is computed. The first four natural modes and frequencies provided in the Ref. [2] are used for the structural equations. To determine the flutter condition at a given Mach number, computations are conducted at several values of dynamic pressure which bracket the flutter point. Changing the dynamic pressure leads to changing the Reynolds number associated with that flow condition. However, the Reynolds number is fixed to 537,700 per wing root chord because the small variation in the Reynolds number doesn't have a significant effect on the flow solution.[12] All computations are carried out on a C-H type grid, which has 20 times root chord length as the outer boundary position. The number of the grid point is $197 \times 67 \times 71$.

Figure 1(a) shows the pressure contours on the upper surface at the steady-state. Pressure contours indicate that a weak shock wave appears on the outboard portion of the wing at about 70 percent of the local chord. In fig. 1(b), the pressure coefficient distributions at each station are in good agreement with the other computed result.[12]

Flutter computations are started from the steady-state result. The time step size is $\Delta t = 0.05$ whose value is the same as that used in other computations.[3][4] For the initial condition of structural equations, a small velocity perturbation, $\dot{q}_1 = 0.0001$ is given. In most cases, the nondimensional dynamic pressure is set to $\bar{Q} = 1.3$ near the flutter point predicted by other computations.[3][4]

Figure 2 shows the responses of the first mode obtained by the first-($\phi = 0$) and second-($\phi = 1/2$)order scheme with the result computed by the small time step size($\Delta t = 0.005$) as a reference response. Three times inner-iteration also used in the other computation [4] is conducted in this case. Time responses obtained by the first-order scheme show divergence behaviors while time responses grow at a very small rate when applying the second-order scheme. As compared with the reference responses, it turns out that the second-order scheme gives better responses than the first-order scheme. However, time responses of the second-order scheme with three times inner-iteration does't yet converge to a accurate solution. This results indicate that the enough number of the inner-iteration more than three is required.

Figure 3 shows the time responses of the first mode obtained by changing the inner-iteration number. It is shown that the amplitude of time responses are decaying by increasing the inner-iteration number. This implies that the effect of the improved time accuracy is to increase the computed flutter speed. Figure 4 shows the change in the amplification factor(AF) behavior. AF is defined as the ratio of the magnitude of a time response peak with the magnitude of the previous peak of corresponding sign. As the inner-iteration number is increased from one to ten, time responses converges to $m = 10$ solution. As a result, flutter points predicted in present computations are $\bar{Q} = 1.576$, $\omega/\omega_\alpha = 0.546$, and 0.506 for the flutter speed index, respectively. The current results lie well within the range of the experiment and computational results presented by other authors.[2][3][4][12]

Time responses obtained by the fully implicit and the loosely coupling method are compared in Fig. 5. In the loosely coupling method, aerodynamic equations are solved by the first-order time accuracy with three times inner-iteration like the fully implicit coupling method and structural equations are solved using the four-stage Runge-Kutta scheme. From obtaining almost the same time responses, it turns out that the one-step time-lag error in a loosely coupling method doesn't affect on time responses for this transonic flow case.

6 Conclusion

The effect of the time accuracy, the inner-iteration number, and coupling issues of aeroelastic solvers are examined through the AGARD445.6 aeroelastic wing computations. It becomes clear that the improved time accuracy and increasing the inner-iteration number lead to damping the amplitude of time responses. The imroved time accuracy in aerodynamic equations plays more improtant role than the fully implicit coupling method and the numerical method of the structural equations.

References

1. Morton, S. A., Meiville, R. B. and Visbal, M. R., *J. of Aircraft*, **35**, (1998), 798-805.
2. Yates, E. C., Jr., AGARD-R-765, (1988).
3. Gordnier R. E. and Melville, R. B., *J. of Aircraft*, **37**, (2000), 872-879.
4. Yang, G. and Obayashi, S., *Proc. of Aerospace Numerical Simulation Symp. 2001*, SP-53, (2001), 291-296.
5. Yoon, S. and Jameson, A., *AIAA J.*, **26**, (1988), 1025-1026.
6. Shima, E. and Jounouchi, T., *Proc. of 14th NAL Symp. on Aircraft Computaional Aerodynamics*, SP-34, (1997), 7-12.
7. Baldwin, B. S. and Lomax, H., AIAA Paper 78-257, (1978).

8. Chakravarthy, S. R., AIAA Paper 84-0165, (1984).
9. Tamura, Y. and Fujii, K., AIAA Paper 93-3365-CP, (1993).
10. Goura, G. S. L., et al., *Aeronaut J.*, **105**, (2001), 215-221.
11. Nakamichi, J., NASA TM-88341, (1986).
12. Lee-Rausch, E. M. and Batina, J. T., *J. of Aircraft*, **33**, (1996), 1139-1147.

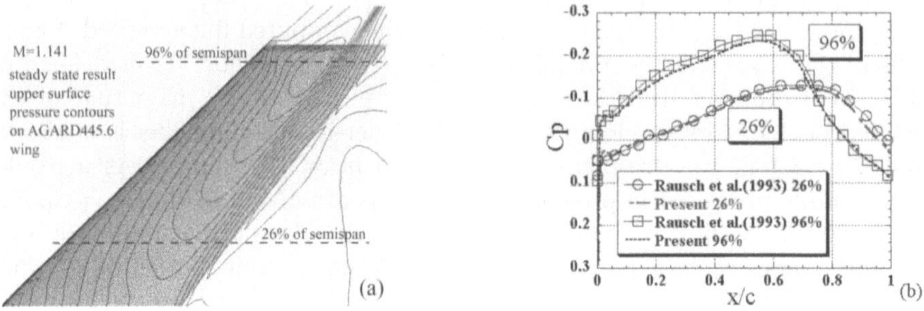

Fig. 1. (**a**) pressure contours on the wing upper surface. (**b**) pressure coefficient distributions at two span stations compared the other result.

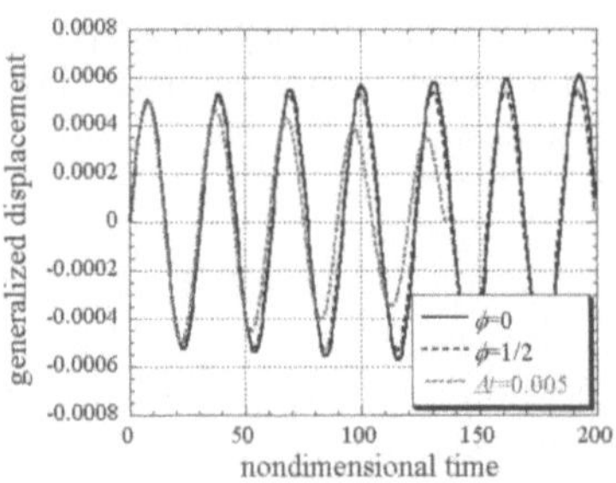

Fig. 2. Effect of the time accuracy.

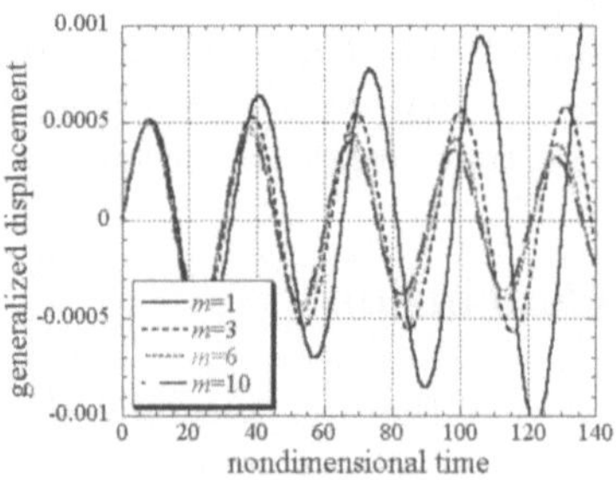

Fig. 3. Effect of maximum number of the inner-iteration.

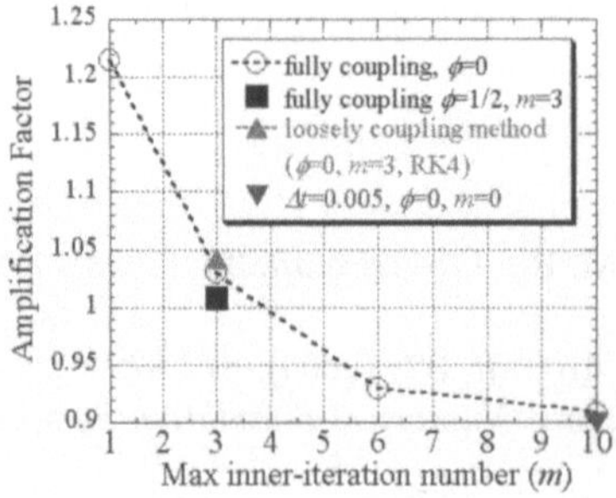

Fig. 4. Amplification factor for various inner-iteration numbers and other methods.

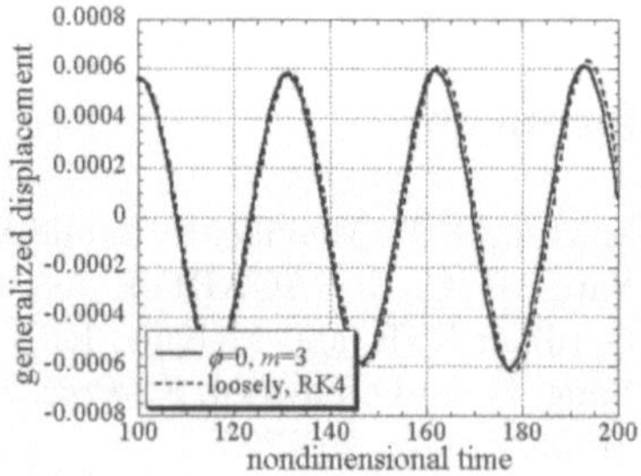

Fig. 5. Effect of coupling methods on mode 1.

Flares of Heat in the Near Wake of a Circular Cylinder in Cross Flow Using a Compact Finite-Difference Scheme

Salem Bouhairie and Vincent H. Chu

McGill University, Montreal, Quebec H3A 2K6, Canada

Abstract. A compact finite-difference scheme is employed to compute the flares of heat from the surface of a circular cylinder in a cross flow. The Poisson equation is solved numerically to the fourth-order of accuracy. Calculations are conducted covering a range of Reynolds numbers from laminar to turbulent flows (Re $= 500 \sim 15550$). The boundary-layer separation, the local and overall heat transfer rates, vortex shedding frequencies and the flares of heat are determined from the numerical calculations.

1 Introduction

The release of heat from a circular cylinder into a cross flow is a highly unsteady process, that is diffusion dominant on the surface and advection dominant in the wake of the cylinder. At a high Reynolds number, heat is trapped in the boundary layer and released by chaotic flares of heat from the layer (Fig. 1). The challenge in the numerical solution of the problem is to resolve of the steep thermal gradient in the boundary layer and at the same time calculate the advection of heat away from the surface of the cylinder with equal accuracy. Calculations of the unsteady heat flux were successful only for flows of relatively low Reynolds numbers . The velocity field nevertheless has been partially computed for a wider range of Reynolds numbers.

Heat transfer calculations of the flow around a circular cylinder were conducted by Dennis et al. [4] at a range of Reynolds numbers up to $Re = 40$ and by Lange et al. [5] up to $Re = 200$. To the best of the author's knowledge, there have not been any previous attempts to calculate the heat transfer rate at higher Reynolds numbers. Since the heat transfer rate is directly proportional to the temperature gradient at the wall, for accurate simulation of the heat transfer rate, the computation grid must have a enough resolution of the heat and flow within the boundary layer. The higher the Reynolds number, the finer is the requirement for the computation grid. In this study, calculations were conducted for Reynolds numbers $Re = 500$, 1000, 5000, 8290 and 15550 with Prandtl number $Pr = 0.7$. The size of the computation grid was selected to be directly proportional to the boundary layer thickness. A fourth-order Hermitian interpolation scheme is used in the present formulation. Only three neighboring points are involved in the interpolation. The accuracy is maintained even in the region very close to the wall.

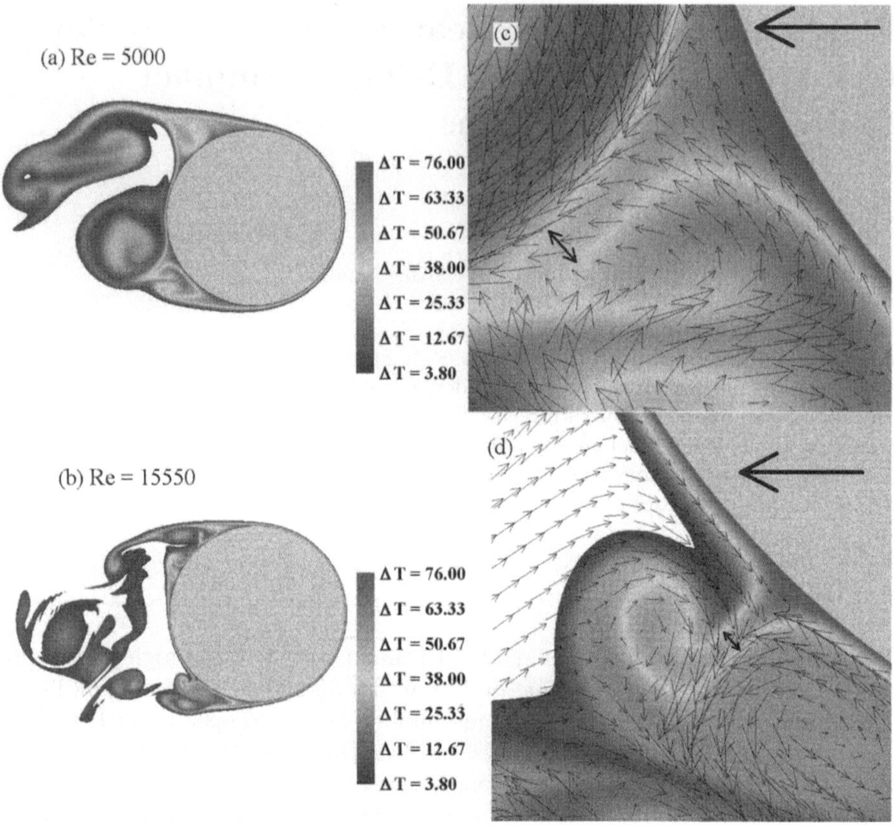

Fig. 1. Flare up of heat from back surface of the cylinder. (**a**) Re = 5000. (**b**) Re = 15550. (**c**)(**d**) Closeup views of the narrow alleyways of heat. The flare thickness as marked by the two-way arrow is $\delta \simeq 4.7/\sqrt{\text{Re}}$

2 Compact Finite Difference Scheme

The numerical solution of the fluid dynamic equations is carried out using a stream-function and vorticity formulation. The finite difference approximation of the Poisson equation, $\nabla^2\psi = \omega$, is supplemented by two Hermitian interpolation formulae:

$$\psi'_{i-1} + 4\psi'_i + \psi'_{i+1} = \frac{3}{h}(\psi_{i+1} - \psi_{i-1}) + O(h^4) \tag{1}$$

$$\psi''_{i-1} + 10\psi''_i + \psi''_{i+1} = \frac{12}{h^2}(\psi_{i+1} - 2\psi_i + \psi_{i-1}) + O(h^4). \tag{2}$$

Unlike other higher-order schemes, the Hermitian interpolation simultaneously takes into account the boundary conditions at the wall not only for the unknown function ψ but also for its first and second derivatives ψ' and ψ''. The fourth-order accuracy is achieved with a three-point approximation because the velocity ψ' and the velocity gradient ψ'' are simultaneously calculated. The higher-order scheme for the Poisson equation provides an accurate velocity field on the surface of the cylinder. A close correlation of the stream function with the vorticity is possible with only three nodal points in the region close to the wall. This is a significant advantage over a conventional high-order scheme because a fewer number of nodal points are involved in the formulation for the boundary condition. Ta Phuoc Loc [8] obtained highly accurate results using the same procedure to calculate the flow around an impulsively started circular cylinder.

3 Heat and Vorticity Equations

The heat and vorticity equations are solved using a number of numerical methods and finite difference approximations. Calculations are conducted using the alternate direction implicit (ADI) method with three different spatial discretizations schemes: a central difference (CD) scheme, a hybrid scheme known as power-law (PL) scheme and a third-order upwind-biased approximation known as QUICK. These implementations will be referred to as CD-ADI scheme, PL-ADI scheme and QUICK-ADI scheme, respectively. The CD discretization is second-order accurate. The PL is a hybrid of the upwind scheme and central difference scheme. The upwind scheme is first-order accurate in space. The QUICK scheme is third-order accurate in space. The ADI schemes are designed to obtained a second order accuracy in time. In addition to these alternate direction implicit schemes, calculations were conducted using an explicit scheme known as the QUICKEST [6] scheme, and the QUICKEST scheme with the application of a universal flux limiter (UFL) [7]. The QUICKEST scheme is third-order accurate in both space and time. These two implementations of the explicit schemes are referred to as QUICKEST and QUICKEST+UFL, respectively.

4 Flares of Heat and Boundary-Layer Detachment

Figure 1 shows the temperature distribution in the wakes of the cylinder obtained by CD-ADI scheme for two Reynolds numbers $Re = 5000$ and 15550. At high Reynolds number, heat is trapped in the thin boundary layer. The greater the Reynolds number, the thinner the boundary layer. The heat is released into the wake at multiple points of boundary-layer separation as fluid and heat lift off from the surface in the form of eruptions. The flares of heat are concentrated in narrow alleyways that are nearly perpendicular to the surface of the cylinder (see the closeup views in Fig. 1c,d). The width of the passages decrease with the Reynolds number. It scales with the boundary layer and is approximately equal to seven times the displacement thickness: $7\delta_1 \simeq 4.7/\sqrt{Re}$. The double arrows in the close-up figures mark the width of the flares.

Temperature and velocity gradients are steep in the boundary layer except in the isolated locations where the heat is erupted from the surface of the cylinder. The flare up of the heat occurs at locations on the surface where the local Nusselt number, Nu, is a minimum. These locations are also the points of boundary-layer detachment where the vorticity vanishes to zero. The loci of zero-vorticity

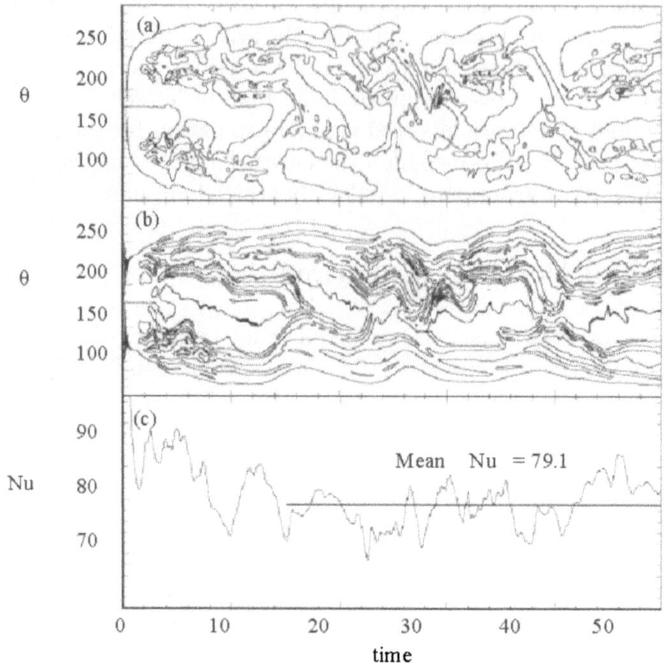

Fig. 2. Results obtained using the QUICKEST+UFL scheme. (**a**) angles of zero vorticity, (**b**) angles of Nusselt-number extrema, and (**c**) circumferentially averaged values of the Nusselt number $\overline{Nu}$, plotted versus the time for flow at $Re = 15550$. The angle θ is measured from the front stagnation point

contours in Fig. 2a, and Nusselt-number minima in Fig. 2a, show the flare-up and detachment to be highly chaotic process.

5 Subgrid Viscosity and Flux Limiter

The results presented in Fig. 2 are obtained using QUICKEST+UFL scheme. We have obtained almost identitical surface heat transfer properties as the QUICKEST+UFL results shown in Fig. 2 using the ADI-CD and ADI-QUICK schemes. The ADI-CD is stable only if a small subgrid viscosity is included in the computation [2]. Subgrid viscosity is not required using ADI-QUICK and ADI-PL and QUICKEST+UFL schemes. With and without a subgrid viscosity, the QUICKEST scheme is not stable even when the calculation was conducted using a very small Courant number.

The application of the flux limiter has made the QUICKEST+UFL a very stable scheme. The computations by the QUICKEST+UFL scheme as shown in Fig. 2 was carried out without the use of the subgrid viscosity. The flux limiter was effective in controlling the unphysical oscillations and at the same time allowed the QUICKEST+UFL scheme to maintain its highest accuracy.

For heat transfer calculation on the surface of the cylinder, both the QUICKEST+UFL scheme and CD-ADI scheme produced accurate results that are in excellent agreement with the available experimental data. Extensive calculations for the surface heat flux have been obtained using the CD-ADI scheme with a subgrid viscosity. Excellent agreement with available experimental data have been obtained. The role of the subgrid viscosity and the validation of CD-ADI scheme for surface heat flux caculations has been reported previously by Bouhairie and Chu [2] [3]. These will not be repeated in this paper.

Central-difference scheme has been reported to be excellent in large-eddy simulation of the turbulent flow in the wake by Beaudan and Moin [1], but the calculations by Beaudan and Moin [1] were for the velocity field only. The central-difference scheme is also known to be affected by unphysical oscillations that is undesirable, particularly in computing the transport of scalar such as temperature and contaminant concentration in turbulent flow. Fig. 3a shows the temperature excess obtained using CD-ADI scheme and Fig. 3b the results obtained using QUICKEST+UFL scheme. The dark-colored wave-like patterns are the unphysical negative undershots. Although the surface heat flux was correctly calculated, huge (nearly a thousand percent) undershots of the temperature excess were observed in the wake data obtained by CD-ADI. However, with the application of the flux limiter, the negative undershots are suppressed. The neg-

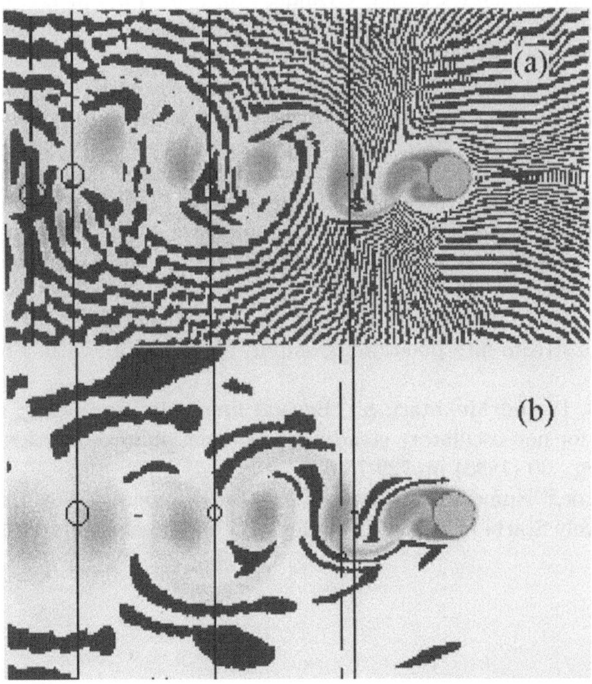

Fig. 3. Negative undershot (dark-colored wave-like patterns) produced by (**a**) CD-ADI scheme and (**b**) QUICKEST+UFL scheme

atives obtained by QUICKEST+UFL are are quite small with values that are less than five percent of the profile maxima.

6 Conclusion

Heat release from the surface of the cylinder in cross flow has been computed for a range of the Reynolds numbers varying from Re = 500 to 15550 using five numerical schemes. The success of the heat flux calculations on the surface of the cylinder is attributed to the use of the compact finite-difference scheme based on the Hermitian interpolation. The flux limiter is effective in suppressing unphysical oscillations. The QUICKEST+UFL scheme, with the help of the flux limiter, has been able to calculate correctly the advection/diffusion of heat both on the surface of the cylinder and in the wake.

References

1. Beaudan, P and Moin, P.: 'Numerical experiments on the flow past a circular cylinder at subcritical Reynolds number'. Dept. of Mechanical Engineering Rept. TF-62. Stanford Univ., California (1994)
2. Bouhairie, S., and Chu, V.H.: 'Heat Transfer Calculations of High-Reynolds-Number Flows around a Circular Cylinder'. In: *Proc. First M.I.T. Conf. Computational Fluid and Solid Mechanics,* Vol. 2 (Massachussetts Institute of Technology, Boston, 2001) pp. 791-794
3. Bouhairie, S. and Chu, V. H.: 'Flare of heat from the surface of a circular cylinder in a cross flow.' *Proc. 2nd Int'l Conf. on Computational Heat and Mass Transfer* (Rio de Janeiro, Brazil, 2001) 8 pp.
4. Dennis, S., Hudson, J., and Smith, N.: 'Steady laminar forced convection from a circular cylinder at low Reynolds numbers'. *Phys. Fluids,* 11:5 (1968) pp. 933-40
5. Lange, C., Durst, F., and Breuer, M.: 'Momentum and Heat Transfer from Cylinders in Laminar Crossflow at $10^{-4} \leq Re \leq 200$'. Int. J. Heat and Mass Transfer, 41 (1998) pp. 3409-3430
6. Leonard, B. P.: 'A stable and accurate convection modelling procedure based on quadratic upstream interpolation'. Comput. Meth. Appl. Mech. Engrg., 19 (1979) pp.59-78
7. Leonard, B. P. and Mokhtari, S.: 'Beyond first-order upwinding: the ultra-sharp alternative for non-oscillatory steady-state simulation of convection'. Int'l J. Num. Meth. Engrg., 30 (1990) pp.729-766.
8. Ta Phuoc Loc.: 'Numerical Analysis of Unsteady Secondary Vortices Generated by an Impulsively Started Circular Cylinder'. J. Fluid Mech., 100:1 (1980) pp. 111-112

Numerical Analysis of the Airfoil's Fluid-Structure Interaction Problems at Large at Large Mean Incidence Angle

Yan Jin[1], Xin Yuan[1], and Byeong Rog Shin[2]

[1] Department of Thermal Engineering, Tsinghua University, Beijing 100084, P. R. China, E-mail: jinyan99@mails.tsinghua.edu.cn
[2] Institute of Fluid Science, Tohoku University, 2-1-1 Katahira, Aoba-ku, Sendai 980-8577, Japan

Abstract. A high accuracy fluid-structure coupling numerical method is developed in the present paper. The complete unsteady Navier-Stokes equations are solved in the fluid zone and the structural equations are solved in the solid zone. The fluid-induced vibration of NACA 0015 aifoil at large attack angle is studied by this numerical method. It is found from the numerical results that "lock-in" will occur at certain M_∞ range near the static stall angle, where the frequency of the vortex shedding will be equal to the natural frequency. The sympathetic oscillation of the vortex and the structure will lead to the stall flutter . The characteristic of "lock-in" can be used to explain all kinds of nonlinear phenomena of the stall flutter.

1 Introduction

When a free stream flows past an airfoil at large attack angle, large scale vortical structure will be constructed around the airfoil, which may lead to airfoil vibration. In Recent years, many aspects of the airfoil flutter have been studied by the structure-fluid coupling approach[1-3]. However, few papers have addressed the issue of airfoil vibration at large attack angle. The difficulties of calculating air stall flutter lies in setting up a mathematical model and numerical algorithms.

The ideal mathematical model is to coupling calculate the Navier-Stokes equations and the structural equations. But it is difficult to solve the this model because of the algorithm and computation time. The idea of structure-fluid coupling has in the past been applied primarily with a simpler aerodynamic analysis. The Euler equations and a two-degree-of-freedom structural dynamic model is the most common aerodynamic model[2-3]. But the non-viscosity assumption breaks down at high angles of attack. The aerodynamic force at large attack can only be analyzed by a more accurate aerodynamic model, Navier-Stokes Equations. A high accuracy numerical method is also needed to calculate the unsteady airload.

In the present paper, a high accuracy fluid-structure coupling numerical method is developed to study the fluid-induced vibration problems of airfoils. A case of deep dynamic stall is calculated to validate the code. Following the code validation, the present techniques are employed to study the fluid-induced vibration of NACA 0015 airfoils at large attack angle. The Characteristic of the airfoil stall flutter are studied from the numerical results.

2 Mathematical Formulation

Governing equations of fluid The Farve-averaged, two-dimensional, Navier-Stokes equations can be written in body-conforming curvilinear coordinate system as follows:

$$\frac{\partial \hat{\boldsymbol{Q}}}{\partial t}+\frac{\partial \hat{\boldsymbol{F}}_i}{\partial \xi_i}+\frac{1}{Re}\hat{\boldsymbol{\mathcal{D}}}+\hat{\boldsymbol{S}}=0 \tag{1}$$

$$\hat{\boldsymbol{Q}}=J\begin{bmatrix}\rho\\ \rho u_1\\ \rho u_2\\ \rho E\\ \rho\vartheta_1\\ \rho\vartheta_2\end{bmatrix}, \hat{\boldsymbol{F}}_i=J\begin{bmatrix}\rho U_i\\ \rho u_1 U_i+\xi_{i,1}p\\ \rho u_2 U_i+\xi_{i,2}p\\ \rho H U_i-\xi_{i,t}p\\ \rho\vartheta_1 U_i\\ \rho\vartheta_2 U_i\end{bmatrix}, \hat{\boldsymbol{\mathcal{D}}}=-\frac{\partial}{\partial \xi_i}J\xi_{i,j}\begin{bmatrix}0\\ \tau_{j1}\\ \tau_{j2}\\ \tau_{jl}u_l+\varpi_j(k)-q_j\\ \varpi_j(\vartheta_1)\\ \varpi_j(\vartheta_2)\end{bmatrix} \tag{2}$$

where, $\xi_{i,j}=\dfrac{\partial \xi_i}{\partial x_j}$, $\xi_{i,t}=\dfrac{\partial \xi_i}{\partial t}$. JU_i can also be chosen as the volume flux in the governing equations of body-conforming curvilinear coordinate system. $\hat{\boldsymbol{S}}$ is the source term.

The low-Reynolds number q-ω turbulence model[6] is applied in the numerical analysis.

Equations of structure The structural dynamics model considered is the PAPA model. The airfoil DOFs are the pitch angle(α), positive nose up, and plunge displacement(φ). The governing equations for structure are given by,

$$\boldsymbol{M}\ddot{Y}+\boldsymbol{D}\dot{Y}+\boldsymbol{K}Y+\boldsymbol{S}=\boldsymbol{F} \tag{3}$$

$$\boldsymbol{M}=\begin{bmatrix}1 & S_\alpha cos(\alpha_0+\alpha)\\ S_\alpha cos(\alpha_0+\alpha) & I_\alpha\end{bmatrix}, Y=\begin{bmatrix}\varphi\\ \\ \alpha\end{bmatrix}, \boldsymbol{K}=\begin{bmatrix}K_\varphi & 0\\ 0 & K_\alpha\end{bmatrix}$$

$$\boldsymbol{D}=\begin{bmatrix}C_\varphi & 0\\ 0 & C_\alpha\end{bmatrix}, \boldsymbol{S}=\begin{bmatrix}-S_\alpha sin(\alpha_0+\alpha)\dot{\alpha}^2\\ \\ 0\end{bmatrix}, \boldsymbol{F}=\begin{bmatrix}-\dfrac{2M_\infty^2 C_l}{\pi\mu}\\ \dfrac{2M_\infty^2 C_m}{\pi\mu}\end{bmatrix}.$$

3 Numerical Scheme

A loose fluid-structure coupling approach has been applied to analyze the flutter of airfoils. The fluid and solid variables are updated alternatively by independent CFD and CSD codes which exchange boundary information at each time step.

An implicit time marching scheme is used to compute the unsteady flowfield. By applying 2nd-order Crank-Nicholson method and Newton subiteration to equation (1), a δ type implicit equation can be obtained as follows:

$$(\boldsymbol{I} + \theta \Delta t \frac{\partial}{\partial \xi_i} \hat{A}_i^n) \delta \hat{\boldsymbol{Q}}^{(m)} = -(\hat{\boldsymbol{Q}}^{(m)} - \hat{\boldsymbol{Q}}^{(n)}) + \frac{1}{2}(\hat{\boldsymbol{R}}^{(n)} + \hat{\boldsymbol{R}}^{(m)}) \tag{5}$$

where $\hat{Q}(m+1) = \hat{Q}(m) + \delta\hat{Q}^m, \hat{Q}(0) = \hat{Q}(n), 0 \leq \theta \leq 1$.

An implicit LU-SGS-GE scheme[5] has been used to solve the left hand of the equation(5), which not only contains all the features of the LU-SGS scheme, but also displays distinctive advantages. Although similarity transformations are used to construct upwind Jacobian matrix of the flux vectors, the block-diagonal matrix inversions are still eliminated, the left hand side operator can be completely vectorized too. As the true Jacobian matrix has been used used, the new implicit scheme leads to a much faster convergency and higher stabilization without improper numerical dissipation and free parameters.

A fourth-order MUSCL TVD scheme[4] is used to solve the left hand of the equation(5) which has a higher accuracy, higher resolution and better stability. It is suited for solving the unsteady flows and catching the complicated vortex shedding process.

The body-fitted C-type computational grids, whose meshes modify their positions with time, are generated with a geometry method. The size of the grids is 361 × 76.

The structural equations are solved by 4th order Runge-Kutta method.

4 Results and Discussion

Prior to study of flutter, a case of deep dynamic stall is was first calculated and then compared with the experimental results to validate the code. The numerical results agree fairly well with the experimental data of Piziali [7] (Fig. 1). The detailed numerical results is in reference[8].

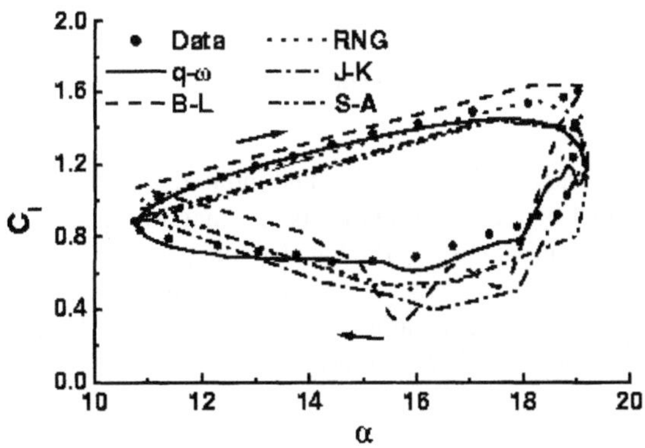

Fig. 1. Numerical results of deep dynamic stall($\alpha = 15° + 4.2° \sin(2kt)$)

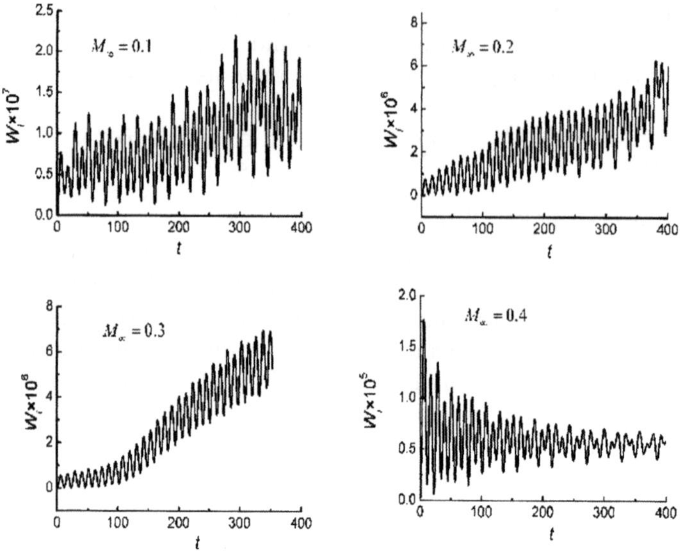

Fig. 2. Influence of M_∞ to the fluid-induced vibration($\alpha_0 = 20°$)

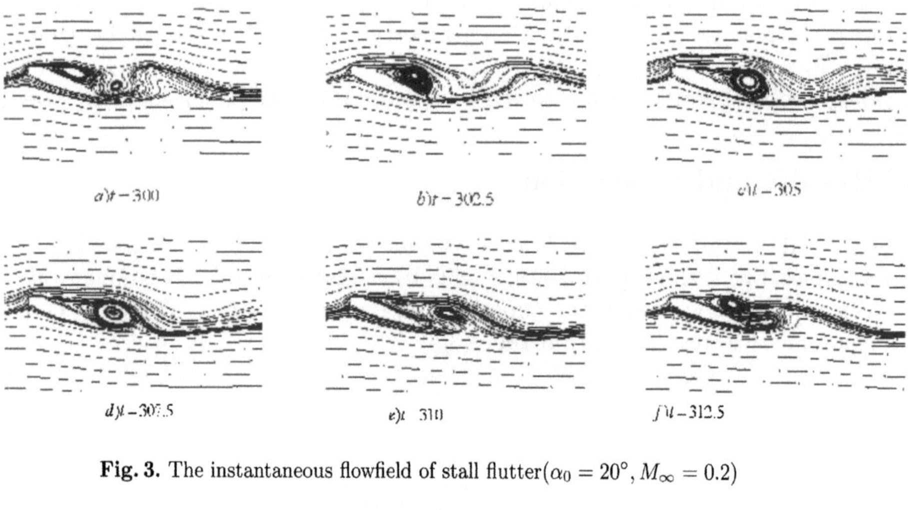

Fig. 3. The instantaneous flowfield of stall flutter($\alpha_0 = 20°, M_\infty = 0.2$)

Following the validation of the code, the fluid-induced vibration of NACA 0015 airfoil is calculated by fluid-structure coupling numerical method. NACA 0015 airfoil has two freedoms of motion, Pitching (α) and plunging (φ). The axis is at 1/4 chord. The natural frequency $\omega_\alpha = 0.05, \omega_\varphi = 0.075$. The influence of free stream Mach number(M_∞) and static attack angle(α_0) to the fluid induced vibration are studied. Figure 2 shows the evolution of the energy transferred

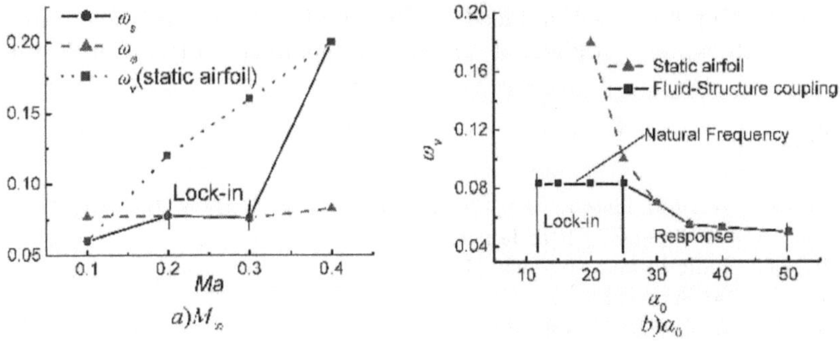

Fig. 4. Influence of M_∞ and α_0 to the oscillating frequency

between the fluid to the airfoil at the attack angle of 20°, from which we can see that the vibration will be unstable when the free stream Mach number $M_\infty = 0.2 \sim 0.3$, while the vibration will remain stable at both small M_∞ and large M_∞. Figure 3 shows the instantaneous flowfield when the stall flutter occur. The large scale vortical structure is generated periodically near the trailing edge of the airfoil. In order to study the the characteristic of the stall flutter, the vortex frequency and the natural frequency are obtained from Fourier analysis. Figure 4 shows the influence of free stream Mach number and attack angle to the oscillating frequency, where ω_s is the vortex shedding frequency considering the fluid-structure interaction, ω_v is the vortex shedding frequency when the airfoil is static and ω_φ is the airfoil's aeroelastic natural frequency. From Fig. 4(a) we can see that ω_s will be influenced by the natural frequency and there is a "lock-in" zone($M_\infty = 0.2 \sim 0.3$), in which ω_s will be equal to ω_φ. The numerical results shows that the sympathetic oscillation of the vortex and the structure will cause the stall flutter.

Figure 4(b) shows the influence of attack angle to the oscillation frequency. A "lock-in" zone is also found near the static stall angle, in which ω_s will be equal to ω_φ. The vortex shedding frequency will be away from the natural frequency at large attack angle. These numerical results can explain the reason why the stall flutter will only occur near the static stall angle and the fluid-induced vibration will be a stable response problem at large incidence angle. The "lock-in" of frequency can also be used to explain all kinds of nonlinear phenomena of stall flutter, such as the stall flutter has oscillating boundary, it contains the characteristic of both self-induced oscillation and forced oscillation.

Conclusions

The fluid-induced vibration at large attack angle is studied by a high accuracy fluid-structural coupling numerical method. It is found from the numerical results that "lock-in" will occur at certain M_∞ range near the static stall angle, where the frequency of the vortex will be equal to the natural frequency and the flutter has the characteristic of self-induced oscillation. The characteristic of "lock-in" can be used to explain all kinds of nonlinear phenomena of the stall flutter.

References

1. Edwards, J.W., and Thomas. J.L., "computational Method for Unsteady Transonic Flows," AIAA Paper 87-0107, Jan. 1987.
2. So R.M.C., Jadic I., and Mignolet M.P., "Oncoming Alternating Vortices," Journal Of Fluids And Structures, Vol.13, No.4, May 1999, pp519-548.
3. Kim D.H., and Lee I., "Transonic and low-supersonic aeroelastic analysis of a two-degree-of-freedom airfoil with a freeplay non-linearity," Journal Of Sound And Vibration, Vol.234, No.5, Jul.2000 pp859-880
4. Daiguji H, Yuan X, Yamamoto S. Stabilization of higher-order high resolution schemes for the compressible Navier-Stokes equations. Int J Numerical Methods for Heat & Fluid Flow. 1997, 7(2,3), pp250-274.
5. Yuan Xin, and Daiguji H., "A Specially Combined Lower-Upper Factored Implicit Scheme for Three-Dimensional Compressible Navier-Stokes Equations," Computers and Fluids, 2001, Vol. 30, No.3, pp339-363.
6. Coakley T. J. Turbulence modeling methods for the compressible Navier-Stokes equations. AIAA Paper, 83-1693.
7. R. A. Piziali An experimental investigation of 2D and 3D oscillating wing aerodynamics for a range of angle of attack including stall. NASA Technical Memorandum 4623(1993)
8. Jin Yan, Yuan Xin. Numerical Study of Unsteady Viscous Flow Past Oscillation Airfoil. Wind Engineering. 2001, 25(4), pp227-237

Part VIII

Incompressible Flows

Application of Two Dimensional Boundary Element Methods to Ship Motion Predictions

D.S. Holloway and M.R. Davis

The University of Tasmania, GPO Box 252-65, Hobart, TAS 7001 Australia

Abstract. This paper outlines and compares two boundary element solutions for the two dimensional hydrodynamic problem of floating ship-like sections, one for periodic motion, and one for arbitrary motion. These solutions are suitable for use in 'strip theory' ship motion predictions. For traditional strip theory, a periodic 2D solutions is required, while for a high speed strip theory developed by the authors a boundary element solution capable of describing arbitrary 2D motion beneath a free surface is required. The boundary element solutions make use of source functions that automatically satisfy the appropriate free surface boundary condition.

1 Background: Strip Theories

The prediction of ship motions in waves requires solution of the hydrodynamic problem governed by the (linearised) equations

$$\frac{\partial^2\phi}{\partial x^2}+\frac{\partial^2\phi}{\partial y^2}+\frac{\partial^2\phi}{\partial z^2} = 0 \quad \text{in the fluid domain} \tag{1}$$

$$\frac{\partial^2\phi}{\partial t^2}+g\frac{\partial\phi}{\partial y} = 0 \quad \text{on the free surface} \tag{2}$$

$$\nabla\phi\cdot\boldsymbol{n} = \boldsymbol{v}\cdot\boldsymbol{n} \quad \text{on the hull surface} \tag{3}$$

$$\nabla\phi \to \nabla\phi_{\mathrm{I}} \quad \text{in the far field,} \tag{4}$$

where ϕ_{I} is the incident wave potential. This is traditionally separated into radiated, incident, and diffracted waves, i.e. $\phi = \phi_{\mathrm{R}} + \phi_{\mathrm{I}} + \phi_{\mathrm{D}}$ say, where each satisfy the same fluid and free surface equations, but with the hull and far field conditions replaced with

$$\nabla\phi_{\mathrm{R}}\cdot\boldsymbol{n} = \boldsymbol{v}\cdot\boldsymbol{n} \quad \text{on the free surface} \tag{5}$$

$$\nabla\phi_{\mathrm{D}}\cdot\boldsymbol{n} = -\nabla\phi_{\mathrm{I}}\cdot\boldsymbol{n} \quad \text{on the free surface} \tag{6}$$

$$\nabla\phi_{\mathrm{R}} \to 0 \quad \text{in the far field} \tag{7}$$

$$\nabla\phi_{\mathrm{D}} \to 0 \quad \text{in the far field.} \tag{8}$$

Strip theory assumes that, due to the slenderness of the ship and the two-dimensionality of the radiated and diffracted wave patterns at high ship speeds, the fluid field equation can be replaced by its two dimensional form

$$\frac{\partial^2\phi}{\partial x^2}+\frac{\partial^2\phi}{\partial y^2} = 0\,, \tag{9}$$

hence the three dimensional problem may be replaced with a set of two dimensional problems corresponding to planes perpendicular to the longitudinal axis of the ship.

In the most widely used theories, those based on [6], all wave fields are taken to be periodic in a reference frame moving with the mean forward speed of the ship, U, hence the $\frac{\partial}{\partial t}$ operator is replaced with $\left(i\omega - U\frac{\partial}{\partial z}\right)$. (The far field condition (7) or (8) also becomes one of outgoing regular waves.) Although this introduces a z dependence in the free surface boundary condition, it is neglected in the interests of retaining the two dimensional formulation. This leads to the requirement that $U/L \ll \omega$, contradicting the high speed assumption implicit in the adoption of a two dimensional Laplace's equation for all but high frequencies. The authors have shown [5] that in the frequency range of significant motions for most ships the assumptions do contradict, although typically traditional strip theory gives acceptable predictions at low speeds (ship Froude number, $U/\sqrt{gL}$, less than about 0.4).

To improve motion predictions at high speed resulting from the above discrepancy of assumptions the authors have proposed a high speed theory [3], [5], in which the hydrodynamic problem is computed in a fixed reference frame, through which the ship passes. This removes the z dependence from the free surface boundary condition, avoiding the low speed restriction, but at the cost of additional complexity due to the requirement of a transient solution (even if the motion in a reference frame moving with the ship is periodic).

The periodic solution described below may thus be used for the evaluation of hydrodynamic forces in a traditional strip theory, while the arbitrary motion solution may be used for the high speed theory of the authors.

2 Boundary Element Solutions in Two Dimensions

2.1 Periodic Motion

The method used is based on that of Doctors [1], which uses a source function that automatically satisfies the appropriate field equation, free surface boundary condition, and far-field radiation condition. Such a source may be obtained from that given by Wehausen and Laitone [7],

$$f(z,\nu) = Q\left\{\frac{1}{2\pi}\left[\left(\ln(z-c) - \ln(z-\bar{c})\right) + 2\mathrm{e}^{-\mathrm{i}\nu(z-\bar{c})}\left(\mathrm{Ei}\left(\mathrm{i}\nu(z-\bar{c})\right) + \delta\pi\mathrm{i}\right)\right]\right.$$
$$\left. + \mathrm{j}\left[\mathrm{e}^{-\mathrm{i}\nu(z-\bar{c})}\right]\right\}, \tag{10}$$

time-complex in j and space-complex in i, where space-complex z and c represent respectively the field and source locations, time-complex Q is a source strength, $\nu = \omega^2/g$ is the wavenumber, and $\delta = 1$ or -1 if $\Re_{\mathrm{i}}(z-c) > 0$ or < 0 respectively. Each ship cross section is discretised into a number of straight line elements, over which these sources are distributed. Source strengths, constant on each element, are adjusted to satisfy the hull boundary condition (5) or (6) on the midpoint of each element, by solving the resulting $n \times n$ system of linear time-complex equations of the form

$$\boldsymbol{A}\boldsymbol{Q} = \boldsymbol{R}, \tag{11}$$

where matrix element $A_{ij} = -\Im_{\mathrm{i}}\left(\mathrm{e}^{\mathrm{i}\alpha} W_{ij}\right)$ (α = slope of element i) is the normal component of fluid velocity W_{ij} at the mid point of element i due to a unit source distributed over element j, Q_j is the unknown source strength on element j, and R_j for the radiation problem is $\boldsymbol{v} \cdot \boldsymbol{n}$ for element j corresponding to (time-complex) cross section motion $\boldsymbol{v}$, or, for the diffraction problem, $-\nabla\phi_{\mathrm{I}} \cdot \boldsymbol{n}$. W_{ij} is given by

$$W_{ij} = \frac{1}{Q_j} \int_{\text{element } j} \frac{\partial f\left(z_i, \nu\right)}{\partial z_i} \mathrm{d}s\,, \tag{12}$$

where $\mathrm{d}s = \mathrm{d}c\,\mathrm{e}^{-\mathrm{i}\beta}$ or $\mathrm{d}\bar{c}\,\mathrm{e}^{\mathrm{i}\beta}$ as appropriate (β = slope of element j).

Finally, forces, or sectional added mass and damping coefficients, are then obtained from the time-real and time-imaginary parts of the hydrodynamic pressures, $p = -\varrho\partial\phi/\partial t = -\varrho \mathrm{j}\omega\phi$, where $\boldsymbol{\phi} = \boldsymbol{B}\boldsymbol{Q}$, and

$$B_{ij} = \Re_{\mathrm{i}}\left\{\frac{1}{Q_j} \int_{\text{element } j} f\left(z_i, \nu\right)\, \mathrm{d}s\right\}. \tag{13}$$

Special treatment of the diagonal terms of $\boldsymbol{A}$ and $\boldsymbol{B}$ is discussed in [3].

2.2 Arbitrary Motion

In this problem it is of no advantage to separate radiated and diffracted waves, hence, given the instantaneous values for $(\boldsymbol{v} - \nabla\phi_{\mathrm{I}}) \cdot \boldsymbol{n}$ on the hull surface at each time step, the source strength distribution is solved, and from these forces are determined, which are integrated to obtain new values of $\boldsymbol{v}$ for the next time step. The method of obtaining source strengths is essentially the same as that for periodic motion, substituting the source function for arbitrary motion [7]

$$\begin{aligned} f\left(z, t\right) = {} & \frac{Q\left(t\right)}{2\pi}\left\{\ln\left(z - c\left(t\right)\right) - \ln\left(z - \bar{c}\left(t\right)\right)\right\} \\ & -\frac{g}{\pi}\int_0^t Q\left(\tau\right)\int_0^\infty \frac{1}{\sqrt{gk}}\mathrm{e}^{-\mathrm{i}k(z-\bar{c}(\tau))} \sin\left[\sqrt{gk}\left(t-\tau\right)\right]\, \mathrm{d}k\, \mathrm{d}\tau \end{aligned} \tag{14}$$

in place of (10). However it is noted that the convolution term in (14) is independent of $Q\left(t\right)$, hence it appears in $\boldsymbol{R}$ rather than $\boldsymbol{A}$, giving in (11)

$$A_{ij} = -\Im\left\{\frac{\mathrm{e}^{\mathrm{i}\alpha}}{2\pi}\left[\mathrm{e}^{-\mathrm{i}\beta}\ln\left(\frac{z - c_1}{z - c_2}\right) - \mathrm{e}^{\mathrm{i}\beta}\left(\frac{z - \bar{c}_1}{z - \bar{c}_2}\right)\right]\right\} \tag{15}$$

and

$$R_i = \boldsymbol{v} \cdot \boldsymbol{n} - \Im\left\{\mathrm{e}^{\mathrm{i}\alpha}\left[\mathrm{i}\sqrt{\frac{g}{\pi}}\sum_{j=1}^{n}\int_0^t \mathrm{e}^{\mathrm{i}\beta(\tau)} Q_j\left(\tau\right)\left[\frac{\mathrm{e}^{w^2}\operatorname{erf} w}{\sqrt{\mathrm{i}\left(z - \bar{c}\left(\tau\right)\right)}}\right]_{w_1}^{w_2} \mathrm{d}\tau\right]\right\}, \tag{16}$$

where $w = \mathrm{i}\,(t-\tau)/2 \times \sqrt{-\mathrm{i}g/\left(z-\bar{c}\,(\tau)\right)}$, w_1 and w_2 correspond to end points of element j, and $\int \ldots \mathrm{d}\tau$ is evaluated numerically using the known discrete values of Q, z, c, and β from previous time steps.

Pressures (hence forces) are obtained at each element i by summing contributions from all elements j of $p_{ij} = -\varrho \frac{\partial \phi_{ij}}{\partial t}$, i.e.

$$
\begin{aligned}
p_{ij} = -\varrho \Re \Bigg\{ & \frac{\mathrm{e}^{-\mathrm{i}\beta}}{2\pi} \left[\left(A_k + (z-c_k)\,B\right) \ln\left(\mathrm{e}^{-\mathrm{i}\gamma}\,(z-c_k)\right)\right]_{k=1}^{2} \\
& -\frac{\mathrm{e}^{\mathrm{i}\beta}}{2\pi} \left[\left(\bar{A}_k + (z-\bar{c}_k)\,\bar{B}\right) \ln\left(z-\bar{c}_k\right)\right]_{k=1}^{2} \\
& +\frac{2\mathrm{i}g}{\sqrt{\pi}} \int_0^t \mathrm{e}^{\mathrm{i}\beta(\tau)} Q\,(\tau) \int_{w_1}^{w_2} \left(\mathrm{e}^{w^2} \operatorname{erf} w + \frac{1}{w\sqrt{\pi}} \right) \mathrm{d}w\,\mathrm{d}\tau \Bigg\}\,, \qquad (17)
\end{aligned}
$$

where $\gamma = \arg\left(z - (c_1 + c_2)/2\right)$ (or $\arg\left(c_2 - c_1\right) + \pi/2$ if $z = (c_1 - c_2)/2$), $A_k = Q\,\mathrm{d}c_k/\mathrm{d}t$, $B_k = -\mathrm{d}Q/\mathrm{d}t + \mathrm{i}Q\,\mathrm{d}\beta/\mathrm{d}t$, all evaluated at time t. Special algorithms were developed to evaluate $\int \left(\mathrm{e}^{w^2} \operatorname{erf} w + 1/w\sqrt{\pi}\right) \mathrm{d}w$ efficiently [4].

3 Validation of Boundary Element Solutions

Extensive validation of the two boundary element methods is given in [3] and [4], against both analytic solutions (e.g. [2]) and a wide variety of numerical solutions to problems of steady oscillation, steady translation, and arbitrary motion (including both Rankine and Green function source methods), showing in all cases excellent agreement. Some results are reported here.

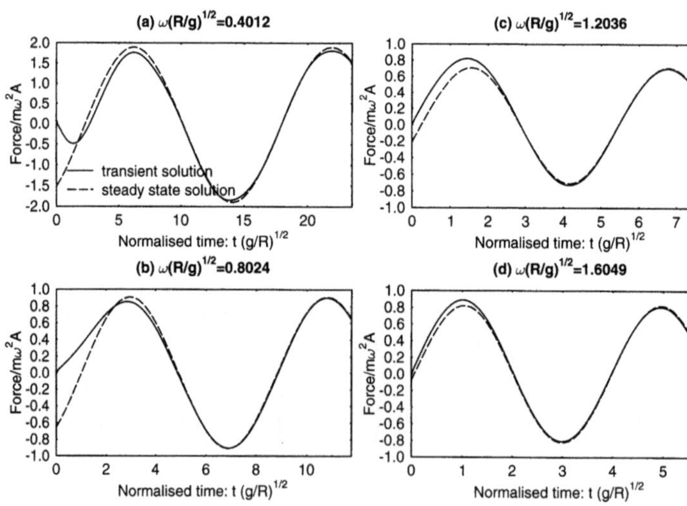

Fig. 1. Comparison between steady periodic and time domain solutions for total hydrodynamic force on a semi-circular cylinder (radius R, displaced mass m) floating on a free surface, oscillating vertically with amplitude A

Figure 1 compares the above two methods, showing agreement in both amplitude and phase of the hydrodynamic force once initial transients have decayed in the time domain solution. The periodic solution experiences irregular frequencies (the lowest being at $\omega\sqrt{R/g} = 1.35$) where the wavelength of internal waves divides integrally into the internal free surface length. These were eliminated by the use of 'lid' elements, forming a closed cross section. For good results using the transient solution no lid was required, but it was found to be necessary to keep the time step small compared with the period of the first irregular frequency (as well as compared with the actual frequency of oscillation).

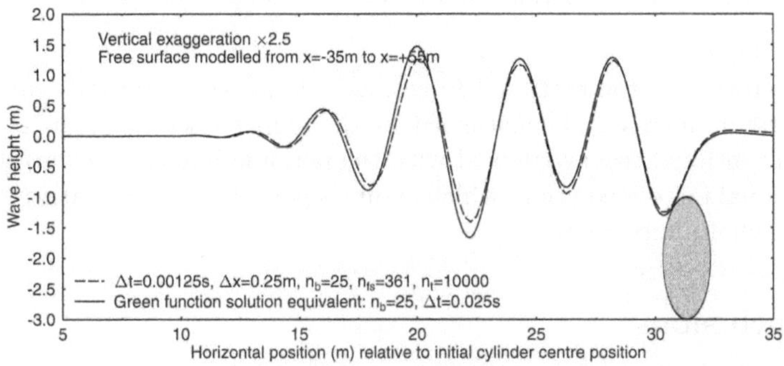

Fig. 2. Comparison of Rankine and Green function time domain solutions for the wave profile generated by a submerged 1 m radius circular cylinder started from rest then translating at constant speed of $Fr = 0.8$ (based on radius), after elapse of 12.5 s ($t\sqrt{g/R} = 39.15$)

Figure 2 compares the present arbitrary motion method to one based on Rankine sources [3], in which a finite length of free surface was represented by elements to which the boundary condition (2) was applied. This represented a severe test, as reasonable computational times in the Rankine method solution limited the combination of element and time step refinement. The time step was found to be more critical in this instance, hence the very small value, while the present solution was able to use a time step 20× larger without significant accuracy penalty. Given that the two methods introduce quite different types of error, the agreement is very good, and comparison with coarser discretisations [3] clearly suggests that the two are converging.

4 Ship Motion Predictions

Figure 3 compares strip theory motion predictions using the conventional theory [6] and the authors' high speed theory. While the improvement for heave motion is pleasing, the improvement for pitch motion is striking. This difference is attributed to the asymmetry of the wedge shaped actual wave pattern. This is predicted well by the high speed theory coupled with the transient boundary

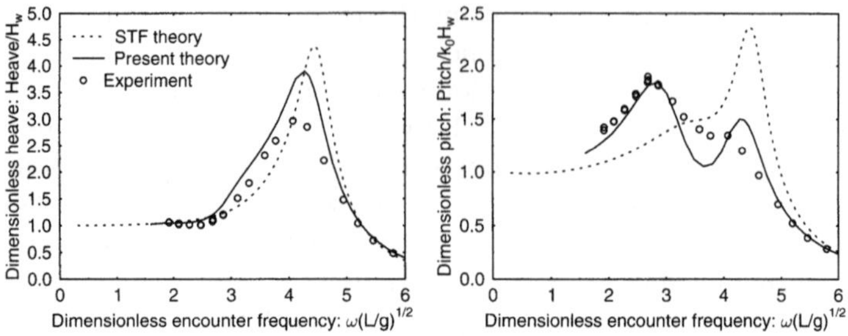

Fig. 3. Comparison of ship motions in regular waves by conventional theory [6] (*STF*) and present theory with experimental response, at Froude number 0.7

element solution to the sectional forces, as it assumes an initially undisturbed strip of water, increasingly disturbed as the ship passes through, as opposed to the conventional theory coupled with the periodic boundary element method force sectional force prediction, which assumes periodic wave radiating to infinity along the whole ship length.

5 Conclusions

A time domain boundary element method for modelling linear water waves is presented, and compared with the corresponding frequency domain method. While it is computationally more involved, because of the multiple evaluations required for the Green function appearing in the convolution integral, it is shown to give significantly better predictions of ship motions when coupled with a strip theory.

References

1. L.J. Doctors: 'Application of the Boundary-Element Method to Bodies Oscillating Near a Free Surface'. In: *In International Symposium on Computational Fluid Dynamics, 1987*, pp. 1–10
2. T.H. Havelock: *The Collected Papers of Sir Thomas Havelock on Hydrodynamics.* ed. by C. Wigley (Office of Naval research, Washington 1963)
3. D.S. Holloway: A High Froude Number Time Domain Strip Theory Applied to the Seakeeping of Semi-SWATHs. PhD Thesis, University of Tasmania, Hobart (1998)
4. D.S. Holloway, M.R. Davis: J. Ship Research **46**, 99 (2002)
5. D.S. Holloway, M.R. Davis: 'Computation of Ship Motions by a High Froude Number Strip Theory', under review for J. Ship Research
6. N. Salvesen, E.O. Tuck, O. Faltinsen: Trans. Society of Naval Architects and Marine Engineers **78**, 250 (1970)
7. J.V. Wehausen, E.V. Laitone: 'Surface Waves'. In: *Handbuch der Physik 9.* ed. by S. Flügge (Springer, Berlin 1960) pp. 445–814

The Characteristic-Based-Split Method for 3D Incompressible Flows on Unstructured Grids

P. Nithiarasu[1] and J. S. Mathur[1,2]

[1] School of Engineering, University of Wales Swansea,
Singleton Park, Swansea, SA2 8PP, U.K.
e-mail: P.Nithiarasu@Swansea.ac.uk; J.S.Mathur@Swansea.ac.uk

[2] Computational & Theoretical Fluid Dynamics Division,
National Aerospace Laboratories, Bangalore 560 017, India
e-mail: jsm@ctfd.cmmacs.ernet.in

Abstract. This paper describes the Characteristic-Based-Split (CBS) algorithm and its application to the solution of three dimensional, incompressible flow problems. Both fully explicit and semi-implicit schemes are discussed. An artificial compressibility is introduced to treat the equations fully explicitly and a Poisson type equation, for pressure, is implicitly solved in the semi-implicit scheme. A stable solution is achieved by using the convective stabilization terms obtained from Characteristic discretization, and no explicit addition of artificial dissipation is necessary.

1 Introduction

The finite element method has been used for the solution of the Navier-Stokes equations for many years now. The standard method is equivalent to a central difference approximation and hence produces a solution with oscillations when applied to convection dominated problems. Such unrealistic solutions are eliminated by using suitable stabilization schemes. However, methods like the Taylor Galerkin or Characteristic Galerkin procedure have shown that the balancing diffusion appears naturally when the equations are discretized in time. In the Characteristic Galerkin method the temporal derivative is discretized along the characteristic resulting in an optimal approximation [1].

In this paper, the finite element based Characteristic-Based-Split (CBS) scheme (CBSflow) is discussed. It is shown that, with appropriate amendments, the artificial compressibility can easily be incorporated into CBSflow. Thus, this method can be used either in its fully explicit or semi-implicit form. Some results are presented to validate the scheme on 3D unstructured meshes.

2 CBSflow Scheme

The Navier-Stokes equations for fluid flow can be written in non-dimensional conservative form as

$$\frac{\partial \boldsymbol{W}}{\partial t} + \frac{\partial \boldsymbol{F}_j}{\partial x_j} - \frac{\partial \boldsymbol{G}_j}{\partial x_j} = \boldsymbol{S} \qquad (1)$$

where

$$\boldsymbol{W} = \begin{pmatrix} \rho \\ \rho u_1 \\ \rho u_2 \\ \rho u_3 \\ \rho T \end{pmatrix}, \boldsymbol{F_j} = \begin{pmatrix} \rho u_j \\ \rho u_1 u_j + p\delta_{1j} \\ \rho u_2 u_j + p\delta_{2j} \\ \rho u_3 u_j + p\delta_{3j} \\ \rho T uj \end{pmatrix}, \boldsymbol{G_j} = \begin{pmatrix} 0 \\ \tau_{1j} \\ \tau_{2j} \\ \tau_{3j} \\ \frac{1}{\mathrm{RePr}}\frac{\partial T}{\partial x_j} \end{pmatrix}, \boldsymbol{S} = \begin{pmatrix} 0 \\ 0 \\ 0 \\ \rho\frac{\mathrm{Gr}}{\mathrm{Re}^2}T \\ 0 \end{pmatrix} \tag{2}$$

These equations represent the conservation of mass, momentum in the three coordinate directions and energy, respectively. Here, ρ is the density, x_j are the coordinate directions, u_j are the velocity components, t is the time, p is the pressure and T is the temperature. Re, Pr and Gr represent the Reynolds, Prandtl and Grashof numbers respectively. The viscous stress is defined by

$$\tau_{ij} = \frac{1}{\mathrm{Re}}\left(\frac{\partial u_i}{\partial x_j} + \frac{\partial u_j}{\partial x_i} - \frac{2}{3}\frac{\partial u_k}{\partial x_k}\delta_{ij}\right) \tag{3}$$

where δ_{ij} is the Kronecker delta which equals 1 when $i = j$ and 0 when $i \neq j$. The force of gravity acts along the x_3 direction as is evident by the inclusion of the buoyancy term in the x_3 momentum equation.

The majority of the schemes for the solution of incompressible flow fall into the categories of either artificial compressibility (AC) or projection (velocity correction) schemes. The AC schemes are simple and easy to employ due to their fully explicit nature, but time step limitations make it difficult to solve some slow flow applications. Velocity correction or projection schemes avoid, to some extent, the time step limitations but introduce the expense of implicit treatment of certain equations. However, a combination of velocity correction and AC has the advantage of being explicit and stable.

We now describe the Characteristic-Based-Split (CBS) algorithm [2] and its application to the solution of three dimensional, incompressible flow problems. We first define the mass flow fluxes $U_j \equiv \rho u_j$ and rewrite the equation for mass conservation in the form

$$\frac{\partial \rho}{\partial t} = \frac{1}{c^2}\frac{\partial p}{\partial t} = -\frac{\partial U_j}{\partial x_j} \tag{4}$$

where c is the speed of sound, and each of the equations for momentum conservation in the form

$$\frac{\partial U_i}{\partial t} = -\frac{\partial}{\partial x_j}(U_i u_j) - \frac{\partial p}{\partial x_i} + \frac{\partial \tau_{ij}}{\partial x_j} + S_i \tag{5}$$

Note that for incompressible flows $c^2 \to \infty$ in (4). However, AC schemes use finite values for c^2 calculated from velocity and temperature fields [3]. Discretization in time using the Characteristic process gives [2]

$$U_i^{n+1} - U_i^n = \Delta t \left[-\frac{\partial}{\partial x_j}(U_i u_j)^n - \frac{\partial p^{n+\theta_2}}{\partial x_i} + \frac{\partial \tau_{ij}^n}{\partial x_j} + S_i^n \right] + \frac{\Delta t^2}{2} u_k \frac{\partial}{\partial x_k} \left(\frac{\partial}{\partial x_j}(U_i u_j)^n + \frac{\partial p^n}{\partial x_i} - S_i^n \right) \tag{6}$$

where the pressure term is not treated explicitly but is evaluated at a time $t^n + \theta_2 \Delta t$ $(0 \leq \theta_2 \leq 1)$ and is given by

$$\frac{\partial p^{n+\theta_2}}{\partial x_i} \equiv \theta_2 \frac{\partial p^{n+1}}{\partial x_i} + (1-\theta_2) \frac{\partial p^n}{\partial x_i} \tag{7}$$

For the solution of these equations, the CBS algorithm uses a fractional step with a split. The four steps can be briefly described as

i. solve momentum equation without pressure terms,
ii. calculate pressure from a Poisson equation,
iii. correct the velocities, and finally
iv. calculate any additional scalar variables such as temperature, concentration etc. from appropriate governing equations.

In the first step, we define a new variable U_i^* such that

$$U_i^* - U_i^n = \Delta t \left[-\frac{\partial}{\partial x_j}(U_i u_j)^n + \frac{\partial \tau_{ij}^n}{\partial x_j} + S_i^n \right] + \frac{\Delta t^2}{2} u_k \frac{\partial}{\partial x_k} \left(\frac{\partial}{\partial x_j}(U_i u_j)^n - S_i^n \right) \tag{8}$$

This represents the first part of the split and is explicit. The corrected velocities can be determined, once the pressure is known, using the equation

$$U_i^{n+1} - U_i^* = \Delta t \frac{\partial p^{n+\theta_2}}{\partial x_i} + \frac{\Delta t^2}{2} u_k \frac{\partial}{\partial x_k} \left(\frac{\partial p^n}{\partial x_i} \right) \tag{9}$$

The solution of this equation is actually the third step in the algorithm. The second step is the determination of the pressure. Referring to (4), we can write

$$\left(\frac{1}{c^2} \right)^n (p^{n+1} - p^n) = -\Delta t \frac{\partial U_i^{n+\theta_1}}{\partial x_i} \tag{10}$$

Now $U_i^{n+\theta_1} \equiv \theta_1 U_i^{n+1} + (1-\theta_1) U_i^n$ where $0.5 \leq \theta_1 \leq 1$, and using (9) we get

$$U_i^{n+\theta_1} = \theta_1 \left[U_i^* - \Delta t \frac{\partial p^{n+\theta_2}}{\partial x_i} \right] + (1-\theta_1) U_i^n \tag{11}$$

and hence (10) can be written as

$$\left(\frac{1}{c^2} \right)^n (p^{n+1} - p^n) = -\Delta t \left[\theta_1 \frac{\partial U_i^*}{\partial x_i} + (1-\theta_1) \frac{\partial U_i^n}{\partial x_i} \right] + \frac{\Delta t^2}{2} \theta_1 \left[\theta_2 \frac{\partial^2 p^{n+1}}{\partial x_i \partial x_i} + (1-\theta_2) \frac{\partial^2 p^n}{\partial x_i \partial x_i} \right] \tag{12}$$

The fourth step of the algorithm involves the calculation of the temperature field using the equation for conservation of energy. Referring to (1) this can be written as

$$\frac{\partial}{\partial t}(\rho T) + \frac{\partial}{\partial x_j}(\rho T u_j) - \frac{\partial}{\partial x_j}\left(\frac{1}{\mathrm{RePr}}\frac{\partial T}{\partial x_j}\right) = 0 \tag{13}$$

and, assuming incompressible flow, Characteristic time discretization gives

$$T^{n+1} - T^n = \Delta t\left[-\frac{\partial}{\partial x_j}(T u_j)^n + \frac{\partial}{\partial x_j}\left(\frac{1}{\mathrm{RePr}}\frac{\partial T^n}{\partial x_j}\right)\right] + \frac{\Delta t^2}{2} u_k \frac{\partial}{\partial x_k}\left[\frac{\partial}{\partial x_j}(T u_j)^n\right] \tag{14}$$

The resulting equations are solved using a standard Galerkin finite element procedure. The computational domain is discretized into a mesh of tetrahedral finite elements. Within each tetrahedron, a variable is approximated by a relation of the form $U_i = \sum_{n=1}^{4} N_n \bar{U}_i|_n$ where n =1–4 represent the four nodes forming the tetrahedron, N_n is the shape function at node n and $\bar{U}_i|_n$ is the value of U_i at node n. This is usually written in the more convenient matrix form $U_i = \mathbf{N}\bar{\mathbf{U}}_\mathbf{i}$ where the matrices are given by

$$\mathbf{N} = \{N_1\ N_2\ N_3\ N_4\} \quad \bar{\mathbf{U}}_\mathbf{i} = \{\bar{U}_i|_1\ \bar{U}_i|_2\ \bar{U}_i|_3\ \bar{U}_i|_4\}^\mathrm{T} \tag{15}$$

Further details on the final matrix form of the equation can be found in [1,2].

2.1 Fully Explicit Scheme

A fully explicit form of the CBS algorithm is obtained by substituting $\theta_2 = 0$ and $\frac{1}{2} \leq \theta_1 \leq 1$ into (9) and (10). Thus it becomes necessary to solve (9) explicitly, for which a finite value for c^2 needs to be evaluated. In general, this is carried out using local velocity and temperature fields [3]. This gives a critical time step $\Delta t = h/(c + |u|)$, where h is the local element size.

2.2 Semi- and Quasi- Implicit Schemes

The CBS algorithm can also be used in its semi-implicit form in which an implicit solution of pressure equation is necessary. The algorithm is conditionally stable if $\frac{1}{2} \leq \theta_1 \leq 1$ and $\frac{1}{2} \leq \theta_2 \leq 1$, with a critical time step $\Delta t = h/(|u|)$. The quasi-implicit scheme is similar to semi-implicit form except that all diffusion terms are also treated implicitly.

3 Results

In this section, the solutions of two 3D incompressible viscous flow problems are presented. The first problem, flow over a sphere, is solved using the fully explicit form and the second problem, the differentially heated thermal cavity, is solved using the semi-implicit form of CBSflow. For the first problem, an unstructured grid containing 475,067 tetrahedral elements has been generated using the PSUE

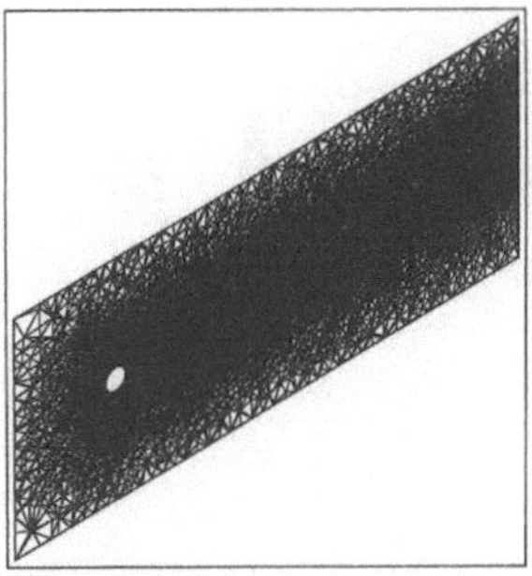

Fig. 1. Unstructured grid, sphere

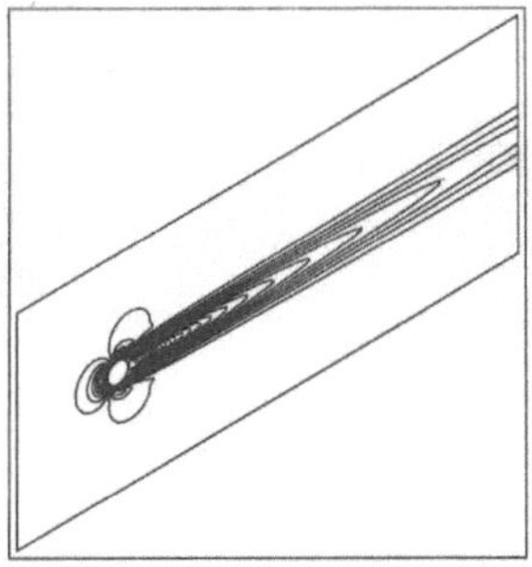

Fig. 2. u_1 contours, Re = 100

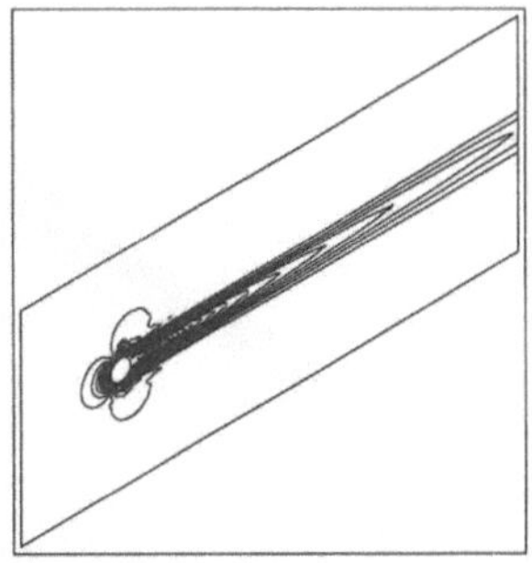

Fig. 3. u_1 contours, Re = 200

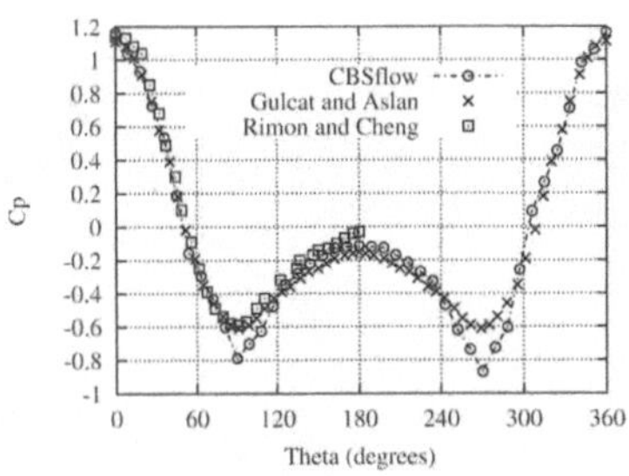

Fig. 4. C_p comparison, Re = 100

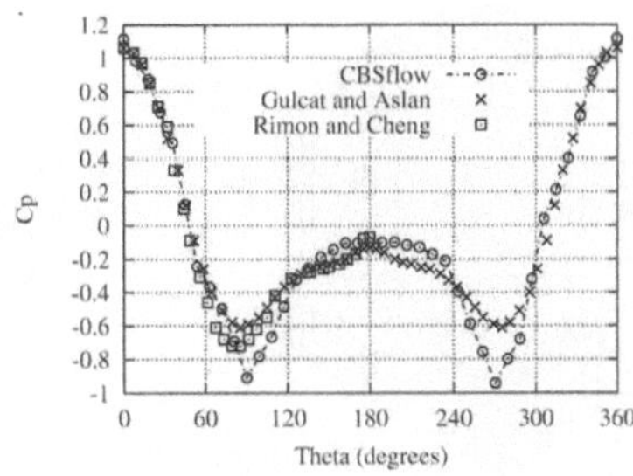

Fig. 5. C_p comparison, Re = 200

software [4]. The computational domain is a rectangular box of length $25D$, where D is the diameter of the sphere, with the down-stream boundary located $20D$ from the centre of the sphere. Figure 1 shows the grid on the symmetry plane and Figs. 2 & 3 show the contours of the u_1 component of the velocity at Reynolds numbers of 100 and 200 respectively. Figures 4 & 5 compare the C_p values on the surface along the symmetry plane with those available in literature [5,6]. The comparisons are generally good and the discrepancies are noted only close to the flow acceleration zones. Figure 6 shows the u_1 contours at a Reynolds number of 600, where the unsteady nature of the wake is apparent and Fig. 7 shows the variation of the u_2 velocity component at a downstream boundary node with real time. The unstructured grid used to solve the differentially heated cavity is shown in Fig. 8. A total of 42,997 tetrahedral elements are used for the computations. Figure 9 shows the computed solution for the flow with a Rayleigh number of 10^6. The temperature contours show the correct flow pattern. Figure 10 shows the variation of u_1 along the centre of the cavity and compares well with the maximum value reported in the benchmark solution [7].

Acknowledgements: This work is sponsored by EPSRC Grant GR/R29321/01. The second author thanks the Director, National Aerospace Laboratories, Bangalore, India and the Council of Scientific & Industrial Research, India, for the grant of sabbatical leave and permission to work in the University of Wales Swansea.

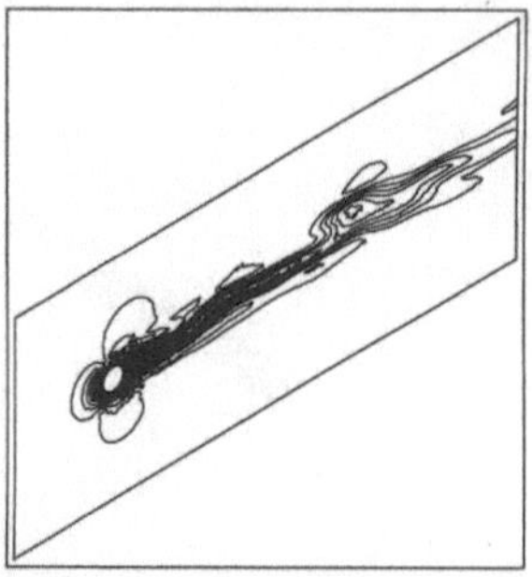

Fig. 6. u_1 contours, Re = 600

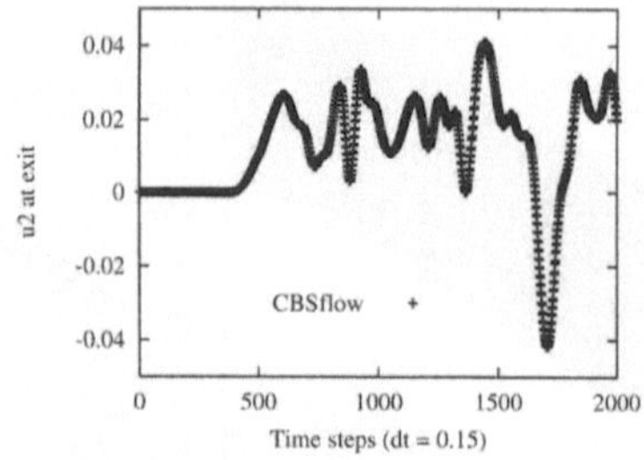

Fig. 7. u_2 variation with time, Re = 600

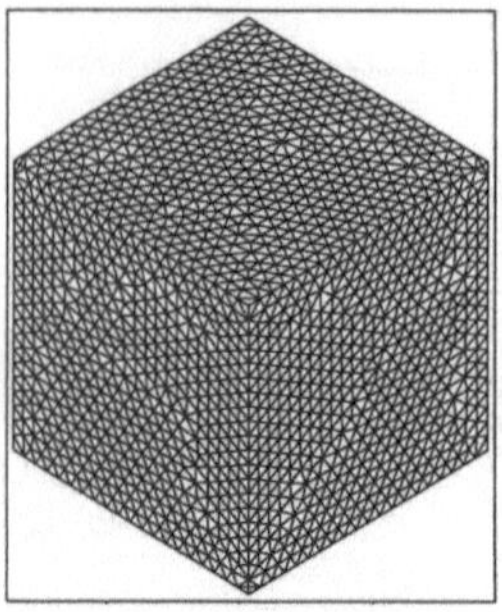

Fig. 8. Unstructured grid, thermal cavity

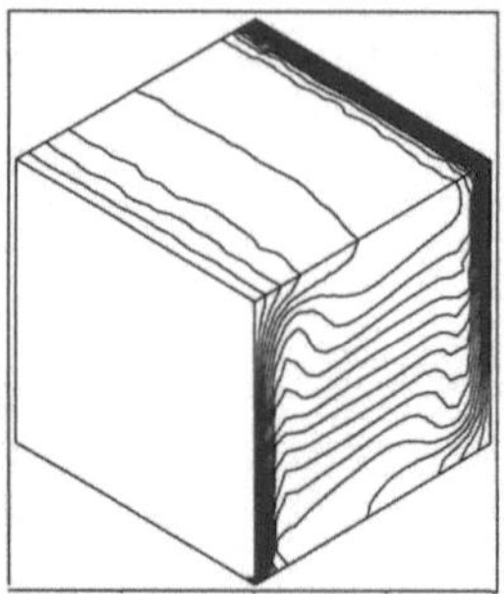

Fig. 9. Temperature contours, Ra = 10^6

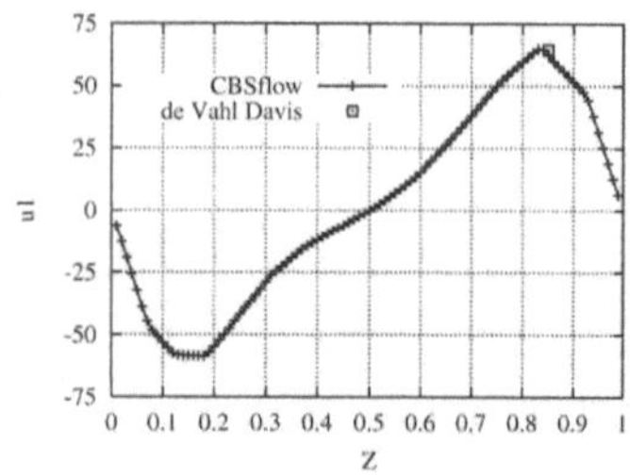

Fig. 10. u_1 velocity, thermal cavity

References

1. O. C. Zienkiewicz, R. L. Taylor: *The finite element method, Vol. 3 fluid dynamics*, 5th edn. (Butterworth-Heinemann, Oxford 2000)
2. O. C. Zienkiewicz, P. Nithiarasu, R. Codina, M. Vázquez, P. Ortiz: Int. J. Numer. Meth. Fluids **31**, 359 (1999)
3. P. Nithiarasu: 'An efficient artificial compressibility (AC) scheme based on the Characteristic Based Split (CBS) method for incompressible flows', submitted to Int. J. Numer. Meth. Engng. (2002)
4. K. Morgan, N. P. Weatherill, O. Hassan, P. J. Brookes, R. Said, J. Jones: Int. J. Numer. Meth. Fluids **31**, 159 (1999)
5. Ü. Gülçat, A. R. Aslan: Int. J. Numer. Meth. Fluids **25**, 985 (1997)
6. Y. Rimon, S. I. Cheng: Phys. Fluids **12**, 949 (1969)
7. G. de Vahl Davis: Int. J. Numer. Meth. Fluids, **3**, 249 (1983)

Spectral Vorticity Method for Computation of Axial and Near-Axial Flow Over Long Circular Cylinders

Milton Woods and Max Bull

Department of Mechanical Engineering,
University of Adelaide, South Australia, 5005

Abstract. A numerical procedure has been developed for simulation of the boundary layer on a long circular cylinder in axial or near-axial flow. The vorticity transport equation is solved on a cylindrical domain using a Fourier-Chebyshev spectral representation for the flow fields and a stiffly-stable time integration scheme. The outer boundary condition for velocity is determined adaptively at each timestep by application of an analytical potential flow solution. The paper presents details of the procedure and initial validation test results.

1 Introduction

Fluid flow over long circular cylinders has been the subject of a great number of investigations, mainly because of the geometrical simplicity of cylinders and their common use in engineering applications. The majority of investigations have been concerned with cylinders whose alignment is perpendicular or nearly perpendicular to the freestream. Particularly in the realm of CFD, far less attention has been directed towards cylinders in axial flow, and even less towards cylinders in near-axial flow. The results of direct numerical simulations of thick boundary layers on cylinders in axial flow are discussed in [1]. No comparable simulation for cylinders in near-axial flow has yet been reported, and it is the aim of the current study to redress this situation.

2 Flow Model

The Prandtl model for analysis of boundary layer flow assumes that viscous effects are limited to boundary layers and wakes while the remainder of the flow is effectively inviscid. When the model is applied to a circular cylinder, the flow is considered to be divided into two distinct regions. The first region, denoted by R_1, lies between the surface of the cylinder, whose radius is r_a, and some larger radius r_b. This region includes the boundary layer, which is associated with a vorticity field $\boldsymbol{\omega}$. The second region, denoted by R_2, extends from r_b to infinite radius. The flow in R_2 is assumed to be irrotational and hence a potential flow.

The simulation procedure presented here confines numerical calculations to the domain R_1. The simple geometry allows the general form of the potential flow on R_2 to be expressed analytically. The velocity boundary condition at r_b is updated during the simulation to ensure consistency between the inner and outer flow fields. The use of a dynamic rather than a fixed velocity boundary condition is preferred for two reasons. The first is that the effect of the boundary layer (and its asymmetry) on the surrounding velocity field is not known in advance except in the case of axial flow. The second reason is that a fixed velocity boundary condition may suppress large scale, unsteady flows across the cylinder of the type reported in [2].

The flow over a long, rigid cylinder is reduced to manageable proportions by the assumption that the flow fields are periodic in the axial direction. This simplification limits the validity of the model to flows in which the mean properties change slowly with axial distance, as is typically the case at positions well downstream from the boundary layer origin.

3 Methodology

3.1 Governing Equations

The principle quantity in the simulation is the vorticity field $\boldsymbol{\omega}$, which is related to the velocity $\boldsymbol{u}$ by the definition

$$\boldsymbol{\omega} = \nabla \times \boldsymbol{u}. \tag{1}$$

The evolution of the vorticity field is governed by the vorticity transport equation,

$$\frac{\partial \boldsymbol{\omega}}{\partial t} = \nabla \times (\boldsymbol{u} \times \boldsymbol{\omega}) + \frac{1}{\mathrm{Re}} \nabla^2 \boldsymbol{\omega}, \tag{2}$$

where Re is the Reynolds number and all quantities are non-dimensional. The constraints on the solution include the continuity equation for incompressible fluid,

$$\nabla \cdot \boldsymbol{u} = 0, \tag{3}$$

and the vector identity

$$\nabla \cdot \boldsymbol{\omega} = \nabla \cdot (\nabla \times \boldsymbol{u}) = 0. \tag{4}$$

3.2 Time Integration

The vorticity field is stepped forward in time by integration of (2). An appropriate stiffly-stable scheme is

$$\frac{\Delta t}{\mathrm{Re}} \nabla^2 \boldsymbol{\omega}_{n+1} - \gamma_0 \boldsymbol{\omega}_{n+1} = -\sum_{i=0}^{I-1} \alpha_i \boldsymbol{\omega}_{n-i} - \Delta t \sum_{i=0}^{I-1} \beta_i \left(\nabla \times (\boldsymbol{u} \times \boldsymbol{\omega})\right)_{n-i}, \tag{5}$$

where Δt is the timestep, n is the current timestep, and I is the desired order of accuracy. The coefficients are given for schemes up to third order in [3]. The non-linear term $\boldsymbol{u} \times \boldsymbol{\omega}$ is treated explicitly for efficiency and the linear term $\nabla^2 \boldsymbol{\omega}$ implicitly for stability.

3.3 Spectral Representation

The periodicity of the flow, natural in the circumferential direction and assumed over a length Z in the axial direction, suggests the use of two-dimensional Fourier series to represent the velocity field. The Fourier series is truncated after a finite number of terms to yield the approximation

$$\boldsymbol{u}[r,\theta,z] \approx \sum_{p=-P}^{P} \sum_{q=-Q}^{Q} \hat{\boldsymbol{u}}_{p,q}[r] \exp(\mathrm{i}p\theta) \exp(\mathrm{i}\beta_q z), \tag{6}$$

where $\beta_q = 2\pi q/Z$. The radial distribution of each Fourier coefficient is approximated by a Chebyshev series. An equivalent series representation is used for vorticity.

Linear partial differential equations involving the flow variables may be separated into independent, ordinary differential equations corresponding to each Fourier mode. Application of the Galerkin method to the modal equations yields banded systems of equations for the unknown Chebyshev coefficients, which may be stored and solved very efficiently.

The majority of calculations required to solve (5) are performed independently for each Fourier mode, allowing them to be distributed across multiple processors with minimal communication overhead. The exception is evaluation of the non-linear term, which involves Fourier transformations to and from the space domain.

3.4 Velocity

Evaluation of the non-linear term of (5) requires that the velocity is calculated from the vorticity at each timestep.

The Fourier modal versions of (1) and (3) may be manipulated to isolate a single velocity component as the unknown in a second order ordinary differential equation, with the remaining components explicitly related to the solution of the equation. The boundary condition for the velocity at radius r_b is not known in advance, so a "trial" solution vector $\boldsymbol{w}$ is obtained using whatever set of boundary conditions is most convenient.

The solution $\boldsymbol{w}$ satisfies the relationship $\boldsymbol{\omega} = \nabla \times \boldsymbol{w}$, although it is not necessarily equal to the velocity $\boldsymbol{u}$. The difference between $\boldsymbol{w}$ and $\boldsymbol{u}$ must be a potential flow, since a potential flow is irrotational by definition. Consequently the velocity field may be expressed as

$$\boldsymbol{u}_1 = \boldsymbol{w} + \boldsymbol{V} + \boldsymbol{v}_1 \tag{7a}$$

$$\boldsymbol{u}_2 = \boldsymbol{V} + \boldsymbol{v}_2, \tag{7b}$$

where the subscripts signify the appropriate region R_1 or R_2. $\boldsymbol{v}$ denotes a distinct potential flow velocity on each region, and $\boldsymbol{V}$ is the freestream velocity common to both regions.

Each Fourier mode of $\boldsymbol{v}$ is related to a potential function by

$$\boldsymbol{v}_{p,q}[r,\theta,z] = \hat{\boldsymbol{v}}_{p,q}[r]\exp(\mathrm{i}p\theta)\exp(\mathrm{i}\beta_q z) = \nabla\phi_{p,q}[r,\theta,z]. \tag{8}$$

The potential function can be written in the general form

$$\phi_{p,q}[r,\theta,z] = (c_1 f[r] + c_2 g[r])\exp(\mathrm{i}p\theta)\exp(\mathrm{i}\beta_q z), \quad (p,q) \neq (0,0). \tag{9}$$

The form for mode $(0,0)$ involves different functions of θ and z. The symbols f and g represent known analytical functions. The constants c need to be determined such that the overall velocity $\boldsymbol{u}$ satisfies the following boundary conditions:

- The radial component of velocity at r_a is zero, since flow through the cylinder surface is not possible.
- The velocity at r_b is continuous between regions R_1 and R_2.
- The velocity at infinite radius equals the freestream velocity $\boldsymbol{V}$.

Equations can be written for each of the above conditions, and an appropriate subset solved simultaneously to yield expressions for the constants in (9).

The potential flow velocity $\boldsymbol{v}_1$ can be expressed as a function of the boundary values of the "trial" solution $\boldsymbol{w}$. As a preprocessing step, the invariant part of $\boldsymbol{v}_1$, which includes the functions f and g, is evaluated and stored. At each timestep, the boundary values of $\boldsymbol{w}$ and hence the potential flow "correction" $\boldsymbol{v}_1$ are evaluated. The velocity $\boldsymbol{u}_1$ follows from (7a).

3.5 Vorticity Boundary Conditions

The cylindrical vector components of (5) are coupled. The components may be decoupled by use of (4) and decomposition into modal equations. Two components of each Fourier mode are obtained by solution of differential equations. The third component is explicitly dependent on the two solutions. Therefore, only two components of the vorticity boundary condition need to be specified.

Since no vorticity is allowed outside region R_1, the boundary condition at r_b is assigned the fixed value of $\boldsymbol{\omega} = \mathbf{0}$. The boundary condition at r_a is not so easily specified. Although the radial component of vorticity must be zero to satisfy the no-slip condition, boundary conditions for the tangential components of vorticity at r_a are initially unknown.

However, the solution can be obtained in several passes because of the linearity of the differential equations. As a preprocessing step, the solution of the differential equations is obtained with homogeneous initial conditions and homogeneous boundary conditions for all except one tangential component of vorticity at r_a. The vorticity and the corresponding velocity are stored. During the simulation, a "trial" solution is obtained for the differential equations with homogeneous boundary conditions and the initial conditions associated with the current timestep. The velocity field is calculated, and a "correction" is applied to the enforce the no-slip condition at the cylinder surface. The correction is achieved by addition of the preprocessed velocity and vorticity after scaling by an appropriate factor, which is determined separately for each Fourier mode.

3.6 Pressure

The pressure field is calculated as a post-processing step. The Navier-Stokes equation can be written in the non-dimensional form

$$\nabla p_0 = \boldsymbol{u} \times \boldsymbol{\omega} + \frac{1}{\mathrm{Re}} \nabla^2 \boldsymbol{u} - \frac{\partial \boldsymbol{u}}{\partial t}, \tag{10}$$

where p_0 is the total pressure. All terms on the right hand side of (10) can be determined from the simulation data. The total pressure relative to the freestream is obtained by integration of the Fourier modes of the gradient from a reference point at infinite radius. Since the potential flow field on region R_2 is an analytical function, the result of integration can be expressed in terms of analytical functions and evaluated at r_b. Inside R_1, the integral is evaluated by use of Chebyshev series.

4 Validation

A code to implement the simulation scheme has been developed. Results are presented for two fundamental flows, laminar axial flow and crossflow, both started impulsively from rest.

For the axial flow case, Fig. 1a shows the computed radial distribution of axial velocity at various times. The exact theoretical solution derived by Carslaw and Jaeger (1947) is discussed in [4]. Figure 1b shows the root mean square difference (error) between the theoretical and simulated velocity profiles for a range of times. As expected, higher order time integration produces better accuracy. Schemes with order greater than 2 are not shown because they are found to be unstable.

A notable feature of Fig. 1b is that error decreases with time over most of the time interval. The initial error is large because the vorticity field is singular

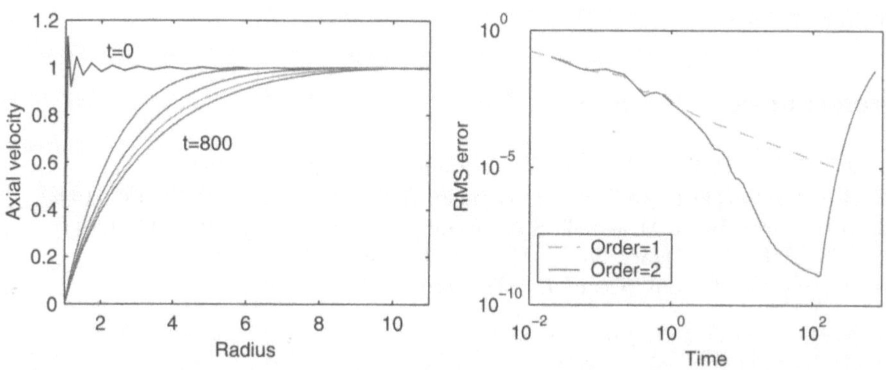

Fig. 1. Cylinder in impulsively started axial flow, $\mathrm{Re_a} = 100$ based on radius. (**a**) Radial distribution of axial velocity for $t = 0, 200, \ldots, 800$. (**b**) Root mean square error for the numerical solution computed with different orders of time integration

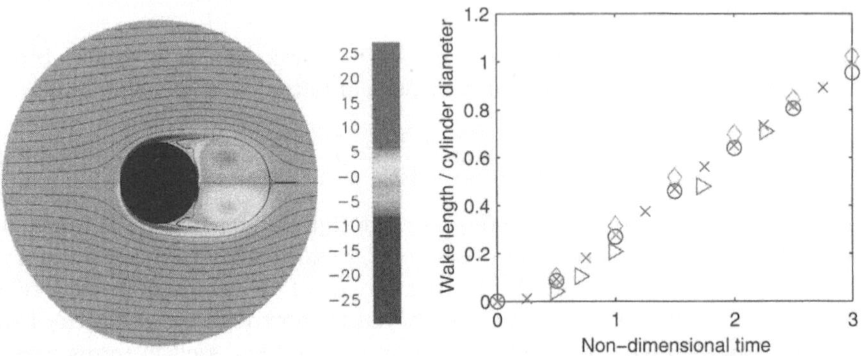

Fig. 2. Cylinder in impulsively started crossflow, $\mathrm{Re_d} = 550$ based on diameter. (**a**) Axial vorticity and streamlines at $t = 3$ with $r_b = 4$. (**b**) Evolution with time of the closed wake length. Present numerical results: $\circ$, $r_b = 4$; $\times$, $r_b = 21$. Experimental data (see [5]): $\triangleright$, Schwabe (1935); $\diamond$, Bouard and Coutanceau (1980)

at the moment of impulse, and it remains difficult to resolve for some time afterwards. As time progresses, the time integration scheme tracks the solution more accurately because the velocity profile changes more slowly. Eventually, the boundary layer grows beyond the limits of the computational domain (at $r = 11$), and the error increases accordingly.

Figure 2a shows the axial vorticity distribution and streamlines for a cylinder in crossflow. Shortly after the impulsive start a recirculation zone forms, whose extent is shown by the closed streamlines behind the cylinder. The streamlines also illustrate that the boundary condition for velocity varies with azimuth as required to accommodate the wake. Figure 2b shows that the evolution of closed wake length with time agrees reasonably well with experimental data, and that the results appear to be independent of r_b.

The initial validation results are promising, and planned future work will extend tests of the simulation procedure to three-dimensional and, ultimately, turbulent flows.

References

1. J.C. Neves, P. Moin, R.D. Moser: J. Fluid Mech. **272**, 349 (1994)
2. R.E. Luxton, M.K. Bull, S. Rajagopalan: Aeronautical Journal **88**, 186 (1984)
3. G.E. Karniadakis, M. Israeli, S.A. Orszag: J. Comput. Phys. **97**, 414 (1991)
4. G.K. Batchelor: Quart. Journ. Mech. and Applied Math. **7**, 179 (1954)
5. R. Bouard, M. Coutanceau: J. Fluid Mech. **101**, 583 (1980)

A Robust and Accurate Scheme for Solving Low Speed Flows

Meng-Sing Liou

Turbomachinery and Propulsion Systems Division, NASA Glenn Research Center at Lewis Field, Cleveland, OH 44135, USA

Abstract. In this paper, we present the ideas and procedure to extend the AUSM-family schemes for solving low speed flows. Specifically, we shall introduce the concept of numerical speed of sound to rescale the Mach number so that the numerical dissipation is uniformly valid over the entire speed regimes. Consequently, the accuracy in the low Mach number is maintained and the convergence enhanced, resulting in a convergence rate insensitive to the Mach numbers. Further, the pressure flux is enhanced by including a velocity diffusion term. As a result, the final scheme eradicate fails found in the previous versions.

1 Introduction

Today, upwind schemes undoubtedly have become a main spatial discretization technique adopted into nearly all major research and commercial codes. In the last decade, the AUSM-family[1] schemes, such as AUSM$^+$ [2,3], AUSMDV[4], and LDFSS[5] have been developed and become a formidable member of the upwind club, providing an attractive alternative scheme.

Deficiencies of using a compressible upwind code for calculating low speed flows have been known : (1) extremely slow or stalled convergence, and (2) the grossly inaccurate solutions. These two phenomena are not related because the first one originates at the continuum level and is tied to the form of governing equations being solved, irrespective of whether the scheme is centered or upwind.

The degraded convergence at low speeds is due to a disparity in eigenvalues which can be characterized by the condition number,

$$\kappa = \frac{|u| + a}{\min(|u|, |u - a|)}. \tag{1}$$

The condition number over a range of speeds as plotted in Fig. 1 reveals several important messages for the nonpreconditioned system: (1) an infinitely large condition number occurs as $M \to 0$ or 1, and (2) a minimum value of 3 is reached at $M = 0.5$ and around which reasonably (low) values are maintained.

However, the second deficiency is inherently tied to the flux scheme employed and is the subject of this paper - we will present some recent progress in developing a robust, accurate and efficient upwind scheme for low speed flows. Extensions to solving multiphase flows are possible, but its presentation is regrettably excluded due to space limitation.

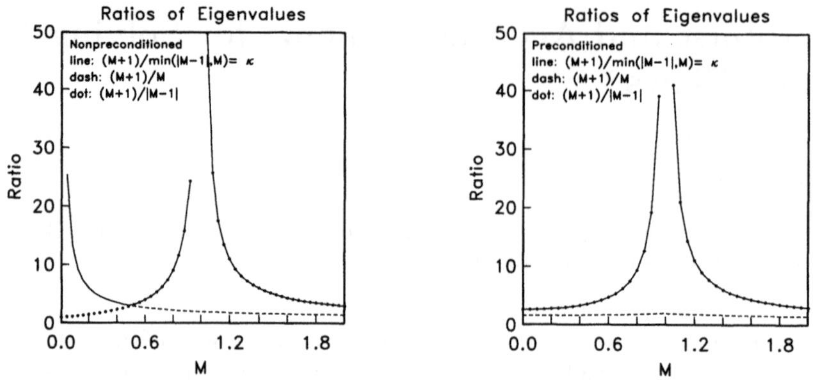

Fig. 1. Ratios of eigenvalues and condition number (shown as upper curves): (L) non-preconditioned system; (R) preconditioned system.

2 Governing Equations

A conditioning matrix $\mathbf{\Gamma}$ is applied to the time-derivative term of conservation laws,

$$\mathbf{\Gamma}\tilde{\mathbf{Q}}_t + \mathrm{div}(\mathbf{F} + \mathbf{F}_{(\mathrm{v})}) = 0, \tag{2}$$

where $\tilde{\mathbf{Q}}$ can take different forms, such as $[p \text{ or } \rho, \boldsymbol{v}, \mathrm{T}]^{\mathrm{T}}$. Several forms of the local preconditioning matrix $\mathbf{\Gamma}$ have been proposed in the literature, e.g., Refs. [6–9]. Robustness is a key issue for designing $\mathbf{\Gamma}$.

We adopted the form given by Weiss and Smith [9], which may be expressed as a rank-one perturbation of the Jacobian matrix $\partial\mathbf{Q}/\partial\tilde{\mathbf{Q}}$ ($\mathbf{Q}$ being the conservative variables). The time-derivative term in the Euler equations is replaced by

$$\mathbf{\Gamma} \equiv (\frac{\partial \mathbf{Q}}{\partial \tilde{\mathbf{Q}}} + \Theta \boldsymbol{u}\boldsymbol{v}^T), \tag{3}$$

where for example,

$$\boldsymbol{u} = [1, u, H]^T, \quad \boldsymbol{v} = [p_\rho, 0, p_T]^T, \quad \Theta = \frac{1}{a^2}(\frac{1}{M_*^2} - 1). \tag{4}$$

The eigenvalues of $\mathbf{\Gamma}^{-1}\partial\mathbf{F}/\partial\tilde{\mathbf{Q}})$ commonly have the following form:

$$u \text{ and } u^{'} \pm a^{'} = \frac{1 + M_*^2}{2}(u \pm a\frac{\sqrt{(1 - M_*^2)^2 M^2 + 4M_*^2}}{1 + M_*^2}). \tag{5}$$

The condition number under this preconditioned system is also shown in Fig. 1. Clearly shown is that this preconditioner essentially rescales the eigenvalues at the low speed range so that the condition number remains bounded. This allows uniform convergence in both low-speed and high-speed flows, see later in Fig. 4.

3 Numerical Flux: AUSM-Family

In this section, we briefly review the basic AUSM scheme, but emphasize instead on the devices required for low speed flows. The reader should consult with previous publications for other details.

As a first step common in the AUSM schemes, we explicitly split the inviscid flux (written in one space-dimension without loss of generality) into the *convective* and *pressure* fluxes:

$$\mathbf{F} = \mathbf{F}^{(c)} + \mathbf{P} = \overset{\bullet}{m}\psi + \mathbf{P}, \tag{6}$$

where

$$\overset{\bullet}{m} = \rho u, \quad \psi = (1, u, H)^T, \quad \mathbf{P} = (0, p, 0)^T. \tag{7}$$

It is possible to write a numerical flux, mimicking the expression at the continuum level, in terms of a common mass flux in a general upwind form.

$$\mathbf{f}_{1/2} = \overset{\bullet}{m}_{1/2}\, \psi_{L/R} + \mathbf{p}_{1/2} = \overset{\bullet}{m}_{1/2} \begin{cases} \psi_L, & \text{if } \overset{\bullet}{m}_{1/2} > 0, \\ \psi_R, & \text{otherwise,} \end{cases} + \mathbf{p}_{1/2}. \tag{8}$$

In what follows, we shall separately describe the mass and pressure fluxes, $\overset{\bullet}{m}_{1/2}$ and $p_{1/2}$.

3.1 Mass Flux

The mass flux is simply determined by upwinding:

$$\overset{\bullet}{m} = \tilde{a}_{1/2} M_{1/2} \begin{cases} \rho_L, & \text{if } M_{1/2} > 0, \\ \rho_R, & \text{otherwise.} \end{cases} \tag{9}$$

Clearly, the main tasks are to define the interface quantities $\tilde{a}_{1/2}$ and $M_{1/2}$.

To properly scale the numerical dissipation with the flow speed, we propose the concept of numerical speed of sound [10]. The numerical speed of sound is conveniently expressed in terms of a scaling function f as:

$$\tilde{a}_{1/2} = f(\bar{M}; M_*) a_{1/2}, \quad a_{1/2} = (a_L + a_R)/2 \ \text{for example,} \tag{10}$$

where the scaling factor may take the form as suggested in Eq. (5),

$$f(\bar{M}; M_*) = \frac{\sqrt{(1 - M_*^2)^2 \bar{M}^2 + 4M_*^2}}{1 + M_*^2}, \quad \bar{M}^2 = (M_L^2 + M_R^2)/2, \tag{11}$$

with the reference Mach number,

$$M_*^2 = \min(1, \max(\bar{M}^2, M_{co}^2)). \tag{12}$$

We note that some other suitably chosen forms for f may be appropriate. The cutoff parameter M_{co} is nominally specified as $M_\infty/2$.

Using the numerical speed of sound $\tilde{a}_{1/2}$ to define

$$\tilde{M}_{L/R} = \frac{1}{2}(u_L + u_R)/\tilde{a}_{1/2} \pm \frac{M_*^2}{2}(u_L - u_R)/\tilde{a}_{1/2}. \tag{13}$$

A simpler version is to replace the coefficient M_*^2 by a unity, yielding fully one-sided formulas. It is found that these blended formulas yield a less diffusive and more robust solution at low Mach numbers.

We can now set the interface Mach number as:

$$M_{1/2} = \mathcal{M}_{(m)}^{+}(\tilde{M}_L) + \mathcal{M}_{(m)}^{-}(\tilde{M}_R) + M_p. \tag{14}$$

The split Mach number $\mathcal{M}_{(m)}^{\pm}$ are polynomial functions of $\tilde{M}_{L/R}$ of degree m, see [3] for detailed expressions. The pressure diffusion term M_p, introduced to enhance calculations of low Mach number and multiphase flows, is suggested as:

$$M_p = -(\frac{1}{M_*^2} - 1)\Delta\mathcal{M}\frac{p_R - p_L}{p_R + p_L}, \tag{15}$$

where

$$\Delta\mathcal{M} = \left[\mathcal{M}_{(4)}^{+}(\tilde{M}_L) - \mathcal{M}_{(1)}^{+}(\tilde{M}_L) - \mathcal{M}_{(4)}^{-}(\tilde{M}_R) + \mathcal{M}_{(1)}^{-}(\tilde{M}_R)\right] \geq 0. \tag{16}$$

As a result of using the numerical speed of sound, the accuracy is restored as it is applied to low Mach flows, as evident in Fig. 2.

Fig. 2. Pressure contours for the shuttle external tank problem for $M_\infty = 0.01$: (L) using the standard AUSM$^+$ at N=6400 time steps; (R) using numerical speed of sound at N=1000 time steps.

3.2 Pressure Flux

In all the AUSM-family schemes, a general Van Leer/Liou - type [3] interface pressure is used as a starting point,

$$p_{1/2} = \mathcal{P}_{(n)}^{+}(\tilde{M}_L)\, p_L + \mathcal{P}_{(n)}^{-}(\tilde{M}_R)\, p_R, \tag{17}$$

where n = 1, 3, or 5 corresponds to the polynomial choices.

Recasting the AUSMDV[4] scheme, we find that the blended momentum flux gives rise to a term proportional to the velocity difference. This interpretation,

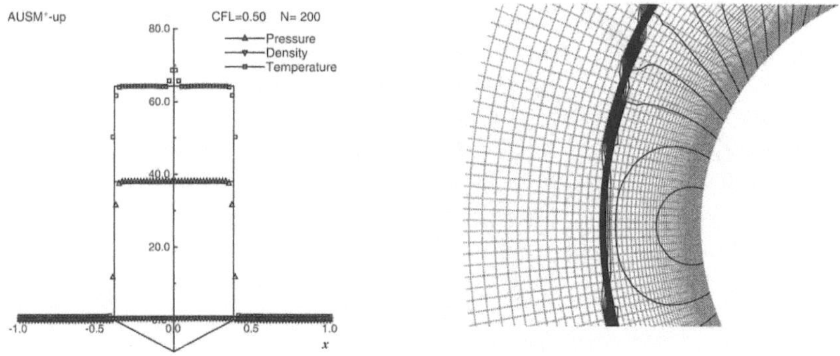

Fig. 3. Effect of the new pressure flux: (L) colliding shocks, $M_L = -M_R = 15$; (R) blunt body problem, $M_\infty = 10$.

however, has a connection to the characteristic equations $dp \pm \rho a du = 0$. Hence, we can beef up the pressure flux with the velocity difference term,

$$
\begin{aligned}
p_{1/2} = & \mathcal{P}^{+}_{(5)}(\tilde{M}_L)p_L + \mathcal{P}^{-}_{(5)}(\tilde{M}_R)p_R \\
& -\mathcal{P}^{+}_{(5)}(\tilde{M}_L)\mathcal{P}^{-}_{(5)}(\tilde{M}_R)\rho_{1/2}\tilde{a}_{1/2}(u_R - u_L),
\end{aligned}
\tag{18}
$$

where $\rho_{1/2} = (\rho_L + \rho_R)/2$ for example.

It's interesting to note that the velocity diffusion is the counterpart of the pressure diffusion in Eq. (14) To denote this new version first appeared in Ref. [1]), we use the suffix "u" (for velocity diffusion) and call it AUSM$^+$-u. Thus, the scheme reads AUSM$^+$-up when M_p is also included in Eq. (14), and AUSM$^+$-aup if the numerical speed of sound is also activated as well.

While the above development is meant for low speed flow, it however has added benefits to the high speed flows as well. The new pressure flux AUSM$^+$-aup scheme completely removes the overshoots, as displayed in Fig. 3.

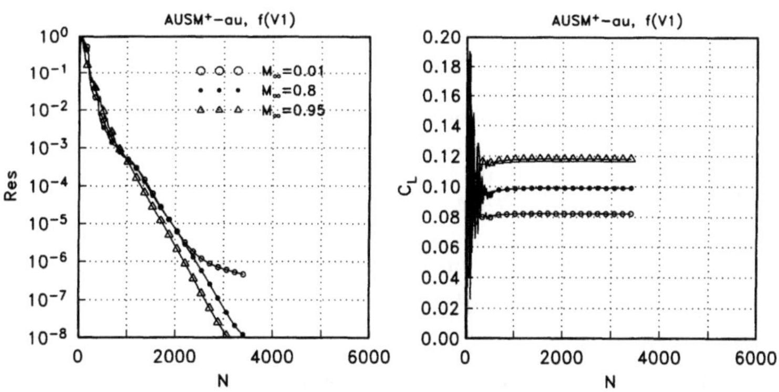

Fig. 4. Convergence history.

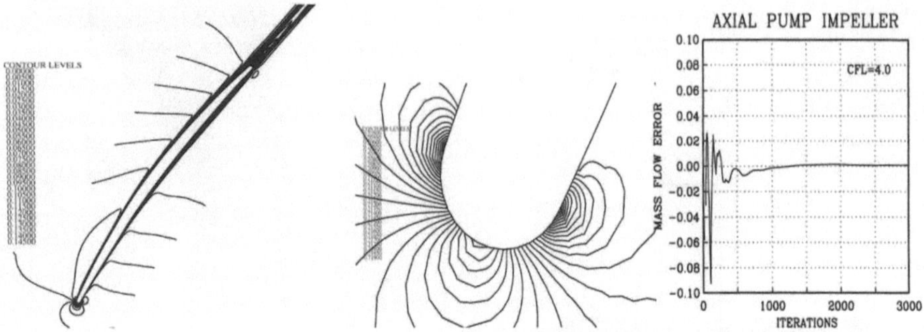

Fig. 5. Low speed pump.

Shown also in Fig. 3 is the pressure contours over a blunt body at a hypersonic speed. Not only is the solution free of carbuncle phenomena, but also the pressure contours are smooth, specifically near the stagnation and sonic line regions.

Moreover, Fig. 4 clearly demonstrates that the convergence rate, in terms of residual and C_L, is nearly equal for all Mach numbers, meeting the objective of preconditioning. This insensitivity holds true (not shown) for different cutoff Mach numbers and time step sizes. However, dependence on time step sizes still exists for the explicit scheme.

For a turbomachinery application, we present turbulent flow results for a low speed pump impeller. Figure 5 shows the overall flow features via Mach number contours, blowup of pressure contours near the leading stagnation point and the mass error history. As above, smooth pressure contours are achieved, together with very small mass error, $O(10^{-5})$ on a 177×49 grid.

4 Concluding Remarks

Modifications of the AUSM-family schemes for calculating flows at all speeds have been outlined in this paper. Numerical results have confirmed shown that the modifications based on numerical speed of sound are effective, yielding accurate solutions at convergence rate rather insensitive to Mach numbers.

References

1. Liou, M.-S., 15th AIAA CFD Conference, Paper 2001-2521-CP, June 11-14 2001.
2. Liou, M.-S. and Steffen, Jr., C. J., *JCP*, 107, 1993, pp. 23–39.
3. Liou, M.-S., *JCP*, 129, 1996, pp. 364–382.
4. Wada, Y. and Liou, M.-S., *SIAM J Scientific and Statistical Computing*, 18, 1997, pp. 633–657.
5. Edwards, J. R., *Computers & Fluids*, 26, 1997, pp. 635–659.
6. Turkel, E., *JCP*, 72, 1987, pp. 277–298.
7. van Leer, B., Lee, W. T., and Roe, P. L., 10th AIAA CFD Conference, Paper 91-1552, June 1991.
8. Choi, Y. H. and Merkle, C. L., *JCP*, 105, 1993, pp. 207–223.
9. Weiss, J. M. and Smith, W. A., AIAA Paper 94-2209, June 1994.
10. Liou, M.-S. and Edwards, J. R., 14th AIAA CFD Conference, 99-3268-CP, 1999.

Part IX

Kinetic Schemes

Implicit Kinetic Schemes for the Euler and the Ideal Magnetohydrodynamics Equations

Ramesh K. Agarwal[1] and H.S.R. Reksoprodjo[2]

[1] Department of Mechanical Engineering,
Washington University, St. Louis, MO 63130, U.S.A.
[2] National Institute for Aviation Research,
Wichita State University, Wichita, KS 67260, U.S.A.

Abstract. The implicit kinetic schemes, namely the Kinetic Flux-Vector Split (KFVS) and the Kinetic Wave/Particle Split (KWPS) schemes are derived for the Euler and the ideal MHD equations. The schemes are applied to compute the flow in a 1-D shock-tube and the 2-D flowfield due to a cylindrical explosion with and without the magnetic field. For the ideal 2-D MHD equations, the homogeneity of the flux-vectors is achieved by employing the approach due to MacCormack, and a Poisson solver is used at every time-step to enforce the solenoidal condition on the magnetic field. To the authors' knowledge, this is the first time that the implicit kinetic schemes have been formulated for the Euler and the ideal MHD equations.

1 Introduction

In recent years, there has been a considerable interest in the kinetic schemes for solving the Euler[1,7], Navier-Stokes[3], and MHD[9,11] equations. However, to date, all the papers on kinetic schemes reported in the literature employ an explicit formulation due to its simplicity. The major drawback of the explicit formulation is the restriction placed on the time-step allowed for the stability of the scheme.

Kinetic schemes are based on the fact that the set of equations governing the motion of fluid flows at the continuum level, i.e., Euler, Navier-Stokes, and Burnett equations, can be obtained by taking the moments of the Boltzmann equation at the molecular level with respect to the collision invariants and the appropriate distribution function. One of the well-known and extensively used kinetic schemes, the Kinetic Flux-Vector Splitting (KFVS) scheme proposed by Mandal & Deshpande[7], splits the flux term in the Boltzmann equation into positive and negative parts based on the sign of the molecular velocity. Taking the moments of this split-flux with respect to the collision invariant vector results in the KFVS algorithm. This scheme, however, requires the evaluation of the computationally expensive error functions. Recently, Agarwal & Acheson[1] have proposed a new flux splitting at Boltzmann level, which they call Kinetic Wave/Particle Splitting (KWPS). They have shown that this new splitting does not lead to the evaluation of error functions, which results in increased computational speed and also simplifies the algebra considerably. By recognizing that the molecular velocity of an individual gas particles can be expressed as the sum of the average fluid velocity of the gas and each particle's thermal (peculiar) velocity, the Boltzmann flux can be split into two components: the convective part and the acoustic part. These two parts of the Boltzmann flux are then upwind discretized and the moments of this discretized equation with the collision invariants then result in the KWPS scheme for the Euler equations. Both the KFVS and KWPS schemes have been investigated using explicit formulations. Only recently, for the first time, implicit formulations of these two kinetic schemes were systematically derived by Reksoprodjo & Agarwal[10] for the Euler equations.

For the ideal 1-D MHD equations, the explicit kinetic schemes have been formulated by Croisille et al[4], Reksoprodjo & Agarwal[9], Tang & Xu[11], and Huba & Lyon[5]. However, the derivation of implicit kinetic schemes for MHD equations requires special consideration due to the nonhomogeneity of the MHD flux-vectors with respect to the state vector. By introducing a "*dummy*" equation, MacCormack[6] has been successful in recovering the homogeneity of the flux-vectors. Using this new expanded system of equations, he derived an implicit scheme based on the Steger-Warming flux-splitting algorithm. It should be noted that Powell's *magnetic divergence wave*[8] is inherently included in this formulation. This methodology has been applied in this paper to derive the implicit formulation of the kinetic schemes for the ideal MHD equations. A Poisson solver is used at every time-step to enforce solenoidal condition on the magnetic field.

2 Implicit Kinetic schemes for the Euler equations

For a gas in a state of collisional equilibrium the collision integral vanishes and the solution of the Boltzmann equation is the Maxwellian probability density distribution function. The Boltzmann equation can be written as

$$\frac{\partial (nf)}{\partial t} + v_i \frac{\partial (nf)}{\partial x_i} = 0 \tag{1}$$

The Maxwellian distribution function is given by

$$f = \frac{1}{\epsilon_0} \exp\left(-\frac{\epsilon}{\epsilon_0}\right) \left(\frac{\beta}{\pi}\right)^{3/2} \exp\left(-\beta c_k^2\right) \tag{2}$$

When moments of the Boltzmann equation (1) with the Maxwellian distribution function (2) are taken with respect to the collisional invariant vector

$$\boldsymbol{\Psi} = \left[m \;\; mv_j \;\; m\epsilon + \tfrac{1}{2}mv_k^2\right]^T,$$

the Euler equations are obtained.

In equations (1) and (2), $\boldsymbol{v}$ is the molecular velocity, $\boldsymbol{c}$ is the thermal velocity, the fluid velocity $\boldsymbol{u} = \boldsymbol{v} - \boldsymbol{c}$, β is the equivalent temperature ($\equiv \frac{1}{2RT}$), and ϵ and $\epsilon_o \equiv \left(\frac{1}{\gamma-1} - \frac{3}{2}\right) RT$ are the internal energy and average internal energy respectively. Also, $\rho \equiv mn$.

To obtain the implicit kinetic scheme for the Euler equations, an implicit algorithm for the Boltzmann equation is formulated. The flux terms are linearized as follows:

$$(v_i nf)^{t_0+\Delta t} \approx (v_i nf)^{t_0} + \Delta t \frac{\partial (v_i nf)}{\partial \mathbf{Q}} \frac{\partial \mathbf{Q}}{\partial t} \approx v_i (nf)^{t_0} + v_i \frac{\partial (nf)}{\partial \mathbf{Q}} \Delta \mathbf{Q}$$

The flux Jacobians are then obtained as

$$\mathbf{A}_i = \left\langle v_i \frac{\partial (nf)}{\partial \mathbf{Q}} \right\rangle$$

where $\mathbf{Q}$ is the state vector of the Euler equations ($\left[\rho\ \rho u_j\ \rho e_t\right]^T$), with e_t defined as the total energy per unit mass.

The KFVS methodology first splits the molecular velocity $\boldsymbol{v}$ into positive and negative spaces, and then takes the moments with respect to the collision invariant vector. Equation (1) is written as:

$$\frac{\partial f}{\partial t} + \frac{\partial}{\partial x_i} \left(\tfrac{v_i \pm |v_i|}{2} f \right) = 0 \tag{3}$$

On the other hand, the KWPS methodology first expresses the molecular velocity as the sum of the *average* fluid velocity and each particle's thermal velocity ($\boldsymbol{v} = \boldsymbol{u} + \boldsymbol{c}$), so that equation (1) becomes:

$$\frac{\partial f}{\partial t} + \frac{\partial}{\partial x_i} \left(\tfrac{u_i \pm |u_i|}{2} f + \tfrac{c_i \pm |c_i|}{2} f \right) = 0 \tag{4}$$

After taking the moments of equations (3) and (4), the split-flux Jacobians are obtained. For KFVS scheme, the split-flux Jacobians are obtained as

$$\mathbf{A}_\xi^\pm = \frac{\|\xi\|}{\mathcal{J}} \left[\tfrac{1 \pm \mathrm{erf}\left(u_\xi \sqrt{\beta}\right)}{2} \left(u_\xi \mathbf{I} + \mathbf{B}_\xi^1\right) \pm \tfrac{\exp\left(-u_\xi^2 \beta\right)}{2\sqrt{\pi\beta}} \left(\mathbf{I} + \mathbf{B}_\xi^2\right) \right] \tag{5}$$

and for KWPS scheme they are

$$\mathbf{A}_\xi^\pm = \frac{\|\xi\|}{\mathcal{J}} \left[\tfrac{u_\xi \pm |u_\xi|}{2} \mathbf{I} + \tfrac{1}{2} \mathbf{B}_\xi^1 \pm \tfrac{1}{2\sqrt{\pi\beta}} \left(\mathbf{I} + \mathbf{B}_\xi^2\right) \right] \tag{6}$$

where

$$\mathbf{B}_\xi^1 = \begin{bmatrix} -u_\xi & \hat{\xi}_l & 0 \\ -u_\xi u_j + \frac{1}{2} u_k^2 \tilde{\gamma} \hat{\xi}_j & u_j \hat{\xi}_l - u_l \tilde{\gamma} \hat{\xi}_j & \tilde{\gamma} \hat{\xi}_j \\ -u_\xi h_t + \frac{1}{2} u_k^2 \tilde{\gamma} u_\xi & h_t \hat{\xi}_l - u_l \tilde{\gamma} u_\xi & \tilde{\gamma} u_\xi \end{bmatrix}, \mathbf{B}_\xi^2 = \begin{bmatrix} 0 & 0 & 0 \\ 0 & 0 & 0 \\ \frac{1}{4} \tilde{\gamma} u_k^2 & -\frac{1}{2} \tilde{\gamma} u_l & \frac{1}{2} \tilde{\gamma} \end{bmatrix}$$

Note that $\tilde{\gamma} \equiv \gamma - 1$, $\mathbf{I} \equiv$identity matrix, $\xi \equiv$generalized coordinate with $\|\xi\| \equiv \sqrt{\xi_k^2}$ and $\hat{\xi} \equiv \frac{\xi}{\|\xi\|}$, $\mathcal{J} \equiv$Jacobian of transformation from Cartesian system to generalized coordinate system, and $h_t \equiv$total enthalpy per unit mass.

3 Kinetic schemes for the ideal MHD equations

By adding a "*dummy*" equation ($\partial_t \hat{a} = 0$) MacCormack[6] has obtained the homogeneity of the ideal MHD flux-vectors, achieving the construction of an implicit upwind split-flux algorithm to be constructed. Since there is no distribution function available from which the complete *8-wave* MHD system can be obtained using the *moment method strategy*, the implicit kinetic schemes for the ideal MHD equations are obtained directly from the explicit formulations. For the KFVS scheme this process is very straightforward. Its split-flux-vectors can be written as

$$\mathbf{F}_\xi^\pm = \frac{\|\xi\|}{\mathcal{J}} \left[\frac{1 \pm \mathrm{erf}\left(u_\xi \sqrt{\beta}\right)}{2} \left(u_\xi \mathbf{G}_\xi^0 + \mathbf{G}_\xi^1\right) \pm \frac{\exp\left(-u_\xi^2 \beta\right)}{2\sqrt{\pi\beta}} \mathbf{G}_\xi^2 \right] \tag{7}$$

The split-flux Jacobians for the KFVS scheme can be easily obtained from the above split-flux-vectors following the procedure employed for the Euler equations. The split-flux-vectors for the KWPS scheme are obtained as

$$\mathbf{F}_\xi^\pm = \frac{\|\xi\|}{\mathcal{J}} \left[\frac{u_\xi \pm |u_\xi|}{2} \mathbf{G}_\xi^0 + \frac{1}{2}\mathbf{G}_\xi^1 \pm \frac{1}{2\sqrt{\pi\beta}} \mathbf{G}_\xi^2 \right] \tag{8}$$

The individual vectors are

$$\mathbf{G}_\xi^0 = \begin{bmatrix} \rho \\ \rho u_j \\ \rho e_t \\ B_j - \left(\frac{B_\xi}{\hat{a}}\right) \hat{\xi}_j \hat{a} \\ 0 \end{bmatrix}, \quad \mathbf{G}_\xi^1 = \begin{bmatrix} 0 \\ \hat{\xi}_j P_o - \left(\frac{B_\xi}{\hat{a}}\right) B_j \\ u_\xi P_o - \left(\frac{B_\xi}{\hat{a}}\right) u_k B_k \\ -\left(\frac{B_\xi}{\hat{a}}\right)\left(u_j - \hat{\xi}_j u_\xi\right) \hat{a} \\ 0 \end{bmatrix}$$

$$\mathbf{G}_\xi^2 = \mathbf{G}_\xi^0 + \begin{bmatrix} 0 & 0 & \frac{1}{2}P_o - \left(\frac{B_\xi}{\hat{a}}\right)\frac{1}{2}B_\xi & 0 & 0 \end{bmatrix}^T$$

Again, the split-flux Jacobians for the KWPS scheme are obtained by following the procedure employed for the Euler equations. In the above equations, $\boldsymbol{B}$ is the magnetic field, $P_o \equiv p + \frac{1}{2}B_k^2$ is the total pressure, and $\hat{a} \equiv 1$ is the "*dummy*" variable. Also, β is redefined to include magnetic field contribution as $\frac{\rho}{2P_o}$. Note that when computing the split-flux Jacobians, the terms involving $\frac{\partial}{\partial \mathbf{Q}}\left(\frac{B_\xi}{\hat{a}}\right)$ can be dropped off since they are homogeneous of degree zero.

4 Numerical results and discussions

The performance of implicit kinetic schemes is evaluated by computing a number of benchmark test cases in 1 and 2 dimensions. In particular, the Euler calculations are performed for 1-D shock-tube and the MHD computations are performed for 1-D MHD shock-tube[2]. In 2-D both Euler and MHD computations are performed for flow field due to a cylindrical explosion. Results are compared with analytical results and solutions from explicit kinetic schemes. It is shown that the implicit kinetic formulations retain the efficiency and robustness of their explicit counterparts without the restrictive time step constraints. This results in an increase of computational speed for steady state computations.

The Euler and MHD shock-tube test cases are computed over a grid of 2000 points divided equally, with density and pressure ratios of 8 and 10 respectively. A time step of $\Delta t = 0.4\,(\Rightarrow CFL \approx 0.877)$ is used in the explicit schemes, and a value of $\Delta t = 8.0$ is used in the implicit schemes. Final solution is obtained at $t = 400$. The initial conditions for the MHD shock-tube test case are the same as in the Euler case, with the addition of a constant longitudinal magnetic field value of $B_x = \frac{3}{4}$, and a tranverse magnetic field having the value of $B_y = +1$ in the high density/pressure region and $B_y = -1$ in the low density/pressure region. Final solution is obtained at $t = 80$. For the Euler computations $\gamma = \frac{7}{5}$, and for the ideal MHD computations $\gamma = 2$.

Figure 1 shows the computations for the Euler shock-tube and figure 2 shows the computations for the ideal MHD shock-tube using the implicit KFVS scheme. Both are obtained using the implicit KFVS scheme

In 2-D, the schemes are employed to compute the flowfield due to a cylindrical explosion with and without the magnetic field. The details of this test case can be found in Tang & Xu[11]. The grid is a 100×100 Cartesian grid. The fluid is initially at rest, with a central circular region $r \leq 0.1L$ of high pressure $p_{\text{high}} = 100 p_{\text{low}}$. The applied magnetic field (case #2) is $B_y = \frac{10}{\sqrt{\pi}}$. Time step is $\Delta t = 0.01$ with final solution is computed at $t = 3.0$. Solutions are obtained using the KWPS scheme.

Figure 3 shows the thermal pressure contours for case #1, and figure 4 show the thermal pressure contours for case #2. Figures 5 and 6 show the magnetic pressure contours and field lines for case #2. From these figures it is clear that the presence of the magnetic field can greatly affect the flow field. Instead of a circular shock front obtained from the Euler computations, the shock front becomes elongated along the direction of the magnetic field while the fluid motion in the normal direction is suppressed. These results show good agreement with those of Tang & Xu[11].

5 Conclusion

Implicit kinetic schemes have been developed for the Euler and the ideal MHD equations. Numerical tests show that the implicit schemes retain accuracy and robustness of the explicit schemes while increasing their efficiency by reducing the total computation time.

References

1. Agarwal, R. K. and Acheson, K. E. (1995) *AIAA 95-2178*
2. Brio, M. and Wu, C. C. (1988) J. Comp. **75**, 400–422
3. Chou, S. Y. and Baganoff, D. (1997) J. Comp. Physics **130**, 217–230

4. Croisille, J.-P., Khanfir, R., and Chanteur, G. (1995) J. Sci. Comput. **10**, 481–492
5. Huba, J. D. and Lyon, J. G. (1998) J. Plasma Physics **61**, 391–405
6. MacCormack, R. W. (1999) *AIAA 99-3609*
7. Mandal, J. C. and Deshpande, S. M. (1994) Computers and Fluids **23**, 447–478
8. Powell, K. G. (1994) *ICASE Report 94-24*
9. Reksoprodjo, H. S. R. and Agarwal, R. K. (2000) *AIAA 2000-0448*
10. Reksoprodjo, H. S. R. and Agarwal, R. K. (2001) *AIAA 2001-2629*
11. Tang, H.-Z. and Xu, K. (2000) J. Comp. Physics **165**, 69–88

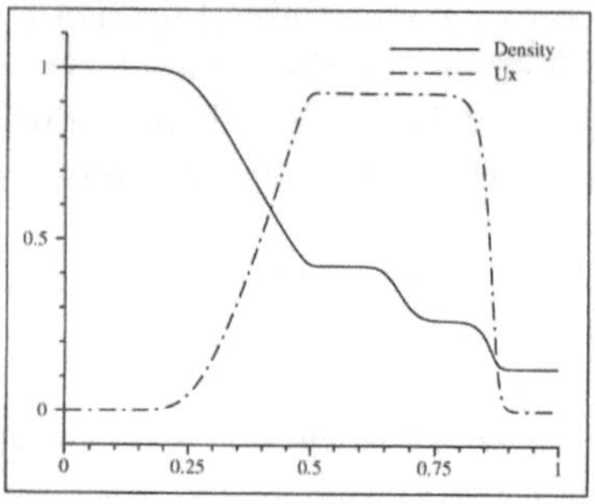

Fig. 1. Euler shock-tube

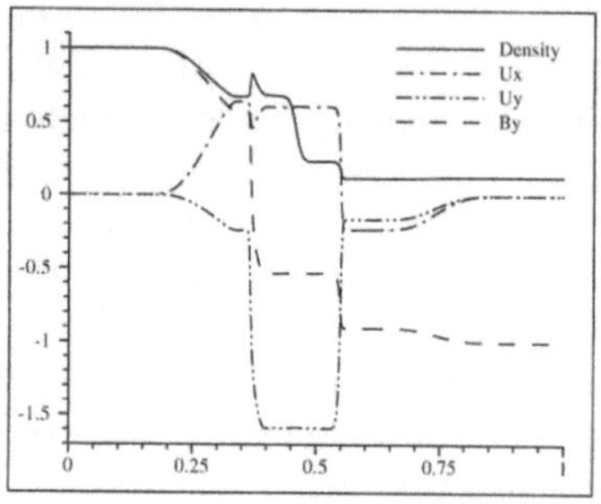

Fig. 2. MHD shock-tube

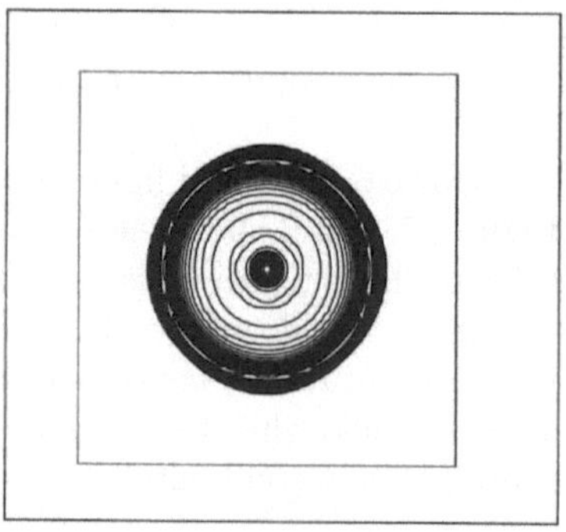

Fig. 3. Thermal pressure (case #1)

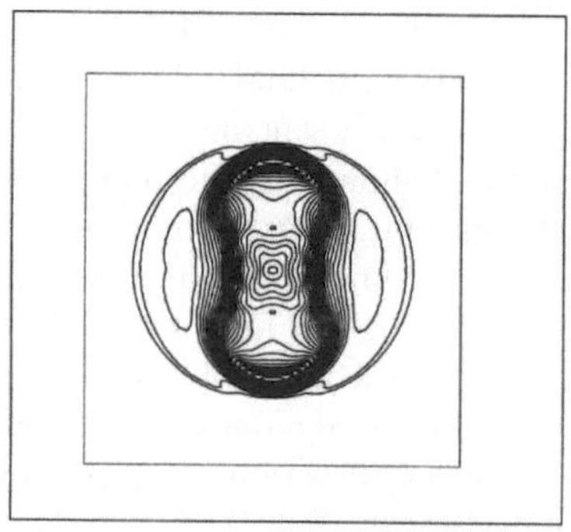

Fig. 4. Thermal pressure (case #2)

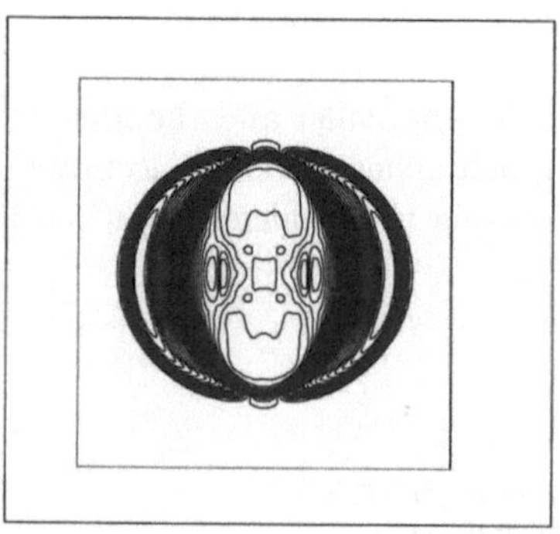

Fig. 5. Magnetic pressure (case #2)

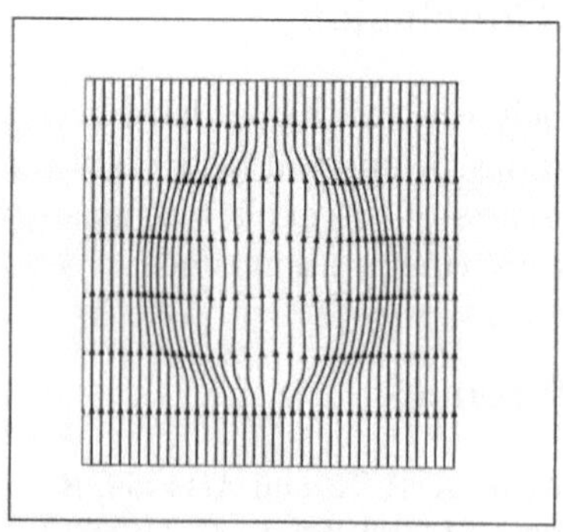

Fig. 6. Magnetic field (case #2)

3-D KFMG Euler Computations for Unsteady Flows Around Oscillating Geometries

R. Krishnamurthy[1], B.S. Sarma[1], and S.M. Deshpande[2]

[1] Defence Research & Development Laboratory, Hyderabad, India
[2] CFD Center, Indian Institute of Science, Bangalore, India

Abstract. A three dimensional Euler solver based on Kinetic Flux Vector Splitting scheme on moving grids(KFMG) is developed. The grid moving algorithm is formulated to tackle moving boundaries based on spring analogy. This KFMG solver has been enhanced with fourth order Runge-Kutta method and high resolution scheme for more time and space accuracies. The boundary conditions to tackle moving inner boundary and fixed outer boundary have also derived. The solver has been applied to a harmonically, oscillating ogive-cylinder-ogive configuration in transonic flow and the unsteady aerodynamic characteristics are presented here.

1 Introduction

Analysis of unsteady flows is a very challenging topic of research. A decade ago, potential flow equations were used to predict unsteady pressures on oscillating bodies. Recognizing the fact that nonlinear aerodynamic formulations are essential to accurately analyze unsteady flows, particularly in transonic and supersonic flows, Euler equations on moving grids have emerged as important CFD tools for unsteady aerodynamics. Numerical solution of these equations based on different upwind schemes such as the ones due to van Leer and Roe have been obtained respectively by J.T. Batina[1] and Ambroci et al.[2] recently. In the present work, Euler computations based on yet another upwind scheme namely Kinetic Flux Vector Splitting (KFVS) scheme due to Deshpande and Mandal[3] which has been proven for its robustness for steady flows, has been extended to formulate a time accurate Euler solutions for problems with moving boundaries(KFMG). The KFVS scheme has inherent simplicity in splitting the flux even on moving grids due to the underlying particle model. 3-D formulations using Kinetic Flux Vector Splitting scheme on Moving Grids (KFMG) have been developed here. The development of a solver(KFMG) is highly useful in dealing with problems of unsteady aerodynamics and computational aeroelasticity[4,5]. The necessary boundary conditions to tackle moving inner boundary and fixed outer boundary have also been developed. The grid movement around the moving inner boundary has been accomplished using a technique called 'spring analogy' which is successfully implemented for 2-D geometries[6].

Numerical results are presented for an ogive-cylinder-ogive configuration for steady flows as well as unsteady flows. The results for steady flows are compared with another Euler solver(VEEMA) and grid variation studies are also presented. Numerical results for unsteady flows compare well with published data.

2 Kinetic Flux vector splitting scheme on Moving Grids (KFMG)

Different upwind schemes are in use to solve Euler equations and in the present study, Kinetic Flux Vector Splitting (KFVS) scheme on stationary grids has been extended for moving grids. The KFVS solver essentially addresses Boltzmann equations of Kinetic theory of gases and is transformed into Euler equations through a moment method strategy. Extensive studies are carried out using 2-D and 3-D formulations of KFVS scheme in different flow regimes. The 1-D KFMG solver based on KFVS scheme developed by Kulkarni and Deshpande[5] revealed that the KFVS scheme has inherent simplicity, especially when applied to moving grid problems. The simplicity stems for the fact that the molecular velocity term present in the governing equations at the Boltzmann level is replaced by the term involving the difference between the molecular velocity and grid velocity. The above aspect has been exploited in deriving the fluxes for moving grids. For 2-D case, the details are presented in Ref.[6]. It is interesting to note that, the split normal fluxes $GN^{\pm}_{MG}$ corresponding to moving grids can be cast in terms of the split normal fluxes $GN^{\pm}_{SG}$ corresponding to stationary grids and the same can be seen from.

$$\begin{pmatrix} GN^{(1)^{\pm}}_{MG} \\ GN^{(2)^{\pm}}_{MG} \\ GN^{(3)^{\pm}}_{MG} \\ GN^{(4)^{\pm}}_{MG} \\ GN^{(5)^{\pm}}_{MG} \end{pmatrix} = \begin{pmatrix} 1 & 0 & 0 & 0 & 0 \\ w_1 & 1 & 0 & 0 & 0 \\ w_2 & 0 & 1 & 0 & 0 \\ w_3 & 0 & 0 & 1 & 0 \\ \frac{w_1{}^2 + w_2{}^2 + w_3{}^2}{2} & w_1 & w_2 & w_3 & 1 \end{pmatrix} \begin{pmatrix} GN^{(1)^{\pm}}_{SG} \\ GN^{(2)^{\pm}}_{SG} \\ GN^{(3)^{\pm}}_{SG} \\ GN^{(4)^{\pm}}_{SG} \\ GN^{(5)^{\pm}}_{SG} \end{pmatrix}$$

where, w_1, w_2 and w_3 represent the grid velocities. From this matrix relation it is obvious that, $GN^{\pm}_{MG}$ is related to $GN^{\pm}_{SG}$ through a matrix dependent only on grid velocity at that point. The grid velocity w_1, w_2 and w_3 at each grid point during the (n+1) time step are calculated using the location of the point in n and (n+1) time steps.

3 Boundary conditions

The flow tangency condition on the moving boundaries requires that a fluid particle on the moving boundary will have normal velocity equal to that of moving surface. In the present work, the boundary conditions are derived on moving and fixing boundaries for finite volume formulation using the concepts of kinetic theory of gases. The kinetic treatment of moving wall boundary condition(KMBC) and far-field boundary condition(KOBC) use specular reflection model of kinetic theory of gases. The KOBC is so general that it can be imposed on far-field boundary which can be subsonic or supersonic.

4 Dynamic Mesh Algorithm

Between two successive time steps, the body changes its orientation in case of oscillation and hence the grid has to be moved or regenerated around the deformed body. Regeneration of grid at each time step is not only laborious but also expensive with respect to CPU time of the computer. Hence, a moving grid strategy given by J.T.Batina[1] based on 'spring analogy' is implemented to move the grid so as to get a good quality grid around the oscillating body. The mesh is moved to conform to the instantaneous position of the body by modeling every edge of each cell by a spring. The spring stiffness for a given edge i-j is taken to be inversely proportional to the length of the edge. Grid points on the outer boundary are held fixed, and the instantaneous locations of the points on the inner boundary (body) are prescribed. At each time step, the static equilibrium equations in the x,y and z directions are solved iteratively at each interior node ijk of the grid for the displacements $\delta_{x_{ijk}}, \delta_{y_{ijk}}$ and $\delta_{z_{ijk}}$. The spring analogy method is coded to obtain the grid around the new orientation of the oscillating airfoil at every time level.

5 Results and Discussions

The capabilities of the dynamic mesh algorithm are demonstrated by considering the grid around ogive-cylinder-ogive configuration. The initial grid is generated by elliptic grid generator ELGRID- TM based on Thomas-Middelcoff procedure[7]. In this example, the geometry is pitched about the center point with an amplitude of 5° and the mesh at a pitch angle of -5° and $+5^o$. The mesh moves smoothly as the configuration pitches, and the procedure is completely general in the sense that it can treat any realistic oscillations.

In the present work, a 3-D cell-centered KFMG Euler code has been developed to operate on structured grids. Fourth order Runge-Kutta method has been incorporated in the code for time accuracy and high resolution scheme has been used for spatial accuracy. Initially, the code is validated against standard shapes such as, transonic flow past a hemisphere cylinder and ONERA M6 wing in steady flows. This code has also been used for predicting hypersonic flow past blunt cone-cylinder-flare configuration for which experimental data is available. The KFMG Euler code has been then applied to predict steady flow around ogive-cylinder-ogive configuration with elliptical cross section(Fig.1). Table-1 shows the aerodynamic coefficients obtained using KFMG code as well as using van Leer method based Euler code VEEMA for this configuration. For the finest grid namely, 90x51x51, the aerodynamic coefficients compare well.

The 3-D KFMG Euler code has been then applied for the ogive-cylinder-ogive configuration oscillating in transonic flow to analyze the unsteady flow behavior. Spring analogy method has been used for generating grids around 3-D moving geometries. Fig.2 shows the grids around ogive-cylinder-ogive geometry at different pitch angles obtained using the spring analogy method. In order to treat the moving and fixed outer boundaries the Kinetic treatment of Moving Boundary

Condition(KMBC) and Kinetic Outer Boundary Condition(KOBC)are formulated and used respectively. Then, the KFMG Euler code has been applied to an oscillating ogive-cylinder-ogive configuration with free stream Mach number 0.9 and pitch angle $15^o + 3^o sin(kt)$. Here, k is the reduced frequency and t is the time. In order to verify the results through inter-code comparison, the unsteady results obtained through reduced Euler formulations due to Nixon et al.[8] have been considered. The unsteady computations have been performed using the KFMG Euler code on a grid (45x41x51) which is very close to the grid chosen in the reference. It has been mentioned in Ref.[8] that, a phase lag of $\pi/2$ was observed in lift coefficients with respect to motion of the configuration for a free stream Mach number of 0.3 with other conditions remaining the same. The unsteady lift coefficients obtained using KFMG code as well as Nixon et al. for Mach 0.9 are plotted in Fig.3. Even though Nixon et al. have not mentioned phase lag for Mach 0.9 in their paper, we have observed the same phase lag of $\pi/2$ for Mach 0.9 case also. Fig.4 shows the phase lag corrected plot of lift coefficient obtained by Nixon et al. The lift coefficient versus time obtained by KFMG code is also shown on the same plot for comparison. It is observed that, the two results compare well except that the peaks are over predicted by KFMG method. The hysteresis loop of lift and moment coefficients confirmed the unsteady behavior during the oscillation of the configuration(Figs. 5 and 6). Two grid sizes are used in this study for the unsteady calculations. This study has proved that, the 3-D KFMG formulations are capable of predicting the unsteady flows satisfactorily. Thus a new capability of 3-D KFMG based Euler formulation for unsteady aerodynamics and computational aeroelasticity has been evolved.

Acknowledgments: The first two authors express their gratitude to Director, DRDL for giving encouragement and support during the course of this work. The first author thanks Head, CFD Division,DRDL for the useful discussions while pursuing this work.

References

1. J.T.Batina: *'Unsteady Euler Airfoil Solutions Using Unstructured Dynamic Meshes'*. In : AIAA Journal, Vol.28 Aug. 1990.
2. Ambrosi D, Gasparini L and Vigevano L (1994), "Full potential and Euler solutions for transonic unsteady flow", Aeronautical Journal
3. Deshpande SM (1996),"Some Recent Developments in Kinetic Schemes", Report no. 96 FM 2, CFD Centre, Department of Aerospace Engineering, Indian Institute of Science, Bangalore.
4. Krishnamurthy,R., Sarma,B.S. and Deshpande,S.M(2001), "KFMG based Euler solver for Computational Aeroelasticity", presented in first International Conference on Computational Fluid Dynamics (ICCFD), held at Kyoto, Japan (June 2000) and published in proceedings of the conferecne
5. Kulkarni PS and Deshpande SM (1996), "Kinetic Flux vector splitting on Moving Grid for Unsteady Aerodynamics", proceedings of third ECCOMAS CFD conference, Paris.
6. Krishnamurthy,R., Sarma,B.S. and Deshpande,S.M, "Kinetic Flux vector splitting scheme for unsteady Euler equations on Moving Grids (KFMG)", presented in 3rd

Asian Computational Fluid Dynamics conference (ACFD3), held at Bangalore, India and published in proceedings of the conference.
7. Krishnamurthy,R.,Anandhanarayanan,K. and Nagarathinam,M. (1998),"Elliptic Grid Around Flight Vehicle Configurations" presented in 4th ECCOMAS Conference on Computational Fluid Dynamics, held at Athens, Greece and published in proceedings of the conference.
8. Nixon D, Reisenthel PH, Torres TO and Klopfer GH (1990), "Prediction of Unsteady Transonic Flow Around Missile Configurations", AIAA paper 90-0934.

Table 1. Inter-code comparison of aerodynamic coefficients

Euler solver	Grid size	*Aerodynamic coefficients*	
		Lift coeff. (C_L)	Drag coeff. (C_D)
KFMG	45x25x27	1.78	0.52
	45x41x51	1.38	0.51
	90x51x51	0.85	0.29
VEEMA	45x25x27	2.06	0.63
	45x41x51	1.10	0.41
	90x51x51	0.82	0.30

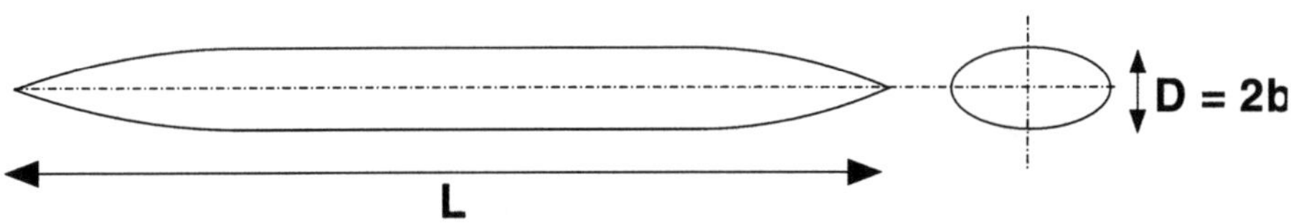

Fig. 1. Ogive-cylinder-ogive configuration

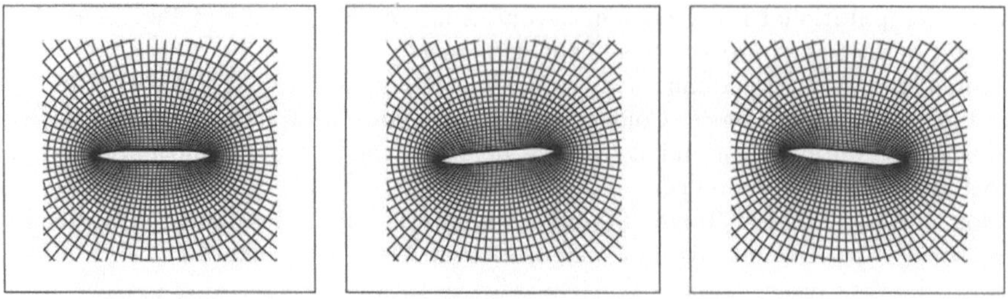

Initial grid Pitch angle = -5 deg. Pitch angle = +5 deg.

Fig. 2. Grids around ogive-cylinder-ogive geometry for different pitch angles

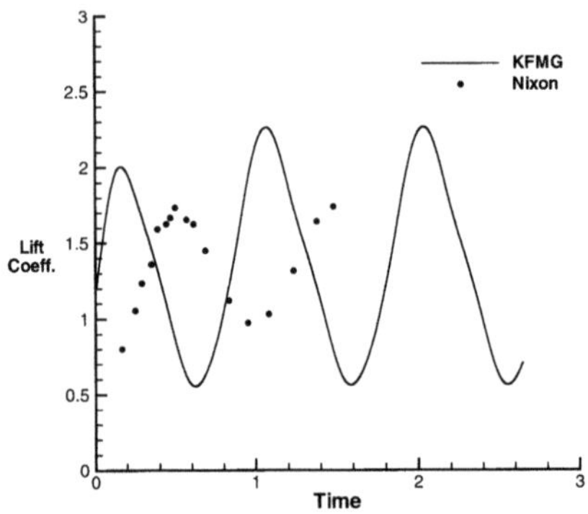

Fig. 3. Variation of unsteady lift coefficient with time

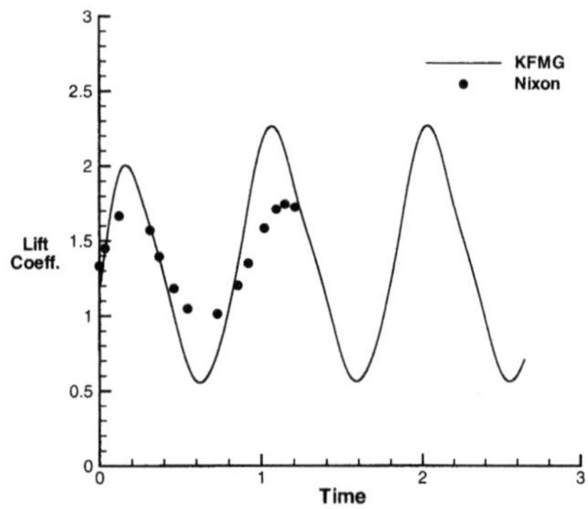

Fig. 4. Variation of unsteady lift coefficient with time

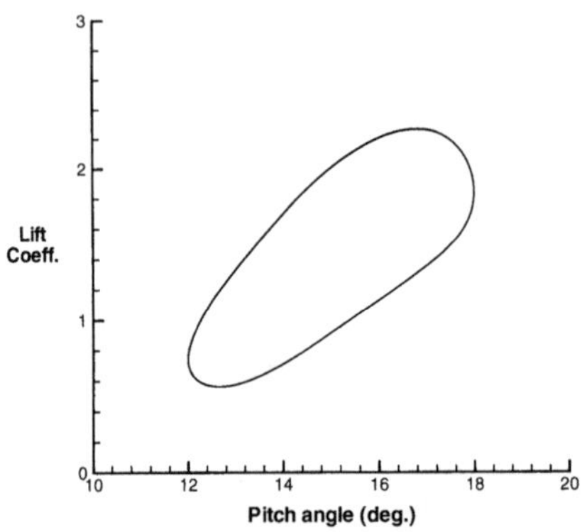

Fig. 5. Variation of unsteady lift coefficient with pitch angle

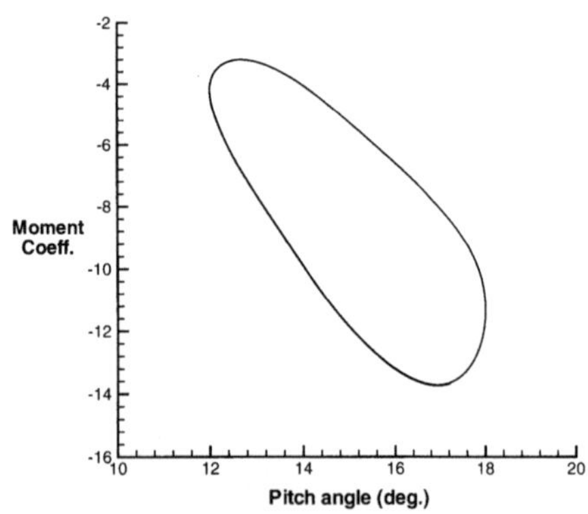

Fig. 6. Variation of unsteady moment coefficient with pitch angle

3D Euler Codes Based on Kinetic Upwind Scheme and Their Verification and Validation Using NAL Wind Tunnel Data

PS Kulkarni

AR & DB CFD centre, Department of Aerospace Engineering, Indian Institute of Science, Bangalore, INDIA

Abstract. The Euler code BHEEMA has been subjected to systematic verification and validation studies. The verified values of computed aerodynamic coefficients C_N and C_M agree very well with the experimental results. The percentage error in prediction of C_A reduces considerably when empirically obtained Skin Friction Coefficient (sfc) is added to computed C_A. The computations of C_N and C_M near zero angle of attack show that the accuracy of experimental measurements near α =0 is doubtful.

1 Introduction

The computation of flow field and aerodynamic coefficients for 3D complex configurations need to be performed quickly and accurately in the design studies. Thoroughly verified and validated CFD codes are required as reliable tools in the design and analysis phase of aerospace vehicles [1]. In accordance with the definition of systematic VVC (Verification, Validation and Certification) studies [2], the verification includes iterative convergence, grid refinement and inter-code comparison. Thereafter a thoroughly verified code should be validated with bench mark experimental results. A systematic VVC study could be taken mainly because of the availability of PACE+24 parallel computer and the availability of in house developed two 3-D Euler codes BHEEMA-KFVS and BHEEMA-PVU and enormous wind tunnel data from NAL for practical configuration is available.

2 Systematic approach to VVC

The inter-relationship of three stages of any CFD code verification, validation and certification is described in the form of flow diagram in Fig 1. The verification stage is concerned mainly with numerical aspects where as validation deals with finding out the degree of approximation of the mathematical model used and therefore needs comparison with experimental data. The certification stage involves operational aspects like realistic configuration, error bands and CPU time per run.

3 Euler codes

The two codes namely BHEEMA-KFVS [3] and BHEEMA-PVU [4] which are based on Kinetic Upwind Method have been considered for the present study. The code BHEEMA-KFVS is based on Kinetic Flux Vector Splitting (KFVS) Method. In case of BHEEMA-KFVS the fluxes are split based on whether the molecular velocity $v < 0$ or $v > 0$ where as in case of BHEEMA-PVU the splitting is done based on whether the molecular velocity v is greater or less than the fluid velocity u. Equivalently the relative velocity $c = v - u$ (*known as Peculiar Velocity*) is used for splitting.

In the kinetic framework the Euler equation are obtained [5] by taking the inner product of moment function ψ and the Boltzmann equation,

$$\left\langle \psi, \frac{\partial \mathrm{F}}{\partial \mathrm{t}} + \overline{v}.\frac{\partial \mathrm{F}}{\partial \overline{\mathrm{x}}} \right\rangle = 0, \text{where } \psi = \begin{bmatrix} 1 \\ \overline{v} \\ 1 + \frac{\overline{\mathrm{v}}^2}{2} \end{bmatrix} \quad \text{and} \tag{1}$$

F is the Maxwellian representing the molecular velocity distribution in the Euler limit. The flux vector G can be evaluated in the closed form in terms of error functions as follows. The 3D Euler equations in the conserved vector form can be written as $\frac{\partial \mathrm{U}}{\partial \mathrm{t}} + \frac{\partial \mathrm{G_1}}{\partial \mathrm{x}} + \frac{\partial \mathrm{G_2}}{\partial \mathrm{y}} + \frac{\partial \mathrm{G_3}}{\partial \mathrm{z}} = 0$

$$\text{where } U = \begin{bmatrix} \rho \\ \rho u_1 \\ \rho u_2 \\ \rho u_3 \\ e \end{bmatrix}, \ G_1 = \begin{bmatrix} \rho u_1 \\ p + \rho {u_1}^2 \\ \rho u_1 u_2 \\ \rho u_1 u_3 \\ (p+e)u_1 \end{bmatrix}, \ G_2 = \begin{bmatrix} \rho u_2 \\ \rho u_2 u_1 \\ p + \rho {u_2}^2 \\ \rho u_2 u_3 \\ (p+e)u_2 \end{bmatrix}, \ G_3 = \begin{bmatrix} \rho u_3 \\ \rho u_3 u_1 \\ \rho u_3 u_2 \\ p + \rho {u_3}^2 \\ (p+e)u_3 \end{bmatrix}$$

The split KFVS fluxes can be written as

$$G^{\pm} = (G)\frac{\mathrm{A}^{\pm}(\mathrm{S})}{2} \pm (GT)\frac{\mathrm{B(S)}}{2} \tag{2}$$

$$\text{where GT} = \text{thermal flux} = \begin{bmatrix} \rho\sqrt{\pi\beta} \\ \rho u_1\sqrt{\pi\beta} \\ \rho u_2\sqrt{\pi\beta} \\ \rho u_3\sqrt{\pi\beta} \\ \left(\frac{(\gamma+1)\mathrm{p}}{2(\gamma-1)} + \frac{1}{2}\rho({u_1}^2 + {u_1}^2 + {u_1}^2)/\sqrt{\pi\beta} \right) \end{bmatrix}$$

$$\text{and} \quad \frac{\mathrm{A}^{\pm}(\mathrm{S})}{2} = \frac{1 \pm \mathrm{erf(S)}}{2}, \ B(S) = \exp(-S^2), \ S = u\sqrt{\beta}$$

The finite volume method based on PVU splitting gives discrete conservation law as $\frac{\partial \mathrm{U}}{\partial \mathrm{t}} + \frac{1}{\mathrm{V_{ij}}} \sum_{k=1}^{6} |S_k| \left[G_k^{(t)} + G_k^{(a)} \right] = 0$ where U is the state vector, S_k is the area of surface k and $G^{(t)}$ and $G^{(a)}$ are the transport and acoustic parts of the flux respectively. The expression for the split PVU fluxes are given by

$$G_k^{(t)} = \left[\frac{\mathrm{u_{n,k}} \pm |\mathrm{u_{n,k}}|}{2}\right] U \quad \text{and} \quad G_k^{(a)} = \begin{bmatrix} \pm\frac{\rho}{2\sqrt{\pi\beta}} \\ n_1\frac{\mathrm{p}}{2}\pm\frac{\rho \mathrm{u_1}}{2\sqrt{\pi\beta}} \\ n_2\frac{\mathrm{p}}{2}\pm\frac{\rho \mathrm{u_2}}{2\sqrt{\pi\beta}} \\ n_2\frac{\mathrm{p}}{2}\pm\frac{\rho \mathrm{u_3}}{2\sqrt{\pi\beta}} \\ \frac{1}{2}pu_n\pm\frac{1}{2\sqrt{\pi\beta}}\left\{\frac{\mathrm{p}}{2}+\rho E\right\} \end{bmatrix} \tag{3}$$

where n_1, n_2 and n_3 are the three direction cosines of the surface normal n_k and ρ, p and E are the density, pressure and internal energy respectively. We can see that PVU split fluxes do not involve error functions and exponentials which appear in KFVS formulation.

4 Candidate Configuration and NAL Wind Tunnel Data

The candidate configuration with ogive-cylinder-wing-tail shown in fig.1 has been considered for the present VVC study because NAL, Bangalore has conducted several careful experiments to measure aerodynamic coefficients such as C_A, C_N and C_M for various Mach numbers and angles of attack for this configuration. We have used nearly 175 points in the $M_\infty - \alpha$ carpet (matrix of points) with M_∞ varying in supersonic region from 1.54 to 3.53 and angle of attack α from -16° to $+16^o$.

5 Methodology

A 2-D elliptic grid generator is used. By connecting corresponding grid points in neighboring cross sections (as shown in Fig 2) a stacked network of finite volumes is formed. We have considered four sets of grids of sizes $1L(97 \times 73 \times 17 = 1,20,377$ finite volumes), $3L(193 \times 73 \times 17 = 2,39,513$ finite volumes), $5L(193 \times 145 \times 17 = 4,75,745$ finite volumes) and $1M(193 \times 145 \times 33 = 9,23,505$ finite volumes) for performing grid convergence (h$\rightarrow$0) study. The codes BHEEMA-KFVS and BHEEMA-PVU have been run for all the cases of $M_\infty - \alpha$ on all the four sets of grids till the iterative convergence (n$\rightarrow\infty$) is achieved.

6 Results & Discussions

Residue fall for grids 5L and 1M is shown in Fig 3. A typical grid convergence result of C_N with α for different grids is shown in Fig 4. The computed results are also compared with whatever a priori knowledge that we have such as, jump conditions and stagnation pressure behind the shocks [6]. Intercode comparison is done by running BHEEMA-PVU for all the Mach numbers and angles of attack and it is observed that there is good agreement with results obtained from BHEEMA-KFVS [3], [4]. The present verification study reveals that the grid of size 5L is good enough for further conducting validation study for this class of flight vehicles. The computed C_A are compared with experimental data (refer Table 1) It can be seen from this table that the maximum % error in prediction of C_A is within 30%. It is very interesting to note that by adding empirically obtained skin friction coefficient (sfc) to the computed C_A the % error reduces to about 9%. The computed values of C_N and C_M have been tabulated in Table 2 for Mach numbers 1.50 and 3.53. The maximum % error in predicting C_N and C_M is less than 8% thus establishing the validity of BHEEMA under the assumption that 10% error is acceptable. It can also be seen from this table that the % error in computed C_N and C_M near zero angle of attack is very large, which are further analysed to find out why experimental measurements while agreeing with CFD values for almost all differ from them near $\alpha = 0$. The slopes C_N and C_M are calculated both for CFD and experimental values using least squares linear fit [1]. The predicted values of C_N and C_M are obtained by using the formula $C_{NM} = \alpha\left(\frac{\mathrm{d}}{\mathrm{d}\alpha}C_{N,M}\right)_{\alpha=0}$. The predicted values of C_N and C_M based on $(C_{N\alpha})$CFD and $(C_{N\alpha})$CFD agree very well with original CFD data (see table 3). The predicted C_N and C_M based on $(C_{N\alpha})$Exptl and $(C_{M\alpha})$Exptl on the other hand differ considerably from original experimental data (see table 4). The analysis of large percentage of error in C_N and C_M near zero angle of attack throws doubt on the experimental measurements of these coefficients at small values of and most probably this is due to the difficulty in measuring small values which requires further investigations.

Acknowledgements: The authors thanks Prof. SM Deshpande for introducing to VVC Study and for useful discussions. Many thanks are due to CFD Division DRDL for their help and Sathyadharma S.G, N. Srinarayana, R. Tyagaraju and Anand Kulkarni for their assistance while preparing this paper .

References

1. PS Kulkarni, Deepak Kumar Singh and SM Deshpande. "Verification and Validation of Euler code BHEEMA using NAL wind tunnel data for generic flight vehicle configuration", FM Report 02FM8, AE, IISc, Bangalore, India
2. Melnik RE, Siclari MJ and Marconi F. "An overview of recent industry effort at CFD code verification", AIAA-95-2229
3. Deshpande SM, Sekar S, M Nagarathinam, R Krishnamurthy, PK Sinha, PS Kulkarni(1992) "3D Upwind Euler solver using KFVS Method". Proceedings, 13 ICNMFD, ITALY Lecture notes in Physics, 414, PP 105-109.
4. Kulkarni PS, SV Raghurama Rao and SM Deshpande(1995) "Application of PVU based 3D Euler Solver BHEEMA-P wing-body-fin configuration", 6th ACFM May 22-26, Singapore
5. Mandal JC and SM Deshpande(1994) "Kinetic Flux Vector Splitting for Euler equations" Computers and Fluids,23(2),447-478
6. Raghurama Rao SV(1994), "New upwind methods based on kinetic theory for inviscid compressible flows", Ph.D Thesis, Department of Aerospace Engineering, IISc, Bangalore, INDIA

Table 1. Comparsion of C_A values with experimental data for $M_\infty = 2.01$

$M_\infty = 2.01$					
A.O.A	Exptl.	KFVS	KFVS+sfc	% error without sfc	% error with sfc
-10.27	0.248	0.186	0.272	24.8	9.7
-2.71	0.251	0.185	0.271	25.9	8.3
0.63	0.250	0.185	0.271	25.7	8.6
2.61	0.250	0.185	0.271	25.6	8.7
15.46	0.250	0.188	0.274	24.5	9.8

Table 2. Comparsion of C_N and C_M values with expt. data on 5L grid

$M_\infty = 1.50$						
A.O.A	C_N		C_M		% error in C_N	% error in C_M
	Computed	Exptl.	Computed	Exptl.	C_N	C_M
4.58	-1.070	-1.1	4.815	4.922	2.6	2.1
0.18	0.040	0.03	-0.184	-0.75	35.9	75.5
0.29	0.999	0.98	-4.508	-4.269	1.9	5.5
$M_\infty = 3.53$						
A.O.A	C_N		C_M		% error in C_N	% error in C_M
	Computed	Exptl.	Computed	Exptl.	C_N	C_M
-4.50	-0.644	-0.69	2.737	2.958	6.6	7.4
0.30	0.042	0.0	-0.177	-0.048	∞	268.7
4.38	0.626	0.62	-2.666	-2.549	0.97	4.3

Table 3. percentage error in predicted C_N

C_N CFD	C_N Exptl.	% error C_N	$C_{N\infty}$ CFD	$C_{N\infty}$ Exptl.	predicted C_N based on $(C_{N\alpha})_{CFD}$	predicted C_N based on $(C_{N\alpha})_{Exptl}$	Corrected % error on C_N
$M_\infty = 1.50 \quad \alpha = 0.18$							
0.040	0.03	35.9	0.235	0.236	0.042	0.042	0.44
$M_\infty = 2.01 \quad \alpha = 0.63$							
0.127	0.13	2.30	0.205	0.204	0.129	0.128	0.31
$M_\infty = 2.54 \quad \alpha = 0.40$							
0.07	0.06	16.6	0.178	0.181	0.071	0.072	1.52
$M_\infty = 3.06 \quad \alpha = 0.64$							
0.099	0.08	23.7	0.158	0.164	0.101	0.105	3.33
$M_\infty = 3.53 \quad \alpha = 0.30$							
0.042	0.00	∞	0.144	0.151	0.043	0.045	4.44

Table 4. percentage error in predicted C_M

C_M CFD	C_M Exptl.	% error C_M	$C_{M\infty}$ CFD	$C_{M\infty}$ Exptl.	predicted C_M based on $(C_{M\alpha})_{CFD}$	predicted C_M based on $(C_{M\alpha})_{Exptl}$	Corrected % error on C_M
$M_\infty = 1.50$ $\alpha = 0.18$							
-0.184	-0.75	75.4	-1.054	-1.051	0.189	0.189	0.24
$M_\infty = 2.01$ $\alpha = 0.63$							
-0.559	-0.514	8.7	-0.913	-0.907	-0.575	-0.571	0.69
$M_\infty = 2.54$ $\alpha = 0.40$							
-0.297	-0.226	31.4	-0.782	-0.789	-0.312	-0.315	0.91
$M_\infty = 3.06$ $\alpha = 0.64$							
-0.425	-0.317	34.0	-0.690	-0.703	-0.441	-0.45	1.84
$M_\infty = 3.53$ $\alpha = 0.30$							
-0.177	-0.048	268.7	-0.617	-0.641	-0.185	-0.192	3.65

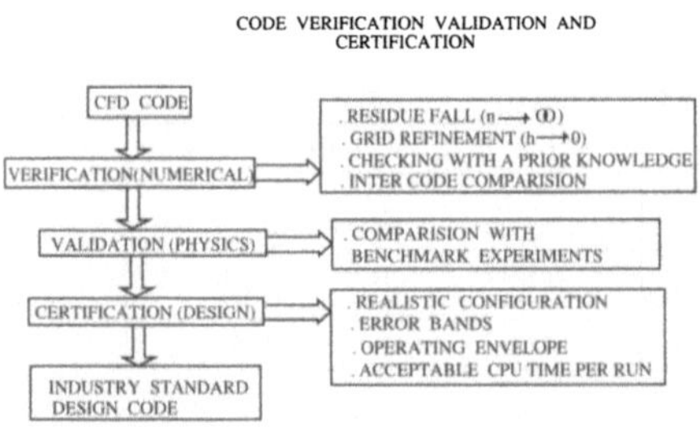

Fig. 1. VVC study of any CFD code

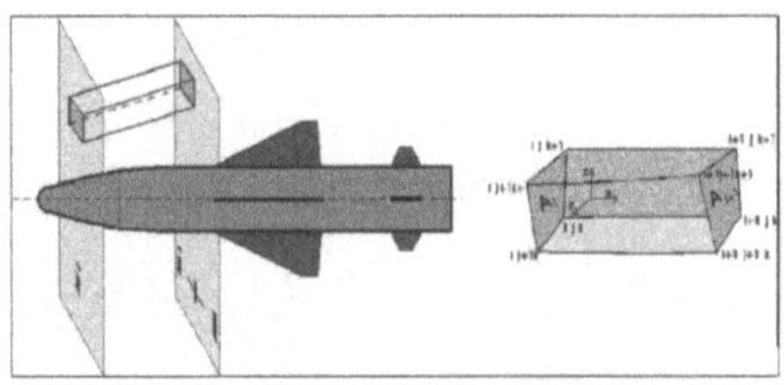

Fig. 2. Stacked network of finite volumes

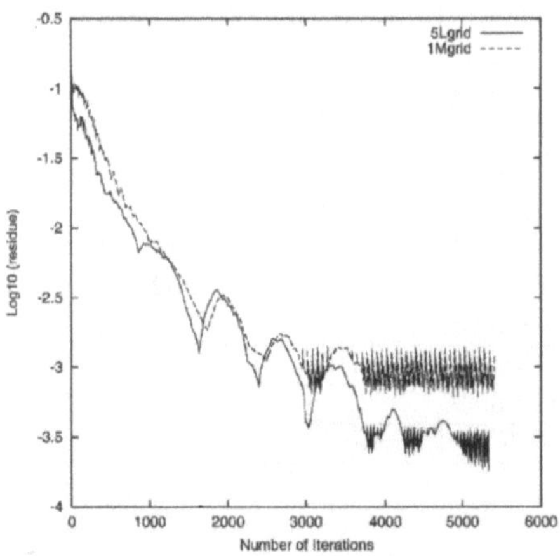

Fig. 3. Iterative Convergence Study ($n \rightarrow \infty$)

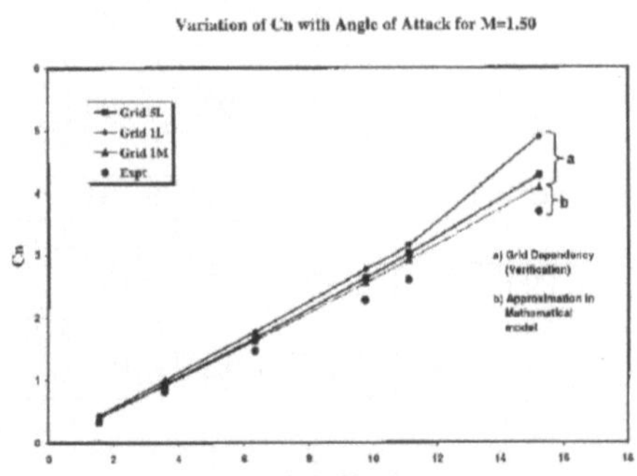

Fig. 4. Grid refinement Study ($h \rightarrow 0$)

Part X

Large Eddy Simulation

A General Approximate Wall Condition for Large Eddy Simulations

P. S. Weng, W. Lo, and C. A. Lin

Department of Power Mechanical Engineering, National Tsing Hua University, Hsinchu 30013, TAIWAN

Abstract. A general approximate wall condition which accounts for the effect of the pressure gradient is proposed. The present formulation does not require spatially averaged velocity in the wall model. For the fully developed channel flow, the present results compare favorably with the wall model of Cabot and Moin, but with less computational effort. The capability of the wall modes are further investigated by the two-dimensional flow with squared sectioned ribs.

1 Introduction

Traditionally the solutions of turbulent flow problems involve the the applications of Reynolds averaging to the equations of motions to obtain Reynolds averaged equations, which describes the evolution of the mean quantities. The effect of the turbulent fluctuations appears in a Reynolds stress $\overline{u_i u_j}$ that must be modeled to close the system. There are many routes to model the unknown stresses, such as eddy-viscosity type model or the Reynolds stress transport models. When the models are applied to flows that are very different from the ones used for calibration, the model may not deliver accurate predictions.

The most straightforward approach to solve the equation without modeling is via direct numerical simulation of turbulence, in which the governing equations are discretised and solved numerically. However, the need to acquire accurate results require the adequate resolutions of different scales presence within the flows.

Large eddy simulation is a technique which computes the large, energy-containing structures to momentum exactly, and only the effects of the smallest scales of turbulence is modeled. The computational cost of a large eddy simulation is therefore substantially reduced over that required by direct numerical simulation. Near the solid wall, on the other hand, the grid resolution has to be very fine to resolve the energy containing structure, which limits the practical applications of LES to high Reynolds number flows.

One remedy is to use approximate wall conditions, where the wall is not resolved and the wall shear stress is determined via the law of the wall. The major drawback is that the law of the wall is strictly speaking applicable to flow in equilibrium conditions. In the present paper, a general approximate wall condition which accounts for the effects of the pressure gradient is proposed. Applications are then applied to the fully developed channel flow with Reynolds number of Re_τ=9000 and the two-dimensional flow with squared sectioned ribs.

2 Governing Equations and Modeling

LES is based on the definition of a filtering operation: a filtered (or resolved, or large-scale) variable, is defined as,

$$f(\tilde{x}) = \int_D f(\tilde{x'})G(\tilde{x}, \tilde{x'})d\tilde{x'} \tag{1}$$

where D is the entire domain and G is the filter function. The filter function determines the size and structure of the small scales.

The governing equations are grid-filtered, incompressible continuity and Navier-Stokes equations,

$$\frac{\partial \rho \overline{u}_i}{\partial x_i} = 0 \tag{2}$$

$$\frac{\partial \rho \overline{u}_i}{\partial t} + \frac{\partial}{\partial x_j}(\rho \overline{u}_i \overline{u}_j) = -\frac{\partial \overline{p}}{\partial x_i} + \frac{\partial}{\partial x_j}[\mu(\frac{\partial \overline{u}_i}{\partial x_j} + \frac{\partial \overline{u}_j}{\partial x_i}) - \tau_{ij}] \tag{3}$$

where

$$\tau_{ij} = \rho \overline{u_i u_j} - \rho \overline{u}_i \overline{u}_j$$

is the sub-grid stress due to the effects of velocities being not resolved by the computational grids.

In the present study, the Smagorinsky model[1] has been used for the sub-grid stress, such that,

$$\tau_{ij} = -(C_s \Delta)^2 \frac{1}{\sqrt{2}} \sqrt{S_{kl} S_{kl}} S_{ij} + \frac{2}{3} \rho k_{sgs} \delta_{ij} \tag{4}$$

where $C_s = 0.1$, $S_{ij} = \frac{\partial \overline{u}_i}{\partial x_j} + \frac{\partial \overline{u}_j}{\partial x_i}$ and $\Delta = (\Delta x \Delta y \Delta z)^{1/3}$. It can be seen that in the present study the mesh size is used as the filtering operator.

If the resolved computation is needed, the length scale has to be damped when approaching the wall. A Van Driest damping function which accounts for the effect of the wall on the subgrid-scale can be adopted and takes the form,

$$l_m = \kappa y[1 - exp(-\frac{y^+}{25})] \tag{5}$$

where y is the distance to the wall and the cell size is redefined as,

$$\Delta = MIN[l_m, (\Delta x \Delta y \Delta z)^{1/3}] \tag{6}$$

Since in the present study, the near wall region is not resolved, the wall shear stress must be determined by wall model. The wall condition is obtained by solving the boundary layer equation within the first near wall cell.

$$\frac{\partial}{\partial y}[(\mu + \mu_t)\frac{\partial \overline{u}_{s,i}}{\partial y}] = \rho \frac{\partial \overline{u}_{m,i}}{\partial t} + \rho \overline{u}_{m,j} \frac{\partial \overline{u}_{m,i}}{\partial x_j} + \frac{\partial \overline{P}_m}{\partial x_i} \tag{7}$$

where s and m stand for the sub-cell and cell quantities and i denotes the directions parallel to the wall.

There are two versions adopted in the present study, namely:

i. Model A

$$\frac{\partial}{\partial y}[(\mu+\mu_t)\frac{\partial \overline{u}_{s,i}}{\partial y}]=0 \tag{8}$$

ii. Model B

$$\frac{\partial}{\partial y}[(\mu+\mu_t)\frac{\partial \overline{u}_{s,i}}{\partial y}]=\frac{\partial \overline{P}_m}{\partial x_i} \tag{9}$$

The above equations can be integrated analytically if the right hand side is treated as a constant and the eddy viscosity is modelled as,

$$\frac{\mu_t}{\mu}=\kappa(y^+-y_v^+) \tag{10}$$

It can be clearly observed that in the near wall region, a viscous sub-layer ($y^+ < Y_v^+$) and a fully turbulent layer ($y^+ > Y_v^+$) has been adopted.

3 Numerical Algorithms

The grid-filtered transport equations are discretised using a staggered finite volume arrangement. The principle of mass-flux continuity is imposed indirectly via the solution of pressure-correction equations according to the SIMPLE algorithm[2].

For the transport equations, the spatial derivatives are approximated using second order central differencing scheme. For the purpose of temporal approximation, the momentum equations may be written as;

$$\frac{\partial \phi}{\partial t}+C_x+C_y+C_z=D_x+D_y+D_z+S \tag{11}$$

where ϕ is the dependent variable, S is the source terms and C and D are the convective and diffusive transport terms, respectively.

The temporal approximation seeks to advance ϕ from any discrete time level (n) to the following level ($n+1$) with some specified order of accuracy. In the present work, a second-order accurate ADI scheme, originally proposed by Brian [3] for 3D heat-conduction, is applied.

4 Results

4.1 Fully developed channel flow

Numerical simulation is first done for a fully developed channel flow of $Re_\tau = 9000$. Periodic boundary conditions are imposed in the streamwise and spanwise directions, while the no-slip boundary is imposed in the transverse direction. The sizes of the computational grid adopted in the simulations are 16x24x16 and

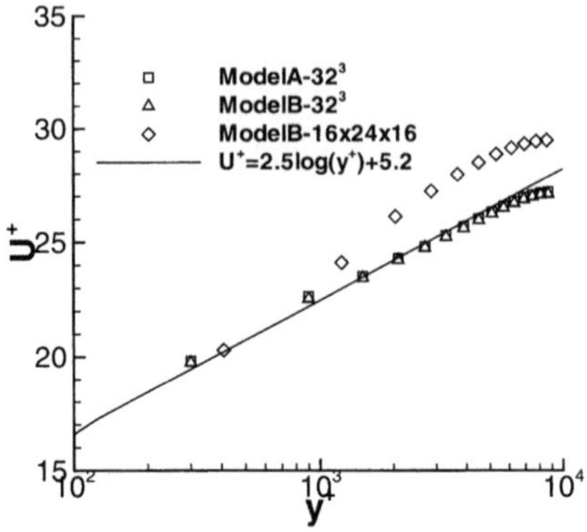

Fig. 1. Effects of wall models and grid density on the axial velocity-Re_τ = 9000

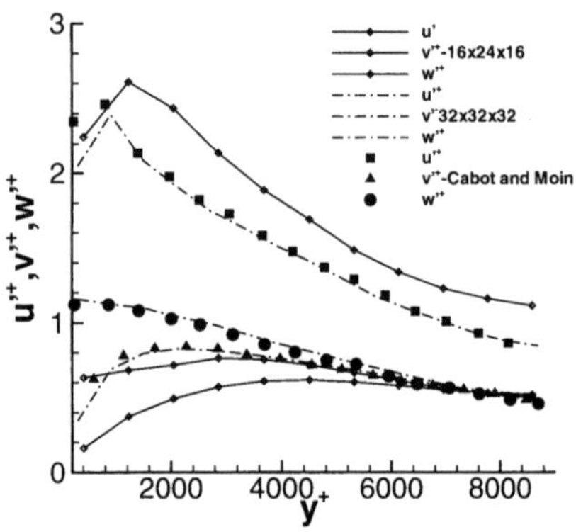

Fig. 2. Effects of wall models (ModelB and Cabot and Moin) and grid density on the turbulence intensities-Re_τ = 9000

32x32x32 in the streamwise, transverse and spanwise directions, respectively. The time step adopted is Δt^+ =0.001.

The time averaged velocity distribution can be examined by looking at the log-law plot, as shown in Figure 1. The first grid is located at y^+ about 300. Because of the wall law is adopted, the viscous sublayer is not resolved. It can be seen that there exists a marginal difference adopting either ModelA or ModelB. This is expected in the fully developed channel flow. The agreement with the empirical formulation is reasonable. However, the figure does show that the coarse grid is not adequate to capture correctly the large scale structure.

The capability of the adopted scheme can be further examined by looking at the predicted turbulence quantities, as shown in Figure 2, where only ModelB results are presented. The anisotropic field of the normal stresses is well represented by the adopted approach. The effect of the grid density on the turbulence quantities is similar to that observed in the axial velocity profiles. The present results compare favorably with the results of Cabot and Moin[4]. Notably, the

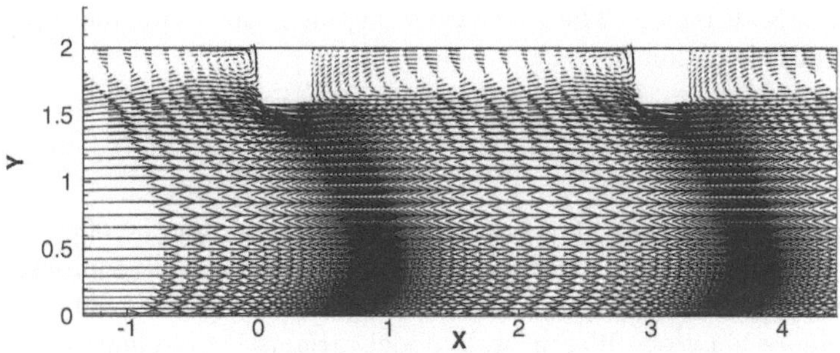

Fig. 3. Velocity vector

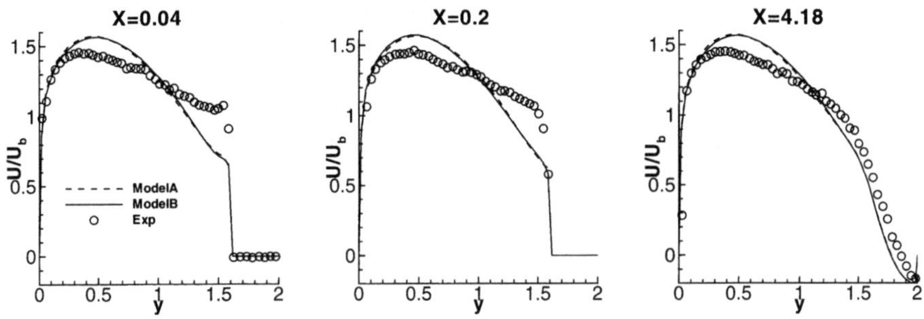

Fig. 4. Axial velocity distributions

Reynolds number of the Cabot and Moin prediction is $Re_\tau = 4000$, which is lower than the present case.

4.2 Two-dimensional flow with periodic squared sectioned ribs

The wall models are further applied to the two-dimensional flow with periodic squared sectioned ribs. The predicted results are contrasted with the measurements of Drain and Martin[5], where the Reynolds number based on the channel height is 3.72×10^4. Since the rib appears periodically, only one periodic domain is adopted. Periodic boundary conditions are imposed in the streamwise and spanwise directions, while the no-slip boundary is imposed in the transverse direction. The size of the computational grid adopted in the simulations is 40x40x30 in the streamwise, transverse and spanwise directions, respectively.

Figure 3 shows the computed velocity vector plot, where the recirculation zone between the ribs can be clearly observed. Notably, the small recirculation bubble right after the rib is predicted by the present approach. The predicted axial velocities at several axial locations are shown in Figure 4. The difference between ModelA and ModelB is surprisingly marginal. The schemes capture correctly the recirculation bubble at x=4.18. However, the comparisons with measured data do show that the shear layer near the rib is not adequately resolved by the present approaches. It might be due to the inadequate modelling of the near wall region. The other cause of this is the inadequate resolution of the grid at this region.

5 Conclusion

A general approximate wall condition which accounts for the effect of the pressure gradient is proposed. The present formulation does not require spatially averaged velocity in the wall model. For the fully developed channel flow Re_τ=9000, the difference between different wall model variants is marginal. The present results compare favorably with the wall model of Cabot and Moin, but with less computational effort. For the two-dimensional flow with sectioned ribs, the comparisons with measured data do show that the shear layer near the rib is not adequately resolved by the present approaches. However, the recirculation bubble is captured correctly by the present schemes. The adopted grid is too coarse, therefore grid refinements should be conducted to examine the grid dependency.

Acknowledgments: This research was supported by National Science Council of Taiwan under grant NSC-90-2212-E-007-058 and the computational facilities were provided by the National Centre for High-Performance Computing of Taiwan which the authors gratefully acknowledge.

References

1. Smagorinsky, J., 1963, "General Circulation Experiments with the Primitive Equations. I. The Basic Experiments," Monthly Weather Review, Vol. 91, pp. 99-164.
2. Patankar, S.V. and Spalding, D.B., "A Calculation Procedure for Heat Mass Momentum Transfer in Three-Dimensional Parabolic Flow", International Journal of Heat and Mass Transfer, Vol 15, October 1972, pp. 1787-1806.
3. Brian, P. L. T., 1961, "A Finite Difference Method of High-Order Accuracy for the Solution of Three-Dimensional Transient Heat Conduction Problem," A. I. Ch. E. Journal, Vol 7, No. 3, pp. 367-370.
4. Cabot, W. and Moin, P., 1999, "Approximate Wall Boundary Condition in the Large-Eddy Simulation of High Reynolds Number Flow," Flow, Turbulence and Combustion, Vol. 63, pp. 269-291.
5. Drain, L. E. and Martin, S., 1985, "Two-Component Velocity Measurements of Turbulent Flow in a Ribbed-Wall Flow Channel;," International Conference on Laser Anemometry-Advances and Application, Manchester.

Large Eddy Simulation of the Purging of Density-Stabilized Basins

Michael Kirkpatrick[1] and Steven Armfield[2]

[1] Center for Turbulence Research, NASA Ames Research Center, CA 94035, USA
[2] The University of Sydney, Sydney, NSW 2006, Australia

Abstract. To date density-stabilized purging flows have proved very difficult to model. While Reynolds-averaging approaches tend to greatly overpredict purging rates, unsteady simulations with no subgrid model tend underpredict it. In this paper we investigate the performance of the large eddy simulation approach for flows of this type. Results are presented for 2D and 3D large eddy simulations using the dynamic Smagorinsky subgrid model and purging rates are compared with experimental results.

1 Introduction

Saline pools, formed by ground water intrusion through the river bed during periods of low flow occur in many of Australia's rivers and can lead to major environmental problems. When the pools are subjected to a sufficiently large fresh overflow, as a result of a rainfall event or the release of water from upstream storage, the resulting mixing and transport purges the pool of saline water. In order to maintain the integrity of the river systems it is necessary to understand the flow dynamics of the water within the pools during the purging flows [1].

Density-stabilized purging flows have proved very difficult to model, however, due to the strongly anisotropic turbulence resulting from buoyancy-driven damping of the vertical scales. Reynolds-averaging approaches using turbulence models such as the $k - \epsilon$ model generally overpredict the purging rate by an order of magnitude due to the assumptions of isotropy inherent in these models [2]. A high-resolution unsteady 2D numerical simulation with no turbulence model by Armfield and Debler [3] gave accurate predictions of early stages of the purging process, where mixing is due mainly to large-scale 2D vortices, but underpredicted the purging rate by an order of magnitude during later stages. Here mixing is dominated by small-scale 3D turbulent eddies at the density interface which were not resolved in the 2D direct simulation.

In this paper we present results obtained using the large eddy simulation code, PUFFIN [4]. A 2D simulation using the dynamic Smagorinsky subgrid model is used to simulate the entire purging process. A 2D simulation cannot, however, be considered a true LES since, in real flows, turbulence must always be three-dimensional. We therefore repeat the simulation of the later stages (in which turbulence is important) in three dimensions.

The non-dimensional parameters used to define the flow are the Reynolds number $Re = \frac{UH_{chan}}{\nu}$ and Rayleigh number $Ra = \frac{gH_{chan}^3 \Delta\varrho}{\nu\kappa}$, where U is the velocity at the free surface of the channel, H_{chan} is the height of the channel,

H_{cav} is the height of the cavity, g is the acceleration due to gravity, ν is the kinematic viscosity of water, and κ the kinematic diffusion coefficient for salt. The density differential between the fresh and salt water is $\Delta\varrho = (\varrho_s - \varrho_f)/\varrho_f$, where ϱ_f and ϱ_s are the initial density of the fresh and salt water. The value of these parameters chosen for the simulation are $Re \approx 7500$, $Ra = 6.5 \times 10^9$.

2 Governing Equations and Numerical Methods

The governing equations are the spatially-filtered equations for conservation of momentum, mass, solute concentration and an equation of state. The filtered equations for conservation of mass and momentum for an incompressible liquid with small variations in density are written

$$\frac{\partial \overline{\varrho}}{\partial t} + \frac{\partial (\overline{\varrho}\,\overline{u}_j)}{\partial x_j} = 0, \tag{1}$$

$$\frac{\partial \overline{\varrho}\,\overline{u}_i}{\partial t} + \frac{\partial (\overline{\varrho}\,\overline{u}_i\,\overline{u}_j)}{\partial x_j} = -\frac{\partial \overline{P}}{\partial x_i} + \frac{\partial}{\partial x_j}\left(2\left(\overline{\mu} + \overline{\mu}_{SGS}\right)\overline{S}_{ij}\right) + \overline{\varrho} g_i. \tag{2}$$

Here the overbar ($^{-}$) is the spatial filtering operator, u_i are the cartesian velocity coordinates, p is the pressure, μ is the molecular viscosity and $\overline{S}_{ij}$ is the large-scale strain rate tensor. The eddy viscosity μ_{SGS} is calculated using the Smagorinsky model [5] with the model coefficient and length-scale calculated using the dynamic procedure of Germano *et al.* [6].

The solute concentration σ in the mixture is normalized by the initial concentration of unmixed saline solution in the cavity σ_o to give the mixture fraction $c = \frac{\sigma}{\sigma_o}$. The filtered transport equation for mixture fraction is written

$$\frac{\partial \overline{\varrho}\,\overline{c}}{\partial t} + \frac{\partial (\overline{\varrho}\,\overline{u}_j\,\overline{c})}{\partial x_j} = \frac{\partial}{\partial x_j}\left[\left(\frac{\overline{\mu}}{Pr} + \frac{\overline{\mu}_{SGS}}{Pr_t}\right)\frac{\partial \overline{c}}{\partial x_j}\right], \tag{3}$$

where the molecular diffusivity has been written in terms of the molecular viscosity μ and a Prandtl number $Pr = 750$ and the eddy diffusivity in terms of the eddy viscosity μ_{SGS} and a turbulent Prandtl number $Pr_t = 1$. The density of the mixture is calculated from the mixture fraction using an equation of state,

$$\overline{\varrho} = \overline{c}\,\overline{\varrho}_s + (1 - \overline{c})\overline{\varrho}_f. \tag{4}$$

The equations are discretized in space using a finite volume formulation on a non-uniform, staggered, Cartesian grid. The upstream boundary condition is an inflow boundary with a $1/7^{th}$ power-law mean velocity profile, zero velocity fluctuations and and zero salt concentration. The downstream boundary uses a zero-normal-gradient outflow boundary condition for velocity and salt concentration. In the 3D simulation, the spanwise boundaries are periodic for all variables. The spanwise dimension of the domain (30mm) was chosen to correspond to the wavelength of the largest flow structures estimated from experimental observations. This dimension also corresponds to $x^+ \approx 100$ where $x^+ = x/\delta_\tau$ and δ_τ is estimated from the mean shear stress of $\tau_w \approx 1.0 \times 10^{-2}\mathrm{N/m}^2$ measured in the numerical simulations on the section of the downstream wall above the interface. The solid boundaries use a power-law wall function [7], while the free-surface of the channel is an impermeable, free-slip boundary. The 3D grid has $15\times140\times147$ cells in the x, y and z directions respectively with $15 \times 100 \times 120$ cells concentrated within the cavity. Cells within the cavity are of uniform size in the x and y directions with $\Delta x = 2$mm and $\Delta y = 1$mm. In the z direction, the cell width is generally $\Delta z = 1$mm except in the upper part of the cavity where the grid is compressed to give $\Delta z = 0.5$mm at the top of the cavity. Outside the cavity the grid is expanded in the y and z directions to reduce computation time. The 2D simulation is performed on the same grid with a single cell in the x direction.

The spatial discretization uses second-order central differences for all terms in the momentum equations and pressure correction equation. Second-order central differences are also used for the diffusion terms in the mixture fraction equation while SHARP [8] is used for the advection terms to prevent numerical oscillations. The solution is advanced in time with one iteration per time-step using a third-order hybrid Adams-Bashforth/Adams-Moulton scheme for the momentum equations and Crank-Nicolson for the mixture fraction equation. The time step is varied automatically to give a Courant number of approximately 0.4. Further description of the numerical scheme may be found in [9,4].

The 2D simulation takes approximately 40 hours to complete on a Compaq DS-10. The 3D simulation, which integrates over the time $t = 100 - 200$s, takes approximately 2 weeks to run and uses approximately 250MB of RAM. Calculations are performed in double-precision. Typical conservation errors per time step for the 3D run calculated as the L_2 norm are: momentum $\sim 5.0 \times 10^{-5}$, mass $\sim 1.0 \times 10^{-9}$, mixture fraction $\sim 5.0 \times 10^{-5}$.

3 Results

The purging process is divided into four stages based on the dominant flow structures which occur at different times during the process. Solute concentration fields representative of each stage are shown in Figure 1.

During Stage 1 ($t = 0 - 5$s) the impulsive start-up pushes a large splash of salt water out of the cavity and generates a vortex at the upstream corner. As the vortex traverses the cavity it forms a breaking wave on the interface.

During Stage 2 ($t = 5 - 20$s) the wave reaches the downstream wall and draws the lighter fresh water down into the cavity. Convective and shear instabilities cause rapid mixing in the upper section of the cavity.

During Stage 3 ($t = 20 - 100$s) Kelvin-Helmholtz type instabilities formed in the shear layer at the top of the cavity are deflected downwards by the downstream wall. The vortices then impinge on the density interface generating waves and overturns which mix the salt and fresh water. The effect of these vortices weakens as the interface level drops.

During Stage 4, ($t = 100 - 200$s) the large clockwise-rotating vortex on the downstream side continues to grow and generates breaking waves on the density

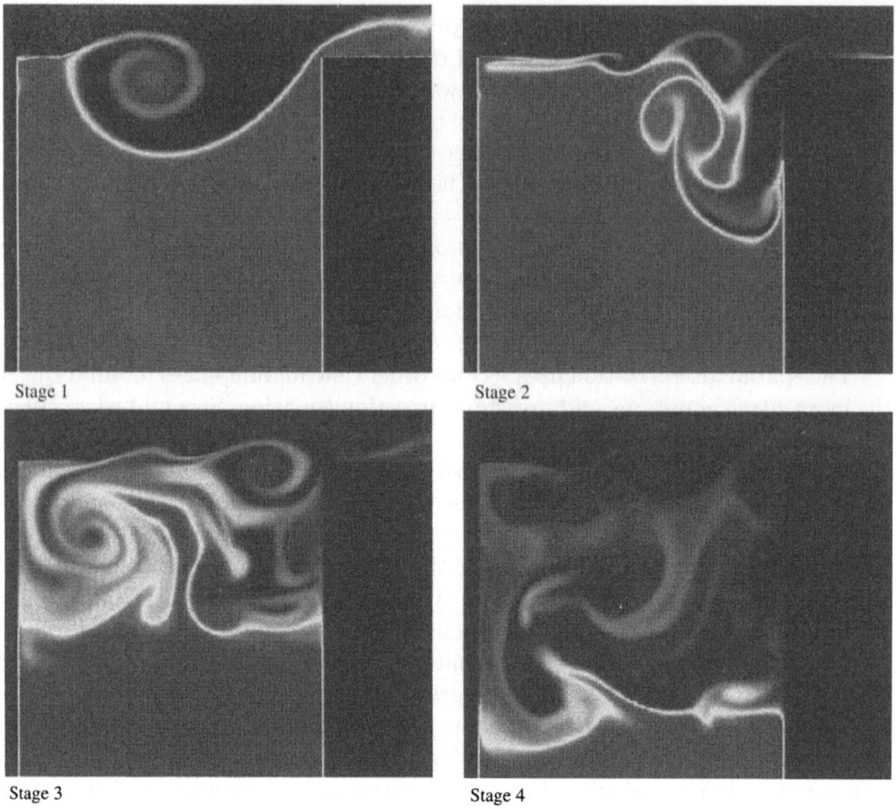

Fig. 1. Four stages which occur during the purging process, flow from left to right

interface. Streamers ejected from the peaks of the waves are the most important mechanism for the transport of salt water across the interface during this stage. While the large vortex generating the waves is essentially two-dimensional, the formation of the streamers is the result of turbulent mixing at the interface. The 2D simulation relies on the subgrid model to give the rate of mixing necessary to correctly generate these streamers.

Figure 2 shows a comparison of numerical results for the minimum interface H height plotted against time with experimental results obtained by the authors for the same flow. There is very good agreement between the numerical and experimental results for Stages 1–3 ($t = 0 - 100s$) with the numerical profile generally overlaying the experimental data points. (The higher interface height in the numerical results during $t = 5 - 25s$ is due to slightly different initial conditions.) During Stage 4 agreement is not as good, with a purging rate predicted by the simulation of 0.16mm/s compared to the experimental result of 0.23mm/s – an error of approximately 30%. This is nevertheless a considerable improvement over results obtained using RANS and 2D direct simulations. From

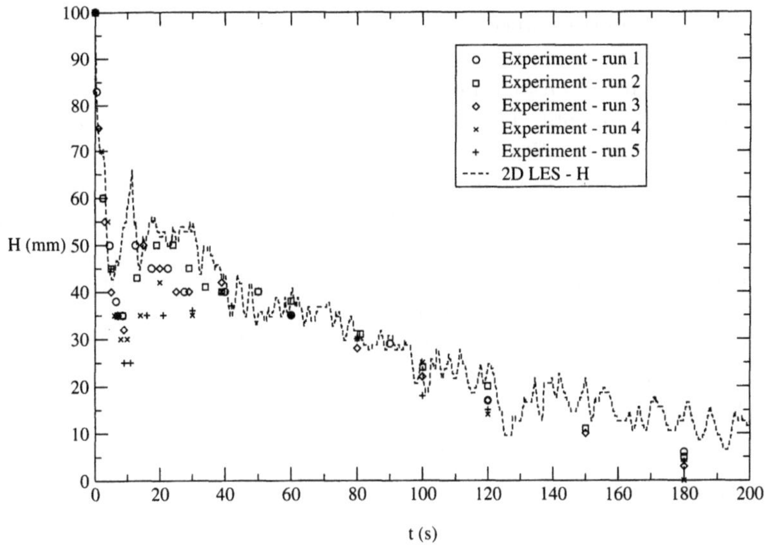

Fig. 2. Comparison of experimental and 2D numerical interface height

this we can conclude that the subgrid model is giving a reasonable prediction of the mixing rate at the surface.

A 2D simulation cannot, however, be considered a true LES since in real flows turbulence must always be three-dimensional. Because interactions between the turbulent velocity field above the interface and the breaking waves are believed to be important in Stage 4, the simulation of this stage was repeated in 3D. Inspection of the 3D results (not shown) indicated significant three-dimensional structure in the flow field above the interface. The most obvious feature is the formation of counter-rotating streamwise vortices in the turbulent boundary layer adjacent to the downstream wall and above the interface. Distortion of the interface by these streamwise vortices adds a spanwise component to the waves generated on the interface. These effects did not enhance the rate of mixing at the surface, however, and in fact the purging rate for the 3D simulation was slightly lower (0.14mm/s) than that obtained in the 2D simulation.

The model coefficient calculated by the dynamic Smagorinsky subgrid model ranges between approximately -0.15 and 0.2 – a range of values typical of turbulent flows. The maximum eddy diffusivity during Stage 4 occurs at the interface and is approximately 650 times the molecular diffusivity. This is a further indication of the importance of the subgrid scales in this simulation.

4 Conclusions

Results have been presented for a flow in which salt water is purged from a square cavity by an overflow of fresh water. The results were used to describe the important transport mechanisms which occur during the purging process and to assess the performance of the large eddy simulation approach for flows of this type. The purging process is divided into four stages based on the dominant flow structures seen at different times during the course of the numerical and physical experiments. The purging rate predicted by the 2D LES was in very good agreement for Stages 1–3. Agreement during Stage 4 is not as good, although the purging rate predicted by the simulation is still within 30% of the experimental result. This is a significant improvement over unsteady RANS simulations which over-predict the purging rate by an order of magnitude, and 2D DNS which under-predicts it by a similar degree. The simulation of Stage 4 was repeated in 3D with results showing that both the interfacial waves and the flow field in the cavity have a complex three-dimensional structure. The 3D simulation however fails to provide any improvement in the prediction of the purging rate during Stage 4.

References

1. J. T. McGuckin: 'Environmental considerations of salinity in the Goulbourn river downstream of Shepparton'. In: *Technical Report Series Nos. 118, Arthur Rylah Institute for Environmental Research*, Dept. of Conservation, Forests and Lands, Victoria, Australia (1991)
2. T. J. Craft, N. Z. Ince and B. E. Launder: Dyn. Atm. and Oceans, **23**, 99 (1996)
3. S. W. Armfield and W. Debler: Int. J. Heat Mass Transfer, **36**, 519 (1993)
4. M. P. Kirkpatrick:A large eddy simulation code for industrial and environmental flows. PhD thesis, The University of Sydney, Sydney (2002)
5. J. Smagorinsky: Monthly Weather Review, **91**, 99 (1963)
6. M. Germano, U. Piomelli, P. Moin, and W. H. Cabot: Phys. Fluids, **A3(7)**, 1760 (1991)
7. H. Werner and H. Wengle: 'Large eddy simulation of turbulent flow over and around a cube in a plate channel'. In: *Eighth Symposium on Turbulent Shear Flows, 1991*
8. B. P. Leonard: 'Sharp simulation of discontinuities in highly convective steady flow'. NASA Technical Memorandum 100240 (1987)
9. M. P. Kirkpatrick, S. W. Armfield, J. H. Kent, and T. Dixon: J. ANZIAM, **42(E)**, 856 (2000)

A Large-Eddy Simulation of Two Side-by-Side Circular Cylinders in Cross Flow

Li Chen[1] and Jiyuan Tu[2]

[1] Maritime Platforms Division, DSTO Melbourne, PO Box 4331, Vic. 3001, Australia,

[2] School of Mechanical and Manufacturing Engineering, RMIT University, Australia

Abstract. The 3D wake flow behind two side-by-side cylinders at Re=750 has been studied numerically using Large Eddy Simulation (LES) at a spacing between two cylinders, T, equal to 3 and 1.7 times the diameter (d) of the cylinder respectively. The numerical simulation reproduce the experimental visualisation with a reasonable accuracy. The wake flows studied in this paper are two-dimensional dominant. It is demonstrated that two anti–phase vortex streets form behind two circular cylinders at a large spacing ($T/d = 3$) due to the interaction between the gap vortexes, where as two in-phase vortex streets are obtained at an intermediary spacing ($T/d = 1.7$). The deflection of the gap flow is a consequence of the gap vortex squeezing and amalgamation.

1 Introduction

Flow over a group of cylinders plays an important role in many industries. This flow depends on the ratio T/d, where T is the center–center cylinder spacing and d is the cylinder diameter. A variety of flow patterns, characterized by the behavior of the wake region, may be discerned as the spacing between two cylinders is changed (e.g. Williamson [5] and Kim and Durbin [1]). For a relatively large spacing $T/d \geq 2$, two distinct vortex streets occur. The two streets are coupled, with a definite phase relationship (e.g. Kim and Durbin [1]), and their frequencies are the same to that found in a single cylinder wake. At an intermediary cylinder spacing, i.e. $1.5 \leq T/d \leq 2$, the interaction between the wakes is expected to intensify. The wake flow behind two cylinders is characterized by a deflected and bistable gap flow, forming a narrow and a wide wake street (e.g.Williamson [5] and Kim and Durbin [1]. Two wake streets demonstrate a different frequency. Recently, Meneghini *et. al* [2] carried out a numerical study of the flow interference between two cylinders at $Re = 200$, but they were unable to identify different shedding frequencies for $T/d = 1.5$. These studies greatly improved our understanding of the flow behind two side–by–side cylinders.

However, a better understanding of the physics associated with the interference between wakes is still required, such as what drives the formation of two predominant symmetrical vortex streets behind two side–by–side cylinders when T/d exceeds 2. The physics behind the formation and stability of a narrow and a wide wake is unclear for an intermediary spacing $1.5 \leq T/d \leq 2$. Furthermore, what triggers the changeover of the gap flow deflection from one side to another?

What is the relationship between the dominant frequency in the narrow wake and in the wide wake? The aim of this numerical study is to address these issues.

Most flows over bluff bodies contain dominant periodic motion associated with the shedding of vortices at distinctive frequencies and the vortices obtain most of their large-scale vorticity from the separated shear layers. Except for a low dispersion, these large–scale motions remain strongly coherent for many diameters downstream. Since the purpose of the Large–Eddy Simulation (LES) approach is full resolution of the large–scale turbulent motion, it is the most appropriate computational approach for wake flows.

In this study, three dimensional turbulent wake flow behind two side-by-side cylinders in a cross flow at $Re = 750$, based on free stream velocity U_∞ and d, was investigated numerically using LES simulation at $T/d = 3$ and $T/d = 1.7$. The choice of this low Reynolds number was based on two considerations: (1) the availability of experimental data, and (2) the expected similarity in the structural dynamics of the near wake with flows over a range of Reynolds number from 300 to 300,000.

2 Numerical method

The governing N-S equation was filtered and discretised using the control volume technique. According to Mittal and Moin [3], a second-order central difference scheme for both convection and diffusion terms, as well as a second-order accuracy in the temporal domain were adopted. The coupling between pressure and velocity was maintained using the SIMPLE technique. Given that the Reynolds number of interest in this study was low and our objective was to provide further insight into the physics of wake interferences, the Smagorinsky-Lilly SGS model was used for the unresolved small scales of eddies.

Before presenting any meaningful results, some numerical considerations need to be addressed. First, a careful grid design, which takes into consideration all essential flow scales, is required. According to many experimental and numerical studies, for $Re = 750$, the typical spanwise length scale of the streamwise eddies is about 0.8d and the structure of the streamwise vortices in the near wake is very ordered. Therefore, cylinders of length $3d$ in the spanwise direction were expected to be sufficient for the LES. A non-uniform grid was used to ensure the first point condition near the wall in the wall unit of $d_y^+ \sim 1$ and enough grid points in the near wake region. Secondly, an adequate resolution of three dimensionality in the near bluff body wake is necessary in order to predict the correct wake dynamics. In our simulation, two cases were simulated using different grid points in the spanwise direction. In case one, 6 grid points were used over the length $3d$ in the spanwise direction to resolve the large scale eddies. This space gave around two grid points for each of the three eddies and a total of 1.53×10^5 mesh points was obtained. In case two, 20 grid points were used over the length of $3.6d$ in spanwise direction, which gave a minimum of four spanwise points for each of the three eddies and a total of 6.2×10^5 mesh points.

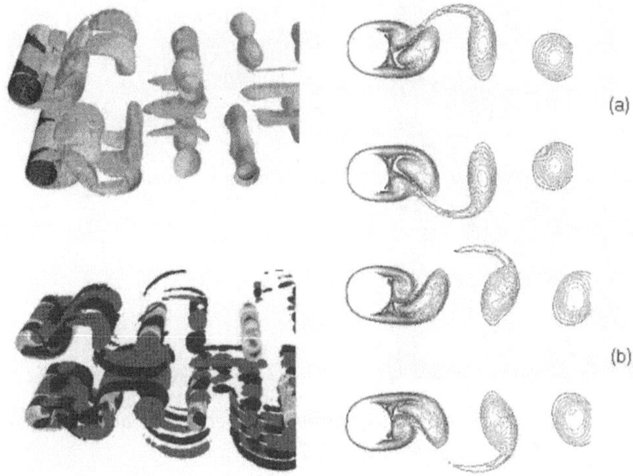

Fig. 1. Iso-surfaces of instantaneous streamwise and spanwise vorticity and contours of spanwise vorticity; (a) for 6 grids over $3d$ and (b) for 20 grids over $3.6d$

The instantaneous iso-surfaces of streamwise vorticity and the contours of the spanwise vorticities predicted using the two meshes are shown in Fig. 1. The fine-scale streamwise vortex structures, termed as 'vortex loops' and 'streamwise vortex pairs', were predicted using both meshes. As one may expect, the coarser mesh smears streamwise eddies and gives an average spanwise wave length $\lambda_c \sim 1.2d$. The case with 20 grid points demonstrates an adequate resolution of the large streamwise eddies and gives $\lambda_c \sim 0.77d$ which closely agrees with the criteria discussed before. However, it is interesting to note from Fig 1 for the contours of the instantaneous spanwise vorticity at $z = 0$, the structure of the spanwise vortices as well as the vortex shedding frequency in term of the Strouhal number were well converged. A further comparison against experimental data confirms that a reasonable resolution of the near wake using the coarser grid is achieved (see Figs. 2. It has been noted that the predicted Strouhal number(0.22), was higher than the experimental data (0.2), but close to other numerical results (e.g. [4]). Therefore, giving the fact that the finner mesh required a significant increase in computing power with a marginal improvement in the solution, it is reasonable to accept the coarse mesh being sufficient to simulate the wake flow reported in this study.

Therefore, the two side-by-side cylinders were finally placed in a computational domain of $22d$ in the streamwise direction (x), $15d$ in the vertical lateral direction (y) and $3d$ in spanwise direction. The upstream of the cylinders consists of $7d$ length from the centerline of the cylinders and $15d$ downstream to exclude the effect of outlet boundary conditions. The $7d$ from the center line of the gap between the cylinders was adopted to implement the free stream boundary condition.

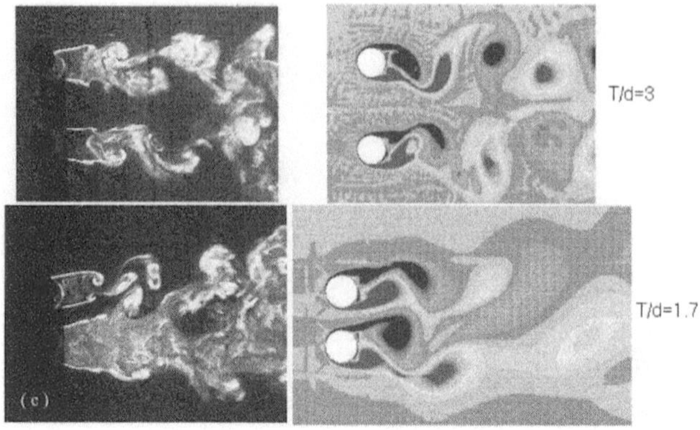

Fig. 2. Experimental visualisation and numerical prediction of instantaneous spanwise vorticity contours for $T/d = 3$ and $T/d = 1.7$

3 Results

The instantaneous spanwise vorticity contours for $T/d = 3$ in Fig. 2 demonstrate two distinct vortex streets and they were dominantly in anti-phase. Based on Von Kámán, with the large spacing, it is impossible to maintain a stable gap vortex street. Since the gap vortices are 'far' apart no squeeze and amalgamation were possible. The two counter-rotating gap vortexes are effectively adjusted each other to maintain two stable vortex streets behind two cylinders. Eventually a symmetrical pressure distribution around the cylinders results leading to two anti-phase vortex streets. A typical pressure distribution on the middle section of two cylinder surfaces and pressure contours at the middle plane of the computational domains are shown in Figs. 3. It can be seen that a mirror-imaged pattern is obtained. This is internally consistent with the generation of the two symmetrical vortex streets. The resultant forces acting on the two cylinders are parallel which closely agrees with the experimental data by Zhou *et. al* [6](not shown). A small variation in pressure distribution on the trailing side of the cylinders is possibly attributed to the wake turbulence. The spectrums of their velocity histories (not shown) demonstrate a pronounced peak at an identical frequency.

However, the experiments by Williamson [5] and Zhou *et. al* [6] observed two unstable in-phase vortex streets. The numerical simulation did not predict this phenomenon. It may suggest that the in-phase mode of the vortex streets is a result of the experimental influence, rather than an intrinsic one.

For $T/d = 1.7$, two asymmetrical vortex streets form in the near wake, marked with a wide wake and narrow wake (see Fig. 4). The gap flow deflects towards one cylinder, and then changes over towards the other with the progress of time. A single vortex street forms further downstream. It can be seen from Fig. 4 that the formation of a stretched gap vortex is a result of amalgamation of gap and outer vortexes. As a consequence, another counter-rotating gap vortex is squeezed and a narrow vortex street results, but this structure is maintained

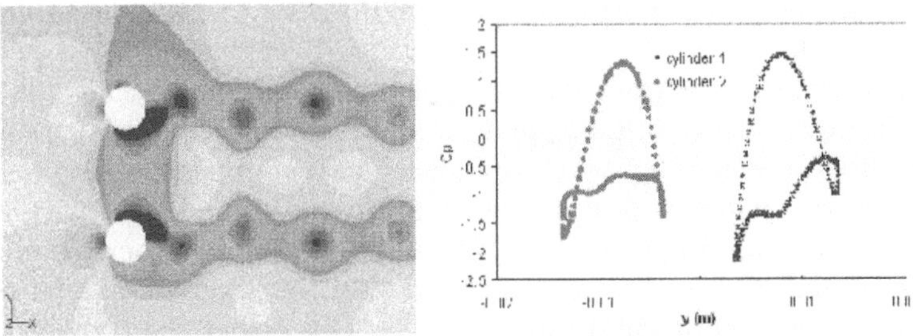

Fig. 3. Pressure distribution around cylinders for $T/d = 3$

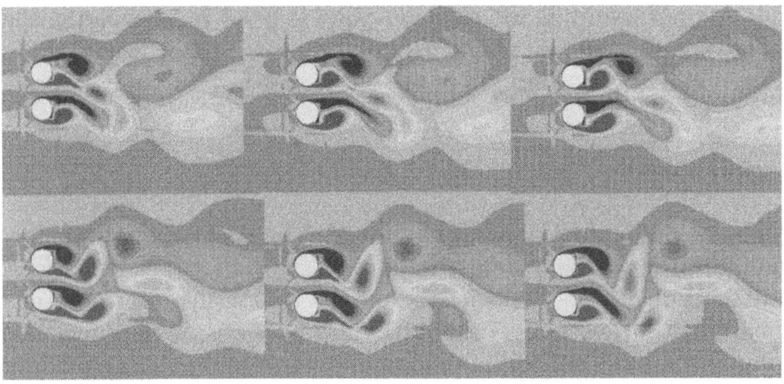

Fig. 4. Evolution of in-phase vortex streets ($T/d = 1.7$)

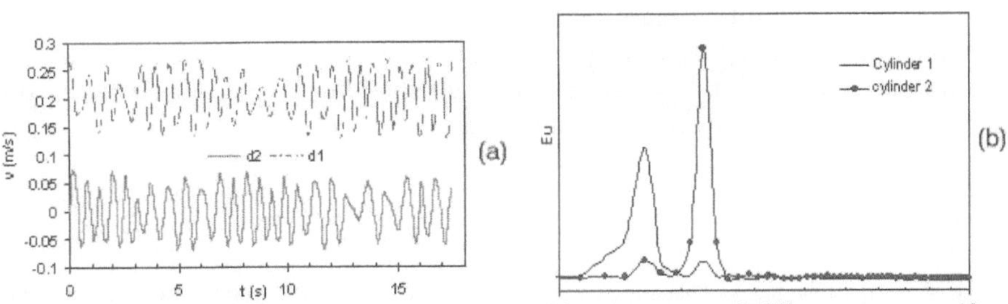

Fig. 5. (a) lateral velocity histories recorded behind two cylinders; (b) Power spectrum of the lateral velocity for $T/d = 1.7$

for a period of a few vortex shedding cycles. It is interesting to note that the gap flow deflection is always following a mature large gap vortex shedding away, but it occurs randomly. This accords with observation of Williamson [5]: the gap flow 'flip' is the consequence of the interference of gap vortices, due to the gap vortexes being squeezed and amalgamated with outer vortex. As may be expected, the narrow vortex street has a lower base pressure and the broad street has a higher base pressure (not shown in this paper). These are in accordance with the results observed experimentally by Williamson [5], Kim and Durbin [1] and Zhou *et al* [6].

The velocity history is tracked at two points: $y = 0.3d$ and $-0.3d$ at $x = 1.5d$ (distance from the center of the cylinder), and is given in Fig. 5(a). Deflection of the gap flow is manifested by the frequency change in the velocity histories, which results two different peak frequencies, $St = 0.16$ and 0.24, with a ratio of 1.5 (shown in Fig. 5(b)). A higher Strouhal number is associated with the narrow wake and a lower Strouhal number with the wide wake. The ratio is different from the experimental value of 3 [6]. Given that the two vortexes in the narrow wake were close to each other, the hot–wise measurement may detect a doubled frequency. It has also been noted that the time scale of gap flow 'flipping' was noticeably larger than the time scale of Karma vortex shedding, but the predicted value was still much lower than the experimental observation by Kim and Durbin [1]. This will be the subject of further study.

4 Conclusion

For $T/d = 3$, the numerical simulation verify the experimental observation of the symmetrical wakes and supports the conclusion that the symmetrical pressure distribution around the cylinders may be responsible for the formation of the symmetrical vortex streets behind two side–by–side cylinders. The numerical simulation did predict the vortex streets of an in-phase mode. This may suggest that the formation of the unstable in-phase vortex streets was due to experimental disturbance. For $T/d = 1.7$, the multiple peaks of shedding frequencies are obtained since the interference of the gap vortexes. The gap flow deflection is the consequence of the squeezing and amalgamation of the vortexes, forming a narrow and a wide vortex street. The ratio of shedding frequencies between the narrow and wide wake was found to be 1.5.

References

1. H. J. Kim and P. A. Durbin:'Investigation of the flow between a pair of circular cylinders in the flopping regime'. In: *J. Fluid Mech.* Vol.196, pp. 431–448, 1988.
2. J. R. Meneghini and F. Saltara:'Numerical simulation of flow interference between two circular cylinders in tandem and side-by-side arrangements'. In: *Journal of Fluid Structure*, vol. 15, pp. 327–350, 2000.
3. R. Mittal and P. Moin:'Suitability of upwind-biased finite-difference schemes for large-eddy simulation of turbulent flows'. In: *AIAA Journal*, Vol. 35, pp.1415–1417, 1997.
4. A. Slaouti and P.K. Stansby: 'Flow around two circular cylinders by the random-vortex method'. In: *Journal of Fluid Structure*, Vol. 6,pp. 641–670, 1992.
5. C. H. Williamson:'Evolution of a single wake behind a pair of bluff bodies'. In: *J. Fluid Mech.*,Vol. 159, pp.1–18, 1985.
6. Y. Zhou, Z.J. Wang, R.M.C. So, S.J. Xu and W. Jin:'Free vibration of two side-by-side cylinders in a cross-flow'. In: *J. Fluid Mech.*, Vol. 443,pp.197–229, 2001.

LES of Turbulent Boundary Layers

Wageeh A. El-Askary, Matthias Meinke, and Wolfgang Schröder

Aerodynamisches Institut, RWTH Aachen
W"ullnerstr. 5-7, 52062 Aachen, Germany

Abstract. The paper presents large-eddy simulations of sub- and supersonic boundary layers. A second-order accurate scheme is used for the simulation of the flow field with non-reflecting boundary conditions and sponge layers at the boundaries of the domain. The LES findings show good agreement with numerical and experimental data of other authors. The boundary layer solutions are then used to provide inflow conditions for the simulation of a subsonic trailing-edge flow and a supersonic ramp flow. For both problems the results are in agreement with experimental and DNS and LES solutions of other works.

1 Introduction

The simulation of turbulent external flows with Large-Eddy Simulation (LES) techniques is both difficult and expensive: First, because such flows usually occur at high Reynolds numbers so that a large number of grid points is necessary for the resolution of the near wall regions. Secondly, because transition from laminar to turbulent flow occurs, which is difficult to predict accurately. The flow phenomena of interest, however, are often located in a limited region, e.g. for trailing edge acoustics in the vicinity of a trailing edge or for hot spots in supersonic flow near shock boundary layer interactions. In such cases it can be sufficient to consider only a small domain of integration just containing that limited region. Then, inflow conditions for a turbulent flow are required, which can be obtained e.g. by a slicing technique from a simultaneously simulated boundary layer. In this paper subsonic and supersonic boundary layers are simulated with LES based on a previously validated method for channel, pipe and jet flows, see [6]. A rescaling technique based on that of Lund et al. [5] and Urbin and Knight [9] is used for the generation of the instantaneous flow variables at the inflow boundary, for which a new rescaling of the inflow temperature profile has been formulated. Then, two examples for sub- and supersonic flows are presented, for which the boundary layer solutions are used as inflow conditions.

2 Method of Solution

A large-eddy simulation is performed to simulate the turbulent compressible flow over a flat plate boundary layer. The discretization of the convective fluxes in the governing equations is a second-order accurate AUSM formulation with a central pressure derivative. The viscous stresses are discretized using central differences of second-order accuracy. An explicit 5-step Runge-Kutta time stepping scheme of second-order accuracy is used for the temporal integration. The scheme is described in more detail in [7], where also results of various validation tests are discussed. The inlet boundary condition for the boundary layer simulation is obtained by extending the method of Lund et al. [5] to compressible flows, which is discussed in the next section. At the surface of the plate the no-slip condition and an adiabatic wall is prescribed, at all outflow boundaries non-reflecting boundary conditions are applied [8]. In addition a sponge layer zone is used, in which the source terms S are computed as a function of the deviation of the instantaneous from the time averaged solution.

The rescaling method of Lund et al. [5] is extended to compressible flows. For the velocity components the procedure proposed by Lund et al. [5] is adopted after a van Driest transformation.

The temperature profile is rescaled as follows. For the mean static temperature T, Walz's equation $T/T_\infty = 1 + A(1 - U^2/U_\infty^2)$ is used, with $A = 0.5(\gamma - 1)rM_\infty^2$, where r represents the recovery factor and M_∞ the freestream Mach number [10]. The temperature fluctuation T' can be calculated assuming negligible total temperature fluctuations $T_t{}'$ compared with the static temperature fluctuation. This assumption is valid for subsonic flows, but for supersonic flows this is only an approximation, since it is known that total temperature fluctuations can attain values in the order of about 10%. Following Bradshaw [2], the static temperature fluctuation is $T'(y,z,t)/T(y,z,t) = -(\gamma - 1)\, M_l^2 u'(y,z,t)/u(y,z,t)$, where M_l is the local Mach number. Using the equation of Walz [10] and formulating the rescaling process similar to the velocity the following equations for the rescaled temperature and its fluctuations are obtained

$$T_{in}^{inner} = \beta_s^2 T_{re}(y_{in}^+) + C_1 T_\infty \quad \text{and} \quad T_{in}^{outer} = \beta_s^2 T_{re}(\eta_{in}) - C_2 \tfrac{U_{re}(\eta_{in})}{U_\infty} T_\infty + C_3 T_\infty$$

$$T'^{inner}_{in} = \beta_s^2 T'_{re}(y_{in}^+, z, t) \ \text{and} \ T'^{outer}_{in} = \beta_s^2 T'_{re}(\eta_{in}, z, t) - C_2 \tfrac{(u')_{re}(\eta_{in}, z, t)}{U_\infty} T_\infty$$

with $C_1 = (1+A)(1-\beta_s^2)$, $C_2 = 2A\beta_s(1-\beta_s)$, and $C_3 = (1-\beta_s)(1+\beta_s+2A\beta_s)$. If the rescaling factor $\beta_s = u_{\tau,in}/u_{\tau,re}$ is assumed to be 1 for the temperature distribution, the formulation that was used by Urbin and Knight [9] is obtained.

The composite distribution that is approximately valid over the entire boundary layer is obtained by forming a weighted average of the inner and outer profile for the velocity and the temperature as was proposed by Lund et al. [5].

The rescaling operation requires the skin friction velocity u_τ and the boundary layer thickness δ, both at the rescaling station and at the inlet. They can be directly determined from the mean velocity profile at the rescaling station, but must be specified at the inlet. It is sufficient to fix δ, whereas u_τ for compressible flow is evaluated via $u_{\tau,in} = u_{\tau,re} \sqrt{\left(\rho_{(wall)_{re}}/\rho_{(wall)_{in}}\right)} \left(\delta_{1re}/\delta_{1in}\right)^{\frac{1}{[2(n-1)]}}$, where δ_1 is the displacement thickness, $\rho_{(wall)}$ is the local density at the wall and the exponent n is set to $n=5$. The above relation can be derived from the standard power-law approximations, $c_f \sim Re_x^{-1/n}$, $\delta_1/x \sim Re_x^{-1/n}$, where c_f is the skin-friction coefficient. Unlike Lund et al. [5] where the momentum thickness is used we incorporate the displacement thickness in the rescaling formulation.

Since the momentum thickness is a non-linear function of the velocity profile, slightly different results are obtained when either the spanwise averaged velocity profile is used to compute the momentum thickness or the momentum thickness is computed locally and then averaged in the spanwise direction.

3 Results

The simulated turbulent boundary layers correspond to the cases conducted by Lund et al. [5] for a Mach number of M_∞=0.4 and Urbin and Knight [9] for a Mach number of M_∞=3.0. All parameters are summarized in table 1.

The mean streamwise velocity profiles are plotted in inner-law scaling in Fig. 1 for the subsonic and in outer scaling for the supersonic case. Both solutions closely follow the analytic form or the reference data taken from the DNS [1]. Fig. 2 illustrates the mean temperature-velocity relationship. The simulation data fits well with the Walz's quadratic profile for an adiabatic wall [10] and other reference data. The turbulence intensities are compared in Fig. 3 with the data of Lund et al. [5] at $Re_\theta = 1850$ for the subsonic flow and with the DNS of Adams [1] and other reference data for the supersonic flow. The good agreement of the present results with experimental data and other LES or DNS for both the sub- and supersonic turbulent boundary layer is clearly visible.

The boundary layer solutions are used for the generation of inflow boundary conditions for a sub- and supersonic flow case. A slicing technique is used in which the instantaneous velocity and temperature distribution at each time level is generated from the simultaneous simulation of a turbulent boundary layer, see Figs. 4,5. The domain of integration comprises only the rear part of the flat plate and the wake region or only a small part of the ramp, respectively. The parameters for the two cases are summarized in table 2.

The vortex structures are visualized with surfaces of constant λ_2 following [4] in Figs. 6,7 for the two different cases considered. Comparisons of the velocity profiles with experimental data, which were obtained with pitot probe measurements by the University of Dresden and Stuttgart for the wake flow, show good agreement in Fig. 8. The velocity profiles for the ramp flow (for brevity not shown here) closely follow the DNS data of Adams [1]. The surface pressure distribution in Fig. 9 also shows good agreement with the reference data.

4 Conclusions

A consistent extension of the rescaling method for incompressible turbulent boundary layers to compressible fluids has been developed. The results for sub- and supersonic turbulent boundary layers indicate that the proposed rescaling of the temperature profile seems to be superior to simpler formulations. LES of the flow past a sharp trailing edge and the simulation of a supersonic ramp flow show good agreement with experimental data and other reference data.

References

1. N. A. Adams. Direct numerical simulation of turbulent compression ramp flow. *J. of Theoretical and Computational Fluid Dynamics*, 12:109–129, 1998.
2. P. Bradshaw. Compressible turbulent shear layers. *Ann. Rev. Fluid Mech*, 9:33–54, 1977.
3. R. Ewert, W. El-Askary, M. Meinke, and W. Schröder. Computation of turbulence related noise on the basis of large-eddy simulation. In H. Krner and J. W. Delfs, editors, *Second Aeroacoustics Workshop*. DLR Braunschweig, Oct. 2000.
4. J. Jeong and F. Hussain. On the identification of a vortex. *JFM*, 285:69–94, 1995.
5. T. S. Lund, X. Wu, and D. Squires. Generation of turbulent inflow data for spatially-developing boundary layer simulations. *JCP*, 140:233–258, 1998.
6. M. Meinke and E. Krause. Application of LES to jets and internal turbulent flows. In R. Peyret and E. Krause, editors, *Advanced Turbulent Flow Simulation*, pages 155–208. Springer, 2000.
7. M. Meinke, W. Schr"oder, E. Krause, and T. Rister. A comparison of second- and sixth-order methods for large-eddy simulations. *Comp. & Fluids*, 31:695–718, 2002.
8. T. J. Poinsot and S. K. Lele. Boundary conditions for direct simulations of compressible viscous flows. *JCP*, 101:104–129, 1992.
9. G. Urbin and D. Knight. Compressible large eddy simulation using unstructured grid: supersonic boundary layer. In *Second AFOSR International Conference on DNS/LES, Rutgers University*, pages 1–16, June 7-9 1999.
10. A. Walz. *Boundary layers of flow and temperature.* MIT Press, 1969.

Table 1. Flow parameters and size of the domain of integration L for the boundary layer simulations

$Re_{\delta o}$	Re_x	Re_θ	M_∞
14400	$5.5 \cdot 10^5$	1400	0.4
L_x	L_y	L_z	grid points
$30\delta_0$	$3.5\delta_0$	$0.64\delta_0$	$2.7 \cdot 10^6$

$Re_{\delta o}$	Re_x	Re_θ	M_∞
20000	$8.25 \cdot 10^5$	1945	3
L_x	L_y	L_z	grid points
$15\delta_0$	$3.5\delta_0$	$0.64\delta_0$	$1.6 \cdot 10^6$

Table 2. Parameters and domain of integration for the wake and ramp flow simulations. Left: wake flow, right: ramp flow. See also Fig. 4 and Fig. 5

Re_L	$Re_{\delta o}$	δ_o/D	M_∞
$5.33 \cdot 10^5$	9869	3.76	0.15
L_1	L_2	L_z	grid points
$13.5\delta_0$	$17\delta_0$	$0.64\delta_0$	$2.25 \cdot 10^6$
Δx^+_{min}	Δx^+_{max}	Δy^+_{min}	Δz^+
2	30	2	25

$Re_{\delta o}$	Re_x	Re_θ	M_∞	α
20000	$8.25 \cdot 10^5$	1945	3	8°
L_1	L_2	L_z	grid points	
$-5.5\delta_0$	$5.5\delta_0$	$0.64\delta_0$	$2.9 \cdot 10^6$	
Δx^+_{min}	Δx^+_{max}	Δy^+_{min}	Δz^+	
0.5	12.4	0.5	7	

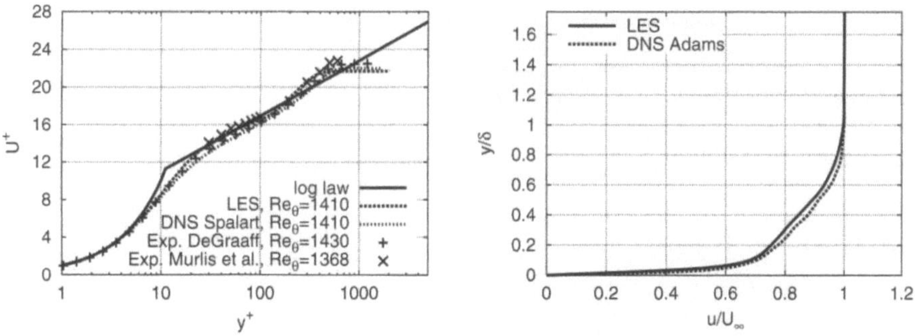

Fig. 1. LES of a turbulent boundary layer. Left: Velocity profiles in inner scaling for the subsonic case, Right: Velocity profiles in outer scaling for the supersonic case

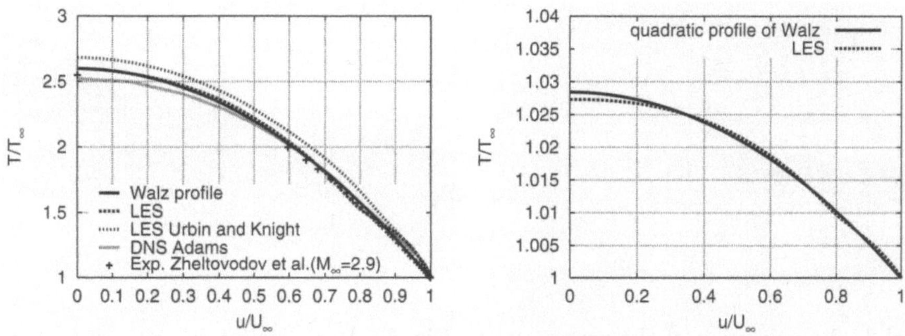

Fig. 2. LES of a turbulent boundary layer. Temperature-velocity relationship compared with the profile of Walz [10] and other reference data. Left: subsonic case, Right: supersonic case

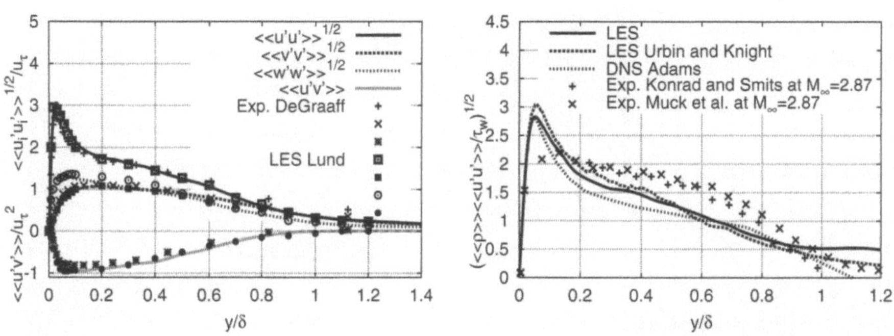

Fig. 3. LES of a turbulent boundary layer. Reynolds stress components in comparison to experimental and other reference data. Left: subsonic case, Right: supersonic case

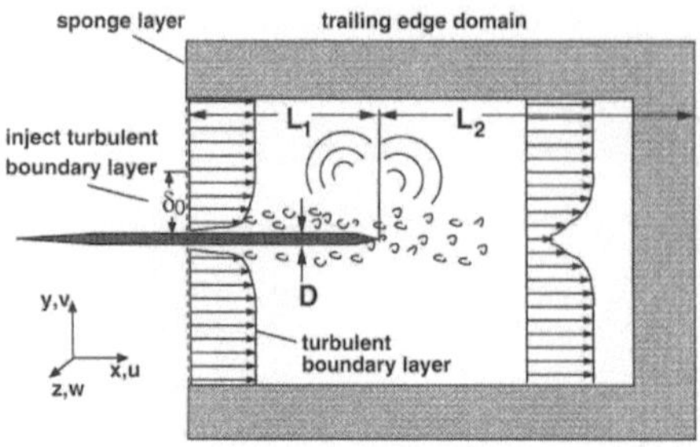

Fig. 4. Sketch of the domain of integration for the trailing edge simulation.

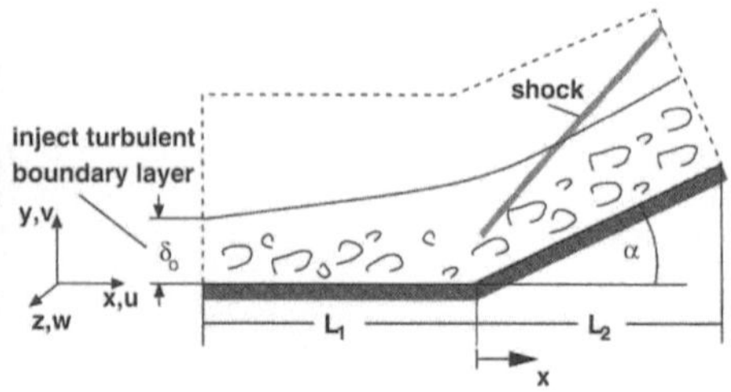

Fig. 5. Sketch of the domain of integration for the ramp simulation.

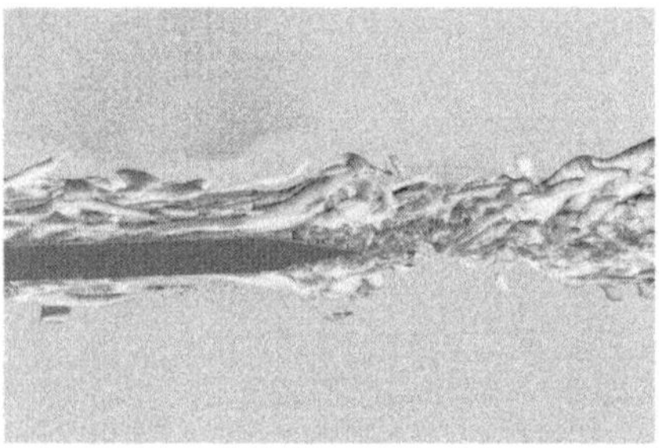

Fig. 6. Visualization of the vortex structures in the boundary layer and in the near wake of the turbulent M_∞=0.15 flow around a sharp trailing edge.

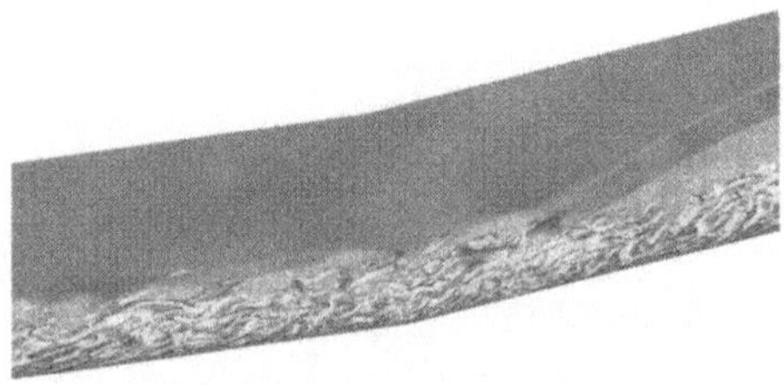

Fig. 7. Visualization of the vortex structures and the shock surface for the ramp flow at a Mach number of M_∞=3 and an angle of 8°.

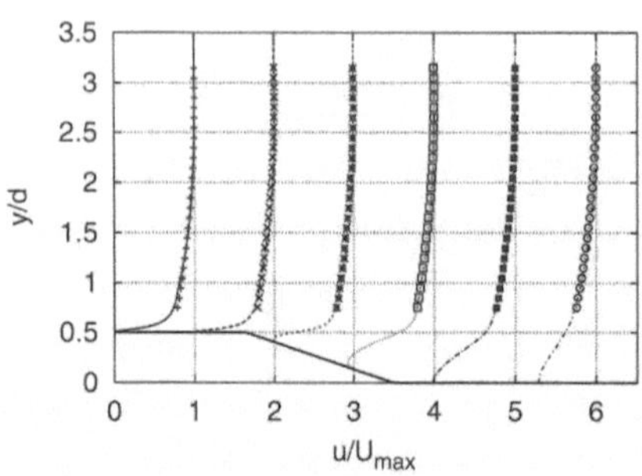

Fig. 8. LES of a trailing-edge flow. Mean velocity profiles in the trailing-edge and the near-wake region compared with experimental data from the University of Dresden and Stuttgart (Symbols).

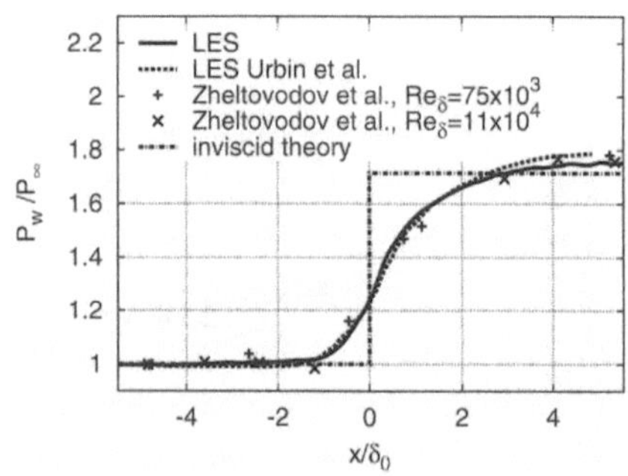

Fig. 9. LES of a supersonic ramp flow Re_x=8.25·10^5 and M_∞=3. Surface pressure distribution.

Part XI

Mesh

An Adaptively-Moving-Grid Finite-Volume Scheme with Boundary-Grid Elimination/Addition

Kenichi Matsuno

Kyoto Institute of Technology, Kyoto 606-8585 JAPAN

1 Introduction

One of interesting problems in CFD is an interaction between fluid motion and body/wall boundary motion. To simulate accurately such a problem with moving boundary, we often encounter two problems. For achieving sufficient resolution of a solution, we have to use enough number of grid points. The increase of grid points, however, means the increase of CPU time. One of the techniques to overcome this problem is to develop an adaptive grid method for moving grid system.

Another one accompanying moving-boundary problem appears when the size of the flow region changes excessively according to the wall-boundary motion. It is, for example, typically seen for the moving piston inside a closed cylinder. As the piston moves toward the end of the cylinder, the flow region becomes narrower and the grid becomes excessively fine. Then the numerical stability becomes worse and results in the use of smaller time step accordingly. On the contrary, for the case that the piston moves away from the end of the cylinder the flow region becomes wide and the grid spacing becomes far wider than the original state. In that case, the accuracy of the computation becomes worse. If we can remove or add the grid points from or to the region, the computation becomes more efficient, numerically stable and accurate.

The objective of this paper is twofold. One is to present a new solution-adaptively-moving-grid finite-volume scheme for non-stationary problems of viscous compressible flows with traveling wall boundaries. The other is to propose a new technique of boundary point elimination/addition procedure and to incorporate it to the present moving-grid finite-volume scheme.

The present scheme combines the moving-mesh finite-volume scheme[1] and the elliptic adaptive-grid generation method[4] into the new adaptively-moving-grid finite-volume scheme with traveling wall boundaries. The present scheme also includes a new boundary-grid elimination/addition procedure in order to keep grid density at proper level for the moving boundary problems. As for the boundary grid treatment, the paper presents the new technique of boundary-grid elimination/addition. The elimination or addition of the boundary grid is performed straightforwardly in the present moving-grid finite-volume scheme.

2 Adaptively-Moving-Grid Finite-Volume Scheme

2.1 Geverning Equation

For a three-dimensional compressible flow, the Navier-Stokes equation may be expressed as

$$\frac{\partial}{\partial t}Q + \frac{\partial}{\partial x}(E - E_V) + \frac{\partial}{\partial y}(F - F_V) + \frac{\partial}{\partial z}(G - G_V) \tag{1}$$

where, x, y and z are coordinates. Here, $Q = (\rho, \rho u, \rho v, \rho w, e)$ is a vector of conservative variables, E, F and G are inviscid fluxes, E_V, F_V and G_V are viscous fluxes. No turbulence model is introduced in this paper and the laminar flow is assumed throughout. Equation(1) can be represented using extended divergent operator $\tilde{\nabla} = (\partial/\partial x, \partial/\partial y, \partial/\partial z, \partial/\partial t)$ and flux tensor $\tilde{\boldsymbol{F}} = (E - E_V, F - F_V, G - G_V, Q)$ as

$$\tilde{\nabla}\tilde{\boldsymbol{F}} = 0. \tag{2}$$

2.2 Moving-Grid Finite-Volume Scheme

Since body-wall boundaries move and change their relative positions, the body-fitted grid system must dynamically change and deform its shape according to the movement of the wall boundary. Moreover, since the flow is non-stationary, the solution-adapted grid is also dynamically changing its shape according to the flow change. As the present method is finite-volume method and we construct a control volume in the frame work of non-stationary grid system. The moving-grid finite-volume scheme was presented in Ref.[1], and here, we present the scheme for three dimensional viscous flow field.

Now let (ξ, η, ζ) be general curvilinear coordinate system. The coordinate system is assumed to be flow-dependent as well as time-dependent. Thus $\xi = \xi(x, y, z, t, Q)$, $\eta = \eta(x, y, z, t, Q)$ and $\zeta = \zeta(x, y, z, t, Q)$. The present scheme is based on a cell-centered finite-volume method and we define flow variables at the cell center (i, j, k, n), where i, j, k, and n are indices of ξ, η, ζ and t directions respectively. Note that the same indices, i, j and k are also used for grid node point, but the position is not the same. Now let the flow and grid system at the time level $n+1$ $(t = t^{n+1})$ be unknown, assuming that the all flow variables and grid system at the time-level up to n $(t = t^n)$ are completely known.

Now we consider a octahedral control volume, of which apexes are the grid points: $\mathbf{r}^n_{i,j,k}$, $\mathbf{r}^n_{i+1,j,k}$, $\mathbf{r}^n_{i+1,j+1,k}$, $\mathbf{r}^n_{i,j+1,k}$, $\mathbf{r}^n_{i,j,k+1}$, $\mathbf{r}^n_{i+1,j,k+1}$, $\mathbf{r}^n_{i+1,j+1,k+1}$, $\mathbf{r}^n_{i,j+1,k+1}$, $\mathbf{r}^{n+1}_{i,j,k}$, $\mathbf{r}^{n+1}_{i+1,j,k}$, $\mathbf{r}^{n+1}_{i+1,j+1,k}$, $\mathbf{r}^{n+1}_{i,j+1,k}$, $\mathbf{r}^{n+1}_{i,j,k+1}$, $\mathbf{r}^{n+1}_{i+1,j,k+1}$, $\mathbf{r}^{n+1}_{i+1,j+1,k+1}$, and $\mathbf{r}^{n+1}_{i,j+1,k+1}$. The control volume stands in unified space-time (x, y, z, t)-domain which is four-dimensional for three-dimensional flows. Now let Ω be the octahedral control volume and $\partial\Omega$ be its boundary, We apply volume integration with respect to Ω to Eq.(2). Then Eq.(2) can be written in surface integral form using the Gauss theorem:

$$\int_{\Omega} \tilde{\nabla}\tilde{\boldsymbol{F}} dV = \int_{\partial\Omega} \tilde{\boldsymbol{F}} \cdot \mathbf{n} dS = 0, \tag{3}$$

where, $\mathbf{n}$ is a unit vector outwardly normal to the control volume surface.

Discretized form of Eq.(3) becomes,

$$\sum_{\ell=1}^{8} \tilde{\boldsymbol{F}}_\ell \cdot \tilde{\mathbf{n}}_\ell = 0. \tag{4}$$

where, $\tilde{\mathbf{n}}_\ell = (n_x, n_y, n_z, n_t)_\ell$ $(\ell = 1, 2, \ldots, 8)$ is the normal vector of the surface of the octahedral control volume, and the length of the vector equals to the area of the surface. When the surfaces, say, $\ell = 7$ and $\ell = 8$, are the surface of the control volume which are perpendicular to the t-axis, they have only n_t component and correspond to the volumes in the (x, y, z)-space at time t^{n+1} and t^n, respectively. Thus, Eq.(4) can be expressed as,

$$\frac{Q^{n+1}_{i,j,k} - Q^n_{i,j,j}\,(V^n_{i,j,k}/\,V^{n+1}_{i,j,k})}{\Delta t} + \frac{1}{\Delta t\, V^{n+1}_{i,j,k}} \sum_{\ell=1}^{6} \tilde{\boldsymbol{F}}^{n+1/2}_\ell \cdot \tilde{\mathbf{n}}_\ell = 0. \tag{5}$$

Here $\tilde{\boldsymbol{F}}^{n+1/2} = (\tilde{\boldsymbol{F}}^{n+1} + \tilde{\boldsymbol{F}}^{n})/2$, and we use $(n_t)_{\ell=7} = V^{n+1}_{i,j,k}$ and $(n_t)_{\ell=8} = V^n_{i,j,k}$ Flux tensor, $\tilde{\boldsymbol{F}}_\ell$ $(\ell = 1, 2, \ldots, 6)$, are evaluated using the Roe flux difference splitting scheme[2] with MUSCL approach for accurate estimation.

Equation (5) has to be coupled with the elliptic grid generation system and must be solved simultaneously. Since Eq.(5) is nonlinear, we have to introduce iterative method of solution. The present scheme employs subiteration strategy with a pseudo-time approach[3] with LDU implicit factorization at every time step.

2.3 Elliptic System for Adaptive Gridding

Now let $\mathbf{r} = (x, y, z, t)$ be a position vector of the grid point at time t and assume that the solution adapted grid at time t^{n+1} is determined by the flow solution Q^{n+1}. Then the elliptic system which generate solution adaptive grid at time t^{n+1} can be expressed by the following elliptic equation[4] $(\boldsymbol{r} = (x, y, z)^T)$:

$$\begin{aligned}
&\alpha(\boldsymbol{r}^{n+1<\nu'+1>}_{\xi\xi} + \phi \boldsymbol{r}^{n+1<\nu'+1>}_{\xi}) + \beta(\boldsymbol{r}^{n+1<\nu'+1>}_{\eta\eta} + \psi \boldsymbol{r}^{n+1<\nu'+1>}_{\eta}) \\
&\quad + \gamma(\boldsymbol{r}^{n+1<\nu'+1>}_{\zeta\zeta} + \omega \boldsymbol{r}^{n+1<\nu'+1>}_{\zeta}) \\
&\quad + 2\kappa \boldsymbol{r}^{n+1<\nu'+1>}_{\xi\eta} + 2\lambda \boldsymbol{r}^{n+1<\nu'+1>}_{\eta\zeta} + 2\mu \boldsymbol{r}^{n+1<\nu'+1>}_{\zeta\xi} = 0,
\end{aligned} \tag{6}$$

Equation(6) is solved iteratively by SLOR and superscript ν' is an iteration counter for this elliptic system. The information of the flow solution is given to the control function ϕ, ψ and ω, see Ref.[4] for detail. Since the flow is non-stationary, the solution adaptive gridding must be performed at every time step. The difficulty is that both the flow solution and the grid system at t^{n+1} are unknown and they are dependent each other and have to be calculated simultaneously. Combination with flow solver will be given in the next section.

2.4 Simultaneous Subiteration

To advance one time step, from t^n to t^{n+1}, there are two types of iterations: for flow solver and grid generation. Both iterations must be converged simultaneously at every time step. The flow-change affects the estimation of the control functions, ϕ, ψ and ω, since the control functions are estimated by the weight functions which are the function of flow solution. While the flow is solved on the grid system which is the solution of the elliptic equations. Thus to get the converged solution the exchange of the influences of each other must be properly performed. Although flow-iteration and grid-iteration are both dependent on each other, each iteration can be performed separately if exchange of the data is properly performed. Thus the present scheme performs each subiteration alternatively. The iteration for elliptic system, however, converged relatively quickly and after few iteration the grid movement terminates. Once the criterion to check the grid movement is satisfied, the iteration for grid generation is terminated and only iteration for flow solver is performed until it is converged. Our experience of the present study showed that the movement of the grid point during the flow subiterations stopped after first 3 or 4 subiteration cycles.

3 Boundary-Grid Elimination/Addition

In this section, we propose a method to eliminate or add the grid point at the wall boundary. The present new method is to add or eliminate grid points from the computational region according to the movement of the wall boundaries in order to keep the total grid density at proper level. The basic idea is to utilize the merit that the control volume of the present moving-grid finite volume scheme is constructed in the unified space-time domain (x,y,z,t). If the grid point is eliminated then we only change the control volume as can be illustrated in Fig. 1.

When we eliminate the grid point, for instance, $(i-1)$ at $t = t^{n+1}$, we consider the surface of the control volume at t^n ($\ell = 8$) to be sum of the $i-1$ and i surface

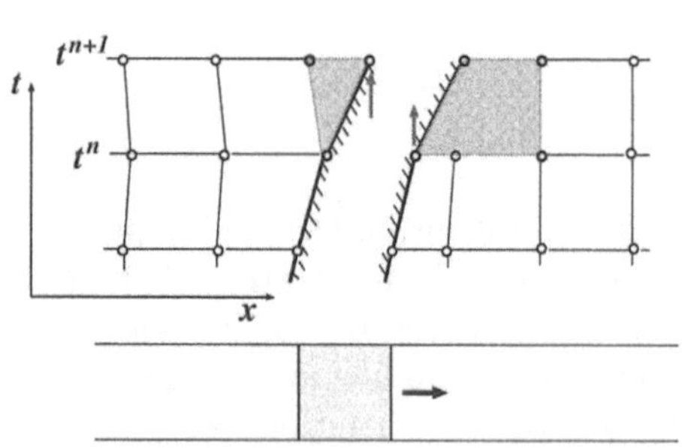

Fig. 1. Control volume for addition and elimination of boundary grid point(cell) for one-dimensional flow

of the octahedron (see Fig.1). Thus Eq.(5) is replaced with

$$\frac{Q^{n+1}_{i,j,k} - \left(Q^{n}_{i-1,j,j}V^{n}_{i-1,j,k} + Q^{n}_{i,j,j}V^{n}_{i,j,k}\right)/V^{n+1}_{i,j,k}}{\Delta t} + \frac{1}{\Delta t\, V^{n+1}_{i,j,k}} \sum_{\ell=1}^{6} \tilde{\boldsymbol{F}}^{n+1/2}_{\ell} \cdot \tilde{\mathbf{n}}_{\ell} = 0. \tag{7}$$

for elimination of the grid point.

When the grid point (i) is added then we put $V^{n}_{i,j,k}$ to be zero and the control volume in the (x, y, z, t) space becomes heptahedral as illustrated in Fig.1. Thus Eq.(5) can be replaced with

$$\frac{Q^{n+1}_{i,j,k}}{\Delta t} + \frac{1}{\Delta t\, V^{n+1}_{i,j,k}} \sum_{\ell=1}^{6} \tilde{\boldsymbol{F}}^{n+1/2}_{\ell} \cdot \tilde{\mathbf{n}}_{\ell} = 0. \tag{8}$$

for addition of the grid point at the wall boundary. The present method treats the grid points such that they appear to go into or through the wall boundary.

4 Numerical Examples

First example is a simulation of the axisymmetric shock tube problem. The cylinder length is 2.5 times of the diameter. Both end walls are stationary and fixed. Figure 2 shows the result of the flow field and adaptive grid.

Next example is a three-dimensional computation of shock-boundary layer interaction in the square duct with moving wall boundary of constant acceleration followed by constant speed. The end wall at $x = 0$ at $t = 0$ moves towards the other end ($x = 1$ fixed). Figure 3 is a result at time $t = 0.9$ with wall position $x = 0.75$ Figure 3 shows Mach number contours on the cut planes of $x = 0.833$ and $y = 0$. Typical interactive flow field can be seen.

The final example is a test problem for the new boundary-grid elimination/addition method applied to the one-dimensional flow of the traveling piston in the closed cylinder. Figure 4 shows results. Without boundary-grid elimination/addition (Fig.4 left), the grid density in the compression side becomes excessively fine as the piston proceeds close to the end of the cylinder, while the

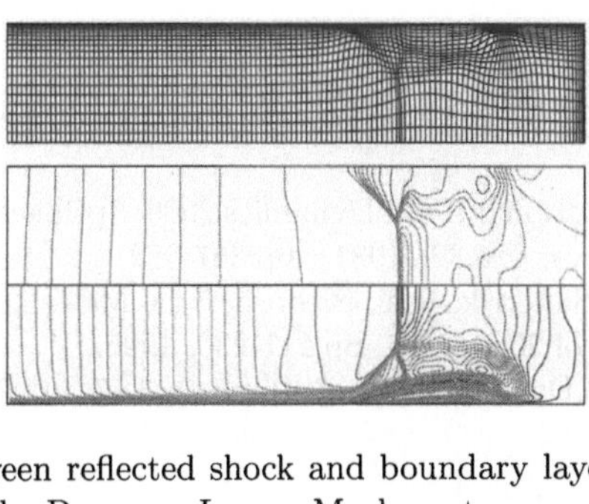

Fig. 2. Interaction between reflected shock and boundary layer in a shock tube. Upper:Adaptive grid. Middle: Pressure. Lower: Mach contours.$p_{41} = \rho_{41} = 10$, Re=5000

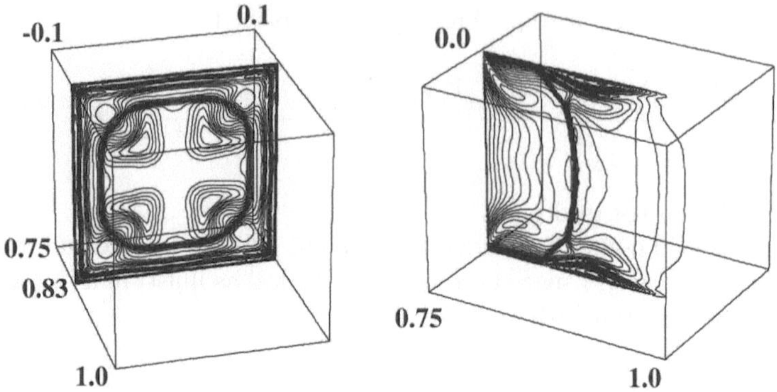

Fig. 3. Shock-boundary layer interaction generated by the traveling wall in a square duct, Mach number contours with constant interval of 0.05.

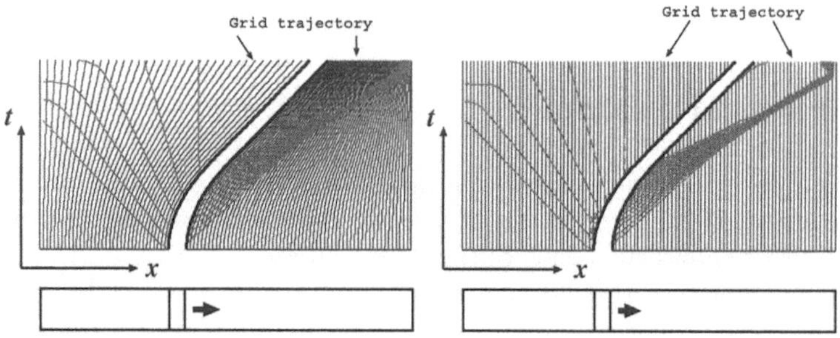

Fig. 4. Illustrative results for one-dimensional piston in a closed tube. Mach number contours in space-time(x-t) domain as well as grid point trajectories. Left: without boundary grid elimination/addition. Right: with boundary grid elimination/addition.

other side of the piston the grid density becomes coarse. On the contrary, with boundary-point elimination/addition procedure(Fig.4, right), the grid density is almost constant at both sides of the piston.

Acknowledgements: This research is supported by the Grant-in-Aide for Scientific Research No.14550152 of the Japan Society for the Promotion of Science.

References

1. K.Matsuno, Computational Fluid Dynamics 2000, Springer July, (2001) pp.705-710.
2. P.L.Roe, J.Comp.Phys. Vol.32 (1981), pp.357-372.
3. C.L. Rumsey, M.D. Sanetrik, R.T. Biedron, N.D. Melson, and E.B. Parlette, Computers and Fluids, Vol.25, (1996), pp.217-236, 1996.
4. K.Matsuno and H.A.Dwyer, J. Comp. Phys., Vol.77, (1988), pp.40-52.

An Adaptive Cartesian Mesh Flow Solver Based on the Tree-data with Anisotropic Mesh Refinement

Takanobu Ogawa

Seikei University, 3-3-1 Kichijoji-Kitamachi, Musashino-shi, Tokyo, JAPAN

Abstract. An adaptive Cartesian mesh flow solver with anisotropic mesh refinement is developed. The tree data structure is utilized to organize the adaptively refined meshes. It is found that the algorithm developed for isotropic mesh refinement can be extended for anisotropic case with little modification. Parallelization of the algorithm is also done in the SPMD paradigm. The domain decomposition technique is used and a tree data structure is split according to the partitioned domain. Parallel efficiency is examined on a PC cluster.

1 Introduction

The adaptive Cartesian mesh approach has been successfully applied to practical flow fields, such as a flow around an airplane[1], [2] and a flow inside a car engine room[3]. In this method, a rectangular mesh is recursively subdivided and mesh adaptation becomes simple and flexible.

For solver efficiency and simple implementation, the data structure to organize the subdivided meshes is important. There are three possible alternatives of the data structure for the adaptive Cartesian mesh approach. The first one is the conventional structured data with the nested mesh technique. Although implementation is relatively easy with this data structure, mesh adaptation becomes less flexible. The second possibility is to use the conventional unstructured data. In this case, the meshes are organized as the conventional unstructured mesh. However, the unstructured data is too flexible for the adaptive Cartesian mesh which has regularity in mesh distribution. The third one is the tree data structure. This data structure is suitable for hierarchical data such as recursively refined mesh.

In contrast to the two other data structures mentioned above, the tree data structure has not yet been common in CFD. However, it is found from the past study[4] that the algorithms developed for the conventional flow solver can be applied for the tree data with little modification, and that the tree data has mixed property between the structured data for its regularity and the unstructured data for its flexibility.

Isotropic mesh refinement in which a mesh is subdivided in all the directions has been used in most of the adaptive Cartesian mesh approaches. However, the isotropic mesh refinement unnecessarily increases the number of meshes particularly when applied to a boundary layer. Thus, anisotropic mesh refinement is efficient in reducing memory consumption[5], [6].

In this paper, the adaptive Cartesian mesh flow solver based on the tree data is extended for anisotropic mesh refinement. Parallelization of the solver is also done with the domain decomposition technique. Computational results for flow fields at low *Re* number are presented with the experimental data and the parallel efficiency is examined on a PC cluster.

2 Numerical Algorithm

2.1 Adaptive Cartesian mesh

In the adaptive Cartesian mesh approach, a rectangular mesh (thereafter called a cell) is recursively and locally subdivided. Figure 1 shows isotropic cell subdivision in which a cell is divided into equally shaped 8 ($= 2^3$) child cells.

In anisotropic case, a cell is split in 1 or 2 directions and 2 ($= 2^1$) or 4 ($= 2^2$) child cells are generated respectively and Fig. 2 shows the anisotropically subdivided cells. When subdivided repeatedly in a specific direction, a cell has a large aspect ratio.

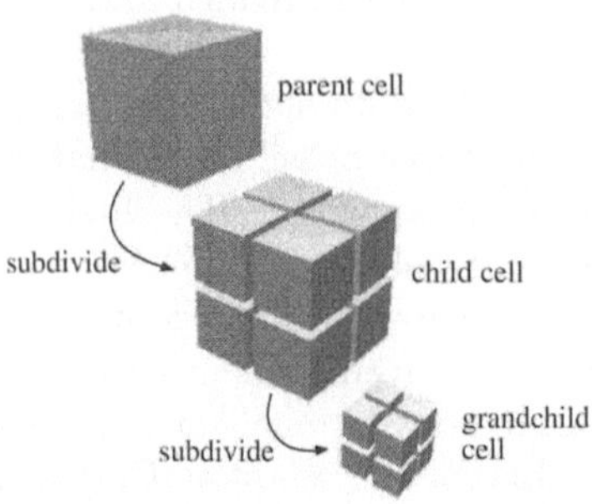

Fig. 1. Recursive cell subdivision

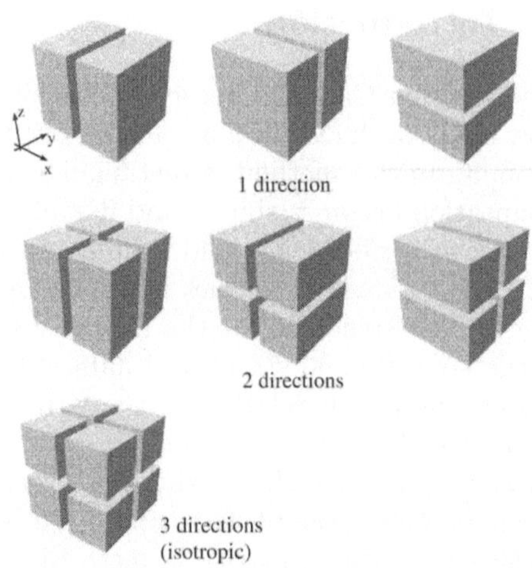

Fig. 2. Anisotropically subdivided cells

2.2 Numerical schemes

The artificial compressibility method [7] is used as numerical method in this study, however, the algorithm presented here can be applied to the other numerical methods based on the conservation law.

The cell-centered finite volume method is used and the semi-discretized form of the basic equations can be formulated as the conventional unstructured mesh approach,

$$\frac{\partial}{\partial t}(QV)_i + \sum_l (F - F_v) \cdot s_l = 0 , \tag{1}$$

where the sum on l is done for the cells surrounding the cell i.

Numerical flux is calculated with Roe's FDS [8]. The physical quantities at the cell interface are extrapolated with the gradient in a cell which is obtained from surrounding cells using the Green-Gauss theorem. The viscous flux is calculated with the velocity gradient averaged at the cell interface.

2.3 The tree data structure

The cells are organized with the tree data structure as shown in Fig. 3. In the tree data, every cell has pointers to its parent and child cells and one can reach any cell from anywhere in a tree by tree traversing. When a cell is anisotropically divided in N directions, the number of child cells is 2^N. Thus, the tree is called a 2^N-tree [5] or N-d tree.

All the cells are generated from the root cell which is at the top of the tree and is the largest cell. A cell at the lowest level is called a leaf cell, which plays the most important role in a flow solver. Only the physical quantities of the leaf cells are used for flow computation except for the case of the Multigrid method in which their ancestor cells are also used. In mesh adaptation, only the leaf cells are divided for refinement and removed for coarsening.

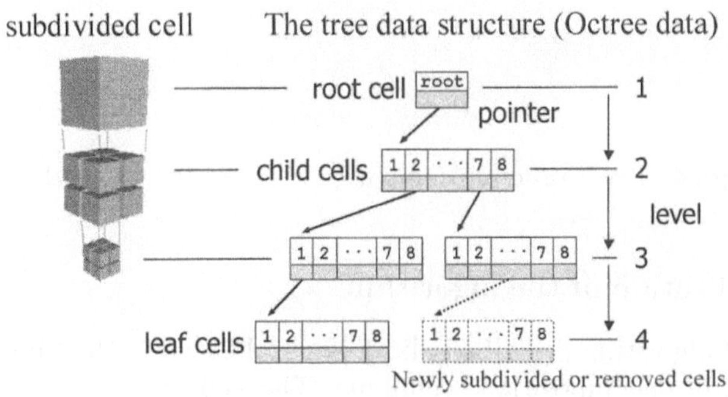

Fig. 3. The tree data structure

2.4 Neighbor cell finding

A neighbor cell is needed to be found by tree traversing for the flux calculation. However, the tree data itself does not have any information about the location of a cell. In the past study [4], the author presented a key named "pedigree" which contains information about the ancestor cells and indicates the location of a cell. This key can be extended to anisotropically refined cells, since "pedigree" is defined uniquely in each coordinate.

2.5 Implicit time integration

For implicit time integration, the LU-SGS method [9],

$$Forward: \quad \Delta Q_i^* = D^{-1}(RHS^n - \textstyle\sum^{lower} sA^+\Delta Q), \tag{2}$$
$$Backward: \quad \Delta Q_i = \Delta Q_i^* + D^{-1}\textstyle\sum^{upper} sA^-\Delta Q, \tag{3}$$

is used. The tree data requires ordering of the cells for the Gauss-Seidel sweep as a conventional unstructured mesh solver. The Z-ordering, which is one of the space filing curves and is obtained on a tree data by a very simple algorithm, can be applied to ordering of anisotropically refined cells as well as isotropically refined cells [4].

Figure 4 shows the Z-ordering curve for sample cell distribution. The forward sweep can be carried out by visiting the leaf cells along the Z-ordering curve. As seen from Fig. 4, the lower cells are always visited before the target cell. For the backward sweep, the curve is traced in the reverse direction.

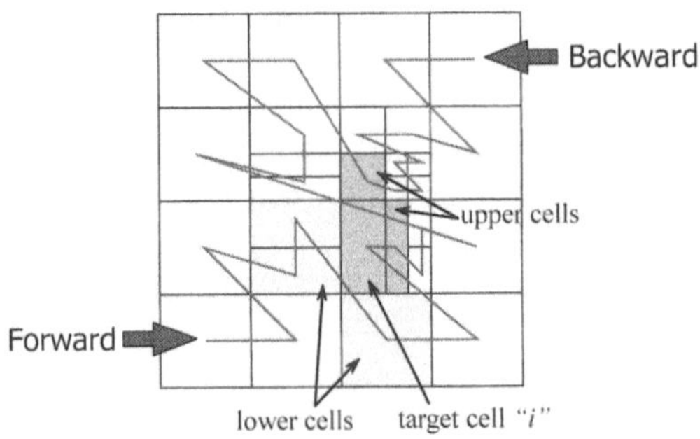

Fig. 4. The Z-ordering curve used for the LU-SGS method

2.6 Parallelization of the algorithm

The numerical algorithm described above is parallelized in the SPMD paradigm with the domain decomposition technique. The cells should be partitioned so that the leaf cells are equally distributed to each processor, since only the leaf cells are used for flow computation.

In partitioning the adaptive Cartesian mesh, the Peano-Hilbert space-filling curve has been successively applied as a partitioner [10]. However, the Peano-Hilbert space-filling curve can not be used for anisotropically refined meshes, since they are no longer recursive. Thus, another appropriate partitioner should be used for an anisotropic refined case.

The partitioning process is illustrated in Fig. 5. Figure 5 (a) shows sample cell distribution partitioned for two processors. The tree data structure for the cells should be split according to the partition and Fig. 5 (b) shows the partitioning line in the tree data. When a tree data is split, some of child cells (indicated with the arrows in Fig. 5 (b)) would lose their parent cells, i.e. they are cut off from the tree. Thus, the same parents cell should be left on the different processors which have their child cells. Figure 5 (c) shows the split tree. The same parent cells, for example, the root cell, exist in both processors so that the child cells do

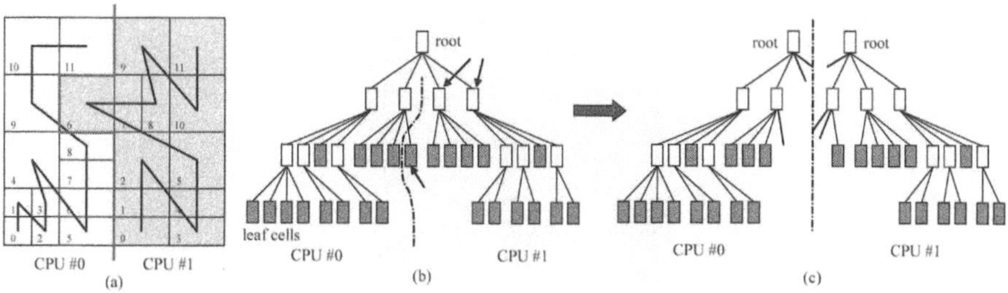

Fig. 5. (a) Cell distribution partitioned for 2 processors and the Z-ordering curve defined in a decomposed domain, (b) the partitioning line in the tree data, (c) the tree data split according to the partition.

not lose their parents. Then, the numerical algorithm same as the serial version can be used in each processor, for example, the Z-ordering curve can be still defined in a decomposed domain.

3 Computational Example

A lid-driven cavity flow at $Re = 400$ is solved as a test case. Figure 6 (a) shows the x-velocity contour plot, and the cell distributions at the steady state is shown in Fig. 6 (b). The cells are anisotropically refined according to the velocity gradient. The cells near the wall are subdivided only along the wall direction. The velocity along the center line is compared with the experimental data in Fig. 6 (c).

This flow field with about 300,000 regularly distributed cells is used to examine the parallel scalability. Parallel computation is carried out on a PC cluster, 8 Intel PentiumIII/1GHz processors connected with a 100BASE-TX switching hub. The SCore [11] is used as a cluster system software. The speed-up is plotted in Fig.7 and the parallel efficiency is about 70% at 8 processors. The current parallel performance is not so good, however, this is the first trial of parallelization and the code has room for performance improvement.

4 Conclusion

An adaptive Cartesian mesh flow solver with anisotropic mesh refinement is developed. The 2^N-tree data structure is used to organize anisotropically subdivided cells. It is found that the numerical algorithms developed for isotropic refinement can be extended to anisotropic refinement with little modification. The algorithm is parallelized in the SPMD paradigm with the domain decomposition technique. A tree data structure is split according to the partitioned domain and the same algorithm as the serial version can be applied in each processor. Parallel efficiency is examined on a PC cluster.

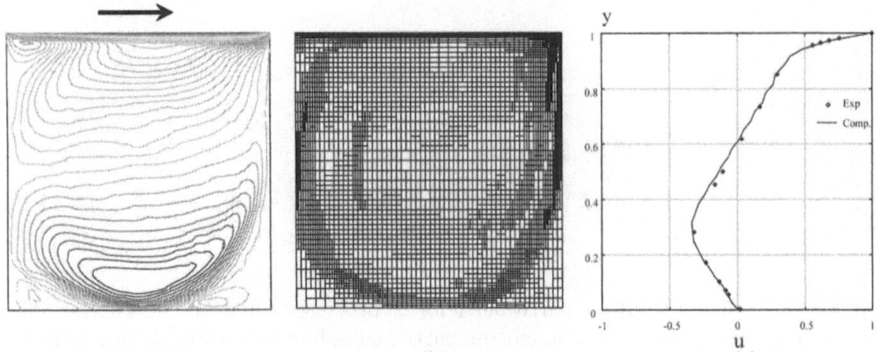

(a) (b) (c)

Fig. 6. A lid-driven cavity flow (a) x-velocity contour plot, (b) mesh distribution, (c) x-velocity plot with the experimental data

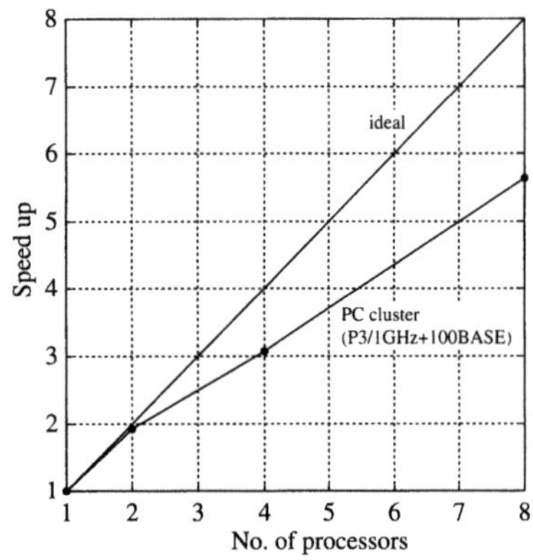

Fig. 7. Parallel scalability on a PC cluster

References

1. Karman, S.L.: AIAA paper 95-0343, (1995)
2. Melton, J., Berger, M. Aftosmis, M. and Wong, M.: AIAA paper 95-0853, (1995)
3. Ono, K., Tomita, N., Fujitani, K. and Himeno, R. : Society of Automotive Engineers of Japan, No.984, pp.165-168 (1998)
4. Ogawa, T.: Proc. of the 1st ICCFD, pp.337-342, (Springer, 2000)
5. Wang, Z.J., Chen, R.F., N. Hariharan, A.J. Przekwas and D. Grove: AIAA paper 99-3300 (1999)
6. Berger, M. and Aftosmis, M.: Proceedings of the 16th ICNMFD, pp.1-12 (Springer, 1998)
7. Chorin, A.J.: J. of comp. phys., No.2, pp.12-26, (1967)
8. Roe, P.L.: J. of Comp. Phys., No.43, pp.357-372 (1981)
9. Yoon, S. and Jameson, A., AIAA J., No.26, pp.1025-1026 (1988)
10. Aftosmis, M.J., Berger, M.J. and Adomavicius, G.: AIAA paper 2000-0808 (2000)
11. http://www.pccluster.org/

A Posteriori Error Determination and Grid Adaptation for AMR and ALE Computational Fluid Dynamics

Giovanni Lapenta

Theoretical Division, Los Alamos National Laboratory, Los Alamos NM87545, USA

Abstract. We discuss grid adaptation for application to AMR and ALE codes. Two new contributions are presented. First, a new method to locate the regions where the truncation error is being created due to an insufficient accuracy: the operator recovery error origin (OREO) detector. The OREO detector is automatic, reliable, easy to implement and extremely inexpensive. Second, a new grid motion technique is presented for application to ALE codes. The method is based on the Brackbill-Saltzman approach but it is directly linked to the OREO detector and moves the grid automatically to minimize the error.

1 Introduction

Grid adaptation can be achieved by grid refinement (i.e. adding more grid points) in some selected areas or by grid motion (i.e. moving grid points to regions of interest from regions of lesser interest). In the first case, the AMR method [1] is obtained. In the second case, the ALE method [2] is obtained. In both cases we need guidance. We need to know what interesting mean. Often, interest is defined based on the knowledge of the solution. If we know the solution, we know where more accuracy is required. However, we do not, in general, know the solution. So we need error detectors which tell us where the error is larger.

The present paper describes a new error detector: the operator recovery error origin (OREO) detector [3].

For AMR codes, the OREO detector provides accurate and automatic determination of where the discretization error is being generated. This knowledge is directly used by the AMR method to refine or to coarsen.

For ALE codes, the knowledge of the error needs to be supplemented by a method to move the grid. Given the error what new grid should we use? To answer this additional problem typical of the ALE method we also present a new technique citelapenta based on the Brackbill-Saltzman approach [4].

Examples are presented to illustrate the new methods.

2 Operator Recovery Computation of the Local Truncation Error

In a previous paper [3], we have proposed a new error origin detector based on the extension of the gradient recovery error estimator [5]. We have named the approach *operator recovery error origin (OREO) detector* since it extends to any operator the method used for the gradient operator by the gradient recovery error estimator. Below, we summarize briefly the procedure involved in its definition and implementation.

For the sake of definiteness, we shall assume a general N-dimensional grid (where one of the dimensions could be time) where a vector field $\mathbf{v}_n$ is node centered. For notation, we label the cells with c and the nodes with n, using further the notation $n(c)$ to indicate the nodes neighboring cell c and $c(n)$ to indicate the cells neighboring node n.

We consider a general multi-dimensional non-linear partial differential operator:

$$\mathbb{O}(q) \tag{1}$$

Equation (1) summarizes the most general operator acting on a function $q(\mathbf{x})$ defined on the multidimensional space $\mathbf{x}$.

Equation (1) is discretized on a grid with N nodes $\mathbf{x}_n$:

$$O_n(q_1, \ldots, q_N) \tag{2}$$

From the discretized field q_n and from the discretized operator X_n applied to q_n defined only on the grid nodes, it is possible to reconstruct two functions defined everywhere in the continuum space $\mathbf{x}$:

$$\begin{aligned} \widetilde{q}(\mathbf{x}) &= \mathcal{L}\{q_n\} \\ \widetilde{O}(\mathbf{x}) &= \mathcal{L}\{O_n(q_1, \ldots, q_N)\} \end{aligned} \tag{3}$$

where $\mathcal{L}$ is the operator representing the multilinear (e.g., linear in 1D, bilinear in 2D) interpolation from the grid points.

The local truncation error is defined as the difference between the linear interpolation of the discretized operator applied to the discretized field $\widetilde{X_q}(\mathbf{x})$ and the exact differential operator applied to the linear interpolation of the discretized field $\widetilde{q}(\mathbf{x})$:

$$e = \widetilde{O}(\mathbf{x}) - \mathbb{O}\widetilde{q}(\mathbf{x}) \tag{4}$$

The average local truncation error on any given cell c is defined as the L_2 norm:

$$e_c = \left(\frac{1}{V_c} \int_{V_c} e^2 dV \right)^{1/2} \tag{5}$$

where e_c is the average local truncation error over cell c and V_c is the cell volume.

3 Variational Grid Adaptation Based on Error Indicators

We have recently proposed a new approach [6] to variational grid adaptation [4] based of the local truncation error defined above. The method can be constructed starting from the following equidistribution theorem proven in Ref. [6]

THEOREM: *In a optimal grid, defined as a grid that minimizes the local truncation error according to the minimzation principle*

$$\int_{\mathcal{V}} | e | \, d^N x \,, \tag{6}$$

the product of the local truncation error in any cell i by the cell volume V_i (given by the Jacobian $J = \sqrt{g}$) is constant:

$$e_i V_i = \text{const} \tag{7}$$

The equidistribution theorem is applied solving the following Euler-Lagrange equations:

$$g^{ij} \frac{\partial}{\partial \xi^i} \left(|e| \frac{\partial x^i}{\partial \xi^j} \right) = 0 \tag{8}$$

This approach creates a grid where $|e_i| V_i$ is constant. Note that the equations above are identical to the equations used by the Brackbill-Saltzman variable diffusion method [4]. The primary innovation is that the monitor function is now directly linked with the local truncation error instead of being left undefined. In the typical implementations of the Brackbill-Saltzman method, the monitor function is defined heuristically by the user. The use of the OREO detector proposed here results in a more accurate scheme [6].

4 ALE Example

We have applied the grid rezoning described above to a standard ALE code [2,7] and we have tested it in a number of problems. Here we limit the discussion to the classic spherical 1D implosion test proposed by Noh [8]. A gas with $\gamma = 5/3$ initially has $\rho = 1$, $e = 10^{-4}$ and uniform velocity $u = -1$ (except in the center where $u(r = 0) = 0$). The problem represents a serious challenge for Lagrangian calculations and the solution is known to suffer from serious wall heating due to the use of artificial viscosity to capture shocks. Note that we are not using artificial heat conduction [8] (a tool to mitigate the wall heating problem) precisely to highlight the trouble of Lagrangian calculations for the present case .

The results of an ALE calculation using the adaptive grid is compared with a reference standard Lagrangian calculation. Figure 1 shows the density at the end of the Lagrangian and ALE calculation. The use of adaptive grid results in a much improved solution. The reason for the improvement is explained by the sharper resolution of the shock achieved by the adaptation. As noted in the original paper by Noh [8], a sharper resolution of the shock also implies a reduction of wall heating, as obseved in Fig. 1 for the ALE case.

The use of grid adaptation based on the OREO detector results in an automatic method to increase the accuracy of the ALE method.

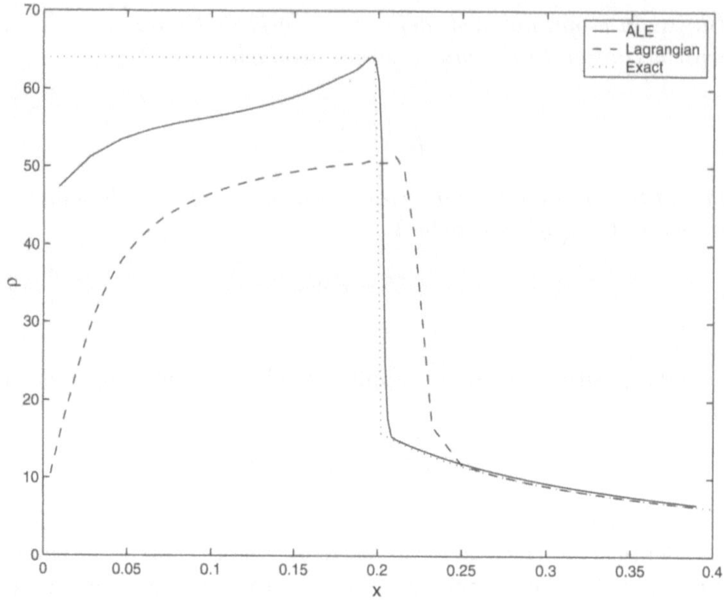

Fig. 1. Noh's spherical benchmark: comparison of the density at the end ($t = 0.6$), for a Lagrangian (dashed) and ALE (solid line) calculation. The exact solution is also shown (dotted line).

4.1 AMR Example

To investigate the performance of the OREO detector in 2D, we have applied it to results obtained with CLAWPACK [10]. CLAWPACK is a publicly available software [11] based on an AMR solution [1] of the conservation laws. We have applied the code to the solution of the gas dynamics equations for the Colella's wedge problem [9]. A planar $M = 10$ shock is incident on an oblique surface; the angle between the shock direction and the surface is $\pi/6$. The actual computed results at time $t = 0.2$ for a 240x120 grid are shown in Fig. 2 where all the expected features [9] can be recognized.

The OREO detector is computed based on the results obtained from CLAWPACK using Algorithm 3. The detector is shown in Fig. 3 for a simulation with a grid 120x60. For comparison we also provide an estimate of the actual error, computed by difference between the solution on a 120x60 grid and the more accurate solution on a 240x120 grid. Clearly the OREO detector is successful in detecting all origins of errors. The shocks are all captured; the slip surface rolling up under the shock is evident. All features are detected.

For reference, Fig. 3-c shows also a similar analysis conducted on another possible candidate for error detection often used in the literature. The detector, which we name *warp indicator* for convenience, measure the local error as the variance among the different values obtained at a node when extrapolating the internal energy from the four directions (backward and forward along x and

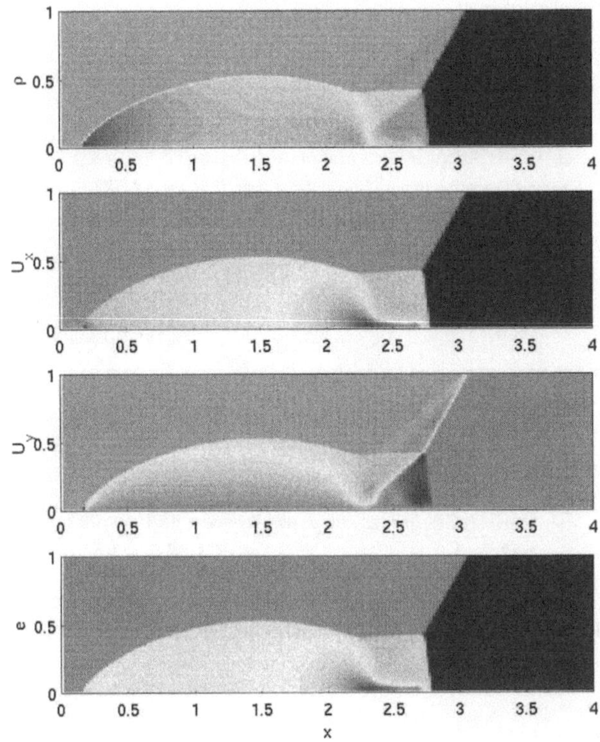

Fig. 2. 2D Gas Dynamics (Eulerian form)- Colella's benchmark on a 240x120grid. Density, velocity and internal energy at the end of a Eulerian calculation ($t = 0.2$). Results obtained using CLAWPACK.

backward and forward along y). The analysis in Fig. 3-c shows that the two rightmost planar shocks are captured well, while the top and bow shocks are barely visible. All the structure inside the rolling up region within the outer shocks is lost: no slip surface is measured and the internal shock is also lost. In practice the warp indicator is often supplemented by other ad hoc detectors to pick up all shocks, but still the rolling up region and the slip surfaces are often left undetected.

The OREO detector does not miss any feature and can be used reliably alone without any other ad hoc detector.

Acknowledgments: The author wishes to acknowledge many useful discussions on the topics reported in the present manuscript with Jerry Brackbill, Doug Kothe, Alex Kurganov, Len Margolin, Tony Scannapieco and Misha Shashkov. This research is supported by the National Nuclear Security Administration under the Advance Strategic Computing Initiative (ASCI) Program.

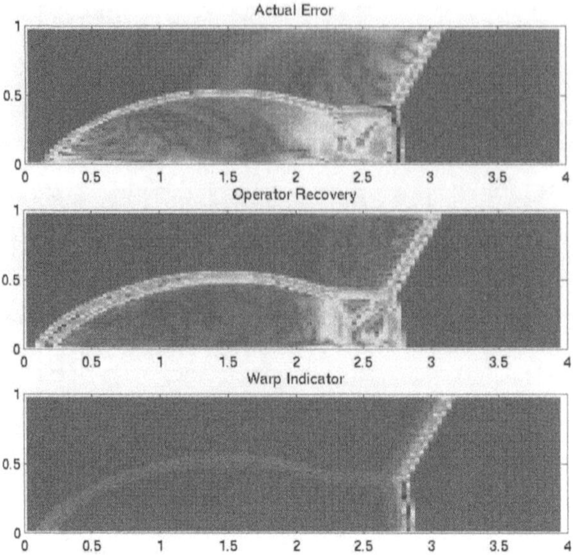

Fig. 3. 2D Gas Dynamics - Colella's benchmark on a 120x60 grid at $t = 0.2$. Comparison of the global truncation error (a) with the OREO detector (b), and the warp indicator (c).

References

1. M.J. Berger, J. Oliger, *J. Comput. Phys.*, **53**, 484(1984)
2. C.W. Hirt, A.A. Amsden, J.L. Cook, *J. Comp. Phys.*, **14**, 227 (1974).
3. G. Lapenta, *Int. J. Num. Meth Engnrg.*, submitted.
4. J.U. Brackbill, *J. Comput. Phys.*, **108**, 38 (1993).
5. M. Ainsworth, J.T. Oden, *A Posteriori Error Estimation in Finite Element Analysis* (Wiley, New York, 2000).
6. G. Lapenta, *J. Comput. Phys.*, submitted.
7. G. Lapenta, "Grid Adaptation and Remapping for Arbitrary Lagrangian Eulerian (ALE) Methods", *Proceedings of the 8th International Conference on Numerical Grid Generation in Computational Field Simulations*, Honolulu, June 2-6, (2002).
8. W.F. Noh, *J. Comput. Phys.*, **72**, 78 (1987).
9. P. Colella, *J. Comput. Phys.*, **87**, 171 (1990)
10. R.J. LeVeque, *CLAWPACK Version 4.0 User's Manual* (University of Washington, 1999).
11. http://www.amath.washington.edu/~claw/

A Posteriori Error Estimates and Adaptive Meshing Applied to Flow in Channels with Corners

Pavel Burda[1], Jaroslav Novotný[2], and Bedřich Sousedík[3]

[1] Department of Mathematics, Faculty of Mechanical Engineering, Czech University of Technology, Karlovo nám. 13, CZ-121 35 Praha 2, Czech Republic
[2] Institute of Thermomechanics AS CR, Dolejškova 5, CZ-18200 Prague 8, Czech R.
[3] Department of Mathematics, Faculty of Civil Engineering, Czech University of Technology, Thákurova 7, CZ-166 29 Praha 6, Czech Republic

Abstract. In the paper we investigate first the problem of a posteriori error estimate for the Navier-Stokes equations to model the flow of incompressible fluid in a 2D domain. We discuss the role of the constant in the a posteriori estimate. Then we apply the a posteriori error estimate with the constant found numerically, to the adaptive mesh refinement. We present numerical results in case of an incompressible flow problem in a domain with corners that cause singularities in the solution. We also show an alternative approach based on the asymptotic behavior of the solution near nonconvex corners.

1 Introduction

At present various a posteriori error estimates for the incompressible flow are available, cf. e.g. references in [2]. In this paper we focus on the aspect of the constant that appears in the estimate – it plays significant role in the adaptive mesh refinement, cf. also [5]. That is why we derive our own a posteriori estimate and trace carefully the role of different constants and their sources, cf. [2] where we derived an a posteriori error estimate for the Stokes problem in a 2-dimensional polygonal domain. In [3] we used similar technique to derive a posteriori error estimates also for 3-dimensional domains.

In the second part we comment on the accurate determination of these constants. In the third part we apply the estimates with the constants found numerically to the adaptive mesh refinement - we solve an incompressible flow problem in a domain with corners that cause singularities in the solution.

2 A Posteriori Estimates for the Steady Navier-Stokes Equations in 2D

Let $\Omega \subset \mathbb{R}^2$ be a bounded domain with boundary $\partial\Omega$, $\mathbf{f} = (f_1, f_2)$ the volumetric loads, and ν is kinematic viscosity coefficient. The steady Navier-Stokes problem for the incompressible fluid consists in finding the velocity $\mathbf{v} = (v_1, v_2)$, and pressure p defined in Ω and satisfying

$$(\mathbf{v}\cdot\nabla)\mathbf{v}-\nu\Delta\mathbf{v}+\nabla p=\mathbf{f} \tag{1}$$

$$\text{div } \mathbf{v} = 0 \tag{2}$$

in Ω together with boundary conditions on disjoint parts of the boundary Γ_{in}, Γ_{wall} and Γ_{out} (meaning, in turn, the inlet, the wall, and the outlet part),

$$\mathbf{v} = \mathbf{g} \text{ on } \Gamma_{in}\cup\Gamma_{wall} \tag{3}$$

$$\nu\frac{\partial\mathbf{v}}{\partial\mathbf{n}} - p\mathbf{n} = 0 \text{ on } \Gamma_{out} \qquad (\text{'do nothing' boundary condition}). \tag{4}$$

For the finite element discretization we use Hood-Taylor elements P2/P1 where velocities are approximated by quadratic shape functions and pressure by linear shape functions. Suppose, that exact solution of the problem is denoted by (v_1, v_2, p) and the approximate finite element solution by (v_1^h, v_2^h, p_h). The exact solution differs from its approximation in the error

$$(e_{v_1}, e_{v_2}, e_p) = (v_1 - v_2^h, v_2 - v_2^h, p - p_h).$$

By the norm of a vector function $\mathbf{u}=(u_1,u_2)$ in Sobolev space $\mathsf{H}^1(\Omega_l)$ we mean

$$\|\mathbf{u}\|_{1,\Omega_l}^2 = \sum_{i=1}^{2}\int_{\Omega_l}\left(u_i^2+\sum_{k=1}^{2}\left(\frac{\partial u_i}{\partial x_k}\right)^2\right)\mathrm{d}\Omega_l.$$

By norm of a vector function $\mathbf{u}=(u_1,u_2)$ in the space $\mathsf{L}^2(\Omega_l)$ we understand

$$\|\mathbf{u}\|_{0,\Omega_l}^2 = \sum_{i=1}^{2}\int_{\Omega_l} u_i^2 \mathrm{d}\Omega_l.$$

For the solution (v_1, v_2, p) we denote

$$\|(v_1,v_2,p)\|_V^2 = \|(v_1,v_2)\|_{1,\Omega}^2 + \|p\|_{0,\Omega}^2$$
$$= \int_\Omega \left(v_1^2+v_2^2+\left(\frac{\partial v_1}{\partial x}\right)^2+\left(\frac{\partial v_1}{\partial y}\right)^2+\left(\frac{\partial v_2}{\partial x}\right)^2+\left(\frac{\partial v_2}{\partial y}\right)^2\right)\mathrm{d}\Omega+\int_\Omega p^2\mathrm{d}\Omega.$$

The estimate in [2] can be generalized to the Navier-Stokes equations:

$$\|(e_{v_1},e_{v_2})\|_{1,\Omega}^2+\|e_p\|_{0,\Omega}^2 \le C\left[h^2\sum_{l\in K}\int_{\Omega_l}(r_1^2+r_2^2)+\sum_{l\in K}\int_{\Omega_l}r_3^2\mathrm{d}\Omega\right], \tag{5}$$

where h denotes the diameter of the finite element Ω_l and r_i stand for the residuals, cf [5]. Let us note that according to our practical experience we use only the element residuals.

It is not possible to determine the constant C by analysis, the contributions to C can be seen in Theorem 1 of [2]. Breefly, the constant C in (5) consists of three parts: the constant C_P from the Poincaré-Friedrichs inequality, the constant C_I coming from interpolation, and the constant C_R coming from regularity. It is important, that this constant doesn't depent on the mesh size and so can be determined experimentally for general situation. In [5] we presented a way how to find the constant C by a numerical experiment.

In our adaptive strategy we use the relative error given by the ratio of absolute norm of the solution error, related to unit area of the element Ω_l, $\frac{1}{|\Omega_l|}\ \mathcal{E}^2(v_1^h, v_2^h, p^h, \Omega_l)$, and the solution norm on the whole domain Ω, related to unit area $\frac{1}{|\Omega|}\ \|(v_1^h, v_2^h, p^h)\|^2_{V,\Omega}$, i.e.

$$\mathcal{R}^2(v_1^h, v_2^h, p^h, \Omega_l) = \frac{|\Omega|\ \mathcal{E}^2(v_1^h, v_2^h, p^h, \Omega_l)}{|\Omega_l|\ \|(v_1^h, v_2^h, p^h)\|^2_{V,\Omega}}. \tag{6}$$

3 Application of Estimates to Adaptive Meshing and Numerical Results

Consider two-dimensional flow of viscous, incompressible fluid described by Navier-Stokes equations in a domain with corner singularity, cf. Fig. 1.

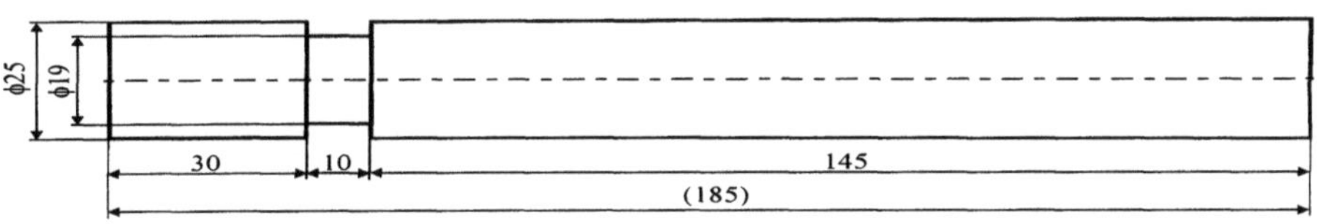

Fig. 1: *Geometry of the channel*

Due to symmetry, we solve the problem only on half of the channel, cf. Figs. 1, 2. On the inflow we consider parabolic velocity profile, at the outflow 'do nothing' boundary condition. On the upper wall, no-slip condition and on the lower wall, condition of symmetry. We consider the following parameters: $\nu = 0.0001\ \mathrm{m}^2/\mathrm{s}$, $v_{in} = 1$ m/s. The initial mesh is in Fig. 2.

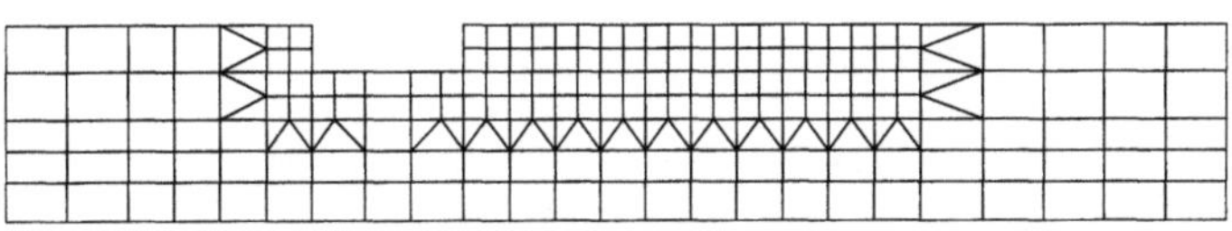

Fig. 2: *Initial finite element mesh*

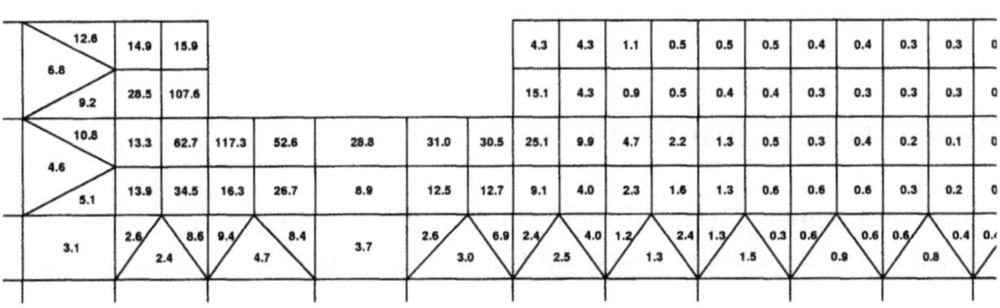

Fig. 3: *Relative errors on elements of initial mesh*

Elements, where the relative error exceeds 3 % (cf. Fig. 3) are refined, and new solution together with new error estimates is computed. The mesh after three successive refinements is shown on Fig. 4.

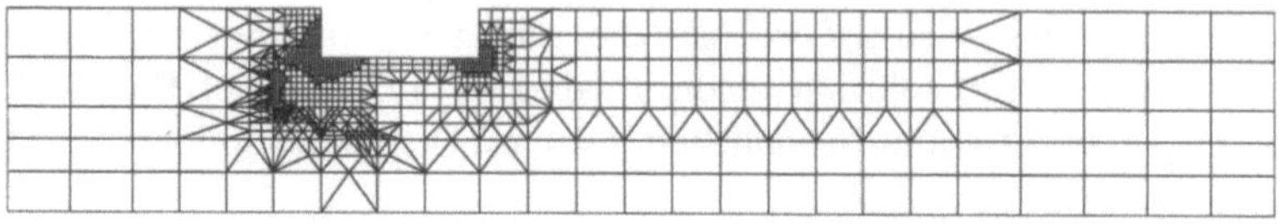

Fig. 4: *Finite element mesh after third refinement*

The results after three refinements are shown on Figures 5 to 8. Streamlines near the corners after third refinement are on Fig. 5. Fig. 6 shows pressure after third refinement. On Figs. 7 and 8 profiles of velocity components v_x and v_y are presented. The singularity is well seen on the v_y component of velocity as well as on the pressure.

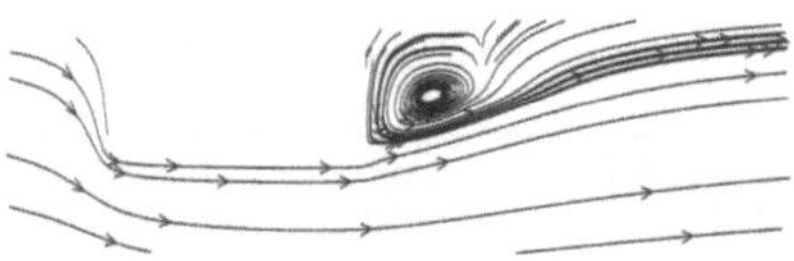

Fig. 5: *Streamlines, 3-rd rfnmt*

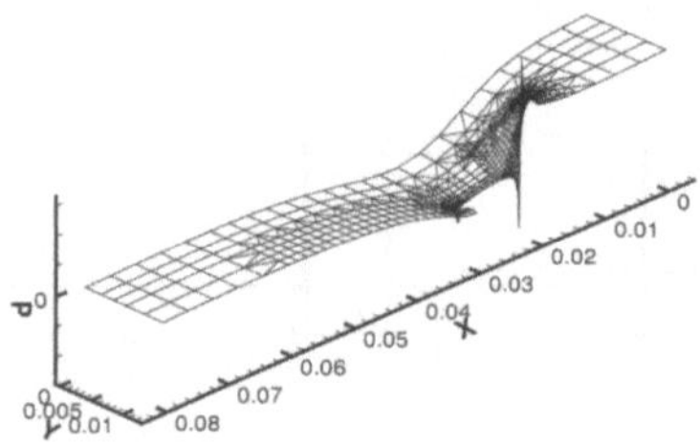

Fig. 6: *Pressure, 3-rd rfnmt*

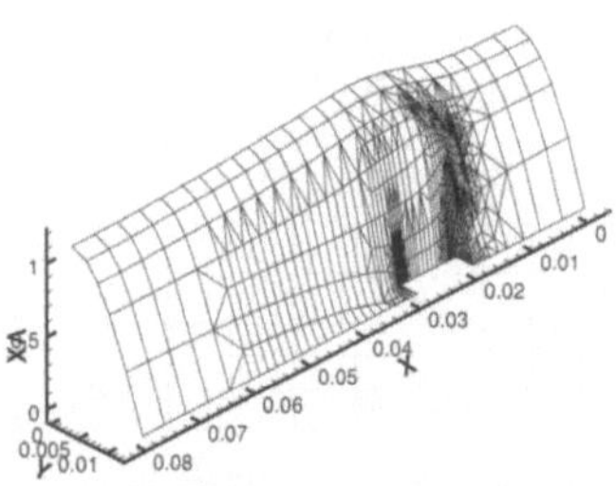

Fig. 7: *Velocity v_x, 3-rd rfnmt*

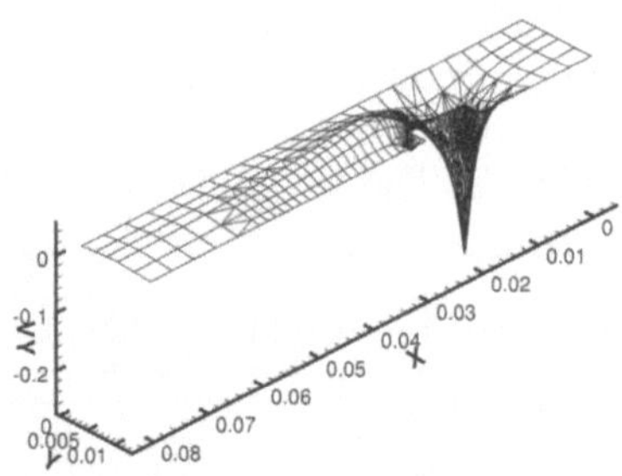

Fig. 8: *Velocity v_y, 3-rd rfnmt*

4 Alternative refinement adjusted to the singularity

Let us note that another way of local mesh refinement near the singularity has been suggested in [1]. This is based on the asymptotic behaviour of the solution

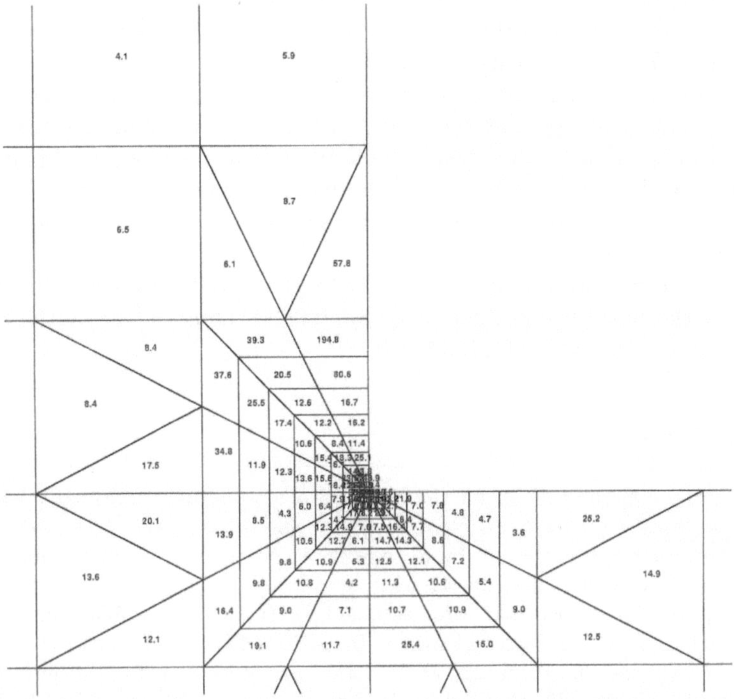

Fig. 9: *Relative errors on adjusted mesh based on quadrilaterals - detail*

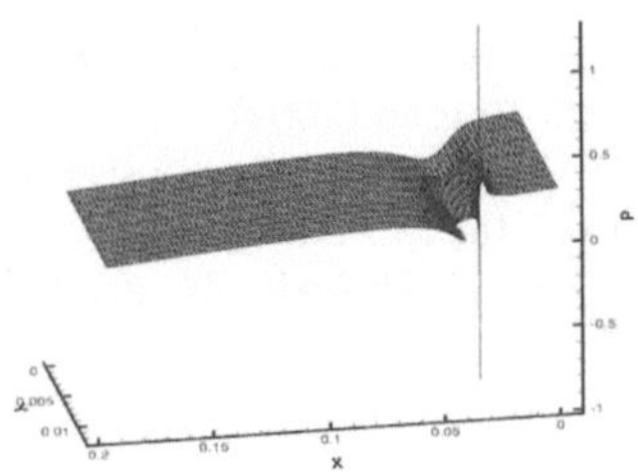

Fig. 10: *Pressure on adjusted mesh*

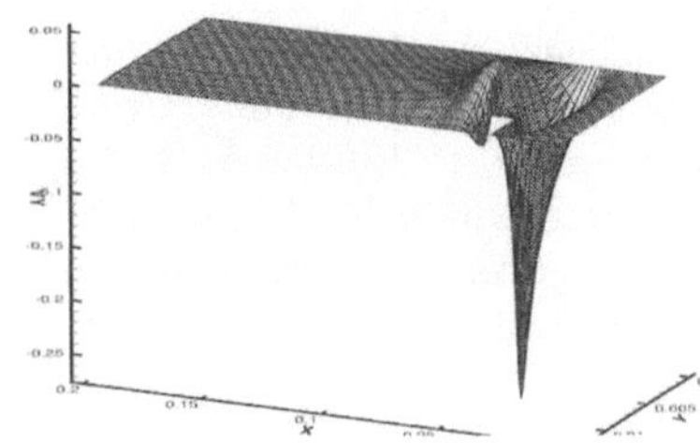

Fig. 11: *Velocity v_y on adjusted mesh*

near the singularity, not on a posteriori error estimates. It has been applied to our model problem in [4]. The refinement of the mesh adjusted to the singularity is shown on Figure 9. Some results are shown in Figures 10 to 11.

5 Conclusions

In the estimate (5) the constant C is in fact composed from the constants C_P, C_R, C_I, corresponding in turn to the Poincaré inequality, the regularity, and the interpolation constants. Of these only the constant C_P can be evaluated directly. The others are not known. But in the application to the adaptive mesh refinement we need the value of the constant C. We developed a technique for calculating the constant with high accuracy.

The alternative approach based on the knowledge of the asymptotic behaviour of the solution near the singularity is, e.g. in our model problem time saving: compared with the usual adaptive refinements based on "halving" the meshsize we do not have to repeat the computation here. But, of course the adaptive strategy based on a posteriori error estimates is much more robust.

Acknowledgements:
This research has been supported partly by the GACR grant No. 101/02/0391, partly by the State Project No. J04/98/210000003 and partly by the State Research Project No. J04/98/210000010.

References

1. P. Burda: 'On the F.E.M. for the Navier-Stokes equations in domains with corner singularities'. In: *Finite Element Methods, Supeconvergence, Post-Processing and A Posteriori Estimates,* ed. by M. Křížek, P. Neittaanmäki, R. Stenberg (Marcel Dekker, New York, 1998) pp. 41–52
2. P. Burda: 'An a posteriori error estimate for the Stokes problem in a polygonal domain using Hood-Taylor elements' In: *ENUMATH 99, Proc. of the 3-rd European Conference on Numerical Mathematics and Advanced Applications, Jyväskylä, Finland, July 26 - 30, 1999,* ed. by P. Neittaanmäki, T. Tiihonen and P. Tarvainen (World Scientific, Singapore, 2000) pp. 448–455
3. P. Burda: 'A posteriori error estimates for the Stokes flow in 2D and 3D domains' In: *Finite Element Methods, 3D Problems, ed. by P. Neittaanmäki, M. Křížek,* GAKUTO Internat. Series, Math. Sci. and Appl. **15** (2001), pp. 34–44
4. M. Jelínek: Applications of Adaptive Finite Element Method to Incompressible Flow Problems, BA Thesis, Czech Univ. Technol., Prague (2002)
5. B. Sousedík, P. Burda, A. Szölös, J. Novotný: 'Application of a posteriori error estimates to fluid flow in a domain with singularity'. In: *Proceedings of 4-th Seminar Euler and Navier-Stokes Equations, ed. by K. Kozel, J. Příhoda, M. Feistauer* (IT AS CR Praha, 2001) pp. 125–128

Part XII

Mixed Topics

Fingering Instabilities in a Miscible Rotating Hele-Shaw Flow

Ching-Yao Chen[1,2], Shu-Wei Wang[1], and Yu-Chia Liu[1]

[1] Department of Mechanical Engineering, Da-Yeh University, Chang-Hua, Taiwan, 515 R.O.C.
[2] email : chingyao@mail.dyu.edu.tw

Abstract. Numerical simulations of interfacial stability for miscible droplet in a rotating Hele-Shaw cell by means of high accurate numerical schemes are presented. The interfacial instability is dominated by two control parameters, such as rotating speed, expressed as a dimensionless Peclet number and viscosity contrast. Great similarities of the fingering patterns with the experimental findings are obtained numerically. More vigorous interfacial fingerings are found at higher rotating speed, and lower viscosity contrast.

1 Introduction

The dynamics of interfacial stability, both for immiscible and miscible fluids, in a Hele-Shaw cell, or porous medium, are investigated intensively since the pioneer works by Hill[4], Saffman and Taylor[8]. Most of the problems studied focus on the evolution of the interface in rectilinear and radial flow geometries where the less viscous fluid displaces more viscous one, cf. the recent reviews by McCloud and Maher [5]. In the opposite scenario, the interfacial instabilities might occur as well when a more viscous fluid displaces a less viscous fluid driven by gravity of centrifugal force. The Hele-Shaw flows driven by centrifugal force have been the subject of recent studies[1-3,9,10], due to their potential applications of in the technology of coating. The rotational effects to the fingering instability differ the radial flows in many ways. Instabilities in radial flows are mainly due to the viscosity contrast, and decay as radial distance increased. Fingerings are triggered if the displacing fluid is less viscous and more vigorous near the origin. However, in the rotational Hele-Shaw cell, the centrifugal force is generated by the density difference, and proportional to the radial distance. Stronger unstable mechanism is expected if the heavier fluid, usually more viscous, is the driving fluid and more pronounced away from the center. The competition of these two counter-effect, viscous stable and density unstable, mechanisms leads to the determination of interfacial stabilities. Carrillo et al.[1] investigated the interfacial stabilities of immiscible fluids in a rotating cell both theoretically and experimentally. Linear growth of two characteristic lengths, the radius of gyration and the radial finger length, were found in the nonlinear regime, i.e. zero injection rate. Experiments of radial displacement of a fluid annulus[2] show the strong dependence of instabilities on the wetting condition of the outer interface. Experimental results are well described by Darcy's law with little capillary effects if the cell is prewetted. In a dry cell, however, a scaling correlation can be formed between capillary number and a dimensionless group that relates to centrifugal force and capillary force. The following up stability analysis[3] shows that the instabilities of two interfaces, which the leading interface is density driven and the trailing is driven by viscosity, are coupled through pressure field. The prediction of stability analysis is conformed by the experimental results. An interesting emission of liquid droplets at finger tip was observed as well. In the present study, evolution of the morphological instability, caused by flow rotation and injection, is simulated by means of highly accurate numerical scheme in a miscible situation. The outline of this paper is as follows. After formulating the physical problem and brief review of the computational technique in § II, Section III will focus on the computational results and their interpretation. Conclusions will be provided in § IV.

2 Physical Problem and Governing Equations

We study the interfacial dynamics of a heavier and more viscous fluid droplet, diameter D_o, that is surrounded by a miscible fluid with less density and viscosity in a rotating Hele-Shaw cell. The governing equations, in a reference frame rotating with the cell, of continuity, momentum of Hele-Shaw equations [9] and concentration conservation take the forms

$$\nabla \cdot \mathbf{u} = 0 \tag{1}$$

$$\nabla p = -\frac{b^2}{12}\mu\mathbf{u} + \rho\omega^2\mathbf{x} + 2\rho\omega \times \mathbf{u} \tag{2}$$

$$\frac{\partial c}{\partial t} + \nabla \cdot (\mathbf{u}c) = D\nabla^2 c \tag{3}$$

Here, $\mathbf{u}$ denotes the velocity, b the Hele-Shaw cell gap spacing, ω the rotational frequency, $\mathbf{x}$ the position vector, c the concentration of lighter fluid, D the diffusion coefficient and p the pressure. Viscosity is indicated by μ.

In order to render the governing equations dimensionless, we take the diameter D_o of droplet as the characteristic length scale and time scale $12\mu_l/b^2\triangle\rho\omega^2$. By furthermore scaling with viscosity μ_l where the subscript l indicates lighter fluid, density difference $\triangle\rho$ and pressure $\triangle\rho\omega^2 {D_0}^2$ associated with the definition of the viscosity-concentration dependence[6,7] and density-concentration forms

$$\mu(c) = e^{R(1-c)}, \quad R = -\frac{1}{\mu}\frac{d\mu}{dc} \tag{4}$$

$$\rho(c) = \frac{\rho_h}{\triangle\rho} + c, \tag{5}$$

the dimensionless momentum and concentration equations are obtained in terms of vorticity (ω) and streamfunction (ψ)

$$\nabla^2\psi = -\omega \tag{6}$$

$$\omega = -R\nabla\psi_{rot}\cdot\nabla c - \frac{1}{\mu}(y\frac{\partial c}{\partial x} - x\frac{\partial c}{\partial y}) - \frac{2Re}{\mu}(u\frac{\partial c}{\partial x} + v\frac{\partial c}{\partial y}) \tag{7}$$

$$\frac{\partial c}{\partial t} + \mathbf{u}\cdot\nabla c = \frac{1}{Pe}\nabla^2 c \tag{8}$$

where the Peclet number Pe and Reynolds number Re take the forms

$$Pe = \frac{\triangle\rho\omega^2 b^2 {D_0}^2}{12\mu_l D}, \quad Re = \frac{\triangle\rho\omega b^2}{12\mu_l} \tag{9}$$

In the present study, Coriolis is neglected ($Re \ll 1$) [1]. The numerical code is largerly identical to one used for earlier investigation in [7]. A Galerkin-type discretization of cosine expansions is used in the stream-wise direction accompanied by sixth order compact finite difference in the normal direction. The code was validated by comparing the growth rates with the values obtained from linear stability theory in a plane front. More details on the implementation of these schemes are provided by [6,7].

3 Results

We start by describing the temporal and spatial evolution of a reference case, in order to identify the dominant mechanisms at work. Subsequently, the values of the governing parameters, such as viscosity parameter R and Peclet number Pe will be varied individually, in order to elucidate their effects.

3.1 Reference case

A representative calculation for $R = 2.5$ and $Pe = 10^4$ is described. The R-value indicates that the droplet is about 12 times more viscous than the environment. Figure 1 displays the time sequences of the concentration fields. At $t = 7.5$, very vigorous interfacial instability is triggered by the strong centrifugal force even the mixing front of droplet is viscously stable. To conserve the mass, opposite fingers of less viscous fluid penetrate toward the droplet origin. Nonlinear tip-splittings are observed as well at this early stage. Due to the nature that the centrifugal force is proportional to radial distance, these fingering instabilities are further enhanced at later time, figure 1(b). However, the nonlinear behaviors, such as tip-splitting and shielding, are not seen more pronounced, c.f. $t = 15.55$. The radial fingers grow comparably in a straight outward trend. This insignificant nonlinear growth of the radial fingers is also observed in the immiscible experiments[3]. The early development of non-linear tip-splitting and evolution of straight fingers at later stage can be understood by the different features of viscous fingerings and gravitational instabilities. At earlier time, the

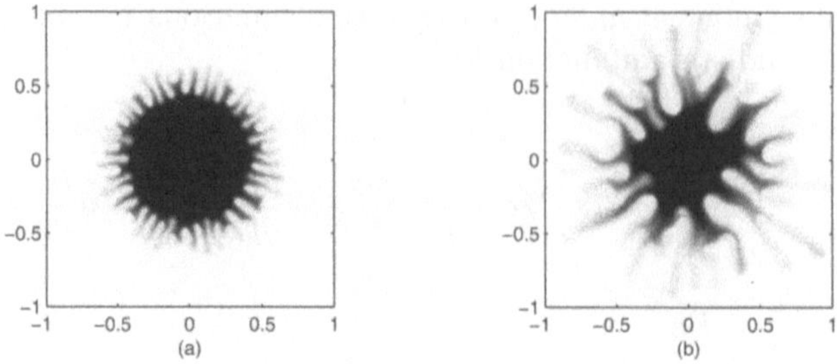

Fig. 1. Reference case $R = 2.5$ and $Pe = 10^4$: Concentration images at (a) $t = 7.5$ and (b) t=15.55

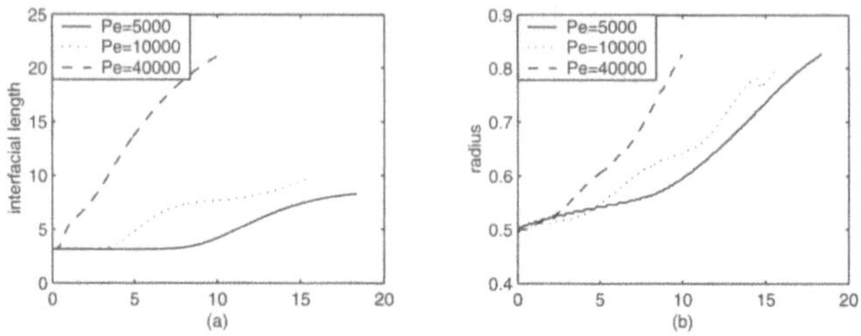

Fig. 2. $R = 2.5$: Evolutions of (a)mixing interfacial length and (b) radius of gyration for various Pe

relatively weaker centrifugal force due to the shorter rotational radius leads to the viscous effects more significantly, and thus the instabilities perform similar to conventional viscous fingering that often leads to non-linear behavior. This viscous effect diminishes as the increase of centrifugal force by the larger rotating radius, and the gravitational instabilities dominate the instability patterns. It explains the appearance of slimmer straight fingers at later stage. Development of fingering instabilities can also be understood quantitatively by the growth of mixing interfacial length, shown in figure 2(a). The interfacial length is seen stretched drastically before a period of latency, which the trend resembles the experimental data in [1]. However, unlike the immiscible case with well-defined interface, the mixing interfacial length relates to the concentration gradient. Milder growth rate is resulted from the strong dispersion after about $t = 8$. Another quantitative characterization that is interesting to be measured is the radius of gyration. The growth of radius is plotted in figure 2(b). Due to diffusive effects, the original radius is slighter greater than 0.5.

3.2 Effect of rotating speed and viscosity contrast

The centrifugal force is directly proportional to the dimensionless Peclet number Pe, which denotes the ratio of radial convection to diffusion. For certain fluids combination, larger Pe implies higher cell rotating frequency. Therefore more vigorous fingerings are expected at higher value of Pe due to the stronger centrifugal force. Figure 3(a) and 3(b) display the concentration images of $Pe = 5 \times$

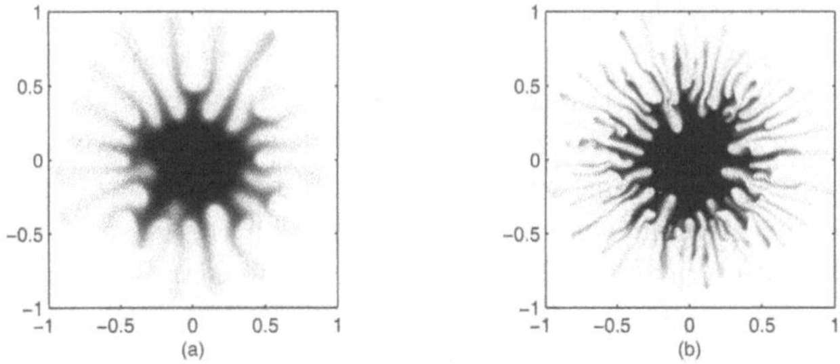

Fig. 3. R=2.5 : Concentration images for (a) $Pe = 5 \times 10^3$ at t=18.4 and (b) $Pe = 4 \times 10^4$ at t=10.0

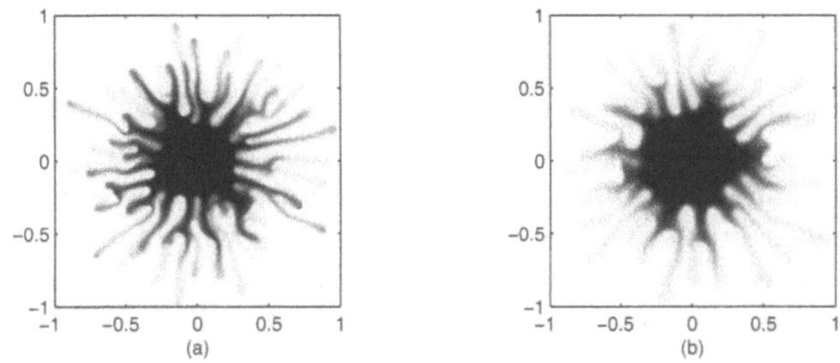

Fig. 4. $Pe = 10^4$: Concentration images for (a) $R = 1$ at t=6.4 and (b) $R = 3$ at t=18.5

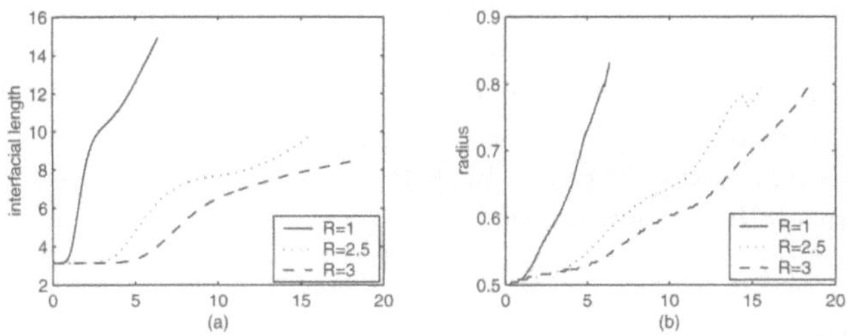

Fig. 5. $Pe = 10^4$: Evolutions of (a)mixing interfacial length and (b) radius of gyration for various R

10^3 and 4×10^4 respectively. While fingerings still develop at lower $Pe = 5 \times 10^3$, the number of fingers have decreased significantly. The fingerings are seen much more vigorous at higher $Pe = 4x10^4$, shown in figure 3(b). The significant reduction of wave number concludes the less unstable pattern for slower rotating speed, which meets the common expectation. Stability analysis of immiscible flow in the absence of injection rate in [1] found that the interface is unconditional unstable and higher most dangerous wave mode at faster rotating speed. Their findings are in the same line of current results. Growth of both interfacial length and radius of gyration, shown in figure 2(a) and 2(b), confirm the less unstable situations if Pe is reduced. Longer period of latency and milder growth are observed for lower Peclet number. A lower viscosity parameter, $R = 1$ shown in figure 4(a), provides less stable viscous damping effects, thus leads to more fingering. Numerous numbers of very slim fingers with tip-splitting are triggered. These slim fingers proceed straight growth outward and reach the computational domain in a very fast path, t =6.4, compared to t =15.55 in reference case. On the other hand, less fingerings are observed at higher viscosity contrast of $R = 3$, as the concentration image shown in figure 4(b). The more stable interface leads to slower fingers' outward speed, arrival to the boundary at about $t = 18.5$. A shorter period of latency as well as a faster growth rate both for the interfacial length, figure 5(a), and radius of gyration, figure 5(b), at lower values of R also reflects the more vigorous fingering on mixing interfaces.

4 Conclusions

We have presented the numerical simulations of interfacial stability for miscible interfaces in a rotating Hele-Shaw cell by means of high accurate numerical schemes. The centrifugal force, if the central fluid is heavier and in general more viscous, provides an unstable outward driving force, while the viscous effect tends to stabilize the contacting front. The interfacial stability is dominated by two dimensionless control parameters, such as Peclet number and viscosity contrast. For a stationary droplet, i.e. with no injection, the occurrence of instability is decided by the competition of unstable centrifugal force and stable viscous damping. As a result, more vigorous interfacial fingerings are found at higher rotating speed, and lower viscosity contrast. Due to the nature of proportionality of centrifugal force and radial distance to the origin, fingerings is more pronounced at later time when the mixing front is further far away from center, which behaves in a opposite way to the radial flow.

Acknowledgments: Support by a ROC NSC Research Grant 90-2212-E-212-013 is gratefully acknowledged.

References

1. Carrillo, L., Magdaleno, F., Casademunt, J. and Ortin, J. 1996 Phys. Rev. E 54,6260-6267.
2. Carrillo, L., Soriano, J. and Ortin, J. 1999 Phys. Fluids 11, 4, 778-785.
3. Carrillo, L., Soriano, J. and Ortin, J. 2000 Phys. Fluids 12, 7, 1685-1698.
4. Hill, S. 1952 Chem. Eng. Sci. 1.
5. McCloud, K. V. and Maher, J. V. 1995 Physics Reports 260, 139.
6. Meiburg, E. and Chen, C.-Y. 2000 SPE J. June 5, 2.
7. Ruith, M. and Meiburg, E. 2000 J. Fluid Mech. 420, 225.
8. Saffman, P. and Taylor, G. 1958 Proc. R. Soc. London. Ser. A, 245, 312.
9. Schwartz, L. 1989 Phys. Fluids A 1, 167.
10. Wang, M. and Chou, F. 2001 J. Electrochemical Society, 148, 5, G283.

A Composite Explicit-Implicit Godunov Method for Unsteady Problems on Highly Stiff Grids

Igor Men'shov and Yoshiaki Nakamura

Nagoya University, Chikusa-ku, Nagoya 464-8603, Japan

Abstract. The present paper is devoted to the development of a composite explicit-implicit method for computing the unsteady Navier-Stokes equations on grids with a high ratio between maximal and minimal grid spacings, here referred to as stiff grids. The method implements a smooth switching between the explicit and implicit modes. In the explicit mode, which is realized in regions of a large grid spacing, the method is represented by a second order Godunov/MUSCL sceme. As a whole, the method satisfies the Max Norm Diminishing (MND) property in the linear case and is stable for any value of the time step in the case of non-linear equations.

1 Introduction

This paper addresses the development of a composite (or hybrid) explicit-implicit numerical method for computing unsteady flows on highly stiff grids. By grid stiffness we mean the ratio between the largest and the smallest grid spacing. The grid stiffness occurs when the problem to be computed involves different spatial scales. Such situation arises, for example, in unsteady high Reynolds number wall-bounded flows. The scale of the boundary layer on the wall is much smaller than a characteristic lenght of the problem. Therefore, a rather fine grid has to be used near the wall along with a moderate cell spacing grid in the base flow off the boundary layer. Another example is adaptive grids, where local grid refinement is performed to capture highly steep gradient solutions, as shocks or contacts. This can also lead to a high grid stiffness.

Spatial stiffness causes temporal stiffness, which makes the use of any explicit scheme almost impossible due to tough time step limitations. Implicit schemes are unconditionally stable, and can formally handle any time step. However, these schemes suffer from excessive numerical diffusion, much larger than that of explicit methods. Another drawback of implicit schemes is the solution of discretized equations, which commonly involves iterative CPU-costly procedures.

An attractive approach to treat unsteady problems on stiff grids is to employ a hybridization between explicit and implicit schemes. The basic idea is to combine an explicit method with an implicit one so that the resultant scheme would be continuously switched from explicit to implicit, and vice versa, depending on local grid spacing and/or the solution to be computed.

A hybrid scheme has been proposed by Collins, Collela, and Glaz [1] for 1D systems of conservation laws, which was then generalized for an advection multidimensional equation by O Rourke and Sahota [2]. It is referred to as the CCG scheme in this paper. This scheme was designed under the following principles:

(a) continuos switching, (b) second-order accuracy in the explicit mode, and (c) MND for all values of the time step. Although all the above design principles are proved for linear equations, the MND property appears to be outstanding, and the scheme works incorrectly, being extended to non-linear equations. To overcome this problem, an additional constraint has to be imposed on the time step. In fact,this is reminiscent of a CFL-type restriction, and therefore advantages originally declared for the CCG scheme are almost cancelledin the non-linear case.

In the present paper we attemp to revise the CCG scheme and develop a new composite explicit-implicit scheme that would ensure the CCG design principles, in particular the unconditional MND property, for non-linear conservation laws, and be easily extended to the unsteady compressible Navier-Stokes equations.

2 Basic principle of hybridization

The basic principle of the proposed hybridization is first considered for a scalar equation

$$\partial_t u + \partial_x [f(u)] = 0 \tag{1}$$

with $s = f'(u) > 0$ and $f''(u) > 0$, which is discretized as

$$u_i^{n+1} = u_i^n - \frac{\Delta t}{\Delta x_i}[F_{i+1/2} - F_{i-1/2}] \tag{2}$$

where $F_{i+1/2}$ is the Godunov's numerical flux at the cell interface $i + 1/2$. This flux is a function of two independent variables $u_{i+1/2}^l$ and $u_{i+1/2}^r$, which represent the values of u on the left- and right-hand sides of the interface, respectively: $F_{i+1/2} = F(u_{i+1/2}^l, u_{i+1/2}^r)$.

The scheme is specified by the values $u_{i+1/2}^l$ and $u_{i+1/2}^r$. When these values are defined as

$$u_{i+1/2}^l = u_i^{n+1}; \quad u_{i+1/2}^r = u_{i+1}^{n+1} \tag{3}$$

an implicit scheme is obtained, which satisfies the MND property for all time steps. However, this scheme shows too much diffusion in calculating discontinuous solutions.

On the other hand, if they are specified as

$$u_{i+1/2}^l = u_i^n - 0.5(\beta_i^n - 1)\delta u_i^n; \quad u_{i+1/2}^r = u_{i+1}^n - 0.5(\beta_{i+1}^n + 1)\delta u_{i+1}^n \tag{4}$$

with $\beta_i = s_i \Delta t / \Delta x_i$, and δu_i^n as a limited difference defined by means of positive coefficients $c_i^{\pm}$ as $\delta u_i^n = c_i^+ \Delta u_{i+1/2} = c_i^- \Delta u_{i-1/2}$, where $\Delta u_{i+1/2} = u_{i+1}^n - u_i^n$, the scheme is converted into a second-order explicit scheme. In this case the MND property holds only under the following conditions:

$$\beta_i^n = \Delta t s_i^n / \Delta x_i \leq 1; \qquad 0 \leq c_{i+1/2}^{\pm} \leq 2 \tag{5}$$

Here $s_i^n = f'(u_i^n)$ denotes the wave speed.

For the proposed hybrid explicit-implicit scheme, the interface values are controlled by parameter ω, $0 \le \omega \le 1$, evaluated for each computational cell, and given by

$$u^l_{i+1/2} = u^\omega_i - 0.5(\omega_i\beta^\omega_i - 1)\delta u^n_i; \quad u^r_{i+1/2} = u^\omega_{i+1} - 0.5(\omega_{i+1}\beta^\omega_{i+1} + 1)\delta u^n_{i+1} \quad (6)$$

where $u^\omega_i = u^n_i + (1-\omega_i)\Delta^n u_i$, and $\Delta^n u_i = u^{n+1}_i - u^n_i$.

A simple interpretation of eq. (6) can be suggested, which will make it possible to extend this hybrid approach to multidimensional cases and fluid dynamics equations in the ensuing section. To compute the interface values, an intermediate time level is first introduced for each computational cell as $t^\omega_i = t^n + (1-\omega_i)\Delta t$, and intermediate values of the solution are defined at this level by interpolating between values at the lower and upper time levels. Then the explicit scheme is launched from the time level t^ω with the time step $\omega\Delta t$, using the intermediate values as the initial data.

Evaluation of the control parameter ω is dictated by two conditions. To suppress strong diffusion inherent in the implicit scheme, it is preferable to take the value of this parameter as close to unity as possible. The other condition is that the scheme must be MND for all time steps. Therefore, the optimal choice is to take the value of ω as large as possible providing the scheme retains the MND property. The following lemma helps us make such a choice.

Lemma 1. *If the parameter ω and constraint factors of limited differences, $c^\pm_{i+1/2}$, satisfy the following inequalities:*

$$\omega_i\beta^\omega_i \le 1; \qquad 0 \le c^\pm_{i+1/2} \le 2 \quad (7)$$

then the hybrid scheme given by eqs. (2) and (6) holds the MND property for all time steps.

In this lemma, β_i is defined by eq. (5), and the superscript ω denotes the value at the intermediate time level, i.e., $\beta^\omega = \beta(u^\omega)$.

In this vein, we can determine the control parameter ω in eq. (6) as

$$\omega_i = \begin{cases} 1 & if\ \beta^\omega_i \le 1 \\ 1/\beta^\omega_i & otherwise \end{cases} \quad (8)$$

It should be noted that the proposed hybridization coincides with the CCG scheme for the scalar equation considered, if β^ω_i in eq. (8) is replaced by β^n_i.

3 Extension to the Navier-Stokes equations

The foregoing method can be easily extended to the multidimensional compressible flow equations. To do this, first we consider a second-order explicit time marching scheme for the compressible Navier-Stokes equations

$$\mathbf{q}^{n+1}_i = \mathbf{q}^n_i + \Delta t L_2(\Delta t, \mathbf{q}^n) \quad (9)$$

where $\mathbf{q}$ is the vector of conservative variables, and L_2 represents a discrete operator that integrates the spatial differential operator of the Navier-Stokes equations over the time step Δt with the second-order of accuracy on a given spatial grid. A common way to construct such an operator is to use the finite-volume method with a MUSCL-type interpolation of the inviscid flux [3].

In accordance with the strategy developed above for the scalar equation, we introduce an intermediate time level t^ω for each computational cell as $t^\omega = t^n + (1-\omega)\Delta t$, which is defined by an appropriate cell-attributed parameter ω_i, $0 \le \omega_i \le 1$, and determine the intermediate solution vector $\mathbf{q}_i^\omega$: $\mathbf{q}_i^\omega = \mathbf{q}_i^n + (1-\omega_i)\left(\mathbf{q}_i^{n+1} - \mathbf{q}_i^n\right)$ With these definitions, the hybrid scheme is determined by means of the operator L_2 as follows:

$$\mathbf{q}_i^{n+1} = \mathbf{q}_i^n + \Delta t L_2\left(\omega_i \Delta t, \mathbf{q}^\omega\right) \tag{10}$$

In other words, the hybrid scheme is thought of as the explicit scheme (9), where the integration is executed over a time step $\omega_i \Delta t$ from the initial data given by the intermediate values.

The hybrid scheme is converted to the explicit second-order scheme of eq. (9) when the intermediate parameter ω takes unity for all cells. Provided that this parameter is zero, the scheme is transformed to the fully implicit (backward Euler) scheme. The latter reveals strong diffusion in numerical solutions. Therefore, it is preferable to make the value of the intermediate parameter as large as possible providing the MND property holds, i.e., $\left\|\mathbf{q}^{n+1}\right\| \le \left\|\mathbf{q}^n\right\|$.

The scheme can be also recast in the following form:

$$\mathbf{q}_i^{n+1} = \mathbf{q}_i^\omega + \omega_i \Delta t L_2\left(\omega_i \Delta t, \mathbf{q}^\omega\right) \tag{11}$$

which exactly coincides with the form of the explicit scheme of eq. (9), where the time integration is executed from the intermediate time level t^ω to the upper time level t^n. Therefore, it will be

$$\left\|\mathbf{q}^{n+1}\right\| \le \left\|\mathbf{q}^\omega\right\| \tag{12}$$

if the time step and the intermediate time level parameter are chosen to comply with the following inequality:

$$\omega_i \Delta t \le \lambda_i(\mathbf{q}^\omega) \tag{13}$$

where $\lambda_i(\cdot)$ denotes a function, which expresses the CFL-condition ensuring the MND property of the explicit scheme.

Since ω varies in the range $0 \le \omega \le 1$, $\left\|\mathbf{q}^\omega\right\| \le \omega\left\|\mathbf{q}^n\right\| + (1-\omega)\left\|\mathbf{q}^{n+1}\right\|$. Combining this inequality with (12), we can infer that the numerical solution of the hybrid scheme also satisfies the inequality $\left\|\mathbf{q}^{n+1}\right\| \le \left\|\mathbf{q}^n\right\|$. Consequently, the condition given by eq. (13) can be treated as a sufficient condition for the hybrid scheme to be MND.

Defining $\beta_i^\omega = \Delta t / \lambda_i(\mathbf{q}^\omega)$ for each computational cell, the intermediate time level parameter ω in the hybrid scheme can therefore be taken as

$$\omega_i = \begin{cases} 1 & if \;\; \beta_i^\omega \le 1 \\ 1/\beta_i^\omega & otherwise \end{cases} \tag{14}$$

The principle of hybridization proposed here provides a rather general approach to construct variable MND explicit-implicit schemes. As a matter of fact, there are no special difficulties to convert any explicit scheme into the hybrid form by means of the way described above to take advantage of relatively large time steps without spoiling the MND property.

4 Numerical results

The hybrid scheme discussed above is realized on the base of a second-order accurate Godunov explicit method, i.e., the operator L_2 in (10) is taken in the form of the Godunov's scheme [4] implemented with the MUSCL second-order accurate reconstraction-evolution procedure for cell interface parameters [3]. The resultant system of discretized equations must be solved regarding the cell values of $\mathbf{q}^{n+1}$ and ω. To obtain this solution, we apply the dual-time (or pseudo-time) approach proposed Jameson [5]. Unsteady in dual time equations are implicitly discretized and solved with successive Newton's iterations to make the unstedy residual vanish for each real time step. A system of linear equations with respect to the iterative increment of the solution vector appeared in this approach is solved with the LU-SGS approximate factorization method of Yoon and Jameson [6].

The convergence of Newton's iterations in the foregoing solution strongly depends on the linearization of numerical flux functions. Exact linearization of the second order inviscid flux-function is extremely expensive, if not impossible. In the hybrid approach this is even more tough as the flux-function depends on the solution at the intermediate time level normally unknown in advance. Therefore we assume two things: only the first-order representation of the inviscid flux is considered for linearization, and the parameter ω is frozen between two sequential iterations. Moreover, the viscous flux is approximately linearized with the use of a diagonal matrix majorizing the viscous Jacobian [7]. Under these assumptions, two methods of linearization of the Godunov flux-function are investigated. One is a conventional approximate linearization (AL) by means of the Jacobian spectral radius [8]. Another is the exact linearization (EL) obtained in [9].

Comparison between the composite scheme and conventional implicit ($\omega = 0$) and explicit ($\omega = 1$) schemes is tested on the solution of the Burgers scalar equation on a stiff grid. The ratio between the time step of the explicit scheme and the hybrid (and implicit) scheme is 0.01. The numerical solutions are displayed in Figs. 1 and 2, which show that the hybrid scheme solution is much more accurate than the solution of the implicit scheme, and even more accurate than the solution of the explicit scheme.

Figures 3 and 4 show the efficiency of the EL implementation versus the AL in realizing the hybrid method. The scheme was applied to compute a boundary layer flow on a flat plate ($M = 0.17,\ Re = 230000$) and an unsteady flow around a Zhukovskii airfoil ($\alpha = 5deg,\ M = 0.5,\ Re = 23000$). The EL implementation essentially accelerates the convergence of the steady residual and also the convergence of the iterative residual in the unsteady flow calculation.

References

1. J. Collins, P. Colella, and H. Glaz: Jour. Comp. Phys. **116**, 195 (1995)
2. P. O'Rourke, M. Sahota: Jour. Comp. Phys. **143**, 312 (1998)
3. B. van Leer: Jour. Comp. Phys **32**, 101 (1979)
4. S. Godunov: Math. Sb. **47**, 3 (1959)
5. A. Jameson: AIAA Tech. report **91-1596** (1991)
6. A. Jameson, S. Yoon: AIAA Journal **25**, 929 (1987)
7. I. Men'shov, Y. Nakamura: Proceed. of 8th Japan Conference on CFD, 13 (1994)
8. A. Jameson, E. Turkel: Mathematics of Computation **37**, 385 (1981)
9. I. Men'shov, Y. Nakamura: Computers and Fluids **29**, 595 (2000)

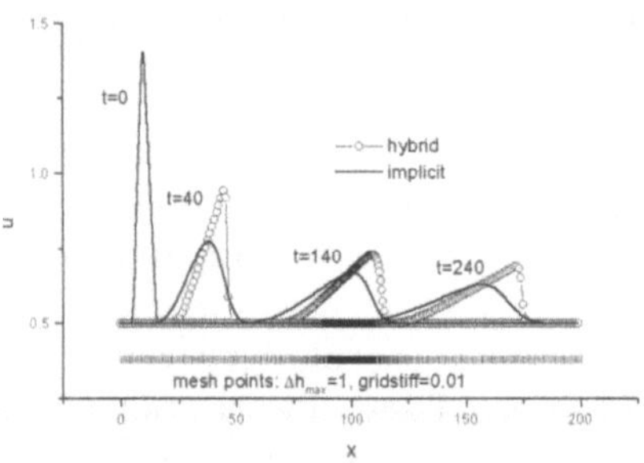

Fig. 1. Hybrid and implicit scheme solutions.

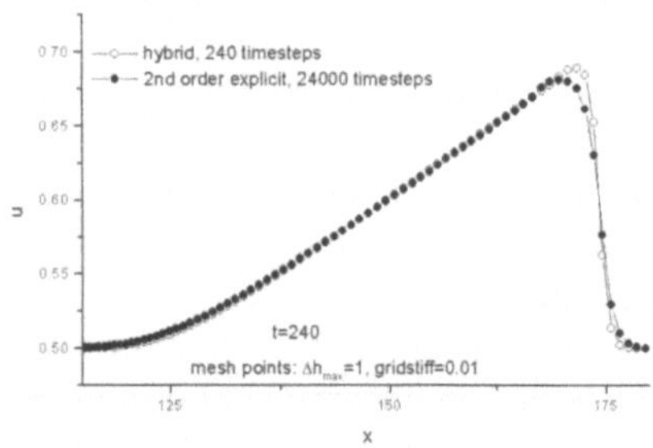

Fig. 2. Hybrid and explicit scheme solutions.

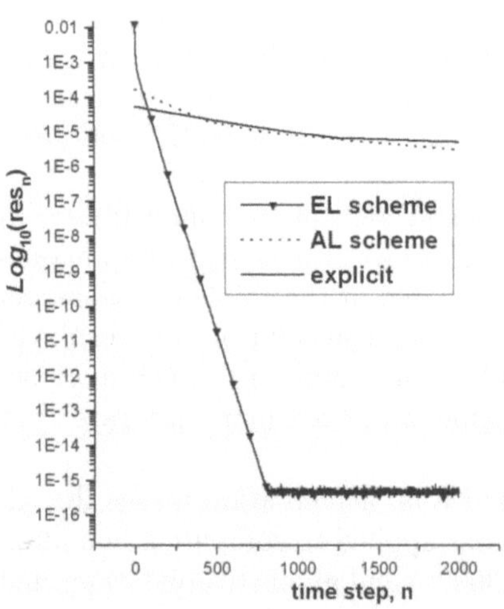

Fig. 3. Residual convergence for the flat plate calculation.

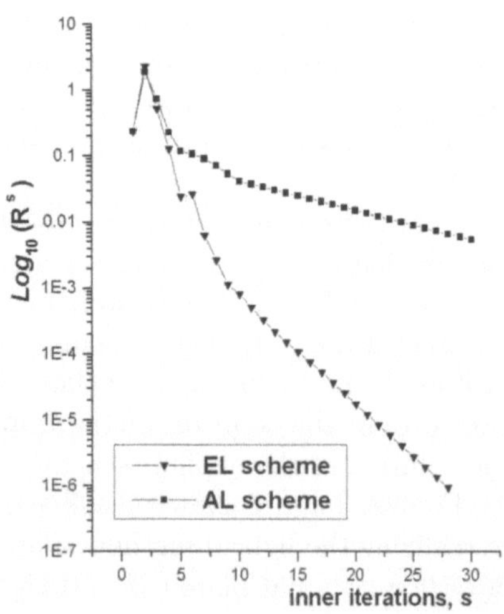

Fig. 4. Inner iterations convergence for unsteady flow past a Zhukovskii airfoil.

Numerical and Experimental Investigation of the Stability of the 3D Flow Between Two Counter–rotating Disks

Olivier Daube[1], Patrick Le Quéré[2], Frédéric Moisy[3], and Marc Rabaud[3]

[1] CEMIF/LME, Université d'Evry, 91020 Evry Cedex, France
[2] LIMSI-CNRS, BP 133, 91403 Orsay Cedex, France
[3] FAST, Bât 502, Université Paris Sud, 91405 Orsay Cedex, France

Abstract. This paper is devoted to a numerical and experimental study of the 3D flow in a cylindrical cavity of large radius-to-height aspect ratio whose top and bottom lids are rotating in opposite directions. This counter rotation gives rise to a new type of instability pattern, referred as negative spirals. Comparisons between numerical and experimental results give evidence that a free shear layer instability is responsible for this new pattern.

1 Introduction

The flows between rotating disks, or von Kármán swirling flows, occur in a variety of situations, and have been studied since a long time. The stability of these flows received considerable interest, mostly in the rotor-stator configuration [1–3] (i.e. between one rotating disk and one stationary disk), and it is only recently that 3D numerical simulations for moderate radius-to-height aspect ratio $\Gamma = R_0/H$ became available [4]. These studies focused on the boundary layer instabilities, leading to spiral or circular waves.

In this study, it is shown that in the case of counter–rotating disks, another type of instability may also occur. The pattern associated to this new instability received the name of negative spirals [5], since they roll up to the center in the direction of the slower disk. Similar patterns have been reported for smaller aspect ratio, $\Gamma = 2$ [6], and were attributed to a free shear layer instability. The influence of the aspect ratio have been recently studied experimentally [7].

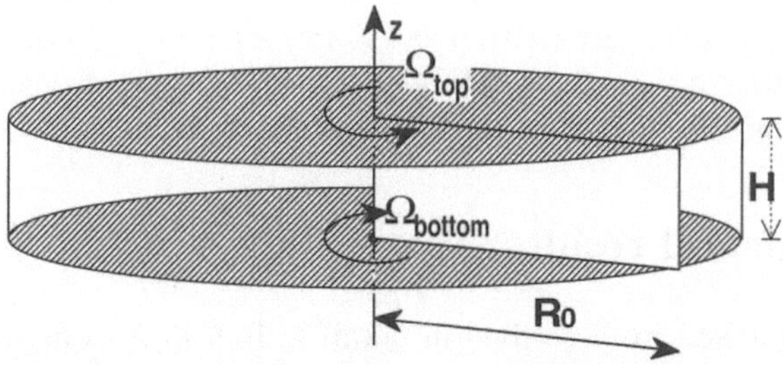

Fig. 1. Geometry of the flow between rotating disks. The lateral endwall rotates with the upper disk.

The configuration is defined in figure 1. The flow is characterized by three non dimensional parameters:

$$\mathrm{Re}_t = \frac{\Omega_t H^2}{\nu} \; ; \; \mathrm{Re}_b = \frac{\Omega_b H^2}{\nu} \; ; \; \Gamma = \frac{R_0}{H}.$$

The set of parameters $(\mathrm{Re}_t, s, \Gamma)$ may be also used for convenience, where $s = \Omega_b/\Omega_t$ is the counter-rotation ratio ($|s| < 1$ in the present study).

2 The numerical method

The numerical method is described in [8] and has the following main ingredients:
Time discretization: A second order time marching procedure is used with an implicit discretization of the linear terms and an explicit Adams-Bashforth type extrapolation of the non linear terms. In the resulting discrete problem to be solved at each time step, the velocity–pressure coupling is handled by means of an incremental projection method [9].
Spatial discretization: In the previously time discretized system, the different unknowns are first expanded in truncated Fourier series over N modes in the azimuthal direction. The coefficients of this expansion are then discretized in the (r, z) planes by means of mimetic finite difference operators [10]:

- A classical staggered uniform grid in cylindrical coordinates (r, θ, z) is used. Noteworthy is the fact that the only unknown located on the axis $r = 0$, is the axial component ζ of the vorticity.
- The space $\mathfrak{H}$ of the discrete vectors and the space H of the discrete scalar functions are equipped with inner products $<,>_{\mathfrak{H}}$ and $<,>_{\mathrm{H}}$ derived from the $\mathbb{L}^2$ inner products of the continuum case,
- The first order differential operators divergence and curl, are discretized by means of the Gauss and Stokes theorems written on elementary cells,
- The discrete gradient operator is defined as the negative adjoint of the discrete divergence operator with respect to the discrete inner products.
- The second order differential operators are built as compounds of first order discrete operators, therefore ensuring that the Laplace operators are self adjoint. This fact, together with the use of a staggered grid, ensures that the discrete problem to be solved at each time step is well posed,
- The non linear terms are written as $(\nabla \times \boldsymbol{v}) \times \boldsymbol{v}$ and discretized in order that $< \boldsymbol{v}\, , (\nabla \times \boldsymbol{v}) \times \boldsymbol{v} >_{\mathfrak{H}} = 0$, therefore ensuring kinetic energy conservation. As usual, they are computed in the physical space with the classical 3/2 rule.

3 Experimental results

The experimental set-up, described in detail in Ref. [5,7], consists in a rotating cylinder of radius $R = 140$ mm, in which a disk of same radius rotates at a different speed. In this paper we are only concerned with the counter-rotation case, where the faster disk is the top one ($|\Omega_t| \geq |\Omega_b|$). The working fluid is a mixture of water and glycerol of kinematic viscosity lying in the range $1.0 \times 10^{-6} < \nu < 4 \times 10^{-6}$ m^2/s at 20°C. Qualitative insight of the flow structure is obtained from visualization of the light reflected by anisotropic flakes seeding the flow (Kalliroscope). The flow is illuminated by a concentric circular light source, and pictures are obtained using a CCD camera located along the disks axis.

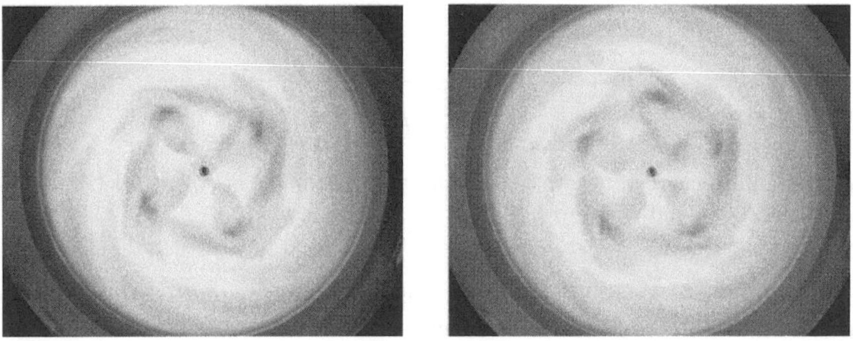

(a) (b)

Fig. 2. Examples of patterns observed between rotating disks. (a) $m = 4$, (b) $m = 5$.

At high enough Reynolds numbers, the counter-rotating flow gives rise to instability patterns such as the ones shown in figure 2. They consist in a circular chain of vortices surrounded by a set of spirals (negative spirals [5]). Depending on the mode and the aspect ratio, only the circular chain of eddies or the negative spirals may be observed. High aspect ratios R/h and/or low mode essentially leads to eddies, while lower aspect ratios and/or higher modes mostly show negative spirals. Intermediate modes, roughly between 4 and 7, usually shows a combination of the two aspects of the pattern.

The stability curve of these patterns is shown in figure 3 for different aspect ratios. These curves are obtained by slowly increasing the bottom disk angular velocity Ω_b at fixed value of Ω_t. No hysteresis is observed within our experimental uncertainty, around 3%. When plotted as functions of the Reynolds numbers $(\Gamma Re_t, \Gamma Re_b)$, the different curves appear to collapse into a single master curve. The fact that the chain of vortices and negative spirals share the same onset curve indicate that they both arise from the same instability mechanism, although the nonlinear saturation leads to very different morphology. We note that at high Reynolds number, the onset is well described by a single non dimensional parameter, the counter-rotation ratio

$$s = \frac{Re_b}{Re_t} \simeq -0.135 \pm 0.010.$$

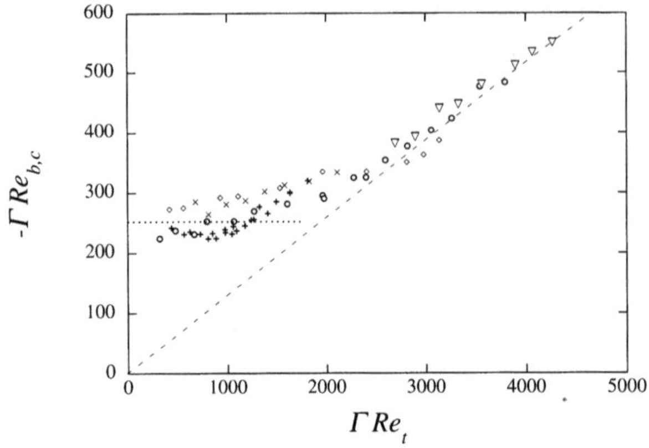

Fig. 3. Stability curve of negative spirals for aspect ratios $\Gamma = R/h = 6.1$ ($\triangledown$), 7.0 (o), 10.8 ($\diamond$), 14.0 ($\times$) and 20.9 ($+$).

However, at lower Reynolds numbers, this linear curve appears to saturate towards a finite bottom Reynolds number,

$$\Gamma Re_b \simeq -230 \pm 20.$$

The collapse of the curves for different aspect ratios gives indication for the mechanism responsible of the instability, because the shear and the boundary layers behave differently when R/h is varied. Assuming that the growth rate is controlled by the shear $S \sim \Omega R/h$, damped by viscous diffusion on a time scale $\tau_v \sim h^2/\nu$, this leads to the natural control parameter $S\tau_v = \Omega Rh/\nu = \Gamma Re$. So the master curve obtained in figure 3 suggests that a shear layer instability is responsible for the different patterns observed in figure 2.

4 Numerical results

The numerical results reported here have been obtained for $\Gamma = 7$ using a $400{\times}100$ grid in the (r, z) variables and 32 Fourier modes in the azimuthal direction. The time step was taken equal to 1/500 revolution time. In figure 4, contours in the plane $z = H/2$ of the axial velocity component are shown for $\Gamma = 7$; $\Gamma \mathrm{Re}_t = 1750$; $\Gamma \mathrm{Re}_b = 260$. The essential features of the experiments are recovered, i.e. the existence of a mode $m = 4$ with surrounding negative spirals. Other calculations for a slightly different Re_b have been performed, showing the existence of a dominant $m = 5$ mode (see figure 5). Additional calculations (not reported here) have shown that for a further decrease of Re_b down to 252, the flow reverts to a steady axisymmetric regime, in excellent agreement with the experimental stability curve in figure 3.

Further insight into the mechanism of this instability can be achieved from these computations. As first noted by [11], at high enough counter-rotation ratio

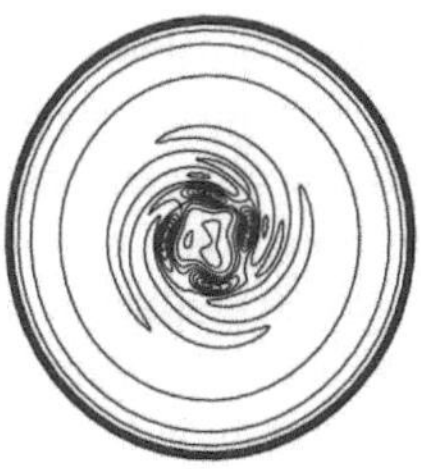

Fig. 4. Contours of the axial velocity component in the plane $z = 0.2H$; $\Gamma = 7$; $\mathrm{Re}_t = 250$; $\mathrm{Re}_b = 37$.

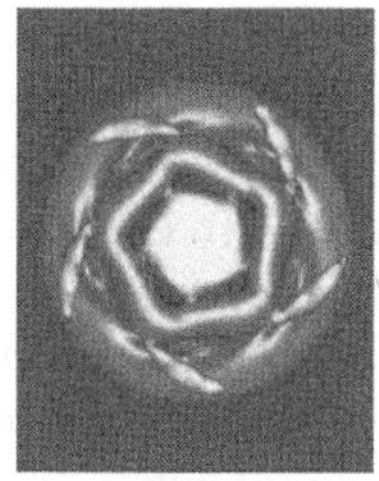

Fig. 5. Isosurface $w = 0.0354$ of the axial velocity ; $\Gamma = 7$; $\mathrm{Re}_t = 250$; $\mathrm{Re}_b = 41$.

(roughly $|s| > 0.1$), the meridional flow gets organized into a two-cell recirculating flow. The competition between the centrifugal effects of each disk leads to the formation of a stagnation ring on the slower disk. This important property of the counter-rotation regime can be shown in figure 6: the sign of the azimuthal vorticity over the slower disk ($z = 0$) changes at $r/R_0 \simeq 0.36$. The presence of this stagnation ring as important consequence on the azimuthal component of the flow, as shown in figure 7. For large radius, the shear layer between the

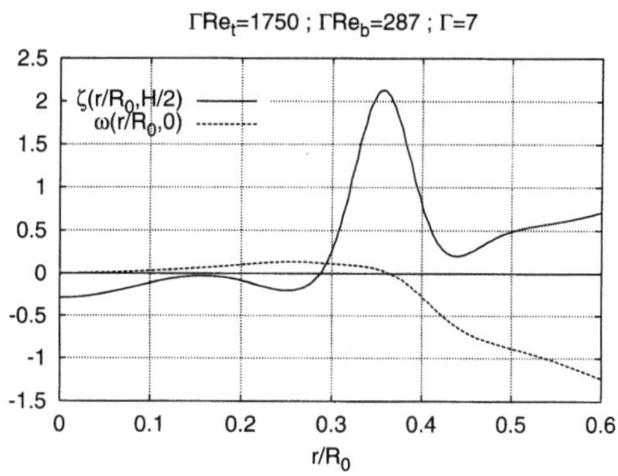

Fig. 6. Azimuthal vorticity along the slower disk $z = 0$ (dashed line) and vertical vorticity at mid-height $z = H/2$ (solid line).

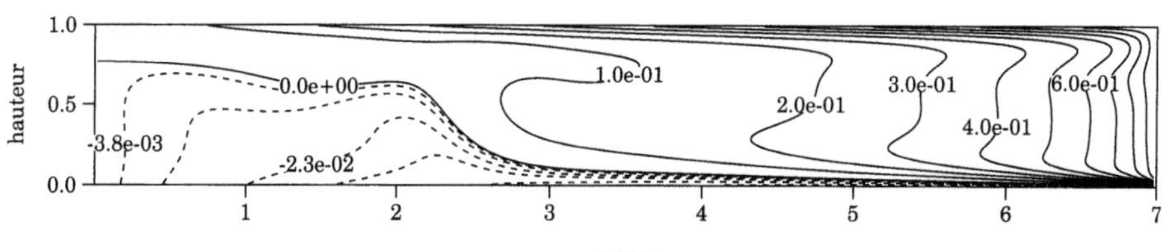

Fig. 7. Contours of the azimuthal velocity in the steady axisymmetric state.

positive velocity (associated to the faster disk) and the negative velocity (slower disk) remains within the inward boundary layer over the slower disk. At smaller radius, $r < 0.25R_0$, this shear layer gets detached because of the stagnation ring, leading to a free shear layer in the bulk of the flow. This phenomenon can be further seen in figure 6, where the location of the stagnation ring appears to coincide with the maximum of vertical vorticity. At high enough Reynolds number, this free shear layer may become unstable, leading to the instability patterns shown above.

Conclusion

The existence and onset of negative spirals have been confirmed both experimentally and numerically. Unlike other instabilities that are reported in the literature for this flow configuration, this new pattern is not associated to a boundary layer instability. Evidence is given that a free shear layer instability is responsible for this new pattern, in agreement with Ref. [6]. Additional experiments and computations show that modes with increasing wave number appear as the radius-to-height ratio is increased.

References

1. A.J. Faller, R.E. Kaylor, *Dynamics of Fluids and Plasmas*, Academic Press (1966).
2. L. Schouveiler, P. Le Gal, M.P. Chauve, J. Fluid Mech. *443*, 329–250 (2001).
3. Gauthier, G., Gondret, P. and Rabaud, M., J. Fluid Mech. *386*, 105–126 (1999).
4. E.Serre, E. Crespo del Arco, P. Bontoux, J. Fluid Mech. *434*, 65–100 (2001).
5. Gauthier, G., Gondret, P., Moisy, F. and Rabaud, M., J. Fluid Mech., in press (2002).
6. J.M. Lopez, J.E. Hart, F. Marques, S. Kittelman, J. Shen, J. Fluid Mech. *462*, 383–409 (2002).
7. Moisy, F., Pasutto, T. and Rabaud, M., to appear in "Nonlinear Processes in Geophysics", Ed. W.G.Früh, (2002).
8. E. Barbosa, O. Daube, A Finite Differences Method in Cylindrical Coordinates for 3D Incompressible Flows, *ECCOMAS CFD Conference*, Swansea, UK (2001).
9. K. Goda, J. Comp. Phys. *30* (1979).
10. J.M. Hyman, M. Shashkov, Comp. Math. Appl. *33*, (1997).
11. Dijkstra, D. and van Heijst, G.J.F., J. Fluid Mech. *128*, 123–154 (1983).

Molecular Dynamics Simulation of Oxygen Recombination on Finite Catalytic Surface

Tomoe Oda[1], Yoshiaki Tamura[2], and Tetuya Kawamura[1]

[1] Ochanomizu University, Tokyo 112-8610, JAPAN
[2] Toyo University, Kawagoe 350-8585, JAPAN

Abstract. In this paper, the behaviors of oxygen atoms on the surface of silica are investigated by classical molecular dynamics simulation for the detailed mechanism on surface catalysis. Three types surface of β-cristobalite are formed with BMH potential and three-body potential. The temperature dependency of mobility and proximity about two bared oxygen atoms is observed and the result supports one of the proposed surface catalysis mechanisms.

1 Introduction

During re-entry of a reusable space vehicle a shock wave forms in front of the vehicle and leads to a very high translational temperature in the flow field. A large number of Nitrogen and Oxygen molecules are dissociated in such a flow field into N and O atoms. These atoms are struck onto the surface of the orbiter and recombined by the catalysis of the surface. The recombination processes generate heats. This heat of recombination causes the increase of the orbiter surface temperature and affects the performance of heat shields of these hypersonic vehicles. This reaction is known as Surface Catalysis. To reduce the reaction, the surface is constructed or coated with materials as little catalytic as possible, such as Silica (SiO_2).

In many papers, the energy transfer recombination coefficient and the reaction rate coefficients are described as a function of the gas temperature[1,2]. Recently, however, it was reported that the recombination rate coefficient depends also on the partial pressure and the partial pressure rates of gas species[3] though the physics of the recombination is still not clear.

The objective of this study is to investigate the behaviors of oxygen atoms on the surface of silica by classical molecular dynamics simulation to obtain more detailed information about catalytic process. Especially the probability of oxygen atom recombination and the temperature dependency of the probability are observed. In the following sections, the model and the method are described.

2 Surface Catalysis

In the literature[4], the heterogeneous catalyst generally has the following process. In the case of recombination of atoms, the process is schematically shown in Fig. 1. (a) The dissociated atoms incident to the surface. (b) Some of the

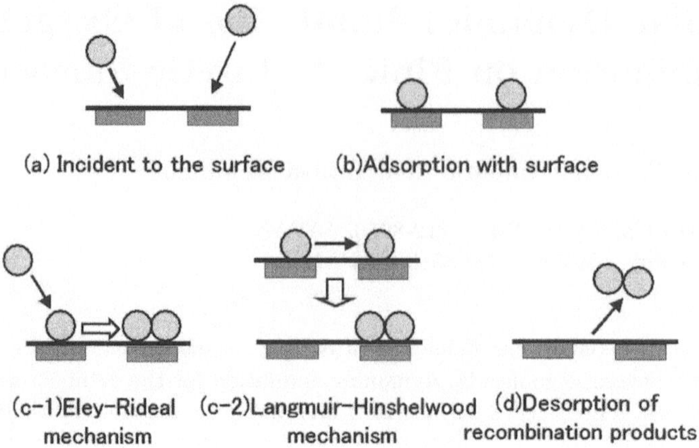

Fig. 1. Elementary steps of recombination process

incident atoms are absorbed at the surface. There are two recombination mechanisms, Eley-Rideal mechanism and Langmuir-Hinshelwood mechanism[4]. (c) An adatom recombine with one more atom. (d) The bond of molecule with surface is broken. Hence, the molecule is desorbed from surface. The neighborhood of the surface is considered to be in the thermal equilibrium.

In this paper, SiO_2, which have low catalysis, is considered as surface and Langmuir-Hinshelwood mechanism is mainly investigated as it has temperature dependency.

3 Molecular Dynamics Method

The motion of atoms is described by the kinetic equation for each particle i.

$$\frac{d^2\boldsymbol{r_i}}{dt^2} = \frac{\boldsymbol{F_i}}{m}, (i = 1, 2, \cdots, N) \tag{1}$$

where $\boldsymbol{r_i}$ is the position of atom i and $\boldsymbol{F_i}$ is the external force acting on atom i.

In this study, only oxygen atoms are considered as gas phase atoms. The β-cristobalite formed at the temperatures involved in the re-entry conditions is adopted. The β-cristobalite[5], shown in Fig. 2, is one of the crystal structures in high temperature of SiO_2. In Fig. 2, the small white particles are Si atoms and the large black particles are O atoms. Because the ionic radius of Si atom is smaller than that of O atom, the Si atoms are indicated as smaller particle in Fig. 2.

The SiO_2 potential, described in [6], consists of a weak three-body interaction and a modified Born-Mayer-Huggins (BMH) ionic, pair interaction. The two-body interaction includes short-range repulsion due to overlap of core electrons and a modified Coulombic term,

$$v_2(r_{ij}) = A_{ij} \exp(-b_{ij} r_{ij}) + \frac{Z_i Z_j e^2}{r_{ij}} \mathrm{erfc}\left(\frac{r_{ij}}{\beta}\right) \tag{2}$$

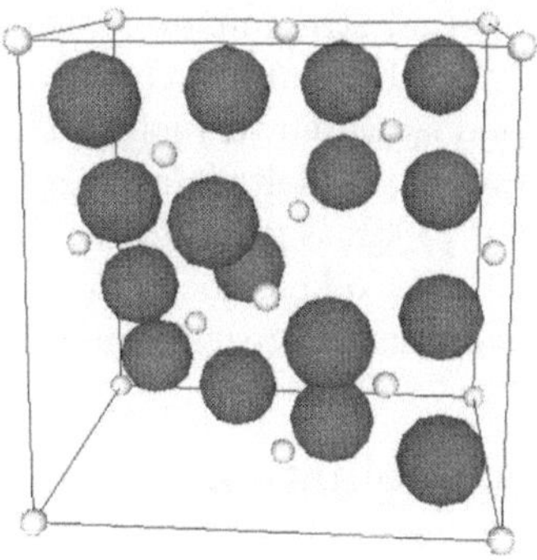

Fig. 2. Crystal structure of β-cristobalite.

where r_{ij} is the distance between atoms, Z_{ie} is the formal ionic charge, and A_{ij}, β_{ij} and ρ are constants. Values of σ, β_{ij}, ρ, A_{ij} have been taken from previous MD simulations using this potential[7]. The first term is the repulsive force between interior shells. The second term is Coulomb term. The erfc(x) is the complementary error function. Since silicon and oxygen are not purely ionic, only the first term of an expansion for the Coulomb distance dependence is used as Coulomb term[6]. Then only the first term in Ewalds Method is solved.

A short-range three-body interaction has been adopted for the O-Si-O and Si-O-Si configurations. This interaction is not adopted all other possible combinations. The three-body term is

$$v_3(r_{ij}, r_{ik}, \theta_{jik}) =$$

$$\begin{cases} \lambda_i \exp\left(\frac{\gamma_i}{r_{ij}-r_i^c} + \frac{\gamma_i}{r_{ik}-r_i^c}\right)\left(\cos\theta_{ijk} + \frac{1}{3}\right)^2 & , r_{ij} < r_i^c \text{ and } r_{ik} < r_i^c , \\ 0 & , \text{ otherwise} . \end{cases} \quad (3)$$

Constants, λ_i, γ_i, r_i^C, have been taken as the same values in [6]. The three-body potential lowers the total binding energy of the system whenever the bond angle θ_{ijk} differs from the perfect tetrahedral angle θ_{ijk}, where $\cos(\theta_{ijk}) = -1/3$. Local order in silicate systems is well defined with the three-body potential. For more details, see [6].

4 Run Procedure

The system contains 648 atoms and a configuration of b-cristobalite positions is given initially. The periodic boundary conditions is employed in the x and y directions. The velocity Verlet algorithm was used for integrating Eq.(1). A time step is 1.0fs. The lowest layer atoms are fix at crystal structure. In order to save the computational time, the bookkeeping method is introduced. In order to remove the internal stress of solid, or to find out the equilibrium positions of atoms, 10000 time steps calculation at 0[K] is carried out firstly. The atoms slightly move due to the internal stress at every time step. Then the velocity of atoms is removed forcibly until the atoms come to the equilibrium positions. After this procedure, the solid is heated to 1500K. The temperature of the solid is expressed by lattice vibrations. A Boltzmann distribution of velocities corresponding to a temperature of 1500K is given to the mobile atoms. The program runs for 3000 time steps with periodic velocity scaling and 7000 time steps at constant energy to allow the system to reach structure relaxation.

5 Evaluation of Recombination

Two surfaces are considered (Fig. 3). At surface A, silica surface is covered with oxygen atoms bonding to underlining silicon atoms. At surface B, the silicon atoms are bared with dangling bonds on the top layer. During the relaxation process at surface B, the silicon atoms in the top layer are being migrated under the oxygen atoms. Then the oxygen atoms are bared. Both surfaces is a model of SiO_2 coating bared in high temperature oxygen atom gas. Surface A simulates a situation that oxygen atoms full cover the SiO_2 surface while Surface B has deficit due to the recombinations of oxygen atoms by either proposed mechanism.

Computed results show that only the top-most oxygen atoms vibrate hard and come close to each other. When two oxygen atoms come close, they may have probability of recombination. In this research, the distance between oxygen atoms is taken as an index of recombination. In Fig. 4, the number of encounters of oxygen atoms versus temperature are plotted with different threshold distances. It hardly occurs that two atoms come closer than 1.7[Å] in both cases. Also the temperature dependency is observed in both cases. However, the probability is much higher in surface B at every temperature. This result is interpreted as follows. The oxygen atoms on surface A are balanced but not on surface B. On surface B, the top-most Si atoms have dangling bonds. To balance a deficit, the oxygen atoms, which are in the same shell (a shell of tetrahedron shape) with a Si, vibrate hard. From the result, it is suggested that once a deficit appears on the surface due to the recombination or any other reason, the unbalance of internal forces enhance the probability of recombination.

Fig. 3. Silica surfaces: left; surface A. right; surface B

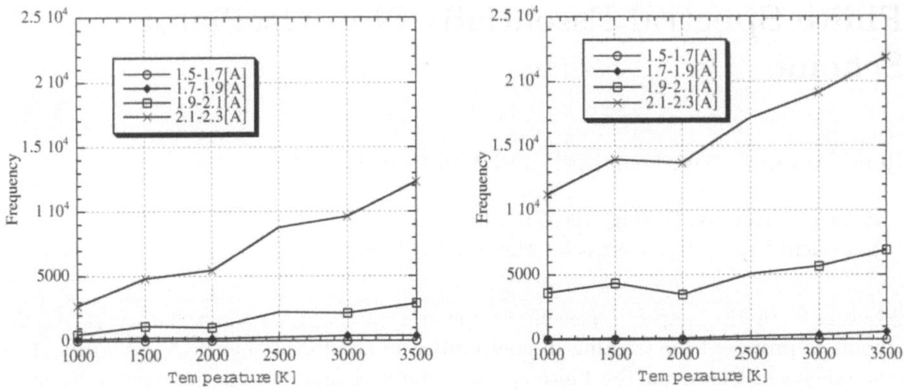

Fig. 4. Frequency of encounters of oxygen atoms: left; surface A. right; surface B

6 Conclusions

The behaviors of oxygen atoms on silica surface are investigated by classical molecular dynamics simulation for detailed information of surface catalysis. The BMH-potential with three-body term is employed and successfully applied to the simulation. As an index of recombination, the distance between two oxygen atoms is measured. From the result, the mechanism that the presence of Si's dangling bonds strengthens the vibration of oxygen atoms and induce the oxygen recombination is suggested.

References

1. M. Barbato, C. Bruno: *Molecular Physics and Hypersonic Flows*, ed. by M. Capitelli, pp. 139-160.
2. O. Deutschmann et al: J. Heat Trans., **117** (1995) pp. 495-501
3. T. Kurotaki: AIAA Paper 2000-2366 (2000).
4. P. W. Atkins: 'Physical Chemistry,' University Lecture and Fellow of Lincoln College, Oxford
5. R. W. G. Wyckoff: 'Crystal Structures,' John Wiley & Sons (1965)
6. B. P. Feuston, S. H. Garofalini: J. Chem. Phys. **91**, 564 (1989)
7. E. B. Webb III, S. H. Grofalini: J. Chem. Phys. **101**, 10101 (1994)

Finite Spectral Essentially Non-Oscillatory Scheme

Hong-Wei Liu[1], Yun-Feng Liu[1], and Jian-Ping Wang[1,2]

[1] Peking University, Beijing, 100871, China
[2] Yokkaichi University, Yokkaichi, Mie 512-8512, Japan

Abstract. In this paper a new finite spectral essentially non-oscillatory (FSENO) scheme is proposed for treating discontinuities in computational fluid dynamics. The new scheme is based on the finite spectral method and the idea of essentially non-oscillatory(ENO). Uniform high-order accuracy in smooth regions of solutions and sharp capturing of shock waves and other discontinuities are achieved by using it. We present numerical results for various one-dimensional Euler equations including Sod's and Lax's test problems. Finally, we simulate a turbulent flow compressed by a shock wave and a two-dimensional detonation.

1 Introduction

In order to deal with complex phenomenons and geometries with high accuracy, flexible and efficient spectral methods are required. Many efforts have been made for decades [1–8]. Localization of spectral methods has been a significant field in those efforts. Domain decomposition techniques [2] have succeeded in reducing the global effects to each subdomain, but they are still not very satisfactory for the need of flexibility in treating those complex problems. Finite spectral method based on non-periodic Fourier transform [7] is the first attempt of dealing with spectral methods pointwise, while more CPU time is required than FFT. Finite spectral method based on non-periodic Fourier integral [8] has been another successful attempt after the former. It can be characterized by local property, non-periodicity, orthogonal relation, efficiency and simplicity. This makes it possible to construct various local spectral schemes with both flexibility and efficiency.

As an application of the finite spectral method, we construct a new finite spectral essentially non-oscillatory (FSENO) scheme for treating discontinuities in computational fluid dynamics. It is based on the finite spectral method [8] and the idea of essentially non-oscillatory(ENO) [9]. The new scheme is applied to the standard Sod' and Lax' test problems [10] and shows itself to be very efficient. Also we use the scheme to simulate a turbulent flow compressed by a shock wave and a two-dimensional detonation.

2 Finite Spectral Method based on Non-Periodic Fourier Integral

The finite spectral method [8] was found when approximating a pulse function in Fourier series. It also can be derived from Fourier integral by truncating the frequency, and then applying the composite trapezoidal rule to the numerical integration in the frequency space. The finite spectral interpolation of an arbitrary function $f(x)$ is represented in a convolution form as

$$f(x) = \sum_{j=N_1}^{N_2} f(x_j) W_N(x - x_j) \tag{1}$$

where $x_j = x_{N_1} + (j - N_1)l/N$, $x_{N_1} \leq x \leq x_{N_2}$, and the basis function $W_N(x)$ is represented as

$$W_N(x) = \frac{1}{2N} \sum_{n=-N}^{N} C_n \exp\frac{\mathrm{i}\pi n x}{l} = \frac{1}{2N} \sum_{n=-N}^{N} C_n \cos\frac{\pi n x}{l} \tag{2}$$

where $-l \leq x \leq l$, and C_n takes 0.5 for $n = \pm N$ but 1 for others. The solution of $W_N(x)$ is showed in Figure 1 (a).

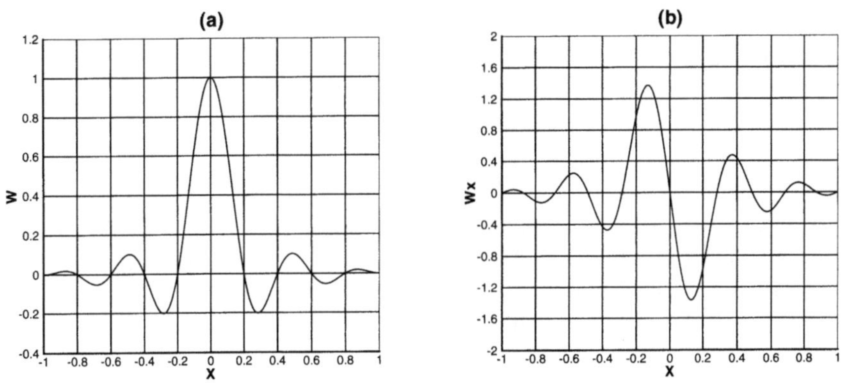

Fig. 1. Basis function and its derivative. (**a**) Basis function(N=5). (**b**) The derivative of basis function(N=5).

The important point to notice is that the periodicity in the original function is no longer required. This is because that the basis functions are local functions for each sample point, which are independent for each other.

Consequently, it is easy to obtain the formulae of calculating derivatives of $f(x)$ at discrete sample points as

$$f_x(x_i) = \sum_{j=N_1}^{N_2} f(x_j) W_N^{(1)}(x_i - x_j) \tag{3}$$

where $W_N^{(1)}(x)$ is the first derivative of $W_N(x)$ and can be represented as

$$W_N^{(1)}(x) = \frac{\mathrm{i}\pi}{2Nl} \sum_{n=-N}^{N} n C_n \exp\frac{\mathrm{i}\pi n x}{l} = -\frac{\pi}{2Nl} \sum_{n=-N}^{N} n C_n \sin\frac{\pi n x}{l} \tag{4}$$

Figure 1 (b) shows the solution of $W_N^{(1)}(x)$. The higher order derivative can be derived in the same way. The above formula gives a finite spectral method which is similar to a finite difference one. $W_N^{(1)}(x_i - x_j)$ can be regarded as the weighted coefficients for discrete sample points, which are calculated previous. Fast Fourier series are no longer required.

3 Finite Spectral Essentially Non-Oscillatory Scheme

The new FSENO scheme is essentially based on the "reconstruction via primitive function" technique [9] and the finite spectral method, and then it will be divided into two processes: in the first process, we can construct the primitive function $H(x)$ and choose the suitable smoothest stencil by "upwinding" and estimating the smoothness of stencils by comparing the Newton divided differences [11,12]; in the second process, the reconstructed numerical flux is obtained by using the finite spectral method to calculate the derivative of $H(x)$ at the corresponding point on the smoothest stencil.

For simplification, we take one-dimensional scalar conservation laws to illustrate this scheme: $u_t(x,t)+f_x(u(x,t)) = 0$. We will concentrate on the discussion of spatial discretization and the time discretization will be discussed latter:

$$u_t(x_i,t) + \frac{1}{\Delta x}(\hat{f}_{i+\frac{1}{2}} - \hat{f}_{i-\frac{1}{2}}) = 0. \tag{5}$$

We emphasize here that the grid must be uniform in this scheme: $\Delta x_i = \Delta x$. Then we have the algorithm of the FSENO scheme as the following steps:

ALGORITHM (FSENO).

Step(1), calculate the primitive function $H(x)$ at $x_{i+\frac{1}{2}}$ by

$$H(x_{i+\frac{1}{2}}) = \Delta x \sum_{k=-\infty}^{i} f(x_k) \tag{6}$$

Step(2), compute the divided difference table of $H(x)$, and identify

$$H[x_{i-\frac{1}{2}}, x_{i+\frac{1}{2}}] = \{H(x_{i+\frac{1}{2}}) - H(x_{i-\frac{1}{2}})\}/(x_{i+\frac{1}{2}} - x_{i-\frac{1}{2}}) = f(x_i) \tag{7}$$

$$H[x_{i-\frac{1}{2}}, x_{i+\frac{1}{2}}, \ldots, x_{i+k-\frac{1}{2}}] = \frac{H[x_{i+\frac{1}{2}}, \ldots, x_{i+k-\frac{1}{2}}] - H[x_{i-\frac{1}{2}}, \ldots, x_{i+k-1-\frac{1}{2}}]}{(x_{i+k-\frac{1}{2}} - x_{i-\frac{1}{2}})},$$
$$(k = 2, 3, \ldots, 2N) \tag{8}$$

Step(3), distinguish the signs of propagation speeds at cell boundaries, for the Euler equations, the characteristic speeds $\bar{c}_{i+\frac{1}{2}}$ ($\bar{u}_{i+\frac{1}{2}}$, $\bar{u}_{i+\frac{1}{2}} \pm \bar{a}_{i+\frac{1}{2}}$) are calculated by using the Roe average [13].

Step(4), choose the smoothest stencil. Let j_{local} be the point on the local coordinate corresponding to the cell boundary point $i+\frac{1}{2}$ on the global coordinate.

$$i_{min}^{(1)} = i,\ j_{local}^{(1)} = -N+1.$$
$$if\ \bar{c}_{i+\frac{1}{2}} < 0,\ \ then\ i_{min}^{(1)} = i+1,\ j_{local}^{(1)} = -N.$$
$$Inductively,$$
$$if\ |H[x_{i_{min}^{(l)}-\frac{1}{2}}, \ldots, x_{i_{min}^{(l)}+l-\frac{1}{2}}]| > |H[x_{i_{min}^{(l)}-1-\frac{1}{2}}, \ldots, x_{i_{min}^{(l)}+l-1-\frac{1}{2}}]|,$$
$$then\ \ i_{min}^{(l)} = i_{min}^{(l-1)} - 1,\ j_{local}^{(l)} = j_{local}^{(l-1)} + 1.$$
$$else\ \ i_{min}^{(l)} = i_{min}^{(l-1)},\ j_{local}^{(l)} = j_{local}^{(l-1)}.$$
$$i_{min} = i_{min}^{(l)},\quad j_{local} = j_{local}^{(l)}.\quad (l = 2, 3, \ldots, 2N).$$

Step(5), interpolate the primitive function $H(x)$ on the smoothest stencil in terms of the finite spectral interpolation method. Replacing H to the local coordinate, we have

$$\tilde{H}(\xi_j^{i+\frac{1}{2}}) = H(x_{i_{min}-\frac{1}{2}+N+j}), \qquad (j = -N, -N+1, \ldots, N) \tag{9}$$

then

$$\tilde{H}(\xi^{i+\frac{1}{2}}) = \sum_{j=-N}^{N} \tilde{H}(\xi_j^{i+\frac{1}{2}}) W_N(\xi^{i+\frac{1}{2}} - \xi_j^{i+\frac{1}{2}}) \tag{10}$$

Step(6), reconstruct the numerical flux by calculating the derivatives of the interpolating function at the cell boundaries as

$$\begin{aligned} \hat{f}_{i+\frac{1}{2}} = \frac{d}{dx} H(x)|_{x=x_{i+\frac{1}{2}}} &= \frac{d}{dx} \tilde{H}(\xi^{i+\frac{1}{2}})|_{\xi^{i+\frac{1}{2}}=\xi_{j_{local}}^{i+\frac{1}{2}}} \\ &= \sum_{j=-N}^{N} \tilde{H}(\xi_j^{i+\frac{1}{2}}) W_N^{(1)}(\xi_{j_{local}}^{i+\frac{1}{2}} - \xi_j^{i+\frac{1}{2}}) \end{aligned} \tag{11}$$

Remarks.

(1) The main difference of the present algorithm from the previous ones [11,12] is the finite spectral treatment in step(5)and(6).

(2) In the step(4), we can shorten the stencil when the error increases. We define $R^{(l)}(x)$ to be the l th error term of Newton interpolation of $f = dH/dx$,

$$then\quad R^{(l)}(x) = |H[x_{i_{min}^{(l)}-\frac{1}{2}}, \ldots, x_{i_{min}^{(l)}+l-\frac{1}{2}}] \prod_{\substack{j = i_{min}^{(l)} \\ j \neq i+1}}^{i_{min}^{(l)}+l-1} (x - x_{j-\frac{1}{2}})| \tag{12}$$

since $R^{(l)}(x_{i+\frac{1}{2}})$ is the approximation of the error, the stencil can be shortened when $kR^{(l)}(x_{i+\frac{1}{2}}) > R^{(l-1)}(x_{i+\frac{1}{2}})$, where k is a selected biasing coefficient which is chosen as $0.5 \leq k \leq 1.5$ in our numerical experiments and shows to be fit.

(3) The scheme will be more robust if a global "flux splitting" is used:

$$f(u) = f^{+}(u) + f^{-}(u) \tag{13}$$

where $f_u^{+}(u) \geq 0$, $f_u^{-}(u) \leq 0$. The smooth Lax-Friedrichs splitting is a good and simple choice:

$$f^{\pm}(u) = \frac{1}{2}(f(u) \pm \alpha u) \tag{14}$$

where α is taken as $\alpha = max_u|f^{'}(u)|$ over the relevant range of u. Therefor we can begin with

$$i_{min}^{(1)} = i,\ j_{local}^{(1)} = -N+1 \quad for \quad f^{+}(u)$$
$$i_{min}^{(1)} = i+1,\ j_{local}^{(1)} = -N \quad for \quad f^{-}(u)$$

in step (4), and then the step (3) is no longer needed.

(4) The treatment of boundary conditions is similar to that in [8,9].

(5) The TVD Runge-Kutta methods [12] are used in the time discretization.

4 Results

Several numerical experiments were carried out with the FSENO scheme. The results indicate that the FSENO scheme is very efficient in our numerical tests. Figure 2 (**a**) and Figure 2 (**b**) illustrate the density distributions of Sod's and Lax's test problems separately. Figure 3 (**a**) shows the density contour of the interaction between turbulence and shock wave. Figure 3 (**b**) shows the pressure contour of two-dimensional detonation.

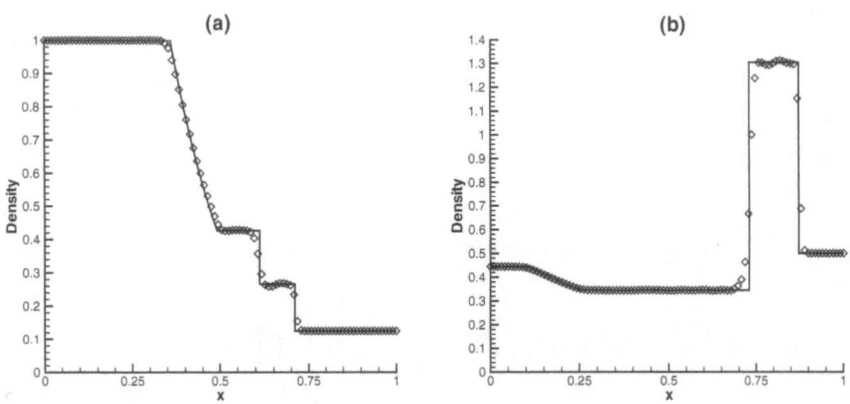

Fig. 2. Results of numerical tests. (**a**) Density distribution of Sod' problem. (**b**) Density distribution of Lax' problem.

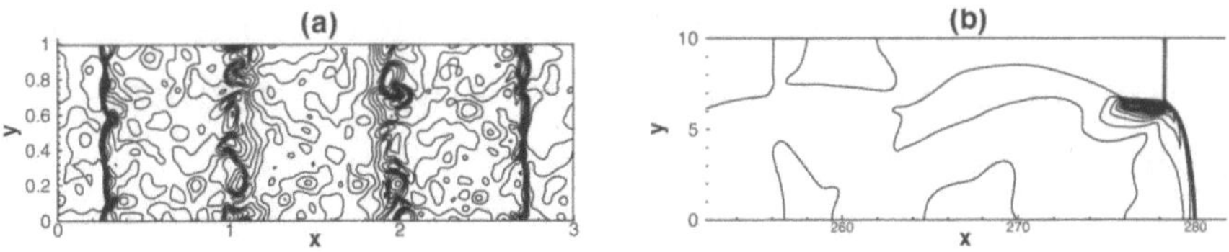

Fig. 3. Results of numerical simulations. (**a**) Density contour of the interaction between turbulence and shock wave. (**b**) Pressure contour of two-dimensional detonation.

5 Concluding Remarks

The new finite spectral essentially non-oscillatory (FSENO) scheme is an efficient and flexible scheme in computational fluid dynamics. The results of the numerical experiments implies that uniform high-order accuracy in smooth regions of solutions and sharp capturing of shock waves and other discontinuities are achieved by using this scheme. The new scheme is another prominent application of the finite spectral method which is a most fundamental methodology and needs further developments in the future.

References

1. S.A. Orszag: J. Comput. Phys. **37**, 70 (1980)
2. A.T. Patera: J. Comput. Phys. **54**, 468 (1984)
3. M.Y. Hussaini, D.A. Kopriva, M.D. Salas, and T.A. Zang: AIAA J. **23**, 64 (1985)
4. W. Cai, D. Gottlieb, and C.W. Shu: Math. Comput. **52**, 389 (1989)
5. W. Cai, C.W. Shu: J. Comput. Phys. **104**, 427 (1993)
6. J.P. Wang, Y. Nakamura, and M. Yasuhara: J. Comput. Phys. **107**, 160 (1993)
7. J.P. Wang: Computers & Fluids **27**, 639 (1998)
8. J.P. Wang: CFD Journal **10**, 569 (2002)
9. A. Harten, B. Engquist, S. osher, and S.R. Chakravarthy: J. Comput. Phys. **71**, 231 (1987)
10. G.A. Sod: J. Comput. Phys. **27**, 1 (1978)
11. C.W. Shu, S. Osher: J. Comput. Phys. **77**, 439 (1988)
12. C.W. Shu, S. Osher: J. Comput. Phys. **83**, 32 (1989)
13. P.L. Roe: A. Rev. Fluid Mech. **18**, 337 (1986)

Numerical Studies on the Reaction Zones in a Detonation Wave with a Detailed Chemical Reaction Model

Xiang Yu Hu[1], Boo Cheong Khoo[1], De Liang Zhang[2], and Zong Lin Jiang[2]

[1] Singapore-MIT Alliance, National University of Singapore, 117576, Singapore
[2] LHD, Institute of Mechanics, Chinese Academy of Science, Beijing, 100080, China

Abstract. This paper is focused on the reaction zones of a detonation wave in $2H_2/O_2/7Ar$ mixture with a detailed chemical reaction model. A regular cellular detonation wave structure is firstly computed, then the characteristics of the reaction zones, such as the ignition mechanisms, the species properties and the vortexes effects caused by the features of the structure, have been studied.

1 Introduction

It is well known that a gaseous detonation wave has cellular structure. The cellular structure is usually complex and involves triple wave configurations, one of which includes an incident shock wave (I), a Mach stem (M) and a transverse wave (T). Thus, instead of the layer-by-layer, one-dimensional shock-chemical induction-exothermic reaction processes described by the ZND model, a variety of burning patterns in the reaction zones is assumed companying with the cellular structure (Fickett and Davis, 1979). From late 1970's, numerical simulations have been employed to study the cellular structure. In the early works, simple one or two-steps chemical reaction model were used (Taki et al, 1978; Gamezo et al, 1999). These works mainly concerned the triple wave configurations. Recently, multi-step, detailed reaction model have been used with some success and new results computed. With this model Lefebvre and Oran (1995) and Oran et al (1998) concluded that, besides primary energy release behind the Mach stem, there is a secondary energy release starting at about two-third of the way through the cell. In this paper, further studies on the nature of reaction zones of a cellular detonation with a detailed reaction model are performed.

2 Physical and numerical models

The governing equations for a two dimensional gaseous detonation with a N species, multi-step chemical reaction model are

$$\frac{\partial \mathbf{U}}{\partial t} + \frac{\partial \mathbf{F}}{\partial x} + \frac{\partial \mathbf{G}}{\partial y} = \mathbf{S}, \tag{1}$$

This set of equations describes the conservation of density ρ, momentum $\rho\mathbf{v} \equiv (\rho u, \rho v)$, total energy density E, density of species $\{\rho_i\}$, where $i = 1, N$, and closed by the equation of states. In this paper, a 9 species, 19 elementary reactions model is used for hydrogen-oxygen combustion (Wilson and MacCormack, 1990). The reacting species are H_2, O_2, H, O, OH, H_2O_2, HO_2, H_2O, and 70% argon dilute is added to the gas mixture.

The finite difference scheme used is based on 3rd order ENO-LLF scheme (Shu and Osher, 1989). For the 2nd and 3rd order terms, a modification is made by which a weighted average is used in place of the two candidate divided differences,

$$D_- = D\left[x_{j-1/2}, ..., x_{j+l-1/2}\right], D_+ = D\left[x_{j+1/2}, ..., x_{j+l+1/2}\right], l = 2, 3 \quad (2)$$

hence ensuring the smoothness of the interpolation (as opposed to the original choosing of the quantity with smaller absolute value). That is

$$D = \frac{D_+W^+ + D_-W_-}{W_+ + W_-}, W_+ = D_-^4, W_- = D_+^4. \quad (3)$$

The time discretization for the fluid dynamic terms is the 3rd order TVD Runge-Kutta method (Shu and Osher, 1989). For the chemical kinetic integrations, the set of coupled source terms is solved by Selected Asymptotic Integration Method (Young, 1979). The time-step splitting scheme is them used to couples the Euler equations to chemical reactions (Oran and Boris, 1987).

3 Computational setup and initialization

The numerical simulation models a detonation propagating in a two dimensional channel from left to right in stoichiometric $H_2 - O_2$ diluted with 70% argon with initial pressure and temperature given as 6.67kPa and 298K, respectively. Firstly, one dimensional steady detonation is initialized by a strong shock wave. Then the steady solution is placed on a two dimensional grid serving as the initial conditions to simulate a detonation wave propagating in a channel. The channel widths is set to $20mm$, and the grid size is $\Delta x = \Delta y = 0.2mm$. The initial condition is perturbed by introducing random disturbances only to the first time-step of chemical kinetic integrations via perturbing the total specific energy. The grid is set to move at about detonation velocity in the positive x-direction. Therefore, the detonation is thus observed for very long physical time and ensured free from initial imposed perturbations.

4 Results and discussions

4.1 Formation of regular cellular structure and detonation cell

The numerical results show that the initial disturbances are very small and one can hardly distinguished the difference from the perturbed pressure contours (see Fig. 1a). However, as the detonation propagates, the disturbances are magnified, and it is observed the existence of transverse waves, Mach stems and incident waves which give rise to the triple wave configurations at about $15\mu s$

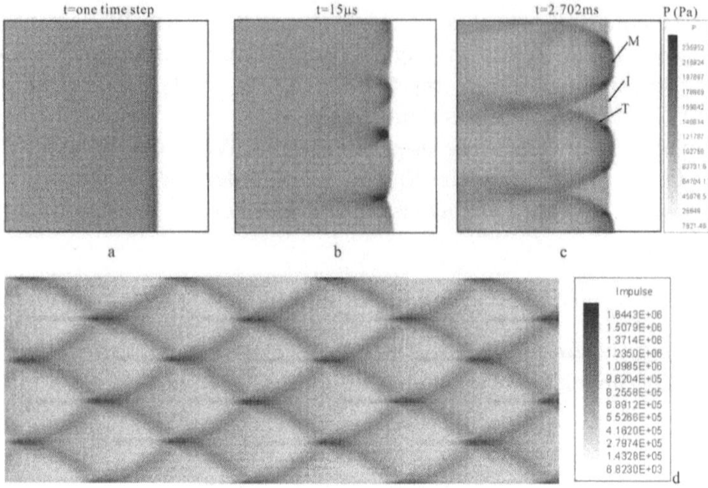

Fig. 1. a,b and c)Pressure contours at 3 times, d)numerical detonation cells

(see Fig. 1b). After a very long time of about $1ms$, the structure becomes very regular structure with less transverse wave number. Once the regular structure is reached, the detonation runs with the same transverse wave equilibrium mode. In the current work, the transverse wave number formed is found to be 5. Figure 1c shows the pressure contours of the regular cellular structure at $2.072ms$.

The quantity, $p|\mathbf{v}|$'s maximum on all grid nodes in time history on each node is used as an analogue of smoke foil tracks of the detonation cells in experiments. The numerical detonation cells are shown in Fig. 1d, whose width/length ratio and other geometric parameters are in full agreement with experimental data for a cell in an ordinary detonation (Fickett and Davis, 1979).

4.2 Ignition mechanisms

From Fig. 2, one can observed that the concentration discontinues can be considered as ignition front approximately. So the induced zones can be defined as the

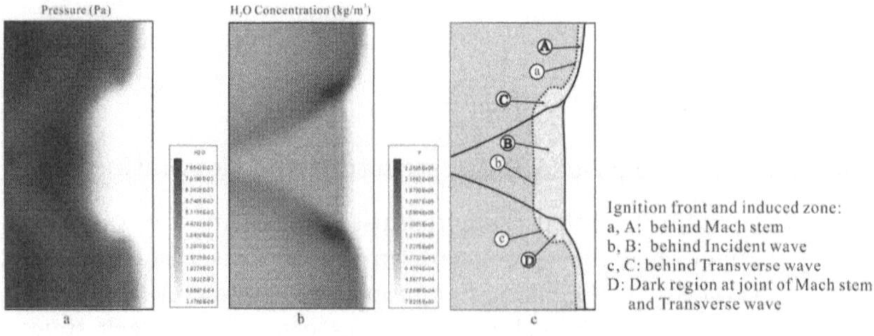

Fig. 2. Pressure, H_2O concentration and ignition front, $t = 2.072ms$,

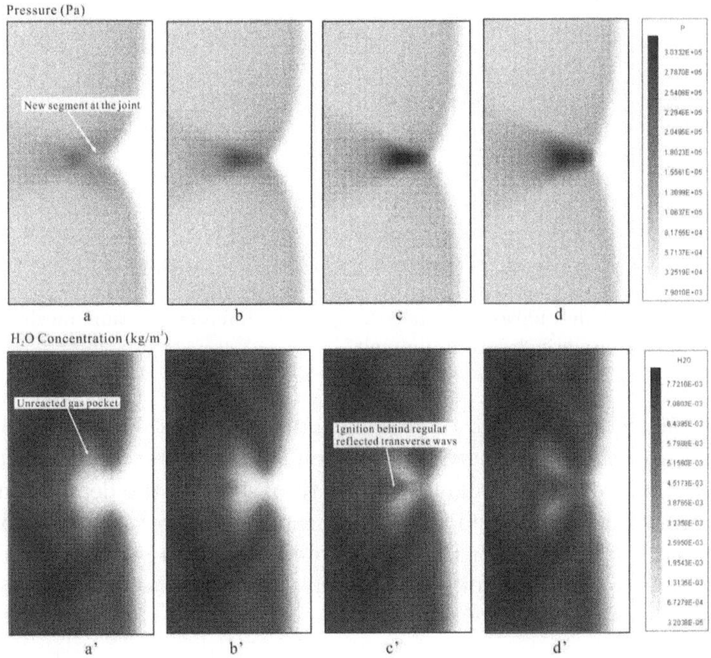

Fig. 3. Pressure and H_2O concentration contours at 4 times through the triple point collision, the interval of each frame is $0.1\mu s$

regions between these discontinues to the leading fronts. Most of the chemical energy releasing activity is completed at ignition front, behind which there are only small concentration fluctuations. The induced length behind the Mach stems is much shorter than that of incident waves showing different ignition mechanisms. While transverse waves sweeping the induced zones behind the incident waves, they are secondary shocked and then followed by ignition. The ignition process is more complex while triple point collision (see Fig. 3). It can be observed that, as the two transverse wave moving closer just before the triple point collision, they collide in the induced zone behind the incident wave, make strong ignition, and show a special ignition pattern characterized two divided unreacted pockets (see Fig. 3). This strong ignition phenomenon is in agreement with the results of Lefebvre and Oran (1995) and Oran (1998). Our further study shows that, in a regular structure, the formed unreacted pockets are mainly ignited by the collided transverse waves. But for irregular structures, the transverse waves collide with each other asymmetricaly and make the collided waves weaker, hence the unreacted pockets may exist for a longer time.

According to the ignition mechanisms above, a detonation cell can be divided into four regions (see Fig. 4). In the first region, about 40 percent of a cell, the gas mixture is ignited by the Mach stem; the gas mixture of the second region, about 25 percent of a cell, is ignited by the incident wave; the two transverse waves ignite the gas mixture in the third region contributing about 22 percent

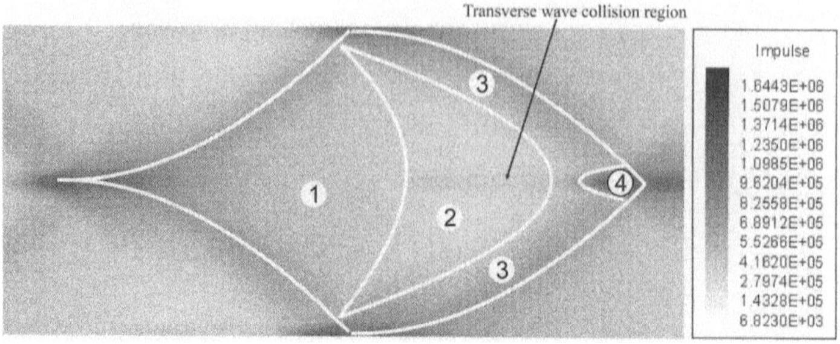

Fig. 4. A divided detonation cell according to 4 different ignition mechanisms, the background is a numerical detonation cell

of a cell; the triple wave collision region comprising about 3 percent of a cell (the fourth region), the gas mixture is ignited by the transverse waves and the triple wave collision process. Mitrofanov (1996) reconstructed the detonation ignition mechanisms from experiments, in which, the incident wave ignition region (region 2) is ignited by so called third shock compression, which is caused by the waves after two transverse waves collide. However, current results shows that, in region 2, ignition occurs before the transverse waves collision (see Fig. 2).

4.3 Species and vortex effects

Fig. 5 gives the concentrations of all species at $2.072ms$. From the current results, all species can be classified into three groups by their magnitude of concentration. The first group, comprising H_2, O_2, H_2O, reacts or produces fast, and then is kept in almost equilibrium concentration at about $0.05mol/m^3$, $0.025mol/m^3$ and $0.46mol/m^3$, respectively. The species HO_2 and H_2O_2 of the second group initially attain their maximum concentrations and then decline monotonically to their equilibrium values very quickly. Especially for H_2O_2, it almost only exists at Mach stems. This indicates different chain reaction mechanism around the Mach stem. For H, O and OH (the third group), they firstly get their maximums at ignition front but then show considerable spatial oscillating characteristics and take the longest time to get to their equilibrium values in the three groups. Fig. 3c, d and e show periodical patterns in the flow field of the species O, H and OH behind the leading shock front. By comparing with the vorticity and the temperature, it can be observed that the regions with higher temperature correspond to higher vorticity, which shows the vortex effects contributing to the spatial oscillating of the concentrations. By studying the structure and its collision process, it can also be observed that the vortexes develop from the triple collision jets and the contact discontinues extending from the triple points.

5 Conclusion remarks

In this paper, the characteristics of reaction zones of a detonation cellular structure in low pressure $H_2/O_2/Ar$ mixture have been studied with a elementary

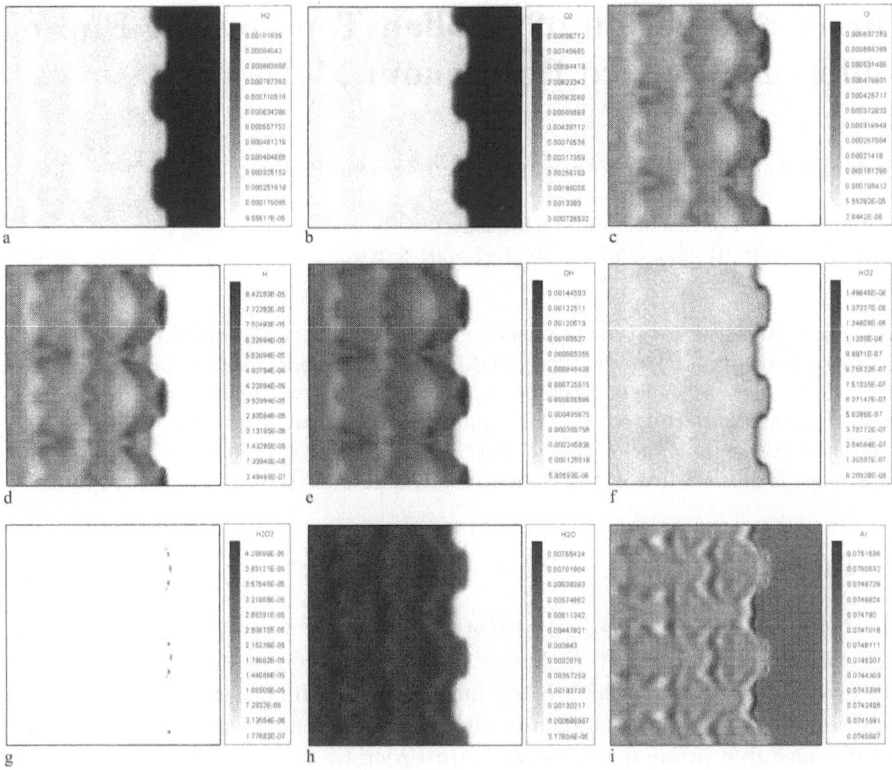

Fig. 5. Concentrations for different species at 2.072ms (kg/m^3), a)H_2, b)O_2, c)O, d)H, e)OH, f)HO_2, g)H_2O_2, h)H_2O, Ar

chemical reaction model. The calculated reaction zones show different ignition mechanisms. The vortexes in reaction zones can also affect the chemical equilibrium of the species O, H and OH considerably, and then affect the detonation states.

References

1. W. Fickett, W.C. Davis: *Detonation.* (University of California Press, Berkeley 1979)
2. V.N. Gamezo, D. Desbordes, E.S. Oran: Combust. Flame. **116**, 154 (1999)
3. M.H. Lefebvre, E.S Oran: Shock Waves. **4**, 277 (1995)
4. V.V. Mitrofanov: 'Modern view of gaseous detonation mechanism' In: *Prog. in Astro. and Aero.* vol 137 (Washington, DC: AIAA) p 327 (1996) **4**, 97 (1984)
5. E.S Oran, J.P. Boris:*Numerical Simulation of Reactive Flow.* (Elsevier, New York 1987)
6. C.W. Shu, S. Osher: J. Comput. Phys. **83**, 32 (1999)
7. S. Taki, T, Fujiwara: AIAA Journal **16**, 73 (1978)
8. G.J. Wilson, R.W. MacCormack: *AIAA-90-2307* (1990)
9. T.R. Young: *AD-A0835545* (1979)

Direct Simulation of Aeolian Tone by the Finite Difference Lattice Boltzmann Method

Michihisa Tsutahara, Makoto Kurita, and Takeshi Kataoka

Graduate School of Science and Technology,
Kobe University, Rokko, Nada, Kobe 657-8501 Japan

Abstract. Acoustic waves in the phenomenon of aeolian tone is simulated by the finite difference lattice Boltzmann method (FDLBM) using a model for thermal compressible fluids and a new scheme to stabilize the calculation. Two-dimensional flow past a circular cylinder is calculated and the sound waves whose pressure is 10^{-3} times smaller than the pressure fluctuation of the flow field are successfully detected. Accuracy of the scheme is also studied in two-dimensional Couette flow.

1 Introduction

The lattice Boltzmann method (LBM) has become a powerful tool of CFD, and the field of application of this method spreads very widely now[1]-[6]. But the models commonly used are for incompressible fluids and the models for compressible and thermal fluids are, although some models have been proposed, usually unstable in computation[7][8]. In order to stabilize the scheme, finite difference methods can be employed[9][10]. In this study, a new scheme of the discrete BGK model for thermal fluids is proposed, and Aeolian tone is calculated successfully.

2 Finite difference lattice Boltzmann method

The discrete BGK equation is

$$\frac{\partial f_i(t,\mathbf{r})}{\partial t} + c_{i\alpha}\frac{\partial f_i(t,\mathbf{r})}{\partial r_\alpha} = -\frac{1}{\phi}\left(f_i(t,\mathbf{r}) - f_i^{(0)}(t,\mathbf{r})\right) \tag{1}$$

where f is the particles distribution function, f_i is the particle density in $i$direction, $f_i^{(0)}$ is the local equilibrium distribution function, α represents the Cartesian coordinates, $\mathbf{c}_i$ is the particle velocity, and ϕ is the relaxation parameter. The macroscopic variables, such as the density, the momentum and the internal energy of the fluid are given by

$$\rho = \sum_i f_i = \sum_i f_i^{(0)} \tag{2}$$

$$\rho\mathbf{u} = \sum_i f_i\mathbf{c}_i = \sum_i f_i^{(0)}\mathbf{c}_i \tag{3}$$

$$\frac{1}{2}\rho u^2 + \rho e = \sum_i \frac{1}{2} f_i c_i^2 = \sum_i \frac{1}{2} f_i^{(0)} c_i^2 \tag{4}$$

In the above equations the terms including the local equilibrium distribution function represent the conservation of the quantities on the LHS at the collision. We propose a new model, instead of (1), as

$$\frac{\partial f_i(t,\mathbf{r})}{\partial t} + c_{i\alpha}\frac{\partial f_i(t,\mathbf{r})}{\partial r_\alpha} - \frac{ac_{i\alpha}}{\phi}\frac{\partial}{\partial r_\alpha}\{f_i(t,\mathbf{r}) - f_i^{(0)}(t,\mathbf{r})\}$$
$$= -\frac{1}{\phi}\left(f_i(t,\mathbf{r}) - f_i^{(0)}(t,\mathbf{r})\right) \tag{5}$$

where the third term of LHS is added. Assuming the same form of the local equilibrium distribution function as the conventional LBM [11]

$$f_{\sigma i}^{(0)} = F_\sigma \rho\Big[1 - 2Bc_{\sigma i\alpha}u_\alpha + 2B^2 c_{\sigma i\alpha}c_{\sigma i\beta}u_\alpha u_\beta + Bu^2$$
$$-2B^2 c_{\sigma i\alpha}u_\alpha u^2 - \frac{4}{3}B^3 c_{\sigma i\alpha}c_{\sigma i\beta}c_{\sigma i\gamma}u_\alpha u_\beta u_\gamma\Big] \tag{6}$$

where

$$F_0 = 1 + \frac{5}{4Bc^2}\left(\frac{17}{96B^2c^4} + \frac{35}{48Bc^2} + \frac{49}{45}\right) \tag{7a}$$

$$F_{11} = -\frac{1}{8Bc^2}\left(\frac{13}{16B^2c^4} + \frac{71}{24Bc^2} + 3\right) \tag{7b}$$

$$F_{12} = \frac{1}{16Bc^2}\left(\frac{5}{16B^2c^4} + \frac{25}{24Bc^2} + \frac{3}{5}\right) \tag{7c}$$

$$F_{13} = -\frac{1}{24Bc^2}\left(\frac{1}{16B^2c^4} + \frac{1}{8Bc^2} + \frac{1}{15}\right) \tag{7d}$$

$$F_{21} = \frac{1}{4B^3c^6}\left(\frac{Bc^2}{3} + \frac{1}{8}\right) \tag{7e}$$

$$F_{22} = -\frac{1}{1536B^3c^6}\left(2Bc^2 + 3\right) \tag{7f}$$

$$B = -\frac{1}{2e} \tag{7g}$$

and using the Chapman-Enskog approach, the Navier-Stokes equations for compressible fluids

$$\frac{\partial \rho}{\partial t} + \frac{\partial}{\partial r_\alpha}(\rho u_\alpha) = 0 \tag{8}$$

$$\frac{\partial}{\partial t}(\rho u_\alpha) + \frac{\partial}{\partial r_\beta}(\rho u_\alpha u_\beta)$$
$$= -\frac{\partial P}{\partial r_\alpha} + \frac{\partial}{\partial r_\beta}\mu\left(\frac{\partial u_\beta}{\partial r_\alpha} + \frac{\partial u_\alpha}{\partial r_\beta}\right) + \frac{\partial}{\partial r_\alpha}\left(\lambda\frac{\partial u_\gamma}{\partial r_\gamma}\right) \tag{9}$$

$$\frac{\partial}{\partial t}\left(\rho e+\frac{1}{2}\rho u^2\right)+\frac{\partial}{\partial r_\alpha}\left(\rho e+P+\frac{1}{2}\rho u^2\right)u_\alpha=\frac{\partial}{\partial r_\alpha}\left(\kappa'\frac{\partial e}{\partial r_\alpha}\right)$$
$$+\frac{\partial}{\partial r_\alpha}\left\{\mu u_\beta\left(\frac{\partial u_\beta}{\partial r_\alpha}+\frac{\partial u_\alpha}{\partial r_\beta}\right)\right\}+\frac{\partial}{\partial r_\alpha}\left(\lambda\frac{\partial u_\beta}{\partial r_\beta}u_\alpha\right) \tag{10}$$

are obtained from the evolution equation (5). The pressure P, the viscosity μ, the second viscosity λ, the heat conductivity k', and the sound speed c_s are given, respectively, by

$$P=\rho e \tag{11a}$$
$$\mu=\rho e\tau\left(\phi-a\right) \tag{11b}$$
$$\lambda=-\rho e\tau\left(\phi-a\right) \tag{11c}$$
$$\kappa=2\rho e\tau\left(\phi-a\right) \tag{11d}$$
$$c_s=\sqrt{2e} \tag{11e}$$

We should note here that the relationship between the viscosity and the relaxation parameter in this model is given by (11b) instead of $\mu=\rho e\tau\phi$ that is for conventional FDLBM model in (1). In the latter case, if the Reynolds number of the flow is very large, ϕ should be very small. On the other hand, from the constraint of the convergence condition of the collision term (RHS term in (1)), $\Delta t/\phi<2$ must be satisfied, where Δt represents the time increment. This means that the time increment must be chosen small for high Reynolds number flows, and the calculation time must be very long. But in newly proposed model, the relationship changes to (11b); therefore, even for high Reynolds number flows, the above stated constraint on Δt can be eliminated if we choose appropriate value to a.

As to the value of a , although even for a very large value the Navier-Stokes equations are recovered, we chose 0.5 only because the relationship between the viscosity and ϕ becomes the same as a conventional LBM.

In this study, the 2D21V model[11] as shown in Fig.1 is used.

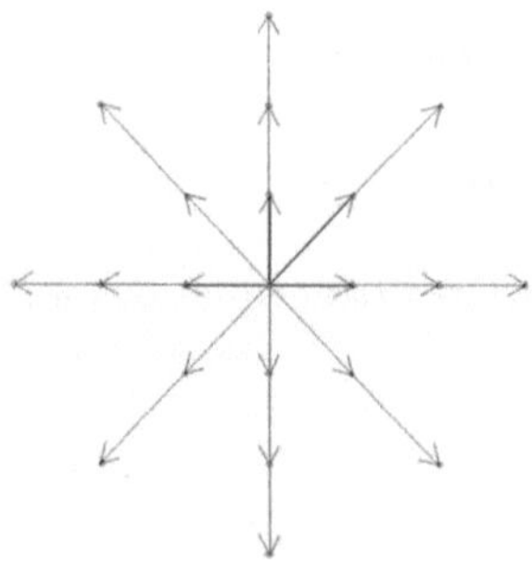

Fig. 1. A compressible lattice Boltzmann model (2D21V)

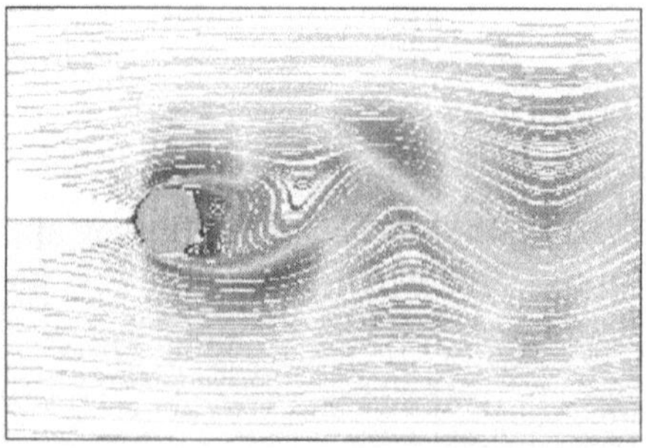

Fig. 2. Karman vortex street

3 Simulation of the Aeolian tone

The calculation is performed for a flow past a circular cylinder in non-uniform cylindrical co-ordinates. The number of lattice is $(r \times \theta) = (151 \times 101)$, and the Reynolds number based on the diameter of the cylinder is 200 and the Mach number of the flow is 0.2. The outer boundary of the computational domain is taken on the circle with radius of 200 times larger of the circular cylinder. The boundary conditions on the outer boundary is the uniform flow and those on the surface of the circular cylinder is non-slip and adiabatic. The space derivative of the distribution function in (5) is formulated by the third order upwind scheme (UTOPIA) and the time marching is performed by the second order Runge-Kutta method.

The stream-lines about the cylinder is shown in Fig.2, and the Karman vortex street appears in the wake.

Here we define also the pressure deviation as

$$\Delta p = (p - p_0)/p_0 \tag{12}$$

where p_0 is the pressure at the outer boundary. In the whole flow-field, the maximum and the minimum values of the pressure deviation is shown in Fig.3 as a function of time. The maximum pressure keeps almost constant value but the minimum pressure is fluctuating. But the difference between the maximum and the minimum pressures $\Delta p_{\max} - \Delta p_{\min}$ is the order of 10^{-1}, and the fluctuation is the order of 10^{-2} . This fluctuation occurs in the vicinity of the cylinder, and the fluctuation is much smaller in the region far from the cylinder.

Figure 4 shows the pressure field within the pressure deviations $\Delta p = \pm 5 \times 10^{-4}$ at the same instance as in Fig.2, and the dipole-like sound, which is synchronized with the vortex shedding, is shown to be emitted. But the high-pressure region concentrates on the upstream side of the cylinder and the lower pressure region on the downstream side. The pressure distribution along the radial direction perpendicular to the flow is shown in Fig.5 and the pressure decreases as $r^{-0.5}$, that agrees well with the theoretical prediction[12]. This sound wave emission is the Aeolian tone, and is successfully simulated by FDLBM.

The pressure deviation of the sound waves is the order of $10^{-5}$and is 10^{-4} times smaller than the pressure difference in the whole flow-field $\Delta p_{\max} - \Delta p_{\min}$.

4 Accuracy of the scheme

In order to simulate the sound waves, the scheme must have enough accuracy[13]-[17]. The accuracy of the proposed model is investigated in two-dimensional unsteady Couette flow in which the top plate suddenly moves with velocity U in x direction. The exact solution of the Navier-Stokes equation is given as [18]

$$u_e(y,t) = U\left(1 - \frac{y}{d}\right) - \frac{2U}{\pi}\sum_{n=1}^{\infty}\frac{1}{n}\exp\left(-n^2\pi^2\frac{\nu t}{d^2}\right)\sin\frac{n\pi y}{d} \tag{13}$$

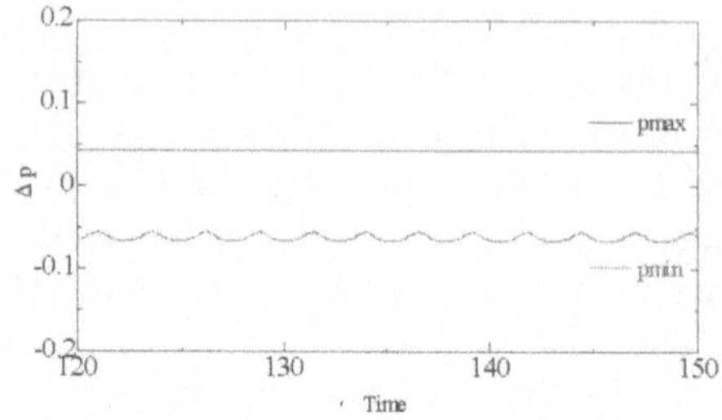

Fig. 3. Time variation of maximum and minimum pressures

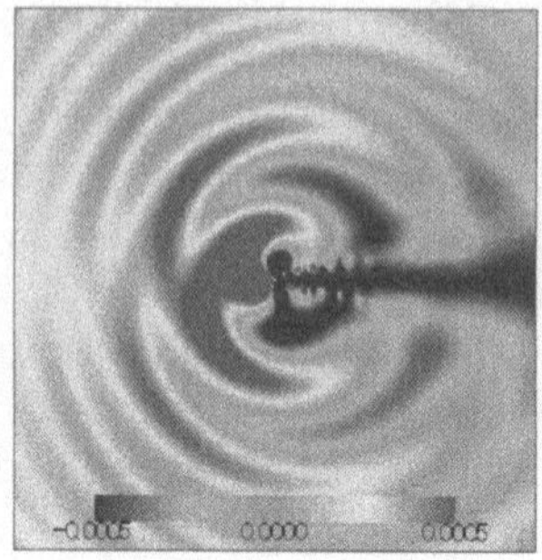

Fig. 4. Pressure field

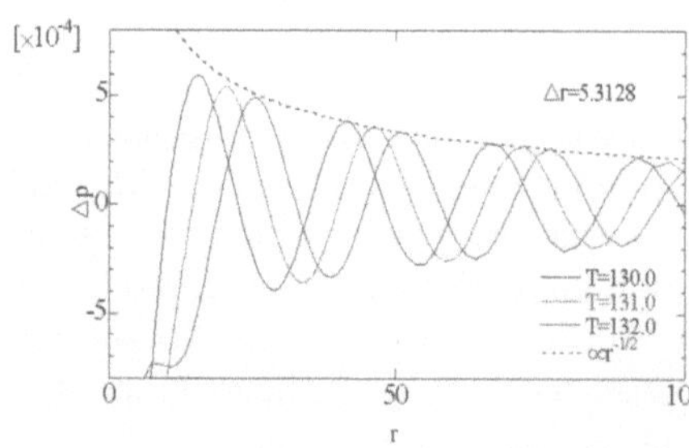

Fig. 5. Relationship between pressure and distance at $\theta = 90°$

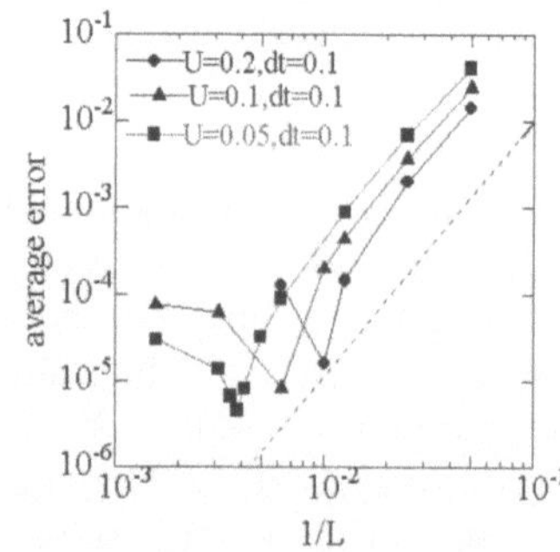

Fig. 6. Error of the scheme

where d is the distance between the top and bottom plates, and y represent the position in the vertical direction. The number of lattice N in the vertical direction changes, and the relative error is defined as

$$E = \sum_{i=1}^{N} |u_i - u_{ei}| \Big/ \sum_{i=1}^{N} |u_{ei}| \tag{14}$$

At the non-dimensional time $t = 5$. the error is shown in Fig.6, in which the dashed line represents the third order accuracy. The accuracy changes from third order to even fifth order depending the number of lattice. The minimum error becomes smaller as the velocity U is smaller because the deviation of this model from the Navier-Stokes equations is $O(u^3)$. The acoustic waves have been successfully simulated by FDLBM with third order upwind scheme because of this high accuracy of this model.

5 Conclusions

Aeolian tone emitted from a two-dimensional circular cylinder is successfully simulated by the finite difference lattice Boltzmann method with a thermal BGK model. The accuracy of this method is shown to be higher than that of the scheme for the evolution equation of the distribution function.

References

1. Qian, Y.H., Succi S. and Orszag S.A.:*Recent Advances in Lattice Boltzmann Computing* , Ann. Rev. of Comp. Phy. III, D. Stauffer ed. World Scientific, (1995) 195.
2. Rothman,D.H. and Zalenski S.:*Lattice-Gas Cellular Automata*, Cambridge U.P. (1997) 73D
3. Chopard, B.. and Droz, M.:*Cellular Automata Modeling of Physical Systems*, Cambridge University Press (1998) 122.
4. Chen S. and Doolen G.D.:*Lattice Boltzmann method for fluid flows*, Ann. Rev. Fluid Mech., Ann. Rev. Inc. (1998) 329.
5. Wolf-Gladrow, D.A.:*Lattice-Gas Cellular Automata and Lattice Boltzmann Models*, Lecture Notes in Mathematics, Springer (2000)159.
6. Succi S.:*The lattice Boltzmann Equation for Fluid Dynamics and Beyond*, Oxford (2001) 51.
7. Alexander F.J. et. al.:Phys. Rev. E, 47 (1993) 2249.
8. Chen Y., et. al.:Phys. Rev. E, 50 (1994) 2776.
9. McNamara G.R.:J. Stat. Phys. (1995) 395.
10. Cao, N., Chen S.J. & Martinez D.:Phy. Rev. E, (1997) R21.
11. Tsutahara, M., Takada, N., Kataoka T.:*Lattice gas and lattice Boltzman methods*, Corona-sha (1999, in Japanese) 111.
12. Lighthill M.J.: *Waves in Fluids*, Cambridge U.P. (1978) 21.
13. Lele S.K.:J. Comp.Phys.103 (1992) 16.
14. Colonius T., Lele S.K., & Moin P.:J.Fluid Mech.260 (1994) 271.
15. Mitchell B.E., Lele S. K. & Moin P.:J.Fluid Mech. 285 (1995) 181.
16. Inoue O. & Hattori Y. :J.Fluid Mech. 380 (1999) 81.
17. Mei R. and Shyy W.:J. Comp. Phys. 143 (1998) 426.
18. Batchelor G.K.:An Introduction to Fluid Dynamics, Cambridge U.P. (1967) 191.

Development and Applications of a Coupled CFD/CSD Methodology Using an Embedded CSD Approach

Joseph D. Baum, Eric L. Mestreau, Hong Luo, Rainald Löhner, Daniele Pelessone, and Charles Charman

SAIC, 1710 SAIC Dr., MS 2-6-9, McLean VA 22102, USA

1 Abstract

This paper describes recent algorithm developments and select applications of a program that couples parallel Computational Fluid Dynamics (CFD) and Computational Structural Dynamics (CSD) methodologies. FEFLO98 is the CFD code used while DYNA3D handles the CSD portion. FEFLO98 solves the time-dependent, compressible Euler and Reynolds-Averaged Navier-Stokes equations on an unstructured mesh of tetrahedral elements. DYNA3D solves explicitly the large deformation, large strain formulation equations on an unstructured grid composed of bricks and hexahedral elements.

While the initial coupled algorithm used the so-called "glued-mesh" approach, where the CFD and CSD faces match identicaly, failure of this approach to model severe structural deformation, as well as crack propagation in steel and concrete, led us to the development of the so-called "embedded-mesh" approach. Here, the CSD objects float through the CFD domain. While each approach has it's own advantages, limitations and deficiencies, the embedded approach was proven to be superior for the class of problems modeled here. Critical applications of both approaches are described, including weapon detonation and fragmentation, fragment/airblast interaction with a steel wall, and an external airblast interaction with a generic steel ship hull.

2 Introduction

In this study, the coupled CFD/CSD methodology is applied to the simulation of detonation wave interactions with structure and weapon fragmentation. These applications constitute a very severe test to the numerical methodology as they require modeling of several complex, interacting physical phenomena: a) Detonation wave initiation and propagation; b) CSD modeling of case expansion and fragmentation; c) Blast wave expansion through the breaking case, and around the flying fragments; and e) Fragments and airblast impact on the structure and the resulting structural deformation. To model the wide variation of physical models we require elasto-plastic material models with rupture criteria (CSD), coupled with either the Euler or the Reynolds-Averaged Navier-Stokes (CFD).

Two approaches can be used to model fluid/structure interaction. The 'tight coupling' approach requires solving both CFD and CSD as one coupled set of equations. It requires the complete rewrite of both solvers. The second approach, called 'loose coupling', decouples the CFD and CSD sets of equations and uses projection methods to transfer interface information between the CFD and CSD domains. We adopted the latter method. By building on pre-existing and well-established codes, a loose-coupled solver can be assembled with minimum modifications to either of the two solvers.

The motion of a solid or a deforming body through the fluid domain can be modeled using two approaches: the glued approach and the embedded approach. The glued approach requires the CFD surface grid to match the CSD Surfaces. The meshes do not need to be identical at the interface. When the CSD moves across the CFD domain, the CFD tracking boundaries also move and deform the volume mesh they are based on. This, in turn, triggers expensive meshing techniques, which are delayed as much as possible by the use of complex mesh movement techniques and ALE formulation.

In the second approach, the embedded CSD mesh floats through the CFD domain without body/surface conformance. The faster Eulerian formulation (rather than the slower ALE) is used while the only extra intensive steps involve the identification of the crossed edges and proper modeling of the boundary conditions. On the negative side, this methods requires the use of new boundary conditions of first and maybe second order, which are more complex.

Over the past several years we have developed and successfully applied the traditional glued approach [Ba98, Ba01]. However, this approach has failed recently for some simulations involving singular points, as well as for cases where initially orthogonal steel plates were deformed such that the core angles became exceedingly small, too small for the contact algorithm from preventing contact and penetration. Moreover, for applications such as crack propagation in steel plates or cased weapons under explosive loading which require frequent CPU-intensive topology reconstruction and fine resolution to properly model the crack, this approach became prohibitively expensive (CPU-wise). These shortcomings of the glued approach are, conversely, the strengths of the embedded approach, where the bodies float through the CFD mesh, as the distance between them, even at contact, poses no difficulty to the CFD solution.

The development of the embedded approach required the following major steps which use the existing 3-D CFD mesh and the wetted CSD faces: 1) Identify CFD edges cut by the CSD faces and mask CFD edges and points; 2) Impose new boundary conditions; 3) Modify the geometry boundary point arrays; 4) Extrapolate the solution near the boundaries; and 5) Develop new display tools. Due to space limitations, only some of these will be discussed here.

3 Numerical Methodology

The CAD definition of both CFD and CSD was generated using FECAD. Mesh generation for both CSD and CFD was performed using FRGEN3D [Lo88]. This unstructured grid generator is based on the advancing front method. The CFD mesh is composed of triangular (surface) and tetrahedral (volume) elements. The CSD mesh includes beams, triangular or quad shells and bricks for the solids. The quads shells are the result of aggregation of triangles while the bricks are the subdivision of tetrahedral elements followed by smoothing techniques.

The flow solver employed is FEFLO98, a 3-D adaptive, unstructured, edge-based hydro-solver based on the Finite-Element Method Flux-Corrected Transport (FEM-FCT) concept [Lo87]. It solves the Arbitrary Lagrangean-Eulerian (ALE) formulation of the Euler and Reynolds-averaged turbulent, Navier-Stokes equations.

Finally, the structural dynamics solver is DYNA3D[Wh91], an unstructured, explicit finite element code. DYNA3D is well suited for modeling large deformations and provides a good base for non-linear materials with elasto-plastic compartmental laws with rupture.

4 Numerical Results

4.1 Weapon Fragmentation Studies

The coupled technology has been applied to the simulation of the detonation and fragmentation of an experimental weapon that hangs tip-down at the center in a reinforced concrete room. The thick-walled steel weapon is top (i.e., base) ignited. The detonation front propagates from the base to the tip at the C-J detonation velocity prescribed in the detonation model (essentially, the program burn model of DYNA3D).

Several CSD meshes of this weapon were tested, using either 8-node hexahedral elements or brick-like parallelepipedal elements, and varying the number of elements from 748 to 8228. 748 brick elements with a single element across the thickness was used for the simulation with the glued approach. The fragment size distribution for the glued simulation was prescribed based on arena test data. Four elements across were used for the embedded, where recently developed first-principles fragmentation model was used to model the break-up of the case.

After ignition, as the detonation wave propagates from the base to the tip, the high-pressure detonation products force the case to expand. The structural elements fail once the element strain computed at the center of each element exceeds 100%. Each failing fragment is then treated as a separate rigid body. A 6 DOF integrator linked to the contact algorithms computes their trajectories. For the glued approach, the failure of bricks triggers automatic remeshing phases to fill up the narrow gaps separating the fragments. The gaps, which should have been of the order of millimeters, are artificially increased to avoid introduction of tiny fluid elements with small integration times. The remeshing (due to topology change) is a CPU intensive process and is allowed to occur only every 5-8 μs. Even the usage of adaptive refinement and automatic placement of sources wherever needed for modeling the small spacing, did not significantly reduce the high computational cost.

Figure 1 shows a comparison between the glued approach (Fig 1a) and the embedded approach (Figs 1b and 1c). The first approach did not allow for the application of a physically-correct crack propagation model, as each crack had to be modeled with at least 4-6 fluid elements across the narrow gap, at a prohibitive computational cost. The fragments had to be pre-defined, hence precluding the modeling of weapons for which there is no arena data. In addition, the break-up of each fragment was a topology-changing event, requiring global remeshing at a fairly high computational cost, even when performed in parallel using a parallelized grid generation scheme. In comparison, the embedded approach allows both the treatment of crack propagation and the formation of small fragments. Thus, the cracking model automatically determines the size of the fragments. The results shown in Figs 1c-e demonstrate: 1) The randomness of the size of the fragments; 2) The narrow cracks between the fragments, some of which are very small; 3) The large number of fragments modeled, a simulation that would have been computationally prohibitive with the glued approach; 4) Mesh adaptation to both the structure (smaller elements within the cracks will yield more accurate pressure relief and detonation products expansion) and density gradients; and 5) Most fragments reach about the same velocity, in agreement with available experimental data.

The fragments achieve their terminal velocity between 120 and 150 μs after the detonation front passage. The final mass-averaged fragment velocity obtained for this simulation, using a strain break-up value of 1.0, was 787.6 m/sec. The experimental measured value was 752 m/sec. To examine the role of the break-up strain value on the final velocity we conducted three more simulations, at break-up strain values of 0.1 (Vf=675.53 m/sec), 0.5 (Vf=741.8 m/sec), and infinity (no break-up, Vf=859.7 m/sec). The results show that the experimental data indicates a break-up strain value of about 0.6, a value that corresponds to an internal expansion of about 90%, and external expansion of about 50%. It must be noted that a recent simulation conducted for a different weapon that used a break-up strain value of 0.6, predicted average fragment velocity of 1282 m/sec, in good agreement with the measured velocity of 1260 m/sec.

4.2 Blast and Fragment Impact on a Steel chamber and a Generic Ship

Here we modeled fragment/airblast interaction with a steel wall chamber. Similar to the weapon case break-up case, three levels of mesh refinement were used in the CSD model. The standard DYNA3D element erosion model was used to eliminate failed CSD elements.

The numerical predictions show that the impacting weapon fragments arrive ahead of the airblast, punching holes through the plate (Fig. 2a). Next, the pressure blast from the detonation tears the weakened plate apart (Fig. 2b). The eroded plate elements were converted into particles that can interact with the rest of the structure. Contact conditions were enforced between all entities to avoid fragment interpenetrations and therefore a failure in meshing.

The last simulation modeled an external blast interaction with a generic ship hull. All dimensions and values for the compartments and materials for the decks, frames and bulkheads were representative of real ships. The large deformation and folding of metal sheets at corners precluded from using the glued approach. The various pictures therefore represent the results obtained by the embedded approach. Fig. 3 depicts the initial conditions (an external blast), deformed CSD surface and CFD pressure contours on a planar cross-section at an early time and later, and the damaged structure at a late time. These results demonstrate that the exterior blast loading resulted in failure of the hull, three interior bulkheads, and significant damage to neighboring bulkheads.

5 Summary and Conclusions

A recently developed loose-coupling algorithm that combines state-of-the-art CFD and CSD methodologies, has been applied to the simulation of weapon detonation and fragmentation and blast and fragment impact on steel and concrete structures. These applications required modeling several complex and interacting physical phenomena. A significant development described in this paper was the failure of the glued mesh approach to model the response of complex-geometry severely-deforming steel-plate structure to a large blast. This forced the development of the embedded mesh approach. The results presented demonstrate the ability of the coupled methodology using the embedded approach to model these processes and yield results that are in good agreement with available experimental data.

References

1. R. Löhner,, C. Yang, J. Cebral, J.D. Baum, H. Luo, D. Pelessone and C. Charman, *Fluid-Structure Interaction Using a Loose Coupling Algorithm and Adaptive Unstructured Grids* AIAA-95-2259, (1995).
2. J.D. Baum, H. Luo, R. Löhner, C. Yang, D. Pelessone and C. Charman, *A Coupled Fluid/Structure Modeling of Shock Interaction with a Truck* , AIAA-96-0795, (1996).
3. J.D. Baum, H. Luo, R. Löhner, *The Numerical Simulation of Strongly Unsteady Flows With Hundreds of Moving Bodies*, AIAA-98-0788, (1998).
4. J.D. Baum et al., *Recent Developments of a Coupled CFD/CSD Methodology*, AIAA-2001-2618, (2001).
5. R. Löhner and P. Parikh, *Three-Dimensional Grid Generation by the Advancing Front Method*, Int. J. Num. Meth. Fluids 8. 1135-1149 (1988).
6. R. Löhner, K. Morgan, J. Peraire and M. Vahdati, *Finite Element Flux-Corrected Transport (FEM-FCT) for the Euler and Navier-Stokes Equations*, Int. J. Num. Meth. Fluids 7, 1093-1109 (1987).
7. R.G. Whirley and J.O. Hallquist, *DYNA3D, A Nonlinear Explicit, Three-Dimensional Finite Element Code for Solid and Structural Mechanics - User Manual*, UCRL-MA-107254 (1991)

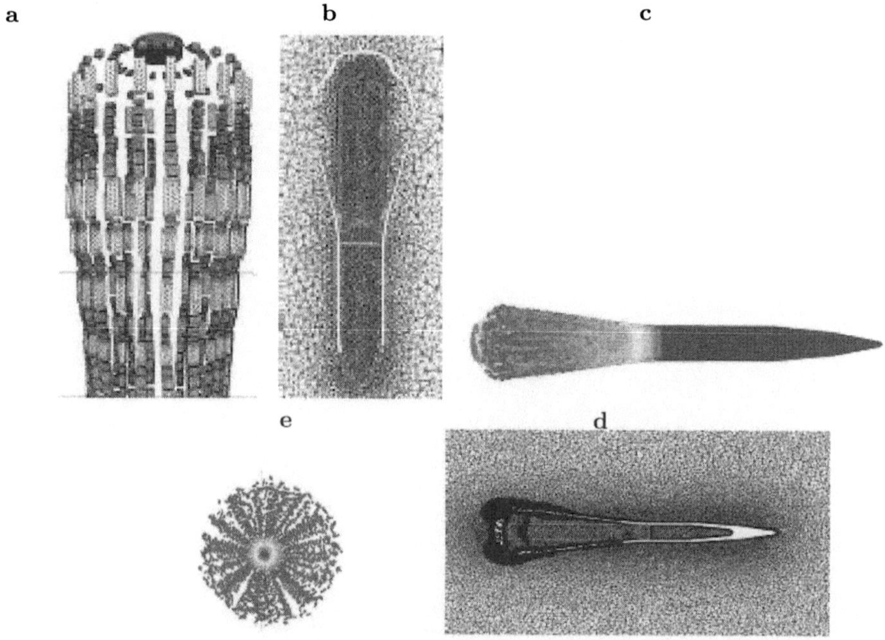

Fig. 1. Comparison of glued and embedded CSD schemes applied to the weapon fragmentation study: **a** CFD surface mesh, **b** pressure contours, **c** CSD velocity contours, **d** adapted pressure contours, and **e** fragment velocity contours.

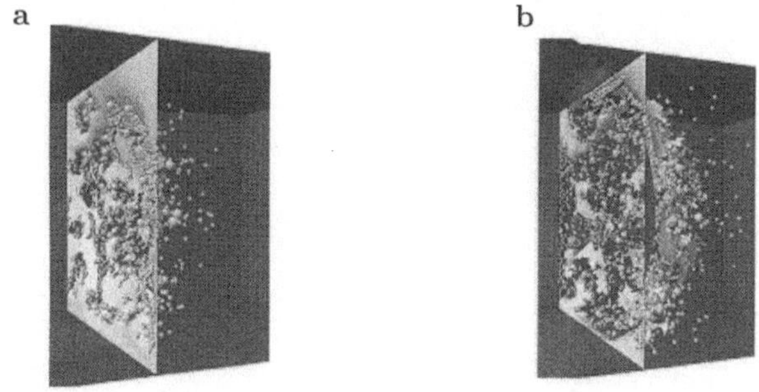

Fig. 2. Airblast/Fragments Impact on a Flat Plate: **a** impacting weapon fragments arriving ahead of the airblast, and **b** blast pressaure from the detonation tearing the weakened plate apart.

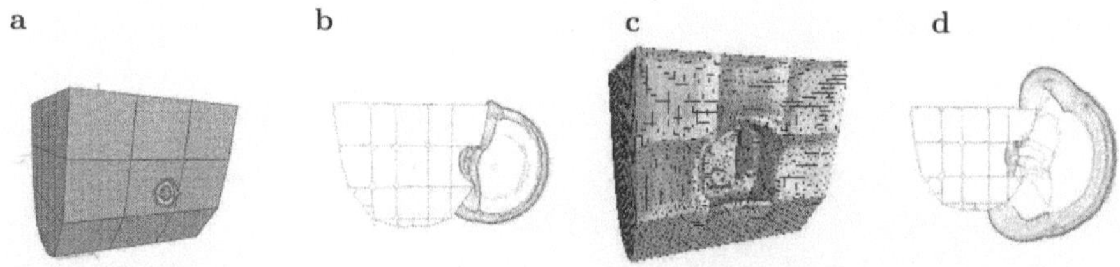

Fig. 3. External Blast and Interaction with Generic Ship Hull: **a** Initial External Blast, **b** Early Pressure Contours, **c** Damaged Ship Hull, and **d** later Pressure Contours.

Part XIII

Multi-phase Flow

Fix for Solution Errors near Interfaces in Two-Fluid Flow Computations

B. Koren[1], E.H. van Brummelen[1], P.W. Hemker[1], B. van Leer[2], and M.R. Lewis[1]

[1] CWI, P.O. Box 94079, 1090 GB Amsterdam, The Netherlands
[2] The University of Michigan, Ann Arbor, MI 48109-2140, USA

Abstract. A finite-volume method is considered for the computation of flows of two compressible, immiscible fluids at very different densities. A level-set technique is employed to distinguish between the two fluids. A simple ghost-fluid method is presented as a fix for the solution errors ('pressure oscillations') that may occur near two-fluid interfaces when applying a capturing method. Computations with it for compressible two-fluid flows with arbitrarily large density ratios yield perfectly sharp, pressure-oscillation-free interfaces. The masses of the separate fluids appear to be conserved up to first-order accuracy.

1 Introduction

A known difficulty of capturing two-fluid interfaces in a conservative formulation of the compressible Euler equations is that $\mathcal{O}(h^0) = \mathcal{O}(1)$ solution errors (in literature often called 'pressure oscillations') may arise near the two-fluid interface. Without remedial intervening, the conservative formulation considered in the present paper also suffers from the pressure-oscillation problem. Fixes for the problem can be found in the literature. We refer to the works of Karni [6,7] and Abgrall [1], their common paper [2], and also to [3,5,9]. In most of the available literature though, the ratio of the two densities at the interface is $\mathcal{O}(1)$–$\mathcal{O}(10^2)$. To our knowledge, only in [3,9] ratios of $\mathcal{O}(10^3)$, typical water-air ratios, are considered. In this paper a fix is proposed, which allows arbitrarily large density ratios. The fix is a simple variant of the ghost-fluid method [3].

2 Flow Model

2.1 Conservation Equations

In 1D, for a sufficiently small control volume Ω, conservation of mass and momentum reads:

$$\int_\Omega \frac{\partial \boldsymbol{q}}{\partial t} \mathrm{d}x + \boldsymbol{f}(\boldsymbol{q})_{\partial\Omega_{\text{right}}} - \boldsymbol{f}(\boldsymbol{q})_{\partial\Omega_{\text{left}}} = 0, \quad \boldsymbol{q} = \begin{pmatrix} \varrho \\ \varrho u \end{pmatrix}, \quad \boldsymbol{f}(\boldsymbol{q}) = \begin{pmatrix} \varrho u \\ \varrho u^2 + p \end{pmatrix}, \tag{1}$$

with ϱ the bulk density: $\varrho = \alpha(x,t)\varrho_w(p) + (1-\alpha(x,t))\varrho_a(p)$, where α is, e.g., the volume-of-water fraction, and where $\varrho_w(p)$ and $\varrho_a(p)$ are the equations of state for water and air, respectively. To balance this system, the equations of state and an equation determining the location of the interface (and hence α) still have to be chosen. An accurate resolution of the interface location(s) is of paramount importance. For this purpose, we follow a level-set approach, to be outlined in the next section.

2.2 Level-Set Equation

To accurately resolve the interface location(s), a level-set approach [10] is more appropriate than the classical volume-of-fluid (VOF) approach [4], because of its better smoothness (and thus accuracy) properties at precisely the point of interest: the interface. Good smoothness of the level-set function is first taken care of in the level-set function's initialization. A common approach is to initialize the level-set function as the signed distance to the initial interface, with the distance positive in, e.g., water and negative in air. During the computation, the level-set function may need to be reinitialized.

Denoting the level-set function by ϕ, in 1D, it is advected: $\frac{\partial \phi}{\partial t} + u\frac{\partial \phi}{\partial x} = 0$. Combined with the bulk-mass conservation equation from (1), this advection equation may be written in the conservative control-volume form

$$\int_{\Omega} \frac{\partial(\varrho\phi)}{\partial t} \mathrm{d}x + (\varrho u \phi)_{\partial\Omega_{\text{right}}} - (\varrho u \phi)_{\partial\Omega_{\text{left}}} = 0 \, . \tag{2}$$

Conservation of $\varrho\phi$ is not physical, there is no conservation law for it. The form (2) is simply practical because it is consistent with system (1).

2.3 Equation of State

In homentropic water-air computations, for both fluids, elegant use can be made of a single equation of state, viz. Tait's: $\frac{p+Bp_{\text{ref}}}{(1+B)p_{\text{ref}}} = \left(\frac{\varrho}{\varrho_{\text{ref}}}\right)^{\gamma}$, where the subscript 'ref' indicates some reference state. The reference pressure p_{ref} is chosen freely but equally for the two fluids. The value of ϱ_{ref} for each of the two fluids corresponds with p_{ref} and is read from standard data bases for fluid properties. Concerning the material constants γ and B, for water it holds: $\gamma = 7$, $B = 3000$ and for air: $\gamma = \frac{7}{5}$, $B = 0$.

3 Ghost-Fluid Method

In [8], it is shown that – unfortunately – the present conservative formulation (1)–(2) also leads to $\mathcal{O}(1)$ pressure errors near two-fluid interfaces. As a remedy against this, we outline a simple variant of the ghost-fluid method.

In updating the finite-volume solutions with a single explicit time step the following is done. Suppose we have an equidistant, cell-centered finite-volume grid Ω_i, $i = 1, 2, \ldots, N$, with cell faces $\partial\Omega_{i+\frac{1}{2}}$, $i = 0, 1, \ldots, N$, where $\partial\Omega_{\frac{1}{2}}$ and $\partial\Omega_{N+\frac{1}{2}}$ are at the domain boundaries. Also suppose that at time level n we have a known, unique solution (u_i^n, p_i^n, ϕ_i^n), $i = 1, 2, \ldots, N$. Then, at first, at the actual time level n the cells and cell faces are classified into different types. For cells, the following three types are distinguished: (i) pure-water cells, (ii) pure-air cells and (iii) cells with one (or two) interface(s). To make this classification, we determine $\phi^n_{i+\frac{1}{2}} = \frac{1}{2}\left(\phi^n_i + \phi^n_{i+1}\right)$, $i = 1, 2, \ldots, N-1$. At the inflow-boundary face, say $\partial\Omega_{\frac{1}{2}}$, we take $\phi^n_{\frac{1}{2}} = \phi^n_b$, with ϕ^n_b the boundary condition, and at the outflow boundary, say $\partial\Omega_{N+\frac{1}{2}}$: $\phi^n_{N+\frac{1}{2}} = \phi_N$. Then, cell Ω_i is: (i) a pure-water cell if $\phi^n_i > 0$, $\phi^n_{i-\frac{1}{2}} > 0$ and $\phi^n_{i+\frac{1}{2}} > 0$, (ii) a pure-air cell if $\phi^n_i < 0$, $\phi^n_{i-\frac{1}{2}} < 0$ and $\phi^n_{i+\frac{1}{2}} < 0$, and – else – (iii) a cell with one or two interfaces. The third type of cells are named ghost cells. This classification is also applied to the cell faces; (i) pure-water, (ii) pure-air and (iii) ghost faces are distinguished. The two faces of a ghost cell are both identified as ghost faces. I.e., if Ω_i is found to be a ghost cell, then both $\partial\Omega_{i-\frac{1}{2}}$ and $\partial\Omega_{i+\frac{1}{2}}$ are ghost faces. A cell face not belonging to a ghost cell is – depending on the sign of ϕ at that face – either a pure-water or a pure-air face. Across the latter two types of faces, the flux is simply computed with the *single-fluid* linearized Godunov scheme $\boldsymbol{F}(\tilde{\boldsymbol{q}}_0, \tilde{\boldsymbol{q}}_1)$, $\tilde{\boldsymbol{q}} \equiv (u, p)$. So, across pure-water faces we get $\boldsymbol{F} = \boldsymbol{F}_w$ and across pure-air faces $\boldsymbol{F} = \boldsymbol{F}_a$. Across the ghost faces *two* fluxes are computed: a water and an air flux, so *both* $\boldsymbol{F}_w(\tilde{\boldsymbol{q}}_0, \tilde{\boldsymbol{q}}_1)$ *and* $\boldsymbol{F}_a(\tilde{\boldsymbol{q}}_0, \tilde{\boldsymbol{q}}_1)$. Applying the forward Euler scheme, the subsequent update of finite-volume solutions reads then:

(i) in pure-water cells: $\boldsymbol{q}_i^{n+1} = \boldsymbol{q}_i^n - \frac{\Delta t}{h}\left((\boldsymbol{F}_w)^n_{i+\frac{1}{2}} - (\boldsymbol{F}_w)^n_{i-\frac{1}{2}}\right)$,

(ii) in pure-air cells: $\boldsymbol{q}_i^{n+1} = \boldsymbol{q}_i^n - \frac{\Delta t}{h}\left((\boldsymbol{F}_a)^n_{i+\frac{1}{2}} - (\boldsymbol{F}_a)^n_{i-\frac{1}{2}}\right)$,

(iii) in ghost cells: $(\boldsymbol{q}_w)_i^{n+1} = \boldsymbol{q}_i^n - \frac{\Delta t}{h}\left((\boldsymbol{F}_w)^n_{i+\frac{1}{2}} - (\boldsymbol{F}_w)^n_{i-\frac{1}{2}}\right)$ and
$(\boldsymbol{q}_a)_i^{n+1} = \boldsymbol{q}_i^n - \frac{\Delta t}{h}\left((\boldsymbol{F}_a)^n_{i+\frac{1}{2}} - (\boldsymbol{F}_a)^n_{i-\frac{1}{2}}\right)$.

So, in ghost cells we are left with two, possibly different updated solutions: $\boldsymbol{q}_w$ and $\boldsymbol{q}_a$. Expressed in $\tilde{\boldsymbol{q}} = (u, p)$-variables, these two new ghost solutions will not differ very much for the flows considered here. In case a solution ambiguity does arise, we proceed as follows. From the updated level-set function, which is updated separately through an advection equation, the volume-of-fluid fraction α_i^{n+1} in the ghost cell is computed. Next, the updated solution in the ghost cell is made unique with

$$\tilde{\boldsymbol{q}}_i^{n+1} = \alpha_i^{n+1}\left(\tilde{\boldsymbol{q}}_w\right)_i^{n+1} + \left(1 - \alpha_i^{n+1}\right)\left(\tilde{\boldsymbol{q}}_a\right)_i^{n+1} . \tag{3}$$

4 Numerical Results Ghost-Fluid Method

4.1 Water Front at Constant Speed and Pressure

Consider a 1D tube with unit length, $x \in [0, 1]$, inflow at $x = 0$, outflow at $x = 1$ and with as initial solution: $u(x, t = 0) = 1$, $p(x, t = 0) = 1$, $(x_{\text{fs}})_{t=0} = 0$, $\varrho(x = 0, t = 0) = \varrho_w(1) = 1$, and $\varrho(x, t = 0)_{x>0} = \varrho_a(1) \ll 1$, where x_{fs} is the location of the free surface, i.e., the water-air interface. The boundary conditions to be imposed are $u(x=0,t)=1$, $\phi(x=0,t)=t$ and $p(x=1,t)=1$. The exact Euler-flow solution reads $u(x,t)=1$, $p(x,t)=1$, $\varrho(x,t)_{x\le t}=1$ and $\varrho(x,t)_{x>t}=\varrho_a(1)$. This simple model flow precisely uncovers the deficiency of capturing methods with regard to material interfaces. Various values are considered for the density ratio $\frac{\varrho_w}{\varrho_a}$. The grids to be used are equidistant. The space discretization is taken first-order accurate and time integration is done with the forward Euler scheme.

Results of several numerical approaches that do not work satisfactorily are given in [8]. The ghost-fluid method performs very well; pressure (not shown) remains constant. In Figure 1 we present the computed bulk-density profiles at $t=0.0,0.1,0.2,\ldots,1.0$. The results are perfect; the interface is captured over a single mesh width h only (thanks to the level-set approach). The method does not break down with increasing density ratio. It works for standard water-air density ratios, $\frac{\varrho_w}{\varrho_a}=\mathcal{O}(10^3)$, as well as for much larger ratios (Figure 2). Reinitialization of the level-set function is not necessary for the running water front.

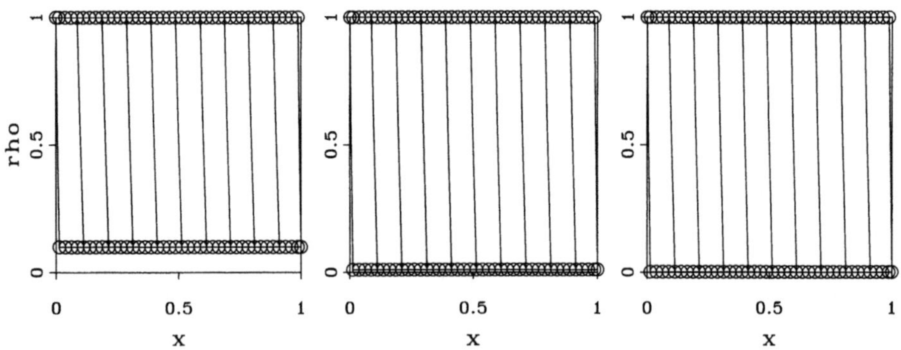

Fig. 1. Bulk-density profiles at $t=0.0,0.1,0.2,\ldots,1.0$, $h=\frac{1}{40}$, density ratios from left to right: 10, 100, 1000

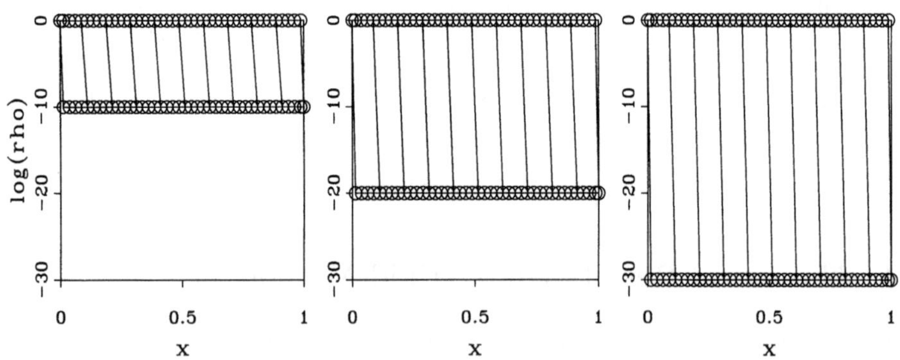

Fig. 2. Bulk-density profiles at $t=0.0,0.1,0.2,\ldots,1.0$, $h=\frac{1}{40}$, density ratios from left to right: $10^{10}, 10^{20}, 10^{30}$

4.2 Oscillating Water Column

In the ghost cells, the conservation laws are applied to virtual amounts of water and air, not to the real physical amounts. Hence, proper conservation of mass and momentum in these cells is not guaranteed. As a second test case, we consider a closed 1D tube (impermeable boundary at the left and right), with the initial condition as sketched in Figure 3. Starting from $t = 0$, the air at the right is

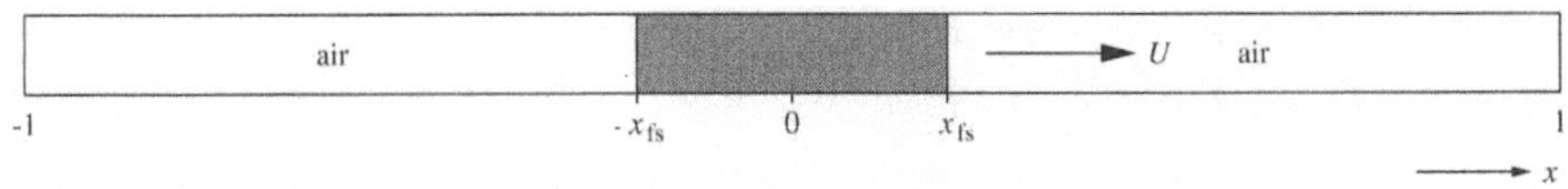

Fig. 3. Initial condition: closed tube with column of water (grey) in between two columns of air, all three columns flowing to the right at constant speed and pressure

compressed by the water and the air at the left expands. The column of water is decelerated until stagnation, and next accelerated to the left. The latter leads to a reverse pressure gradient across the water column, which redirects the flow from left to right again, and so on: the water column starts to oscillate.

We present numerical results obtained through the ghost-fluid method, taking $\frac{\varrho_w}{\varrho_a} = 1000$ and $(x_{\text{fs}})_{t=0} = 0.1$. An equidistant grid with $h = \frac{1}{40}$ is applied. The space discretization is again first-order accurate and time integration is done again with the forward Euler scheme. The level-set function is taken as the signed-distance function. For this test case, the level-set function is reinitialized. (The reinitialization is done after each time step.)

In the left graph of Figure 4 we give the time evolution of the relative mass error $\mathcal{M}(t) \equiv \frac{m_a(t)-m_a(0)}{m_a(0)}$, where $m_a(t)$ is the total mass of air in the tube at time t. The error appears to be composed of two components: one oscillating and the other behaving linearly in time. The total mass of air is slowly decreasing; air is converted into water. Fortunately, the orders of both the oscillatory and the linear error component are close to the computational method's order of accuracy, which is first-order. To show the latter, in the right graph of Figure 4 the time evolution of the relative air-mass error is also given for a grid and time step twice as fine. (The orders of accuracy of the oscillatory and linear error component – in going from $h = \frac{1}{40}$ to $h = \frac{1}{80}$ – appear to be 0.78 and 0.90, respectively.)

5 Conclusions

To avoid large solution errors near interfaces, a problem for many conservative capturing methods, a simple ghost-fluid fix has been proposed. For density ratios of the order 1000 (typical water-air ratio) the simple ghost-fluid technique performs perfectly. Even the computation of fronts running into vacuum is expected to be possible. Extension of the method to higher-order accuracy is straightforward through the use of, e.g., a MUSCL approach and a multi-stage

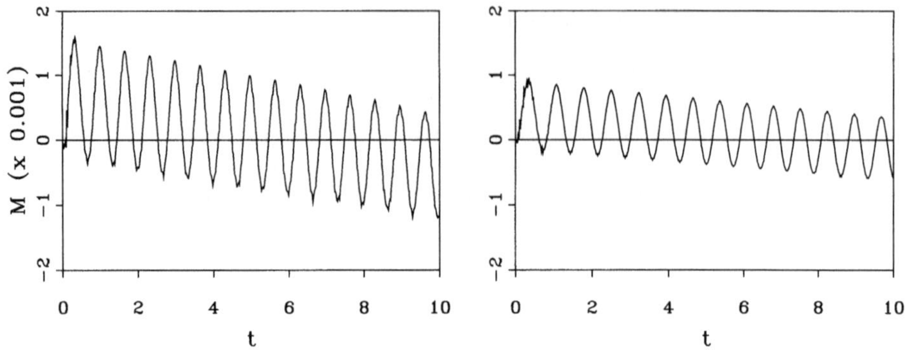

Fig. 4. Time evolution of relative error in total mass of air in closed tube, left: $h = \frac{1}{40}$, right: $h = \frac{1}{80}$

time integrator. Concerning the extension to higher dimensions, no principal difficulties exist.

Acknowledgements: This research was supported by the Dutch Technology Foundation STW.

References

1. R. Abgrall: J. Comput. Phys. **125**, 150 (1996)
2. R. Abgrall, S. Karni: J. Comput. Phys. **169**, 594 (2001)
3. R.P. Fedkiw, T. Aslam, B. Merriman, S. Osher: J. Comput. Phys. **152**, 457 (1999)
4. C.W. Hirt, B.D. Nichols: J. Comput. Phys. **39**, 201 (1981)
5. P. Jenny, B. Müller, H. Thomann: J. Comput. Phys. **132**, 91 (1997)
6. S. Karni: J. Comput. Phys. **112**, 31 (1994)
7. S. Karni: SIAM J. Sci. Comput. **17**, 1019 (1996)
8. B. Koren, M.R. Lewis, E.H. van Brummelen, B. van Leer: Report MAS-R0112, CWI, Amsterdam (2001)
9. R. Saurel, R. Abgrall: SIAM J. Sci. Comput. **21**, 1115 (1999)
10. J.A. Sethian: *Level Set Methods: Evolving Interfaces in Geometry, Fluid Mechanics, Computer Vision, and Materials Science* (Cambridge University Press, Cambridge, 1996)

Numerical Analysis of Particle Behavior Penetrating into Liquid by Level Set Method

Tsuyoshi Oda[1], Nobuyuki Satofuka[2], and Hidetoshi Nishida[2]

[1] Kobe Steel,LTD., Takatsukadai, Nishi-ku, Kobe, 651-2271 JAPAN
[2] Kyoto Institute of Technology, Matsugasaki, Sakyo-ku, Kyoto 606-8585 JAPAN

Abstract. A numerical analysis technique using the level set method has been developed to calculate the critical velocity of a particle penetrating into liquid. The contact angle, a dominant factor in this phenomenon, is given by imaginary gas-liquid interfaces inside a solid wall. A Cartesian grid method with immersed boundary is also applied in order to accurately represent a boundary of a moving particle. The current results agree with other calculated and experimental results in contact-angle problems. Finally, calculations of a particle impinging on a liquid surface have been carried out. Current results of the critical velocity agree with experimental and theoretical results.

1 Introduction

A powder injection refining is used for dephosphorization and desulfurization of hot metal in steel making processes. It is very important whether or not the injected powder can penetrate into hot metal and contribute to a chemical reaction. A basic model experiment with a single particle impinging on a liquid surface has been carried out and the necessary conditions of the particle have been investigated.

In this paper, we apply the numerical simulation with the level set method to the model experiment with a single particle. A Cartesian grid method with immersed boundaries is used to accurately represent the boundaries of a moving particle and the surface tension with an effect of contact angle is taken into consideration.

2 Numerical methods

2.1 Governing equations

The 2-dimensional incompressible Navier-Stokes equations can be written as:

$$\nabla \cdot \mathbf{u} = 0 \tag{1}$$

$$\frac{\partial \mathbf{u}}{\partial t} + \nabla \cdot (\mathbf{u}\mathbf{u}) = \frac{1}{\rho}\{-\nabla p + \nabla \cdot \tau + \sigma\kappa\delta(\phi)\mathbf{n}\} + \mathbf{g} \tag{2}$$

where $\mathbf{u}$,t,ρ,p,τ,σ,κ,$\mathbf{n}$,$\mathbf{g}$ are velocity, time, density, pressure, viscous stress tensor, surface tension, curvature, normal vector to liquid interface and gravity, respectively. An advection equation of a level set function ϕ is written as:

$$\frac{\partial \phi}{\partial t} + (\mathbf{u} \cdot \nabla)\,\phi = 0 \tag{3}$$

The finite volume method with a co-location grid is used. The 2nd-order ENO, the 2nd-order central difference, the Adams-Bashforth and the Crank-Nicolson are used in a spatial discretization of advection terms and diffusion terms, and in a time integration of advection terms and diffusion terms, respectively. A turbulence model is not used because of laminar flow.

2.2 Cartesian grid method with immersed boundaries

A Cartesian grid method with immersed boundaries proposed by T.Ye [1] is employed in this paper. In the method, a cell area divided by a boundary is absorbed by another cell on the opposite side of the boundary, as shown in Fig. 1.

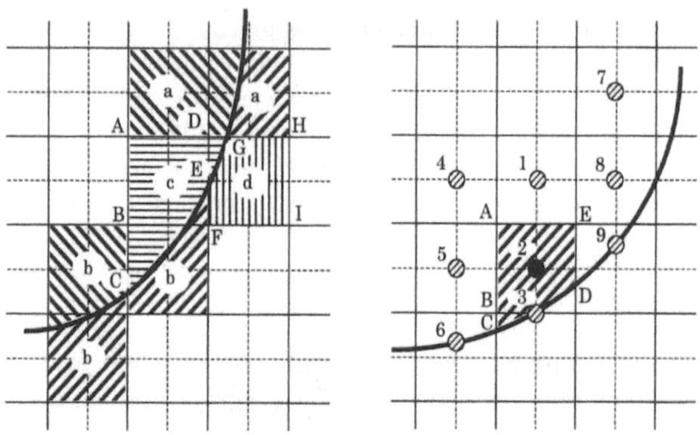

Fig. 1. Transition between x-pairing cells and y-pairing cells(left), and 9 points used for interpolation(right). 1st-order interpolation along x-direction and 2nd-order interpolation along y-direction are used in the right hand side case.

2.3 Surface tension and contact angle

The CSF [2] model is used for the surface tension. A weighted distribution function δ is a function of the level set function ϕ and is defined as:

$$\left.\begin{array}{ll} |\phi| < \alpha & \delta(\phi) = \frac{1}{2}\frac{(1+\cos(\pi\phi/\alpha))}{\alpha} \\ \text{else} & \delta(\phi) = 0 \end{array}\right\} \quad (4)$$

A grid interval Δx or Δy is used as a smearing width α.

In this paper, we propose a unique technique which generates imaginary gas-liquid interfaces which have the same gradient with the contact angle inside a wall. Fig. 2 shows an example of a droplet falling on an inclined plane. Black circles represent cells within a smearing width of gas-liquid interfaces and white circles represent the other cells. The level set function $\hat{\phi}$ of a solid wall is also defined. A unit vector normal to the gas-liquid interface, a velocity vector, a unit vector normal to the wall and a unit vector tangential to the wall at the ternary contact point of the gas, liquid and solid phases are represented by those vectors, $\mathbf{n}_A = (\phi_x, \phi_y)_AC\mathbf{u}_A = (u, v)_AC\hat{\mathbf{n}}_A = (\hat{\phi}_x, \hat{\phi}_y)_A$ and $\hat{\mathbf{s}}_A = (-\hat{\phi}_y, \hat{\phi}_x)_A$ at the nearest gas-phase cell A from the ternary contact point, respectively.

Corrections by equation(5) are made in hatched cells B which exist inside the wall near the imaginary interfaces in order to generate the imaginary interface.

$$\begin{array}{llll} \hat{\mathbf{s}}_A \cdot \mathbf{n}_A \geq 0 : & \phi_x & = \cos(\hat{\theta} + \theta_d) \\ & \phi_y & = \sin(\hat{\theta} + \theta_d) \\ & V_{CL} & = \hat{\mathbf{s}}_A \cdot \mathbf{u}_A \\ \text{else} \qquad : & \phi_x & = \cos(\hat{\theta} - \theta_d) \\ & \phi_y & = \sin(\hat{\theta} - \theta_d) \\ & V_{CL} & = -\hat{\mathbf{s}}_A \cdot \mathbf{u}_A \end{array} \tag{5}$$

where $\hat{\phi}_x = \cos\hat{\theta}, \hat{\phi}_y = \sin\hat{\theta}$, θ_d : the contact angle, V_{CL} : the velocity of the ternary contact point used in a dynamic contact angle model [3] shown in Fig. 3.

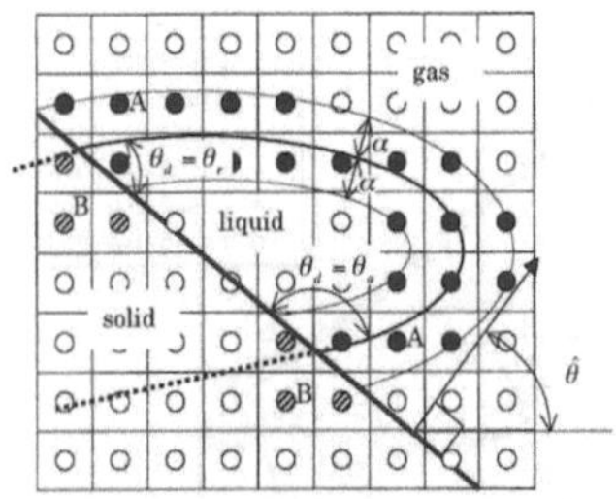

Fig. 2. Schematic around a droplet on an incline. Gradients of the level set function at the cell B are corrected according to the imaginary interfaces inside the wall.

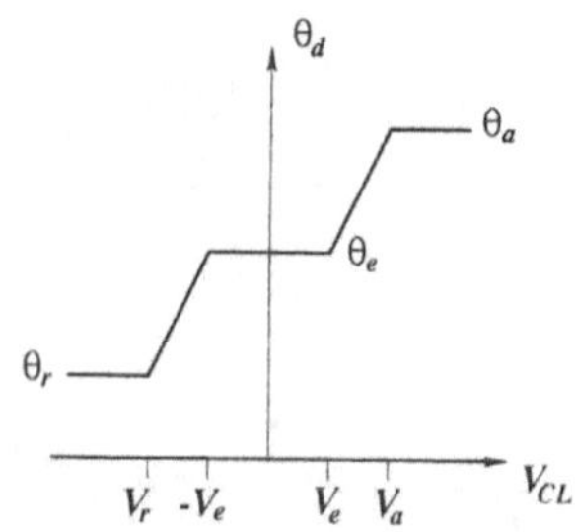

Fig. 3. Model of dynamic contact angle [3]. $V_a = -V_r = 2V_e = 0.1$(m/s) is used in this paper.

3 Numerical tests of contact angle

Numerical tests of a droplet impacting on a plate with a variety of contact angles are presented. Those results are compared with experimental results by J.Fukai [4]. The experimental conditions are shown in Table 1 where r_0 is the initial radius of the droplet and v_0 is the impact velocity.

Axi-symmetric 2-D numerical simulations with a central y-axis are carried out. A numerical domain with 11.2mm in x-width and 7.75mm in y-height is divided into 112 cells at regular intervals along the x-direction and into 128 cells at irregular intervals along the y-direction, respectively.

Figs. 4‘6 show comparisons of dimensionless splat radius and height. Fig. 7 shows comparisons of droplet shapes between the experimental results(left) and the calculated results(right). J.Fukai et al [4]. describe that a Eulerian approach with fixed grid makes it difficult to discretize accurately the normal and tangential stress balance at the free surface. Very similar results with J.Fukai et al [4] are obtained because the level set method can provide a gradient of the interface accurately though it is a Eulerian approach.

Table 1 Experimental conditions

case	1	2	3	4	5
r_0 (mm)	1.88	1.85	1.86	1.86	1.87
v_0 (m/s)	1.50	2.09	1.49	1.58	3.76
Re	3010	4130	3320	3130	7390
We	58.4	111	57.5	64.1	364
Bo	0.478	0.463	0.472	0.468	0.472
θ_a (deg)	60	60	70	92	92
θ_r (deg)	22	22	37	60	60
θ_e (deg)	34	34	54	76	76

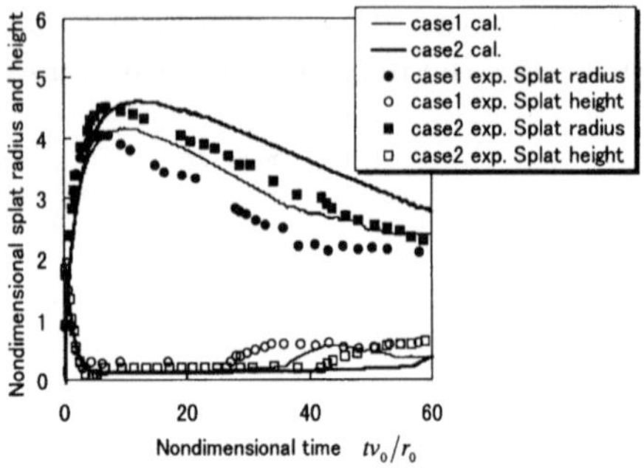

Fig. 4. Comparison of splat radius and height between current calculation and experiment [4] for case 1 and case 2

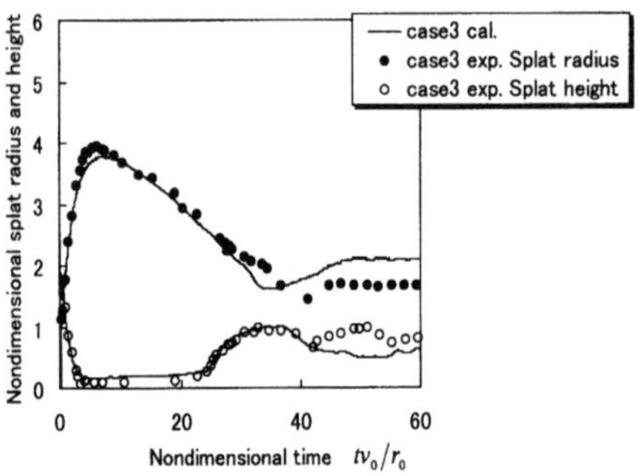

Fig. 5. Comparison of splat radius and height between current calculation and experiment [4] for case 3

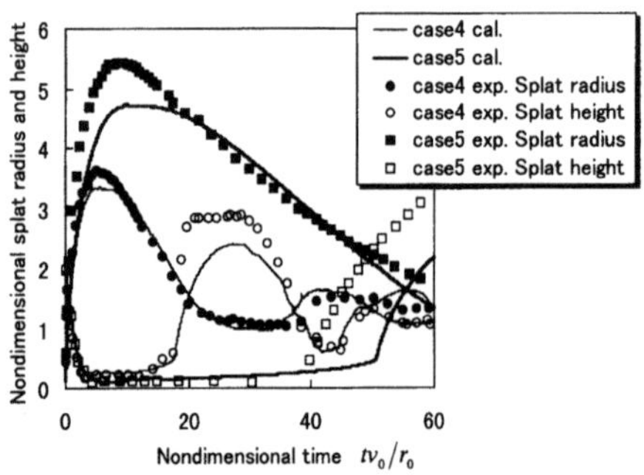

Fig. 6. Comparison of splat radius and height between current calculation and experiment [4] for case 4 and case 5

4 Numerical tests of a particle impinging on liquid

Engh [6] derived a critical Weber number We_c for a particle to overcome the drag and the surface tension of fluid.

$$We_c = 8\Big(1 - \exp\big(\frac{3}{4\rho^*}\big)\Big)\Big(\cos\theta_e - 1 + \frac{8}{3}\rho^*\Big) + 16 \tag{6}$$

Y.Ozawa et al.[5] modified the above equation(6) from model experiments, as:

$$We_c = \frac{1}{0.44}\Bigg(\Big(1 - \exp\big(\frac{0.66}{\rho^* + 0.25}\big)\Big) \Big(\cos\theta_e - 1 + \frac{\rho^* + 0.25}{0.33}\Big) + 2\Bigg) \tag{7}$$

where ρ_l:density of liquid, ρ_s:density of solid(particle), v_c:critical velocity for a particle to penetrate into liquid, r_s:radius of a particle. The critical Weber number We_c and the density ratio ρ^* are defined as follows:

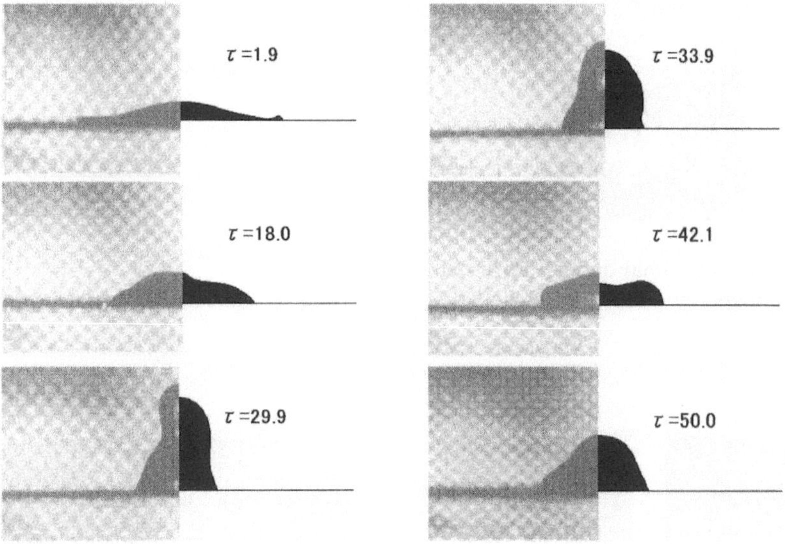

Fig. 7. Comparison of droplet shape between current calculation(right) and experiment [4](left) for case 4 ($\tau = tv_0/r_0$)

$$We_c = \frac{\rho_l v_c^2 r_s}{\sigma}, \qquad \rho^* = \frac{\rho_s}{\rho_l} \tag{8}$$

K.Ogawa et al.[7] confirmed agreement of the equations(6) and (7) by experiment.

In this paper, the experiment by K.Ogawa et al.[7] is taken as an example. The particle is made of polystyrene and its density is 1020kg/m^3, its spherical diameter is 65μm. The gas is air and the liquid is water. The equilibrium contact angle between the particle and water is 87°. The particle shape is represented precisely by the immersed boundaries.

Axi-symmetric 2-D numerical simulations with a central y-axis are carried out. A numerical domain with 325μm in both of x-width and y-height is divided into regular 100×100 cells shown in Fig. 8. The movement of the particle is calculated solving an equation of motion with pressure and shear stress from the surrounding fluid and is represented by coordinate transformation.

The critical velocities obtained from the equations(6) and (7) are 1.16m/s and 1.09m/s, respectively. Therefore, the two velocities of 1.00m/s and 1.20m/s are taken as an initial condition of the particle. Fig. 9 shows histories of the two particle velocities and Figs. 10 and 11 show gas-liquid interfaces and particle positions under the two initial conditions. As a result, the same critial velocity can be obtained with the equations(6), (7) and K.Ogawa et al.[7]. On the other hand, Numerical instability is observed especially in the case of the initial velocity of 1.20m/s. Further remedies for stability need to be done for versatile use.

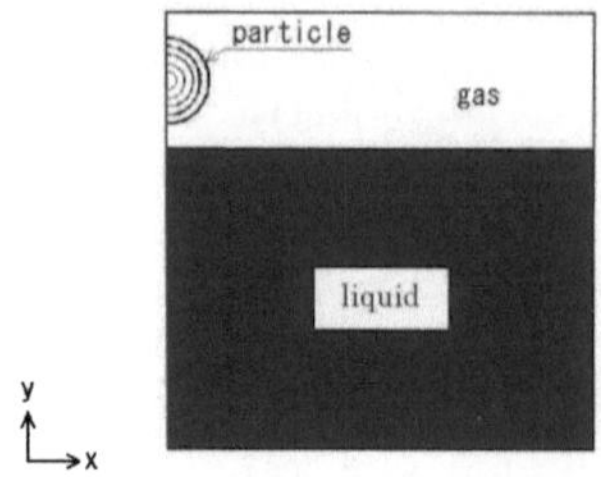

Fig. 8. Computational domain for a particle impinging on a liquid surface

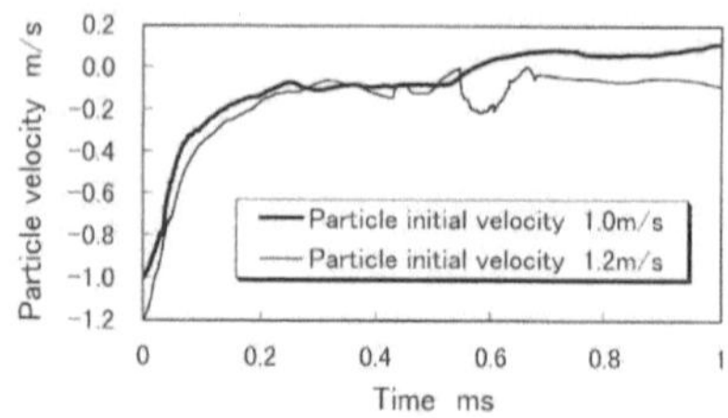

Fig. 9. Particle velocity histories

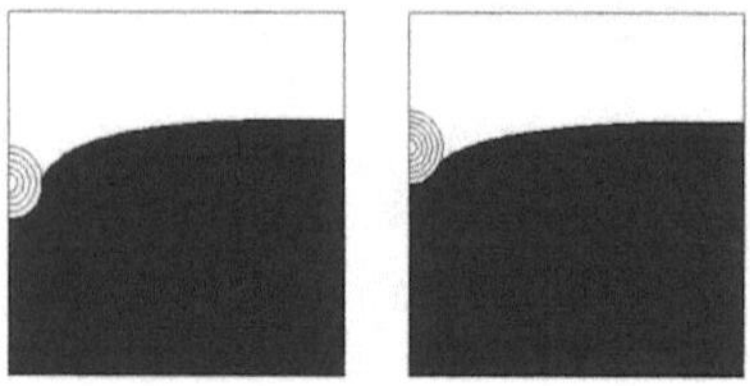

Fig. 10. Location of a particle and a liquid interface for particle initial velocity of 1.0(m/s) left:t=0.6(ms), right:t=1.0(ms)

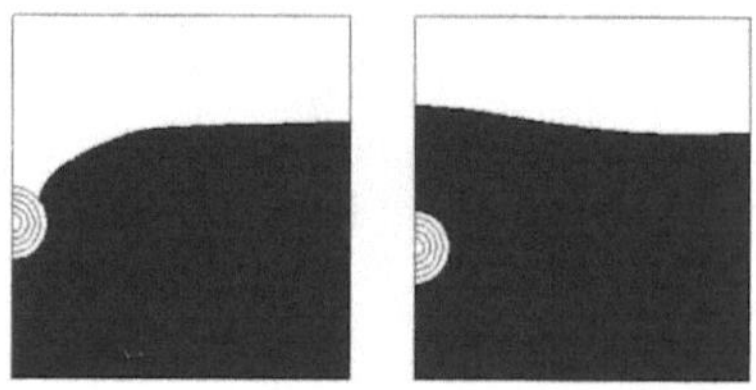

Fig. 11. Location of a particle and a liquid interface for particle initial velocity of 1.2(m/s) left:t=0.6(ms), right:t=1.0(ms)

5 Conclusions

i. Numerical results of droplet shapes impinging on a flat plate by the level set method with the unique treatment of a contact angle are found to be equivalent to a conventional Lagrangian approach.
ii. Numerical results of a particle critical velocity penetrating into liquid are found to be equivalent to experimental and theoretical results.
iii. Further remedies for stability need to be done for versatile use because numerical instability with immersed boundaries and surface tension is observed.

References

1. T. Ye, R. Mittal, H.S. Udaykumar and W. Shyy, J.Comput.Phys., 156,209-240,1999
2. J.U. Brackbill and D.B. Kothe and C.Zemach, J.Comput.Phys., 100,335-354,1992
3. M. Bussmann and J. Mostaghimi and S. Chandra, Physics of Fluids, 11,6,1406-1417,June 1999
4. J. Fukai, Y. Shiiba, T. Yamamoto and O. Miyatake, Physics of Fluids, 7,2,236-247,February 1995
5. K. Mori, Y. Ozawa, K. Suzuki, 19th committee of steelmaking reaction council, 19-10438, reaction-546, 28th October 1982 (in Japanese)
6. T.A. Engh, H. Sandberg, A. Hultkvist and L.G. Norberg, Scand. J. Met. of Metal, 1,103-114,1972
7. K. Ogawa, H. Matsumoto, CAMP-ISIJ, 10(1997)-80 (in Japanese)

A New Algorithm for Cavitating Flow Simulations Based on the Bubble Dynamics

Yoshiaki Tamura[1] and Yoichiro Matsumoto[2]

[1] Toyo University, Kawagoe 350-8585, JAPAN
[2] University of Tokyo, Tokyo 113-8656, JAPAN

Abstract. A numerical method for simulating cavitating flows is developed. The cavitation is modeled with the dynamics of bubbles which radii change with Rayleigh-Plesset equation. The bubbles are also allowed to have slip velocities so that the bubble accumulation could be simulated. Pseudocompressibility concept for incompressible flow simulation is extended for the present systems of equations for efficient and robust computations while the divergence free condition is practically satisfied. Computations of two-dimensional cavitating flows verify the present method and show its capability to the practical applications of cavitating flows.

1 Introduction

Computational Fluid Dynamics (CFD) has been progressing for decades and now it is applied to many kinds of flow fields. However, there is still large application area that the current CFD technology is not sufficiently applied to. One is cavitating flow. The difficulty for numerical simulations of cavitating flow may come from two facts. One is that the flow contain two phases, liquid and gas, and their individual acoustic speed of sound as well as that of the mixture differs. For instance, the speed of sound of water is about $1500m/s$, while that of the mixture of water and vapor is $25m/s$ at the minimum. Another is scale difference. The size of cavitation bubble is the order of micro- to millimeter while the scale of the global flow field is the order of meter. These scale problems must make numerical approaches difficult and time consuming. In the literature, there are several papers treating the cavitating flow simulation [1–6]. As seen in the references, however, many of them do use rather empirical cavitation models (pressure-density-void fraction relation, for instance) than physical modeling of cavitation or cavitation bubbles. The authors have derived the governing equations of cavitating flow including individual bubbles and their dynamics [7,8]. They have developed two algorithms for solving these equations. The results have shown qualitatively good agreement with the experiments. However, these algorithms still have the problem in stability and accuracy. In the present paper, yet another solution algorithm for the governing equations of cavitating flows is presented. The computed results are compared with the experiment to show its stability and accuracy.

2 Governing equations

2.1 Assumptions

Followings are assumed to construct the governing equations.

- Liquid phase is incompressible, gas phase is compressible.
- Density and momentum of gas phase is sufficiently small compared to liquid phase.
- Gas phase consists of bubbles. All bubbles are spherical and no collision and coalescence occurs.
- Mass change due to the phase change is much smaller than the mass of liquid.
- Bubbles are filled with vapor and non-condensable gas. The pressure inside a bubble is uniform. The vapor pressure is constant. The pressure of non-condensable gas is modeled to simulate vaporization and condensation at the interface.
- The heat capacity of liquid is large enough to keep the temperature constant.

Most critical assumption is that the gas phase consists of groups of bubbles. As no collision and coalescence is allowed, high void fraction cannot be treated or, even possible, the solution around such area is not reliable. Other assumptions are physical and sufficiently reliable solutions are expected.

2.2 Governing Equations

According to the assumptions in 2.1, following equations are derived.

1) conservation of the number density of bubbles

$$\frac{\partial n_G}{\partial t} + J\frac{\partial (n_G U_{Gi}/J)}{\partial \xi_i} = 0\,. \tag{1}$$

2) translational motion of a bubble due to the forces acting on the bubble

$$F_{Ai} + F_{Pi} + F_{Di} + F_{Li} = 0. \tag{2}$$

F_{Ai} is an added mass force and denotes,

$$F_{Ai} = \frac{4}{3}\pi\beta\left\{\left(\frac{\partial (r_G^3 \rho_L u_{Gi})}{\partial t} + U_{Gj}\frac{\partial (r_G^3 \rho_L u_{Gi})}{\partial \xi_j}\right)\right.$$
$$\left. - \left(\frac{\partial (r_G^3 \rho_L u_{Li})}{\partial t} + U_{Lj}\frac{\partial (r_G^3 \rho_L u_{Li})}{\partial \xi_j}\right)\right\}. \tag{3}$$

β is taken as 1/2 for a spherical bubble. The subscripts L and G denote the liquid and gas phase, respectively. U is the contravariant velocity.

F_{Pi} is the force by the acceleration of surrounding fluid and,

$$F_{Pi} = -\frac{4}{3}\pi r_G^3 \rho_L \left(\frac{\partial u_{Li}}{\partial t} + U_{Lj} \frac{\partial u_{Li}}{\partial \xi_j} \right) . \tag{4}$$

Drag and lift forces are given as,

$$F_{Di} = \frac{1}{2}\pi r_G^2 \rho_L C_D |\boldsymbol{u_G} - \boldsymbol{u_L}| (u_{Gi} - u_{Li}) , \tag{5}$$

$$F_{Li} = \frac{1}{2}\pi r_G^2 \rho_L C_L |\boldsymbol{u_G} - \boldsymbol{u_L}| \varepsilon_{ijk} \omega_{Lk} (u_{Gi} - u_{Li}) / |\boldsymbol{\omega_L}| . \tag{6}$$

C_D and C_L are given as written in [7,8].

3) volumetric motion of a bubble (Rayleigh-Plesset Equations)

$$r_G \frac{D^2 r_G}{Dt^2} + \frac{3}{2} \left(\frac{D r_G}{Dt} \right)^2 = \frac{p_B - p_L}{\rho_L} + \frac{1}{4} (u_{Li} - u_{Gi})(u_{Li} - u_{Gi}) . \tag{7}$$

p_B is a pressure inside of a bubble.

$$p_B = p_v + p_G - \frac{2T}{r_G} - 4\mu \frac{1}{r_G} \frac{D r_G}{Dt} . \tag{8}$$

T is the surface tension and p_v the vapor pressure and both are constant. p_G is the pressure of non-condensable gas and modeled [7,8] as,

$$\begin{aligned} p_G r_G^3 &= const. \ (D r_G / Dt > 0) , \\ p_G r_G^{3\kappa} &= const. \ (D r_G / Dt < 0) . \end{aligned} \tag{9}$$

κ is the specific heat ratio and taken to be 1.4.

4) conservation of the volumetric fraction of liquid phase

$$\frac{\partial f_L}{\partial t} + J \frac{\partial (f_L U_{Lj} / J)}{\partial \xi_j} = 0 . \tag{10}$$

f_L is the volumetric fraction of liquid phase.

5) conservation of momentum of the mixture

$$\frac{\partial (\rho_L f_L u_{Li})}{\partial t} + J \frac{\partial (\rho_L f_L u_{Li} U_{Lj} / J)}{\partial \xi_j} = -\nabla_i p + \nabla_j (\mu \nabla_j u_{Li}) + \frac{1}{3} \nabla_i (\mu \nabla_k u_{Lk}) . \tag{11}$$

The density of gas is neglected. μ is the effective viscosity of bubbly flow and given as,

$$\mu = \left\{ 1 + f_G \left(\frac{\mu_L + 5/2 \mu_G}{\mu_L + \mu_G} \right) \right\} \mu_L \cong \{ 1 + f_G \} \mu_L . \tag{12}$$

f_G is the void fraction.

6) constraint of volumetric fractions

$$f_G + f_L = 1 = \frac{4}{3}\pi r_G^3 n_G + f_L\,. \tag{13}$$

This closes the system of equations.

2.3 Numerical algorithm

In [7], the governing equations were solved based on SMAC-like method. It works well for non-cavitation region (as the method is originally for incompressible flows) but tends to be unstable in the cavitation region. In [8], pseudocompressibility is introduced to relieve the stiffness of the equations instead of Eq. (10) as,

$$\frac{1}{c^2}\frac{\partial p}{\partial t} + \nabla_j f_L u_{Lj} + \nabla_j f_G u_{Gj} - 4\pi r_G^2 n_G \frac{Dr_G}{Dt} = 0 \tag{14}$$

where c^2 is a coefficient. Equation (14) is further time-split into,

$$\frac{1}{c^2}\frac{\partial p}{\partial t} + \nabla_j f_L u_{Lj} + \nabla_j f_G u_{Gj} = 0\,, \tag{15}$$

$$\frac{1}{c^2}\frac{\partial p}{\partial t} = 4\pi r_G^2 n_G \frac{Dr_G}{Dt}\,. \tag{16}$$

The method is more stable but has the accuracy problem for the divergence free condition. To overcome this problem, Eq. (10) is solved as well as Eq. (14) in the present method to satisfy the condition. The present procedure is summarized as follows.

- advection equations, Eqs. (1), (10), (11) and (15), are solved.
- Rayleigh-Plesset equation, Eq. (7), and Eq. (16) is solved as Equation (16) is directly related to the radius change of bubbles.
- the error of divergence free is calculated as,

$$err = 1 - \left(f_L^* + \frac{4}{3}\pi r_G^{*3} n_G^*\right) \tag{17}$$

  where $(\cdot)^*$ denotes the intermediate value of sub-iteration. Then f_L and r_G is modified so that,

$$\begin{aligned} f_L^{**} &= f_L^* - 0.5err\,, \\ r_G^{**} &= \left\{(1 - f_L^{**})\big/\left(\tfrac{4}{3}\pi n_G^*\right)\right\}^{1/3}. \end{aligned} \tag{18}$$

  n_G is not modified as the bubble number density is rather free from this problem.
- repeat these steps till the pressure converges.

3 Results

Results of two-dimensional hydrofoil of Clark-Y 11.4% are only presented here. The flow conditions are as follows.

angles-of-attack	: 2.0°
initial radius of bubble nucleus	: 10μm
uniform flow velocity	: 10m/s
Reynolds number	: 6$\times$10^5

Figure 1 shows the pressure distribution around and on the hydrofoil with two cavitation numbers defined as $\sigma = (p_\infty - p_v)/(0.5\rho_\infty u_\infty^2)$. Typical pressure distributions with and without cavitation are observed. Comparison with the experiments are made in Fig. 2 with different initial void fractions. The computed results shows good agreement with experiment and little dependency on initial void fractions.

Figure 3 shows lift coefficients with different initial bubble radii. The effect of initial bubble radius is larger than that of initial void fraction, but the result is concluded not to be sensitive to the initial conditions of bubbles.

CPU time required for one case of computation with 281$\times$81 grid points is about 20 hours of Pentium III PC.

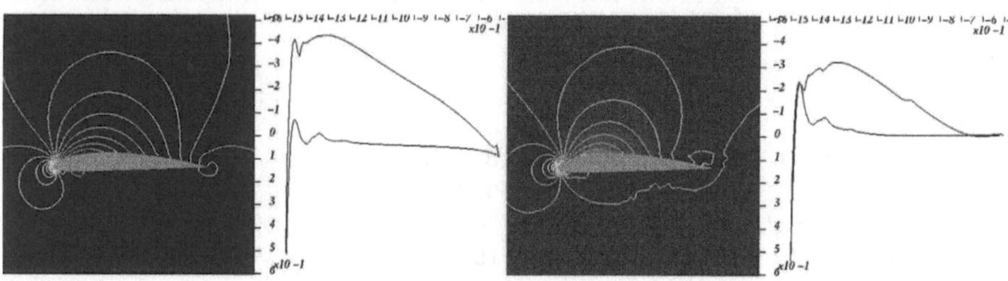

Fig. 1. Pressure distribution around Clark-Y 11.4%. left; σ=2.0. right; σ=0.6.

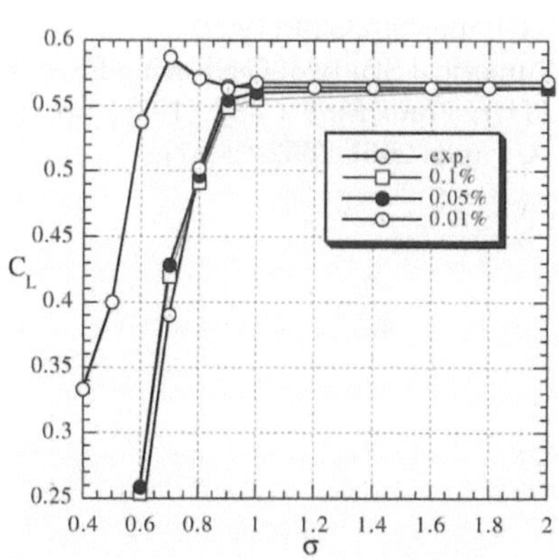

Fig. 2. Lift coefficient of Clark-Y 11.4% with different initial void fractions.

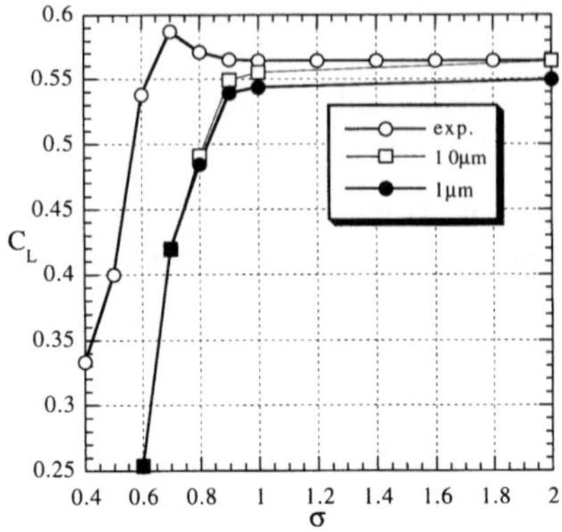

Fig. 3. Dependency of initial bubble radius.

4 Conclusions

A solution algorithm for the present systems of equations are developed. It is robust and efficient for the future extension to three-dimensional practical applications. The effect of initial conditions of bubbles are also examined and the computed results are found out not sensitive to the initial conditions of bubbles.

References

1. Y. Chen, S. D. Heister: J. Fluid Eng. **118**, (1996)
2. Y. Delaunay, J. L. Kueny: 'Cavity Flow Predictions Based on the Euler Equations'. In: *ASME Cavitation and Multi-Phase Flow Forum*, **109** (1990) pp. 153-158
3. H. Lemonnier, A. Rowe: J. Fluid Mech. **195** (1998) pp. 557-580
4. A. Kubota et al: J. Fluid Mech. **240** (1992) pp. 59-96
5. C. L. Merkle et al: 'Computational Modeling of the Dynamics of Sheet Cavitation'. In: *3rd Int. Symp. Cavitation* (1998)
6. R. F. Kunz et al: AIAA Paper 99-3329 (1999)
7. Y. Matsumoto et al: 'Numerical Study of Cavitating Flow Structure on a Hydrofoil'. In: *Proc. 4th KSME-JSME Fluid Eng. Conf.* (1998) pp. 249-252.
8. Y. Tamura et al: AIAA Paper 2001-2652 (2001)

Free-Lagrange Simulations of Shock/Bubble Interaction in Shock Wave Lithotripsy

Ahmad R. Jamaluddin[1], Graham J. Ball[2], and Timothy G. Leighton[3]

[1] School of Engineering Sciences, University of Southampton, Highfield, SO17 1BJ, UK
[2] Atomic Weapons Establishment, Aldermaston, Reading, RG7 4PR, UK
[3] Institute of Sound and Vibration Research, University of Southampton, Highfield, SO17 1BJ, UK

Abstract. A Free-Lagrange CFD code is used to model the near-field interactions between a lithotripter shock wave and a single spherical air bubble in water in axisymmetric form. The shock wave interaction causes the bubble to collapse and undergo jetting which consequently generates a blast wave into the surrounding water due to a liquid-liquid impact. Following impact, the bubble attains a toroidal shape and undergoes subsequent expansion and collapse. A detailed description of the shock/bubble interaction event is given, which includes both primary and secondary collapse. Near-field pressure-time histories are presented.

1 Introduction

Studies of SWL have shown that cavitation bubbles are induced *in vivo* near the lithotripter focus by the tensile stress of lithotripter shock wave pulses [1]. An idealised lithotripter shock profile consists of a leading shock front with peak positive pressure, P^+, up to $100\,MPa$, followed by a tensile wave with a peak negative pressure, P^-, down to $-10\,MPa$, and a total pulse duration of 3 to $7\,\mu s$ [2]. The collapse of such cavitation bubbles near a solid surface [3] or by an incident shock [4] can produce a liquid jet. The mechanical stresses generated by the shock-bubble interaction and subsequent jet impact on the kidney stone have been identified as a possible mechanism of kidney stone fragmentation during lithotripsy [5]. However, the violent collapse of cavitation bubbles also causes collateral damage to the kidneys as well as the surrounding tissue [6]. The costs and benefits of the treatment are therefore highly dependent on whether the focal point of the lithotripter generator is accurately targeted on the stone.

Despite these findings, current commercial lithotripters are not equipped with any means to assess qualitatively the cavitation activity in patients during clinical lithotripsy. One means of detecting the presence of cavitation bubbles is to measure their acoustic emissions [7,8]. In an attempt to explore the feasibility of assessing inertial cavitation *in vivo* during SWL based on acoustic measurements, Zhong *et al.* [9] studied the dynamics of cavitation *in vitro* using high-speed photography and measured the associated acoustic emission in water emanating from the focus of an electrohydraulic shock wave lithotripter. A clear correlation between the dynamics of lithotripsy-induced cavitation bubbles and resultant acoustic emission was identified. On the other hand, Cunningham *et al.* [8] used time-frequency analysis to quantify cavitation activity *in vivo* and used the time of the detected acoustic emissions to infer a value for the radius of stable bubble. However, in most experimental work, understanding of the fluid mechanics involved is incomplete owing to the limited temporal and spatial resolution of available experimental diagnostics.

The main objective of the present work is to simulate the cavitation events of shock-induced bubble collapse by lithotripter shock wave using the Free-Lagrange method in axisymmetric form and to predict the acoustic emission in the near field.

2 Numerical method

The numerical simulation is performed using the Free-Lagrange (FL) code, *Vucalm*, developed by Ball [10]. The code solves the unsteady, inviscid and compressible Euler equations, in axisymmetric form. The flow solver is of Godunov-type, and nominal 2nd order spatial accuracy is achieved using a piece-wise linear reconstruction of conserved variables. In order to ensure monotonicity at steep gradients, a MUSCL slope limiter is used.

The FL method uses a fully unstructured Lagrangian mesh in which the connectivity is allowed to change freely as the flow evolves, and is thereby suitable for highly deforming flows. The mesh is constructed using a Voronoi diagram. The working fluid is divided into discrete cells, each containing a single particle which carries information of fluid type, properties, coordinates and flow conditions. The mass and fluid type of each mesh cell is assigned from the start of the simulation and never changes. There are no mixed cells and hence the material interfaces are always sharply resolved. The code employs a simple interface smoothing algorithm which acts as a form of artificial surface tension [11]. This prevents numerically seeded Richtmyer-Meshkov instability occurring on the the interface when strongly shocked. The water is represented by the Tait equation of state, while the ideal gas equation is used for air. The code uses the HLLC approximate Riemann solver at gas/gas interfaces and an exact solver at gas/water and water/water cell interfaces [12].

3 Problem specification

The problem studied in the present work comprises a single spherical air bubble immersed in water (Fig. 1). The initial density for air and water are 1.2246 kgm^{-3} and 1000 kgm^{-3} respectively while the initial temperature and pressure for both fluids are 0.1 MPa and 288.15 K. The initial air bubble radius R_0 = 0.06 mm [7], which was chosen to be typical of a secondary stable bubble[1]. A

[1] A secondary stable bubble is a bubble which has been formed as a result of the interaction of a preceding lithotripter pulse with a cavitation nucleus, and which has reached a state of mechanical equilibrium with the surrounding fluid.

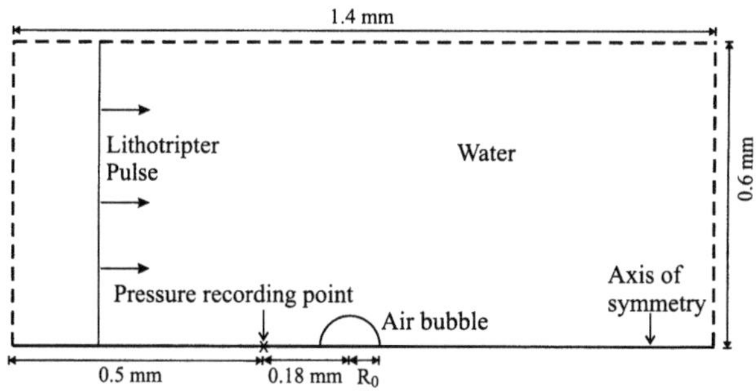

Fig. 1. The geometry of the problem

planar lithotripter pulse, with $P^+ = 90\,MPa$ and $P^- = -10\,MPa$, propagates through the water from left to right. Only half of the problem is simulated. The lower domain boundary represents the axis of symmetry. All elapsed times are measured from the first shock/bubble impact.

The lithotripter pulse is introduced by imposing a time-dependent pressure boundary condition on the left boundary. The top and right boundaries are non-reflecting at all times. A pressure recording point is positioned on the axis of rotation, $0.18\,mm$ from the initial bubble centre in order to register pressure pulses produced by cavitation event. A mesh of approximately 4×10^4 cells has been used.

4 Results and discussion

As a result of the profound acoustic impedance mismatch, a relatively weak shock is transmitted into the air bubble when the lithotripter shock (IS) hits the left bubble wall, whilst a strong expansion fan is produced in the water, running leftwards and upwards (EX in Fig. 3(a)). The high particle velocity behind the incident shock causes the bubble wall to deform to the right. At $t = 0.07\,\mu$s, the incident shock has traversed almost the full bubble width (Fig. 2(a)). The interaction between the shock and expansion waves originating at the bubble surface results in weakening and curvature of the shock. The air shock propagates more slowly and decouples from the incident shock. The deformation of the upstream bubble wall continues after the incident shock has passed because of the inertia of the water.

Interaction of the lithotripter pulse with the bubble causes it to collapse rapidly. The pressure gradient in the water near the upstream bubble increases as time progresses. It is clear from Fig. 2 that the collapse is asymmetric as the downstream bubble wall remains stationary up to about $t = 0.11\,\mu$s (Fig. 2(b)). At about $t = 0.20\,\mu$s (Fig. 2(c)), the upstream bubble wall starts to involute to form a distinct jet of liquid running to the right along the symmetry axis. The motion of the bubble during this phase is controlled almost exclusively by the inertia of the water. The liquid jet continues to accelerate and hits the downstream wall at about $t = 0.22\,\mu$s, isolating a lobe of trapped and highly compressed gas

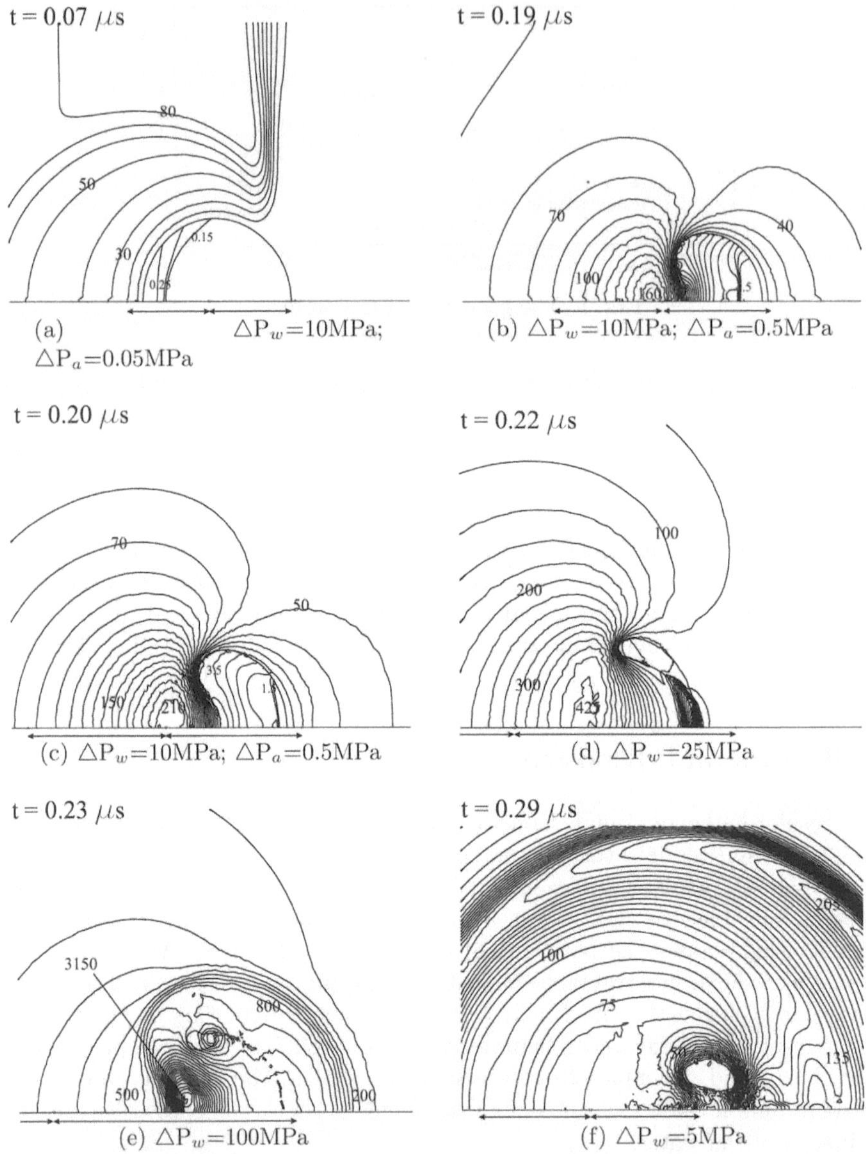

Fig. 2. Pressure contours for an air bubble impacted by a lithotripter shock with $P^{+} = 90\,MPa$ and $P^{-} = 10\,MPa$. Horizontal arrows indicate initial position and size of bubble. $\triangle P_w$ and $\triangle P_a$ indicate the increments between contours in the water and air respectively

which form a toroid in three dimensions (Fig. 2(d)). The variation of jet velocity with time is shown in Fig. 3(b). The jet continues to accelerate as it pierces the bubble, reaching a maximum of over $1200\,ms^{-1}$ immediately prior to jet impact.

It is believed that high-speed jets of this type play a primary role in cavitation erosion [13] as well as formation of circular pits and indentation on metal foils [1]. It is clear that the motion of the primary collapse is driven solely by the compressive component of the lithotripter pulse as the bubble does not encounter the tensile portion of the pulse before the primary collapse is complete.

The impact of the jet on the downstream bubble wall produces an intense blast wave in the surrounding water. It also leads to the creation of bubble fragments (Fig. 2(e)). These fragments may coalesce with the main cavity or act as nuclei for further cavitation events. The peak overpressure exceeds $1.0\,GPa$. As a result of the high velocity of the jet fluid, the blast wave advances relatively slowly to the left below the bubble. Consequently, the blast front is asymmetric. The interaction between the high-momentum liquid jet and the downstream low-momentum water produces a strong vortex flow.

In Fig. 2(e), the air cavity is drawn into the vortex core while the blast wave continues to propagate outwards radially from the bubble. The blast wave produces a sharp peak (BW) on the pressure-time history curve recorded at the pressure point (Fig. 3(a)). The strength of the blast wave decreases as it propagates into the surrounding water. The radiated blast wave could explain the large pressure spikes recorded by Zhong *et al.* [9] near primary collapse in their experimental studies.

The time history of the cavity volume is shown in Fig. 3(a). The volume reduces linearly with time from shock-bubble impact, until the first minimum at $t \approx 0.22\,\mu s$. The end of the linear phase correlates with the liquid jet impact. At this time the internal pressure greatly exceeds that of the surrounding water, and therefore the bubble begins to expand, entering an oscillatory state with two further cycles of expansion and collapse. Our simulation was halted after the third collapse, however the Gilmore-Akulichev model predicts that when the bubble encounters the tensile portion of the lithotripter pulse it will enter a phase

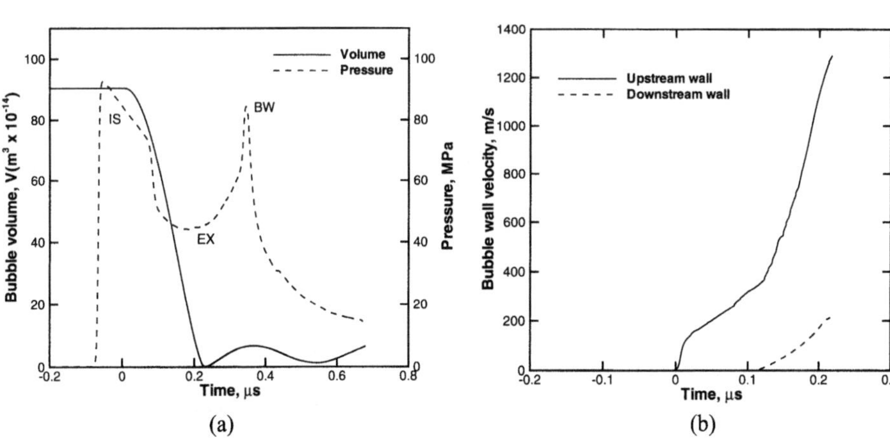

Fig. 3. (a) Pressure and bubble volume time history. Pressure is measured at point 'x' on Fig. 1. IS - Lithotripter shock, EX - Expansion waves, BW - Blast wave **(b)** Liquid jet velocity history

of prolonged expansion, followed by a series of lower-frequency oscillations [2]. This stage of the bubble behaviour will be examined in our future work.

5 Conclusion

The simulation of the near field interaction of a single air bubble with a lithotripter pulse, in axisymmetric form, has been performed using the Free-Lagrange code *Vucalm*. The results showed that the method allows sharp capture of the bubble boundary at all times and successfully predicts many details of the shock/bubble interaction. The impact of the shock on the upstream bubble wall causes it to involute and form a jet of liquid. The jet penetrates the interior of the bubble and strikes the downstream wall, generating a strong near-spherical blast wave into the surrounding fluid. Successive cycles of rebound and collapse occur prior to the long expansion phase, each collapse of the bubble emitting weak pressure waves into the surrounding water. This is predicted by the Gilmore-Akulichev model but has been overlooked by other workers. It is postulated that the liquid jet and strong spherical blast wave may assist in the fragmentation of kidney stones during clinical lithotripsy. Further work has been carried out to examine the influence on initial bubble size, shock strength and the proximity of a solid surface on the behaviour of the bubble. However, due to limited space, this work can not be included in the current paper.

References

1. A.J. Coleman, J.E. Saunders, L.A. Crum, M. Dyson: Ultrasound Med. Biol. **13/2**, 69–76 (1987)
2. C.C. Church: J. Acoust. Soc. Am. **86/1**, 215–227 (1989)
3. A. Phillip, W. Lauterborn: J. Fluid Mech. **361**, 75–116 (1998)
4. N.K. Bourne, J.E. Field: J. Fluid Mech. **244**, 225–240 (1992)
5. L.A. Crum: J. Urol. **140**, 1587–1590 (1988)
6. D. Howard, B. Sturtevant: Ultrasound Med. Biol. **23** 1107–1122 (1997)
7. A.J. Coleman, M.J. Choi, J.E. Saunders, T.G. Leighton: Ultrasound Med. Biol. **18/3**, 267–281 (1992)
8. K.B. Cunningham, A.J. Coleman, T.G. Leighton, P.R. White: Acoustics Bulletin **26/5** 10–16 (2001)
9. P. Zhong, I. Cioanta, F.H. Cocks, G.M. Preminger: J. Acoust. Soc. Am. **101/5**, 2940–2950 (1997)
10. G.J. Ball: Shock Waves **5**, 311–325 (1996)
11. B.P. Howell, G.J. Ball: Shock Waves **10**, 253–264 (2000)
12. G.J. Ball, B.P. Howell, T.G. Leighton, M.J. Schofield: Shock Waves **10**, 265–276 (2000)
13. T.B Benjamin, A.T. Ellis: Phil. Trans. R. Soc. Lond. **A260** 221–240 (1966)

Effects of Differencing Schemes on Simulation of Dense Gas-Particle Two-Phase Flows

S. J. Zhang[1], P. Zulli[2], Y. Q. Feng[3], X. F. Dong[3], and A. B. Yu[3]

[1] Engineering Sciences, Inc., 1900 Golf Road, Suite D, Huntsville, AL 35802, USA
[2] BHP Steel Research Laboratories, P.O. Box 202, Port Kembla, NSW 2505, Australia
[3] School of Materials Science and Engineering, the University of New South Wales, Sydney NSW 2052, Australia

Abstract. This work concerns the effects of differencing schemes on the simulation of dense gas-particle two-phase flows. The first-order upwind difference, the central difference, the second-order upwind difference, the central difference plus artificial dissipation, the deferred correction, the quadratic upstream interpolation (for convective kinematics) and the monotone upstream-centered schemes for conservation laws with different forms of flux limiters were all accomplished to simulate the multi-phase flows. It was found that differencing schemes may significantly affect the solution accuracy and numerical convergence. The monotone upstream-centered schemes for conservation laws are the best choice of all, because they can effectively suppress the nonphysical oscillations with the introduction of adaptive numerical dissipation into numerical solutions.

1 Introduction

Gas fluidization is widely encountered in engineering practice, and accurate prediction of its performance under different operational conditions has been a goal in multi-phase computational fluid dynamics for many years [1-3]. Two fundamentally different approaches to solve such dispersed multi-phase flows have evolved in the past. Their conceptual difference lies in the treatment of the dispersed phases, where one is based on Lagrangian of computational particles [4,5] and the other is based on the idea of interpenetrating continua and the Eulerian frame of reference, i.e. multiphase model, leading to a set of conservation equations [6,7]. The former is naturally suited to dilute flows and the latter meets its theoretical justification for denser flows [8]. When the conservation equations are arranged for numerical solutions, it is a common practice to divide the equations into three descriptive parts: a transient term, a total flux term (including both convection and diffusion) and a linearized source term. Unlike the straightforward treatment of transient, diffusive and source terms, the discretization of convective term is quite intrinsic and has been one of toughest tasks in computational fluid dynamics. Comparison and evaluation of many schemes for convective terms such as First-Order Upwind (FOU), Central Difference (CD), Second-Order Upwind (SOU), Central Difference plus Artificial Dissipation (CD+AD), Deferred Correction (DC), Quadratic Upstream Interpolation (for Convective Kinematics) (QUICK) and Monotone Upstream-Centered

Schemes for Conservation Laws (MUSCL) have been performed extensively in the past. However, all previous works were focused on linear or nonlinear model equations, Euler or Navier-Stokes equations. The effects of differencing schemes for multiphase flows, in which two sets of coupling N-S equations are employed and solved will be addressed in this paper.

2 Differencing Schemes

The governing equations employed herein are based on Gidaspow's hydrodynamics model A which is widely used to describe the motion of gas-particle two-phase flow, the detailed derivation and the constitutive relationships that close the model can be found in [9], The two sets of conservation equations, closed with the constitutive equations and supplemented with initial and boundary conditions, have to be numerically solved. To date there are two common solution algorithms for sequential iterative multi-phase control volume solver. They are the InterPhase Slip Algorithm (IPSA) [10] and the Implicit MultiField (IMF) method [11], of which the treatment here is based on the more implicit IPSA algorithm enhanced with Partial Elimination Algorithm (PEA) [10]. The treatment of the convective terms in conservation equations determines the accuracy of the solutions of the model equations describing the motion of multi-phase flows. Proper treatment is not straightforward because the more accurate schemes tend to less robust or slower. Because of this, many discretized methods have been developed to accommodate the different requirements in terms of flow conditions such as FOU, CD, CD+AD, SOU, QUICK, MUSCL and DC. To illustrate the convective terms and derive their discretized formula, we will consider the control volume shown in Fig. 1. The accomplishments of these schemes are listed below, taking an example of west face (w).

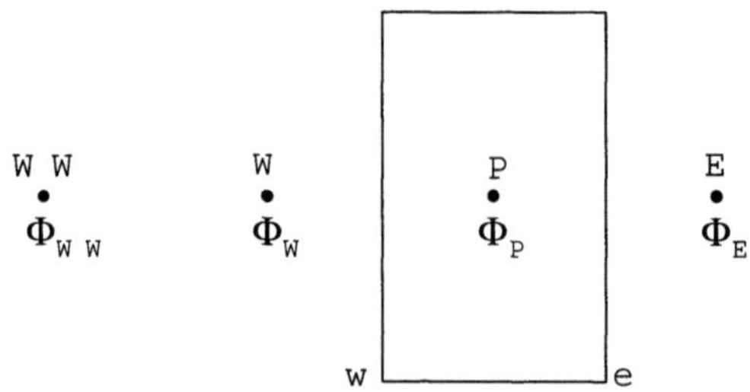

Fig. 1. Control volume notation

In FOU scheme, the convective value of variable ϕ at the west face of the normal volume is taken to be ϕ_W, i.e. $\phi_w = \phi_W$. In CD scheme, ϕ_w is: $\phi_w = \frac{1}{2}(\phi_W + \phi_P)$. SOU is an upwind scheme producing second-order accuracy by extrapolating to the face from the two upwind points. Thus ϕ_w is given by $\phi_w = \frac{3}{2}\phi_W - \frac{1}{2}\phi_{WW}$.

The compact form is: $\phi_w = \phi_W + \frac{1}{2}(\phi_W - \phi_{WW})$. QUICK is an upwind scheme which is third-order accurate and uses two upstream points and one downstream point. Thus ϕ_w is derived: $\phi_w = \frac{3}{8}\phi_P + \frac{3}{4}\phi_W - \frac{1}{8}\phi_{WW}$, To retain the matrix structure, this is written as: $\phi_w = \frac{5}{8}\phi_W + \frac{3}{8}\phi_P + \frac{1}{8}(\phi_W - \phi_{WW})$. MUSCL and TVD schemes are modifications to the higher-order upwind scheme with flux limiters to ensure boundedness and monotonicity of the solution. They have been found to be particularly useful in high Mach number flows with shocks. There are three different MUSCL schemes according to different flux limiters. These are Min-Mod, Van Leer and Superbee. These schemes only differ through their flux limiter. On the other hand, in terms of approximate accuracy, there are two types of MUSCL schemes, one is second-order accurate, the other third-order accurate. In this study only is the second-order one used. For second-order MUSCL schemes, ϕ_W is given by: $\phi_w = (1 + \frac{1}{2}\psi)\phi_W - \frac{1}{2}\psi\phi_{WW}$. Here ψ is the flux limiter which is given in terms of the ratio $r = \frac{\phi_P - \phi_W}{\phi_W - \phi_{WW}}$. The different schemes are corresponding to the different flux limiters, i.e., Min-Mod vs $\psi = \max(0, \min(1, r))$, Van Leer vs $\psi = (r + |r|)/(1 + |r|)$, and Superbee vs $\psi = \max(0, \min(2r, 1), \min(r, 2))$. DC scheme can be expressed as: $\phi_w = \phi_w^L + \beta(\phi_w^H - \phi_w^L)^{old}$, ϕ_w^Lstands for the approximation by some lower-order scheme (an upwind method is often used) and ϕ_w^H is the higher-order approximation. Th term in brackets is evaluated using values from the previous iteration, as indicated by the superscript 'old'. It is normally small compared to the implicit part, so its explicit treatment does not affect the convergence significantly. The term in brackets may also be multiplied by a factor $0 \leq \beta \leq 1$, thus blending the two schemes. In this study β is set to 0.5.

3 Results and Discussion

This work is primarily concerned with the numerical methods, aiming to assess the effects of differencing schemes on the simulation of gas-particle two-phase flows. Therefore, to reduce the computational effort, a simplest but representative one-dimensional condition is considered. The conditions and the physical properties of gas and particle used in the computation are listed in Table 1. The time step size is $1. \times 10^{-4}$s. All computational results are presented in terms of solid volume fraction to demonstrate the effects of numerical dissipation and dispersion on the solutions. When Rhie-Chow interpolation [12] which was initially proposed for non-staggered grid to prevent the solutions from odd-even decoupling is employed for a very small time step or existing of steep solution gradients such as shock waves, strongly swirling flows, free surfaces and discontinuities in porosity, the solution accuracy is severely deteriorated. Even though some efforts have been made to overcome this drawback no general desired remedy method without the loss of accuracy is available. Therefore staggered grid with 292 cells are used here.

Fig. 2(a) shows the FOU results without any nonphysical oscillation but serious numerical diffusion is introduced, conforming to many other theoretical analyses and observations [13]. While in Fig. 2(b) strongest oscillation near the

Table 1. Data for the numerical computation

Bed height = 0.292 m	Bed initial height = 0.1 m
Grid size = 1. × 10^{-3} m	Initial freeboard pressure = 101,325.0 Pa
Gas density = 1.2 kg/m^3	Gas viscosity = 1.8 × 10^{-5} kg/m·s
Particle diameter = 0.0005 m	Particle density = 2500 kg/m^3
Uniform gas inlet velocity = 0.4 m/s	Solid phase viscosity = 1. Kg/m·s
Minimum fluidizing porosity = 0.402	Minimum fluidizing velocity (u_{mf}) = 0.25 m/s

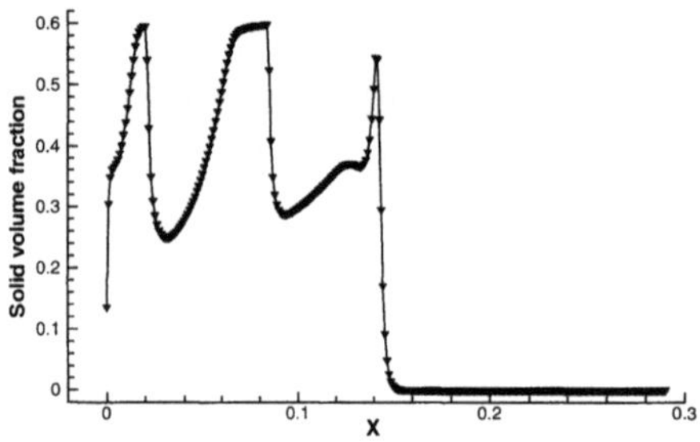

(a) FOU

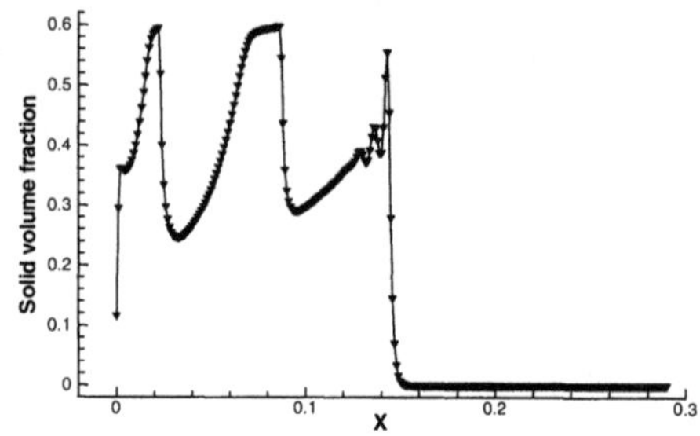

(b) CD

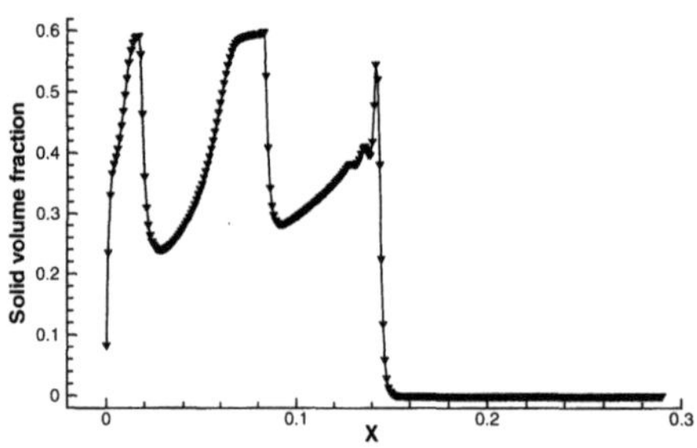

(c) CD+AD(0.01)

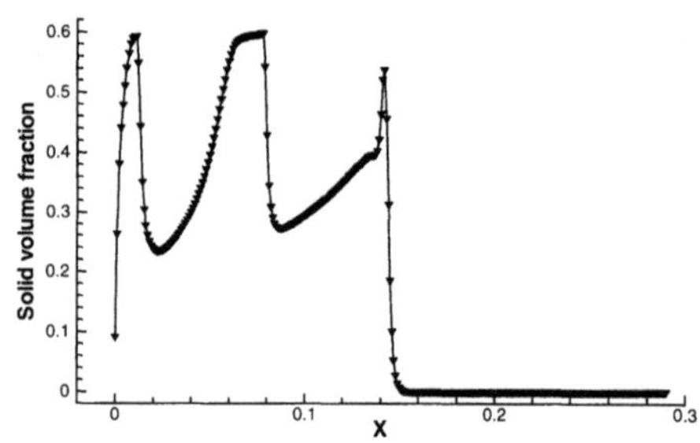

(d) CD+AD(0.025)

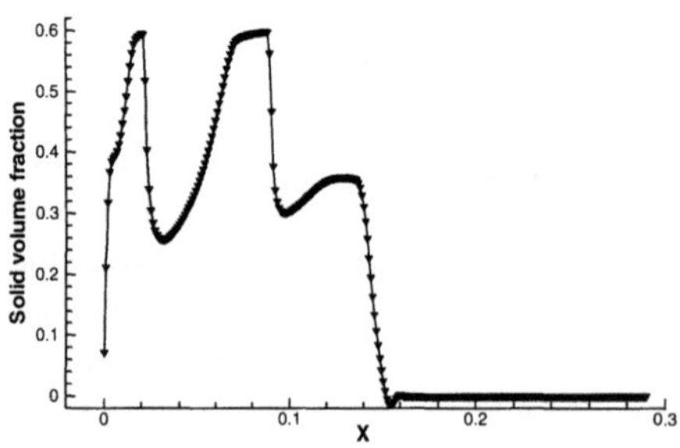

(e) SOU

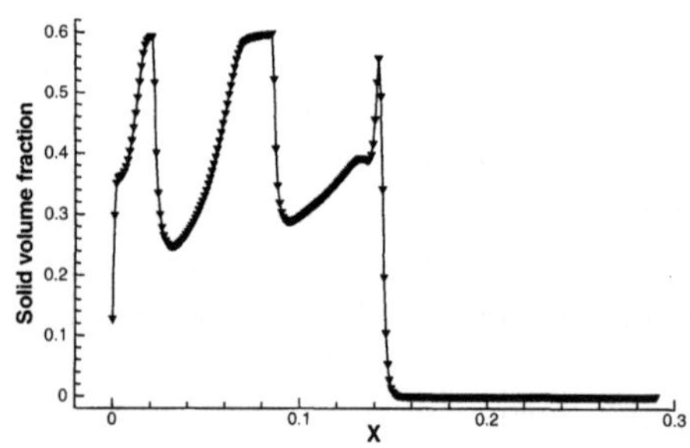

(f) DC

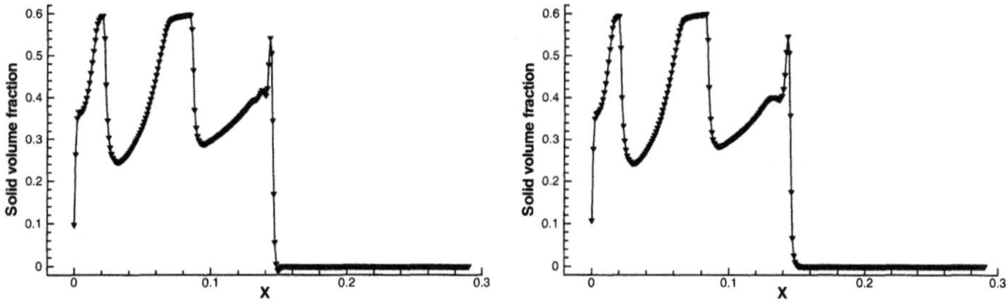

(g) QUICK

(h) MUSCL (min-mod)

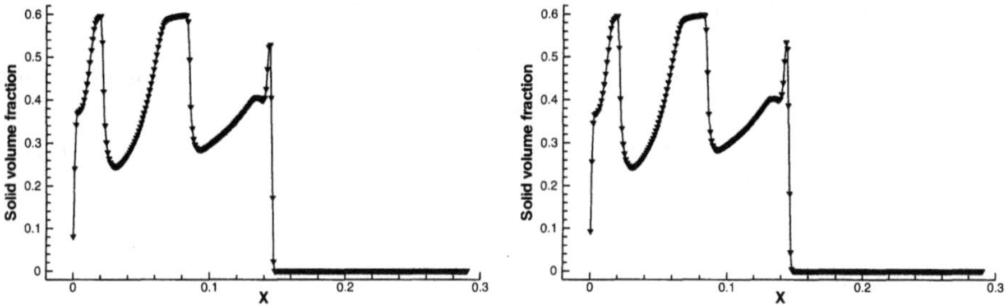

(i) MUSCL (super-bee)

(j) MUSCL (van-leer)

Fig. 2. Computational results with different schemes (at t = 0.25 s)

porous interfaces can be seen since CD often leads to numerical dispersion errors [13]. The classical method to remedy such a non-dissipative scheme is to add a proper amount of artificial dissipation. Figs. 2(c) and 2(d) show the results based on CD+AD with different constant dissipation values. The artificial dissipation equal to 0.01 is not large enough to suppress nonphysical maxima and minima, so that a small magnitude oscillation is still observed. When the artificial dissipation is set to 0.025, the numerical oscillation is effectively eliminated.

The computational representation of step profiles associated with discontinuities was difficult at the initial CFD development. FOU scheme introduces excessive diffusion which smoothes away the step. CD scheme is inherently without dissipation leading to ripples phenomena [13]. These undesired features of FOU and CD schemes have led to the development of other schemes such as SOU, DC, QUICK etc. Figs. 2(e) and 2(f) illustrate the results with SOU and DC, respectively. It can be seen both schemes have moderated numerical dissipation and dispersion. In Fig. 2(g), QUICK results are illustrated, very small maxima and minima occur. Because these schemes do not basically solve numerical dissipation and dispersion. Figs. 2(h)-2(j) show the MUSCL results with three forms of flux limiters, i.e. min-mod, super-bee and van-leer. It can be seen that the computations with the uses of these three limiters all produce good results without serious dissipation gradients or nonphysical oscillations near large gradients. Super-bee demonstrates the best performance in three.

4 Conclusions

The effects of differencing schemes on the simulation of dense gas-particle two-phase flows with reference of one-dimensional condition have been investigated. The first-order upwind difference, the central difference, the second-order upwind difference, the central difference plus artificial dissipation, the deferred correction, the quadratic upstream interpolation (for convective kinematics) and the monotone upstream-centered schemes for conservation laws were all accomplished to simulate the multi-phase flows. It was found that differencing schemes may significantly affect the solution accuracy and numerical convergence. The monotone upstream-centered schemes for conservation laws are the best choice in all, because they can effectively suppress the nonphysical oscillations with the introduction of adaptive numerical dissipation into numerical solutions.

References

1. Sinclair, J.L. and Jackson, R., Gas-Particle Flow in a Vertical Pipe with Particle-Particle Interactions, AIChE Journal, Vol. 35, 1989, p. 1473.
2. Ding, J. and Gidaspow, D., A Bubbling Fluidization Model Using Kinetic Theory of Granular Flow, AIChE Journal, Vol. 36, 1990, p. 523.
3. Zhang, S. J. and Yu, A. B., Computational Investigation of Slugging Behavior in Gas Fluidized Beds, Powder Technology, Vol. 123, 2002, p. 147.
4. Gosman, A.D. and Ioannides, E., Aspects of Computer Simulation if Liquid-Fuelled Combustors, AIAA paper-81-0323, 1981.
5. Faeth, G.M., Recent Advances in Modeling Particle Transport Properties and Dispersion in Turbulent Flow, in: Proc. ASME-JSME Thermal Engineering Conference, New York: ASME, Vol. 2, 1983, p.517.
6. Soo, S.L., Multiphase Fluid Dynamics, Science Press, Beijing, 1990.
7. Drew, D.A., Mathematical Modeling of Two-Phase Flow, Ann Rev Fluid Mech, Vol. 15, 1983, p. 261.
8. Crowe, C.T., REVIEW-Numerical Models for Dilute Gas-Particle Flows, J Fluid Eng, Vol. 104, 1982, p. 297.
9. Gidaspow, D., Multiphase Flow and Fluidization, Academic Press, Boston, 1994.
10. Spalding, D.B., Numerical Computation of Multi-Phase Fluid Flow and Heat Transfer, in: Taylor, C, Morgan, K., editors, Recent Advances in Numerical Methods in Fluids, Swansea: Pineridge Press, Vol. 1, 1980, p. 139.
11. Harlow, F.H., and Amsden, A.A., Numerical Calculation of Multiphase Fluid Flow, J Comp Phys, Vol. 17, 1975, p. 19.
12. Rhie, C. M. and Chow, W. L., Numerical Study of the Turbulent Flow Past an Airfoil with Trailing-Edge Separation, AIAA Juornal, Vol. 21(11), 1983, p. 1525.
13. Hirsch, C., Numerical Computation of Internal and External Flows, Vol. 2, chap. 21, John Wiley, New York, 1988.

Numerical Modelling of Multiple Three Dimensional Gas Bubbles Rising in a Quiescent Liquid Using a Volume of Fluid Method

Matt Gaston[1], John Reizes[1] and Geoff Evans[2]

[1] Faculty of Engineering, University of Technology, Sydney,Broadway, NSW, 2007, Australia.

[2] Centre for Multiphase Processes, Department of Chemical Engineering University of Newcastle, Callaghan, NSW, 2308 Australia

Abstract. This paper presents numerical investigation of the interaction of multiple three dimensional gas bubbles rising in a square cylinder using the volume of fluid method (VOF) including surface tension effects. Bubble trajectories, dynamics, shapes, and wake interaction are discussed.

1 Introduction

The extent and rate of mixing in bubble columns is directly related to the motion of bubbles and the quantity of liquid that is transported with them. Mixing is also promoted via the stretching, breaking, coalescing, and rolling action of the bubbles caused by fluid vorticity acting upon the gas-liquid interface. Numerical modeling of the dynamics of gas bubbles in multiphase systems has been the subject of a number of investigations, most of which have been based on simplifying assumptions which relied on empirical correlations [1]. The Volume of Fluid (VOF) Method [1] uses the fewest assumptions, and can simulate a mixture of two incompressible, immiscible fluids and the effects of surface tension. The VOF method is used to study the motion of, one, two, and six bubbles released simultaneously in a square cylinder in various relation to the centre line and each other.

2 Theory

Computations were performed with a three-dimensional, multi-domain, commercial flow solver, CFD-ACE(U) 2002. This is a pressure-based code that employs a cell-centred control volume approach [3]. In the VOF method, the fractions of the local volume occupied by each phase are represented by scalar functions F and *(1-F)* respectively. The variation in F is represented by the additional transport equation,

$$\frac{\partial F}{\partial t} + \nabla \cdot \boldsymbol{V} F = 0, \tag{1}$$

where $\boldsymbol{V}$ is the velocity vector. Equation (1) is solved together with the equations of conservation of mass and momentum. The interface between the gas and liquid is reconstructed so surface tension forces can be determined. A piecewise linear interface construction (PLIC) was used to reconstruct the surface [3]. The code was validated against Chen et al [2].

The computations were performed with second order spatial and first order temporal differencing schemes. An automatic time stepping was used to maintain a CFL criterion equal to 0.1, which is based on the motion of the liquid-gas interface only. Air and water at a constant temperature of 20°C were modelled. Two square cylinder computational domains were used with dimensions of 30x30x75 mm and 30x30x120 mm respectively. The domains were modelled with a uniform grid 0.75 mm cubed. The gas bubbles were initially spherical. Three sets of interactions were investigated; single bubble, two bubbles both with a 5 mm initial radius and six bubbles. Two single bubble cases were investigated with one bubble placed on the centre line of the cylinder and the other 5mm off centre towards one face. Both were positioned 10 mm from the bottom of the cylinder. Table 1 presents the initial coordinates of the two bubble interactions. The final case involved the interaction of six bubbles. The initial positions of the centre of each bubble and its radius are given in Table 2.

Table 1. Initial co-ordinates of the two bubble cases (mm).

Case	*No.*	*X*	*Y*	*Z*
Two Horizontal	1	7.5	10.0	15.0
	2	22.5	10.0	15.0
Two Vertical	1	15.0	25.0	15.0
	2	15.0	10.0	15.0
Two Diagonal	1	9.6967	20.6066	15.0
	2	20.3033	10.0	15.0

Table 2. Initial co-ordinates of the six bubble case (mm).

No.	*Radius*	*X*	*Y*	*Z*
1	5.0	7.5	10.0	22.5
2	5.0	22.5	30.0	7.5
3	3.75	7.5	30.0	7.5
4	3.75	7.5	20.0	15.0
5	2.5	15.0	20.0	7.5
6	2.5	22.5	10.0	22.5

3 Results and Discussion

3.1 Single Bubble

The motions of the single centered and off-centre (OC) bubble under the action of buoyancy forces are shown in Fig. 1. Both bubbles initially roll up into a mushroom shape, with a ring vortex around its rim. As the bubbles continue to rise they form into to an ellipsoidal bubble which is consistent with observations from previous studies [6]. In both cases the bubbles pulsate in the vertical direction. The frequency of the pulsations was 13.3 Hz in both cases, although the amplitude of the oscillations of the OC bubble was slightly less than that of the centered bubble; perhaps because of its rocking motion with a frequency of ≈8 Hz. With each vertical pulsation a ring vortex is shed in the wake of the bubble. The OC bubble moved away from the nearest wall towards the centre of the cylinder. The trajectory had a complicated repeating pattern due to the combination of vertical pulsation and rocking. The centered bubble appears to reach its terminal velocity of 0.216 m/s after ≈0.1 s. This velocity compares favorably with the experimentally reported unbounded terminal velocity of 0.2 m/s [7]. The terminal velocity of the OC bubble is 0.21 m/s which is slightly less than that of the centered bubble due to the closer proximity to the wall.

3.2 Two Bubbles

Shown in Fig. 2 are the three dimensional iso-surfaces depicting the motion of the two vertical, two horizontal and two diagonal bubble cases. Like the single bubble each bubble, shown in Fig. 2 initially rolls up into a mushroom shape, with a ring vortex around each rim. The two vertical bubbles pulsate in unison in the vertical direction for only one cycle then just prior to coalescence the

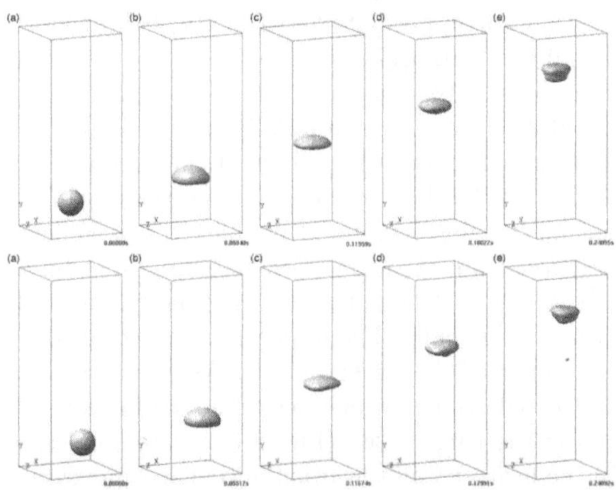

Fig. 1. Single bubble iso-surfaces; Centered (top) and Off-centered bubble (bottom), (a) 0.0s, (b) 0.06s, (c) 0.12s, (d) 0.18s, and (e) 0.24s.

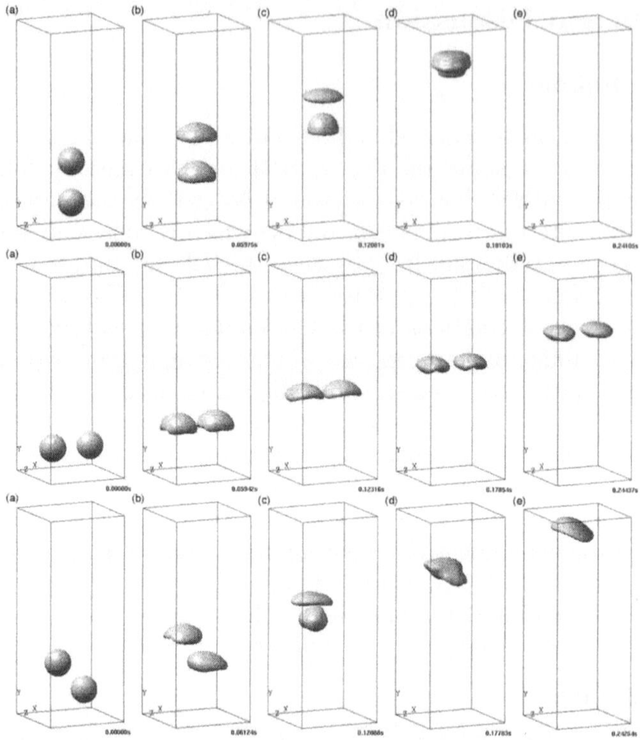

Fig. 2. Two bubble iso-surfaces; Vertical (top), Horizontal (centre) and Diagonal (bottom), (a) 0.0s, (b) 0.06s, (c) 0.12s, (d) 0.18s, and (e) 0.24s.

motion of the lower bubble appears to be out of phase with that of the upper bubble. The lower bubble does not reach its terminal velocity before coalescence, however, the velocity of the upper bubble reaches 0.228 m/s which is higher than that of the single bubble. This appears to be due to the ring vortex of the lower bubble driving the upper bubble faster as shown in Fig. 3(a).

The two horizontal bubbles (centre Fig. 2) have a similar motion to that of the single centered bubble, pulsating in the vertical direction but with a higher frequency of 14.7 Hz. The bubbles reach a terminal velocity of 0.179 m/s, which is $\approx$17 % slower than the single centered bubble. This is believed to be caused by the flow restrictions caused by the two bubbles being on the same cross-section (Figure 3(b)). The motion of the two diagonal bubbles (bottom Fig. 2) is the most interesting dynamically. The upper bubble's trajectory is only slightly affected by the presence of the lower bubble. As it moves towards the upper bubble just prior to coalescence, which occurs at $\approx$0.135 s it is pulled towards the upper bubble by the ring vortex (Fig. 3(c)). After coalescence the bubble moves away from the nearest wall towards the centre of the cylinder in the same manner as the OC bubble. A terminal velocity of 0.219 m/s is reached after $\approx$0.16 s. This value is slightly higher than that of the single bubble due top the

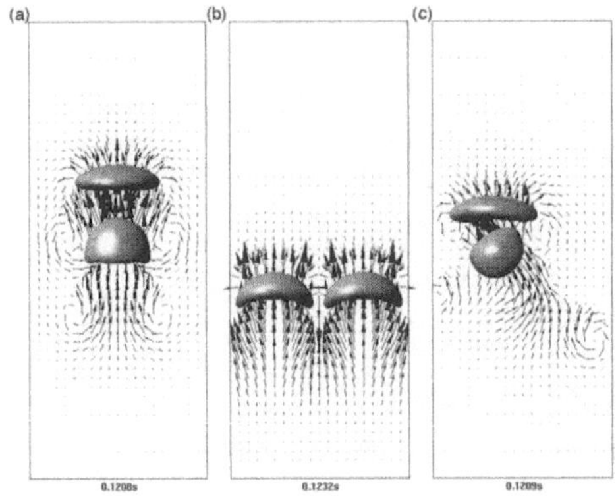

Fig. 3. Two bubble iso-surfaces with centre plane velocity vectors; (a) Vertical, (b) Horizontal and (c) Diagonal.

flow-field induced by the lower bubble. Figure 3 shows the two bubble iso-surfaces with centre plane velocity vectors taken at 0.12 s. Figure 3(a) clearly shows that the motion of the two vertical bubbles is out of phase just prior to coalescence and the induced velocity flow-field of the lower bubble on the upper. Another flow feature shown is the shed of the ring vortices below the lower bubble. The ring vortex generated by the two horizontal bubbles (Fig. 3(b)) are constricted by the side walls which causes an increase in frequency of the vertical pulsations and a reduction in the amplitude by more than a factor of two compared with that of the single centred bubble. The two diagonal bubble flow-field shown in Fig. 3(c) is somewhat more complex. Like the two vertical bubbles the lower bubble affects the flow on the upper bubble. Also shown is a shed vortex from the lower bubble which has moved to the wall and is effectively stationary.

3.3 Six Bubbles

Figure 4 shows the motion of six bubbles with various initial radii (Table 2). The motion of the bubbles is complex and a full description would require more space than is available in this paper. Briefly each bubble can be described as having a motion similar to that of either of the two horizontal, vertical, or diagonal bubbles or a combination of them. However, to understand the bubble dynamics fully a more detailed analysis is necessary, requiring the identification of all attached vortices and their cores. For example a small bubble can be seen approximately halfway down which appears not to be moving. On closer inspection it can be shown that this bubble was trapped within a core of a vortex, similar to that shown in Fig. 3(c), which due to its initial motion and the induced velocities of the other vortices is prevented from moving.

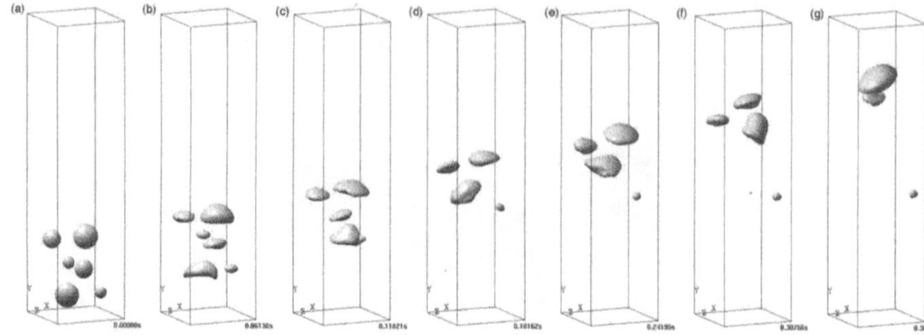

Fig. 4. Six bubble iso-surfaces; (a) 0.0s, (b) 0.06s, (c) 0.12s, (d) 0.18s, (e) 0.24s, (f) 0.30s, and (g) 0.36s.

4 Conclusions

A numerical investigation of the interaction of multiple three dimensional gas bubbles in a square cylinder utilizing the volume of fluid method including surface tension effects has been presented. The modeling of the interaction of two gas bubbles has been shown to assist in the understanding of the dynamic behavior of a larger numbers of gas bubbles. With increases in computer processor power reaching nearly a factor of two every year, the ability to model very large numbers of gas bubbles such as that present in gas column systems, without needing experimentally determined correlations and in a reasonable time frame, is becoming more practical.

References

1. Deen, N.G., Solberg T. and Hjertager B.H.,(2001),' Large eddy Simulation of the Gas-Liquid Flow in a Square Cross-Sectioned Bubble Column', *6th World Cong. of Chem. Eng. Melb., Aust. 23-27 Sept. 2001.*
2. Hirt, C.W. and Nichols B.D., 'Volume of fluid method for the dynamics of free boundaries', *J. Comput. Phys.*, **39**, 201-225, (1981).
3. CFDRC, 'CFD-ACE(U) module manual version 2002', CFD Research Corporation, Huntsville, AL, USA, (2002).
4. Chen, L., Garimella, S.V., Reizes, J.A. and Leonardi, E., 'The Development of a Bubble Rising in a Viscous Liquid', *J. Fluid Mech.* **387**, 61-96, (1999).
5. Kothe, D.B., Rider, W.J., Mosso, S.J. and Brock, J.S., 'Volume Tracking of Interfaces having Surface Tension in Two and Three Dimensions', AIAA, Paper 96-0859, (1996).
6. Aybers, N.M. and Tapucu, A., *Wrme-Stoffbertrag*, **2**, 171-177, (1969).
7. Sclift, R., Grace, J.R. and Weber, M.E., 'Bubbles, Drops and Particles', Academic Press, New York, 172, (1978).

Numerical Investigation of Flow Instability in a Low-Pressure Subcooled Boiling Channel

Y.Z. Li[1], G. H. Yeoh[2], and J. Y. Tu[3]

[1] School of Energy and Power Engineering, Xi'an Jiaotong University, Xi'an,710049, P. R. China
[2] Australian Nuclear Science and Technology Organisation (ANSTO), PMB 1, Menai, NSW 2234, Australia
[3] School of Aerospace, Mechanical and Manufacturing Engineering, RMIT University, Vic. 3083, Australia

Abstract. A three-dimensional two-fluid model to predict pressure drop and void fraction in a subcooled boiling flow at low pressure is adopted to investigate the flow instability due to the presence of two-phase flow in a heated channel. Good agreement on the overall pressure drop against experimental measurements was achieved. Predicted results of void fraction provided useful information towards a more fundamental understanding of the occurrence of incipient boiling, onset of significant voiding and onset of flow instability. The phenomenon of boiling onset oscillations observed in the experiment was also revealed through the use of the present two-fluid model.

1 Introduction

Flow instability is an important consideration in the design of nuclear reactors. The present investigation addresses the static flow instability in a heated channel. Static flow instability may arise in narrow cooling channels due to steam formation in the case of loss of coolant accidents. The formation of steam in a heated channel can have a significant effect on the overall pressure drop along the channel. For fully developed single-phase flow, the pressure drop is comprised primarily of effects due to wall friction, associated with surface roughness condition. With a reduction of the flow velocity, two-phase effects appear in a single-phase liquid at the point of incipient boiling (IB) and become more pronounced with the onset of significant voiding (OSV) as demonstrated in Fig. 1. With the presence of vapour in the bulk liquid during boiling, it causes an increase of frictional drag, and also acts to affect the pressure drop through acceleration and buoyancy effects. At some intermediate velocity, the increase in pressure drop due to boiling completely offsets the decrease in pressure drop due to channel frictional components. Further velocity reductions cause the pressure drop to rise which results in a minimum point, the onset of flow instability (OFI) point, in the pressure drop versus velocity curve (Fig. 1). If parallel flow paths exist, such as in fuel element designs, this increase in pressure drop in one channel may cause flow to be diverted to alternate channels, destabilising the system, and resulting in excursive or Ledinegg [1] stability.

Extensive validation against experimental data covering a wide range of conditions with significant improvements incorporated within the computational

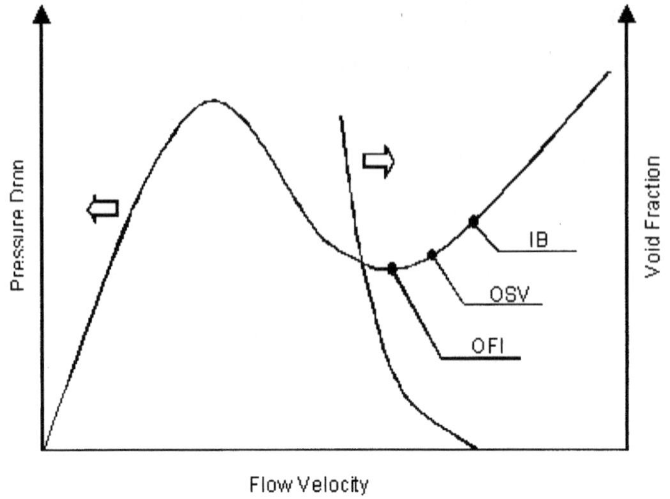

Fig. 1. Presence of Incipient Boiling (IB), Onset of Significant Voiding (OSV) and Onset of Flow Instability (OFI) along the pressure drop and void fraction versus flow velocity curves

fluid dynamics (CFD) two-phase boiling flow model in a generic computer code CFX4.4 has been recently performed in Tu and Yeoh [2], Yeoh *et. al.* [3] and Yeoh and Tu [4]. We further extend our investigation in this study to demonstrate the feasibility of employing this boiling model to predict the pressure drop along an upward boiling flow in a vertical heated channel and its relationship with the void fraction distribution. An important consideration is the identification of the velocity at which the OFI point occurs and the capability of the model to predict this point is assessed. The locations of IB and OSV points can also be obtained through the void fraction distribution along the heated channel. These results are also reported.

2 Two-Phase Boiling Flow Model

A two-fluid model formulated in terms of two sets of governing equations for the balance of mass, momentum and energy of each phase is used to predict the boiling process. A standard k-e model is employed to account for the turbulent boiling flow. The interaction of these two phases is effected through closure relationships such as inter-phase drag, heat and mass transfer terms in the field equations. These terms have been highlighted elsewhere (Tu and Yeoh [2]) and will not be repeated here.

2.1 Mean Bubble Diameter

The correlation developed by Zeitoun and Shoukri [5] is adopted for determining the mean bubble diameter because of its applicability for subcooled boiling flow at low pressures. They correlated their experimental data with:

$$\frac{D_s}{\sqrt{\sigma/\triangle\rho}} = \frac{0.0683(\rho_l/\rho_g)^{1.326}}{Re^{0.324}\left(Ja + \frac{1.492(\rho_l/\rho_g)^{1.326}}{Bo^{0.487}Re^{1.6}}\right)} \tag{1}$$

where ρ_l and ρ_g are the liquid and vapour densities and $\triangle\rho = \rho_l - \rho_g$. Here D_s is the mean Sauter diameter; σ is the surface tension; and g is the gravitational acceleration. The non-dimensional parameters in the above equation are the flow Reynolds number,$Bo = Q_w/Gh_{fg}$, where G is the mass flux and Dh is the hydraulic diameter; the boiling number , where Q_w is the wall heat flux and h_{fg} is the latent heat; and the Jakob number $Ja = \rho_l C_{pl}(T_{sat} - T_l)/\rho_g h_{fg}$, where T_{sat} is the saturation temperature. The mean bubble diameter, d, is taken to be equivalent to the mean Sauter diameter D_s.

2.2 Bubble Departure Diameter

For the bubble departure diameter dbw at the heated wall, the correlation of Kocamustafaogullari and Ishii [6] is employed:

$$d_{bw} = 2.64 \times 10^{-5}\theta\left(\frac{\sigma}{g\triangle\rho}\right)^{0.5}\left(\frac{\triangle\rho}{\rho_g}\right)^{0.9} \tag{2}$$

where $\triangle\rho = \rho_l - \rho_g$ and the equilibrium contact angle θ is taken to be at $80°C$ as suggested in Rogers and Li [7].

2.3 Wall Heat Partition Model

Various experimental and theoretical investigations for low-pressure subcooled boiling flow (Judd and Hwang [8]) suggest that the wall heat flux can be divided into three components: (i) heat transferred by conduction to the superheated layer next to the wall (nucleate boiling or surface quenching), Q_q; (ii) heat transferred by evaporation or vapour generation, Q_e; and (iii) heat transferred by turbulent convection, Q_c.

These are discussed further in [2].

3 Results and Discussion

The two-fluid CFD boiling model predictions are compared against a series of comprehensive experiments performed at CEA-Grenoble in a vertical rectangular channel. The height, width and thickness of the channel flow are 600 mm, 38 mm and 3.6 mm respectively. Heat was applied uniformly around the outer test section wall. More details regarding the experimental setup, instrumentation and measurement uncertainties are reported in the two CEA-Grenoble reports [9] [10].

The conservation equations were discretised using the control volume technique. The velocity-pressure linkage was handled through the SIMPLE [11] procedure. The discretised equations were solved using Stone's Strongly Implicit Procedure [12]. Since the wall heat flux was applied uniformly throughout the outer test section wall and taking advantage of the rectangular geometrical shape, only a quarter of the rectangular channel was considered as the domain for simulation. Convergence was achieved within 4000 iterations when the mass residual dropped below a reasonably small value $\sim 1 \times 10^{-7}$.

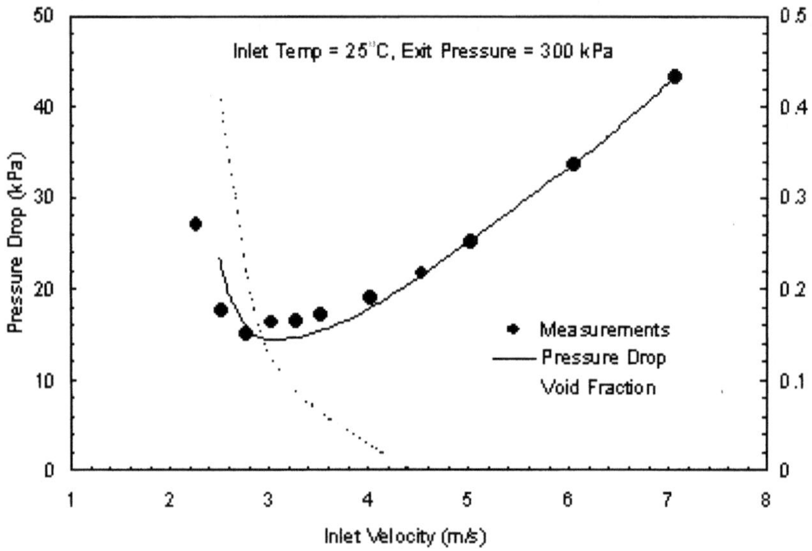

Fig. 2. Pressure drop and void fraction versus inlet flow velocity profile for heat flux, exit pressure and inlet subcooled temperature of $3MW/m2$, $300kPa$ and $25\circ C$ respectively

Figure 2 shows the comparison of pressure drop versus the inlet velocity in the rectangular channel for a surface heat flux of $3MW/m2$, exit pressure of 300 kPa and an inlet subcooling temperature of $25°C$. Accompanying the pressure drop profile is also the outlet void fraction distribution. Excellent agreement between the experimental results and model prediction was achieved. From Fig. 2, incipient boiling (IB) occurred at around an inlet velocity of $4.4m/s$. Further velocity reductions triggered the onset of static flow instability by the sharp increase of the pressure drop accompanying also by the sharp increase of the outlet void fraction. The predicted OFI point corresponding to the minimum pressure drop versus inlet velocity curve was $3.1m/s$, which was in very good agreement with experimental measured OFI point at $2.9m/s$.

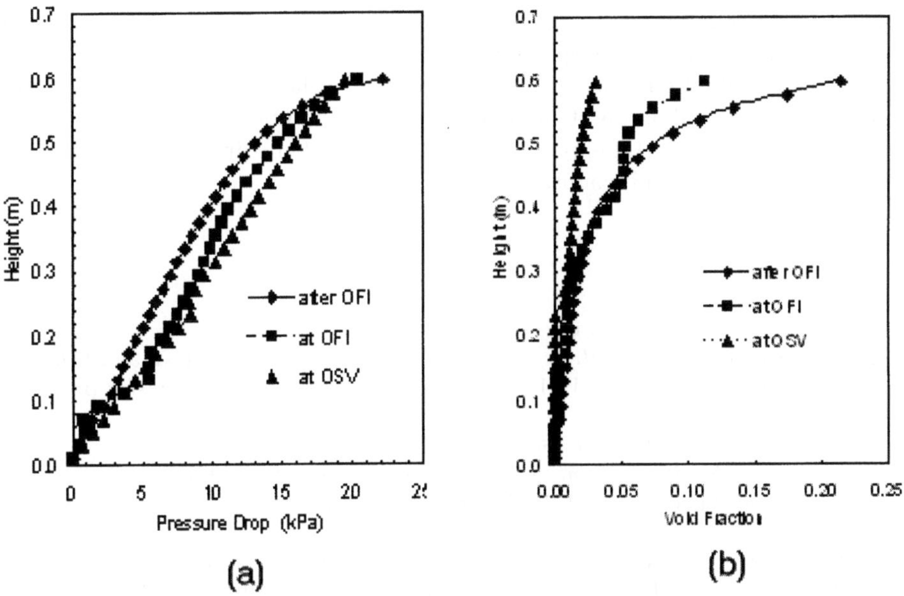

Fig. 3. Axial distribution along the channel height: (a) Pressure drop and (b) Void fraction for the for heat flux, exit pressure and inlet subcooled temperature of $3MW/m2$, $300kPa$ and $45°C$ respectively.

Figures 3(a) and 3(b) reveal the axial distributions of the pressure drop and void fraction along the channel height. The results corresponded to an inlet subcooling temperature of $45°C$, exit pressure of 300 kPa and a wall heat flux of $3MW/m2$. The profiles are recorded at three different situations during the reduction of inlet velocity - at OSV, OFI and the point after OFI. The increasing irregularity or waviness of the profiles for the axial pressure drop and void fraction at the OFI point as the boiling passes through the OSV regime clearly illustrated the presence of the phenomenon due to boiling onset oscillations. The flow became stable again after it passed through beyond the OFI point as evidenced by the smooth axial profiles of the pressure drop and system pressure, which was also similarly reported in Wang *et. al* [13].

4 Conclusion

The presence of bubbles as a consequence of heating flow through a vertical channel has a significant effect on the overall pressure drop along the channel. A three-dimensional two-fluid CFD model is adopted to investigate the two-phase flow and heat transfer characteristics in the heated channel. Comparisons with experimental data for a variety of operating conditions show good agreement with the predicted results. The OFI velocity is accurately determined through the two-fluid CFD model and the predicted results of void fraction supplement a more fundamental understanding especially the occurrence of various points along the pressure drop versus the velocity curve regarding IB, OSV and OFI. The two-fluid CFD model also demonstrates the capability to predict the phenomenon of boiling onset oscillations within the OSV regime as confirmed through Wang *et. al.* [13] experiments. at the end of the line that is to have no number, and set the sub-counter to zero.

References

1. M. Ledinegg: *Die Wrme*, 61, (1938) pp. 891-898.
2. J. Y. Tu, G. H. Yeoh: *Int. J. Heat Mass Transf.*, 45, (2002a) pp. 1197-1209.
3. G. H. Yeoh, J. Y. Tu, T. H. Lee TH, G. C. Park: *Num. Heat Transfer* (in press) (2002b).
4. G. H. Yeoh, J. Y. Tu: *Nuc. Sci. Eng.* 140, (2002c) pp. 181-188.
5. O. Zeitoun, M. Shoukri: *Int. J. Heat Mass Transfer*, 40, (1997) pp. 867-879.
6. G. Kocamustafaogullari, M. Ishii: *Int. J. Heat Mass Transfer*, 38, (1995) pp. 481-493.
7. J. T. Rogers, J. H.Li: *ASME J. Heat Transfer*, 116, (1994) pp. 1049-1053.
8. R. L. Judd, K. S. Hwang: *ASME J. Heat Transfer*, 98, (1976) pp. 623-662.
9. M. Courtaud, G. Coulon, F. Mazzili: Compte rendu d'essais boucle casimir traces de courbes ens, Centre D'etudes Nucleaires de Grenoble, Rapport TT/66-2-B/MC-GC/FM (1966).
10. M. Courtaud, K. Schleisiek: Pertes de pression en ebullition locale a basse pression dans des canaux de section rectangulaire, Centre D'etudes Nucleaires de Grenoble, Rapport T.T. n 84 (1968).
11. S. V. Patankar: *Numerical heat transfer and fluid flow*, Hemisphere Publ Corp, New York (1980).
12. H. L. Stone: *SIAM J. Num. Analysis*, 5, (1968) pp. 530-558.
13. Q. Wang, X. J. Chen, S. Kakac, Y. Ding: *Int. J. Heat Fluid Flow*, 17, (1996) pp. 418-423.

Numerical Simulation of Particle Generation in Gas Phase Reactors

Rainer Kaiser[1], Dieter Hänel[1], Einar Kruis[2], and Roland Vilsmeier[1]

[1] Inst. of Combustion and Gasdynamics (IVG), [2] Process and Aerosol Measurement Technology, University of Duisburg, D-47048 Duisburg, Germany

Abstract. The present paper is concerned with the numerical simulation of the processes in gas phase reactors for generating nano-size particle. A numerical method for compressible gas flow is extended to the simulation of suspensions of agglomerating particle. A discrete-sectional model for the size distribution function is chosen, yielding large systems in the order of 100 unknowns per grid-node. The numerical approach is semi-implicit with a preconditioned explicit transport formulation for gas and particle phase and an implicit treatment of particle interaction source terms. The development is based on the finite volume library MOUSE for the simulation of partial differential equations on arbitrary grids.

1 Introduction

Nano particle technology is an emerging field in mechanical, chemical and electro-engineering, opening up essential new fields for future developments and industrial products. For technical or research purposes, nano-particles can be generated in arbitrary ways, for instance via electro-sprays, partial evaporation of liquid droplets with dissolved agents, laser ablation or, as considered here, in gas phase reactors. Above mentioned methods have their specific assets and drawbacks. Concerning gas phase reactors, the main advantage is that relatively large amounts can be generated, while the distribution of size and morphological quantities of the particles generated can be critical.

Upon gaseous educts fed into a reactor, precursor reactions yield the later particle substance under consideration. At first, very small primary particles are produced in a nucleation process. They form agglomerates of different size and shape until they are deposited or collected.

Current nano-size particle research is mostly dominated by experimental methods and obviously, due to their small size, morphological variations and large amounts, corresponding experimental analysis are difficult and expensive. Numerical methods may thus be of essential benefit. Microscopic simulations on molecular level may help to understand formation processes and access basic physical data. Macroscopic simulations may predict nano-particle size and/or morphological distribution in space and time, enabling optimization of technical and research equipment.

Due to the large systems arising by modelling of agglomeration, most macroscopic methods are related to 0-D or 1-D simulations in time, [1], [2]. The methods to describe the distribution functions in space and time differ essentially.

2 Governing equations

The two phase mixture of gas and particles, including chemical reactions, particle transport and coagulation is formulated in a system of conservation laws. The system can be arranged in three types representing gas dynamics (*gas*), gaseous species transport and chemical reactions (*chem*) and particle phase (*part*):

$$\int_V \frac{\partial \boldsymbol{Q}_{type}}{\partial t}\, \mathrm{d}V + \oint_A \boldsymbol{H}_{type} \cdot \boldsymbol{n}\, \mathrm{d}A = \int_V \boldsymbol{S}_{type}\, \mathrm{d}V \quad ; \qquad type \in \{gas, chem, part\} \tag{1}$$

where $\boldsymbol{Q}$ is the set of conservation quantities, $\boldsymbol{H}$ is the flux vector and $\boldsymbol{S}$ a source term. V is a volume, A is its surface with the local normal vector $\boldsymbol{n}$ and t the time. In above system, the conservative quantities read:

$$\boldsymbol{Q}_{gas} = (\rho, \rho\boldsymbol{v}, \rho E)^T \,;\, \boldsymbol{Q}_{chem} = (\rho y_{g1}, \ldots)^T \,;\, \boldsymbol{Q}_{part} = (\rho_p, \rho_p\boldsymbol{v}_p, \rho_p E_p, \rho_p y_{p1}, \ldots)^T$$

with ρ the density, $\boldsymbol{v}$ the velocity vector, E the total internal energy and ρy the partial density. The indices g and p refer gaseous and particle phase species respectively. So above subsystems of equations are coupled:

$$\boldsymbol{H} = \boldsymbol{H}(\boldsymbol{Q}_{gas}, \boldsymbol{Q}_{chem}, \boldsymbol{Q}_{part}) \qquad \text{and} \qquad \boldsymbol{S} = \boldsymbol{S}(\boldsymbol{Q}_{gas}, \boldsymbol{Q}_{chem}, \boldsymbol{Q}_{part}).$$

3 Basic numerical concept

The developments shown here are based on the object oriented finite volume library MOUSE [3]. This open source library is in continuous development at the site of the authors and serves as basis for several research projects.
The system of equations (1) is solved on arbitrary grids employing a nodal finite volume method, Figure 1. For a discrete control volume V_d the above system reads:

$$\left.\frac{\Delta Q}{\Delta t}\right|_{V_d} + \boldsymbol{Res}^t_{\Delta,V_d} = \boldsymbol{Res}^s_{\Delta,V_d} \tag{2}$$

where t and s refer the transport and the source terms respectively. The transport residual $\boldsymbol{Res}^t_{\Delta,V_d}$ is composed of a sum of edge-wise contributions:

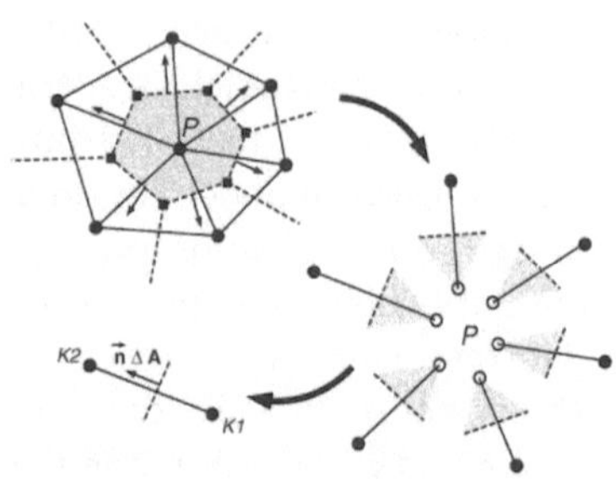

Figure 1: Control volume

$$\boldsymbol{Res}^t_{\Delta,V_d} = \frac{1}{V_{V_d}} \sum_{k=1}^{nk} (\boldsymbol{H}_k \cdot \boldsymbol{n}_k\, \Delta A_k) \tag{3}$$

with the index k referring to the edges appended to node P. Specific methods for gas dynamics (*gas*) [4] and chemistry (*chem*) [5] can be found in literature and are thus not repeated in detail here. For the convective particle transport a simple upwind formulation is chosen. A particle diffusion model based on concentration gradients and gas kinetic mobility is added. Note, that the particle phase equations (*part*) contain a proprietary momentum and energy variable, instead of being directly coupled to the gas flow: First, a decoupling of the particle speed from the speed of the carrier fluid is possible (particle deposition) and second, the structure of a given flow solver (*gas*) and (*chem*) can be preserved. However, the inertia and heat capacity cannot be ignored. For the present simulations full accommodation between gas and particle phase speed and temperature is assumed, while conservation is respected:

$$\boldsymbol{v}^* = \boldsymbol{v}_p^*; \quad T^* = T_p^*; \quad \text{with:} \quad q^* + q_p^* = q + q_p; \quad q \in \{\rho \boldsymbol{v}, \rho E\} \tag{4}$$

The time scales for the source terms, for chemistry as well as for particle formation, can vary by many orders of magnitude, yielding very stiff systems. The time integration process used is thus semi-implicit, with explicit transport and implicit sources. With linearisation of the implicit part one obtains:

$$\left(\frac{\boldsymbol{I}}{\Delta t} - \frac{\partial \boldsymbol{Res}^s_{\Delta,V_d}}{\partial \boldsymbol{Q}} \right) \Delta \boldsymbol{Q}|_{V_d} = -\boldsymbol{Res}^t_{\Delta,V_d}(\boldsymbol{Q}^m) + \boldsymbol{Res}^s_{\Delta,V_d}(\boldsymbol{Q}^m) \tag{5}$$

where m refers to the previous time level. The treatment of source terms for the particle phase is described in the next chapter. Due to large amounts of unknowns for chemistry and particle phase, the solution of the systems repeated once per node and partial time step is the essential computational difficulty.

4 Discrete-sectional Model for particle size distribution

Nano-particles are formed in a huge manifold of diversity of different sizes and shapes. This behavior is a consequence of the complex physics involved. After a nucleation process the particles agglomerate by random inter-particle collisions. The agglomerates formed are of arbitrary shape, depending on their individual history. This process is blended further on by evaporation, coalescence and heterogenous chemical reactions on their surfaces. As consequence, an individual particle can be characterised by a set of parameters $\boldsymbol{P_p}$, like volume and surface or further properties depending on the actual physics. For simplicity and due to large computational resources needed, current approaches are restricted to the particle sizes, while a constant fractal dimension of the surface is assumed.

In contrast to Monte Carlo type methods, particles are not treated as individuals. Their amounts (concentrations) are described by distribution functions. Several approaches exist, [6]. In the current project a discrete-sectional model is used. The concentrations of small particles are described by discrete size classes for multiples of monomers. Larger particles are grouped in sections.

The physical process of agglomeration is based on collisions arising from a gas kinetic random motion of particles. The collision frequency depends on collision cross section, concentration and mobility (speed). In a discrete and sectional model for particle density distribution, colliding particles vanish from their previous size classes and reappear as a single larger particle in another [1] size class, Figure 2. Consider i, j and k as indices for three discrete size classes, each one containing particles of i, j or k molecules respectively.

For the kth class the ballance term then reads:

$$\frac{\partial n_k}{\partial t} = \underbrace{\frac{1}{2}\sum_{i+j=k} \beta_{i,j} n_i n_j}_{\text{production}} + \underbrace{n_k \sum_{j=1}^{N_p} \beta_{k,j} n_j}_{\text{loss}} \tag{6}$$

where n_i and n_j are number-concentrations in the corresponding classes and N_p is the amount of particle classes considered. β is the collision probability:

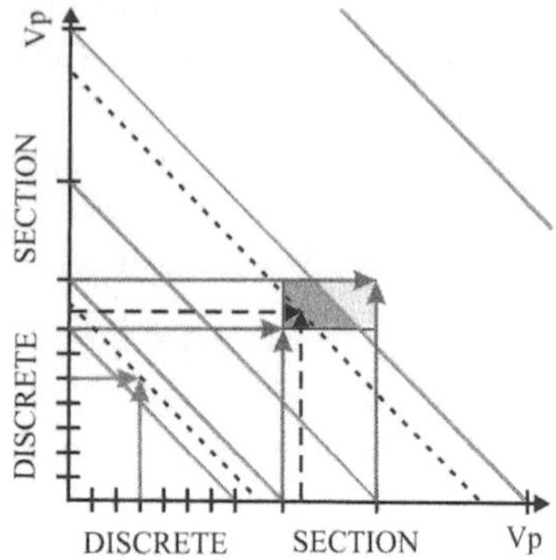

Figure 2: Discrete-sectional model: Agglomeration.

$$\beta_{i,j} := \beta(v_i, v_j) = \sqrt{\frac{6k_bT}{\rho_p^0}} \left(\frac{3}{4\pi}\right)^{\frac{2}{D_f}-\frac{1}{2}} r_o^{2\left(1-\frac{3}{D_f}\right)} \left(\frac{1}{v_i}+\frac{1}{v_j}\right)^{\frac{1}{2}} \left(v_i^{\frac{1}{D_f}} + v_j^{\frac{1}{D_f}}\right)^2 \tag{7}$$

where k_b is the Boltzmann constant, ρ_p^0 is the density of the particles material, v the particle volume, r the radius of a particle and T the temperature. The index $_o$ refers to primary particles. D_f is introduced as a fractal dimension [2], allowing an approximate description of the particles collision cross section:

$$\frac{v_i}{v_o} = \left(\frac{r_i}{r_o}\right)^{D_f} \tag{8}$$

To describe the agglomeration among sectional classes, a distribution function of the particles within these must be assumed. In the present work, volume based equipartition and a constant material density ρ_p^0 is chosen:

$$n_i(\mathbf{v}, t) = \frac{\rho_{p_i}}{\rho_p^0 \mathbf{v}(v_i^{max} - v_i^{min})} \tag{9}$$

where v_i^{max} and v_i^{min} are the upper and lower volume limits per particle in class i. With this distribution, the collision probability for sections read:

$$\hat{\beta}_{i,j} := \frac{\rho_{p_i}\rho_{p_j}}{(\rho_p^0)^2} \cdot \frac{1}{(v_i^{max} - v_i^{min})(v_j^{max} - v_j^{min})} \cdot \int_{v_i^{min}}^{v_i^{max}} \int_{v_j^{min}}^{v_j^{max}} \frac{\beta(\mathbf{u}, \mathbf{v})}{\mathbf{uv}} \mathbf{F}(\mathbf{u}, \mathbf{v}) \mathrm{d}\mathbf{u}\mathrm{d}\mathbf{v} \tag{10}$$

and the function $\mathbf{F}(\mathbf{u}, \mathbf{v})$ rules the production and loss in sections i, j and k.

$$\mathbf{F}(\mathbf{u}, \mathbf{v}) = \begin{cases} -\mathbf{v} & \text{loss in class } i \\ -\mathbf{u} & \text{loss in class } j \\ \mathbf{u}+\mathbf{v} & \text{production in a class } k \end{cases} \tag{11}$$

[1] It is possible, that an agglomerated particle reappears in a previous size class.

Note that not all possible agglomerating partners from classes i and j yield necessarily a production in the same class k. Figure 2 illustrates that, as the rectangle of production is located in several sections.

5 Sample results

As a first test case, a simple one-dimensional flow problem with particle agglomeration is considered. Ten discrete and 50 sectional classes with the upper limits

$$v_j^{max} = (10{+}0.5)v_o{\cdot}2^j\ ;\ j = 1,\ldots,50$$

were used. The entry Mach-number is $Ma_\infty = 0.01$. Particles are passively transported and agglomerate. Figure 3 shows the partial density of the 10th sectional class over the distance from the entry in dependence of the fractal dimension D_f.

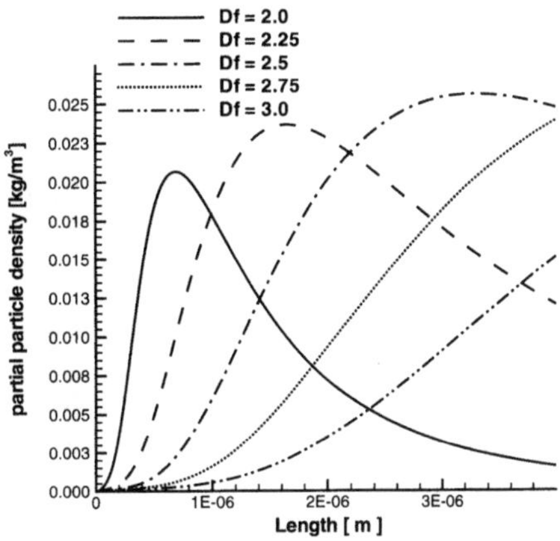

Figure 3: Particle density in 10th section.

In a second test case, the model of a reactor is studied. As inflow condition, three parabolic profiles are set, Figure 4 with $Ma_\infty^{max} = 0.4$ and $Re_d^{max} = 5000$. The central profile contains primary particles as additive, the outer flow consists of pure gas to prevent wall contact. 10 discrete and 30 sectional particle classes were used. The flow behaviour is cyclic. Figure 5 shows colors of constant partial particle density of the 10th sectional class near the entry of the reactor. Figure 6 shows a snapshot of the total particle density. A detail of the primary particle density near the entry and colors of constant geometric standard deviation of the particle size are shown in Figures 7 and 8 respectively.

6 Further applications and future developments

Other than in production devices, nano-size particles, especially those generated in the gas phase, might be of severe environmental impact due to bio-chemical activity. For example, such undesired particles are generated in combustion engines, and below a reasonable size limit filtration is nearly impossible. Efficient numerical simulation methods may assist product optimization.

In the present study, a mono-dimensional distribution function was used assuming all particles, disregard their history, can be considered with a simple relation between size and collision cross section. Microscopic analysis reveal that this assumption is too simple in many cases, as morphological transitions occur in dependence of time and temperature. In these cases a multidimensional distribution function is required, allowing the description of coalescence processes. The number of unknowns per node and computational resources needed rise further on, calling for optimized efficiency and very large parallel computers.

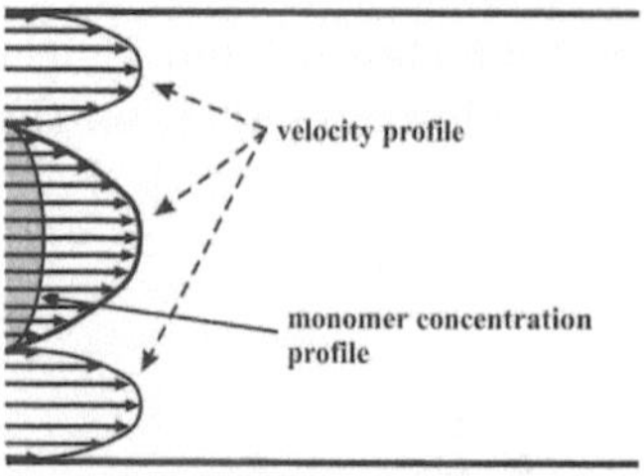

Figure 4: Inflow conditions for a modelled reactor.

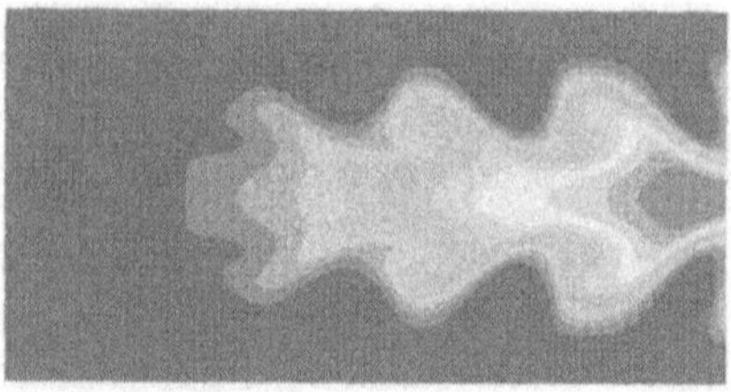

Figure 5: Colors of constant partial particle density of the 10th discrete class near the entry.

Figure 6: Colors of constant total particle density.

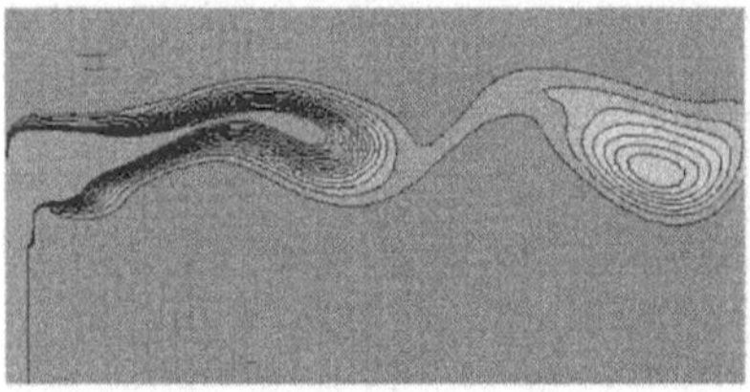

Figure 7: Concentration of primary particles at entry, border of core flow.

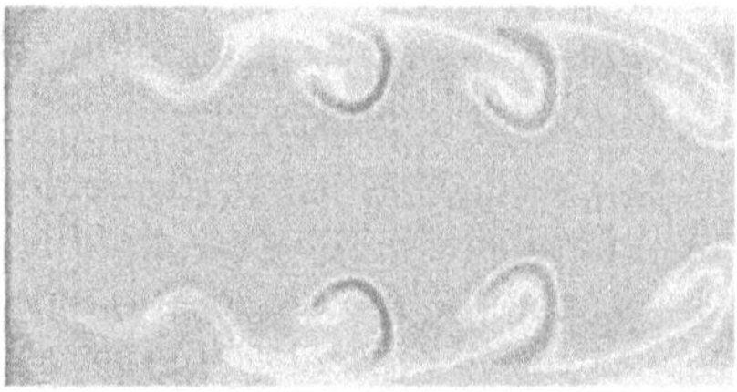

Figure 8: Colors of constant standard deviation of particle sizes.

References

1. Landgrebe, J.D. and Pratsinis, S.E., J. Colloid Interf. Sci., **39**, 63, (1990)
2. Matsoukas, T. und Friedlander, S.K., J. Colloid Interf. Sci., **146**, 495, (1991)
3. O. Gloth and R. Vilsmeier and D.Hänel, Homepage of Project MOUSE, *Online: http://www.vug.uni-duisburg.de/MOUSE* (periodically updated)
4. E. Turkel,*J. Comput. Phys.* **72**, 277-298, (1987)
5. Elmahi, I.,Benkhaldoun, F., Vilsmeier, R., Gloth, O., Patschull A. and Hänel D., J. of Comput. and Appl. Mathematics, **103**, 187, (1999).
6. E. Otto and H. Fissan, Advanced Powder Technol., **10**, (1), 1-20, (1999)

Comparison Between Experimental and Numerical Results of a Mould Filling with Viscous Fluid

Jérôme Breil[1*], Jean Paul Caltagirone[2], Eric Giraud[3], Hideaki Aiso[1], and Tadayasu Takahashi[1]

[1] CFD Technology Center, National Aerospace Laboratory of Japan, (NAL) Jindaiji-Higashi-machi 7-44-1, Chofu, Tokyo 182-8522, Japan
E-mail:breil@nal.go.jp

[2] Université Bordeaux I, Laboratoire MASTER-ENSCPB, Av. pey Berland, BP 108, 33402 Talence Cedex, France

[3] Groupe SNPE, Service Développement Procédés, BP 57, Av. Gay Lussac, 33160 Saint-Médard en Jalles, France

1 Introduction

Dramatically viscous fluids are required to fill missiles, rockets and the boosters of the Ariane V Rocket. Such a rocket engine is called solid propellant engine, since the dynamic viscosity of the casted material increases in the course of time. At the end of the operation, the propellant becomes solid. Since the geometries we deal with are relatively complex, we solve the three-dimensional Navier-Stokes equations together with the transport of the fluid interface. The rheological behavior of the fluid involved, (the propellant, modelled as a fluid) is Newtonian.

A mathematical modeling of mould filling, which has the possibility to handle very viscous fluids, will be presented. The Navier-Stokes equations resolution is ensured by an algorithm composed by an Augmented Lagrangian method [3], and corrected by a vectorial projection method [2] while the interface transport is ensured by following a voluminal function on a fixed grid using the VOF PLIC method (Piecewise Linear Interface Calculation) [1].

Firstly, the method is validated on an academic problem to show good agreement between the computed and existent studies. This will follow by an experimental model filling [5]. We then, compare the obtained experimental data with those predicted by our numerical method on the identical problem. Obviously, the experimental model is chosen to be representative of a typical industrial problem that we have to solve.

2 VOF PLIC Method

As previously explained, the advection of the interface is carried out using the VOF PLIC method [4]. This method can be broken down into three distinct

* Supported by an STA Postdoctoral Fellowship under ID No. 100071.

stages. Initially, we use the volumine fraction θ, present on each node, to reconstruct the interface surfaces, which are lines in two-dimensal cases or planes in three-dimensional cases. In the first part of the reconstruction process, a normal direction to the interface is estimated by a finite difference formula: $\boldsymbol{m} = \nabla\theta$. The determination of the normal direction and the volumine fraction present in the cell gives us the position of the interface. We can then easily apply the speed field previously calculated on the mesh. Using the new position new volume fraction value can be evaluated based on the method of an area calculation. Obviously, in order to determine the new volume fraction, it is necessary to calculate the contribution of the cells upstream and downstream to obtain the final voluminal fraction, as shown in Fig. 1. In three dimensions the procedure pro-

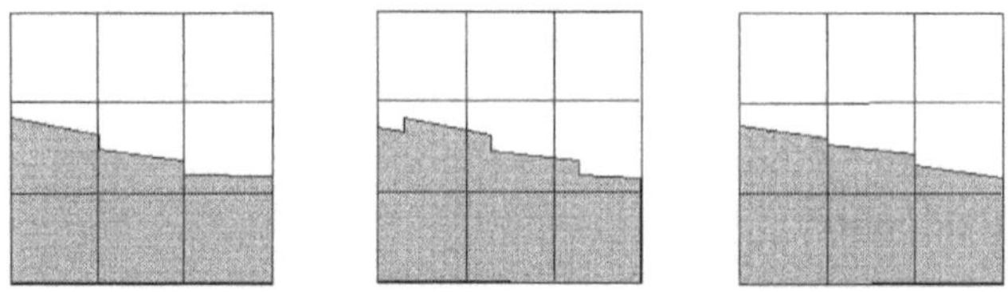

Fig. 1. The three stages of the VOF-PLIC method.

grammed for the advection in the x direction is applied to the y and z direction as well. This algorithm using fractional steps simplifies the programming of the method by reducing the number of possible cases in the final calculation of area. Our computer code is validated using a strong shear speed field. The analytical speed field (Fig. 2b) $\mathbf{V}(x,y) = u(x,y)\mathbf{x} + v(x,y)\mathbf{y}$ is given by :

$$\begin{aligned} u(x,y) &= K\cos(\pi(x-x_0))\,\sin(\pi(y-y_0)) \\ v(x,y) &= -K\sin(\pi(x-x_0))\,\cos(\pi(y-y_0)) \end{aligned} \tag{1}$$

where K is a fraction of π. During this simulation, the initial circle (Fig. 2a) is deformed during N iterations. We then reverse the field speed, and start simulation again for N new iterations. By carrying out N iterations in the opposite direction, we should obtain the initial solution again. At the end of N=1200 iterations, the return of the initial state is carried out satisfactorily (Fig. 2d). For N=4200 iterations, the initial circle is completely deformed in the shape of a long filament(Fig. 2e). When we return to the initial position in this case we obtain a form which approaches a slightly embossed circle (Fig. 2f).

3 Flow Resolution

In the first approximation, the flow is assumed to be incompressible for all fluids considered, and the surface tension is neglected. The Navier-Stokes equations are solved in several stages. In the first stage (Eqn. 2a), we seek to solve a prediction of the field speed $\boldsymbol{V}^*$ starting from the $\boldsymbol{V}^n$ field using the Augmented

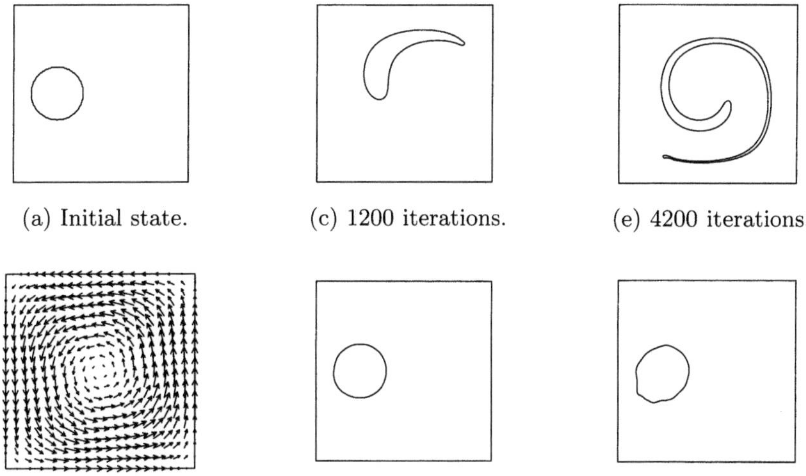

(a) Initial state. (c) 1200 iterations. (e) 4200 iterations.

(b) Speed field. (d) Back to initial. (f) Back to initial.

Fig. 2. Validation of VOF PLIC in a strong shear rate speed field.

Lagrangian method [3]. The parameter dr, introduced by this method, is a coupling constraint term on the velocity field. The pressure field is obtained as usual in this method (Eqn. 2d),

$$\rho\left[\frac{\boldsymbol{V}^*-\boldsymbol{V}^n}{\Delta t}+(\boldsymbol{V}^n.\nabla)\,\boldsymbol{V}^*\right]-dr\nabla(\nabla.\boldsymbol{V}^*)=-\nabla p^n+\boldsymbol{f}+\nabla.\left[\mu(\nabla\boldsymbol{V}^*+\nabla^t\boldsymbol{V}^*)\right] \quad (2a)$$

$$\nabla\left(\nabla.\boldsymbol{V}'\right)=-\nabla\left(\nabla.\boldsymbol{V}^*\right), \quad (2b)$$

$$\boldsymbol{V}^{n+1}=\boldsymbol{V}^*+\boldsymbol{V}', \quad (2c)$$

$$p^{n+1}=p^n-dr\nabla.\boldsymbol{V}^*. \quad (2d)$$

A correction speed $\boldsymbol{V}'$ is then calculated using $\boldsymbol{V}^*$ by a vectorial projection method (2b) [2]. The $\boldsymbol{V}^{n+1}$ fields obtained (2c) is then null divergence, which is of primary importance for calculating free surface flow.

To validate our method, we simulate the instability of a 3D jet. Tomé et al. [5] give a oscillation criterion of a jet running up against a plate based on experimental tests and theoretical study. This criterion indicates that the viscous jet will oscillate if the following are satisfied:

$$Re<0.56, \qquad \frac{H}{D}>3\pi, \quad (3)$$

where $Re=\dfrac{\rho U_0 D}{\mu}$ is the Reynolds number based on the injector diameter D, the injection speed U_0 of the fluid, the fluid density ρ and viscosity μ and the height between the injector and the plate, H. For the first simulation we use $Re=0.72$ and $\frac{H}{D}=8.33$. Figure 3 shows the filling of the cavity. The filling is

t = 0.56 s. t = 2.4 s. t = 4.9 s.

Fig. 3. Non-oscillating jet on a plane.

t = 0.61 s. t = 3.1 s. t = 4.9 s..

Fig. 4. Oscillating jet on a plane.

uniform and no oscillations during the filling were observed. Figure 4 shows the second simulation, and here we use $Re = 0.36$ and $\frac{H}{D} = 10$. In this case, after the impact, the jet become unstable and began to oscillate.

4 Comparison with Experimental Filling

Simulations are now commonly used in the industrial process to improve the quality of castings. The purpose of this section is to present the validation we have done comparing experimental and numerical results for different mould fillings.

4.1 Complex Geometry Mould

The experiment we conducted here has been simulated using the same model geometry and fluid property. We compare the experimental data with those predicted by our numerical method. Obviously, the experimental model is chosen to be representative of a typical industrial problem that we have to solve. Figures 6 and 7 depicted the results for the filling of a mould of complex geometry with one fluid. We can see that the experimental and simulation results are in good agrement.

4.2 Filling of a Mould with Four Fluids

Due to some technical reasons, the filling of big rockets requires successive castings. Because the dynamic viscosity of the material increases with time, the loading, at the end, is composed of a stacking of fluid of different viscosity. Therefore, it is necessary to follow the interfaces between fluids of different viscosity. We validate here a numerical modeling of mould filling which make possible to take into account several fluids. Figures 5a and 5b show the results at the end of the casting. Figure 5c shows the fluid stacking comparison inside the model. They show that the modeling gives a good prediction on the fluid stacking.

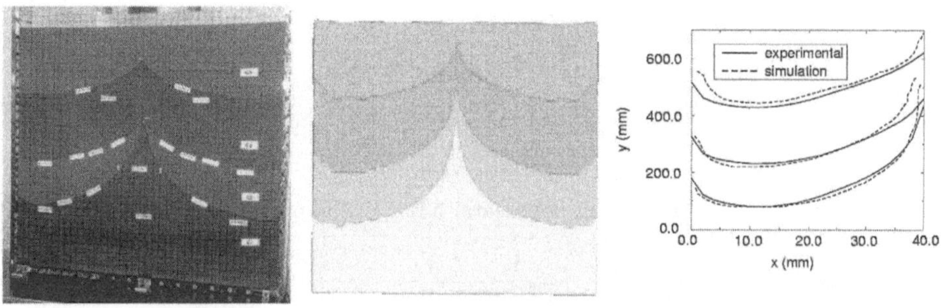

(a) Experimental. (b) Simulation. (c) Interface comparison inside the mould.

Fig. 5. Comparison between experimental result and simulation of the fluids stacking at the end of the casting.

References

1. Breil J. (2001) Modélisation du remplissage en propergol de moteur à propulsion solide *PhD. thesis*, Université de Bordeaux I.
2. Caltagirone J. P., Breil J. (1999) Sur un méthode de projection vectorielle pour la résolution des équations de Navier-Stokes. *C. R. Acad. Sci. Paris*, t. 327, Série II b, p. 1179-1184.
3. Fortin M., Glowinski R. (1982) Méthodes de Lagrangien Augmenté; Application à la résolution numérique de problèmes aux limites, *Méthodes Mathématiques de l'Informatique*, Dunod, Paris.
4. Zaleski S., Li J., Succi S., Scardovelli R. and Zanetti G. (1995) Direct Numerical Simulation of Flows with Interfaces. *Proceedings of the 2nd International Conference on Multiphase Flow*, Kyoto, Apr. 3-7, 1995, v. 2, PT2-1-PT2-12.
5. Tomé M.F., Duffy B. and McKee S. (1996) A numerical technique for solving unsteady non-newtonian free surface flows. *J. Non-Newtonian Fluid Mech*, vol.62, pages 9-34, 1996.

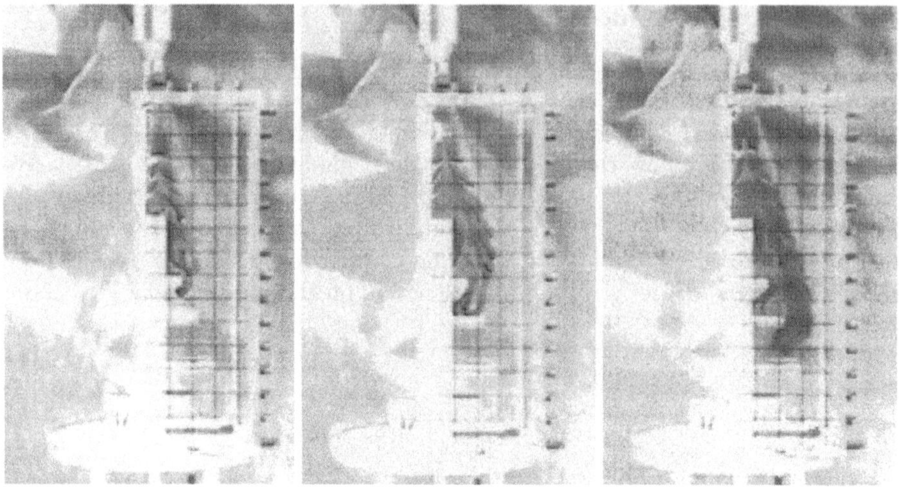

t = 8 s. t = 14 s. t = 18 s..

Fig. 6. Experimental filling of the mould.

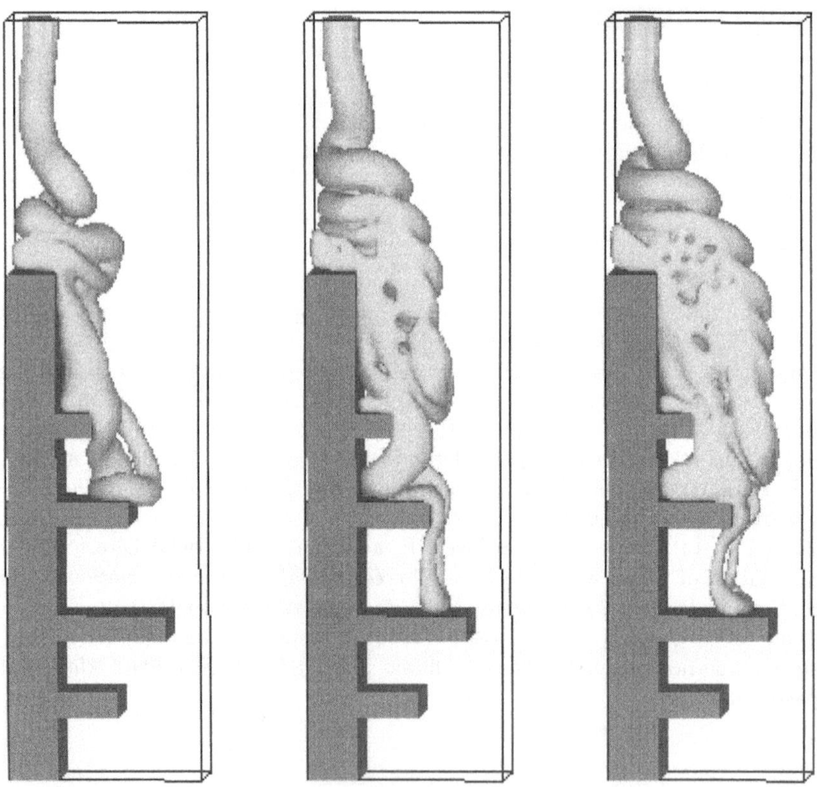

t = 8.1 s. t = 15 s. t = 18.6 s.

Fig. 7. Numerical simulation of the mould filling.

On the Computation of Multi-Material Flows Using ALE Formulation

Hong Luo[1], Joseph D. Baum[1], and Rainald Löhner[2]

[1] SAIC, 1710 SAIC Dr., MS 2-6-9, McLean VA 22102, USA
[2] George Mason University, Fairfax VA 22030, USA

1 Introduction

Development of numerically accurate and computationally efficient algorithms for multi-material flow simulations remains one of the unresolved issues in computational fluid dynamics. The most remarkable feature of such flow problems is the existence of material interfaces. Two general options exist for solving the interface problems: interface tracking and interface capturing. The former explicitly tracks the interface, and treats it like a boundary. An arbitrary Lagrangian-Eulerian formulation is required in this context, since the interface boundaries are moving and deforming continuously. The compelling feature of this approach is its clean definition of the material interface, which ensures an unambiguous separation of different phases during the calculation. The main drawback of the interface tracking approach is its difficulty to handle very complex free surface problems, especially those involving interface topological changes. In contrast, the interface capturing does not require the explicit calculation of the free surface locations. The free surfaces appear as slightly "smeared" interfaces, similar to the way shock waves are "captured" by many Eulerian, compressible flow codes. This makes interface capturing especially attractive and promising to solve very complex free surface problems involving interface topological changes. Unfortunately, maintaining pressure equilibrium among fluid composition across material interfaces still remains a problem to be solved, especially when the equations of state for different materials are drastically different. In this paper, the computation of compressible multi-material flow problems is discussed using the interface tracking approach. Both HLLC[1] and approximate Riemann Solver of Collela[2] are presented and implemented in the context of ALE formulation for solving the unsteady compressible Euler equations. The developed methods have been used for computing a variety of multi-material interface problems. Numerical results indicate that both ALE HLLC and Godunov schemes demonstrate their simplicity and robustness for solving such multi-phase flow problems.

2 Governing Equations

The unsteady compressible Euler equations for a moving control volume can be expressed in integral form as

$$\frac{\partial}{\partial t}\int_{\Omega(t)} \mathbf{U} d\Omega + \int_{\Gamma(t)} \mathbf{F} d\Gamma = 0, \tag{2.1}$$

where $\Omega(t)$ is the moving control volume, and $\Gamma(t)$ its boundary. The flow variable vector $\mathbf{U}$ and the inviscid flux vector $\mathbf{F}$ are defined by

$$\mathbf{U} = \begin{pmatrix} \rho \\ \rho\mathbf{v} \\ \rho E \end{pmatrix}, \mathbf{F} = \begin{pmatrix} (\mathbf{v}-\dot{\mathbf{x}})\cdot\mathbf{n} & \rho \\ (\mathbf{v}-\dot{\mathbf{x}})\cdot\mathbf{n} & \rho\mathbf{v}+p\mathbf{n} \\ (\mathbf{v}-\dot{\mathbf{x}})\cdot\mathbf{n} & \rho E+p\mathbf{v}\cdot\mathbf{n} \end{pmatrix}, \tag{2.2}$$

where ρ, p, and E are the density, pressure, and specific total energy of the fluid, respectively, and $\mathbf{v}$ is the velocity vector of the flow. $\mathbf{n}$ denotes the unit outward normal vector to the moving boundary $\Gamma(t)$, whose velocity is denoted by $\dot{\mathbf{x}}$. In the following, $\mathbf{u}$ is used to denote the fluid velocity relative to the interface velocity, i.e., $\mathbf{u} = \mathbf{v} - \dot{\mathbf{x}}$. This set of equations is completed by the addition of an equation of state, which is taken to be of the form $p = p(\rho, e)$, where the specific internal energy, e, is related to the specific total energy by $E = e + \mid \mathbf{v} \mid^2 /2$.

When grids are moving and/or deforming, the conservation property may be violated unless the following geometric conservation law (GCL) is satisfied

$$\frac{\partial \Omega}{\partial t} - \int_{\Gamma(t)} \dot{\mathbf{x}} \cdot \mathbf{n} d\Gamma = 0. \tag{2.3}$$

The GCL can be easily derived by assuming uniform velocity and density fields in the continuity equation. Clearly, GCL will ensure free stream capturing with moving and/or deforming grids.

3 Numerical Methods

Over the last two decades characteristic-based upwind methods have established themselves as the methods of choice for prescribing the numerical fluxes for compressible Euler equations. The two methods considered here, HLLC[1], and approximate Godunov[2] share the same feature that they all can easily and simply accommodate general equation of state: a necessary requirement for the problems to be solved.

3.1 HLLC scheme on moving grids

The particular version of the HLLC flux[1] used in this work is defined by

$$\mathbf{F}_{ij}^{\mathrm{HLLC}} = \begin{cases} \mathbf{F}_i, & \text{if } S_i > 0 \\ \mathbf{F}(\mathbf{U}_i^*), & \text{if } S_i \le 0 < S_M \\ \mathbf{F}(\mathbf{U}_j^*), & \text{if } S_M \le 0 \le S_j \\ \mathbf{F}_j, & \text{if } S_j < 0 \end{cases} \tag{3.1}$$

where

$$\mathbf{U}_i^* = \begin{pmatrix} \rho_i^* \\ (\rho\mathbf{v})_i^* \\ (\rho E)_i^* \end{pmatrix} = \frac{1}{S_i - S_M} \begin{pmatrix} (S_i - u_{n_i}) & \rho_i \\ (S_i - u_{n_i}) & (\rho\mathbf{v})_i + (p^* - p_i)\mathbf{n} \\ (S_i - u_{n_i}) & (\rho E)_i - p_i u_{n_i} + p^* S_M \end{pmatrix}, \tag{3.2}$$

$$\mathbf{U}_j^* = \begin{pmatrix} \rho_j^* \\ (\rho\mathbf{v})_j^* \\ (\rho E)_j^* \end{pmatrix} = \frac{1}{S_j - S_M} \begin{pmatrix} (S_j - u_{n_j}) & \rho_j \\ (S_j - u_{n_j}) & (\rho\mathbf{v})_j + (p^* - p_j)\mathbf{n} \\ (S_j - u_{n_j}) & (\rho E)_j - p_j u_{n_j} + p^* S_M \end{pmatrix}, \tag{3.3}$$

$$\mathbf{F}_i^* \equiv \mathbf{F}(\mathbf{U}_i^*) = \begin{pmatrix} S_M & \rho_i^* \\ S_M & (\rho\mathbf{v})_i^* + p^*\mathbf{n} \\ S_M & (\rho E)_i^* + (S_m + \dot{\mathbf{x}} \cdot \mathbf{n})p^* \end{pmatrix}, \tag{3.4}$$

$$\mathbf{F}_j^* \equiv \mathbf{F}(\mathbf{U}_j^*) = \begin{pmatrix} S_M & \rho_j^* \\ S_M & (\rho\mathbf{v})_j^* + p^*\mathbf{n} \\ S_M & (\rho E)_j^* + (S_m + \dot{\mathbf{x}} \cdot \mathbf{n})p^* \end{pmatrix}, \tag{3.5}$$

$$p^* = \rho_i(u_{n_i} - S_i)(u_{n_i} - S_M) + p_i = \rho_j(u_{n_j} - S_j)(u_{n_j} - S_M) + p_j, \tag{3.6}$$

and $u_n = \mathbf{u} \cdot \mathbf{n}$. S_M is defined as

$$S_M = \frac{\rho_j u_{n_j}(S_j - u_{n_j}) - \rho_i u_{n_i}(S_i - u_{n_i}) + p_i - p_j}{\rho_j(S_j - u_{n_j}) - \rho_i(S_i - u_{n_i})}. \tag{3.7}$$

Signal velocities S_i, and S_j are defined as $S_i = \min[u_{n_i} - c_i, (\hat{\mathbf{v}} - \dot{\mathbf{x}}) \cdot \mathbf{n} - \hat{c}], S_j = \max[u_{n_j} + c_j, (\hat{\mathbf{v}} - \dot{\mathbf{x}}) \cdot \mathbf{n} + \hat{c}]$ with $\hat{\mathbf{v}}$ and $\hat{c}$ being Roe's average variables for velocity and speed of sound.

3.2 Approximate Godunov scheme on moving grids

The approximate Riemann problem solver for a general equation of state, developed by Colella[2], starts with the calculation of (p^*, u_n^*) by linearization:

$$p^* = \frac{W_j p_i + W_i p_j + W_i W_j (u_{ni} - u_{nj})}{W_i + W_j}, u_n^* = \frac{W_i u_{ni} + W_j u_{nj} + p_i - p_j}{W_i + W_j}, \tag{3.8}$$

where $W = \rho c$. The remainder of the $*$-state calculation proceeds as follows:

$$\rho_{i,j}^* = \rho_{i,j} + \frac{p^* - p_{i,j}}{c_{i,j}^2}, \tag{3.9}$$

$$(c_{i,j}^*)^2 = \frac{p^* \rho_{i,j} c_{i,j}^2}{p_{i,j} \rho_{i,j}^*}, \tag{3.10}$$

$$(\rho e)_{i,j}^* = (\rho e)_{i,j} + (p^* - p_{i,j})(\frac{e + p/\rho}{c^2})_{i,j}, \tag{3.11}$$

$$u_{t\,i,j}^* = u_{t\,i,j}, \tag{3.12}$$

where $u_t = \mathbf{u} \cdot \mathbf{t}$ is the tangential velocity with $\mathbf{t}$ being the unit tangential vector. The (approximate) Godunov state $\mathbf{Q}_G$ is defined to be the value of $\mathbf{Q} = (\rho, \mathbf{u}, \rho e, p)^t$ along the ray $x/t = 0$ and is computed in two steps. First, we set

$$\mathbf{Q}, \mathbf{Q}^* = \begin{cases} \mathbf{Q}_i, \mathbf{Q}_i^*, & \text{if } u_n^* > 0; \\ \mathbf{Q}_j, \mathbf{Q}_j^*, & \text{otherwise,} \end{cases} \tag{3.13}$$

and $\lambda, \lambda^* = c - su_n, c^* - su_n^*$, where $s = sign(u_n^*)$. Also, an approximate shock speed $\sigma = (\lambda + \lambda^*)/2$ is computed. Second, if $\lambda^* > \lambda$, then

$$\mathbf{Q}_G = \begin{cases} \mathbf{Q}, & \text{if } \sigma < 0; \\ \mathbf{Q}^*, & \text{otherwise}; \end{cases} \tag{3.14}$$

otherwise, if $\lambda^* < \lambda$, then

$$\mathbf{Q}_G = \begin{cases} \mathbf{Q}, & \text{if } \lambda < 0; \\ \mathbf{Q}^*, & \text{if } \lambda^* > 0; \\ \alpha\mathbf{Q}^* + (1-\alpha)\mathbf{Q}, & \text{otherwise}. \end{cases} \tag{3.15}$$

Here, $\alpha = \frac{1}{2}(1 + \frac{\lambda+\lambda^*}{\lambda-\lambda^*})$. The conservative variable vector at the interface is then computed from the Godunov variable vector as

$$\mathbf{U}_{ij} = \begin{pmatrix} \rho_{ij} \\ (\rho\mathbf{v})_{ij} \\ (\rho E)_{ij} \end{pmatrix} = \begin{pmatrix} \rho_G \\ \rho_G(\mathbf{u}_G + \dot{\mathbf{x}}) \\ (\rho e)_G + 0.5\rho_G \mid \mathbf{u}_G + \dot{\mathbf{x}} \mid^2 \end{pmatrix}. \tag{3.16}$$

Finally, the fluxes for flow equations at the interface can be evaluated from the Godunov state as follows:

$$\mathbf{F}_{ij} = \begin{pmatrix} \mathbf{u}_G & \rho_{ij} \\ \mathbf{u}_G & (\rho\mathbf{v})_{ij} + p_G\mathbf{n} \\ \mathbf{u}_G & ((\rho E)_{ij} + p_G) + p_G\dot{\mathbf{x}}\cdot\mathbf{n} \end{pmatrix}. \tag{3.17}$$

It is worth noting: (1) At the Lagrangian contact where $p_i = p_j, \mathbf{u} = \mathbf{0}$, the dissipation vanishes for both schemes. This fact easily explains why the contact discontinuities can be sharply resolved by the moving grid methods; (2) In the case of uniform free stream flow conditions, both schemes lead to the geometric conservation law.

Numerical Example As an illustrative example, the developed two schemes were used to compute a benchmark 1D spherically symmetric underwater explosion problem. This represents a serious challenge to any multi-material computation methods, as the materials are drastically different, and the EOS for water is very stiff. Any oscillations in the material interface will lead to invalid EOS (negative pressure), resulting in either collapse or meaningless of calculation. The computational conditions are: detonation of 300 g of TNT spherical charge (radius of 3.5287 cm) at a depth of 91.4 m (ambient pressure of 1.0E+07 d/cm^2). The problem can be essentially regarded as a spherical two-phase shocktube problem where the initial conditions in cgs units are given by $\rho = 1.63000000, u = 0, e = 4.29E+10$, for $0.0 \leq x \leq 3.5287$; $\rho = 1.00037984, u = 0, p = 1.00E+07$, for $3.5287 < x \leq 25$. The Jones-Wilkins-Lee (JWL) and Tait equations of state are used for the detonation-products gas and liquid water, respectively:

$$\text{Tait}: \qquad p = B[(\frac{\rho}{\rho_0})^\gamma - 1] + A,$$

where $B = 3.31E + 09, \rho_0 = 1.0, A = 1.0E + 06, \gamma = 7.15$

$$\text{JWL}: \qquad p = A(1 - \frac{\omega\rho}{R_1\rho_0})e^{-\frac{R_1\rho_0}{\rho}} + B(1 - \frac{\omega\rho}{R_2\rho_0})e^{-\frac{R_2\rho_0}{\rho}} + \omega\rho e,$$

where $A = 7.712E + 12, B = 0.0323E + 12R_1 = 4.15, R_2 = 0.95, \omega = 0.3, \rho_0 = 1.63, e_0 = 0.0429E + 12$. Note the speed of sound for Tait EOS is computed as $c^2 = \gamma\frac{p+A}{\rho}$. The computation was performed on a series of 4 grids in order to get a mesh converged solution. These grids contain 751, 1501, 3001, and 6001 uniform placed points, respectively. It is found that the mesh, that features 3001 points with 424 points in the bubble and 2577 grid points in the water, is adequate to provide a grid converged solution. A Lagrangian interface is established at the material interface by constraining the grid velocity to match the fluid velocity at the interface. The computation was performed using both HLLC and Godunov, respectively. It is found that both HLLC and Godunov methods produced the virtually identical results. Unless stated otherwise, the numerical results to be presented here are obtained using ALE Godunov scheme on the 3001 grid-points mesh. Figures 1-2 show the evolution of the computed density and pressure through different phases of spherically symmetric underwater explosion problem: the initial phase, the shock and free-surface interaction phase, incompressible phase, and bubble collapse phase. The bubble radius and bubble-water interface pressure time history for all four grids are compared in Fig. 3. The grid density has significant impact on the bubble period, however the maximum bubble radius differs very little among those four grids. Table 1 shows a detailed comparison among the four grids and the experimental data of Swift and Decius[3]. The difference of the solutions between the last two grids is insignificant, indicating a converged solution can be considered as a grid converged solution on the third grid, which has 3001 grid points with 424 grid points in the bubble.

4 Conclusion

HLLC and Godunov schemes have been presented and implemented in the context of ALE formulation. The developed methods have been used for computing a variety of multi-material interface problems. Computational results show that both ALE HLLC and Godunov schemes are able to offer accurate and robust solutions for capturing strong shock, contact, and phase discontinuities on arbitrarily moving grids.

References

1. P. Batten, M. A. Leschziner, and U. C. Goldberg, J. Comp. Phy., 137, (1997).
2. P. Colella and H. M. Glaz, *Multifluid Algorithm for Eulerian Finite Difference Methods*, Private Communication, (1992).
3. E. Swift and J. Decius, *Measurement of Bubble Pulse Phenomena*, Navord Report, (1946).

a

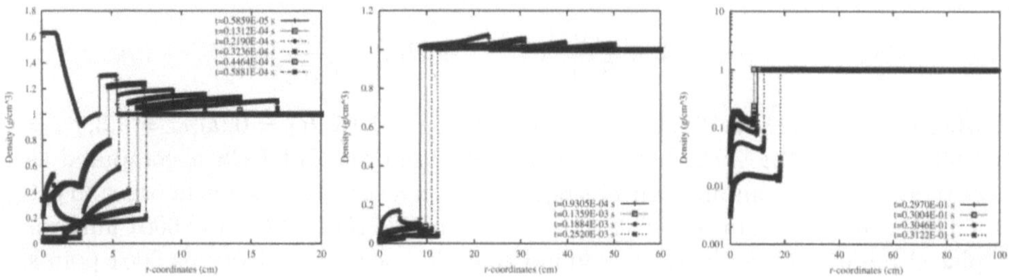

b

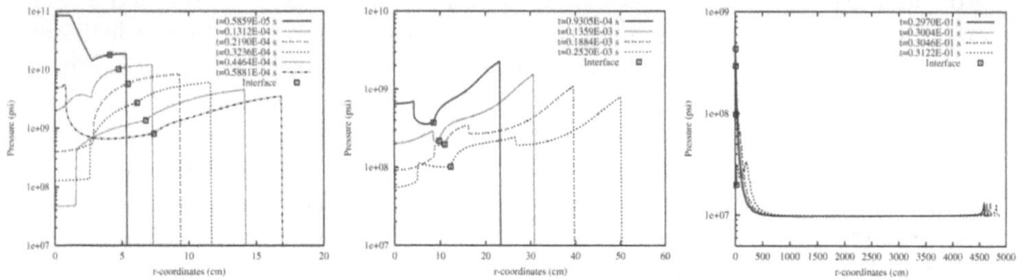

Fig. 1. Computed **(a)** density and **(b)** pressure distributions for different phases of a spherically symmetric underwater explosion problem: initial phase, shock/material-surface interaction phase, and bubble collapse phase

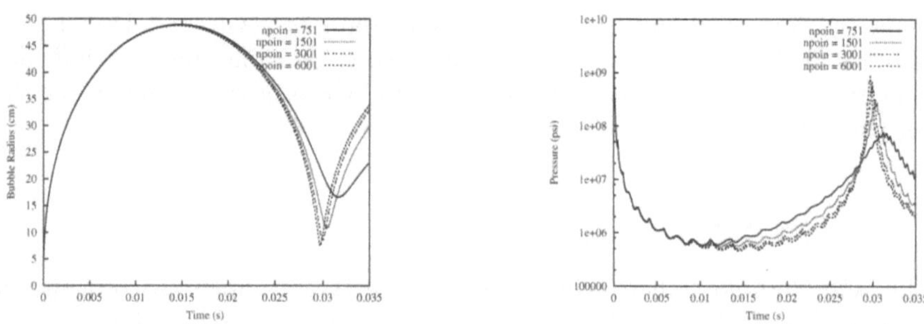

Fig. 2. Comparison of bubble radius time and interface pressure history for different mesh size solutions of the spherically symmetric underwater explosion problem.

Table 1. Computation and experiment for underwater explosion bubble

Grid Convergence Study				
Grid Size	Period (ms)	Relative Error	R_{max} (cm)	Relative Error
751 (106)	31.65596	6.2%	48.92498	5.45%
1501 (212)	30.39513	2.0%	48.88203	5.35%
3001 (424)	29.94145	0.47%	48.83771	5.25%
6001 (847)	29.70043	-0.33%	48.75348	5.07%
Experiment	29.8	0%	46.4	0%

Part XIV

Optimisation

Turbomachinery Airfoil Design Optimization Using Differential Evolution

Nateri K. Madavan

NASA Advanced Supercomputing Division,
NASA Ames Research Center, Moffett Field, CA 94035-1000, USA

Abstract. An aerodynamic design optimization method that is based on an evolutionary algorithm known at Differential Evolution (DE) is described. DE is a simple, fast, and robust evolutionary strategy that has been proven effective in determining the global optimum for several difficult optimization problems. A Navier-Stokes solver is used to evaluate the various intermediate designs and provide inputs to the DE optimizer. The method is implemented on distributed parallel computers so that new designs can be obtained within reasonable turnaround times. A hybrid version of the method that uses the DE algorithm in conjunction with a neural network is also developed and shown to reduce overall computing time requirements. Results are presented for the inverse design of a turbine airfoil from a modern jet engine. The capability of the method to search large design spaces and obtain the optimal airfoils in an automatic fashion is demonstrated.

1 Introduction

Remarkable progress has been made in recent years in the ability to design turbomachinery airfoil shapes that are optimal with regard to certain desired characteristics. This progress has been achieved by combining improved methods for predicting the complicated flow fields in turbomachinery with efficient numerical optimization techniques and by harnessing the powerful capabilities of modern computers. Both steady and unsteady Euler and Navier-Stokes solvers have been combined with various gradient-based and other traditional optimization techniques to obtain optimal airfoil designs.

More recently, there has been considerable interest in the development of turbomachinery airfoil design optimization techniques that are based on non-traditional approaches such as evolutionary algorithms and neural networks. Various approaches based on neural networks [1], [2]), neural networks in conjunction with response surfaces [3], [4], genetic algorithms [5], [6], and genetic algorithms in conjunction with neural networks [7],[8], among others, have been reported in the literature. These techniques offer several advantages over traditional optimization methods.

This paper deals with the development of a turbomachinery airfoil design optimization procedure that is based on the Differential Evolution (DE) algorithm [9]. DE is conceptually simple and possesses good convergence properties that have been demonstrated in a variety of applications [10]. Its application in aeronautics, however, has been rather limited. It has been used in the predictive control of aircraft dynamics [11]. It has been used in conjunction with a potential flow solver in the inverse design of turbomachinery airfoils [12]; the same authors have also presented a hybridized version [13] that combines DE with a local search method to minimize the number of objective function evaluations using the potential flow solver.

The turbomachinery airfoil design optimization method described in this paper uses the DE algorithm in conjunction with a Navier-Stokes solver that performs the objective function evaluations. The method is implemented on a distributed parallel computer in a straightforward manner. A hybrid DE-NN method that uses a neural network that is trained on the computational fluid dynamics (CFD) simulation data and acts as a surrogate for the Navier-Stokes solver in the objective function evaluations is also described.

2 Design Optimization Method Features

Differential Evolution DE uses a population of real-valued parameter vectors of design variables that is usually initialized in a random fashion. The population size is maintained constant throughout the optimization. The key ingredient of DE is mutation. New parameter vectors for the subsequent generation are formed using weighted differences between two (or more) parameter vectors selected randomly from the current population to provide appropriately scaled perturbations that modify another parameter vector (or, comparison vector) selected from the same population. This can be implemented in various forms [9]. A discrete recombination strategy is also used in addition to the mutation operator. The selection scheme is deterministic and based on local competition only, with the child trial vector competing against one population member and the survivor entering the new population.

Constraint Handling An efficient constraint handling mechanism has been incorporated into the algorithm. It is a parameter-less approach that helps steer the algorithm away from infeasible regions of the design space. Physical constraints (e.g. maximum airfoil thickness) are imposed, as well as aerodynamic constraints (e.g. wavy surfaces). Airfoil geometries that do not violate the constraints but for which the CFD solver fails to converge are also deemed infeasible and treated accordingly by the constraint handling mechanism.

Airfoil Geometry Parametrization Geometry parametrization and prudent selection of design variables are critical aspects of any shape optimization procedure. Here, the airfoil geometry parametrization method described in [3] that uses a total of 13 parameters to define the turbine airfoil geometry is used. Different airfoil geometries are obtained by varying these parameters.

CFD Solver A two-dimensional Navier-Stokes solver is used to perform the flow simulations (objective function evaluations) that serve as inputs to the optimization process.

Parallel Implementation Details The results in this article were obtained on the SGI Origin 3000 and the Cray SV1 distributed parallel computers at NASA Ames Research Center. Parallel implementation of the method is quite straightforward and relies on the simultaneous computation of multiple, independent aerodynamic simulations on separate processors. A script-based procedure is used that invokes a variable number of processors depending on processor availability and the size of the population used. The number of processors can also be adjusted as the design proceeds. The current setup is based on a "master-slave" arrangement, with the master handling the tasks of setting up the simulations, neural network training for the hybrid method, and farming out of the aerodynamic computations to the other "slave" processors. Since the aerodynamic computations are independent of each other, no communication between the processors is required until the computations are completed. The slave processors then communicate their results to the master which then performs the necessary calculations to determine the members of the next population.

Hybrid Differential Evolution–Neural Network (DE-NN) Approach While the DE algorithm is quite efficient and effective in exploring the entire design space in search of the optimal solution, the algorithm has a tendency to slow down as it approaches the vicinity of the optimal solution. Many subsequent function evaluations are then required to obtain the exact optimum. The computational effort required by these function evaluations in the vicinity of the optimum can be nearly eliminated using a hybrid approach that combines the DE algorithm with a neural network that is trained on the CFD simulation data. The trained neural network is then used to evaluate the objective function with negligible computational effort and expense instead of using the CFD solver. While hybridization is always useful, it must be done with caution. Here the neural network is used only in the latter stages after the entire population has evolved to the general vicinity of the optimal solution. Thus the neural network is used as a "local" response surface with validity only in a small region of the design space. This makes it easier to train the neural network and improves its generalization abilities. For the inverse design optimization problem, the neural network is trained on the sum-of-squares error between the actual pressure computed by the CFD solver and the target pressure at various points on the airfoil.

3 Results

The design method was used in the inverse design of a turbine airfoil with a specified pressure distribution. The target pressure distribution was obtained at the midspan of a turbine vane from a modern Pratt and Whitney jet engine. Several flow and geometry parameters were also supplied and used in the design process. The design objective function was formulated as the equally-weighted sum-of-squares error between the target and actual pressure obtained during the optimization process at 45 locations on the airfoil.

The initial design space was chosen to be quite large to allow a wide range of airfoil shapes to be explored. A sampling of some of the initial airfoil geometries is shown in Fig. 1. In order to hold the CFD function evaluations to a reasonable number, 6 (instead of 13) design variables were used in the initial stages of the design. The population of 50 members were then evolved using the DE algorithm.

Figure 2 shows the pressure distribution for the optimal airfoil that represents the best airfoil obtained after 13 generations using the DE optimization method with 6 design parameters. The algorithm is able to approach the target distribution within about 650 function evaluations. Note that in the early stages of evolution several of the airfoil geometries were infeasible and hence were not evaluated by the CFD solver. After 13 generations, the number of design variables is increased to 13 and the population evolved further. The optimal pressure distribution shown in Fig. 2 was obtained after an additional 13 generations and agrees well with the target distribution. The airfoil geometries corresponding to the optimal designs with both 6 and 13 design parameters are shown in Fig. 3.

The convergence history of the algorithm is shown in Fig. 4 which plots the population mean and variance as a function of the number of generations. Both the mean and the variance decrease with generation number except for the slight increase corresponding to the switch from 6 to 13 design parameters. The figure also shows that convergence of the algorithm is slower in the vicinity of the optimal solution. This is typical of many other evolutionary algorithms and highlights the need for a hybrid approach that combines the global search and exploration capabilities of the DE algorithm with other algorithms that can converge rapidly to an optimum when initialized in the local neighborhood. As described earlier, a different hybrid approach to reducing overall design time was adopted here in the DE-NN method where the populations after the first few generations (using 13 design parameters) were used to train a neural network. Figure 5 shows the envelope of pressure data around the target pressure distribution that was used to train the neural network. Note, however, that the network was trained directly on the sum-square-error and not on the individual pressure data. The data envelope in Fig. 5 represents the variation of the pressure distributions for the range of airfoil geometries in the neural network training set and is meant to convey the local nature of the neural network response surface in the vicinity of the optimal solution. The pressure distribution for the optimal airfoil design obtained by the DE-NN approach is shown in Fig. 6 and compares well with the target and optimal DE design pressure distributions.

4 Summary and Conclusions

An aerodynamic design optimization method that is based on Differential Evolution is described. The method is combined with a Navier-Stokes solver that evaluates the various intermediate designs and provides inputs to the optimization procedure. Results are presented for the inverse design of a turbine airfoil from a modern jet engine. The method can search large design spaces and obtain the optimal airfoils with a reasonable number of CFD function evaluations in an automatic fashion. The airfoil geometry parametrization and constraint handling procedure help limit the number of CFD function evaluations required by disallowing airfoil geometries that are infeasible from a physical or aerodynamic standpoint. Substantial reductions in the overall computing time requirements are achieved by the hybrid DE-NN algorithm that uses a neural network trained on the CFD data to evaluate the objective function in the latter stages of the design evolution. Parallel implementation of the method also helps reduce design turnaround time.

It is noted that the inverse design problem was chosen simply to demonstrate the general capabilities of the method. In fact, this problem is quite straightforward and can be solved more efficiently (with far fewer function evaluations) using other optimization techniques, e.g. [1], [3]. Applications of the method to more appropriate aerodynamic problems involving multimodal landscapes, multiobjective or multipoint design, etc. are currently being investigated.

References

1. M. M. Rai: 'A Rapid Aerodynamic Design Procedure Based on Artificial Neural Networks'. *AIAA Paper No. 2001-0315, Jan. 2001*
2. S. Pierret, R. A. V. Braembussche: Journal of Turbomachinery, **121**, 326 (1999)
3. M. M. Rai, N. K. Madavan: AIAA Journal, **38**, 173 (2000)
4. N. Papila, W. Shyy, L. Griffin, D. J. Dorney: 'Shape optimization of supersonic turbines using response surface and neural network methods'. *AIAA Paper No. 2001-1065, Jan. 2001*
5. S. Obayashi, S. Takanashi: AIAA Journal, **34**, 881 (1996)
6. B. H. Dennis, I. N. Egorov, Z. X. Han, G. S. Dulikravich, C. Poloni: 'Multi-Objective Optimization of Turbomachinery Cascades for Minimum Loss, Maximum Loading, and Maximum Gap-to-Chord Ratio'. *AIAA Paper No. 2000-4876, Sept. 2000*
7. M. Uelschen, M. Lawerenz: 'Design of Axial Compressor Airfoils with Artificial Neural Networks and Genetic Algorithms'. *AIAA Paper No. 2000-2546, June 2000*
8. C. Poloni, A. Giurgevich, L. Onesti, V. Pediroda: Comp. Meth. Appl. Mech. Engr., **186**, 403 (2000)
9. R. Storn, K. Price: Dr. Dobb's Journal, **22**, 18 (1997)
10. J. Lampinen: A Bibliography of Differential Evolution Algorithm. Technical Report. Lappeenranta University of Technology, Department of Information Technology, Laboratory of Information Processing, Lappeenranta, Finland (2001)
11. K. Nho, R. K. Agarwal: 'Fuzzy Logic Model-Based Predictive Control of Aircraft Dynamics Using ANFIS'. *AIAA Paper 2001-0316, Jan. 2001*
12. T. Rogalsky. R. W. Derksen, S. Kocabiyik: Can. Aero. Space Inst. Jour., **46**, 183 (2000)
13. T. Rogalsky, R. W. Derksen: 'Hybridization of Differential Evolution for Aerodynamic Design'. *Proceedings 8th Ann. Conf. Comp. Fluid Dyn. Soc. Can., pp. 729–736, June 11–13, 2000*

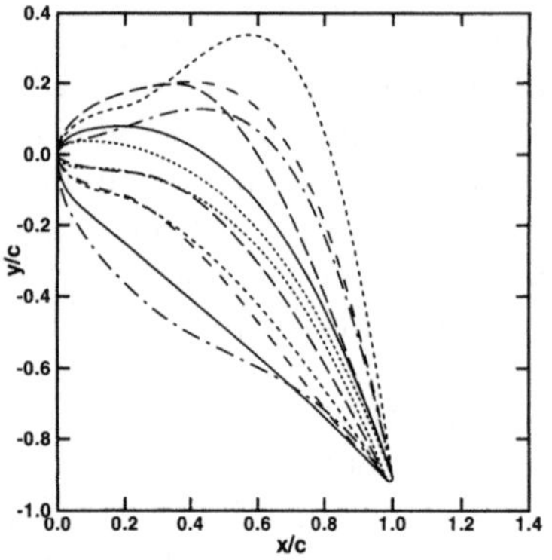

Fig. 1. Sampling of initial airfoil geometries

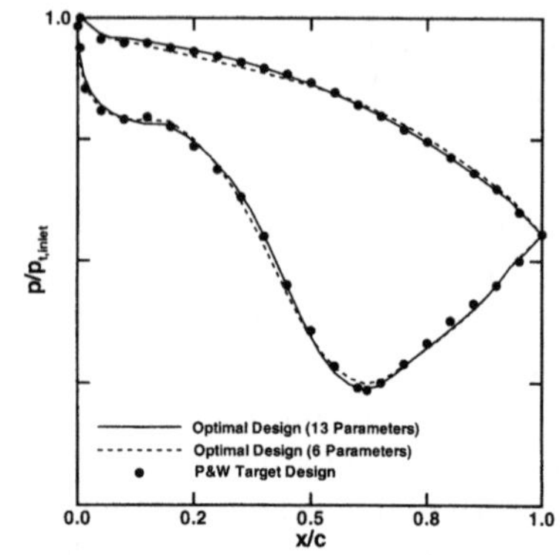

Fig. 2. Airfoil pressure loading for the optimal design

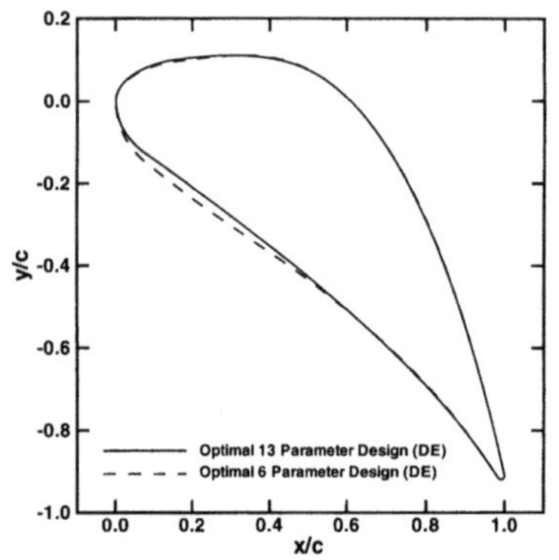

Fig. 3. Airfoil geometry for the optimal design

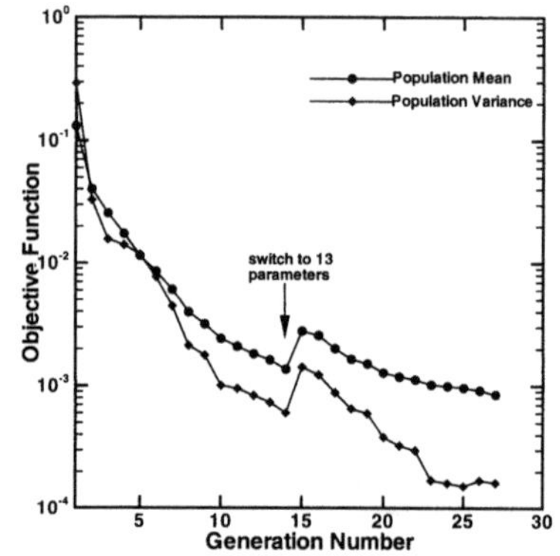

Fig. 4. Convergence history

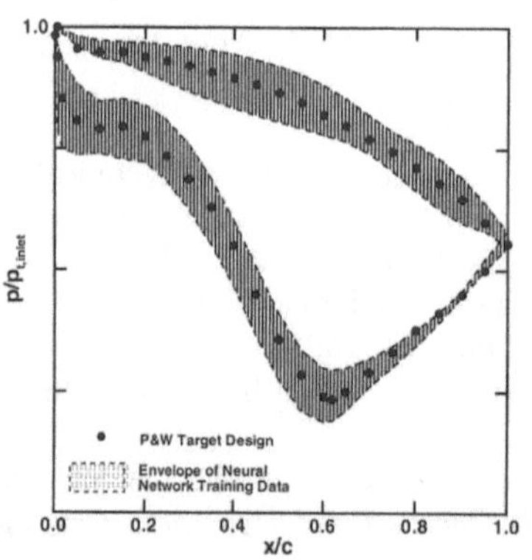

Fig. 5. Envelope of airfoil loadings used to train the neural network for the DE-NN approach

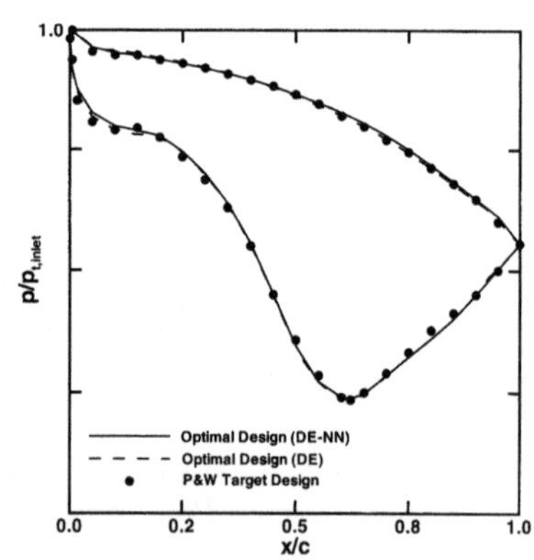

Fig. 6. Airfoil pressure loadings for the optimal design using the DE-NN approach

Design Optimization of Fluid Dynamic Problems Using Cartesian Grids

Andrea Dadone[1] and Bernard Grossman[2]

[1] Politecnico di Bari, Dip. Ing. Meccanica e Gestionale, 70126 Bari, Italy
[2] Virginia Tech, Aerospace & Ocean Eng., Blacksburg, VA 24061, USA

Abstract. An efficient formulation for the robust design optimization of inviscid fluid flow problems is presented. The approach is based upon a new Cartesian grid technique developed by the authors which is suited for design applications. The methodology has a number of essential ingredients which contribute to its efficiency and robustness including: an accurate Cartesian grid flow solver, approximate design sensitivities from a discrete adjoint formulation based on a dissipative flow solver, progressive optimization, whereby a sequence of operations, containing a partially converged flow solution, followed by an adjoint solution followed by an optimization step is performed and the use of progressively finer grids. Test problems for two-dimensional inverse design indicate that the method is accurate, robust and efficient, with converged design optimization results produced in no more than the amount of computational work to perform one flow analysis.

1 Introduction

Fluid dynamic design problems require the calculation of many computationally expensive flow analyses. One impediment to the use of CFD in the conceptual design process is the need to determine detailed body-fitted grids over wide classes of different geometric topologies. The grid generation process is often laborious and is not always as automated as desired. For conceptual design, where inviscid methods may be adequate, procedures involving Cartesian grid methods, particularly in combination with grid adaptation have proved to be effective, *e. g.*, [1].

Recently, the authors have introduced a new Cartesian grid technique called *immersed body methodology* [2] for the inviscid flow about bodies using strictly Cartesian grids with no cut cells. The methodology is based on the utilization of *ghost cells* and highly accurate surface boundary conditions for inviscid flows [3]. The aim of the present paper is to develop a design optimization procedure for fluid dynamic problems by using flow analyses performed on Cartesian grids with the immersed body methodology and by computing the corresponding adjoint equations on the same Cartesian grid.

The robust and efficient design methodology that we utilize was originally developed for body-fitted grids [4] and is adapted here for Cartesian grids. In the following, we will briefly present the immersed body methodology, followed by the progressive optimization procedure. Then the optimization will be applied to airfoil inverse designs in subsonic and transonic flow conditions.

2 Immersed Body Methodology and Flow Solver

In this approach, the body wall geometry is represented pointwise and independent of mesh. A Cartesian mesh is overlayed on the body, and a first row of ghost cell centers closest to the body (interior) are determined. These cell centers are represented by circles in Fig. 1, which presents a part of the Cartesian mesh and the curved body profile. Then, a second row of ghost cell centers located at the interior of the body and adjacent to the first row is singled out (triangles in Fig. 1). Finally, points symmetric to the previous ghost points with respect to the body surface and located inside the flow field (body exterior) are determined (dark squares in Fig. 1). As an example, the point B and the cell center A in Fig. 1 are symmetric with respect to the body. The fluid dynamic properties at these symmetric points are evaluated by a bilinear interpolation involving the four surrounding cell centers.

The fluid dynamic conditions at the ghost cell centers are determined using an assumed inviscid flow model in the vicinity of the wall [3], which is a vortex flow of constant entropy and total enthalpy, which satisfies the normal momentum equation and the non-penetration condition at the wall. The physical variables at the cell center A are given by:

$$p_A = p_B - \varrho_B \frac{\tilde{u}_B^2}{R} \Delta n \,, \qquad \varrho_A = \varrho_B \left(\frac{p_A}{p_B} \right)^{1/\gamma} , \tag{1}$$

$$\tilde{u}_A^2 = \tilde{u}_B^2 + \frac{2\gamma}{\gamma - 1} \left(\frac{p_B}{\varrho_B} - \frac{p_A}{\varrho_A} \right) , \qquad \tilde{v}_A = -\tilde{v}_B \,. \tag{2}$$

In (1), Δn is the distance between points A and B, while R is the signed local radius of curvature of the body.

In order to accurately capture shock waves, a finite-volume, flux-difference-splitting method has been used as the basic flow solver. For the present two-dimensional flow applications, the computational flow field will be divided into

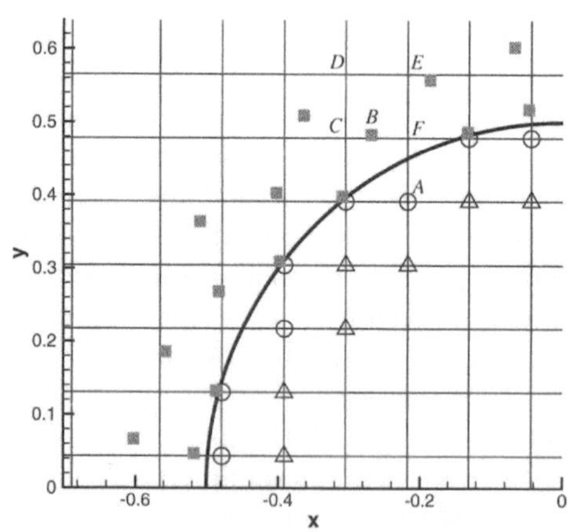

Fig. 1. Cartesian grid: cell center net, cell centers close to the body and symmetric points

an $N \times M$ finite-volume mesh and stagnation flow conditions are enforced at cell centers which are interior to the body and different from the previously outlined ghost cell centers.

3 Inverse Design

Inverse design problems for airfoils generally consist of finding the geometric contours whose pressure distribution along the surface s, matches a target pressure distribution. A discrete objective function, I, may be defined as:

$$I(\xi) = \frac{1}{2\,N_b} \sum_{i=1}^{N_b} [p_i - \hat{p}_i]^2 \,, \tag{3}$$

where the body length has been subdivided into N_b intervals, whose center locations are s_i $(i = 1, \ldots, N_b)$. The discrete computed pressures are $p_i = p(s_i, \xi)$ and the discrete target pressures are $\hat{p}_i = \hat{p}(s_i)$, where ξ represents the design parameters.

We utilize a simplified, dissipative auxiliary flow solver for the computation of smooth sensitivity derivatives, as described in [4]. The sensitivity derivatives appear in the discrete adjoint formulation [4] described below:

$$\frac{\partial I}{\partial \xi_j} = \left(\frac{\partial I}{\partial \xi_j}\right)_{\mathbf{u}} - \mathbf{\Lambda}_k^T \left(\frac{\partial \mathbf{w}_k}{\partial \xi_j}\right)_{\mathbf{u}} \,, \tag{4}$$

where the adjoint variables $\mathbf{\Lambda}_k$ are evaluated according to the following conditions:

$$\frac{\partial \mathbf{\Lambda}_i}{\partial t} + \left(\frac{\partial \mathbf{w}_k}{\partial \mathbf{u}_i}\right)_{\xi}^{T} \mathbf{\Lambda}_k = \left(\frac{\partial I}{\partial \mathbf{u}_i}\right)_{\xi}^{T} \,, \tag{5}$$

for $i = 1, \ldots, N \times M$ and $j = 1, \ldots, n$, with n the number of design parameters. In the previous equations, $\mathbf{u}_i$ corresponds to the vector of the physical variables at each cell center, i, and $\mathbf{w}_k(\mathbf{u}, \xi) = 0$ represents the discretized system of governing partial differential equations corresponding to the auxiliary flow solver.

The terms $(\partial \mathbf{w}_k / \partial \mathbf{u}_i)_{\xi}$ at the cell centers away from the body can be evaluated directly from the system of discretized flow equations and depend on the solution of the flow field. At the cell centers nearest to the body, they are numerically evaluated following the procedure suggested in [5], which determines these values by perturbing u_i and by determining the corresponding variations of the residuals at the cell center and its neighbors. Then $(\partial \mathbf{w}_k / \partial \mathbf{u}_i)_{\xi}$ is approximated by the ratio between the variation of the residuals and the corresponding variation of u_i. In this computation, the corresponding changes of the boundary conditions at ghost cells are also taken into account. The terms $(\partial \mathbf{w}_k / \partial \xi_j)_{\mathbf{u}}$ are numerically evaluated by perturbing the design parameters ξ, by evaluating the corresponding variations of the fluid dynamic properties at the row of cell centers surrounding the body on the interior and by determining the corresponding variations of the residuals at the cell centers surrounding the body and located in the flow field. The terms $(\partial I/\partial \mathbf{u}_i)_\xi$ are analytically evaluated. The terms $(\partial I/\partial \xi_j)_\mathbf{u}$ are evaluated by perturbing the design parameters and by evaluating the corresponding variation of the objective function.

The fluid dynamic variables computed with the accurate, basic flow solver are used to compute the terms $(\partial I/\partial \mathbf{u}_i)_\xi$ and $(\partial \mathbf{w}_k/\partial \mathbf{u}_i)_\xi$. Then, the adjoint equations are solved for the adjoint variables, $\mathbf{\Lambda_k}$. During this computation, the far-field boundaries are systematically moved closer to the body, in order to significantly reduce the CPU time required to evaluate the adjoint variables. In addition, the adjoint variables for cell centers located inside the body are set to zero. Finally, (4) is used to evaluate $\partial I/\partial \xi_j$.

The goal of the progressive convergence approach is to simultaneously determine the global minimum of the objective function, while reaching a converged solution for the flow field and the adjoint equations. The progressive algorithm is as follows:

i. Define the coarsest mesh level and several finer levels. Each finer mesh is obtained from the corresponding coarser mesh by doubling the number of intervals in each direction. Then start with an initial set of design variables and start the flow computations on the coarsest grid.
ii. Advance the flow solver for several iterations. Then evaluate the adjoint equations by advancing the adjoint solver for several iterations, with the condition that the residuals drop at least one order of magnitude with respect to the value it assumes at the beginning of the procedure. Then compute the objective function gradient, ∇I and update the design variables according to

$$\xi_j^{l+1} = \xi_j^l - a_j \left(\partial I/\partial \xi_j\right) , \tag{6}$$

where a_j are positive parameters [4].
iii. Repeat steps 2 until the gradient of the objective function is sufficiently decreased (one half order of magnitude), with the condition that the residuals of the flow and of the adjoint equations drop progressively.
iv. Move to the corresponding finer mesh and interpolate the computed flow solution to the finer grid. Then repeat steps 2 until the gradient of the objective function is decreased one order of magnitude more.
v. Repeat steps 4 until the finest grid is reached. Then repeat steps 2 until the objective function has sufficiently decreased.

In the computed results, we have used three grid levels and we have considered the objective function sufficiently decreased when the norm of its gradient has decreased 2.5 orders of magnitude with respect to the value it assumes after the first global step.

The inverse design of airfoils in transonic flow conditions has been considered with the airfoil surface shape described as:

$$Y(x/c) = \sum_{i=1}^{N_a} \xi_i \, Y_i(x/c) , \tag{7}$$

where $Y = y/c$, with y being the airfoil ordinate and c the airfoil chord length. The shape functions $Y_i(x/c)$ are the ordinates of the N_a specified airfoils, which depend on x/c, and the weighting terms ξ_i are the design parameters.

First, we considered the subsonic flow about an $NACA$ 0012 with $M_\infty = 0.63$ at $\alpha = 2°$. The flow solution on the finest mesh converged to machine zero was used to define the target pressure distribution. This airfoil optimization test was performed with only one design variable. The target airfoil is the $NACA$ 0012 airfoil, corresponding to $\xi = 1$. We assumed an initial airfoil shape of $\xi = 2.5$. The optimization process was performed with three mesh levels made by 32×29, 64×59, 128×119 cells. Figure 2 shows the convergence history of the logarithm of the objective function, $log_{10}(I)$, and of $log_{10}|\nabla I|$. The abscissa in Fig. 2 is the required work, with a unit of work considered as the computational time required to run a single analysis solution to convergence on the finest mesh. We see that $log_{10}(I)$ and $log_{10}|\nabla I|$ converged to machine zero and the design parameter reached its target value, equal to 1, with a residual error of 10^{-6}.

Then, we considered the transonic flow about an $NACA$ 0012 with $M_\infty = 0.80$ at $\alpha = 1.25°$. The airfoil optimization test was performed with $N_a = 4$ and Y_1–Y_4 as the shape of four specified airfoils, $NACA$ 0012, $NACA$ $64_1 - 412$, $NACA$ 64_2A215, $NACA$ $65_2 - 415$. The target airfoil is the $NACA$ 0012 airfoil so that the target set of coefficients is $\xi = (1, 0, 0, 0)$. We assumed an initial airfoil shape corresponding to the set of coefficients $\xi = (1.5, 0.5, 0.5, 0.5)$. Figure 3 shows the convergence history of the $log_{10}(I)$, and of $log_{10}|\nabla I|$. The results indicate that the work required to converge the gradient of the objective function by 2.5 orders of magnitude is equal to 1.0 work units. In our progressive optimization strategy, large gains in efficiency have been obtained since most of the geometry changes are performed at the coarsest mesh level at a very low computational cost. Figure 4 shows that the target pressure distribution on the airfoil surface is practically coincident with the converged pressure distribution. Figure 5 shows the initial airfoil shape (dashed line) and the optimized airfoil shape (symbols) nearly coincident with the target airfoil shape (continuous line).

In order to analyze the influence of the mesh size, the previous optimization process has been repeated by employing three finer meshes. The corresponding convergence history shows a small reduction in the work required. This result indicates that refining the mesh does not impair the rapid convergence of the present technique.

Furthermore, a sensitivity analysis has been performed by changing the typical values [4] of the coefficients a_j in (6) by a factor of 2. Also, the level of convergence of the optimization procedure at each mesh level has been increased by up to 50%. These sensitivity studies showed that the computational work increases by a maximum of 40% with respect to the computational work with typical parameter values [4].

Acknowledgements: The research of the first author has been supported by the Italian Agency MIUR. The research of the second author was partially supported by AFOSR throught grant F49620-96-1-0329.

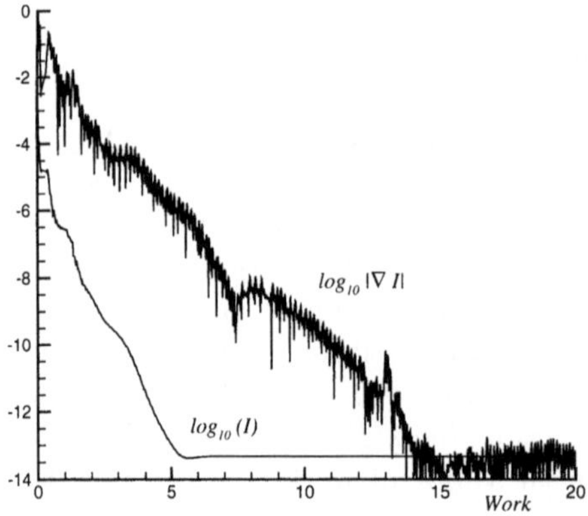

Fig. 2. Convergence history for subsonic NACA0012 airfoil. $M_\infty = 0.63$, $\alpha = 2.°$

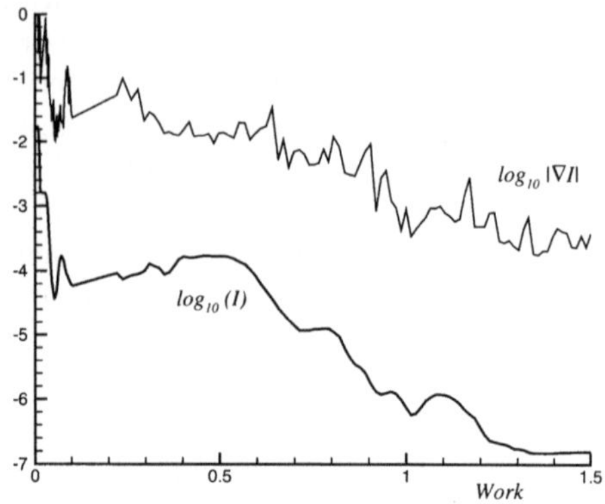

Fig. 3. Convergence history for transonic NACA0012 airfoil. $M_\infty = 0.8$, $\alpha = 1.25°$

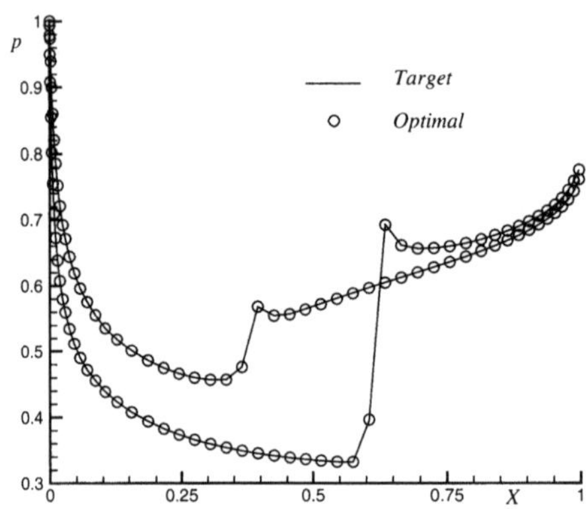

Fig. 4. Transonic NACA0012 airfoil. Target and optimized pressure distribution on airfoil surface

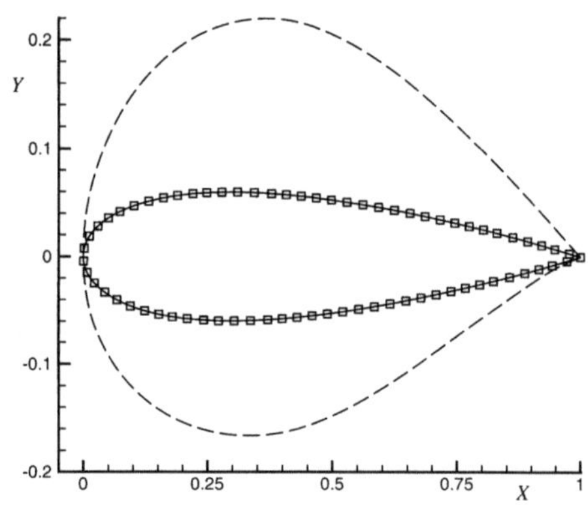

Fig. 5. Transonic NACA0012 airfoil. Initial, target and optimized airfoil shape

References

1. J.E. Melton, M.J. Berger, M.J. Aftosmis, M.D. Wong: '3D Applications of a Cartesian Grid Euler Method'. AIAA Paper 95-0853 (1995).
2. A. Dadone, B. Grossman: 'An Immersed Body Methodology for the Computation of Inviscid Flows on Cartesian Grids'. AIAA Paper 2002-1059 (2002).
3. A. Dadone, B. Grossman: 'Surface Boundary Conditions for the Numerical Solution of the Euler Equations'. AIAA J. **32**, 285 (1994).
4. A. Dadone, B. Grossman: 'Fast Convergence of Inviscid Fluid Dynamic Design Problems'. To appear in Computer and Fluids.
5. L.A. Catalano, A. Dadone: 'Progressive Optimization for the Efficient Design of 3D Cascades", AIAA Paper 2001-2578 (2001).

Multi-Criteria Aerodynamic Shape Design Problems in CFD Using a Modern Evolutionary Algorithm on Distributed Computers

E. Whitney[1], L. Gonzalez[1], K. Srinivas[1], and J. Périaux[2]

[1] School of Aerospace, Mechanical and Mechatronic Engineering, The University of Sydney, NSW 2006, Australia.
[2] Dassault Aviation, 78 Quai Marcel Dassault, 99214 Saint-Cloud, France.

Abstract. A modern Evolutionary Algorithm (EA) for multi–criteria aerodynamic shape design optimisation is presented. This algorithm attempts to overcome some of the drawbacks associated with earlier evolutionary algorithms, the most important being that of large computational time.

We propose the use of a Hierarchical Asynchronous Parallel Evolution Algorithm (HAPEA) where the search for the best solution takes place successively in separate hierarchical layers comprising different CFD solvers or resolutions. Each of the layers can be independently tuned to suit a given problem or CFD solver. Other strategies employed to reduce optimisation time are parallel computing and asynchronous evaluation.

We show that when applied to multi–criteria optimisation problems the use of a mix of fast and simple models together with slower and more detailed models on an asynchronous parallel architecture results in the same design quality as that obtained using a classical EA which would normally utilise only a complex model and involve much larger computational expense.

Results indicate that overall, the algorithm is easy to implement, robust and flexible when applied to CFD problems. It provides possibilities for future applications in Multi–disciplinary Design Optimisation (MDO).

1 Introduction

Evolutionary algorithms are based on Darwinian evolution, where populations evolve on a search space and adapt to the environment by the use of different mechanisms such as mutation, crossover and selection. EAs iteratively map multiple points in the design space simultaneously and require no fitness gradient information of any kind to proceed. EAs have the capability of finding globally optimum solutions, are easy to process in parallel and can be easily adapted to CFD codes without major modification.

The main drawback associated with using EAs, is that they are inherently slow; a typical optimisation may involve hundreds or thousands of function evaluations, thus limiting their application in aeronautical design problems where a time consuming CFD code is necessary to evaluate the fitness of each individual. Research has been undertaken to identify strategies to overcome this drawback and Hierarchical Asynchronous Parallel Evolution Algorithms (HAPEA) seem to be a promising approach. Some of the major advantages of EAs are their robustness and the fact that they can handle multi–criteria problems easily.

The paper gives an introduction to the features of the method, a brief description of a general multi–criteria problem and describes the strategies to increase the speed and efficiency of the optimisation. These concepts are then illustrated in two CFD problems, namely, a two objective nozzle reconstruction problem using a 2D Navier–Stokes solver and constrained single element aerofoil design.

2 Multi–Criteria Hierarchical Parallel Evolution Algorithm

2.1 Hierarchical Topology and Multiple Models

Hierarchical Genetic Algorithms (HGAs), are a particular approach developed by Sefrioui [3]. They use a hierarchical topology for the layout of the sub–populations. The main feature of this approach is the interaction between the given layers. The best solutions progress from the bottom layer to the top layer where they are refined. This circulation of solutions up and down the tree becomes even more interesting if we keep in mind that each node can be handled by a different EA code, and has specific parameters that can be tuned (figure 1).

In other words, the nodes of each layer can have a different purpose, defined by their associated EA:

i. The top layer concentrates on refining solutions. That can be achieved by tuning the EA in a way that makes mutation take very small steps. This layer incorporates the most precise formulation of the solver.
ii. The intermediate layer is a compromise between exploitation and exploration. This layer incorporates an intermediate approximation of the solver.
iii. The bottom layer can be entirely devoted to exploration. That means that the EA can make large leaps in the search space via a broad mutation span. This layer includes the most coarse approximation of the solver.

2.2 Multi–Criteria Optimisation

Often aeronautical design problems require a simultaneous optimisation of conflicting objectives and associated number of constraints, unlike single objective optimisation problems, the solution is a set of points known a Pareto optimal set. As EAs evaluate multiple populations of points, they are capable of finding a number of solutions in a Pareto set. Pareto selection ranks the population and selects the non dominated individuals for the Pareto fronts. Other aspects of the method can be found in [5].

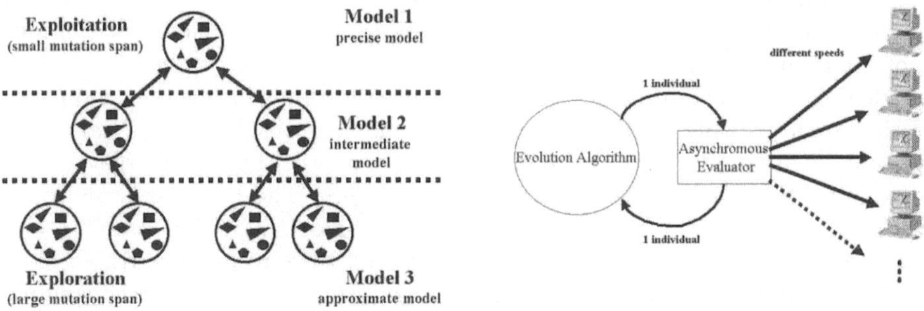

Fig. 1. Hierarchical Topology and Asynchronous Solver Flow Diagram.

2.3 Parallel Computing and Asynchronous Evaluation

EAs are particularly adaptable to parallel computing, especially when implemented using the master–slave approach. In the optimisation described in this paper the master computer carries on the optimisation process while the remote machines compute the solver code. The remote solvers do not need to run at the same speed (synchronise) or even on the same local network. Solver nodes can be added or deleted dynamically during the execution. A schematic of such implementation is shown in figure 1.

The parallel implementation requires modifications to the canonical Evolution Strategy (ES) [2] which ordinarily evaluates entire populations simultaneously. The distinctive method of an asynchronous approach is that it generates only one candidate solution at a time and only re–incorporates one individual at a time, rather than an entire population at every generation as is usual with traditional EAs. Consequently solutions can be generated and returned *out of order*. This allows the implementation of an asynchronous fitness evaluation giving the method its name. This is an extension of the work by Wakunda and Zell [6].

With an asynchronous approach there is no waiting time (or bottleneck) for individuals to return. As soon as a solution is available, it is incorporated back into the optimisation process.

3 Applications

3.1 Two Objective–Two Dimensional Nozzle Inverse Optimisation Problem.

Problem definition The first case considers the fitting of two different shapes to two converging–diverging nozzles. For two different pre–existing target nozzles, the corresponding pressure distribution is computed using a quasi–steady two dimensional approximation for the flow. The equations are solved by a time marching technique using a CUSP scheme with an iterative solver [4].

The objective of the optimisation task for the EA is to rebuild the target nozzles by finding the Pareto set of nozzles between the two pressure distributions that approximately fit the two target pressure distributions. For each candidate nozzle the CFD solver provides the pressure distribution which is then compared to the target pressure distribution in a least-squares sense. The individual nozzle wall shapes are implemented using a Bézier curve of order five. The target nozzles are simply taken as two different quadratic shapes. The boundary conditions are fixed at $M_{outlet} = 0.7$, $Re = 10000$, with inlet stagnation pressure and enthalpy imposed and all other variables treated by zero order extrapolation. Each solution case is assumed to be converged after the 2–norm of the residual falls below 10^{-3}.

Implementation The solution to this problem has been implemented for both single population and hierarchical methods.

- Single Population EA (EA–SP): Population size 20, the computational grid is 75×37 points equally spaced along the nozzle.
- Hierarchical Asynchronous Parallel EA (HAPEA): Multiple models with different models in different layers, the computational grids are 25×12, 50×25 and 75×37 points in the bottom, intermediate and top layers respectively, run asynchronously.

Results The CPU time required for the single and multiple models is given in table 1. Evaluations correspond to the average number of evaluations over twenty runs. CPU time denotes the average CPU time before reaching the stopping precision of 10^{-3}, which is the average sum of the minimum distance from each point of the Pareto set to a precomputed perfect Pareto set. As can be seen, the HAPEA is three times faster when compared to the EA–SP.

Table 1. CPU Comparison.

Model	Evals	CPU time
EA–SP	2311±224	152 ± 20 m
HPEA	504 ± 490 (-78%)	48 ± 24 m (-68%)

3.2 Constrained Single Element Aerofoil Design.

Problem definition Transport aircraft are often designed for cruise at single flight conditions while meeting other design constraints. The penalty for operating at off design conditions can be significant. Aerodynamic performance might be improved if a multi–criteria multi point optimisation can be developed that considers numerous separate design points. One dual point design procedure is described here to find the Pareto set of aerofoils for minimum drag at these

Table 2. Flight Conditions.

Property	Flt. Con. 1	Flt. Con 2
$MachNo.$	0.75	0.75
$ReynoldsNo.$	9×10^6	9.0×10^6
c_l	0.65	0.715

two design points. The flow conditions for the two points analysed are listed in table 2.

The objective functions to be minimised are therefore:

$f_1 = Cd_{FlightCondition_1} + Penalty$, $f_2 = Cd_{FlightCondition_2} + Penalty$

Two Bézier curves represent the aerofoil, with six points on the mean line curve and ten points on the thickness distribution curve. Constraints are imposed on minimum thickness ($t/c \geq 12.11\%$) and position of maximum thickness ($20\% \leq x_t/c \leq 55\%$). If any of these constraints are violated both fitnesses are linearly penalised to ensure an unbiased Pareto set.

CFD Solver and Implementation The two conditions are evaluated using an Euler + boundary layer flow solver, using a structured quadrilateral mesh which is coupled to an integral boundary layer based on a multi layer velocity profile representation (for more details refer to [1]). On the top, middle and bottom levels grids of 215×36, 99×16 and 71×12 are used respectively.

Results The algorithm was allowed to run for 10000 functions evaluations of the head node and took approximately twelve hours to run. The final population (including the Pareto optimal front) is shown in figure 2. It can be seen how the hierarchical approach gives a uniformly distributed front.

Figure 2 also shows the aerofoils obtained from the Pareto set. The bottom left aerofoil corresponds to Pareto member one which is best suited for flight condition one, the top right aerofoil corresponds to Pareto member eighteen which is best suited for flight condition two. Results are presented in table 3.

Table 3. Aerofoil Comparison.

Aerofoil	c_d $[c_l = 0.65]$	c_d $[c_l = 0.715]$
RAE 2822	0.0147	0.0185
Nadarajah	0.0098 (-33.3%)	0.0130 (-29.7%)
HAPEA opt	0.0094 (-36.1%)	0.0108 (-41.6%)

4 Conclusions

The HAPEA method presented is shown to be promising for direct and inverse design optimisation problems. A wide variety of optimisation problems including

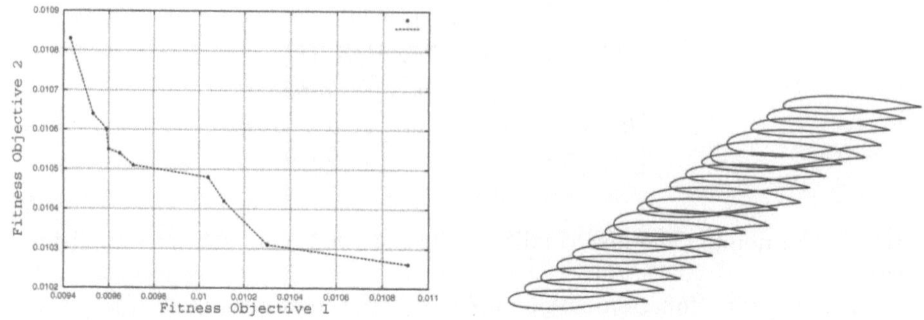

Fig. 2. A sampling of the Pareto set of aerofoils and Pareto front obtained for the constrained transonic aerofoil design problem.

Multi–disciplinary Design Optimisation (MDO) problems can be solved using the same evolution algorithm coding with minor implementation effort providing an appropriate function evaluation mechanism is available.

The benefits of using parallel computing, hierarchical optimisation and evolution algorithms to provide solutions for multi–criteria problems has been demonstrated. This is of significant importance when dealing with multi-criteria optimisation problems where the flowfield solution involves lengthy computing time.

References

1. M. Drela. *A Users Guide to MSES 2.95.* MIT Computational Aerospace Sciences Laboratory, 1996.
2. Z. Michalewicz. *Genetic Algorithms + Data Structures = Evolution Programs.* Artifical Intelligence. Springer-Verlag, 1992.
3. M. Sefrioui and J. Périaux. A hierarchical genetic algorithm using multiple models for optimization. In M. Schoenauer, K. Deb, G. Rudolph, X. Yao, E. Lutton, J. J. Merelo, and H-P. Schwefel, editors, *Parallel Problem Solving from Nature, PPSN VI*, pages 879–888. Springer, 2000.
4. K. Srinivas. Computation of cascade flows by a modified cusp scheme. In *Computational Fluid Dynamics Journal*, volume 2, pages 285–295, 1999.
5. N. Srinivas and K. Deb. Multiobjective optimisation using nondominated sorting genetic algorithm. In *Evolutionary Computation*, volume 2(3), pages 221–248, 1995.
6. J. Wakunda and A. Zell. Median-selection for parallel steady-state evolution strategies. In *Proceedings of the 6th Internation Conference "Parallel Problem Solving from Nature - PPSN VI*, pages 405–414, 2000.

Comparison of Simultaneous Perturbation Stochastic Approximation and Simulated Annealing in Transonic Airfoil Design Optimization

X.Q.Xing and M.Damodaran

Singapore-Massachusetts Institute of Technology Alliance (SMA) and
School of Mechanical and Production Engineering,
Nanyang Technological University
50 Nanyang Avenue, Singapore, 639798

Abstract. Constrained and unconstrained transonic airfoil design optimization problems are investigated by applying the simultaneous perturbation stochastic approximation (SPSA) method to objective functions which are estimated using a finite volume computational fluid dynamics (CFD) flow solver solving the Navier-Stokes equations during each design cycle . The performance and advantages of SPSA method over Simulated Annealing, another global optimization method are assessed. The computed results show that the SPSA method is robust in reaching optimal aerodynamic shapes, is easy to implement and efficient in decreasing the computational cost compared with Simulated Annealing.

1 Introduction

Aerodynamic shape design optimization problems have been extensively investigated in recent years using computational fluid dynamics (CFD) and a number of design optimization techniques. Besides deterministic methods, stochastic methods such as genetic algorithm (GA) and simulated annealing (SA) algorithm etc have recently found applications in practical engineering design optimization problems. SA and GA algorithms are stochastic in nature and are easily implemented in robust CFD codes as compared with deterministic methods. They have the advantage of yielding a global minimum and overcoming the limitations of deterministic gradient-based search methods which have a tendency of getting trapped in local minima. However, SA and GA methods require large number of function evaluations and long computation time especially in the case of complex design optimization problems. One approach to reduce computational time would be to use parallel GA and parallel SA as outlined in Wang and Damodaran[1]. While parallel SA and GA can speedup the computation, they still require enormous computational resources. An attractive alternative to SA and GA is the Simultaneous Perturbation Stochastic Approximation (SPSA) method outlined in Spall[2,3]. It only uses objective function measurements and does not require direct measurements of the gradient of the objective function. Only two function measurements are required to approximate the gradient independent of the number of parameters being optimized. The SPSA method has attracted considerable range of applications in many different areas such as statistical parameter estimation, feedback control, simulation-based optimization, signal and image processing, etc. Based on SPSA method's successful use on wing design problem outlined in Xing and Damodaran[4], the feasibility of SPSA in transonic airfoil optimization problems (including constrained and unconstrained problems) is considered in this study. The performance and advantages of SPSA method over SA method are also assessed in this study.

2 Aerodynamic Shape Design Problems

Problems concerning the design of transonic airfoil shape to satisfy desired aerodynamic behavior fall into constrained and unconstrained design optimization problems. The optimization of selected aerodynamic parameters such as aerodynamic lift, drag or pitching-moment subject to some imposed design constraints generally belong to the class of constrained optimization problems. An example of unconstrained design problem is the inverse design problem of determining the airfoil shape which will support a target airfoil surface pressure distribution. This shape is determined by minimizing the discrepancy between the target and the evolving airfoil surface pressure distribution corresponding to the designed airfoil. The present study focuses on the application of SPSA method for inverse and constrained aerodynamic design of transonic airfoil shapes. During the optimization process, the objective function is evaluated using a compressible Navier-Stokes flow solver. The objective function is then minimized using SPSA.

2.1 Inverse Airfoil Shape Design

Inverse design problems require a baseline airfoil configuration to initiate the design process. In this study, a baseline NACA 0012 airfoil is first chosen and a steady flow field around it at a Mach number of 0.73 and angle of attack of 2.78° is computed for starting the design cycle iterations. The airfoil shape is updated by adding smooth perturbations $\Delta y_k(x)$ defined as a linear combination of a family of smooth curves over the range $0 < x < 1$ as follows:

$$\Delta y_k(x) = \sum_{k=1}^{K} \delta_k f_k(x) \tag{1}$$

where x is the normalized chord-wise position of the coordinates defining the airfoil contour, δ_k are the design variables which will change during the design iterations and K is the number of basis functions $f_k(x)$ and $K = 14$ in this study. The leading edge and the trailing edge are fixed at the origin and a chord length away from the origin respectively along the x-axis. For inverse design problems a typical objective function $F(X)$ to be minimized is defined as follows:

$$F(X) = \left[\sum_{m=1}^{M} (Cp_{t_m} - Cp_{b_m})^2 \Delta S_m / \sum_{m=1}^{M} \Delta S_m \right]^{\frac{1}{2}} \tag{2}$$

where Cp_{t_m} is the pressure distribution of the target airfoil which for this example corresponds to that around a RAE 2822 airfoil at the same Mach number and angle of attack as that of the baseline airfoil, Cp_{b_m} is the pressure distribution of the designed airfoil which evolves after each design iteration, ΔS_m is the length of the airfoil surface element and the summation is done for the M coordinate points defining the airfoil contour. Details of this process are outlined in Lee and Damodaran[5].

2.2 Transonic Airfoil Drag Minimization

The aim of constrained design problem is to improve the aerodynamic shape of a baseline airfoil so that the drag can be minimized while the cross sectional area and the lift of the airfoil are maintained during the optimization process. The constrained design optimization problem is stated as:

$$MinmizeF(X) = C_d \tag{3}$$

Subject to:

$$g_1(X) = 1 - \frac{C_l}{C_{l_0}} \leq 0 \tag{4}$$

$$g_2(X) = 1 - \frac{A}{A_0} \leq 0 \tag{5}$$

where X is the vector of k design variables , C_l and C_d are the aerodynamic lift and drag coefficients respectively. A is the airfoil area, A_0 is the area of the baseline airfoil and C_{l_0} denotes the initial value of the lift coefficient. The baseline airfoil to initiate this design optimization process in this study is the RAE 2822 airfoil immersed in transonic flow at a Mach number $M = 0.726$, angle of attack $\alpha = 2.44°$ and Reynolds number $Re = 6.5\times10^6$.Here an external penalty function method is used to incorporate the constraints so that the composite function to be minimized can be defined as:

$$F(X) = C_d + \sum [max(0, g_j(X)]^2, j = 1, 2. \tag{6}$$

3 Optimization and Simulation

The CFD model used to evaluate the objective function is based on the finite volume formulation of the unsteady two-dimensional compressible Navier-Stokes equations. Specific details of the flow modeling are outlined in Martinelli and Jameson[6]. The objective function is minimized using SPSA which is an algorithm based on a "simultaneous perturbation" gradient approximation. The "simultaneous perturbation" approximation uses only two function measurements independent of the number of parameters being optimized. The details of the implementation of the SPSA algorithm are outlined in Spall[2].

4 Results and Discussion

4.1 Inverse design problem

Fig. 1 shows the convergence of the minimization of the objective function with design cycles using SPSA method and compares the initial, designed and target airfoil shapes and the corresponding the surface pressure distributions . From Fig.1(a) it can be seen that SPSA method requires about 800 design cycles or 2400 objective function evaluations to reach convergence. The designed airfoil shape and the distribution of the pressure coefficients are identical to that of the target. Comparing the results of SPSA method with that obtained using SA method as in Lee and Damodaran[5], and shown in Fig.2, it can be seen that SPSA method requires fewer function evaluations to reach the target airfoil than Simulated Annealing. Although SA method requires 3000 accepted function evaluations to converge to the desired optima and this SA method actually requires about 6000 objective function evaluations to get the convergence results because only about half of objective function evaluations are accepted using SA method.

4.2 Transonic airfoil drag minimization

Fig.3 shows the convergence of the minimization of the objective function with design cycles and compares the initial and final design which results from the constrained optimization problem using SPSA method. From the surface pressure coefficient distribution, it can be observed that the decrease of the drag was achieved mainly by weakening the shock waves as a result of the changed shape. The convergence history of the objective function shows that the airfoil shape obtained from the SPSA method yields a reduction in C_d by 9.78% (from 0.01800 to 0.01624) while contraints defined by equations 4 and 5 are satisfied. This is consistent with the results shown in Lee and Damodaran[5] using SA algorithm as shown in Fig.4. The SPSA method requires 1489 iterations or 4467 function evaluations to get the convergence results using SPSA method, while it required 4500 accepted objective function evaluations (or 9000 total function evaluations) to get the same levels of convergence using SA method. Compared with SA method, SPSA method has some marginal gain in efficiency in terms of computational cost for this optimization problem.

5 Conclusion

SPSA method has been successfully implemented to address aerodynamic shape design optimization problems covering both inverse and constrained airfoil design in this study. Comparisons of results from SPSA and SA method show that SPSA method provides a reasonable computational advantage over SA method. This results from the fact that SPSA method requires only two measurements of the objective function to approximate the gradient regardless of the dimensions of the design space corresponding to the optimization problem and consequently the cost of optimization decreases. Examination of the trends in convergence of the objective functions with the number of design iterations SA shows a smooth convergence while SPSA shows an oscillatory convergence with decreased amplitude to optimal values. This implies that the SPSA method could be combined with a local optimization method when the objective function reaches an oscillating convergence value so that the global optimum can be found by SPSA method while the accuracy will be improved by the local optimization method.

References

1. X. Wang and M. Damodaran, AIAA Journal of Aircraft, Vol. 37, No.5 (2000) pp.929-932.
2. J. C. Spall, Vol.19, No.4(1998) pp.482-492
3. J. C. Spall, Vol. 34, No.3, July(1998) pp.817-823
4. X.Q. Xing and M. Damodaran, AIAA, Journal of Aircraft, Vol.39, No.2 (2002) pp.379-381
5. S. L. Lee, and M. Damodaran, AIAA Paper 2000-0782 (2000)
6. L. Martinelli, A. Jameson and F. Grasso, AIAA Paper 86-0208 (1986)

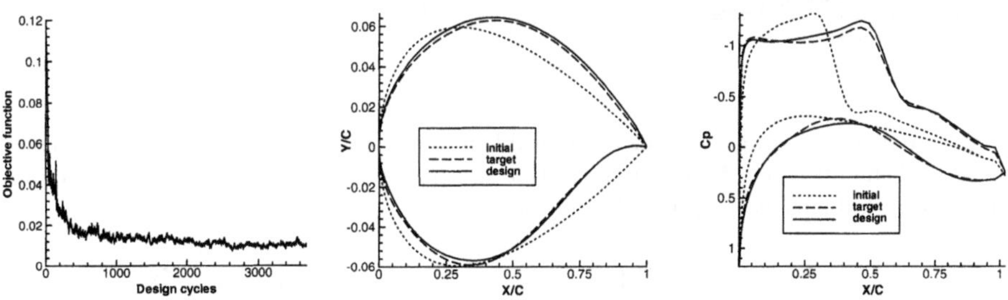

(a) (b) (c)

Fig. 1. Results of inverse design problem using SPSA method (a) convergence history of the objective function (b) initial, designed and target airfoil shapes (c) initial, designed and target airfoil surface pressure distributions

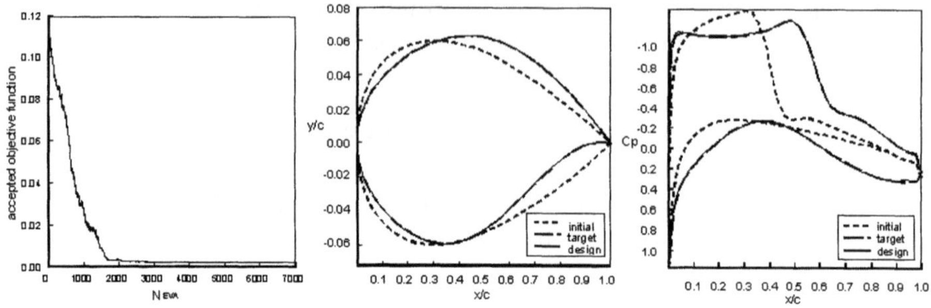

(a) (b) (c)

Fig. 2. Result of inverse design problem using SA method (a) convergence history of the objective function (b) initial, designed and target airfoil shapes (c) initial, designed and target airfoil surface pressure distributions

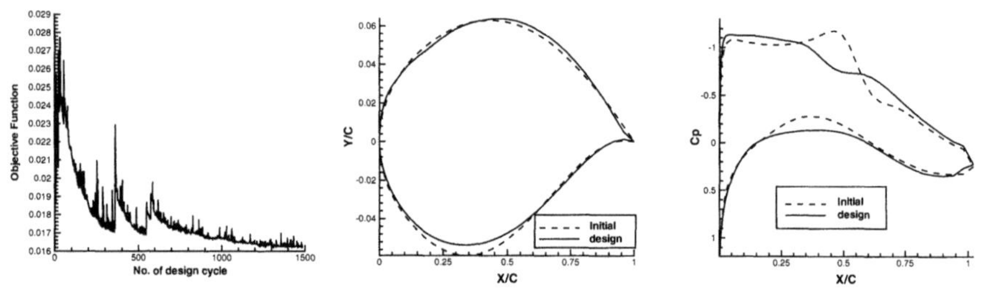

(a) (b) (c)

Fig. 3. Results of drag minimization problem using SPSA method (a) convergence history of the objective function (b) initial and designed airfoil shapes (c) initial and designed airfoil surface pressure distributions

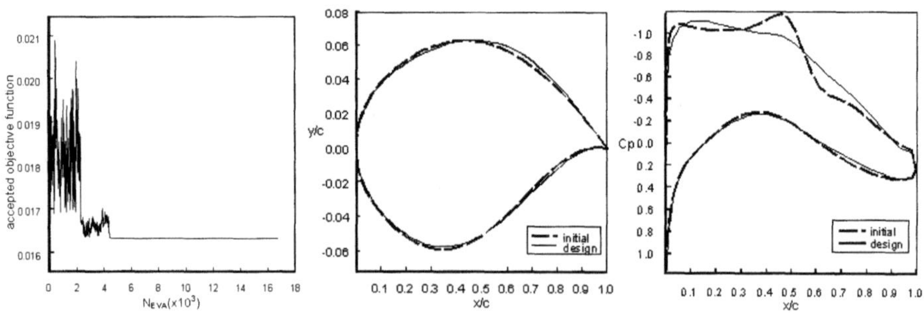

(a) (b) (c)

Fig. 4. Results of drag minimization problem using SPSA method (a) convergence history of the objective function (b) initial and designed airfoil shapes (c) initial and designed airfoil surface pressure distributions

Part XV

Supersonic Flows and Shock Waves

Numerical Simulation of Fluctuations in a Shock Tunnel Flow

Richard J. Goozée[1], David R. Buttsworth[2], and Peter A. Jacobs[1]

[1] The University of Queensland, Brisbane Qld 4072, Australia
[2] University of Southern Queensland, Toowoomba Qld 4350, Australia

Abstract. Shock tunnel experiments and simulations were performed to explore the mechanisms leading to the generation of noise in shock tunnel flows. The shock tunnel was operated at two different conditions (one with a nitrogen driver, the other with a helium driver), and simulations were performed using an axisymmetric compressible Navier-Stokes solver. The nozzle reservoir region and nozzle exit pitot pressures are simulated with reasonable accuracy at both conditions, and the magnitude of fluctuations at frequencies between 10 and 50 kHz in the nozzle exit flow is also accurately simulated.

1 Introduction

It is well known that free stream fluctuations have a strong influence on laminar-turbulent transition in supersonic and hypersonic flow and that noise levels measured in conventional supersonic wind tunnels are often more than an order of magnitude larger than those present during flight [9]. To address this noise issue, a number of 'quiet' supersonic and hypersonic wind tunnels have been developed; however, it is also known that enthalpy can have a significant effect on boundary layer transition [1] and that these quiet facilities do not correctly simulate the high enthalpy associated with hypersonic flight.

Free piston shock tunnels can simulate the high enthalpy conditions associated with hypersonic flight and, therefore, should be a good source of data for high-speed transitional flows. Although a number of transition experiments have been performed in these facilities [6],[1], very little data is available on their free stream disturbances [9].

Quasi-one dimensional simulations [7] provide a computationally efficient method of predicting the performance of shock tunnels; however, noise in the flow is generated by multi-dimensional phenomena and so axisymmetric or three dimensional simulations are necessary in order to study these effects.

2 Shock Tunnel – Instrumentation and Operation

The shock tunnel facility used in this work is located at the University of Queensland and is illustrated in Fig. 1. Additional facility details are available in [4]. Two piezoelectric pressure transducers were used: 1) in the nozzle reservoir region of the shock tube; and 2) close to the exit of the Mach 4 conical nozzle, as illustrated in Fig. 1.

Two operating conditions were studied: 1) a nitrogen driver gas condition (3.25 MPa of N_2 in the driver section and 30kPa of N_2 in the shock tube section); and 2) a helium driver gas condition (5.60 MPa of He in the driver section and 61.4 kPa of N_2 in the shock tube section). The tunnel was fired by rupturing the primary diaphragm (1 mm aluminium plate) with the piercing device (Fig. 1).

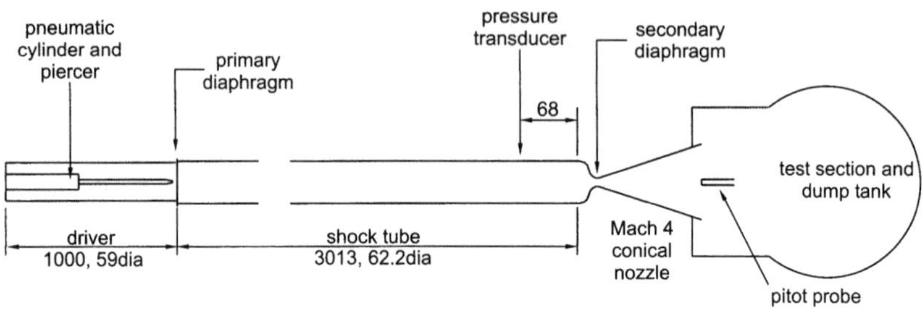

Fig. 1. Illustration of the shock tunnel and instrumentation

3 Shock Tunnel Simulations

3.1 Compressible Navier-Stokes Solver

The simulations were performed using the multi-block CFD code, MB_CNS [8], which is based on a finite-volume formulation of the compressible Navier-Stokes equations. It has a shock-capturing capability through the use of a limited reconstruction scheme and an adaptive flux calculator that switches from AUSM-D to the Equilibrium Flux Method (EFM) where large compressive velocity gradients are detected.

Multiple gas components are modelled by solving additional equations for conservation of each of the species. This allows the movement of the driver and driven gases in the facility to be tracked over time. Data can be recorded at the geometric location of sensors in the real facility and can be used for comparison.

In order to make the use of large simulations practical, the code has been parallelised using OpenMP (shared memory) and MPI (message passing). The implementation of the parallel code in these simulations is discussed in more detail in [5].

3.2 Modelling of the Facility

Previous attempts to model reflected shock tunnel flows have typically simulated only the end section of the shock tube using an inflow condition derived from theoretical analyses; however, for simulation of the nozzle reservoir region, it is more accurate to model the development of the flow along the entire length of the facility.

A body fitted mesh covered the whole facility, from the driver section to the test section and dump tank. The mesh was decomposed into 24 blocks in order to represent the geometry of the facility and to allow parallelisation of the solution. The mesh was refined towards the wall to focus cells in the boundary layers and towards the end of the tube to increase resolution in the nozzle reservoir region. Mesh refinement studies were conducted by uniformly refining the mesh from 80,850 to 181,980 and 323,400 cells. In these simulations, the turbulent boundary layer and its interaction with the reflected shock may not be fully resolved. The axial resolution of the mesh along the shock tube has little effect on the flow.

The simulations were initiated with the instantaneous removal of the primary diaphragm between the driver and driven gas. The mechanics of a real diaphragm rupturing introduces significant fluctuations into the flow which are not modelled by this ideal removal. The secondary diaphragm is also removed instantaneously after a specified pressure is reached in the cells immediately upstream of the diaphragm.

Mass flow into the boundary layer on the shock tube wall attenuates the shock and accelerates the contact surface, resulting in a reduction of the test time. For the present conditions, it is expected that these boundary layers rapidly undergo transition to turbulence. The Baldwin-Lomax eddy viscosity model [2] is used to model the effect of turbulence in the boundary layers. Coefficients used in the model were approximated from the literature and were adjusted using the experimental results. For nitrogen driving nitrogen, modified values of k=0.18 for the inner layer and the combination of C_{cp}=6.4 and C_{kleb}=0.3 for the outer layer were used; for helium driving nitrogen, values of k=0.12, C_{cp}=1.6 and C_{kleb}=0.3 were used.

4 Results

4.1 Stagnation pressure

Figure 2 provides a comparison of the measured and simulated shock tube pressures. In the N_2 driver case, the laminar simulation over estimates the incident and the reflected shock pressures, whereas the calibrated turbulent simulation computed both of these pressure levels accurately. Also, the arrival time of the compression waves from the driver gas interface, is over estimated by the laminar simulation. This means that insufficient test gas is entrained into the shock tube boundary layers by this model. In the He driver case, the laminar simulation again over estimated the incident and reflected shock pressures and duration of the stagnated region, whereas, although significant fluctuations are evident in the stagnated region of the trace, the turbulent simulation correctly estimated the average pressures and the duration. The remainder of the simulations discussed in this article use the calibrated turbulence model.

The most significant difference between the turbulent simulation and experimental traces is at the arrival of the reflected shock. The foot of the reflected

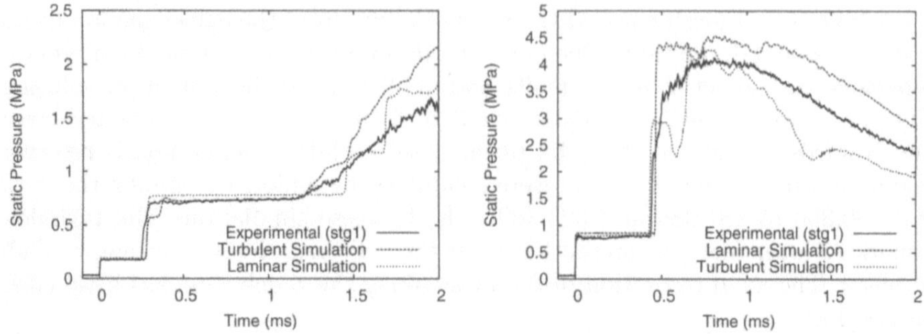

Fig. 2. Shock tube pressure measurements and simulations. **(left)** N_2 driver. **(right)** He driver

shock has bifurcated due to its interaction with the incident boundary layers. The pressure histories are recorded, in both the simulation and experiment, inside the boundary layer making these traces very sensitive to this phenomenon. The passage of this shock structure over the sensor appears in the experimental trace as a small step, followed by a gradual rise in pressure. In the simulated pressure trace, the step is significantly more defined.

4.2 Reflected Shock Interaction

The development of flow features in the stagnation region is illustrated in Fig. 3 which shows sequences of images of the end region of the shock tube and the nozzle entry. Numerical schlieren images are shown in the upper half of each frame and driver gas mass fraction images are shown in the lower half.

The shock reflection from the contoured nozzle entry results in the generation of a nonuniform nozzle reservoir region. As the reflected shock moves back upstream, the energy deficient boundary layers do not have enough energy to cross the normal shock. Boundary layer material builds up at the foot of the shock and is carried with it; this causes the foot of the shock to bifurcate. This interaction appears to be responsible for introducing additional disturbances into the test flow and can cause jetting of material towards the nozzle. These disturbances can be seen in the numerical schlieren plots in the region between the reflected shock and the nozzle.

The interaction between the reflected shock and the incident contact surface can also be seen in these frames. This interaction causes significant amounts of vorticity to be generated in the fluid around the interface. This induced vorticity is combined with the jetting through the foot of the reflected shock and, in both the N_2 and He driver cases, results in the early contamination of the test flow with driver gas.

In the N_2 driver case, driver gas at the interface is pushed outwards into the shock tube walls through its interaction with the reflected shock. This gas then folds back inwards towards the nozzle entry ahead of much of the test gas. In the He driver case, the driver gas is similarly pushed outwards; however, the head

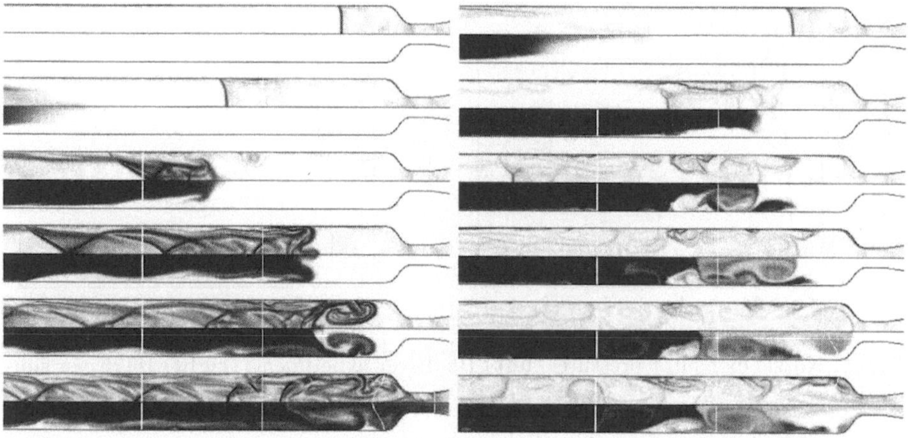

Fig. 3. Sequence of simulated schlieren images (upper half of frames) and driver gas mass fraction images (lower half of frames) showing the development of the nozzle reservoir region. (**left**) N_2 driver (at 400μs intervals). (**right**) He driver (at 300μs intervals)

of the interface forms a vortex ring which accelerates along the centreline of the tube towards the nozzle. This appears to be a new mechanism, distinct from wall jetting, that can cause early contamination of the test gas.

4.3 Nozzle exit fluctuations

Figure 4 illustrates the simulated and experimental pitot pressure at the nozzle exit for the He driver case. Restricting attention to frequencies between 10 and 50 kHz, the measured RMS fluctuation in the He driver case is 1.4 % over the period from about 0.5 to 1 ms on the time scale in Fig. 4 (see [3] for further details). The simulation yields RMS pitot pressure fluctuations of 1.2 % over the same period and frequency band. This result suggests that the dominant source of the measured pitot pressure fluctuations is the nozzle reservoir region rather than aerodynamic noise radiating from the nozzle boundary layers since the nozzle boundary layer contribution is not simulated in the present work.

5 Conclusions

The compressible Navier-Stokes solver provides reasonable estimates for the nozzle reservoir pressure and the nozzle exit pitot pressure for the present shock tube operating conditions. In the present simulations, an algebraic turbulence model was used. Improvements in the shock tube simulations may be achieved through implementation of a high order turbulence model. Nevertheless, the magnitude of the pitot pressure fluctuations at the nozzle exit has been accurately simulated. This suggests that the various shock reflection and interaction processes

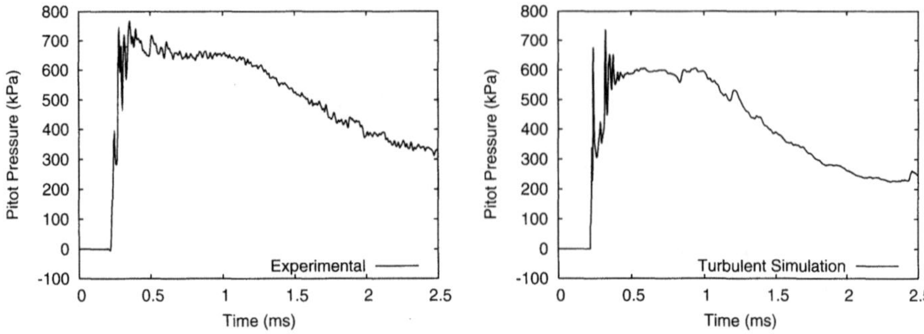

Fig. 4. Nozzle exit pitot pressure. (**left**) He driver experiment. (**right**) He driver simulation

in the nozzle reservoir region are primarily responsible for the measured pitot pressure fluctuations.

References

1. P.H. Adam, H.G. Hornung: J. Spacecraft Rockets **34**, 614 (1997)
2. B.S. Baldwin, H. Lomax: 'Thin layer approximation and algebraic model for separated turbulent flows'. Paper 78-257 AIAA 16th Aerospace Sciences Meeting, Washington D.C., January, 1978
3. D.R. Buttsworth, P.A. Jacobs: 'Measurement of fluctuations in a Mach 4 shock tunnel nozzle flow'. In: *7th Australasian Heat and Mass Transfer Conference*, Townsville, Australia, July, 2000, ed. by G.B. Brassington, J.C. Patterson (Chalkface Press, Western Australia, 2000) pp. 53-59
4. C.S. Craddock: Operational instructions for the small shock tunnel at UQ. Report 8/98, Dept. Mech. Eng., University of Queensland (1998)
5. R.J. Goozée, P.A. Jacobs: 'Parallelism for simulations of an entire shock tunnel facility'. In: *ICCFD2*, Sydney, Australia, July, 2002
6. Y. He, R.G. Morgan: Aero. J. **98**, 25 (1994)
7. P.A. Jacobs: AIAA J. **32**(1), 137 (1994)
8. P.A. Jacobs: MB_CNS: A computer program for the simulation of transient compressible flows. Report 10/96, Dept. Mech. Eng., University of Queensland (1996)
9. S.P. Schneider: J. Spacecraft Rockets **38**, 323 (2001)

Numerical Simulations of Sonic Booms in Non-uniform Flows

Mohamed Hafez[1] and Essam Wahba[1]

University of California, Davis CA 95616, USA

Abstract. Sonic booms can be simulated based on a small disturbance potential formulation which leads to a nonlinear mixed type equation similar to that of transonic flows [1]. Numerical results using Murman's scheme are given in [2] for 2-D and 3-D problems. In the present work, the governing equations are written as a system of first-order equations in terms of the velocity components. The resulting system is a generalized form of Cauchy/Riemann equations with the vorticity as a forcing function. Using Crocco's relation, the vorticity can be related to the gradients of entropy and total enthalpy in general. For some applications, the entropy variation can be neglected and the isentropic flow model is adequate even with variable total enthalpy.

The Cauchy-Riemann equations are solved numerically as in [4]. Artificial time dependent terms are introduced leading to a symmetric hyperbolic system, which is solved using standard numerical methods. Centered schemes are used with explicit artificial viscosity terms added to the equations for numerical stability. The artificial viscosity is constructed via a least squares procedure; as in [3], and additional dissipation is needed for shock capturing. Explicit time integration, which is fully parallelizable, is implemented to obtain steady state solution. Typical numerical results are shown for linear and parabolic velocity profiles of the main flow, with shock reflection in the field and smooth signatures on the ground.

1 Governing Equations and Boundary Conditions for the Sonic Boom Problem

In [1], Cheng derived a nonlinear equation describing the sonic boom problem in terms of a small disturbance potential (ϕ), which is of mixed type and similar to the transonic flow equation except the similarity parameter is linearly dependent on the altitude. For three-dimensional flows, the equation reads,

$$(K - \phi_x)\,\phi_{xx} + \phi_{yy} + \phi_{zz} = 0 \tag{1}$$

or, in conservation form,

$$\left(K\phi_x - \frac{1}{2}\phi_x^2\right)_x + \phi_{yy} + \phi_{zz} = 0 \tag{2}$$

or,

$$f_x + \phi_{yy} + \phi_{zz} = 0$$

where,

$$f\,(x) = K\phi_x - \frac{1}{2}\phi_x^2$$

where K is a function of y
The boundary conditions for the 3D problem are:

a) Pressure Signature, $\phi = x(x-1)$ at $z = z_M$ and $(0 < x < 1)$, and $\phi = 0$ otherwise.

b) On the ground $(z = z_o)$, $\phi_z = 0$.

c) Inflow at $x = x_o$, $\phi_x = 0$. For supersonic flow $\phi = \phi_x = 0$.

d) Outflow at $x = x_M$, only for subsonic flow $\phi_x = 0$.

The limits of the computational domain are $x_M = 8$ and $y_M = 3$

The First Order and Second Order schemes of Engquist and Osher [5], [6] are used for simulating the sonic boom problem.

2 Alternative Formulation for the Sonic Boom Problem

In order to apply standard higher order schemes, the governing equations can be written as a first order system. In particular, the transonic small disturbance equation can be written as a symmetric hyperbolic system in the form

$$u_t = (K-u)\,u_x + v_y + w_z, \qquad v_t = u_y - v_x, \qquad w_t = u_z - w_x \tag{3}$$

or, in a matrix form

$$\begin{pmatrix} u \\ v \\ w \end{pmatrix}_t = \begin{bmatrix} (K-u) & 0 & 0 \\ 0 & -1 & 0 \\ 0 & 0 & -1 \end{bmatrix} \begin{pmatrix} u \\ v \\ w \end{pmatrix}_x + \begin{bmatrix} 0\,1\,0 \\ 1\,0\,0 \\ 0\,0\,0 \end{bmatrix} \begin{pmatrix} u \\ v \\ w \end{pmatrix}_y + \begin{bmatrix} 0\,0\,1 \\ 0\,0\,0 \\ 1\,0\,0 \end{bmatrix} \begin{pmatrix} u \\ v \\ w \end{pmatrix}_z \tag{4}$$

The main flow is assumed almost aligned with the x-axis. The only nonlinear term is uu_x which can be approximated by a conservative discretization. The Lplacian of the velocity components will provide only first order artificial viscosity terms. Outside the shock region, higher order viscosity based on least squares including the nonlinear term uu_x can be used. The artifical viscosity terms are obtained via a least squares procedure minimizing the functional,

$$I(u,v,w) = \int\!\!\int\!\!\int \left[(K-u)\,u_x + v_y\right]^2 + \left[-u_y + v_x\right]^2 + \\ \left[-w_x + u_z\right]^2 + \left[w_y - v_z\right]^2 dxdydz \tag{5}$$

To allow for a general flow direction (relative to the coordinate system) with an arbitrary free stream Mach number, the full potential equation must be used as in [7].

In figures (1) and (3), comparison of the solutions of the potential and Cauchy/Riemann equations for a main flow with a parabolic velocity profile is given in terms of the velocity distributions. The pressure contours are gievn in figures (2) and (4).

Application of Lerat's third order scheme [8] produces sharper shocks with some wiggles. Comparison with other higher order schemes will be presented elsewhere.

3 Wave Propagation in Non-uniform Flows

A more general formulation for 2D problems can be based on the continuity equation and the definition of vorticity

$$\begin{aligned}(\rho u)_x + (\rho v)_y &= 0 \\ v_x - u_y &= \omega\end{aligned} \tag{6}$$

The first equation can be rewritten as

$$u_x + v_y = \frac{-(u\rho_x + v\rho_y)}{\rho} = S \tag{7}$$

The artificial viscosity is obtained via minimizing the functional

$$f(u,v) = \int\int \left[(u_x + v_y - S)^2 + [u_y - v_x + \omega]^2\right] dxdy \tag{8}$$

The density and vorticity must be given as functions of velocity There are two distinct cases of non-uniform flows to be considered:

i) The main flow is isentropic (S=constant) with constant thermodynamic variables (P, ρ and T). The main flow is parallel to the x-axis (i.e. $v = 0$) while $u = u(y)$. The total enthalpy is a function of y, namely

$$H = \frac{\gamma}{\gamma - 1}\frac{P_\infty}{\rho_\infty} + \frac{1}{2}q_\infty^2(y) \tag{9}$$

In this case, the density is given by

$$\rho = \left[1 - \frac{\gamma - 1}{2}M_o^2\left(q^2 - q_\infty^2(y)\right)\right]^{\frac{1}{\gamma-1}} \tag{10}$$

where $M_o = \frac{u_o}{a_o}$ is a reference Mach number. In terms of a potential superimposed on the main shear flow, the equation becomes

$$\left(\rho\left[\phi_x + u_\infty(y)\right]\right)_x + \left(\rho\phi_y\right)_y = 0 \tag{11}$$

In figure (5), comparison of the solutions of the full potential equation and Cauchy/Riemann equations is given in terms of the velocity distribution. The pressure contours are given in figure (6).

ii) The main flow is isoenergetic (H=constant) where the pressure is constant but the density and the temperature vary with y. Again, $v = 0$ and $u = u(y)$. The entropy, S, is a function of y also. There is a relationship however between the variations of the entropy and of the velocity, namely,

$$-uu_y = T\Delta S_y \tag{12}$$

In this case, the density is given by $\rho = \rho_i K$ where
$K = e^{-\frac{\Delta S}{R}}$ and $\rho_i = \left[1 - \frac{\gamma-1}{2}M_\infty^2\left(q^2 - 1\right)\right]^{\frac{1}{\gamma-1}}$

A similar case was studied analytically by Lighthill in [9].

The pressure contours for the case of parabolic shear flow is given in figure (7).In the proper limit, the formulation can be reduced to a small disturbance equation in the form

$$\left[\left(1-M_l^2\right)-\frac{\phi_x u_\infty}{a_\infty^2}\left(2+(\gamma-1)\,M_l^2\right)\right]\phi_{xx}+\phi_{yy}=0 \tag{13}$$

where $M_l = u_\infty(y)/a_\infty(y)$. If the entropy and vorticity are allowed to change due to the disturbance, the formulation becomes equivalent to Euler Equations for adiabatic inviscid flows with no further approximations.

Acknowledgements: The authors would like to thank Prof. H.K. Cheng of USC for many interesting discussions of the sonic boom problem.

References

1. H. K. Cheng and M. M. Hafez: *Analysis and Numerical Simulation of the Superboom Problem* (Frontiers of Computational Fluid Dynamics, World Scientific 1998)
2. W. H. Guo and M. M. Hafez: *Numerical Simulation of the Super Boom problem* (Contemporary Research in Engineering Science, Springer Verlag 1995)
3. T. J. R. Hughes and L. P. Franca and G. M. Hulbert: *A new finite element formulation for Computational Fluid Dynamics: viii the Galerkin/Least Squares Method for advective diffusive equations* (Methods in Applied Mechanics and Engineering 73, pp 173-189, 1989)
4. J. Roy and M. M. Hafez and J. J. Chattot: *Explicit Methods for the solution of the generalized Cauchy-Riemann Equations and Simulation of Inviscid Rotational Flows* (Computers and Fluids 2001)
5. B. Engquist and S. Osher: *Stable and Entropy satisfying approximations for transonic flow calculations* (Mathematics of Computation, 34, pp 45-75, 1980)
6. B. Engquist and S. Osher: *One-sided difference approximations for nonlinear conservation laws* (Mathematics of Computation, 36, pp 321-352, 1981)
7. M. M. Hafez and W. H. Guo: *Wake Capturing in potential flow calculations* (Acta Mechanica, 143 No.1-2, pp 47-56, 2000)
8. A. Lerat and C. Core: *Residual-Based Compact Schemes for multidimensional hyperbolic systems of conservation laws* (Computers and Fluids, 31, No. 4-7, pp 639-662, 2002)
9. M. J. Lighthill: *Reflection at a Laminar Boundary Layer of a weak steady disturbance to a supersonic stream neglecting viscosity and heat conduction* (Quart. J. Mech. App. Math, 3, 1950)

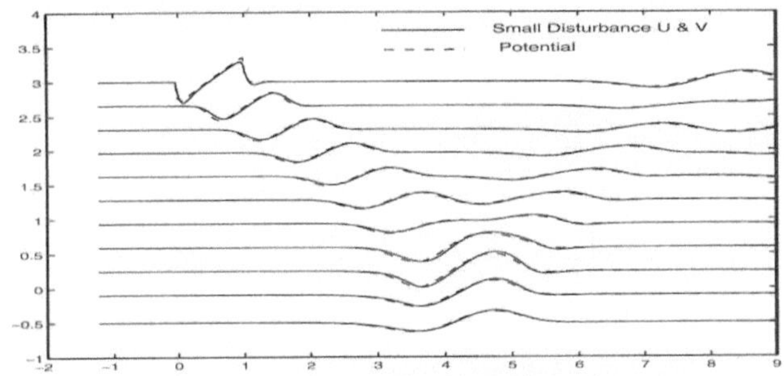

Fig. 1. U-velocity Distribution for 2D case

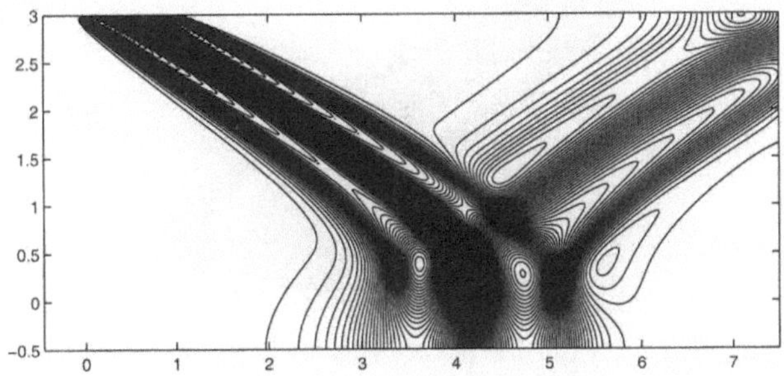

Fig. 2. Pressure Contour for 2D case - Potential Method

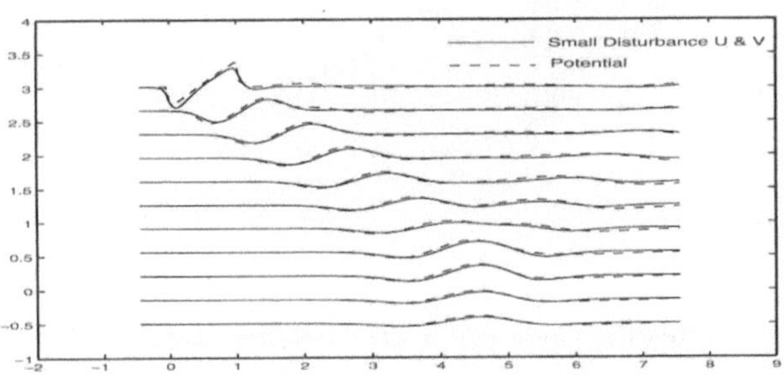

Fig. 3. U-velocity Distribution for 3D case

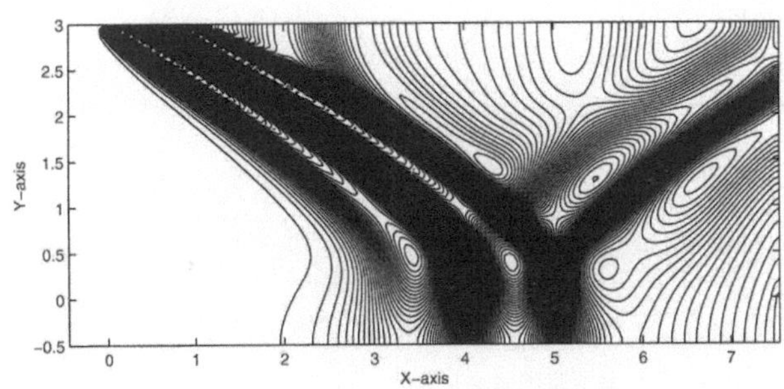

Fig. 4. Pressure Contour for 3D case - Symmetry Plane

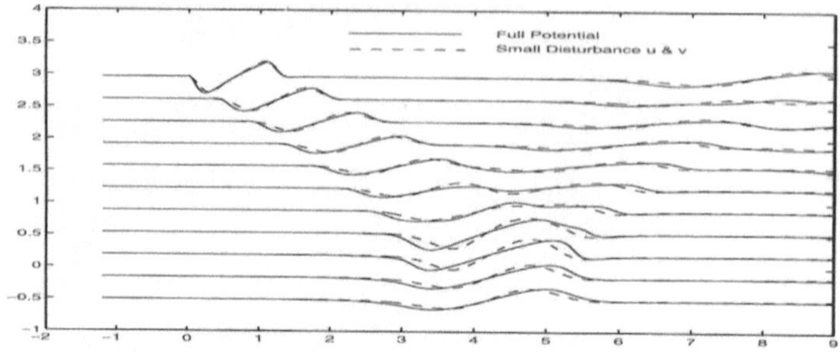

Fig. 5. U-velocity Distribution for 2D case

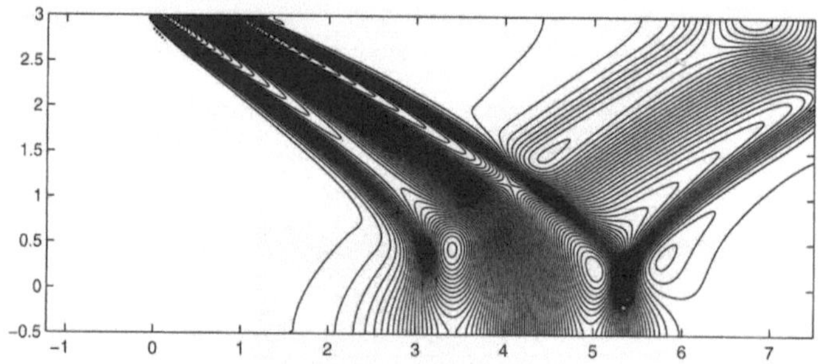

Fig. 6. Pressure Contour for 2D case - Isentropic Main Flow

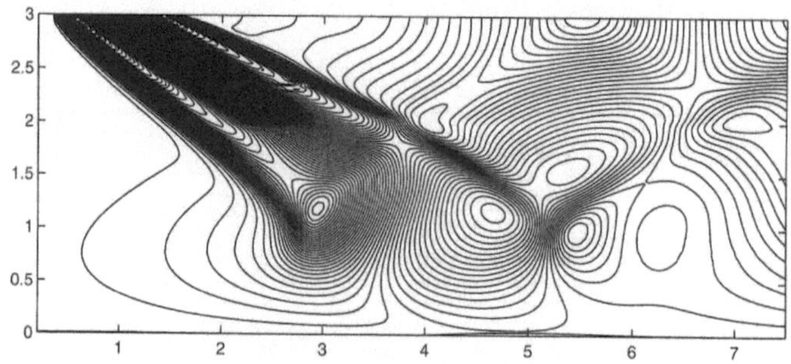

Fig. 7. Pressure Contour for 2D case - Isoenergetic Main Flow

Computing Jet Screech: A Complex Aeroacoustic Feedback System

Ching Y. Loh[1] and Lennart S. Hultgren[2]

[1] Taitech Inc., Brook Park, OH 44135, USA
[2] NASA Glenn Research Center, Brook Park, OH 44135, USA

Abstract. The space-time conservation-element and solution-element method is employed to numerically study the near-field axisymmetric screech-tone noise of a typical underexpanded circular jet issuing from a sonic nozzle. For the computed case, corresponding to a fully expanded Mach number of 1.19, the self-sustained feedback loop is established without artificial means. The computed shock-cell structure, acoustic wave length, screech tone frequency, and sound pressure levels are in good agreement with existing experimental results.

1 Introduction

Underexpanded supersonic jets radiate mixing noise, broadband shock-associated noise, as well as very loud discrete screech tones under certain conditions. Screech is of particular interest not only because of general noise-reduction concerns, but also because of potentially destructive structural interaction leading to sonic fatigue. Many experimental and theoretical investigations of jet noise have been carried out, see the review papers [1,2] for a comprehensive discussion. Mixing noise is directly associated with large-scale structures, or instability waves, in the jet shear layer; whereas, broadband shock-associated noise and screech tones are associated with the interaction of these waves with the shock-cell structure in the jet core. The screech tones arise because part of the acoustic waves generated by the wave/shock-cell interaction propagate upstream and re-generate the instability waves at, or in the vicinity of, the nozzle lip. This feedback loop leading to distinct screech tones is sensitive to small changes in the system conditions, and its explanation is to date mostly based on experimental observations, see [1,2].

Jet noise is a challenging topic in computational aeroacoustics – in particular, near-field noise computation in the presence of shock cells in the jet core – and reliable direct numerical simulation of jet screech noise has up to quite recently not been feasible. It imposes stringent requirements that the scheme must be able to: (i) handle shock waves, (ii) resolve acoustic waves with low dispersion and dissipation errors, (iii) resolve the instability waves in the jet shear layers and their interaction with the core shock-cell structure and (iv) have an effective non-reflecting boundary condition. Successful direct computation of screech for circular jets have been carried out by Shen and Tam [3,4] using the well-known DRP scheme and by the present authors [5] using the recent space time conservation element and solution element method, CE/SE, [6,7]. Further direct simulation of screech is presented below using the axisymmetric CE/SE Navier-Stokes scheme. The method is described in detail in [5,6].

2 Numerical Method

In general, the CE/SE method systematically solves a set of discretized (second-order accurate in both space and time) *integral* equations derived directly from the physical conservation laws and naturally captures shocks and other discontinuities in the flow. Both dependent variables and their derivatives are computed simultaneously. As a consequence, the flow vorticity can be obtained without reduction in accuracy. Non-reflecting boundary conditions (NRBCs) are also easily implemented because of the flux-conservation formulation.

With an unstructured grid, the CE/SE procedure is easily adapted to complicated geometries. Here, an unstructured triangle grid is used. Figure 1 shows a typical triangle cell, ΔABC, with O being its center and D, E, F being the centers of its three neighboring cells. The flow variables at the current time level

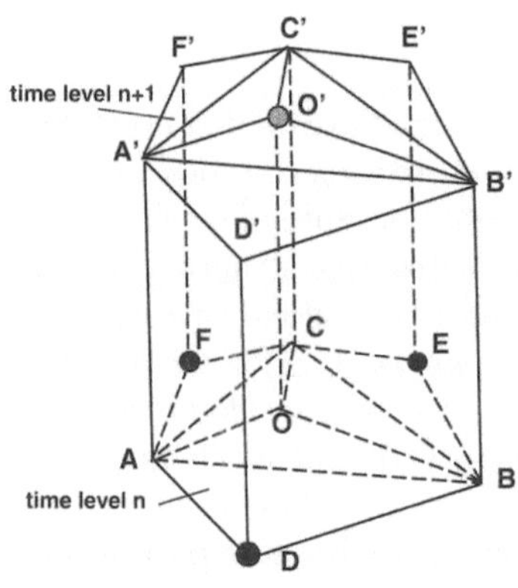

Fig. 1. CE/SE unstructured grid: hexagon cylinder ADBECFA-A'D'B'E'C'F'A' containing three CEs OADBO-O'A'D'B'O', OBECO-O'B'E'C'O', OCFAO-O'C'F'A'O'

n are stored at the center of each triangle cell. Three quadrilateral cylinders (conservation elements, CEs) are formed by the edges that connect the vertices and the center of the triangle and its three neighbors. The solution elements, SEs, are the interfaces between the CEs. An integral form of the conservation laws is applied to the hexagon cylinder that consists of these three quadrilateral cylinder conservation elements and explicitly determines the full solution at O' at the new time level $n+1$ using only information stored at D, E, and F (no extrapolations/interpolations across a stencil of cells are needed). Discontinuities are allowed to occur in a conservation element. Details about the unstructured CE/SE method can be found in [8]. The weighted a-ϵ CE/SE scheme is used with the weighting parameter $\alpha = 1$ and $\epsilon = 0.5$ in the present computations.

3 Computation and Comparison with Experiment

Figure 2(a) shows the geometry of the convergent nozzle in Panda's experiment [9]. The flow is choked (i.e. Mach number $M = 1$) at the nozzle exit. The jet Mach number which represents the plenum/ambient pressure ratio for the simulation is $M_j = 1.19$ [9]. The computational domain, see Fig. 2(b), spans between $-8.3D \leq x \leq 6D$ and $0 \leq r \leq 11.7D$, with x and r being the streamwise and radial coordinates and D being the jet nozzle diameter. The flow inside the nozzle is not computed, rather the steady flow conditions are prescribed at the nozzle exit which is located at $x = 0$. This inflow plane is recessed by two cells so as not to numerically restrict or influence the feed-back loop. There are totally 230,000 triangle cells. Non-reflecting boundary conditions are applied to the upper and outflow boundaries and a symmetry condition is applied at the center axis. The last 10 streamwise cells have exponentially growing size and serve as a buffer, or sponge, zone to essentially eliminate any small remaining numerical reflection from the downstream outflow boundary. Initially, the entire flow is at rest and at ambient conditions; the jet flow is then impulsively started. The Reynolds number $Re = Da_o/\nu = 570,000$, where a_o is the ambient speed of sound (used as velocity scale) and ν is the kinematic viscosity at the ambient conditions. Figure 3(a) displays numerical pressure contours and numerical Schlieren contours, outside and inside of the jet core respectively, well after the start-up transients has passed out of the computational domain. Distinct screech waves are observed to emit from the 3rd to the 5th shock-cell and are reflected at the flange/nozzle body. The screech wavelength ($1.6D$) and the shock-cell structure ($0.8D$ spacing) agree well with the experiment [9]; for the latter see Fig. 3(b). Spectral analysis yields a computed screech frequency of 8513 Hz, which agrees well with the experimental value of 8525 Hz. The sound pressure level (SPL) along an inclined line at the outer edge of the shear layer is shown in Fig. 4. This figure shows Panda's [9] data for the A_2 axisymmetric screech mode, the corresponding result from the simulation as well as the computed total SPL and subharmonic

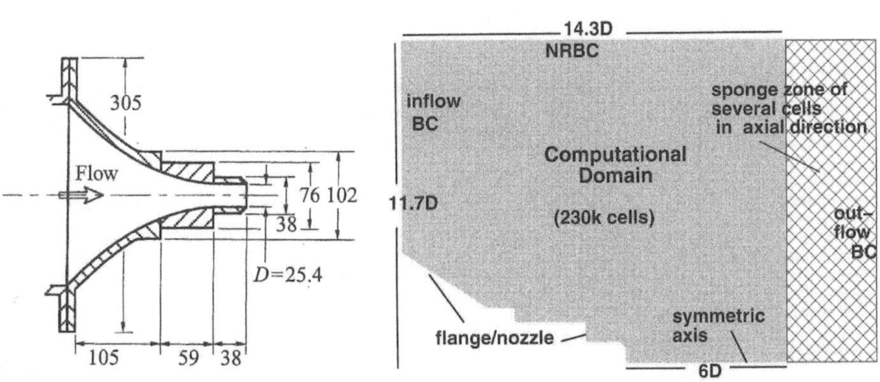

Fig. 2. (**a**) Geometry of the convergent nozzle and flange in Panda's experiment (dimensions in mm). (**b**) Computational domain

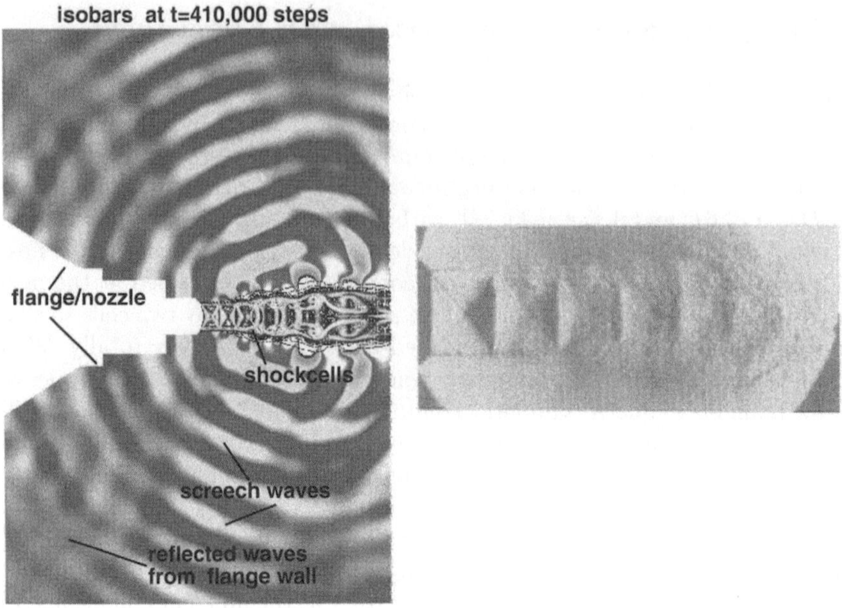

Fig. 3. (**a**) Isobars at t=410,000 steps, showing screech waves and shock cell structures. (**b**) Experimental Schlieren picture showing the shock-cell structure [9]

of the A_2 mode. Even though the A_2 SPL level in the vicinity of the nozzle lip is too low, its early streamwise growth rate is, however, well predicted indicating that the jet shear layer is in general well resolved. Nonlinearity in the shear layer limits the amplitude that can be obtained and once the SPL level 'catches up' the agreement is very good, in particular the SPL level in the streamwise region where the backward radiating acoustic waves are generated is well predicted. The

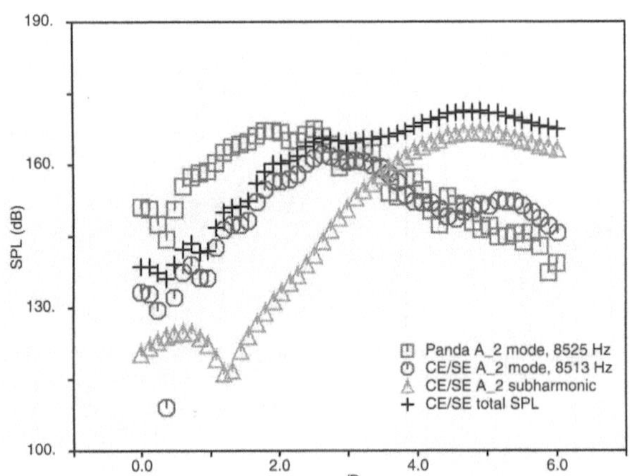

Fig. 4. Comparison of computed and experimental SPL along the shear layer edge

strong subharmonic that appears further downstream is due to (axisymmetric) vortex pairing in the shear layer and as a consequence a second streamwise peak at the A_2 frequency occurs due to nonlinear effects. However, 3-D effects are most likely to have come into play in the experiment at these streamwise locations leading to suppression of the subharmonic. The reason for the low initial SPL values in the simulation is currently being pursued.

Figure 5 shows computed SPL contours for the A_2 mode. As in the experiment [9], a standing wave structure can be observed along the edge of the jet shear layer. The results are in general agreement with the experimental ones except, as pointed out earlier, that the computed SPL levels are too low in the vicinity of the nozzle lip and the existence of a second, or extended, region of elevated values further downstream ($x/D > 4$).

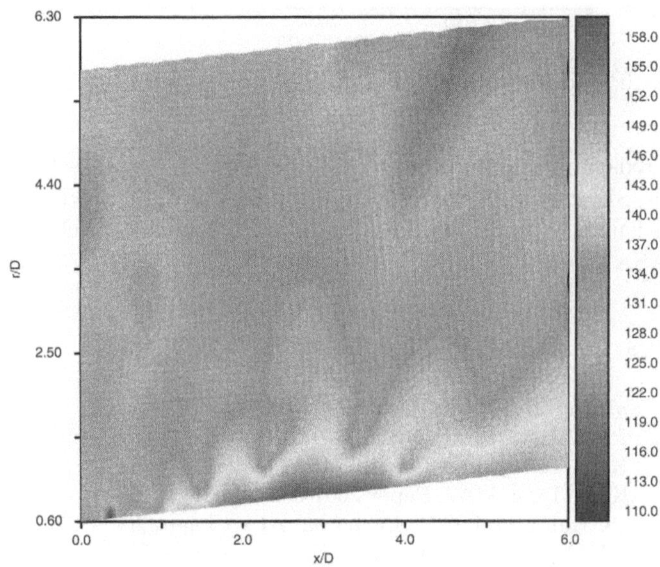

Fig. 5. Computed A_2-mode SPL levels for Panda's experiment

Figure 6 displays the computed SPL at the nozzle exit lip wall, $x/D = 0$, $r/D = 0.6$. The SPL shows a distinct spike corresponding to the A_2-mode screech tone observed in Panda's [9] experiment. Note that in our previous work [5], which had a much simplified description of the nozzle external geometry, the A_1-mode screech tone (not observed in [9]) was also obtained at this condition. This illustrates the sensitivity of the screech phenomenon to geometry changes.

4 Summary

It is concluded that the simulation shows a reasonably good agreement with experimental data in the streamwise region where the flow is expected to be pre-

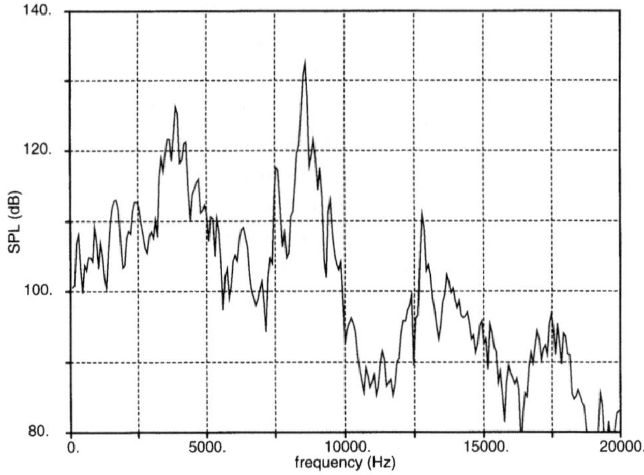

Fig. 6. SPL at nozzle exit ($x/D = 0$, $r/D = 0.6$), 75 Hz digital binwidth

dominantly axisymmetric and, hence, that jet screech is successfully simulated using the CE/SE scheme. **Acknowledgements:** The authors wish to thank Dr. J. Panda for fruitful discussions. The Supersonic Propulsion Technology Project Office of NASA Glenn Research Center supported this work.

References

1. C.K.W. Tam: Ann. Rev. Fluid Mech. **27**, 17-43 (1995)
2. G. Raman: Prog. Aerospace Sci. **34**, 45-106 (1998)
3. H. Shen, C.K.W. Tam: AIAA Paper 98-0283 (1998)
4. H. Shen, C.K.W. Tam: AIAA Paper 2001-0820 (2001)
5. C.Y. Loh, L.S. Hultgren, P.C.E. Jorgenson: AIAA Paper 2001-2252 (2001).
6. S.-C. Chang, X.-Y. Wang, Chow, C.-Y. Chow: J. Comp. Phys. **156**, 89-136 (1999)
7. C.Y. Loh, L.S. Hultgren, S.-C. Chang: *AIAA J.*, **39**, 794-801 (2001)
8. X.-Y. Wang, S.-C. Chang: C.F.D. J. **8**, 309-325 (1999)
9. J. Panda: *J. Fluid Mech.* **378**, 71-96 (1999)

Direct Simulation of Transitional Shock Boundary-Layer Interactions at Mach 2

Yufeng Yao and Neil Sandham

University of Southampton, Southampton SO17 1BJ, United Kingdom

Abstract. Direct simulation of transitional shock boundary-layer interactions at Mach 2 has been carried out by solving the Navier-Stokes equations. The undisturbed two-dimensional base flow was simulated and validated first, followed by a linear stability analysis to identify the most unstable modes. The linear stability predictions are compared with results from two- and three-dimensional unsteady simulations. The overall amplification factors seen in theory and simulations are consistent with the observations of transition in laboratory experiments. This indicates that parallel flow inviscid stability analysis is a useful tool to predict transition in these flows. Simulation of the early nonlinear stages of transition show the presence of streamwise vortical structures.

1 Introduction

Shock boundary-layer interaction at Mach 2 has been studied for several decades by experiment [1] and numerical computations [2] [3] [4]. The main features of the interaction between the laminar boundary-layer and an impinging oblique shock are well-understood and are illustrated in figure 1(a). The shock strength is represented by the impinging shock angle (ϕ). Providing the shock is strong enough, the viscous boundary-layer separates upstream and re-attaches downstream of the shock impingement point, forming a closed recirculation bubble.

Hakkinen *et al.* [1] carried out experimental studies at Mach 2 with Reynolds numbers of approximately 3×10^5 (based on freestream quantities and the distance from the leading edge to the shock impingement point). They found that for weak/moderate interactions the flow is fully laminar. As the shock strength increased, however, the flow exhibited transitional behaviour, as shown by the large increase of the skin friction after flow re-attachment. The presence of mean flow three-dimensionality in the experiments is a complicating factor, so in this contribution we consider the transition process in a flow that is two-dimensional in the mean.

2 Numerical Method

The computation uses a three-dimensional compressible direct numerical simulation code [5], in which 4th-order central finite differences were employed for

[1] The authors acknowledge the financial support by the UK Engineering and Physical Science Research Council (EPSRC) through grant GR/M 84336.

evaluating spatial derivatives with an explicit multi-stage compact Runge-Kutta algorithm for time advancement. Additional features include entropy splitting for the Euler terms and a stable higher order boundary treatment. A shock-capturing method of the total variation diminishing family, combined with the artificial compression method, is used to detect flow discontinuities. Code validation has been carried out for a wide range of flow problems [6], including laminar/turbulent boundary-layer and channel flows. Results are in agreement with analytical solutions and other computations.

3 Results and Discussions

3.1 Problem definition

We consider a computational domain of $L_x = 400, L_y = 115$ with lengths normalized by inflow boundary-layer displacement thickness (δ_1^*). The shock impinges approximately at the centre of the lower boundary. The computational mesh is generated analytically, resulting in a grid of $N_x = 151, N_y = 128$ points, uniformly spaced in the streamwise direction (x) and non-uniformly spaced in the wall-normal direction (y) (see figure 1(b)). The inflow laminar boundary-layer at Mach 2 is produced by a separate self-similar compressible boundary-layer solution. The Reynolds number is 2.96×10^5 based on the distance from the leading edge to the shock impinging point. Systematic domain dependency and grid refinement studies have been carried out (see Yao *et al.* [4] for details) to minimize the numerical influence on the solution. It has been confirmed that the current domain size and grid resolution are sufficient for this study.

3.2 Undisturbed 2D base flow

Simulation of an undisturbed 2D base flow has been carried out previously [4]. The oblique shock was introduced on the upper boundary at $x \simeq 40.0$ with angles $\phi = 31.67^o, 32.58^o, 34.05^o$ for cases $1-3$, respectively. Case 1 has a weak interaction with incipient flow separation. Case 2 has a moderate interaction with an asymmetrical separation bubble. Case 3 has a strong interaction with extensive flow separation and large reverse flow (the reverse velocity is up to 5% of freestream velocity). For this case, a secondary vortex forms inside the primary vortex.

Figure 2(a) shows the skin friction (C_f) distributions for cases $1-3$. The zero pressure gradient simulation and compressible boundary-layer solution are also plotted for reference. It is clear that the separation bubble increases in size as the shock strength increases. Validations have been done for case 1 and case 2 by comparison with references [1] [2] [3]. As an example, figure 2(b) shows the C_f comparisons for case 2. All computations captured the secondary minimum in C_f just ahead of a rise associated with re-attachment. Computations using different numerical approaches agree with each other and also match fairly well with the experiment except for the lower recovery of C_f downstream after re-attachment.

Differences after the flow re-attachment may be due to measurement difficulties, transition or (most likely) the three-dimensional mean flow effects (influence of tunnel side walls). Given the lack of a comparable simulation of case 3 at the moment, a careful study has been done to confirm that the computed solution is grid-independent. The ratio of separation bubble length to the undisturbed boundary-layer displacement thickness at the impingement position, l_B/δ_o^*, is estimated to be 130, in agreement with an extrapolated value based on the study of Katzer [2].

3.3 Linear stability analysis

Parallel flow instability analysis has been carried out with a base flow taken from the numerical solution given in the previous section. The inviscid equations are linearised by introducing a decomposition into a base flow and a small perturbation

$$(\rho, u_{i,}, T) = (\overline{\rho}, \overline{u}_{i,}, \overline{T}) + (\rho', u'_{i,}, T'). \tag{1}$$

The perturbations are prescribed by normal modes

$$(\rho', u'_{i,}, T') = (\hat{\rho}, \hat{u}_{i,}, \hat{T}) \exp[i(\alpha x - \omega t + \beta z)], \tag{2}$$

where α and β are streamwise and spanwise wavenumbers respectively, ω is a frequency and hatted variables are eigenfunctions that are a function only of the normal coordinate y. We consider the case of spatial instability, where ω and β are real and α is complex. To build up a complete stability map the stability problem is first solved at some streamwise location inside the separation bubble. Then the solution is marched both upstream and downstream, using the previously converged solution as the initial guess and continuing until the mode becomes neutral. An n-factor is computed as

$$n = \ln\left(\frac{A}{A_0}\right), \tag{3}$$

where A is the wave amplitude and a subscript zero denotes the upstream neutral point. The process is repeated for many ω and β until an overall maximum is found, representing the mode most likely to cause transition. Figure 3(a) shows the n-factor variation for the three shock impingement cases, showing the most dangerous mode (highest overall n-factor) in each case. The peak n-factor increases from around 3 for the lowest impingement angle to 10 at the highest angle. A common transition criterion is n=9, implying that transition is expected at the highest shock impingement angle. There is some evidence from Hakkinen [1] that transition is indeed a feature of the flow close to this case.

Stability of high speed shear layers depends on the convective Mach number M_c (Papamoschou and Roshko [7]), with significant compressibility effects expected for $M_c > 0.3$. For the present flow we can take the wall as the lower limit, giving a convective Mach number as

$$M_c = \frac{U_e}{a_e + a_w}, \tag{4}$$

where a is the sound speed and subscripts e and w refer to local freestream and wall conditions respectively. Convective Mach numbers defined in this way were computed to be in the range 0.75 to 0.85. The most unstable modes are at angles of between 50° and 60°, consistent with the local convective Mach numbers being in excess of 0.6 (Sandham and Reynolds [8]).

In the following section we compare n-factors computed from linear stability with those from two- and three-dimensional DNS. This will help assess the usefulness of inviscid parallel flow stability results for predicting linear disturbance growth in transitional shock boundary layer interaction problems at moderate Mach numbers.

3.4 2D/3D flow with small disturbances

We consider the strong interaction (case 3) to evaluate the linear disturbance growth in the simulation. The external forcing disturbance takes the form

$$v' = a_f \exp[-b_f(x - c_f)^2] \sin(\omega t) \cos(\beta z), \tag{5}$$

which is applied on the lower boundary. Constants are $b_f = 0.03125, c_f = 40$. For the two-dimensional simulation we take $a_f = 1.0 \times 10^{-3}, \omega = 0.1, \beta = 0.0$, and for the three-dimensional simulation $a_f = 1.0 \times 10^{-6}, \omega = 0.06, \beta = 0.18$. The spanwise computational box length is chosen as 35 with 17 grid points used.

The n-factor from the simulations is derived by calculating the amplitude of the normal velocity v signal at about $50T_\omega$ ($T_\omega = 2\pi/\omega$) and at maximum $\partial u/\partial y$ locations. A reference location is taken at $x = 140$ for the 2D/3D simulations. At this location the simulated growth factor is matched to the linear stability result. Figures 3(b) and 3(c) show comparisons of simulations using the small external forcing disturbances given in (5) with the linear stability analysis. The results show better agreement upstream of the shock impingement point ($x \simeq 200$). Differences appear after the shock, with higher growth predicted by simulation, indicating that the shock plays an additional role in the amplification of disturbances. Other factors that could cause the difference are non-parallel flow effects, curvature, and viscous effects in the simulation.

3.5 3D transitional flow simulation

After the validation in the linear region, a 3D simulation of case 3 using larger disturbances ($a_f = 2.5 \times 10^{-6}$) was carried out up to time $60T_\omega$. Figure 4(a) shows the skin friction (C_f) distribution. It can be seen that C_f remains broadly unchanged upstream of the shock impingement, but shows a significant increase after re-attachment, similar to that found in [1] at $\phi = 34.5^o$. Figure 4(b) shows the instantaneous vorticity contours (ω_y) at $x = 340$ downstream of re-attachment. Figure 5 shows a top-view of vorticity (ω_y) at $y = 1$. The flow becomes fully three-dimensional with counter-rotating vortex pairs appearing.

Effects of curvature for current Mach 2 simulation (case 3) were also assessed. In the region where streamline curvature is destabilising, the ratio of radius of curvature to shear layer vorticity thickness is in excess of 100, indicating that the Görtler mode of transition probably has only a minor effect in our study.

4 Conclusions

Direct simulation of transitional shock boundary-layer interaction at Mach 2 has been carried out together with linear stability analysis to identify the most unstable modes. For the strong shock interaction, the growth (n) factors agree reasonably well with those from simulations using small disturbances. A three-dimensional simulation using larger disturbances has been carried out showing that the early stage of flow transition occurs after re-attachment with a higher skin friction value and well-developed three-dimensional vorticity field. The Görtler mode of transition probably has only a minor effect due to a large radius of streamline curvature. Future work will include higher Mach number flow and heat transfer effects.

References

1. R.J. Hakkinen, I. Greber, L. Trilling and S.S. Abarbanel: *The interaction of an oblique shock wave with a laminar boundary layer*, NASA Memo 2-18-59W, (1959)
2. E. Katzer: Journal of Fluid Mechanics, **206**, 477-496 (1989)
3. B. Wasistho: Spatial Direct Numerical Simulation of Compressible Boundary Layer Flow. Ph.D thesis, University of Twente, Netherlands (1998)
4. Y.F. Yao, R.O. Bura, N.D. Sandham and G.T. Roberts: 'Time-accurate simulation of supersonic and hypersonic laminar shock/boundary-layer interactions'. In: *Proceedings of CEAS Aerospace Aerodynamics Research Conference, Cambridge, United Kingdom, June 10-13, 2002.*
5. N.D. Sandham, Q. Li and H.C. Yee: Journal of Computational Physics, **178**, 307-322 (2002)
6. Y.F. Yao, A.A. Lawal, N.D. Sandham, I.C. Wolton, M. Ashworth and D.R. Emerson: 'Massively parallel simulation of shock/boundary-layer interactions'. In: *Proceedings of International Conference of Applied Computational Fluid Dynamics at Beijing, China, October 17-20, 2000*, ed. by J.H. Wu and Z.J. Zhu, pp. 728-735
7. D. Papamoschou and A. Roshko: Journal of Fluid Mechanics, **197**, 453-477 (1988)
8. N.D. Sandham and W.C. Reynolds: AIAA Journal **28(4)**, 618-624 (1990)

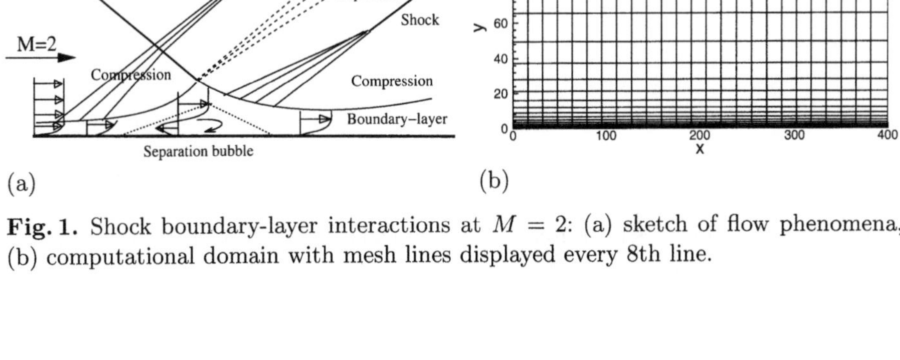

(a) (b)

Fig. 1. Shock boundary-layer interactions at $M = 2$: (a) sketch of flow phenomena, (b) computational domain with mesh lines displayed every 8th line.

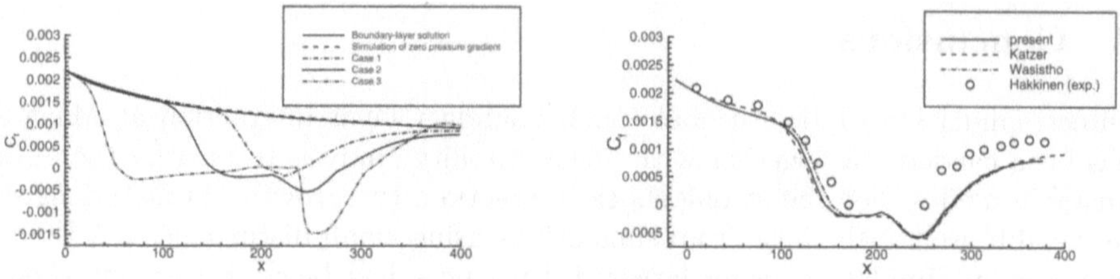

(a) (b)

Fig. 2. Simulation of undisturbed 2D base flow: (a) skin friction distributions for cases 1 − 3, (b) skin friction distributions of case 2 compared with experiment and other computations.

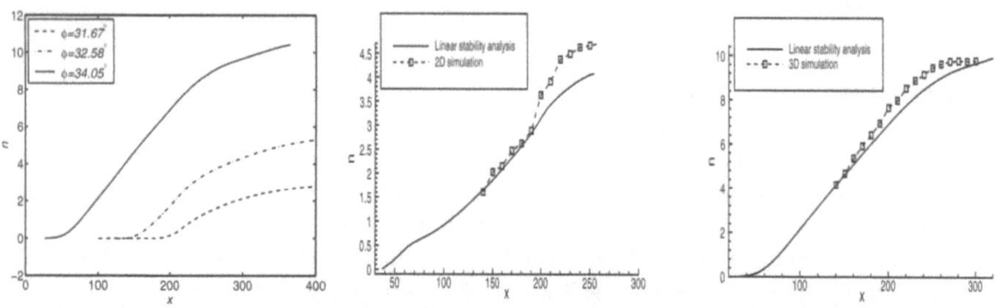

(a) (b) (c)

Fig. 3. Comparisons of linear growth (n) factors: (a) linear stability analysis for cases 1-3, (b) 2d simulation, (c) 3d simulation.

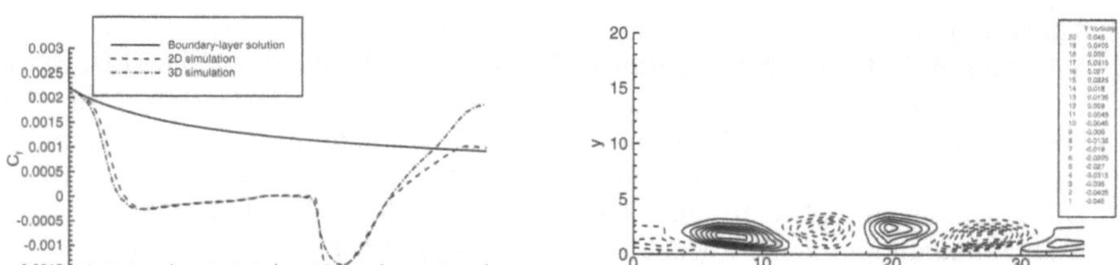

(a) (b)

Fig. 4. 3D transitional simulation (case 3) with large disturbances: (a) skin friction distributions, (b) vorticity contours (ω_y) in $y - z$ plane at a location of $x = 340$.

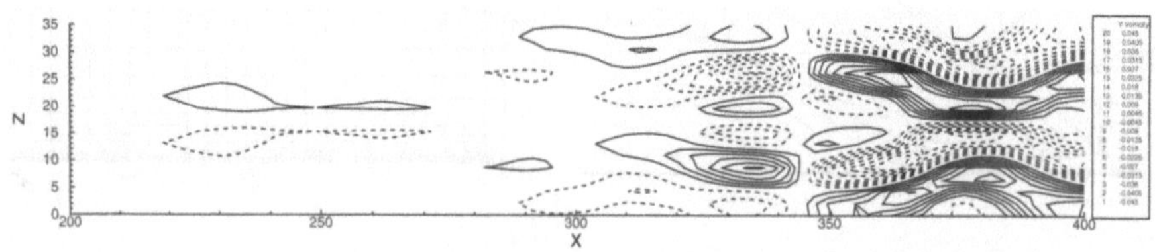

Fig. 5. Vorticity contours (ω_y) from 3D transitional simulation (case 3) in $x - z$ plane at a location of $y = 1$.

Numerical Simulation of Shock-Interface Interaction With Modified Compact Scheme

Fu Dexun and Ma Yanwen

Lab. of Nonlinear Mechanics, Institute of Mechanics, CAS,
Beijing 100080, China, fudx@lnm.imech.ac.cn

Abstract. A modified fourth order accurate compact scheme with group velocity control (GVC) is developed. The scheme is high order accurate with less stencil, simple, less computer consuming, and can capture the shocks with high cross pressure ratio and the contact discontinuity with high density ratio. The developed method is used to simulate Richtmyer-Meshkov (R-M) instability problem driven by a cylindrical shock. The behavior of R-M instability is studied.

1 Introduction

For solving shock-interface interaction with high cross density ratio and pressure ratio it is required that the numerical method can capture both shock and contact discontinuity well. There have been a lot of activities geared towards constructing efficient schemes with high resolution of the shock. These include TVD and ENO type schemes. Development of the shock capturing methods of TVD and ENO types is mainly from the viewpoint of mathematics. In the most existing TVD and ENO type schemes the physical reason of oscillation production is not considered directly. In [2] the relevance of group velocity in numerical solution to the behavior of the solution of finite difference schemes was considered. Schemes were divided into three groups with the group velocity of wave packets: slower (SLW), faster (FST) and mixed (MXD). For the SLW schemes the numerical wave packets are propagated slower than the physical ones, and the oscillations in numerical solutions may appear behind the shock. For the FST schemes the numerical wave packets are propagated faster than the physical ones and the oscillations may appear in front of the shocks. For the MXD schemes in some range of low and middle wave numbers the schemes are FST, but are SLW in the range of high wave numbers. In order to improve the numerical solutions the method of group velocity control (GVC) is presented in [2]. The basic idea of the GVC scheme is that the scheme used must be reconstructed so that it is FST/MXD behind the shock, and is SLW in front of the shock. According to the basic idea of the GVC method a modified fourth order accurate compact scheme is developed. The scheme is used to simulate Richtmyer-Meshkov (R-M) instability problem. The numerical results for interaction between the cylindrical shock and the cylindrical interface with small disturbance are presented. The behavior of R-M instability is studied.

2 Method Development

2.1 Scheme description

Consider the following model equation and its semi-discrete approximation

$$\frac{\partial u}{\partial t}+\frac{\partial f}{\partial x}=0, f=cu, c=const.>0 \tag{1}$$

$$\frac{\partial u_j}{\partial t}+\frac{F_j}{\Delta x}=0 \tag{2}$$

where $F_j/\Delta x$ is an approximation of $\partial f/\partial x$ and F_j is obtained from

$$\frac{1}{6}F_{j+1}+\frac{2}{3}F_j+\frac{1}{6}F_{j-1}+H_{j+1/2}-H_{j-1/2}=h_{j+1/2}-h_{j-1/2} \tag{3}$$

$$H_{j+1/2}=(\varepsilon_2-\varepsilon_4)_{j+1/2}(F_{j+1}+F_j)$$

$$h_{j+1/2}=(f_j+f_{j+1})/2-2\varepsilon_{4,j+1/2}(f_{j+1}-f_j) \qquad \varepsilon_{2,j+1/2}=\left[g_{j+1/2}(\rho)\right]^3$$

$$\varepsilon_{4,j+1/2}=\left[1+\gamma_0 S_{j+1/2}(\rho)\right]g_{j+1/2}(\rho), \quad S_{j+1/2}(\rho)=\text{sign}\left[\partial\rho/\partial x, \partial^2\rho/\partial x^2\right]$$

$$g_{j+1/2}(\rho)=[g_j(\rho)+g_{j+1}(\rho)]/2$$

$$g_j(\rho)=\left[|\rho_{j+1}-2\rho_j+\rho_{j-1}|/(\rho_{j+1}+2\rho_j+\rho_{j-1}\right]^{1/2}$$

The function $g(\rho)$ is used to control the accuracy and dissipativity, $S(\rho)$ is used to control the group velocity. The 3-stage TVD R-K method is used for time discretization. Here not any limiter and minmod function are used.

2.2 Accuracy and behavior analysis of numerical solution

According to the analysis in Ref.[2] for improvement of shock resolution the scheme should be constructed so that it is SLW in front of the shock and FST or MXD behind the shock. Fourier analysis shows that scheme (3) is dissipative for $\varepsilon_2>0$ and $\varepsilon_4>0$, SLW for small ε_2 and ε_4, and MXD for $\varepsilon_4>\varepsilon_2$ and $\varepsilon_4>\varepsilon_0$ (ε_0 is a positive constant). In fig.1 is given variation of $\partial k_i(\alpha)/\partial\alpha$ related to the group velocity of wave packet and $k_r(\alpha)$ related to the numerical dissipativity as function of α. The accuracy of the scheme is $O(\Delta x\varepsilon_2+\Delta x^3\varepsilon_4)$. $D(\alpha)$ is the group velocity of wave packet and $\alpha=k\Delta x$, k is wave number. It is obvious that the constructed scheme is fourth order accurate in the smooth region. Numerical experiments show that the scheme has high resolution of the shock with high cross pressure ratio and high density difference across the interface.

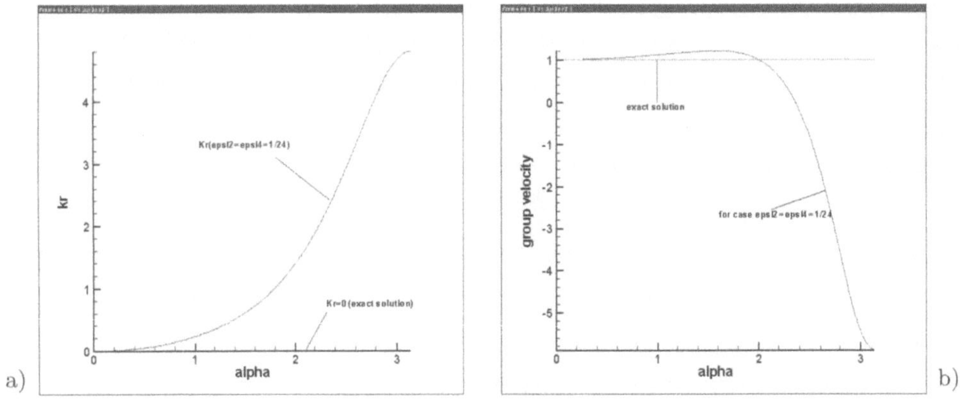

Fig. 1. (a) Variation of $K_r(\alpha)$ for $\varepsilon_2 = \varepsilon_4 = \frac{1}{24}$. (b) Variation of $dk_i/d\alpha$

3 Numerical Experiments

3.1 Model problems

The scheme is used to solve the Euler equations for simulation of 1-D shock tube and Riemann problems. Numerical results of 1-D shock tube with steady shock for incoming Mach number M_a=30 are given in fig.2. Numerical results for 1-D Riemann problems at $t = 0.4$ with initial condition $(t = 0) : u_1 = u_2 = 0, p_1/p_2 = 10, \rho_1/\rho_2 = 800$ and 8 are given in fig.3. u, p and ρ are velocity, pressure and density, respectively, the lower index 1 and 2 related to parameters in two sides of discontinuity. From fig.3.2 it can be seen that numerical results agree well with exact solutions.

For 2-D Riemann problem the initial conditions are taken as shown in fig.4: lower index corresponds to the sub domain. The computational region is $-0.5 \leq x \leq 0.5, -0.5 \leq y \leq 0.5$ with grid points 400×400. Numerical results at $t = 0.6$ are given in fig.5 for density counters and in fig.6 for pressure counters. From fig.5 and fig.6 it can be seen that the double Mach reflection is formed, and obtained results including contact discontinuity, position of the mushroom cap and fine structures agree well with results obtained in ref. [1]. It should be noted that in our results there is not nonphysical discontinuity produced due to nonphysical oscillations.

3.2 Richtmyer-Meshkov instability

The developed scheme with three stage R-K method is used to solve the 2-D N-S equations for simulation of Richtmyer-Meshkov(R-M) instability problem. A cylindrical shock collides a cylindrical interface with small sinusoid disturbances. The shock wave implodes from heavy fluid to light fluid. The initial interface is located as follows

$$r = 1 + a_0 cos\,(n\theta)$$

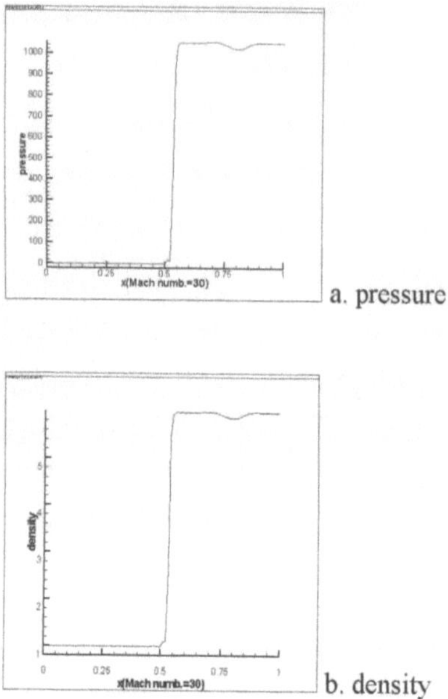

Fig. 2. Shock tube $Ma_\infty = 30$

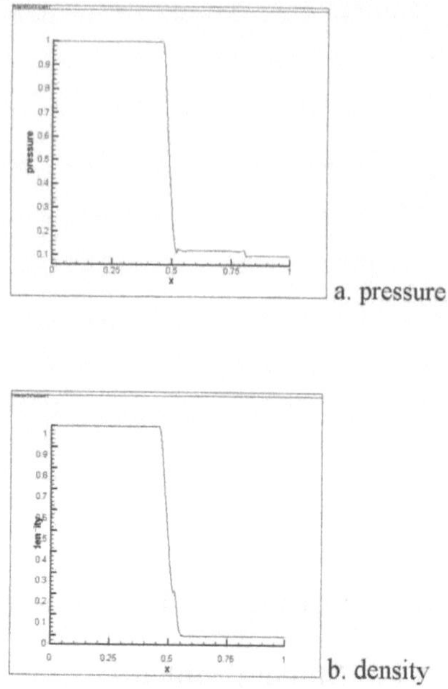

Fig. 3.1. Results of Sod problem $\rho_1/\rho_2 = 800$

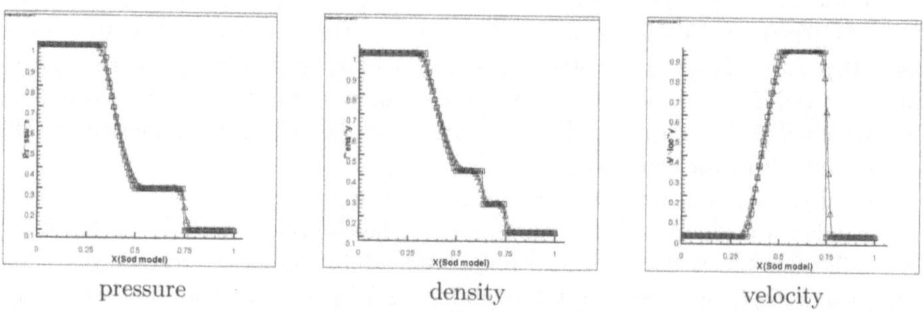

Fig. 3.2 Results of Sod model with $\rho_1/\rho_2 = 8$ and comparison with exact solution

where $r = \bar{r}/R_0$, $a = \bar{a}/R_0$ are dimensionless radius and perturbation amplitude, R_0 is the mean radius of the interface at $t = 0.0$ (the radius of the unperturbed interface). In present computation the initial dimensionless perturbation amplitude at the interface is a_0=0.033, and the Mach number of the shock is Ma=1.2. The ratio between outside and inside densities of the interface is ρ_1/ρ_2=5 and 20. The Reynolds number is Re=5000. For getting fine structures in computation non-uniform mesh grid system near the disturbed region is used. The density counters at different times are given in fig.7 for ρ_1/ρ_2=5, the pressure and the density distributions along a radial direction of 45 degrees at the corresponding times are also given. From figures it can be seen when the incident shock collides the interface at t =0.034 (where t is dimensionless time defined as

$t = \bar{t}\, R_0/U$, and U is the speed of the incident shock), the shock bifurcates into a transmitted shock which is moving radially inward, two reflected rarefaction waves which are moving radially outward. After bifurcation the phase inversion at the interface is produced due to interaction, and the interface is flattened. There are a series of wave interactions behind the transmitted shock. At $t = 1.0$ the phase inversion is completed, and the transmitted shock is reflected back from the center of the cylinder. The interface is re-shocked twice by the reflected back shock, second bifurcation is produced, and the interface is moving toward to the center. For second bifurcation the incoming shock from the center explodes from light fluid to heavy, so there is no phase inversion at the interface, and the reflected wave is reflected shock, which will move away from the center. In this case there are several re-shocks at the interface. From above results it can be seen that the numerical method used in present paper can capture shocks, reflected waves, interface and fine structures well.

References

1. M. Brio, A. R. Zakharian and G. M. Webb, JCP 167:177-195, 2001
2. Fu Dexun & Ma Yanwen, JCP 134: 1-15, 1997
3. Q. Zhang, M. J. Graham, Phys. Fluid, Vol.10, No4, 1998

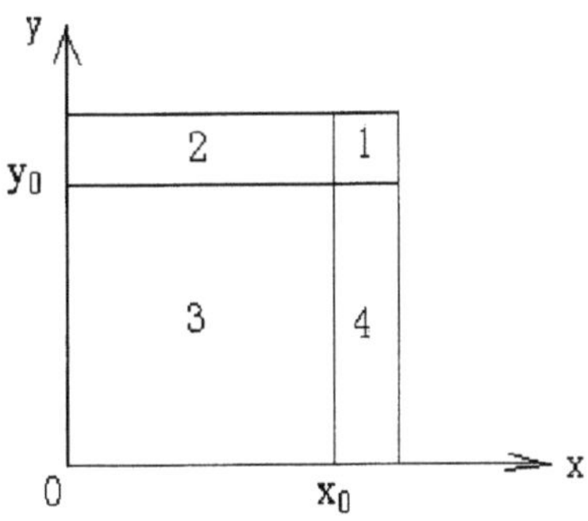

$$\rho_1 = 1.5, p_1 = 1.5, u_1 = v_1 = 0.0$$
$$\rho_2 = 0.5323, p_2 = 0.3, u_2 = 1.206, v_2 = 0.0$$
$$\rho_3 = 0.1379, p_3 = 0.029, u_3 = 1.206, v_3 = 1.206$$
$$\rho_4 = 0.5323, p_4 = 0.3, u_4 = 0.0, v_4 = 1.206$$

Fig. 4. Initial conditions ($x_0 = y_0 = 0.8$)

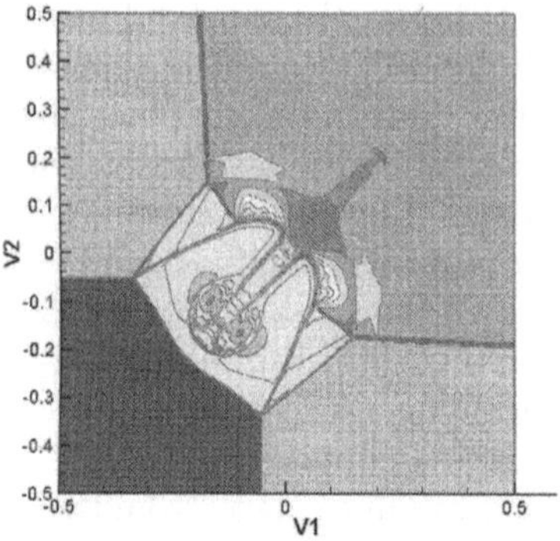

Fig. 5. Density contours

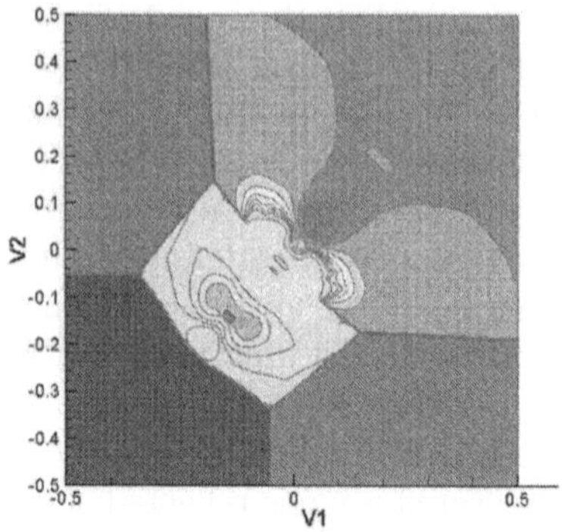

Fig. 6. Pressure contours

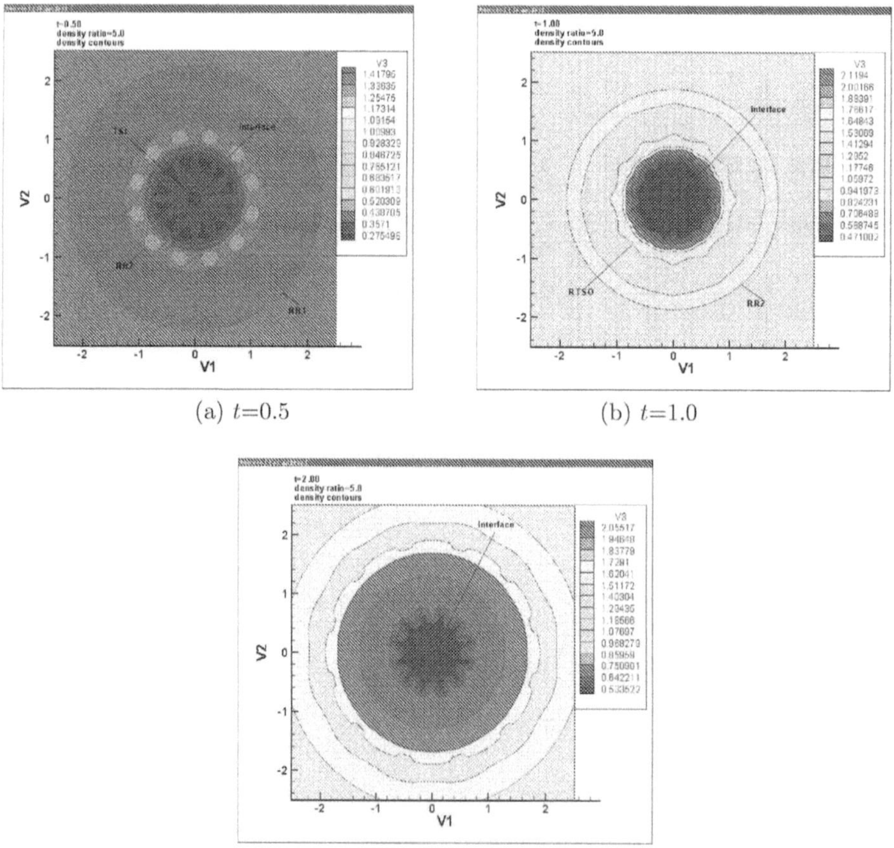

(a) t=0.5

(b) t=1.0

(c) t=2.0

Fig. 7. Density contours at different time for ρ_1/ρ_2=5, RR-Reflected rarefaction wave; TSI-Transmitted shock moving inward; RTSO-Reflected transmitted shock moving outward.

Direct Molecular Simulation (Monte-Carlo Method) of Weak Mach Shock Reflection

D.J. Auld

School of Aerospace, Mechanical and Mechatronic Engineering,
The University of Sydney 2006, Australia

Abstract. The problem of predicting the behaviour of reflected shock waves has been an ongoing one over many years. The use of Rankin-Hugoniot theory [1] to predict gas property changes for normal and oblique shocks and the case of has been shown to be extremely accurate. However in the case of weak Mach reflection of shocks there has been reported that a discrepancy exists between observations and the usual application of this theory [2][3].

Previous experimental results [4] appear to indicate that an enthalpy dissipation term is required to the standard application of Rankin-Hugoniot theory so as to produce correct predictions when there is a confluence of three shocks as is the case of the weak Mach reflection.

As a means of independently verifying the flow properties for this case, DSMC simulations have been carried out. A range of shock reflection cases have been examined covering the standard regular reflection solution region through to transition to weak Mach reflection. Simulation runs show excellent agreement with Rankin-Hugoniot theory except in the vicinity of the shocks themselves.

In these regions, the simulations indicate that there may be enthalpy transfer between the temperature states of the gas. For the case of weak Mach reflection this may lead to an apparent reduction in total enthalpy downstream of the shock system.

The behaviour of strong and weak oblique shock waves produced by disturbances in supersonic flow can be predicted by the Rankin-Hugoniot equations [1]. Here the flow of an assumed perfect gas can be predicted using the principles of conservation of mass, momentum and energy. Disturbances due to flow deflection are assumed to produce infinitely thin oblique shocks across which flow properties change discontinuously. Over a certain range of free stream conditions this leads to the possibility of two distinct solutions, strong and weak wave solutions. Real physical flow will conform to the solution requiring minimum energy change to the flow.

For the case of oblique wave reflecting at a plane of symmetry the solution satisfies energy conditions that take the complex form of a Mach reflection. This is usually the case for weak wave reflection solutions. In this solution, the original oblique wave does not reflect from the plane of symmetry but reflects from a subsonic zone that is formed on the plane of symmetry and bounded by an upstream normal shock and shear surface between two distinct gas zones.

The behaviour of the Mach reflection solution is relatively complex, especially when the refected oblique wave is extremely weak and produces a region of subsonic flow above the shear layer. The point of confluence of the leading normal and oblique shocks with the reflected shock is called the triple point and has been an area of intense study.

Previous experiments [4] have concentrated on measuring the gas properties just downstream of the triple point. In some of these studies, measurements of stagnation temperature indicated that for the case of the very weak Mach reflection a loss of total enthalpy is observed.

This study will attenmpt to validate these results using a DSMC simulation method applied to the experimental geometry, Figure 1.

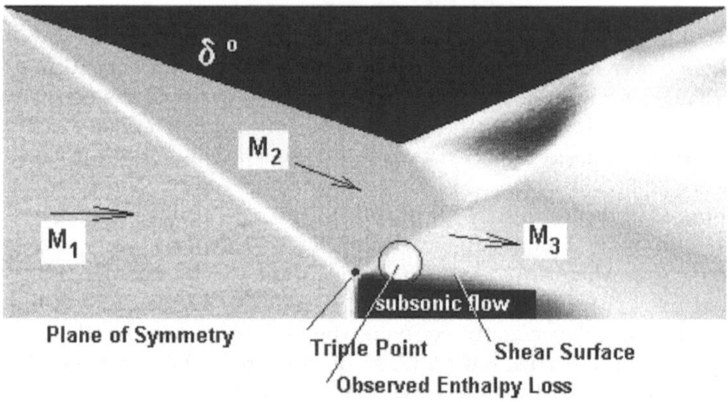

Fig. 1. Mach Reflection showing region of observed enthalpy loss.

1 DSMC Simulation

DSMC is a gas flow simulation technique, originally designed for the solution of problems in rarefied gas dynamics, but now being used in many applications for near-continuum flows and for high density flows with micro-scale geometry [5].

The technique is a computer simulation of the governing Boltzmann equation [6] for gas flow. A statistical approach is used with representative computer molecules being tracked through a flow field and undergoing intermolecular collisions and boundary interactions. The simulation employs two independent solution algorithms.

Firstly, a Lagrangian time stepping technique is used to move simulated molecules over small time steps of 10^{-6} sec. Secondly, these simulated molecules are processed for binary-pair, nearest neighbour collisions that would occur during the time step. This algorithm is based on collision probability and the collision rate required for small volumes of the flow field.

The gas used for this current simulation is ideal air ($80O_2$). Molecule characteristics are modelled using the fundamental laws of physics [5]. The simulation employs a Variable Hard Sphere (VHS) molecular model with a viscosity-temperature index for representing translational energy effects. It employs a Larsen-Borgnakke model for molecule rotational degrees of freedom.

Gas/surface interactions are modelled directly as part of the movement algorithm using a specular reflection model for all solid boundaries including the plane of symmetry. This removes the influence of any low Reynolds number effects due the boundaries of the flow field.

By its very nature the simulation is an unsteady flow solver. The prediction of flow properties is obtained by the statistical averaging of molecular behaviour. Time averaging of flow properites is used to predict the final steady flow behaviour of the gas.

Both translational (kinetic) and rotational components of temperature are obtained from the molecular sampling. Overall temperature, T_{ov}, is calculated as the average of these separate effects. Stagnation temperature, T_O, is then calculated by applying the standard equations of gasdynamics, $T_O = T_{ov} \times (1 + \frac{\Gamma-1}{2} M^2$, where Γ is the ratio of specific heats for air, 1.4, and M is the local flow Mach number.

2 Results

The simulation of gas flow in the small nozzle shown in Figure 1, has been run for a flow of Mach numbers in the range 1.5 to 2.0 for a flow deflection angle of 10°. Predictions of total temperature versus position in the nozzle have been obtained and are shown as contours in the following Figures. In all cases stagnation temperature variations are found to be no more than a few degrees.

The most pronounced variation in stagnation temperature occurs internally within the shock waves. For both the regular reflection, Figure 2, and the Mach reflecton, Figure 3, the translation component of stagnation temperature drops rapidly while the rotational component rises. The overall stagnation temperature drops slightly as can be seen in the contour graphs. The greatest deficit occurs inside the normal shock stem of the Mach reflection. Behind the shock waves, the balance is quickly restored and the original value of stagnation temperature is recovered.

This result matches standard gasdynamic theory with the reservation that theory treats the shocks as instantaneous discontinuities without internal detail.

The other region of interest where a change in stagnation temperature seems to have been recorded are the sub-sonic regions behind the normal and reflected shocks for the Mach reflection case, Figure 3. In this case the stagnation temperature seems to be slightly reduced behind the normal shock and slightly increased behind the reflected wave. It should be noted however that these variations are much less than those inside the waves and are difficult to quantify due to the stastistical flucuations in the measured temperature.

The other regions where stagnation temperature is varying are along the walls where the flow is strongly influenced by boundary conditions. Although these effects could be ignored there may be some influence on the flow region the reflected wave as the flow here is subsonic, slightly less than Mach 1.

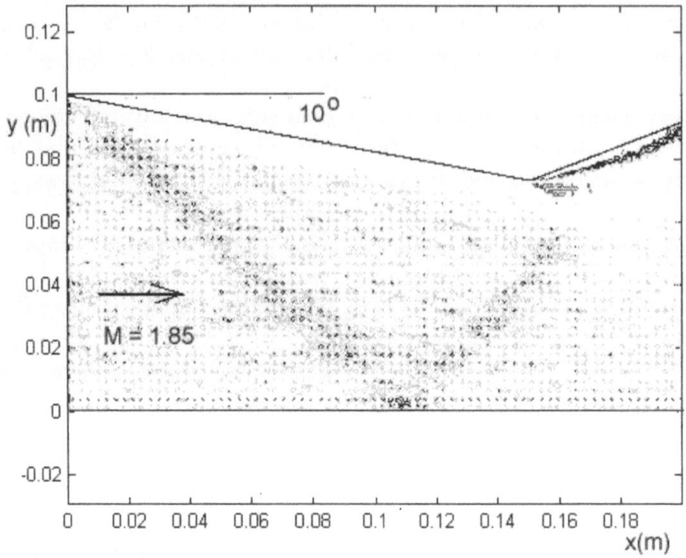

Fig. 2. Stagnation Temperature Contours, Regular Reflection, (480 to 500 Kelvin, steps of 2 K)

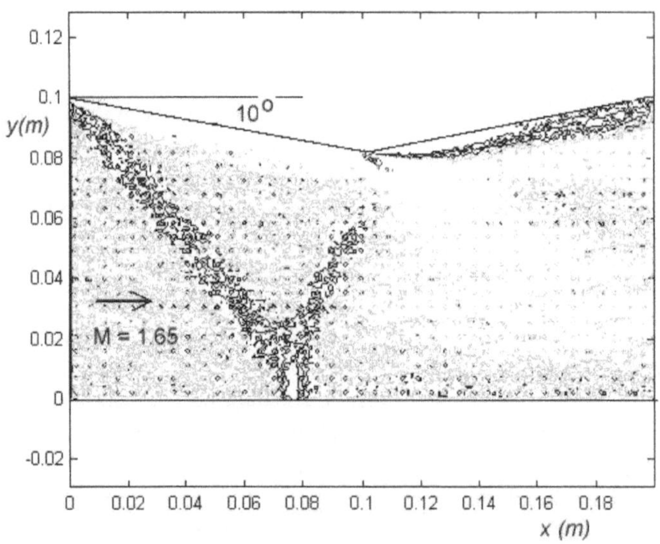

Fig. 3. Stagnation Temperature Contours, Mach Reflection, (440 to 450 Kelvin, steps of 1 K)

3 Discussion

DSMC simulations as described above have indicated the possibility of variation in stagnation temperature with in a gas flow. The mechanism inside a shock wave for transfer of energy between translational and rotational modes is complex and not yet fully investigated. The mechanism by which stagnation temperature is different between the two downstream Mach reflection regions is most likely a cause of the differing internal structures of a normal shock and an oblique shock. At the triple point these differences must interact hence giving rise to the possibility of a final inbalance of stagnation temperature being shed downstream.

While not conclusive, the simulation does indicate that there are possible mechanisms in the molecular behaviour of a simulated gas which may lead to the previously observed reductions in stagnation temperatures of real physical flow. The simulation algorithm is now at a level of accuracy where absolute temperature variations of just a few degrees can be observed. This translates to accuracy levels of less than

References

1. Ben-Dor G. Shockwave reflection phenomena *Springer Verlag*(1992)
2. Tabak, E.G. Rosales, R.R. Focusing of weak shock waves and the von Neumann paradox of oblique shock reflection *Phys of Fluids*(1994) **6**, 1875–1982.
3. Zakharian, A.R. et al The von Neumann paradox in weak shock reflection. *J. Fluid Mech*(2000) **422**, 193–205.
4. Siegenthaler, A. & Madhani, J. Experimental verification of the departure from Rankin-Hugoniot behaviour in weak Mach reflection *13th Australian Fluid Mechanics Conf, Monash Uni , Melbourne* (1998), 549–554.
5. Bird G.A. Molecular Gas Dynamics and the Direct Simulation of Gas Flows. *Oxford Uni Press*(1994).
6. Bird, G.A. DSWT Simulation Software *http://www.gab.com.au*(1999).

Part XVI

Turbomachinery Flow

Numerical Study of the Flow Around the High-torque Wind Turbine of Vertical Axis Type

Tetuya Kawamura and Yuko Sato

Ochanomizu University, 2-1-1 Otsuka Bunkyo-ku, Tokyo, Japan

Abstract. Numerical simulations of two-dimensional flows around a vertical axis type wind turbine are investigated. The turbines studied in this paper are the Savonius turbine and the cross-flow turbine of twelve blades. The governing equations are the incompressible Navier-Stokes equations. A rotational coordinate system, which rotates at the same speed of the turbine, is used in order to simplify the boundary conditions on the blades. Additionally, a boundary fitted coordinate system is employed in order to express the shape of the blades precisely. The fractional step method is used to solve the basic equations. A third order upwind scheme is chosen for the approximation of non-linear terms. The results are compared with experiments and agreements are reasonable.

1 Introduction

Recently, wind energy has received much attention as a natural energy source. A wind turbine is one of the best candidates for getting wind energy. While there are many kinds of wind turbines, they can be classified into two types, i.e. the horizontal axis type and vertical axis type. The typical example of the former is the propeller turbine, which is the most widely used for generating electricity. Disadvantage of this type is that the performance of the turbine is greatly affected by the direction of the wind. On the other hand, the vertical axis type is free from this disadvantage, i.e. it can be rotated by the wind of any direction. Moreover, the structure of the turbine is simple, therefore, it can be produced easily. Another classification of the wind turbine is also possible when we look at the driving force acting on the blades of the turbine, i.e. the turbine is called lift-type or drag-type when the turbine is rotated by the lift or the drag, respectively. The rotational speed of the former is higher than the latter while the torque of the latter is larger than the former. Therefore, the drag-type turbine is suitable for pumping water. Since we are interested in irrigation in arid land, we focus on the Savonius turbine [1,2] and the cross-flow turbine with many blades [3], both of which are drag-type wind turbines of vertical axis, and simulate the flow fields around them in order to obtain the basic data for the design of these turbines.

2 Numerical Method

Since the drag-type wind turbine rotates rather slowly, the flow around it can be analyze by solving the incompressible Navier-Stokes equations. The rotational coordinate, which rotates at the same speed of the blades of the turbine, is suitable for the present computation due to the simplicity of the boundary conditions on the blades. Therefore, the basic equation becomes

$$\frac{\partial \boldsymbol{V}}{\partial t} + (\boldsymbol{V} \cdot \nabla) \boldsymbol{V} = -\frac{1}{\varrho} \nabla p - \boldsymbol{\Omega} \times (\boldsymbol{\Omega} \times \boldsymbol{r}) - 2\boldsymbol{\Omega} \times \boldsymbol{V} + \nu \nabla^2 \boldsymbol{V}, \tag{1}$$

where Ω is the constant angular velocity. This equation is solved by the fractional step method [4]. This method is summarized briefly for the present cases as follows. At first, a temporal velocity is computed by using the following equation

$$\boldsymbol{V}^* = \boldsymbol{V} + \Delta t \left(-(\boldsymbol{V} \cdot \nabla) \boldsymbol{V} - \boldsymbol{\Omega} \times (\boldsymbol{\Omega} \times \boldsymbol{r}) - 2\boldsymbol{\Omega} \times \boldsymbol{V} + \nu \nabla^2 \boldsymbol{V} \right). \tag{2}$$

The right-hand-side of (2) is calculated by using the velocity at 'n'th time step. Then, the pressure is determined from the Poisson equation:

$$\frac{1}{\varrho} \nabla^2 p = \frac{\nabla \cdot \boldsymbol{V}^*}{\Delta t}. \tag{3}$$

The new velocity at '$n+1$'th time step is calculated from

$$\boldsymbol{V}^{n+1} = \boldsymbol{V}^* - \Delta t \nabla p. \tag{4}$$

This procedure is continued until it reaches the prescribed maximum time step. In order to impose the boundary conditions on the blades (no-slip in rotational coordinate system), a boundary-fitted coordinate system is employed. The transformed equations are solved by the finite difference method. All the spatial derivatives except non-linear terms are approximated by central differences. Non-linear terms are approximated by third order upwind differences since it provides a stable solution without any turbulence models even at considerably high Reynolds number [5]. In this study, the Savonius turbine and the cylindrical cross-flow turbine with 12 blades are investigated. These are shown in Fig.1. The shape of the cross section of the blade for each turbine is an arc. The thickness of the blade is assumed to be zero in this computation.

3 Results

3.1 Savonius turbine

Figure 2 shows the grid used for the computation of the flow around the Savonius turbine. As shown in this figure, the region is divided into small regions and they are connected to each other. The total grid points is 81×65. The transfinite interpolation is used to generate the grid in each region. The Reynolds number of the main computations is 2000. Figure 3 shows the velocity vectors and the pressure contours at a typical time. In this case, the flow comes from the left side of the figure. Figure 3(**a**) shows the results for S=0, where S is the gap between the blades, and Fig.3(**b**) shows those for S=0.2D, where D is the diameter of the blade. Figures 4(**a**) and (**b**) display the effect of the tip speed ratio (=tip speed of the turbine / wind speed) on the power coefficient for S=0 and S=0.2D, respectively. Higher coefficients are obtained for S=0.2D.

3.2 Cross flow turbine

Figure 5 is the grid system for the cross-flow turbine. Considering the shape of the turbine, the grid is first generated over one-twelfth of the region. The grid is generated algebraically with grid concentration enforced near the blade. Then, the whole grid is obtained from this grid by rotating 30 (=360/12) degrees 12 times. The total number of grid points is 92×184 for the main computations. Figure 6 shows the pressure contours and velocity vectors at a typical time. The flow comes from the left side of the figure. The Reynolds number of the main computation is 40000. Figure 7 is the time history of the torque coefficient. The interval between each peak of the curve corresponds to the 1/12 rotation of the rotor. In order to get high performance of the turbine, some accessories are proposed. One of such accessories is the guide vane shown in Fig.1(**c**), which weakens the drag acting on the opposite direction of the rotation. Figure 8 shows the effect of the guide vane on the power coefficient at various tip speed ratio. The power coefficient of the turbine with the guide vane is almost 1.5 times as large as that without the guide vane. The height of the actual turbine is finite and the effect of the end-plate cannot be ignored. Therefore, we also compute the three-dimensional flow fields. The total number of grid points is $92 \times 184 \times 21$. Figure 9 shows the results obtained for one of such computations.

4 Summary

The flows around the drag-type wind turbine of vertical axis are analyzed by solving the incompressible Navier-Stokes equations numerically. A rotational coordinate system is employed in order to simplify the boundary conditions on the rotating blades. The Savonius turbine and the cross-flow turbine are chosen in this study. Complex flow fields due to the rotation of the turbine are obtained. For the Savonius turbine, the power coefficients are compared with the experiments and the agreement between them is reasonable. For the cross-flow turbine, it is found that the guide vane works well even if it is quite simple. Moreover, the three-dimensional flows around the cross-flow turbine are computed.

References

1. K.Ishimatsu, T.Shinohara and F.Takuma : *Numerical Simulation for Savonius Rotors (Running Performance and Flow Fields).* Trans. JSME (B),60(569)(1994) pp154-160
2. B.F. Blackwell R.E. Sheldahl and L.V. Feltz : *Wind Tunnel Performance Data for two- and three-bucket Savonius Rotors.* J.Energy (1978)
3. M.J. Holgate, et al. : *A Cross Flow Wind Turbine.* Int. Symp. Wind Energy System (1976)
4. Yanenko,N.N. : *The Method of Fractional Steps.* (Springer, New York 1971)
5. T.Kawamura, K.Kuwahara and H.Takami : *Computation of High Reynolds number flow around a Circular Cylinder with Surface Roughness.* Fluid Dynamics Research, Vol.1 (1986) pp.145-162

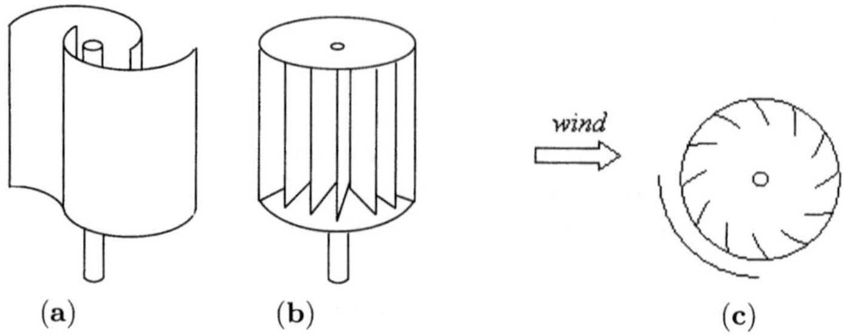

Fig. 1. Two kinds of the drag-type wind turbines of the vertical axis. (**a**) Savonius, (**b**) Cross-flow, (**c**) Cross-flow with a guide vane (one cross section of the turbine)

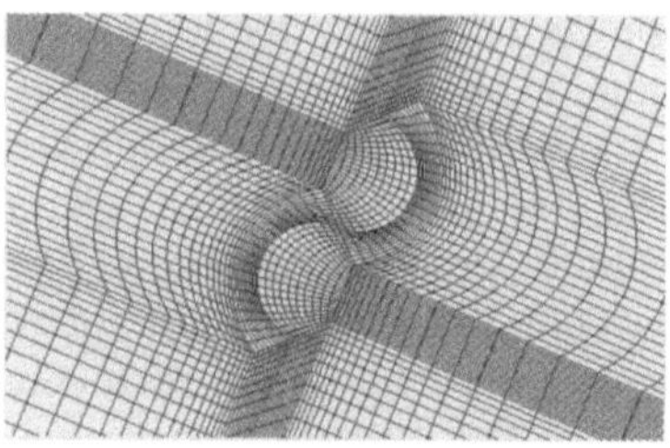

Fig. 2. Grids for the Savonius turbine

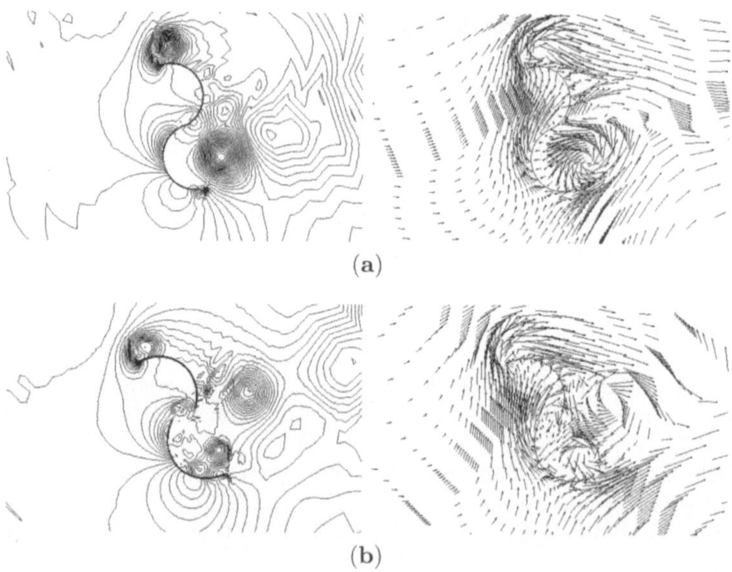

Fig. 3. Example of the pressure contours and velocity vectors around the Savonius turbine. (**a**) S=0, (**b**) S=0.2D

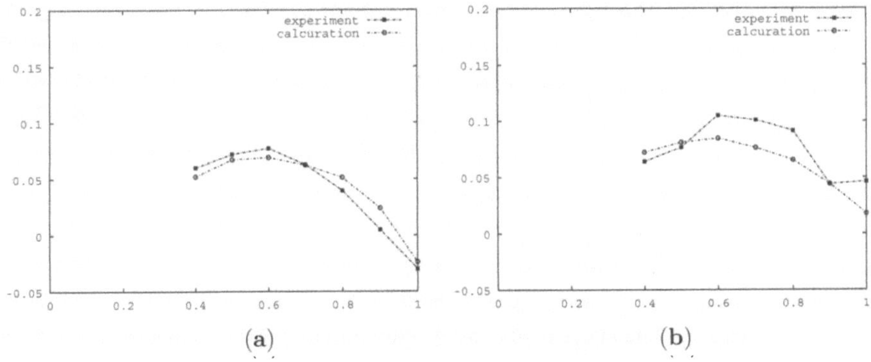

Fig. 4. Comparison of the power coefficient with experiment. (**a**) S=0, (**b**) S=0.2D

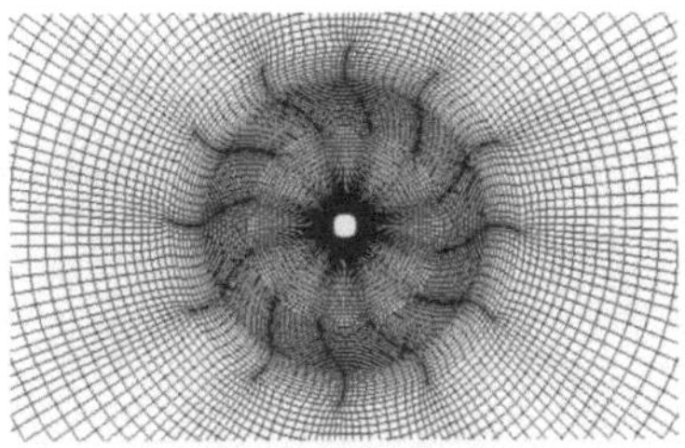

Fig. 5. Grids for the cross-flow turbine

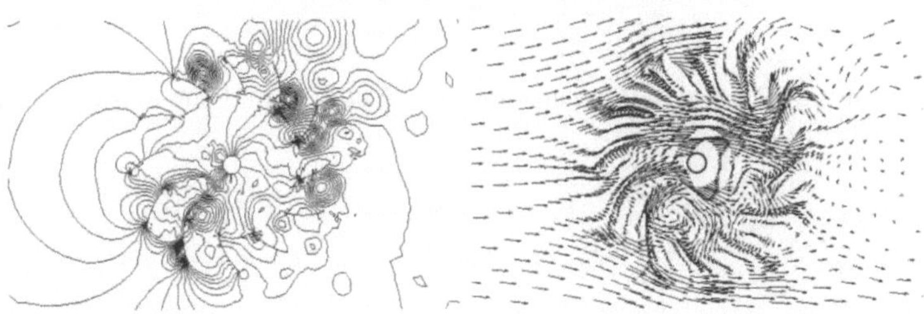

Fig. 6. Example of the pressure contours and velocity vectors around the cross-flow turbine

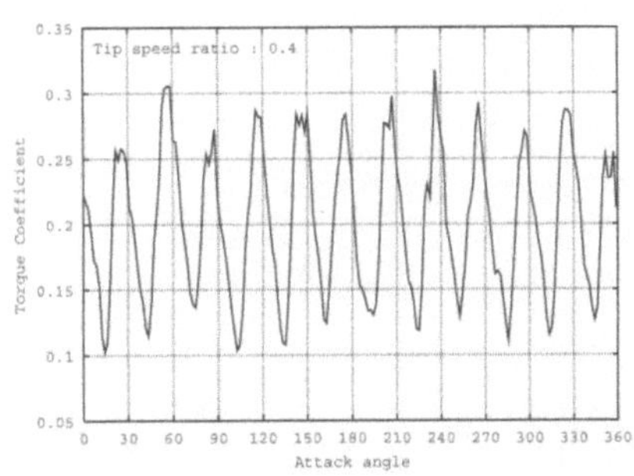

Fig. 7. Time history of the torque coefficient

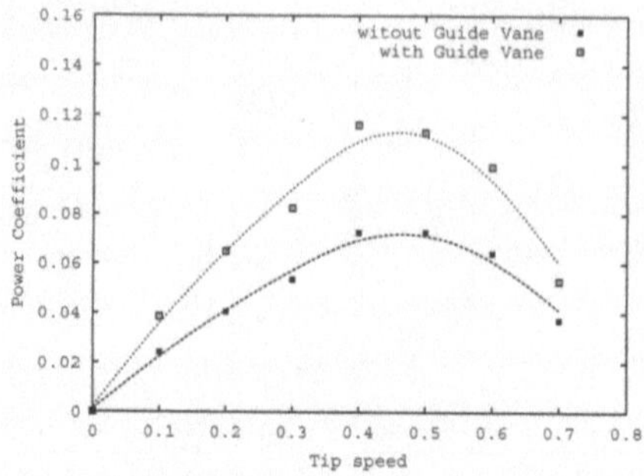

Fig. 8. The effect of the guide vane on the power coefficient

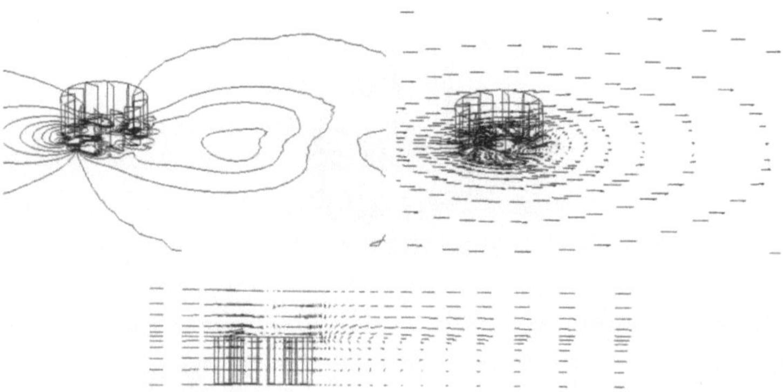

Fig. 9. Example of the three-dimensional flow fields around the cross-flow turbine

Numerical Simulation of Steady and Unsteady Turbomachinery Flows Using a Navier-Stokes Solver

J. J. Liu

Institute of Engineering Thermophysics, Chinese Academy of Sciences
P.O. Box 2706, Beijing 100080, CHINA

Abstract. This paper describes the approach used for the numerical simulation of steady and unsteady viscous flows in turbomachinery cascades or blade-rows. The flow solver is based upon the finite-volume TVD Lax–Wendroff scheme and the Baldwin–Lomax turbulence model. For steady flow calculations, the local time–steps and multigrid method are used to speed up the convergence to a steady state. For unsteady flow calculations, a dual–time step method is introduced in which an implicit discretization in the physical time and an explicit discretizatoin in the pseudo time are utilized. The convergence–acceleration methods of local time–steps and multigrid method are still used to drive the residuals in the pseudo time to zero at each physical time–step. Sample applications of the present approach are presented.

1 Introduction

CFD is of great importance in the flow analysis and aerodynamic design of turbomachinery. Flow solvers for the solution of Reynolds averaged Navier-Stokes equations incorporating turbulence closures are widely applied to the flow prediction and aerodynamic optimization of 2D cascades and 3D blade rows. The assumption of steady flow is appropriate for an isolated cascade/blade row or for blade rows with relatively large axial spacing between rotor and stator. Because of the relative motion between adjacent blade rows, the turbomachinery flow is inherently unsteady. The unsteadiness needs to be taken into account for the blade rows with compact axial spacing in modern turbomachinery. It is useful to develop a flow solver that is capable of simulating both steady and unsteady flows.

A 3D Navier-Stokes solver based on the TVD Lax-Wendroff scheme has been reported in [2]. This paper describes the extension of the solver to the simulation of steady and unsteady turbomachinery flows. After this introduction, the numerical approach for the solution of steady and unsteady flows are described. The application of the present method to steady flows in the VKI gas turbine cascade, NASA guide vane and unsteady flow in a 1.5 stage steam turbine is presented.

2 Numerical Approach

2.1 Governing Equations

The Reynolds averaged Navier-Stokes equations can be written in the following nondimensional form,

$$\frac{\partial U}{\partial t}+\frac{\partial F}{\partial x}+\frac{\partial G}{\partial y}+\frac{\partial H}{\partial z}=\frac{1}{Re_\infty}\left(\frac{\partial F_v}{\partial x}+\frac{\partial G_v}{\partial y}+\frac{\partial H_v}{\partial z}\right)+Q \tag{1}$$

where U is the vector of conversation variables, F, G, H are the vectors of convection fluxes, F_v, G_v, H_v are the vectors of diffusion fluxes, Q is the vector of source terms. Re_∞ is Reynolds number based on the reference state and length scale.

The turbulence eddy viscosity is estimated by using the Baldwin–Lomax model[2]. In this model a turbulent boundary layer is divided into an inner and an outer region. In the inner region, the eddy viscosity is given by,

$$(\mu_t)_{inner}=\rho l^2 \mid \omega \mid \tag{2}$$

where ρ is density, l is a length scale proportional to the distance from the wall (scaled by a damping factor D), and ω is the vorticity. In the outer region, the eddy viscosity is given by,

$$(\mu_t)_{outer}=C_{outer}\rho F_{wake}F_{kleb} \tag{3}$$

where C_{outer} is a constant, F_{kleb} is the intermittency factor, F_{wake} is a function of d_{max}, F_{max} and V_{dif} (the difference between the maximum and minimum velocity in the profile). F_{max} is the maximum value of the function $F=d\mid\omega\mid D$, and d_{max} is the distance d (from the wall or wake center) at which this maximum occurs. For a complicated flow field, F may exhibit multiple local maximum. The one closest to the wall is chosen in the present implementation. Standard coefficients due to Baldwin and Lomax are used.

In practice, both inner–layer viscosity and outer–layer viscosity are computed for the entire boundary and the smaller one is taken as the actual eddy viscosity.

For 3D turbomachinery flows, two solid walls (i.e., the blade surface A and the endwall B) intersect. The eddy viscosity in this case is first computed according to each of the two walls and then weighted by the distance from the wall to form the effective eddy viscosity,

$$\mu_t=\frac{d_B^2(\mu_t)_A+d_A^2(\mu_t)_B}{d_A^2+d_B^2} \tag{4}$$

2.2 Solution Method

Equation 1 is solved numerically by using the explicit TVD Lax-Wendroff scheme incorporated with Roe's approximate Riemann solver[3]. The V-cycle multigrid method[4] is adopted to speed up the convergence to a steady state. For *steady flow* simulations, the explicit time-marching scheme applied can be written in the following form,

$$\frac{U^{n+1}-U^{n}}{\Delta t}=-R(U^{n}),\quad -\Delta t\cdot R(U^{n})=\delta U_{h}+\sum_{m=1}^{N}T_{2mh}^{h}\delta U_{2mh} \tag{5}$$

where Δt is time–step, R is residual, δU_h is correction on the fine mesh, δU_{2mh} is correction on the successive coarse meshes, T is the interpolation operator from coarse mesh to fine mesh, N is the total level of coarse meshes. When $R(U^n)$ approaches zero, U^{n+1} approaches U^n and a steady solution is obtained. The discretization details can be found in [2].

For *unsteady flow* simulations, a three-point backward implicit discretization in time is applied,

$$\frac{3U^{n+1}-4U^{n}+U^{n-1}}{2\Delta t}=-R(U^{n+1}) \tag{6}$$

Equation 6 can be solved by any of the implicit methods. Alternatively, the explicit dual–time step method due to Jameson[5] is adopted to solve the above equation. By introducing a pseudo time–step $\Delta\tau$, equation 6 can be re-written as,

$$\frac{U^{(q+1)}-U^{(q)}}{\Delta\tau}+\frac{3U^{(q)}-4U^{n}+U^{n-1}}{2\Delta t}=-R(U^{(q)}) \tag{7}$$

It is noted that U^{n+1} in equation 6 is replaced by $U^{(q)}$ in equation 7. Equation 7 can be re-written in the form similar to equation 5, i.e.,

$$\frac{U^{(q+1)}-U^{(q)}}{\Delta\tau}=-R^{*}(U^{(q)}),\quad R^{*}(U^{(q)})=\frac{3U^{(q)}-4U^{n}+U^{n-1}}{2\Delta t}+R(U^{(q)}) \tag{8}$$

At each physical time–step Δt, an inner iteration in the pseudo time τ, similar to the previous explicit time–marching procedure used for steady flow simulations, is carried out to drive the residual R^* to zero such that $U^{(q+1)} \rightarrow U^{(q)} \rightarrow U^{n+1}$, and equation 6 is thus solved. The initial value of $U^{(q)}$ for the inner iteration can be set as,

$$U^{(0)}=U^{n}+\frac{3U^{n}-4U^{n-1}+U^{n-2}}{2} \tag{9}$$

2.3 Boundary Conditions

Five types of numerical boundary exist in the computational domain, namely, inlet, outlet, solid wall, bladerow interface and periodicity. For subsonic axial flow, distributions of total pressure, total temperature, swirl angle and pitch angle are specified at the domain inlet, and averaged static pressure is specified at the domain outlet. A slip condition and the wall function due to Spalding are used at the solid wall. The wall is also assumed to be adiabatic. At the blade interface, a frozen–rotor technique is used for steady flow simulations and interpolation is used at the sliding interface for unsteady flow simulations. The periodic boundary can be treated as interior by shifting the corresponding blade passage.

3 Results

VKI Cascade The geometry and experimental data are obtained from [6]. The mesh of 177×41 used for the simulation is shown in Figure 1(a). At the inlet, uniform total pressure, total temperature and flow angle are specified. The averaged static pressure is specified at the outlet and the static pressure is allowed to vary in the tangential direction according to the local flow.

The computed Mach contours for the test case of M_{2is}=0.99 are shown in Figure 1(b). The computed and measured isentropic Mach number distributions are presented in Figure 1(c), the agreement between the computed result and the experimental data is good.

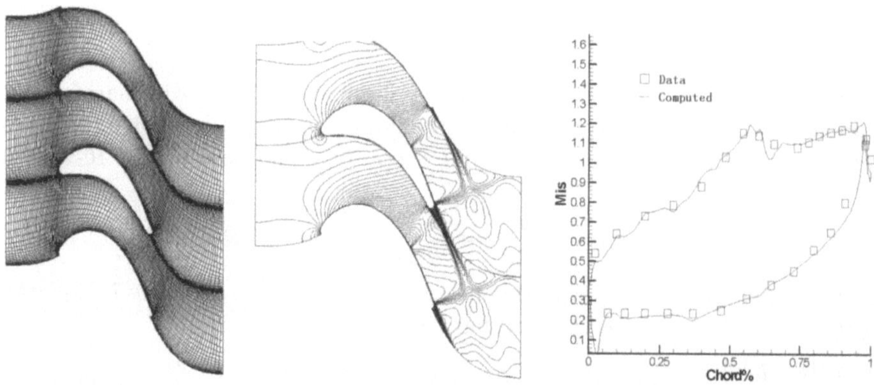

(a) (b) (c)

Fig. 1. VKI Cascade: (a) Mesh, (b) Computed Mach Contours, (c) Isentropic Mach Number Distributions

NASA Guide Vane The geometry and experimental data are obtained from [7]. The 3D mesh of $137 \times 33 \times 21$ used for the simulation is shown in Figure 2(a). Uniform total pressure, total temperature, swirl angle and pitch angle are specified at the inlet. Static pressure at the hub of the outlet is specified and static pressure at other radial locations of the outlet is obtained by using radial equilibrium equation.

The computed and measured isentropic Mach number distributions at blade hub, mid-span and tip sections are presented in Figure 2(b), 2(c) and 2(d), respectively. The agreement between the computed result and the experimental data is quite good.

1.5 Stage Turbine The mesh used for the stator–rotor–stator unsteady interactions is shown in Figure 3(a). The geometry represents the mid-span section of a high–pressure steam turbine. The original blade number ratio is 92:122:92, which is scaled to 90:120:90 (i.e., 3:4:3) in the simulation. The mesh dimensions of the three rows of cascade are $3 \times (145 \times 49)$, $4 \times (137 \times 49)$ and $3 \times (145 \times 49)$. A rotor passing period is divided into 64 physical time–steps, 30 to 40 multigrid

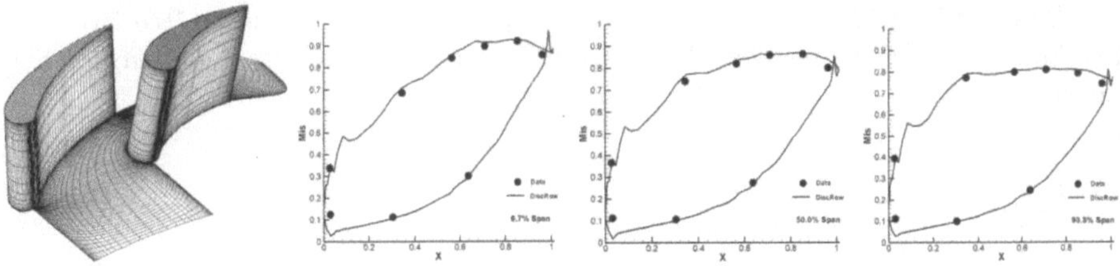

(a) (b) (c) (d)

Fig. 2. NASA Guide Vane: (a) 3D Mesh, and Isentropic Mach Number Distributions (b) at Hub, (c) at Mid-span and (d) at Tip Sections

cycles in the pseudo time are carried out at each physical time–step to obtain the time accurate solution of equation 6.

The lift coefficient variation of the rotor is presented in Figure 3(b). The envelope of unsteady surface pressure distributions of the rotor, the steady and time-averaged surface pressure distributions are presented in Figure 3(c).

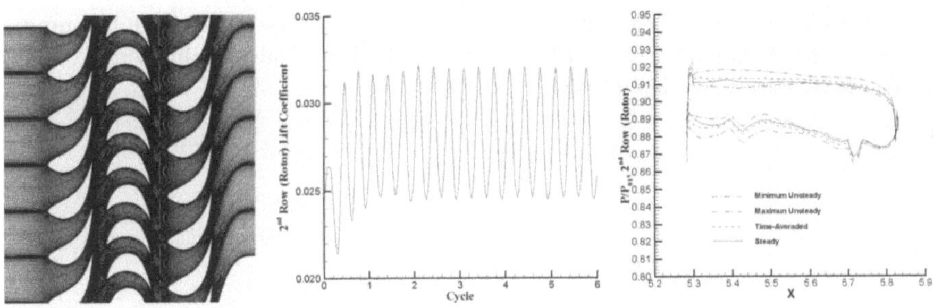

(a) (b) (c)

Fig. 3. Stator–Rotor–Stator Unsteady Interactions: (a) Mesh, (b) Lift Coefficient Variation of the Rotor, (c) Surface Pressure Distributions of the Rotor

4 Concluding Remarks

The numerical approach used for the simulation of steady and unsteady flows in 2D cascades and/or 3D blade rows has been presented. The relation between the steady and unsteady solution method is emphasized. By using the dual–time step technique, an explicit steady flow solver can be extended to the simulation of unsteady flows with implicit discretization in the physical time. Numerical results for the steady flow in the VKI cascade and NASA guide vane, and unsteady flow in a 1.5 stage turbine are presented to demonstrate the capability of the present method.

Acknowledgements: This work is supported by the National 973 Program of China and the Knowledge Innovation Program of Chinese Academy of Sciences.

References

1. J.J. Liu, T.P. Hynes: 'A Navier-Stokes Solver using Edge-Based Smoothing'. In: *Computational Fluid Dynamics 2000, Proceedings of the First International Conference on Computational Fluid Dynamics, Kyoto, Japan, July 10-14 2000,* ed. by N. Satofuka (Springer 2001) pp.313-318
2. B.S. Baldwin, H. Lomax: 'Thin Layer Approximation and Algebraic Model for Separated Turbulent Flows'. AIAA paper No.78-257 (1978)
3. P.L. Roe: J. Comput. Phys., **43**, 357, (1981)
4. R.H. Ni: 'Prediction of 3D Multi-Stage Turbine Flow Field Using a Multiple-Grid Euler Solver'. AIAA paper No.89-0203 (1989)
5. A. Jameson: 'Time Dependent Calculation Using Multigrid with Applications to Unsteady Flow Past Airfoils and Wings'. AIAA paper No.91-1956 (1991)
6. C.H. Sieverding: 'Experimental Data on Two Transonic Turbine Blade Sections and Comparison with Various Theoretical Method'. VKI Report LS59 (1973)
7. L. J. Goldman, R.G. Seasholtz, 'Laser Anemometer Measurements and Computations in an Annular Cascade of High Turning Turbine Core Turbine Vanes'. NASA Technical Paper No.3252 (1992)

Numerical Simulation of Windturbine for Interference Between Rotors and Tower

Makoto Iida and Chuichi Arakawa

The University of Tokyo, 7-3-1, Hongo, BUnkyo-ku, Tokyo, 113-8656, JAPAN

Abstract. It is very important to analyze the unsteady phenomena such as various wind speeds, their directions, influence of other wind turbines, the effects of tower and so on. The authors simulate the unsteady flow field around the horizontal axis wind turbine of WINDMELV with their own code of pseudo-compressibility and moving overset grid method. The computational results are compared with the blade element and momentum method. When the blade overlaps with the tower, the thrust coefficient is instantaneously reduced in 20% of averaged one.

1 Introduction

It is very important that wind turbines achieve energy supply stability. One way of solving this problem is to analyze the unsteady phenomenon on a wind turbine blade. Unsteady phenomena on wind turbines include load fluctuations of the blades by change of wind speed and directions (it means an influence of Reynolds number changes and the angle of attack). Also these unsteady phenomenons include load fluctuations of the effect of wakes and influence of other wind turbines as well as the tower. It has been studied how the flow depends on the Reynolds number and the angle of attack although the transition that is unsolved in the study of fluid dynamics. It has been said that Blade Element and Momentum theory (BEM) are useful tools to analyze and design the wind turbines because of good computational efficiency and the accuracy for the design condition. However, it is very difficult for BEM to analyze and design for the unsteady phenomenon.

In this paper, the authors describe the unsteady flow around the wind turbine, and study the effect of the tower with numerical simulation. Currently, CFD has been used in many fields because of recent improvement in computer power. There fore, not only BEM but also CFD is used for the analysis of the flow around the wind turbines [1], [2], [3], [4].

The authors have been researching about and developing the MEL airfoil family for the wind turbines in the wind tunnel [5]. Then we developed the horizontal axis wind turbine WINDMELV on the basis of this knowledge. On the other hand, we analyzed the small wind turbine in steady flow condition and we could get useful results that agree with the experimental data. In this paper, we describe the computational evaluation of performance of WINDMELVwhich has no corning angle (not real). We compute the unsteady flow around the wind turbine, and study the effect of the tower.

2 WINDMELV

Figure 1 shows WINDMEL V that is upwind type. It has two blades, with a diameter of 15m, power of 15kW, rated speed of 81rpm and pitch control systems. WINDMEL V is constructed of the MEL Airfoil family, which has high and stable performance compared with other airfoil families for low Reynolds numbers region (: 10^5-10^6) and over a wide range of angles of attack [6]. To be concrete, this airfoil family does not have sudden decrease of lift for low Reynolds numbers and has gentle-stall characteristics for large angle of attack. It is very important that a wind turbine has these characters, and it means that performance of the wind turbine changes gently when the tip speed ratio varies.

Figure 2 shows that performance of WINDMELV computed by BEM, where x-axis is tip speed ratio, y-axis is power coefficient and the symbols show different cases for pitch angles. This figure also shows that the power curve changes gently for variation in the tip speed ratio and pitch angles.

Fig. 1. Outlook of WINDMELV

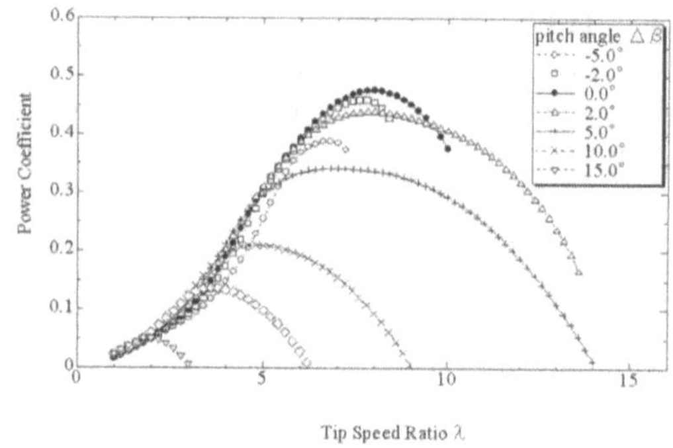

Fig. 2. Performance curve of WINDMELV

The wind turbine performance depends on the constructed aifoils. Because the MEL airfoil family has stable characteristic for low Reynolds numbers, WINDMELV has the stable operation without rapidly power reduction on low tip speed ratio, see Figure 2.

3 Methodology

In this study, we compute the flow with the pseudo-compressible method that Chorin [7] developed. The pseudo-compressible method is originally developed technique to solve for steady phenomenon. We improve it in order to solve the unsteady flow field using δ^2-Scheme which Matuno developed [8]. I used the third order upwind difference, TVD [9] for discretion of convection term, and second order central difference for viscosity term. The time difference is Euler implicit method for pseudo-time and second order imlicit method(Crank-Nicorson) for

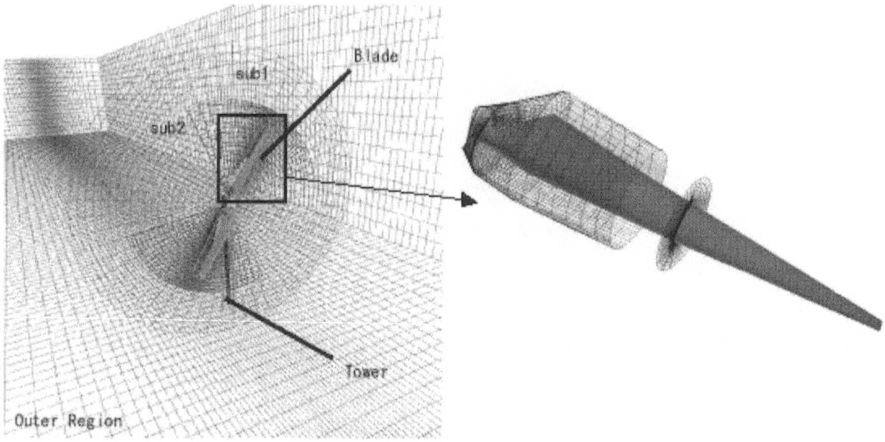

Fig. 3. Computational grid (Left: outline Right: detail of blade)

physical-time. Low Reynolds type turbulent model (Launder-Sharma [10]) is used.

Figure 3 illustrates the computational grid. A total of seven computational domains are used. There is one outer region where we solve the outer flow (Fig. 3 Outer Region), a region for flow around the tower (Fig. 3 Tower), two grid regions around the blade (Fig. 3 Blade), and intermediate domains (Fig. 3 sub1, sub2) to interpolate the physical data smoothly. The least square method is used to interpolate every physical time step. The size of the Outer Region is 15D (5D is in front and upper, 10D is wake of wind turbine). In addition, this computational wind turbine does not have any corning angle. It is different from the real wind turbine.

We use 5 nodes in the super computer HITACHI SR8000 that implements 8CPU, 8G flops and memory of 8GB in 1Node. It takes 120 hours to compute five rotations of the wind turbine.

4 Results

4.1 Performance curve

Because the WINDMELV is currently being, tested experimental data has not been obtained so far. In this paper, the numerical results are compared with the results of Blade Element and Momentum Theory (BEM). Figure 4 shows the power curve calculated by CFD and BEM, where x-axis is tip speed ratio and y-axis is power coefficients of WINDMELV. The CFD results for both λ =8 and 7.5 are smaller than BEM. There are two reasons why the CFD results are smaller than BEM. At first, the loss of CFD computations is larger than BEM because BEM does not include influence of viscosity in this computation.

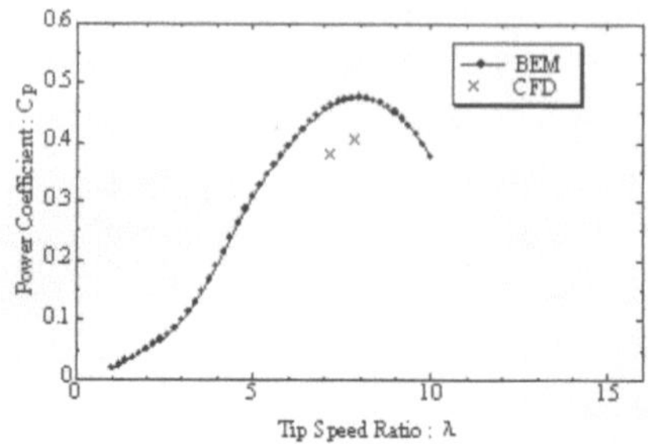

Fig. 4. CFD result compare with BEM for power coefficient

Second, CFD can evaluate the effect of three dimensions, tip vortex and the loss at the root, while BEM include those as experienced low.

4.2 Effects of the tower

Figure 5 shows the definition of the azimuth angle Ψ. When the blades lie a long the horizontal axis, the azimuth angle Ψ is taken 0 deg. The azimuth angle increases the clockwise direction. The blades in overlap with the tower at Ψ =90 deg and 270 deg.

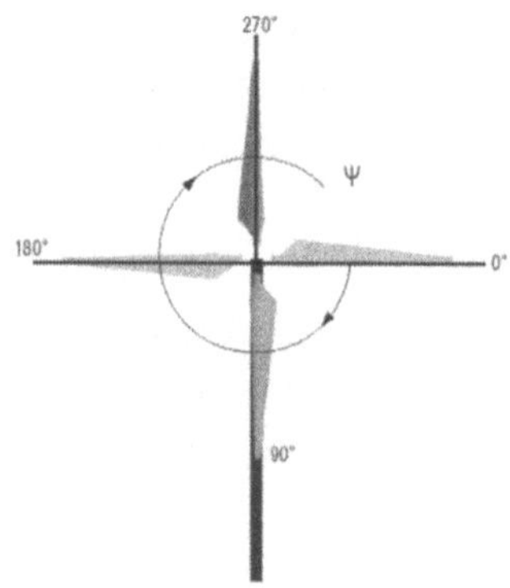

Fig. 5. Definition of azimuth angle

Figure 6 and 7 illustrate the contour of pressure distribution around the tower. Figure 6 is a result for Ψ =0 deg and figure 8 is a result for Ψ =100 deg. The color of the contours shows the strength of pressure, and a line shows an isobar of pressure. For Ψ =0 deg, the pressure contour of the flow is constant along the tower. On the other hand, for Ψ =100 deg, because the interval between the blade and tower is small and the chord length is large at the root, the pressure of the stagnation around the tower is strong at the root side rather than at the tip side. It can be seen in Figure 5 that the propagation of the pressure fluctuation is

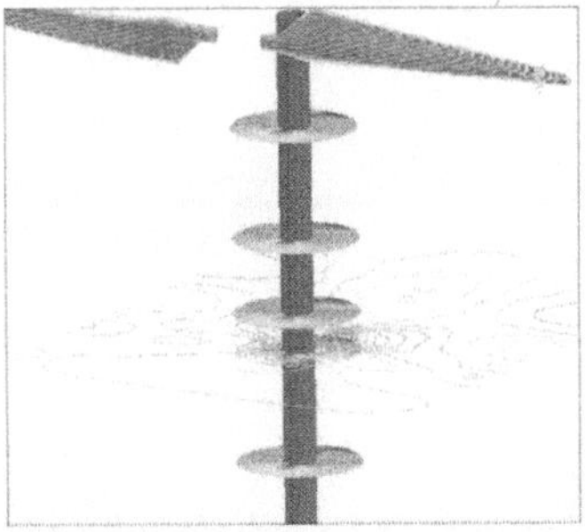

Fig. 6. Pressure distribution on section for Ψ= 0deg

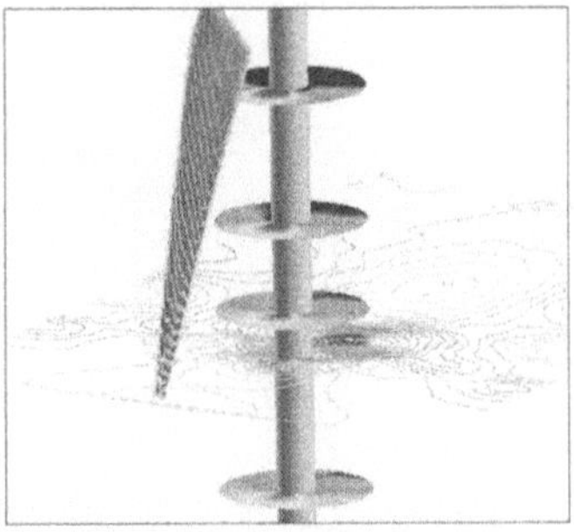

Fig. 7. Pressure distribution on section for Ψ= 100deg

much stronger at the root side than at the tip side. Thus, CFD can model appear the influence of the tower. We have obtained the effects of the different chord length and of the distance between tower and blade on the pressure distribution.

Figure 8 shows the definition of the thrust coefficient CT and CT- Azimuth angle curve. The thrust coefficient is the force of the flow direction against the blade. Figure shows that the thrust coefficient decreases by 20 percent of the average value for Ψ =90 deg, 270 deg that is when the blade overlaps with the tower. It means that the blade receives the opposite force to a flow. The thrust coefficient fluctuates for other azimuth angles. Figure 9 shows the pressure distribution on each section for Ψ =0 deg, 100 deg. When the blade overlap the tower, the pressure level is lower for the Ψ=0 degthan Ψ=100 degat all sections, which means abrupt increase of thrust appear for the upside in the Fig. 9.

5 Conclusions

In this paper, we compute the unsteady three-dimensional flow around horizontal wind turbine WINDMELVusing CFD. When we analyze and design, the wind turbine, it is very difficult to evaluate the effect of the tower by conventional analysis technique. We have analyzed the flow around the wind turbine detail using overset grid method. In this paper, we can analyse the influence of the tower using both overset grid method and moving grid system method. Concerning the power coefficient, the results of BEM calculation are larger than CFD. It means that the loss of three-dimensional effects and the viscosity is too small in BEM compared to CFD. Both of CFD and BEM show gentle trend in the performance for off-design a conditions. It means that characteristics of airfoils that consist of WINDMEL are stable for a wide range of Reynolds numbers and angles of attack. When the blade overlaps with the tower, the thrust coefficient decreases by 20 percent of the average value because the pressure level is decreased.

References

1. Butterfield, C.P., Musial, W.P. and Simms, D.A.: NREL/TP-**257-4655**,(1992)

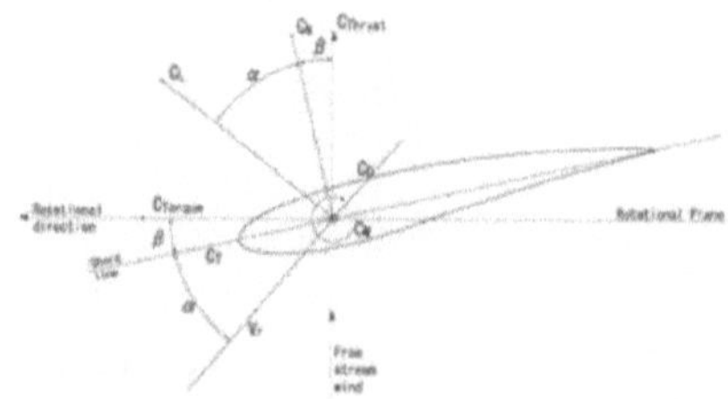

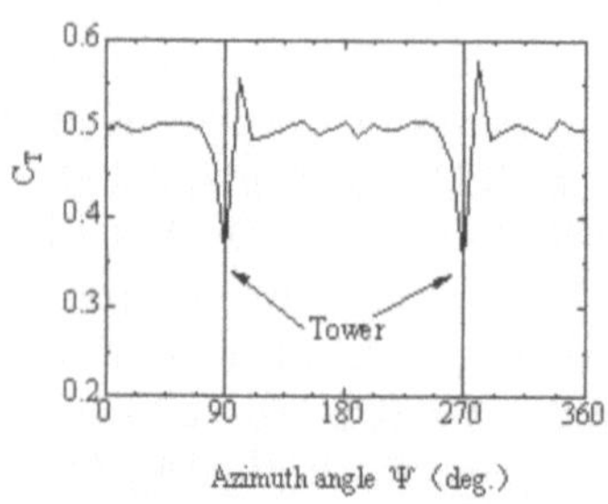

Fig. 8. Thrust coefficient for each Azimuth angle

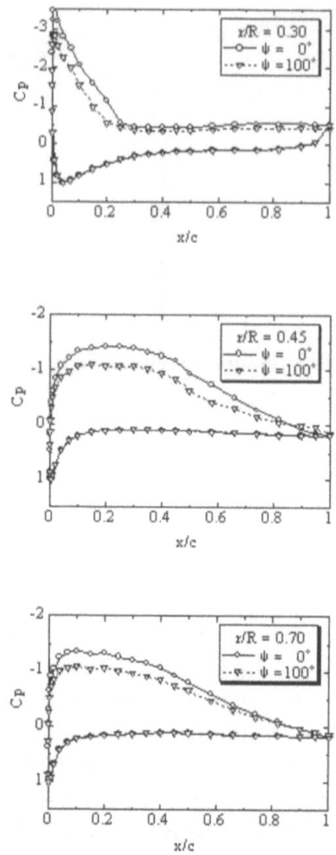

Fig. 9. Pressure distribution on section for Ψ = 0deg and 100deg

2. Butterfield, C.P., Musial, W.P., and Simms, D.A.: NREL TP-**422-4807**, (1992)
3. Earl P.N. Duque, Wayne Johnson et al.: AIAA-**2000-0038**, (2000)
4. Makoto Iida, Chuichi Arakawa, Hikaru Matsumiya: AIAA-**2001-0058**, (2001)
5. Hikaru Matsumiya: Wind energy symposium **11**, (1990) pp.8-17
6. Giancarlo Chiatti et al.: Wind Engineering **21**, 5, (1997) pp.295-306.
7. Chorin, A. J.: J. of Computational Physics, **2**, (1967) pp.12-26.
8. K. Matsuno: Theoritical and Applied Mechanics **37**, 424, (1988), pp.91-99
9. Yee, H.C., Warming, R.F. and Harten, A.: **22**, (1985) pp.357-377.
10. Launder, B.E. and Sharma, B.I.: Letters in Heat Mass Transfer, **1**, (1974) pp.131-138.

Unsteady Flow Simulation of Turbine Stage

Francesco Martelli, Elisabetta Belardini, and Paolo Adami

Energetics Dept. 'S. Stecco' - University of Florence

Abstract. The prediction of unsteady flow field in turbine blades is required by the reduced margin for increasing efficiency, stability and life of propulsion components. Advanced numerical tools are available to run within reasonable time 3D unsteady calculation for full stage. New techniques on the computation and parallel computer allow the improvements of results in terms of cost and accuracy; the physical modelling is sometimes still a challenge and the huge amount of data extracted from experiments require care and skill to become useful design tools. A full-3D and parallel unstructured solver has been developed and applied to the simulation of an unsteady turbine stage.

1 The governing equations and numerical method

The basic numerical scheme of HybFlow code solves the compressible Navier-Stokes equations cast in strong conservative form [1] and extended to cope with a moving reference frame [2]. The spatial discretization is based on a finite volume approach for hybrid unstructured grids. Roe's approximate method is used for the upwind scheme. A linear reconstruction of the solution inside the elements provides a second order discretisation, and monotonicity is ensured through the TVD slope limiter proposed by Barth [3]. The dual-time stepping approach used for time accuracy. More details can be found in in [1],[2]

Turbulence is accounted using the traditional low Reynolds two equation k-w model proposed by Wilcox [6]. The turbulence model also incorporates an extra algebraic equation which enforces a realizability constraint [5] and helps solving the problem of extra turbulence production in stagnation points and in the developing boundary layer.

1.1 The Parallel Multi-Block Approach for Stage Computations

A significant memory saving and reduction of the overall CPU cost has been obtained by means of different numerical tools. Firstly a multi-block strategy implementation allowed a significant reduction of memory usage, thanks to the reduced dimensions of the linear system matrix during the implicit marching of the solution. The parallel version of the code, based on the standard MPI standard allowed a further relevant decrease of computational costs. The load distribution strategy has been studied to guarantee almost the same number of elements inside each block and a quite uniform distribution of blocks to different Processors, thus allowing a global load balancing.

1.2 Grid Sequencing Strategy

In a CFD simulation usually an initial guess solution is used to start the computation. This rough solution is updated till the solver reaches the final exact numerical solution of the physical problem. When a complex domain and different physical aspects are involved, rarely this initial guess solution can be chosen close to the real one. Stability problems may arise while several time steps can be required before the final convergence is obtained. An unsteady inviscid computation can be performed on the coarser grid to capture the fundamental flow physics or the basic frequency spectrum of the unsteady pressure fluctuations which are both almost independent from viscous phenomena. Once the solution is obtained on the coarse grid a viscous and turbulent computation can be carried out on the second refined grid to capture the effect boundary layers development on the performance and efficiency of the stage. The initial guess solution can be represented by the inviscid solution obtained on the coarse grid which can be transferred to the refined simulation .

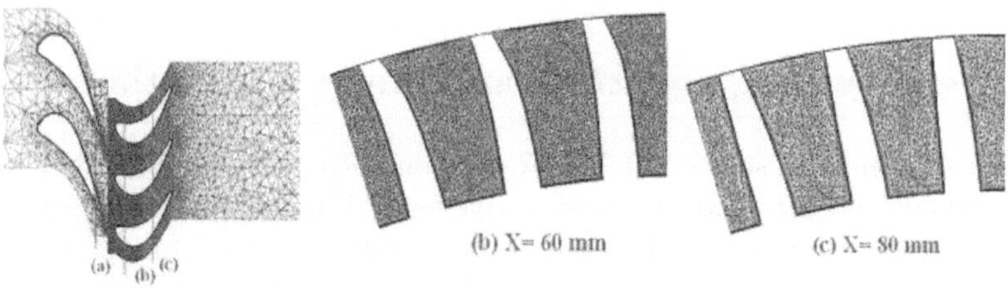

Fig. 1. Refined Grid for Viscous Computations

A further refinement step can be performed using the viscous grids shown in Figures 1. Grid refinement is evident in the radial section at midspan (Figure 1) where 10 prismatic layers have been introduced all around the NGV and rotor blade surfaces. Node elements have been also added in the rotor vane passage flow and mainly in the coolant injection area (slide (a) in figure 1) of the NGV and all over the pitch-wise direction ahead the rotor blade (slide (b)) in order to accurately capture the effect of the NGV wake for all relative positions of the two rows.

2 Application to the 3D VKI Annular Stage

2.1 Results and discussion

The time-averaged pressure profiles on the rotor blade are reported in Figure 2(a), Figure 2(b) and Figure 2(c), for three radial sections at 50%, 15% and 85% of the blade span. The agreement with experiments [4]is satisfactory on the pressure side, while on the suction side and close to s/smax=0.2, the flow acceleration is slightly under predicted.

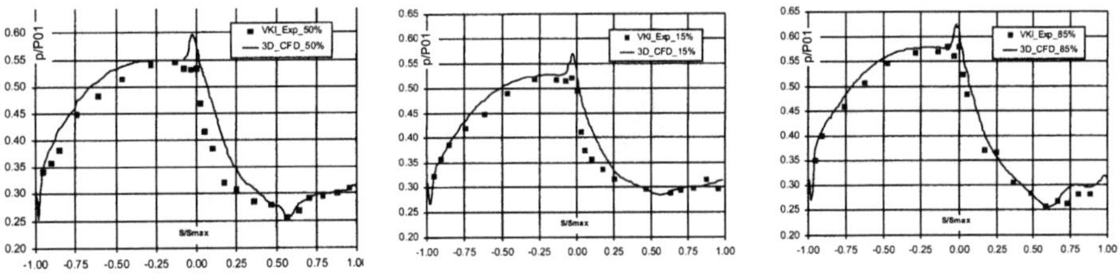

(a) Averaged rotor pressure - 50% span

(b) Averaged rotor pressure - 15% span

(c) Averaged rotor pressure - 85% span

Fig. 2. Time averaged Pressure Profiles

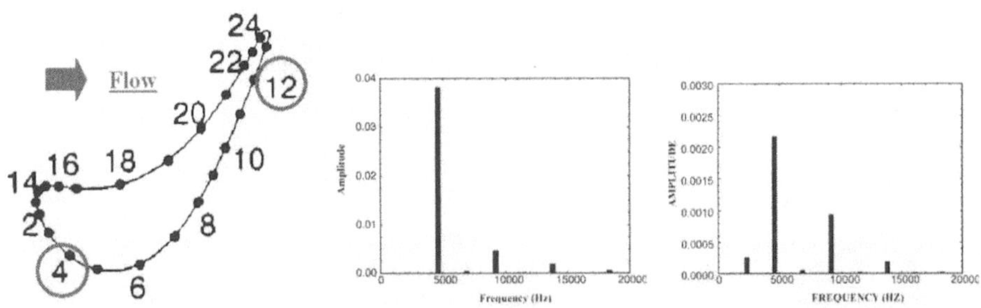

(a) Rotor Gauge Position

(b) Rotor LE gauge 4

(c) Rotor TE gauge 12

Fig. 3. Frequency Spectrum

The frequency analysis for two points located near LE and TE of rotor blade on suction surface are reported in Figure 3. For both gauges the main harmonic observed is the stator blades passing frequency.

These basic frequencies of the spectrum (Figures 3(b) and 3(c) can be computed from the speed rotation (6500 RPM) and the blade count ratio (2:3) leading to a main peaks at f=4.66 kHz and the corresponding multiples (9.32 kHz, 13.98 kHz ...). The remarkable amplitude of the first harmonic at LE is due both to the not uniform exit flow field from the stator blades (mainly due to the presence of vane TE shock impinging on the rotor) and to the small gap existing between the stator TE and the rotor LE. The amplitude of the pressure discontinuity varies considerably depending on the location on the rotor blade. The maximum amplitude is at gauge 4 because directly heat by the shock wave. Flowing toward the TE, the pressure fluctuations are strongly reduced (two orders) by the action of the vane acceleration effect, as can be observed in figure 3(c). Figure 4(a) and Figure 4(b) shows the unsteady pressure fluctuations $\Delta p/P01 = (p - p_{averaged})/P01$ for two rotor blade locations at mid-span. The

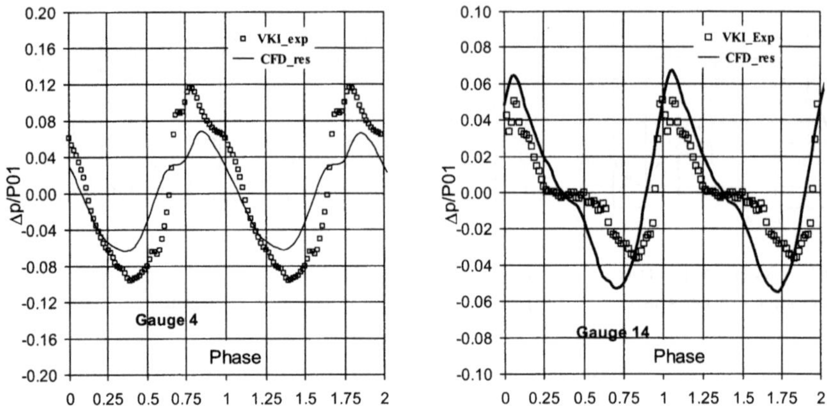

(a) Gauge Location 4 (b) Gauge Location 14

Fig. 4. Unsteady Pressure Fluctuations

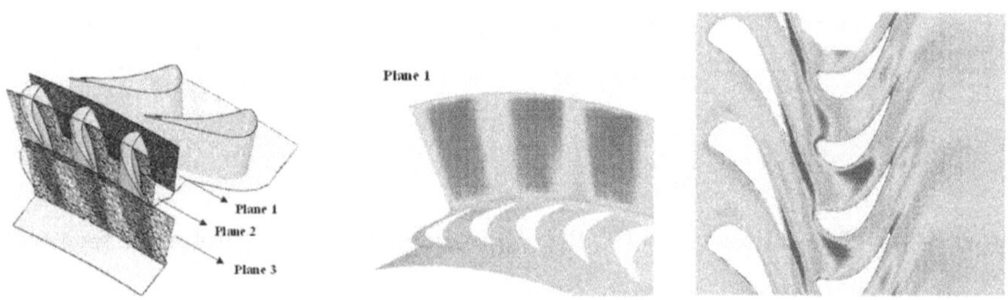

(a) (b) Entropy Contours – Plane 1 (c) Entropy Contours – Plane Midspan

Fig. 5. Entropy Contours

two selected points lay on the suction side at approximately 20% axial chord (again point 4 in Figure 3) and on the leading edge (point 14). The basic feature of the instantaneous signal are reproduced by the numerical results even if peak amplitudes of pressure fluctuations are under predicted mainly at gauge 4. In Figure 5(b) the entropy contours at the stator exit are shown in axial plane 1, situated at 10 % of the axial chord downstream the stator TE. The stator wake is well outlined by the high loss region in the middle of the area. As can be noticed the mid-span flow is essentially 2D, most of the loss being confined in the blade wake.

In Figure 5(c) a turbulent viscosity contours at 50% span is reported. The NGV TE wakes can be clearly identified by the high turbulent regions impinging on the rotor LE. The turbulence viscosity level is a fundamental parameter for heat transfer prediction in view of the strong impact on the Nusselt number

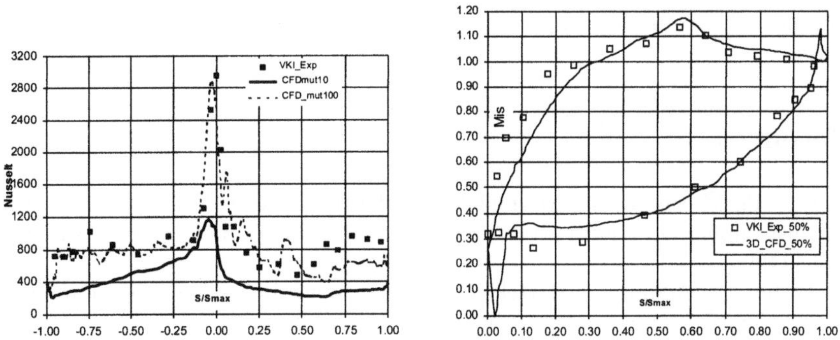

(a) Time Averaged Nusselt Number

(b) Time Averaged Rotoric Mis

Fig. 6. Nusselt Number Comparisons

and thus on the convection heat transfer coefficient. With low inlet turbulent (figure 6(a)) viscosity levels the experimental Nusselt numbers are largely underestimated all over the blade profile. The mach number distribution of figure 6(b) can help the understanding of the boundary layer development on the suction and pressure side and explain some inaccuracies of the numerical results.

Around 60% of the blade the isentropic mach number in figure 6(b) decreases, indicating a deceleration of the flow maybe promoting a transitional state. The agreement between experimental and numerical results is reduced in this blade part following a boundary layer laminar to turbulent transition. This is due to the present limitations of the turbulence closure implemented which is known to fail for the detection of the transition onset. An improvement of the numerical results in this blade region could be obtained with the addition of a proper transition model. On PS the accuracy of the averaged numerical results is better than on the SS still with some inaccuracies.

3 Conclusions

To accurately reproduce heat transfer experimental data stringent requirements are imposed on the grid cell dimension. The grid dimensions typical of a realistic turbulent computation (about 4.0 ML elements) created difficulties for the management and storage of the input and output data. Efficient interpolation techniques and a rationale use of the I/O management were necessary to reduce the length of the grid sequencing technique and all the post processing phases of the results management. In conclusion great effort have been successfully carried out on the different aspects of the calculation to face correctly and profitable the challenge to compute and investigate unsteady flow in a full turbine stage. The Nusselt numerical averaged and unsteady distribution, discussed against experiments showed that further transition modelling and grid sensitivity analysis should help to further improve the accurate description of heat transfer flow features.

References

1. Adami P., Michelassi V., Martelli F.: 'Performances.of a Newton-Krylov scheme against implicit and multi-grid solvers for inviscid flows'. AIAA paper 98-2429-257-4655,(1998)
2. Belardini E., Adami P., Martelli F.: 'Development of an Unsteady Parallel Approach for 3D Stator-Rotor Interaction'. IMechE Journal of Power and Energy2001-Vol.215. n.A6, (2001)
3. Barth T.J.: 'Numerical Aspects of Computing Viscous High-Reynolds Number Flows on Unstructured Meshes'. AIAA Paper 91-0721, Jan, (1991)
4. Denos R., Arts T., Paniagua G., Michelassi, V. and Martelli F.: 'Investigation of the unsteady Rotor Aerodinamics in a Transonic Turbine Stage'.2000-GT-435.**422-4807**, (2000)
5. Durbin P.A.: 'On the k-e Stagnation Point Anomaly'.Int. J. Heat and Fluid Flow, 17 (1996)
6. Wilcox D.C.: 'Simularion of transition with a two-equation turbulence model'.AIAA J. Vol. 32,No.2a, (1994)

A Three-Dimensional Parallel Time-Accurate Turbopump Simulation Procedure Using Overset Grid Systems

C. Kiris, W. Chan, and D. Kwak

NAS Applications Branch, M.S, T27B-1 NASA-Ames Research Center, Moffett Field, CA 94035 kiris@nas.nasa.gov wchan@nas.nasa.gov dkwak@mail.arc.nasa.gov

Abstract. The objective of the current effort is to provide a computational framework for design and analysis of the entire fuel supply system of a liquid rocket engine, including high-fidelity unsteady turbopump flow analysis. This capability is needed to support the design of pump sub-systems for advanced space transportation vehicles that are likely to involve liquid propulsion systems. To date, computational tools for design/analysis of turbopump flows are based on relatively lower fidelity methods. An unsteady, three-dimensional viscous flow analysis tool involving stationary and rotational components for the entire turbopump assembly has not been available for real-world engineering applications. The present effort provides developers with information such as transient flow phenomena at start up, non-uniform inflows, and its impact on system vibration and structures. In the proposed paper, the progress toward the capability of complete simulation of the turbo-pump for a liquid rocket engine is reported. The Space Shuttle Main Engine (SSME) turbo-pump is used as a test case for evaluation of the hybrid MPI/Open-MP and MLP versions of the INS3D code. CAD to solution auto-scripting capability is developed for tubopump applications. The relative motion of the grid systems for the rotor-stator interaction was obtained using overset grid techniques. Unsteady computations for the SSME turbo-pump, which contains 114 zones with 34.5 million grid points, are carried out on Origin 3000 systems at NASA Ames Research Center. Results from these time-accurate simulations with moving boundary capability are presented along with the performance of parallel versions of the code.

1 Introduction

Until recently, the high performance pump design process was not significantly different from that of 30 years ago. During the past 30 years a vast amount of experimental and operational experience has demonstrated that there are many important features, which are not accounted for in the semi-empirical design process. Pump design processes are no more technologically advanced than those designed for the Space Shuttle Main Engine (SSME). During that same time span huge strides have been made in computers, numerical algorithms, and physical modeling. In order to increase the role of Computational Fluid Dynamics (CFD) in the rocket pump design process, the CFD analysis tools must be evaluated and validated so that designers gain confidence in their use. As a part of extending the CFD technology to pump design work, INS3D code has been validated for pump component analysis using data from a rocket pump inducer, and pump impellers [5]. When rotating and stationary parts are included, unsteady interactions between stationary and rotating components need to be included. To resolve the complex geometry in relative motion, an overset grid approach is employed in the current effort. In this approach, a geometrically complex body is decomposed into a number of simple grid components. In order to solve the entire configuration including inlet guide vanes, impeller blades, and diffuser blades, an overset grid system was constructed using 34.3 Million grid points with 114 zones. Since neighboring grids may overlap arbitrarily, these grids can be created independently from each other. Thus it is easier to generate high quality and nearly orthogonal grids using this approach. Connectivity between neighboring grids is established by interpolation at the grid outer boundaries. Addition of new components to the system and simulation of arbitrary relative motion between multiple bodies are achieved by establishing new connectivity without disturbing the existing grids.

2 Numerical Method

The present computations are performed utilizing the INS3D computer code, which solves the incompressible Navier-Stokes equations for both steady-state and unsteady flows. The numerical solution of the incompressible Navier-Stokes equations requires special attention in order to satisfy the divergence-free constraint on the velocity field because the incompressible formulation does not yield the pressure field explicitly from the equation of state or through the continuity equation. One way to avoid the numerical difficulty originated by the elliptic nature of the problem is to use an artificial compressibility method, developed by Chorin [2]. The artificial compressibility algorithm, which introduces a time-derivative of the pressure term into the continuity equation, transforms the elliptic-parabolic type partial differential equations into the hyperbolic-parabolic type. An incompressible flow solver, INS3D, has been developed [9] based on this algorithm. Since the convective terms of the resulting equations are hyperbolic, upwind differencing can be applied to these terms. The current version uses Roe's flux-difference splitting [8]. The third and fifth-order upwind differencing used here are implementations of a class of high-accuracy flux-differencing schemes for the compressible flow equations. To obtain time-accurate solutions, the equations are iterated to convergence in pseudo-time for each physical time step until the divergence of the velocity field has been reduced below a specified tolerance value. The total number of sub-iterations required varies depending on the problem, time step size and the artificial compressibility parameter used, and typically ranges from 10 to 30 sub-iterations. The matrix equation is solved iteratively by using a non-factored Gauss-Seidel type line-relaxation scheme [7], which maintains stability and allows a large pseudo-time step to be taken. Details of the numerical method can be found in [9]. The GMRES scheme has also been utilized for the solution of the resulting matrix equation. Computer memory requirement for the flow solver INS3D with line-relaxation is 35 times the number of grid points in words, and with GMRES-ILU(0) scheme is 220 times the number of grid points in words. When a fast converging scheme, such as a GMRES-ILU(0) solver, was implemented into the artificial compressibility method, reasonable agreement was obtained between computed results and experimental data [6].

3 Turbopump Scripts

For each component of a turbopump (inlet guide vanes, impeller and diffuser), a script was developed to accelerate the simulation process prior to the flow solver call. The starting point is the geometry definition of a component which consists of the hub and shroud profile curves and the surface panels of each distinct blade. Functions provided by the script include overset surface and volume grid generation, creation of object X-ray maps for hole-cutting and creation of overset domain connectivity inputs. After running the domain connectivity program OVERFLOW-D/DCF [1] with the script-created input file, the resulting grids and iblank connectivity information are used as input for the INS3D flow solver. Figure 1 illustrates the procedures automatically handled by the scripts. Important attributes of each component are parameterized in the scripts. For example, parameterized geometric attributes include the number of blades for the inlet guide vanes and the diffuser; and the number of distinct blades per section and the number of sections for the impeller. Parameterized grid attributes include hub and shroud surface grid spacings, leading and trailing edge spacings, viscous wall spacing for the volume grids, stretching ratios and marching distances for the hyperbolic surface and volume grids. All parameters are controlled via an input file and the script can be rapidly re-run to generate different configurations. For example, the user can input different geometry definitions for the hub, shroud and blades, and try different number of blades and grid resolutions. A high level scripting language Tcl is utilized which provides modularity as well as floating point arithmetic capability (unlike basic shell languages such as C-shell). Although extra time and expertise are needed to construct the scripts, the pay-off is well worth the effort. The advantages include the ability to rapidly re-run the entire process in just a few minutes versus one or more days of manual

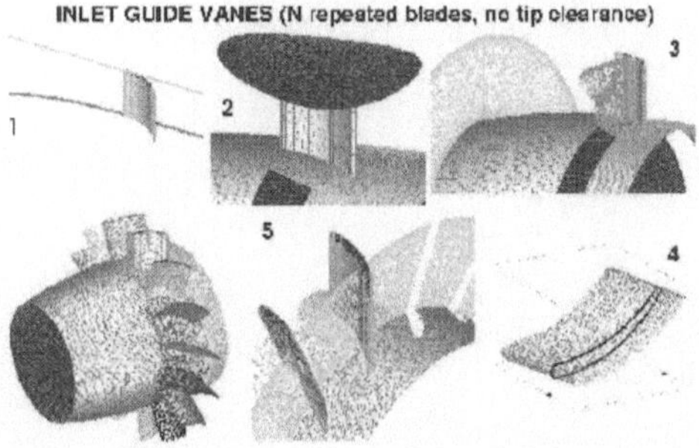

Fig. 1. Illustration of turbopump scripts for inlet guide vanes.

effort. Tedious repetitive error-prone inputs are avoided with the scripts while they also provide a documentation of the entire grid generation procedure. Grid refinement and parameter studies can also be easily and efficiently performed. For example, the process time from geometry definition to domain connectivity inputs for a 6 million point Inlet Guide Vanes grid system required about 1 day of manual effort without the scripts, versus 43 seconds using the scripts on an R12000 300MHz Silicon Graphics workstation. A similar time reduction was achieved on a 15 million point Impeller grid system from 2 weeks of manual effort to just over 5 minutes of CPU time with the scripts.

4 Parallel Computations

Parallel computing strategies vary depending on computer architecture such as memory arrangement relative to processing units. Two approaches have been implemented in the present study: the first approach is hybrid MPI / OpenMP and the second one is Multi-Level Parallelism (MLP) developed at NASA-Ames Research Center. The first approach is obtained by using message-passing interface (MPI) for inter-zone parallelism [3], and by using OpenMP directives for intra-zone parallelism [4]. INS3D-MPI [3] is based on the explicit message-passing interface across MPI groups and is designed for coarse grain parallelism. The primary strategy is to distribute the zones across a set of processors. During the iteration, all the processors would exchange boundary condition data between processors whose zones shared interfaces with zones on other processors. A simple master-worker architecture was selected because it is relatively simple to implement and it is a common architecture for parallel CFD applications. All I/O was performed by the master MPI process and data was distributed to the workers. After the initialization phase is complete, the program begins its main iteration loop. Figure 2(a) shows the MPI/OpenMP scalability for SSME impeller computations using 19.2 Million grid points. The second approach in

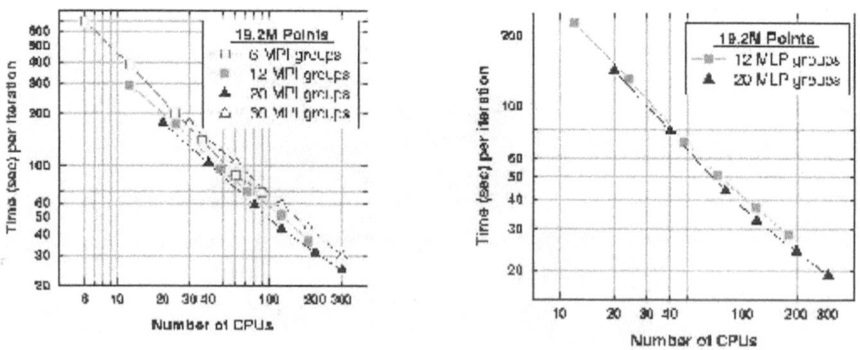

Fig. 2. (a) Performance of INS3D-MPI/OpenMP and (b) IND3D-MLP on SGI Origin platform.

Multi-Level Parallelism (MLP) is obtained by using NAS-MLP [10] routines. The shared memory MLP technique developed at NASA Ames Research Center has been incorporated into the INS3D code. This approach differs from the MPI/OpenMP approach in a fundamental way in that it does not use messaging at all. All data communication at the coarsest and finest level is accomplished via direct memory referencing instructions. This approach also provides a simpler mechanism for converting legacy code, such as INS3D, than MPI. For shared memory MLP, the coarsest level parallelism is supplied by spawning off independent processes via the standard UNIX fork. The advantage of the UNIX fork over MPI procedure is that the user does not have to change the initialization section of the large production code. A library of routines is used to initiate forks, to establish shared memory arenas, and to provide synchronization primitives. The boundary data for the overset grid system is archived in the shared memory arena by each process. Other processes access the data from the arena as needed. Figure 2(b) shows the MLP scalability for SSME impeller computations using 19.2 Million grid points. Using the INS3D-MLP parallel implementation, time-accurate computations for SSME-rig1 configuration have been carried out on SGI Origin 2000 and 3000 platforms. Instantaneous snapshots of particle traces and pressure surfaces from these computations are shown in Figure 3. This procedure provides flow-field details not readily available from experiments. The initial conditions for these simulations imposed stationary flow with an impulsive start-up for the impeller. Three full impeller rotations were completed in the simulations using 34.3 million grid points. Using 128 SGI Origin 3000 CPUs, one impeller rotation was completed in less than 3.5 days wall-clock time. When 480 processors were used, one impeller rotation was completed in less than 1.5 days. In 1999, one impeller rotation required 42 days of wall-clock time to compute. More than 30 times speed-up has been obtained over 1999 capabilities.

Fig. 3. Snapshots of particle traces and pressure surfaces from unsteady turbopump computations (first rotation and the end of third rotation).

5 Summary

An incompressible flow solver in steady and time-accurate formulations has been utilized for parallel turbopump computations. Numerical procedures and turbopump scripts for rapid grid system generation and for overset connectivity input generation are outlined. Performance of the INS3D-MPI/ OpenMP and INS3D-MLP versions of the flow solver for the SSME impeller model are presented. Using INS3D-MLP parallel implementation, time-accurate computations for the SSME-rig1 model, which includes 114 zones with 34.5 million grid points, have been carried out on SGI Origin platforms. More than 30 times speed-up has been obtained over 1999 capability for these computations.

References

1. W. M. Chan, R. L. Meakin, and M. A. Potsdam. CHSSI software for geometrically complex unsteady aerodynamic applications. AIAA Paper 2001–0593, Jan 2001. Reno, Nevada.
2. A. J. Chorin. A numerical method for solving incompressible viscous flow problems. *Journal of Computational Physics*, 2:12–26, 1967.
3. T. Faulkner and J. Mariani. MPI parallelization of the incompressible Navier–Stokes solver (INS3D). http://www.nas.nasa.gov/faulkner/home.html.
4. H. Jin, M. Frumkin, and J. Yan. Automatic generation of OpenMP directives and its application to computational fluid dynamics codes. In *the Proceedings of the Third International Symposium on High Performance Computing, Tokyo, Japan Oct. 16-18*, 2000.
5. C. Kiris, L. Chang, D. Kwak, and S. E. Rogers. Incompressible Navier–Stokes computations of rotating flows. AIAA Paper No. 93–0678, 1993.
6. C. Kiris and D. Kwak. Aspects of unsteady incompressible flow simulation. *Computers and Fluids Journal*, 31(4–7):627–638, 2002.
7. R. W. MacCormack. Current status of numerical solutions of the Navier–Stokes equations. AIAA Paper No. 85–0032, 1985.
8. P. L. Roe. Approximate riemann solvers, parameter vectors, and difference schemes. *Journal of Computational Physics*, 43(1):357–372, 1981.
9. S. E. Rogers, D. Kwak, and C. Kiris. Numerical solution of the incompressible Navier-Stokes equations for steady and time-dependent problems. *AIAA Journal*, 29(4):603–610, 1991.
10. J. R. Taft. Performance of the OVERFLOW–MLP and LAURA–MLP CFD codes and the NASA Ames 512 CPU origin systems. In *HPCC/CAS 2000 Workshop, NASA Ames Research Center*, 2000.

Part XVII

Turbulence

Cycle-resolved Computations of Compressible Vortical Flow in Cylinders of Engines

Ken Naitoh[1], Masashi Ono[2], Kunio Kuwahara[3], and Egon Krause[4]

[1] Yamagata University, 4-3-16 Jyonan, Yonezawa, Yamagata 992-8510, Japan
[2] Nissan Motor Co., Ltd, Kanagawa, Japan (Yamagata University until March, 2002)
[3] Institute of Space and Astronautical Science, Yoshinodai, Kanagawa 229-8510, Japan
[4] Aerodynamisches Institut der RWTH Aachen, Germany

Abstract. Cycle-by-cycle differences of the flow in engine cylinders, that is, the cyclic variations, are analyzed in a fine grid system of about 1,000,000 grid points. A computational method without grid motions is employed on the basis of the multi-level formulation. We apply this method to a square-piston engine. First, we can see that the computational results obtained in an engine cycle agree well with the experiments using freon as the working fluid under the condition with a Mach number in excess of 0.5. Drastic transition to turbulence near TDC is also grasped by the computations and experiments. Next, the calculations are performed during three continuous cycles of the engine. The flow structure around the center of engine cylinder, that is, around the ignition point, varies cycle by cycle. The present cyclic variations of flow field may lead to combustion instability. It should be stressed that the present method clarifies the cyclic variations, although a lot of researches have been devoted for ensemble-averaged computations.

1 Introduction

Cyclic variations, the cycle-by-cycle differences of the flow field in engines, are analyzed in detail with a fine grid system of about 1,000,000 points. We employ the highly-accurate numerical method [1] [2] [3] [4] and the way without grid motions.

A lot of efforts have been devoted in order to perform the computations of cycle-averaged flows in engine cylinders (cycle-averaged computations) [5] [6]. Most of the cycle-averaged computations are based on ensemble-averaged turbulence models. In principle, the cycle-averaged computations are difficult to analyze the cyclic variations and combustion instability of engine. Numerical codes are expected, which can perform the cycle-resolved computations.

In the previous reports [1] [2] [3] [4], we proposed the numerical method for calculating the compressible turbulence with high accuracy. The flows in an engine having square piston and slit type of inlet were simulated, because the detailed experimental photographs were taken for the square-piston engine by the group of RWTH Aachen [2]. The computational results during an engine cycle agree well with the experiments using freon as the working fluid under the condition with a Mach number in excess of 0.5. Drastic transition to turbulence near TDC is also grasped by the computations.

In the present report, the detailed calculations are performed during three continuous cycles of the engine. The flow structure around the center of engine cylinder, that is, around the ignition point, varies cycle by cycle. The present cyclic variations of flow field may bring combustion instability.

2 Formulation

2.1 Basic equation and multi-level formulation

The compressible Navier-Stokes (N-S) equation [7] involving the mass, momentum, and energy conservation laws, Eq. (1), is employed as the basic equation.

$$\frac{D\boldsymbol{q}}{Dt} = \boldsymbol{rhs}. \tag{1}$$

where the vector $\boldsymbol{q}$ denotes the thermo-fluid dynamic quantities of density, velocity, and enthalpy and the vector $\boldsymbol{rhs}$ denotes the right hand side in the N-S equation, while D/Dt is the substantial derivative.

The basic equation is transformed to the special formulation (the multi-level formulation [1] with three levels, which involves the usual N-S equation, the higher derivatives, and the form integrated over the entire analytical domain). (See Eqs. 2,3, and 4.)

$$\frac{D\boldsymbol{q}}{Dt} = \boldsymbol{rhs}, \tag{2}$$

$$\frac{D\boldsymbol{Q}}{Dt} = \boldsymbol{RHS}, \tag{3}$$

$$\frac{D\boldsymbol{q'}}{Dt} = \boldsymbol{rhs'}, \tag{4}$$

where the vectors Q and q' denote the spatially-integration and spatial derivative of q, respectively, while RHS and rhs' are the spatial integral and derivative of rhs.

This transformation is necessary, because only the discretization of the usual N-S equation (Eq.1) brings us some numerical difficulties while computing the unsteady compressible vortical flows in engines. Equation 2 is discretized by the finite difference method, which is combined with the third-order upwind scheme for convection. Details of formulation are shown in References [1] [2] [3] [4].

2.2 Gridless approach

Moving arbitrary coordinate systems have been previously used to calculate engine flows, because of the piston motions. However, the approach with moving grids leads us to a disadvantage, that grid size varies temporally. (Spatial accuracy varies temporally.)

In the present research, the piston is moved in the fixed Cartesian coordinate system, that is, the homogeneous and orthogonal grid system without the

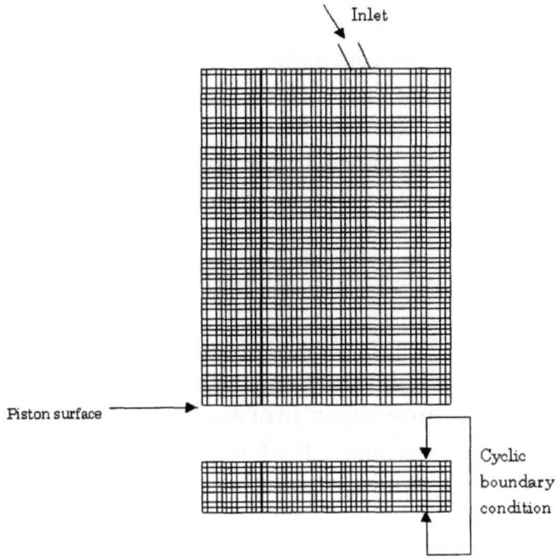

Fig. 1. Grid system (202 x 151 x 31 points)

motions of grid points. This is one of the gridless methods. Figure 1 show the engine geometry and grid system. The engine analyzed has a square piston and a slit type of inlet. (The inlet also becomes the outlet in the exhaust process.) Computations are performed in three-dimensional space. Engine parameters and conditions are shown in Table 1.

Table 1. Engine geometry and operating conditions

Parameters	*Values*
Compression ratio	3.9:1
Bore x Stroke	100 mm x 100 mm
Engine speed	998 rpm
Working fluid	Freon-12

3 Results

3.1 Comparison between computations and experiments

The multi-level formulation (Eqs. 2,3, and 4) with "moving grid" systems calculated the compressible turbulence in engine accurately [1] [2] [3].

Here, we examine whether the "gridless" method also computes the engine flows with high accuracy. Figures 2 and 3 show the computed density contours in the intake and compression processes obtained without grid motions and the corresponding experimental photographs. A large vortex in the cylinder and small vortices generated at the edge of inlet are simulated well during the intake stroke. The drastic transition to turbulence is also observed near the end of the compression stage.(See Figs. 3 and 4.) The number of density contours in the calculated large vortex agrees with that of the experiment fairly well.

3.2 Cyclic variations

Computations in the intake, compression, expansion, and exhaust processes are performed during three continuous cycles of engine. Figure 5 shows the density contours at the end of the compression process in each cycle. The flow structure around the center of engine cylinder, that is, around the ignition point, varies cycle by cycle. The present cyclic variations of flow field may lead us to the instability of combustion.

4 Conclusions

The present numerical method involving the multi-level formulation [1] [2] [3] and the gridless approach on the fixed Cartesian coordinate makes it possible to perform the cycle-resolved computations. Small disturbance at the initiation of intake process leads to large difference at the end of compression stage.

References

1. K. Naitoh and K. Kuwahara, Large eddy simulation and direct simulation of compressible turbulence and combusting flows in engines based on the BI-SCALES method, J. of Fluid Dynamics Research., 10, (1992), 299-325.
2. K. Naitoh, K. Kuwahara, M. Jeschke, and E. Krause, Numerical Simulation of the Small Vortices in the Intake and Compression Processes of an Engine, JSME International J., Series II, Vol. 35, No.4 (1992)., 549-558.
3. K. Naitoh, Y. Takagi, and K. Kuwahara, Cycle-resolved computation of compressible turbulence and premixed flame in an engine, Computers and Fluids, Vol. 22, No.4/5 (1993), pp.623-648.
4. K. Naitoh, H. Fujii, Y. Takagi, and K. Kuwahara, Numerical simulation of the detailed flow in engine ports and cylinders, Trans. of SAEpaper900256 (1990).
5. A. Amsden, P.J. O'Rourke, and T.D. Butler, KIVA-II, a computer program for chemically-reactive flows with sprays, Report LA-11560-MS (1989).
6. R. Taghavi and A. Dupont, Assessment of chamber and inlet port designs using multidimensional flow modeling, IMechE, (1989).
7. R. Aris, Vectors, tensors, and basic equations of fluid mechanics, Prentice-Hall, (1962).

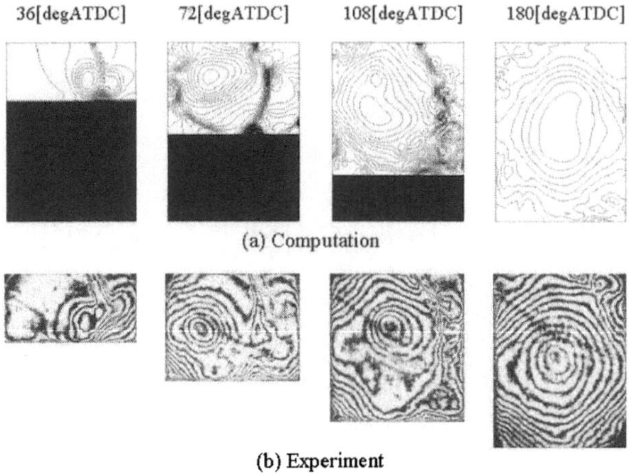

Fig. 2. Computed density contours in the intake process of first cycle and the corresponding experimental photograph (Mach-Zehnder Interferometry [2]).

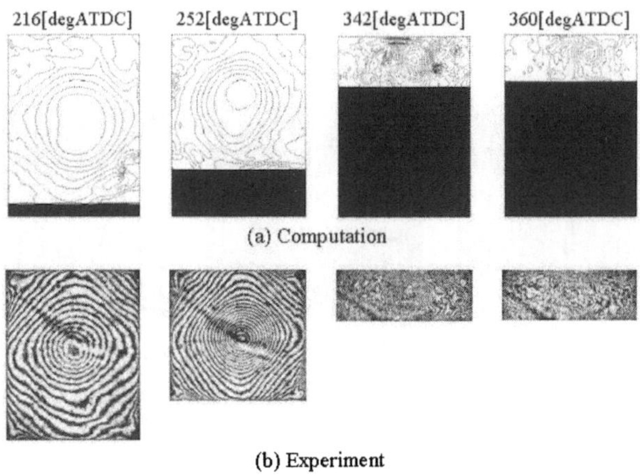

Fig. 3. Computed density contours in the compression process of first cycle and the corresponding experimental photograph (Mach-Zehnder Interferometry [2]).

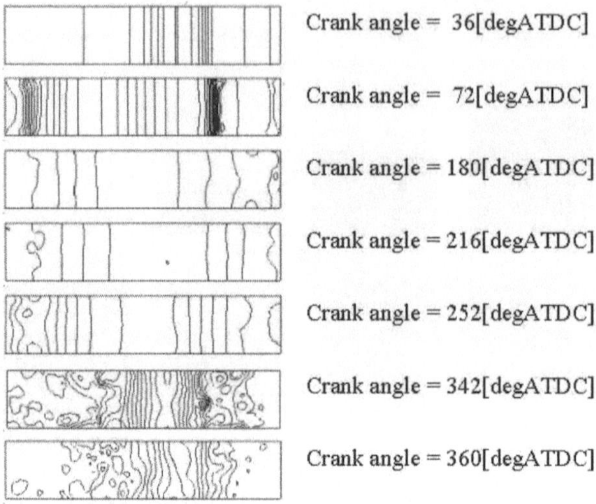

Fig. 4. Computed density contours in the depth of the square-piston engine.

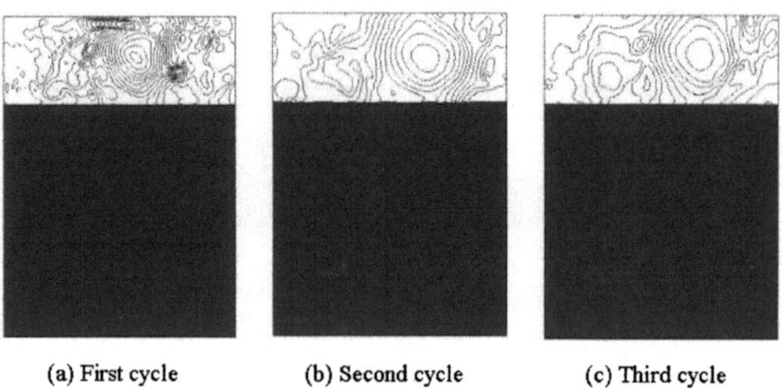

Fig. 5. Calculated density contours during three cycles (the end of compression stage).

Turbulent Pipe Flow of Non-Newtonian Fluids

M. Rudman and H. M. Blackburn

CSIRO Manufacturing and Infrastructure Technology,
PO Box 56, Highett 3190, Australia

Abstract. Direct numerical simulation of the weakly turbulent flow of non-Newtonian fluids with shear-thinning rheology are undertaken. Results agree reasonably well with the experimentally determined logarithmic layer correlation presented by Clapp (1961) and with other previously published experimental work. As the power law index becomes smaller for the same Reynolds number, the flow deviates further from the Newtonian profile and the results suggest that transition is delayed. Use of this technique shows promise in understanding transition and turbulence in non-Newtonian fluids.

1 Introduction

The flow of non-Newtonian fluids in pipes occurs in a wide range of practical applications in the process industries. Transitional and turbulent flows are of significant interest, although if the fluid has a high yield stress, or if its effective viscosity is high, industrially relevant flow rates may occur in the laminar flow regime. While some experimental work has appeared on the transitional and turbulent flow of non-Newtonian fluids, little is known of how these differ in detail from corresponding Newtonian flows. There is evidence that the radial and azimuthal turbulence intensities can be lower by 20–40% for a power law fluid compared to a Newtonian fluid, whereas the axial intensities are marginally higher, consistent with experimental results [6,9].

Computational modelling of non-Newtonian flows using direct numerical simulation (DNS) shows promise in understanding transition and turbulence in these fluids. For a wide range of important materials, the non-Newtonian rheology is primarily of a shear-thinning nature and apart from [12,13] there have been few published DNS investigations of turbulent flows of shear-thinning non-Newtonian fluids. Previous studies [1,5] have largely concentrated on non-Newtonian rheologies in which visco-elastic effects were the dominant feature. The aim of the present study is to examine the effect of model parameters on turbulent pipe flows of power law shear thinning fluids.

Comparisons between DNS of power law fluids and experimental data were presented in [12,13]. It was seen there that a significant disagreement existed, with the predicted superficial velocities for a given pressure gradient being 10–15% too low and the velocity profiles having a different shape. This was believed to be a result of elastic effects in the 0.5 wt% carboxymethylcellulose (CMC) solution used as the experimental fluid [7]. The discrepancy between experiment and DNS observed in [12,13] is consistent with the drag reduction reported for example in [7]. Although the discrepancy highlighted the difficulty in approximating power law fluids using polymer solutions, it underlines the fact that once a CFD model can be correctly validated for these flows, it can be reliably used to model the behaviour of power law fluids without the added complication of elastic and elongational effects that influence the experiments.

2 Rheology Model

The viscosity η is described using the power law model

$$\eta = K\dot{\gamma}^{n-1}\,, \tag{1}$$

where $\dot{\gamma}$ is the shear rate, K is the consistency and n is the flow index. For $n < 1$ the fluid is shear-thinning, $n = 1$ for Newtonian, and $n > 1$ for shear-thickening fluids. The shear rate is computed from the rate-of-strain tensor $\boldsymbol{S}$ as

$$\dot{\gamma} = (2\boldsymbol{S}:\boldsymbol{S})^{1/2}\,. \tag{2}$$

A generalised Reynolds number, Re_{G}, is based on the pipe diameter D, superficial velocity, fluid density ϱ, and a wall viscosity, η_{w}, which is derived from (1) using the mean wall shear stress τ_{w} as the basis for estimating a shear rate at the wall.

3 Numerical Method

The spatial discretisation employs a spectral element–Fourier formulation, which allows arbitrary geometry in the (x, y) plane, but requires periodicity in the z (axial, or out-of-plane) direction. A second-order-in-time mixed explicit–implicit technique is employed for time integration of the incompressible momentum equations, which for a spatially variable viscosity η read

$$\partial_t \boldsymbol{u} + \boldsymbol{N}(\boldsymbol{u}) = -\frac{1}{\varrho}\nabla p + \frac{1}{\varrho}\nabla\cdot\left[\eta\left\{\nabla\boldsymbol{u} + (\nabla\boldsymbol{u})^T\right\}\right]\,, \quad \text{with} \quad \nabla\cdot\boldsymbol{u} = 0\,. \tag{3}$$

To allow a semi-implicit treatment of the viscous terms, the non-Newtonian viscosity is decomposed into a spatially constant component, η_R, and a spatially varying component $\eta - \eta_R$. The spatially varying component is treated with a second-order explicit formulation and the constant component is treated implicitly. Because the power law rheology has a singular viscosity at zero shear rate, a 'cut-off' value is used, below which the shear rate is assumed to be constant when computing the viscosity. This value is chosen to be 10^{-5} times the mean shear rate; in practice the limit is never tripped. In order to drive the flow in the axial (z) direction, a body force per unit mass equivalent to the pressure gradient measured in the experiments is applied to the z-momentum equation.

The planar/Fourier representation of three spatial dimensions leads naturally to parallel implementations in which for most of the timestep each process carries a subset of two-dimensional complex modes. The non-linear terms $\boldsymbol{N}(\boldsymbol{u})$

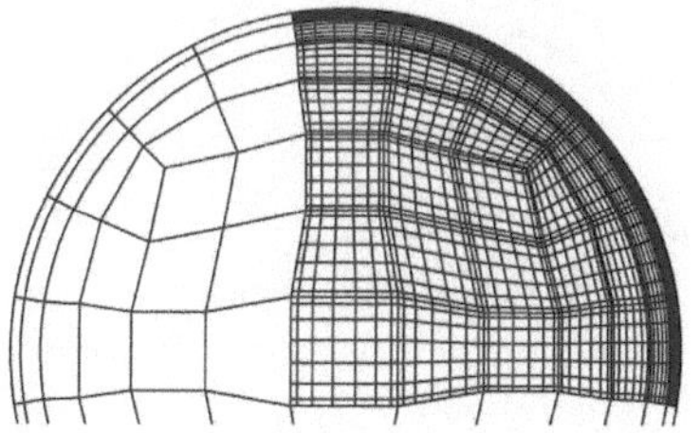

Fig. 1. Upper part of a two-dimensional projection of the simulation mesh

Table 1. Simulation parameters: 'length' is the axial periodic length of the domain, 'modes' is the number of Fourier modes employed in the pipe-axis direction, and $\varrho^{-1}\partial p/\partial z$ is the body force per unit mass that drives the flow in the axial direction

n	K	$\varrho^{-1}\partial p/\partial z$	Re_{G}	length	modes
0.5	1.527	1441	5339	$5\pi D$	128
0.69	0.483	1695	5501	$5\pi D$	108
0.75	0.320	1745	5514	$4\pi D$	108

are formed pseudospectrally with the aid of interprocess memory exchanges. The underlying numerical code has been validated for both DNS and LES of pipe and channel flow [2,11]. The implementation of the power-law non-Newtonian viscosity was validated against laminar pipe flow and axisymmetric Taylor–Couette flow of power-law fluids, both of which have analytic solutions. In all cases, numerical and theoretical velocity profiles agreed to within 0.01%.

The computational domain consists of 105 8th-order elements in the pipe cross section (see Fig. 1) and 80–128 Fourier modes (160–256 data planes) in the axial direction, with domain lengths of 3–5 π. In terms of wall units (based on the standard Newtonian analysis with η_{w} taking the place of the Newtonian viscosity), the near-wall mesh spacing is $r^+ \approx 0.5$, $R\theta^+ \approx 8$ and $z^+ \approx 35$.

4 Results

Results from three simulations are presented here for power law fluids at a generalised Reynolds number of approximately 5500. Simulation parameters are presented in Table 1. The effect of modifying the power law index is of fundamental interest, so to maintain the generalised Reynolds number for the simulations as close to 5500 as practicable as n was changed, the consistency, K, and the axial body force were altered to maintain constant wall viscosity and superficial velocity.

The mean axial velocity for the three simulations at $Re_{\mathrm{G}} = 5500$ for $n = 0.5$, 0.69 and 0.75 are shown in Fig. 2 a plotted with standard 'Law of the wall' nondimensionalisation and compared to the standard (low Reynolds number)

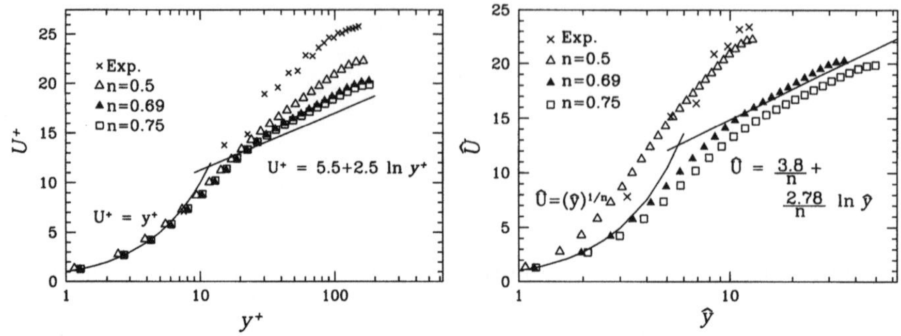

Fig. 2. Velocity profiles for three different power law fluids for $Re_{\mathrm{G}} = 5500$, with values of $n = 0.5$, 0.69 and 0.75, plotted using standard Newtonian non-dimensionalisation (a, left) and Clapp's non-dimensionalisation (b, right). Shown for comparison are experimental results obtained in a CMC solution with $n = 0.69$ and$Re_{\mathrm{G}} = 5780$

Newtonian velocity profile for pipe flow [4]. As the index n increases, the profiles for the power law fluids approach the Newtonian profile, as expected. Results obtained from experiment [12] (with $n = 0.69$) are also plotted and fall well above the other curves; as discussed in the Introduction, this discrepancy is thought to result from viscoelastic effects in the CMC solution.

Shown in Fig. 2 b, the axial velocity profiles are compared to the experimentally derived correlation of [3]. Based on dimensional arguments and experimental measurements, Clapp determines that the logarithmic velocity profile for power law fluids is a function of the power law index, n and satisfies

$$\widehat{U} = \frac{3.8}{n} + \frac{2.78}{n} \ln \widehat{y}\,, \quad \text{where} \quad \widehat{y} = \left[(\varrho^n \tau_{\mathrm{w}}^{2-n})^{1/2} / K \right] y\,. \tag{4}$$

Clearly seen is that the CFD results for $n = 0.69$ agree quite well with Clapp's correlation, however the profile for $n = 0.75$ falls slightly below the correlation and those for $n = 0.5$ falls well above and are suggestive of undeveloped turbulent flow. Of interest is the close agreement between the results for $n = 0.5$ and the experimental results for a CMC solution with $n = 0.69$, suggesting that the experimental flow was not fully developed. This conclusion is supported by dye visualisation in related physical experiments that demonstrated turbulent 'puff' type behaviour was occurring for this flow.

Instantaneous viscosity contours are shown in Fig. 3. The contour scales are equivalent. The cross sections have been taken in regions of the flow that are active and show the degree of unsteadiness as well as the high shear rates (and hence low viscosities) in the wall regions. Of note is the large range of viscosities, and higher viscosities in the central regions for the case of $n = 0.5$ – this is expected because the fluid is more shear thinning.

Turbulence intensities and Reynolds stresses are plotted in Fig. 4. For both the axial turbulence intensities and the Reynolds stresses, the results for the power law fluids are close to the Newtonian results (DNS at $Re = 5000$ [8]), however for both radial and azimuthal velocity fluctuations, the values for the

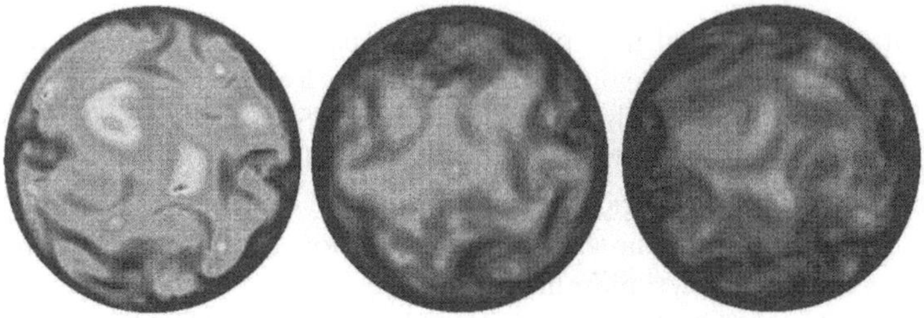

Fig. 3. Instantaneous contours of viscosity for power law fluids, $Re_G = 5500$ and $n = 0.5$ (left), 0.69 (centre) and 0.75 (right). (Red is high, blue low for all contours)

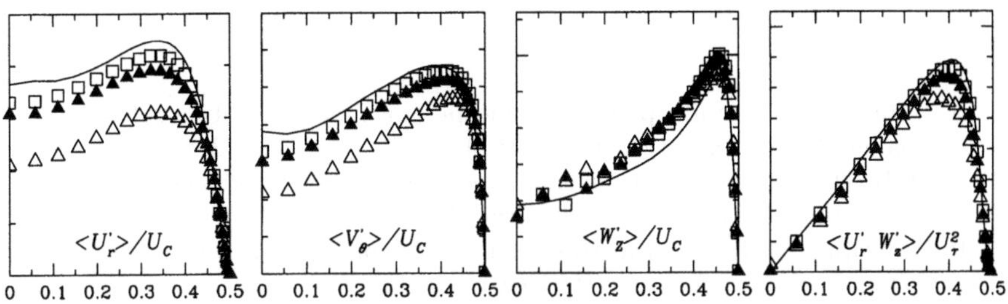

Fig. 4. Turbulence intensities and Reynolds stress as a function of r/D. Radial velocity (top left), azimuthal velocity (top right), axial velocity (lower left) and Reynolds stress (lower right). (*Solid line* for Newtonian DNS, power law fluids are $n = 0.5$ (△), $n = 0.69$ (▲) and $n = 0.75$ (□)

power law fluids are significantly lower than the Newtonian case. This has been observed experimentally [7,9] although currently there is no clear understanding of why this is the case. As n approaches unity, the results approach the Newtonian correlation, as expected.

5 Summary

The applicability of Clapp's scaling for power law fluids is backed up by the DNS results for $n = 0.69$ and 0.75 although the Reynolds numbers used here are lower than those used to derive the experimental correlation. The results for $n = 0.5$ are significantly different and are suggestive of transitional flow. The results also suggest that as the power law index (n) is decreased, and the deviation from Newtonian rheology increases, the value of Re_G at which transition occurs also increases.

It appears that pipe flow of power law fluids make the transition to turbulence via intermittency and turbulent events like the slugs and puffs observed in Newtonian flow. Although the results are not conclusive because of the insufficient domain length of the simulations, they are believed to be qualitatively correct. These unsteady structures are potentially able to re-suspend small settling particles in particulate flows.

Difficulties encountered in experimentation as a result of using real polymer solutions to approximate idealised rheologies can lead to problems of interpretation and understanding. The application of DNS to flows of non-Newtonian fluids with certainty of the rheology being studied has the potential to enable the effect of different rheological parameters to be correctly quantified and understood. This is possibly the greatest contribution that DNS can bring to the study of flows of non-Newtonian fluids.

Acknowledgments: We gratefully acknowledge AMIRA, BHP–Billiton, De Beers, Rio Tinto, Western Mining Corp and Warman International for partial sponsorship of this work.

References

1. A. N. Beris, C. D. Dimitropoulos, Pseudospectral simulation of turbulent viscoelastic channel flow, Comput. Methods Appl. Mech. Eng., **180**, 365–392 (1999)
2. H. M. Blackburn, 'LES of turbulent channel flow using spectral elements'. In: *13th A/Asian Fluid Mech. Conf.* (Melbourne, 1998)
3. R. M. Clapp, 'Turbulent heat transfer in pseudoplastic non-Newtonian fluids'. In: *Int. Dev. Heat Transfer part III*, (New York, 1961)
4. J. M. J. den Toonder, F. T. M. Nieuwstadt, Reynolds number effects in a turbulent pipe flow for low to moderate *Re*, Phys. Fluids **9** (11), 3398–3409 (1997)
5. C. D. Dimitropoulos, R. Sureshkumar, A. N. Beris, Direct numerical simulation of viscoelastic turbulent channel flow exhibiting drag reduction: effect of the variation of rheological parameters, J. Non-Newtonian Fluid Mech. **79**, 433–468 (1998)
6. M. P. Escudier, F. Presti, Pipe flow of a thixotropic liquid, J. Non-Newtonian Fluid. Mech. **62**, 291–306 (1996)
7. M. P. Escudier, F. Presti, S. Smith, Drag reduction in the turbulent pipe flow of polymers, J. Non-Newtonian Fluid Mech. **81**, 197–213 (1999)
8. S. Schmidt, D. M. McIver, H. M. Blackburn, M. Rudman, G. J. Nathan, 'Spectral element based simulation of turbulent pipe flow'. In: *14th A/Asian Fluid Mech. Conf.* (Adelaide, 2001)
9. F. T. Pinho, J. H. Whitelaw, Flow of non-Newtonian fluids in a pipe, J. Non-Newtonian Fluid Mech. **34**, 129–144 (1990)
10. L. Pullum, L. J. W. Graham, 'A new high concentration pipeline test loop facility'. In: *Hydrotransport 14* (Maastrict, 1999)
11. M. Rudman, H. M. Blackburn, 'Large eddy simulation of turbulent pipe flow'. In: *2nd Int. Conf. on CFD in the Minerals and Process Ind.* (Melbourne, 1999)
12. M. Rudman, H. M. Blackburn, L. J. W. Graham, L. Pullum, 'Weakly turbulent pipe flow of a power law fluid'. In: *14th A/Asian Fluid Mech. Conf.* (Adelaide, 2001)
13. M. Rudman, L. J. W. Graham, H. M. Blackburn, L. Pullum, 'Non-Newtonian turbulent and transitional pipe flow'. In: *Hydrotransport 15.* (Banff, 2002)

Some Aspects of Shocklets in Isotropic Compressible Turbulence

Fenggan Zhuang,[1] Xinliang Li[2]

[1] China aerospace science and technology corporation, Beijing, China
[2] SMC Center, Department of Mechanics, Tsinghua University, Beijing, China

Abstract. To overcome the difficulty in the study of shocklets in isotropic compressible turbulence at high turbulent Mach number, a new difference scheme called GVC8 is developed. We have succeeded in direct numerical simulation up to turbulent Mach number 0.95. The statistical quantities thus obtained at lower turbulent Mach number agree well with those from previous authors start with same initial conditions, but they are limited to simulate at lower turbulent Mach numbers due to the so-called start-up problem. In addition with our scheme we have good capability for shock wave capturing and resolution of small scale. Evidence indicates that the extended self-similarity holds in decaying isotropic compressible turbulence despite of the occurrence of shocklets.

1 Introduction

In the study of numerical simulation of isotropic incompressible turbulence much attention has been paid to the problem of aliasing error when using spectral method. It should be pointed out that aliasing error also exists implicitly in the finite difference scheme. The use of upwind scheme may yield a heavy damping in the high wave number region thus limiting the growth of those wave components. We are not going to discuss these aliasing errors further in our paper as many techniques developed for incompressible flow can be used. Particular attention paid here is to study the appearance of shocklets. In compressible turbulent flow, when turbulent Mach number is high enough, those shock waves in compressible turbulent flow pose severe challenges to DNS [1]. According to reference [1], most difference schemes designed for steady aerodynamic flow are too dissipative, and even ENO schemes also seemed too dissipative. But if non-dissipative schemes are used, the non-physical oscillations near shock waves will cause in many cases the appearance of negative temperatures or pressures in the flow, if the turbulent Mach number is high enough.

Samtaney et al (2001) [2] performed DNS of decaying compressible turbulence using the 10th Pade scheme. Limited by "start-up" problem, the initial turbulent number (M_t) is limited to less than 0.5. Li et al (2002) [3] performed the DNS of the same flow using 7th order upwind-biased difference and 8th order center difference schemes. The maximum initial turbulent number is improved to 0.7. But when the turbulent Mach number is further increased, the "start-up" problem occurs again. Fu and Ma (1996,1997)[4,5] advanced a group velocity control theory for shock-capture. According to the theory, a scheme must be "fast type" and "slow type" at the left and right side of the discontinuity (assume the discontinuity propagates toward right). Zhang and Zhuang (1996) [6] advanced a "suppressing oscillations principle", which let the coefficient of the odd term of modified equations be negative and positive at the left and right side of the discontinuity. The suppressing oscillations principle and group control theory are essentially equivalent.

2 The GVC8 schemes

Consider the model equation: $\frac{\mathcal{P}_{(5,\alpha)}u}{\mathcal{P}_{(5,\alpha)}t}+a\frac{\mathcal{P}_{(5,\alpha)}u}{\mathcal{P}_{(5,\alpha)}x}=0$, assume $a>0$ Construct the following upwind-biased scheme:

$$\begin{aligned} u' = (&a_1u_{j+4}+a_2u_{j+3}+a_3u_{j+2}+a_4u_{j+1}+a_5u_j+a_6u_{j-1}\\ &+a_7u_{j-2}+a_8u_{j-3}+a_9u_{j-4}+a_{10}u_{j-5})/\Delta \end{aligned} \tag{1}$$

where Δ is the grid size. The following control equations of $a_j(j=1,2,\cdots,10)$ can be got from the Taylor expansion:

$$\sum_{j=1}^{10}(5-j)^k a_j = S_k \qquad (k=0,1,\cdots,K_{\max}) \tag{2}$$

where $S_1=1$, $S_k=0(k\neq 1)$. (1) has $K_{\max}$-th order accuracy.

Let $K_{\max}=8$, and let a_{10} be free coefficient. When a_{10} is given, other coefficients can be obtained by solving equations (2).

The numerical wave number associated with dispersion [6] is $k_i=-a_1\sin 4\alpha-a_2\sin 3\alpha-a_3\sin 2\alpha-a_4\sin\alpha+a_6\sin\alpha+a_7\sin 2\alpha+a_8\sin 3\alpha+a_9\sin 4\alpha+a_{10}\sin 5\alpha$. According to group velocity control theory, $\frac{dk_i}{d\alpha}>1$ at the left of discontinuity, $\frac{dk_i}{d\alpha}<1$ at the right of discontinuity. Figure 1 shows the variation of k_i with α for $a_{10}=-0.008$, -0.006, -0.004, -0.002, 0, 0.002, which shows that when $a_{10}=-0.008,-0.006$ or -0.004 $\frac{dk_i}{d\alpha}>1$ and when $a_{10}=-0.002, 0$ or 0.002 $\frac{dk_i}{d\alpha}<1$ in the wave number region of interest. We can choose different a_{10} at the left and right of discontinuity.

Following is GVC8 scheme for detail. Let

$$u_j'=(F_{j+1/2}-F_{j-1/2})/\Delta, \tag{3}$$

where $F_{j+1/2}=b_1u_{j+4}+b_2u_{j+3}+b_3u_{j+2}+b_4u_{j+1}+b_5u_j+b_6u_{j-1}+b_7u_{j-2}+b_8u_{j-3}+b_9u_{j-4}$, and $b_1=a_1, b_k=b_{k-1}+a_k(k=2,3,\cdots,9)$. In general we may choose:

$a_{10}=a_{10}^{(1)}$ when $|\Delta u_{j-1/2}|<|\Delta u_{j+1/2}|$ (at the left of discontinuity),

$a_{10}=a_{10}^{(2)}$ when $|\Delta u_{j-1/2}|>|\Delta u_{j+1/2}|$ (at the right of discontinuity),

where $\Delta u_j=u_{j+1/2}-u_{j-1/2}$.

When $a_{10}^{(1)}$ and $a_{10}^{(2)}$ are given, the coefficients can be obtained by solving equations (2) . The expression (3), is the 8th order group velocity control scheme (GVC8) when $a > 0$. We can construct the GVC8 schemes by using the same method when $a < 0$. The GVC8 schemes have 8th order accuracy at the smooth area, and have 7th order at some special points. This scheme is very efficient, and the computing work of this scheme is as much as an ordinary 9th upwind difference scheme and much less than those of WENO schemes[7].

There are two free coefficients ($a_{10}^{(1)}$ and $a_{10}^{(2)}$) in the GVC8 scheme. We will select the two coefficients to improve the shock wave capture capability and the resolution of small scales. Consider one dimensional SOD shock tube problem [7] with the following initial conditions:

$$\begin{cases} u = 0, \rho = \rho_s, p = p_s & x < 1/2 \\ u = 0, \rho = 0.125, p = 0.1 & x \geq 1/2 \end{cases} \quad \text{at } t = 0$$

Let $p_s = \rho_s$. The higher the p_s denotes the stronger the shock waves. Define $p_{s\,\max}$ is the maximum of p_s which the method (GVC8+3rd R-K, mesh size: 128) can simulate (if $p_s > p_{s\,\max}$ the computation will diverge).

Table 1. The maximum p_s of different GVC8 schemes

$(a_{10}^{(1)}, a_{10}^{(2)})$	$p_{s\,\max}$	$(a_{10}^{(1)}, a_{10}^{(2)})$	$p_{s\,\max}$
(−0.00128,0.0042)	8.5	(−0.006,−0.004)	3.5
(−0.006,0.001)	17.0	(−0.008,−0.003)	5.0
(-0.006,0.002)	12.0	(−0.006,−0.006)	2.5
(−0.006,0.0015)	18.5	(−0.0015873,−0.0015873)	1.5

Table 1 shows the $p_{s\,\max}$ of different GVC8 schemes. From Table 1 if we consider when $(a_{10}^{(1)}, a_{10}^{(2)}) = (-0.006, 0.001)$ or $(-0.006, 0.0015)$ the corresponding $p_{s\,\max}$ is high enough, then these values $(a_{10}^{(1)}, a_{10}^{(2)})$ here may be a good choose.

The resolution of small scale is an important problem in the simulation of turbulence. Schemes will be too dissipative if the resolution of small scale is not good enough. Let $u = \sin 16x, x \in [0, 2\pi]$, u'_j is the differential approximation of $\frac{\partial u}{\partial x}$ by using GVC8 scheme. Define the error: $err = \max|u'_j - 16\cos 16x_j|$ and let the mesh size is 128. Figure 4 shows the error varies with $a_{10}^{(1)}$ and $a_{10}^{(2)}$. From the figure we can see, in the range of this figure, the error decreases as $a_{10}^{(1)}$ increases, and the error increases as $a_{10}^{(2)}$ increases. Considering the capability of shock-capture and the resolution of small scales, we choose $(a_{10}^{(1)}, a_{10}^{(2)}) = (-0.006, 0.001)$.

3 DNS of compressible decaying isotropic turbulence at relative high turbulent Mach numbers

We perform DNS of compressible decaying turbulence [2] by using the GVC8 schemes. Flux vector splitting is adopted for the non-viscous terms, the GVC8 scheme is used to discretize it, 10th order center difference scheme is used to discretize viscous terms, 3 stage TVD type Runge-Kutta method [5] is used for time advance. We choose $(a_{10}^{(1)}, a_{10}^{(2)}) = (-0.006, 0.001)$ in the GVC8 scheme. The velocity starts with a specified spectrum which is initially divergence free and the energy spectrum is $E(k) = Ak^4 \exp(-2k/k_0^2)$, the coefficient A and k_0 are chosen the same as those in Ref [2]. All initial thermodynamic quantities (pressure, density and temperature) are constant. Periodic boundary conditions and uniform meshes are used in x-, y- and z- directions. Table 2 shows the DNS cases we studied. From which we see that the initial turbulent Mach number is increased to 0.95. All the Reynolds numbers (based on Taylor scale and RMS velocity fluctuation) are 72. For comparison, we performed DNS with initial turbulent number $M_t = 0.5$ by using 5th order WENO scheme [8] (5th WENO for non-viscous terms and 10th center difference for viscous terms). The program was coded by using MPI Fortran77 . In the program, long messages ware cut into short messages to avoid dead lock and improve the efficiency.

Figure 3 is the time history of normalized average kinetic energy $K(t)/K_0$ and velocity derivative skewness [2] for Case D1. The horizontal ordinate of Figure 3 denotes normalized time t/τ, where τ is large-eddy-turnover time at t=0 [2]. Symbols in figure3 show the results of Ref. [2], which agree well with those of this paper. This means that the dissipative error of the GVC8 schemes can be ignored in the mesh resolution (the result does not seem to be over-dissipative). Velocity derivative skewness, which is a high-order statistical quantity, agrees well with those of Ref [2], which verified our results. Figure 4 shows the time history of normalized turbulent Mach number and normalized RMS of density fluctuations[2]. Symbols in the figure denote the corresponding results of Ref. [2], which agree well with those of this paper. The results of case D5 (by using WENO5) are also shown in figure 3 and figure 4, which seemed too dissipative. Figure 5 shows the time history of normalized turbulent kinetic energy of case D3 (mesh size:128^3)and D6(mesh size:256^3), we can see the result of case D3 agree well with those of D6, shows our result is reliable. Figure 6 shows the time history of normalized turbulent kinetic energy of D1-D4, we can see the turbulent kinetic energy decays faster as initial turbulent Mach number increases, which means the more kinetic energy will be turned into intrinsic energy as compressibility effect increases.

Shocklets will occur in the flow when turbulent Mach number is higher enough [2]. Figure 7 is a snapshot of normalized pressure fluctuation p' at

Table 2. The computational conditions of each DNS cases

CASE	Re_λ	Mt	Scheme	Mesh size
D1	72	0.5	GVC8	128^3
D2	72	0.8	GVC8	128^3
D3	72	0.9	GVC8	128^3
D4	72	0.95	GVC8	128^3
D5	72	0.5	WENO5	128^3
D6	72	0.9	GVC8	256^3

the middle section of flow fluid of computation case D6 (at $t/\tau = 1$), where $p' = \frac{p-\langle p\rangle}{\langle p\rangle}$. From this figure the steep gradient of p'can be used to identify possible locations of the shocklets in the isotropic compressible turbulent. According to the computation data, there we have very high local Mach number in some area even when the averaged one is not so high. For example, In the computation case D6, when $t/\tau = 1$, the averaged Mach number is 0.653, but the maximum local Mach number of this flow fluid is 3.276, the maximum local Mach number is far larger than the averaged Mach number. This is probably because of the intermittency of turbulence, so the authors conjecture that the intermittency of turbulence is an importance reason for the creation of shocklets.

Extended Self-Similarity (ESS) is an important development in the measurement of the scaling exponents. It states that the velocity structure function of any order p depends on the structure functions of order q (usually 3) with a much better power law: $< |\delta u_l|^p > \sim < |\delta u_l|^q >^{\zeta_{p,q}}$, $\zeta_{p,q} = \zeta_p/\zeta_q$ is called the relative scaling exponent. Figure 8 shows the plot of $LARGE < |\delta u_l|^p >$ as a function of $LARGE < |\delta u_l|^3 >$ in flow flied of case D6 at $t/\tau = 1$ (in this flow flied, $Re_\lambda = 33.05$, $M_t = 0.715$). For all p, $LARGE < |\delta u_l|^p >$ is shown as a strict linear function of $LARGE < |\delta u_l|^3 >$. This figure shows clearly the existence of ESS in this compressible turbulence fluid in spite of the existence of shocklets. The slope rate of lines in this figure is the relative scaling exponent ζ_p/ζ_3. The numerical values of the relative scaling exponents obtained here agree well with those of SL theory [8]. This shows that compressibility has little effect on scaling exponents when turbulent Mach number is not very high.

4 Conclusions

From the DNS cases we can see GVC8 is an effective scheme for the DNS of compressible turbulence with shocklets. Evidence is obtained that scaling laws and ESS hold in the compressible turbulent flow in spite of the presence of shocklets, and compressibility has little effect on scaling exponents.

This work is supported by NKBRSF (CG199032805) and NNSFC (19872069). The authors thank Professor Zhang Linbo for his help in programming, Professors Chen Shiyi and He Guowei for setting up the initial conditions, and LSEC for providing computer time.

References

1. Moin P, Mahesh K. 1998. Annu. Fluid Mech. 1998,30:539-78.
2. Samtaney R, Pullin DI, Kosovic B. Phys. Fluids. 2001,13(5),1415-1430.
3. Li XL, Fu DX, Ma YW. Science in China, serial A (to appear).
4. Fu DX, Ma YW, Kobayashi T, et al. CFD JOURNAL. Vol 4, no. 4, 1996,427-450.
5. Fu DX, Ma YW. J. of Comp. Phys., 134, 1997, 1-15.
6. Zhang HX, Zhuang FG. Analysis and numerical simulation of vortex(Book 1) 1996.75-81.
7. Liu X D, Osher S, Chan T. J. Comp. Phys., 1994, 115: 200-212
8. She ZS. 1998. Prog. Ther. Phys. 1998, S130:87-102.

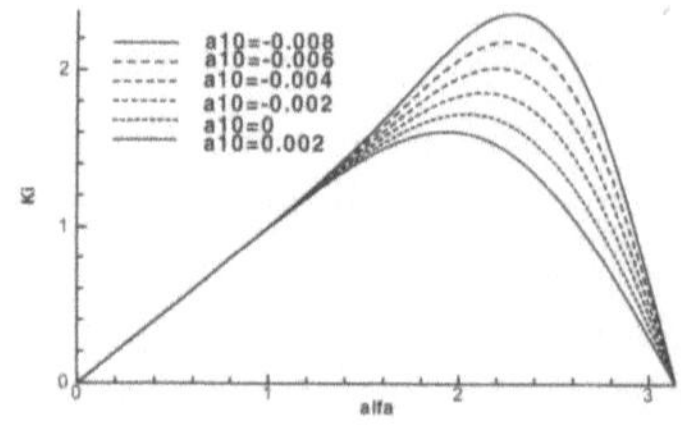

Figure 1. k_i varies as a function α for different a_{10}

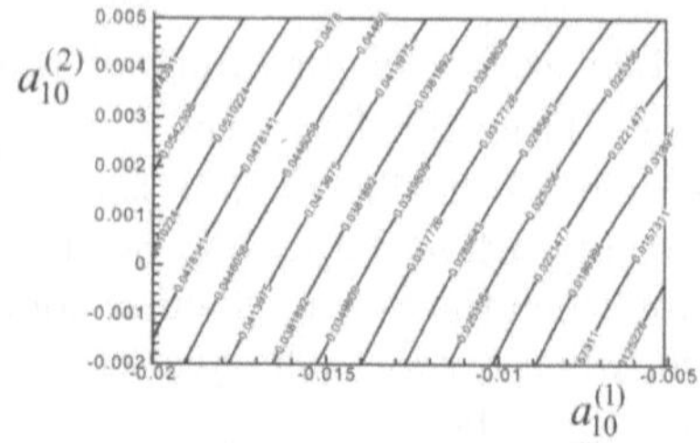

Figure 2. Error varies with $a_{10}^{(1)}$ and $a_{10}^{(2)}$

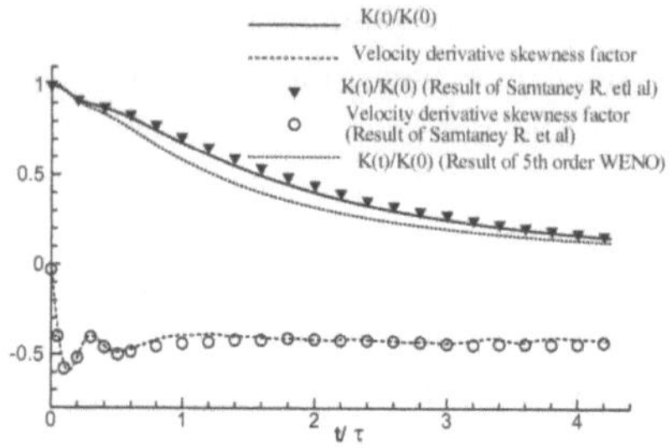

Figure 3. Normalized turbulent energy and velocity derivative skewness (Mt=0.5)

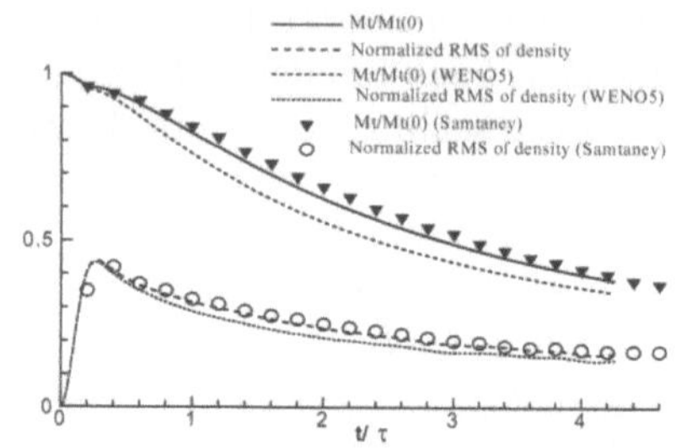

Figure 4. Normalized turbulent Mach number and normalized RMS of density (Mt=0.5)

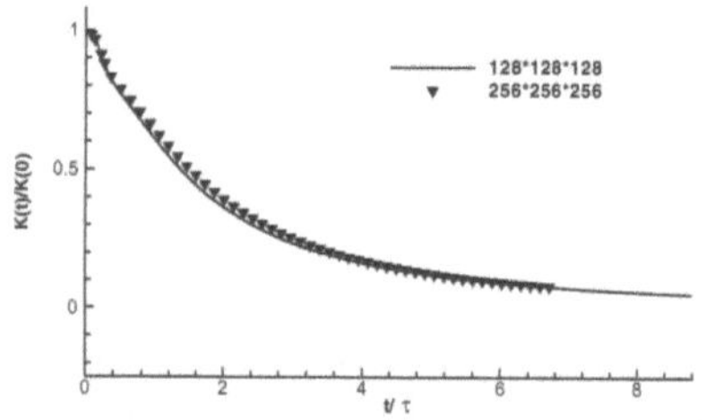

Figure 5. Time history of normalized turbulent kinetic energy of D3 and D6

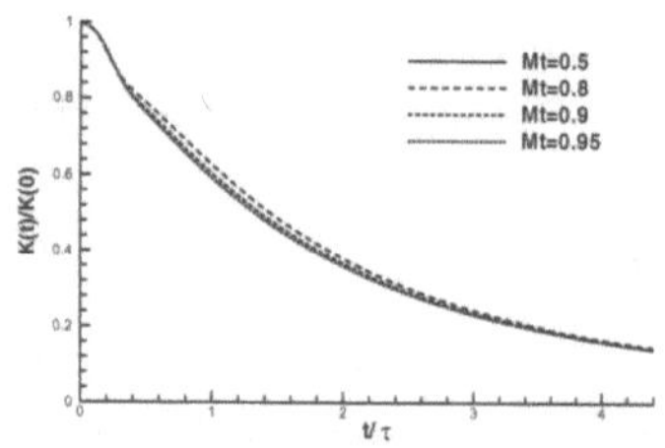

Figure 6. Time history of normalized turbulent kinetic energy of D1-D4

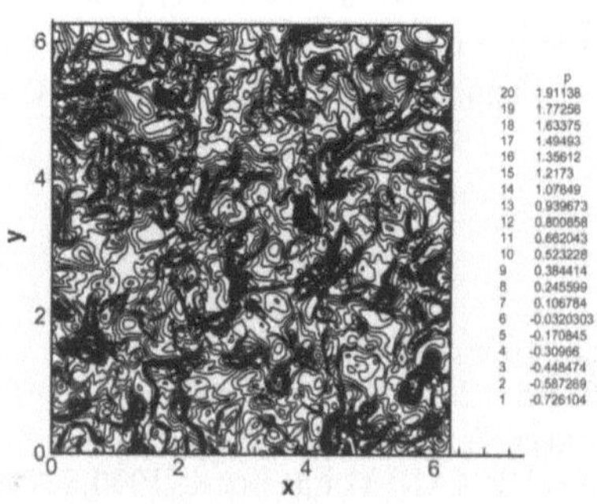

Figure 7. Pressure fluctuations in the middle section of the fluid $t/\tau=1$ of D6

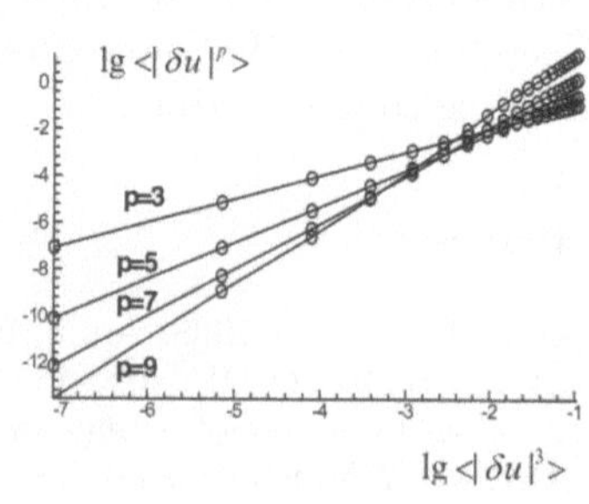

Figure 8. ESS of the flow field of case D6 at $t/\tau=1$

Part XVIII

Unsteady Flow

Numerical Simulation of Unsteady Viscous Flow Around a Flapping Wing

Koji Isogai

Department of Aeronautics and Astronautics
Kyushu University
6-10-1 Hakozaki, Higashiku, Fukuoka 812-8581, Japan

Abstract. In order to investigate the effects of dynamic stall phenomenon on the propulsive efficiency, thrust and lift of flapping wing of a bird, unsteady viscous flow simulations have been performed by using a 3D Navier-Stokes code. As a result of the simulations about a typical pigeon wing which has thick-large-cambered airfoil sections and a peculiar nose-up twist distribution, it is shown that the occurrence of the flow separation is unavoidable around the tip region to maintain the high lift and appropriate thrust, wherefore the propulsive efficiency remains less than 0.5. This is in sharp contrast with the results of the existing potential flow analyses that predict the very high propulsive efficiency of the order of $0.8 \sim 0.9$.

1 Introduction

Unlike aircraft, a bird should generate lift and thrust at the same time by a flapping motion of a wing. In order to investigate the aerodynamic mechanism of generating lift and thrust by a flapping wing, many theoretical [1,2] and experimental studies [3,4] have been conducted so far. However, the mechanism has not yet been well understood. As far as the three-dimensional (3D) wing is concerned, all the theoretical studies presented so far are based on the potential flow assumption. As you can see in the flying birds, however, the amplitude of the flapping motion of the wing is very large (for example, the flapping angle of a pigeon is about 50 deg as shown later in the example). In such a large amplitude of the flapping motion, the induced angle of attack (due to the flapping motion of a wing) exceeds easily stalling angle. This means that the discussions based on the potential flow assumption become meaningless. In the present study, the numerical simulation of unsteady viscous flow around a flapping wing has been conducted by using a 3D Navier-Stokes code and the role of the flow separation, especially, the effect of dynamic stall on the propulsive efficiency, thrust and lift of a flapping wing has been investigated.

2 Method of Analysis

As is well known, the basic motion of a bird wing is a coupled oscillation of flapping and feathering (twisting motion). As defined in Fig. 1, the vertical

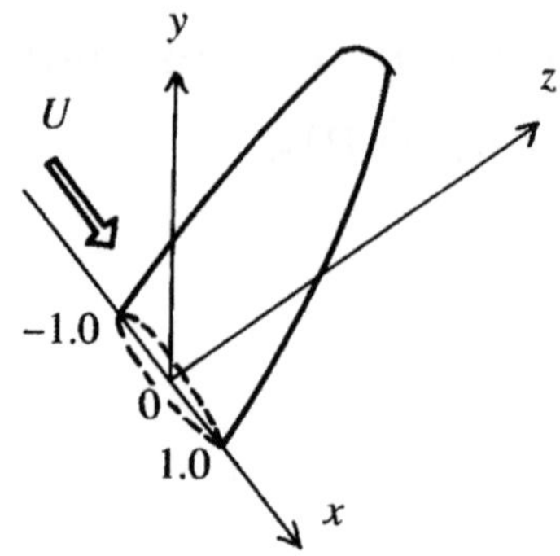

Fig. 1. Definitions of coordinate and wing motion

displacement of the arbitrary point of the upper/lower surfaces of the wing can be expressed as

$$y(x,z,t) = y_s(x,z) + f(x,z,t) \tag{1}$$

where y_s is the time-mean displacement of the upper/lower surfaces, and where f is the displacement of the upper/lower surfaces from the time-mean displacement y_s. $f(x,z,t)$ in Eq.(1) can be given by

$$f(x,z,t) = (h_r + \phi_0 z)\sin(kt) - (\theta_r + b_\theta z)(x - x_p(z))\sin(kt + \phi) \tag{2}$$

where k is the reduced frequency based on the root semichord b_r and is defined by $k = (b_r\omega)/U$, where ω is the circular frequency of the wing and U is the free-stream velocity. In Eqs. (1) and (2), all the physical quantities are nondimensionalized by b_r. In Eq.(2), t is the dimensionless time, and h_r is the amplitude of the heaving oscillation at the root station, and ϕ_0 is the amplitude of the flapping oscillation, and θ_r is the amplitude of the pitching oscillation at the root station. The amplitude of the pitching oscillation is assumed to increase linearly toward tip station with the slope of b_θ, and x_p is the x coordinate of the pitch axis, and ϕ is the phase-advance angle of the pitching oscillation ahead of the flapping oscillation. In order to compute the unsteady viscous flow around the oscillating wing whose motion is given by Eq. (1), a 3D compressible Navier-Stokes code is applied. The grid used is C-H type structured grid, 240 (chord wise) $\times$ 31 (normal to surfaces) $\times$ 19 (span wise). In the computations, Mach number and Reynolds number are assumed to be 0.30 and 10^5, respectively, and Baldwin & Lomax's algebraic turbulence model [5] is employed.

3 Results and Discussion

3.1 Code Validation

Since, as far as we concern a 3D flapping wing, there is no existing NS computation nor reliable experimental data that are published so far, we will compare our NS computations with those of the 3D potential theories. The case we have selected is a rectangular wing of aspect ratio 8 oscillating in a coupled heaving and pitching, for which the potential flow computation by Lan [1], who employed Unsteady Quasi-Vortex Lattice Method (UQVLM), exists. The details of the wing motion are as follows: $k = 0.25$, $h_r = 2.0$, $\phi_0 = 0$ deg, $\theta_r = 23$ deg, $x_p = 0.5$, $\phi = 90$ deg. The wing section is NACA0012 and the mean angle of attack is 0 deg. The results of the present NS simulation for this case are as follows: propulsive efficiency $\eta_p = 0.70$ ($\eta_p = TU/W$, T : time mean thrust, W : time mean rate of work), thrust coefficient $C_T = 0.0623$ ($C_T = T/(1/2\rho U^2 S)$). The results obtained by Lan [1] for this case by using UQVLM are $\eta_p = 0.94$ and $C_T = 0.0623$. Since no viscous effect is included in those obtained by Lan, those of our present NS simulation seem to be reasonable.

3.2 Numerical Example and Discussion

In the following examples, we will show that the occurrence of flow separation reduces the propulsive efficiency and thrust considerably. The wing studied is a straight tapered wing of aspect ratio 7.04 and the taper ratio 0.42. The wing section is NACA0012 and the mean angle of attack is assumed to be 0 deg. b_r and U are assumed to be 0.0905 m and 11 m/s, respectively. For this wing, we have computed two cases of different oscillation mode. The reduced frequency k is 0.30. The first case is for a pure flapping motion ($\phi_0 = 33$ deg) without twist. In Fig. 2, the flow pattern (iso-vorticity distribution) at $kt = \pi$ where the maximum effective angle of attack induced by the flapping motion becomes 31 deg at 75 % semispan station is shown. As seen in the figure, the large scale flow separation can be seen from 50 % semispan station to the tip station. η_p and C_T for this case are 0.111 and 0.0374, respectively. The second case is for a coupled flapping and feathering motion, namely, $\phi_0 = 33$ deg, $\theta_r = 0$ deg, $b_\theta = 0.0931$ (the pitch amplitude at 75 % semispan station is about 20 deg), $\phi = 90$ deg. In this case, the maximum effective angle of attack at 75% semispan station is about 11 deg. The flow pattern at $kt = \pi$ where the effective angle of attack becomes maximum is shown in Fig. 3. As seen in the figure, there is no large scale flow separation. η_p and C_T for this case are 0.619 and 0.106, that are considerably higher than those of the first case. In Fig. 4, the variations of thrust and rate of work for the two cases are shown. As seen in the figure, the second case (the coupled flapping and feathering motion) generates considerably higher thrust by consuming a smaller amount of energy compared with the first case (the pure flapping motion without twist).

As an example of more realistic case where thrust and lift are generated at the same time, the flow around a pigeon wing shown in Fig. 5 is computed. In Fig. 5, the planform, wing section and the geometrical twist distribution, that are measured by Nachtigall and Wieser [6], are shown. The detailed values of the wing geometry are as follows: the aspect ratio is 7.2, full span length is 0.66 m, $b_r = 0.055$ m, full span wing area is 0.062 m^2, mass is 0.39 kg. As to the wing motion, the following values at the cruise condition[2] are taken, that is, $U = 11$m/s, $\omega = 50.24$ rad/s, $k = 0.25$, $h_r = 0.0$, $\phi_0 = 50$ deg, $\theta_r = 0$, $b_\theta =$

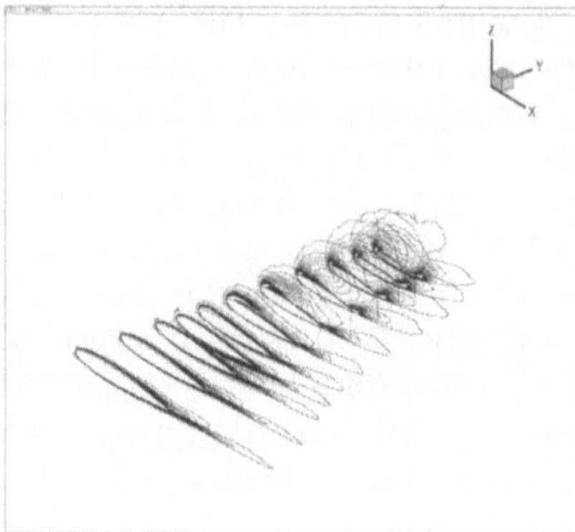

Fig. 2. Flow pattern at $kt = \pi$ (pure flapping)

Fig. 3. Flow pattern at $kt = \pi$ (coupled flapping and pitching)

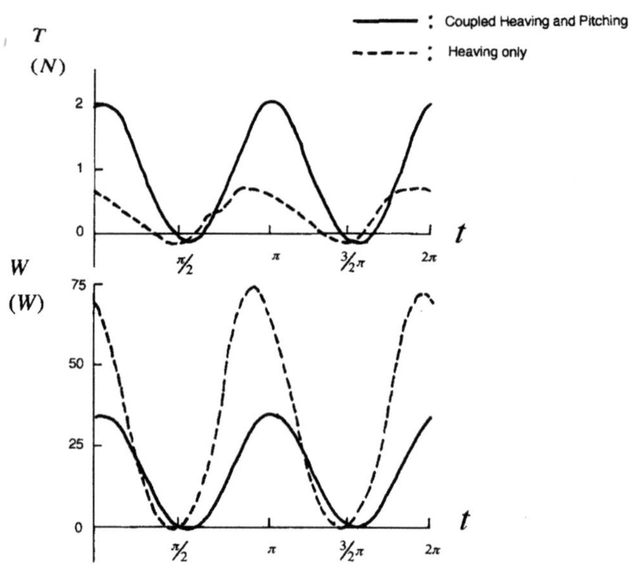

Fig. 4. Variations of thrust and rate of work

0.129 (pitching amplitude is 30.4 deg at 75 % semi-span station), $\phi = 90$ deg, $x_p = -0.50$.

The results of the numerical simulation for this case are as follows: propulsive efficiency $\eta_p = 0.42$, thrust coefficient $C_T = 0.12$, lift coefficient $C_L = 0.72$. Vest and Katz [2] have computed the same case by using a panel method, obtaining $\eta_p = 0.64$, $C_T = 0.13$, $C_L = 0.85$. In our computations, the large scale flow separation is observed, as shown in Fig.6 (iso-vorticity in z direction contour), on the upper surface from 70 % semispan to tip station at the instant of $kt = \pi$ when the induced angle of attack becomes maximum. Since the occurrence of

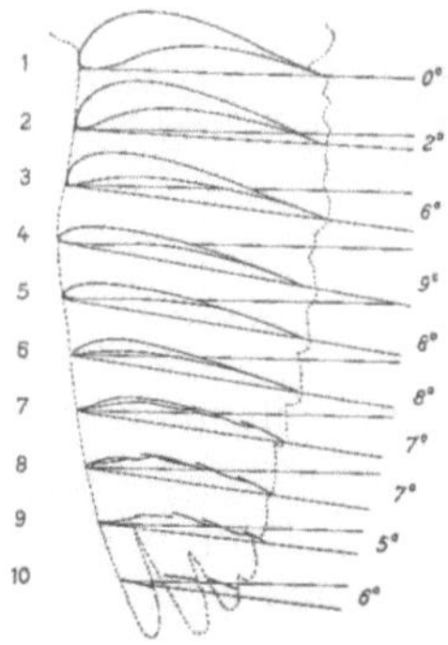

Fig. 5. Planform and wing sections of a pigeon wing (Nachtigall and Wieser [6])

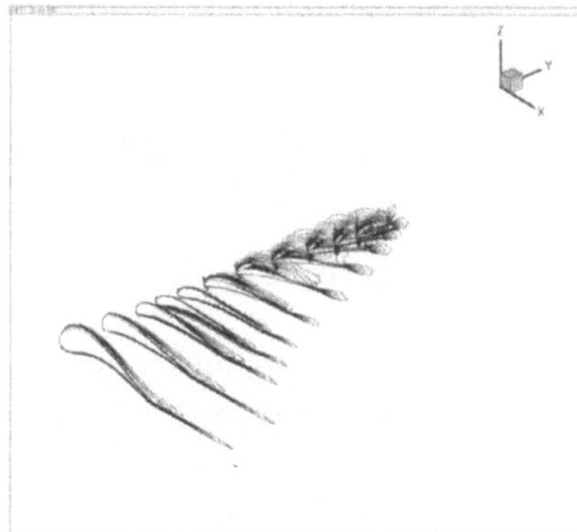

Fig. 6. Flow pattern at $kt = \pi$

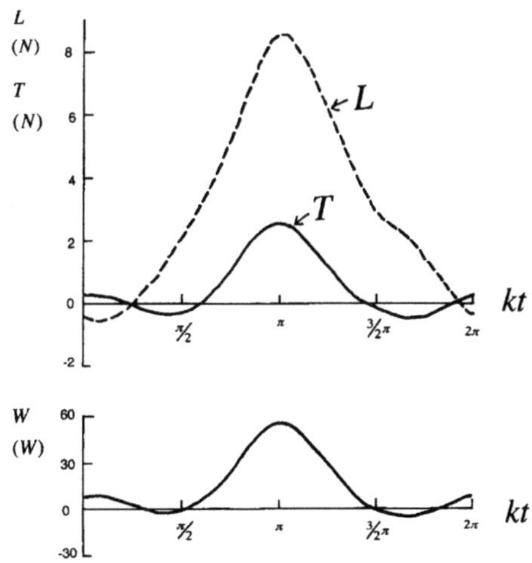

Fig. 7. Variations of lift, thrust and rate of work during one cycle of oscillation

the flow separation reduces the propulsive efficiency considerably [7], the results we have obtained seem to be quite reasonable compared with those obtained by Vest and Katz [2] by using the panel method. In Fig.7, the variations of lift, thrust and rate of work during one cycle of oscillation are shown. In the figure, it can be seen that most of the lift and thrust are generated during the process of down stroke ($kt = \pi/2 - 3/2\pi$).

By the present numerical simulations of the unsteady viscous flow around the pigeon wing, it is disclosed that the requirements of obtaining high lift and high propulsive efficiency at the same time are contradictory. That is, in order to obtain the high mean lift such as the present pigeon case (C_L required to sustain its weight at $U = 11$ m/s is 0.83), the maximum effective angle of attack in the process of down stroke, which is induced by the flapping motion, easily exceeds the stalling angle, and that results in considerable reduction of the propulsive efficiency. It is conjectured, however, that the large cambered airfoil sections and the peculiar built-in twist distribution of the present pigeon wing seem to be the device to attain these contradictory requirements as high as possible.

4 Conclusion

Numerical simulations of unsteady viscous flow around a flapping wing have been conducted by using a 3D Navier-Stokes code, and the following results have been obtained. 1) The occurrence of flow separation reduces the propulsive efficiency and thrust considerably. 2) The propulsive efficiency and thrust can be improved considerably by adding the appropriate twisting motion to the flapping motion, that prevents the flow separation. 3) The numerical simulation of a flapping pigeon wing which has large cambered airfoil sections and a peculiar built-in twist distribution has revealed that the flow separation is unavoidable around the tip region to maintain the high lift and appropriate thrust, wherefore the propulsive efficiency remains less than 0.5. 4) Although requirements of obtaining high lift and high propulsive efficiency at the same time are contradictory, the large cambered airfoil sections and the peculiar built-in twist distribution of the pigeon wing seem to be the device to attain these contradictory requirements as high as possible.

References

1. Lan, C.E., J.: Fluid Mechanics, Vol. 93, Part 4, pp. 747–765 (1979)
2. Vest, M.S. and Katz, J.: AIAA Journal, Vol. 34, No. 7, pp. 1435–1440 (1996)
3. Fejtek, I. and Nehera, J.: Aeronautical Journal, Vol. 84, pp. 28–33 (1980)
4. Vest, M.S. and Katz, J.: AIAA 99-0994, Jan (1999)
5. Baldwin, B.S. and Lomax, H.: AIAA Paper 78-257 (1978)
6. Nachtigall, W. und Wieser, J.: Zeitschrift fur vergleichende Physiologie 52, pp. 333–346 (1966)
7. Isogai, K. and Shinmoto, Y. and Watanabe, Y.: AIAA Journal, Vol. 37, No. 10, pp. 1145–1151 (1999)

Validation of Unsteady Turbulent Flow Predictions for Over-expanded Rocket Nozzle

Richard Schwane[1], Lars Torngren[2], and Henry Wong[3]

[1] European Space Agency, ESTEC
[2] Swedish Defense Research Agency, FOI
[3] AtosOrigin, The Netherlands

Abstract. The present paper reports on a validation activity, where CFD solutions for the coupling of fluctuating ambient pressure with flow separation in over-expanded rocket nozzles are validated by means of a dedicated experiment in a hypersonic facility. The experimental and CFD methodology is discussed and the related data is compared. A simplified quasi-one-dimensional CFD model has been designed to identify the physical processes that govern the interaction. The methodology is then applied to the present experimental configuration and conclusions are drawn on the severity of the observed coupling for the design of space applications.

1 Introduction

The most important requirement of present research for access to space is the reduction of launch costs. For a propulsion system, the engine and the expansion nozzle, reduction of costs is equivalent to the increase of performance with at least maintained reliability and simplicity.

The most significant parameter for engine performance is the area ratio of the expansion nozzle that is constrained by the need for attached flow conditions for all ambient pressure levels, occuring during ascent of a launcher. Increase of the area ratio would result in flow separation for see level operation and, consequently, destruction of the nozzle due to excessive side loads.

All presently existing launchers have been designed based mainly on experimental methods. Such methods allow accurate conclusions for discrete points of operation on the trajectory. Development costs, however, increase strongly with the number of degrees of freedom that have to be considered for the design. The transition from fully flowing nozzles to nozzles that permit controlled separation during ascent is a typical example for design that requires reliable CFD tools to optimize the large number of design parameters without creating prohibitive development costs.

The present paper describes applications of advanced CFD tools for research on altitude adaptive expansion nozzle concepts, including the experimental validation for discrete points of operation. For a nozzle with a Truncated Ideal Contour (TIC) the numerically predicted turbulent steady and unsteady separated flow fields are validated with experimental data. Special attention is paid to the generation of side loads due to coupling of unsteady flow in the base region of launchers (base buffeting) with the separation in the nozzle.

2 Experimental activities

Using CFD methods for the design of propulsion systems requires thorough validation of all components that contribute to the numerical solution, e.g. computational grids, numerical solvers, turbulence and real gas modeling. For the present activity a dedicated experiment has been performed in which steady state and time dependent experimental data have been collected. This data will be used to validate the numerical method and the physical modeling.

2.1 Test Facility

The HYP500 hypersonic wind tunnel at FOI was used as a test facility. It has been modified to simulate the time dependent ambient flow field that a launcher nozzle would experience due to base buffeting. Details of the experimental set-up can be found in [7].

3 Numerical Methodology

Numerical computations are performed with different time-accurate two or three dimensional Navier-Stokes Computational Fluid Dynamics (CFD) codes. The three-dimensional viscous flow solver EURANUS, developed by FFA, Sweden [1] is used for flows under perfect gas or equilibrium assumptions with high temporal accuracy requirements. Different numerical schemes are available in the code, for the present computations the symmetric Harten / Yee scheme has been used. The time integration employs the pseudo-time approach (Jameson 91) [2] which relieves the CFL restriction of explicit schemes and allows to choose the time step based on physical considerations, e.g. a suitable resolution of unsteady processes in the flow field. A number of turbulence models can be selected, in the present paper the algebraic Baldwin-Lomax (BL) model and the Chien low Reynolds k-epsilon ($\kappa - \epsilon$) model coupled with the Sarkar correction that accounts for compressibility effects have been employed [3].

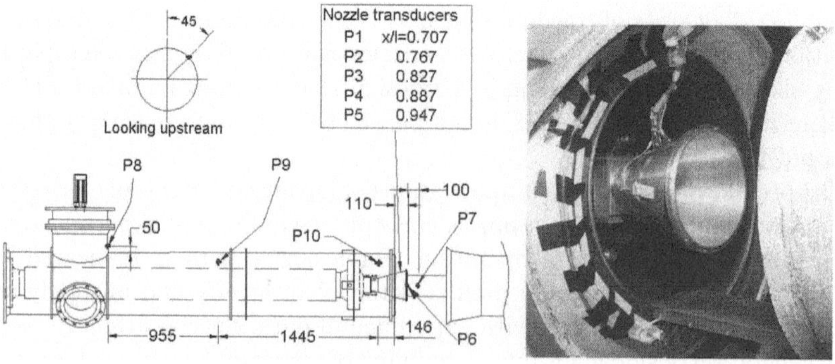

Fig. 1. HYP500 - Experimental setup

The commercially available CFDRC-FASTRAN code [5] was used in the fully implicit mode with a higher order spatial discretization and Roe's approximative Riemann solver. A large variety of turbulence models is available with this code that have been thoroughly validated for super- and hypersonic flows [6].

Finally, these institutional or commercial codes have been compared to in-house developed Navier Stokes codes with higher order time accuracy, e.g. for the numerical solution for the quasi-one-dimensional flows in section 5.

4 Steady State Validation

In a first test campaign steady flow conditions with different stagnation to ambient pressure ratios between 14.5 and 59.7 have been investigated, leading to steady separated flow fields in the nozzle. The experimental results have been numerically rebuilt by axisymmetric turbulent Navier Stokes computations and used to 'fine tune' the computational grid, the numerical scheme and the turbulence models. Figure 2 shows the wall pressure and Mach number distribution for a pressure ratio of $\frac{p_c}{p_a} = 21$, that matches the experimental results in terms of the location of the over-expansion shock and the pressure levels.

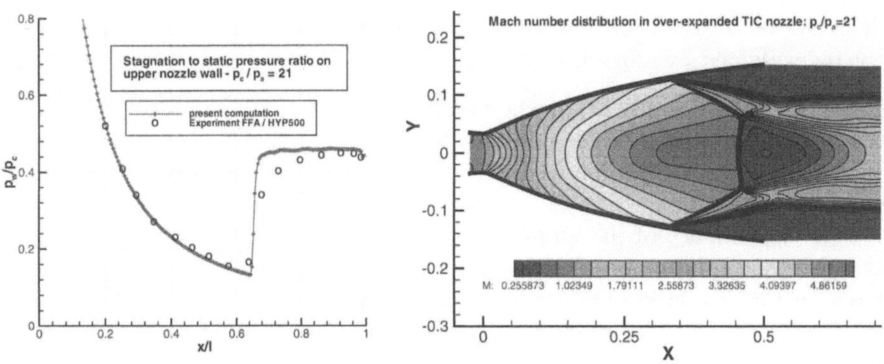

Fig. 2. Experimental and numerical steady wall pressure and Mach number distributions

5 Time Dependent Validation

The steady state experimental set-up of section 2.1 is now extended to unsteady flows by modulation of the ambient flow past the nozzle exit with a periodic pressure fluctuation created by means of fast revolving slotted disk upstream at sensor P8, as sketched in Figure 1.

Time Dependent Experiments Time dependent experimental data include the pressure levels, ambient to the nozzle exit and at five positions inside the nozzle.

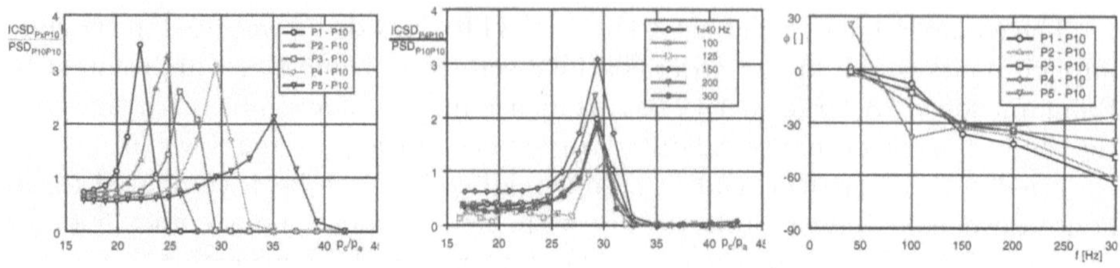

Fig. 3. Amplification and Phasing: internal versus external pressure

Figure 3 shows amplification rates, e.g. the ratio of the cross-correlation between the inner (transducer P1-P5 in the nozzle, see Figure 1) and outer pressure (transducer P10), normalized by the power-spectra of the ambient pressure signal for different stagnation pressure levels. The fact that the ratios in Figure 3-left assume values larger than one, does not by itself indicate dynamic amplification, since the pressure jump over the over-expansion shock and the fluctuation of outer pressure are not directly correlated.

Evidence for dynamic behavior, e.g. resonance in the experiment, is given in Figure 3 (center and right). The amplification ratio at transducer P4 and the phase angle of all internal sensors indicate a dependency of the response of the shock system in the nozzle on the frequency of the external excitation that can not be explained by a quasi-steady approach.

Time Dependent CFD In the subsequent section the results of the CFD analysis are presented. Given the presently not completely understood physics in the experiment, CFD has primarily been used to analyze and investigate the basic flow physics.

The dependency of the amplification ratio on the excitation frequency hints at a resonance phenomenon. Resonance, however, requires that the compression shock in the nozzle has properties of an oscillator. Consequently, it should be possible to assign the characteristic parameters of a damped oscillation , e.g. a mass, a spring stiffness, a damping constant and a related eigenfrequency to a shock in an over-expanded expansion nozzle.

Quasi-one-dimensional Numerical Experiments The time-accurate computation of a normal shock wave in a 2d-conical nozzle (quasi-one-dimensional, inviscid flow) has been selected as a simplified model for the considered flow problem. The computation starts from a steady condition, where the ratio between stagnation pressure and ambient pressure determines the position of a nearly normal shock wave in the nozzle.

At t=0 the ambient pressure is suddenly increased creating a subsonic pressure wave that moves from the nozzle exit towards the steady normal shock wave. When the pressure wave reaches the shock wave it causes the shock to move upstream towards its new steady state condition for the modified ambient pressure. However, it overshoots this location and performs a damped oscillation until it settles to a new steady state condition after a large non-dimensional time ($t > 8$). This is also demonstrated in Figure 4-right where the pressure distribution

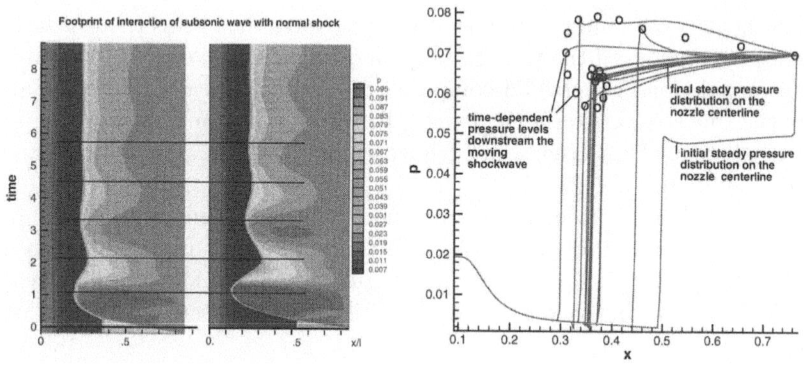

Fig. 4. Time dependent shock positions for the simplified numerical model

on the centerline of the nozzle is plotted versus the nozzle length for different elapsed time levels. Additionally, the symbols show the pressure levels immediately downstream of the moving shock, demonstrating the transient behavior of the shock wave.

Obviously, the shock behaves similar to a damped oscillator and exposes a characteristic time scale, e.g. the time interval between two extrema of the shock location. The horizontal lines in Figure 4-left, showing the pressure distribution along the nozzle center line versus the elapsed time as a color plot, indicate that the oscillation frequency is approximatively constant over time in both computations, even if different amplitudes of the initial excitation by a subsonic wave and different initial shock locations in the conical nozzle are assumed. This is a main characteristic of a linear oscillator.

Since the shock behaves similar to a linear oscillator and, in particular, has an inherent time scale nearly independent of the amplitude of the excitation, an eigenfrequency can be assigned to it. Excitation with the eigenfrequency leads to a shock movement, where the amplitude of the response is only depending on the damping. The main mechanism of the damping is the distribution (wave transport) and dissipation of the local maximum of pressure and total enthalpy that is created behind the moving shock. The transport is directed into the sub-sonic regime, towards the end of the nozzle, as can also be observed in Figure 4.

Application to the S6 Nozzle Based on the results from the numerical experiment in the previous section, the same CFD methodology has been applied to the experimental configuration, as sketched in Figure 1. Axisymmetric, turbulent and time accurate Navier Stokes computations [4] are performed to explain and validate the experimental observations.

For an excitation frequency close to the eigenfrequency of the shock system in the nozzle (here 200Hz), the response, e.g. the pressure immediately downstream of the shock is in resonance with the excitation. Unlike the quasi-steady case, where the pressure distribution in the separated region forms a plateau (see Figure 2), in case of resonance the time-dependent pressure distribution downstream the shock can be interpreted as a standing wave that leads to alternating positive/negative slopes of the wall pressure and an increased effect of the shock movement on side load generation compared to the quasi-steady case. The amplitude of the shock-excursion is then strongly enlarged and only depends on the damping coefficient of the oscillator, if linear properties are assumed. Detailed numerical results are published in [8].

Damping, as already discussed for the one-dimensional simplified model and explained by wave transport and dissipation of the pressure and enthalpy peak behind the fast moving shock wave (see Figure 4) in section 5 will have to be quantified in more detail for an improved in-depth assessment of side load generation.

6 Conclusions

CFD solutions for the coupling of ambient fluctuating flow with separated flow in over-expanded rocket nozzles have been validated by means of a dedicated experiment. The design of a simplified quasi-one-dimensional flow model allows for the analysis of the physical principles that govern the interaction. The existence of resonance between ambient fluctuation and internal flow has been demonstrated, however, the quantitative results from the CFD simulation have not yet been confirmed by the experiment. This will be the objective of a new test campaign that is presently under preparation.

Acknowledgments: This work has been motivated by the activities of the Flow Separation Control Device (FSCD) group with CNES, Astrium, SNECMA, Volvo Aero, DLR, LEA Onera and ESTEC as members, managed by the French Space Agency CNES. The authors are grateful to all members of the FSCD group for fruitful discussions.

References

1. Eliasson, P. & Nordström, J. : FFA TN 1995-39
2. Jameson, A. : AIAA-91-1596
3. Wong, H.; Schwane, R.: "Numerical Investigation of Transition in Flow Separation in a Dual-bell Nozzles" Proc. 4th European Symp. on Aerothermodynamics for Space Vehicles, Oct. 2001,Italy.
4. Herget, S. : TUM-FLM-2000/11 2000.
5. Coirier, W.J.;Jiang, Y.; Fricker, D.M.; Haq, I.; Przekwas, A.J.;Harrand,V.J.; Singhal, A.K.: An Unstructured Parallel Solution Adaptive Simulation System for CFD and Multidisciplinary Analysis AIAA 14th CFD Conference, July 1999.
6. Schwane, R.;Hagemann,G.;Reijasse, P.: AIAA 2002-0585
7. Torngren, L.: Correlation between outer flow and internal nozzle pressure fluctuations, Proc. 4th European Symp. on Aerothermodynamics for Space Vehicles, Oct. 2001,Italy.
8. Schwane, R.; Wong, H., Toerngren, L.: Experimental and Numerical Investigation of Side load Generation in Over-expanded Rocket Nozzles. Accepted for Publication. 4th European Colloquium for Launcher Technologies, Liege, Belgium, Dec. 2002.

Gust Acoustics Computation with a Space-Time CE/SE Parallel 3D Solver

X.Y. Wang[1], A. Himansu[2], S.C. Chang[3] and P.C.E. Jorgenson[4]

[1,2]Taitech Inc., NASA Glenn Research Center, Cleveland, Ohio
[3,4]NASA Glenn Research Center, Cleveland, Ohio
email: [1]*wangxy@turbot.grc.nasa.gov*, [2]*fshiman@suspiria.grc.nasa.gov*
email: [3]*sin-chung.chang@grc.nasa.gov*, [4]*jorgenson@grc.nasa.gov*

Abstract. The benchmark Problem 2 in Category 3 of the Third Computational Aero-Acoustics (CAA) Workshop is solved using the space-time conservation element and solution element (CE/SE) method. This problem concerns the unsteady response of an isolated finite-span swept flat-plate airfoil bounded by two parallel walls to an incident gust. The acoustic field generated by the interaction of the gust with the flat-plate airfoil is computed by solving the 3D Euler equations in the time domain using a parallel version of a 3D CE/SE solver. The effect of the gust orientation on the far-field directivity is studied. Numerical solutions are presented and compared with analytical solutions, showing a reasonable agreement.

1 Introduction

The noise generated by the interaction of vortical disturbances originating upstream with propeller or turbomachinery blades has been of great interest in noise studies. A model problem regarding a one-dimensional vortical wave interaction with an isolated finite-span swept flat-plate airfoil bounded by two parallel walls was posed as a benchmark problem in Category 3 of the Third CAA Workshop [1]. The effect of different gust orientations on the far-field directivity is investigated here.

This problem is solved numerically by solving the unsteady 3D Euler equations in the time domain using the space-time conservation element and solution element (CE/SE) method. The CE/SE method is an innovative numerical method [2–5]. It applies flux conservation to finite space-time volumes, and achieves second-order accuracy in both space and time for uniform space-time meshes. Its salient properties are summarized briefly as follows. First, both local and global flux conservations are enforced in space and time instead of in space only. Second, all the dependent variables and their spatial derivatives are considered as individual unknowns to be solved for simultaneously at each grid point. Third, every CE/SE scheme is based upon a non-dissipative scheme with addition of fully controllable numerical dissipation. This results in very low numerical dissipation. Fourth, the flux-based specification of the CE/SE schemes gives rise in a natural fashion to a simple yet generally effective non-reflecting boundary condition, which is an important issue in CAA. A detailed description of this method and accompanying analysis can be found in [2–7]. Applications of this method to CAA problems reveal that the CE/SE method can produce accurate results in a simple way [8–11].

In this paper, a parallel version of the 3D CE/SE Euler solver is used to compute the acoustic field generated by the interaction of a vortical gust with an isolated swept flat-plate airfoil bounded by two parallel walls. Different spanwise wavenumbers of the gust are used to investigate the far-field directivity.

In the following, the description of the benchmark problem is given first, which is followed by the boundary and initial conditions used in the calculation, and numerical results and discussion.

2 Gust - Flat-Plate Airfoil Problem

Consider an isolated finite-span swept flat-plate airfoil bounded by two parallel walls shown in Fig. 1. The x-axis is aligned with the chord of the airfoil, the y-axis is perpendicular to it and the z-axis is normal to the bounding walls. The normal distance between walls, h, is $2.6c$, where c is the chord length of the flat plate. The sweep angle of the flat-plate airfoil is $15°$.

The mean flow is assumed to be uniform and aligned with the x-axis. The mean flow variables are inflow velocity, U_0, static density, ρ_0, and static pressure, p_0. The inflow Mach number, M_0, is 0.5. Flow variables are non-dimensionalized by using a_0(the speed of sound) as the velocity scale, c as the length scale, c/a_0 as the time scale, ρ_0 as the density scale and $\rho_0 a_0^2$ as the pressure scale. Thus the mean flow is described in dimensionless variables as

$$\bar{\rho} = 1, \quad \bar{u} = 0.5, \quad \bar{v} = 0, \quad \bar{w} = 0, \quad \bar{p} = 1/\gamma \tag{1}$$

where γ is the specific heat ratio and assumed to be 1.4.

The incident gust carried by the mean flow has x, y, and z velocity components given by

$$u' = 0, \quad v' = v_G \cos(k_x x + k_z z - \omega t), \quad w' = 0 \tag{2}$$

respectively, where $v_G = 0.05$ and $k_x = 5.5, k_z = 3.6m$, with $m = 0, 1, 2$, and $\omega = M_0 k_x$, respectively. The corresponding period of the gust wave is $T = 2\pi/\omega$. It is assumed that the gust is "frozen" and convected by the uniform mean flow. Thus the gust satisfies the linearized Euler equations. This implies that $\rho' = p' = 0$.

The gust is convected by the mean flow and interacts with the swept flat-plate airfoil. Thus, acoustic waves are generated which propagate both upstream and downstream. Let $(u^*, v^*, w^*, p^*, \rho^*)$ represent the unsteady solution due to the propagating acoustic waves, and define

$$\hat{u} = \bar{u} + u^*, \quad \hat{v} = \bar{v} + v^*, \quad \hat{w} = \bar{w} + w^*, \quad \hat{p} = \bar{p} + p^*, \quad \hat{\rho} = \bar{\rho} + \rho^*, \tag{3}$$

which represent the mean flow and the acoustic waves. Thus, we have

$$u = \hat{u} + u', \quad v = \hat{v} + v', \quad w = \hat{w} + w', \quad p = \hat{p} + p', \quad \rho = \hat{\rho} + \rho', \tag{4}$$

where (u, v, w, p, ρ) is the solution of the total flowfield, including the mean flow, the gust, and the acoustic waves.

In the current computation, the mean flow and the acoustic waves $(\hat{u}, \hat{v}, \hat{w}, \hat{p}, \hat{\rho})$ are computed by solving the nonlinear Euler equations. The convection of the gust is not simulated here.

3 Initial and Boundary Conditions

At $t = 0$, the time-marching variables in the whole domain are defined using the mean flow variables, and their spatial derivatives are set as zero. At the inlet, outlet, and the open boundary in y- direction, the time-marching variables and their spatial derivatives at the nth time level, are specified using the $(n - 1/2)$th time level value of the corresponding variables at the interior spatial point of its immediate neighbor. It allows the flux to "stream" out of the spatial domain smoothly with minimal reflection [7].

On the top and bottom walls bounding the flat-plate airfoil and the flat-plate airfoil surface, it is assumed that the velocity (u, v, w) is tangent to the wall/airfoil surface. By using Eq. (4), this implies that

$$n_x\hat{u} + n_y\hat{v} + n_z\hat{w} = -(n_x u' + n_y v' + n_z w') \tag{5}$$

where (n_x, n_y, n_z) is any unit vector normal to the wall/airfoil surface at the point under consideration.

On the symmetric plane at $y = 0$, the anti-symmetric condition is used for the solution of the acoustic waves as follows:

$$p^*(x, -y, z) = -p^*(x, y, z), \quad \rho^*(x, -y, z) = -\rho^*(x, y, z), \quad u^*(x, -y, z) = -u^*(x, y, z), \quad v^*(x, -y, z) = v^*(x, y, z), \quad w^*(x, -y, z) = -w^*(x, y, z) \tag{6}$$

No grid points are located at the flat-plate airfoil leading and trailing edges, in order to avoid the singular flow behavior.

4 Numerical Results and Discussion

A structured 281x71x41 hexahedral grid is generated by using an algebraic transformation in the computational domain of $-7 \leq x \leq 7$, $0 \leq y \leq 7$, and $0 \leq z \leq h$ first. A slice of the grid in x-z planes is shown in Fig. 2. Each hexahedron is cut into six tetrahedrons, which results in 4704000(280x70x40x6) tetrahedral cells in the computation. The parallel version of the 3D Euler solver is used with $\alpha = 0$ and $\epsilon = 0.5$, where α and ϵ are the parameters connected with the specification of the numerical dissipation in the solver [5]. A detailed description of the 3D Euler solver is referred to in [5] and the parallelization part is given in [6]. In the current computation, $\Delta t = T/168$ is used.

The spanwise wavenumber $k_z = 3.6m$ with $m = 0, 1, 2$, corresponding to three different gust orientations, are used here to study its effect on the far-field directivity. The analytical solution of the RMS acoustic pressure on the circle of radius 5 centered at $(0, 0, h/2)$ on the middle plane between the two bounding walls is available to compare with the corresponding numerical result. In all figures, the acoustic pressure p^* non-dimensionalized by v_G is plotted. The time history of the acoustic pressure at point $(-5, 0, h/2)$, which is located at the upstream on the specified circle, is shown in Fig. 3 for the three different values of m. It can be seen that (i) the solution is fully converged by $t = 50T$ for $m = 0$; (ii) a slight variation in the maximum absolute value of the acoustic pressure is still observed for $m = 1$ and 2; and (iii) the absolute amplitude of the acoustic pressure at the upstream decreases when the value of m increases. The solution convergence is also checked by comparing the RMS acoustic pressure on the specified circle at $t = 40T$ and $t = 50T$, which shows to be identical.

The computed result of the RMS acoustic pressure on the specified circle normalized by the maximum RMS pressure value at $m = 0$ is plotted in Fig. 4, while the corresponding analytical solution taken from [1] is reproduced in Fig. 5. It can be seen that both computed and analytical solutions show that the RMS pressure decreases when the spanwise wavenumber increases. The computed far-field radiation pattern is similar to those shown in the analytical solution. However some discrepancies are also observed, which is possibly caused by the non-reflecting boundary conditions used in the computation. Some reflections are created at the outer boundaries. A larger computational domain can be used to reduce the reflection errors from the outer boundary. However, a larger domain involves extensive mesh points and the computational cost is too expensive.

The parallel 3D CE/SE code was run on Origin2000 clusters. For a mesh of 4704000 cells, it takes around 35 hrs of wall-clock time to reach $t = 50T$ (16800 time iterations) using 32 CPUs.

5 Conclusion

The Problem 2 in Category 3 of the Third CAA Workshop has been solved by using the 3D parallel CE/SE Euler solver. The numerical results of the acoustic RMS pressure on the circle of radius 5 centered at $(0, 0, h/2)$ on the middle plane between two bounding walls are compared with the analytical solution for three different gust orientations, showing a reasonable agreement.

Acknowledgments: This work was supported by NASA Glenn Research Center through Contract NAS3-97186. The first author would like to thank Dr. Edmane Envia for his valuable help in this work.

References

1. The Proceeding of the Third Computational Aeroacoustics (CAA) Workshop on Benchmark Problems, NASA/CP-2000-209790, August, 2000.
2. S.C. Chang, "The Method of Space-Time Conservation Element and Solution Element – A New Approach for Solving the Navier-Stokes and Euler Equations," *J. Comput. Phys.*, **119,** pp. 295-324, (1995).

3. S.C. Chang, X.Y. Wang, and C.Y. Chow, "The Space-Time Conservation Element and Solution Element Method – A New High-Resolution and Genuinely Multidimensional Paradigm for Solving Conservation Laws," *J. Comput. Phys.*, **156**, pp. 89–136, (1999).
4. X.Y. Wang and S.C. Chang, "A 2D Non-splitting Unstructured-triangular-mesh Euler Solver based on the Method of Space-Time Conservation Element and Solution Element," Vol. 8, No. 2, pp. 326-340, 1999, *Computational Fluid Dynamics JOURNAL.*
5. X.Y. Wang and S.C. Chang, "A 3D Non-splitting Structured/Unstructured Euler Solver based on the Method of Space-Time Conservation Element and Solution Element," AIAA Paper 98-3278, Norfolk, Virginia, June, 1998.
6. A. Himansu, P. Jorgenson, X.Y. Wang and S.C. Chang, "Parallel CE/SE Computations via Domain Decomposition," in *Proceedings of 1st International Conference of Computational Fluid Dynamics*, Kyoto, June, 2000.
7. S.C. Chang, A. Himansu, C.Y. Loh, X.Y. Wang, S.T. Yu and P.C.E. Jorgenson, "Robust and Simple Non-Reflecting Boundary Conditions for the Space-Time Conservation Element and Solution Element Method," AIAA Paper 97-2077, June 29-July 2, 1997, Snowmass, CO.
8. C.Y. Loh, L.S. Hultgren and S.C. Chang, "Waves Computations in Compressible Flow Using the Space-Time Conservation Element and Solution Element method," *AIAA Journal*, Vol.39, No.5, pp. 794–801, May 2001.
9. X.Y. Wang, C.Y. Chow and S.C. Chang, "Numerical Simulation of Gust Generated Aeroacoustics in a Cascade Using the Space-Time Conservation Element and Solution Element Method," AIAA Paper 98-0178, January 12-15, 1998, Reno, NV.
10. X.Y. Wang, A. Himansu, P. Jorgenson, and S.C. Chang, "Gust Acoustic Response of a Swept Rectilinear Cascade Using the Space-Time CE/SE Method," FEDSM 2001-18134, in *Proceedings of FEDSM'01 2001 Fluids Engineering Summer Meeting*, May 29–June 1, 2001.
11. X.Y. Wang, S.C. Chang, A. Himansu, and P. Jorgenson, "Gust Acoustic Response of a Single Airfoil Using the Space-Time CE/SE Method," AIAA 2002-0801, Reno, NV, January, 2002.

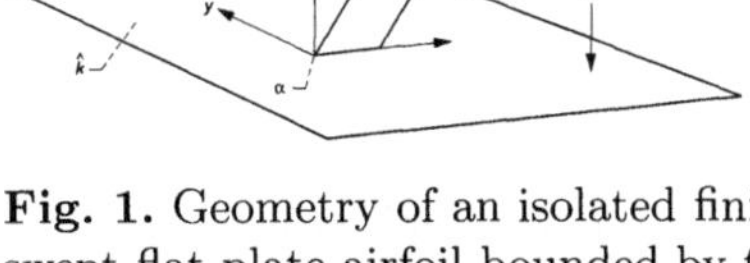

Fig. 1. Geometry of an isolated finite-span, swept flat-plate airfoil bounded by two walls.

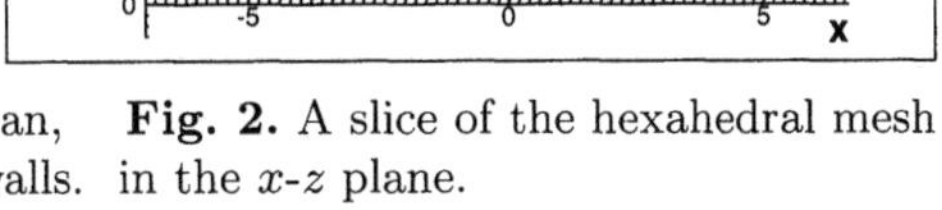

Fig. 2. A slice of the hexahedral mesh in the x-z plane.

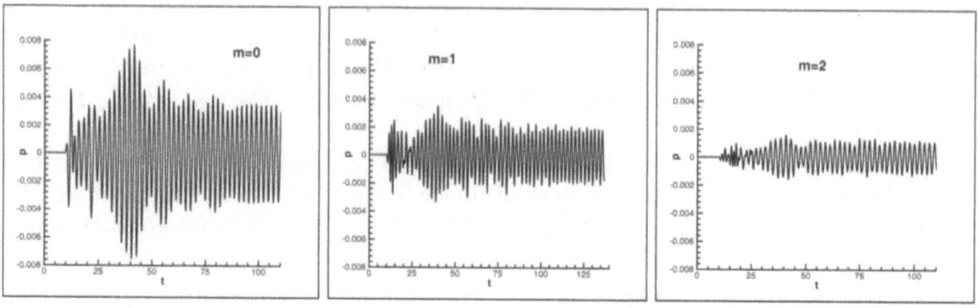

Fig. 3. The time history of the acoustic pressure at point $(-5, 0, h/2)$ for the three different values of m..

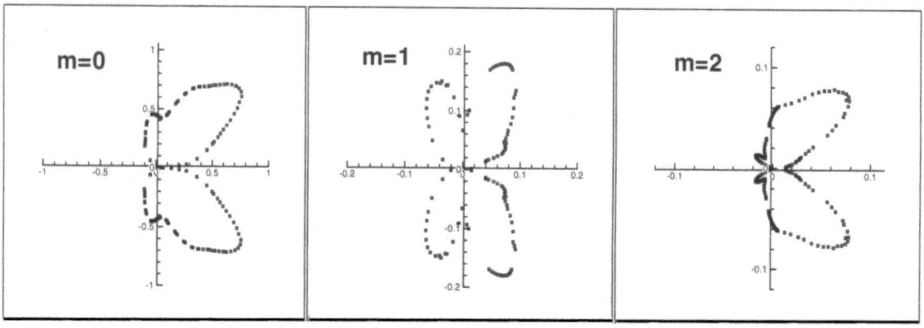

Fig. 4. The computed directivity pattern (RMS pressure on a circle of radius 5 centered at point $(0, 0, h/2)$ on the plane of $z = h/2$) for the three different values of m.

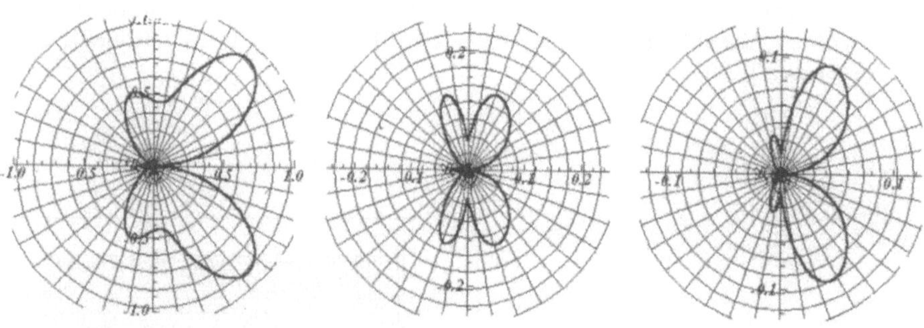

Fig. 5. The corresponding directivity pattern exhibited by the analytical solution for the three different values of m.

Unsteady Flow and Noise Characteristics of Wing in Ground Effect at Close Proximity

Young J. Moon[1], Sung-Ryong Koh[1], Jung-Hee Seo[1], Yong-Hyun Yoon[2], and Jin-Soo Cho[3]

[1] Dept. of Mechanical Engineering, Korea University, Seoul, Korea
[2] Air Force Academy, Chungwon, Korea
[3] Dept. of Mechanical Engineering, Hanyang University, Seoul, Korea

Abstract. The unsteady turbulent flow characteristics of NACA4406 airfoil at close proximity to the ground are numerically investigated. Attention is focused on finding a noise generation mechanism near the blunt trailing edge. The unsteady two-dimensional compressible Navier-Stokes equations are solved by the 6th-order compact scheme and the 4th-order Runge-Kutta method with a Spalart-Allmaras turbulence closure model. The computation shows noise generation by a feedback mechanism at the blunt trailing edge, where an acoustic-fluidic coupling occurs between the wall-reflected acoustic waves and the periodically disturbed turbulent shear layer.

1 Introduction

An eco-friendly, aero-levitated, high-speed ground-surface transport vehicle [1] is designed to cruise at a Mach number of 0.4 (or 500 km/hr) under the influence of ground effect, and it demands aerodynamic safety, operation stability, and aeroacoustic quiescence.

This vehicle generates not only lift increase by the ground effect but also unstable wake due to its closeness to the ground. An isolated airfoil is generally known to produce broadband and/or tonal noises, depending upon the thickness of the turbulent boundary layers and the trailing edge bluntness [2]. An airfoil with ground effect, however, may have noise generation by an acoustic-fluidic coupling between the wall-reflected acoustic waves and the unstable wake at the trailing edge. A gap tone by a feedback mechanism has been investigated by Tam and Pastouchenko [3] for a wall jet, where the acoustic-fluidic coupling occurs between the wall-reflected sound waves and the discrete vortex shedding at the tip (zero thickness).

In the present study, the unsteady turbulent flow characteristics of NACA4406 airfoil at close proximity to the ground are numerically investigated. Our attention is focused on finding a noise generation mechanism near the blunt trailing edge of the airfoil; more specifically investigating an acoustic-fluidic coupling mechanism between the wall-reflected acoustic waves and the unstable turbulent wake at the blunt trailing edge. The unsteady two-dimensional compressible Navier-Stokes equations are solved by the 6th-order compact scheme and the 4th-order Runge-Kutta method with a Spalart-Allmaras turbulence closure model for direct acoustic numerical simulation.

2 Numerical Method

2.1 Governing Equations

With the Favre averaging and the assumption of Boussinesq eddy viscosity approximation, the two-dimensional governing equations in generalized curvilinear coordinates (ξ, η) are written as

$$\frac{\partial \boldsymbol{Q}}{\partial t}+\frac{\partial \boldsymbol{E}}{\partial \xi}+\frac{\partial \boldsymbol{F}}{\partial \eta} = Re^{-1}J^{-1}\left(\xi_x\frac{\partial \boldsymbol{E_v}}{\partial \xi}+\xi_y\frac{\partial \boldsymbol{F_v}}{\partial \xi}+\eta_x\frac{\partial \boldsymbol{E_v}}{\partial \eta}+\eta_y\frac{\partial \boldsymbol{F_v}}{\partial \eta}\right) \quad (1)$$

where $\boldsymbol{Q} = J^{-1}[\varrho \;\; \varrho u \;\; \varrho v \;\; e]^T$ is a dependent variable vector. The inviscid flux vectors, $\boldsymbol{E}$ and $\boldsymbol{F}$, are given in a strong conservative form with the contravariant velocity components, and the viscous flux vectors, $\boldsymbol{E_v}$ and $\boldsymbol{F_v}$, are expressed with the laminar stress tensor, the Reynolds stress tensor and the turbulent heat flux. A one-equation Spalart-Allmaras model [4] is used as a turbulence closure model.

2.2 Numerical Schemes

The governing equations are integrated in time explicitly with a four-stage Runge-Kutta method, and are spatially discretized with a sixth-order compact finite difference scheme [5]. The tridiagonal system of the difference equations about the first and second derivatives of a scalar quantity ϕ at point i are expressed as

$$\phi'_{i-1}+3\phi'_i+\phi'_{i+1} = \frac{\phi_{i+2}+28(\phi_{i+1}-\phi_{i-1})-\phi_{i-2}}{12\Delta s}$$

$$2\phi''_{i+1}+11\phi''_i+2\phi''_{i-1} = 12\,\frac{\phi_{i+1}-2\phi_i+\phi_{i-1}}{\Delta s^2}+\frac{3}{4}\,\frac{\phi_{i+2}-2\phi_i+\phi_{i-2}}{\Delta s^2} \quad (2)$$

where Δs is a grid spacing. In order to prevent the grid to grid oscillations by a successive application of the first derivative operator, a non-conservative formation is adopted for evaluating the viscous fluxes that have the second derivatives. For the scalar quantity ϕ, the second derivative $\frac{\partial^2\phi}{\partial x^2}$ is written through a chain-rule differentiation as

$$\frac{\partial^2\phi}{\partial x^2} = \xi_x^2\frac{\partial^2\phi}{\partial \xi^2}+\eta_x^2\frac{\partial^2\phi}{\partial \eta^2}+2\xi_x\eta_x\frac{\partial^2\phi}{\partial\xi\partial\eta}+\xi_{xx}\frac{\partial\phi}{\partial\xi}+\eta_{xx}\frac{\partial\phi}{\partial\eta} \quad (3)$$

where the Jacobian transformation metrics ξ_{xx} and η_{xx} are expressed as

$$\begin{aligned}\xi_{xx} &= -J^3[y_\eta^3 x_{\xi\xi}-y_\eta^2(2x_{\xi\eta}y_\xi+x_\eta y_{\xi\xi})+\\ &\qquad y_\xi^2(x_{\eta\eta}y_\eta-x_\eta y_{\eta\eta})+2x_\eta y_\xi y_\eta y_{\xi\eta}]\,, &(4a)\\ \eta_{xx} &= J^3[y_\xi^3 x_{\eta\eta}-y_\xi^2(2x_{\xi\eta}y_\eta+x_\xi y_{\eta\eta})+\\ &\qquad y_\eta^2(y_\xi x_{\xi\xi}-y_{\xi\xi}x_\xi)+2x_\xi y_\xi y_\eta y_{\xi\eta}]\,. &(4b)\end{aligned}$$

In practical aspect, application of high-order compact scheme to the stretched meshes has numerical instability arising from the numerical truncation or failure of capturing the high wave number phenomena. Therefore, a tenth-order spatial filtering [6] is used for enhancing the numerical stability with high frequency cutoff.

3 Computational Results and Discussion

3.1 Computational Details

A NACA4406 airfoil with a blunt trailing edge of thickness $0.005c$ is considered in the present computation. Schematics of the wing-ground configuration are shown in Fig. 1 with the airfoil trailing edge details. A free stream Mach number of the present aero-levitated vehicle is 0.4, Re_c is $2\text{x}10^6$, the angle of attack is $2°$, and a gap between the trailing edge and the ground h/c is 0.2. The computational domain is set to a distance $3c$ away from the airfoil with a buffer zone of $1c$. Here c indicates the airfoil chord length. The computational meshes of approximately 73,000 points are used for the 3 block H-type grids. At the solid wall, an adiabatic no-slip condition is used, while, for the inflow and outflow boundaries, Freund's buffer zone boundary conditions [7] are applied not to generate spurious waves. At the far-field boundary, one-dimensional characteristic relations [8] are also used so that out-going acoustic waves escape the boundary without generating undesirable reflected waves.

3.2 Feedback Noise Generation

A typical aerodynamic characteristic of a wing in ground effect (WIG) is demonstrated for the present NACA4406 airfoil in Fig. 2. Experimental data for the wing of aspect ratio(AR)=2 show that the lift coefficient non-linearly increases as h/c becomes less than 0.15 and the computed C_L, corrected to the case of AR=2 by a Lanchester-Prandtl theory, quite closely matches the data for the h/c=0.2 case.

Figure 3 shows the turbulent boundary layers and wake profiles near the blunt trailing edge. The boundary layer thickness is $0.045c$ and $0.025c$ at the upper and lower surfaces of the trailing edge and δ^*/δ is 0.4 and 0.173, respectively. According to Brooks and Hodgeson [2], if $h_T/\delta^* < 0.3$ (h_T: trailing edge thickness), there is no Karman-type vortex shedding at the blunt trailing edge. From the present computation, h_T/δ^* is 0.27 and 1.14 at the upper and lower surfaces of the trailing edge. So it is expected that there will be no formation of Karman vortex sheddings and the associated tonal noise generation, while the shear layer at the lower surface could be in an unstable mode.

Even though no discrete vortex sheddings at the trailing edge, a direct acoustic numerical simulation result still shows an acoustic pressure field closely related to a tone generated at the trailing edge (see Fig. 4). In Fig. 5(a1)-(a4), one can clearly see the acoustic wave generation, reflection, and upstream propagation between the lower surface of the airfoil and the ground. More specifically, the acoustic waves generated at the trailing edge are fortified by the ground and are reflected back to disturb the turbulent shear layer at the trailing edge (see (b4)→(b1)). The sequential process of this acoustic-fluidic coupling is also demonstrated in Fig. 5(b1)-(b4). Fig. 5(b1) is an instant when a reflected acoustic wave impinges onto the trailing edge and deflects the shear layer downward. This local action guides an extended nozzle effect and makes the pressure drop at the trailing edge. The expansion process continues to (b2), while, at (b3), the shear layer is deflected upward and makes an local compression. The shear layer oscillations cause the pressure to periodically fluctuate at the trailing edge and generate acoustic waves. This acoustic-fluidic coupling between the wall-reflected acoustic waves and the unstable shear layer creates a feedback noise generation mechanism at the blunt trailing edge. Figure 6 (density field) also clearly shows the unstable turbulent shear layer developed at the lower surface of the trailing edge.

If this acoustic wave is a gap tone[3] by a feedback mechanism, the frequency must be determined by a period of acoustic wave traveling the gap between the airfoil and the ground, and its non-dimensional form is expressed by

$$St = \frac{f}{a/c} = n\frac{\sqrt{1-M^2}}{2h/c}, \qquad n = 1, 2, 3, ... \tag{5}$$

where a is a speed of sound and n is the mode. For the present case of $M = 0.4$ and $h/c = 0.2$, St becomes 2.29 for mode one. Figure 7 shows a time history of acoustic pressure fluctuations near the trailing edge and its fast Fourier transformed spectrum. The computed solution clearly shows the frequency of St=2.3, and it confirms that the present tonal noise is generated by a feedback mechanism between the wall-reflected acoustic waves and the periodically disturbed turbulent shear layer at the blunt trailing edge.

4 Conclusions

A direct acoustic numerical simulation of unsteady turbulent flow of NACA4406 airfoil at close proximity to the ground reveals a tonal noise generation (St=2.3) by a feedback mechanism at the blunt trailing edge, and the acoustic-fluidic coupling occurs between the wall-reflected acoustic waves and the periodically disturbed turbulent shear layer.

References

1. J.S. Cho, C.H. Han, Y.J. Moon, J.H. Baek: AIAA Paper 2001-2435 (2001)
2. T.F. Brooks, T.H. Hodgeson: J. Sound Vib. **78** (1981)
3. C.K. Tam, N. Pastouchenko: AIAA J. **39**, 8 (2001)
4. P.R. Spalart, S.R. Allmaras: AIAA Paper 92-0439 (1992)
5. S.K. Lele: J. Comput. Phys. **103**, 16 (1992)
6. M.R. Visbal, D. Gaitonde: AIAA J. **37**, 10 (1999)
7. J.B. Freund: AIAA J. **35** (4) (1997)
8. T.J. Poinsot, S.K. Lele: J. Comput. Phys. **101**, 104 (1992)

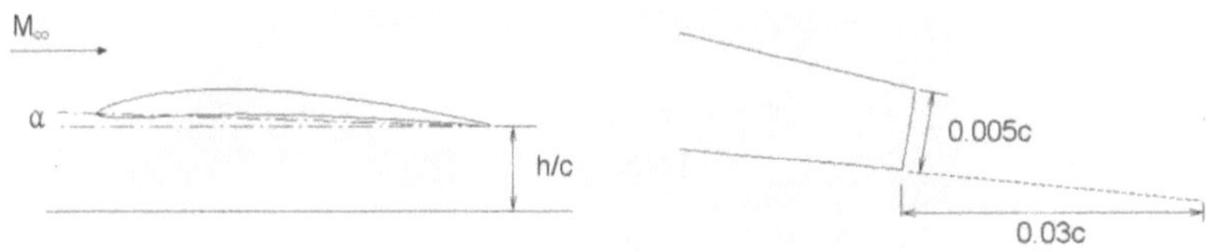

Fig. 1. NACA4406 airfoil with a blunt trailing edge

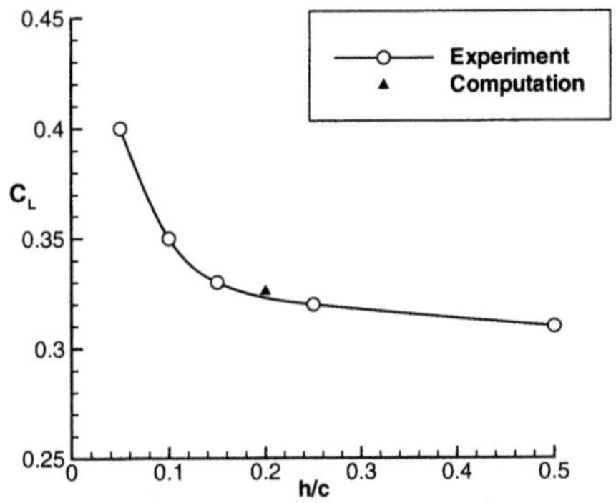

Fig. 2. Lift coefficient

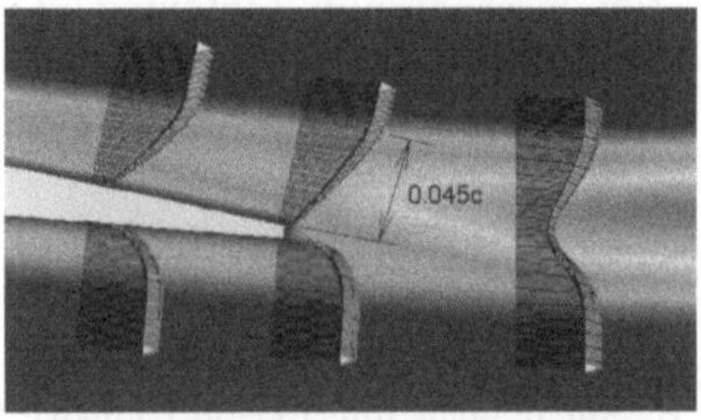

Fig. 3. Turbulent boundary layers and wake profiles

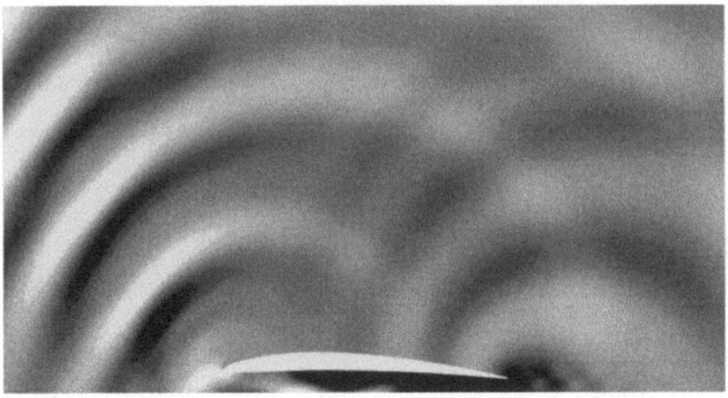

Fig. 4. Acoustic pressure field

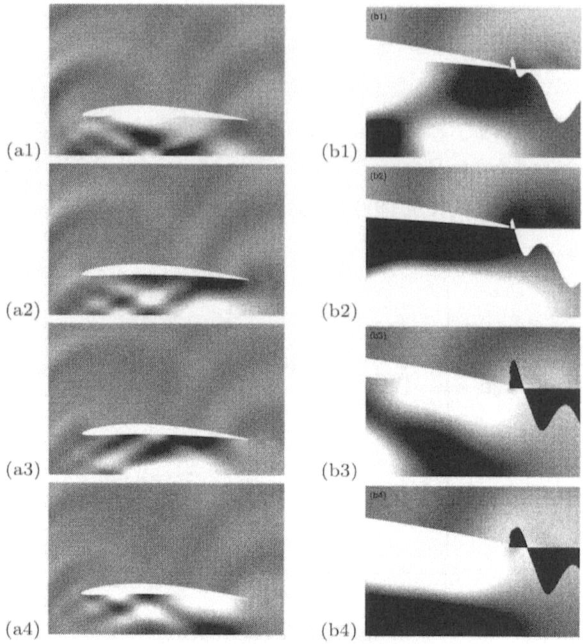

Fig. 5. Noise generation mechanism; p' contours and v vectors(magnified)

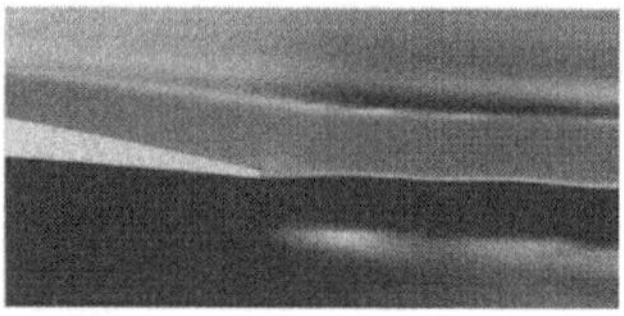

Fig. 6. Shear layer fluctuation (density field)

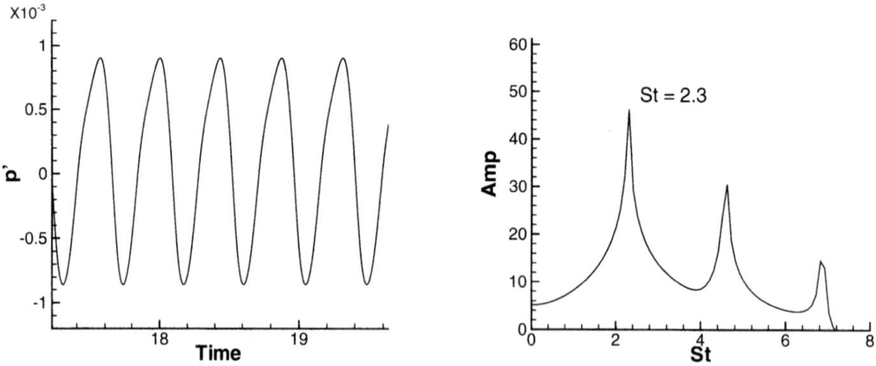

(a) Acoustic pressure fluctuations (b) FFT spectrum

Fig. 7. Frequency characteristics

An Accurate and Efficient Technique for Unsteady Viscous Flows in Two Dimensions

G. Pascazio, A. Grimaldi, and M. Napolitano

DIMeG, Sez. Macchine ed Energetica, & CEMeC,
Politecnico di Bari, Via Re David 200, 70125 Bari, Italy

Abstract. In the study of the physics of vortex dominated flows, vorticity plays a fundamental role, its dynamics being the primary tool which enables one to understand the time evolution of vortical structures; therefore, in the numerical solution of the time-dependent incompressible Navier–Stokes equations, it is convenient to select vorticity as one of the dependent variables. In order to be able to solve two-dimensional as well as three-dimensional flows, the velocity vector is used as the second dependent variable, as originally proposed by Fasel [1]. For such a formulation of the incompressible Navier–Stokes equations the authors have recently developed a numerical method for solving steady flows in two-dimensional general boundary-fitted curvilinear coordinates [2]. The governing equations are written as a system of a scalar vorticity transport equation and a second-order equation for the velocity vector, obtained by combining the vorticity definition and the incompressibility condition. The equations are discretized in space by a staggered-grid second-order-accurate finite-volume scheme, the main merit of the method being its remarkable accuracy, achieved by satisfying the discrete counterparts of the vorticity definition and of the continuity equation, *exactly*, *i.e.*, within machine accuracy. The method has been then extended to the case of unsteady flows, by solving the vorticity transport equation by means of an alternating direction implicit technique and the elliptic equations for the velocity components by means of a multigrid line-Gauss–Seidel relaxation procedure [3]. Such an approach requires an inner iteration to evaluate the vorticity boundary values at each new time level in order to achieve second-order accuracy in space and time. Here a new method is proposed, which achieves the same accuracy at a markedly lower computational cost. The vorticity transport equation is discretized in time by means of a fully implicit three-level scheme coupled with the velocity vector equations and the resulting large linear system is solved by a dual-time stepping technique: at each fictitious time step the vorticity and velocity fields are iterated to convergence by means of an efficient multigrid line-Gauss–Seidel relaxation scheme.

In the following, the numerical method is described in detail; then, it is used to compute a test problem with an exact solution, the vortex shedding phenomenon around a circular cylinder, for Reynolds number equal to 100, and the impulsively started flow past a circular cylinder for Reynolds number equal to 3000.

1 Numerical Method

The two-dimensional incompressible vorticity–velocity Navier–Stokes equations are written in conservative form as:

$$\frac{\partial \omega}{\partial t} + \boldsymbol{\nabla} \cdot (\boldsymbol{u}\omega) - \frac{1}{Re}\nabla^2 \omega = 0, \tag{1}$$

$$\nabla \cdot \boldsymbol{u} = 0, \tag{2}$$

$$\nabla \times \boldsymbol{u} = \omega \boldsymbol{k}, \tag{3}$$

where $\boldsymbol{u}$ is the velocity vector, ω is the only component of $\boldsymbol{\omega}$, $\boldsymbol{k}$ is the unit vector in the direction normal to the plane of the flow, t is the time, and Re is the Reynolds number. The problem stated in equations (1)-(3) is written in the so-called *div-curl* formulation and can be solved numerically as it is. However, from a computational standpoint, it is convenient to obtain a second-order elliptic PDE for the velocity vector in place of the two first-order scalar equations (2) and (3), so as to use standard relaxation methods, while satisfying equations (2) and (3) exactly in the discrete. Following [2], one subtracts the gradient of equation (2) from the *curl* of equation (3), to obtain:

$$\nabla \times \nabla \times \boldsymbol{u} - \nabla(\nabla \cdot \boldsymbol{u}) = \nabla \times \omega \boldsymbol{k}. \tag{4}$$

By virtue of Gauss' theorem, equations (1) and (4) are discretized in space as:

$$\mathcal{V}\frac{\partial \omega}{\partial t} + \sum_l \left(\boldsymbol{u}\omega - \frac{1}{Re}\nabla \omega \right) \cdot \boldsymbol{S}^l = 0, \tag{5}$$

$$\sum_l (\nabla \times \boldsymbol{u} - \omega \boldsymbol{k}) \times \boldsymbol{S}^l + \sum_l (\nabla \cdot \boldsymbol{u}) \boldsymbol{S}^l = 0, \tag{6}$$

where $\mathcal{V}$ and $\boldsymbol{S}^l$ indicate the cell control volume and its face area vectors, respectively, the summation being extended to all faces of $\mathcal{V}$.

In the present staggered-grid finite-volume discretization, the vorticity is defined at the nodes, whereas the discrete velocity field is approximated by means of the flow rates across the faces of the cells. With reference to the elementary cell, depicted in figure 1, one as:

$$U = \boldsymbol{u} \cdot \boldsymbol{S}^{\xi}, \quad V = \boldsymbol{u} \cdot \boldsymbol{S}^{\eta}. \tag{7}$$

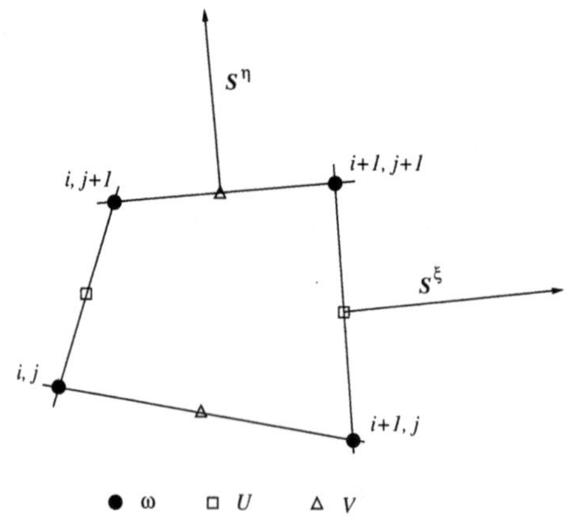

Fig. 1. Elementary cell.

In equation (7), $\boldsymbol{S}^\xi$ and $\boldsymbol{S}^\eta$ are the vectors having magnitude equal to the face area and direction normal to it, so that the velocity vector can be expressed as:

$$\boldsymbol{u} = U\boldsymbol{S}_\xi + V\boldsymbol{S}_\eta, \qquad \boldsymbol{S}^m \cdot \boldsymbol{S}_n = \delta^m_n, \tag{8}$$

δ^m_n being the Kronecker delta.

The vorticity transport equation, equation (5), is rewritten as:

$$\mathcal{V}_\omega \frac{d\omega_{i,j}}{dt} = R_{i,j} = -\mathcal{C}_{i,j}(U,V,\omega) + \frac{1}{Re}\mathcal{L}_{i,j}(\omega), \tag{9}$$

where $\mathcal{V}_\omega$ is the vorticity control volume, whereas $\mathcal{C}_{i,j}$ and $\mathcal{L}_{i,j}$ indicate the centered second-order accurate discrete convective and diffusive operators, as given in [2]. The vorticity time derivative is approximated using a three-points backward difference,

$$\frac{d\omega_{i,j}}{dt} = \frac{3\,\omega^{n+1}_{i,j} - 4\,\omega^n_{i,j} + \omega^{n-1}_{i,j}}{2\,\Delta t}.$$

Then, the dot product of the vector equation (6) times each face area vector, $\boldsymbol{S}^\xi$, $\boldsymbol{S}^\eta$, provides two scalar equations for the dependent variables U and V, respectively. Adding a pseudo-time derivative to each governing equation and discretizing it in pseudo-time using a two-level implicit scheme and the delta form of Beam and Warming [4], one obtains:

$$\mathcal{V}_\omega \frac{\Delta\omega_{i,j}}{\Delta t} + \mathcal{C}^u_{i,j}(U,V,\Delta\omega) - \frac{1}{Re}\mathcal{L}_{i,j}(\Delta\omega) = -\mathcal{V}_\omega \frac{d\omega_{i,j}}{dt} + R_{i,j}. \tag{10}$$

$$\mathcal{V}_U \frac{\Delta U_{i,j+\frac{1}{2}}}{\Delta\tau} - \mathcal{L}^\xi_{i,j+\frac{1}{2}}(\Delta U, V) = \mathcal{L}^\xi_{i,j+\frac{1}{2}}(U,V) - \mathcal{D}^\xi_{i,j+\frac{1}{2}}(\omega), \tag{11}$$

$$\mathcal{V}_V \frac{\Delta V_{i+\frac{1}{2},j}}{\Delta\tau} - \mathcal{L}^\eta_{i+\frac{1}{2},j}(U, \Delta V) = \mathcal{L}^\eta_{i+\frac{1}{2},j}(U,V) - \mathcal{D}^\eta_{i+\frac{1}{2},j}(\omega). \tag{12}$$

In the equations above, $\mathcal{V}_U$ and $\mathcal{V}_V$ are the velocity-component control volumes, $\Delta\tau$ is the pseudo-time step, $\Delta\omega$, ΔU, ΔV are the unknowns, namely, the variations of the dependent variables at the new physical time level $(n+1)$ from the old (k) to the new $(k+1)$ pseudo-time level, and the right-hand-sides (*rhs*) are the physical solutions at $(n+1,k)$. Moreover, in equation (10), $\mathcal{C}^u_{i,j}$ indicates the first-order-accurate upwind operator, according to the deferred correction strategy [7], whereas $\mathcal{L}^\xi$, $\mathcal{D}^\xi$, $\mathcal{L}^\eta$ and $\mathcal{D}^\eta$ are shorthand notations for the diffusion-like operators and the vorticity coupling terms for the U and V equations, respectively. The operators in equations (11) and (12) are discretized using a second-order finite volume centered scheme, so as to produce solutions satisfying conditions (2) and (3) exactly in the discrete [2]. The vorticity at the boundary is computed using the discrete form of its definition, equation (3).

The three uncoupled equations (10), (11), and (12) are solved sequentially by means of a forward-backward line-Gauss–Seidel relaxation procedure combined with a FAS multigrid scheme [5] using full weighting collection and bilinear prolongation operators, see [6]. After driving the *rhs* unsteady residuals to zero, the solution is advanced to the new physical time step.

2 Results

The well known Taylor problem, having the analytical solution:

$$u = -\cos x \,\sin y \, e^{-2t/Re},$$
$$v = \sin x \,\cos y \, e^{-2t/Re},$$
$$\omega = 2\cos x \,\cos y \, e^{-2t/Re},$$

has been used at first to verify the accuracy of the proposed numerical scheme. The problem has been solved in a square domain $[0, \pi/2]^2$, for $0 \leq t \leq 1$ and $Re = 10$. Numerical results have been obtained on uniform grids with $(N+1)\times(N+1)$ points and N_t time steps, for several values of N, N_t/N being constant. The numerical errors at the final time, $t = 1$, have been computed as the L^1 norms of the differences between the computed and exact solutions. The corresponding orders of accuracy are then evaluated according to

$$p = \frac{\ln(\text{error}_H/\text{error}_h)}{\ln(H/h)}.$$

Table 1 provides the values of the errors and of the corresponding orders of accuracy for four values of N and N_t, obtained driving the L^1 norm of the continuity equation to 10^{-6} at each pseudo-time step. The results in the table are comparable to those of [3], but have been obtained at a seven times lower computational cost. In particular, the CPU time for the finest-mesh computation is about six minutes on a single processor AlphaServer DS20E.

The non symmetric vortex shedding over a circular cylinder for $Re = 100$ has been computed using two grids with 129×129 grid-points and $\Delta t = 0.1$ and 193×193 grid-points and $\Delta t = 0.067$. For such a case a stretched cylindrical grid is employed ($r = r_{\text{out}} + (r_{\text{out}} - r_{\text{cil}}) \tanh\left[2(\eta - \eta_{max})/(\eta_{max} - 1)\right]/\tanh 2$, $\theta = \pi(1 - a^{\xi - 1})/(1 - a^{\xi_{mid} - 1})$, a being equal to 1.025 and 1.015 for the coarse

Table 1. Taylor's problem: accuracy study.

	ω		U		V	
N, N_t	error	p	error	p	error	p
8, 10	0.5404 (−2)		0.6025 (−3)		0.5396 (−3)	
		1.880		2.053		2.003
16, 20	0.1468 (−2)		0.1452 (−3)		0.1346 (−3)	
		1.982		2.045		2.027
32, 40	0.3715 (−3)		0.3519 (−4)		0.3302 (−4)	
		2.032		2.050		2.049
64, 80	0.9083 (−4)		0.8496 (−5)		0.7978 (−5)	

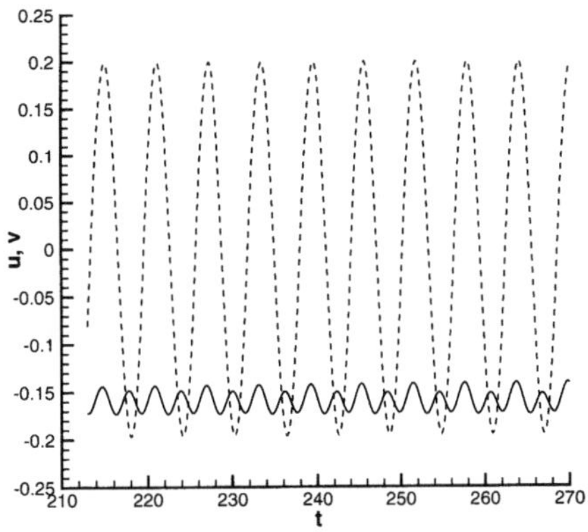

Fig. 2. Time history of the horizontal (u, solid line) and vertical (v, dashed line) velocity components at $(1,0)$.

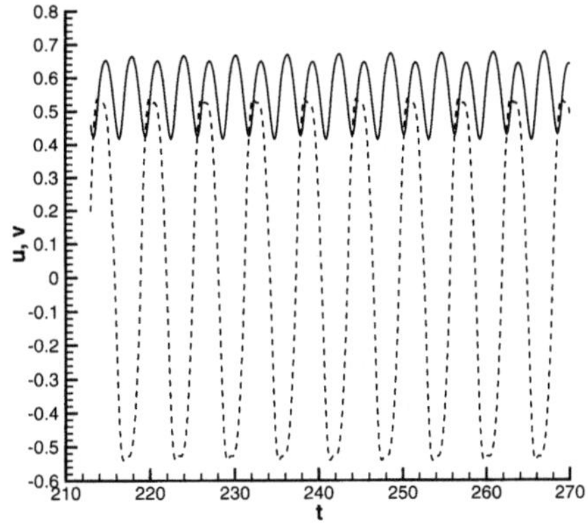

Fig. 3. Time history of the horizontal (u, solid line) and vertical (v, dashed line) velocity components at $(4,0)$.

and fine mesh, respectively), the outer boundary being at a distance of 30 diameters away from the cylinder. Figures 2 and 3 provide the time variation of the horizontal and vertical velocity components at points $(1,0)$ and $(4,0)$, re-

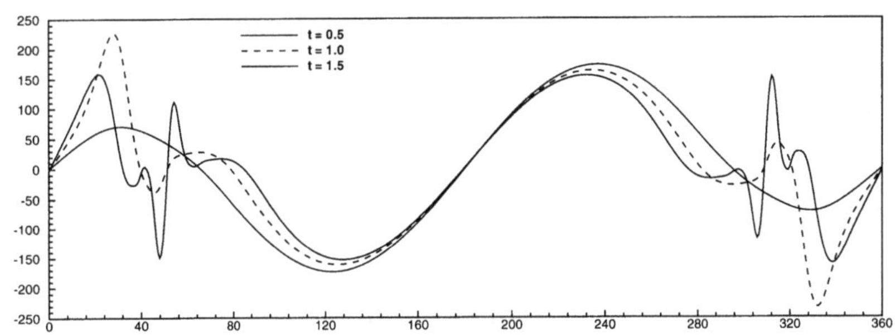

Fig. 4. Vorticity distributions for $Re = 3000$.

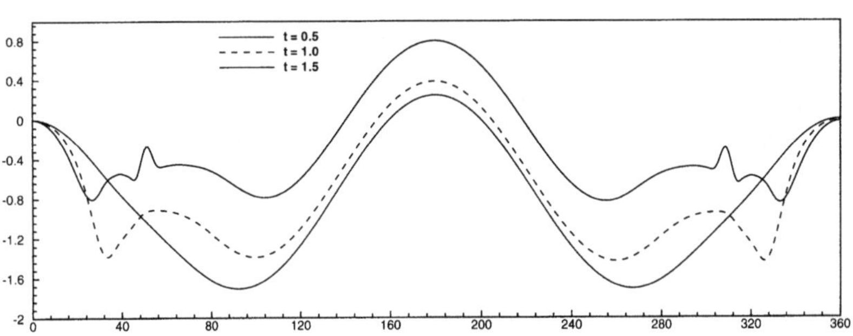

Fig. 5. Pressure distributions for $Re = 3000$.

spectively, obtained with the finer grid. They are practically grid-converged and are in good agreement with those of [8]. Moreover, the computed grid-converged Strouhal number is equal to 0.161, which compares well with the value 0.164 given by the interpolation of experimental data provided in [9]. Concerning efficiency, 30 and 60 multigrid cycles are required to drive the L^1 norm of the continuity equation to 10^{-5} on the coarse and fine grids, respectively.

Finally, the impulsively started symmetric flow past a circular cylinder has been considered, for $Re = 3000$. For such a flow the outer boundary has been located at 6 diameters away from the cylinder and the computations have been performed until $t = 1.5$ using a tangentially uniform grid with 513×257 grid-points and $\Delta t = 0.005$ ($r = r_{\rm out} + (r_{\rm out} - r_{\rm cil}) \tanh[1.4(\eta - \eta_{max})/(\eta_{max} - 1)]/\tanh 2$). The vorticity and pressure distributions along the cylinder at three different times are provided in figures 4 and 5, respectively; the present results are in good agreement with the very accurate numerical solution of [10], but are not yet grid-converged.

3 Conclusions

An accurate and efficient method has been developed to compute incompressible unsteady viscous flows in two space dimensions using the vorticity–velocity formulation and applied successfully to well known test-cases. In the near future, the proposed method will be used to provide benchmark solutions for flows of interest and then extended to three-dimensional flows.

References

1. H. Fasel: J. Fluid Mech. **78** 355 (1976).
2. G. Pascazio and M. Napolitano: Comput. Fluids **25**, 433 (1996).
3. P. De Palma, G. Pascazio, and M. Napolitano: Int. J. Num. Meth. Heat and Fluid Flow **11**, 286 (2001).
4. R. M. Beam and R. F. Warming: AIAA J. **16**, 393 (1978).
5. A. Brandt: 'Guide to multigrid development'. *Lecture Notes in Mathematics* **960**, pp. 220-312 (1982).
6. M. Napolitano and G. Pascazio: Comput. Fluids **19**, 489 (1991).
7. P. K. Khosla and S. G. Rubin: Comput. Fluids **2**, 207 (1974).
8. M. S. Engelman and M.-A. Jamnia: Int. J. Num. Meth. Fluid **11**, 985 (1990).
9. U. Fey, M. König, and H. Eckekmann: Phys. Fluids **10**, 1547 (1998).
10. C. R. Anderson and M. B. Reider: J. Comput. Phys. **125**, 207 (1996).

A Second-order-accurate Monotone Implicit Fluctuation Splitting Scheme for Unsteady Problems

P. De Palma, G. Pascazio, G. Rossiello, and M. Napolitano

DIMeG, Sez. Macchine ed Energetica, & CEMeC,
Politecnico di Bari, Via Re David 200, 70125 Bari, Italy

Abstract. This paper provides a novel implicit fluctuation splitting scheme which achieves second-order accuracy in both space and time, while guaranteeing monotone solutions. The method is based on a dual-time stepping approach which allows to separate the spatial and temporal discretizations. A novel distribution scheme of the unsteady-fluctuation term, coupled with a new limiting procedure for hampering spurious extrema, are the key-ingredients leading to the remarkable accuracy of the new method. This is applied with success to both scalar advection problems and unsteady inviscid flows in two space dimensions. In both applications the scheme is markedly superior to current state-of-the-art explicit fluctuation splitting schemes.

1 Introduction

Consider the linear advection equation for the scalar variable u with velocity $\boldsymbol{\lambda} = a\,\boldsymbol{i} + b\,\boldsymbol{j}$ and a discretization of the computational domain by triangular elements. If the variable u is assumed to vary bilinearly over each triangle, the discrete fluctuation, namely the flux balance over the cell, can be evaluated exactly as:

$$\phi_T = -\sum_{j=1}^{3} k_j u_j, \qquad k_j = \frac{1}{2}\boldsymbol{\lambda} \cdot \boldsymbol{n}_j l_j, \tag{1}$$

$\boldsymbol{n}_j$ and l_j being the inward unit vector normal to the edge opposing node j and its length, respectively. Explicit Fluctuation Splitting (FS) schemes are obtained by two main steps [1]: i) the residual is distributed among the nodes j of each triangle; ii) the solution at each node i of the computational domain is obtained by summing up all nodal contributions from the triangles $\triangle_i$ sharing node i as:

$$u_i^{n+1} = u_i^n + \frac{\Delta t}{S_i} \sum_{T \in \cup \triangle_i} \beta_{T,i}\phi_T = u_i^n + \frac{\Delta t}{S_i} \sum_{T \in \cup \triangle_i} \phi_{T,i}. \tag{2}$$

Two FS schemes are of interest as the basic tools for constructing a second-order-accurate monotone scheme, namely, the first-order-accurate monotone N scheme and the second-order-accurate non-monotone *Upwind Control Volume* (UCV) scheme, see [1]. In principle, it is impossible to use such schemes to obtain an unsteady solver with temporal accuracy higher than one. For this reason, current state-of-the-art FS schemes are based on the Lax–Wendroff (LW) scheme, limited with a Flux Corrected Transport (FCT) procedure [2,3]. The main drawback of such schemes is their unsatisfactory phase accuracy. In order to obtain a more accurate scheme, one needs to employ a dual-time stepping procedure and achieve second-order accuracy for the unsteady residual. This goal is obtained here using a novel residual distribution for the unsteady term, coupled with a UCV scheme for the spatial fluctuations, as described below.

2 Implicit scheme

The proposed scheme is based on a second-order accurate three-level discretization of the time derivative and a dual-time-stepping technique [4]. Each step in the fictitious time can be written as:

$$u_i^{n+1,k+1} = u_i^{n+1,k} - \frac{\Delta\tau}{S_i} \sum_{T \in \cup \Delta_i} (\alpha_{T,i}\phi_T^t + \beta_{T,i}\phi_T)^{n+1,k}, \tag{3}$$

where

$$\phi_T^t = \int_T \frac{\partial u}{\partial t} dS = \frac{S_T}{3} \sum_{i=1}^{3} \left(\frac{\partial u}{\partial t}\right)_i, \quad \left(\frac{\partial u}{\partial t}\right)_i = \frac{3u_i^{n+1} - 4u_i^n + u_i^{n-1}}{2\Delta t}.$$

The $\beta_{T,i}$ in equation (3) are those of the UCV scheme. Concerning the discretization of the critical unsteady term, ϕ_T^t, the distribution coefficients $\alpha_{T,i}$ are given as:

$$\alpha_{T,i} = \frac{1}{\phi_T^t} \int_{A_i} \frac{\partial u}{\partial t} dS, \quad A_i = \beta_{T,i} S_T. \tag{4}$$

The integral in the equation above is evaluated analytically, the time derivative varying linearly over the triangle, as:

$$\int_{A_i} \frac{\partial u}{\partial t} dS = A_i \sum_{j=1}^{3} \gamma_{i,j} \left(\frac{\partial u}{\partial t}\right)_j, \tag{5}$$

where

$$\gamma_{i,i} = \frac{2 - \beta_{T,i}}{3}, \qquad \gamma_{i,j} = \frac{1 - \beta_{T,j}}{3}, \; j \neq i. \tag{6}$$

Such coefficients guarantee that in the *two-target case* [1] the unsteady term is distributed only between the downstream nodes according to equations (4), A_i being the two triangular areas obtained by dividing the cell using the advection vector through the upwind node.

The proposed scheme is second-order accurate in both space and time, and monotone, when limited using the novel limiting procedure based on the N scheme, described later; as such it will be referred as the FS-UCV-N scheme.

3 Limiting procedure

For simplicity, the limiting procedure using the lower-order N scheme is described below for the case of the explicit FS-LW scheme; a new explicit FS-LW-N scheme is thus obtained, which will be used in the results section, for comparison.

i. For each node i of the domain, compute the lower-order and higher-order solutions:

$$u_i^{lo} = u_i^n + \frac{\Delta t}{S_i} \sum_{T \in \cup \Delta_i} \phi_{T,i}^N, \quad u_i^{ho} = u_i^n + \frac{\Delta t}{S_i} \sum_{T \in \cup \Delta_i} \phi_{T,i}^{LW},$$

and the quantities: $u_i^{\pm} = \left\{ \begin{matrix} \max \\ \min \end{matrix} \right. (u_i^{lo}, u_i^n)$.

ii. For each element T compute the quantities: $u_T^{\pm} = \left\{ \begin{matrix} \max \\ \min \end{matrix} \right. u_j^{\pm}, \quad \forall j \in T.$

iii. For each node i compute the quantities:

$$\bar{u}_i^{\pm} = \left\{ \begin{matrix} \max \\ \min \end{matrix} \right. u_T^{\pm}, \quad \forall T \in \cup \Delta_i,$$

$$\Delta u_{i,max}^{\pm} = \begin{matrix} \max \\ \min \end{matrix} [0, \bar{u}_i^{\pm} - u_i^{ho}], \quad \rightarrow \quad \Delta f_{i,max}^{\pm} = \Delta u_{i,max}^{\pm} \frac{S_i}{\Delta t},$$

$$\Delta u_i^{\pm} = \begin{matrix} \max \\ \min \end{matrix} [0, u_i^{ho} - \bar{u}_i^{\pm}], \quad \rightarrow \quad \Delta f_i^{\pm} = \Delta u_i^{\pm} \frac{S_i}{\Delta t},$$

iv. For each node i:
if $\Delta f_i^+ > 0$ $(\Delta f_i^- < 0)$ then

$$S_{i,max}^{\pm} = \sum_j \Delta f_{j,max}^{\pm}, \; j \in \cup \Delta_i, \; j \neq i,$$

if $S_{i,max}^+ \geq \Delta f_i^+$ $(S_{i,max}^- \leq \Delta f_i^-)$ then

$$\Delta f_{ij}^{\pm} = \Delta f_i^{\pm} \frac{\Delta f_{j,max}^{\pm}}{S_{i,max}^{\pm}},$$

else

$$\Delta f_{ij}^{\pm} = \Delta f_{j,max}^{\pm} \qquad R_i^{\pm} = \pm \Delta f_i^{\pm} \mp S_{i,max}^{\pm}.$$

For the rare situations in which the quantity $R_i^{\pm}$ is not equal to zero, it is distributed among the neighboring points according to the same criteria.

Update the solution as:

$$\tilde{u}_i = u_i^{ho} \mp \Delta f_i^{\pm} \frac{\Delta t}{S_i}, \quad \tilde{u}_j = u_j^{ho} \pm \Delta f_{ij}^{\pm} \frac{\Delta t}{S_i}, \; j \in \cup \Delta_i, \; j \neq i,$$

where $\tilde{u} = u^{n+1}$, when the entire domain has been swept.
Notice that $\Delta f_{i,max}^{\pm}$ needs to be updated throughout every sweep as:

$$\Delta f_{j,max}^{\pm} = \Delta f_{j,max}^{\pm} \mp \Delta f_{ij}^{\pm}, \; j \in \cup \Delta_i, \; j \neq i.$$

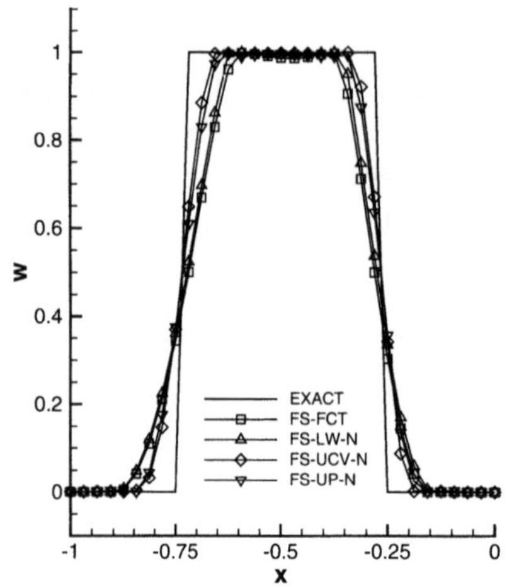

Fig. 1. Rotating cylinder: horizontal cut at $t = 1$.

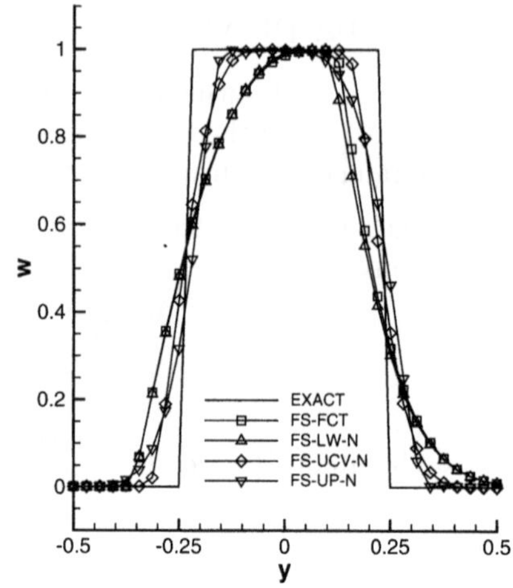

Fig. 2. Rotating cylinder: vertical cut at $t = 1$.

It is noteworthy that the key feature of the proposed limiting procedure is its global character, namely, it is applied to all nodes of the domain, rather than to those of each cell.

4 Extension to hyperbolic systems

The proposed implicit FS scheme is extended to the Euler equations, by generalizing equations (4)-(6) as follows:

$$\Phi^t_{T,i} = S_T\,\boldsymbol{\beta}_{T,i} \sum_{j=1}^{3} \Gamma_{i,j} \left(\frac{\partial U}{\partial t}\right)_j , \tag{7}$$

with

$$\Gamma_{i,i} = \frac{1}{3}\,(2\,I - \boldsymbol{\beta}_{T,i}), \qquad \Gamma_{i,j} = \frac{1}{3}\,(I - \boldsymbol{\beta}_{T,j}),\; j \neq i, \tag{8}$$

where $\Phi^t_{T,i}$ is the unsteady contribution sent to node i, U is the vector of the conservative variables, and $\boldsymbol{\beta}_{T,i}$ is the distribution matrix to node i. The limiting procedure described above is applied to each equation of the Euler system.

5 Results

5.1 Scalar advection equation

The circular advection of a cylinder in the square domain $[-1,1]^2$ has been computed to verify the monotonicity of the proposed schemes. The cylinder is

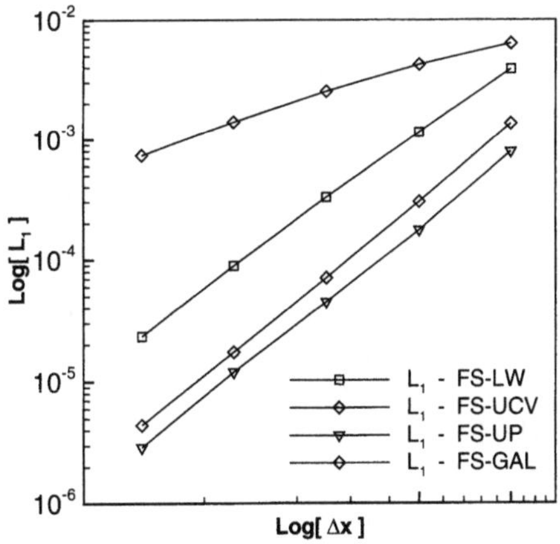

Fig. 3. Euler equations: accuracy study.

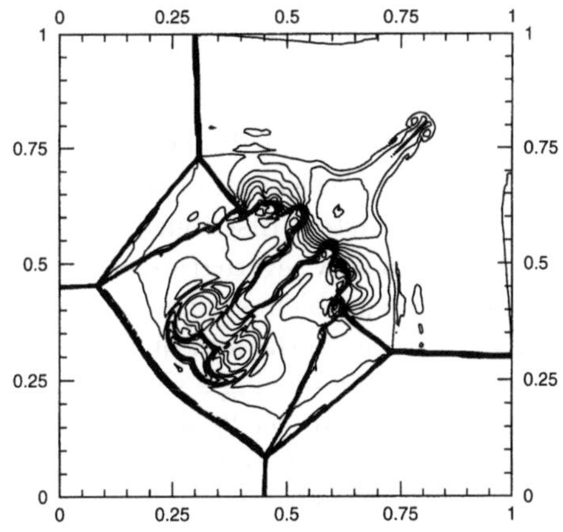

Fig. 4. Two-dimensional Riemann problem: density contours ($\rho_{min} = 0.135, \rho_{max} = 1.761, \Delta\rho = 0.08$).

defined as:

$$u = \begin{cases} 1 \text{ for } r \leq 0.25 \\ 0 \text{ for } r > 0.25 \end{cases},$$

where $r^2 = (x+0.5)^2 + y^2$ and $\boldsymbol{\lambda} = (-2\pi y,\, 2\pi x)^T$. A uniform Cartesian grid has been employed with 64×64 quad-cells, divided into triangles by means of the right-running diagonals. Figures 1 and 2 provide the horizontal and vertical solution cuts through the exact centre of the cylinder $(-0.5, 0)$ at $t = 1$ ($\Delta t = 0.0025$) using the proposed FS-LW-N and FS-UCV-N schemes, the FS-FCT scheme of [2], and a fourth implicit scheme, proposed in [5], obtained employing $\alpha_{T,i} = \beta_{T,i}$ in equation (3) and denoted here as FS-UP-N scheme; the exact solution is also provided, for comparison. All schemes provide monotone solutions, the implicit ones showing smaller dissipative and dispersive errors. It is noteworthy that the proposed FS-UCV-N scheme enjoys the lowest dispersive error.

5.2 Euler equations

The accuracy of the proposed schemes has been verified by computing the test case given in [6], namely, the smooth, unsteady, two-dimensional, inviscid flow of a perfect gas, initially at rest, driven by radially symmetric smooth initial data. The two-dimensional computations are performed on the unit square $[0,1]^2$ up to $t = 0.5$, on a sequence of five grids, the coarsest one using $\Delta x = \Delta y = 1/20$ and $\Delta t = 0.025$. The numerical results are compared with the *exact* solution obtained by solving the one-dimensional Euler equations with a source term for radial symmetry on a much finer grid [6]. In figure 3 the L_1 norm of the errors of the continuity equation, obtained with the FS-LW scheme, and the implicit FS-UCV and FS-UP schemes without limiting procedure, are plotted versus the mesh size. All schemes are second-order accurate, the implicit ones providing smaller errors. Furthermore, a fourth scheme is considered, named FS-GAL, which is obtained by distributing the steady term with the UCV coefficient matrix and the unsteady term with a Galerkin-type approach (see [5] for details). Such a scheme is clearly only first-order accurate, demonstrating the fundamental role of a *consistent* distribution of the steady and unsteady terms.

Finally, a very severe test-case has been considered, namely, the two-dimensional Riemann problem studied in [6]. The initial condition consists of four different constant values in the four quadrants of the square domain $[0,1]^2$, such that each pair of data gives a single shock wave, the interaction at the corner producing a complex structure, see [6] for details. Only the FS-LW-N, FS-FCT, and FS-UCV-N schemes were able to provide solutions at $t = 0.8$ ($\Delta t = 0.0016$) on a grid with 200×200 quad-cells, divided into triangles by means of the right-running diagonals. Moreover, only the FS-UCV-N scheme well captures the Kelvin-Helmholtz instability of the slip lines on this relatively coarse grid, thanks to its lowest dissipative error (see figure 4; the other results are not reported for brevity).

6 Conclusions

A novel implicit fluctuation splitting scheme is proposed which achieves second-order accuracy in both space and time, while guaranteeing monotone solutions. The method provides lower phase error than previously developed implicit fluctuation splitting schemes; more importantly, it is the first monotone scheme with such an accuracy, thanks to a very effective new limitation procedure. More extensive results will be provided in a future more detailed paper.

References

1. R. Struijs, H. Deconinck, and P. L. Roe: 'Fluctuation splitting schemes for the 2D Euler equations'. *VKI LS 1991-01*, von Karman Institute, Belgium (1991).
2. P. De Palma, G. Pascazio, and M. Napolitano: 'A second-order-accurate fluctuation splitting scheme for unsteady hyperbolic problems'. *NNFM*, **78**, ed. by M. Hafez, K. Morinishi and J. Periaux (Springer Verlag, New York, 2001), pp. 103–116.
3. M. E. Hubbard and P. L. Roe: 'Multidimensional upwind fluctuation distribution schemes for time dependent problems'. In: *Numerical Methods for Fluid Dynamics*, **6**, pp. 363-369 (1998).
4. A. Jameson: 'Time dependent calculations using multigrid with applications to unsteady flows past airfoils and wings'. *AIAA Paper 91-1596*, 1991.
5. D. A. Caraeni: 'Development of a multidimensional residual distribution solver for large eddy simulation of industrial turbulent flows'. Ph. D. Thesis, Lund Institute of Technology, 2000.
6. R. J. Le Veque: J. of Comp. Phys. **131**, 327 (1997).

Part XIX

Vortical Flows

Direct Navier–Stokes Simulation of Acoustic Waves Radiated by Vortex–Body Interaction

Nozomu Hatakeyama, Osamu Inoue, and Tomohiro Irie

Institute of Fluid Science, Tohoku University, 2-1-1 Katahira, Aoba-ku, Sendai 980-8577, Japan

Abstract. Acoustic waves radiated by the interaction of a single vortex colliding head-on with a body in a uniform flow are simulated numerically. The two-dimensional, unsteady, compressible Navier–Stokes equations are solved by a finite difference method. The results show that the dipole sounds are generated basically three times according to two times oscillations of lift force due to the vortex–body collision itself. It is also found that the third wave varies qualitatively according to the azimuthal Mach number of the interaction vortex.

1 Introduction

Computational aeroacoustics (CAA) is one of the rapidly developing fields in CFD. Engineering fields to which CAA is applicable are also expanding rapidly. The noise problems caused by vortex–body interaction can be seen widely in the field of engineering; for example, noise from helicopter blades, airfoils, compressors, electric fans, and so on [1]. However, to the authors' knowledge, direct numerical simulations (DNS) of acoustic waves generated by the interaction between a vortex and a body have not yet been reported so far, although theoretical, experimental and computational researches were carried out by using the acoustic analogy equation [2–4]. Our purpose in this paper is to study, by using DNS, acoustic waves generated by the head-on collision of a single vortex with a body. At an initial instant, a Taylor-type vortex is introduced upstream of a body in a uniform flow. With increased time, the vortex collides with the body and then passes through it; during the process, acoustic waves are generated. DNS makes it possible to investigate the sounds from generation to propagation quantitatively, and to increase our understanding of the basic features.

2 Numerical Procedure

A schematic diagram of the flow model is illustrated in Fig. 1. The coordinates parallel and normal to the free stream are denoted by x and y respectively, and a body with the characteristic length L is fixed as the leftmost point on the surface lies at the origin. The polar coordinates (r, θ) are also defined and the azimuthal angle θ is taken from upstream in the clockwise direction. The Mach number of the uniform flow is given by $M = U_\infty / c_\infty$, where U_∞ denotes the velocity of the uniform flow and c_∞ denotes the speed of sound at infinity. The Reynolds number is defined as $Re = U_\infty L/\nu_\infty$ where ν_∞ is the kinematic viscosity. The length, time, and density are made dimensionless by L, L/c_∞, and the ambient density ρ_∞, respectively.

The two-dimensional, unsteady, compressible Navier–Stokes equations for a perfect gas are solved by a finite difference method, where the Prandtl number is assumed to be 0.75 and the ratio of specific heats to be $\gamma = 1.4$. A sixth-order-accurate compact Padé scheme [5] is adopted for spatial derivatives. The fourth-order-accurate Runge–Kutta scheme is used for time integration. Adiabatic and no-slip conditions are applied on the body surface, and non-reflecting boundary conditions [6] are used on the outer boundary so that wave-reflections from computational boundaries are removed.

In the present study, two different bodies are considered; an elliptic cylinder with the thickness ratio of 0.5, and a NACA0012 airfoil. The body is placed parallel to the free stream without any attack angle, and the Reynolds number is prescribed to be $Re = 150$. Then the flow becomes steady and two standing vortices appear behind the body symmetrically. This is convenient to study acoustic waves radiated by a vortex–body interaction itself apart from the sounds caused by the periodic shedding of vortices such as aeolian tones for a circular cylinder [7]. The Mach number of a uniform flow is also fixed to be $M = 0.2$.

At an initial instant a vortex is superposed upstream at $(x, y) = (x_0, 0)$ to collide head-on with the body. The vortex is assumed to be axisymmetric at the tangential velocity of the Taylor-type, $u_\theta(r) = M_v(r/R_c)\exp[(1 - r^2/R_c^2)/2]$, where R_c denotes the core radius at which the velocity reaches its maximum, M_v [8]. The initial distributions of the pressure and the density are uniquely calculated by the adiabatic condition [9]. The influence of the vortex vanishes around $r = 4R_c$, hence $x_0 = -2$ and $R_c = 0.5$ are chosen. The Mach number of the vortex is prescribed to be M_v = 0.1 to 0.5.

We adopt a O-grid system around the elliptic cylinder and a C-grid system around the NACA0012 airfoil. The former is almost the same as the previous study on a circular cylinder [7]. The latter C-grid is constructed as follows. It is sufficient to consider only the region of $y \geq 0$ because of the x-axis symmetry. Let ξ denote the coordinate tangential to the airfoil surface measured from the origin ($x = y = 0$) in the clockwise direction, and η denotes the normal coordinate from the surface. Here the spacing in each direction, $\Delta\xi$ and $\Delta\eta$, is taken to identify that in the physical coordinate system (x, y). The ξ-spacings must be varied according to η while the η-grids can be distributed with little dependence on ξ, so the region in the ξ-direction will be expressed with ξ_0, the coordinate at $\eta = 0$. The domain is divided into three regions: a surface region $[0 \leq \xi_0 \leq x_{\text{surface}}]$ and $[0 \leq \eta \leq r_{\text{surface}}]$, a sound region $[0 \leq \xi_0 \leq x_{\text{sound}}]$ and $[r_{\text{surface}} \leq \eta \leq r_{\text{sound}}]$, a buffer region $[x_{\text{sound}} \leq \xi_0 \leq x_{\text{buffer}}]$ or $[r_{\text{sound}} \leq \eta \leq r_{\text{buffer}}]$, where $x_{\text{surface}} = 1.5$, $r_{\text{surface}} = 0.5$, $x_{\text{sound}} = 110$, $r_{\text{sound}} = 90$, $x_{\text{buffer}} = r_{\text{buffer}} = 1000$. The spacings are determined to be 0.005 in the surface region and to increase to 0.3 in the sound region. The largest spacing in the buffer region is 28. The point number of the whole grid system results in 1769 (ξ-direction) $\times$ 750 (η-direction), in which 127 points are distributed in the buffer region in each direction.

3 Results

Typical examples of computational results are presented in Figs. 2 to 6 for the case of an NACA0012 airfoil. The Reynolds number is $Re = 150$ and the Mach number of the uniform flow is $M = 0.2$. The sense of rotation of the vortex is prescribed to be anticlockwise. Without the introduced vortex, the flow shows no periodic shedding of vortices, and thus the so-called Kármán vortex street is not observed.

The development of the vorticity field is shown in Fig. 2a–d when the vortex of the Mach number $M_v = 0.1$ is initially, at $t = 0$, placed at $(x, y) = (-2, 0)$. The vortex collides head-on with the cylinder (Fig. 2a,b) and is divided into two parts; one passes on the upper side of the cylinder and the other passes on the lower side (Fig. 2c,d). The similar division of the interaction vortex has been also observed in a PIV experiment [10]. After the vortex passed through the airfoil, the flow restores the initial steady state and no periodic vortices are shed from the airfoil. For the case of an elliptic cylinder instead of the airfoil, weak vortices are excited by the interaction and shed successively to form a Kármán vortex street downstream.

Figure 3a shows a time history of the lift coefficient on the airfoil for three different Mach numbers, $M_v = 0.1, 0.25, 0.5$. Due to the anticlockwise sense of the vortex rotation, first the airfoil is lifted upward during the period of $Mt = 1$–2 (compare with Fig. 2a,b), then forced downward during $Mt = 3$–4 (Fig. 2c,d). For the strong vortex of $M_v = 0.5$, an additional lift-oscillation is observed around $Mt = 4$–5, which is exerted by the clockwise vortex uprolling from the shear layer on the upper side (see Fig. 2e,f). In comparison with the lift variation, three pressure pulses are observed at $r = 20, \theta = 90°$, as seen in Fig. 3b. Here the sound pressure Δp is defined by $\Delta p = p - p_\infty$ where $p_\infty = \gamma^{-1}$ denotes the ambient pressure at infinity. The generated sound is well approximated by the two-dimensional Curle's equation modified as includes the Doppler effect; $\Delta p_{\mathrm{Curle}}(r, \theta, t) = (1/\sqrt{8\pi^2 c_\theta r}) \int_{-\infty}^{\tau} [-F'_\theta(\tau')/\sqrt{\tau - \tau'}] \mathrm{d}\tau'$, where $\tau = t - r/c_\theta$, $c_\theta = c_\infty(1 - M\cos\theta)$, and $F'_\theta(t)$ denotes the time derivative of the force on the body in the θ-direction [7]. When $\theta = 90°$, F'_θ may be replaced by the time-derivative of the lift on the body, and $\Delta p_{\mathrm{Curle}}(20, 90°, t)$ gives a negative-positive-negative variation in response to the positive-negative variation of the lift.

The time series of a fluctuation pressure field during generation of the three waves are shown in Fig. 4. Here the fluctuation pressure is defined by $\Delta\tilde{p} = \Delta p - \Delta p_0$ where Δp_0 denotes the initial pressure field without an interacting vortex. At first glance, we can easily see the dipolar nature of the sounds. Figure 4a shows that a negative pulse is generated from the upper side and a positive pulse is generated from the lower side as the first sound wave, till the vortex center reaches the airfoil front (see Fig. 2b). Then the second wave is generated in antiphase with the first wave while the vortex is divided by the airfoil about $Mt = 2$–3, as seen from Fig. 4b in comparison with Fig. 2c. Figure 4c shows that the third wave is generated again in phase with the first wave but much smaller than the preceding two waves.

The effect of the mean pressure with respect to the total pressure is shown in Fig. 5. It is found that the third wave on the lower side is invisible from Fig. 5a of the total pressure Δp at $t=25$ after the three waves were radiated. This is because Δp includes the time-independent component, identified to the initial pressure Δp_0, in the nature of dipole toward upstream shown in Fig. 5b. Hence the fluctuation $\Delta\tilde{p}$ is better to analyze the wave propagation, as seen in Fig 5c. Note that the propagation of the peaks directs slightly upstream from $\theta=\pm 90°$ due to the Doppler effect. This direction θ_p depends on M with the relation of $\theta_p = \cos^{-1}M$, if the drag variation is much smaller than the lift variation [7].

Figure 6 shows that the three waves are propagated as acoustic sounds in two dimensions. The radial distributions of the pressure at the angle of $\theta = 90°$ are presented at three different times $t = 30, 40, 50$ in Fig. 6a. We can see that the propagation speed of the wave approximately equals unity. The decay of the peaks like $r^{-1/2}$ is clearly seen in Fig. 6b which is a logarithmic plot of the peak amplitudes during the propagation.

4 Concluding Remarks

By using the direct Navier-Stokes simulation in two dimensions, the acoustic waves radiated by the vortex-body interactions have been computed. The dipole sounds are radiated three times during the head-on collision of a vortex with a NACA0012 airfoil, where the third wave is much smaller than the preceding waves. The generation of these sounds are closely related with the lift force fluctuation exerted by the vortex motion passing through the body. The three acoustic waves are propagated nearly at the speed of c_∞ toward the vertical direction to the free stream and decay in proportion to $r^{-1/2}$ in the far acoustic field. For the case of an elliptic cylinder, however, weak acoustic waves are generated successively after the vortex passed through the body, because the successive vortex shedding is exerted by the vortex-body interaction.

We thank Mr. Sakari Onuma for his technical assistance. Computations were performed on the NEC SX-5 at the Institute of Fluid Science, Tohoku University.

References

1. D. Rockwell: Ann. Rev. Fluid Mech. **30**, 199 (1998)
2. M.S. Howe: *Acoustics of Fluid-Structure Interactions* (Cambridge University Press 1998), Sect. 3.3
3. E.R. Booth Jr.: AIAA J. **28**, 1353 (1990)
4. J.M. Gallman: AIAA J. **32**, 232 (1994)
5. S.K. Lele: J. Comput. Phys. **103**, 16 (1992)
6. T.J. Poinsot, S.K. Lele: J. Comput. Phys. **104**, 104 (1992)
7. O. Inoue, N. Hatakeyama: J. Fluid Mech. (2002, in press)
8. G.I. Taylor: Aero. Res. Comm., R. & M. 598 (1918)
9. O. Inoue, Y. Hattori: J. Fluid Mech. **380**, 81 (1999)
10. M.B. Horner, R.A.M. Galbraith, F.N. Coton, J.N. Stearts, I. Grant: AIAA J. **34**, 1188 (1996)

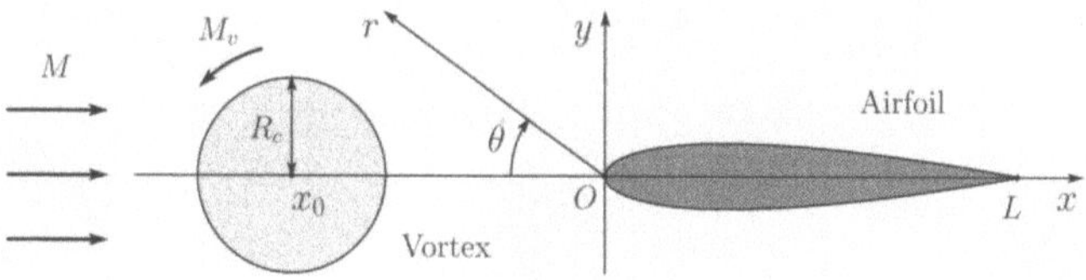

Fig. 1. Schematic diagram of vortex-body interaction

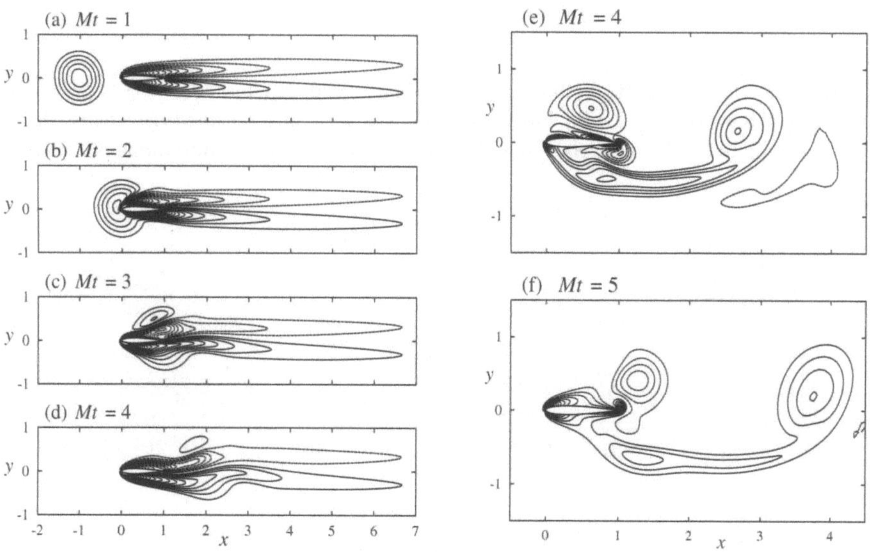

Fig. 2. Time development of vorticity field. The contours of $\omega > 0$ (*solid lines*) and $\omega < 0$ (*dashed lines*) are shown individually. $Re = 150, M = 0.2, x_0 = -2, R_c = 0.5$, (**a–d**) $M_v = 0.1$, with a contour level of $\Delta\omega = 0.1$, (**e,f**) $M_v = 0.5$, with $\Delta\omega = 0.3$

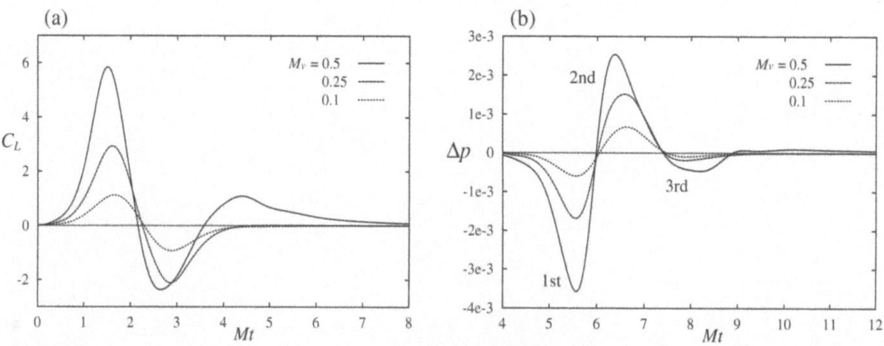

Fig. 3. (**a**) Time history of lift coefficient C_L. (**b**) Time history of sound pressure $\Delta p = p - p_\infty$ measured at $r = 20, \theta = 90°$. $Re = 150, M = 0.2, x_0 = -2, R_c = 0.5$

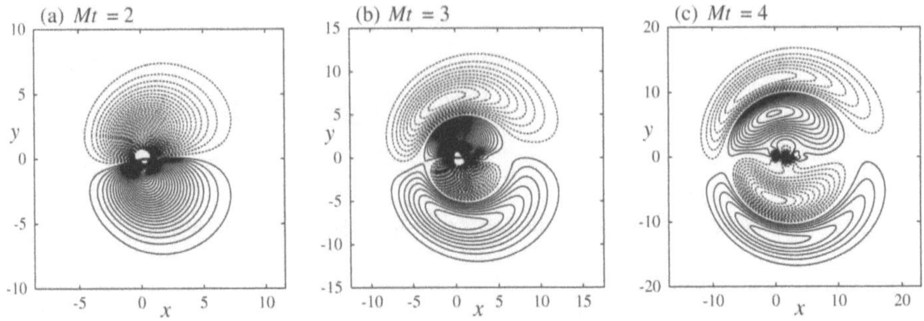

Fig. 4. Time development of fluctuation pressure field $\Delta\tilde{p}$ for $M_v = 0.1$. The contours of positive pressure (*solid lines*) and negative pressure (*dashed lines*) are shown individually. The levels are from -4×10^{-3} to 4×10^{-3} with an increment of 1×10^{-4}

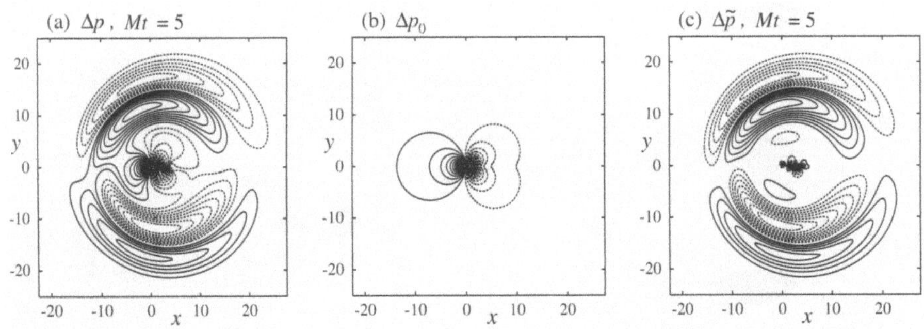

Fig. 5. Effect of mean pressure. $Mt = 5, M_v = 0.1$. (**a**) Total pressure Δp. (**b**) Initial pressure without a colliding vortex, Δp_0. (**c**) Fluctuation pressure $\Delta\tilde{p} = \Delta p - \Delta p_0$. The contour levels are the same as in Fig. 4

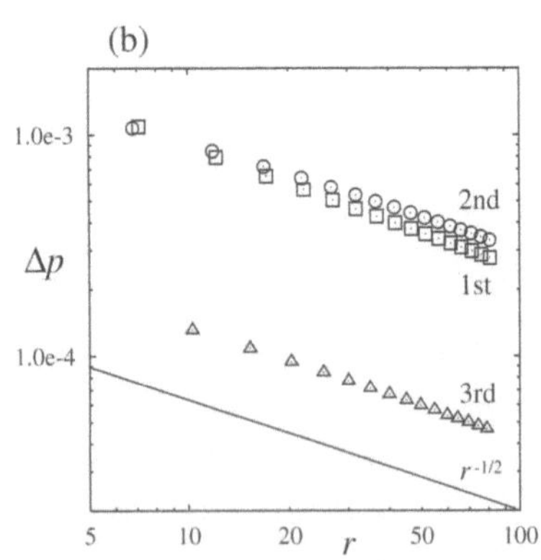

Fig. 6. Propagation and decay of acoustic waves in the direction of $\theta = 90°$ for $M_v = 0.1$. (**a**) Radial distributions of Δp at three different times. (**b**) Decay of the pressure peaks on a logarithmic scale

Simulation of Streamwise Vortices on Turbulent Hypersonic Ramps

Heinrich Lüdeke[1] and Erich Schülein[2]

[1] German Aerospace Center, Institute of Aerodynamics and Flow Technology Lilienthalplatz 7, D-38108 Braunschweig
[2] Bunsenstraße 10, D-37073 Göttingen

Abstract. Purpose of this investigation is the simulation of separated turbulent ramp flow as a simple generic model for re-entry vehicle flaps under three dimensional disturbances. Parameter variations of cross flow, disturbance intensity and wave length of the disturbances are tested. Comparison with wind tunnel data of the Ludwieg tube facility in Göttingen (RWG) was used to validate the numerical code and getting a deeper insight into the flow topology. The dependence of the vortices from surface disturbances and a tendency of circular vortex shape at reattachment has been shown experimentally and numerically in good agreement. Finally the damping influence of cross flow in the separation region was investigated.

1 Introduction

The influence of streamwise vortices in separated boundary layers on the heat flux at reattachment was shown in different studies during the last decades [1]. These vortices initiate spanwise heat flux oscillations at the reattachment line and lead to significant heat flux peaks in comparison with the undisturbed flow. This effect is of special interest for the design of deflected control surfaces at re-entry vehicles which is driven by the requirement of minimizing the overall weight of thermal protection [2]. For practical three dimensional cases the only efficient way to predict vortex influences on deflected flaps are combined CFD and wind tunnel investigations. In the following a basic study of different influencing parameters on these structures will be presented for a turbulent 20° ramp flow at $M = 3$, $Re = 12 \cdot 10^6$. In a first step the dependence of streamwise vortices on different disturbances, excited by thin pasted foils on the surface, was shown. Furthermore different spanwise vortex diameters and vortex damping by cross flow influences was investigated. Studies of the simulated flow field allow visualizations of the perturbation and the vortex behaviour at reattachment.

2 Numerical tools and grids

Numerical investigations have been carried out using the DLR-CEVCATS-N code which was validated for a wide range of sub- trans- and hypersonic flow cases [3]. CEVCATS-N is a multiblock finite volume flow solver for the steady Euler and Navier-Stokes equations in the integral form. Spacial discretization is done by a modified AUSM scheme that ensures high accuracy as well as a sharp

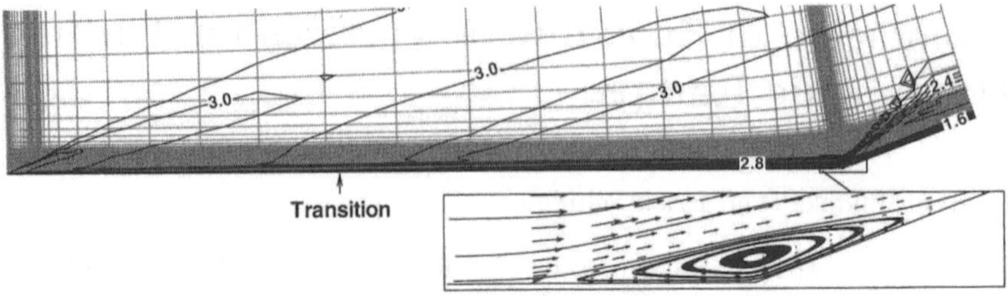

Fig. 1. Separated flow field of the 20° ramp. $M_\infty = 3$, $Re = 12 \cdot 10^6$

shock resolution [4]. Second order accuracy is obtained by MUSCL extrapolation. Turbulent separated boundary layers are modelled by a compressible extended Spalart Allmaras model.

The ramp grids were generated by conform mappings of cartesian basic ones. This non iterative technique has the advantage of global orthogonality and gives an approximation of the inviscid flow field by the grid lines. Line clustering was included in regions with high velocity gradients (Fig. 1). For 3D cases these grids were staggered in spanwise direction with a vortex resolution of 8 cells per vortex. In wall normal direction, a fine discretization was chosen all over the boundary layer. In the standard grid 5 vortex pairs were covered, it has 112x64 cells in streamwise and wall normal direction and 80 cells spanwise.

3 The DLR Ludwieg tube facility

The experimental part of this work was carried out at the Ludwieg-Tube facility in Göttingen. This supersonic/hypersonic Ludwieg-Tube (RWG) covers a Mach number range of $3 \leq M \leq 7$ and a unit Reynolds number range from $5 \cdot 10^6 m^{-1}$ to $80 \cdot 10^6 m^{-1}$. the facility consists of an 80-meter long tube used as pressure reservoir which is separated from the nozzles and the test section by a quick-acting gate valve. Hypersonic flow is generated by individual nozzle blocks, the

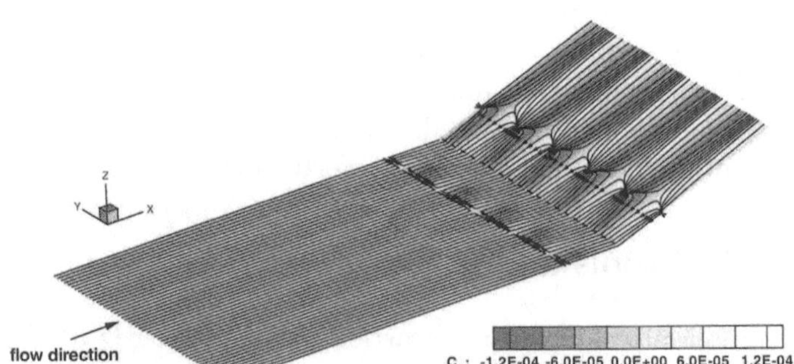

Fig. 2. Vortex influence on the attachment line. $M_\infty = 3$, $Re = 12 \cdot 10^6$

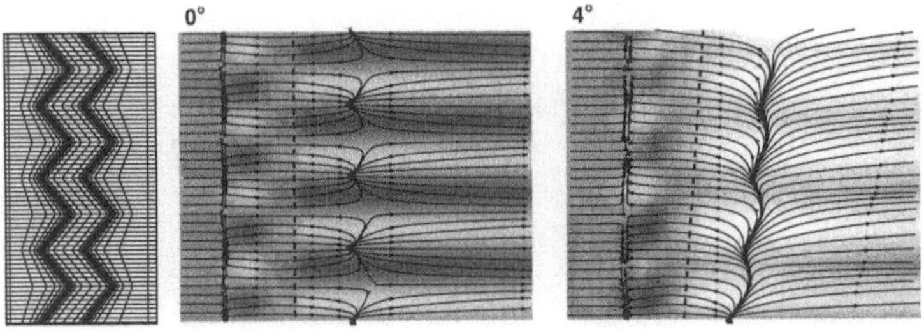

Fig. 3. Simulated streamlines for various yaw angles. $M_\infty = 3$, $Re = 12 \cdot 10^6$

test section for the Mach numbers $M = 3$ and 4 has a cross section of $0.5x0.5 m^2$. It was used for present investigations of the 20° ramp configuration.

4 Investigation of streamwise vortices

For the experimental setup a 20° ramp model of 200 mm width and 60 mm compression surface was mounted on a flat plate with a distance of 346 mm between the sharp leading edge (LE) and the ramp. In the following this dimension is used as reference length. To generate artificial disturbances a foil in form of a zigzag band (a so called "turbulator") of 0.05 mm thickness was pasted 10 mm downstream the LE [5]. These turbulators were included also in the 3D computational grids. The distance between the turbulator cusps was varied from 3 mm to 15 mm. For investigations of cross flow influences the ramp was turned by a yaw angle of 2° to 8°. The flow conditions of $M_\infty = 3$ and $Re = 12 \cdot 10^6$ were chosen as an approximation of those at deflected flaps. For simulations the ramp-width was minimized to save computational time, the configuration is treated infinite in spanwise direction by adequate boundary conditions. Laminar-turbulent transition was set at an $x = const$ line in the first third of the plate (Fig. 1), neglecting influences of the turbulator on the transition line.

4.1 Numerical and experimental results

As a first step for the computations an undisturbed 2D ramp flow was simulated (Fig. 1). Independence of the solution from numerical influences was shown by

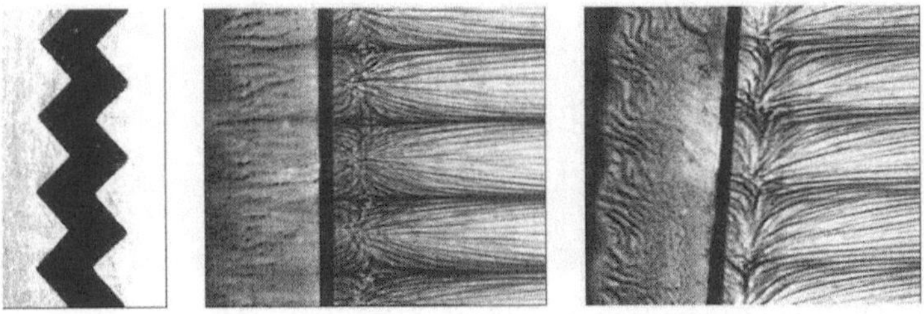

Fig. 4. Oil flow pictures for various yaw angles. Left: 0°, right: 4°.

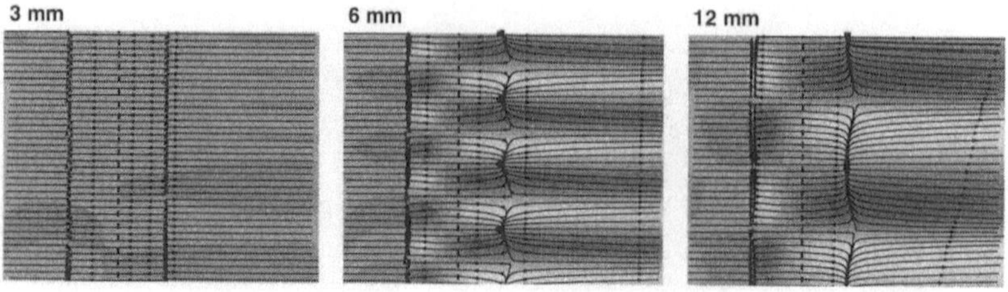

Fig. 5. Simulated streamlines for various vortex diameters. $M_\infty = 3$, $Re = 12 \cdot 10^6$

grid studies [6]. In the following the two dimensional flow field was staggered in spanwise direction as a basic solution for 3D computations.

Initiated by the turbulators counter rotating streamwise vortices appear in the separation region, resulting in a redistribution of the attachment line on the ramp (Fig. 2). The vortex influence leads to a periodic sequence of nodes and saddle points, and after reattachment to alternating converging and diverging streamlines. This is a typical pattern of the perturbed flow field, which was investigated several times under different flow conditions [1]. Nodes and saddle points were visualized numerically and experimentially in good agreement (Fig. 3 - 4). Streamwise vortices with different intensity, generated by turbulators of varying thickness 10 mm downstream the LE, have shown a negligible influence on the reattachment line as long as the turbulators are smaller than 0.025 mm [5].

Former investigations on the flaps of the X-38 re-entry vehicle have shown a weakening of streamwise structures under cross flow influence. To generate cross flow in the separation region a yaw angle of the ramp between 2° and 8° was introduced. Increasing yaw angles lead to a merging of nodes and saddle points resulting in a continuous reattachment line with a wavy character that decreases with the yaw angle (Fig. 3 - 4). This behaviour was simulated in good agreement with the oil flow pictures.

To check the influence of different vortex cross sections the spanwise dimensions of the vortices were varied by the distance of turbulator cusps. A pronounced influence was found for nearly circular vortices at the reattachment line. For smaller or larger cusp distances the influence weakened. Although this tendency was shown numerically as well as experimentially, the resulting attachment lines are different. While in the simulation only structures of the given

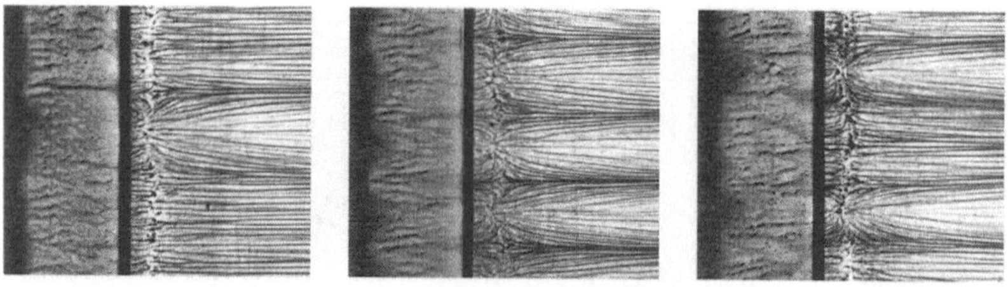

Fig. 6. Oil flow pictures for various vortex diameters. $M_\infty = 3$, $Re = 12 \cdot 10^6$

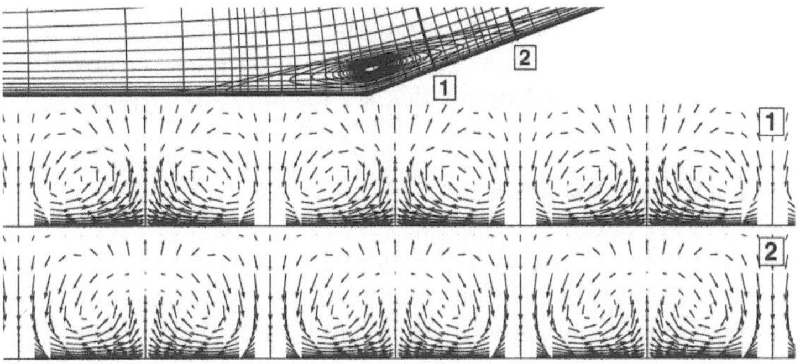

Fig. 7. Perturbation flow in two cross-sections. $M_\infty = 3$, $Re = 12 \cdot 10^6$

distance are visible with different impact on the reattachment line (Fig. 5), oil flow pictures show vortices of circular form for all cases with nearly vortex free zones in between (Fig. 6). These discrepancies may result by an imperfect manufacturing of the turbulators since any disturbance of the non-circular vortices could generate stronger ones of circular shape. On the other hand a lack of modelling of the wind tunnel conditions is possible, since influences of turbulators on the transition line were neglected by the simulation, so the deformation of this line by turbulators and the influence on streamwise vortices would be an interesting approach for further studies.

To visualize the vortex structures in the separation region, the perturbation flow field, where the vortices appear, had to be derived from the numerical results. To do so the perturbed 3D flow without yaw angle, shown in Fig. 2, was subtracted from the undisturbed basic flow. Two cross sections of the resulting velocity field are shown in Fig. 7. Beside the counter rotating structure also a wall normal vortex groth downstream is visible. This is a typical picture for streamwise vortices in growing boundary layers. 3D stream traces for the perturbed flow field in the separation region are shown in Fig. 8. The complex flow topology results from the wall normal components in the 2D separation bub-

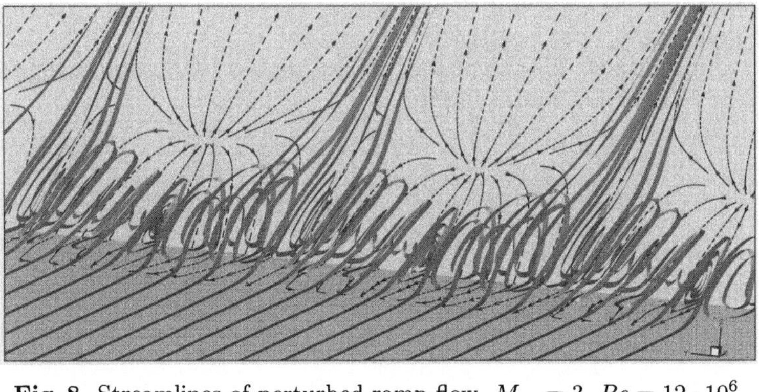

Fig. 8. Streamlines of perturbed ramp flow. $M_\infty = 3$, $Re = 12 \cdot 10^6$

ble. The stream traces show a converging behaviour downstream reattachment similar to the wall stream traces. This leads to a periodic outflow at the saddle points in the reattachment region which is typical for 3D separations.

5 Concluding remarks

Purpose of the present study was the investigation of streamwise counter rotating vortices at the reattachment line of a separated 20°-ramp flow. The influence of different parameters on the vortices, excited by turbulators near the leading edge, was shown. The generated vortices lead to a redistribution of the attachment line on the ramp into a periodic sequence of nodes and saddle points. Parameters like turbulator thickness and cusp distance were varied numerically as well as experimentially with good results. For vortices with non circular diameter at reattachment a different behaviour was found in the wind tunnel in comparison with the simulation. Suggested reasons were imperfections in the pasted turbulators or a lack of numerical transition modelling. Cross flow influences on the ramp have shown weakening effects on vortex influences in wind tunnel as well as numerical studies. From these investigations a strategy for reentry vehicles can be derived to prevent vortex generation and accompanying heat flux peaks on deflected flaps.

References

1. J.J. Ginoux: *Streamwise Vortices in Reattaching High-Speed Flows: A Suggested Approach* AIAA Journal Vol.9, No. 4 (1971), pp. 759-760.
2. H. Lüdeke: *Computation of Görtler Vortices in Separated Hypersonic Flows* Computational Fluid Dynamics 2000, Springer Verlag (2001), pp. 171-176.
3. S. Brück, R. Radespiel, J.M.A. Longo: *Comparison of Nonequilibrium Flows past a simplified Space-Shuttle Configuration* AIAA Paper 97-0275 (1997).
4. N. Kroll, R. Radespiel: *An Improved Flux Vector Split Discretization Scheme for Viscous Flows* DLR-FB 93-53 (1993).
5. E. Schülein: *Messungen der Wandschubspannung und des Wärmeflusses in einigen turbulenten Strömungen mit Ablösungen im Überschall* DLR IB 223 - 01 A 06, Göttingen (2002), to be published.
6. S. Nassif, H. Lüdeke: *Investigation of Görtler vortices on Hypersonic Ramp configurations* DLR IB 129-2001/3 (2000).

Numerical Simulation of the Interaction of Two Adjacent Synthetic Jet Actuators

Dahai Guo, Andrew W. Cary, and Ramesh K. Agarwal

Washington University, St. Louis, MO 63130, USA

Abstract. The interaction of two adjacent synthetic jet actuators is simulated. The computational results compare well with the experiment of Smith and Glezer[1]; the computed vortex structures in the flow-field are in general agreement with the experimental data. The influence of relative phase $\Delta\phi$ between the two synthetic jets on the vectoring of the merged jet is studied.

1 Introduction

Smith and Glezer[1] investigated the performance of a rectangular synthetic jet actuator for flow control. The synthetic jet is created by driving one side of a cavity using a piston or piezoelectric diaphragm in a periodic manner. Thus it requires no fluid ducting, only mechanical or electrical power. Figure 1 illustrates the cross-section of an actuator operating in still air. Two synthetic jets issuing side by side can be used to form a larger jet, which is depicted in Figure 2. The resultant merged jet can be manipulated by controlling the relative phase of the driving wave form. Figure 3 shows the effect of varying the phase angle between the driving signals of adjacent jets. If the two jets are in phase (left photo), the merged jet just moves away from the wall. When the left jet is leading in phase, the merged jet is vectored towards the actuator leading in phase. The merged jet becomes attached to the exit plane at a relative phase angle of 130° (right photo).

In this paper, the incompressible flow solver, INS2D, developed by Rogers and Kwak[5] is used for the simulations. The unsteady, two-dimensional and incompressible RANS equations are solved to simulate the interaction of two adjacent synthetic jet actuators. Cases with different relative phases between the adjacent jets are simulated. Comparison is made with the experiments of Smith and Glezer[1].

2 Results and Analysis

The computational parameters are matched with the experimental parameters of Smith and Glezer[1]. The Reynolds number $Re_{U_0} = 300$ and the space-averaged velocity amplitude at the exit of the orifice is 29.6 m/s. The oscillation frequency of the jet is 600 Hz. The left synthetic jet is considered to lead in phase. The synthetic jet actuator is simulated with the following approach[2–4]. At the exit of the orifice, the velocity of both the left(L) and right(R) jets is periodic, given by

$$\boldsymbol{V}_{O_L}(x=0,y,t) = U_A \sin(\omega t + \phi_L)\boldsymbol{i} \quad (1)$$
$$\boldsymbol{V}_{O_R}(x=0,y,t) = U_A \sin(\omega t + \phi_R)\boldsymbol{i} \quad (2)$$

with $\Delta\phi = \phi_L - \phi_R > 0$. The cavity geometry is included in the second approach and the actuator boundary conditions using Eqs. (1) and (2) are applied at the bottom of the two cavities.

2.1 Comparsion of Numerical Simulations with the Experiments[1]

The comparisons with the experiments focus on three cases: $\Delta\phi = 0°, 60°$ and $130°$. Figure 4 shows the phase-difference schematic of velocity for the three cases.

The temporal evolutions of vorticity for the three cases with and without the cavity are shown in Figures 5, 6, and 7. The main vortex structures in the simulations compare well with the experiments of Smith *et al.*[1]. When $\Delta\phi = 0°$, the two jets interact with each other and are merged into a large jet (Figure 5), moving away from the exit of the jets. When the two jets blow, each generates a vortex pair. The inner two vortices are weaker than the external two because of the small distance between them. At the end of the blowing($t^* = 0.5$), the outside vortex pair becomes stronger while the two inner vortices become weaker. At the peak suction, the outside vortex pair separates from the exit of the orifices and the inner vortex pair dissipates and disappears. The outside vortices generate induced velocity upon each other and travel away from the wall. When $\Delta\phi = 60°$, the vortex pair of the phase-leading jet(left) is generated first. At the same time, the suction stroke of the right jet attracts this pair to the right side. When the vortex pair of the right jet is generated, the two pairs interact with each other and the merged jet is vectored to the left (Figure 6). In the case $\Delta\phi = 130°$, the merged jet is attached to the left exit plane (Figure 7). The vortex pairs generated by the blowing strokes of the two jets are both vectored towards the left wall even before they separate from the jets, and travel away from the synthetic jets along the wall and dissipate gradually.

2.2 Time- and space-averaged pressure coefficient $\bar{\bar{C}}_{p_o}$

The time- and space-averaged pressure coefficient at the exits of the synthetic jets is defined as

$$\bar{\bar{C}}_{p_o} = \frac{1}{Th}\int_0^T \int_0^h C_p dx dt \quad (3)$$

It provides the general distribution of pressure at the exits of the synthetic jets. Figure 8 shows the variation of $\bar{\bar{C}}_{p_o}$ with $\Delta\phi$. $\bar{\bar{C}}_{p_o}$ of the phase-lagging jet (right) varies between -0.61 and -0.55. However, $\bar{\bar{C}}_{p_o}$ of the phase-leading jet (left) changes rapidly with $\Delta\phi$. When $\Delta\phi$ changes from $0°$ to $110°$, $\bar{\bar{C}}_{p_o}$ of the left jet reduces from -0.55 to -0.84. After $\Delta\phi > 110°$, $\bar{\bar{C}}_{p_o}$ of the left jet recovers and at $\Delta\phi = 160°$, it is about -0.59, even higher than the value of the right jet (-0.61). Although when $\Delta\phi > 70°$, the merged jet is always attached to the wall, $\Delta\bar{\bar{C}}_{p_o} = \bar{\bar{C}}^R_{p_o} - \bar{\bar{C}}^L_{p_o}$ between the two jets is different at different $\Delta\phi$. Since at $\Delta\phi = 110°$ $\Delta\bar{\bar{C}}_{p_o}$ is the largest, this suggests that the merged flow at $\Delta\phi = 110°$ is the most stable; in other words, the ability to resist the change of the direction of the merged flow is the strongest.

2.3 Effect of Large Phase Difference $\Delta\phi$ Between the Jets

As shown in Figure 8, $\bar{\bar{C}}^R_{p_o}$ is even lower than $\bar{\bar{C}}^L_{p_o}$ when $\Delta\phi = 160°$. This suggests that when the relative phase $\Delta\phi$ is very large, the flow direction of the merged jet can be very different depending on the initial conditions. The computations show that the flow transition occurs for a $\Delta\phi$ of about 160° if the cavities are not included in the computations. Figure 9 shows the vorticity distribution at time $t^* = 0.75$ with different initial phases of the two jets for $\Delta\phi = 155°, 160°, 170°$ and 180°. When $\Delta\phi = 155°$, the merged jet always goes to the left side; however, when $\Delta\phi \geq 160°$, the flow direction of the merged jet begins to depend on the initial phases of the two jets. For the cases $\Delta\phi = 160°$ and 170°, when $\phi_L = 0$ or $7\pi/4$, the merged jets are vectored to the right side, while for all other values of ϕ_L, they are vectored to the left side.

$\Delta\phi = 180°$ represents a special case since

$$\sin(\omega t + \phi_R) = \sin(\omega t + \phi_L - \pi) = -\sin(\omega t + \phi_L), \tag{4}$$

thus the jets are completely out of phase with each other. Figure 9(d) shows that when $3\pi/4 \leq \phi_L \leq 3\pi/2$, the merged jet is vectored to the left side while it is vectored to the right side when ϕ_L takes on other values. This phase difference of $\Delta\phi = 180°$ represents a critical point for the merged jet since it changes its flow direction from left to right. When $180° < \Delta\phi < 360°$,

$$\sin(\omega t + \phi_R) = \sin(\omega t + \phi_L + 2\pi - \Delta\phi) \tag{5}$$

and thus the right jet leads in phase. Therefore, for a small range of $\Delta\phi$ near 180°, the vectoring direction of the merged jet depends sensitively on the initial phases of the jets.

3 Summary

The interaction of two adjacent synthetic jet actuators is simulated. The computational results compare well with the experiment of Smith and Glezer[1]; the computed vortex structures in the flow-field are in general agreement with the experimental data. The influence of relative phase $\Delta\phi$ between the two synthetic jets on the vectoring of the merged jet is studied.

References

1. Smith, B. L. and Glezer, A., 'The Formation and Evolution of Synthetic Jets,' *Physics of Fluids*, Vol. 10, No. 9, September 1998.
2. Kral, L. D., Donovan, J. F., Cain, A. B., and Cary, A. W., 'Numerical Simulation of Synthetic Jet Actuators,' *AIAA* 97-1824, *4th AIAA Shear Flow Control Conference*, Snowmass, CO, June 1997.
3. Kral, L. D. and Guo, D.- H., 'Characterization of Jet Actuators for Active Flow Control', AIAA 99-3573, *30th AIAA Fluid Dynamics Conference*, Norfolk, VA, June 1999.
4. Guo, D.- H., Kral, L. D. and Cary, A. W., 'Numerical Simulation of the Interaction of Adjacent Synthetic Jet Actuators', AIAA 2000-2565, *Fluid 2000 Conference*, Denver, CO, June 2000.
5. Rogers, S. E. and Kwak, D., 'An Upwind Differencing Scheme for the Time-Accurate Incompressible Navier-Stokes Equations', AIAA Journal, Vol. 28, No. 2, pp. 253-262, 1990.

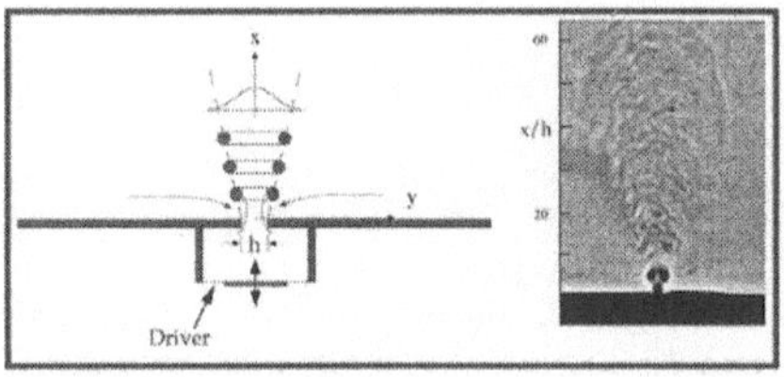

Fig. 1. Schematic of a synthetic jet actuator and Schlieren flow visualization from the experiments of Smith and Glezer[1].

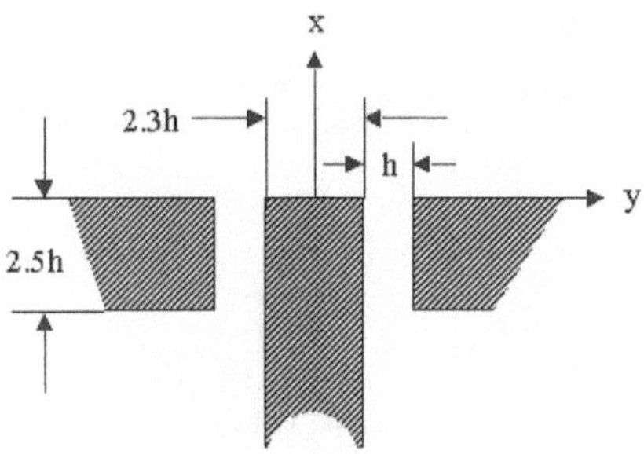

Fig. 2. Geometry schematic of two rectangular adjacent synthetic jets.

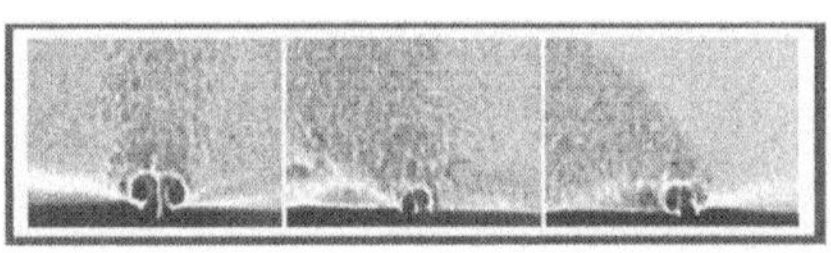

Fig. 3. Schlieren images of two adjacent synthetic jets which demonstrate the effect of varying the phase angle between the two driving signals. (from experiments of Smith and Glezer[1])

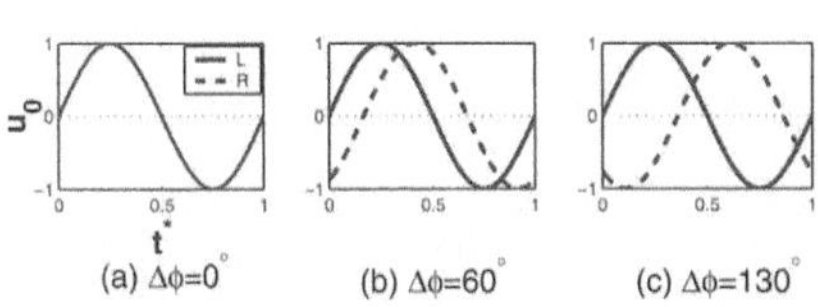

Fig. 4. Velocity wave form for relative phase differences $\Delta\phi$ between the two synthetic jets.

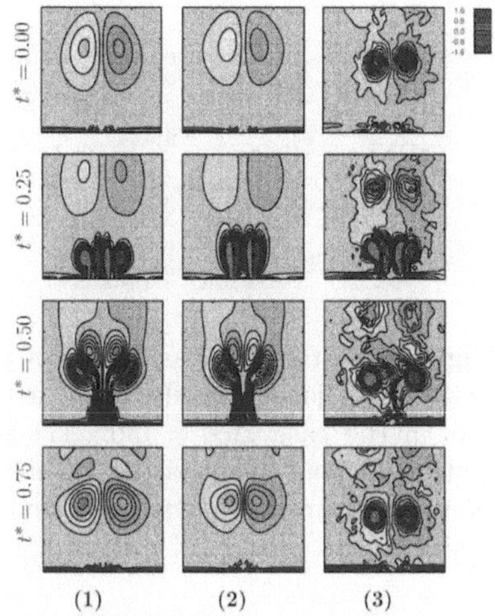

Fig. 5. The temporal evolution of vorticity for adjacent synthetic jet actuators in the case $\Delta\phi = 0°$, (1) simulation without the cavities (2) simulation with the cavities (3) experiments of Smith *et al.*[1].

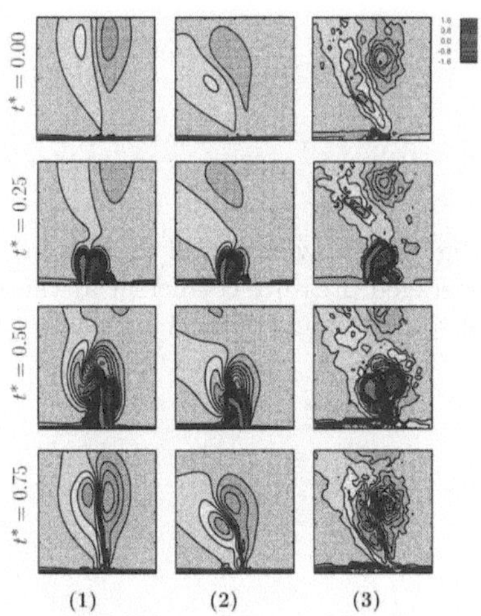

Fig. 6. The temporal evolution of vorticity for adjacent synthetic jet actuators in the case $\Delta\phi = 60°$, (1) simulation without the cavities (2) simulation with the cavities (3) experiments of Smith *et al.*[1].

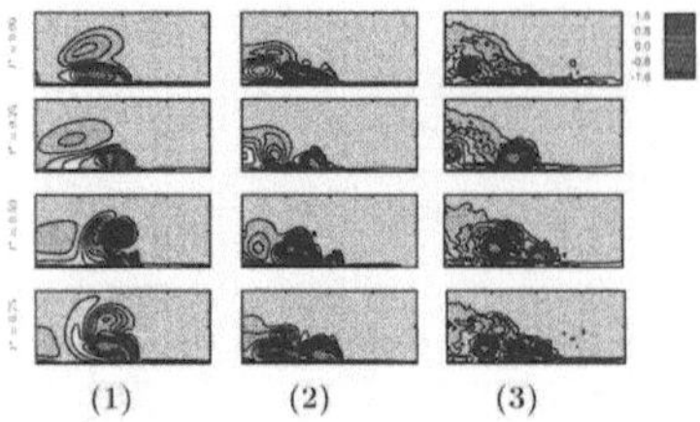

Fig. 7. The temporal evolution of vorticity for adjacent synthetic jet actuators in the case $\Delta\phi = 130°$, (1) simulation without the cavities (2) simulation with the cavities (3) experiments of Smith *et al.*[1].

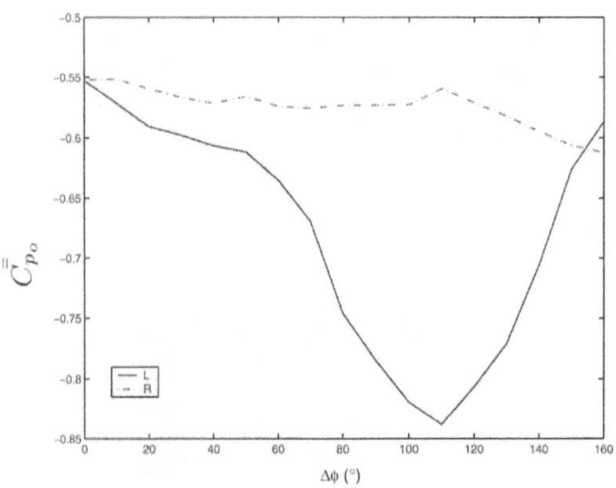

Fig. 8. The variation of time and space averaged pressure coefficient at the exit of synthetic jets with $\Delta\phi$.

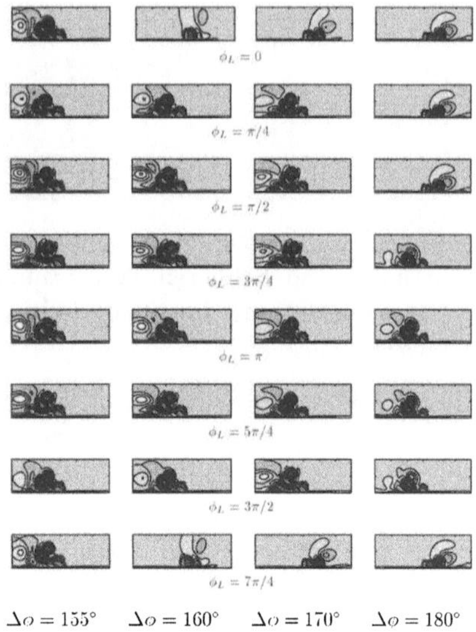

Fig. 9. Vorticity distributions at $t^* = 0.75$ for different values of the initial phase ϕ_L for $\Delta\phi = 155°, 160°, 170°$ and $180°$.

Numerical Investigation of Spiral and Wavy Vortices in Spherical Gaps

Li Yuan

Institute of Computational Mathematics and LSEC, Academy of Mathematics and System Sciences, Chinese Academy of Sciences, Beijing 100080, P. R. CHINA

Abstract. The supercritical flow states of the spherical Couette flow between two concentric spheres with the inner sphere rotating are studied numerically by solving the unsteady three-dimensional Navier-Stokes equations. Narrow to medium gap widths are chosen with clearance ratio $\sigma = 0.06, 0.14$ and 0.18 respectively, and a wide range of Reynolds number is encompassed in the present study. The numerical method used is the artificial compressibility method with a dual-time stepping technique for obtaining time accuracy. A third-order upwind compact finite difference is used for the advective terms. The numerical results show that spiral and wavy Taylor-Görtler vortices are formed in supercritical high Reynolds number range, and multiple time-dependent solutions exist for the medium gap case $\sigma = 0.18$.

1 Introduction

The flow in spherical shells is an important problem in fluid dynamics and geophysics [8]. A parade example for studying a wide spectrum of flow structures is the spherical Couette flow where the flow is driven by the rotation of the inner sphere. There are two control parameters that determine various flow regimes: a Reynolds number $Re = \Omega_1 R_1^2/\nu$ and a clearance ratio $\sigma = (R_2 - R_1)/R_1$, with ν the kinematic viscosity, Ω_1 the angular velocity of the inner sphere, and R_1 and R_2 the radii of the inner and outer sphere. Geometrically, a spherical shell can be considered as a combination of two other simpler systems with parallel disks in the pole region and cylindrical annulus near the equator. In narrow to medium gaps, different numbers of axisymmetric toroidal vortices (Taylor-Görtler vortex, TG) will be generated in the equatorial region if $Re \geq Re_c$, where Re_c is the critical Reynolds number for occurrence of the TG vortex. At higher Re numbers secondary instabilities develop in the form of spiral and wavy TG vortices [4]. In wide gaps, the first instability appears as nonaxisymmetric spiral vortices at the equator [9]. Numerical simulations of the spiral TG vortices were conducted [1,6], yet most were restricted to only two or three Re numbers with one specific gap width. The objective of present study is to investigate the multiple flow states for a wider range of Re numbers, especially the multiple unsteady solutions that received little attention before. To show the robustness of such multiple solutions, clearance ratios of $\sigma = 0.06, 0.14$ and 0.18 respectively, are chosen, which correspond to narrow ($\sigma < 0.12$) and medium ($0.12 \leq \sigma < 0.24$) gap widths as divided by Marcus [3].

The artificial compressibility method implemented in conjunction with a dual-time stepping technique for obtaining second-order time accuracy is used to solve the unsteady 3D incompressible Navier-Stokes equations. A third-order upwind compact finite difference scheme is used for the advective terms. The resulting discretized equation is solved by a diagonalized ADI scheme.

2 Mathematical Formulation and Numerical Method

Consider an annulus between two concentric spheres filled with an incompressible Newtonian fluid of constant density ρ and kinematic viscosity ν. The inner sphere is constrained to rotate about the vertical axis with a prescribed angular velocity Ω_1. The governing equations are written in conservative form in generalized coordinate system with pseudo-time derivatives added:

$$\frac{\partial \hat{\boldsymbol{Q}}}{\partial \tau}+\boldsymbol{I}_m\frac{\partial \hat{\boldsymbol{Q}}}{\partial t}+\frac{\partial(\hat{\boldsymbol{E}}-\hat{\boldsymbol{E}}_\nu)}{\partial \xi}+\frac{\partial(\hat{\boldsymbol{F}}-\hat{\boldsymbol{F}}_\nu)}{\partial \eta}+\frac{\partial(\hat{\boldsymbol{G}}-\hat{\boldsymbol{G}}_\nu)}{\partial \zeta}=0, \tag{1}$$

where $\hat{\boldsymbol{Q}}=\boldsymbol{Q}/J=(p,u,v,w)^{\mathrm{T}}/J$, $\boldsymbol{I}_m=\mathrm{diag}(0,1,1,1)^{\mathrm{T}}$, J is the Jacobian of the coordinate transformation, τ and t are the pseudo-time and physical time, respectively, and the detailed formulations of the inviscid flux vectors $\hat{\boldsymbol{E}},\hat{\boldsymbol{F}},\hat{\boldsymbol{G}}$ and viscous flux vectors $\hat{\boldsymbol{E}}_\nu,\hat{\boldsymbol{F}}_\nu,\hat{\boldsymbol{G}}_\nu$ can be found elsewhere [5,10]. Note that a subiteration is needed at each physical time step to drive the artificial compressibility terms toward zero to satisfy the continuity equation.

Applying an implicit backward difference to the pseudotime derivatives , a second-order, three point backward difference to the physical time derivatives, a first-order flux-difference splitting to the advective terms of the LHS, and the diagonalized Beam-Warming scheme, one obtain [10]

$$\begin{aligned}&\boldsymbol{T}_\xi\left[\boldsymbol{D}+\delta_\xi^-\boldsymbol{\Lambda}_\xi^++\delta_\xi^+\boldsymbol{\Lambda}_\xi^- -\delta_\xi\boldsymbol{A}_\nu'\bar{\delta}_\xi\right]\boldsymbol{T}_\xi^{-1}\boldsymbol{D}^{-1}\boldsymbol{T}_\eta\left[\boldsymbol{D}+\delta_\eta^-\boldsymbol{\Lambda}_\eta^++\delta_\eta^+\boldsymbol{\Lambda}_\eta^- -\delta_\eta\boldsymbol{B}_\nu'\bar{\delta}_\eta\right]\\&\boldsymbol{T}_\eta^{-1}\boldsymbol{D}^{-1}\boldsymbol{T}_\zeta\left[\boldsymbol{D}+\delta_\zeta^-\boldsymbol{\Lambda}_\zeta^++\delta_\zeta^+\boldsymbol{\Lambda}_\zeta^- -\delta_\zeta\boldsymbol{C}_\nu'\bar{\delta}_\zeta\right]\boldsymbol{T}_\zeta^{-1}\Delta\boldsymbol{Q}^{n+1,m}=-\left[\partial_\xi\hat{\boldsymbol{E}}+\partial_\eta\hat{\boldsymbol{F}}+\right.\\&\left.\partial_\zeta\hat{\boldsymbol{G}}-\delta_\xi\hat{\boldsymbol{E}}_\nu-\delta_\eta\hat{\boldsymbol{F}}_\nu-\delta_\zeta\hat{\boldsymbol{G}}_\nu\right]^{n+1,m}-\frac{\boldsymbol{I}_m}{\Delta t}\left(1.5\hat{\boldsymbol{Q}}^{n+1,m}-2\hat{\boldsymbol{Q}}^{n}+0.5\hat{\boldsymbol{Q}}^{n-1}\right),\end{aligned} \tag{2}$$

where $\boldsymbol{D}=\frac{1}{J}\left(\frac{1}{\Delta\tau}+\frac{1.5}{\Delta t}\right)\boldsymbol{I}$, $\boldsymbol{I}$ is the identity matrix, δ^+,δ^- and δ ($\bar{\delta}$) are forward, backward and central (midpoint central) differences, respectively, $\boldsymbol{A}_\nu',\boldsymbol{B}_\nu'$, and $\boldsymbol{C}_\nu'$ are the modified orthogonal part of viscous terms, $\boldsymbol{T}_\xi^{-1},\boldsymbol{T}_\xi$ and $\boldsymbol{\Lambda}_\xi$ are the left and right eigenvector matrices and eigenvalues of the Jacobian $\boldsymbol{A}\equiv\partial\hat{\boldsymbol{E}}/\partial\boldsymbol{Q}$, and $\boldsymbol{\Lambda}_\xi^\pm=\frac{1}{2}(\boldsymbol{\Lambda}_\xi\pm|\boldsymbol{\Lambda}_\xi|)$, cf. [5]. The third-order upwind compact scheme is applied to the discretization of the advective terms in the RHS of Eq.(2) dimension by dimension:

$$\partial_\xi\hat{\boldsymbol{E}}+\partial_\eta\hat{\boldsymbol{F}}+\partial_\xi\hat{\boldsymbol{G}}=\boldsymbol{A}\partial_\xi\boldsymbol{Q}+\boldsymbol{B}\partial_\eta\boldsymbol{Q}+\boldsymbol{C}\partial_\zeta\boldsymbol{Q}, \tag{3}$$

where $\boldsymbol{A}\partial_\xi\boldsymbol{Q}=\boldsymbol{A}^+(\partial_\xi\boldsymbol{Q})^++\boldsymbol{A}^-(\partial_\xi\boldsymbol{Q})^-$, $\boldsymbol{A}^\pm=\boldsymbol{T}_\xi\boldsymbol{\Lambda}_\xi^\pm\boldsymbol{T}_\xi^{-1}$, and $(\partial_\xi\boldsymbol{Q})^\pm$ can be computed by a third-order upwind compact finite difference scheme [11].

In the calculation, the computational domain of the spherical shell is divided into a number of grids in the radial (r), meridional(θ) and azimuthal (ϕ) directions, respectively. The grids are uniform in θ and ϕ directions, but are clustered near the wall of two spheres in the radial direction. A grid with $21 \times 361 \times 89$ grid points is used for narrow gap ($\sigma = 0.06$), and that with $31 \times 361 \times 129$ is used for medium gaps ($\sigma = 0.14$ and 0.18). For subcritical Reynolds numbers ($Re < Re_c$), the calculations were initiated from the Stokes flow and marched in pseudo-time only. For higher Reynolds numbers, however, low-Re flow fields were used for initialization in order to avoid excessive computational times for initial transients. For all unsteady simulations the physical time step is $\Delta t = 0.01$. The subiteration for unsteady computation is thought as converged when L_2 norm of the residuals has been reduced 10^{-2} from its initial value [10]. After the initial transients have died out, this level of convergence is achieved within 5-15 subiteration per physical time step.

Non-slip conditions are applied to velocities, and the pressure is obtained by the momentum equation in the radial direction on both spheres. A reference pressure is taken at a specified point in the interior domain.

3 Results

For comparison, three gap widths $\sigma = 0.06, 0.14$ and 0.18 have been selected to be the same as those in [1], [6] and [7], respectively. A wider range of Reynolds number was used in this study compared with those in previous studies.

For $\sigma = 0.06$, the range $2667 \leq Re \leq 5000$ is studied. Firstly, the basic flow state with zero vortex is obtained at $Re = 2667$ by steady-state computation. Then the Reynolds number was increased abruptly to $Re = 3100$. In the initial 300 time steps, two artificial dividing lines were symmetrically set on the northern and southern sides of the equator, and a "symmetric" condition is used about each line. A "crude" 1-vortex flow state is thus produced between each dividing line and the equator. After each line is removed, the flow settles down to steady-state 1-vortex flow (with a pair of toroidal TG vortices). By changing Re number with increment of 5, it is found that the supercritical 1-vortex flow exists stably for $2970 \leq Re < 3085$, while the supercritical 2-vortex flow exists stably for $3080 \leq Re < 3130$ once it is directly computed at $Re = 3100$. Therefore, there is a small overlap Re regime where both 1- and 2-vortex flow modes coexist, see Fig. 1, which is consistent with the results for $\sigma = 0.18$ [3,11]. The 2-vortex flow becomes periodic, axially symmetric but equatorially asymmetric for $3130 \leq Re < 3240$, but becomes fully symmetric for $3240 \leq Re < Re_s$. Here Re_s denotes the Reynolds number for the occurrence of spiral TG vortices. The structure of the spiral TG vortex flow is shown in Fig. 2. It can be seen that three pairs of spiral TG vortices are generated on both sides of the equator at $Re = 3300$ (see Fig. 2(a) and (d)). The TG vortex near the equator becomes wavy at $Re = 4000$, and the spiral TG vortices are more inclined in the azimuthal direction in the middle-latitude region. At $Re = 5000$, there are more threads of spiral TG vortices, and the flow becomes more chaotic.

For $\sigma = 0.18$, multiple unsteady flow solutions are found to appear in high supercritical Re range. Fig. 3 shows formation process of 8 pairs of spiral vortices at $Re = 8000$, which is evolved from steady 2-vortex flow at $Re = 1200$. Fig. 4 shows 1-vortex and 0-vortex flows with spiral vortex-like 3D disturbances at $Re = 8000$, which are evolved from steady 1-vortex flow at $Re = 750$, and Stovkes flow at $Re = 600$, respectively. Note that the spiral vortices are inclined opposite to that in 2-vortex case.

For $\sigma = 0.14$, the range $900 \leq Re \leq 1600$ with $\Delta Re = 10$ was studied. The supercritical 1-vortex flow exists for $920 \leq Re \leq 1330$, and 2-vortex flow for $1000 \leq Re \leq 1600$. There is also an overlap Re regime between the two flow modes. We tried to simulate the spiral TG vortex flow at $Re = 1100$ with the same initial condition and angular acceleration as those in [6], but only obtained steady-state 1-vortex flow. It is noticed that Junk [2] gave $Re_s = 1245$ for $\sigma = 0.498$. One might expect that Re_s should be far above $Re = 1100$ for $\sigma = 0.14$ from Junk's stability analysis. This contradiction calls for further study.

4 Conclusions

Numerical solutions of the unsteady, three-dimensional Navier-Stokes equations were obtained to study the high Reynolds number spherical Couette flow. Narrow to medium gap cases with clearance ratio $\sigma = 0.06, 0.14$ and 0.18 were studied. The numerical results show that there is a range of Reynolds number where steady-state 1- and 2- vortex flow modes coexist for all the three gap widths chosen. Spiral and wavy Taylor-Görtler vortices appear at higher Reynolds numbers. It is also found that multiple unsteady flow modes with 2-,1- and 0-vortex, respectively, exist stably for $\sigma = 0.18$ at higher Reynolds numbers.

Acknowledgements: This work was supported by state key program for developing basic sciences (G1999032801) and National Natural Science Foundation of China (G10172089).

References

1. G. Dumas and A. Leonard: J. Comput. Phys. **111**, 205-219 (1994)
2. M. Junk, R. Hollerbach and C. Egbers: 'Numerical investigation of the flow in wide spherical gaps'. In: *12th International Couette-Taylor Workshop at Evanston, IL USA, September 6-8, 2001*
3. P. Marcus and L. Tuckerman: J. Fluid Mech. **185**, 1 (1987)
4. K. Nakabayashi: J. Fluid Mech. **132**, 209-230 (1983)
5. S. Rogers, J. Chang, D. Kwak: J. Comput. Phys. **73**, 364-379 (1987)
6. W. Sha and K. Nakabayashi: J. Fluid Mech. **431**, 323-345 (2001)
7. M. Wimmer: J. Fluid Mech. **78**, 317 (1976)
8. M. Wimmer: Prog. Aerosp. Sci. **25**, 43 (1988)
9. P. Wulf, C. Egbers and H. Rath: Phys. Fluids **11**, 1359-1372 (1999)
10. L. Yuan: J. Comput. Phys. **177**, 134-155 (2002)
11. L. Yuan, D.X. Fu and Y.W. Ma: Sci. China Ser. A **39**, 187 (1996)

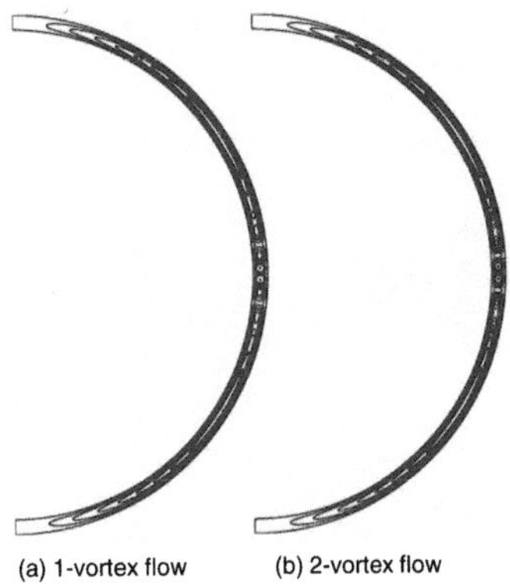

Fig. 1. Contours of meridional streamfunctions for steady 1- and 2-vortex flows, $\sigma = 0.06$, $Re = 3080$.

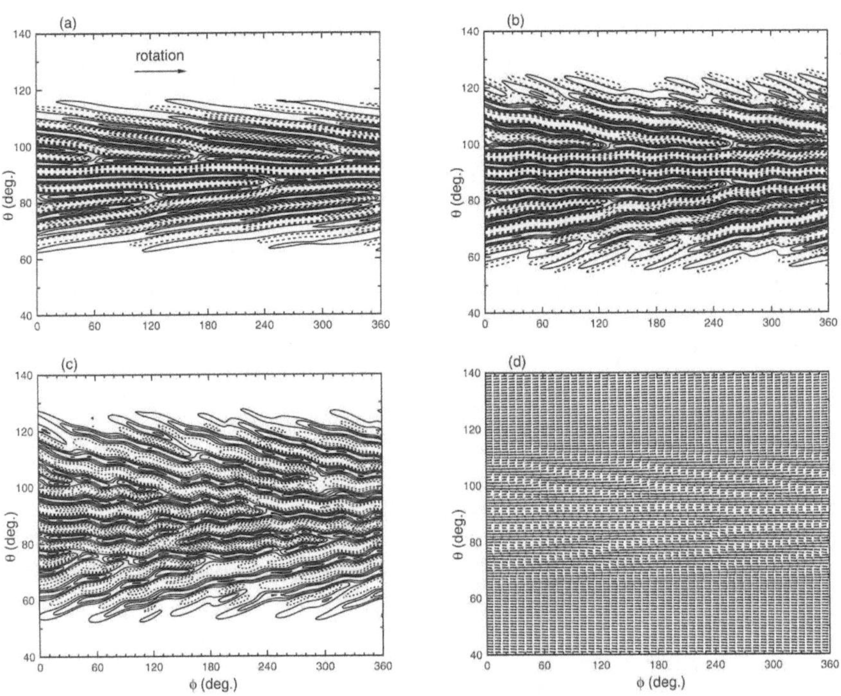

Fig. 2. Radial velocity contours in the $(\phi - \theta)$ spherical surface at the radial position $r = R_1 + 0.5\sigma$ for $\sigma = 0.06$: (a) $Re = 3300$; (b) $Re = 4000$; (c) $Re = 5000$;(d) v_ϕ and v_θ vectors for $Re = 3300$. In (a-c) solid lines are for $v_r \geq 0$, while dashed lines for $v_r < 0$.

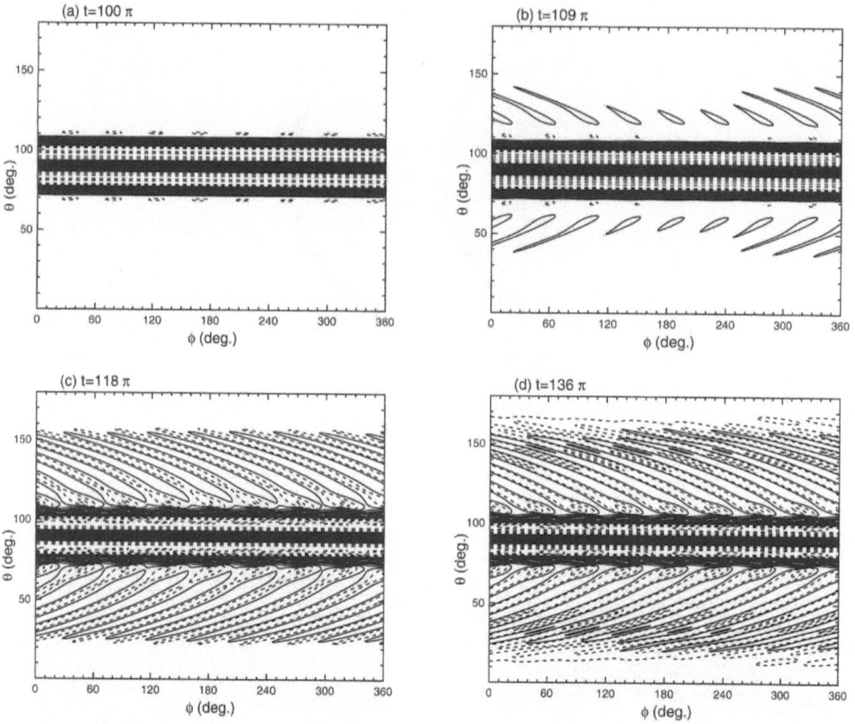

Fig. 3. Radial velocity contours in the $(\phi - \theta)$ spherical surface at the radial position $r = R_1 + 0.7\sigma$. The developed flow field has two TG vortex and 8 pairs of spiral vortices. $Re = 8000, \sigma = 0.18$.

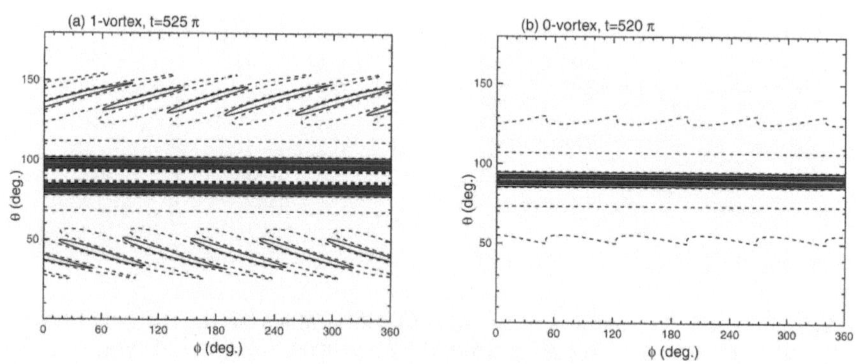

Fig. 4. Radial velocity contours in the $(\phi - \theta)$ spherical surface at the radial position $r = R_1 + 0.7\sigma$. The initial flow field are steady 1- and 0- vortex flows in (a) and (b), respectively. There are 5 pairs of spiral vortices in (a). $Re = 8000, \sigma = 0.18$.

Part XX

Technical Notes

A Numerical Model of the Dry Spinning Process

Joyce M Aitchison

Applied Mathematics & Operational Research Group, Cranfield University,
RMCS Shrivenham, Swindon SN6 8LA, United Kingdom.

1 Introduction

The dry spinning process is a standard method for the industrial production of synthetic fibres. The polymer is dissolved in a volatile solvent and the solution is pumped through a spinneret with many holes. After the filaments leave the spinneret, the liquid polymer is solidified by evaporating the solvent in a stream of hot air or inert gas. The fibres are then collected on a take-up wheel with a specified speed which also causes them to stretch.

This continuous production process can be modelled as a steady state flow. In this paper we describe a mathematical model and associated numerical algorithm for the calculation of the behaviour of a single axisymmetric polymer filament subject to evaporation and extension, representing that part of the flow between the exit of the spinneret and the take-up wheel.

The first mathematical model of this process is due to Ohzawa *et al* [3], who gave a detailed derivation of equations for each separate mechanism. However a significant number of simplifying assumptions were then made to the model. Partial solutions were calculated using variables which had been averaged over the cross section of the filament, and so only considered variations in the axial direction. Subsequently Ishihara *et al* [2] extended this model to allow the solvent concentration to vary with radial position, but all other variables were still radially averaged. This same level of simplification was also used by Kotani *et al* [4].

Brazinsky *et al* [1] used the same underlying model but introduced a radial dependence for solvent concentration and temperature. However, like [3], they ignored the radial velocity of the spin line and the axial component velocity was assumed to be flat. When this model is applied to realistic data representing cellulose acetate in acetone, there is a significant lack of conservation of polymer.

2 Revised model

In this paper we describe an extension of the model in [1], which is conservative. The diffusion of solvent and conduction of heat are modelled as non-linear partial differential equations in terms of the radial and axial position (r, z). Evaporation of the solvent takes place at the surface of the filament which is defined by the radius $R(z)$. The modelling of the problem as a steady state flow means that $R(z)$ is an unknown free surface rather than a moving interface, and must be determined as part of the solution.

The mechanics of the process are described by a constitutive equation, involving a highly nonlinear viscosity, and the conservation of mass and momentum. The system is modelled by assuming that both the axial velocity of the filament and the tension are functions of z only. A radial component of velocity is included in the model but then formally eliminated. This leads to a system of ordinary differential and algebraic equations which are linked to the diffusion equations above through the axial velocity, $V(z)$, and the radius, $R(z)$. The axial velocity is specified at both the inlet (extrusion from the spinneret) and the outlet (take-up wheel), but there are no boundary conditions for the tension.

3 Numerical solution

The previous studies [1,2,4] do not discuss the computational issues involved. However since those models reduced most features to one dimension, the computation would be fairly straightforward. In this paper we consider the computational problems associated with a more detailed model, in particular concentrating on the development of a robust but flexible algorithm with appropriate conservation properties.

The system of linked algebraic, ordinary and partial differential equations form a boundary value problem in the z-direction, with an unbalanced set of boundary conditions. Such a system frequently occurs when a continuous manufacturing process is described by a steady state model. Here we use a shooting method, solving an appropriate sequence of initial value problems, which aims to match the take-up speed at the outlet.

All the differential equations are approximated by finite differences in the z-direction, and the integration proceeds forward in a time-like manner, in the direction of flow. The variable radius of the filament occurring in the diffusion equations is handled by a coordinate transformation $\xi = r/R(z)$. Hence the interval $0 < r < R(z)$ becomes the fixed range $0 < \xi < 1$. The transformed equations are approximated by finite differences in the ξ-direction. However this does not resolve the difficulty that $R(z)$ is unknown.

In practice, the computation is not carried out on the whole length of the filament. Near the spinneret both $R(z)$ and $V(z)$ vary significantly (incidentally demonstrating the need to respect the conservation equations). However as the solvent evaporates and the filament solidifies they become effectively constant. Therefore the solution can be stopped at some point $z = L$ where no further changes take place. Typically $L \sim 2$ cm whereas $R(z) \sim 30$ μm.

References

1. I Brazinsky, A G Williams, H L LaNieve: Polymer Eng & Sci **15**, 834 (1975)
2. H Ishihara, K Tani, S Hayashi, H Ikeuchi: J Polymer Engineering **6**, 237 (1986)
3. Y Ohzawa, Y Nagano, T Matsuo: J Applied Polymer Science **13** 257 (1969)
4. T Kotani, S Hayashi, H Ishihara, T Matsui: J Macromol. Sci.-Phys **B31**, 65 (1992)

Transient Sulfur Dioxide Uptake by an Aerosol Droplet in a Convective Flow

Wei-Hsin Chen

Department of Environmental Engineering and Sanitation
Foo Yin University
Taliao, Kaohsiung Hsien, Taiwan 831, R.O.C.

Studies of sulfur dioxide absorbed by an aerosol droplet command both scientific and practical interest. On the scientific aspect the atmosphere is filled with aerosols in which the liquid water droplets dominate a considerable amount. For instance, when these droplets are small and suspended in the sky, they are called clouds or fogs. In contrast, if the droplets grow to a great extent, they fall to form rains or snows. For these water droplets, they can absorb atmospheric trace gases such as SO2 and NH3 [1,2], as well as collect and scavenge small solid particles [3,4]. Thus, they provide an important mechanism for cleaning the atmosphere. On the practical side it is known that spray towers and wet scrubbers have been widely applied in industrial processes to remove air pollutants, especially for sulfate. In those of the mentioned equipment, water is an important medium that is commonly utilized to wash flue gas, thereby transferring hazardous components from gas phase to liquid phase. The absorbing medium is frequently dispersed

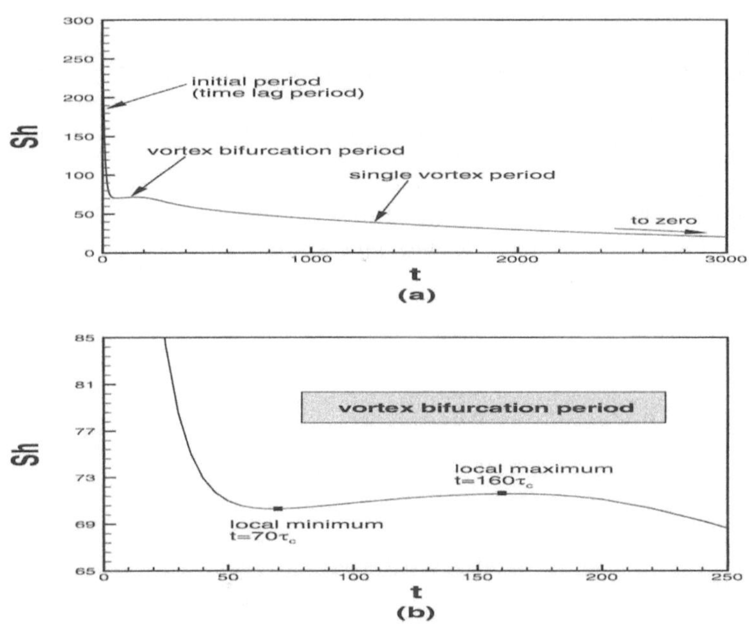

Fig. 1. Time distribution of the average Sherwood number (a) with increasing time and (b) within the vortex bifurcation period.

by means of spray in which water is in the form of droplets. In other words, the basic unit of the mass transfer between air pollutants and water is droplet absorption. Hence a detailed recognition on sulfur dioxide uptake into a water droplet also provides an engineering application. In this study, an analysis of transient sulfur dioxide uptake by an aerosol droplet, which suddenly exposes to an atmospheric flow, is performed numerically. Particular emphasis is placed on the interfacial momentum and mass transport phenomiena. When a water droplet is under a specific Reynolds number, say, Re=120, the numerical result indicates that there is a drastic interaction on the flow field between the gas phase and the liquid phase. Specifically, with increasing time the droplet internal flow proceeds from a single vortex, a double vortex, and then to a single vortex again. For the double vortex it is constructed by a primary and a secondary vortexes. Accordingly, the entire absorption process is divided into there sub-periods, consisting of the initial period, the vortex bifurcation period, and the single vortex period. Furthermore, it is found that the flow interaction between the two phases has a dramatic influence on the interfacial mass transfer rate, notably in the vortex bifurcation period. As a whole, on account of progressive decrease in droplet internal concentration gradient, the distribution of average Sherwood number decreases with time, except in the vortex bifurcation period in which a slightly undulate profile is characterized, as shown in Fig.1.

References

1. Seinfeld, J.H. 1986 Atmospheric Chemistry and Physics of Air Pollution. John Wiley & Sons, Chapter 5.
2. Seinfeld, J.H. and Pandis, S.N. 1998 Atmospheric Chemistry and Physics. Wiley, New York.
3. Ebert, P., Kibler, M., Mainka, A., Tenberken, B., Baechmann, K., Frank, G. & Tschiersch, J. 1998 A field study of particle scavenging by raindrops of different sizes using monodisperse trace aerosol, J. Aerosol Sci. 29, 173-186.
4. Hinds, W.C. 1999 Aerosol Technology: Properties, Behavior, and Measurement of Airborne Particles. John Wiley & Sons, pp.272-275.

Parallelism for Simulations of an Entire Shock Tunnel Facility

Richard J. Goozée and Peter A. Jacobs

The University of Queensland, Brisbane Qld 4072, Australia

Shock Tunnel Modelling

During the operation of a shock tunnel the growth of the viscous boundary layer on the tube wall causes attenuation of the shock and acceleration of the contact surface. This significantly reduces the test time and, because the boundary layer is often turbulent, results in noise being introduced into the test flow. The interaction of the reflected shock with the boundary layer and contact surface also introduces noise and can cause driver gas to be accelerated downstream, prematurely contaminating the test flow.

Simulations are being performed with the aim of identifying and modelling the sources of convected temperature fluctuations in a shock tunnel [1]. The multi-block CFD code MB_CNS [3], based on a finite-volume formulation of the compressible Navier-Stokes equations, is being used.

Until recently, the only practical method of modelling an entire shock tunnel was to use a quasi-one dimensional model. This type of model assumes that the flow only varies along the length of the tube and a mass loss submodel must be used to include the effect of the boundary layers on the core flow [2]. Axisymmetric simulations remove this assumption and allow investigations of noise generation mechanisms; however, the flow in these facilities is complex and significant resolution is required in the computational mesh.

Although the whole facility must be modelled, the mesh can be refined in order to focus the computational effort where it is needed. This is most important towards the walls of the tube to model the boundary layer with fewer cells across the rest of the tube.

Even with the high speed processors that are available, the requirement for large, very fine meshes means that parallel computation is desirable. MB_CNS has been parallelised using the shared memory approach of OpenMP and the message passing approach of MPI. Being a multi-block code, individual processors can be assigned to particular blocks, and only require information from the boundaries of adjacent blocks.

Three meshes were used for refinement studies: a coarse mesh with 80,850 cells, a medium mesh with 181,890 cells and a fine mesh with 323,400 cells. Full simulations running sequentially on the APAC National Facility used 56 hours, 119 hours and 225 hours of CPU time respectively. The performance of the code running in parallel, using OpenMP, was investigated by solving segments of the full simulations. The parallel efficiency of the code on an SGI Origin 3000 and a Compaq Alphaserver SC are shown in Figure 1.

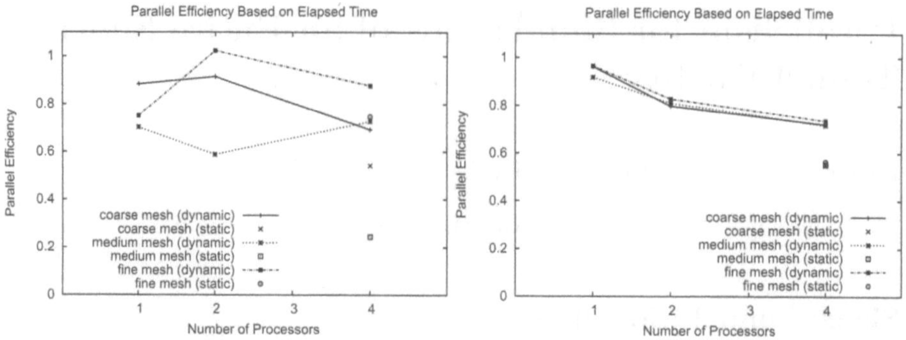

Fig. 1. Parallel efficiency of the OpenMP version of MB_CNS on an SGI Origin 3000 (left) and a Compaq Alphaserver SC (right)

Using OpenMP for multi-block CFD simulations is relatively straight-forward. Since all threads are reading from the same memory space, they can read the boundaries of the blocks being solved by other threads directly from the shared memory. All processors reading from the same memory space can place limits on the number of processors that can be used efficiently. The high interconnection speeds on newer shared memory computers lessen this effect.

In the MPI version of MB_CNS each block is solved in a separate memory space. Fluid properties required by neighbouring blocks must be sent and received explicitly between the processes; however, the volume of this data is quite small as only the data for the two row of cells along each neighbouring boundary is transferred. By not relying on physically accessing the same memory, for larger numbers of processors, MPI can become significantly more efficient than OpenMP [4]. Future development of the MPI version will allow computation and communication to be performed concurrently. This can be done by updating the properties of cells in the centre of blocks while the boundary conditions from the previous time step are communicated.

References

1. D. Buttsworth, P. Jacobs and R. Goozée: 'Numerical Simulations of Fluctuations in a Shock Tunnel'. In: *ICCFD2*, Sydney, Australia, July 15-18, 2002
2. C. J. Doolan and P. A. Jacobs: AIAA Journal **34**, 6 (1996)
3. P. A. Jacobs: Department of Mechanical Engineering Report 10/1996, The University of Queensland, Australia
4. R. J. Goozée and P. A. Jacobs: Department of Mechanical Engineering Report 08/2000, The University of Queensland, Australia

Numerical Analysis of Transpiration Cooling in Carbon Dioxide Atmosphere at Hypersonic Mach Numbers

P.S. Kulkarni, V.N. Kulkarni, B.R. Ravi and K.P.J. Reddy

Department of Aerospace Engineering Indian Institute of Science Bangalore-560 012, India Email: psk@aero.iisc.ernet.in

The 2-D Navier-Stokes equations coupled with species continuity equations are solved to compute flow fields along the flat plate in CO_2 atmosphere in the presence of transpiration cooling using helium gas. Variation of skin friction, heat transfer and growth of boundary layer along the flat plate have been estimated.

1 Introduction

The aerodynamic heating is a major design concern for vehicles flying at hypersonic Mach numbers. The cold fluid is injected into the boundary layer in transpiration cooling. Many experimental [1] as well as analytical and numerical studies have been reported in the literature [2]- [5]. The focus in recent times has been on the planet Mars. This paper presents numerical analysis of transpiraton cooling in carbon dioxide atmosphere which is dominant in Mars.

2 Governing Equations and Numerical Algorithm

The modified non-dimensional N-S equations in the conservative ingeral form are written as:

$$\frac{\partial \bar{U}}{\partial t} + \frac{1}{A} \oint_s \boldsymbol{Fn} dS = 0 \tag{1}$$

Where, $\bar{U} = \frac{1}{A} \int \int U dA$ is the average value of the column vector of conserved variables $U = [\rho, \rho u, \rho v, \rho E, \rho m_1]^T$ in the control volume and $\boldsymbol{F}$ is the flux vector defined in Ref. [4]. Here, ρ is the density, u and v are the velocity components in the x and y directions, E is the total energy of the mixture, A and S are the area and boundary of the control volume, m_1 is the mass fraction of the species 1 and $\boldsymbol{n}$ is the unit outward normal. The viscosity coefficient μ for the gas mixture is expressed as a funtion of temperature based on either Sutherland's law or Wilkes mixture rule [3].

3 Results And Discussions

The code has been validated for a flat plate at $M_\infty = 5.75$ in IISc hypersonic Shock Tunnel (HST2). The code is used to compute flow fields at $M_\infty = 6$ and $Re = 10^6$. The variation of normalized skin friction coefficient along the flat plate for different injecting gases is shown in Fig 1 1. The CO_2 is found to more effective in reducing the skin friction. The variation of temperature in the normal direction is shown in Fig 2 2. The results clearly show that the thermal boundary layer gets affected significantly due to blowing rate of Skin friction, Wall temperature and Boundary layer.

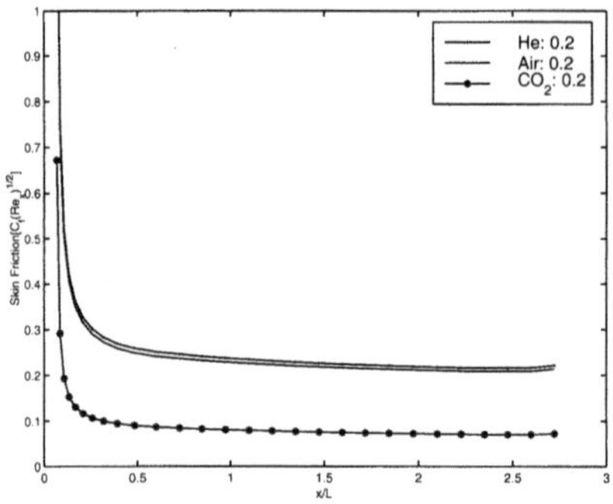

Fig. 1. Variation of skin friction for injection of different gases for M_∞=6.0, Re=10^6 m^{-1}

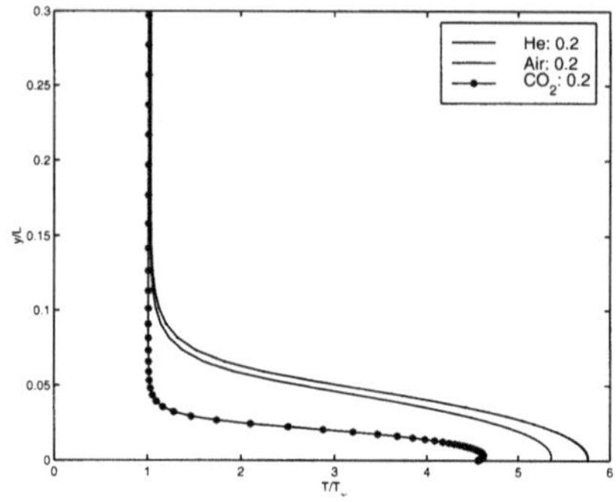

Fig. 2. Variation of temperature profiles for injection of different gases for M_∞=6.0, Re=10^6 m^{-1}

References

1. Libby PA, Cresci RJ. Experimental investigation of the downstream influence of stagnation point mass transfer. Jounal of Aeronautical Sciences, 1961,28:51-64
2. Hartnett JP, Eckert ERG. Mass transfer cooling in a laminar boundary layer with constant fluid properties. Trans. Amer. Society for Mechanical Engineers, 1957,79:247
3. Sreekanth, Reddy NM.Transpiration cooling analysis including binary diffusion using 2-D Navier-Stokes equations. AIAA Paper 94-2075(1994)
4. Ravi BR. Transpiration cooling analysis including binary diffusion using 2-D Navier-Stokes equations at hypersonic Mach numbers. MS dissertation, Dept. of Aero Engg., IISc, Bangalore, India (1997)
5. PS Kulkarni, Ravi BR and KPJ Reddy. Two dimensional Navier-Stokes solutions for transpiration cooling at hypersonic Mach numbers. 9th ACFM, Iran

Numerical Analysis of Unsteady Effects in Two-Phase Nozzles

U. Maatje and E. von Lavante

Institute of Turbomachinery, University of Essen, Germany

Pressure swirl atomizers are quite common due to their favorable atomization characteristics, low clogging tendency and geometrical simplicity: The fluid to be atomized enters through one, two or more tangential inlet ports, passes through the swirl chamber with a conical convergent part and exits through a cylindrical orifice. The high swirl velocity as well as the acceleration of the liquid through a convergent exit part produce an hollow conical liquid sheet as well as a gaseous core extending from the outlet through the swirl chamber mostly up to the top lid. After some distance downward the nozzle exit, the sheet becomes unstable and breaks up to form a spray of droplets. Although the geometrical characteristics are relatively simple, the inner flow pattern is rather complex. It has been observed repeatedly in experiments and computations that the corresponding flow in the swirl chamber and, consequently, in the liquid sheet as well, displays a significant unsteady component. With this background in mind, the present authors decided to carry out a detailed phenomenological numerical study of several models of typical pressurized simplex nozzles operated using water or glycerol/water mixture sraying into air. The present numerical computations were performed on conventional clusters of PCs using the commercial Navier-Stokes solvers **COMET** and **FLUENT**.

The numerical simulations were performed using two-dimensional axisymmetric and three-dimensional models of the simplex nozzles investigated by [1] and [2]. The swirl (tangential velocity component) was taken into account. The free-surface between the liquid and gaseous pahases has been modeled using an interface-tracking algorithm of VOF-type.

In Fig. 1, typical contours of the axial velocity obtained in the numerical simulations are shown. Regions of higher velocity in the liquid can be found near the air core and at the wall. Interestingly, approximately one half of the total flow passes through these areas, being in good agreement with the corresponding experimental observations by Cooper [2]. The liquid flows pass a concave wall, giving rise to hydrodynamic instabilities that result in the formation of Görtler vortices. Figure 1 displays a detailed view of one of such Görtler vortices, as visualized by pathlines coloured by the magnitude of the axial velocity.

In both nozzles, the interface between the liquid and the gas becomes unsteady, displaying waves of small amplitude along its surface (Fig. 1). The waves originate at the nozzle top, at the stagnation point on the lid, and propagate toward the exit. Their amplitude seems to decrease with increasing axial distance. After the exit, however, it increases again, causing at times a violent movement of the liquid sheet. For comparison purposes, the mean spray angle as resulting

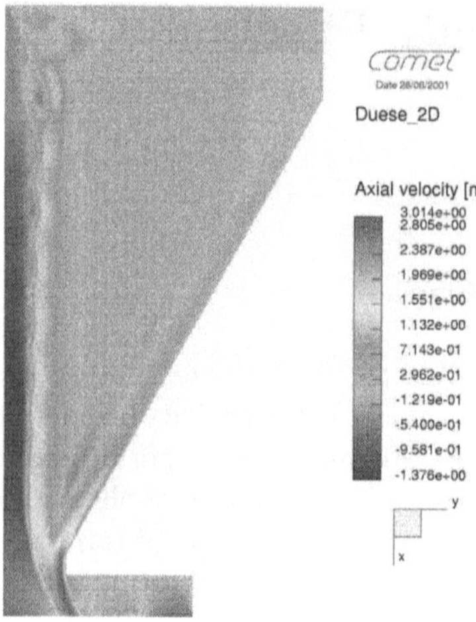

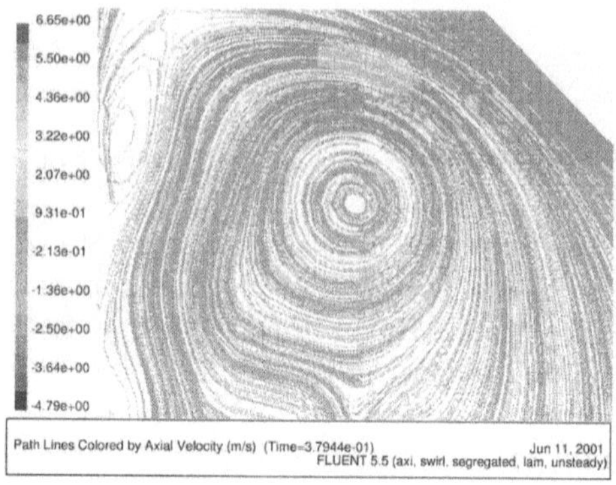

Fig. 1. Example of instantaneous contours of axial velocity within the large nozzle. **Fig.2.** [right] Görtler vortices in the swirl chamber of the small nozzle

from the computations was also determined. For the large nozzle, it was 84° - an exactly same value as mesuread. In the case of the small nozzle, the computed spray angle was 99.5°, the experimentally determined spray angle amounted to 100.8° . Again, the agreement can be considered good.

Conclusions The flow simulation in a large and a small model nozzles was carried out first in two dimensions using the assumption of axisymmetric flow. Subsequently, the simulation was repeated for the corresponding three-dimensional configurations. For this purpose, two commercially available computer codes were used, **FLUENT** and **COMET**. Both delivered very similar results. The flow at the air core interface has been shown to be unsteady, with waves generated at the nozzle top end and propagating towards the open end. From spectral analysis, we found out frequency values that are typical for observations made by [2]. The mean spray angles agreed also well with experimental data.

Acknowledgements: The authors thank gratefully the German Research Foundation (DFG) for its financial support under the grant Pe-130/13-3.

References

1. Yule, A.J. and Chinn, J.J:, The internal flow and exit conditions of pressure swirl atomizers, Atomization & Sprays, 2000.
2. Cooper, D., Chinn, J.J. and Yule, A.J., Experimental measurements and computational predictions of the internal flow field in a pressure swirl atomizer, ILASS-Europe 1999

Numerical Simulation of Three-Dimensional Flow around a Rectangular Cylinder

Zhiliang Ou and Liang Cheng

Department of Civil and Resource Engineering, University of Western Australia, Stirling Highway, Crawley, Western Australia 6009

1 Introduction

A vast amount of investigations has been carried out for the flow around a circular cylinder. While there are very few were concerned about flow around rectangular cylindrical structures. For flow around a rectangular cylinder, experimental results showed that flow patterns and hydrodynamical forces such as lift, drag and St are strongly dependent on both Reynolds number (Re) and the side ratio of the cylinder's crosssection. In the range of Re between 10^2 and 2.0×10^3, the St of a square cylinder shows a slight and continuous change around a constant value. Whereas for flow around a rectangular cylinder with $B/A = 2$, St varies with Re. Re has little influence on St when it's greater than 2000 [1,3]. Some researchers have conducted 2-D numerical simulations of flow around a rectangular cylinder in the past, 3-D simulation, however has not been developed. It has been found that a 2-D model can provide accurate results only for Re up to 300 [2,5]. Therefore 3-D simulations are necessary for flow around a rectangular cylinder at higher Re.

This paper presents a direct 3-D numerical simulation of flow around a rectangular cylinder in Re between 10^2 and 10^3. The model uses a finite difference method with a modified second-order upwind scheme to solve three-dimensional Navier-Stokes equations in a body-fitted coordinate system [4]. The flow domain used in the calculations is $22A \times 16A \times 3A$. The cylinder is located $7A$ from the inlet flow. A mesh of $130 \times 61 \times 40$ points in x, y, and z directions was used with the minimum radial grid size of $0.01A$ in $x - y$ plane for all the computations.

2 Numerical results

Fig. 1(a) shows the dependence of St on Re for a square cylinder. It exhibits that the St varies around a constant value ($St = 0.13$) for Re up to 1000. Flow around a square cylinder behaves in a very similar way to the flow around a circular cylinder. The main difference is the separation points are at the corners (both front and rear) of the cylinder. For lower Re, the flow separates at the rear corners of the cylinder while for higher Reynolds number the separation points are at the front corners. The St versus Re for flow around a rectangular cylinder with $B/A = 2$ is shown in Fig. 1(b). At lower Re such as $Re \leq 400$, the flow behaves similar to that around a square cylinder. There is a single sharp peak in the power spectrum of lift coefficient. At Re around 500, there is a discontinuity in the $St-Re$ curve. This is believed due to the flow reattachment on the cylinder surface. With further increase of Re, the separated flows detach themselves from the surface while the reattached flow forms an eddy and trails off the cylinder. Therefore the lower-frequency component becomes a dominating one while the higher-frequency component gradually disappeared in the power spectrum.

From the computation results described above, the following conclusions can be obtained.

1. St depends on the B/A ratio of rectangular cylinders.

2. The discontinuity in $St-Re$ curve for $B/A = 2$ as reported in experimental studies, is successfully simulated.

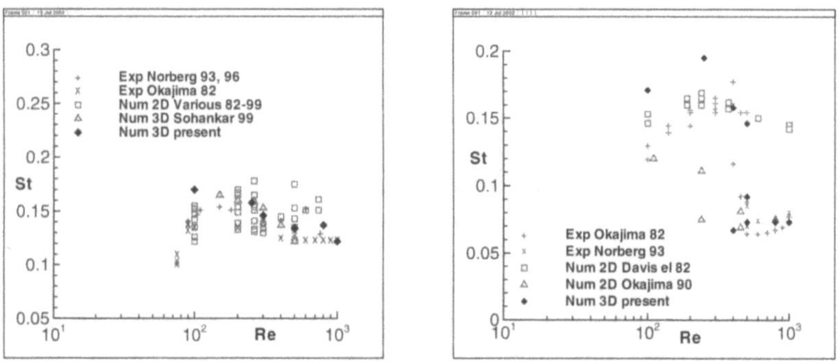

(a) (b)

Fig. 1.

References

1. T. Igarashi, Characteristics of flow around rectangular cylinders, Bulletin of JSME, **28**(242), 1690-1696 (1985)
2. C. Lei, L. Cheng, K. Kavanagh, Spanwise length effects on three-dimensional modelling of flow over a circular cylinder, Computer Methods in Applied Mechanics and Engineering, **190**(22-23), 2909-2923 (2001)
3. A. Okajima, Strouhal number of rectangular cylinders, Journal of Fluid Mechanics, **123**, 379-398 (1982)
4. Z. Ou, L. Cheng, Numerical simulation of three-dimensional flow around a vertical cylinder, The Proceedings of The First Asian and Pacific Coastal Engineering Conference, APACE 2001, **2**, pp. 711-720 (2001)
5. A. Sohankar, C. Norberg, L. Davidson, Numerical simulation of flow past a square cylinder, Proceeding of FEDSM99, 3rd ASME/JSME Joint Fluids Engineering Conference, July 1999, San Francisco, California, USA, FEDSM99-7172, pp. 1-6 (1999)

Investigation of Flow Conditioning in Pipes

G. Moniz Pereira[1], B. Mickan[1], E. von Lavante[2], and R. Kramer[1]

[1] Department of Fluid Mechanics, Physikalisch-Technische Bundesanstalt, Germany
Tel.:+49531 5921336, Fax: +49531 592691336, e-mail: gabriel.moniz-pereira@ptb.de
[2] Institute of Turbomachinery, University of Essen, Germany

Gas flow meters used in gas metering stations and in industry react differently to flow conditions in the inlet pipe section. Various pipe configurations such as bends, headers, T-fittings and armatures cause disturbed velocity profiles which can have an enormous unfavourable influence on meters. A research program initiated at the *Physikalisch-Technische Bundesanstalt (PTB)* investigates systematically the flow profiles of pre-defined installation configurations.

These investigations are not only of significant scientific interest, but also due to the huge volume of gas sales by the corresponding industry economically important. The measurement of the effects of various pipe configurations on the flow meter accuracy usually requires a set of custom-made steel pipes, making it sumptuary and financially an intensive undertaken. To simplify the investigations of the metrological behaviour of volumetric flow meters, a perturbation plate has been developed at the *PTB*, consisting of non-uniformly holes drilled at an angle to the axial direction to produce a circumferencial component.

To produce such a perforated plate, having certain properties such as generating swirl and deformation of the axial flow profile, a detailed knowledge of the physics in the perturbation-plate as well as local flow conditions is important.

Therefore the present authors have decided to gain knowledge of the complete flow field in the above configuration and determined to use a CFD numerical simulation, carried out using `ACHIEVE`, a parallelised Navier-Stokes CFD-Code developed at the Institute of Turbomachinery at the *University of Essen*.

This finite-volume upwind solver uses the Roe Flux-Difference Splitting procedure to carry out the spatial discretization of the governing equations. The interpolation slope of the primitive variables is limited by the modified van Albada limiter, according to a previously published *MUSCL* type procedure to obtain spatial discretization of second and third order.

The numerical simulation is validated by the highly accurate measurements of the flow profiles, realised at the test facility of installation effects at the PTB. In the configurations considered in the present study a detailed study of the phenomena of the flow was carried out.

The detailed measurements were very helpful for developing new empirical turbulence models in the present numerical simulation. Here, the emphasis was put on the correct simulation of highly separated flow regions, followed by low-pressure zone. Also the correct simulation of all the shear and boundary layers was of importance. The present study starts with a simple configuration that was gradually developed to the full configuration.

It was decided to proceed in three steps:

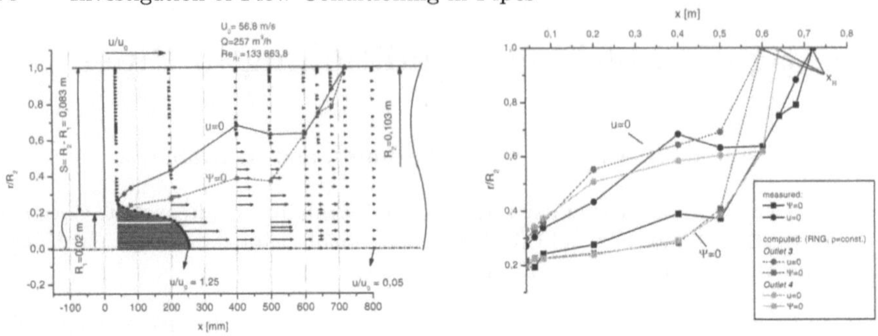

Fig. 1. Velocity profiles, reattachment, streamline $\Psi = 0$ and axial velocity $u = 0$

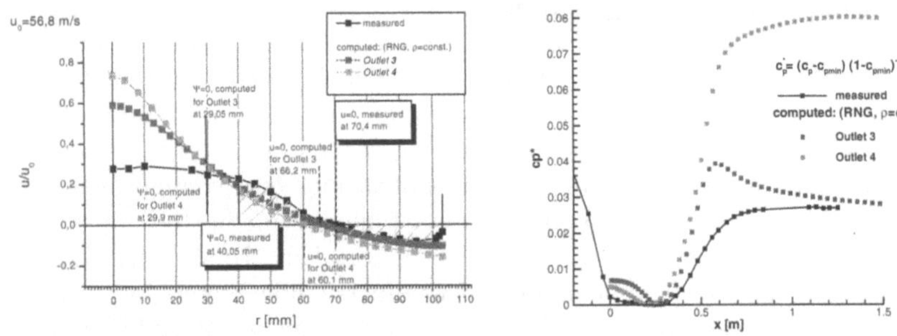

Fig. 2. Axial velocity profiles at 400mm downstream of the outlet and wall pressure

In the first step of this work the flow behaviour downstream of a step-like enlargement of a tube having a diameter of 40mm exhausting into a larger tube having a diameter of 206mm was examined.

In the second step the behaviour of the tube configuration from the first part will be combined using a bundle of two and four tubes. This way the interaction of multiple jets each issuing from a separate tube will be studied in detail. Here the main emphasis will be put on the interaction of the single jets with the shear layer.

In the last part the successive reduction of the tube diameter will produce a configuration that agrees with the physics of the original perturbation-plate.

In the issue the flow phenomena appearing downstream of the single tube will be discussed in detail, including formation of production of turbulence, reduction of the diameter of the turbulent jet and the fully developed flow profile in the large tube.

Conclusions. The present results show the necessity of improvement of turbulence modelling. The discrepancies concerning computed results depend on used turbulence models and mesh edge geometries. Whereas rounded edge geometry leans towards better agreement with measured results. Thus an improvement of numerical results is aimed by modelling fine scale turbulence using various subgrid-scale models by means of *LES*.

Outlet 3: mesh with 45° angle at the outlet, Outlet 4: equal to Outlet 3 but with rounded edges

A Combined SPH-experimental Systematic Approach to Resonant Free Surface Movements

Antonio Souto and Rosario Bravo

Naval Architecture Department,
Universidad Politécnica de Madrid. 28040 Madrid, Spain.

We describe the work done in the experimental tests and numerical simulation of the behavior of water in tanks. These tanks are subjected to a rolling motion around an axis. We vary systematically the amplitude of the rolling motion, the rolling frequency and the water level. With this oscillation motion we try to simulate the disturbance produced in the fluid by a vibrating structure because our intention is to develop a computational tool aimed at designing tuned liquid dampers for structures with vibration problems. We measure the moment produced by the water motion, and we analyze its harmonic components. We also record the images of the motion in order to compare free surface deformations.

To perform the simulations we use the Smoothed Particle Hydrodynamics (SPH) method [3,5]. With SPH it is feasible to model large free surface deformations and wave breaking. These phenomena appear in the experimental tests and for a conventional grid-type numerical technique it is not easy to cope with such large deformations, but SPH can try to model it as can be seen in figure 1. The most interesting free surface shapes occur when the rolling axis is far away from the tank. In these cases the motion can be very violent, even for low frequencies.

In SPH the fluid is represented by a set of particles which follow the fluid motion and advect fluid quantities such as mass and moment. For the time integration we employ a first-order Euler scheme. The particle positions are advected with the particle velocities, and the velocities with the particle accelerations obtained from the moment conservation equation. We use this first order scheme because it makes it easier to control the collisions with the tank walls. The collision time partially defines the time step and with an Euler scheme it is simple to reduce the time step in order not to collide with the limits. Since the time steps required for the collision criteria are very small, it is not worth having a higher order integration scheme.

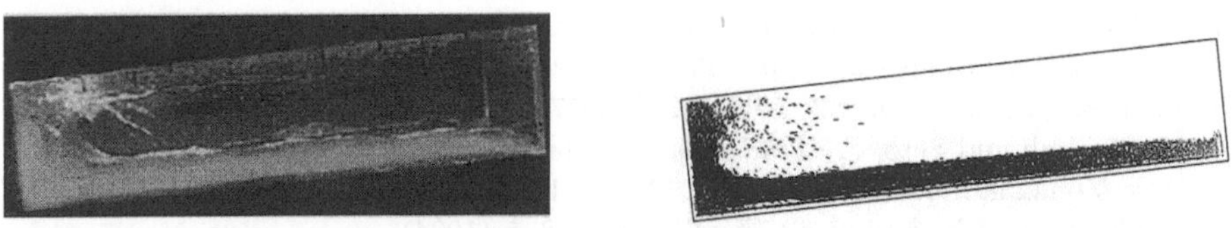

Fig. 1. Rectangular tank without baffles. Maximum Amplitude. Experimental test and simulation. 5E4 particles.

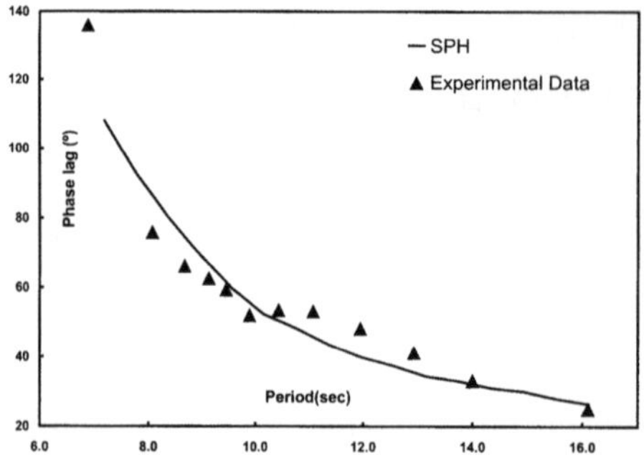

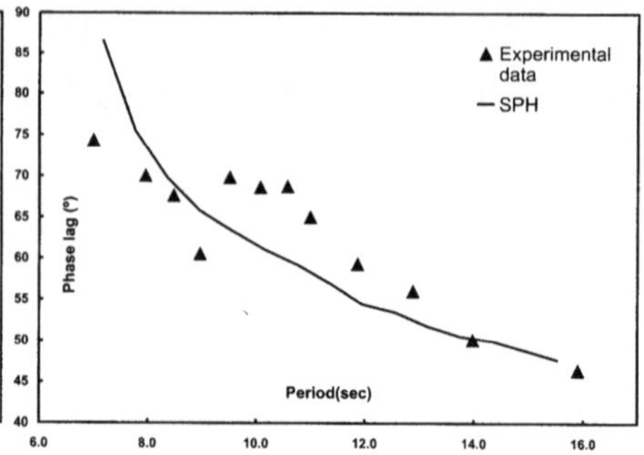

Fig. 2. Phase lag, rectangular tank without and with baffles, water level 0.75 m.

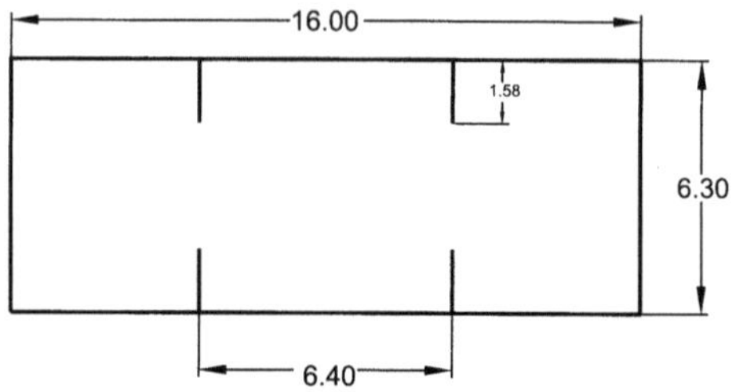

Fig. 3. Rectangular tank with baffles ground section.

We use for our experiments a rectangular tank. Many papers have reported on the sloshing phenomena in a rigid rectangular tank. Theoretical and experimental research has been carried out by means of the shallow water wave theory[1,4,6], boundary element methods[2], finite element methods, etc., but not much has been done regarding large free surface deformations, wave breaking, and relevant three dimensional effects. In our work, we deal not only with rectangular tanks, but with rectangular tanks with obstacles, see fig. 3, which can not be modelled by two dimensional approximations. In these cases strong non linearities appear in the water moment curves. We study these non linearities with detail, and try to simulate them. We compare phase lags and amplitudes of the first harmonic component (see fig. 2) with experimental data obtained in our facilities. Results are promising but further work has yet to be done.

References

1. T. Ikeda and N. Nakagawa, *J. Sound Vib.*, **201**, 1 (1997).
2. Z. Liu and Y. Huang, *J. Sound Vib.*, **175**, 2 (1994).
3. L. B. Lucy, *Astron. J.*, **82**, 1013-1024, (1977).
4. M. L. Modi and Seto, *J. Wind Eng. Ind. Aerod.*, **67-68**, 1-3 (1997).
5. J. J. Monaghan, *J. Comput. Phys*, **110**, 2 (1994).
6. L. M. Sun and Y. Fujino, *J. Fluid. Struct.*, **8**, 5 (1994).

Development of a Simple 2-D Steady State Model of the Flow from a Precessing Jet Nozzle

Elizabeth Smith, Bassam Dally, and Graham Nathan

The University of Adelaide, Adelaide 5005, Australia

Rotary kilns are used in the production of cement clinker, alumina, lime, and nickel. These kilns require high heat energy, with temperatures up to 1500^0C, usually supplied from the combustion of fossil fuels. Rotary kilns are a relatively inefficient process and there is much room for improvement in their design. One method for improving the combustion within a rotary kiln is the use of the precessing jet nozzle developed at The University of Adelaide [2]. The precessing jet nozzle is licensed under the name Gyro-Therm^{TM} and has been successfully applied to rotary kilns producing cement clinker, alumina, lime and zinc oxide, [1] and [5]. When firing natural gas in rotary kilns, a precessing jet provides a more luminous flame, lower peak temperatures, high flame stability and about a 50 percent less NOx emissions than comparable swirling and non-swirling flames.

To further the application of precessing jet (PJ) flows in industry, an improved understanding of the dominant mechanisms by which this complex flow influences a flame is required. Experimental work has given insight into the mixing and combustion characteristics of these flows. A CFD model of the precessing jet would allow assessment of the benefits and optimisation of an installation of PJ flows for a given application. However at present no CFD models of PJ flows exist. The present research program arises from the need for a CFD model of the precessing jet for industrial application.

The initial flow from a precessing jet nozzle is highly three-dimensional (3-D) and unsteady. Its initial rates of spreading and decay are much greater than those of simple jet flows. However most of the combustion occurs downstream from this highly unsteady region, where the unsteadiness in the flow is comparable with that in a simple jet. Modelling of the entire nozzle and flame in a rotary kiln using a 3-D, time dependent model would not presently be justified as an industrial design tool, due to the prohibitive computational expense and numerical difficulties. A more desirable means of modelling large industrial flows is to apply a robust and relatively simple turbulence model, such as a two-equation model, to simulate the flows. Assuming symmetry and using a two-dimensional (2-D) approximation further simplifies the model, making it more attractive to industry, and allows application of existing modelling techniques. Although this will involve some inevitable simplification of the initial region it is hoped it will provide a useful model of at least key aspects of the flow and flame.

It is known that the 2 equation k-ε model over-predicts the spreading and decay rate of a round jet flow by 40 percent [4]. To improve the accuracy of the k-ε model for solving round jet flows the turbulence constants $C_{\epsilon 1}$ and $C_{\epsilon 2}$ of the dissipation term may be modified, as they are responsible for the generation/destruction of the dissipation term. This technique of varying the turbulence constants to alter the spreading and decay rate of a round simple jet may be applied to the analysis of the precessing jet. That is, $C_{\epsilon 1}$ and $C_{\epsilon 2}$ can be varied to seek to match the spreading and decay rates of the precessing jet flow. The present study examines the possibility of matching the gross features of the flow emerging a PJ nozzle by applying this technique of varying the turbulence constants and modelling the flow as a two-dimensional steady state problem.

The approach chosen requires the use of experimental data with which to calibrate the choice of turbulence constants. The constants will be selected to meet one criterion and then assess the validity of their predictions with regard to other turbulence statistics and conditions. The experimental data selected for comparison was collected by [3] and provides scalar planer data of a confined PJ flow in a co-flow. The commercially available CFD program CFX 4.4 is used for all calculations. CFX uses a finite volume formulation over a structured mesh. The CFX program is commonly used throughout industry and accommodates user routines to enable changes to the default settings to be made.

Various combinations of turbulence constants were examined and since the flow has different rates of spread and decay in the near and far fields, different constants are required for each of these regions. Thus two sets of constants are required, one set for the near field and a different set for the far field. A user routine is incorporated to gradually introduce the transition from one set of constants to the other, in the region $x/D = 1$ to 1.35 (where D is the PJ nozzle diameter, $D = 38$ mm). The location of this region is consistent for most PJ flows.

The results obtained thus far show that a simple 2-D steady state round jet k-ε model can be adapted to provide reasonable prediction of the mean flow characteristics of the highly unsteady 3-D PJ flow. Further evaluation is required to assess the agreement of the higher order statistics and its usefulness in combustion applications.

References

1. C. Manias, A. Balendra, D. Retallack: 'New Combustion Technology for Lime Production' In: *World Cement*, December (1996)
2. G.J. Nathan, S.J. Hill, R.E. Luxton: *J. Fluid Mech.* **370** (1998)
3. J.J. Parham: Control and Optimisation of Mixing and Combustion from a Precessing Jet Nozzle. Ph.D. Thesis, The University of Adelaide, Australia (2000)
4. S.B. Pope: *AIAA Journal* **16**, 3 (1978)
5. D. Rapson, B. Stokes, S. Hill: 'Kiln Flame Shape Optimisation using a Gyro-Therm Gas Burner' In: *World Cement*, **26**, 7 (1995)

The Practical Application of Polyhedral Finite Volume Methodology to Problems with Large Scale Discrepancies

Ron Cresswell[1] and Paul Croaker[2]

[1] James Hardie Research Pty Ltd, 10 Colquhoun St, Rosehill, NSW, 2142
[2] MSC.Software Australia, Level 13 309 Pitt St, Sydney, NSW, 2000

Abstract. The application of polyhedral finite volumes to the study of buoyancy driven flow in a large warehouse will be presented. In this case, a hot storage pile provides both the topography and driving force required to generate upslope air currents and recirculate the air within the shed. Some air escapes through openings in the walls of a far smaller scale than the shed itself. The modelling of these openings with sufficient resolution to adequately measure total egress of air from the shed is made possible through the use of arbitrary interfaces where dissimilar meshes are matched at the junction through the subdivision of element faces. This results in elements with an arbitrary number of sides, but one to one matching of faces across the interface.

1 Summary

The problem of large scale discrepancies where high resolution is required at small scales within the context of a far larger scale flow pattern is one that may be tackled in a number of different ways. One method is to use nested grids where a high resolution mesh is imbedded within a low resolution larger scale mesh. This method in general requires the use of interpolation from the larger scale mesh to the smaller scale mesh, and it usually does not solve on the two meshes concurrently. In this paper we will discuss a method appropriate to unstructured meshes where the scale separation issue is addressed without interpolation. The commercial CFD code CFD-ACE is an unstructured finite volume code based on the finite volume method. All variables are held at cell centres resulting in considerable freedom in element choice. One practical implication of this is that cells may have any number of sides. This paper will discuss the application of the method of arbitrary interfaces, where dissimilar meshes may be brought together and non-matching faces sub-divided until 1:1 matching is achieved, thus allowing flow through the interface without the need for interpolation. Although intended to be used for ease of mesh generation, vastly differing scales can be incorporated into the mesh which would otherwise require an impractical number of elements to resolve both scales simultaneously. Buoyancy driven flow in a large building will be presented where the focus is on calculation of the total volume of flow leaving the building through small openings in the walls. The building considered is a warehouse approximately 185m x 20m x 30m with a large scale natural convection flow recirculating in the interior. The gaps in the walls vary in size from 1m x 2m down to narrow openings 10cm in width. The natural convection flow in the building sets up a vertical gradient of static pressure which drives flow out of openings towards the top of the building and draws flow into the building through openings lower down. The junction between the building mesh and the meshes representing the openings is handled using arbitrary interfaces and the mass flow is shown to be accurately calculated despite the large scale discrepancy. As can be seen in Fig. 1, the arbitrary interfaces perform a "cookie-cutter" operation on the adjoining faces. In this case, one triangular face of a tetrahedral cell becomes 17 new faces and the tetrahedral cell becomes a 20 sided polyhedral cell.

As the faces match 1:1 after the operation, the momentum equations are fully conservative across the interface. Each hexahedral cell in the smaller mesh (which is used to model the gap in the wall) sees the local pressure at the cell centre of the larger tetrahedral cell. This correctly predicts the momentum flux through the small cell, and the total mass flux of gas leaving the larger cell is also correctly accounted for. Hence the modification to the local pressure in the tetrahedral cell is also likely to be correctly predicted.

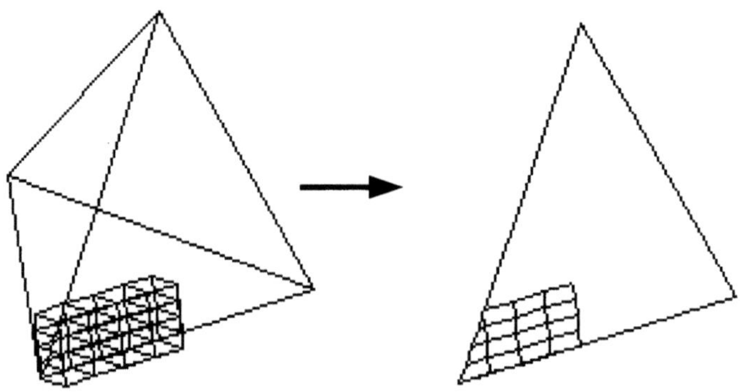

Fig. 1. The effect of arbitrary interfaces is to perform a "cookie-cutter" operation on the adjoining faces to achieve 1:1 face matching.

Transition of Strouhal Frequency Associated with Drag and Lift across Pitch to Diameter Ratio 1.8 of Square-Pitched Cylinder Array in Cross Flow

Hisashi Hishida[1], Takashi Adachi[1], and Hirotoshi Hishida[2]

[1] Saitama Institute of Technology, Okabe, Saitama 369-0293, JAPAN
[2] Nippon Steal Corporation, Nagoya R&D Laboratories, Tokai, Aichi 476-8686, JAPAN

1 Background of the current numerical work

Transient characteristics of the lift and the drag with the associated Strouhal frequencies have numerically been evaluated for 4×4 square-pitched circular cylinder array elastically supported inline to the cross flow as shown in Fig.1a with pitch P to diameter D ratios P/D =1.35, 1.5, 1.8, 2.1 and 2.5 under the Reynolds number Re=1000. The following numerical results were obtained [1]. (1) Between cylinders, gap mass flow rate defined in Fig.1b showed periodicity. The associated Strouhal frequency component of $St \cong 0.09$ was found dominant for the array with P/D <1.8 and $St \cong 0.2$ for P/D >2.0, respectively. For the array with P/D =1.8, duality in frequency components of $St \cong 0.09$ and 2.0 was seen . (2) The time-variation in drag and lift on a cylinder was found to be subject to the periodicity associated with the mass flow rate in gaps. (3) From cylinders in the array with P/D =2.1 and 2.5, vortex shedding was observed. (4) For cylinders in the array with P/D =1.8, vortices were observed. However, they remained inside the array until being replaced for the newly generated vortices. (5) No vortex was generated inside the array with P/D =1.35 and 1.5. (6) Cylinders in the outermost rows or the trailing column were affected by the peripheral pressure change with $St \cong 0.2$ due to vortices generated under the periodic switch in biasing direction of the outflow jet.

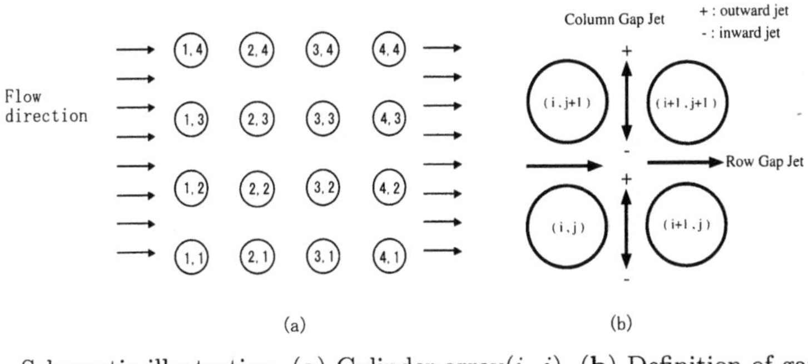

Fig. 1. Schematic illustration. (**a**) Cylinder array(i, j), (**b**) Definition of gap jets.

2 Objectives and results of the present work

The present numerical analyses are aimed to investigate how the Strouhal frequency associated with the non-dimensional gap mass flow rate undergoes transition across P/D =1.8 of the same square-pitched circular cylinder array under Re =1000. From figures 2a and 2b, it is seen that the Strouhal frequencies associated with row gap jet between cylinders (3,2) and (3,3) gradually shifts from $St \cong 0.09$ to 0.17 as P/D increases. It was numerically shown that vortices started to be created between the adjacent cylinders inside array with P/D =1.6 remaining almost still. As P/D increased, the periodic disappearance and creation was observed leading to vortex shedding at P/D =1.8 as shown in figures 3a and 3b, in consequence of which the transition in the Strouhal frequency associated with the non-dimensional gap mass flow rate as well as the drag and the lift was observed.

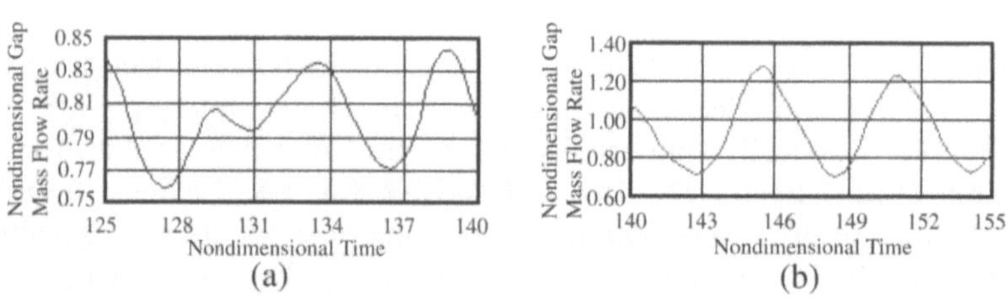

Fig. 2. Mass flow rate through gap between cylinders (3,2) and (3,3). (**a**) P/D =1.7, (**b**) P/D =1.9.

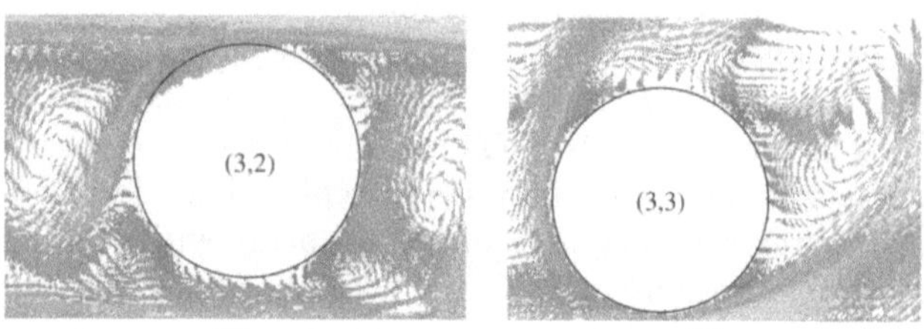

Fig. 3. Flow vector field. (**a**) t =133.5, P/D =1.7, (**b**) t =144.5, P/D =1.9.

References

1. H. Hishida, T. Adachi, S. Koike: Jpn. J. Comp. Fluid. Dyn. **11**, No.1, 112 (2002)

Investigation of Scalar Transport Closures Using Direct Numerical Simulation

Anthony Keating[1], Matthew Bilson[1], Klaus Bremhorst[1], and Srdjan Nesic[2]

[1] Department of Mechanical Engineering, The University of Queensland, Australia
[2] Department of Chemical Engineering, Ohio University, USA

1 Introduction

Direct numerical simulation (DNS) of turbulent scalar transport in fully developed, incompressible channel and pipe flow is used to test various turbulence modelling approaches. Three scalar transport closures are considered: the eddy diffusivity model, the generalized gradient diffusion hypothesis and the generalized equation for scalar variance.

Fully developed, incompressible, turbulent channel (pipe) flow, with scalar transport is computed using DNS. The Reynolds number is 2800 (2650), based on bulk velocity and channel half width (pipe radius). The Prandtl number is 0.71 (corresponding to heat transfer in air). Computational conditions are given in Table 1.

The Navier-Stokes equations are integrated in time using a fractional step technique and discretized in space with a fully conservative second-order finite volume scheme. An implicit Crank-Nicholson method is used for wall normal diffusion, while other terms are advanced using a second-order explicit Adams-Bashforth method. The flow is driven by a pressure gradient that is modified to keep a constant mass flux, while a constant wall flux is used for the scalar transport.

2 Results

Statistics are shown in Figure 1 and are compared to Eggels [1], Kawamura et al. [2] and Moser et al. [4]. Coefficients for scalar transport closures (see equations below) and comparison with commonly used values [3] are shown in Figure 2. The results show that all coefficients vary significantly in space.

$$\nu_t = C_\mu \frac{k^2}{\epsilon} \qquad \overline{u_i\phi} = -\frac{\nu_t}{Pr_t}\frac{dT}{dx_i}$$

$$\overline{u_i\phi} = -C_\theta \frac{k}{\epsilon}\overline{u_i u_j}\frac{\partial T}{\partial x_j} \qquad \overline{\phi\phi} = -C_t \frac{k}{\epsilon}\overline{u_i\phi}\frac{\partial T}{\partial x_i}$$

References

1. J.M. Eggels: PhD Thesis, Delft University of Technology, The Netherlands (1994)
2. H. Kawamura, K. Ohsaka, H. Abe, K. Yanamoto: Int. J. H. Fl. Flow **19**, 482 (1998)
3. B.E. Launder: J. Heat Trans. **110**, 1112 (1988)
4. R.D. Moser, J. Kim, N.N. Mansour: Phys. Fluids **11**, 943 (1999)

Table 1. Computational conditions

	Channel	Pipe
Re_b, Re_τ	$2800, 177.4$	$2650, 181.5$
Pr	0.71	0.71
Computational volume (x, y, z)	$4\pi\delta \times 2\delta \times 4/3\pi\delta$	$4\pi R \times R \times 2\pi$
Grid	$256 \times 128 \times 196$	$256 \times 68 \times 128$
Spatial resolution $(\Delta x^+, \Delta z^+, \Delta y^+_{min})$	8.8, 3.8, 0.12	7.0, 8.8, 0.12

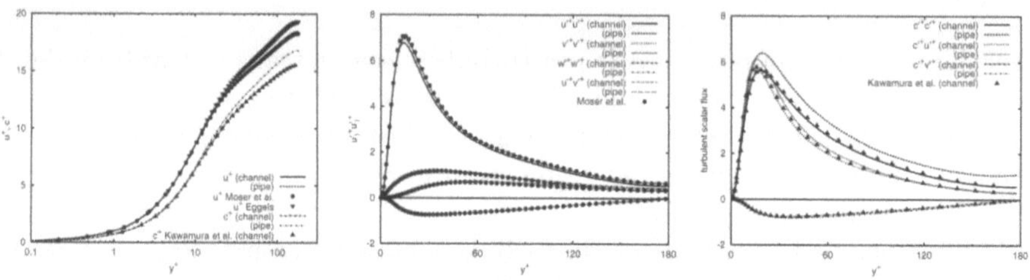

Fig. 1. First and second order statistics for velocity and scalar fields

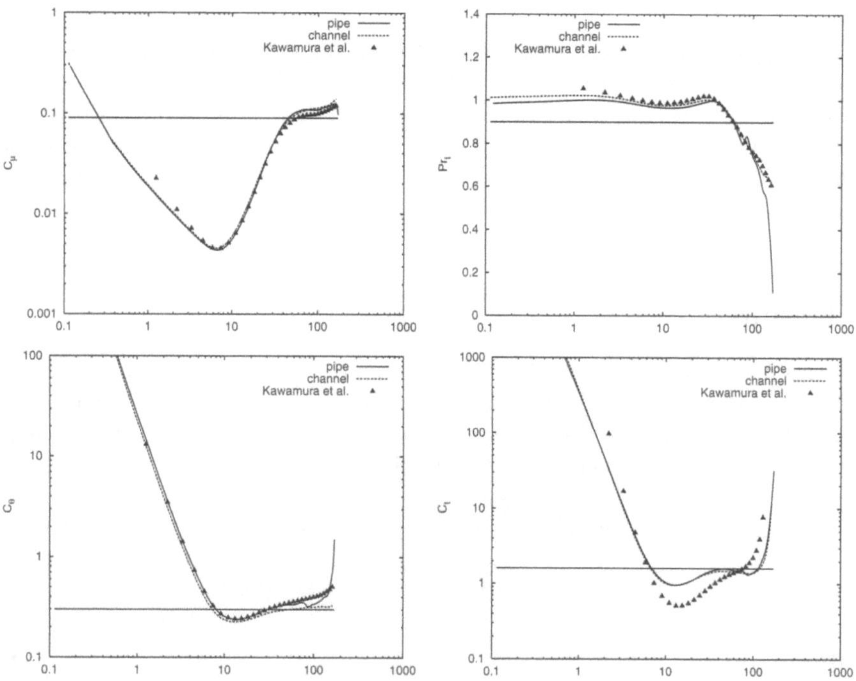

Fig. 2. Closure coefficients C_μ, Pr_t, C_θ and C_t

Implementation of Multiple Mapping Conditioning for Single Conserved Scalar

Andrew P. Wandel[1], Nigel S.A. Smith[2], and A.Y. Klimenko[1]

[1] Division of Mechanical Engineering,
The University of Queensland, St Lucia Qld 4072, Australia
[2] Aeronautical and Maritime Research Laboratory,
Defence Science and Technology Organisation, Melbourne Vic. 3207, Australia

Abstract. The Multiple Mapping Conditioning (MMC) model is a combination of probability density function (pdf) modelling and Conditional Moment Closure (CMC) and applicable to a wide range of turbulent flow and combustion situations. We are the first to implement the model and favourably compare the mixture fraction probability density function at various locations for an inhomogeneous mixing layer with those observed in experiments. The extension to multiple scalar flows will be future work.

Two models developed for turbulent combustion, probability density function (pdf) models [4] and Conditional Moment Closure (CMC) [2], can be applied to a wide range of flows. The former provide extensive data on the complete range of scales of turbulence and combustion at considerable computational expense, while the latter sacrifices some of this information, recovering some of this cost. Multiple Mapping Conditioning (MMC) [3] incorporates the ideas of both models by treating major species with pdf modelling techniques, while minor species are computed by CMC. The resulting model elegantly returns a single equation.

Pdf models are eminently suited for turbulence since velocity pdfs are joint-normal for homogeneous turbulence, and close to joint-normal in the inhomogeneous case. Species' properties are often highly non-Gaussian, so the Mapping Closure has been used with some success in homogeneous turbulence by modelling a Gaussian reference variable which is subsequently mapped to obtain the mixture fraction. Although it has been extended to multi-variate spaces, they were forced to be ordered, while MMC is applicable to non-homogeneous flows without differentiation between the dimensions. The implementation of MMC for all but the simplest cases necessitates a Lagrangian reference frame; the closure difficulty is transferred from the computation of complex double derivatives to a mixing process. Mixing models for particles almost exclusively select pairs from arbitrary points in the flow and average their properties with varying degrees of sophistication. We here instead utilise the expanding cloud process [5], where each particle represents a swarm of particles that diffuse over the mixing timescale, transporting the parent particle's properties and blending them with neighbouring particles in reference space. Only those particles which are physically located close together are allowed to mix; since particles do not diffuse in physical space, there is no mathematical error introduced by discretising physical space for determination of neighbours.

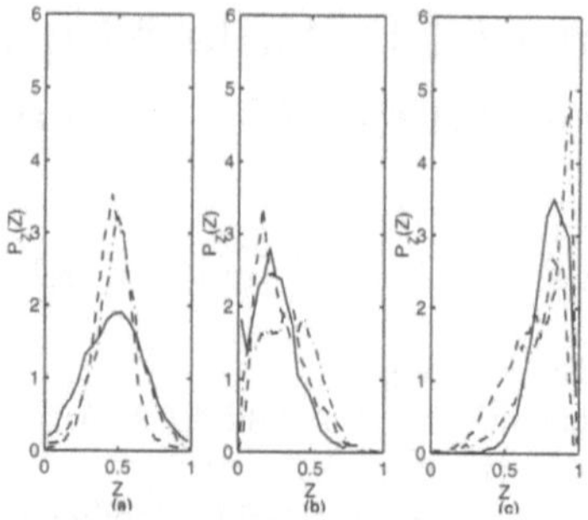

Fig. 1. Comparison of numerical and experimental [1] mixture fraction pdfs for various cross-stream positions; downstream location: 21 grid spacings. (a) $y/\delta \approx 0$, (b) $y/\delta \approx -0.3$, (c) $y/\delta \approx 0.45$. —, MMC; – –, $N_D = 0.3$; –·, $N_D = 1.81$

The testing of the model considered here is for the single-scalar case; this is the first implementation of a model of this type. Once it is validated for single-scalars, the model will be implemented for multiple scalars. The Lagrangian simulation of single scalars is inefficient, however it becomes significantly more efficient with each additional dimension considered. Tests were favourably performed on a homogeneous case with uniform pdf and constant dissipation. The model was then utilised to simulate a non-homogeneous mixing layer behind a turbulence-generating grid [1]. The standard Reynolds stress model was used to compute the mean and covariance properties of the flow, the Boussinesq approximation for the scalar statistics. Particles were injected into the flow downstream of the recirculations caused by the grid. The domain was meshed and averages defined for cells; particles' properties were then assigned. This is numerically more efficient than ensemble averaging of particles and similar to a particle-mesh algorithm previously studied [6].

Figure 1 compares the mixture fraction pdf computed by MMC with those determined experimentally (Fig. 12 [1]) for various cross-stream locations (δ is the cross-stream distance between mixture fraction values of 0.1 and 0.9). The MMC data is for 1000 particles per physical cell and ensemble averaged over three simulations. The centre-line MMC pdf has a higher variance than those measured, while the off-axis pdfs are narrower. This is likely due to the numerical pdfs taking non-zero boundary values.

We have successfully tested MMC's ability to produce reasonable distributions of the mixture fraction by using large numbers of particles. Future simulations will incorporate chemistry utilising an order of magnitude fewer particles.

References

1. R.W. Bilger, L.R. Saetran, L.V. Krishnamoorthy: J. Fluid Mech. **233**, 211 (1991)
2. A.Y. Klimenko, R.W. Bilger: Progr. Energy Combust. Sci. **25**, 595 (1999)
3. A.Y. Klimenko, S.B. Pope: Phys. Fluids (submitted) (2002)
4. S.B. Pope: Progr. Energy Combust. Sci. **11**, 119 (1985)
5. S.B. Pope: *Private Communication* (2002)
6. J. Xu, S.B. Pope: J. Comput. Phys. **152**, 192 (1999)

Micro-VOF: An Improved Free Surface Tracking Algorithm for Unsteady Free Surface Flow Problems

J.R. Whelan[1], M.R. Davis[1], and D.S. Holloway[1]

The University of Tasmania, GPO Box 252-65, Hobart, TAS 7001 Australia

An improved free surface tracking method (micro-VOF) is presented for use in numerical solutions of unsteady, incompressible free surface flow problems. Problems of this sort are governed by the Navier- Stokes and continuity equations. These equations maybe solved using an Eulerian mesh and appropriate finite difference approximations to the terms of the equations [1]. The evolution of the free surface must also be solved for as its spatial and temporal location is not known a priori. The method is validated against various test cases.

There are several methods of solving for the evolution of the free surface. These include marker particles [1] [2], a method which tracks the location of the fluid by the use of marker particles that move with the fluid. The drawback of this method is the large memory and computational requirements to store and flux the marker particles. Also it is difficult to reconstruct the exact location of the free surface. Height functions [3] [4] track the height of the free surface above some reference line as a function of location and time. This method is limited to problems in which the free surface is single valued. Line segment methods [5] use an ordered list of line segments to identify the free surface. These methods experience difficulties when two surfaces interact in an arbitrary manner and the list needs to be reordered. Surface marker and micro cell method [6] combine the ideas of marker particles and area flagging. Marker particles are placed on the free surface and are used to flag mesh cells (micro cells) that are finer than the computational mesh used for the solution of the pressure and velocity fields. This method only allows for the free surface pressure boundary condition to be applied to within the accuracy of the micro cells. The volume of fluid method [7] [8] uses a scalar function F that represents the fractional volume of fluid in a mesh cell. A mesh cell with $F = 1.0$ is considered full, with $F = 0.0$ is considered empty and if $0.0 < F < 1.0$ the mesh cell may contain a free surface. The scalar variable F moves with the fluid:

$$\frac{\partial F}{\partial t} + u\frac{\partial F}{\partial x} + v\frac{\partial F}{\partial y} = 0. \qquad (1)$$

The volume of fluid method has the advantages that the memory storage requirements are not excessive, the interaction of free surfaces is easily dealt with, and the method can be generalised to three dimensions. However considerable numerical diffusion occurs when standard finite difference schemes are used to model the spatial convection terms in equation 1. This leads to inaccurate results. [8].

The authors propose a method of solving equation 1 which uses third order upwind differencing to approximate the convection terms and integrates equation 1 over a mesh that is three times finer than the mesh used to solve for the pressure and velocity fields. The numeric diffusion of the free surface is limited to 1.5 pressure and velocity mesh cells in total. The free surface is located by linear interpolation within the F mesh. Correct application of the free surface pressure boundary condition is also possible and this adds to the accuracy of the solution method [5].

The method is validated against the collapsing column of water problem [7]. (Figure 1). The column of water is initially $1 \times a$ units wide and $2 \times a$ units high. Validation is also carried out against the run up of a Laitone solitary wave on a vertical wall [5] (figure 2). The still water depth is d, wave crest height H and run up height R. Results of both the test problems compare well with experimental results and other numerical results for these problems.

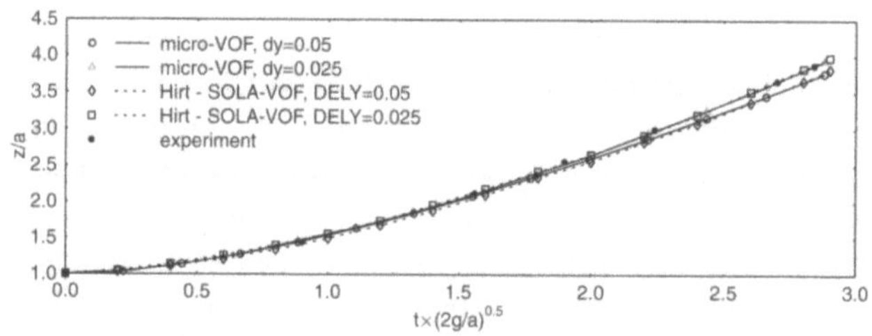

Fig. 1. Comparison of experimental and computer generated data for location of the surge front of a collapsing column of water

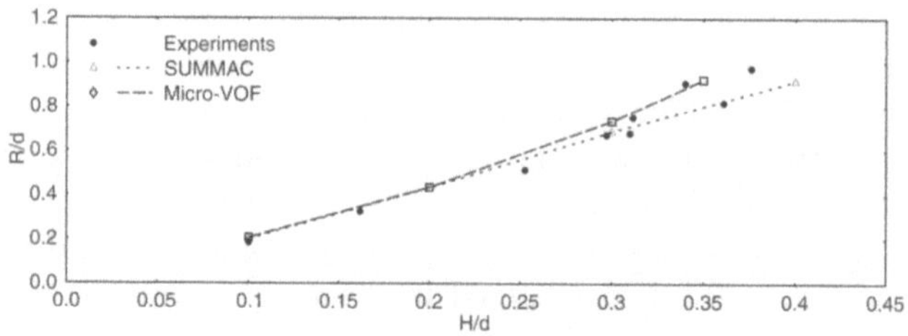

Fig. 2. Wave run up on a vertical Wall (R/d)

References

1. F.H. Harlow, E.J. Welch: The Physics of Fluids **8(12)**, 2182 (1965)
2. A.A. Amsden, F.H. Harlow: J. Comp. Phys. **6**, 322 (1970)
3. B.D. Nichols, C.W. Hirt: J. Comp. Phys. **12**, 234 (1976)
4. R.W. Lardner, Y. Song: Int. J. Numerical methods in Fluids **14**, 109 (1992)
5. R.K.-C. Chan, R.L. Street: J. Comp. Phys. **6**, 68 (1970)
6. P.E. Raad, S. Chen, D.B. Johnson: J. Fluids Engg. **117**, 683 (1995)
7. C.W. Hirt, B.D. Nichols: J. Comp. Phys. **39**, 201 (1981)
8. K.S. Varyani, R.M. Gatiganti: Ocean Engg. **27**, 729 (2000)

Performance of Geometry-Independent Second-Moment Closures in Application to Recirculating Flows

Sergey N. Yakovenko[1,2], C.H. Hsu[1], and K.C. Chang[1]

[1] Department of Aeronautics and Astronautics, National Cheng Kung University, 1 University Road, Tainan, Taiwan 70101, R.O.C.

[2] Lab of Turbulent Flows Modelling, Institute of Theoretical and Applied Mechanics, Siberian Branch of RAS, 4/1 Institutskaya Street, Novosibirsk 630090, Russia

Application of the traditional near-wall second-moment closure models to flows with complex geometry may encounter difficulties since they usually contain the wall-normal units. A number of geometry-independent models was developed recently (e.g. [1–3]) to overcome this problem. In this work, a backward-facing step flow with the developed turbulent boundary layer at inlet is chosen to examine applicability of several selective turbulence models through the comparison with the DNS data [4]. Tested models include the basic high-Reynolds-number Reynolds-stress model (RSM) [5,6] (denoted as LRR and SSG) which serve as a baseline for comparison, and two modified low-Reynolds-number RSM [1–3] containing no wall-normal units. The computation results show (Fig. 1,2) that the RSM developed by Shima [1] gives (except the skin friction coefficient after reattachment cross-section) the worst predictions since it was initially formulated for flows with smooth walls producing no recirculation, whereas the NNWRS model of So and Yuan [2] and SSG [6] yield, by and large, better results. The more careful assessment of NNWRS is performed through the several flow configurations and shows that two NNWRS versions (SY (1998) [2] and SY (1999) [3]) are still not able to yield the satisfactory predictions. On the other hand, the modified NNWRS model (SY-c) using the new coefficients $\alpha^* = -0.34$ and $\beta^* = 0$ which is proposed in this work can improve the predictions of either simple flows, such as a plane channel flow and a flat-plate boundary layer, or recirculating flow, such as the backstep flow, especially in the regions upstream of sharp edges or in the boundary-layer recovery regions (Fig. 1,2).

References

1. N. Shima: Int. J. Heat and Fluid Flow **19**, 549 (1998)
2. R.M.C. So, S.P. Yuan: Int. J. Engng Sci. **36**, 283 (1998).
3. R.M.C. So, S.P. Yuan: Int. J. Engng Sci. **37**, 33 (1999).
4. H. Le, P. Moin, J. Kim: J. Fluid Mech. **330**, 349 (1997).
5. B.E. Launder, P.G. Reece, W. Rodi: J. Fluid Mech. **68**, 537 (1975).
6. C.G. Speziale, S. Sarkar, T.B. Gatski: J. Fluid Mech. **227**, 245 (1991).

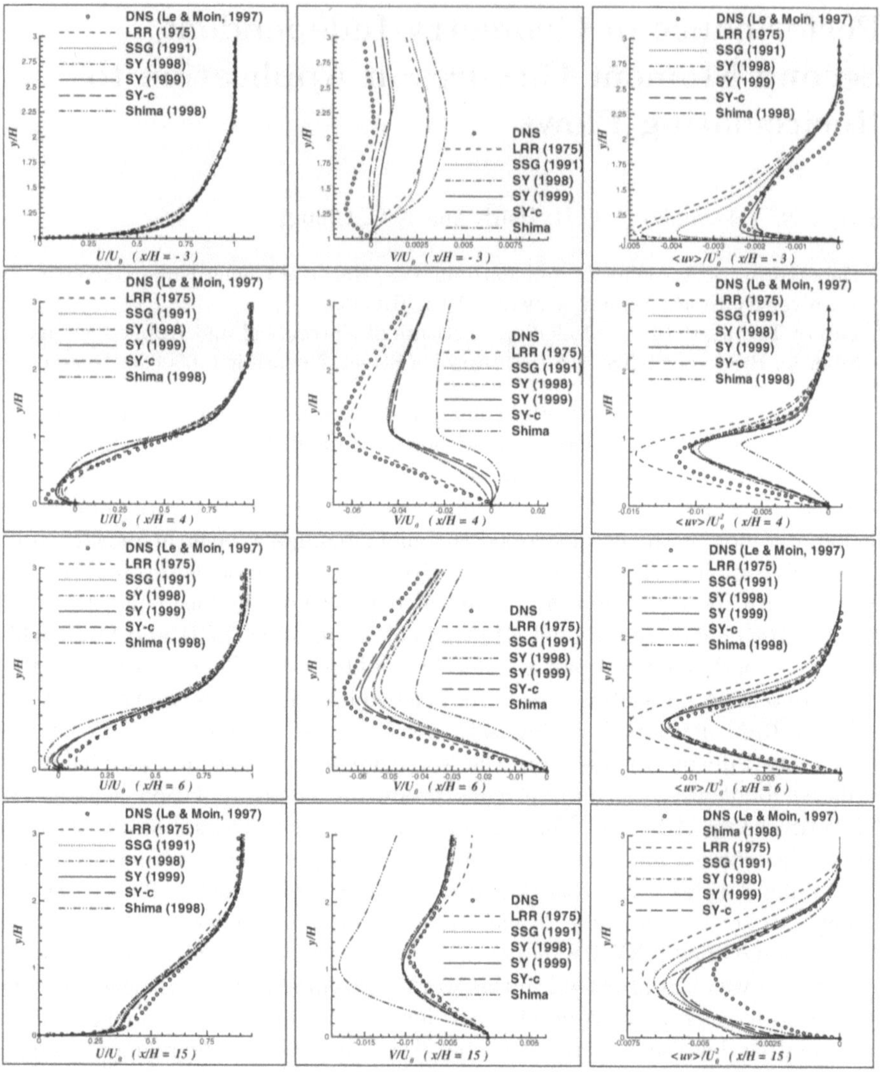

Fig. 1. Profiles upstream of step, in recirculation, reattachment and recovery zones

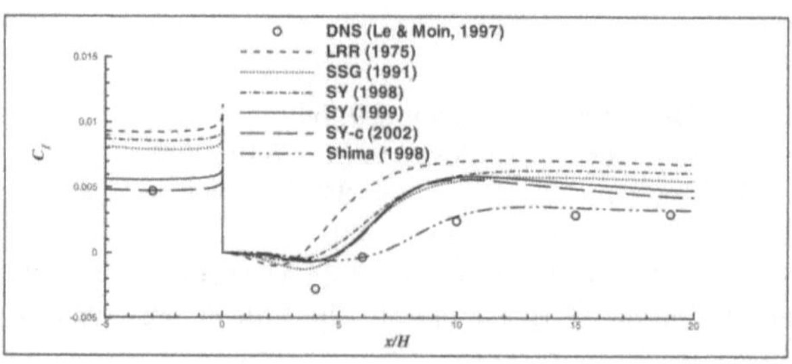

Fig. 2. Skin friction coefficients (C_f) computed in the backstep flow

Difference Approximation on Non-Uniform Mesh and Applications

Ma Yanwen,[1] Fu Dexun[2]

[1] Lab. of High Temperature Dynamics and
[2] Lab. of Nonlinear Mechanics, Institute of Mechanics, CAS, Beijing 100080, fudx@lnm.imech.ac.cn

Abstract. Some difference approximations on non-uniform mesh for traditional, compact and upwind compact schemes, and super compact schemes are presented.

1 Compact Difference Approximation

For solving practical problems in fluid dynamics usually physical parameters have large gradient in a smaller narrow subdomain. For this kind of problems it is better to use refined mesh for capturing the fine structures. There are two ways to obtain the schemes on refined mesh, the method of coordinate transformation and construction of schemes on non-uniform mesh. Analysis and practices show that it is better to use high order accurate scheme on non-uniform mesh near the wall.

Define

$$h_j = x_{j+1} - x_j, \theta_j = h_{j-1}/h_j, F_j = \left(\partial f/\partial x\right)_j, \delta_x^{\pm} f_j = \mp\left(f_j - f_{j\pm1}\right),$$

we can construct traditional and compact schemes on non-uniform mesh. Some compact schemes on non-uniform mesh are listed as follows

$$\alpha_j F_{j-1} + F_j + \gamma_j F_{j+1} = \left(a_j \delta_x^+/h_j + b_j \delta_x^-/h_{j-1}\right) f_j \tag{1}$$

4^{th}order central compact difference

$$\alpha_j = 1/\left[2\left(1+\theta_j\right)\right], \gamma_j = \theta_j/\left[2\left(1+\theta_j\right)\right], a_j = 3\theta_j/\left[2\left(1+\theta_j\right)\right], b_j = 3/\left[2\left(1+\theta_j\right)\right]$$

3rd order upwind compact difference(for $c > 0$)

$$\alpha_j = 1/\left(1+\theta_j\right), a_j = \theta_j^2/\left(1+\theta_j\right)^2, b_j = \left(2+3\theta_j\right)/\left(1+\theta_j\right)^2 \quad \gamma_j = 0$$

We can also construct 5th order accurate upwind compact difference approximation and Super Compact Finite Difference approximation[1,2]. One of the simpler super compact difference approximation is

$$-L\left(1/\theta_{j-1}\right)\overline{F}_{j-1}/2 + \left[L(1) + A\left(\theta_j\right)\right]\overline{F}_j - L\left(\theta_{j+1}\right)\overline{F}_{j+1}/2 = \left(\alpha\delta_x^+ + \beta\delta_x^-\right) f_j e_1/2 \tag{2}$$

where $\overline{F_j} = \left[f^{<1>}, f^{<2>}, \cdots, f^{<N>}\right]_j^T$, α and β are free parameters, L and A are $N \times N$ matrices, $\overline{F}$ and e_1 are N-dimensional vectors.$f_j^{<k>}/h_j^k$ approximates $\left(\partial^k f/\partial x^k\right)_j$with accuracy of order $N - K + 1$.

2 Analysis

Analysis shows that if variation of the mesh size from point to point is not very large for lower accurate schemes with small stencil the accuracy of solutions obtained by the scheme on non-uniform mesh and by coordonate transformation is almost the same. If the mesh can not be obtained by analitical transformation, or if high order accurate schemes with large stencil are used, especially near the wall, it is better to use schemes on non-uniform mesh. Suppose we have a function $u(x) = \sin(\pi x)$ defined on $[-1, 1]$ from which $\partial u/\partial x$ can be calculated. The non-uniform mesh is generated by $x_j = \tanh(b\eta_j)/\tanh(b)$ where $b = 3$, η_j=jH/N-1, H=2, N=160. The error is defined by $\varepsilon_j = F_j - \pi cos(\pi x_j)$. In Tab.1 the errors for 5th order upwind compact scheme based on non-uniform mesh and by using coordinate transformation near the boundary are given. We see that the results obtained by scheme construction on non-uniform mesh is much better.

Table 1. (5th Order Scheme)

ε_j	Coor. Trans.	Non-uniform Mesh
j=2	$+0.275^{-2}$	-0.108^{-6}
j=3	-0.153^{-4}	-0.972^{-12}
j=4	$+0.780^{-7}$	-0.435^{-15}

3 DNS of Fully Developed Compressible Channel Flow

A 5th order upwind compact difference based on non-uniform mesh is used to approximate the convection terms of the N-S equations, a 6th order central compact difference is used to approximate the viscous terms, and a 3 stage R-K method is used in advance of time[1]. In computational region $[2\pi, 2, 2\pi]$ $101\times140\times101$ points are used. The physical parameters used in computation are $Ma = 0.8$, $Re = 3300$. Density-weighted mean-stream velocity and RMS velocity fluctuation are very close to those of computed data for incompressible flow(Kim et al 1987) and compressible flow($Ma = 1.5$, $Ma = 3$, Coleman et al 1995). The skewness and flatness factors of velocity fluctuation in compressible turbulence are given. From the computed results we can see the difference of streamwise velocity distribution near the wall between the compressible and incompressible flows. The reason is that the compressibility effect makes a part of kinetic energy absorbed by pressure-dilatation and turned into inner energy which makes streamwise velocity streaks more smooth.

Acknowledgements: The work is supported by NKBRSF (G1999032805), NNSFC-10135010, and NSAF-10176033.

References

1. Li Xinlian, Ma Yanwen, and Fu Dexun, CFD J., 8(4): 536-543, 2000
2. Ma Yanwen, Fu Dexun, CFD J. 5(2): 259-276, 1995

Thermal Layer Instability in a Shallow Wedge Subject to Solar Radiation

Chengwang Lei[1], John C. Patterson[1], and Duncan E. Farrow[2]

[1] School of Engineering, James Cook University, Townsville, Australia 4811
[2] Mathematics and Statistics, Murdoch University, Perth, Australia 6150

1 Introduction

A model for radiation-induced natural convection in a water-filled wedge is described in [3]. With this model there is the potential for a thermal instability [2]. Flow visualisation revealed the primary form of the instability [5], which was confirmed by numerical simulations [4]. A corresponding scaling analysis has identified the critical condition and the controlling parameters as the Prandtl number Pr, the Grashof number Gr, and the cavity aspect ratio A [6].

Here, the stability properties are examined using the method of direct stability analysis [1]. A small amplitude perturbation is added to the heat flux condition on the sloping bottom as follows (after normalisation):

$$\frac{\partial T}{\partial n} = -e^{-A\eta x} + \varepsilon[\text{rand}(0,1) - 0.5] \quad (y = -Ax), \tag{1}$$

where the first term on the RHS quantifies the residual radiation and the second term is the artificial perturbation; ε specifies the intensity of the perturbation; rand(0, 1) generates numbers between 0 and 1. The governing equations are solved numerically. Results are obtained for $A = 0.1$ and $Pr = 7$ with $\varepsilon = 1.0\%$.

2 Numerical results

A transverse temperature profile in the thermal boundary layer indicates the flow becomes 3-D after the onset of the instability. The strength of the instability is quantified by the standard deviation of the temperature. A time series of the standard deviation provides an indication of the growth of the perturbation. An example is given in Fig. 1(a), showing three distinct regions: a constant-response region, an exponential-growth region and a connecting transitional region. In the first region, the perturbation is not amplified, indicating that the flow is stable. In the growth region, the standard deviation is represented by

$$T_d = ae^{c(t-t_B)} \tag{2}$$

where a is the amplification, c is the growth rate and t_B is the time for the onset of the instability.

Results are obtained for different Gr ranging from 10^3 to 2.5×10^6. No growth of the perturbation has been detected at $Gr < 5 \times 10^3$. At $Gr = 10^4$,

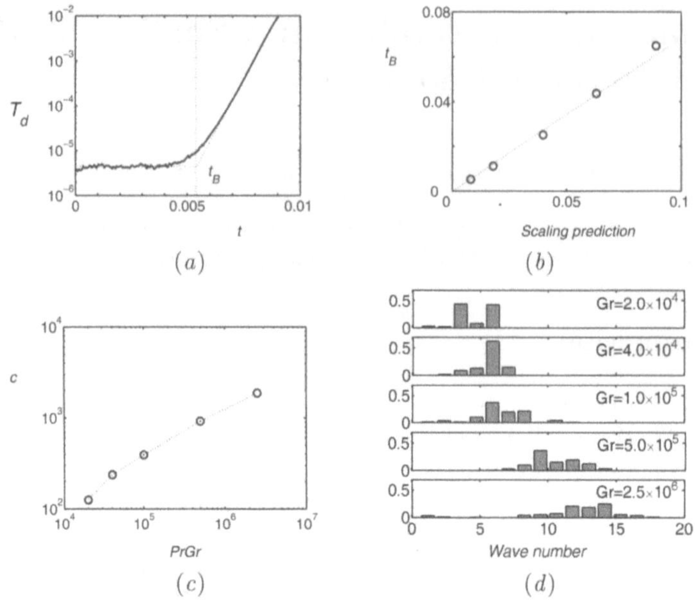

Fig. 1. (*a*) Growth of perturbation for $Gr = 2.5 \times 10^6$. (*b*) Critical time vs scaling prediction. (*c*) Growth rates. (*d*) Distribution of dominant wave numbers.

weak growth of the perturbation is detected. For $Gr > 2 \times 10^4$, the growth of the perturbation becomes clear and increasingly strong as Gr increases. These observations indicate that the critical Gr lies between 5×10^3 and 10^4, consistent with the previous scaling analysis [6].

For all Gr above the critical value, the instability properties can be obtained. Fig. 1(*b*) plots the calculated critical time against the scaling prediction [6], showing a linear dependence of the data set, which confirms the scaling prediction. The growth rates of the perturbation for different Gr are presented in Fig. 1(*c*). This figure clearly demonstrates that the growth rate increases with the Gr. Fig. 1(*d*) plots the distribution of the dominant wave numbers for different Gr, which approximately indicates the preferred mode of the convective instability. It is seen in Fig. 1(*d*) that the dominant wave number decreases with Gr.

References

1. S. Armfield, R. Janssen: Int J Heat Fluid Flow **17**, 539 (1996)
2. D.E. Farrow, J.C. Patterson: Int J Heat Mass Transfer **36**, 89 (1993)
3. D.E. Farrow, J.C. Patterson: Int J Heat Mass Transfer **37**, 1957 (1994)
4. C. Lei, J.C. Patterson. In: *Proc of The 14th Australasian Fluid Mech Conf, Australia, Dec 10-14, 2001*, ed. by B.B. Dally (Adelaide University 2001), pp. 359–362
5. C. Lei, J.C. Patterson: Exp Fluids **32**, 590 (2002)
6. C. Lei, J.C. Patterson: J Fluid Mech **460**, 181 (2002)

Turbulence Modeling Using Artificial Compressibility with Implicit LU-SGS Method

Lee Thong See, Liao Wei, and Low Hong Tong

National University of Singapore, Singapore 119260

Abstract. A numerical method is developed to solve steady and unsteady incompressible turbulent flows in a curvilinear coordinate system on basis of [1]. The solution procedure is based on the method of artificial compressibility and uses a decoupled approach to solve the Reynolds-averaged Navier-Stokes(RANS)equations and k-ω turbulence model equations [2]. The third-order flux-difference splitting upwind differencing scheme, which is built to satisfy TVD condition, is used to discretize the convective terms in order to deal with the steep gradients existing in the k, ω fields. A time-accurate scheme for unsteady terms is achieved by using an implicit real time discretization and a dual-time approach, which introduces pseudo-unsteady terms into both RANS equations and k-ω model equations. An efficient implicit algorithm LU-SGS is developed for the pseudo-compressibility formulation of the incompressible RANS equations and turbulence model equations for both steady and unsteady flows. To validate the present method, numerical solutions for steady flows in a channel and in a circular tube with a severe constriction and for unsteady flow in a circular pipe are presented and compared with other numerical results or experimental data [3] [4].

1. Fully developed steady channel flow for $Re_{\mathbf{H}} = 13750$

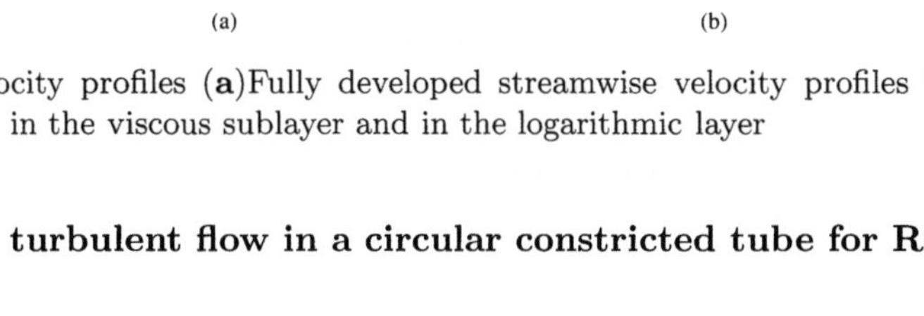

Fig. 1. Velocity profiles (**a**)Fully developed streamwise velocity profiles (**b**)Velocity distribution in the viscous sublayer and in the logarithmic layer

2. Steady turbulent flow in a circular constricted tube for Re=15000

Table 1. Separation and reattachment locations

	Separation, zs/r0	Reattachment, zr/r0
Experiment, Deshpande et al. [3]	0.4-0.5	4.0-4.5
Present, 221×41	0.41	4.35

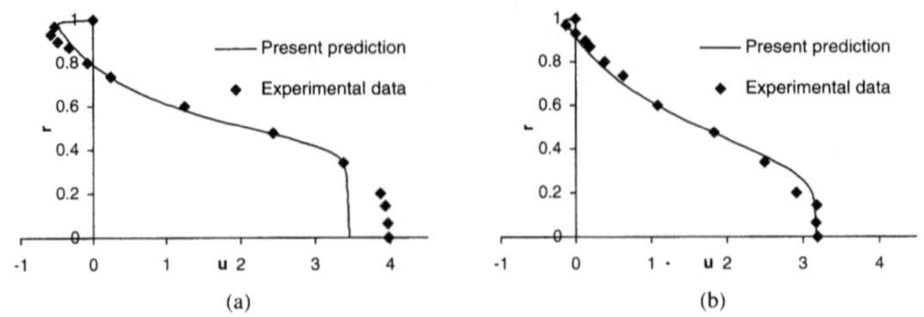

Fig. 2. Comparison of computed and measured [3]axial velocity profiles at different z-positions (**a**)z=2.5 (**b**)z=4

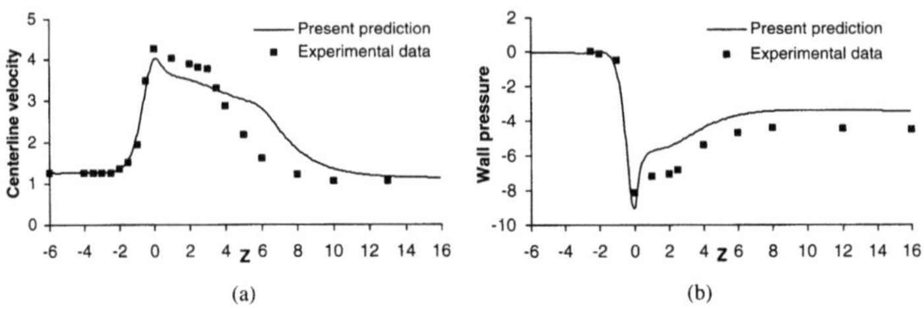

Fig. 3. Comparison of computed and measured [3]results (**a**) Distribution of centerline velocity (**b**) Wall static pressure through the constriction

3. Oscillating turbulent flow in a pipe for Re=19300 and Va=272

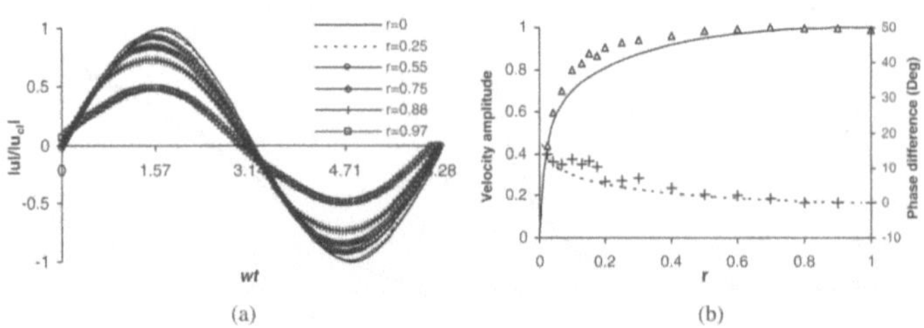

Fig. 4. (**a**) Turbulent axial velocity distribution versus crank angle during a cycle (**b**) Profile of the velocity amplitude and phase difference versus radius. Velocity amplitude—(*solid line*)present numerical data, (△)experimental data [4]; Phase difference—(*dashed line*) present numerical data, (+) experimental data [4]

References

1. T.S. Lee, W. Liao, H.T. Low: Int. J. Comp. Fluids Dynamics **15**, pp.197-208 (2001)
2. D.C. Wilcox: *Turbulence Modeling for CFD*, (DCW Industries, Inc., 1998)
3. M. D. Deshpande, D.P. Giddens: J. Fluid Mech. **97**, pp.65-89 (1980)
4. M. Ohmi, M. Iguchi, I. Urahata: Bulletin of JSME **25**, pp.536-543 (1982)

A Stable Numerical Method for the Time Dependent Navier-Stokes Equations and Its Application in Street Canyon Flows

Jiyang Xia[1] and Dennis Y.C. Leung[2]

[1] School of Computational Science and Information Technology, Florida State University, Tallahassee, Florida 32306-4210, USA
[2] Department of Mechanical Engineering, The University of Hong Kong, China

The flow over street canyon with different configurations is simulated using the time dependent Navier-Stokes equations. A three-step fractional method [1] is used to solve the velocity field and the pressure field separately from the governing equations. In the first step of the fractional method, the intermediate velocity components, which do not satisfy the continuity equation, are firstly computed from the momentum equations with the pressure terms dropped. Next, the pressure field is obtained from the pressure Poisson equation. In the third step, the velocities are updated based on the pressure obtained from the second step. An eight-node isoparametric element with bi-quadratic interpolation for velocity and bi-linear interpolation for pressure over an element is used to discretise the space domain. The convective terms are discretised using the second order Adams-Bashforth scheme. The Streamline Upwind Petrov-Galerkin (SUPG) technique is introduced to prevent numerical oscillations for flows at high Reynolds numbers. The wind field model is validated by simulating a backward-facing step flow within a wide range of Reynolds numbers. The dynamical characteristics of the simulated flow have excellent agreement with the experimental data [2] at Reynolds numbers lower than 600 and higher than 1500. For Reynolds numbers between 600 and 1500, it is generally difficult to predict precisely in a numerical way due to the complexity of the transitional flow.

The wind field model is then applied to simulate more complex flow phenomena in street canyons. Two typical building configurations, i.e. two identical buildings and multiple non-identical buildings, are used in the simulations. Three interesting findings are obtained from the present study, which worth noticing for dispersion modellers. 1) The flow at rooftop of buildings may not be assumed parallel to the ground as some numerical modellers did. A counter-clockwise rotating vortex may be found in street canyons even in a two-building configuration with an inflow from the left to right. This is because of the fact that the incident flow may separate at the front corner of the upstream building, generating a clockwise rotating vortex in the wake area above the street canyon, driving a counter-clockwise rotating vortex in the street canyon. Spectrum analysis confirms this hypothesis. The velocity frequency components in the street canyon is the same as those at the roof top while the magnitude of velocity in the street canyon is only half of that at the roof top. This implies that the flow regime in the street canyon is significantly affected by that at the roof top. 2) Increasing building height can increase velocity fluctuations in the street canyon under certain circumstances [3], which facilitate pollutant dispersion. This phenomenon is not reported in previous wind tunnel experiments largely because they focused on the effect of the height and width of street canyons on flow patterns. The buildings in most of experiments were set identical and may not reflect the complex building configurations in real world. It reveals a fact that the target street canyon cannot be isolated from the surrounding building settings when a dispersion model in urban street canyons is concerned. 3) At high Reynolds numbers, the flow regimes in street canyons do not change with inflow velocity. A strong vortex is formed in the street canyon in simulations when inflow velocities U_0 are set at 1.5 m/s and 6.0 m/s, respectively. The normalised maximum wind velocities U/U_0 in street canyons maintain at a constant value (0.26 ± 0.05) indicating that wind velocities in street canyons increases proportionally with the inflow velocity.

References

1. J. Xia and D.Y.C. Leung: J. Environ. Eng, **127**, pp. 369-376 (2001)
2. B.F. Armaly, F. Durst, J.C.F. Pereira, and B. Schönung: J. Fluid Mech. **127**, pp. 473-496 (1983)
3. J. Xia and D.Y.C. Leung: Atmospheric Environment, **35**, pp. 2033-2043 (2001)

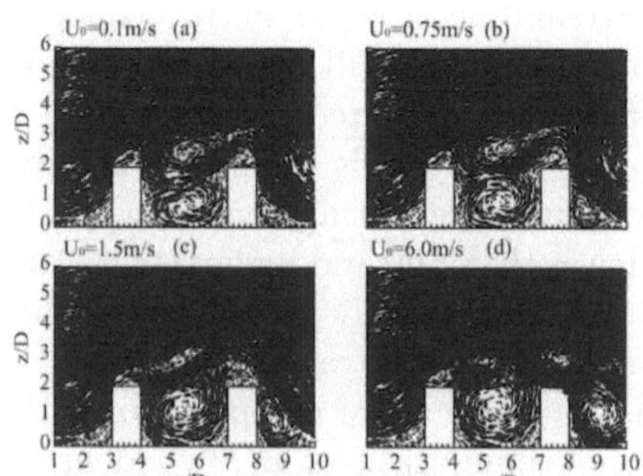

Fig. 1. Wind field around two identical buildings at different inflow velocities.

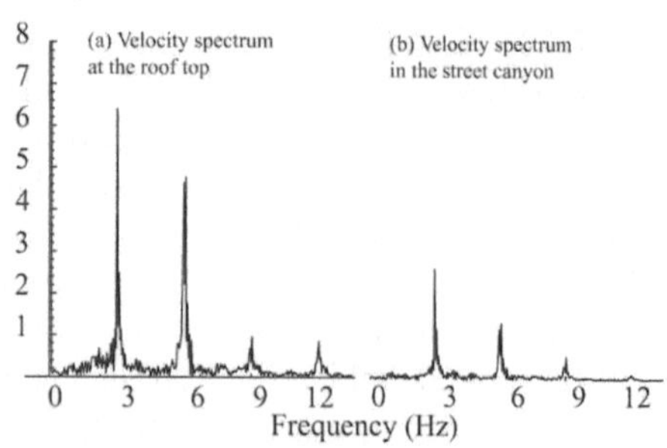

Fig. 2. Velocity spectrum at the roof top and in the street canyon

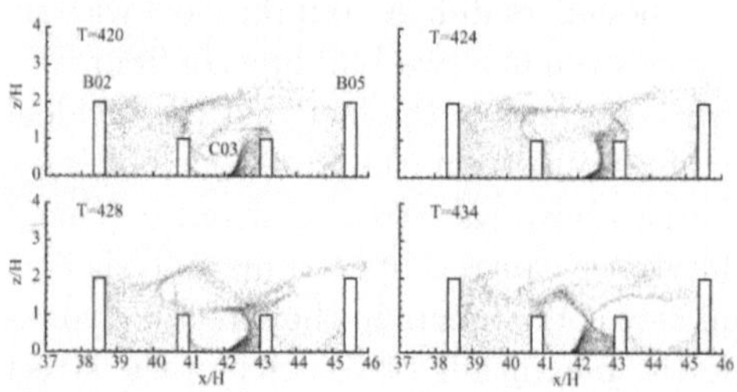

Fig. 3. Pollutant dispersion in a street canyon surrounded by high buildings.

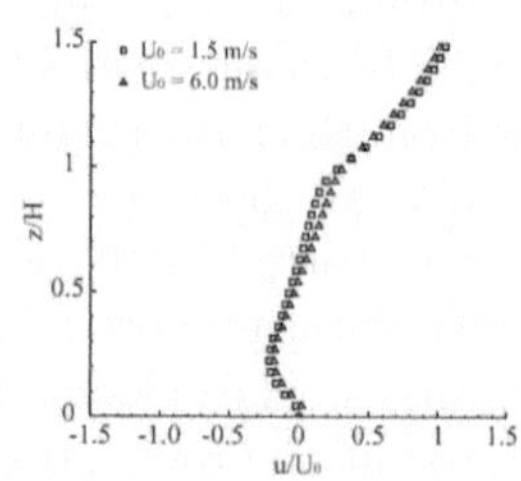

Fig. 4. Wind profile in a street canyon with different inflow velocities

A Finite Volume Least-Squares Method

Darryl Whitlow* and Jean-Jacques Chattot**

University of California, Davis

Abstract. A finite volume method is derived and applied to the solution of problems of incompressible flow. In particular, external inviscid flows and boundary-layer flows are examined. The method is a vertex-centered least-squares method and is second order accurate. Its quality is investigated for several types of inviscid flow problems and to solve the Prandtl boundary-layer equations over a flat plate.

1 Inviscid Incompressible Flow

This section concerns the numerical solution of Laplace's equation. Namely,

$$\Delta\phi = 0$$

for some potential ϕ. This is the governing equation of inviscid incompressible flow. Using $\phi_x = u$ and $\phi_y = v$ it can be written as the following equations:

$$\begin{aligned} 0 &= u_x + v_y \\ 0 &= v_x - u_y \end{aligned} \tag{1}$$

which are also known as the Cauchy-Riemann equations. The functional over the domain is

$$\mathcal{I} = \sum_T \frac{1}{2}\left((u_x + v_y)^2 + (v_x - u_y)^2\right) \Omega_T \tag{2}$$

where T indexes the triangular elements of the domain. A linear approximation is used for the unknowns u, v and the derivatives u_x, u_y, v_x, v_y are constant on each element. The functional is minimized to solve for the flow over a sinusoidal wall.

2 Boundary-Layer Flow

In dimensionless form, the Prandtl boundary-layer equations are:

$$u_x + v_y = 0 \tag{3}$$

$$u u_x + v u_y - \omega_y - \mathcal{U}\mathcal{U}_x = 0 \tag{4}$$

$$u_y - \omega = 0 \tag{5}$$

* e-mail:whitlow@math.ucdavis.edu

** e-mail:jjchattot@ucdavis.edu

The functional for the boundary-layer equations is

$$\mathcal{I} = \sum_T \frac{1}{2} \left((u_x + v_y)^2 + (uu_x + vu_y - \omega_y - \mathcal{U}\mathcal{U}_x)^2 \right.$$
$$\left. + (u_y - \omega)^2 \right) \Omega_T \tag{6}$$

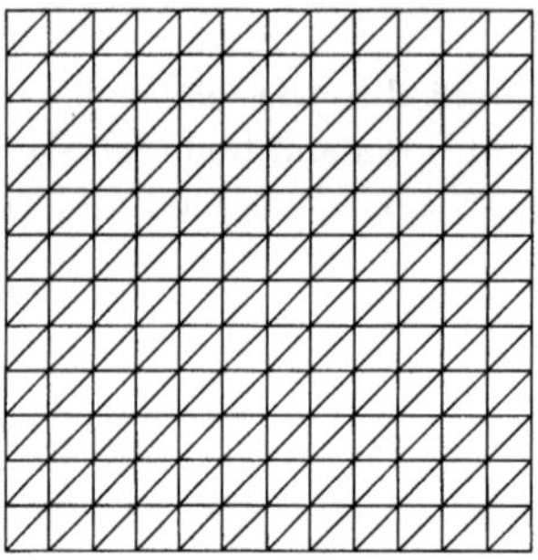

Fig. 1. Initial grid

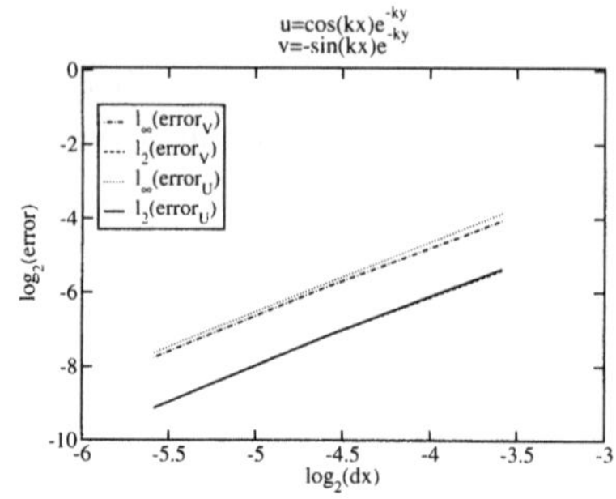

Fig. 2. Errors for the sinusoidal wall

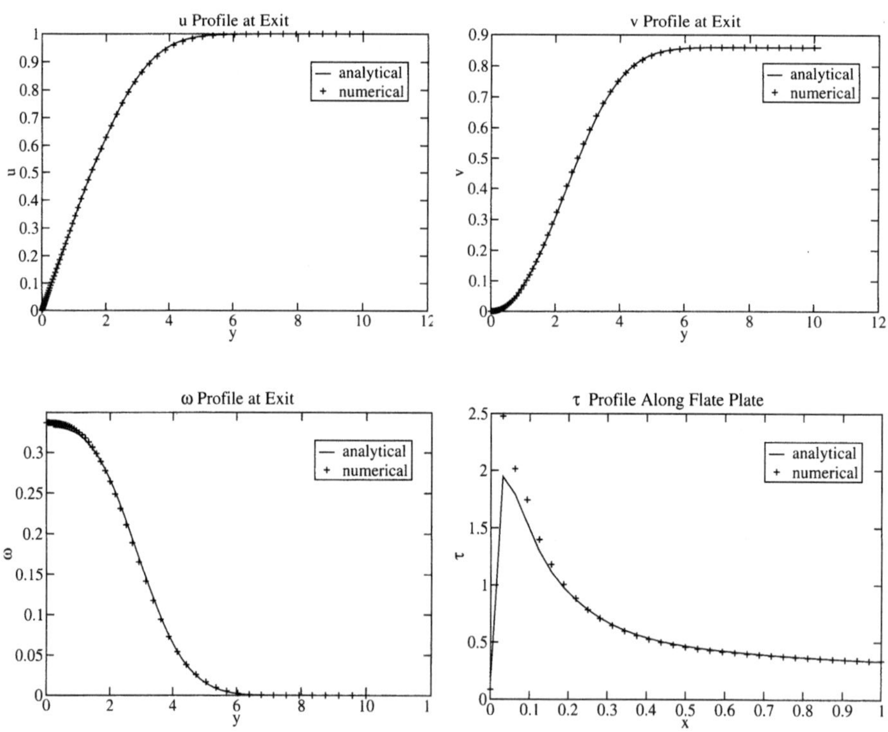

Fig. 3. Exit profiles for $u(y), v(y), \omega(y)$ and the wall shear stess $\tau(x)$

Development of A 3-D Laminar Compressible Navier-Stokes Solver

R. Balasubramanian[1] and N. Balakrishnan[2]

[1] Scientist, CFD Division, Defence Research & Development Laboratory, Hyderabad, India.
[2] Asst. Professor, Dept. of Aerospace Engineering, IISc., Bangalore, India.

1 Introduction

It is conventional to apply upwind methods for inviscid terms and central differencing schemes for the viscous terms for numerical simulation of compressible viscous flows. Differing from this, the KFVS [1] based discretization procedure in the form of Chapman-Enskog fluxes [2], treats both the inviscid and viscous terms by upwinding. Recently, a 2-D flowsolver has been developed and validated [3]. In the present work, the above approach has been extended to develop a 3-D Navier-Stokes solver which has been validated for a blunt cone-cylinder-flare configuration and the results are compared with results of another viscous code MB-EURANIUM [6]. The flowsolver CERANS works in conjunction with a versatile preprocessor which provides the necessary data to the solver in an unstructured format. Higher order spatial accuracy has been obtained using linear reconstruction of primitive variables. Convergence acceleration is obtained using LU-SGS [5].

2 Results and Discussions

The grid for this configuration (Fig.1) has been generated using AUTOELGRID [4] with two different grid sizes, namely, 97x41x45 (0.18 million) and 111x45x50 (0.25 million). The following test cases are considered for validation of the solver L-CERANS.

- *Case 1:*M_∞=8.0, $\alpha = 2^o$, Re=1.67 x 10^5, T_∞= 263.15 K, Adiabatic wall.
- *Case 2:*M_∞=8.0, $\alpha = 2^o$, Re=1.47 x 10^6, T_∞= 231.35 K, Adiabatic wall.

The C_p distribution (Fig.2) for test case 1 for both the grids are quite close except near the leeward side of the cylinder-flare junction where a separation bubble is seen. The overall results for both the test cases for both the grids are comparable except that the separation bubble size is smaller for grid 1 (Fig.3, Fig.5) and larger for grid 2 (Fig.4, Fig.6) but only slightly smaller than that obtained by MB-EURANIUM. The aerodynamic coefficients are compared with MB-EURANIUM as shown in Table 1. Since the size of the laminar separation bubble seems to grow as the grid size is increased, further grid refinement has to be made to obtain a better grid-independent results.

References

1. Mandal, J.C., and Deshpande, S.M., "Kinetic Flux Vector Splitting for Euler Equations",Computers and Fluids, Vol.23, No.2,pp 447-478,1996.
2. Manoj Kumar, B., Raghurama Rao, S.V., Balasubramanian, R. and Deshpande S.M., "Kinetic Flux Vector Splitting for the Navier-Stokes Equations based on Chapman-Enskog Distribution", FM report, FM9, IISc., Bangalore, India, 1997.
3. Anandhanarayanan, K., Nagarathinam, M. and Deshpande, S.M., "A KFVS based Navier-Stokes solver", 3rd CFD Symp. of AeSI, Bangalore,2000.
4. Krishnamurthy, R., and Anandhanarayanan, K., "Elliptic Grid Generation around Complex Configurations", ADA/DRDO-DERA Workshop on CFD, Bangalore,1997.
5. Sharov, D., and Nakahashi, K., "Reordering of Hybrid Unstructured Grids for Lower-Upper Symmetric Gauss-Siedel Computations", AIAAJ,36,No.3,1998.
6. Manoj T. Nair and Saxena, S.K., "Computation of Supersonic / Hypersonic Flow Past a Blunt Cone-Cylinder-Flare Configuration", PD-CF 0106, NAL, Bangalore, India,2001.

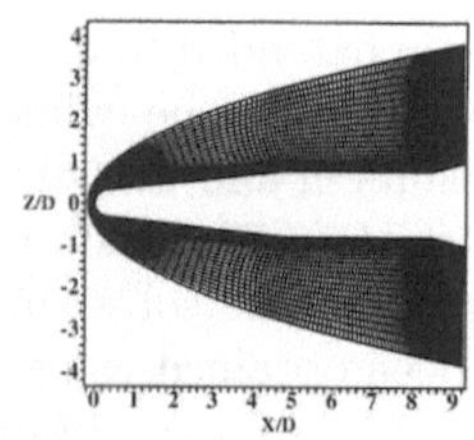

Fig. 1. Grid in Pitch Plane

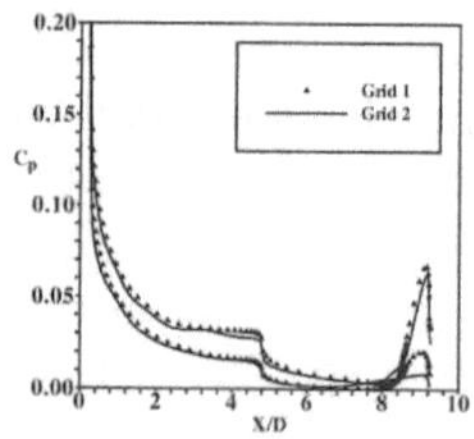

Fig. 2. C_p plot(Case 1)

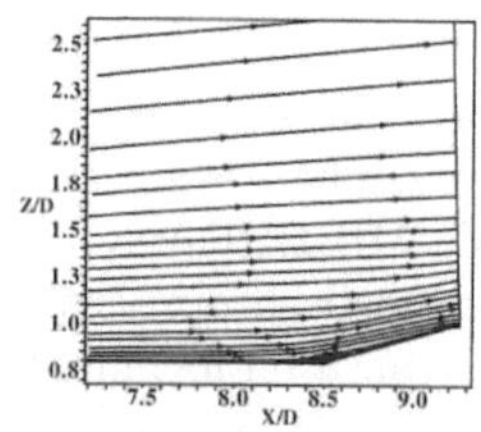

Fig. 3. Streamlines(Case 1,Grid 1)

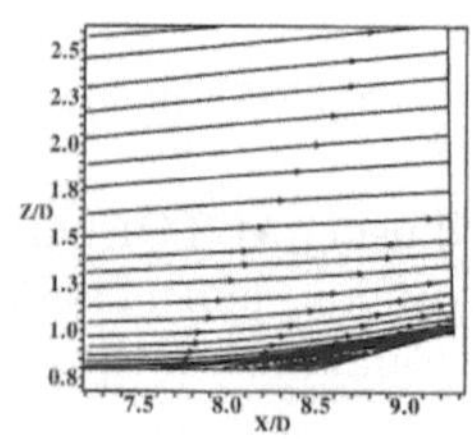

Fig. 4. Streamlines (Case 1,Grid 2)

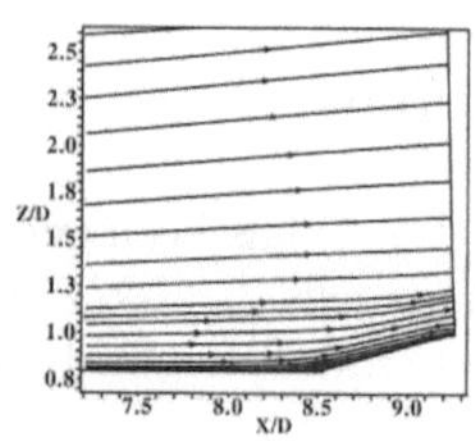

Fig. 5. Streamlines (Case 2,Grid 1)

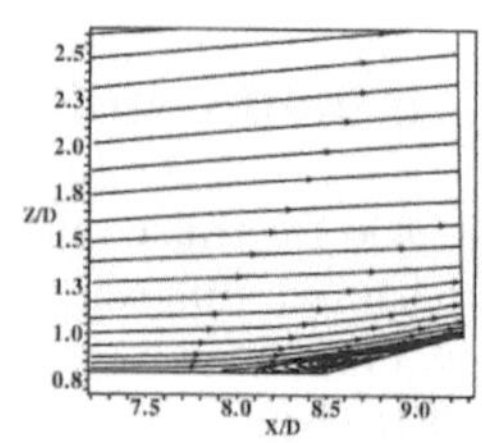

Fig. 6. Streamlines (Case 2,Grid 2)

Table 1. Comparison of aerodynamic coefficients

	Test case 1: Re=1.67x10^5			Test case 2: Re=1.47x10^6		
PARAMETER	Grid 1	Grid 2	Ref.[6]	Grid 1	Grid 2	Ref.[6]
C_A	0.1507	0.1394	0.1280	0.1430	0.1326	0.1205
C_N	0.0425	0.0418	0.0395	0.0441	0.0429	0.0485
C_m	-0.1074	-0.1060	-0.0952	-0.1163	-0.1124	-0.1288
X_{cp}/D	2.5278	2.5373	2.4105	2.6366	2.6205	2.8096

Energy Methods for Fluid-Structure Interaction Simulation

Pénélope Leyland[1], Bernd Hagenah[2],, and Volker Carstens[3]

[1] Institut d'Energie, STI, EPFL, CH-1015 Lausanne, Switzerland
[2] Phonak Systems, CH-8000 Zürich, Switzerland;
[3] DLR, D-37073 Göttingen, Germany

1 Introduction

Aeroelasticity is investigated in the time domain (non-linear) and in the frequency domain (linear). The algorithms for the Fluid-Structure Interaction procedure are analysed using an energy analysis. Accurate prediction of the interaction is shown to depend upon the consistency of interface boundary conditions with the time integration of the equations, i.e. the evaluation of the interaction as a fully coupled system allows less non-physical energy dissipation. This is important for obtaining confidence in for instance flutter boundary prediction. When time integration is pushed further, limit cycle behaviour can be found.

Frequency domain techniques are based on linearised equations and assume a linear relationship between aerodynamic loads and structural deformation. They are particulary adapted for mistuning investigations and subsonic flow regimes. When non-linear effects are important as for example vibrating shocks in transonic regimes or nonlinearities within the structure properties, then time domain methods are prefered.

2 Algorithms

The Fluid and the Structure systems can be solved separately or together, with appropriate interface conditions. The geometrical representation and the numerical techniques for CFD and CSM are not usually the sane and coupling algorithms of the separate methods are necessary. These algorithms, and the individual methods fro the fluid and teh structure, should

- maintain energy transfer conservation
- be time consistent
- be time accuracy

There are two main methods : firstly, the so-called **Staggered:** technique, whereby the fluid and the structure are solved separately, updated at specific points and subcycling within one of the (usuallly fluid) solvers can be done. However, this leads to a time delay at the moment of updates and can give non-physical energy dissipation, [1].

To be accurate and competitive (larger time steps), implicit techniques are necessary, this allows also to consider **direct coupled methods:** whereby the coupling takes place at the same time level. Then the energy dissipation of the method should be null. The drawbacks of simultaneous resolution are

i. Structure and fluid are integrated on same time scale but scales may be completely different.
ii. Specific methods of time integration for Structure or Fluid can no longer be applied directly.

By combining implicit time-staggered methods in such a way that globally the fluid and the structure are evaluated at the same time level which allow individual solvers for each part to be used. This is the best compromise between accuracy, methodology and implementation possibilities, For these techniques the displacement vector $\boldsymbol{q_s}$ of the structure, or the fluid state vector $\mathbf{W}$ are *predicted* at time step $n+1$ while ($\mathbf{W}$ or $\boldsymbol{q_s}$) are integrated implicitly by its own specific method. The decoupled update of $\boldsymbol{q_s}$ and $\mathbf{W}$ - allows application of different time steps in the individual structure and fluid parts. An overall time step for advancing $\boldsymbol{q_s}$ and $\mathbf{W}$ from n to $n+1$ is prescribed. There are many versions of this, see for instance [2].

The strong coupling algorithm applied to more complex situations than generic problems in this paper is obtained by applying several corrector steps to a staggered algorithm, in such a way that the structure is integrated at the same level as the fluid in time.

References

1. P. Leyland, V. Carstens : Fully Coupled Fluid-Structure Algorithms for Aeroelasticity and Forced Vibration Induced Flutter. In book edited by A. Dervieux "Fluid-Structure Interaction", ed. HERMES, Paris Dec. 2000.
2. Piperno S., Farhat C.: Design of Efficient Partitioned Procedures for the Transient Solution of Aeroelastic Problems, *Revue Européenne des Eléments Finis*, Dec. 2000.

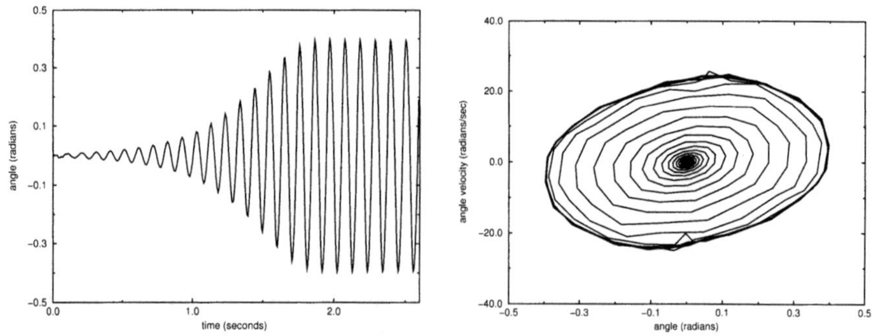

Angle and Phase plot

Fig. 1. Non-linear limit cycle behaviour for an unstable mode of aeroelastic behaviour of a NACA0012 profile at reduced speed $U^* = 6.928$ - when the shock reaches trailing edge there is a phase shift between airfoil motion and pressure response. The phase shift is responsible for flutter at transonic flow conditions

CFD Modelling of Vehicle A-Pillar Aerodynamics

Nurul M. Murad[1], Jamal Naser[1], Firoz Alam[2], and Simon Watkins[2]

[1] Swinburne University of Technology, School of Engineering and Science, PO Box 218 Hawthorn, Victoria 3122 Australia
[2] RMIT University, Department of Mechanical and Manufacturing Engineering, Bundoora East Campus, PO Box 71 Bundoora, Victoria 3083 Australia

1 Introduction

Vortices formed around a vehicle A-pillar region can lead to aerodynamic noise generation [5]. Noise generated from the A-Pillar region can be transferred to the passenger cabin causing annoyance and much discomfort to the vehicle occupants [4]. At present there is no specific method or procedure outlining the best way to model A-Pillar aerodynamic noise using CFD. Most of the CFD approaches used in the literatures focuses on obtaining results using various CFD codes and different vehicle geometry configurations [5].

The geometry configurations used to structure the vehicle such as the A-pillar height, slant angle and the windshield radius geometry plays an important part in the outcome of the flow past the A-Pillar [1,2]. However, only changes to the A-Pillar slant angle and windshield geometry are the ones that need attention because the different magnitude of separation they will generate. This investigation focuses on the effect of airflow with the changing of windshield radius geometry.

The idea behind this investigation is to demonstrate that as the windshield radius becomes rounder, the A-Pillar airflow will lead to attachments and will be easier to model using CFD. Only when the A-Pillar radius starts to open up, separation starts to build up which will lead to stronger vortex generation [4]. CFD modelling will in turn becomes much more complex. This investigation will be the foundation work that can hopefully leads to a method of CFD modelling of different A-Pillar windshield radiuses.

2 Vehicle Geometry, Boundary Conditions and Numerics

Experimental data for correlation was obtained from the PhD thesis of Alam [1]. In his thesis, Alam used a simplified model of a vehicle and test it using different windshield radiuses at different speed. For this paper, FLUENT by Fluent Inc. was used as the CFD processor. Two types of radiuses were used (Semi Circular and Ellipsoidal) for the simulation with the boundary conditions at the velocity inlet were set at 60, 100 and 140 km/h respectively.

Unstructured tetrahedral meshes were used in the simulation with the initial discretization convection scheme set first order upwind and later changed to second order upwind. The pressure-velocity coupling discretization scheme used was the SIMPLE scheme and the turbulence model used was standard $k - \epsilon$ model with non-equilibrium wall function.

3 Results and Discussion

The result obtained from the simulation in general is within close agreement with the experimental results obtained from Alam . The standard k-ϵ turbulence model with the non-equilibrium wall function successfully modelled the coefficient of static pressure (Cp) values. This shows that very small separation and vortex are being generated by the airflow making it convenient for CFD user to generate grid on this type of geometry radius. However, this will only apply to a more rounded A-Pillar windshield radius. More work will have to be done in order to know the transition radius of the windshield when the airflow that goes past the A-Pillar starts to separate. Different meshing strategy will then have to be applied to accomodate requirements needed by the turbulence and near wall model in order to model a larger vortex generated area with higher degree of accuracy.

4 Concluding Remarks

A simplified vehicle body with different windshield radius and at different velocity was modeled using FLUENT. Unstructured tetrahedral meshes were used with standard k-ϵ turbulence model and non-equilibrium wall model. The results obtained were in close agreement with the experimental results.

References

1. F. Alam: The effects of Car A-Pillar and Windshield Geometry and Yaw Angles on Local Flow and Noise. PhD Thesis, RMIT University, Melbourne, Australia (2000)
2. R.H. Barnard: *Road Vehicle Aerodynamic Design-An Introduction*, 2nd edn. (Longman, Cambridge 1996)
3. Fluent Incorporated: *Fluent User's Guide*, Version 6.0 (USA 1998)
4. A. George: *Automobile Aerodynamic Noise*, SAE 900315 (1990)
5. W.H. Hucho ed.: *Aerodynamics of Road Vehicles*, 2nd edn.(Butterworths, London 1998)

An Unsplit Relaxation Scheme for Hyperbolic Conservation Laws

S.V. Raghurama Rao[1] and Sachin Khosla[2]

[1] Assistant Professor, Centre of Excellence for Aerospace CFD
Department of Aerospace Engineering, Indian Institute of Science
Bangalore- 560 012, India. e-mail : raghu@aero.iisc.ernet.in
[2] B.Tech. Student, Department of Mechanical Engineering
Punjab Engineering College, Chandigarh-160012, India

Jin and Xin [1] proposed new schemes called **Relaxation Schemes** to solve hyperbolic equations. For a hyperbolic system given by

$$\frac{\Delta u}{\Delta t} + \frac{\Delta g(u)}{\Delta x} = 0 \tag{1}$$

the Relaxation System can be written as

$$\left.\begin{aligned} \tfrac{\Delta u}{\Delta t} + \tfrac{\Delta v}{\Delta x} &= 0 \\ \tfrac{\Delta v}{\Delta t} + \lambda^2 \tfrac{\Delta u}{\Delta x} &= -\tfrac{1}{\epsilon}\left[v - g\left(u\right)\right] \end{aligned}\right\} \text{Relaxation System} \tag{2}$$

where v is a new variable with the initial condition $v(t=0) = g(u(t=0))$. This system can also be interpreted as a Discrete Boltzmann equation [2] as

$$\frac{\Delta f}{\Delta t} + \Lambda \frac{\Delta f}{\Delta x} = \frac{1}{\epsilon}\left(F - f\right) \tag{3}$$

$$f = \begin{bmatrix} f_1 \\ f_2 \end{bmatrix}, = \begin{bmatrix} \frac{12}{u} - \frac{v}{2\lambda} \\ \frac{12}{u} + \frac{v}{2\lambda} \end{bmatrix} F = \begin{bmatrix} F_1 \\ F_2 \end{bmatrix} = \begin{bmatrix} \frac{12}{u} - \frac{g(u)}{2\lambda} \\ \frac{12}{u} + \frac{g(u)}{2\lambda} \end{bmatrix} \text{ and } \Lambda = \begin{bmatrix} -\lambda & 0 \\ 0 & \lambda \end{bmatrix} \tag{4}$$

The linearity of the left hand side makes this system very easy to solve, as there is no necessity to use any Riemann solvers or complicated flux splittings. The Relaxation System is usually solved by *Splitting Method* in which convection and relaxation parts are solved in separate steps. Because of the inherent *splitting error* in this approach, the scheme is always first order accurate in time, even if convection step is solved to second order accuracy. *Strang Splitting* is usually used as a remedy, but even then the method reduces to formal first order accuracy in the stiff limit. Jin proposed a correction to this defect, which was used in [1]. Prendergast and Kun Xu [3] presented a way to solve the BGK form of the classical Boltzmann equation in an *unsplit way*, while developing a Kinetic Scheme for Euler equations. They integrate the BGK form of the Boltzmann equation fully and give suitable approximations to get the solution of the distribution function. Using a similar strategy, we have solved the relaxation system in an unsplit way. This way of approximating and solving the equations removes the stiffness from the original system. Higher order accuracy is achieved by simple reconstruction of the state variables. The details of the method are available in [4,5]. The results for bench-mark test cases demonstrate that the new method is accurate and resolves the flow features well.

References

1. S. Jin and Z. Xin : Comm. Pure Appl. Math., 48, 235 - 277 (1995).
2. Denise Aregba-Driollet and Roberto Natalini, SIAM J. Num. Anal. 37, 6, 1973-2004 (1998).
3. K.H. Prendergast and Kun Xu : J. Comp. Phy., 109, 53-66 (1993).
4. S.V. Raghurama Rao and Sachin Khosla : Fluid Mechanics report, Department of Aerospace Engg., Indian Institute of Science (2002).
5. S.V. Raghurama Rao : berichte 249, Arbeitsgruppe Technomathematik, Universitaet Kaiserslautern, Germany (2002).

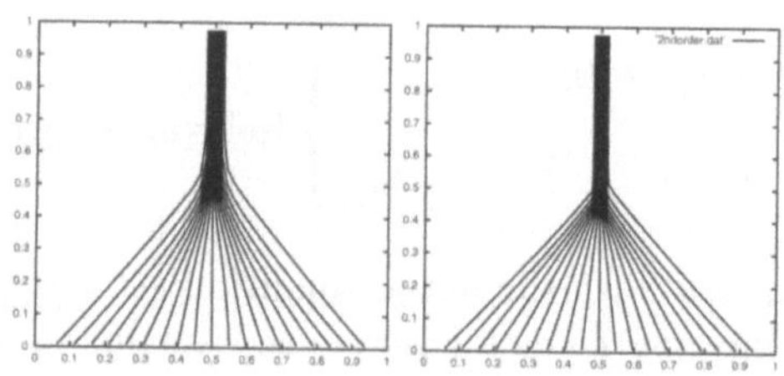

Fig. 1. I order and II order accurate results for 2-D Burgers equation

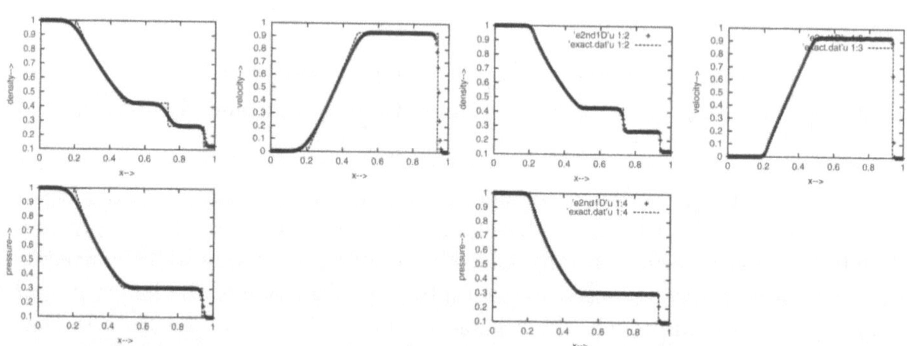

Fig. 2. I order and II order accurate results for 1-D Euler equations

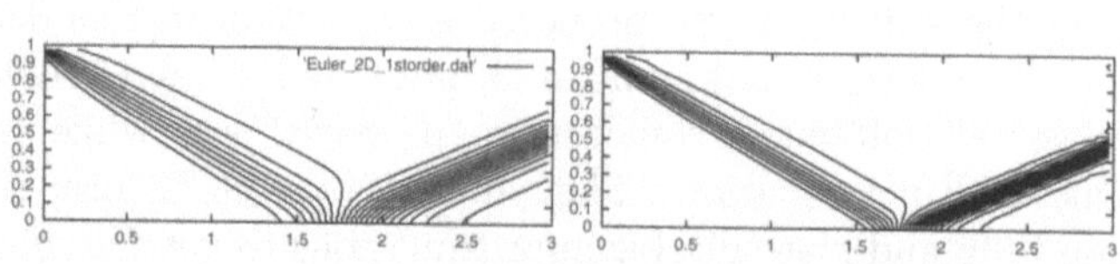

Fig. 3. I order and II order accurate results for 2-D Euler equations

A Computational Study of Combustion of Pulverised Coal and Softwood Blends

Chatphol Meesri and Behdad Moghtaderi

CRC for Coal in Sustainable Development, Discipline of Chemical Engineering, The University of Newcastle, Callaghan, NSW 2308, Australia

Abstract. This work aimed to numerically investigate the impact of blending ratio and particle size of pulverised wood on the combustion characteristics of a pulverised fuel combustor co-firing fine coal and softwood. The predicted results indicated that the increase in the blending ratio and the size of wood particles adversely affected the combustion behaviours.

1 Introduction

This study presents computational fluid dynamics (CFD) modelling of pulverised coal (bituminous) and softwood (pine sawdust) co-firing under conditions pertinent to PF boilers. A commercially available CFD code known as FLUENT was used to carry out the calculations. The model was validated against two sets of benchmarking data [1] and [2] from the International Flame Research Foundation (IFRF).

2 Results and Discussion

The predictions were obtained by numerical solution of the conservation equations for the gas and the particle phases. Turbulence of the gas phase was modelled using a standard $k-\varepsilon$ model. The gas phase combustion was treated using the mixture fraction approach. In order to account for the variability of off-gases from coal and woody biomass, volatiles from each solid fuel were tracked separately using two mixture fractions. The oxidation reactions of both fine coal and wood were modelled using a kinetic/diffusion-limited rate model by assuming that the solid fuel particles possess a spherical shape and constant diameters. The kinetic parameters were extracted from experimental measurements. The system considered was a 10x1 m two-dimensional axis-symmetric furnace. The inlet of the furnace was split into two streams: (1) a high-speed stream (50m/s) near the centre of the furnace (spans 0.125 m from the symmetry axis); and (2) a low-speed stream (15 m/s) spanning 0.375 m.

The predicted contour plots of the furnace temperatures as a function of blending ratio and the plots of the centre-line temperatures as a function of the furnace length, blending ratio, and the particle size for coal/sawdust co-firing are shown in Figures 1 (a) and (b), respectively. It can be seen from Fig.1 (a) that, for the finer sawdust (0.10 mm), the flame moves towards the injection port as the

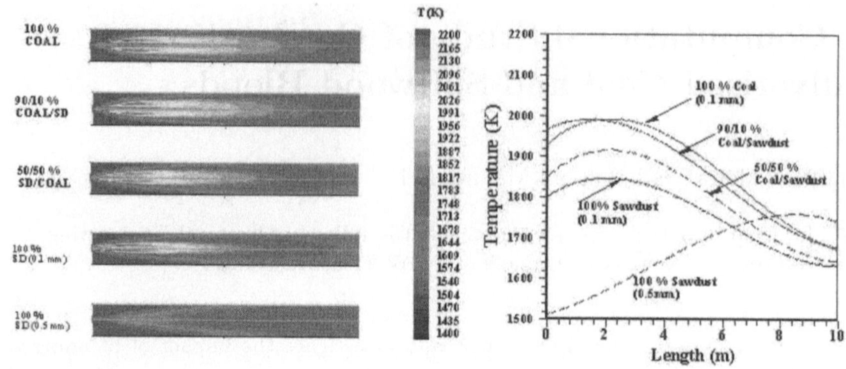

1(a) 1(b)

Fig. 1. (**a**) Contour plots of the furnace temperatures (**b**) profiles of the centre-line temperatures

sawdust content increases. This is mainly due to the rapid evolution of volatiles of the sawdust compared to that of coal, which consequently causes the overall ignition and combustion reactions of the blend to take place closer to the burner mouth. In addition, Figs. 1 (a) and (b) show that the flame width and length, as well as the furnace centre-line temperature decrease as the sawdust proportion increases. These are chiefly attributed to the impact of the lower heating value of the sawdust. However, these effects are perceived to be small in the case of 10 % w/w blending ratio. For the combustion of the unblended sawdust, it can be observed clearly that the bigger-sized sawdust (0.50 mm) produces no visible flame (Fig. 1 (a), the bottom contour) as well as significantly decreases the furnace centre-line temperatures (Fig. 1 (b), the bottom plot). This is due to the longer heat-up times required for the larger sawdust particles, which may lead to the incomplete combustion of these particles.

3 Conclusions

The predicted results imply that the blending ratio and the particle size of the wood component adversely affect the combustion behaviours of coal/wood blends in typical PF boilers. They suggest that utilising low proportions of pulverised wood (eg. 10 % w/w blending ratio of sawdust) in the blends could avoid such impacts.

References

1. Peters AAF. and Weber R., *Combust. Sci. Tech.*, (1997), **122**, 131-182.
2. Van De Kamp WL. and Morgan DJ. *Combust. Sci. Tech.*, (1996), **121**, 317-332.
3. Meesri C., Moghtaderi B., Gupta RP., Rezaei HR. and Wall TF., Co-firing of Biomass with Coal: Combustion Issues, *Eighteenth Annual International Pittsburgh Coal Conference, Newcastle, Australia*(2001), Session 42.

Numerical Simulations of Aortic Flow Problems Based on A Parallelized Artificial Compressibility Solver

Yang-Yao Niu[1], Wei-Kuang Chu[1], Shou-Cheng Tcheng[2], Lung-Cheng Lee[2], and Hsi-Yu Yu[3]

[1] Institute of Mechanical & Aerspace Engineeringcs, Chung Hua University Hsin-Chu, Taiwan, ROC
[2] National Center for the High-Performance Computing Center Hsin-Chu, Taiwan, ROC
[3] National Taiwan University Hospital Taipei, ROC

During the past years, some fundamental difficulties in the diagnosis and treatment of vascular disorders such that the re-stenosis and no-reflow phenomena of arteries after balloon angioplasty are still the unsolved problems and required in a further study. It is well known that blood flow characteristics in arteries are altered significantly by arterial disease, such as stenosis and aneurysm. In general, the prestenotic flow is usually laminar, flow in the post-stenotic region could become highly disordered or even turbulent depending upon the flow conditions and severity of the stenosis. The post-stenotic flow became unstable during this vortex generation phase of the cycle, resulting in intense fluctuations in the formation of wall shear stresses [1]. The study of physiologically realistic pulsatile flow through stenosis could have profound implications for the above-mentioned post-stenotic flow phenomena. In the current study, a dual-time stepping method based on a three-order upwind type approximated Riemann solvers [2,3] is used to solve the stationary and transient Navier- Stokes equations for analysis of separated blood flows in artery. A four-stage Runge-Kutta scheme is used in the pseudo-time iteration, also combining with MPI parallelized tool. The test results also display very promising potential of parallel processing. Newtonian flows through one curved stenosis model, one health aorta model and one aortic dissection model are performed to predict the wall stress distributions. The curved stenosis model with Reynolds numbers of 100, 200, 400, and 600 are constructed. It is noted that geometric effects on flow features such as velocity profiles, pressure and wall shear stress distributions in the post-stenotic region are significant. The location of maximum wall shear stress is found to relate to the geometric effect much than the Reynolds number effect. The current parallelized CFD solver is also demonstrated the capability on the simulation of the wall shear stresses for a health aorta and an aortic dissection cases.

References

1. Cheng Tu and Michel Deville, "Pulsatile Flow of Non-Newtonian Fluids through Arterial Stenoses", J. Biomechanics, Vol.29, No.7, pp. 2899-908, 1996

2. S. E. Roger and D.Kwak," Upwind differencing scheme for the time-accurate Incompressible Navier-Stokes Equations", AIAA Journal 28(2) pp. 253-262,1990
3. 3. D. Pan and S. R. Chakravarthy," Unified Formulation for Incompressible Flows", AIAA paper 89-0122, Jan., 1989.

Multi-block Grid with Overlapping for Implicit Parallel Computation of Flow Field within Turbomachinery

Zhirong Lin and Xin Yuan

Department of Thermal Engineering, Tsinghua University, Beijing 100084, P.R. China

1 Introduction

A accurate and fast numerical solver of 3D compressible viscous flow within turbomachinery is very helpful for blade optimization and design. Parallelization based on overlapping mesh technique is introduced to accelerate the solver.

2 Computational model and method

Compressible Navier-Stokes equations and the low-Reynolds-number two equation q-ω turbulence model are considered in governing equations set. To solve this set, one of the present authors has proposed a specially combined lower-upper factored implicit scheme based on the lower-upper symmetric-Gauss-Seidel implicit scheme in conjunction with diagonalization and Gaussian elimination (LU-SGS-GE) [1], and the modified higher-order MUSCL TVD scheme [2]. The most natural way to accelerate this process by parallel is utilizing domain decomposition. This is easy for explicit schemes because they are defined pointwisely and inherently parallel. For implicit schemes, time-lagging interface conditions is introduced. Previous study shows that this method is very effective [3].

3 Results

The typical results for a cluster of 16-processor in Table 1 shows good speed-up.

Linear, real and ideal speed-up coefficient are compared in Fig. 1. The idea speed-up coefficient presume that the network data transfer speed is infinite in

Table 1. Results on 251×31×51 grid and 100 steps with overlapping width 4

N_p	Time(S)	Speed-up (N_p)	Comm.(S)	Ratio(%)	Ideal speed-up(N_p)
1	1673.86	1.00	0.00	0.00	1.00
8	199.45	8.39	3.96	1.99	8.56
12	144.90	11.55	7.84	5.41	12.21
16	118.73	14.10	7.39	6.22	15.03

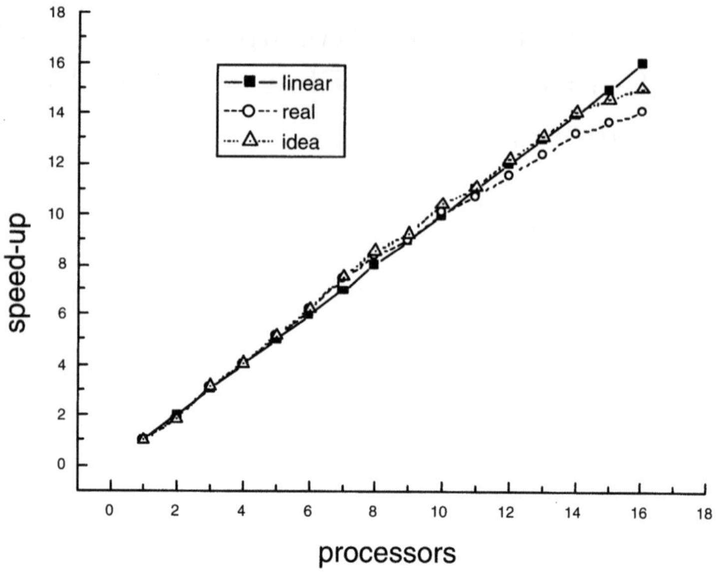

Fig. 1. The parallel speed-up coefficient

order to show the potential of the present parallel computation method. The super-linear speed-up is observed from 3 to 10 processors.

4 Conclusions

We can conclude as follows: the present numerical scheme and parallel method are suitable for solving the compressible, three-dimensional Navier-Stokes equations and the low-Reynolds-number two-equations q-ω turbulence model. The numerical results show the present solver is one of the most effective implicit solver for the large-scale problem by using cheap cluster computer.

References

1. Yuan Xin, Daiguji H., A Specially Combined Lower-Upper Factored Implicit Scheme For Three-Dimensional Compressible Navier-Stokes Equations, Computers and Fluids, 2001, Vol. **30**, No. **3**, pp. 339-363.
2. Daiguji H., Yuan Xin and Yamamoto S., Stabilization of Higher-Order High Resolution Schemes for the Compressible Navier-Stokes Equations, Int. J. of Numerical Methods for Heat & Fluid Flow, 1997, Vol.**7**, No.**2/3**, pp.159-182.
3. Wu Z. N., Zou H., Grid overlapping for implicit parallel computation of compressible flows, Journal of Computational Physics, 2000, Vol. **157**, No. **1**, pp. 2-43.

Numerical Investigation of Oscillating Flow around a Circular Cylinder

Lee Woon Chew[1] Liang Cheng[1] and Bing Chen[2]

[1] Department of Civil and Resource Engineering, University of Western Australia, Stirling Highway, Crawley, Western Australia 6009
[2] Department of Civil Engineering, Dalian University of Technology, Dalian, 116024, P. R. China

1 Introduction and Numerical Methods

Most numerical investigations on oscillating flow around a circular cylinder reported in literature chose computation domain and grid sizes based on the experience of investigators, which may results uncertainty and inaccuracy. It was shown for steady flows around a circular cylinder that calculation results strongly on the sizes of calculation domain and grid sizes [2]. Hence, the main objective of this paper is to investigate the effects of calculation domain and grid sizes on hydrodynamic forces. The hydrodynamic forces studied here includes drag coefficient C_{D}, lift coefficient C_{L}, and inertial coefficient C_{M}. A combination of in-line force, C_{X}, is also compute to investigate the combination effect of both C_{D} and C_{M}.

A three-step finite element method has been used to solve the 2D Navier-Stokes equations for oscillating flow around a circular cylinder. For details of the method, please refer to [3]. It should be noted, however, that an upwind scheme is implemented in to obtain a better numerical stability [1].

2 Results and Discussions

The calculations are carried out on a circular domain with an O-type grid. The optimum choice of the radius R of the outer boundary should be the one that gives the minimum R and satisfies the accuracy requirement of solution. Hence, calculations with R=20D, 30D, 40D, 53D and 69D are carried out at Reynolds number, Re=2000 and Keulegan Carpenter number, KC=11.7.

The calculation results of four force coefficients are generally not very sensitive to the domain sizes for R>30D Fig. 1(a). The only exception is $(C_{\mathrm{L}})_{\mathrm{rms}}$ where the trend does not converge as R increases. This may be due to the fact that the meshes used are not the optimum and the lift is more sensitive to the change of the mesh. Furthermore, it is known that Morison's equation may not express the lift well in the range of 7<KC<15. From this domain size study, it is apparent that R=40D is large enough to get domain size independent results at this particular Re and KC number.

It is well known that the accuracy of numerical solution also depends on the grid resolution. In this study, the effects of the number of nodes on the cylinder surface (Nc), the number of nodes in the radial direction (Nr) and the minimum radial spacing between the first circumferential grid line and the cylinder surface (dr) on numerical results are tested on a calculation domain of R=40D. All three parameters are varied simultaneously to achieve an optimum grid. There are five cases which are: case G1 with Nr=100, Nc=70 and dr=0.05 (total nodes=7000), case G2 with Nr=120, Nc=140, dr=0.02 (total nodes=16800), case G3 with Nr=136, Nc=200, dr=0.01 (total nodes=27200), case G4 with Nr=142, Nc=250, dr=0.0075 (total nodes=35500), and case G5 with Nr=152, Nc=300, dr=0.005 (total nodes=45600). Fig. 1(*b*) shows that the coefficients examined show slight variations with relatively coarse meshes but approach steady values as the mesh is refined. Based on the considerations of the computation time and accuracy, G3 appears to be the optimum mesh at the current Re and KC numbers.

It should be noted that the present study is at its preliminary stage and future work should include a wider range of Re and KC numbers.

As a conclusion, for oscillating flow around a circular cylinder at Re=2000 and KC=11.7, a calculation domain of 40D, Nc=200, Nr=136 and dr=0.01 is sufficient. However, further study is required to investigate the dependence of the present conclusions on Re and KC number.

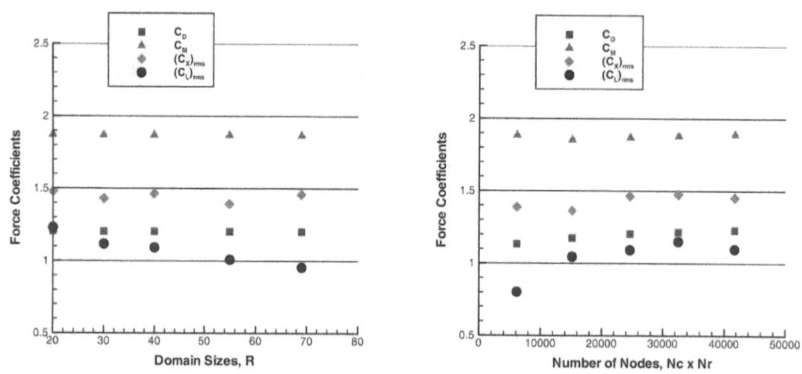

(*a*) (*b*)

Fig. 1.

References

1. A. N. Brooks, and T. J. R. Hughes, Streamline upwind/ Petrov-Galerkin Formulations for Convection dominated flows with particular emphasis on the incompressible Navier-Stokes equation, Computer methods in Applied Mechanics and Engineering, **32**, pp199-259 (1982).
2. C. Lei, Experimental and numerical studies of vortex shedding flow over a circular cylinder near a wall, PhD thesis, University of Western Australia, Australia (1999).
3. Y. Li, B. Chen and G. Lai, The numerical simulation of wave forces on seabed pipeline by three-step finite element method and large eddy simulation, Proceeding of the seventh (1997) International Offshore Polar Engineering Conference, Honolulu, USA, **II**, pp273-277 (1997).

A Two Dimensional CFD Code for Teaching

H. Siddique[1,2] and J. Naser[1,2]

[1] Swinburne University of Technology, School of Engineering and Science, PO Box 218 Hawthorn, Victoria 3122 Australia

[2] Victorian Partnership for Advanced Computing Limited Carlton South, Victoria, Australia

1 Introduction

This paper is concerned with the solution of the 2-D Navier-Stokes equation and the heat equation employing finite-volume method and is targeted for beginners in CFD. The code is able to solve channel flow and couette flow between 2 parallel plates. It is also possible to introduce an obstacle in the middle of the domain. In this paper, the authors aim to provide a simple, satisfying and motivational code for readers who are learning CFD for the first time.

MS-Excel is a Microsoft Office application, which comes along with the operating system of most personal computers. The present code utilises Excel macro feature and the program is capable of producing tabular and graphical output e.g., chart, vector plot and colour contour. The code has been prepared using standard procedures and numerical methods outlined by Patankar [1]. Subroutines have been included for data validation and data integrity.

2 GUI Interface

The 2-D code resides in an Excel file; user enters preliminary data e.g., solution domain, grid arrangement, fluid properties and boundary conditions, in designated cells of Sheet 1 of the worksheet. All data are in SI units. After macro is invoked, user is presented with the GUI presented in Figure 1.

The top 2 rows of command buttons request user for different controls. DOMAIN is the default frame displayed anytime macro is invoked. FLOW button asks for flow type, e.g., normal flow between plates or couette flow. The program will not run unless a flow type is selected. EQUATION button prompts for the transport equations to solve, e.g., x-momentum, y-momentum, pressure correction and energy.

The frame displayed when GRAPHICS button is pressed is shown in Figure 2. Graphical representation of results include X-Y plot, colour contour plot and vector plot. The system of algebraic equations derived from the discretised equations are solved by indirect method (iterative method). User has the control over the number of iteration, TDMA implies line by line iteration (inner loop) while MAIN is for the number of loops around the transport equations (outer loop). After entering all data, user can click on SOLVE button, which starts solution of transport equations and prints output at predefined locations of the worksheet.

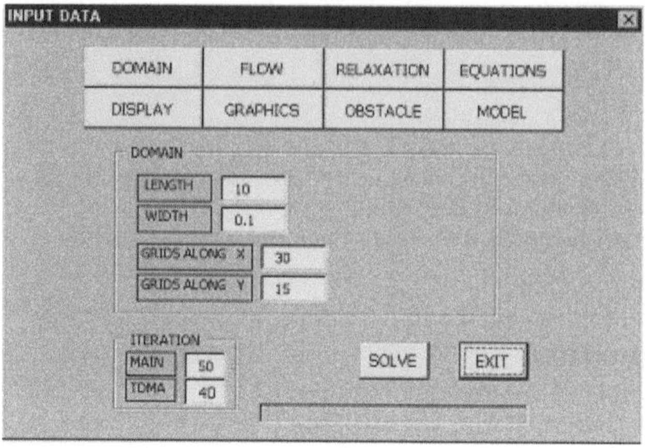

Fig. 1. GUI Interface when macro is invoked

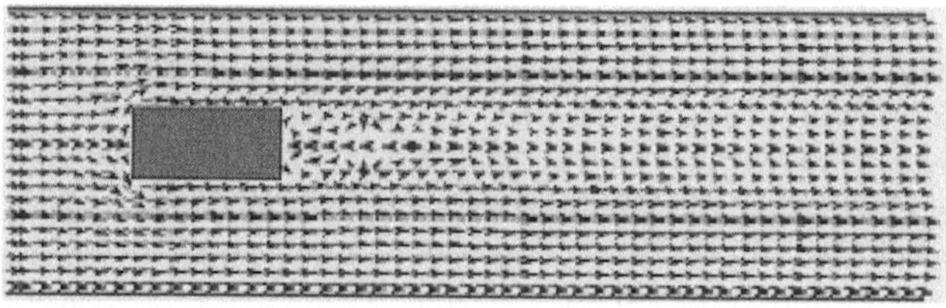

Fig. 2. velocity vector in the channel produced by the program

Acknowledgements: This project was funded by Victorian Partnership of Advanced Computing.

References

1. S.V Patankar: *Numerical Heat Transfer and Fluid Flow*, (Hemisphere Publishing Corporation, New York 1980)

Evolutionary Algorithms, Game Theory and Hierarchical Models in CFD

J. Périaux[1], M. Sefrioui[1], E. J. Whitney[2], L. Gonzalez[2], K. Srinivas[2], and J. Wang[3]

[1] Dassault Aviation, 78 Quai Marcel Dassault, 99214 Saint–Cloud, France.
[2] School of Aerospace, Mechanical and Mechatronic Engineering, The University of Sydney, NSW 2006, Australia.
[3] College of Aerospace Engineering, NUAA. 210016 Nanjing, P.R. China

Abstract. This paper details some current extensions to Evolutionary Algorithms (EAs) using concepts from game theory, as well as multiple–model and asynchronous computing. Some optimization case–studies are presented, which indicate the robustness and broad applicability of the method. These cases are 2D multi–element aerofoil reconstruction using Nash games, multiple 1D nozzle reconstruction, 2D nozzle reconstruction using multiple models and asynchronous evaluation, and 2D multiobjective nozzle reconstruction using Pareto techniques. In each case it is shown that application of the extensions provide often significant gains in speed when compared to the traditional EA framework.

1 Introduction

Practical uses of Computational Fluid Dynamics in the 21st century include automatic design and optimization for multidisciplinary challenges in aeronautics. Within this effort groups utilising conventional optimizers are achieving very good results, however due to the often multimodal and approximate nature of the solution tools available, the search for a global optimum solution to most problems must eventually take a different path. To this end we have focussed our efforts on Evolutionary Algorithms (EAs), which can often find a solution to a problem that no deterministic tool can provide.

In this paper, we discuss some aspects of the design tools we have devised, in particular the new approaches using game theory, hierarchical optimization and asynchronous computing. Finally, we give four examples where these techniques are applied to representative design problems which may face an aerospace company today.

2 Concepts from Game Theory, Hierarchical Topology, Asynchronous Computing

The evolutionary algorithms utilised in the present work are based on canonical forms widely used today, for details refer to [1,5,3]. We introduce in the following some novel concepts that greatly improve both the performance and the applicability of the method.

2.1 Nash games

A well known concept from game theory which is very promising for fast multiobjective or multidisciplinary optimization is the Nash equlibirum [7,6]. The Nash equlibirum is determined by n players competing symmetrically for n criteria, where each player optimizes a unique subset of the optimization variables, and all other variables are determined by the other players. For example, for player i the vector of problem variables is $\boldsymbol{X} = (\overline{\boldsymbol{x_1}}, \overline{\boldsymbol{x_2}}, \ldots, \boldsymbol{x_i}, \overline{\boldsymbol{x_{i+1}}}, \ldots, \overline{\boldsymbol{x_n}})$ where all subsets $\overline{\boldsymbol{x_j}}, j \neq i$ are 'locked' by the other players. Player i is interested only in one objective, namely $f_i = f_i(\boldsymbol{X})$ where $\boldsymbol{F(X)} = (f_1(\boldsymbol{X}), f_2(\boldsymbol{X}), \ldots, f_n(\boldsymbol{X}))$ is the entire multiobjective problem. We implement this using one EA for each player, as can be seen in fig 1a, whereby information is exchanged between the EAs after every generation.

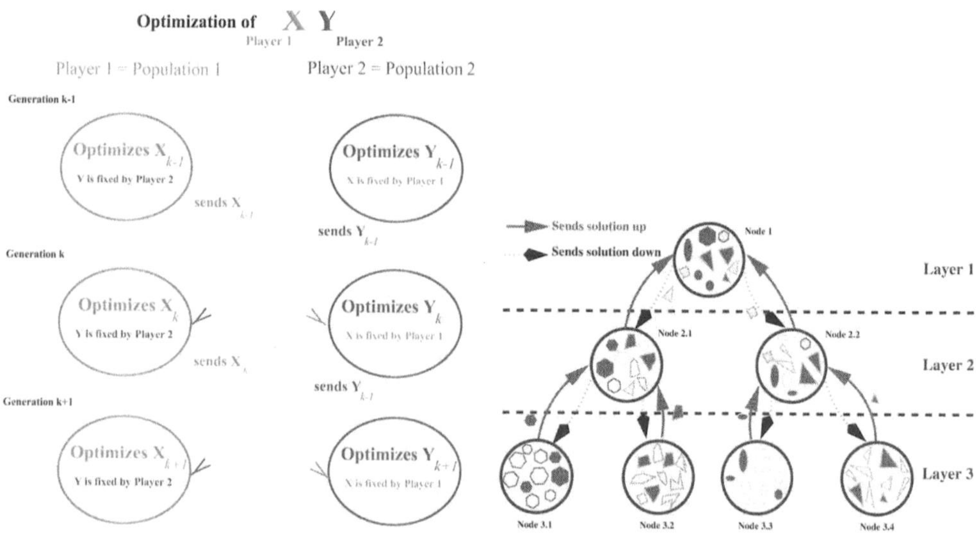

(a) Nash EA (b) Hierarchical EA

Fig. 1. Nash and hierarchical layout.

2.2 Pareto optimization

Optimization of multiobjective problems where it is difficult to determine *a priori* the importance of one objective over the other can be performed by calculating the optimum Pareto front. The Pareto front is composed of non–dominated (Pareto optimal) solutions. The solution x^* is non–dominated when all other solutions x satisfy: $f_i(x^*) \leq f_i(x)$ for all $i = 1 \ldots M$ and $f_j(x^*) < f_j(x)$ for one $j = 1 \ldots M$ where M is the number of objectives. The Pareto set of solutions to a given multiobjective problem will form a surface in the M dimensional space, and in general may have concave or convex components as well as discontinuous (as well as disconnected) regions. In general the conversion of a multiobjective problem into a single objective one using weighting will cause a failure to find any portion of the Pareto set using a conventional optimizer. In addition, more sophisticated deterministic methods may fail to locate discontinous regions and will fail to find disconnected regions. A population based EA has no such difficulty finding these regions.

2.3 Hierarchical approach

The hierarchical approach to EAs can be viewed conceptually as a kind of multi–grid or multi–physics optimization. We remove the onerous task of having to compute the fitness function of each variant solution using the most precise model available, by instead searching for initially promising solutions using relatively inexpensive methods. Good candidates are then passed on to solution methods of increasing fidelity, and when a more precise model finds a promising solution it is in turn based down to the less precise solver as a 'seed' for further exploration. The layout chosen in these study is a binary–tree hierarchical one as shown in fig 1b. Details of this hierarchical approach can be found in ref [8].

2.4 Asynchronous computing

The asynchronous evaluation method exploits heterogeneous parallel cluster computing and the fact that solvers may take varying time to compute a solution for a given problem. A normal parallel approach is where an EA sends an entire generation of individuals (perhaps 5 to 50) to be computed simultaneously. However, if one of the solutions takes much longer to compute than the others, the evolution is suspended while the fitness of this last individual is determined; this is called a *bottleneck* in the process (a rate determining step). Our approach based on the work of Wakunda and Zell [10], whereby only one individual is sent and/or received at any one time, and these individuals can return out of order. Therefore, there is no imperative that the parallel solution nodes wait for each other to finish.

3 Application One: Inverse Three Element Aerofoil Reconstruction

This section treats the reconstruction of a three–element aerofoil, both for the displacements as well as geometrical components of the slat and flap. The reconstruction is evaluated using the following techniques:

- Genetic Algorithm using a single high–fidelity model (GA).
- Nash GA using a single high-fidelity model (N-GA).
- Hierarchical Nash GA using a two-fidelity model LF-HF (HN-GA).

3.1 Solver and Definition of Models

Two-level unstructured meshes with different grid density, are used as different fidelity models for the reconstruction problem. This is performed using the software PLTMG (Piecewise Linear Triangulation Multi–Grid) developed by R.E. Bank [2]. This software package uses modules for unstructured mesh generation and adaptation using the domain decomposition method for compressible potential flow solvers in the subsonic regime. The mesh adaptation and refinement on a coarse mesh are realized based on the distribution of an 'a posteriori' error estimate to generate a fine mesh. This process can also be executed on sub meshes, with the final fine mesh constructed by gluing all sub meshes. For a given multi-element aerofoil configuration, the adaptation and refinement steps are:

i. Generation of the coarse mesh (global mesh).
ii. Resolution of flowfield and calculation of *a posteriori* the error estimate.
iii. Decomposition of the global mesh to sub-meshes.
iv. Refinement of a sub mesh based on the error estimate, with others non–refined.
v. Glue all sub meshes to give the fine mesh.

In a hierarchical genetic approach, this process is executed for each individual to generate meshes with different density. A two–fidelity model is used in this paper, where the coarse mesh is used as low–fidelity model (LF) and the fine mesh as high–fidelity model (HF).

3.2 Shape Representation

The variables of conception for the problems resolved in this paper are:

- Shape: The upper surface of the slat and lower surface of the flap are modeled by Bézier spline of order 11, each has 10 variables (control points).
- Position: The position of slat and flap given by translation and rotation. Each element has 3 variables.

In the Nash approach we split these variables between two players, with player one optimizing the slat upper surface and slat position ($\boldsymbol{V}_1$), while player two optimizes the flap lower surface and flap position ($\boldsymbol{V}_2$) so each player has 13 variables.

3.3 Problem Formulation

The problem is formulated as a least squares minimisation over the surface pressures of the candidate vs. the target solution. For a given candidate geometry C and target T: $f(\boldsymbol{V}_1, \boldsymbol{V}_2) = \sum_{i=1}^{N} (P_i^C - P_i^T)^2$ where N is the number of surface points and may vary depending on whether LF or HF models are used. In the Nash game each player minimises $f(\boldsymbol{V}_1, \boldsymbol{V}_2)$ according to their variable set.

3.4 Results

The speedup properties of hierarchical algorithms are tested with sequential algorithms (GA, N-GA and HN-GA). Figure 2a shows the comparison of convergence of three algorithms. Figure 2b shows an example of a final adapted mesh. A comparison of speedup is shown in table 1, which illustrates that the hierarchical Nash GA has much better convergence properties.

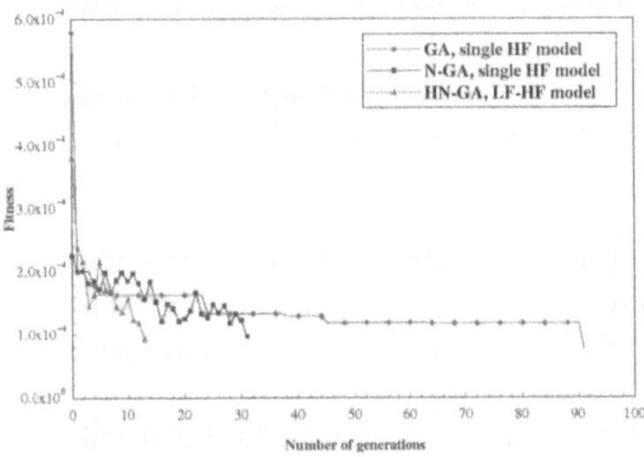

(a) Convergence for GA, N-GA and HN-GA.

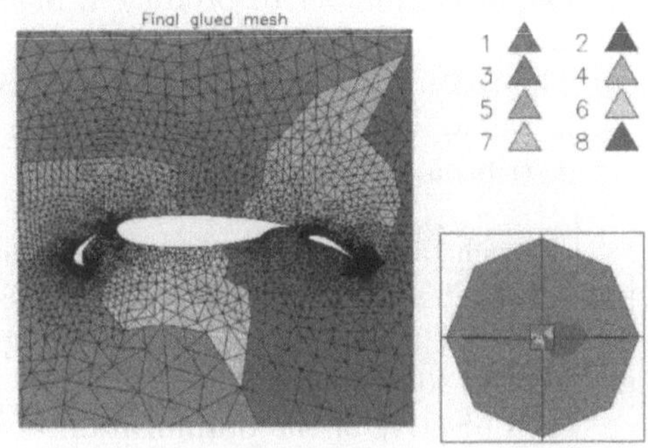

(b) Example of final glued mesh.

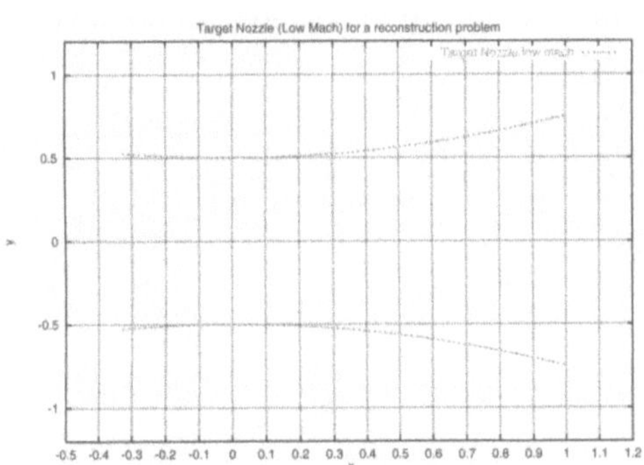

(c) Low Mach.

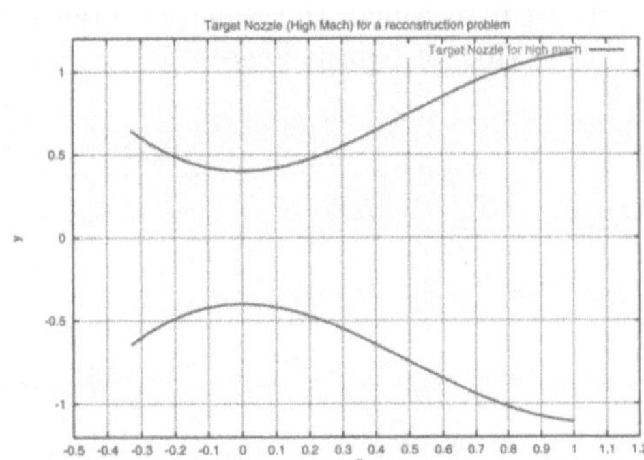

(d) High Mach.

Fig. 2. Target nozzles, low and high Mach number.

Table 1. Speedup for GA, N–GA and HN–GA.

	GA	*N-GA*	*HN-GA*
Time	437 m	293 m	89 m
Speedup	1.0	1.5	4.9

4 Application Two: Nozzle Optimization with Multiple Objectives

The Nash strategy is tested via the optimization of the shape of a nozzle for a transonic and a subsonic flow involving a shock. The problem is a reconstruction problem involving two target nozzles at two different Mach numbers:

i. The first target nozzle (Nozzle One) is shown in figure 2a. It is used with a low Mach number to produce a Mach number distribution shown in figure 3.
ii. The second target nozzle (Nozzle Two) is shown in figure 2b and is used in a flow with a high Mach number to produce the second Mach number distribution shown in figure 3.

For each target nozzle, the corresponding Mach number distribution is computed, using a quasi-steady one-dimensional approximation for the flow. The equations are solved by a time marching technique using a CUSP scheme with an iterative solver [4,9].

The objective of the optimization task for the EA is to try to find a nozzle that will be the best compromise between the two targets:

i. At low Mach, the Mach number distribution should be as close as possible to that of Nozzle One.
ii. At high Mach, the Mach number distribution should be as close as possible to that of Nozzle Two.

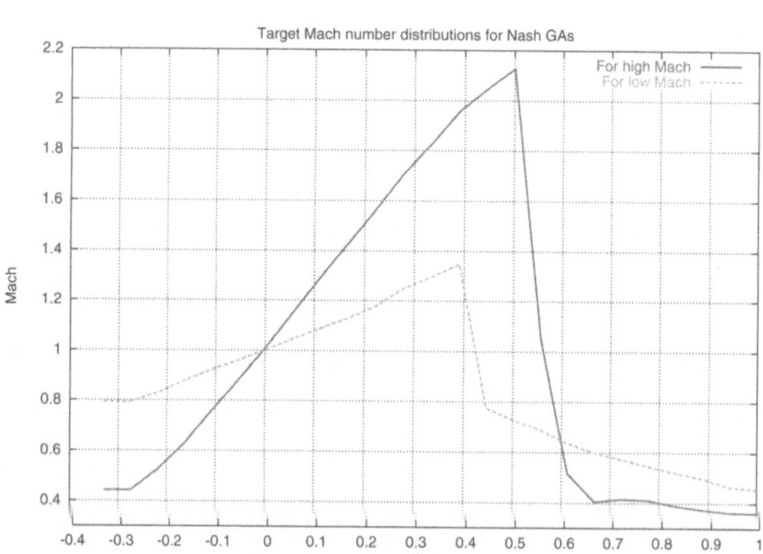

Fig. 3. Target Mach distributions for high Mach and low Mach.

In other words, for each candidate nozzle, the CFD solver provides a Mach number distribution at low or high Mach, and it is then compared to the relevant target distribution.

4.1 Nozzle Shape Parameterization

The optimization we present is done by a real-coded Nash EA, with two players having a population of size 20 each. As the nozzle is symmetric, we optimize only the upper half of it. The shape of the nozzle is defined by a Bézier curve of order 8 (i.e. a polynomial curve of order 7). The variables of the optimization are the y coordinates of the control points, $P_y^1, \cdots, P_y^8$. For the target Mach number distribution, we use 25 points along the nozzle for the grid. Let M_i be the value of the Mach number distribution of the candidate nozzle for point i. The fitness function is based on a least square method and depends on the player:

i. For player 1, the solver uses fitness $f_1 = \sum_{i=1}^{25}(M_i - M_i^{\text{Noz 1}})^2$
ii. For player 2, the solver uses fitness $f_2 = \sum_{i=1}^{25}(M_i - M_i^{\text{Noz 2}})^2$

In the case presented, Player 1 moves the Bézier control points of the left hand side of the nozzle $(P_y^1 \dots P_y^4)$, and Player 2 moves the right hand points $(P_y^5 \dots P_y^8)$.

4.2 Results

Figure 4 shows the convergence of both players in the fitness space. It is quite clear that each of them has reached a Nash equilibrium.

The solution nozzle found by Nash EA is displayed in figure 5a.

To have a better idea of how the shape evolved compared to the two target shapes, figure 5b shows the upper half of Nozzle One, Nozzle Two and the solution nozzle. The most interesting point is that the solution nozzle is by no means a compromise of the two shapes but rather an entirely novel shape. This is quite promising, as the fitness was not based on the similarity to a given shape, but rather on the similarity to a Mach number distribution.

From the computations described above, it is clear that Nash EAs offer a fast and robust alternative when it comes to multi-objective optimization. Another point of interest is that they offer solutions that are an equilibrium between two players, not necessarily a compromise. That means that new solutions can arise that are quite different from either target objectives.

5 Application Three: Nozzle Reconstruction with Asynchronous Parallel Computing

A reconstruction of an internal 'bump' geometry is tested using EAs with synchronous and asynchronous parallel techniques. The asynchronous approach has been developed in response to the demand of computing larger problems in parallel using possible heterogenous clusters of computers, with solvers that take varying amounts of time to compute solutions.

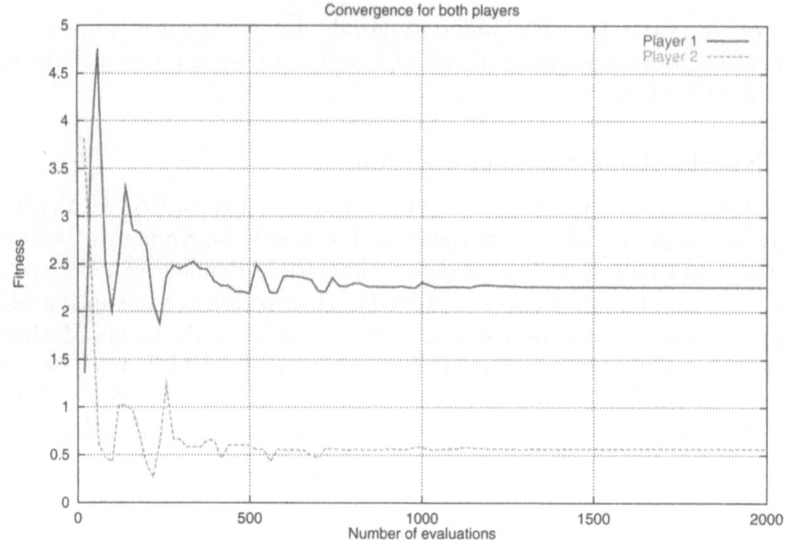

Fig. 4. Convergence for both players.

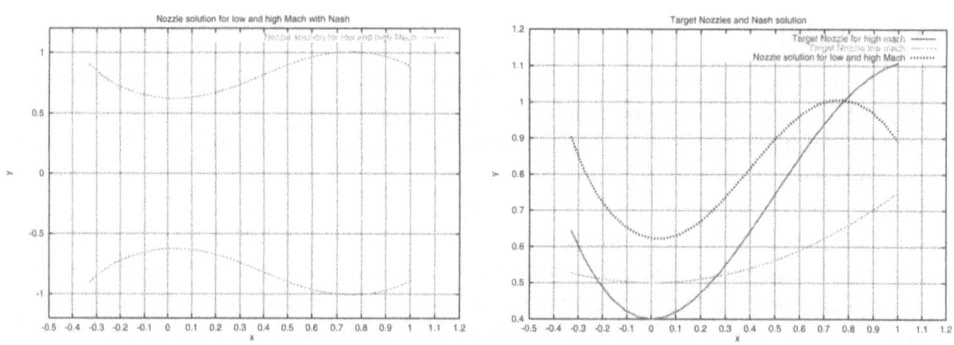

(a) Nash solution.

(b) Target nozzles and Nash solution.

Fig. 5. Nash solution and target nozzles.

5.1 Problem definition

The problem is to find the original semicircular bump geometry using a Bezier curve with 7 control points, with their vertical positions y_i as the problem unknowns (with the first and last fixed, this gives a problem in 5 dimensions). The target geometry is shown in figure 6. The wall pressures are used to find the original geometry, so that the objective is to minimise: $f = \sum_{i=1}^{G}(P_i - P_i^{Target})^2$ where G is the number of streamwise grid points. The target geometry wall is given by $y = h_t(1 - 4x^2)$ for $x \in [-0.5, +0.5]$ and $y = 0$ for the remainder. The flowfield extends from $x \in [-1.0, 1.5]$ and $y \in [-0.5, 0]$. For this case, we take $h_t = 0.1$.

5.2 Flow solver and boundary conditions

The flow is computed using the laminar Navier–Stokes equations and a CUSP solver by Srinivas [9] using laminar viscosity, on a regular grid of $G_X \times (\frac{G}{2})_Y$ points. The flow conditions are $M_{exit} = 0.95$ and $Re = 10000$; the flow enters at high subsonic speeds, becomes supersonic over the bump, and finally passes through a shock to become subsonic at the exit. The upper boundary is treated with a symmetry condition to represent internal nozzle flow and the remainder of the variables are treated with a zero–order extrapolation. The boundary layer is approximately captured with five points in the thinnest area.

Fig. 6. Target 2–D nozzle.

5.3 Results

Both synchronous and asynchronous runs were performed using the hierarchical topology as laid out by Sefrioui [8]. A three level binary tree arrangement is used with $G = 50$ on the top layer, $G = 30$ on the middle layer and $G = 15$ on the bottom layer and all populations have 10 members. An example of the convergence of a run is given in figure 7. Five runs were performed in each case, with the performance increase shown in table 2. There is approximately a 10% improvement with asynchronism, and this is at the low end of the improvement scale; All machines used in this case were identical Intel Pentium IV machines running at 2.0 GHz, and the solver execution time only varied within a range of approximately 5%. The approach appears to be even more promising when executed on a more heterogeneous cluster, where the speed–to–cost tradeoff will be much better.

Table 2. Synchronous and asynchronous performace.

Technique	*Execution Time*
Synchronous	774s ± 95s
Asynchronous	693s ± 243s

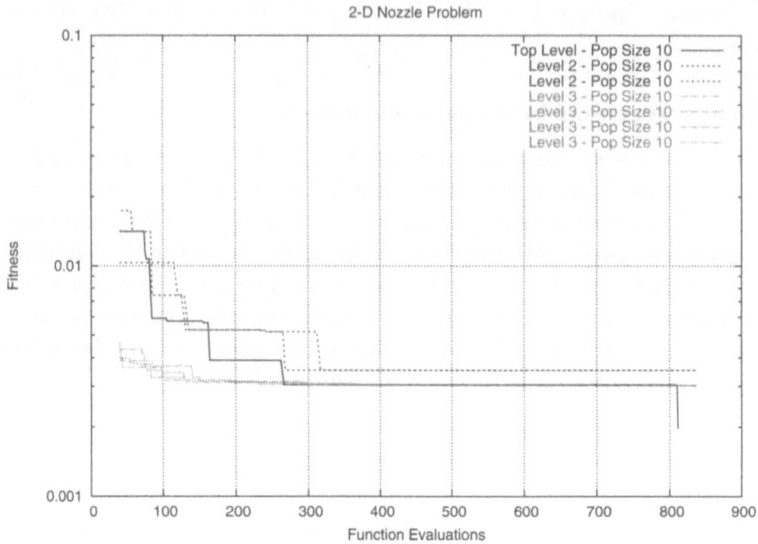

Fig. 7. Optimization progress.

6 Application Four: Multiobjective Pareto Nozzle Reconstruction

This section treats the reconstruction of two separate nozzle via the application of the Pareto methodology. In addition the hierarchical optimization topology is used to accelerate the process.

6.1 Problem definition

The problem is similar to that given in section 5, where the least–squares wall pressure difference is minimised. In this section, this is duplicated to produce two fitness functions; $f_1 = \sum_{i=1}^{G}(P_i - P_i^{Target1})^2$ and $f_2 = \sum_{i=1}^{G}(P_i - P_i^{Target2})^2$. The flow solver is the same as used in section 5. The two targets are represent by the earlier equation, this time taking $h_1 = 0.1$ and $h_2 = 0.15$. The first target flow is shown in the previous figure 6, and the second target flow is shown in figure 8.

6.2 Results

The Pareto set found at the end of the optimization process is shown in figure 9. It is seen that a well–defined Pareto optimal front is captured, and that the solutions are well spread along it. In an industrial optimization situation this would allow the designer to select a final implementation from one of many of these solutions. Individuals from the left, right and approximate centre of the

Fig. 8. Target two, 2–D nozzle.

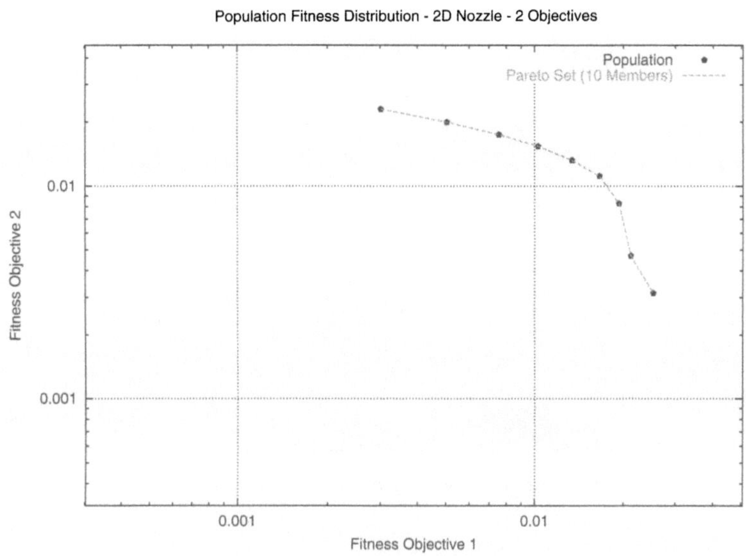

Fig. 9. Final Pareto front.

Pareto front are shown in figures 10, 11 and 12 respectively. It can be seen that the left and right Pareto points fit the two targets almost exactly, and the compromise solution has traits of both targets. Similar experiments regarding the Pareto front computation of multi–point designs can be found in ref [11].

7 Conclusion

The promising use of extended Evolution Algorithms for practical problems in aeronautics has been presented. In particular, the Nash approach is seen to enable the splitting of large optimization problems into smaller sections with both single and multiple objectives. The direct splitting of the original aerofoil reconstruction problem into a Nash game results in a speedup by a factor of 1.5 and a factor of 4.9 when multiple models are used. In fact, the hierarchical approach has appeared very promising when applied to problems so far encountered, giving a speedup of at least 3.0 in all situations encountered. The asynchronous method

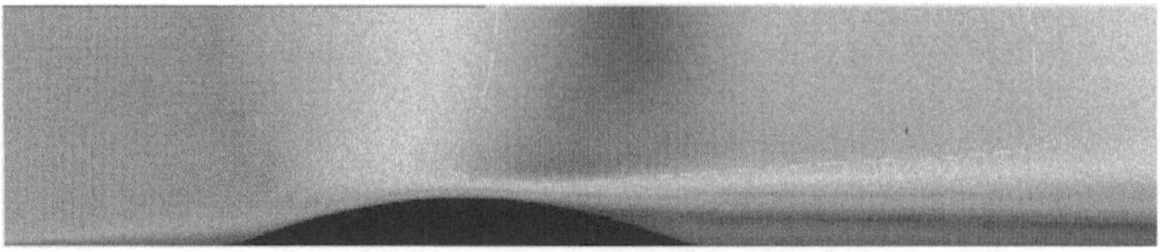

Fig. 10. First objective optimal solution.

Fig. 11. Compromise optimal solution.

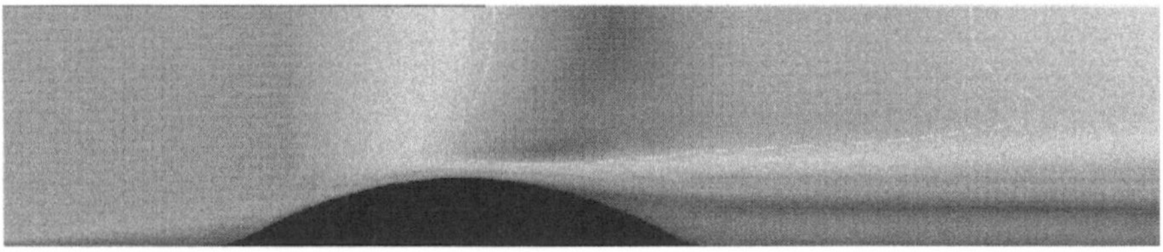

Fig. 12. Second objective optimal solution.

of fitness function evaluation has enabled the future use of highly heterogenous clusters, and has improved convergence speed by 10% on a homogenous cluster. In conclusion it seems that Evolutionary Algorithms are poised to take a much more important role in future optimization studies in CFD with one or many criteria.

Acknowledgements The authors are grateful to R.E. Bank for fruitful comments and suggestions and for providing the PLTMG software. The third author would like to thank Dassault Aviation for partially supporting his PhD research at The University of Sydney.

References

1. T. Bäck, U. Hammel, R. Lewandoski, M. Mandischer, B. Naujoks, S. Rolf, M. Schutz, H.-P. Schwefel, J. Sprave, and S. Theis. Evolutionnary algorithms: Applications at the informatik center dortmund. Trieste, Italy, November 1997. EuroGen 97, John Wiley.
2. E.D. Bank. *PLTMG: A Software Package for Solving Elliptic Partial Differential Equations, Users' guide 8.0.* University of California at San Diego, La Jolia California, 1998.

3. D.E. Goldberg. *Genetic Algorithms in Search, Optimization, and Machine Learning*. Addison-Wesley, Reading, Mass., 1989.
4. A. Jameson. Analysis and design of numerical schemes for gas dynamics, 2: Artificial diffusion and discrete shock structure. *Int. J. Comput. Fluid Dynamics*, 5:1–38, 1995.
5. Z. Michalewicz. *Genetic algorithms + data structures = evolution programs*. Artificial Intelligence. Springer-Verlag, New York, 1992.
6. J. F. Nash. Equilibrium points in n-person games. *Proc. Nat. Acad. Sci. U.S.A*, 36:46–49, 1950.
7. J. F. Nash. Noncooperative games. *Annals of Mathematics*, 54:289, 1951.
8. M. Sefrioui and J. Périaux. A hierarchical genetic algorithm using multiple models for optimization. In *Parallel Problem Solving from Nature - PPSN 6*, Paris, September 2000. Springer.
9. K. Srinivas. Computation of cascade flows by a modified cusp scheme. *Computational Fluid Dynamics Journal*, 8(2):285–295, 1999.
10. J. Wakunda and A. Zell. Median-selection for parallel steady-state evolution strategies. In *Proc. of Parallel Problem Solving From Nature (PPSN-VI)*, pages 405–414, 2000.
11. E. J. Whitney, L. F. Gonzalez, K. Srinivas, and J. Périaux. Multi–criteria aerodynamic shape design problems in cfd using a modern evolutionary algorithm on distributed computers. In *ICCFD2 Proceedings*, 2002.

4. [illegible]
[illegible] 1995.
5. [illegible]
[illegible] 199[illegible]
6. [illegible]
[illegible] 199[illegible]
7. [illegible]
8. [illegible]
[illegible]
[illegible]

Index